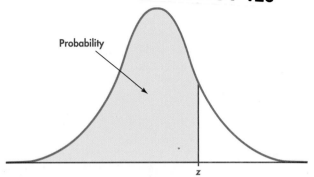

Table entry for z is the area under the standard normal curve to the left of z.

TABLE A Standard normal probabilities *(continued)*

z	.00	.01	.02	.03	.04	.05	.06	.07	.08	.09
0.0	.5000	.5040	.5080	.5120	.5160	.5199	.5239	.5279	.5319	.5359
0.1	.5398	.5438	.5478	.5517	.5557	.5596	.5636	.5675	.5714	.5753
0.2	.5793	.5832	.5871	.5910	.5948	.5987	.6026	.6064	.6103	.6141
0.3	.6179	.6217	.6255	.6293	.6331	.6368	.6406	.6443	.6480	.6517
0.4	.6554	.6591	.6628	.6664	.6700	.6736	.6772	.6808	.6844	.6879
0.5	.6915	.6950	.6985	.7019	.7054	.7088	.7123	.7157	.7190	.7224
0.6	.7257	.7291	.7324	.7357	.7389	.7422	.7454	.7486	.7517	.7549
0.7	.7580	.7611	.7642	.7673	.7704	.7734	.7764	.7794	.7823	.7852
0.8	.7881	.7910	.7939	.7967	.7995	.8023	.8051	.8078	.8106	.8133
0.9	.8159	.8186	.8212	.8238	.8264	.8289	.8315	.8340	.8365	.8389
1.0	.8413	.8438	.8461	.8485	.8508	.8531	.8554	.8577	.8599	.8621
1.1	.8643	.8665	.8686	.8708	.8729	.8749	.8770	.8790	.8810	.8830
1.2	.8849	.8869	.8888	.8907	.8925	.8944	.8962	.8980	.8997	.9015
1.3	.9032	.9049	.9066	.9082	.9099	.9115	.9131	.9147	.9162	.9177
1.4	.9192	.9207	.9222	.9236	.9251	.9265	.9279	.9292	.9306	.9319
1.5	.9332	.9345	.9357	.9370	.9382	.9394	.9406	.9418	.9429	.9441
1.6	.9452	.9463	.9474	.9484	.9495	.9505	.9515	.9525	.9535	.9545
1.7	.9554	.9564	.9573	.9582	.9591	.9599	.9608	.9616	.9625	.9633
1.8	.9641	.9649	.9656	.9664	.9671	.9678	.9686	.9693	.9699	.9706
1.9	.9713	.9719	.9726	.9732	.9738	.9744	.9750	.9756	.9761	.9767
2.0	.9772	.9778	.9783	.9788	.9793	.9798	.9803	.9808	.9812	.9817
2.1	.9821	.9826	.9830	.9834	.9838	.9842	.9846	.9850	.9854	.9857
2.2	.9861	.9864	.9868	.9871	.9875	.9878	.9881	.9884	.9887	.9890
2.3	.9893	.9896	.9898	.9901	.9904	.9906	.9909	.9911	.9913	.9916
2.4	.9918	.9920	.9922	.9925	.9927	.9929	.9931	.9932	.9934	.9936
2.5	.9938	.9940	.9941	.9943	.9945	.9946	.9948	.9949	.9951	.9952
2.6	.9953	.9955	.9956	.9957	.9959	.9960	.9961	.9962	.9963	.9964
2.7	.9965	.9966	.9967	.9968	.9969	.9970	.9971	.9972	.9973	.9974
2.8	.9974	.9975	.9976	.9977	.9977	.9978	.9979	.9979	.9980	.9981
2.9	.9981	.9982	.9982	.9983	.9984	.9984	.9985	.9985	.9986	.9986
3.0	.9987	.9987	.9987	.9988	.9988	.9989	.9989	.9989	.9990	.9990
3.1	.9990	.9991	.9991	.9991	.9992	.9992	.9992	.9992	.9993	.9993
3.2	.9993	.9993	.9994	.9994	.9994	.9994	.9994	.9995	.9995	.9995
3.3	.9995	.9995	.9995	.9996	.9996	.9996	.9996	.9996	.9996	.9997
3.4	.9997	.9997	.9997	.9997	.9997	.9997	.9997	.9997	.9997	.9998

THE PRACTICE OF STATISTICS
FOR BUSINESS AND ECONOMICS

THE PRACTICE OF STATISTICS
FOR BUSINESS AND ECONOMICS

THIRD EDITION

David S. Moore
Purdue University

George P. McCabe
Purdue University

Layth C. Alwan
University of Wisconsin, Milwaukee

Bruce A. Craig
Purdue University

William M. Duckworth
Creighton University

W. H. Freeman and Company
New York

Publisher: Ruth Baruth

Acquisitions Editor: Karen Carson

Senior Developmental Editor: Bruce Kaplan

Associate Editor: Katrina Wilhelm

Media Editor: Laura Capuano

Assistant Media Editor: Catriona Kaplan

Executive Marketing Manager: Jennifer Somerville

Head of Strategic Market Development: Steven Rigolosi

Editorial Assistant: Lauren Kimmich

Photo Editor: Cecilia Varas

Senior Project Editor: Mary Louise Byrd

Design Manager: Vicki Tomaselli

Cover Designer: Brian Sheridan

Text Designer: Cambraia Fernandes

Production Coordinator: Paul W. Rohloff

Composition: MPS Limited, a Macmillan Company

Printing and Binding: RR Donnelley

Library of Congress Control Number: 2010928451

ISBN-13: 978-1-4292-3281-4
ISBN-10: 1-4292-3281-1

W. H. Freeman and Company
41 Madison Avenue, New York, NY 10010
Houndmills, Basingstoke RG21 6XS, England
www.whfreeman.com

BRIEF TABLE OF CONTENTS

The Core book includes Chapters 1–14. Chapters 15–17 are individual optional Companion Chapters and can be found on the CD-ROM that accompanies this text or on the companion Web site (www.whfreeman.com/psbe3e).

TABLE OF CONTENTS

CHAPTER 12 Statistics for Quality: Control and Capability   633

PART IV Optional Companion Chapters

CHAPTER 15   Two-Way Analysis of Variance

Introduction

15.1 The Two-Way ANOVA Model
Advantages of two-way ANOVA
The two-way ANOVA model
Main effects and interactions
Section 15.1 Summary

15.2 Inference for Two-Way ANOVA
The ANOVA table for two-way ANOVA
Carrying out a two-way ANOVA
Case 15.1 Discounts and Expected Prices

TO INSTRUCTORS: *ABOUT THIS BOOK*

Statistics is the science of data. ***The Practice of Statistics for Business and Economics (PSBE)*** is an introduction to statistics for students of business and economics based on this principle. We present methods of basic statistics in a way that emphasizes working with data and mastering statistical reasoning. *PSBE* is elementary in mathematical level but conceptually rich in statistical ideas. After completing a course based on our text, students should be able to think objectively about conclusions drawn from data and use statistical methods in their own work.

In *PSBE,* we combine attention to basic statistical concepts with a comprehensive presentation of the elementary statistical methods that students will find useful in their work. We believe that you will enjoy using *PSBE* for several reasons:

1. *PSBE* examines the nature of modern statistical practice at a level suitable for beginners. We focus on the production and analysis of data as well as the traditional topics of probability and inference.

2. *PSBE* has a logical overall progression, so data production and data analysis are a major focus, while inference is treated as a tool that helps us to draw conclusions from data in an appropriate way.

3. *PSBE* presents data analysis as more than a collection of techniques for exploring data. We emphasize systematic ways of thinking about data. Simple principles guide the analysis: always plot your data; look for overall patterns and deviations from them; when looking at the overall pattern of a distribution for one variable, consider shape, center, and spread; for relations between two variables, consider form, direction, and strength; always ask whether a relationship between variables is influenced by other variables lurking in the background. We warn students about pitfalls in clear cautionary discussions.

4. *PSBE* uses real examples and exercises from business and economics to illustrate and enforce key ideas. Students learn the technique of least-squares regression and how to interpret the regression slope. But they also learn the conceptual ties between regression and correlation—the importance of looking for influential observations.

5. *PSBE* is aware of current developments both in statistical science and in teaching statistics. Brief, optional "Beyond the Basics" sections give quick overviews of topics such as density estimation, scatterplot smoothers, data mining, nonlinear regression, and meta-analysis.

Themes of This Book

Look at your data is a consistent theme in *PSBE.* Rushing to inference—often automated by software—without first exploring the data is the most common source of statistical errors that we see in working with users from many fields. A second theme is that *where the data come from matters.* When we do statistical inference, we are acting as if the data come from properly randomized sample or experimental designs. A basic understanding of these designs helps students grasp how inference works. The distinction between observational and experimental data helps students understand the truth of the mantra that "association does not imply causation." Moreover, managers need to understand the use of sample surveys for market research and customer satisfaction and of statistically designed experiments for product development, as in clinical trials of pharmaceuticals.

Another strand that runs through *PSBE* is that *data lead to decisions in a specific setting.* A calculation or graph or "reject H0" is not the conclusion of an exercise in statistics. We encourage students to state a conclusion in the specific problem context, even though quite simple, and we hope that you will require them to do so.

Finally, we think that a first course in any discipline should focus on the essentials. *PSBE* equips students to use statistics (and learn more statistics as needed) by presenting the major concepts and most-used tools of the discipline. Longer lists of procedures "covered" tend to reduce student understanding and ability to use any procedures to deal with real problems.

What's New in the Third Edition

- **Title** We have added economics to the title because of the number of examples and applications that are economics-oriented and because the book can be used in the statistics course for economics. Hence, the title is now *The Practice of Statistics for Business and Economics.*

- **New Co-author** We are delighted to welcome Bruce Craig to the author team. Bruce currently is professor of statistics and director of the Statistical Consulting Service at Purdue University. Bruce also is a co-author of *Introduction to the Practice of Statistics,* sixth edition.

- **Structural changes** There is no change in the chapter order, but the order of topics in three chapters has been changed significantly: Chapter 5 on probability theory, Chapter 12 on quality control, and Chapter 13 on time series forecasting. This change within chapters improves topic flow.

- **New cases and examples** We have taken extra effort to bring in new or updated cases and examples that make the material as interesting and business oriented as possible. Here are a few of them:
  - Worldwide data on the time needed to start a business (Case 1.2)
  - Predicting a movie's U.S. box office revenue based on opening weekend results (Case 11.2)
  - Unemployment rates in the United States and Canada (Chapter 1)
  - The business of the NFL (Chapter 5)
  - Turnaround lab testing time from a hospital emergency room (Case 12.1)

- **Exercises** Over 25% of the exercises are new, and over 25% include updated data sets, making over half of the exercises different from the previous edition. We have placed additional emphasis on making the business or economics relevance of the exercises clear to the reader.

- **Increased emphasis on computing** Software usage has been updated and given additional emphasis. For example, in Chapter 3, we now explain how to perform randomization for samples and experiments using spreadsheets rather than relying exclusively on tables of random numbers.

- **Data file names** The names of data files now suggest their content rather than the example, exercise, or table where they are used. The names are given with a marginal icon for examples and with highlighted text for exercises.

Content and Style

PSBE adapts to the business and economics statistics setting the approach to introductory instruction that was inaugurated and proved successful in the best-selling general statistics text *Introduction to the Practice of Statistics* (sixth edition, Freeman 2009). *PSBE* features the use of real data in examples and exercises and emphasizes statistical thinking as well as mastery of techniques. As the continuing revolution in computing automates most tiresome details, an emphasis on statistical concepts and on insight from data becomes both more practical for students and teachers and more important for users who must supply what is not automated.

Chapters 1 and 2 present the methods and unifying ideas of data analysis. Students appreciate the usefulness of data analysis, and that they can actually do it relieves a bit of their anxiety about statistics. We hope that they will grow accustomed to examining data and will continue to do so even when formal inference to answer a specific question is the ultimate goal. Note in particular that Chapter 2 gives an extended treatment of correlation and regression as descriptive tools, with attention to issues such as influential observations and the dangers posed by lurking variables. These ideas and tools have wider scope than an emphasis on inference (Chapters 10 and 11) allows. We think that a full discussion of data analysis for both one and several variables before students meet inference in these settings both reflects statistical practice and is pedagogically helpful.

Teachers will notice some nonstandard ideas in these chapters, particularly regarding the Normal distributions—we capitalize "Normal" to avoid suggesting that these distributions are "normal" in the usual sense of the word. We introduce density curves and Normal distributions in Chapter 1 as models for the overall pattern of some sets of data. Only later (Chapter 4) do we see that the same tools can describe probability distributions. Although unusual, this presentation reflects the historical origin of Normal distributions and also helps break up the mass of probability that is so often a barrier that students fail to surmount. We use the notation $N(\mu, \sigma)$ rather than $N(\mu, \sigma^2)$ for Normal distributions. The traditional notation is in fact indefensible other than as inherited tradition. The standard deviation, not the variance, is the natural measure of scale in Normal distributions, visible on the density curve, used in standardization, and so on. We want students to think in terms of mean and standard deviation, so we talk in these terms.

In Chapter 3, we discuss random sampling and randomized comparative experiments. The exposition pays attention to practical difficulties, such as nonresponse in sample surveys, that can greatly reduce the value of data. An understanding of such broader issues is particularly important for managers who must use data but do not themselves produce data. Discussion of statistics in practice alongside more technical material is part of our emphasis on data leading to practical decisions. We include a section on data ethics, a topic of increasing importance for business managers.

Chapter 3 also uses the idea of random sampling to motivate the need for statistical inference (sample results vary) and probability (patterns of random variation) as the foundation for inference. Chapters 4 and 5 then present probability. We have chosen an unusual approach: Chapter 4 contains only the probability material that is needed to understand statistical inference, and this material is presented quite informally. Chapter 5 presents additional probability in a more traditional manner. Chapter 4 is required to read the rest of the book, but Chapter 5 is optional. We suggest that you consider omitting Chapter 5 unless your students are well prepared or have some need to know probability beyond an understanding of basic statistics. One reason is to maintain content balance—less time spent on formal probability allows full attention to data analysis without reducing coverage of inference. Pedagogical concerns are more compelling. Experienced teachers recognize that students find probability difficult. Research on learning confirms our experience. Even students who can do formally posed probability problems often have a very fragile conceptual grasp of probability ideas. Formal probability does not help students master the ideas of inference (at least not as much as we teachers imagine), and it depletes reserves of mental energy that might better be applied to essentially statistical ideas.

The remaining chapters present statistical inference, still encouraging students to ask where the data come from and to look at the data rather than quickly choosing a statistical test from an Excel menu. Chapter 6, which describes the reasoning of inference, is the cornerstone. Chapters 7 and 8 discuss one-sample and two-sample procedures, which almost any first course will cover. We take the opportunity in these core "statistical practice" chapters to discuss practical aspects of inference in the context of specific examples. Chapters 9, 10, and 11 present selected more advanced topics in inference: two-way tables and simple and multiple regression. Chapters 12, 13, and 14 present additional advanced topics in inference: quality control, time series forecasting, and one-way analysis of variance. Instructors who wish to customize a single-semester course or to add a second semester will find a wide choice of additional topics in the Companion Chapters that extend *PSBE*:

- Chapter 15 Two-Way Analysis of Variance
- Chapter 16 Nonparametric Tests
- Chapter 17 Logistic Regression

Companion Chapters can be found on the book's companion Web site: `www.whfreeman.com/psbe3e`.

Accessible Technology

Any mention of the current state of statistical practice reminds us that quick, cheap, and easy computation has changed the field. Procedures such as our recommended two-sample t and logistic regression depend on software. Even the mantra "look at your data" depends in practice on software, as making multiple plots by hand is too tedious when quick decisions are required. Also, automating calculations and graphs increases students' ability to complete problems, reduces their frustration, and helps them concentrate on ideas and problem recognition rather than mechanics.

We therefore strongly recommend that a course based on PSBE be accompanied by software of your choice. Instructors will find using software easier because all data sets for PSBE can be found in several common formats both on the Web (`www.whfreeman.com/psbe3e`) and on the CD-ROM that accompanies each copy of the book.

The Microsoft Excel spreadsheet is by far the most common program used for statistical analysis in business. Our displays of output therefore emphasize Excel, though output from several other programs also appears. *PSBE* is not tied to specific software. However, appendices found at the end of most chapters provide general instructions for doing statistical procedures in Excel and Minitab. Even so, one of our emphases is that a student who has mastered the basics of, say, regression can interpret and use regression output from almost any software.

We are well aware that Excel lacks many advanced statistical procedures. More seriously, Excel's statistical procedures have been found to be inaccurate, and they lack adequate warnings for users when they encounter data for which they may give incorrect answers. There is good reason for people whose profession requires continual use of statistical analysis to avoid Excel. But there are also good practical reasons why managers whose work is not purely statistical prefer a program that they regularly use for other purposes. Excel appears to be adequate for simpler analyses of the kind that occur most often in business applications.

Some statistical work, both in practice and in *PSBE*, can be done with a calculator rather than software. Students should have at least a two-variable statistics calculator with functions for correlation and the least-squares regression line as well as for the mean and standard deviation. Graphing calculators offer considerably more capability. Because students have calculators, the text doesn't discuss "computing formulas" for the sample standard deviation or the least-squares regression line.

Technology can be used to assist learning statistics as well as doing statistics. The design of good software for *learning* is often quite different from that of software for *doing*. We want to call particular attention to the set of statistical applets available on the *PSBE* Web site (www.whfreeman.com/psbe3e). These interactive graphical programs are by far the most effective way to help students grasp the sensitivity of correlation and regression to outliers, the idea of a confidence interval, the way ANOVA responds to both within-group and among-group variation, and many other statistical fundamentals. Exercises using these applets appear throughout the text, marked by a distinctive icon. We urge you to assign some of these, and we suggest that if your classroom is suitably equipped, the applets are very helpful tools for classroom presentation as well.

Carefully Structured Pedagogy

Few students find statistics easy. An emphasis on real data and real problems helps maintain motivation, and there is no substitute for clear writing. Beginning with data analysis builds confidence and gives students a chance to become familiar with your chosen software before the statistical content becomes intimidating. We have adopted several structural devices to aid students. Major settings that drive the exposition are presented as cases with more background information than other examples. (But we avoid the temptation to give so much information that the case obscures the statistics.) A distinctive icon ties together examples and exercises based on a case.

CASE

The *exercises* are structured with particular care. Short "Apply Your Knowledge" sections pose straightforward problems immediately after each major new idea. These give students stopping points (in itself a great help to beginners) and also tell them that "you should be able to do these things right now." Each numbered section in the text ends with a substantial set of exercises, and more appear as review exercises at the end of each chapter. Finally, each chapter ends with a few Case Study Exercises that are suitable for individual or group projects. Case Study Exercises are more ambitious, offer less explicit guidance, and often use large data sets.

Acknowledgments

We are grateful to the many colleagues and students who have provided helpful comments about *Introduction to the Practice of Statistics* and *The Basic Practice of Statistics*. They have contributed to improving *PSBE* as well. In particular, we would like to thank the following colleagues who, as reviewers, offered specific comments on *PSBE*, third edition:

Nimer Alrushiedat,
  California State University, Fullerton
Robert J. Banis,
  University of Missouri-St. Louis
Matthew Bognar,
  University of Iowa
Dave Bregenzer,
  Utah State University

Si Chen,
  Murray State University
Joan M. Donohue,
  University of South Carolina
Paramjit Gill,
  The University of British Columbia, Okanagan

Betsy Greenberg,
   University of Texas, Austin
Bob Hammond,
   North Carolina State University
Abigail Jager,
   Kansas State University
Ronald V. Kalafsky,
   The University of Tennessee-Knoxville
Michael Kriley,
   Webster University
Gregory LaBlanc,
   University of California, Berkeley
Leigh Lawton,
   University of St. Thomas
Carolyn H. Monroe,
   Baylor University
Robert M. Nauss,
   University of Missouri-St. Louis
Thomas J. Page Jr.,
   Michigan State University
Joseph A. Petry,
   University of Illinois
   at Urbana-Champaign
Cathy D. Poliak,
   University of Wisconsin-Milwaukee
Deborah J. Rumsey,
   The Ohio State University

Martin J. Sabo,
   Community College of Denver
Mosen Sahebjame,
   California State University,
   Long Beach
Said E. Said,
   East Carolina University
Fati Salimian,
   Salisbury University
Bonnie Schroeder,
   The Ohio State University
Rose Sebastianelli,
   The University of Scranton
Pali Sen,
   The Ohio State University
Laura Shick,
   Clemson University
Brian E. Smith,
   McGill University
Rafael Solis,
   California State University, Fresno
Kenneth C. Strazzeri,
   University of Virginia
Elizabeth J. Wark,
   Worcester State College
Morty Yalovsky,
   McGill University

Also, we would like to thank reviewers who offered specific comments on *PSBE*, second edition:

Radha Bose,
   Florida State University
Bruce C. Brown,
   Cal State Polytechnic Univ., Pomona
Howard R. Clayton,
   University of Georgia
James Czachor,
   Fordham University–Lincoln Center
Nicholas E. Flores,
   University of Colorado
Dr. Guldem Gokcek,
   New York University
Dr. Michael W. Hero,
   University of Wisconsin–Milwaukee
Marlynne Beth Ingram,
   The University of Iowa
Morgan Jones,
   Kenan-Flagler Business School,
   UNC–Chapel Hill
Elias T. Kirche,
   Florida Gulf Coast University
Paul Daniel Martin,
   Pasadena City College
Elaine McGivern,
   University of Pittsburgh

Scott Nicholson,
   Stanford University
Thomas J. Page, Jr.,
   Ohio State University
Dr. Dawn C. Porter,
   Georgetown University
Krishan C. Rana,
   Virginia State University
Deborah Rumsey-Johnson,
   The Ohio State University
Lori E. Seward,
   University of Colorado at Boulder
Paul Shaman,
   The University of Pennsylvania,
   The Wharton School
Xiaofeng (Charlie) Shi,
   Diablo Valley College
Alexandre Borges Sugiyama,
   University of Arizona
Mark Tendall,
   Stanford University
Elwin Tobing,
   The University of Iowa
James S. Weber,
   University of Illinois at Chicago

We also would like to thank the participants in an extremely useful focus group held at the University of Tennessee, Knoxville, and offer a special note of thanks to William Seaver for spearheading this focus group:

Kimberly Dawn Cooper,
University of Tennessee
Charles Cweik,
University of Tennessee
Jennifer Golek,
University of Tennessee
James Michael Lanning,
University of Tennessee

Derek M. Norton,
University of Tennessee
William L. Seaver,
University of Tennessee
Janna L. Young,
University of Tennessee
Russell Zaretzki,
University of Tennessee

We also want to thank the following colleagues who reviewed or class-tested the first edition of *PSBE:*

Mohamed H. Albohali,
Indiana University of Pennsylvania
Mary Alguire,
University of Arkansas
Andrew T. Allen,
University of San Diego
Djeto Assane,
University of Nevada–Las Vegas
Lynda L. Ballou,
Kansas State University
Ronald Barnes,
University of Houston–Downtown
Paul Baum,
California State University–Northridge
Vanessa Beddo,
University of California–Los Angeles
Dan Brick,
University of St. Thomas,
St. Paul–Minnesota
Alan S. Chesen,
Wright State University
Siddhartha Chib,
Washington University
Judith Clarke,
California State University–Stanislaus
Lewis Coopersmith,
Rider University
Jose Luis Guerrero Cusumano,
Georgetown University
Frederick W. Derrick,
Loyola College, Maryland
Zvi Drezner,
California State University–Fullerton
Abdul Fazal,
California State University–
Stanislaus
Yue Feng,
University of Oregon

Paul W. Guy,
California State University–Chico
Robert Hannum,
University of Denver
Erin M. Hodgess,
University of Houston–Downtown
J. D. Jobson,
University of Alberta
Howard S. Kaplon,
Towson University
Mohyeddin Kassar,
University of Illinois–Chicago
Michael L. Kazlow,
Pace University
Nathan R. Keith,
Devry University
John F. Kottas,
College of William and Mary
Linda S. Leighton,
Fordham University
Ramon V. Leon,
University of Tennessee–Knoxville
Ben Lev,
University of Michigan–Dearborn
Vivian Lew,
University of California–Los Angeles
Gene Lindsay,
St. Charles Community College
Richard N. Madsen,
University of Utah
Roberto S. Mariano,
University of Pennsylvania
David Mathiason,
Rochester Institute of Technology
B. D. McCullough,
Drexel University
John D. McKenzie, Jr.,
Babson College

Tim Novotny,
Mesa State College

Tom Obremski,
University of Denver

J. B. Orris,
Butler University

Steve Ramsier,
Florida State University

Stephen Reid,
BC Institute of Technology

Ralph Russo,
University of Iowa

Neil C. Schwertman,
California State University–Chico

Carlton Scott,
University of California–Irvine

Thomas R. Sexton,
SUNY–Stony Brook

Tayyeb Shabbir,
University of Pennsylvania

Anthony B. Sindone,
University of Notre Dame

John Sparks,
University of Illinois–Chicago

Michael Speed,
Texas A&M University

Debra Stiver,
University of Nevada–Reno

Sandra Strasser,
Valparaiso University

David Thiel,
University of Nevada–Las Vegas

Milan Velebit,
University of Illinois–Chicago

Raja Velu,
Syracuse University

Robert J. Vokurka,
Texas A&M University–Corpus Christi

Art Warburton,
Simon Fraser University

Jay Weber,
University of Illinois–Chicago

Rodney M. Wong,
University of California–Davis

Elaine Zanutto,
University of Pennsylvania

Zhe George Zhang,
Western Washington University

The professionals at W. H. Freeman and Company, in particular Mary Louise Byrd, Bruce Kaplan, Terri Ward, Laura Capuano, Katrina Wilhelm, Catriona Kaplan, and Karen Carson have contributed greatly to the success of *PSBE*. Most of all, we are grateful to the many people in varied disciplines and occupations with whom we have worked to gain understanding from data. They have provided both material for this book and the experience that enabled us to write it. What the eminent statistician John Tukey called "the real problems experience and the real data experience" has shaped our view of statistics. It has convinced us of the need for beginning instruction to focus on data and concepts, building intellectual skills that transfer to more elaborate settings and remain essential when all details are automated. We hope that users and potential users of statistical techniques will find this emphasis helpful.

For Students

STATS P◯RTAL

www.yourstatsportal.com (Access code or online purchase required.) *StatsPortal* is the digital gateway to *The Practice of Statistics for Business and Economics,* Third Edition, designed to enrich the course and enhance students' study skills through a collection of Web-based tools. *StatsPortal* integrates a rich suite of diagnostic, assessment, tutorial, and enrichment features, enabling students to master statistics at their own pace. *StatsPortal* is organized around three main teaching and learning components:

- **Interactive eBook** integrates a complete and customizable online version of the text, with all of its media resources. Students can quickly search the text, and they can personalize the eBook just as they would the print version, with highlighting, bookmarking, and note-taking features. Instructors can add, hide, and reorder content; integrate their own material; and highlight key text.

- **Resources** for *PSBE* 3e are all organized into one location for ease of use. These resources include the following:

  - **NEW! Statistical Video Series** consists of StatClips, StatClips Examples, and Statistically Speaking "Snapshots." Animated lecture videos, whiteboard lessons, and documentary-style footage illustrate key statistical concepts and help students visualize statistics in real-world scenarios.
  - **StatTutor Tutorials** offer audio-multimedia tutorials, including videos, applets, and animations.
  - **Statistical Applets** offer 17 interactive applets to help students master key statistical concepts.
  - **CrunchIt! 2.0® Statistical Software** allows users to analyze data from any Internet location. Designed with the novice user in mind, the software is not only easily accessible but also easy to use. CrunchIt! 2.0® offers all the basic statistical routines covered in introductory statistics courses and more.
  - **Stats@Work Simulations** put students in the role of consultants, helping them better understand statistics within the context of real-life scenarios.
  - **EESEE Case Studies,** developed by The Ohio State University Statistics Department, teach students to apply their statistical skills by exploring actual case studies, using real data, and answering questions about the study.
  - **Data sets** are available in ASCII, Excel, JMP, Minitab, TI, SPSS[†], and S-Plus formats.
  - **Student Solutions to Odd-Numbered Exercises** include explanations of crucial concepts and detailed solutions to odd-numbered problems with step-by-step models of important statistical techniques.
  - **Statistical Software Manuals** for Excel, TI-83/84, Minitab, JMP, and SPSS provide instruction, examples, and exercises using specific statistical software packages.

[†]SPSS was acquired by IBM in October 2009.

- **Interactive Table Reader** allows students to use statistical tables interactively to seek the information they need.
- **Tables** reproduce the statistical tables found at the back of the textbook.
- **Companion Chapters:** Chapter 15 – Two-Way Analysis of Variance, Chapter 16 – Nonparametric Tests, and Chapter 17 – Logistic Regression are available as downloadable pdfs.

Resources for Instructors only:

- **Instructor's Guide with Full Solutions** includes worked-out solutions to all exercises, teaching suggestions, and chapter comments.
- **Test Bank** contains hundreds of multiple choice questions.
- **Lecture PowerPoint slides** offer a detailed lecture presentation of statistical concepts covered in each chapter of *PSBE* 3e.
- **NEW! SolutionMaster** is a Web-based version of the solutions in the *Instructor's Guide with Full Solutions*. This easy-to-use tool lets instructors generate a solution file for any set of homework exercises. Solutions can be downloaded in PDF format for convenient printing and posting. For more information or a demonstration, contact your local W. H. Freeman sales representative.
- **Assignments** organizes assignments and guides instructors through an easy-to-create assignment process and provides access to questions from the Test Bank, Web Quizzes, and Exercises from the text, including many algorithmic problems.

Online Study Center 2.0: www.whfreeman.com/osc/psbe3e (Access code or on-line purchase required.) The Online Study Center helps students identify content areas that are particularly difficult so that they can focus their study time where it is most needed. The OSC offers all the resources available in *StatsPortal*, except the eBook and Assignment Center.

Companion Web Site: www.whfreeman.com/psbe3e For students, this site serves as a FREE 24/7 electronic study guide, and it includes such features as statistical applets, data sets, and self-quizzes.

Interactive Student CD-ROM Included with every new copy of *PSBE* 3e, the CD contains access to all the content available on the Companion Web site. CrunchIt! 2.0® Statistical Software and EESEE case studies are available via an access-code-protected Web site. (Access code is included with every new text.)

NEW! Video Tool Kit www.whfreeman.com/statvtk (Access code or online purchase required.) This new Statistical Video Series consists of three types of videos that illustrate key statistical concepts and help students visualize statistics in real-world scenarios:

- **StatClips** lecture videos, created and presented by Alan Dabney, PhD, Texas A&M University, are innovative visual tutorials that illustrate key statistical concepts. In 3-5 minutes, each StatClips video combines dynamic animation, data sets, and interesting scenarios to help students understand the concepts in an introductory statistics course.
- **StatClips Examples,** which are linked to the StatClips videos, are also created and presented by Alan Dabney. Each walks students through step-by-step examples related to the StatClips lecture videos to reinforce the concepts through problem solving.

- **SnapShots** videos are abbreviated, student-friendly versions of the **Statistically Speaking** video series *Against All Odds*. SnapShots present new and updated documentary footage and interviews of real people using data analysis to make important decisions in their careers and in their daily lives. From business to medicine, from the environment to understanding the Census, **SnapShots** focus on why statistics is important for students' careers and how statistics can be a powerful tool to understand their world.

Printed Student Solutions Manual to Odd-Numbered Exercises This Solutions Manual provides solutions to odd-numbered text exercises, along with summaries of the key concepts needed to solve the problems. ISBN: 1-4292-4250-7

Software Manuals Software manuals covering Excel, Minitab, SPSS, TI-83/84, and JMP are offered within StatsPortal and the Online Study Center. These manuals are available in printed versions through custom publishing. They serve as basic introductions to popular statistical software options and guides for their use with *PSBE* 3e.

Special Software Packages Student versions of JMP, Minitab, S-PLUS, and SPSS are available on a CD-ROM packaged with the textbook. This software is not sold separately and must be packaged with the text. Contact your W.H. Freeman representative for information or visit www.whfreeman.com.

For Instructors

The **Instructor's Web site,** www.whfreeman.com/psbe3e, requires user registration as an instructor and features all of the student Web materials, plus:

- Instructor version of **EESEE** (Electronic Encyclopedia of Statistical Examples and Exercises), with solutions to the exercises in the student version.
- **PowerPoint slides** containing all textbook figures and tables.
- **Lecture PowerPoint slides** offering a detailed lecture presentation of statistical concepts covered in each chapter of *PSBE* 3e.

Printed Instructor's Guide with Solutions

This printed guide includes full solutions to all exercises and provides additional examples and data sets for class use, Internet resources, and sample examinations. It also contains brief discussions of the *PSBE* approach for each chapter.
ISBN: 1-4292-4245-0

Test Bank

The test bank contains hundreds of multiple-choice questions to generate quizzes and tests. This is available in print as well as electronically on CD-ROM (for Windows and Mac), where questions can be downloaded, edited, and re-sequenced to suit each instructor's needs.
   Printed Version, ISBN: 1-4292-4251-5
   Computerized (CD) Version, ISBN: 1-4292-4252-3

Enhanced Instructor's Resource CD-ROM: Allows instructors to **search** and **export** (by key term or chapter) all the material from the student CD, plus:

- All text images and tables
- *Instructor's Guide with Full Solutions*

- PowerPoint files and lecture slides
- Test bank files

  ISBN: 1-4292-4244-2

Course Management Systems W. H. Freeman and Company provides courses for Blackboard, WebCT (Campus Edition and Vista), Angel, Desire2Learn, Moodle, and Sakai course management systems. They are completely integrated courses that you can easily customize and adapt to meet your teaching goals and course objectives. Contact your local sales representative for more information

i-clicker

Developed for educators by educators, i-clicker is the easiest-to-use and most flexible classroom response system available.

TO STUDENTS: *WHAT IS STATISTICS?*

Statistics is the science of collecting, organizing, and interpreting numerical facts, which we call *data*. We are bombarded by data in our everyday lives. The news mentions movie box-office sales, the latest poll of the president's popularity, and the average high temperature for today's date. Advertisements claim that data show the superiority of the advertiser's product. All sides in public debates about economics, education, and social policy argue from data. A knowledge of statistics helps separate sense from nonsense in this flood of data.

The study and collection of data are also important in the work of many professions, so training in the science of statistics is valuable preparation for a variety of careers. Each month, for example, government statistical offices release the latest numerical information on unemployment and inflation. Economists and financial advisors, as well as policy makers in government and business, study these data to make informed decisions. Doctors must understand the origin and trustworthiness of the data that appear in medical journals. Politicians rely on data from polls of public opinion. Business decisions are based on market research data that reveal consumer tastes and preferences. Engineers gather data on the quality and reliability of manufactured products. Most areas of academic study make use of numbers and therefore also make use of the methods of statistics. This means it is extremely likely that your undergraduate research projects will involve, at some level, the use of statistics.

Learning from Data

The goal of statistics is learn from data. To learn, we often perform calculations or make graphs based on a set of numbers. But to learn from data, we must do more than calculate and plot because data are not just numbers; they are numbers that have some context that helps us learn from them.

Two-thirds of Americans are overweight or obese according to the Center for Disease Control and Prevention (CDC) Web site (www.cdc.gov/nchs/nhanes.htm). What does it mean to be obese or to be overweight? To answer this question, we need to talk about body mass index (BMI). Your weight in kilograms divided by the square of your height in meters is your BMI. A person who is 6 feet tall (1.83 meters) and weighs 180 pounds (81.65 kilograms) will have a BMI of $81.65/(1.83)^2 = 24.4$ kg/m^2. How do we interpret this number? According to the CDC, a person is classified as overweight or obese if their BMI is 25 kg/m^2 or greater, and as obese if their BMI is 30 kg/m^2 or more. Therefore, two-thirds of Americans have a BMI of 25 kg/m^2 or more. The person who weighs 180 pounds and is 6 feet tall is not overweight or obese, but if he or she gains 5 pounds, the BMI would increase to 25.1, and he or she would be classified as overweight. What does this have to do with business and economics? Obesity in the United States costs about \$147 billion per year in direct medical costs!

When you do statistical problems, even straightforward textbook problems, don't just graph or calculate. Think about the context and state your conclusions in the specific setting of the problem. As you are learning how to do statistical calculations and graphs, remember that the goal of statistics is not calculation for its own sake but gaining understanding from numbers. The calculations and graphs can be automated by a calculator or software, but you must supply the understanding. This book presents only the most common specific procedures for statistical analysis. A thorough grasp of the principles of statistics will enable you to quickly learn more advanced methods as needed. On the other hand, a fancy computer analysis carried out without attention to basic principles will often produce elaborate nonsense. As you read, seek to understand the principles as well as the necessary details of methods and recipes.

The Rise of Statistics

Historically, the ideas and methods of statistics developed gradually as society grew interested in collecting and using data for a variety of applications. The earliest origins of statistics lie in the desire of rulers to count the number of inhabitants or measure the value of taxable land in their domains. As the physical sciences developed in the seventeenth and eighteenth centuries, the importance of careful measurements of weights, distances, and other physical quantities grew. Astronomers and surveyors striving for exactness had to deal with variation in their measurements. Numerous measurements should be better than a single measurement, even though they vary among themselves. How can we best combine many varying observations? Statistical methods that are still important were invented to analyze scientific measurements.

By the nineteenth century, the agricultural, life, and behavioral sciences also began to rely on data to answer fundamental questions. How are the heights of parents and children related? Does a new variety of wheat produce higher yields than the old, and under what conditions of rainfall and fertilizer? Can a person's mental ability and behavior be measured just as we measure height and reaction time? Effective methods for dealing with such questions developed slowly and with much debate.

As methods for producing and understanding data grew in number and sophistication, the new discipline of statistics took shape in the twentieth century. Ideas and techniques that originated in the collection of government data, in the study of astronomical or biological measurements, and in the attempt to understand heredity or intelligence came together to form a unified "science of data." That science of data—statistics—is the topic of this text.

The Organization of This Book

Part I of this book, "Data," concerns data analysis and data production. The first two chapters deal with statistical methods for organizing and describing data. These chapters progress from simpler to more complex data. Chapter 1 examines data on a single variable, Chapter 2 is devoted to relationships among two or more variables. You will learn both how to examine data produced by others and how to organize and summarize your own data. These summaries will be first graphical, then numerical, and then, when appropriate, in the form of a mathematical model that gives a compact description of the overall pattern of the data. Chapter 3 outlines arrangements (called designs) for producing data that answer specific questions. The principles presented in this chapter will help you design proper samples and experiments for your research projects and to evaluate other such investigations in your field of study.

Part II, "Probability and Inference," consisting of Chapters 4 to 8, introduces statistical inference—formal methods for drawing conclusions from properly produced data. Statistical inference uses the language of probability to describe how reliable its conclusions are, so some basic facts about probability are needed to understand inference. Probability is the subject of Chapters 4 and 5. Chapter 6, perhaps the most important chapter in the text, introduces the reasoning of statistical inference. Effective inference is based on good procedures for producing data (Chapter 3), careful examination of the data (Chapters 1 and 2), and an understanding of the nature of statistical inference as discussed in Chapter 6. Chapters 7 and 8 describe some of the most common specific methods of inference, for drawing conclusions about means and proportions from one and two samples.

The six shorter chapters in Part III, "Topics in Inference," introduce somewhat more advanced methods of inference, dealing with relations in categorical data, regression and correlation, and analysis of variance. Supplementary chapters, available from the text Web site, present additional statistical topics.

What Lies Ahead

The Practice of Statistics for Business and Economics is full of data from many different areas of life and study. Many exercises ask you to express briefly some understanding gained from the data. In practice, you would know much more about the background of the data you work with and about the questions you hope the data will answer. No textbook can be fully realistic. But it is important to form the habit of asking "What do the data tell me?" rather than just concentrating on making graphs and doing calculations.

You should have some help in automating many of the graphs and calculations. You should certainly have a calculator with basic statistical functions. Look for keywords such as "two-variable statistics" or "regression" when you shop for a calculator. More advanced (and more expensive) calculators will do much more, including some statistical graphs. You may be asked to use software as well. There are many kinds of statistical software, from spreadsheets to large programs for advanced users of statistics. The kind of computing available to learners varies a great deal from place to place—but the big ideas of statistics don't depend on any particular level of access to computing.

Because graphing and calculating are automated in statistical practice, the most important assets you can gain from the study of statistics are an understanding of the big ideas and the beginnings of good judgment in working with data. Ideas and judgment can't (at least yet) be automated. They guide you in telling the computer what to do and in interpreting its output. This book explains the most important ideas of statistics, instead of just teach methods. Some examples of big ideas that you will meet are "always plot your data," "randomized comparative experiments," and "statistical significance."

You learn statistics by doing statistical problems. "Practice, practice, practice." Be prepared to work problems. The basic principle of learning is persistence. Being organized and persistent is more helpful in reading this book than knowing lots of math. The main ideas of statistics, like the main ideas of any important subject, took a long time to discover and take some time to master. The gain will be worth the pain.

INDEX OF CASES

INDEX OF DATA TABLES

BEYOND THE BASICS INDEX

David S. Moore is Shanti S. Gupta Distinguished Professor of Statistics, Emeritus, at Purdue University and was 1998 president of the American Statistical Association. He received his A.B. from Princeton and his Ph.D. from Cornell, both in mathematics. He has written many research papers in statistical theory and served on the editorial boards of several major journals. Professor Moore is an elected fellow of the American Statistical Association and of the Institute of Mathematical Statistics and an elected member of the International Statistical Institute. He has served as program director for statistics and probability at the National Science Foundation.

In recent years, Professor Moore has devoted his attention to the teaching of statistics. He was the content developer for the Annenberg/Corporation for Public Broadcasting college-level telecourse *Against All Odds: Inside Statistics* and for the series of video modules *Statistics: Decisions through Data,* intended to aid the teaching of statistics in schools. He is the author of influential articles on statistics education and of several leading texts. Professor Moore has served as president of the International Association for Statistical Education and has received the Mathematical Association of America's national award for distinguished college or university teaching of mathematics.

George P. McCabe is the Associate Dean for Academic Affairs in the College of Science and a Professor of Statistics at Purdue University. In 1966 he received a B.S. degree in mathematics from Providence College and in 1970 a Ph.D. in mathematical statistics from Columbia University. His entire professional career has been spent at Purdue with sabbaticals at Princeton; the Commonwealth Scientific and Industrial Research Organization (CSIRO) in Melbourne (Australia); the University of Berne (Switzerland); the National Institute of Standards and Technology (NIST) in Boulder, Colorado; and the National University of Ireland in Galway. Professor McCabe is an elected fellow of the American Association for the Advancement of Science and of the American Statistical Association; he was 1998 Chair of its section on Statistical Consulting. From 2008 to 2010, he served on the Institute of Medicine Committee on Nutrition Standards for the National School Lunch and Breakfast Programs. He has served on the editorial boards of several statistics journals. He has consulted with many major corporations and has testified as an expert witness on the use of statistics in several cases.

Professor McCabe's research interests have focused on applications of statistics. Much of his recent work has focused on problems in nutrition, including nutrient requirements, calcium metabolism, and bone health. He is author or coauthor of more than 160 publications in many different journals.

Layth C. Alwan is Associate Professor of Business Statistics and Operations Management, Sheldon B. Lubar School of Business, University of Wisconsin-Milwaukee. He received a B.A. in mathematics, B.S. in statistics, M.B.A., and Ph.D. in business statistics/operations management, all from the University of Chicago, and an M.S. in computer science from DePaul University.

Professor Alwan is an author of many research articles related to statistical process control and business forecasting. He has consulted for many leading companies on statistical issues related to quality, forecasting, and operations/supply chain management applications. On the teaching front, he is focused on engaging and motivating business students on how statistical thinking and data analysis methods have practical importance in business. He is the recipient of several teaching awards, including Business School Teacher of the Year and Executive MBA Outstanding Teacher of the Year.

Bruce A. Craig is Professor of Statistics and Director of the Statistical Consulting Service at Purdue University. He received his B.S. in mathematics and economics from Washington University in St. Louis and his Ph.D. in statistics from the University of Wisconsin-Madison. He is an active member of the American Statistical Association and was chair of its section on Statistical Consulting in 2009. He also is an active member of the Eastern North American Region of the International Biometrics Society and was elected by the voting membership to the Regional Committee from 2003 to 2006. Professor Craig has served on the editorial board of several statistical journals and has been a member of several data and safety monitoring boards, including Purdue's IRB.

Professor Craig's research interest focus is on the development of novel statistical methodology to address research questions in the life sciences. Areas of current interest are protein structure determination, diagnostic testing, and animal abundance estimation. In 2005, he was named Purdue University Faculty Scholar.

William M. Duckworth specializes in statistics education, business applications of statistics, and design of experiments. He holds a B.S. and an M.S. from Miami University (Ohio) in mathematics and statistics and a Ph.D. in statistics from the University of North Carolina at Chapel Hill. His professional affiliations include the American Statistical Association (ASA), the International Association for Statistical Education (IASE), and the Decision Sciences Institute (DSI). He currently serves on the editorial board of the *Journal of Statistics Education* and has served as the ASA Editor for Statistics Education Web Content. Professor Duckworth was also a member of the Undergraduate Statistics Education Initiative (USEI) which developed curriculum guidelines for undergraduate programs in statistical science that were officially adopted by the ASA.

Professor Duckworth has published research papers and been invited to speak at professional meetings and at company training workshops. During his tenure in the Statistics Department at Iowa State University his main responsibility was coordinating, teaching, and improving introductory business statistics courses for over 1000 business students a year. He received the Iowa State University Foundation Award for Early Achievement in Teaching, based in part on his improvements to introductory business statistics. Professor Duckworth now teaches business statistics in the College of Business Administration at Creighton University.

PART I

Data

© Photodisc/AgeFotostock

Data are being collected and stored every day. Data can tell us interesting and important things about the world around us. A major challenge is to extract useful information from data. We begin our study of statistics by exploring some simple methods for accomplishing this goal.

© CORBIS

Examining Distributions

An iPod can hold thousands of songs. Apple has developed the iTunes playlist to organize data about the songs on an iPod. Example 1.1 discusses how these data are organized.

Introduction

Statistics is the science of data. Data are numerical facts. In this chapter, we will master the art of examining data.

A statistical analysis starts with a set of data. We construct a set of data by first deciding what *cases* or units that we want to study. For each case, we record information about characteristics that we call *variables*.

Cases, Labels, Variables, and Values

Cases are the objects described by a set of data. Cases may be customers, companies, subjects in a study, or other objects.

A **label** is a special variable used in some data sets to distinguish the different cases.

A **variable** is a characteristic of a case.

Different cases can have different **values** for the variables.

CHAPTER OUTLINE

1.1 Displaying Distributions with Graphs

1.2 Describing Distributions with Numbers

1.3 Density Curves and the Normal Distributions

EXAMPLE 1.1   Over 5 Billion Sold

Apple's music-related products and services generated $1.05 billion in the first quarter of 2008 and accounted for 13% of the company's revenue. Since Apple started marketing iTunes in 2003, they have sold over 5 billion songs. Let's take a look at this remarkable product. Figure 1.1 is part of an iTunes playlist named PSBE. The four songs shown are cases. They are numbered from 1 to 4 in the first column. These numbers are the labels that distinguish the four songs. The following five columns give the name (of the song), time (the length of time it takes to play the song), artist, album, and genre.

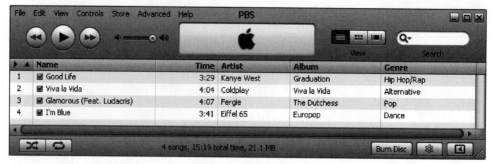

FIGURE 1.1 Part of an iTunes playlist, for Example 1.1.

Some variables, like the name of a song and the artist simply place cases into categories. Others, like the length of a song, take numerical values for which we can do arithmetic. It makes sense to give an average length of time for a collection of songs, but it does not make sense to give an "average" album. We can, however, count the numbers of songs for different albums, and we can do arithmetic with these counts.

Categorical and Quantitative Variables

A **categorical variable** places a case into one of several groups or categories.

A **quantitative variable** takes numerical values for which arithmetic operations such as adding and averaging make sense.

The **distribution** of a variable tells us what values it takes and how often it takes these values.

EXAMPLE 1.2   Categorical and Quantitative Variables in the iTunes Playlist

The PSBE iTunes playlist contains five variables. These are the name, time, artist, album, and genre. The time is a quantitative variable. Name, artist, album, and genre are categorical variables.

An appropriate label for your cases should be chosen carefully. In our iTunes example, a natural choice of a label would be the name of the song. However, if you have more than one artist performing the same song, or the same artist performing the same song on different albums, then the name of the song would not uniquely label each of the songs in your playlist.

A quantitative variable such as the time in the iTunes playlist requires some special attention before we can do arithmetic with its values. The first song in the playlist has time equal to 3:29—that is, 3 minutes and 29 seconds. To do arithmetic with this variable, we should first convert all of the values so that they have a single unit of measurement. We could convert to seconds; 3 minutes is 180 seconds, so the total time is 180 + 29, or 209 seconds. An alternative would be to convert to minutes; 29 seconds is .483 minutes, so the time calculated in this way is 3.483 minutes.

APPLY YOUR KNOWLEDGE

1.1 Time in the iTunes playlist. In the iTunes playlist, do you prefer to convert the time to seconds or minutes? Give a reason for your answer.

In practice, any set of data is accompanied by background information that helps us understand the data. When you plan a statistical study or explore data from someone else's work, ask yourself the following questions:

1. **Who?** What **cases** do the data describe? **How many** cases appear in the data?
2. **What?** How many **variables** do the data contain? What are the **exact definitions** of these variables? In what **unit of measurement** is each variable recorded?
3. **Why?** **What purpose** do the data have? Do we hope to answer some specific questions? Do we want to draw conclusions about cases other than the ones we actually have data for? Are the variables that are recorded suitable for the intended purpose?

EXAMPLE 1.3 Data for Students in a Statistics Class

Figure 1.2 shows part of a data set for students enrolled in an introductory statistics class. Each row gives data on one student. The values for the different variables are in the columns. This data set has eight variables. ID is an identifier, or label, for each student. Exam1, Exam2, Homework, Final, and Project give the points earned, out of a total of 100 possible, for each of these course requirements. Final grades are based on a possible 200 points for each exam and the final, 300 points for Homework, and 100 points for Project. TotalPoints is the variable that gives the composite score. It is computed by adding 2 times Exam1, Exam2, and Final, 3 times Homework, and 1 times Project. Grade is the grade earned in the course. This instructor used cutoffs of 900, 800, 700, etc. for the letter grades.

	A	B	C	D	E	F	G	H
1	ID	Exam1	Exam2	Homework	Final	Project	Total Points	Grade
2	101	89	94	88	87	95	899	A
3	102	78	84	90	89	94	866	B
4	103	71	80	75	79	95	780	C
5	104	95	98	97	96	93	962	A
6	105	79	88	85	88	96	861	B

FIGURE 1.2 Spreadsheet for Example 1.3.

APPLY YOUR KNOWLEDGE

1.2 Who, what, and why for the statistics class data. Answer the Who, What, and Why questions for the statistics class data set.

1.3 Read the spreadsheet. Refer to Figure 1.2. Give the values of the variables Exam1, Exam2, and Final for the student with ID equal to 103.

1.4 Calculate the grade. A student whose data do not appear on the spreadsheet scored 88 on Exam1, 85 on Exam2, 77 for Homework, 90 on the Final, and 80 on the Project. Find TotalPoints for this student and give the grade earned.

spreadsheet   The display in Figure 1.2 is from an Excel **spreadsheet.** Spreadsheets are very useful for doing the kind of simple computations that you did in Exercise 1.4. You can type in a formula and have the same computation performed for each row.

Note that the names we have chosen for the variables in our spreadsheet do not have spaces. For example, we could have used the name "Exam 1" for the first exam score rather than Exam1. In some statistical software packages, however, spaces are not allowed in variable names. For this reason, when creating spreadsheets for eventual use with statistical software, it is best to avoid spaces in variable names. Another convention is to use an underscore (_) where you would normally use a space. For our data set, we could use Exam_1, Exam_2, and Final_Exam.

EXAMPLE 1.4   Cases and Variables for the Statistics Class Data

The data set in Figure 1.2 was constructed to keep track of the grades for students in an introductory statistics course. The cases are the students in the class. There are 8 variables in this data set. These include an identifier for each student and scores for the various course requirements. There are no units of measurement for ID and grade; they are categorical variables. The other variables all are measured in "points"; since it makes sense to do arithmetic with these values, these variables are quantitative variables.

EXAMPLE 1.5   Statistics Class Data for a Different Purpose

Suppose the data for the students in the introductory statistics class were also to be used to study relationships between student characteristics and success in the course. For this purpose, we might want to use a data set that includes other variables such as Gender, PrevStat (whether or not the student has taken a statistics course previously), and Year (student classification as first, second, third, or fourth year). ID is a categorical variable, total points is a quantitative variable, and the remaining variables are all categorical.

In our example, the possible values for the grade variable are A, B, C, D, and F. When computing grade point averages, many colleges and universities translate these letter grades into numbers using $A = 4, B = 3, C = 2, D = 1$, and $F = 0$. The transformed variable with numeric values is considered to be quantitative because we can average the numerical values across different courses to obtain a grade point average.

Sometimes, experts argue about numerical scales such as this. They ask whether or not the difference between an A and a B is the same as the difference between a D and an F. Similarly, many questionnaires ask people to respond on a 1 to 5 scale with 1 representing strongly agree, 2 representing agree, etc. Again we could ask whether or not the five possible values for this scale are equally spaced in some sense. From a practical point of view, the averages that can be computed when we convert categorical scales such as these to numerical values frequently provide a very useful way to summarize data.

APPLY YOUR KNOWLEDGE

1.5 Apartment rentals for students. A data set lists apartments available for students to rent. Information provided includes the monthly rent, whether or not cable is included free of charge, whether or not pets are allowed, the number of bedrooms, and the distance to the campus. Describe the cases in the data set, give the number of variables, and specify whether each variable is categorical or quantitative.

Knowledge of the context of data includes an understanding of the variables that are recorded. Often the variables in a statistical study are easy to understand: height in centimeters, study time in minutes, and so on. But each area of work also has its own special variables. A psychologist uses the Minnesota Multiphasic Personality Inventory (MMPI), and a physical fitness expert measures "VO2 max," the volume of oxygen consumed per minute while exercising at your maximum capacity. Both of these variables are measured with special **instruments.** VO2 max is measured by exercising while breathing into a mouthpiece connected to an apparatus that measures oxygen consumed. Scores on the MMPI are based on a long questionnaire, which is also an instrument. Part of mastering your field of work is learning what variables are important and how they are best measured. Because details of particular measurements usually require knowledge of the particular field of study, we will say little about them.

instrument

Be sure that each variable really does measure what you want it to. A poor choice of variables can lead to misleading conclusions. Often, for example, the **rate** at which something occurs is a more meaningful measure than a simple count of occurrences.

rate

EXAMPLE 1.6 Insurance for Passenger Cars and Motorcycles

Should insurance rates be higher for passenger cars than for motorcycles or should they be lower? Part of the answer to this question can be found by examining accidents for these two types of vehicles. The government's Fatal Accident Reporting System says that 22,856 passenger cars were involved in fatal accidents in 2007. Only 5306 motorcycles had fatal accidents that year.[1] Does this mean that motorcycles are safer than cars? Not at all—there are many more cars than motorcycles, so we expect cars to have a higher *count* of fatal accidents.

A better measure of the dangers of driving is a *rate,* the number of fatal accidents divided by the number of vehicles on the road. In 2007, passenger cars had about 16.6 fatal accidents for each 100,000 vehicles registered. There were about 74.3 fatal accidents for each 100,000 motorcycles registered. The rate for motorcycles is more than three times the rate for cars. Motorcycles are, as we might guess, much more dangerous than cars.

1.1 Displaying Distributions with Graphs

Statistical tools and ideas help us examine data to describe their main features. This examination is called **exploratory data analysis.** Like an explorer crossing unknown lands, we want first to simply describe what we see. Here are two basic strategies that help us organize our exploration of a set of data:

exploratory data analysis

- Begin by examining each variable by itself. Then move on to study the relationships among the variables.

- Begin with a graph or graphs. Then add numerical summaries of specific aspects of the data.

We will follow these principles in organizing our learning. This chapter presents methods for describing a single variable. We study relationships among two or more variables in Chapter 2. Within each chapter, we begin with graphical displays, then add numerical summaries for a more complete description.

Categorical variables: bar graphs and pie charts

The values of a categorical variable are labels for the categories, such as "Yes" and "No." The **distribution of a categorical variable** lists the categories and gives either the **count** or the **percent** of cases that fall in each category.

distribution of a categorical variable

EXAMPLE 1.7   GPS Market Share

The Global Positioning System (GPS) uses satellites to transmit microwave signals that enable GPS receivers to determine the exact location of the receiver. Here are the market shares for the major GPS receiver brands sold in the United States.[2]

Company	Percent
Garmin	47
TomTom	19
Magellan	17
Mio	7
Other	10

Company is the categorical variable in this example, and the values of this variable are the names of the companies that provide GPS receivers in this market.

Note that the last value of the variable Company is "Other," which includes all receivers sold by companies other than the four listed by name. For data sets that have a large number of values for a categorical variable, we often create a category such as this that includes categories that have relatively small counts or percents. Careful judgment is needed when doing this. You don't want to cover up some important piece of information contained in the data by combining data in this way.

When we look at the GPS market share data set, we see that Garmin dominates the market with almost half of the sales. By using graphical methods, we can easily see this information and other characteristics of the data easily. We now examine two graphical ways to do this.

EXAMPLE 1.8   Bar Graph for the GPS Market Share Data

bar graph

Figure 1.3 displays the GPS market share data using a **bar graph.** The heights of the five bars show the market shares for the four companies and the "Other" category.

FIGURE 1.3 Bar graph for the GPS data in Example 1.8.

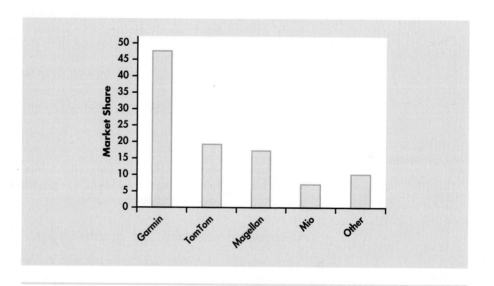

The categories in a bar graph can be put in any order. In Figure 1.3, we ordered the companies based on their market share, with the "Other" category coming last. For other data sets, an alphabetical ordering or some other arrangement might produce a more useful graphical display. You should always consider the best way to order the values of the categorical variable in a bar graph. Choose an ordering that will be useful to you. If you are uncertain, ask a friend whether your choice communicates what you expect.

DATA FILE
GPS

EXAMPLE 1.9   Pie Chart for the GPS Market Share Data

pie chart

The **pie chart** in Figure 1.4 helps us see what part of the whole each group forms. Even if we did not include the percents, it would be very easy to see that Garmin has about half of the market.

FIGURE 1.4 Pie chart for the GPS data in Example 1.9.

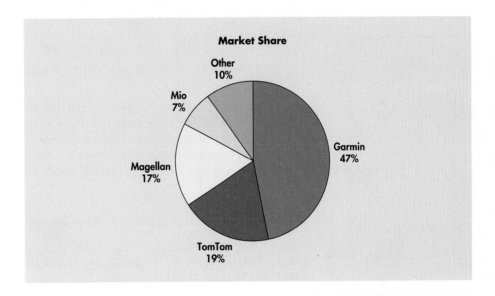

To make a pie chart, you must include all the categories that make up a whole. A category such as "Other" in this example can be used, but the sum of the percents for all of the categories should be 100%.

Bar graphs are more flexible. For example, you can use a bar graph to compare the numbers of students at your college majoring in biology, business, and political science. A pie chart cannot make this comparison, because not all students fall into one of these three majors.

We use graphical displays to help us learn things from data. Here is another example.

EXAMPLE 1.10   The Cost Is $164 Billion!

DATA FILE
CRASHES

Auto accidents cost $164 billion each year.[3] How can this enormous burden on the economy be reduced? Let's look at some data.[4] Figure 1.5 is a bar graph that gives the percents of auto accidents for each day of the week. What do we learn from this graph? The highest percent is on Saturday, about 17%, and the lowest is on Monday, about 10%. If we were to seek government funding for a program to reduce accidents, we might do some research on the Saturday accidents.

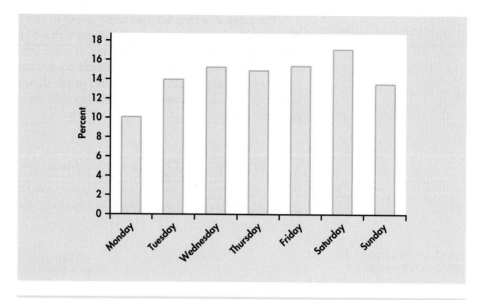

The categories in Figure 1.5 are ordered by the days of the week, Monday through
Sunday. In exploring what these data tell us about accidents, we focused on the day of the
week with the highest percent of accidents. Let's pursue this idea a little further and order
the categories from highest percent to lowest percent. A bar graph whose categories are
ordered from most frequent to least frequent is called a **Pareto chart.**[5]

Pareto chart

CRASHES

EXAMPLE 1.11 Pareto Chart for Automobile Accidents

Figure 1.6 displays the Pareto chart for the automobile accident data. Here it is easy to see that
Saturday is the highest. Friday, Wednesday, and Thursday are also relatively high. Tuesday and
Sunday are a bit lower. Monday is the lowest.

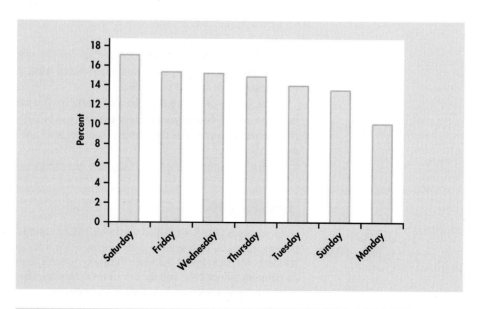

Pareto charts are frequently used in quality control settings. Here, the purpose is often to identify common types of defects in a manufactured product. Deciding upon strategies for corrective action can then be based on what would be most effective. Chapter 12 gives more examples of settings where Pareto charts are used.

Bar graphs, pie charts, and Pareto charts help an audience grasp a distribution quickly. When you prepare them, keep in mind this purpose. We will move on to quantitative variables, where graphs are essential tools.

CANADIAN POPULATION

APPLY YOUR KNOWLEDGE

1.6 Population of Canadian provinces and territories. Here are populations of 13 Canadian provinces and territories based on the 2006 census:[6]

Province/territory	Population
Alberta	3,290,350
British Columbia	4,113,487
Manitoba	1,148,401
New Brunswick	729,997
Newfoundland and Labrador	505,469
Northwest Territories	41,464
Nova Scotia	913,462
Nunavut	29,474
Ontario	12,160,282
Prince Edward Island	135,851
Quebec	7,546,131
Saskatchewan	968,157
Yukon	30,372

(a) Display these data in a bar graph using the alphabetical order of provinces and territories in the table.
(b) Use a Pareto chart to display these data.
(c) Compare the two graphs. Which do you prefer? Give a reason for your answer.

GPSEUROPE

1.7 GPS market share in Europe. In Examples 1.7 to 1.9 (pages 8 to 9), we examined the U.S. market share of several companies that sell GPS receivers. Here is a similar table for the European market:[7]

Company	Market share (%)
TomTom	38
Other	26
Garmin	19

(a) Display the data in a bar graph. Be sure to choose the ordering for the companies carefully. Explain why you made this choice.

(b) Compare this graph with the bar graph in Figure 1.3. Garmin has its world headquarters in Olathe, Kansas, while TomTom's registered address is Amsterdam, the Netherlands. Explain how this information helps you to understand the differences between the two bar graphs.

Quantitative variables: histograms

histogram

Quantitative variables often take many values. A graph of the distribution is clearer if nearby values are grouped together. The most common graph of the distribution of a single quantitative variable is a **histogram.**

TBILLRATES

CASE 1.1

Treasury Bills Treasury bills, also known as T-bills, are bonds issued by the U.S. Department of the Treasury. You buy them at a discount from their face value, and they mature in a fixed period of time. For example, you might buy a $1000 T-bill for $980. When it matures, six months later, you would receive $1000—your original $980 investment plus $20 interest. This interest rate is $20 divided by $980, which is 2.04% for six months. Interest is usually reported as a rate per year, so for this example the interest rate would be 4.08%. Rates are determined by an auction that is held every four weeks. The data set contains the interest rates for T-bills for each auction from December 12, 1958, to October 3, 2008.[8]

Our data set contains 2600 cases. The two variables in the data set are the date of the auction and the interest rate. To learn something about T-bill interest rates, we begin with a histogram.

EXAMPLE 1.12   A Histogram of T-bill Interest Rates

CASE 1.1

classes

To make a histogram of the T-bill interest rates, we proceed as follows.

Step 1. Divide the range of the interest rates into **classes** of equal width. The T-bill interest rates range from 0.85% to 15.76%, so we choose as our classes

TBILLRATES

$$0.00 \leq \text{rate} < 2.00$$
$$2.00 \leq \text{rate} < 4.00$$
$$\vdots$$
$$14.00 \leq \text{rate} < 16.00$$

Be sure to specify the classes precisely so that each case falls into exactly one class. An interest rate of 1.98% would fall into the first class, but 2.00% would falls into the second.

Step 2. Count the number of cases in each class. Here are the counts:

Class	Count	Class	Count
$0.00 \leq$ rate < 2.00	178	$8.00 \leq$ rate < 10.00	235
$2.00 \leq$ rate < 4.00	575	$10.00 \leq$ rate < 12.00	64
$4.00 \leq$ rate < 6.00	951	$12.00 \leq$ rate < 14.00	58
$6.00 \leq$ rate < 8.00	501	$14.00 \leq$ rate < 16.00	38

Step 3. Draw the histogram. Mark on the horizontal axis the scale for the variable whose distribution you are displaying. The variable is "interest rate" in this example. The scale runs from 0 to 16 to span the data. The vertical axis contains the scale of counts. Each bar represents a class.

The base of the bar covers the class, and the bar height is the class count. Notice that the scale on the vertical axis runs from 0 to 1000 to accommodate the tallest bar, which has a height of 951. There is no horizontal space between the bars unless a class is empty, so that its bar has height zero. Figure 1.7 is our histogram.

FIGURE 1.7 Histogram for T-bill interest rates, for Example 1.12.

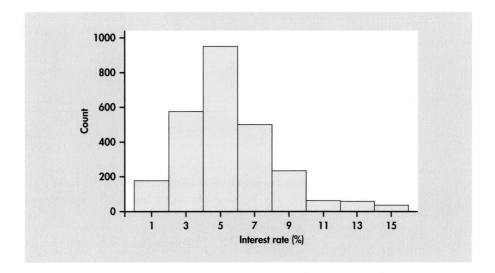

Our eyes respond to the *area* of the bars in a histogram.[9] Because the classes are all the same width, area is determined by height and all classes are fairly represented. There is no one right choice of the classes in a histogram. Too few classes will give a "skyscraper" graph, with all values in a few classes with tall bars. Too many will produce a "pancake" graph, with most classes having one or no observations. Neither choice will give a good picture of the shape of the distribution. You must always use your judgment in choosing classes to display the shape. Statistics software will choose the classes for you. The computer's choice is usually a good one. Sometimes, however, the classes chosen by software differ from the natural choices that you would make. Usually, options are available for you to change them. The next example illustrates a situation where the wrong choice of classes will cause you to miss a very important characteristic of a data set.

EXAMPLE 1.13   Calls to a Customer Service Center

CALLCENTER80

Many businesses operate call centers to serve customers who want to place an order or make an inquiry. Customers want their requests handled thoroughly. Businesses want to treat customers well, but they also want to avoid wasted time on the phone. They therefore monitor the length of calls and encourage their representatives to keep calls short.

We have data on the length of all 31,492 calls made to the customer service center of a small bank in a month. Table 1.1 displays the lengths of the first 80 calls.[10]

Take a look at the data in Table 1.1. In this data set the *cases* are calls made to the bank's call center. The *variable* recorded is the length of each call. The *units of measurement* are seconds. We see that the call lengths vary a great deal. The longest call lasted 2631 seconds, almost 44 minutes. More striking is that 8 of these 80 calls lasted less than 10 seconds. What's going on?

TABLE 1.1   Service times (seconds) for calls to a customer service center

77	289	128	59	19	148	157	203
126	118	104	141	290	48	3	2
372	140	438	56	44	274	479	211
179	1	68	386	2631	90	30	57
89	116	225	700	40	73	75	51
148	9	115	19	76	138	178	76
67	102	35	80	143	951	106	55
4	54	137	367	277	201	52	9
700	182	73	199	325	75	103	64
121	11	9	88	1148	2	465	25

We started our study of the customer service center data by examining a few cases, the ones displayed in Table 1.1. It would be very difficult to examine all 31,492 cases in this way. We need a better method. Let's try a histogram.

EXAMPLE 1.14   Histogram for Customer Service Center Call Lengths

CALLCENTER

Figure 1.8 is a histogram of the lengths of all 31,492 calls. We did not plot the few lengths greater than 1200 seconds (20 minutes). As expected, the graph shows that most calls last between about 1 and 5 minutes, with some lasting much longer when customers have complicated problems. More striking is the fact that 7.6% of all calls are no more than 10 seconds long. It turns out that the bank penalized representatives whose average call length was too long—so some representatives just hung up on customers in order to bring their average length down. Neither the customers nor the bank were happy about this. The bank changed its policy, and later data showed that calls under 10 seconds had almost disappeared.

FIGURE 1.8 The distribution of call lengths for 31,492 calls to a bank's customer service center, for Example 1.14. The data show a surprising number of very short calls. These are mostly due to representatives deliberately hanging up in order to bring down their average call length.

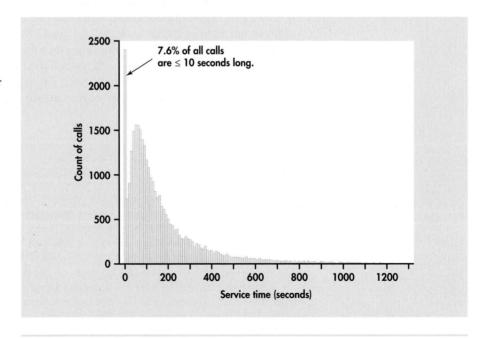

, The choice of the classes is an important part of making a histogram. Let's look at the customer service center call lengths again.

EXAMPLE 1.15 Another Histogram for Customer Service Center Call Lengths

CALLCENTER

Figure 1.9 is a histogram of the lengths of all 31,492 calls with class boundaries of 0, 100, 200, etc. seconds. Statistical software made this choice as a default option. Notice that the spike representing the very brief calls that appears in Figure 1.8 is covered up in the 0 to 100 seconds class in Figure 1.9.

FIGURE 1.9 The "default" histogram produced by software for the call lengths, for Example 1.15. This choice of classes hides the large number of very short calls that is revealed by the histogram of the same data in Figure 1.8.

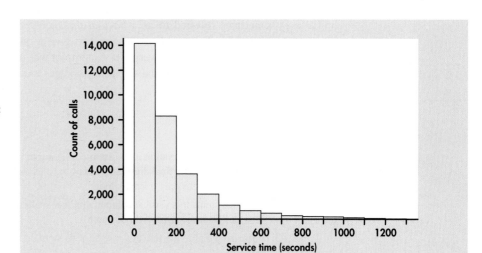

If we let software choose the classes, we would miss one of the most important features of the data, the calls of very short duration. We were alerted to this unexpected characteristic of the data by our examination of the 80 cases displayed in Table 1.1. Beware of letting statistical software do your thinking for you. Example 1.15 illustrates the danger of doing this. To do an effective analysis of data, we often need to look at data in more than one way. For histograms, looking at several choices of classes will lead us to a good choice. Fortunately, with software, examining choices such as this is relatively easy.

APPLY YOUR KNOWLEDGE

1.8 Exam grades in a statistics course. The table below summarizes the exam scores of students in an introductory statistics course. Use the summary to sketch a histogram that shows the distribution of scores.

Class	Count
$60 \leq$ score < 70	11
$70 \leq$ score < 80	36
$80 \leq$ score < 90	57
$90 \leq$ score < 100	29

1.9 Suppose some students scored 100. No students earned a perfect score of 100 on the exam described in the previous exercise. Note that the last class included only scores that were greater than or equal to 90 and *less than* 100. Explain how you would change the class definitions for a similar exam on which some students earned a perfect score.

Quantitative variables: stemplots

Histograms are not the only graphical display of distributions of quantitative variables. For small data sets, a *stemplot* is quicker to make and presents more detailed information. It is sometimes referred to as a *back-of-the-envelope* technique. Popularized by the statistician John Tukey, it was designed to give a quick and informative look at the distribution of a quantitative variable. A stemplot was originally designed to be made by hand, although many statistical software packages include this capability.

> **Stemplot**
>
> To make a **stemplot:**
>
> 1. Separate each observation into a **stem** consisting of all but the final (rightmost) digit and a **leaf,** the final digit. Stems may have as many digits as needed, but each leaf contains only a single digit.
>
> 2. Write the stems in a vertical column with the smallest at the top, and draw a vertical line at the right of this column.
>
> 3. Write each leaf in the row to the right of its stem, in increasing order out from the stem.

EXAMPLE 1.16   A Stemplot of T-bill Interest Rates

CASE 1.1

TBILLRATES50

The histogram that we produced in Example 1.12 to examine the T-bill interest rates used all 2600 cases in the data set. To illustrate the idea of a stemplot, we will take a simple random sample of size 50 from this data set. We will learn more about how to take such samples in Chapter 3. Here are the data:

7.2	5.7	6.0	5.0	12.8	7.8	11.6	4.6	2.7	4.9
5.8	13.8	1.5	4.6	3.7	8.3	7.0	3.2	5.8	1.0
7.2	8.0	3.2	7.5	5.4	5.3	6.9	5.8	5.0	9.4
10.4	4.3	6.8	1.0	5.5	5.1	4.6	6.6	4.7	6.1
5.7	1.0	3.8	7.3	6.5	3.0	3.9	8.0	3.0	7.9

The original data set gave the interest rates with two digits after the decimal point. To make the job of preparing our stemplot easier, we first rounded the values to one place following the decimal.

Figure 1.10 illustrates the key steps in constructing the stemplot for these data. How does the stemplot for this sample of size 50 compare with the histogram based on all 2600 interest rates that we examined in Figure 1.7 (page 13)?

rounding

You can choose the classes in a histogram. The classes (the stems) of a stemplot are given to you. When the observed values have many digits, it is often best to **round** the numbers to just a few digits before making a stemplot, as we did in Example 1.16.

FIGURE 1.10 Steps in creating a stemplot for the sample of 50 T-bill interest rates, for Example 1.16. (a) Write the stems in a column, from smallest to largest, and draw a vertical line to their right. (b) Add each leaf to the right of its stem. (c) Arrange each leaf in increasing order out from its stem.

(a)	(b)	(c)
1	1 | 5 0 0 0	1 | 0 0 0 5
2	2 | 7	2 | 7
3	3 | 7 2 2 8 0 9 0	3 | 0 0 2 2 7 8 9
4	4 | 6 9 6 3 6 7	4 | 3 6 6 6 7 9
5	5 | 7 0 8 8 4 3 8 0 5 1 7	5 | 0 0 1 3 4 5 7 7 8 8 8
6	6 | 0 9 8 6 1 5	6 | 0 1 5 6 8 9
7	7 | 2 8 0 2 5 3 9	7 | 0 2 2 3 5 8 9
8	8 | 3 0 0	8 | 0 0 3
9	9 | 4	9 | 4
10	10 | 4	10 | 4
11	11 | 6	11 | 6
12	12 | 8	12 | 8
13	13 | 8	13 | 8

splitting stems

You can also **split stems** to double the number of stems when all the leaves would otherwise fall on just a few stems. Each stem then appears twice. Leaves 0 to 4 go on the upper stem and leaves 5 to 9 go on the lower stem. Rounding and splitting stems are matters for judgment, like choosing the classes in a histogram. Stemplots work well for small sets of data. When there are more than 100 observations, a histogram is almost always a better choice.

Special considerations apply for very large data sets. It is often useful to take a sample and examine it in detail as a first step. This is what we did in Example 1.16. Sampling can be done in many different ways. A company with a very large number of customer records, for example, might look at those from a particular region or country for an initial analysis.

Interpreting histograms and stemplots

Making a statistical graph is not an end in itself. The purpose of the graph is to help us understand the data. After you make a graph, always ask, "What do I see?" Once you have displayed a distribution, you can see its important features as follows.

> **Examining a Distribution**
>
> In any graph of data, look for the **overall pattern** and for striking **deviations** from that pattern.
>
> You can describe the overall pattern of a histogram by its **shape, center,** and **spread.**
>
> An important kind of deviation is an **outlier,** an individual value that falls outside the overall pattern.

We will learn how to describe center and spread numerically in Section 1.2. For now, we can describe the center of a distribution by its *midpoint,* the value with roughly half the observations taking smaller values and half taking larger values. We can describe the spread of a distribution by giving the *smallest and largest values.*

EXAMPLE 1.17   The Distribution of T-bill Interest Rates

TBILLRATES50

Let's look again at the histogram in Figure 1.7. There appear to be some relatively large interest rates. The largest is 15.76%. What do we think about this value? Is it so extreme relative to the other values that we would call it an *outlier?* To qualify for this status an observation should stand apart from the other observations either alone or with very few other cases. A careful examination of the data indicates that this 15.76% does not qualify for outlier status. There are interest rates of 15.72%, 15.68%, and 15.58%. In fact, there are 15 auctions with interest rates of 15% or higher.

The distribution has a *single peak* at around 5%. The distribution is somewhat *right-skewed*—that is, the right tail extends farther from the peak than does the left tail.

When you describe a distribution, concentrate on the main features. Look for major peaks, not for minor ups and downs in the bars of the histogram. Look for clear outliers, not just for the smallest and largest observations. Look for rough *symmetry* or clear *skewness.*

> **Symmetric and Skewed Distributions**
>
> A distribution is **symmetric** if the right and left sides of the histogram are approximately mirror images of each other.
>
> A distribution is **skewed to the right** if the right side of the histogram (containing the half of the observations with larger values) extends much farther out than the left side. It is **skewed to the left** if the left side of the histogram extends much farther out than the right side. We also use the term **"skewed toward large values"** for distributions that are skewed to the right. This is the most common type of skewness seen in real data.

EXAMPLE 1.18   IQ Scores of Fifth-Grade Students

IQ

Figure 1.11 displays a histogram of the IQ scores of 60 fifth-grade students. There is a single peak around 110 and the distribution is approximately symmetric. The tails decrease smoothly as we move away from the peak. Measures such as this are usually constructed so that they have *nice* distributions like the one shown in Figure 1.11.

FIGURE 1.11 Histogram of the IQ scores of 60 fifth-grade students, for Example 1.18.

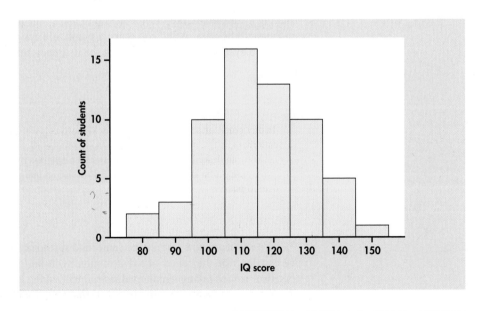

The overall shape of a distribution is important information about a variable. Some types of data regularly produce distributions that are symmetric or skewed. For example, data on the diameters of ball bearings produced by a manufacturing process tend to be symmetric. Data on incomes (whether of individuals, companies, or nations) are usually strongly skewed to the right. There are many moderate incomes, some large incomes, and a few very large incomes. Do remember that many distributions have shapes that are neither symmetric nor skewed. Some data show other patterns. Scores on an exam, for example, may have a cluster near the top of the scale if many students did well. Or they may show two distinct peaks if a tough problem divided the class into those who did and didn't solve it. Use your eyes and describe what you see.

APPLY YOUR KNOWLEDGE

1.10 Make a stemplot. Make a stemplot for a distribution that has a single peak, approximately symmetric with one high and two low outliers.

1.11 Make another one. Make a stemplot of a distribution that is skewed toward large values.

Time plots

Many variables are measured at intervals over time. We might, for example, measure the cost of raw materials for a manufacturing process each month or the price of a stock at the end of each day. In these examples, our main interest is change over time. To display change over time, make a *time plot*.

> **Time Plot**
>
> A **time plot** of a variable plots each observation against the time at which it was measured. Always put time on the horizontal scale of your plot and the variable you are measuring on the vertical scale. Connecting the data points by lines helps emphasize any change over time.

More details about how to analyze data that vary over time are given in Chapter 13, "Time Series Forecasting." For now, we will examine how a time plot can reveal some additional important information about T-bill interest rates.

EXAMPLE 1.19   A Time Plot for T-bill Interest Rates

CASE 1.1   The Web site of the Federal Reserve Bank of St. Louis provided a very interesting graph of T-bill interest rates.[11] It is shown in Figure 1.12. A time plot shows us the relationship between two variables, in this case interest rate and the auctions that occurred at four-week intervals. Notice how the Federal Reserve Bank included information about a third variable in this plot. The third variable is a categorical variable that indicates whether or not the United States was in a recession. It is indicated by the shaded areas in the plot.

APPLY YOUR KNOWLEDGE

CASE 1.1   **1.12 What does the time plot show?** Carefully examine the time plot in Figure 1.12.
(a) How do the T-bill interest rates vary over time?
(b) What can you say about the relationship between the rates and the recession periods?

FIGURE 1.12 Time plot for the
T-bill interest rates, for
Example 1.19.

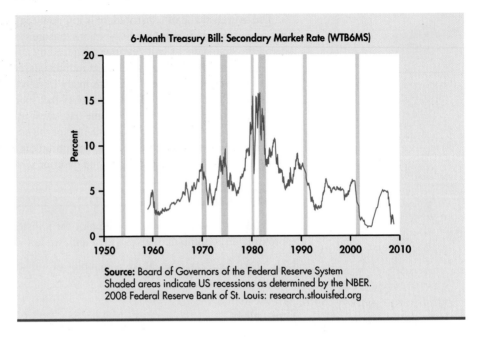

In Example 1.12 (page 12) we examined the distribution of T-bill interest rates for the period December 12, 1958, to October 3, 2008. The histogram in Figure 1.7 showed us the shape of the distribution. By looking at the time plot in Figure 1.12, we now see that there is more to this data set than is revealed by the histogram. This scenario illustrates the types of steps used in an effective statistical analysis of data. We are rarely able to completely plan our analysis in advance, set up the appropriate steps to be taken, and then click on the appropriate buttons in a software package to obtain useful results. An effective analysis requires that we proceed in an organized way, use a variety of analytical tools as we proceed, and exercise careful judgment at each step in the process.

SECTION 1.1 Summary

- A data set contains information on a number of **cases.** Cases may be people, animals, or things. For each case, the data give values for one or more **variables.** A variable describes some characteristic of an individual, such as a person's height, gender, or salary. Variables can have different **values** for different cases.

- Some variables are **categorical** and others are **quantitative.** A categorical variable places each case into a category, such as male or female. A quantitative variable has numerical values that measure some characteristic of each case, such as height in centimeters or salary in dollars per year.

- **Exploratory data analysis** uses graphs and numerical summaries to describe the variables in a data set and the relations among them.

- The **distribution** of a variable describes what values the variable takes and how often it takes these values.

- To describe a distribution, begin with a graph. **Bar graphs** and **pie charts** describe the distribution of a categorical variable, and **Pareto charts** identify the most important categories for a categorical variable. **Histograms** and **stemplots** graph the distributions of quantitative variables.

- When examining any graph, look for an **overall pattern** and for notable **deviations** from the pattern.

- **Shape, center,** and **spread** describe the overall pattern of a distribution. Some distributions have simple shapes, such as **symmetric** and **skewed.** Not all distributions have a simple overall shape, especially when there are few observations.

- **Outliers** are observations that lie outside the overall pattern of a distribution. Always look for outliers and try to explain them.

- When observations on a variable are taken over time, make a **time plot** that graphs time horizontally and the values of the variable vertically. A time plot can reveal interesting patterns in a set of data.

SECTION 1.1 Exercises

For Exercise 1.1, see page 4; for 1.2 to 1.4, see page 5; for 1.5, see page 6; for 1.6 and 1.7, see pages 11–12; for 1.8 and 1.9, see pages 15–16; for 1.10 and 1.11, see page 19; and for 1.12, see page 19.

1.13 Employee application data. The personnel department keeps records on all employees in a company. Here is the information that they keep in one of their data files: employee identification number, last name, first name, middle initial, department, number of years with the company, salary, education (coded as high school, some college, or college degree), and age.
(a) What are the cases for this data set?
(b) Identify each item kept in the data files as a label, a quantitative variable, or a categorical variable.
(c) Set up a spreadsheet that could be used to record the data. Give appropriate column headings and five sample cases.

1.14 Where should you locate your business? You are interested in choosing a new location for your business. Create a list of criteria that you would use to rank cities. Include at least eight variables and give reasons for your choices. Classify each variable as quantitative or categorical.

1.15 Survey of students. A survey of students in an introductory statistics class asked the following questions: (a) age; (b) do you like to dance? (yes, no); (c) can you play a musical instrument (not at all, a little, pretty well); (d) how much did you spend on food last week? (e) height; (f) do you like broccoli? (yes, no). Classify each of these variables as categorical or quantitative and give reasons for your answers.

1.16 What questions would you ask? Refer to the previous exercise. Make up your own survey questions with at least six questions. Include at least two categorical variables and at least two quantitative variables. Tell which variables are categorical and which are quantitative. Give reasons for your answers.

1.17 Study habits of students. You are planning a survey to collect information about the study habits of college students. Describe two categorical variables and two quantitative variables that you might measure for each student. Give the units of measurement for the quantitative variables.

1.18 What color should you use for your product? What is your favorite color? One survey produced the following summary of responses to that question: blue, 42%; green, 14%; purple, 14%; red, 8%; black, 7%; orange, 5%; yellow, 3%; brown, 3%; gray, 2%; and white, 2%.[12] Make a bar graph of the percents and write a short summary of the major features of your graph.
FAVORITECOLORS

1.19 Least-favorite colors. Refer to the previous exercise. The same study also asked people about their least-favorite color. Here are the results: orange, 30%; brown, 23%; purple, 13%; yellow, 13%; gray, 12%; green, 4%; white, 4%; red, 1%; black, 0%; and blue, 0%. Make a bar graph of these percents and write a summary of the results.   LEASTFAVORITECOLORS

1.20 Market share doubles in a year. The market share of iPhones doubled from 5.3% to 10.8% between the first quarter of 2008 and the first quarter of 2009.[13] One of the attractions of the iPhone is the Web browser, which they market as the most advanced Web browser on a mobile device. Users of iPhones were asked to respond to the statement "I do a lot more browsing on the iPhone than I did on my previous mobile phone." Here are the results:[14]

Response	Percent
Strongly agree	54
Mildly agree	22
Mildly disagree	16
Strongly disagree	8

(a) Make a bar graph to display the distribution of the responses.
(b) Display the distribution with a pie chart.
(c) Summarize the information in these charts.
(d) Do you prefer the bar graph or the pie chart? Give a reason for your answer.   BROWSING

1.21 What did the iPhone replace? The survey in the previous exercise also asked iPhone users what phone, if any, did the iPhone replace. Here are the responses:

Response	Percent	Response	Percent
Motorola Razr	23.8	Blackberry	13.0
Symbian	3.9	Windows Mobile	13.9
Sidekick	4.1	Replaced nothing	10.0
Palm	6.7	Other phone	24.5

Make a bar graph for these data. Carefully consider how you will order the responses. Explain why you chose the ordering that you did. 🌀 PHONEREPLACEMENT

1.22 Garbage is big business. The formal name for garbage is "municipal solid waste." In the United States, approximately 254 million tons of garbage are generated in a year. Below is a breakdown of the materials that made up American municipal solid waste in 2007.[15] 🌀 GARBAGE

Material	Weight (million tons)	Percent of total
Food scraps	31.7	12.5
Glass	13.6	5.3
Metals	20.8	8.2
Paper, paperboard	83.0	32.7
Plastics	30.7	12.1
Rubber, leather, textiles	19.4	7.6
Wood	14.2	5.6
Yard trimmings	32.6	12.8
Other	8.2	3.2
Total	254.1	100.0

(a) Add the weights. The sum is not exactly equal to the value of 254.1 million tons given in the table. Why?
(b) Make a bar graph of the percents. The graph gives a clearer picture of the main contributors to garbage if you order the bars from tallest to shortest.
(c) Also make a pie chart of the percents. Comparing the two graphs, notice that it is easier to see the small differences among "Food scraps," "Plastics," and "Yard trimmings" in the bar graph.

1.23 Market share for search engines. The following table gives the market share for the major search engines.[16] 🌀 SEARCHENGINES

Search engine	Market share	Search engine	Market share
Google-Global	79.9%	Microsoft Live Search	1.6%
Yahoo-Global	11.3%		
MSN-Global	3.4%	Ask-Global	1.2%
AOL-Global	2.4%	Other	0.2%

(a) Use a bar graph to display the market shares.
(b) Summarize what the graph tells you about market shares for search engines.

1.24 Market share for computer operating systems. The following table gives the market share for the major computer operating systems.[17] 🌀 OPERATINGSYSTEMS

Operating system	Market share	Operating system	Market share
Windows	90.29%	Playstation	0.03%
Mac	8.23%	SunOS	0.01%
Linux	0.91%	Other	0.21%
iPhone	0.32%		

(a) Make a bar graph of this market share data.
(b) Write a short paragraph summarizing these data.

1.25 Your Facebook app can generate a million dollars a month. A report on Facebook suggests that Facebook apps can generate large amounts of money, as much as one million dollars a month.[18] The market is international. The following table gives the numbers of Facebook users by country for the top 20 countries (excluding the United States) as of September 29, 2008.[19] 🌀 FACEBOOKBYCOUNTRY

Country	Facebook users (in millions)	Country	Facebook users (in millions)
United Kingdom	11.39	Venezuela	1.01
Canada	9.51	South Africa	0.97
Turkey	3.50	Hong Kong	0.91
Australia	3.36	Egypt	0.80
Colombia	2.69	Denmark	0.79
Chile	2.46	Spain	0.77
France	2.45	India	0.77
Norway	1.14	Germany	0.70
Sweden	1.14	Israel	0.61
Mexico	1.01	Italy	0.57

(a) Use a bar graph to describe these data.
(b) Describe the major features of your chart in a short paragraph.

1.26 Facebook use increases, by country. Facebook use has been increasing rapidly. Data are available on the increases between February 8, 2008, and September 29, 2008.[20] 🌀 FACEBOOKINCREASES The table below gives the percent increase in the numbers of Facebook users for the same 20 countries that we studied in the previous exercise. Note that there is no entry for Hong Kong, because the number of users as of February 8, 2008, is not reported.

Country	Increase in Facebook users	Country	Increase in Facebook users
United Kingdom	31%	Venezuela	683%
Canada	9%	South Africa	33%
Turkey	23%	Hong Kong	
Australia	43%	Egypt	31%
Colombia	246%	Denmark	92%
Chile	2197%	Spain	132%
France	92%	India	42%
Norway	7%	Germany	44%
Sweden	4%	Israel	42%
Mexico	69%	Italy	139%

(a) Summarize the data by carefully examining the table. Are there any extreme outliers? Which ones would you classify in this way?

(b) Use a stemplot to describe these data. You can list any extreme outliers separately from the plot.

(c) Describe the major features of these data using your plot and your list of outliers.

(d) How effective is the stemplot for summarizing these data? Give reasons for your answer.

1.27 U.S. unemployment rates. An unemployment rate is the number of people who are not working but who are available for work divided by the total number of people in the workforce, expressed as a percent. Table 1.2 gives the U.S. unemployment rates for each state as of August 2008.[21] **UNEMPLOYMENT**

(a) Construct a histogram of these rates.

(b) Prepare a stemplot of the rates.

(c) Discuss the advantages and disadvantages of (a) and (b). Which do you prefer for this set of data? Explain your answer.

1.28 Unemployment rates in Canadian provinces. Here are 2007 unemployment rates for 10 Canadian provinces:[22] **UNEMPLOYMENTCANADA**

Province	Unemployment rate
Alberta	3.5%
British Columbia	4.2%
Manitoba	4.4%
New Brunswick	7.5%
Newfoundland and Labrador	13.6%
Nova Scotia	8.0%
Ontario	6.4%
Prince Edward Island	10.3%
Quebec	7.2%
Saskatchewan	4.2%

(a) Construct a histogram of these rates.

(b) Prepare a stemplot of the rates.

(c) Discuss the advantages and disadvantages of (a) and (b). Which do you prefer for this set of data? Explain your answer.

TABLE 1.2   Unemployment rates by state, August 2008

State	Rate	State	Rate	State	Rate
Alabama	4.9	Louisiana	4.7	Ohio	7.4
Alaska	6.9	Maine	5.5	Oklahoma	4.0
Arizona	5.6	Maryland	4.5	Oregon	6.5
Arkansas	4.8	Massachusetts	5.3	Pennsylvania	5.8
California	7.7	Michigan	8.9	Rhode Island	8.5
Colorado	5.4	Minnesota	6.2	South Carolina	7.6
Connecticut	6.5	Mississippi	7.7	South Dakota	3.3
Delaware	4.9	Missouri	6.6	Tennessee	6.6
Florida	6.5	Montana	4.4	Texas	5.0
Georgia	6.3	Nebraska	3.5	Utah	3.7
Hawaii	4.2	Nevada	7.1	Vermont	4.9
Idaho	4.6	New Hampshire	4.2	Virginia	4.6
Illinois	7.3	New Jersey	5.9	Washington	6.0
Indiana	6.4	New Mexico	4.6	West Virginia	4.1
Iowa	4.6	New York	5.8	Wisconsin	5.1
Kansas	4.7	North Carolina	6.9	Wyoming	3.9
Kentucky	6.8	North Dakota	3.6		

1.29 Vehicle colors. Vehicle colors differ among types of vehicle. Here are data on the most popular colors in 2007 for luxury cars and for intermediate-price cars in North America:[23]

VEHICLECOLORS

Color	Luxury car (%)	Intermediate-price car (%)
Black	22	10
Silver	16	25
White Pearl	14	4
Gray	12	12
White	11	8
Blue	7	13
Red	7	10
Yellow/Gold	6	4
Other	5	14

(a) Make a bar graph for the luxury car percents.

(b) Make a bar graph for the intermediate-price car percents.

(c) Now, be creative: make *one* bar graph that compares the two vehicle types as well as comparing colors. Arrange your graph so that it is easy to compare the two types of vehicle.

1.30 Procter & Gamble sales. The 2007 annual report of the Procter & Gamble Company (P&G) states that global net sales were over $76 billion. The sales information is organized into global segments. The following summary gives the net sales for each global segment of P&G:[24]   PANDGSALES

Segment	Net sales ($ millions)
Beauty	22,981
Health care	8,964
Fabric care and home care	18,971
Baby care and family care	12,726
Snacks, coffee, and pet care	4,537
Blades and razors	5,229
Duracell and Braun	4,031

Summarize these data graphically and write a paragraph describing the net sales of P&G.

1.31 Products for senior citizens. The market for products designed for senior citizens in the United States is expanding. Here is a stemplot of the percents of residents aged 65 and older in the 50 states, for 2006, as estimated by the U.S. Census Bureau.[25] The stems are whole percents and the leaves are tenths of a percent.   POPOVER65BYSTATE

```
 7 | 0
 8 | 8
 9 | 9
10 | 01
11 | 0177889
12 | 122225567799
13 | 0000112333345556699
14 | 033678
15 | 25
16 |
17 | 0
```

(a) There is an outlier: Florida has the highest percent of residents aged 65 and older and clearly stands out. Alaska has the lowest percent, but it is at the end of a relatively flat tail on the low end of the distribution. What are the percents for these two states?

(b) Describe the shape, center, and spread of this distribution.

1.32 U.S. population 65 and older. Make a stemplot of the percent of residents aged 65 and older in the states other than Alaska and Florida by splitting stems 8 to 15 in the plot from the previous exercise. Which plot do you prefer? Why?   POPOVER65BYSTATE

1.33 The Canadian market. Refer to Exercise 1.31. Here are similar data for the 13 Canadian provinces and territories:[26]   CANADIANPOPULATION

Province/Territory	Percent over 65
Alberta	10.7
British Columbia	14.6
Manitoba	14.1
New Brunswick	14.7
Newfoundland and Labrador	13.9
Northwest Territories	4.8
Nova Scotia	15.1
Nunavut	2.7
Ontario	13.6
Prince Edward Island	14.9
Quebec	14.3
Saskatchewan	15.4
Yukon	7.5

(a) Display the data graphically and describe the major features of your plot.

(b) Explain why you chose the particular format for your graphical display. What other types of graph could you have used? What are the strengths and weaknesses of each for displaying this set of data?

1.34 Left-skew. Sketch a histogram for a distribution that is skewed to the left. Suppose that you and your friends emptied your pockets of coins and recorded the year marked on each coin.

The distribution of dates would be skewed to the left. Explain why.

1.35 Is the supply adequate? How much oil the wells in a given field will ultimately produce is key information in deciding whether to drill more wells. Here are the estimated total amounts of oil recovered from 64 wells in the Devonian Richmond Dolomite area of the Michigan basin, in thousands of barrels:[27] 🌀 **OILWELLS**

21.7	53.2	46.4	42.7	50.4	97.7	103.1	51.9	43.4	69.5
156.5	34.6	37.9	12.9	2.5	31.4	79.5	26.9	18.5	14.7
32.9	196.0	24.9	118.2	82.2	35.1	47.6	54.2	63.1	69.8
57.4	65.6	56.4	49.4	44.9	34.6	92.2	37.0	58.8	21.3
36.6	64.9	14.8	17.6	29.1	61.4	38.6	32.5	12.0	28.3
204.9	44.5	10.3	37.7	33.7	81.1	12.1	20.1	30.5	7.1
10.1	18.0	3.0	2.0						

Graph the distribution and describe its main features.

1.36 The changing age distribution of the United States. The distribution of the ages of a nation's population has a strong influence on economic and social conditions. Table 1.3 shows the age distribution of U.S. residents in 1950 and 2075, in millions of people. The 1950 data come from that year's census, while the 2075 data are projections made by the Census Bureau. 🌀 **USPOPULATION**
(a) Because the total population in 2075 is much larger than the 1950 population, comparing percents in each age group is clearer than comparing counts. Make a table of the percent of the total population in each age group for both 1950 and 2075.
(b) Make a histogram with vertical scale in percents of the 1950 age distribution. Describe the main features of the distribution.

In particular, look at the percent of children relative to the rest of the population.
(c) Make a histogram with vertical scale in percents of the projected age distribution for the year 2075. Use the same scales as in (b) for easy comparison. What are the most important changes in the U.S. age distribution projected for the years between 1950 and 2075?

1.37 Reliability of household appliances. Always ask whether a particular variable is really a suitable measure for your purpose. You are writing an article for a consumer magazine based on a survey of the magazine's readers on the reliability of their household appliances. Of 13,376 readers who reported owning Brand A dishwashers, 2942 required a service call during the past year. Only 192 service calls were reported by the 480 readers who owned Brand B dishwashers.
(a) Why is the count of service calls (2942 versus 192) not a good measure of the reliability of these two brands of dishwashers?
(b) Use the information given to calculate a suitable measure of reliability. What do you conclude about the reliability of Brand A and Brand B?

1.38 A multimillion-dollar business is threatened. Bristol Bay of Alaska, has typically produced more wild-caught sockeye salmon, *Oncorhynchus nerka,* than any other region in the world. In good years, the runs typically exceed 50 million fish. The sockeye salmon industry here provides thousands of jobs and generates millions of dollars per year.[28] Here are the numbers of sockeye salmon in runs at Bristol Bay between 1988 and 2007:[29] 🌀 **BERINGSEAFISH**

© Natalie Fobes/Corbis

TABLE 1.3	Age distribution in the United States, 1950 and 2075 (in millions of persons)	
Age group	**1950**	**2075**
Under 10 years	29.3	53.3
10–19 years	21.8	53.2
20–29 years	24.0	51.2
30–39 years	22.8	50.5
40–49 years	19.3	47.5
50–59 years	15.5	44.8
60–69 years	11.0	40.7
70–79 years	5.5	30.9
80–89 years	1.6	21.7
90–99 years	0.1	8.8
100–109 years	0.0	1.1
Total	151.1	403.7

Year	Runs (millions)	Year	Runs (millions)	Year	Runs (millions)	Year	Runs (millions)
1988	22.9	1993	52.7	1998	18.1	2003	26.5
1989	44.5	1994	50.3	1999	39.5	2004	43.5
1990	47.1	1995	60.8	2000	28.4	2005	39.3
1991	42.0	1996	37.0	2001	22.0	2006	43.1
1992	45.6	1997	18.8	2002	17.2	2007	44.3

(a) Make a graph to display the distribution of salmon run size, then describe the pattern and any striking deviations that you see.

(b) Make a time plot of run size and describe its pattern. As is often the case with data measured at specific time intervals, a time plot is needed to understand what is happening.

1.39 Watch those scales! The impression that a time plot gives depends on the scales you use on the two axes. If you stretch the vertical axis and compress the time axis, data appear to be more variable. Compressing the vertical axis and stretching the time axis make variations appear to be smaller. Make two time plots of the data in the previous exercise to illustrate this idea. Make one plot that makes variability appear to be larger and one plot that makes variability appear to be smaller. The moral of this exercise is: pay close attention to the scales when you look at a time plot.

 BERINGSEAFISH

1.2 Describing Distributions with Numbers

In the previous section, we used the shape, center, and spread as ways to describe the overall pattern of any distribution for a quantitative variable. In this section, we will learn specific ways to use numbers to measure the center and the spread of a distribution. The numbers, like the graphs of Section 1.1, are aids to understanding the data, not "the answer" in themselves.

Dejan Patic/Getty Images

CASE 1.2

DATA FILE
TIMETOSTART24

Time to Start a Business An entrepreneur faces many bureaucratic and legal hurdles when starting a new business. The World Bank collects information about starting businesses throughout the world. It has determined the time, in days, to complete all of the procedures required to start a business.[30] Data for 195 countries are included in the data set. For this section we will examine data for a sample of 24 of these countries. Here are the data:

23	4	29	44	47	24	40	23	23	44	33	27
60	46	61	11	23	62	31	44	77	14	65	42

EXAMPLE 1.20   The Distribution of Business Start Times

CASE 1.2

DATA FILE
TIMETOSTART24

The stemplot in Figure 1.13 shows us the *shape, center,* and *spread* of the business start times. The stems are tens of days, and the leaves are days. As is often the case when there are few observations, the shape of the distribution is irregular. There are peaks in the 20s and the 40s. The values range from 4 to 77 days, with a center somewhere in the middle of these two extremes. There do not appear to be any outliers.

```
0 | 4
1 | 1 4
2 | 3 3 3 4 7 9
3 | 1 3
4 | 0 2 4 4 4 6 7
5 |
6 | 0 1 2 5
7 | 7
```

FIGURE 1.13 Stemplot for sample of 24 business start times, for Example 1.20.

Measuring center: the mean

A description of a distribution almost always includes a measure of its center. The most common measure of center is the ordinary arithmetic average, or *mean*.

> **The Mean $\bar{x}$**
>
> To find the **mean** of a set of observations, add their values and divide by the number of observations. If the n observations are $x_1, x_2, \ldots, x_n$, their mean is
>
> $$\bar{x} = \frac{x_1 + x_2 + \cdots + x_n}{n}$$
>
> or, in more compact notation,
>
> $$\bar{x} = \frac{1}{n} \sum x_i$$

The $\sum$ (capital Greek sigma) in the formula for the mean is short for "add them all up." The subscripts on the observations x_i are just a way of keeping the n observations distinct. They do not necessarily indicate order or any other special facts about the data. The bar over the x indicates the mean of all the x-values. Pronounce the mean $\bar{x}$ as "x-bar." This notation is very common. When writers who are discussing data use $\bar{x}$ or $\bar{y}$, they are talking about a mean.

EXAMPLE 1.21   Mean Time to Start a Business

CASE 1.2

TIMETOSTART24

The mean time to start a business is

$$
\begin{aligned}
\bar{x} &= \frac{x_1 + x_2 + \cdots + x_n}{n} \\
&= \frac{23 + 4 + \cdots + 42}{24} \\
&= \frac{897}{24} = 37.375
\end{aligned}
$$

The mean time to start a business for the 24 countries in our data set is 37.4 days. Note that we have rounded the answer. Our goal in using the mean to describe the center of a distribution is not to demonstrate that we can compute with great accuracy. The additional digits do not provide any additional useful information. In fact, they distract our attention from the important digits that are meaningful. Do you think it would be better to report the mean as 37 days?

In practice, you can key the data into your calculator and hit the Mean key. You don't have to actually add and divide. But you should know that this is what the calculator is doing.

APPLY YOUR KNOWLEDGE

TIMETOSTART25

CASE 1.2   **1.40 Include the outlier.** The complete business start time data set with 195 countries has a few with very large start times. In constructing the data set for Case 1.2 a random sample of 25 countries was selected. This sample included the South American country of Suriname, where the start time is 694 days. This case was deleted for Case 1.2. Reconstruct the original random sample by including Suriname. Show that the mean has increased to 64 days. (This is a rounded number. You should report the mean with two digits after the decimal.)

Exercise 1.40 illustrates an important fact about the mean as a measure of center: it is sensitive to the influence of one or more extreme observations. These may be outliers, but a skewed distribution that has no outliers will also pull the mean toward its long tail. Because the mean cannot resist the influence of extreme observations, we say that it is *not* a **resistant measure** of center.

resistant measure

APPLY YOUR KNOWLEDGE

 CALLCENTER80

1.41 Calls to a customer service center. The service times for 80 calls to a customer service center are given in Table 1.1 (page 14). Use these data to compute the mean service time.

1.42 Find the mean of the first-exam scores. Here are the scores on the first exam in an introductory statistics course for 10 students:

STATCOURSE

$$80 \quad 73 \quad 92 \quad 85 \quad 75 \quad 98 \quad 93 \quad 55 \quad 80 \quad 90$$

Find the mean first-exam score for these students.

Measuring center: the median

In Section 1.1, we used the midpoint of a distribution as an informal measure of center. The *median* is the formal version of the midpoint, with a specific rule for calculation.

> **The Median M**
>
> The **median M** is the midpoint of a distribution, the number such that half the observations are smaller and the other half are larger. To find the median of a distribution:
>
> 1. Arrange all observations in order of size, from smallest to largest.
>
> 2. If the number of observations n is odd, the median M is the center observation in the ordered list. Find the location of the median by counting $(n + 1)/2$ observations up from the bottom of the list.
>
> 3. If the number of observations n is even, the median M is the mean of the two center observations in the ordered list. The location of the median is again $(n + 1)/2$ from the bottom of the list.

Note that the formula $(n + 1)/2$ does *not* give the median, just the location of the median in the ordered list. Medians require little arithmetic, so they are easy to find by hand for small sets of data. Arranging even a moderate number of observations in order is very tedious, however, so that finding the median by hand for larger sets of data is unpleasant. Even simple calculators have an $\bar{x}$ button, but you will need software or a graphing calculator to automate finding the median.

EXAMPLE 1.22   Median Time to Start a Business

CASE 1.2

 TIMETOSTART24

To find the median time to start a business for our 24 countries, we first arrange the data in order from smallest to largest:

$$4 \quad 11 \quad 14 \quad 23 \quad 23 \quad 23 \quad 23 \quad 24 \quad 27 \quad 29 \quad 31 \quad 33$$
$$40 \quad 42 \quad 44 \quad 44 \quad 44 \quad 46 \quad 47 \quad 60 \quad 61 \quad 62 \quad 65 \quad 77$$

The count of observations $n = 24$ is even. The median, then, is the average of the two center observations in the ordered list. To find the location of the center observations, we first compute

$$\text{location of } M = \frac{n+1}{2} = \frac{25}{2} = 12.5$$

Therefore, the center observations are the 12th and 13th observations in the ordered list. The median is

$$M = \frac{33 + 40}{2} = 36.5$$

Note that you can use the stemplot directly to compute the median. In the stemplot the cases are already ordered and you simply need to count from the top or the bottom to the desired location.

APPLY YOUR KNOWLEDGE

 TIMETOSTART25

CASE 1.2 **1.43 Include the outlier.** Include Suriname, where the start time is 694 days, in the data set and show that the median is 40 days. Note that with this case included, the sample size is now 25 and the median is the 13th observation in the ordered list. Write out the ordered list and circle the outlier. Describe the effect of the outlier on the median for this set of data.

 CALLCENTER80

1.44 Calls to a customer service center. The service times for 80 calls to a customer service center are given in Table 1.1 (page 14). Use these data to compute the median service time.

 STATCOURSE

1.45 Find the median of the first-exam scores. Here are the scores on the first exam in an introductory statistics course for 10 students:

80   73   92   85   75   98   93   55   80   90

Find the median first-exam score for these students.

Comparing the mean and the median

Exercises 1.40 and 1.43 illustrate an important difference between the mean and the median. Suriname pulls the mean time to start a business up from 37 days to 64 days. The increase in the median is a lot less, from 36 days to 40 days.

The median is more *resistant* than the mean. If the largest starting time in the data set was 1200 days, the median for all 25 countries would still be 40 days. The largest observation just counts as one observation above the center, no matter how far above the center it lies. The mean uses the actual value of each observation and so will chase a single large observation upward.

The best way to compare the response of the mean and median to extreme observations is to use an interactive applet that allows you to place points on a line and then drag them with your computer's mouse. Exercises 1.68 to 1.70 use the *Mean and Median* applet on the Web site for this book, www.whfreeman.com/psbe, to compare mean and median.

The mean and median of a symmetric distribution are close together. If the distribution is exactly symmetric, the mean and median are exactly the same. In a skewed distribution, the mean is farther out in the long tail than is the median.

Consider the prices of existing single-family homes in the United States. The mean price in 2007 was $266,200 while the median was $217,900. This distribution is strongly skewed to the right. There are many moderately priced houses and a few very expensive mansions. The few expensive houses pull the mean up but do not affect the median.

Reports about house prices, incomes, and other strongly skewed distributions usually give the median ("midpoint") rather than the mean ("arithmetic average"). However, if you are a tax assessor interested in the total value of houses in your area, use the mean. The total is the mean times the number of houses, but it has no connection with the median. The mean and median measure center in different ways, and both are useful.

APPLY YOUR KNOWLEDGE

GDP12

1.46 Gross domestic product. The success of companies expanding to developing regions of the world depends in part on the prosperity of the countries in those regions. Here are World Bank data on the growth of gross domestic product (percent per year) for the period 2000 to 2004 in countries in Asia (not including Japan):

Country	Growth
Bangladesh	5.2
China	9.4
Hong Kong	3.2
India	6.2
Indonesia	4.6
Korea (South)	4.7
Malaysia	4.4
Pakistan	4.1
Philippines	3.9
Singapore	2.9
Thailand	5.4
Vietnam	7.2

(a) Make a stemplot of the data. Note the high outlier.
(b) Find the mean and median growth rates. How does the outlier explain the difference between your two results?
(c) Find the mean and median growth rates without the outlier. How does comparing your results in (b) and (c) illustrate the resistance of the median and the lack of resistance of the mean?

Measuring spread: the quartiles

A measure of center alone can be misleading. Two nations with the same median household income are very different if one has extremes of wealth and poverty and the other has little variation among households. A drug with the correct mean concentration of active ingredient is dangerous if some batches are much too high and others much too low. We are interested in the *spread* or *variability* of incomes and drug potencies as well as their centers. **The simplest useful numerical description of a distribution consists of both a measure of center and a measure of spread.**

One way to measure spread is to give the smallest and largest observations. For example, the times to start a business in our data set that included Suriname ranged from

4 to 694 days. Without Suriname, the range is 4 to 77 days. These largest and smallest observations show the full spread of the data and are highly influenced by outliers.

We can improve our description of spread by also giving several percentiles. The

percentile **pth percentile** of a distribution is the value such that p percent of the observations fall at or below it. The median is just the 50th percentile, so the use of percentiles to report spread is particularly appropriate when the median is our measure of center.

The most commonly used percentiles other than the median are the *quartiles*. The first quartile is the 25th percentile, and the third quartile is the 75th percentile. That is, the first and third quartiles show the spread of the middle half of the data. (The second quartile is the median itself.) To calculate a percentile, arrange the observations in increasing order and count up the required percent from the bottom of the list. Our definition of percentiles is a bit inexact because there is not always a value with exactly p percent of the data at or below it. We will be content to take the nearest observation for most percentiles, but the quartiles are important enough to require an exact recipe. The rule for calculating the quartiles uses the rule for the median.

The Quartiles Q_1 and Q_3

To calculate the **quartiles:**

1. Arrange the observations in increasing order and locate the median M in the ordered list of observations.

2. The **first quartile Q_1** is the median of the observations whose position in the ordered list is to the left of the location of the overall median.

3. The **third quartile Q_3** is the median of the observations whose position in the ordered list is to the right of the location of the overall median.

Here is an example that shows how the rules for the quartiles work for both odd and even numbers of observations.

EXAMPLE 1.23 Finding the Quartiles

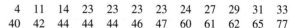

CASE 1.2

Here is the ordered list of the times to start a business in our sample of 24 countries:

| 4 | 11 | 14 | 23 | 23 | 23 | 23 | 24 | 27 | 29 | 31 | 33 |
| 40 | 42 | 44 | 44 | 44 | 46 | 47 | 60 | 61 | 62 | 65 | 77 |

TIMETOSTART24

The count of observations $n = 24$ is even, so the median is at position $(24 + 1)/2 = 12.5$, that is, between the 12th and the 13th observation in the ordered list. There are 12 cases above this position and 12 below it. The first quartile is the median of the first 12 observations, and the third quartile is the median of the last 12 observations. Check that $Q_1 = 23$ and $Q_3 = 46.5$.

Notice that the quartiles are resistant. For example, Q_3 would have the same value if the highest start time was 770 days rather than 77 days.

There are slight differences in the methods used by software to compute percentiles. However, the results will generally be quite similar except in cases where the sample sizes are very small.

Be careful when several observations take the same numerical value. Write down all the observations and apply the rules just as if they all had distinct values.

The five-number summary and boxplots

The smallest and largest observations tell us little about the distribution as a whole, but they give information about the tails of the distribution that is missing if we know only Q_1, M, and Q_3. To get a quick summary of both center and spread, combine all five numbers. The result is the *five-number summary* and a graph based on it.

> **The Five-Number Summary and Boxplots**
>
> The **five-number summary** of a distribution consists of the smallest observation, the first quartile, the median, the third quartile, and the largest observation, written in order from smallest to largest. In symbols, the five-number summary is
>
> $$\text{Minimum} \quad Q_1 \quad M \quad Q_3 \quad \text{Maximum}$$
>
> A **boxplot** is a graph of the five-number summary.
>
> - A central box spans the quartiles.
> - A line in the box marks the median.
> - Lines extend from the box out to the smallest and largest observations.
>
> Boxplots are most useful for side-by-side comparison of several distributions.

You can draw boxplots either horizontally or vertically. Be sure to include a numerical scale in the graph. When you look at a boxplot, first locate the median, which marks the center of the distribution. Then look at the spread. The quartiles show the spread of the middle half of the data, and the extremes (the smallest and largest observations) show the spread of the entire data set. We now have the tools for a preliminary examination of the customer service center call lengths.

EXAMPLE 1.24 Service Center Call Lengths

CALLCENTER80

Table 1.1 (page 14) displays the customer service center call lengths for a random sample of 80 calls that we discussed in Example 1.13. The five-number summary for these data is 1.0, 54.4, 103.5, 200, 2631. The distribution is highly skewed. The mean is 197 seconds, a value that is very close to the third quartile. The boxplot is displayed in Figure 1.14. The skewness of the distribution is the major feature that we see in this plot. Note that the mean is marked with a "+" and appears very close to the upper edge of the box.

FIGURE 1.14 Boxplot for sample of 80 service center call lengths, for Example 1.24.

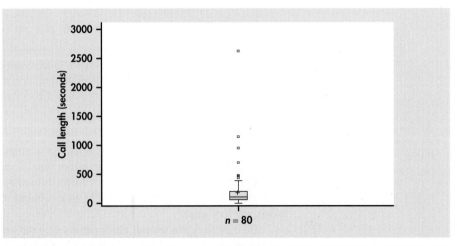

Because of the skewness in this distribution, we selected a software option to plot extreme points individually in Figure 1.14. This is one of several different ways to improve the appearance of boxplots for particular data sets. These variations are called

modified boxplots   *modified boxplots.*

Boxplots can show the symmetry or skewness of a distribution. In a symmetric distribution, the first and third quartiles are equally distant from the median. This is not what we see in Figure 1.14. Here, the distribution is skewed to the right. The third quartile is farther above the median than the first quartile is below it. The extremes behave the same way. Boxplots do not always give a clear indication of the nature of a skewed set of data. For example, the quartiles may indicate right-skewness while the whiskers indicate left-skewness.

Boxplots are particularly useful for comparing several distributions. Here is an example.

EXAMPLE 1.25   Fuel Efficiency Sells Cars

Fuel efficiency has become a major issue for people thinking about buying a new car. The Environmental Protection Agency provides data on the fuel efficiencies of vehicles sold in the United States each year.[31] Figure 1.15 gives side-by-side boxplots of the miles per gallon (mpg) for four vehicle classes: convertibles, pickup trucks, SUVs, and small cars. Small cars appear to have better efficiency than the other three classes. Pickup trucks show less variation than the other classes; the range of mpg values is less, and the first and third quartiles are closer together. The distributions for SUVs and small cars show some skewness, with some vehicles having particularly good fuel efficiency. However, note that the mean (marked with a "+") and the median are very close for all four classes.

FIGURE 1.15 Side-by-side boxplots of fuel efficiency for selected model year 2009 vehicle classes, for Example 1.25.

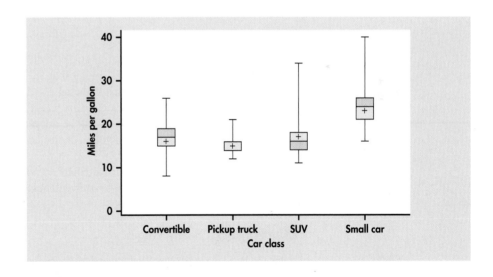

APPLY YOUR KNOWLEDGE

CASE 1.2 **1.47 Time to start a business.** Refer to the data on times to start a business in 24 countries described in Case 1.2 on page 26. Use a boxplot to display the distribution. Discuss the features of the data that you see in the boxplot, and compare it with the stemplot in Figure 1.13. Which do you prefer? Give reasons for your answer.

STATCOURSE

1.48 First-exam scores. Here are the scores on the first exam in an introductory statistics course for 10 students:

$$80 \quad 73 \quad 92 \quad 85 \quad 75 \quad 98 \quad 93 \quad 55 \quad 80 \quad 90$$

Display the distribution with a boxplot. Discuss whether or not a stemplot would provide a better way to look at this distribution.

Measuring spread: the standard deviation

The five-number summary is not the most common numerical description of a distribution. That distinction belongs to the combination of the mean to measure center and the *standard deviation* to measure spread. The standard deviation measures spread by looking at how far the observations are from their mean.

The Standard Deviation s

The **variance s^2** of a set of observations is essentially the average of the squares of the deviations of the observations from their mean. In symbols, the variance of n observations $x_1, x_2, \ldots, x_n$ is

$$s^2 = \frac{(x_1 - \overline{x})^2 + (x_2 - \overline{x})^2 + \cdots + (x_n - \overline{x})^2}{n - 1}$$

or, more compactly,

$$s^2 = \frac{1}{n - 1} \sum (x_i - \overline{x})^2$$

The **standard deviation s** is the square root of the variance s^2:

$$s = \sqrt{\frac{1}{n - 1} \sum (x_i - \overline{x})^2}$$

Notice that the "average" in the variance s^2 divides the sum by 1 less than the number of observations, that is, $n - 1$ rather than n. The reason is that the deviations $x_i - \overline{x}$ always sum to exactly 0, so that knowing $n - 1$ of them determines the last one. Only $n - 1$ of the squared deviations can vary freely, and we average by dividing the

degrees of freedom   total by $n - 1$. The number $n - 1$ is called the **degrees of freedom** of the variance or standard deviation. Many calculators offer a choice between dividing by n and dividing by $n - 1$, so be sure to use $n - 1$.

In practice, use software or your calculator to obtain the standard deviation from keyed-in data. Doing an example step-by-step will help you understand how the variance and standard deviation work, however.

EXAMPLE 1.26   Hourly Wages

BLSWAGES

Planning to be a lawyer or other legal professional? The Bureau of Labor Statistics lists average hourly wages for 9 categories of law-related occupations (OCC Code 23-0000) (the units are dollars per hour):[32]

$$75 \quad 38 \quad 27 \quad 48 \quad 23 \quad 23 \quad 20 \quad 20 \quad 26$$

First find the mean:

$$\overline{x} = \frac{75 + 38 + 27 + 48 + 23 + 23 + 20 + 20 + 26}{9}$$

$$= \frac{300}{9} = 33.33 \text{ dollars per hour}$$

We organize the rest of the arithmetic in a table. This is a good way to do calculations such as this when you need to work through all the details.

Observations x_i	Deviations $x_i - \bar{x}$	Squared deviations $(x_i - \bar{x})^2$
75	$75 - 33.33 = \quad 41.67$	$(41.67)^2 = 1736.39$
38	$38 - 33.33 = \quad 4.67$	$(4.67)^2 = \quad 21.81$
27	$27 - 33.33 = \quad -6.33$	$(-6.33)^2 = \quad 40.07$
48	$48 - 33.33 = \quad 14.67$	$(14.67)^2 = \quad 215.21$
23	$23 - 33.33 = -10.33$	$(-10.33)^2 = \quad 106.71$
23	$23 - 33.33 = -10.33$	$(-10.33)^2 = \quad 106.71$
20	$20 - 33.33 = -13.33$	$(-13.33)^2 = \quad 177.69$
20	$20 - 33.33 = -13.33$	$(-13.33)^2 = \quad 177.69$
26	$26 - 33.33 = \quad -7.33$	$(-7.33)^2 = \quad 53.73$
	sum $= \quad 0.03$	sum $= 2636.01$

The variance is the sum of the squared deviations divided by 1 less than the number of observations:

$$s^2 = \frac{2636.01}{8} = 329.5$$

The standard deviation is the square root of the variance:

$$s = \sqrt{329.5} = 18.15 \text{ dollars per hour}$$

More important than the details of hand calculation are the properties that determine the usefulness of the standard deviation:

- s measures spread about the mean and should be used only when the mean is chosen as the measure of center.

- $s = 0$ only when there is *no spread*. This happens only when all observations have the same value. Otherwise, s is greater than zero. As the observations become more spread out about their mean, s gets larger.

- s has the same units of measurement as the original observations. For example, if you measure wages in dollars per hour, s is also in dollars per hour.

- Like the mean $\bar{x}$, s is not resistant. Strong skewness or a few outliers can greatly increase s.

APPLY YOUR KNOWLEDGE

TIMETOSTART24

TIMETOSTART25

CASE 1.2 **1.49 Time to start a business.** Verify the statement in the last bullet above using the data on the time to start a business. First, use the 24 cases from Case 1.2 (page 26) to calculate a standard deviation. Next, include the country Suriname, where the time to start a business is 694 days. Show that the inclusion of this single outlier increases the standard deviation from 19 to 133.

You may rightly feel that the importance of the standard deviation is not yet clear. We will see in the next section that the standard deviation is the natural measure of spread for an important class of symmetric distributions, the Normal distributions. The usefulness of many statistical procedures is tied to distributions with particular shapes. This is certainly true of the standard deviation.

Choosing measures of center and spread

How do we choose between the five-number summary and $\bar{x}$ and s to describe the center and spread of a distribution? Because the two sides of a strongly skewed distribution have different spreads, no single number such as s describes the spread well. The five-number summary, with its two quartiles and two extremes, does a better job.

> **Choosing a Summary**
>
> The five-number summary is usually better than the mean and standard deviation for describing a skewed distribution or a distribution with extreme outliers. Use $\bar{x}$ and s only for reasonably symmetric distributions that are free of outliers.

APPLY YOUR KNOWLEDGE

STATCOURSE

1.50 First-exam scores. Below are the scores on the first exam in an introductory statistics course for 10 students. We found the mean of these scores in Exercise 1.42 (page 28) and the median in Exercise 1.45 (page 29).

80   73   92   85   75   98   93   55   80   90

(a) Make a stemplot of these data.
(b) Compute the standard deviation.
(c) Are the mean and the standard deviation effective in describing the distribution of these scores? Explain your answer.

CALLCENTER80

1.51 Calls to a customer service center. We displayed the distribution of the lengths of 80 calls to a customer service center in Figure 1.14 (page 32).
(a) Compute the mean and the standard deviation for these 80 calls (the data are given in Table 1.1, page 14).
(b) Find the five-number summary.
(c) Which summary does a better job of describing the distribution of these calls? Give reasons for your answer.

BEYOND THE BASICS: Risk and Return

A central principle in the study of investments is that taking bigger risks is rewarded by higher returns, at least on the average over long periods of time. It is usual in finance to measure risk by the standard deviation of returns, on the grounds that investments whose returns show a large spread from year to year are less predictable and therefore more risky than those whose returns have a small spread. Compare, for example, the approximate mean and standard deviation of the annual percent returns on American common stocks and U.S. Treasury bills over a fifty-year period starting in 1950:

Investment	Mean return	Standard deviation
Common stocks	14.0%	16.9%
Treasury bills	5.2%	2.9%

Stocks are risky. They went up 14% per year on the average during this period, but they dropped almost 28% in the worst year. The large standard deviation reflects the fact that stocks have produced both large gains and large losses. When you buy a Treasury bill,

FIGURE 1.16(a) Stemplot of the annual returns on Treasury bills for 50 years. The stems are percents.

```
 0 | 9
 1 | 2 5 5 6 6 8
 2 | 1 5 7 7 9
 3 | 0 1 1 5 5 8 9 9
 4 | 2 4 7 7 8
 5 | 1 1 2 2 2 5 6 6 8
 6 | 2 4 5 6 9
 7 | 2 7 8
 8 | 0 4 8
 9 | 8
10 | 4 5
11 | 3
12 |
13 |
14 | 7
```

(a) T-bills

FIGURE 1.16(b) Stemplot of the annual returns on common stocks for 50 years. The stems are percents.

```
-2 | 8
-1 | 9 1 1 0 0
-0 | 9 6 4 3
 0 | 0 0 0 1 2 3 8 9 9
 1 | 1 3 3 4 4 6 6 6 7 8
 2 | 0 1 1 2 3 4 4 4 5 7 7 9 9
 3 | 0 1 1 3 4 6 7
 4 | 5
 5 | 0
```

(b) Stocks

on the other hand, you are lending money to the government for one year. You know that the government will pay you back with interest. That is much less risky than buying stocks, so (on the average) you get a smaller return.

Are $\bar{x}$ and s good summaries for distributions of investment returns? Figures 1.16(a) and 1.16(b) display stemplots of the annual returns for both investments. You see that returns on Treasury bills have a right-skewed distribution. Convention in the financial world calls for $\bar{x}$ and s because some parts of investment theory use them. For describing this right-skewed distribution, however, the five-number summary would be more informative.

Remember that a graph gives the best overall picture of a distribution. Numerical measures of center and spread report specific facts about a distribution, but they do not describe its entire shape. Numerical summaries do not disclose the presence of multiple peaks or gaps, for example. **Always plot your data.**

SECTION 1.2 Summary

- A numerical summary of a distribution should report its **center** and its **spread** or **variability.**
- The **mean** $\bar{x}$ and the **median** M describe the center of a distribution in different ways. The mean is the arithmetic average of the observations, and the median is the midpoint of the values.
- When you use the median to indicate the center of the distribution, describe its spread by giving the **quartiles.** The **first quartile** Q_1 has one-fourth of the observations below it, and the **third quartile** Q_3 has three-fourths of the observations below it.

- The **five-number summary** consisting of the median, the quartiles, and the high and low extremes provides a quick overall description of a distribution. The median describes the center, and the quartiles and extremes show the spread.

- **Boxplots** based on the five-number summary are useful for comparing several distributions. The box spans the quartiles and shows the spread of the central half of the distribution. The median is marked within the box. Lines extend from the box to the extremes and show the full spread of the data.

- The **variance** s^2 and especially its square root, the **standard deviation** s, are common measures of spread about the mean as center. The standard deviation s is zero when there is no spread and gets larger as the spread increases.

- A **resistant measure** of any aspect of a distribution is relatively unaffected by changes in the numerical value of a small proportion of the total number of observations, no matter how large these changes are. The median and quartiles are resistant, but the mean and the standard deviation are not.

- The mean and standard deviation are good descriptions for symmetric distributions without outliers. They are most useful for the Normal distributions, introduced in the next section. The five-number summary is a better exploratory summary for skewed distributions.

SECTION 1.2 Exercises

For Exercises 1.41 and 1.42, see page 28; for 1.43 to 1.45, see page 29; for 1.46, see page 30; for 1.47 and 1.48, see pages 33–34; for 1.49, see page 35; and for 1.50 and 1.51, see page 36.

1.52 Gross domestic product growth in 120 countries. The gross domestic product (GDP) of a country is the total value of all goods and services produced in the country. It is an important measure of the health of a country's economy. For this exercise, you will analyze the growth in GDP, expressed as a percent, for 120 countries.[33] 📁 COUNTRIES120

(a) Compute the mean and the standard deviation.
(b) Which two countries are outliers for this variable?
(c) Recompute the mean and standard deviation without the outliers. Explain how the mean and standard deviation changed when you deleted the outliers.

1.53 Use the resistant measures for GDP. Repeat parts (a) and (c) of the previous exercise using the median and the quartiles. Summarize your results and compare them with those of the previous exercise. 📁 COUNTRIES120

1.54 Trade balance for 120 countries. Trade balance is another important variable that describes a country's economy. It is defined as the difference between the value of a country's exports and its imports. A negative trade balance occurs when a country imports more than it exports. Similarly, the trade balance will be positive for a country that exports more than it imports. Note that values of this variable are missing for five countries. In this data set, missing values are coded as a periods. 📁 COUNTRIES120

(a) Describe the distribution of trade balance using the mean and the standard deviation.
(b) Do the same using the median and the quartiles.
(c) Using only the information from parts (a) and (b), give a description of the data.
Do not look at any graphical summaries or other numerical summaries for this part of the exercise.

1.55 What do the trade balance graphical summaries show? Refer to the previous exercise. 📁 COUNTRIES120

(a) Use graphical summaries to describe the distribution of the trade balance for these countries.
(b) Give the names of the countries that correspond to extreme values in this distribution.
(c) Reanalyze the data without the outliers.
(d) Summarize what you have learned about the distribution of the trade balance for these countries. Include appropriate graphical and numerical summaries as well as comments about the outliers.

1.56 U.S. unemployment rates. Refer to Exercise 1.27 and Table 1.2 (page 23) for the U.S. unemployment rates for each of the 50 states. 📁 UNEMPLOYMENT

(a) Find the mean and the standard deviation.
(b) Find the five-number summary.
(c) Draw a boxplot.
(d) How do you prefer to summarize these data? Include numerical and graphical summaries and explain the reasons for your preference.

1.57 Canadian unemployment rates. Unemployment rates for 10 Canadian provinces are given in Exercise 1.28 (page 23). Answer the questions in the previous exercise for these data. The U.S. data set has 50 cases while the Canadian data set has 10 cases. Discuss how this difference influences the way in which you summarize the data. 🖱 UNEMPLOYMENTCANADA

1.58 Compare U.S. and Canadian unemployment rates. Refer to the previous two exercises. 🖱 UNEMPLOYMENT, UNEMPLOYMENTCANADA

(a) Use side-by-side boxplots to give a graphical summary of the two sets of unemployment rates.

(b) Use a *back-to-back stemplot* to compare the two sets of rates. A back-to-back stemplot has a single stem with leaves on the left for one group and leaves on the right for the other.

(c) Summarize the major differences and similarities between the two sets of unemployment rates.

(d) Which graphical comparison do you prefer? Give reasons for your answer.

1.59 Recoverable oil. The estimated amounts of recoverable oil from 64 oil wells in the Devonian Richmond Dolomite area of Michigan are given Exercise 1.35 (page 25). 🖱 OILWELLS

(a) Find the mean and the standard deviation.

(b) Find the five-number summary.

(c) Draw a boxplot.

(d) How do you prefer to summarize these data? Include numerical and graphical summaries and explain the reasons for your preference.

1.60 Variability of an agricultural product. A quality product is one that is consistent and has very little variability in its characteristics. Controlling variability can be more difficult with agricultural products than with those that are manufactured. The following table gives the weights, in ounces, of the 25 potatoes sold in a 10-pound bag. 🖱 POTATOES

7.8	7.9	8.2	7.3	6.7	7.9	7.9	7.9	7.6	7.8	7.0	4.7	7.6
6.3	4.7	4.7	4.7	6.3	6.0	5.3	4.3	7.9	5.2	6.0	3.7	

(a) Summarize the data graphically and numerically. Give reasons for the methods you chose to use in your summaries.

(b) Do you think that your numerical summaries do an effective job of describing these data? Why or why not?

(c) There appear to be two distinct clusters of weights for these potatoes. Divide the sample into two subsamples based on the clustering. Give the mean and standard deviation for each subsample. Do you think that this way of summarizing these data is better than a numerical summary that uses all the data as a single sample? Give a reason for your answer.

1.61 The value of brands. A brand is a symbol or images that are associated with a company. An effective brand identifies the company and its products. Using a variety of measures, dollar values for brands can be calculated.[34] The most valuable brand is Coca-Cola, with a value of $66,667 million. Coke is followed by IBM, at $59,031 million; Microsoft, at $59,007 million; GE, at $53,086 million; and Toyota, at $34,050 million. For this exercise you will use the brand values, reported in millions of dollars, for the top 100 brands. 🖱 BRANDS

(a) Graphically display the distribution of the values of these brands.

(b) Use numerical measures to summarize the distribution.

(c) Write a short paragraph discussing the dollar values of the top 100 brands. Include the results of your analysis.

1.62 The alcohol content of beer. Brewing beer involves a variety of steps that can affect the alcohol content. A Web site gives the percent alcohol for 86 domestic brands of beer.[35] 🖱 BEER

(a) Use graphical and numerical summaries of your choice to describe these data. Give reasons for your choice.

(b) The data set contains an outlier. Explain why this particular beer is unusual and how its outlier status is related to how it is marketed.

1.63 An outlier for alcohol content of beer. Refer to the previous exercise. 🖱 BEER

(a) Calculate the mean with and without the outlier. Do the same for the median. Explain how these values change when the outlier is excluded.

(b) Calculate the standard deviation with and without the outlier. Do the same for the quartiles. Explain how these values change when the outlier is excluded.

(c) Write a short paragraph summarizing what you have learned in this exercise.

1.64 Calories in beer. Refer to the previous two exercises. The data set also gives the calories per 12 ounces of beverage. 🖱 BEER

(a) Analyze the data and summarize the distribution of calories for these 86 brands of beer.

(b) In Exercise 1.62 you identified one brand of beer as an outlier. To what extent is this brand an outlier in the distribution of calories? Explain your answer.

(c) The distribution of calories suggests that there may be two groups of beers that might be marketed differently. Examine the data file carefully and explain the characteristics of the two groups.

1.65 Create a data set. Create a data set for which the median would change by a large amount if the smallest observation is deleted.

1.66 Salaries of the Chicago Cubs. The mean salary of the players on the 2008 Chicago Cubs baseball team is $5,274,108, while the median salary is $4,350,000. What explains the difference between these two measures of center?

1.67 Discovering outliers. Whether an observation is an outlier is a matter of judgment. It is convenient to have a rule for identifying suspected outliers. The *1.5 × IQR rule* is in common use:

1. The *interquartile range IQR* is the distance between the first and third quartiles, $IQR = Q_3 - Q_1$. This is the spread of the middle half of the data.

2. An observation is a suspected outlier if it lies more than $1.5 \times IQR$ below the first quartile Q_1 or above the third quartile Q_3.

The stemplot in Exercise 1.31 (page 24) displays the distribution of the percents of residents aged 65 and older in the 50 states. Stemplots help you find the five-number summary because they arrange the observations in increasing order.

 POPOVER65BYSTATE

(a) Give the five-number summary of this distribution.

(b) Does the $1.5 \times IQR$ rule identify Alaska and Florida as suspected outliers? Does it also flag any other states?

The following three exercises use the *Mean and Median* applet available at www.whfreeman.com/psbe to explore the behavior of the mean and median.

1.68 Mean = median? Place two observations on the line by clicking below it. Why does only one arrow appear?

1.69 Extreme observations. Place three observations on the line by clicking below it, two close together near the center of the line and one somewhat to the right of these two.

(a) Pull the rightmost observation out to the right. (Place the cursor on the point, hold down a mouse button, and drag the point.) How does the mean behave? How does the median behave? Explain briefly why each measure acts as it does.

(b) Now drag the rightmost point to the left as far as you can. What happens to the mean? What happens to the median as you drag this point past the other two (watch carefully)?

1.70 Don't change the median. Place 5 observations on the line by clicking below it.

(a) Add 1 additional observation *without changing the median*. Where is your new point?

(b) Use the applet to convince yourself that when you add yet another observation (there are now 7 in all), the median does not change no matter where you put the 7th point. Explain why this must be true.

1.71 $\bar{x}$ and s are not enough. The mean $\bar{x}$ and standard deviation s measure center and spread but are not a complete description of a distribution. Data sets with different shapes can have the same mean and standard deviation. To demonstrate this fact, find $\bar{x}$ and s for these two small data sets. Then make a stemplot of each and comment on the shape of each distribution. ABDATA

Data A:	9.14	8.14	8.74	8.77	9.26	8.10
	6.13	3.10	9.13	7.26	4.74	
Data B:	6.58	5.76	7.71	8.84	8.47	7.04
	5.25	5.56	7.91	6.89	12.50	

CASE 1.1 **1.72 Returns on Treasury bills.** Figure 1.16(a) (page 37) is a stemplot of the annual returns on U.S. Treasury bills for fifty years. (The entries are rounded to the nearest tenth of a percent.) TBILLRATES50

(a) Use the stemplot to find the five-number summary of T-bill returns.

(b) The mean of these returns is about 5.19%. Explain from the shape of the distribution why the mean return is larger than the median return.

1.73 Salary increase for the owners. Last year a small accounting firm paid each of its five clerks $30,000, two junior accountants $65,000 each, and the firm's owner $355,000.

(a) What is the mean salary paid at this firm? How many of the employees earn less than the mean? What is the median salary?

(b) This year the firm gives no raises to the clerks and junior accountants, while the owner's take increases to $455,000. How does this change affect the mean? How does it affect the median?

1.74 A skewed distribution. Sketch a distribution that is skewed to the left. On your sketch, indicate the approximate position of the mean and the median. Explain why these two values are not equal.

1.75 A standard deviation contest. You must choose four numbers from the whole numbers 10 to 20, with repeats allowed.

(a) Choose four numbers that have the smallest possible standard deviation.

(b) Choose four numbers that have the largest possible standard deviation.

(c) Is more than one choice possible in either (a) or (b)? Explain.

1.76 Imputation. Various problems with data collection can cause some observations to be missing. Suppose a data set has 20 cases. Here are the values of the variable x for 10 of these cases: IMPUTATION

17   6   12   14   20   23   9   12   16   21

The values for the other 10 cases are missing. One way to deal with missing data is called *imputation*. The basic idea is that missing values are replaced, or imputed, with values that are based on an analysis of the data that are not missing. For a data set with a single variable, the usual choice of a value for imputation is the mean of the values that are not missing. The mean for this data set is 15.

(a) Verify that the mean is 15 and find the standard deviation for the 10 cases for which x is not missing.

(b) Create a new data set with 20 cases by setting the values for the 10 missing cases to 15. Compute the mean and standard deviation for this data set.

(c) Summarize what you have learned about the possible effects of this type of imputation on the mean and the standard deviation.

1.77 A different type of mean. The *trimmed mean* is a measure of center that is more resistant than the mean but uses more of the available information than the median. To compute the 5% trimmed mean, discard the highest 5% and the lowest 5% of the observations and compute the mean of the remaining 90%. Trimming eliminates the effect of a small number of outliers. Use the data on the values of the top 100 brands that we studied in Exercise 1.61 (page 39) to find the 5% trimmed mean. Compare this result with the value of the mean computed in the usual way. BRANDS

1.3 Density Curves and the Normal Distributions

We now have a kit of graphical and numerical tools for describing distributions. What is more, we have a clear strategy for exploring data on a single quantitative variable:

1. Always plot your data: make a graph, usually a histogram or a stemplot.
2. Look for the overall pattern (shape, center, spread) and for striking deviations such as outliers.
3. Calculate a numerical summary to briefly describe center and spread.

Here is one more step to add to this strategy:

4. Sometimes the overall pattern of a large number of observations is so regular that we can describe it by a smooth curve.

Density curves

mathematical model

A density curve is a **mathematical model** for a distribution. Mathematical models are idealized descriptions. They allow us to easily make many statements in an idealized world. The statements are useful when the idealized world is similar to the real world. The density curves that we will study give a compact picture of the overall pattern of data. They ignore minor irregularities as well as outliers. For some situations, we are able to capture all of the essential characteristics of a distribution with a density curve. For other situations, our idealized model misses some important characteristics. As with so many things in statistics, your careful judgment is needed to decide what is important and how close is good enough.

MPG2009

| **EXAMPLE 1.27** | Gas Mileage |

Figure 1.17 is a histogram of the city gas mileage achieved by all 1140 motor vehicles (2009 model year) listed in the government's annual fuel economy report.[36] Superimposed on the histogram is

FIGURE 1.17 Histogram of fuel efficiency (miles per gallon) of 1140 autos (model year 2009), for Example 1.27. The smooth curve shows the overall shape of the distribution.

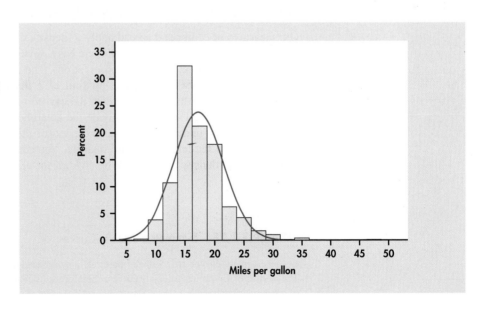

a density curve. The histogram shows that there are a few vehicles with very good fuel efficiency. These are high outliers in the distribution. The distribution is somewhat skewed to the right, reflecting the successful attempts of the auto industry to produce high-fuel-efficiency vehicles. There is a single peak around 15 miles per gallon. Both tails fall off quite smoothly. The density curve in Figure 1.17 is close to the histogram in many places but fails to capture some important characteristics of the distribution displayed by the histogram.

If we use a density curve that ignores vehicles that are outliers, we would capture the main features of the distribution of fuel efficiency for 2009 vehicles. On the other hand, we would miss the fact that some of these vehicles have been engineered to give excellent fuel efficiency. A marketing campaign based on this outstanding performance could be very effective for selling vehicles in an economy with high fuel prices. Be careful about how you deal with outliers. They may be data errors or they may be the most important feature of the distribution. Computer software cannot make this judgment. Only you can.

Here are some details about density curves. We need these basic ideas to understand the rest of this chapter.

Density Curve

A **density curve** is a curve that

- is always on or above the horizontal axis and

- has area exactly 1 underneath it.

A density curve describes the overall pattern of a distribution. The area under the curve and above any range of values is the proportion of all observations that fall in that range.

The median and mean of a density curve

Our measures of center and spread apply to density curves as well as to actual sets of observations. The median and quartiles are easy. Areas under a density curve represent proportions of the total number of observations. The median is the point with half the observations on either side. So **the median of a density curve is the equal-areas point,** the point with half the area under the curve to its left and the remaining half of the area to its right. The quartiles divide the area under the curve into quarters. One-fourth of the area under the curve is to the left of the first quartile, and three-fourths of the area is to the left of the third quartile. You can roughly locate the median and quartiles of any density curve by eye by dividing the area under the curve into four equal parts.

EXAMPLE 1.28 Symmetric Density Curves

Because density curves are idealized patterns, a symmetric density curve is exactly symmetric. The median of a symmetric density curve is therefore at its center. Figure 1.18(a) shows the median of a symmetric curve.

FIGURE 1.18(a) The median
and mean of a symmetric
density curve, for
Example 1.28.

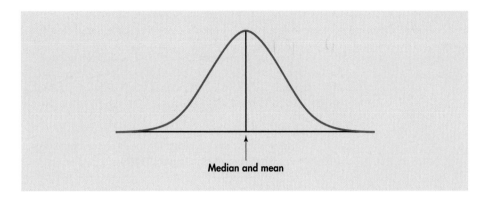

The situation is different for skewed density curves. Here is an example.

EXAMPLE 1.29   Skewed Density Curves

It isn't so easy to spot the equal-areas point on a skewed curve. There are mathematical ways of finding the median for any density curve. We did that to mark the median on the skewed curve in Figure 1.18(b).

FIGURE 1.18(b) The median
and mean of a right-skewed
density curve, for
Example 1.29.

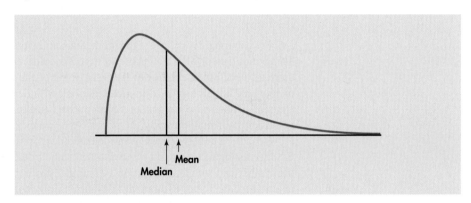

1.78 Another skewed curve. Sketch a curve similar to Figure 1.18(b) for a left-skewed density curve. Be sure to mark the location of the mean and the median.

What about the mean? The mean of a set of observations is their arithmetic average. If we think of the observations as weights strung out along a thin rod, the mean is the point at which the rod would balance. This fact is also true of density curves. **The mean is the point at which the curve would balance if made of solid material.**

EXAMPLE 1.30   Mean and Median

Figure 1.19 illustrates this fact about the mean. A symmetric curve balances at its center because the two sides are identical. **The mean and median of a symmetric density curve are equal,** as in Figure 1.18(a). We know that the mean of a skewed distribution is pulled toward the long tail. Figure 1.18(b) shows how the mean of a skewed density curve is pulled toward the long tail more than is the median. It's hard to locate the balance point by eye on a skewed curve. There are

FIGURE 1.19 The mean is the
balance point of a density
curve.

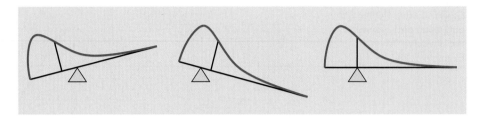

mathematical ways of calculating the mean for any density curve, so we are able to mark the mean as well as the median in Figure 1.18(b).

Median and Mean of a Density Curve

The **median** of a density curve is the equal-areas point, the point that divides the area under the curve in half.

The **mean** of a density curve is the balance point, at which the curve would balance if made of solid material.

The median and mean are the same for a symmetric density curve. They both lie at the center of the curve. The mean of a skewed curve is pulled away from the median in the direction of the long tail.

We can roughly locate the mean, median, and quartiles of any density curve by eye. This is not true of the standard deviation. When necessary, we can once again call on more advanced mathematics to learn the value of the standard deviation. The study of mathematical methods for doing calculations with density curves is part of theoretical statistics. Though we are concentrating on statistical practice, we often make use of the results of mathematical study.

Because a density curve is an idealized description of the distribution of data, we need to distinguish between the mean and standard deviation of the density curve and the mean $\overline{x}$ and standard deviation s computed from the actual observations. The usual

mean μ
standard deviation σ

notation for the mean of an idealized distribution is μ (the Greek letter mu). We write the standard deviation of a density curve as σ (the Greek letter sigma).

APPLY YOUR KNOWLEDGE

1.79 A symmetric curve.   Sketch a density curve that is symmetric but has a shape different from that of the curve in Figure 1.18(a).

uniform distribution

1.80 A uniform distribution.   Figure 1.20 displays the density curve of a **uniform distribution.** The curve takes the constant value 1 over the interval from 0 to 1 and is

FIGURE 1.20 The density curve
of a uniform distribution, for
Exercise 1.80.

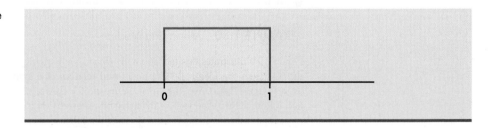

FIGURE 1.21 Three density curves, for Exercise 1.81.

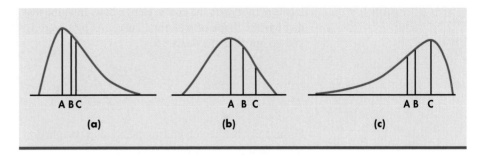

(a)   **(b)**   **(c)**

0 outside that range of values. This means that data described by this distribution take values that are uniformly spread between 0 and 1. Use areas under this density curve to answer the following questions.

(a) Why is the total area under this curve equal to 1?

(b) What percent of the observations lie above 0.8?

(c) What percent of the observations lie below 0.6?

(d) What percent of the observations lie between 0.25 and 0.75?

(e) What is the mean μ of this distribution?

1.81 Three curves. Figure 1.21 displays three density curves, each with three points marked. At which of these points on each curve do the mean and the median fall?

Normal distributions

Normal distributions

One particularly important class of density curves has already appeared in Figure 1.18(a). These density curves are symmetric, single-peaked, and bell-shaped. They are called *Normal curves,* and they describe **Normal distributions.** All Normal distributions have the same overall shape. The exact density curve for a particular Normal distribution is described by giving its mean μ and its standard deviation σ. The mean is located at the center of the symmetric curve and is the same as the median. Changing μ without changing σ moves the Normal curve along the horizontal axis without changing its spread. The standard deviation σ controls the spread of a Normal curve. Figure 1.22 shows two Normal curves with different values of σ. The curve with the larger standard deviation is more spread out.

The standard deviation σ is the natural measure of spread for Normal distributions. Not only do μ and σ completely determine the shape of a Normal curve, but we can

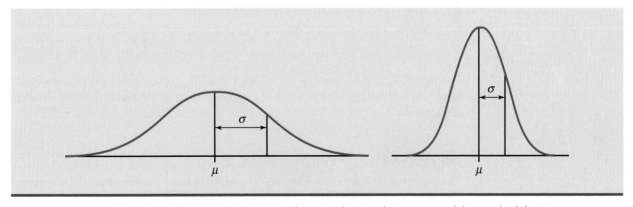

FIGURE 1.22 Two Normal curves, showing the mean μ and the standard deviation σ.

locate σ by eye on the curve. Here's how. Imagine that you are skiing down a mountain that has the shape of a Normal curve. At first, you descend at an ever-steeper angle as you go out from the peak:

Fortunately, before you find yourself going straight down, the slope begins to grow flatter rather than steeper as you go out and down:

The points at which this change of curvature takes place are located along the horizontal axis at distance σ on either side of the mean μ. Remember that μ and σ alone do not specify the shape of most distributions, and that the shape of density curves in general does not reveal σ. These are special properties of Normal distributions.

Why are the Normal distributions important in statistics? Here are three reasons. First, Normal distributions are good descriptions for some distributions of *real data*. Distributions that are often close to Normal include scores on tests taken by many people (such as GMAT exams), repeated careful measurements of the same quantity (such as measurements taken from a production process), and characteristics of biological populations (such as yields of corn). Second, Normal distributions are good approximations to the results of many kinds of *chance outcomes,* such as tossing a coin many times. Third, and most important many of the *statistical inference* procedures that we will study in later chapters are based on Normal distributions.

The 68–95–99.7 rule

Although there are many Normal curves, they all have common properties. In particular, all Normal distributions obey the following rule.

> **The 68–95–99.7 Rule**
>
> In the Normal distribution with mean μ and standard deviation σ:
>
> - **68%** of the observations fall within σ of the mean μ.
> - **95%** of the observations fall within 2σ of μ.
> - **99.7%** of the observations fall within 3σ of μ.

Figure 1.23 illustrates the 68–95–99.7 rule. By remembering these three numbers, you can think about Normal distributions without constantly making detailed calculations.

EXAMPLE 1.31   Using the 68–95–99.7 Rule

The distribution of weights of 9-ounce bags of a particular brand of potato chips is approximately Normal with mean $\mu = 9.12$ ounces and standard deviation $\sigma = 0.15$ ounce. Figure 1.24 shows what the 68–95–99.7 rule says about this distribution.

FIGURE 1.23 The 68–95–99.7 rule for Normal distributions.

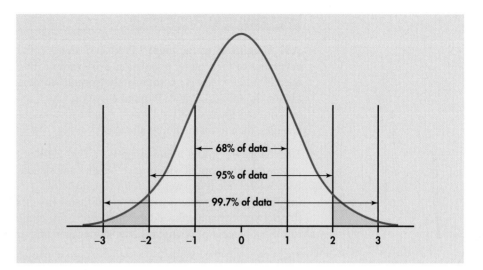

FIGURE 1.24 The 68–95–99.7 rule applied to the distribution of weights of bags of potato chips, for Example 1.31.

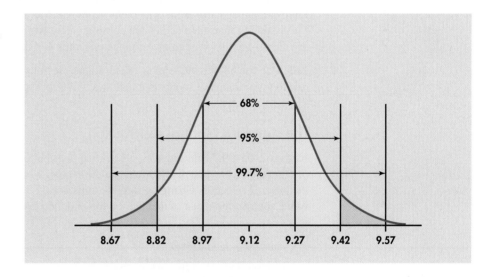

Two standard deviations is 0.3 ounces for this distribution. The 95 part of the 68–95–99.7 rule says that the middle 95% of 9-ounce bags weigh between $9.12 - 0.3$ and $9.12 + 0.3$ ounces, that is, between 8.82 ounces and 9.42 ounces. This fact is exactly true for an exactly Normal distribution. It is approximately true for the weights of 9-ounce bags of chips because the distribution of these weights is approximately Normal.

The other 5% of bags have weights outside the range from 8.82 to 9.42 ounces. Because the Normal distributions are symmetric, half of these bags are on the heavy side. So the heaviest 2.5% of 9-ounce bags are heavier than 9.42 ounces.

The 99.7 part of the 68–95–99.7 rule says that almost all bags (99.7% of them) have weights between $\mu - 3\sigma$ and $\mu + 3\sigma$. This range of weights is 8.67 to 9.57 ounces.

Because we will mention Normal distributions often, a short notation is helpful. We abbreviate the Normal distribution with mean μ and standard deviation σ as $N(\mu, \sigma)$. For example, the distribution of weights in the previous example is $N(9.12, 0.15)$.

1.82 Heights of young men. Product designers often must consider physical characteristics of their target population. For example, the distribution of heights of men aged 20 to 29 years is approximately Normal with mean 69 inches and standard deviation 2.5 inches. Draw a Normal curve on which this mean and standard deviation are correctly located. (*Hint:* Draw the curve first, locate the points where the curvature changes, then mark the horizontal axis.)

1.83 More on young men's heights. The distribution of heights of young men is approximately Normal with mean 69 inches and standard deviation 2.5 inches. Use the 68–95–99.7 rule to answer the following questions.
(a) What percent of these men are taller than 74 inches?
(b) Between what heights do the middle 95% of young men fall?
(c) What percent of young men are shorter than 66.5 inches?

1.84 Test scores. Many states have programs for assessing the skills of students in various grades. The Indiana Statewide Testing for Educational Progress (ISTEP) is one such program.[37] In a recent year, 76,531, tenth-grade Indiana students took the English/language arts exam. The mean score was 572 and the standard deviation was 51. Assuming that these scores are approximately Normally distributed, $N(572, 51)$, use the 68–95–99.7 rule to give a range of scores that includes 95% of these students.

1.85 Use the 68–95–99.7 rule. Refer to the previous exercise. Use the 68–95–99.7 rule to give a range of scores that includes 99.7% of these students.

The standard Normal distribution

As the 68–95–99.7 rule suggests, all Normal distributions share many common properties. In fact, all Normal distributions are the same if we measure in units of size σ about the mean μ as center. Changing to these units is called *standardizing*. To standardize a value, subtract the mean of the distribution and then divide by the standard deviation.

> **Standardizing and z-Scores**
>
> If x is an observation from a distribution that has mean μ and standard deviation σ, the **standardized value** of x is
>
> $$z = \frac{x - \mu}{\sigma}$$
>
> A standardized value is often called a **z-score.**

A z-score tells us how many standard deviations the original observation falls away from the mean, and in which direction. Observations larger than the mean are positive when standardized, and observations smaller than the mean are negative when standardized.

EXAMPLE 1.32   Standardizing Potato Chip Bag Weights

The weights of 9-ounce potato chip bags are approximately Normal with $\mu = 9.12$ ounces and $\sigma = 0.15$ ounce. The standardized weight is

$$z = \frac{\text{weight} - 9.12}{0.15}$$

A bag's standardized weight is the number of standard deviations by which its weight differs from the mean weight of all bags. A bag weighing 9.3 ounces, for example, has *standardized* weight

$$z = \frac{9.3 - 9.12}{0.15} = 1.2$$

or 1.2 standard deviations above the mean. Similarly, a bag weighing 8.7 ounces has standardized weight

$$z = \frac{8.7 - 9.12}{0.15} = -2.8$$

or 2.8 standard deviations below the mean bag weight.

If the variable we standardize has a Normal distribution, standardizing does more than give a common scale. It makes all Normal distributions into a single distribution, and this distribution is still Normal. Standardizing a variable that has any Normal distribution produces a new variable that has the *standard Normal distribution*.

> **Standard Normal Distribution**
>
> The **standard Normal distribution** is the Normal distribution $N(0, 1)$ with mean 0 and standard deviation 1.
>
> If a variable x has any Normal distribution $N(\mu, \sigma)$ with mean μ and standard deviation σ, then the standardized variable
>
> $$z = \frac{x - \mu}{\sigma}$$
>
> has the standard Normal distribution.

APPLY YOUR KNOWLEDGE

1.86 SAT versus ACT. Eleanor scores 680 on the Mathematics part of the SAT. The distribution of SAT scores in a reference population is Normal, with mean 500 and standard deviation 100. Gerald takes the American College Testing (ACT) Mathematics test and scores 27. ACT scores are Normally distributed with mean 18 and standard deviation 6. Find the standardized scores for both students. Assuming that both tests measure the same kind of ability, who has the higher score?

Normal distribution calculations

Areas under a Normal curve represent proportions of observations from that Normal distribution. There is no easy formula for areas under a Normal curve. To find areas of interest, either software that calculates areas or a table of areas can be used. The table and most software calculate one kind of area: **cumulative proportions.** A cumulative proportion is the proportion of observations in a distribution that lie at or below a given value. When the distribution is given by a density curve, the cumulative proportion is the area under the curve to the left of a given value. Figure 1.25 shows the idea more clearly than words do.

cumulative proportion

The key to calculating Normal proportions is to match the area you want with areas that represent cumulative proportions. Then get areas for cumulative proportions. The following examples illustrate the methods.

FIGURE 1.25 The *cumulative proportion* for a value *x* is the proportion of all observations from the distribution that are less than or equal to *x*. This is the area to the left of *x* under the Normal curve.

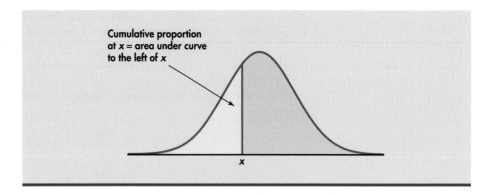

Cumulative proportion at *x* = area under curve to the left of *x*

x

EXAMPLE 1.33   The NCAA Standard for SAT Scores

The National Collegiate Athletic Association (NCAA) requires Division I athletes to get a combined score of at least 820 on the SAT Mathematics and Verbal tests to compete in their first college year. (Higher scores are required for students with poor high school grades.) The scores of the 1.4 million students in the class of 2003 who took the SATs were approximately Normal with mean 1026 and standard deviation 209. What proportion of all students had SAT scores of at least 820?

Here is the calculation in pictures: the proportion of scores above 820 is the area under the curve to the right of 820. That's the total area under the curve (which is always 1) minus the cumulative proportion up to 820.

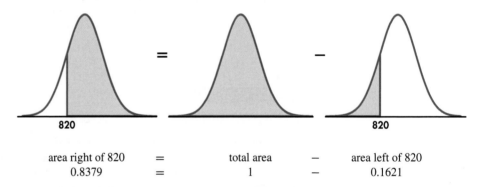

820   820

| area right of 820 | = | total area | − | area left of 820 |
| 0.8379 | = | 1 | − | 0.1621 |

That is, the proportion of all SAT takers who would be NCAA qualifiers is 0.8379, or about 84%.

There is *no* area under a smooth curve and exactly over the point 820. Consequently, the area to the right of 820 (the proportion of scores > 820) is the same as the area at or to the right of this point (the proportion of scores ≥ 820). The actual data may contain a student who scored exactly 820 on the SAT. That the proportion of scores exactly equal to 820 is 0 for a Normal distribution is a consequence of the idealized smoothing of Normal distributions for data.

EXAMPLE 1.34   NCAA Partial Qualifiers

The NCAA considers a student a "partial qualifier" eligible to practice and receive an athletic scholarship, but not to compete, if the combined SAT score is at least 720. What proportion of all students who take the SAT would be partial qualifiers? That is, what proportion have scores between 720 and 820? Here are the pictures:

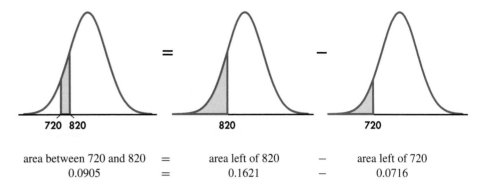

area between 720 and 820	=	area left of 820	–	area left of 720
0.0905	=	0.1621	–	0.0716

About 9% of all students who take the SAT have scores between 720 and 820.

How do we find the numerical values of the areas in Examples 1.33 and 1.34? If you use software, just plug in mean 1026 and standard deviation 209. Then ask for the cumulative proportions for 820 and for 720. (Your software will probably refer to these as "cumulative probabilities." We will learn in Chapter 4 why the language of probability fits.) If you make a sketch of the area you want, you will rarely go wrong.

You can use the *Normal Curve* applet on the text CD and Web site to find Normal proportions. The applet is more flexible than most software—it will find any Normal proportion, not just cumulative proportions. The applet is an excellent way to understand Normal curves. But, because of the limitations of Web browsers, the applet is not as accurate as statistical software.

If you are not using software, you can find cumulative proportions for Normal curves from a table. That requires an extra step, as we now explain.

Using the standard Normal table

The extra step in finding cumulative proportions from a table is that we must first standardize to express the problem in the standard scale of z-scores. This allows us to get by with just one table, a table of *standard Normal cumulative proportions*. Table A in the back of the book gives cumulative proportions for the standard Normal distribution. Table A also appears on the inside front cover. The pictures at the top of the table remind us that the entries are cumulative proportions, areas under the curve to the left of a value z.

EXAMPLE 1.35   Find the Proportion from z

What proportion of observations on a standard Normal variable Z take values less than $z = 1.47$?

Solution: To find the area to the left of 1.47, locate 1.4 in the left-hand column of Table A, then locate the remaining digit 7 as .07 in the top row. The entry opposite 1.4 and under .07 is 0.9292. This is the cumulative proportion we seek. Figure 1.26 illustrates this area.

Now that you see how Table A works, let's redo the NCAA Examples 1.33 and 1.34 using the table.

FIGURE 1.26 The area under the standard Normal curve to the left of the point $z = 1.47$ is 0.9292, for Example 1.35.

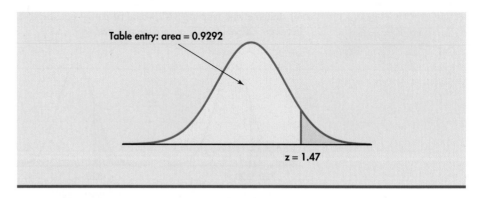

Table entry: area = 0.9292

$z = 1.47$

EXAMPLE 1.36   Find the Proportion from x

What proportion of all students who take the SAT have scores of at least 820? The picture that leads to the answer is exactly the same as in Example 1.33. The extra step is that we first standardize in order to read cumulative proportions from Table A. If X is SAT score, we want the proportion of students for whom $X \geq 820$.

Step 1. *Standardize.* Subtract the mean, then divide by the standard deviation, to transform the problem about X into a problem about a standard Normal Z:

$$X \geq 820$$

$$\frac{X - 1026}{209} \geq \frac{820 - 1026}{209}$$

$$Z \geq -0.99$$

Step 2. *Use the table.* Look at the pictures in Example 1.33. From Table A, we see that the proportion of observations less than -0.99 is 0.1611. The area to the right of -0.99 is therefore $1 - 0.1611 = 0.8389$. This is about 84%.

The area from the table in Example 1.36 (0.8389) is slightly less accurate than the area from software in Example 1.33 (0.8379) because we must round z to two places when we use Table A. The difference is rarely important in practice.

EXAMPLE 1.37   Proportion of Partial Qualifiers

What proportion of all students who take the SAT would be partial qualifiers in the eyes of the NCAA? That is, what proportion of students have SAT scores between 720 and 820? First, sketch the areas, exactly as in Example 1.34. We again use X as shorthand for an SAT score.

Step 1. Standardize.

$$720 \leq X < 820$$

$$\frac{720 - 1026}{209} \leq \frac{X - 1026}{209} < \frac{820 - 1026}{209}$$

$$-1.46 \leq Z < -0.99$$

Step 2. Use the table.

$$\text{area between } -1.46 \text{ and } -0.99 = (\text{area left of } -0.99) - (\text{area left of } -1.46)$$

$$= 0.1611 - 0.0721 = 0.0890$$

As in Example 1.34, about 9% of students would be partial qualifiers.

Sometimes we encounter a value of z more extreme than those appearing in Table A. For example, the area to the left of $z = -4$ is not given directly in the table. The z-values in Table A leave only area 0.0002 in each tail unaccounted for. For practical purposes, we can act as if there is zero area outside the range of Table A.

APPLY YOUR KNOWLEDGE

1.87 Find the proportion. Use the fact that the ISTEP scores from Exercise 1.84 (page 48) are approximately Normal, $N(572, 51)$. Find the proportion of students who have scores less than 600. Find the proportion of students who have scores greater than or equal to 600. Sketch the relationship between these two calculations using pictures of Normal curves similar to the ones given in Example 1.33.

1.88 Find another proportion. Use the fact that the ISTEP scores are approximately Normal, $N(572, 51)$. Find the proportion of students who have scores between 600 and 650. Use pictures of Normal curves similar to the ones given in Example 1.34 to illustrate your calculations.

Inverse Normal calculations

Examples 1.33 to 1.36 illustrate the use of Normal distributions to find the proportion of observations in a given event, such as "SAT score between 720 and 820." We may instead want to find the observed value corresponding to a given proportion.

Statistical software will do this directly. Without software, use Table A backward, finding the desired proportion in the body of the table and then reading the corresponding z from the left column and top row.

EXAMPLE 1.38   How High for the Top 10%?

Scores on the SAT Verbal test in recent years follow approximately the $N(505, 110)$ distribution. How high must a student score in order to place in the top 10% of all students taking the SAT?

Again, the key to the problem is to draw a picture. Figure 1.27 shows that we want the score x with area above it 0.10. That's the same as area below x equal to 0.90.

Statistical software has a function that will give you the x for any cumulative proportion you specify. The function often has a name such as "inverse cumulative probability." Plug in mean 505, standard deviation 110, and cumulative proportion 0.9. The software tells you that $x = 645.97$. We see that a student must score at least 646 to place in the highest 10%.

FIGURE 1.27 Locating the point on a Normal curve with area 0.10 to its right, for Example 1.38. The result is $x = 646$, or $z = 1.28$ in the standard scale.

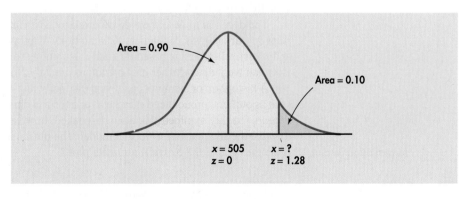

Without software, first find the standard score z with cumulative proportion 0.9, then "unstandardize" to find x. Here is the two-step process:

1. *Use the table.* Look in the body of Table A for the entry closest to 0.9. It is 0.8997. This is the entry corresponding to $z = 1.28$. So $z = 1.28$ is the standardized value with area 0.9 to its left.

2. *Unstandardize* to transform the solution from z back to the original x scale. We know that the standardized value of the unknown x is $z = 1.28$. So x itself satisfies

$$\frac{x - 505}{110} = 1.28$$

Solving this equation for x gives

$$x = 505 + (1.28)(110) = 645.8$$

This equation should make sense: it finds the x that lies 1.28 standard deviations above the mean on this particular Normal curve. That is the "unstandardized" meaning of $z = 1.28$. The general rule for unstandardizing a z-score is

$$x = \mu + z\sigma$$

APPLY YOUR KNOWLEDGE

1.89 What score is needed to be in the top 5%?   Consider the ISTEP scores, which are approximately Normal, $N(572, 51)$. How high a score is needed to be in the top 5% of students who take this exam?

1.90 Find the score that 60% of students will exceed.   Consider the ISTEP scores, which are approximately Normal, $N(572, 51)$. Sixty percent of the students will score above x on this exam. Find x.

$X \sim N(572, 51)$

$0.6 = P(X < x)$

Assessing the Normality of data

The Normal distributions provide good models for some distributions of real data. Examples include the miles per gallon ratings of vehicles, average payrolls of Major League Baseball teams, and statewide unemployment rates. The distributions of some other common variables are usually skewed and therefore distinctly non-Normal. Examples include personal income, gross sales of business firms, and the service lifetime of mechanical or electronic components. While experience can suggest whether or not a Normal model is plausible in a particular case, it is risky to assume that a distribution is Normal without actually inspecting the data.

The decision to describe a distribution by a Normal model may determine the later steps in our analysis of the data. Calculations of proportions, as we have done above, and statistical inference based on such calculations follow from the choice of a model. How can we judge whether data are approximately Normal?

A histogram or stemplot can reveal distinctly non-Normal features of a distribution, such as outliers, pronounced skewness, or gaps and clusters. If the stemplot or histogram appears roughly symmetric and single-peaked, however, we need a more sensitive way to judge the adequacy of a Normal model. The most useful tool for assessing Normality is another graph, the **Normal quantile plot.**[*]

Normal quantile plot

---

*Some software calls these graphs *Normal probability plots*. There is a technical distinction between the two types of graphs, but the terms are often used loosely.

Here is the idea of a simple version of a Normal quantile plot. It is not feasible to make Normal quantile plots by hand, but software makes them for us, using more sophisticated versions of this basic idea.

1. Arrange the observed data values from smallest to largest. Record what percentile of the data each value occupies. For example, the smallest observation in a set of 20 is at the 5% point, the second smallest is at the 10% point, and so on.

Normal scores

2. Find the same percentiles for the Normal distribution using Table A or statistical software. Percentiles of the standard Normal distribution are often called **Normal scores.** For example, $z = -1.645$ is the 5% point of the standard Normal distribution, and $z = -1.282$ is the 10% point.

3. Plot each data point x against the corresponding Normal score z. If the data distribution is close to standard Normal, the plotted points will lie close to the 45-degree line $x = z$. If the data distribution is close to any Normal distribution, the plotted points will lie close to some straight line.

Any Normal distribution produces a straight line on the plot because standardizing turns any Normal distribution into a standard Normal distribution. Standardizing is a linear transformation that can change the slope and intercept of the line in our plot but cannot turn a line into a curved pattern.

> **Use of Normal Quantile Plots**
>
> If the points on a Normal quantile plot lie close to a straight line, the plot indicates that the data are Normal. Systematic deviations from a straight line indicate a non-Normal distribution. Outliers appear as points that are far away from the overall pattern of the plot.

Figures 1.28 to 1.31 are Normal quantile plots for data we have met earlier. The data x are plotted vertically against the corresponding Normal scores z plotted horizontally. For small data sets, the z axis extends from -3 to 3 because almost all of a standard Normal curve lies between these values. With larger sample sizes, values in the extremes are more likely, and the z axis will extend farther from zero. These figures show how Normal quantile plots behave.

EXAMPLE 1.39   IQ Scores Are Normal

In Example 1.18 we examined the distribution of IQ scores for a sample of 60 fifth-grade students. Figure 1.28 gives a Normal quantile plot for these data. Notice that the points have a pattern that is pretty close to a straight line. This pattern indicates that the distribution is approximately Normal. When we constructed a histogram of the data in Figure 1.11 (page 18), we noted that the distribution has a single peak, is approximately symmetric, and has tails that decrease in a smooth way. We can now add to that description by stating that the distribution is approximately Normal.

Figure 1.28 does, of course, show some deviation from a straight line. Real data almost always show some departure from the theoretical Normal model. It is important to confine your examination of a Normal quantile plot to searching for shapes that show *clear departures from Normality*. Don't overreact to minor wiggles in the plot. When we discuss statistical methods that are based on the Normal model, we will pay attention to the sensitivity of each method to departures from Normality. Many common methods work well as long as the data are reasonably symmetric and outliers are not present.

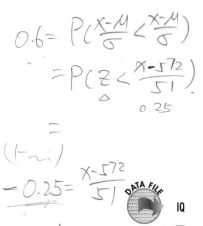

FIGURE 1.28 Normal quantile plot for the IQ data, for Example 1.39. This pattern indicates that the data are approximately Normal.

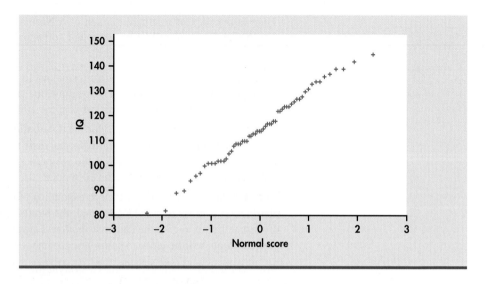

EXAMPLE 1.40   T-bill Interest Rates Are Not Normal

We made a histogram for the distribution of interest rates for T-bills in Example 1.12 (page 12). A Normal quantile plot for these data is shown in Figure 1.29. This plot shows some interesting features of the distribution. First, in the central part, from about $z = -2$ to $z = 1$, the points fall approximately on a straight line. This suggests that the distribution is approximately Normal in this range. Then there is the region from slightly above $z = 1$ to slightly above $z = 2$, where the points also fall approximately on a straight line. This line, however, has a different slope. Combined, these features suggest that the distribution of interest rates may actually be a mixture or a combination of two Normal populations. Finally, in both the lower and the upper extremes the points flatten out. This occurs at an interest rate of around 1% for the lower tail and at 15% for the upper tail. There may be some marked considerations that restrain interest rates from going outside these bounds.

FIGURE 1.29 Normal quantile plot for the T-bill interest rates, for Example 1.40. These data are not approximately Normal.

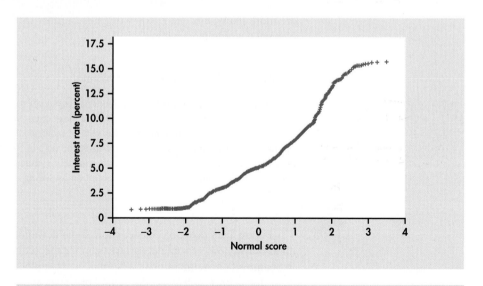

The idea that distributions are approximately Normal within a range of values is an old tradition. The remark "All distributions are approximately Normal in the middle" has been attributed to the statistician Charlie Winsor.[38]

APPLY YOUR KNOWLEDGE

TIMETOSTART25

CASE 1.2 **1.91 Length of time to start a business.**   In Exercise 1.40 we noted that the sample of times to start a business from 25 countries contained an outlier. For Suriname, the reported time is 694 days. This case is the most extreme in the entire data set, which includes 195 counties. Figure 1.30 shows the Normal quantile plot for these data with Suriname excluded.

(a) These data are skewed to the right. How does this feature appear in the Normal quantile plot?

(b) Compare the shape of the upper portion of this Normal quantile plot with the upper portion of the plot for the T-bill interest rates in Figure 1.29, and with the upper portion of the plot for the IQ scores in Figure 1.28. Make a general statement about what the shape of the upper portion of a Normal quantile plot tells you about the upper tail of a distribution.

CALLCENTER

1.92 Customer service center call lengths.   Figure 1.31 is a Normal quantile plot for the customer center call lengths. We looked at these data in Example 1.14, and we examined the distribution using a histogram in Figure 1.8 (page 14). There are clearly some very large outliers. In making the Normal quantile plot, we eliminated all calls that lasted longer than 2 hours (7200 seconds). This distribution is strongly skewed to the right. How does this show up in the Normal quantile plot?

BEYOND THE BASICS: Density Estimation

A density curve gives a compact summary of the overall shape of a distribution. Figure 1.17 (page 41) shows a Normal density curve that summarizes the distribution of miles per gallon ratings for 1140 vehicles. It captures some characteristics of the distribution but misses others.

Many distributions do not have the Normal shape. There are other families of density curves that are used as mathematical models for various distribution shapes.

density estimation

Modern software offers a more flexible option: **density estimation.** A density estimator does not start with any specific shape, such as the Normal shape. It looks at the data and draws a density curve that describes the overall shape of the data.

FIGURE 1.30 Normal quantile plot for the length of time required to start a business, for Exercise 1.91. Suriname, with a time of 694 days, has been excluded.

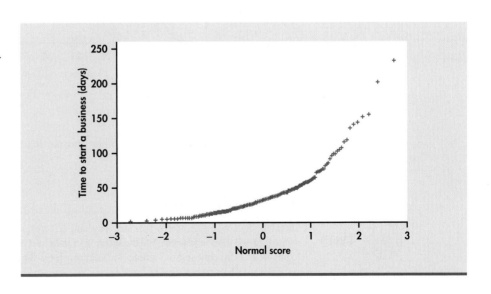

FIGURE 1.31 Normal quantile plot for the customer service center call lengths, for Exercise 1.93. Data for calls lasting more than 7200 seconds (2 hours) have been excluded.

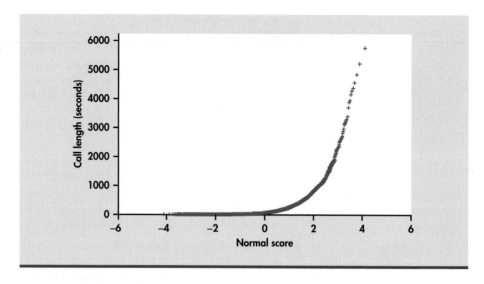

EXAMPLE 1.41   Fuel Efficiency Data

MPG2009

Figure 1.32 gives the histogram of the miles per gallon distribution with a density estimate produced by software. Compare this figure with Figure 1.17 (page 41). Notice how the density estimate captures more of the unusual features of the distribution than the Normal density curve does.

FIGURE 1.32 Histogram of fuel efficiency for 1140 vehicles, with a density estimate, for Example 1.41.

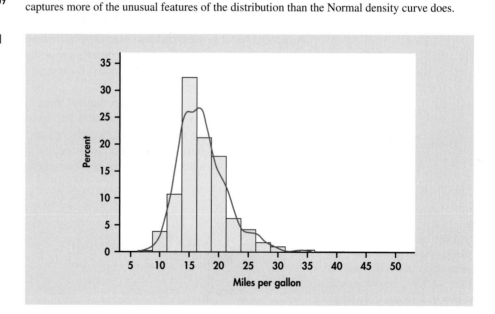

Density estimates can capture other unusual features of a distribution. Here is an example.

EXAMPLE 1.42   StubHub!

STUBHUB

StubHub! is a Web site where fans can buy and sell tickets to sporting events. Ticket holders wanting to sell their tickets provide the location of their seats and the selling price. People wanting to buy tickets can choose from among the tickets offered for a given event.[39]

On Saturday, October 18, 2008, the eleventh-ranked Missouri football team was scheduled to play the first-ranked Texas team in Austin. On Thursday, October 16, 2008, StubHub! listed 64 pairs of tickets for the game. One pair was offered at $883 per ticket. It was noted that these seats were in a suite and that food and bar were included. We discarded this outlier and examined the distribution of the price per ticket for the remaining 63 pairs of tickets. The histogram with a density estimate is given in Figure 1.33. The distribution has two peaks, one around $160 and another around $360. This is the identifying characteristic of a **bimodal distribution.** Since the stadium has upper- and lower-level seats, we suspect that the difference in price between these two types of seats is responsible for the two peaks. (Texas won 56 to 31.)

bimodal distribution

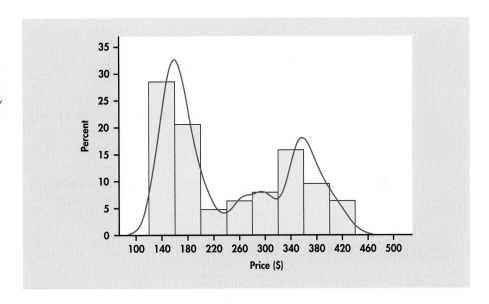

FIGURE 1.33 Histogram of StubHub! price per seat for tickets to the Missouri-Texas football game on October 18, 2008, with a density estimate, for Example 1.42. One outlier, with a price per seat of $883, was deleted.

Example 1.42 reminds us of a continuing theme for data analysis. We looked at a histogram and a density estimate and saw something interesting. This led us to speculate. Additional data on the type and location of the seats may explain more about the prices than we see in Figure 1.33.

SECTION 1.3 Summary

- We can sometimes describe the overall pattern of a distribution by a **density curve.** A density curve has total area 1 underneath it. An area under a density curve gives the proportion of observations that fall in a range of values.

- A density curve is an idealized description of the overall pattern of a distribution that smooths out the irregularities in the actual data. We write the mean of a density curve as μ and the standard deviation of a density curve as σ to distinguish them from the mean $\overline{x}$ and standard deviation s of the actual data.

- The mean, the median, and the quartiles of a density curve can be located by eye. The **mean** μ is the balance point of the curve. The **median** divides the area under the curve in half. The **quartiles** and the median divide the area under the curve into quarters. The **standard deviation** σ cannot be located by eye on most density curves.

- The mean and median are equal for symmetric density curves. The mean of a skewed curve is located farther toward the long tail than is the median.

- The **Normal distributions** are described by a special family of bell-shaped, symmetric density curves, called **Normal curves.** The mean μ and standard deviation σ completely specify a Normal distribution $N(\mu, \sigma)$. The mean is the center of the curve, and σ is the distance from μ to the change-of-curvature points on either side.

- To **standardize** any observation x, subtract the mean of the distribution and then divide by the standard deviation. The resulting **z-score**

$$z = \frac{x - \mu}{\sigma}$$

  says how many standard deviations x lies from the distribution mean.

- All Normal distributions are the same when measurements are transformed to the standardized scale. In particular, all Normal distributions satisfy the **68–95–99.7 rule,** which describes what percent of observations lie within one, two, and three standard deviations of the mean.

- If x has the $N(\mu, \sigma)$ distribution, then the **standardized variable** $z = (x - \mu)/\sigma$ has the **standard Normal distribution $N(0, 1)$** with mean 0 and standard deviation 1. Table A gives the proportions of standard Normal observations that are less than z for many values of z. By standardizing, we can use Table A for any Normal distribution.

- The adequacy of a Normal model for describing a distribution of data is best assessed by a **Normal quantile plot,** which is available in most statistical software packages. A pattern on such a plot that deviates substantially from a straight line indicates that the data are not Normal.

SECTION 1.3 Exercises

For Exercise 1.78, see page 43; for 1.79 to 1.81, see pages 44–45; for 1.82 to 1.85, see page 48; for 1.86, see page 49; for 1.87 and 1.88, see page 53; for 1.89 and 1.90, see page 54; and for 1.91 and 1.92, see page 57.

1.93 Sketch some Normal curves.
(a) Sketch a Normal curve that has mean 10 and standard deviation 3.
(b) On the same x axis, sketch a Normal curve that has mean 20 and standard deviation 3.
(c) How does the Normal curve change when the mean is varied but the standard deviation stays the same?

1.94 The effect of changing the standard deviation.
(a) Sketch a Normal curve that has mean 10 and standard deviation 3.
(b) On the same x axis, sketch a Normal curve that has mean 10 and standard deviation 1.
(c) How does the Normal curve change when the standard deviation is varied but the mean stays the same?

1.95 Know your density. Sketch density curves that might describe distributions with the following shapes.
(a) Symmetric, but with two peaks (that is, two strong clusters of observations).
(b) Single peak and skewed to the left.

1.96 Gross domestic product. Refer to Exercise 1.52, where we examined the gross domestic product of 120 countries.
COUNTRIES120

(a) Compute the mean and the standard deviation.
(b) Apply the 68–95–99.7 rule to this distribution.
(c) Compare the results of the rule with the actual percents within one, two, and three standard deviations of the mean.
(d) Summarize your conclusions.

1.97 Do women talk more? Conventional wisdom suggests that women are more talkative than men. One study designed to examine this stereotype collected data on the speech of 42 women and 37 men in the United States.[40]
(a) The mean number of words spoken per day by the women was 14,297 with a standard deviation of 9065. Use the 68–95–99.7 rule to describe this distribution.
(b) Do you think that applying the rule in this situation is reasonable? Explain your answer.
(c) The men averaged 14,060 words per day with a standard deviation of 9056. Answer the questions in parts (a) and (b) for the men.
(d) Do you think that the data support the conventional wisdom? Explain your answer. Note that in Section 7.2 we will learn formal statistical methods to answer this type of question.

1.98 Data from Mexico. Refer to the previous exercise. A similar study in Mexico was conducted with 31 women and 20 men. The women averaged 14,704 words per day with a standard deviation of 6215. For men the mean was 15,022 and the standard deviation was 7864.
(a) Answer the questions from the previous exercise for the Mexican study.

(b) The means for both men and women are higher for the Mexican study than for the U.S. study. What conclusions can you draw from this observation?

1.99 Total scores. Below are the total scores of 10 students in an introductory statistics course: 🔲 **STATCOURSE**

> 68   54   92   75   73   98   64   55   80   70

Previous experience with this course suggests that these scores should come from a distribution that is approximately Normal with mean 70 and standard deviation 10.

(a) Using these values for μ and σ, standardize the scores of these 10 students.

(b) If the grading policy is to give a grade of A to the top 15% of scores based on the Normal distribution with mean 70 and standard deviation 10, what is the cutoff for an A in terms of a standardized score?

(c) Which students earned an A for this course?

1.100 Assign more grades. Refer to the previous exercise. The grading policy says that the cutoffs for the other grades correspond to the following: the bottom 5% receive an F, the next 10% receive a D, the next 40% receive a C, and the next 30% receive a B. These cutoffs are based on the $N(70, 10)$ distribution.

(a) Give the cutoffs for the grades in terms of standardized scores.

(b) Give the cutoffs in terms of actual scores.

(c) Do you think that this method of assigning grades is a good one? Give reasons for your answer.

1.101 Selling apartment buildings. Owning an apartment building can be very profitable, as can selling an apartment building. Data for this exercise are selling prices (in dollars) and building square footages for 18 apartment buildings sold in a particular city during 2005.[41] 🔲 **APARTMENTS**

(a) Use statistical software to obtain histograms and Normal quantile plots of selling prices and building square footages.

(b) Do either of these variables appear to be Normally distributed? Explain in what way the plots match (or don't match) what you would expect to see for Normally distributed data.

(c) One apartment building appears to be an outlier with respect to both selling price and square footage. Report the selling price and square footage for this apartment building.

1.102 Selling apartment buildings. Continue with the data from the previous exercise. Create a new variable (call it Sale Price Per Sqft) by dividing the selling price for each apartment building by the square footage for each apartment building. 🔲 **APARTMENTS**

(a) When plotting selling prices or building square footages, one apartment building stands out as an outlier. Does this same apartment building stand out in terms of the new variable you created for this exercise? Explain your response clearly.

(b) Use statistical software to obtain a histogram and a Normal quantile plot of the new variable Sale Price Per Sqft.

(c) Does the distribution of Sale Price Per Sqft appear to be Normal? Describe precisely what about the histogram and the Normal quantile plot leads you to your conclusion.

1.103 Selling apartment buildings. Continue with the variable Sale Price Per Sqft created in the previous exercise. 🔲 **APARTMENTS**

(a) Calculate the mean and standard deviation of the Sale Price Per Sqft values.

(b) Calculate the intervals $\bar{x} \pm s$, $\bar{x} \pm 2s$, and $\bar{x} \pm 3s$.

(c) Create a table that allows one to easily compare the distribution of Sale Price Per Sqft with the 68–95–99.7 rule for the three intervals calculated in part (b).

(d) Does your table from part (c) provide a clear indication of Normality (or non-Normality) for the data values?

1.104 Exploring Normal quantile plots.

(a) Create three data sets: one that is clearly skewed to the right, one that is clearly skewed to the left, and one that is clearly symmetric and mound-shaped. (As an alternative to creating data sets, you can look through this chapter and find an example of each type of data set requested.)

(b) Using statistical software, obtain Normal quantile plots for each of your three data sets.

(c) Clearly describe the pattern of each data set in the Normal quantile plots from part (b).

The table below contains data on a random sample of 22 telecom stocks—companies that specialize in telecommunication products. For each company, trading volume and revenue growth (over the last year) have been reported. Exercises 1.105 to 1.108 concern these data.[42]

Ticker symbol	Trading volume	Revenue growth
AATK	68,654	0.0482
ALLN	3,500	−0.0300
ATGN	5,650	0.1514
AVCI	68,482	0.2580
AXE	85,900	0.0739
CGN	100	−0.1098
COVD	2,410,204	−0.0166
CTV	254,600	−0.0437
CYBD	6,900	0
ETCIA	1,741	−0.2391
GCOM	27,392	0.4337
HLIT	690,026	−0.1765
PCTU	6,500	−0.2898
PTSC	314,680	−0.556
QCOM	6,696,185	0.2001
SRTI	2,000	−0.0006
TCCO	1,100	0.0856
TKLC	246,101	0.0009
VERA	25,000	0.0081
WJCI	59,408	−0.3544
XXIA	1,750,027	0.1930
ZOOM	21,295	−0.1298

1.105 Telecom shares traded. 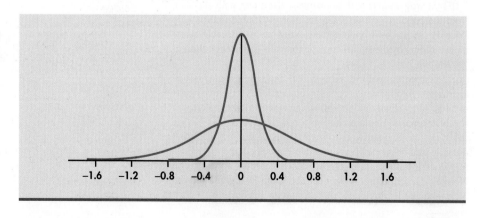 **TELECOMSTOCKS**
(a) Calculate the mean and standard deviation of the 22 trading-volume values.
(b) Calculate $\bar{x} \pm 3s$.
(c) Clearly explain why your calculations in part (b) show that the distribution of trading volume is not symmetric and mound-shaped.

1.106 Telecom revenue growth. **TELECOMSTOCKS**
(a) Calculate the mean and standard deviation of the 22 revenue growth values.
(b) Calculate the ranges $\bar{x} \pm s$, $\bar{x} \pm 2s$, and $\bar{x} \pm 3s$.
(c) Determine the percent of revenue growth values that fall into each of the three ranges that you calculated in part (b). How do these percents compare with the 68–95–99.7 rule?

1.107 Telecom shares traded. **TELECOMSTOCKS**
(a) Use statistical software to create a histogram of the trading volumes for these 22 telecom stocks.
(b) The histogram shows that these data are clearly right-skewed. Sketch what you think a Normal quantile plot of these data will look like.
(c) Use statistical software to create a Normal quantile plot of these data. How well does your sketch from part (b) match the plot generated by your software?

1.108 Telecom revenue growth. **TELECOMSTOCKS**
(a) Construct a stemplot of the revenue growth for these 22 telecom stocks. You will need a 0 and a −0 on the stem. Use the tenths place of these values on the stem and the hundredths place as the leaves. For example, −0.556 rounds to −0.56 and would appear as −5|6 in the stemplot.
(b) Describe the distribution of these revenue growth values. Sketch what you think a Normal quantile plot of these data will look like.
(c) Use statistical software to create a Normal quantile plot of these data. How well does your sketch from part (b) match the plot generated by your software?

1.109 Visualizing the standard deviation. Figure 1.34 shows two Normal curves, both with mean 0. Approximately what is the standard deviation of each of these curves?

1.110 Length of pregnancies. Some health insurance companies treat pregnancy as a "preexisting condition" when it comes to paying for maternity expenses for a new policyholder. Sometimes the exact date of conception is unknown, so the insurance company must count back from the expected due date to judge whether or not conception occurred before or after the new policy began. The length of human pregnancies from conception to birth varies according to a distribution that is approximately Normal with mean 266 days and standard deviation 16 days. Use the 68–95–99.7 rule to answer the following questions.
(a) Between what values do the lengths of the middle 95% of all pregnancies fall?
(b) How short are the shortest 2.5% of all pregnancies?
(c) How likely is it that a woman with an expected due date 218 days after her policy began conceived the child after her policy began?

1.111 Use Table A. Use Table A to find the proportion of observations from a standard Normal distribution that falls in each of the following regions. In each case, sketch a standard Normal curve and shade the area representing the region.
(a) $z \leq -2.30$
(b) $z \geq -2.30$
(c) $z > 1.70$
(d) $-2.30 < z < 1.70$

1.112 Use Table A. Use Table A to find the value of z for each of the situations below. In each case, sketch a standard Normal curve and shade the area representing the region.
(a) Ten percent of the values of a standard Normal distribution are greater than z.
(b) Ten percent of the values of a standard Normal distribution are greater than or equal to z.
(c) Ten percent of the values of a standard Normal distribution are less than z.
(d) Fifty percent of the values of a standard Normal distribution are less than z.

1.113 Use Table A. Consider a Normal distribution with mean 100 and standard deviation 10.
(a) Find the proportion of the distribution with values 90 and 105. Illustrate your calculation with a sketch.

FIGURE 1.34 Two Normal curves with the same mean but different standard deviations, for Exercise 1.109.

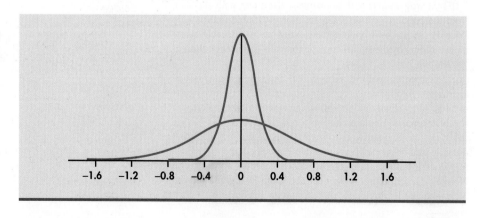

(b) Find the values of x_1 and x_2 such that the proportion of the distribution with values between x_1 and x_2 include the central 85% of the distribution. Illustrate your calculation with a sketch.

1.114 Length of pregnancies. The length of human pregnancies from conception to birth varies according to a distribution that is approximately Normal with mean 266 days and standard deviation 16 days.
(a) What percent of pregnancies last fewer than 240 days (that's about 8 months)?
(b) What percent of pregnancies last between 240 and 270 days (roughly between 8 and 9 months)?
(c) How long do the longest 25% of pregnancies last?

1.115 Quartiles of Normal distributions. The median of any Normal distribution is the same as its mean. We can use Normal calculations to find the quartiles for Normal distributions.
(a) What is the area under the standard Normal curve to the left of the first quartile? Use this to find the value of the first quartile for a standard Normal distribution. Find the third quartile similarly.
(b) Your work in (a) gives the Normal scores z for the quartiles of any Normal distribution. What are the quartiles for the lengths of human pregnancies? (Use the distribution given in the previous exercise.)

1.116 Deciles of Normal distributions. The *deciles* of any distribution are the 10th, 20th, . . . , 90th percentiles. The first and last deciles are the 10th and 90th percentiles, respectively.
(a) What are the first and last deciles of the standard Normal distribution?
(b) The weights of 9-ounce potato chip bags are approximately Normal with mean 9.12 ounces and standard deviation 0.15 ounce. What are the first and last deciles of this distribution?

1.117 Normal random numbers. Use software to generate 100 observations from the standard Normal distribution. Make a histogram of these observations. How does the shape of the histogram compare with a Normal density curve? Make a Normal quantile plot of the data. Does the plot suggest any important deviations from Normality? (Repeating this exercise several times is a good way to become familiar with how Normal quantile plots look when data actually are close to Normal.)

1.118 Uniform random numbers. Use software to generate 100 observations from the distribution described in Exercise 1.80 (page 44). (The software will probably call this a "uniform distribution.") Make a histogram of these observations. How does the histogram compare with the density curve in Figure 1.20? Make a Normal quantile plot of your data. According to this plot, how does the uniform distribution deviate from Normality?

STATISTICS IN SUMMARY

Data analysis is the art of describing data using graphs and numerical summaries. The purpose of data analysis is to describe the most important features of a set of data. This chapter introduces data analysis by presenting statistical ideas and tools for describing the distribution of a single variable. The Statistics in Summary figure below will help you organize the big ideas. The question marks at the last two stages remind us that the usefulness of numerical summaries and models such as Normal distributions depends on what we find when we examine the data using graphs. Here is a review list of the most important skills you should have acquired from your study of this chapter.

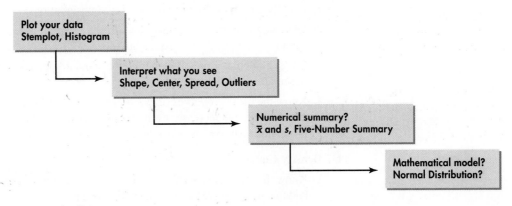

A. Data

1. Identify the cases and variables in a set of data.

2. Identify each variable as categorical or quantitative. Identify the units in which each quantitative variable is measured.

B. Displaying Distributions

1. Make a bar graph, pie chart, and/or Pareto chart of the distribution of a categorical variable. Interpret bar graphs, pie charts, and Pareto charts.

2. Make a histogram of the distribution of a quantitative variable.

3. Make a stemplot of the distribution of a small set of observations. Round leaves or split stems as needed to make an effective stemplot.

C. Inspecting Distributions (Quantitative Variable)

1. Look for the overall pattern and for major deviations from the pattern.

2. Assess from a histogram or stemplot whether the shape of a distribution is roughly symmetric, distinctly skewed, or neither. Assess whether the distribution has one or more major peaks.

3. Describe the overall pattern by giving numerical measures of center and spread in addition to a verbal description of shape.

4. Decide which measures of center and spread are more appropriate: the mean and standard deviation (especially for symmetric distributions) or the five-number summary (especially for skewed distributions).

5. Recognize outliers.

D. Time Plots

1. Make a time plot of data, with the time of each observation on the horizontal axis and the value of the observed variable on the vertical axis.

2. Recognize patterns in a time plot.

E. Measuring Center

1. Find the mean $\bar{x}$ of a set of observations.

2. Find the median M of a set of observations.

3. Understand that the median is more resistant (less affected by extreme observations) than the mean. Recognize that skewness in a distribution moves the mean away from the median toward the long tail.

F. Measuring Spread

1. Find the quartiles Q_1 and Q_3 for a set of observations.

2. Give the five-number summary and draw a boxplot; assess center, spread, symmetry, and skewness from a boxplot.

3. Using a calculator or software, find the standard deviation s for a set of observations.

4. Know the basic properties of s: $s \geq 0$ always; $s = 0$ only when all observations are identical and increases as the spread increases; s has the same units as the original measurements; s is pulled strongly up by outliers or skewness.

G. Density Curves

1. Know that areas under a density curve represent proportions of all observations and that the total area under a density curve is 1.

2. Approximately locate the median (equal-areas point) and the mean (balance point) on a density curve.

3. Know that the mean and median both lie at the center of a symmetric density curve and that the mean moves farther toward the long tail of a skewed curve.

H. Normal Distributions

1. Recognize the shape of Normal curves and be able to estimate by eye both the mean and the standard deviation from such a curve.

2. Use the 68–95–99.7 rule and symmetry to state what percent of the observations from a Normal distribution fall between two points when the points lie one, two, or three standard deviations on either side of the mean.

3. Find the standardized value (z-score) of an observation. Interpret z-scores and understand that any Normal distribution becomes standard Normal $N(0, 1)$ when standardized.

4. Given that a variable has the Normal distribution with a stated mean μ and standard deviation σ, calculate the proportion of values above a stated number, below a stated number, or between two stated numbers.

5. Given that a variable has the Normal distribution with a stated mean μ and standard deviation σ, calculate the point having a stated proportion of all values above it. Also calculate the point having a stated proportion of all values below it.

6. Assess the Normality of a set of data by inspecting a Normal quantile plot.

CHAPTER 1 Review Exercises

1.119 Identify the histograms. A survey of a large college class asked the following questions:
(a) Are you female or male? (In the data, male = 0, female = 1.)
(b) Are you right-handed or left-handed? (In the data, right = 0, left = 1.)
(c) What is your height in inches?
(d) How many minutes do you study on a typical weeknight?
Figure 1.35 shows histograms of the student responses, in scrambled order and without scale markings. Which histogram goes with each variable? Explain your reasoning.

1.120 How much does it cost to make a movie? Making movies is a very expensive activity and many cost more than they earn. On the other hand, enormous profits are also a possibility. For this exercise you will analyze the budgets for 160 films made between 2003 and 2007.[43] BOXOFFICE160
(a) Examine the distribution of the budgets for these 160 films graphically. Describe key features of the distribution.
(b) Plot the budgets versus time. Describe any patterns that you see.
(c) Provide appropriate numerical summaries for the budgets of these 160 films.
(d) Write a summary of what you learned from these data that would be useful to someone who would like to invest in making movies.

1.121 Customers' home state. A sample of 1095 customers entering a retail store were asked to fill out a brief survey. One question on the survey asked each person to identify his or her current state of residency. The data from this question are summarized in the table below. IOWA
(a) The state in which the retail store resides is easily deduced from the table. In which state is this store located?

(b) One way to make a pie chart of these data would be to use one slice in the pie chart for each state in the table. Give at least one reason why this would not result in a useful pie chart.
(c) Group all customers from states other than Iowa (IA) into a category called Other and make a pie chart with an Other slice. Be sure to include the percent or count for each slice of your pie chart.

State	Count	State	Count
AR	1	MI	2
AZ	1	MO	2
CA	2	MS	2
CO	1	NE	3
FL	1	NY	1
GA	2	OH	2
IA	1053	OK	1
ID	2	OR	5
IL	6	TN	1
KS	1	TX	1
LA	1	UT	1
MA	1	WI	2

1.122 Help-wanted advertising in newspapers. One source of revenue for newspapers is printing help-wanted ads for companies that are looking for new employees. For this exercise we will use monthly data on help-wanted advertising in newspapers from January 1951 to April 2005. The time series uses an index value with 1987 as the base year. That is, the monthly average for 1987 is taken to be 100, so a month with an index value of 50 had only half as much help-wanted advertising in newspapers as the monthly average for 1987, while a month with an index value of

FIGURE 1.35 Match each
histogram with its variable, for
Exercise 1.119.

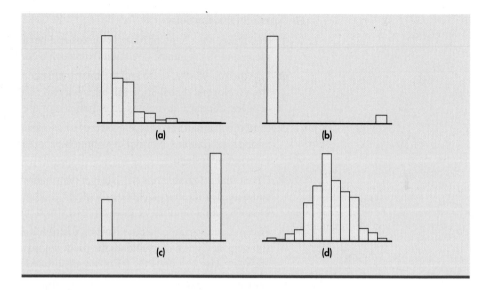

140 had 40% more help-wanted advertising in newspapers than the monthly average for 1987.[44] 🔘 **HELPWANTED**

(a) Using statistical software, obtain a time plot of the index values for help-wanted advertising in newspapers. Add a horizontal line to your time plot at the value of $\bar{x}$ for these data.

(b) What do you notice about the beginning years of the time series relative to the overall average of the time series? Which month in the time series is the first to be greater than the overall average?

(c) Describe the trend of the index values beginning in January 2000. Which month is the last month to be greater than the time series average?

(d) Propose at least one reasonable explanation for the observed trend in help-wanted advertising in newspapers since January 2000.

1.123 A closer look at customer refunds. A retail store specializing in children's clothing and toys has a relatively strict "no refunds" policy. Exceptions to this policy are sometimes granted in specific cases as determined by management. The store would like to look at refund activity for the year 2005. Data recorded include the date, amount, and item count for all refund transactions in 2005. Of the 10,939 transactions conducted between the store and customers during 2005, only 103 of these transactions were refunds (less than 1%). 🔘 **REFUNDS**

(a) Using statistical software, calculate the five-number summary for refund amounts. (*Note:* All refunds are recorded as negative numbers.)

(b) What percent of all refunds in 2005 were $10 or less?

(c) Construct a boxplot of the refund amounts based on your five-number summary.

(d) What does your boxplot indicate about the skewness of these data?

1.124 A closer look at customer refunds. Continue with the data on refunds described in the previous exercise. Upon inspec-

tion of the item counts for the refunds, we see that 83 of the 103 refunds were for one item. Using only this information and without using software or a calculator, answer the following questions. 🔘 **REFUNDS**

(a) Provide the first four numbers of the five-number summary for the item counts. (You cannot determine the maximum item count using only the information given in this exercise.)

(b) Construct a boxplot for the item counts using 14 as the maximum item count. How long is the box in your boxplot? Explain why this makes sense, given the data on item counts.

(c) What does your boxplot indicate about the skewness of these data?

1.125 Telecom revenue growth. The data on revenue growth for a random sample of telecommunications companies displayed before Exercise 1.105 (page 62) closely follow a Normal distribution with a mean of -0.0224 and a standard deviation of 0.2180. Take as a model for telecom revenue growth the $N(-0.0224, 0.2180)$ distribution and answer the following questions. 🔘 **TELECOMSTOCKS**

(a) Calculate $\mu + 3\sigma$ for the model for telecom revenue growth.

(b) From the population of all telecom companies, what percent should we expect to have revenue growth greater than $\mu + 3\sigma$? Explain how you arrived at your response.

(c) What percent of the telecom companies in our sample have revenue growth greater than $\mu + 3\sigma$? Is this percent different from your response to part (b)? Clearly explain why these two percents being different is not inconsistent with our assumption of a Normal distribution for the model for telecom revenue growth.

1.126 Telecom revenue growth. Take the $N(-0.0224, 0.2180)$ distribution as the model for telecom revenue growth as described in the previous exercise and answer the following questions. 🔘 **TELECOMSTOCKS**

(a) What percent of telecom companies had negative revenue growth over the past year? Show your work.

(b) What does "negative revenue growth" mean for a company?

(c) What percent of telecom companies had revenue growth greater than 0.50 (50%)? Show your work.

(d) In terms of revenue growth, the top 25% of all telecom companies had revenue growth greater than what value? Show your work.

1.127 What influences buying? Product preference depends in part on the age, income, and gender of the consumer. A market researcher selects a large sample of potential car buyers. For each consumer, she records gender, age, household income, and automobile preference. Which of these variables are categorical and which are quantitative?

1.128 Evaluating the improvement in a product. Corn is an important animal food. Normal corn lacks certain amino acids, which are building blocks for protein. Plant scientists have developed new corn varieties that contain these amino acids. To test a new corn as an animal food, a group of 20 one-day-old male chicks was fed a ration containing the new corn. A control group of another 20 chicks was fed a ration that was identical except that it contained normal corn. Here are the weight gains (in grams) after 21 days:[45] 🍲 CORN

Normal corn				New corn			
380	321	366	356	361	447	401	375
283	349	402	462	434	403	393	426
356	410	329	399	406	318	467	407
350	384	316	272	427	420	477	392
345	455	360	431	430	339	410	326

(a) Compute five-number summaries for the weight gains of the two groups of chicks. Then make boxplots to compare the two distributions. What do the data show about the effect of the new corn?

(b) The researchers actually reported means and standard deviations for the two groups of chicks. What are they? How much larger is the mean weight gain of chicks fed the new corn?

1.129 Fuel efficiency of hatchbacks and large sedans. Let's compare the fuel efficiencies (mpg) of model year 2009 hatchbacks and large sedans.[46] 🍲 MPGHATCHLARGE Here are the data:

Hatchbacks
30 29 28 27 27 27 27 27 26 25 25 25 24 24 24
24 24 23 23 22 22 21 21 21 21 21 21 21 20 20
20 20 20 20 20 20 19 19 19 18 16 16

Large sedans
19 19 18 18 18 18 17 17 17 17 17 17 17 17 17
17 16 16 16 16 16 16 16 16 15 15 13 13

Give graphical and numerical descriptions of the fuel efficiencies for these two types of vehicle. What are the main features of the distributions? Compare the two distributions and summarize your results in a short paragraph.

1.130 How much oil? How much oil the wells in a given field will ultimately produce is key information in deciding whether to drill more wells. The table below gives the estimated total amount of oil recovered from 64 wells in the Devonian Richmond Dolomite area of the Michigan basin.[47] 🍲 OILWELLS

21.7	53.2	46.4	42.7	50.4	97.7	103.1	51.9	43.4	69.5
156.5	34.6	37.9	12.9	2.5	31.4	79.5	26.9	18.5	14.7
32.9	196.0	24.9	118.2	82.2	35.1	47.6	54.2	63.1	69.8
57.4	65.6	56.4	49.4	44.9	34.6	92.2	37.0	58.8	21.3
36.6	64.9	14.8	17.6	29.1	61.4	38.6	32.5	12.0	28.3
204.9	44.5	10.3	37.7	33.7	81.1	12.1	20.1	30.5	7.1
10.1	18.0	3.0	2.0						

(a) Graph the distribution and describe its main features.

(b) Find the mean and median of the amounts recovered. Explain how the relationship between the mean and the median reflects the shape of the distribution.

(c) Give the five-number summary and explain briefly how it reflects the shape of the distribution.

1.131 The 1.5 × IQR rule. Exercise 1.67 (page 39) describes the most common rule for identifying suspected outliers. Find the interquartile range IQR for the oil recovery data in the previous exercise. Are there any outliers according to the $1.5 \times IQR$ rule?

1.132 Grading managers. Some companies "grade on a bell curve" to compare the performance of their managers. This forces the use of some low performance ratings, so that not all managers are graded "above average." A company decides to give A's to the managers and professional workers who score in the top 15% on their performance reviews, C's to those who score in the bottom 15%, and B's to the rest. Suppose that a company's performance scores are Normally distributed. This year, managers with scores less than 25 received C's and those with scores above 475 received A's. What are the mean and standard deviation of the scores?

1.133 The Statistical Abstract of the United States. Find in the library or at the U.S. Census Bureau Web site (www.census.gov) the most recent edition of the annual *Statistical Abstract of the United States*. Look up data on (a) the number of businesses started ("business starts") and (b) the number of business failures for the 50 states. Make graphs and numerical summaries to display the distributions, and write a brief description of the most important characteristics of each distribution. Suggest an explanation for any outliers you see.

1.134 Canada's balance of international payments. Visit the Web page www40.statcan.ca/l01/cst01/econ01a.htm, which provides data on Canada's balance of international payments. Select some data from this Web page and use the methods that you learned in this chapter to create graphical and numerical summaries. Write a report summarizing your findings that includes supporting evidence from your analyses.

1.135 Canadian government revenue and expenditures by province and territory. Visit the Web pages www40.statcan.ca/l01/cst01/govt08a.htm, www40.statcan.ca/l01/cst01/govt08b.htm, and www40.statcan.ca/l01/cst01/govt08c.htm. You need to look at the three pages to obtain data for all provinces and territories. Select some data from these Web pages and use the methods that you learned in this chapter to create graphical and numerical summaries. Write a report summarizing your findings that includes supporting evidence from your analyses.

1.136 Simulated observations. Most statistical software packages have routines for simulating values of variables having specified distributions. Use your statistical software to generate 25 observations from the $N(30, 5)$ distribution. Compute the mean and standard deviation $\bar{x}$ and s of the 25 values you obtain. How close are $\bar{x}$ and s to the μ and σ of the distribution from which the observations were drawn?

Repeat 19 more times the process of generating 25 observations from the $N(30, 5)$ distribution and recording $\bar{x}$ and s. Make a stemplot of the 20 values of $\bar{x}$ and another stemplot of the 20 values of s. Make Normal quantile plots of both sets of data. Briefly describe each of these distributions. Are they symmetric or skewed? Are they roughly Normal? Where are their centers? (The distributions of measures like $\bar{x}$ and s when repeated sets of observations are made from the same theoretical distribution will be very important in later chapters.)

CHAPTER 1   Case Study Exercises

CASE STUDY EXERCISE 1: What colors sell? Vehicle colors differ among types of vehicle in different regions. Here are data on the most popular colors in 2007 for several different regions of the world:[48]   **VEHICLECOLORSBYCOUNTRY**

Color	North America (percent)	South America (percent)	Europe (percent)	China (percent)	South Korea (percent)	Japan (percent)
Silver	19	26	28	24	21	27
White	16	11	4	16	18	24
Gray	13	14	16	3	19	12
Black	13	20	24	19	20	16
Blue	11	8	13	17	9	10
Red	11	10	6	9	6	3
Brown	7	7	4	1	6	2
Other	10	4	5	11	1	6

Use the methods you learned in this chapter to compare the vehicle color preferences for the regions of the world presented in this table. Write a report summarizing your findings with an emphasis on similarities and differences across regions. Include recommendations related to marketing and advertising of vehicles in these regions.

CASE STUDY EXERCISE 2: The business of health. The Behavioral Risk Factor Surveillance System (BRFSS) conducts a large survey of health conditions and risk behaviors in the United States.[49] The BRFSS data set contains data on 29 demographic factors and risk factors for each state. Pick three or more variables from this data set and summarize the distributions graphically and numerically. Write a report describing your summary. Include a discussion of business opportunities that you would consider on the basis of your analysis. **BRFSS**

CHAPTER 1 Appendix

Using Software for Statistical Analysis

Good statistical analysis relies heavily on interactive statistical software. In this Appendix, we discuss the use of Minitab and Excel for conducting statistical analysis. As a specialized statistical package, Minitab is one of the most popular software choices both in industry and in colleges and schools of business. As an all-purpose spreadsheet program, Excel provides a limited set of statistical analysis options in comparison to Minitab, or to any other statistics package for that matter. However, given its pervasiveness and wide acceptance in industry and the computer world at large, we believe it is important to give Excel proper attention. It should be noted that for users who want more statistical capabilities but want to work in an Excel environment, there are a number of commercially available add-on packages.

Even though basic guidance for using Minitab and Excel is provided in this and subsequent Appendices,

it should be emphasized that we are not bound to these software programs. Because computer output from statistical packages is very similar, you can feel quite comfortable using any one of a number of excellent statistical packages.

Getting Started with Minitab

In this section, we provide a basic overview of Minitab Release 15. For more instruction, Minitab provides a number of Help features found under the **Help** selection on the toolbar (see Figure App. 1.1). The **Tutorials** option, for example, introduces the user to basic Minitab features and walks the user through some example Minitab sessions. In addition, at Minitab's Web site, www.minitab.com, you can search through its knowledge base of customer support questions and their answers.

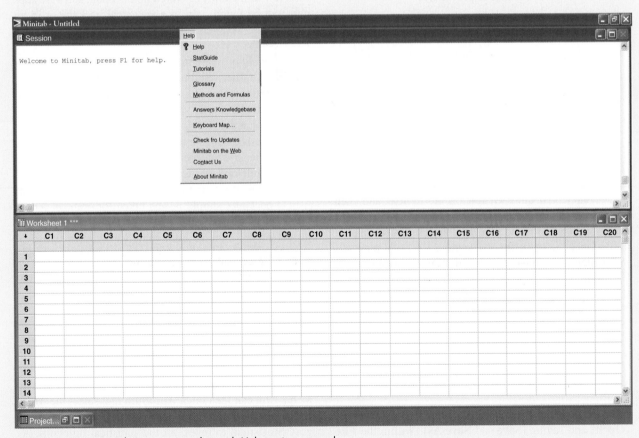

FIGURE App. 1.1 Minitab open screen shot with Help option opened.

Minitab Windows

Upon entering Minitab, you will find the display partitioned into two windows, as seen in Figure App. 1.1. The **Session window** is the area where all nongraphical statistical output and Minitab commands generating statistical output (graphical and nongraphical) are displayed. The **Data window** displays a spreadsheet environment (known as a worksheet) where the data can be directly entered and edited. Each column represents a variable to be analyzed. Unlike Excel, cells in a Minitab worksheet are not active in that formulas cannot be embedded within the cells. A Minitab worksheet is simply an environment for data to reside within.

There is a third window, which is minimized upon entering Minitab, known as the **Project Manager window.** This window allows you to do a variety of housekeeping tasks such as keeping track of all commands issued or seeing the basic attributes of the worksheet.

Invoking Statistical Procedures

There are two ways to invoke procedures:

1. You can type session commands in the Session window. To do so, the command language must be enabled, which will in turn produce an "MTB>" prompt in the Session window. At this prompt, you can then type desired commands. For more details on enabling session commands, refer to Minitab's Help options.

2. Users can make a sequence of selections from a series of menus that all begin in the toolbar menu. For example, in this chapter, we produced a graph known as a boxplot. To create this graph, you would click **Graph** on the toolbar and then select **Boxplot.** In this book, such a sequence of selections will be presented as Graph → Boxplot. Once the sequence of selections has been made, dialog and/or option boxes will be encountered that allow you to indicate which variable(s) will be part of the analysis, along with other information. If further help is needed, you can click the **Help** button that appears with every pop-up box. Once all appropriate information is provided, click the **OK** button to get the desired output.

Minitab Files

Minitab provides standard file options for retrieving (**Open**) and saving (**Save** and **Save As**). Within the **File** menu, you will notice that files can be opened or saved as worksheets or as projects. Worksheet files (.MTW extension) simply store the data found in the Data window, while project files (.MPJ extension) store all the current work, including the data, Session window output, and graphs. Thus, if you save a project prior to exiting Minitab and open the project at a later time, you can resume from where you last left off. Minitab files for selected examples and exercises provided on this book's CD are worksheet files.

Getting Started with Excel

In this section, we provide a basic overview of the statistical analysis options in Excel 2007. We assume that the reader is familiar with the basic layout and usage of Excel. As with all Microsoft products, Excel provides comprehensive support for the user in terms of the general use of its software or the more specific details of a particular procedure. As noted earlier, Excel provides a number of standard statistical analysis procedures but is not as comprehensive as a stand-alone statistical package. Therefore, for a few of the topics covered in this book, software support will be found only in a statistical package or in an enhanced add-on version of Excel rather than in standard Excel.

It should be noted that the accuracy of statistical procedures in earlier versions of Excel (2002 and earlier) has been called into question. Some of the problems revolved around Excel's use of shortcut formulas for certain statistical computations. A number of these problems have been addressed with the newest version of Excel, although a comprehensive independent study of the software has not been released at the time of the publication of this book. It is worth noting that reliability of established statistical packages should not be taken for granted. Albeit less serious than Excel's earlier problems, inaccuracies have been reported for even some well-known statistical packages.[50]

Built-in Statistical Functions and Charts

Excel has a variety of built-in statistical functions that can be used to compute many common descriptive statistics for a given set of data or to compute probabilities from a number of well-known statistical distributions. To find these functions, select the **Formulas** tab found in the main menu. You can then click **AutoSum** and select the **More Functions** option, which allows you to select the category **Statistical** to reveal all the statistical functions. As an alternative to clicking **AutoSum,** you can click **More Functions** and then move the cursor to your **Statistical Functions** menu choice.

In addition to the built-in statistical functions, a number of graphing options are available that may prove useful for data analysis. The available charts are found by selecting the **Insert** tab found in the main menu. One then finds a variety of graphing options in the **Charts** group. A few statistical options (for example, regression fitting) can be implemented in conjunction with the charts.

Installing Analysis ToolPak

Excel's built-in statistical functions can be useful for isolated computations. However, attempting to do a more complete statistical analysis with a collection of "raw" functions can be a laborious and clumsy process. Excel provides an add-on known as **Analysis ToolPak** that enables you to perform a more integrative statistical analysis. This add-on is not loaded with the standard installation of Excel. To install this add-on, click the **Microsoft Office Button,** click **Excel Options,** click **Add-Ins,** and then, in the **Manage** box, choose "Excel Add-ins" and click **Go.** At this point, select **Analysis ToolPak** in the **Add-ins available** box and finally click **OK.**

Invoking Analysis ToolPak Procedures

Once the **Analysis ToolPak** is installed, the statistical analysis routines are found by first selecting the **Data** tab found on the main toolbar. You will then see the **Data Analysis** command in the **Analysis** group. Figure App. 1.2 shows a blank Excel spreadsheet with the **Data Analysis** command invoked, resulting in the appearance of the **Data Analysis** menu box.

Within the **Data Analysis** menu box, there are 19 menu choices. When you select one of the menu choices, a box specific to the statistical routine will appear that calls for you to indicate where the data reside and where you want the output to be displayed. In particular, to indicate where the data for analysis reside, you specify the range of cells for the data in the **Input Range** box. This can be

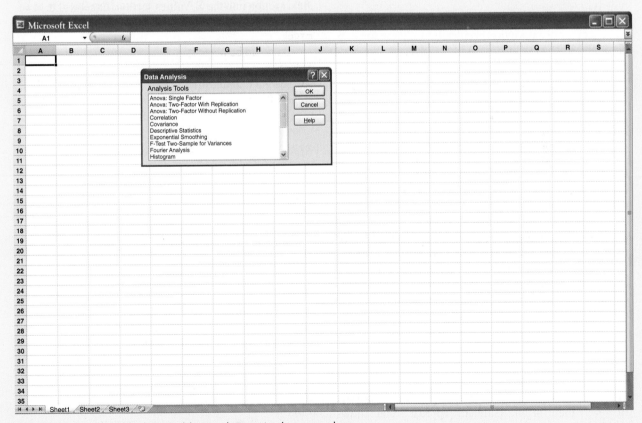

FIGURE App. 1.2 Excel blank spreadsheet with Data Analysis menu box.

accomplished by first clicking the cursor in the **Input Range** box and then typing in the cell range, or more easily you can highlight the data by clicking and dragging the mouse over the cell range. The statistical output can be placed either in the current worksheet (placement indicated with **Output Range** box), in a new worksheet tabbed with the current workbook (**New Output Ply** option), or in an entirely new workbook (**New Workbook** option).

Excel Data Files

As noted, we assume that you are familiar with the basics of Excel, including how to save and open files. It should be noted that files saved by Excel 2007 as an Excel Workbook cannot be opened by earlier versions of Excel. There is, however, an option to save workbooks as an Excel 97-2003 Workbook. Excel 2007 is backward compatible in terms of opening workbooks of older versions. Data files for selected examples and exercises provided on this book's CD are compatible with all versions of Excel.

Using Minitab and Excel for Examining Distributions

Now that we have provided a general overview of Minitab and Excel, we discuss more specifically how these software programs can be used to create the graphs and numerical summaries presented in this chapter.

Bar Graphs

Minitab:

Graph ➤ Bar Chart

If the frequencies have been pretabulated, select "Values from a table" from the **Bars represent** menu. If the frequencies have not been tabulated and you want Minitab to make the counts, select "Counts of unique values" from the **Bars represent** menu. Select "Simple" for the type of bar graph, then click **OK.** For pretabulated frequencies, click-in the data column into the **Graph variables** box and click-in the column that has the names of the categories into the **Categorical variables** box. If the frequencies have not been pretabulated, click-in the column that has data on the categorical names that need to be counted into the **Categorical variables** box. Click **OK.**

Excel:

There are a few ways to create bar graphs in Excel. However, there is one particular approach that allows you to create bar graphs based on providing in the spreadsheet the total counts of each category or having

Excel make the counts. For pretabulated frequencies, the spreadsheet should have two columns of information. With a column name in the top row, one column should have the names of the distinct categories. The other column, with its column name in the top row, should have the total counts of each category. If Excel needs to make the counts, there should be a column, with a column name in the top row, that has the data on the names of the categories that need to be counted. Once the one or two columns have been created, all the cells should be selected by dragging the mouse. Then click the **Insert** tab and click **PivotTable** in the **Tables** group and finally click **PivotChart.** You will then notice that Excel will produce a **PivotTable Field List** box. You will find that the column name(s) that you highlighted will be listed as fields. Select the field(s) presented to you by clicking a checkmark next to the name(s). For pretabulated frequencies, a bar graph will be created automatically. When you have only one column that requires counting, you will find that the field name appears in a section titled **Axis Fields (Categories).** You want to also have this field name in the section titled Σ **Values.** To do so, click and hold the field name and then drag the field from the field section into the Σ **Values** section. Excel will then automatically make the counts and create a corresponding bar graph.

Pie Charts

Minitab:

Graph ➤ Pie Chart

Making a pie chart is quite similar to making a bar graph. If the frequencies have been pretabulated, select the **Chart values from a table** option. If the frequencies have not been tabulated, select the **Chart counts of unique values** option. For pretabulated frequencies, click-in the data column into the **Summary variables** box, and click-in the column that has the names of the categories into the **Categorical variables** box. If the frequencies have not been pretabulated, click-in the column that has data on the categorical names that need to be counted into the **Categorical variables** box. If you wish to have the pie slices labeled by categorical names and have percents reported (as in Figure 1.4), click the **Label** button and then click the **Slice Labels** tab and finally place checkmarks next to the desired labels.

Excel:

To make a pie chart, you should follow the exact steps for making a bar graph. You want to now simply change the created bar graph into a pie chart. To do so, click the **Design** tab and then click the **Change Chart Type** in the **Type** group and finally select the **Pie** chart type. Alternatively, you can right-click on the bar graph and find the **Change Chart Type** option. To add labels to the pie slices, first right-click on one of the pie slices and then choose the **Add Data Labels** option. Once labels have been added, right-click again on one of the pie slices and then choose the **Format Data Labels** option and finally place checkmarks next to the desired labels.

Pareto Charts

Minitab:

Stat ➤ Quality Tools ➤ Pareto Chart

If the frequencies have been pretabulated, select the **Chart defects table** option. If the frequencies have not been tabulated, select the **Chart defects data in** option. For pretabulated frequencies, click-in the data column into the **Labels in** box and click-in the column that has the names of the categories into the **Frequencies in** box. If the frequencies have not been pretabulated, click-in the column that has data on the categorical names that need to be counted into the topmost box next to the **Chart defects data in** option. An alternative way to create a Pareto chart is to follow the steps for creating a bar graph but then click the **Chart Options** button and select the **Decreasing Y** option and place a checkmark next to the **Show Y as Percent** option.

Excel:

As a first step, create a bar graph as already described. You will find in the spreadsheet a PivotTable report made up of two columns: (1) a column labeled "Row Labels" and (2) a column with the frequencies. Highlight the contents of the report (that is, the cells with the category names and the cells with the frequencies). Now click the **Data** tab and then click **Sort** in the **Sort & Filter** group. At this point, choose the **Descending (Z to A)** option and select the column associated with the frequency numbers in the menu box found immediately below the option. We now want to convert the counts into percents. To do so, click the field name found in the Σ **Values** section, select the **Value Field Setting** option, click the **Show values as** tab, finally select "% of total" from the **Show values as** menu and then click **OK.**

Histograms

Minitab:

Graph ➤ Histogram

Select "Simple" for the type of histogram, then click **OK.** Click-in the data column into the **Graph Variables** box and then click **OK.** If you wish to change the automatically selected classes, double-click on the horizontal axis to make the **Edit Scale** box appear. Now, click the **Binning** tab and then choose the **Midpoint/Cutpoint positions** option found in the **Interval Definition** section. Depending on whether you choose the **Interval type** as "Midpoint" or "Cutpoint," you then give the desired values of the midpoints (that is, the middle values of the classes) or the cutpoints (that is, lower and upper values of the classes).

Excel:

Select "Histogram" in the **Data Analysis** menu box and click **OK.** Enter the cell range of the data into the **Input Range** box. If you want Excel to automatically select the classes, leave the **Bin Range** box empty. Place a checkmark next to the **Chart Output** option. Click **OK.** Excel will then create a histogram with gaps between the data bars. To remove these gaps, right-click on any one of the bars and then select the **Format Data Series** option. You will then have the opportunity to set the gap width to 0%. With the bars now closed up to each other, it is a good idea to border the bars with line edges. Before closing the **Format Data Series** box, click the **Border Color** option and select the **Solid line** option and finally click **Close.** If you wish to change the automatically selected classes, enter upper values for each class into the spreadsheet and input their cell range in the **Bin Range** box.

Stemplots

Minitab:

Graph ➤ Stem-and-Leaf

Click-in the data column into the **Graph Variables** box and then click **OK.**

Excel:

Stemplots are available in neither standard Excel nor the enhanced add-on version of Excel.

Time Plots

Minitab:

Graph ➤ Time Series Plot

Select "Simple" for the type of time series plot, then click **OK.** Click-in the data column into the **Series** box. In default mode, Minitab will label the time periods as "1," "2,"

"3," and so on. If you wish to label the time periods by year, as in Figure 1.12, then click the **Time/Scale** button, select the **Calendar** option, select the desired time periods (for example, "Year") from the adjacent menu, and click **OK** to close the pop-up. Click **OK** to produce the plot.

Excel:

Click and drag the mouse to highlight the cell range of the data you wish to time plot (include the column name if you wish it to appear as a chart label). With the cell range highlighted, click the **Insert** tab and then click **Line** in the **Charts** group. Within the **2-D Line** choices, you can choose whether to have data symbols at the data values or not.

Numerical Summaries of Distribution

Minitab:

Stat ➤ Basic Statistics ➤ Display Descriptive Statistics

Click-in the data column(s) for which you want to get numerical summaries into the **Variables** box. To choose what numerical summaries you want reported, click the **Statistics** button, place checkmarks next to all desired measures, and then click **OK** to close the pop-up. Click **OK** to have the summaries reported in the Session window.

Excel:

Select "Descriptive Statistics" in the **Data Analysis** menu box and click **OK.** Enter the cell range of the data into the **Input Range** box. Place a checkmark next to the **Chart Output** option. Click **OK.** You will find that the first and third quartiles are not reported. If you wish to compute these quartiles, click an empty cell in the spreadsheet and then proceed to the **Statistical** function menu as described in the overview section of this Appendix. Scroll down the list of functions and double-click on the **QUARTILE** function choice. In the **Array** box, input the cell range of the data. In the **Quart** box, input the value "1" to get the first quartile or the value "3" to get the third quartile and then click **OK.**

Boxplots

Minitab:

Graph ➤ Boxplot

If you have only one variable, select "One Y Simple" for the type of boxplot, then click **OK.** If you have multi-ple boxplots that you want to display together, as in Figure 1.15, select "Multiple Y's Simple" for the type of boxplot, then click **OK.** In either case, click-in the data column(s) for which you want to construct boxplots into the **Graph variables** box. Click **OK.**

Excel:

Boxplots are not available in standard Excel, but they are available in the enhanced add-on version of Excel.

Normal Distribution

Minitab:

Graph ➤ Probability Distribution Plot

This pull-down sequence will allow you to visualize areas under the Normal curve. Select "View Probability" and then click **OK.** The standard Normal distribution is the default distribution. You can change the values for the mean and/or standard deviation. Now click the **Shaded Area** tab. If you want to find the area under the curve associated with a specified value, select the **X Value** option. You can choose to find the area to the left or right of that specified value or even between two values by clicking the appropriate picture. You then enter the specified value(s) in the **X value** box. Click **OK.** As an exercise, you should be able to reproduce Examples 1.35, 1.36, and 1.37 (pages 51–52). To do inverse Normal calculations, select the **Probability** option rather than the **X Value** option. Depending on whether you are considering the area to the left or to the right of a value, enter the desired area in the **Probability** box and click **OK.** If more accurate reporting of numbers is desired, then you can consider the following pull-down sequence:

Calc ➤ Probability Distributions ➤ Normal

Choose the **Cumulative probability** option if you wish to find the area to the left of a specified value. Choose the **Inverse cumulative probability** option if you wish to find the value associated with a specified area to the left of that value. You can then select the **Input constant** option. In the box next to this option enter the specified value of x or z or enter the specified area. Click **OK** to find the results reported in the Session window.

Excel:

Excel does not provide a means to visualize areas under the Normal curve, but it can compute areas under the Normal curve or work backward. In either case, click an empty cell in the spreadsheet and then proceed to the

Statistical function menu as described in the overview section of this Appendix. If you wish to find the area to the left of a specified value under the standard Normal curve, then scroll down the list of functions and double-click on the **NORMSDIST** function choice. Type the value of z in the **Z** box and click **OK.** To do inverse standard Normal calculations, double-click on the **NORMSINV** function choice. Type the specified area in the **Probability** box and click **OK.**

Normal Quantile Plots

Minitab:

Stat ➤ Basic Statistics ➤ Normality Test

This pull-down sequence will produce a Normal probability plot. As noted in this chapter, there is a bit of a technical distinction between a Normal quantile plot and a Normal probability plot. However, the interpretation is the same in that the closer the data points plot to a straight line, the closer is the conformity to the Normal distribution. Upon doing the noted pull-down sequence, click-in the data column of interest into the **Variable** box and then click **OK.**

Excel:

Neither Normal quantile plots nor Normal probability plots are available in standard Excel, but Normal probability plots are available in the enhanced add-on version of Excel.

CHAPTER 2

Examining Relationships

Spam can destroy your ability to use your computer. This $140 billion problem is discussed in Case 2.1.

Introduction

A marketing study finds that men spend more money online than women. An insurance group reports that heavier cars have fewer deaths per 10,000 vehicles registered than do lighter cars. These and many other statistical studies look at the relationship between two variables. To understand such a relationship, we must often examine other variables as well. To conclude that men spend more online, for example, the researchers had to eliminate the effect of other variables such as annual income. Our topic in this chapter is relationships between variables. One of our main themes is that the relationship between two variables can be strongly influenced by other variables that are lurking in the background.

To study a relationship between two variables, we measure both variables on the same individuals. Often, we take the view that one of the variables explains or influences the other.

> **Response Variable, Explanatory Variable**
>
> A **response variable** measures an outcome of a study. An **explanatory variable** explains or influences changes in a response variable.

You will often find explanatory variables called **independent variables,** and response variables called **dependent variables.** The idea behind this language is that the response variable depends on the explanatory variable. Because the words "independent" and "dependent" have other meanings in statistics that are unrelated to the explanatory-response distinction, we prefer to avoid those words.

It is easiest to identify explanatory and response variables when we actually set values of one variable to see how it affects another variable.

independent variable
dependent variable

EXAMPLE 2.1 The Best Price?

Price is important to consumers and therefore to retailers. Sales of an item typically increase as its price falls, except for some luxury items, where high price suggests exclusivity. The seller's profits for an item often increase as the price is reduced, due to increased sales, until the point at which lower profit per item cancels rising sales. A retail chain therefore introduces a new DVD player at several different price points and monitors sales. The chain wants to discover the price at which its profits are greatest. Price is the explanatory variable, and total profit from sales of the player is the response variable.

When we don't set the values of either variable but just observe both variables, there may or may not be explanatory and response variables. Whether there are depends on how we plan to use the data.

EXAMPLE 2.2 Inventory and Sales

Jim is a district manager for a retail chain. He wants to know how the average monthly inventory and monthly sales for the stores in his district are related to each other. Jim doesn't think that either inventory level or sales explains or causes the other. He has two related variables, and neither is an explanatory variable.

Sue manages another district for the same chain. She asks, "Can I predict a store's monthly sales if I know its inventory level?" Sue is treating the inventory level as the explanatory variable and the monthly sales as the response variable.

In Example 2.1, price differences actually *cause* differences in profits from sales of DVD players. There is no cause-and-effect relationship between inventory levels and sales in Example 2.2. Because inventory and sales are closely related, we can nonetheless use a store's inventory level to predict its monthly sales. We will learn how to do the prediction in Section 2.3. Prediction requires that we identify an explanatory variable and a response variable. Some other statistical techniques ignore this distinction. Do remember that calling one variable explanatory and the other response doesn't necessarily mean that changes in one *cause* changes in the other.

Most statistical studies examine data on more than one variable. Fortunately, statistical analysis of several-variable data builds on the tools we used to examine individual variables. The principles that guide our work also remain the same:

- First plot the data, then add numerical summaries.
- Look for overall patterns and deviations from those patterns.
- When the overall pattern is quite regular, use a compact mathematical model to describe it.

APPLY YOUR KNOWLEDGE

2.1 Relationship between worker productivity and sleep. A study is designed to examine the relationship between how effectively employees work and how much sleep they get. Think about making a data set for this study.

(a) What are the cases?

(b) Would your data set have a label variable? If yes, describe it.

(c) What are the variables? Are they quantitative or categorical?

(d) Is there an explanatory variable and a response variable? Explain your answer.

2.2 Price versus size. You visit a local Starbucks to buy a Mocha Frappuccino©. The barista explains that this blended coffee beverage comes in three sizes and asks if you want a Tall, a Grande, or a Venti. The prices are $3.50, $4.00, and $4.50, respectively.

(a) What are the variables and cases?

(b) Which variable is the explanatory variable? Which is the response variable? Explain your answers.

(c) The Tall contains 12 ounces of beverage, the Grande contains 16 ounces, and the Venti contains 20 ounces. Answer parts (a) and (b) with ounces in place of the names for the sizes.

2.1 Scatterplots

CASE 2.1

Spam Botnets: A $140 Billion Problem A botnet is a remotely and silently controlled collection of networked computers. Botnets are illicitly created through the use of viruses, Trojans, and other malware to assimilate computers, or bots, into the botnet, generally without the knowledge of the computer owner. Some botnets can grow to many thousands of bots located all over the world. A botnet that is used to send unwanted commercial emails, called spam, is called a spam botnet.[1] About 120 billion spam messages are sent per day, and the cost of dealing with spam messages is estimated to be $140 billion per year.[2] Here is some information about 10 large botnets:

Botnet	Bots (thousands)	Spams per day (billions)	Botnet	Bots (thousands)	Spams per day (billions)
Srizbi	315	60	Grum	50	2
Bobax	185	9	Ozdok	35	10
Rustock	150	30	Nucrypt	20	5
Cutwail	125	16	Wopla	20	0.6
Storm	85	3	Spamthru	12	0.35

The variables are the number of bots operated by the botnet and the number of spam messages per day produced by these bots. The first botnet listed is a botnet called Srizbi, which was discovered on June 30, 2007.[3] Srizbi has 315,000 bots that generate 60 billion spams per day.

APPLY YOUR KNOWLEDGE

CASE 2.1 2.3 Make a data set.

(a) Create a spreadsheet that contains the spam botnet data.

(b) How many cases are in your data set?

(c) Describe the labels, variables, and values that you used.

(d) Which columns give quantitative variables?

CASE 2.1 2.4 Use your data set. Using the data set that you created in the previous exercise, find graphical and numerical summaries for bots and spam messages per day.

The most common way to display the relation between two quantitative variables is a *scatterplot*. Figure 2.1 shows an example of a scatterplot.

FIGURE 2.1 Scatterplot of spams per day (in billions) versus bots (in thousands), for Example 2.3.

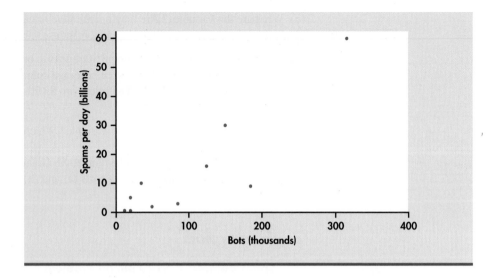

EXAMPLE 2.3   Bots and Spam Messages

We think that a netbot that has a large number of bots would be capable of generating a large number of spam messages relative to a netbot that has a smaller number of bots. Therefore, we think of the number of bots as an explanatory variable and the number of spam messages as a response variable. We begin our study of this relationship with a graphical display of the two variables.

Figure 2.1 is a scatterplot that displays the relationship between the response variable, spam messages per day, and the explanatory variable, number of bots. Notice that 6 of the 10 botnets are clustered in the lower-left part of the plot, with relatively low values for both bots and spam messages per day. On the other hand, the botnet Srizbi stands out with the highest values for both variables.

Scatterplot

A **scatterplot** shows the relationship between two quantitative variables measured on the same individuals. The values of one variable appear on the horizontal axis, and the values of the other variable appear on the vertical axis. Each individual in the data appears as the point in the plot fixed by the values of both variables for that individual.

Always plot the explanatory variable, if there is one, on the horizontal axis (the x axis) of a scatterplot. As a reminder, we usually call the explanatory variable x and the response variable y. If there is no explanatory-response distinction, either variable can go on the horizontal axis. The time plots in Section 1.1 (page 19) are special scatterplots where the explanatory variable x is a measure of time.

APPLY YOUR KNOWLEDGE

2.5 Make a scatterplot.

(a) Make a scatterplot similar to Figure 2.1 for the spam botnet data.
(b) Mark the location of the botnet Bobax on your plot.

2.6 Change the units.

(a) Create a spreadsheet with the spam botnet data using the actual values. In other words, for Srizbi use 315,000 for the number of bots and 60,000,000,000 for the number of spam messages per day.

(b) Make a scatterplot for the data coded in this way.

(c) Describe how this scatterplot differs from Figure 2.1.

Interpreting scatterplots

To interpret a scatterplot, apply the strategies of data analysis learned in Chapter 1.

> **Examining a Scatterplot**
>
> In any graph of data, look for the **overall pattern** and for striking **deviations** from that pattern.
>
> You can describe the overall pattern of a scatterplot by the **form, direction,** and **strength** of the relationship.
>
> An important kind of deviation is an **outlier,** an individual value that falls outside the overall pattern of the relationship.

linear relationship   The scatterplot in Figure 2.1 shows a clear *form:* the data lie in a roughly straight-line, or **linear,** pattern. To help us see this relationship, we can use software to put a straight line through the data. (We will show how this is done in Section 2.3.)

CASE 2.1   **EXAMPLE 2.4**   Scatterplot with a Straight Line

cluster

Figure 2.2 plots the botnet data along with a fitted straight line. There is a **cluster** of points in the lower left. Although Srizbi might be an outlier, it lies roughly in the same linear pattern as the other botnets.

FIGURE 2.2 Scatterplot of spams per day (in billions) versus bots (in thousands) with a fitted straight line, for Example 2.4.

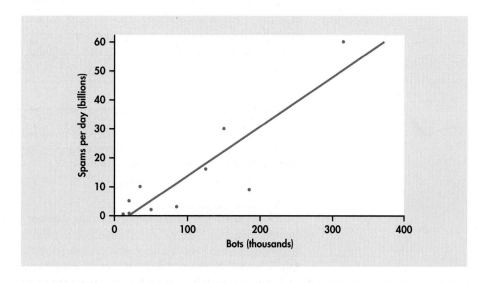

The relationship in Figure 2.2 also has a clear *direction:* botnets with more bots generate more spam messages than botnets that have fewer bots. This is a *positive association* between the two variables.

> **Positive Association, Negative Association**
>
> Two variables are **positively associated** when above-average values of one tend to accompany above-average values of the other, and below-average values also tend to occur together.
>
> Two variables are **negatively associated** when above-average values of one tend to accompany below-average values of the other, and vice versa.

The *strength* of a relationship in a scatterplot is determined by how closely the points follow a clear form. The strength of the relationship in Figure 2.1 is moderate. Botnets with similar numbers of bots have a fair amount of scatter in the number of spam messages per day that they produce. Here is an example of a stronger linear relationship.

EXAMPLE 2.5   Debt for 24 Countries

DEBT

The amount of debt owed by a country is a measure of its economic health. The Organization for Economic Co-operation and Development (OECD) collects data on the central government debt for many countries. One of their tables gives the debt for 30 countries for the years 1998 to 2007.[4] Since there are a few countries with a very large amount of debt, let's concentrate on the 24 countries with debt less than US$1 trillion in 2006. The six countries excluded are the United Kingdom, France, Germany, Italy, the United States, and Japan. The data are given in Figure 2.3.

Figure 2.4 is a scatterplot of the central government debt in 2007 versus the central government debt in 2006. The scatterplot shows a strong positive relationship between the debt in these two years.

The relationship in Figure 2.4 appears to be linear. Some statistical software packages provide a tool to help us make this kind of judgment. These use computer-intensive *algorithms* methods, called **algorithms,** that calculate a smooth curve that gives an approximate *smoothing* fit to the points in a scatterplot. This is called **smoothing** a scatterplot. Usually, these methods use a smoothing parameter that determines how smooth the fit will be. You can vary the parameter until you have a fit that you judge suitable for your data. Here is an example.

EXAMPLE 2.6   Debt for 24 Countries with a Smooth Fit

DEBT

Figure 2.5 gives the scatterplot that we examined in Figure 2.4 with a smooth fit superimposed. Notice that the smooth curve fits almost all of the points. The curve is too wavy and therefore does not provide a good summary of the relationship.

Our first attempt at smoothing the data was not very successful. This scenario is common when we use data analysis methods to learn something from our data. Don't be discouraged when your first attempt at summarizing data produces unsatisfactory results. Take what you learn and refine your analysis until you are satisfied that you have found a good summary. It is your last attempt, not your first, that is most important.

FIGURE 2.3 Central government debt in 2006 and 2007 for 24 countries with debt in 2006 less than US$1 trillion, in US$ billions, for Example 2.5.

	A	B	C
1	Country	Debt2006	Debt2007
2	Luxembourg	0.65	0.78
3	Iceland	4.04	4.92
4	Slovak Republic	18.44	22.79
5	New Zealand	21.44	27.82
6	Norway	43.03	49.05
7	Czech Republic	38.44	49.36
8	Australia	43.91	49.45
9	Ireland	47.30	55.29
10	Finland	77.58	82.54
11	Hungary	76.76	90.24
12	Denmark	94.29	92.74
13	Switzerland	101.28	107.49
14	Portugal	142.97	166.06
15	Sweden	185.01	182.11
16	Poland	164.41	205.97
17	Mexico	195.66	216.76
18	Austria	204.51	231.56
19	Turkey	244.89	286.96
20	Korea	295.07	309.34
21	Netherlands	278.67	316.07
22	Greece	297.93	352.80
23	Canada	347.68	389.83
24	Belgium	366.91	420.74
25	Spain	427.06	464.99
26			

Microsoft Excel – GovDebt200...

E25

FIGURE 2.4 Scatterplot of debt in 2007 (US$ billions) versus debt in 2006 (US$ billions) for 24 countries with less than US$1 trillion debt in 2006, for Example 2.5.

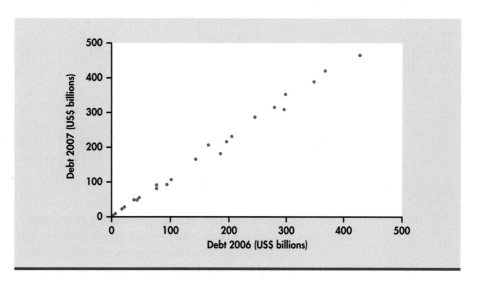

83

FIGURE 2.5 Scatterplot of debt in 2007 (US$ billions) versus debt in 2006 (US$ billions) for 24 countries with less than US$1 trillion debt in 2006, with a smooth curve fitted to the data, for Example 2.6. This smooth curve fits the data too well and does not provide a good summary of the relationship.

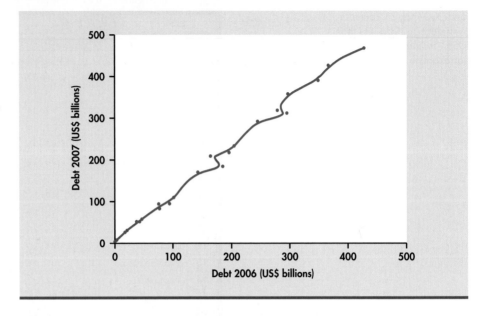

 DEBT

EXAMPLE 2.7   A Better Smooth Fit for the Debt Data

By varying the smoothing parameter, we can make the curve more or less smooth. Figure 2.6 gives the same data in the previous two figures with a smooth fit that is smoother. The smooth curve is very close to a straight line. In this way we have confirmed our original impression that the relationship between these two variables is approximately linear.

FIGURE 2.6 Scatterplot of debt in 2007 (US$ billions) versus debt in 2006 (US$ billions) for 24 countries with less than US$1 trillion debt in 2006, with a smooth curve fitted to the data, for Example 2.7. This smooth curve gives a good summary of the relationship, which is approximately linear.

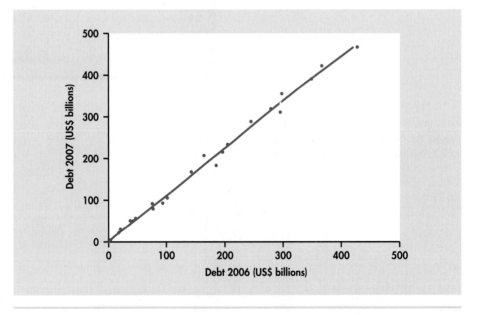

It is tempting to conclude that the strong linear relationship that we have found between the central government debt in 2006 and 2007 for the 24 countries is evidence that the amount of debt for each country is approximately the same in the two years. The first exercise below asks you to explore this temptation.

DATA FILE
DEBT

DATA FILE
DEBT29

2.7 Are the debts in 2006 and 2007 approximately the same? Use the methods you learned in Chapter 1 to examine whether or not the central government debts in 2006 and 2007 are approximately the same. (*Hint:* Think about creating a new variable that would help you to answer this question.)

2.8 What about the countries with very large debts. In Exercise 2.7 we excluded six countries. The original data file did not list a value for Japan's debt in 2007. Here are the debts, in US$ billions, for the other five countries:

Country	2006 debt	2007 debt
United Kingdom	1168	1231
France	1240	1454
Germany	1252	1409
Italy	1892	2167
United States	4848	5055

Add the data for these five countries to your data set and make a scatterplot that includes the data for all 29 countries. Summarize the relationship. Do the additional data change the relationship? Explain your answer.

Of course, not all relationships are linear. Here is an example where the relationship is described by a nonlinear curve.

EXAMPLE 2.8   Forbes.com Best Countries for Business

DATA FILE
BESTCOUNTRIES

Forbes.com analyzes business climates in 120 countries and creates an ordered list of these countries called the Best Countries for Business.[5] Let's look at two of the variables that they use to determine the ranks in their list: gross domestic product (GDP) per capita, in US$ per person, and unemployment rate. We exclude data from a few countries that have very extreme values or that are missing values for one or more of the variables used in the rankings. Figure 2.7 is a scatterplot of GDP per capita versus unemployment rate for 99 countries.

FIGURE 2.7 Scatterplot of gross domestic product (US$) per capita versus unemployment rate (percent) for 99 countries, with a smooth curve, for Example 2.8. There is a negative nonlinear relationship between these two variables.

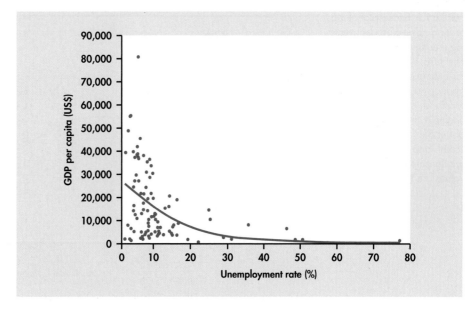

The smooth curve suggests that there is a negative relationship between these two variables. The relationship is approximately linear for small values of unemployment, particularly for values less than about 15%. However, after that point, the curve decreases more slowly; so overall, we have a nonlinear relationship.

Look at Figure 2.7 carefully. Notice that the two variables that we are examining are both skewed toward large values. This shows up in the plot, where most of the data points are at the left and in the lower part of the scatterplot. Notice that a relatively large proportion of the data clusters in the lower-left corner of the scatterplot. Skewed data are quite common in business applications of statistics, particularly when the measured variable *transformation* is some kind of count or amount. In these situations, we often apply a **transformation** to the data. This means that we replace the original values by the transformed values and then use the transformed values for our analysis.

The log transformation

log transformation   The most important transformation that we will use is the **log transformation.** This transformation can be used only for variables that have positive values. Occasionally, we use it when there are zeros, but in this case we first replace the zero values by some small value, often one-half of the smallest positive value in the data set.

You have probably encountered logarithms in one of your high school mathematics courses as a way to do certain kinds of arithmetic. Usually, these are base 10 logarithms. Logarithms are a lot more fun when used in statistical analyses. For our statistical applications, we will use natural logarithms. Statistical software and statistical calculators generally provide easy ways to perform this transformation.

EXAMPLE 2.9   Gross Domestic Product per Capita and Unemployment with Logarithms

BESTCOUNTRIES

Figure 2.8 is a scatterplot of the log of the gross domestic product per capita versus the log of the unemployment rate. Notice how the data now fill up much of the central part of the scatterplot,

FIGURE 2.8 Scatterplot of log gross domestic product versus log unemployment for 99 countries, with a smooth curve, for Example 2.9. There appears to be essentially no relationship for low unemployment values, up to about 5% (1.6 on the log scale), and a negative relationship for unemployment values greater than 5%.

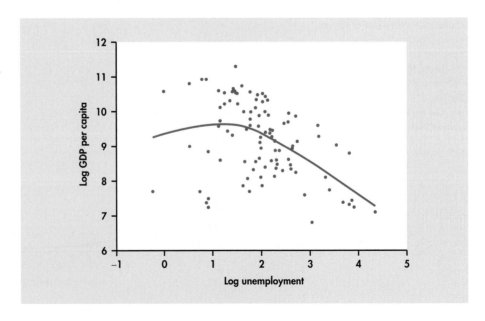

in contrast to the clustering that we noticed in Figure 2.7. Here we see that the relationship is essentially flat for values of log unemployment that are less than about 1.6. This point corresponds to an unemployment rate of about 5%. So for low unemployment rates, there appears to be little or no relationship between unemployment and GDP per capita. On the other hand, for values of unemployment greater than 5%, we see an approximately linear negative relationship with GDP per capita. High unemployment is associated with low GDP per capita.

The interpretation of scatterplots, including knowing when to use transformations, is an art that requires judgment and knowledge about the variables that we are studying. Always ask yourself if the relationship that you see makes sense. If it does not, then additional analyses are needed to understand the data.

Adding categorical variables to scatterplots

In the previous two examples, we looked at two of the variables used by Forbes.com to construct their Best Countries for Business list. They use several more variables, but do not give the details about exactly how their list is constructed. Let's take a look at how the two variables we examined relate to whether or not a country ranks high or low on the list.

EXAMPLE 2.10 Gross Domestic Product per Capita and the Rankings

We start by creating a categorical variable that indicates whether or not a country ranks in the top half of the Best Countries for Business list. In our scatterplot, we will use the symbol "H" for countries that rank in the top half of the list and "L" for countries that rank in the bottom half. Examine the scatterplot in Figure 2.9 carefully. Notice that the countries in the top part of the plot, those with relatively high GDP per capita, tend to rank high in the Best Countries for Business list. On the other hand, countries in the bottom part, those with relatively low GDP per capita, tend to rank low. What about unemployment? No clear pattern is evident. Although the countries with the six highest unemployment rates are all ranked low, they also have low values for GDP per capita.

FIGURE 2.9 Scatterplot of log gross domestic product per capita versus log unemployment for 106 countries using different plotting symbols, for Example 2.10. Countries in the top half of the Best Countries for Business rankings are plotted with the symbol "H" and those in the bottom half are plotted with the symbol "L." Countries with high GDP tend to be rated high, while the unemployment rate does not appear to be a major factor in determining high or low ranking.

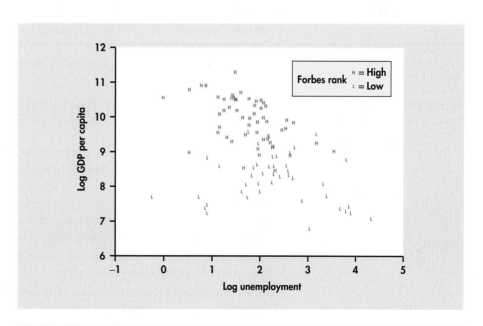

In this example, we used a quantitative variable, rank in the Best Countries for Business list, to create a categorical variable that indicated whether a country ranked high or low in the list. Of course, if the variable that you want to plot is categorical, no conversion is necessary. Careful judgment is needed in applying this graphical method. Don't be discouraged if your first attempt is not very successful. In performing a good data analysis, you will often produce several plots before you find the one that is most effective in describing the data.[6]

APPLY YOUR KNOWLEDGE

 BESTCOUNTRIES

2.9 Change the plotting symbol. In Example 2.10 we used the plotting symbols "H" and "L" to distinguish countries that ranked high and low on the Best Countries for Business list. Let's see if we can learn anything more about these variables by refining our categorical variable further. Define a new categorical variable that has three distinct values corresponding to a rank in the top third (ranks 1 to 40), the middle third (ranks 41 to 80), and the bottom third (ranks 81 to 120). Choose appropriate plotting symbols and make a scatterplot similar to Figure 2.9 using this categorical variable. Describe your scatterplot and compare what you can learn from it with what we learned in Example 2.10.

SECTION 2.1 Summary

- To study relationships between variables, we must measure the variables on the same group of individuals.

- If we think that a variable x may explain or even cause changes in another variable y, we call x an **explanatory variable** and y a **response variable.**

- A **scatterplot** displays the relationship between two quantitative variables measured on the same individuals. Mark values of one variable on the horizontal axis (x axis) and values of the other variable on the vertical axis (y axis). Plot each individual's data as a point on the graph.

- Always plot the explanatory variable, if there is one, on the x axis of a scatterplot. Plot the response variable on the y axis.

- Plot points with different colors or symbols to see the effect of a categorical variable in a scatterplot.

- In examining a scatterplot, look for an overall pattern showing the **form, direction,** and **strength** of the relationship, and then for **outliers** or other deviations from this pattern.

- **Form: Linear relationships,** where the points show a straight-line pattern, are an important form of relationship between two variables. Curved relationships and **clusters** are other forms to watch for.

- **Direction:** If the relationship has a clear direction, we speak of either **positive association** (high values of the two variables tend to occur together) or **negative association** (high values of one variable tend to occur with low values of the other variable).

- **Strength:** The **strength** of a relationship is determined by how close the points in the scatterplot lie to a simple form such as a line.

- A **transformation** uses a formula or some other method to replace the original values of a variable with other values for an analysis. The transformation is successful if it helps us to learn something about the data.

• The **log transformation** is frequently used in business applications of statistics. It tends to make skewed distributions more symmetric, and it can help us to better see relationships between variables in a scatterplot.

SECTION 2.1 Exercises

For Exercises 2.1 and 2.2, see pages 78–79; for 2.3 and 2.4, see page 79; for 2.5 and 2.6, see pages 80–81; for 2.7 and 2.8, see page 85; and for 2.9, see page 88.

2.10 What's wrong? Explain what is wrong with each of the following:
(a) A boxplot can be used to examine the relationship between two variables.
(b) In a scatterplot we put the response variable on the y axis and the explanatory variable on the x axis.
(c) If two variables are positively associated, then high values of one variable are associated with low values of the other variable.

2.11 Make some sketches. For each of the following situations, make a scatterplot that illustrates the given relationship between two variables.
(a) A strong negative linear relationship.
(b) No apparent relationship.
(c) A weak positive relationship.
(d) A more complicated relationship. Explain the relationship.

2.12 Financing a college education. How well does the income of a college student's parents predict how much the student will borrow to pay for college? We have data on parents' income and college debt for a sample of 1200 recent college graduates. What are the explanatory and response variables? Are these variables categorical or quantitative? Do you expect a positive or negative association between these variables? Why?

2.13 Is the cost too high? Because it is so costly, many individuals and families cannot afford to purchase health insurance. The Current Population Survey collected data on the characteristics of the uninsured.[7] Below are the numbers of uninsured and the total number of people classified by age. The units are thousands of people. 🖥 HEALTHINSURANCE

Age group	Number uninsured	Total number
Under 18 years	8,661	74,101
18 to 24 years	8,323	28,405
25 to 34 years	10,713	39,868
35 to 44 years	8,018	42,762
45 to 64 years	10,738	75,653
65 years and older	541	36,035

(a) Plot the number of uninsured versus age group.

(b) Find the total number of uninsured persons, and use this total to compute the percent of the uninsured who are in each age group.
(c) Plot the percents versus age group.
(d) Explain how the plot you produced in part (c) differs from the plot that you made in part (a).
(e) Summarize what you can conclude from these plots.

2.14 Which age groups have higher percents of uninsured? Refer to the previous exercise. Let's take a look at the data from a different point of view. 🖥 HEALTHINSURANCE
(a) For each age group calculate the percent that are uninsured using the number of uninsured persons and the total number of persons in each group.
(b) Make a plot of the percent uninsured versus age group.
(c) Summarize the information in your plot and write a short summary of what you conclude from your analysis.

2.15 Compare the two percents. In the previous two exercises, you computed percents in two different ways and generated plots versus age group. Describe the difference between the two ways with an emphasis on what kinds of conclusions can be drawn from each. 🖥 HEALTHINSURANCE

2.16 Numbers of uninsured in 2006 and 2007. In the previous three exercises, we looked at data on uninsured persons for 2006. Here are similar data for 2007: 🖥 UNINSURED

Age group	Number uninsured	Total number
Under 18 years	8,149	74,703
18 to 24 years	7,911	28,398
25 to 34 years	10,329	40,146
35 to 44 years	7,717	42,132
45 to 64 years	10,784	77,237
65 years and older	686	36,790

(a) Make a scatterplot of the number of uninsured persons in 2007 versus the number in 2006. Note that each point in your scatterplot corresponds to an age group.
(b) Describe the relationship in your scatterplot.

2.17 Can you conclude that the numbers are similar? Examine the scatterplot that you produced in the previous exercise. 🖥 UNINSURED
(a) Can you conclude from the strength of the relationship that the numbers of uninsured in these age groups remained approximately the same from 2006 to 2007? Explain your answer.

FIGURE 2.10 Scatterplot of carbohydrates (grams) versus percent alcohol for 86 brands of beer, for Exercise 2.18.

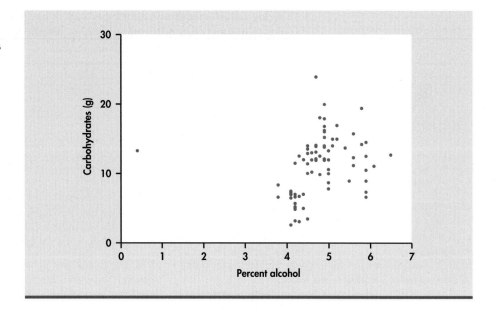

(b) Use graphical and numerical summaries to address the question of whether or not the numbers of uninsured were approximately the same in 2006 and 2007.

2.18 Brand-to-brand variation in a product. Beer100.com advertises itself as "Your Place for All Things Beer." One of their "things" is a list of 86 domestic beer brands with the percent alcohol, calories per 12 ounces, and carbohydrates (in grams).[8] **BEER**
(a) Figure 2.10 gives a scatterplot of carbohydrates versus percent alcohol. Give a short summary of what can be learned from the plot.
(b) One of the points is an outlier. Use the data file to find the outlier brand of beer. How is this brand of beer marketed as compared with the other brands?
(c) Remove the outlier from the data set and generate a scatterplot of the remaining data.
(d) Describe the relationship between carbohydrates and percent alcohol based on what you see in your scatterplot.

2.19 More beer. Refer to the previous exercise. **BEER**
(a) Make a scatterplot of calories versus percent alcohol using the data file without the outlier.
(b) Describe the relationship between these two variables.

2.20 Marketing in Canada. Many consumer items are marketed to particular age groups in a population. To plan such marketing strategies, it is helpful to know the demographic profile for different areas. Statistics Canada provides a great deal of demographic data organized in different ways.[9] Figure 2.11 gives the percent of the population over 65 and the percent under 15 for each of the 13 Canadian provinces and territories. Figure 2.12 is a scatterplot of the percent of the population over 65 versus the percent under 15. Write a short paragraph explaining

what the plot tells you about these two demographic groups in the 13 Canadian provinces and territories. **CANADIANPOPULATION**

2.21 Compare the provinces with the territories. Refer to the previous exercise. The three Canadian territories are the Northwest Territories, Nunavut, and the Yukon Territories. All of the other entries in Figure 2.11 are provinces. **CANADIANPOPULATION**
(a) Generate a scatterplot of the Canadian demographic data similar to Figure 2.12 but with the points labeled "P" for provinces and "T" for territories.
(b) Use your new scatterplot to write a new summary of the demographics for the 13 Canadian provinces and territories.

2.22 World Bank data on living longer and using the Internet. The World Bank collects data on many variables related to development for countries throughout the world. Two of these variables are Internet use (in number of users per 100 people) and life expectancy (in years).[10] Figure 2.13 is a scatterplot of life expectancy versus Internet use. **INTERNETANDLIFE**
(a) Describe the relationship between these two variables.
(b) A friend looks at this plot and concludes that using the Internet will increase the length of your life. Write a short paragraph explaining why the association seen in the scatterplot does not provide a reason to draw this conclusion.

2.23 Let's look at Europe. Refer to the previous exercise. Figure 2.14 is a scatterplot of only the 48 European countries in the data set. Compare this figure with Figure 2.13, which plots the data for all 181 countries in the data set. Write a paragraph summarizing the relationship between life expectancy and Internet use for European countries with an emphasis on how the European countries compare with the entire set of 181 countries. Be sure to take into account the fact that the software used here to create the scatterplots automatically chooses the range of values

FIGURE 2.11 Percent of the population over 65 and percent of the population under 15 in the 13 Canadian provinces and territories, for Exercise 2.20.

	A	B	C
	ProvinceOrTerritory	PercentUnder15	PercentOver65
1	ProvinceOrTerritory	PercentUnder15	PercentOver65
2	Newfoundland and Labrador	15.5	13.9
3	Prince Edward Island	17.7	14.9
4	Nova Scotia	16.0	15.1
5	New Brunswick	16.2	14.7
6	Quebec	16.6	14.3
7	Ontario	18.2	13.6
8	Manitoba	19.6	14.1
9	Saskatchewan	19.4	15.4
10	Alberta	19.2	10.7
11	British Columbia	16.5	14.6
12	Yukon Territories	18.8	7.5
13	Northwest Territories	23.9	4.8
14	Nunavut	33.9	2.7

Microsoft Excel – CanadianPopulation.xls   F14   fx

FIGURE 2.12 Scatterplot of percent of the population over 65 versus percent of the population under 15 for the 13 Canadian provinces and territories, for Exercise 2.20.

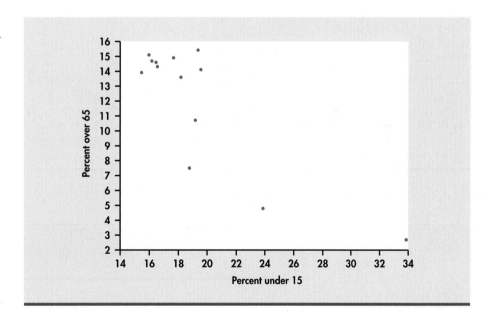

for each axis so that the space in the plot is used efficiently. In this case, the range of values for Internet use is the same for both scatterplots, but the range of values for life expectancy is quite different. 🌐 INTERNETANDLIFEE

2.24 How would you make a better plot? In the previous two exercises, we looked at the relationship between life expectancy and Internet use. First, we examined a scatterplot for all 181 countries in the data set. Then we examined one for the subset of 48 European countries. Explain how you would construct a single plot to compare the European countries with the other countries in the data set. (*Optional:* Make the plot if you have software that can do what you need.) 🌐 INTERNETANDLIFE

FIGURE 2.13 Scatterplot of life expectancy (in years) versus Internet users (per 100 people) for 181 countries, for Exercise 2.22.

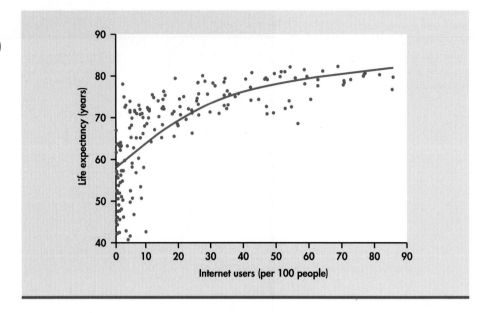

FIGURE 2.14 Scatterplot of life expectancy (in years) versus Internet users (per 100 people) for 48 European countries, for Exercise 2.23.

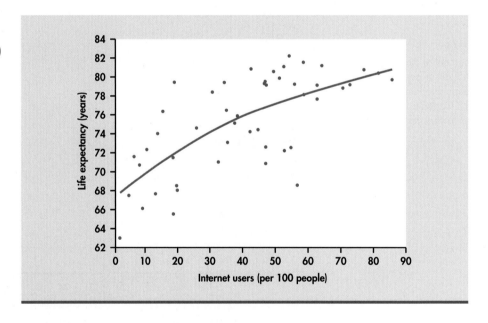

2.2 Correlation

A scatterplot displays the form, direction, and strength of the relationship between two quantitative variables. Linear relations are particularly important because a straight line is a simple pattern that is quite common. We say a linear relation is strong if the points lie close to a straight line, and weak if they are widely scattered about a line. Our eyes are not good judges of how strong a linear relationship is. The two scatterplots in Figure 2.15 depict exactly the same data, but the lower plot is drawn smaller in a large field. The

FIGURE 2.15 Two scatterplots of the same data. The straight-line pattern in the lower plot appears stronger because of the surrounding open space.

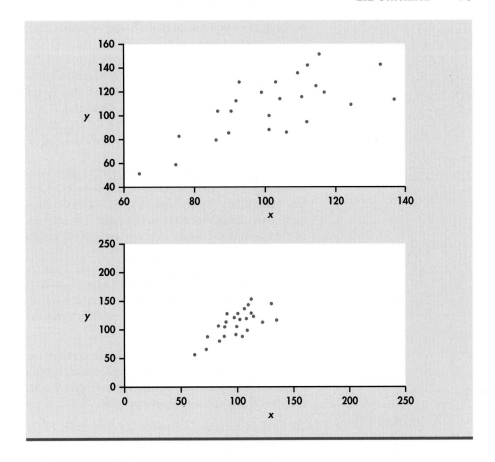

lower plot seems to show a stronger linear relationship. Our eyes can be fooled by changing the plotting scales or the amount of white space around the cloud of points in a scatterplot.[11] We need to follow our strategy for data analysis by using a numerical measure to supplement the graph. *Correlation* is the measure we use.

The correlation r

> **Correlation**
>
> The **correlation** measures the direction and strength of the linear relationship between two quantitative variables. Correlation is usually written as r.
>
> Suppose that we have data on variables x and y for n individuals. The values for the first individual are x_1 and y_1, the values for the second individual are x_2 and y_2, and so on. The means and standard deviations of the two variables are $\overline{x}$ and s_x for the x-values, and $\overline{y}$ and s_y for the y-values. The correlation r between x and y is
>
> $$r = \frac{1}{n-1} \sum \left(\frac{x_i - \overline{x}}{s_x} \right) \left(\frac{y_i - \overline{y}}{s_y} \right)$$

As always, the summation sign $\sum$ means "add these terms for all the individuals." The formula for the correlation r is a bit complex. It helps us to see what correlation is, but in practice you should use software or a calculator that finds r from keyed-in values of two variables x and y.

The formula for r begins by standardizing the observations. Suppose, for example, that x is height in centimeters and y is weight in kilograms and that we have height and weight measurements for n people. Then $\overline{x}$ and s_x are the mean and standard deviation of the n heights, both in centimeters. The value

$$\frac{x_i - \overline{x}}{s_x}$$

is the standardized height of the ith person, familiar from Chapter 1. The standardized height says how many standard deviations above or below the mean a person's height lies. Standardized values have no units—in this example, they are no longer measured in centimeters. Standardize the weights also. The correlation r is an average of the products of the standardized height and the standardized weight for the n people.

APPLY YOUR KNOWLEDGE

 SPAM

CASE 2.1 **2.25 Spam botnets.** In Exercise 2.3 you made a data set for the botnet data. Use that data set to compute the correlation between the number of bots and the number of spam messages per day.

 SPAM

CASE 2.1 **2.26 Change the units.** In the previous exercise bots were given in thousands and spam messages per day were recorded in billions. In Exercise 2.6 you created a data set using the actual values. For example, Srizbi has 315,000 bots and generates 60,000,000,000 spam messages per day.

(a) Find the correlation between bots and spam messages using this data set.
(b) Compare this correlation with the one that you computed in the previous exercise.
(c) What can you say in general about the effect of changing units in this way on the size of the correlation?

Facts about correlation

The formula for correlation helps us see that r is positive when there is a positive association between the variables. Height and weight, for example, have a positive association. People who are above average in height tend also to be above average in weight. Both the standardized height and the standardized weight are positive. People who are below average in height tend also to have below-average weight. Then both standardized height and standardized weight are negative. In both cases, the products in the formula for r are mostly positive and so r is positive. In the same way, we can see that r is negative when the association between x and y is negative. More detailed study of the formula gives more detailed properties of r. Here is what you need to know to interpret correlation.

1. Correlation makes no distinction between explanatory and response variables. It makes no difference which variable you call x and which you call y in calculating the correlation.

2. Correlation requires that both variables be quantitative, so that it makes sense to do the arithmetic indicated by the formula for r. We cannot calculate a correlation between the incomes of a group of people and what city they live in, because city is a categorical variable.

3. Because r uses the standardized values of the observations, r does not change when we change the units of measurement of x, y, or both. Measuring height in inches

rather than centimeters and weight in pounds rather than kilograms does not change the correlation between height and weight. The correlation r itself has no unit of measurement; it is just a number.

4. Positive r indicates positive association between the variables, and negative r indicates negative association.

5. The correlation r is always a number between -1 and 1. Values of r near 0 indicate a very weak linear relationship. The strength of the linear relationship increases as r moves away from 0 toward either -1 or 1. Values of r close to -1 or 1 indicate that the points in a scatterplot lie close to a straight line. The extreme values $r = -1$ and $r = 1$ occur only in the case of a perfect linear relationship, when the points lie exactly along a straight line.

6. Correlation measures the strength of only a linear relationship between two variables. Correlation does not describe curved relationships between variables, no matter how strong they are.

7. Like the mean and standard deviation, the correlation is not resistant: r is strongly affected by a few outlying observations. Use r with caution when outliers appear in the scatterplot.

The scatterplots in Figure 2.16 illustrate how values of r closer to 1 or -1 correspond to stronger linear relationships. To make the meaning of r clearer, the standard deviations of both variables in these plots are equal and the horizontal and vertical scales are the same. In general, it is not so easy to guess the value of r from the appearance of a scatterplot. Remember that changing the plotting scales in a scatterplot may mislead our eyes, but it does not change the correlation.

Do remember that **correlation is not a complete description of two-variable data,** even when the relationship between the variables is linear. You should give the means

FIGURE 2.16 How correlation measures the strength of a linear relationship. Patterns closer to a straight line have correlations closer to 1 or -1.

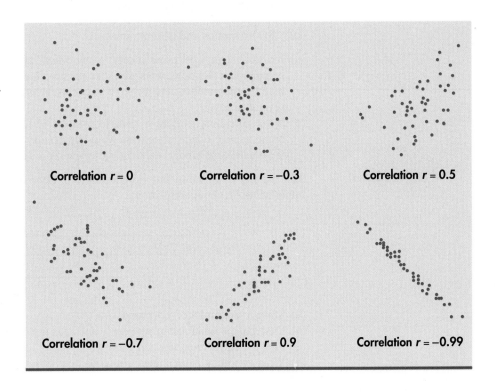

Correlation $r = 0$

Correlation $r = -0.3$

Correlation $r = 0.5$

Correlation $r = -0.7$

Correlation $r = 0.9$

Correlation $r = -0.99$

and standard deviations of both x and y along with the correlation. (Because the formula for correlation uses the means and standard deviations, these measures are the proper choice to accompany a correlation.) Conclusions based on correlations alone may require rethinking in the light of a more complete description of the data.

EXAMPLE 2.11 Forecasting Earnings

Stock analysts regularly forecast the earnings per share (EPS) of companies they follow. EPS is calculated by dividing a company's net income for a given time period by the number of common stock shares outstanding. We have two analysts' EPS forecasts for a computer manufacturer for the next 6 quarters. How well do the two forecasts agree? The correlation between them is $r = 0.9$, but the mean of the first analyst's forecasts is 3 dollars per share lower than the second analyst's mean.

These facts do not contradict each other. They are simply different kinds of information. The means show that the first analyst predicts lower EPS than the second. But because the first analyst's EPS predictions are about 3 dollars per share lower than the second analyst's *for every quarter,* the correlation remains high. Adding or subtracting the same number to all values of either x or y does not change the correlation. The two analysts agree on which quarters will see higher EPS values. The high r shows this agreement, despite the fact that the actual predicted values differ by 3 dollars per share.

DEBT

APPLY YOUR KNOWLEDGE

2.27 Correlation for debt. Figure 2.6 (page 84) is a scatterplot of 2007 debt versus 2006 debt for 24 countries. Is the correlation r for these data near -1, clearly negative but not near -1, near 0, clearly positive but not near 1, or near 1? Explain your answer.

2.28 Brand names and generic products.

(a) If a store always prices its generic "store brand" products at 90% of the brand name products' prices, what would be the correlation between the prices of the brand name products and the store brand products? (*Hint:* Draw a scatterplot for several prices.)

(b) If the store always prices its generic products $1 less than the corresponding brand name products, then what would be the correlation between the prices of the brand name products and the store brand products?

CORRELATION

2.29 Strong association but no correlation. Here is a data set that illustrates an important point about correlation:

x	20	30	40	50	60
y	10	30	50	30	10

(a) Make a scatterplot of y versus x.
(b) Describe the relationship between y and x. Is it weak or strong? Is it linear?
(c) Find the correlation between y and x.
(d) What important point about correlation does this exercise illustrate?

SECTION 2.2 Summary

- The **correlation** r measures the strength and direction of the linear association between two quantitative variables x and y. Although you can calculate a correlation for any scatterplot, r measures only straight-line relationships.

- Correlation indicates the direction of a linear relationship by its sign: $r > 0$ for a positive association and $r < 0$ for a negative association.

- Correlation always satisfies $-1 \le r \le 1$ and indicates the strength of a relationship by how close it is to -1 or 1. Perfect correlation, $r = \pm 1$, occurs only when the points on a scatterplot lie exactly on a straight line.

- Correlation ignores the distinction between explanatory and response variables. The value of r is not affected by changes in the unit of measurement of either variable. Correlation is not resistant, so outliers can greatly change the value of r.

SECTION 2.2 Exercises

For Exercises 2.25 and 2.26, see page 94; and for 2.27 to 2.29, see page 96.

2.30 Best countries for business. Figure 2.7 (page 85) is a scatterplot of the gross domestic product per capita versus the unemployment rate for 99 countries. 🌐 BESTCOUNTRIES
(a) Do you think that the correlation is a good statistic to describe the strength of the relationship between these two variables? Explain your answer.
(b) Find the correlation between these two variables.
(c) Comment on the following statement: "If we use statistical software to perform statistical calculations, then the results are correct, meaningful, and useful."

2.31 Best countries for business with logs. Figure 2.8 (page 86) is a scatterplot of the log gross domestic product per capita versus the log unemployment rate for 99 countries. 🌐 BESTCOUNTRIES
(a) Do you think that the correlation is a good statistic to describe the strength of the relationship between these two variables? Give a reason for your answer.
(b) Find the correlation between these two variables.
(c) Would your answer to part (a) change if you restricted attention to the countries with log unemployment rate greater than 1.5? Explain your answer.

2.32 First test and final exam. How strong is the relationship between the score on the first test and the score on the final exam in an elementary statistics course? Here are data for eight students from such a course: 🌐 STATCOURSE8

First-test score	153	144	162	149	127	158	158	153
Final-exam score	145	140	145	170	145	175	170	160

(a) Do you think that one of these variables should be an explanatory variable and the other a response variable? Give reasons for your answer.

(b) Make a scatterplot and describe the relationship.
(c) Find the correlation.
(d) Give some possible reasons why this relationship is so weak.

2.33 Second test and final exam. Refer to the previous exercise. Here are the data for the second test and the final exam for the same students: 🌐 STATCOURSE8

Second-test score	158	162	144	162	136	158	175	153
Final-exam score	145	140	145	170	145	175	170	160

(a) Explain why you should use the second-test score as the explanatory variable.
(b) Make a scatterplot and describe the relationship.
(c) Find the correlation.
(d) Why do you think the relationship between the second-test score and the final-exam score is stronger than the relationship between the first-test score and the final-exam score?

2.34 Add an outlier. Refer to the previous exercise. Add a ninth student whose scores on the second test and final exam would lead you to classify the additional data point as an outlier.
(a) Highlight the outlier on your scatterplot.
(b) Find the correlation and describe the effect of the outlier on the correlation.
(c) Describe the performance of the student on the second exam and final exam and why that leads to the conclusion that the result is an outlier. Give a possible reason for the performance of this student.

2.35 Alcohol and carbohydrates in beer. Figure 2.10 (page 90) is a scatterplot of carbohydrates versus percent alcohol in 86 brands of beer. Compute the correlation for these data. 🌐 BEER

2.36 Alcohol and carbohydrates in beer revisited. Refer to the previous exercise. The data that you used to compute the correlation includes an outlier. 🌀 **BEER**
(a) Remove the outlier and recompute the correlation.
(b) Write a short paragraph about the possible effects of outliers on a correlation, using this example to illustrate your ideas.

2.37 Marketing in Canada. In Exercise 2.20 (page 90) you examined the relationship between the percent of the population over 65 and the percent under 15 for the 13 Canadian provinces and territories. Figure 2.12 is a scatterplot of the data. Find the value of the correlation r. Does this numerical summary give a good indication of the strength of the relationship between these two variables? Explain your answer. 🌀 **CANADIANPOPULATION**

2.38 Nunavut. Refer to the previous exercise and to the data given in Figure 2.11 (page 91). 🌀 **CANADIANPOPULATION**
(a) Do you think that Nunavut is an outlier? Explain your answer.
(b) Find the correlation without Nunavut. Using your work from the previous exercise, summarize the effect of Nunavut on the correlation.

2.39 Compare the provinces with the territories. Refer to the previous exercises. The three Canadian territories are the Northwest Territories, Nunavut, and the Yukon Territories. All the other entries in Figure 2.11 are provinces. 🌀 **CANADIANPOPULATION**
(a) Generate a scatterplot of the Canadian demographic data similar to Figure 2.12 with the points labeled "P" for provinces and "T" for territories. Use your plot from Exercise 2.21 if you completed that exercise.
(b) Find the correlation using the data from the territories only.
(c) Do the same for the provinces.
(d) Summarize your analysis and conclusions in a short paragraph.

2.40 World Bank data on living longer and using the Internet. Figure 2.13 is a scatterplot of life expectancy versus the number of Internet users per 100 people for 181 countries. In Exercise 2.22 you described this relationship. Make a plot of the data similar to Figure 2.13 and report the correlation. 🌀 **INTERNETANDLIFE**

2.41 Let's look at Europe. Refer to the previous exercise. Figure 2.14 (page 92) is a scatterplot of the same data for the 48 European countries in the data set. 🌀 **INTERNETANDLIFE**
(a) Make a plot of the data similar to Figure 2.14.
(b) Report the correlation.
(c) Summarize the differences and similarities between the relationship for all 181 countries and the relationship that you found in this exercise for the European countries only.

2.42 Prices paid to producers and additional land for production. Coffee is a leading export from several developing countries. When coffee prices are high, farmers often clear forest to plant more coffee trees. Here are data for five years on prices paid to coffee growers in Indonesia and the rate of deforestation in a national park that lies in a coffee-producing region:[12] 🌀 **COFFEE**

Price (cents per pound)	Deforestation (percent)
29	0.49
40	1.59
54	1.69
55	1.82
72	3.10

(a) Make a scatterplot. Which is the explanatory variable? What kind of pattern does your plot show?
(b) Find the correlation r step-by-step. That is, find the mean and standard deviation of the two variables. Then find the five standardized values for each variable and use the formula for r. Explain how your value for r matches your graph in (a).
(c) Next, enter these data into your calculator or software and use the correlation function to find r. Check that you get the same result as in (b), up to roundoff error.

2.43 Prices and land. Coffee is currently priced in dollars. If it were priced in euros, and the dollar prices in the previous exercise were translated into the equivalent prices in euros, would the correlation between coffee price and percent deforestation change? Explain your answer. 🌀 **COFFEE**

2.44 Prices and land. Refer to the previous two exercises. These data show that deforestation in a national park in Indonesia goes up when high prices for coffee encourage farmers to clear forest in order to plant more coffee. There are only 5 observations, so we worry that the apparent relationship may be just chance. 🌀 **COFFEE**
(a) Make a scatterplot and find the correlation r.
(b) Assume that we know that coffee prices and deforestation rates are unrelated. Add at least 20 points (by hand) to the scatterplot you made in part (a) to create a scatterplot that corresponds to this assumption. Use a different marking symbol for these new points than that used for the 5 data points in part (a).
(c) What would be the approximate correlation of your points in part (b)?

2.45 Match the correlation. The *Correlation and Regression* applet at www.whfreeman.com/psbe allows you to create a scatterplot by clicking and dragging with the mouse. The applet calculates and displays the correlation as you change the plot. You will use this applet to make scatterplots with 8 points that have correlation close to 0.8. The lesson is that many patterns can have the same correlation. Always plot your data before you trust a correlation.
(a) Stop after adding the first 2 points. What is the value of the correlation? Why does it have this value?
(b) Make a lower-left to upper-right pattern of 8 points with correlation about $r = 0.8$. (You can drag points up or down to adjust r after you have 8 points.) Make a rough sketch of your scatterplot.
(c) Make another scatterplot with 7 points in a vertical stack at the right of the plot. Add 1 point far to the left and move it

until the correlation is close to 0.8. Make a rough sketch of your scatterplot.

(d) Make yet another scatterplot with 8 points in a curved pattern that starts at the lower left, rises to the right, then falls again at the far right. Adjust the points up or down until you have a quite smooth curve with correlation close to 0.8. Make a rough sketch of this scatterplot also.

2.46 Mutual fund performance. Many mutual funds compare their performance with that of a benchmark, an index of the returns on all securities of the kind that the fund buys. The Vanguard International Growth Fund, for example, takes as its benchmark the Morgan Stanley Europe, Australasia, Far East (EAFE) index of overseas stock market performance. Here are the values of a $10,000 investment in the Vanguard Fund made in 1998 and the hypothetical value of a $10,000 investment in the benchmark (EAFE):[13] 🔷 **MUTUALFUNDS**

Year	EAFE	Fund	Year	EAFE	Fund
1998	10,000	10,000	2003	10,029	9,615
1999	12,110	11,787	2004	12,739	12,465
2000	11,440	10,939	2005	13,867	13,523
2001	9,365	8,846	2006	17,879	17,336
2002	8,378	7,740	2007	21,835	20,335

Make a scatterplot suitable for predicting fund returns from EAFE returns. Is there a clear straight-line pattern? How strong is this pattern? (Give a numerical measure.) Are there any extreme outliers?

2.47 What is the correlation? Suppose that women always married men 2 years older than themselves. Draw a scatterplot of the ages of 5 married couples, with the wife's age as the explanatory variable. What is the correlation r for your data? Why?

2.48 Stretching a scatterplot. Changing the units of measurement can greatly alter *the appearance* of a scatterplot. Consider the following data: 🔷 **STRETCH**

x	−4	−4	−3	3	4	4
y	0.5	−0.6	−0.5	0.5	0.5	−0.6

(a) Draw x and y axes each extending from −6 to 6. Plot the data on these axes.

(b) Calculate the values of new variables $x^* = x/10$ and $y^* = 10y$, starting from the values of x and y. Plot y^* against x^* on the same axes using a different plotting symbol. The two plots are very different in appearance.

(c) Find the correlation between x and y. Then find the correlation between x^* and y^*. How are the two correlations related? Explain why this isn't surprising.

2.49 Make a plot and find the correlation. The following 20 observations on y and x were generated by a computer program. 🔷 **GENERATEDDATA**

y	x	y	x
34.38	22.06	27.07	17.75
30.38	19.88	31.17	19.96
26.13	18.83	27.74	17.87
31.85	22.09	30.01	20.20
26.77	17.19	29.61	20.65
29.00	20.72	31.78	20.32
28.92	18.10	32.93	21.37
26.30	18.01	30.29	17.31
29.49	18.69	28.57	23.50
31.36	18.05	29.80	22.02

(a) Make a scatterplot and describe the relationship between y and x.

(b) Find the correlation between y and x.

2.50 Add an outlier. Refer to the previous exercise. Add an additional observation with $y = 50$ and $x = 30$ to the data set. 🔷 **GENERATEDDATA21A**

(a) Make a scatterplot of the 21 observations and explain why the additional observation is an outlier.

(b) Compute the correlation and compare it with the correlation that you computed for the original data set.

2.51 Add a different outlier. Refer to the previous two exercises. Add an additional observation with $y = 29$ and $x = 50$ to the original data set. 🔷 **GENERATEDDATA21B**

(a) Make a scatterplot of the 21 observations and explain why this additional observation is an outlier.

(b) Compute the correlation and compare it with the correlation that you computed for the original data set.

(c) In this exercise and in the previous one, you added an outlier to the original data set and recomputed the correlation. Write a short summary of the changes in correlations that can result from different kinds of outliers.

2.52 CEO compensation and stock market performance. An academic study concludes, "The evidence indicates that the correlation between the compensation of corporate CEOs and the performance of their company's stock is close to zero." A business magazine reports this as "A new study shows that companies that pay their CEOs highly tend to perform poorly in the stock market, and vice versa." Explain why the magazine's report is wrong. Write a statement in plain language (don't use the word "correlation") to explain the study's conclusion.

2.53 Investment reports and correlations. Investment reports often include correlations. Following a table of correlations among mutual funds, a report adds, "Two funds can have perfect correlation, yet different levels of risk. For example, Fund A and Fund B may be perfectly correlated, yet Fund A moves 20% whenever Fund B moves 10%." Write a brief explanation, for someone who knows no statistics, of how this can happen. Include a sketch to illustrate your explanation.

2.54 Sloppy writing about correlation. Each of the following statements contains a blunder. Explain in each case what is wrong.

(a) "There is a high correlation between the make of an automobile and its quality."

(b) "In an introductory statistics course, there is a very high correlation ($r = 1.2$) between total points earned for homework and the final grade."

(c) "The correlation between y and x is $r = 0.4$, but the correlation between x and y is $r = -0.4$."

2.3 Least-Squares Regression

Correlation measures the direction and strength of the straight-line (linear) relationship between two quantitative variables. If a scatterplot shows a linear relationship, we would like to summarize this overall pattern by drawing a line on the scatterplot. A *regression line* summarizes the relationship between two variables, but only in a specific setting: when one of the variables helps explain or predict the other. That is, regression describes a relationship between an explanatory variable and a response variable.

> **Regression Line**
>
> A **regression line** is a straight line that describes how a response variable y changes as an explanatory variable x changes. We often use a regression line to predict the value of y for a given value of x.

EXAMPLE 2.12   World Financial Markets

 FINMARK

The World Economic Forum studies data on many variables related to financial development in the countries of the world. They rank countries on their financial development based on a collection of factors related to economic growth.[14] Two of the variables studied are gross domestic product per capita and net assets per capita. Here are the data for 15 countries that ranked high on financial development:

Country	GDP	Assets	Country	GDP	Assets	Country	GDP	Assets
United Kingdom	43.8	199	Switzerland	67.4	358	Germany	44.7	145
Australia	47.4	166	Netherlands	52.0	242	Belgium	47.1	167
United States	47.9	191	Japan	38.6	176	Sweden	52.8	169
Singapore	40.0	168	Denmark	62.6	224	Spain	35.3	152
Canada	45.4	170	France	46.0	149	Ireland	61.8	214

In this table, GDP is gross domestic product per capita in thousands of dollars, and assets is net assets per capita in thousands of dollars. Figure 2.17 is a scatterplot of the data. There is a moderately weak linear relationship between the score on the second test and the score on the final exam. The correlation is $r = 0.76$. The scatterplot includes a regression line drawn through the points.

Suppose we want to use this relationship between GDP per capita and net assets per capita to predict the net assets per capita for a country that has a GDP per capita of

FIGURE 2.17 Scatterplot of GDP per capita and net assets per capita for 15 countries that rank high on financial development, for Example 2.12. The dashed line indicates how to use the regression line to predict net assets per capita for a country with a GDP per capita of 50.

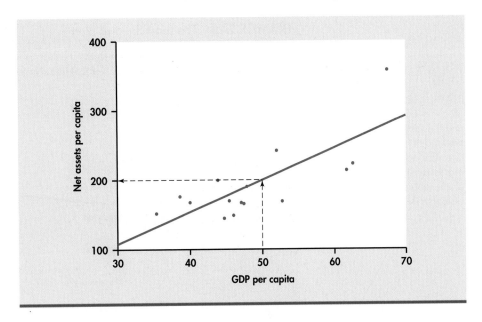

prediction

50 thousand dollars. To **predict** the net assets per capita (in thousands of dollars), first locate 50 on the x axis. Then go "up and over" as in Figure 2.17 to find the GDP per capita y that corresponds to $x = 50$. We predict that a country with a GDP per capita of 50 thousand dollars will have net assets per capita of about 200 thousand dollars.

The least-squares regression line

Different people might draw different lines by eye on a scatterplot. We need a way to draw a regression line that doesn't depend on our guess as to where the line should be. We will use the line to predict y from x, so the prediction errors we make are errors in y, the vertical direction in the scatterplot. If we predict net assets per capita of 177 and the actual net assets per capita are 170, our prediction error is

$$\text{error} = \text{observed } y - \text{predicted } y$$
$$= 170 - 177 = -7$$

The error is $-\$7000$.

APPLY YOUR KNOWLEDGE

2.55 Find a prediction error. Use Figure 2.17 to estimate the net assets per capita for a country that has a GDP per capita of $50 thousand. If the actual net assets per capita are $280, find the prediction error.

2.56 Positive and negative prediction errors. Examine Figure 2.17 carefully. How many of the prediction errors are positive? How many are negative?

No line will pass exactly through all the points in the scatterplot. We want the *vertical* distances of the points from the line to be as small as possible.

EXAMPLE 2.13   The Least-Squares Idea

Figure 2.18 illustrates the idea. This plot shows the data, along with a line. The vertical distances of the data points from the line appear as vertical line segments.

FIGURE 2.18 The least-squares idea. For each observation find the vertical distance of each point from a regression line. The least-squares regression line makes the sum of the squares of these distances as small as possible.

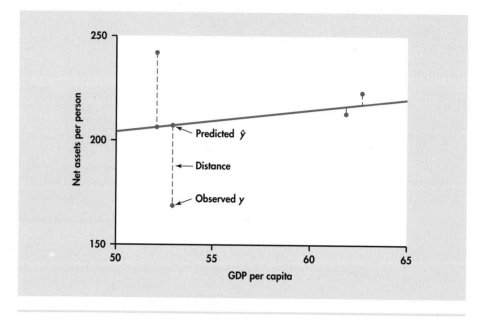

There are several ways to make the collection of vertical distances "as small as possible." The most common is the *least-squares* method.

> **Least-Squares Regression Line**
>
> The **least-squares regression line** of y on x is the line that makes the sum of the squares of the vertical distances of the data points from the line as small as possible.

One reason for the popularity of the least-squares regression line is that the problem of finding the line has a simple solution. We can give the recipe for the least-squares line in terms of the means and standard deviations of the two variables and their correlation.

> **Equation of the Least-Squares Regression Line**
>
> We have data on an explanatory variable x and a response variable y for n individuals. From the data, calculate the means $\bar{x}$ and $\bar{y}$ and the standard deviations s_x and s_y of the two variables, and their correlation r. The least-squares regression line is the line
>
> $$\hat{y} = b_0 + b_1 x$$
>
> with **slope**
>
> $$b_1 = r\frac{s_y}{s_x}$$
>
> and **intercept**
>
> $$b_0 = \bar{y} - b_1\bar{x}$$

We write $\hat{y}$ (read "y hat") in the equation of the regression line to emphasize that the line gives a *predicted* response $\hat{y}$ for any x. Because of the scatter of points about the line, the predicted response will usually not be exactly the same as the actually *observed* response y. In practice, you don't need to calculate the means, standard deviations, and correlation first. Statistical software or your calculator will give the slope b_1 and intercept b_0 of the least-squares line from keyed-in values of the variables x and y. You can then concentrate on understanding and using the regression line. Be warned—different software packages and calculators label the slope and intercept differently in their output, so remember that the slope is the value that multiplies x in the equation.

EXAMPLE 2.14   The Equation for Predicting GDP

FINMARK

The line in Figure 2.17 is in fact the least-squares regression line for predicting net assets per capita from GDP per capita. The equation of this line is

$$\hat{y} = -27.17 + 4.500x$$

slope

The **slope** of a regression line is almost always important for interpreting the data. The slope is the rate of change, the amount of change in $\hat{y}$ when x increases by 1. The slope $b_1 = 4.5$ in this example says that each additional thousand dollars of GDP per capita is associated with an additional \$4500 in net assets per capita.

intercept

The **intercept** of the regression line is the value of $\hat{y}$ when $x = 0$. Although we need the value of the intercept to draw the line, it is statistically meaningful only when x can actually take values close to zero. In our example, $x = 0$ occurs when a country has zero GDP. Such a situation would be very unusual, and we would not include it within the framework of our analysis.

EXAMPLE 2.15   Predict the GDP

prediction

The equation of the regression line makes **prediction** easy. Just substitute a value of x into the equation. To predict the net assets per capita for a country that has a GDP per capita of \$50 thousand, we use $x = 50$:

$$\hat{y} = -27.17 + 4.500x$$
$$= -27.17 + (4.500)(50)$$
$$= -27.17 + 225.00 = 198$$

The predicted net assets per capita is \$198 thousand.

plotting a line

To **plot the line** on the scatterplot, you can use the equation to find $\hat{y}$ for two values of x, one near each end of the range of x in the data. Plot each $\hat{y}$ above its x and draw the line through the two points. As a check, it is a good idea to compute y for a third value of x and verify that this point is on your line.

APPLY YOUR KNOWLEDGE

2.57 A regression line. A regression equation is $y = 10 + 20x$.

(a) What is the slope of the regression line?
(b) What is the intercept of the regression line?
(c) Find the predicted values of y for $x = 10$, for $x = 20$, and for $x = 30$.
(d) Plot the regression line for values of x between 0 and 50.

FINMARK

EXAMPLE 2.16 GDP and Assets Results Using Software

Figure 2.19 displays the selected regression output for the world financial markets data from Minitab and Excel. The complete outputs contain many other items that we will study in Chapter 10.

Minitab

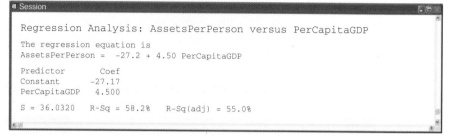

Excel

		Regression Statistics				
Multiple R			0.763116			
R Square			0.582347			
Standard Error			36.03196			
Observations			15			
		Coefficients				
Intercept		−27.16823305				
PerCapitaGDP		4.4998956				

FIGURE 2.19 Selected least-squares regression output for the world financial markets data. (a) Minitab. (b) Excel.

Let's look at the Minitab output first. The first entry is the regression equation. Below that is a table giving the regression intercept and slope under the column heading "Coef." This is an abbreviation for **coefficient,** a generic term that refers to the quantities that define a regression equation. Note that the intercept is labeled "Constant," and the slope is labeled with the name of the explanatory variable. In the table Minitab reports the intercept as −27.17 and the slope as 4.500.

coefficient

Excel provides the same information in a slightly different format. Here the intercept is reported as −27.16823305, and the slope is reported as 4.4998956.

How many digits should we keep in reporting the results of statistical calculations? The answer depends on how the results will be used. For example, if we are giving a description of the equation, then rounding the coefficients and reporting the equation as $y = -27 + 4.5x$ would be fine. If we will use the equation to calculate predicted values, we should keep a few more digits and then round the resulting calculation as we did in Example 2.15.

FINMARK

APPLY YOUR KNOWLEDGE

2.58 Predicted values and residuals for GDP and assets. Refer to the world financial markets data in Example 2.12.

(a) Use software to compute the coefficients of the regression equation. Indicate where to find the slope and the intercept on the output, and report these values.

(b) Make a scatterplot of the data with the least-squares line.

(c) Find the predicted value of assets for each country.

(d) Find the difference between the actual value and the predicted value for each country.

Facts about least-squares regression

Regression is one of the most common statistical settings, and least squares is the most common method for fitting a regression line to data. Here are some facts about least-squares regression lines.

Fact 1. There is a close connection between correlation and the slope of the least-squares line. The slope is

$$b_1 = r\frac{s_y}{s_x}$$

This equation says that along the regression line, **a change of one standard deviation in x corresponds to a change of r standard deviations in y.** When the variables are perfectly correlated ($r = 1$ or $r = -1$), the change in the predicted response $\hat{y}$ is the same (in standard deviation units) as the change in x. Otherwise, because $-1 \le r \le 1$, the change in $\hat{y}$ is less than the change in x. As the correlation grows less strong, the prediction $\hat{y}$ moves less in response to changes in x.

Fact 2. The least-squares regression line always passes through the point $(\overline{x}, \overline{y})$ on the graph of y against x. So the least-squares regression line of y on x is the line with slope rs_y/s_x that passes through the point $(\overline{x}, \overline{y})$. We can describe regression entirely in terms of the basic descriptive measures $\overline{x}$, s_x, $\overline{y}$, s_y, and r.

Fact 3. The distinction between explanatory and response variables is essential in regression. Least-squares regression looks at the distances of the data points from the line only in the y direction. If we reverse the roles of the two variables, we get a different least-squares regression line.

EXAMPLE 2.17　Bots and Spam Messages per Day

CASE 2.1

SPAM

Figure 2.20 is a scatterplot of the spam botnet data described in Case 2.1 (page 79). There is a positive linear relationship.

FIGURE 2.20 Scatterplot of spams per day versus the number of bots. The two lines are the least-squares regression lines: using bots to predict spams per day (solid) and using spams per day to predict bots (dashed), for Example 2.17.

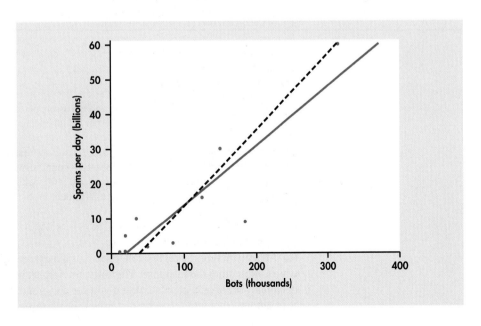

The two lines on the plot are the two least-squares regression lines. The regression line for using bots to predict spam messages per day is solid. The regression line for using spam messages per day to predict bots is dashed. *Regression of spams per day on bots and regression of bots on spams per day give different lines.* In the regression setting, you must decide which variable is explanatory.

Interpretation of r^2

The square of the correlation r describes the strength of a straight-line relationship. Here is the basic idea. Think about trying to predict a new value of y. With no other information than our sample of values of y, a reasonable choice is $\overline{y}$.

Now consider how your prediction would change if you had an explanatory variable. If we use the regression equation to predict, we would use $\hat{y} = b_0 + b_1x$. This prediction takes into account the value of the explanatory variable x.

Let's compare our two choices for predicting y. With the explanatory variable x, we use $\hat{y}$; without this information, we use $\overline{y}$. How can we compare these two choices? When we use $\overline{y}$ to predict, our prediction error is $y - \overline{y}$. If, instead, we use $\hat{y}$, our prediction error is $y - \hat{y}$. The use of x in our prediction changes our prediction error from is $y - \overline{y}$ to $y - \hat{y}$. The difference is $\hat{y} - \overline{y}$. Our comparison uses the sums of squares of these differences $\sum(y - \overline{y})^2$ and $\sum(\hat{y} - \overline{y})^2$. The ratio of these two quantities is the square of the correlation:

$$r^2 = \frac{\sum(\hat{y} - \overline{y})^2}{\sum(y - \overline{y})^2}$$

The numerator represents the variation in y that is explained by x, and the denominator represents the total variation in y.

> **Percent of Variation Explained by the Least-Squares Equation**
>
> To find the percent of variation explained by the least-squares equation, square the value of the correlation and express the result as a percent.

EXAMPLE 2.18 Using r^2

The correlation between GDP per capita and net assets per capita in Example 2.12 (page 100) is $r = 0.76312$, and so $r^2 = 0.58234$. GDP per capita explains about 58% of the variability in net assets per capita.

On the other hand, in Example 2.5 (page 82), we saw that there was a strong linear association between debt in 2007 and debt in 2006 for the 24 countries with debt less than US$1 trillion. The correlation between these two variables is $r = 0.9971$, which means that $r^2 = 0.9943$. Debt in 2006 explains over 99% of the variation in 2007 debt for these 24 countries.

When you report a regression, give r^2 as a measure of how successful the regression was in explaining the response. The software outputs in Figure 2.19 include r^2, either in decimal form or as a percent. When you see a correlation (often listed as R or Multiple R in outputs), square it to get a better feel for the strength of the association.

2.59 The "January effect." Some people think that the behavior of the stock market in January predicts its behavior for the rest of the year. Take the explanatory variable x to be the percent change in a stock market index in January and the response variable y to be the change in the index for the entire year. We expect a positive correlation between x and y because the change during January contributes to the full year's change. Calculation based on 38 years of data gives

$$\bar{x} = 1.75\% \qquad s_x = 5.36\% \qquad r = 0.596$$
$$\bar{y} = 9.07\% \qquad s_y = 15.35\%$$

(a) What percent of the observed variation in yearly changes in the index is explained by a straight-line relationship with the change during January?
(b) What is the equation of the least-squares line for predicting full-year change from January change?
(c) The mean change in January is $\bar{x} = 1.75\%$. Use your regression line to predict the change in the index in a year in which the index rises 1.75% in January. Why could you have given this result (up to roundoff error) without doing the calculation?

2.60 Is regression useful? In Exercise 2.45 (page 98) you used the *Correlation and Regression* applet to create three scatterplots having correlation about $r = 0.8$ between the horizontal variable x and the vertical variable y. Create three similar scatterplots again, after clicking the "Show least-squares line" box to display the regression line. Correlation $r = 0.8$ is considered reasonably strong in many areas of work. Because there is a reasonably strong correlation, we might use a regression line to predict y from x. In which of your three scatterplots does it make sense to use a straight line for prediction?

Residuals

A regression line is a mathematical model for the overall pattern of a linear relationship between an explanatory variable and a response variable. Deviations from the overall pattern are also important. In the regression setting, we see deviations by looking at the scatter of the data points about the regression line. The vertical distances from the points to the least-squares regression line are as small as possible in the sense that they have the smallest possible sum of squares. Because they represent "leftover" variation in the response after fitting the regression line, these distances are called *residuals*.

> **Residuals**
>
> A **residual** is the difference between an observed value of the response variable and the value predicted by the regression line. That is,
>
> $$\text{residual} = \text{observed } y - \text{predicted } y$$
> $$= y - \hat{y}$$

CASE 2.1

EXAMPLE 2.19 Spam Botnets

SPAM

Figure 2.21 is a scatterplot showing the number of spam messages per day versus the number of bots for the 10 botnets that we studied in Case 2.1 (page 79). Included on the scatterplot is the least-squares line. The points for the botnets Srizbi, Bobax, and Ozdok are marked individually.

The equation of the least-squares line is $\hat{y} = -3.4192 + 0.17048x$ where y represents spam messages per day and x represents the number of bots.

Let's look carefully at the data for the botnet Srizbi, $y = 60$ and $x = 315$. The predicted number of spam messages per day for a botnet with 315 bots is

$$\hat{y} = -3.4192 + 0.17048(315)$$
$$= 50.3$$

The residual for Srizbi is the difference between the observed number of spam messages per day (y) and this predicted value.

$$\text{residual} = y - \hat{y}$$
$$= 60 - 50.3$$
$$= 9.7$$

FIGURE 2.21 Scatterplot of spams per day versus the number of bots for 10 botnets, with the least-squares line and selected points labeled, for Example 2.19.

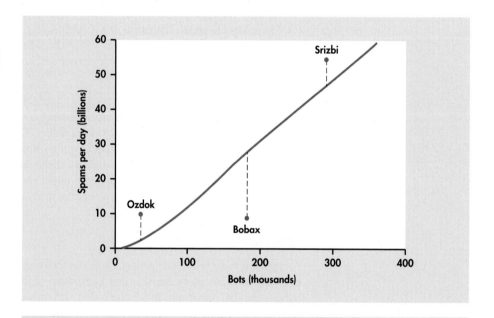

On the scatterplot, the residual of this observation is shown as a dashed vertical line between the observation and the least-squares line.

APPLY YOUR KNOWLEDGE

2.61 Residual for Bobax. The botnet Bobax has 185 bots that send 9 (billion) spam messages per day.

(a) Find the predicted number of spam messages per day for Bobax.
(b) Find the residual for Bobax.
(c) Which botnet, Srizbi or Bobax, has a greater deviation from the regression line?

There is a residual for each data point. Finding the residuals with a calculator is a bit unpleasant, because you must first find the predicted response for every x. Statistical software gives you the residuals all at once.

EXAMPLE 2.20   Spam Botnet Residuals

CASE 2.1   Here are the residuals for the 10 botnets:

9.7	−19.1	7.8	−1.9	−8.1	−3.1	7.4	5.0	0.6	1.7

Because the residuals show how far the data fall from our regression line, examining the residuals helps us assess how well the line describes the data. Although residuals can be calculated from any model fitted to data, the residuals from the least-squares line have a special property: **the mean of the least-squares residuals is always zero.**

APPLY YOUR KNOWLEDGE

CASE 2.1   **2.62 Sum the botnet residuals.** What is the sum of the residuals for the spam botnet data? Is there a deviation from the expected sum due to roundoff error?

As usual, when we perform statistical calculations, we prefer to display the results graphically. We can do this for the residuals.

> **Residual Plots**
>
> A **residual plot** is a scatterplot of the regression residuals against the explanatory variable. Residual plots help us assess the fit of a regression line.

CASE 2.1   **EXAMPLE 2.21**   Residual Plot for Spam Botnets

 SPAM   Figure 2.22 gives the residual plot for the botnet data. The horizontal line at zero in the plot helps orient us.

FIGURE 2.22 Residual plot for the botnet data, for Example 2.21.

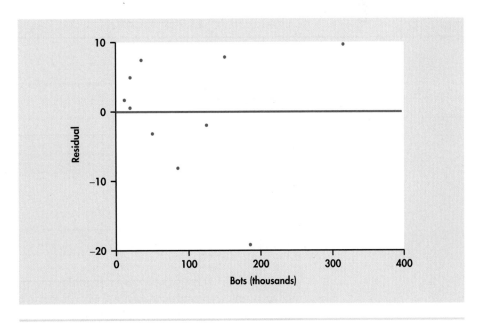

SPAM

CASE 2.1 **2.63 Identify the three botnets.** In Figure 2.21, three botnets are identified by name: Ozdok, Bobax, and Srizbi. The dashed lines in the plot represent the residuals.

(a) Make a sketch of the residual plot in Figure 2.22, and write in the names of the botnets Ozdok, Bobax, and Srizbi.
(b) Explain how you were able to identify these three points on your sketch.

If the regression line captures the overall relationship between x and y, the residuals should have no systematic pattern. The residual plot will look something like the pattern in Figure 2.23(a). That plot shows a scatter of points about the fitted line, with no unusual individual observations or systematic change as x increases. Here are some things to look for when you examine a residual plot:

- **A curved pattern** shows that the relationship is not linear. Figure 2.23(b) is a simplified example. A straight line is not a good summary for such data. The residuals for Figure 2.13 (page 92) would have this form.

- **Increasing or decreasing spread about the line** as x increases. Figure 2.23(c) is a simplified example. Prediction of y will be less accurate for larger x in that example.

FIGURE 2.23 Idealized patterns in plots of least-squares residuals. Plot (a) indicates that the regression line fits the data well. The data in plot (b) have a curved pattern, so a straight line fits poorly. The response variable y in plot (c) has more spread for larger values of the explanatory variable x, so prediction will be less accurate when x is large.

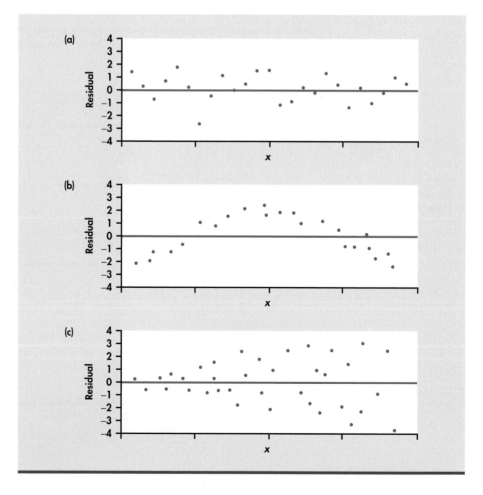

- **Individual points with large residuals,** like the botnet Bobax in Figures 2.21 and 2.22. Such points are outliers in the vertical (y) direction because they lie far from the line that describes the overall pattern.

- **Individual points that are extreme in the x direction,** like Srizbi in Figures 2.21 and 2.22. Such points may or may not have large residuals, but they can be very important. We address such points next.

Influential observations

Bobax and Srizbi are both unusual. They are unusual in different ways. Bobax lies far from the regression line. For a botnet with 185 bots, it produces relatively few spams per day.

EXAMPLE 2.22   Srizbi Is Influential

CASE 2.1

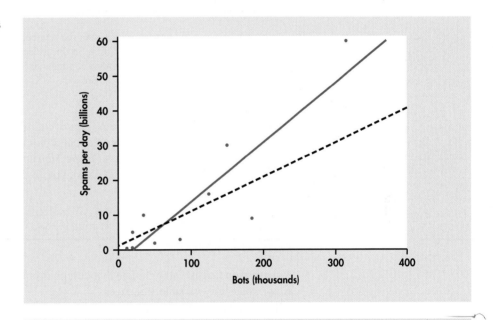
SPAM

Srizbi is closer to the line but very far out in the x direction. It has 315 bots, while the next highest value is 185 bots for Bobax. *Because of its extreme number of bots, this point has a strong influence on the position of the regression line.* Figure 2.24 adds a second regression line, calculated after leaving out Srizbi.

FIGURE 2.24 Two least-squares lines for the botnet data, for Example 2.22. The solid line is calculated using all of the data. The dashed line leaves out the data for Srizbi, an influential botnet. Srizbi is influential because leaving it out moves the line quite a bit.

You can see that removing Srizbi causes the line to move quite a bit. We call points that do this *influential.*

> **Outliers and Influential Observations in Regression**
>
> An **outlier** is an observation that lies outside the overall pattern of the other observations. Points that are outliers in the y direction of a scatterplot have large regression residuals, but other outliers need not have large residuals.
>
> An observation is **influential** for a statistical calculation if removing it would markedly change the result of the calculation. Points that are extreme in the x direction of a scatterplot are often influential for the least-squares regression line.

Bobax and Srizbi are both quite far from the regression line in Figure 2.21. Srizbi influences the slope of the least-squares line because it is far out in the x direction. Bobax is extreme in the y direction. It has less influence on the regression line because the many other points with similar values of x anchor the line well above the outlying point.

APPLY YOUR KNOWLEDGE

 SPAM

CASE 2.1   **2.64 The influence of Bobax.** Make a plot similar to Figure 2.24 giving regression lines with and without Bobax. Explain how the influence of Bobax differs from that of Srizbi.

Influential observations may have small residuals, because they pull the regression line toward themselves. That is, you can't always rely on residuals to point out influential observations. Influential observations can change the interpretation of data. For a linear regression, we compute a slope, an intercept, and a correlation. An individual observation can be influential for one or more of these quantities.

EXAMPLE 2.23   Effects on the Correlation

CASE 2.1

 SPAM

The correlation between the number of bots and the number of spam messages per day for the 10 bots in our sample is $r = 0.88$. If we drop Srizbi, it decreases to 0.66. If, instead, we drop Bobax, it increases to 0.96. Thus, both points are influential for the correlation.

The best way to grasp the important idea of influence is to use an interactive animation that allows you to move points on a scatterplot and observe how correlation and regression respond. The *Correlation and Regression* applet on the companion Web site for this book, whfreeman.com/psbe, allows you to do this. Exercises 2.77 and 2.78 guide the use of this applet.

SECTION 2.3 Summary

- A **regression line** is a straight line that describes how a response variable y changes as an explanatory variable x changes.

- The most common method of fitting a line to a scatterplot is least squares. The **least-squares regression line** is the straight line $\hat{y} = b_0 + b_1 x$ that minimizes the sum of the squares of the vertical distances of the observed points from the line.

- You can use a regression line to **predict** the value of y for any value of x by substituting this x into the equation of the line.

- The **slope** b_1 of a regression line $\hat{y} = b_0 + b_1 x$ is the rate at which the predicted response $\hat{y}$ changes along the line as the explanatory variable x changes. Specifically, b_1 is the change in $\hat{y}$ when x increases by 1.

- The **intercept** b_0 of a regression line $\hat{y} = b_0 + b_1 x$ is the predicted response $\hat{y}$ when the explanatory variable $x = 0$. This prediction is of no statistical use unless x can actually take values near 0.

- The least-squares regression line of y on x is the line with slope $b_1 = r s_y / s_x$ and intercept $b_0 = \overline{y} - b_1 \overline{x}$. This line always passes through the point $(\overline{x}, \overline{y})$.

- **Correlation and regression** are closely connected. The correlation r is the slope of the least-squares regression line when we measure both x and y in standardized units. The square of the correlation r^2 is the fraction of the variance of one variable that is explained by least-squares regression on the other variable.

- You can examine the fit of a regression line by studying the **residuals,** which are the differences between the observed and predicted values of y. Be on the lookout for outlying points with unusually large residuals and also for nonlinear patterns and uneven variation about the line.

- Also look for **influential observations,** individual points that substantially change the regression line. Influential observations are often outliers in the x direction, but they need not have large residuals.

SECTION 2.3 Exercises

For Exercises 2.55 and 2.56, see page 101; for 2.57, see page 103; for 2.58, see page 104; for 2.59 and 2.60, see page 107; for 2.61, see page 108; for 2.62, see page 109; for 2.63, see page 110; and for 2.64, see page 112.

2.65 What is the equation for the selling price? You buy items at a cost of x and sell them for y. Assume that your selling price includes a profit of 10% plus a fixed cost of $5.00. Give an equation that can be used to determine y from x.

2.66 Production costs for cell phone batteries. A company manufactures batteries for cell phones. The overhead expenses of keeping the factory operational for a month—even if no batteries are made—total $600,000. Batteries are manufactured in lots (1000 batteries per lot) costing $8000 to make. In this scenario, $600,000 is the *fixed* cost associated with producing cell phone batteries, and $8000 is the *marginal* (or *variable*) cost of producing each lot of batteries. The total monthly cost y of producing x lots of cell phone batteries is given by the equation

$$y = 600{,}000 + 8000x$$

(a) Draw a graph of this equation. (Choose two values of x, such as 0 and 10, to draw the line and a third for a check. Compute the corresponding values of y from the equation. Plot these two points on graph paper and draw the straight line joining them.)
(b) What will it cost to produce 20 lots of batteries (20,000 batteries)?
(c) If each lot cost $12,000 instead of $8000 to produce, what is the equation that describes total monthly cost for x lots produced?

2.67 Inventory of DVD players. A local consumer electronics store sells exactly 6 DVD players of a particular model each week. The store expects no more shipments of this particular model, and they have 96 such units in their current inventory.
(a) Give an equation for the number of DVD players of this particular model in inventory after x weeks. What is the slope of this line?

(b) Draw a graph of this line between now (Week 0) and Week 10.
(c) Would you be willing to use this line to predict the inventory after 25 weeks? Do the prediction and think about the reasonableness of the result.

2.68 Compare the cell phone payment plans. A cellular telephone company offers two plans. Plan A charges $25 a month for up to 100 minutes of airtime and $0.50 per minute above 100 minutes. Plan B charges $30 a month for up to 200 minutes and $0.45 per minute above 200 minutes.
(a) Draw a graph of the Plan A charge against minutes used from 0 to 250 minutes.
(b) How many minutes a month must the user talk in order for Plan B to be less expensive than Plan A?

2.69 Predict one characteristic of a product using another characteristic. Figure 2.10 (page 90) is a scatterplot of carbohydrates versus percent alcohol in 86 brands of beer. In Exercise 2.35 you calculated the correlation between these two variables. Find the equation of the least-squares regression line for these data. 🗃 BEER

2.70 Carbohydrates and alcohol in beer revisited. Refer to the previous exercise. The data that you used to compute the least-squares regression line includes an outlier. 🗃 BEER
(a) Remove the outlier and recompute the least-squares regression line.
(b) Write a short paragraph about the possible effects of outliers on a least-squares regression line using this example to illustrate your ideas.

2.71 Mutual fund performance. Refer to Exercise 2.46, where you examined the relationship between returns for the Vanguard International Growth Fund and the Morgan Stanley Europe, Australasia, Far East (EAFE) index of overseas stock market performance. 🗃 MUTUALFUNDS
(a) Make a scatterplot suitable for predicting fund returns from EAFE returns.

(b) Find the equation of the least-squares line and add the line to your plot.

(c) What is the residual for the year 2007? Show this residual graphically on your plot.

(d) Plot the residuals versus year and describe the pattern.

(e) What proportion of the variation in EAFE is explained by year?

2.72 Data generated by software. The following 20 observations on y and x were generated by a computer program. GENERATEDDATA

y	x	y	x
34.38	22.06	27.07	17.75
30.38	19.88	31.17	19.96
26.13	18.83	27.74	17.87
31.85	22.09	30.01	20.20
26.77	17.19	29.61	20.65
29.00	20.72	31.78	20.32
28.92	18.10	32.93	21.37
26.30	18.01	30.29	17.31
29.49	18.69	28.57	23.50
31.36	18.05	29.80	22.02

(a) Make a scatterplot and describe the relationship between y and x.

(b) Find the equation of the least-squares regression line and add the line to your plot

(c) Plot the residuals versus x.

(d) What percent of the variability in y is explained by x?

(e) Summarize your analysis of these data in a short paragraph.

2.73 Add an outlier. Refer to the previous exercise. Add an additional observation with $y = 50$ and $x = 30$ to the data set. Repeat the analysis that you performed in the previous exercise and summarize your results, paying particular attention to the effect of this outlier. GENERATEDDATA21A

2.74 Add a different outlier. Refer to the previous two exercises. Add an additional observation with $y = 29$ and $x = 50$ to the original data set. GENERATEDDATA21B

(a) Repeat the analysis that you performed in the first exercise and summarize your results, paying particular attention to the effect of this outlier.

(b) In this exercise and in the previous one, you added an outlier to the original data set and reanalyzed the data. Write a short summary of the changes in correlations that can result from different kinds of outliers.

2.75 Monitoring the water quality near a manufacturing plant. Manufacturing companies (and the Environmental Protection Agency) monitor the quality of the water near their facilities.

Measurements of pollutants in water are indirect—a typical analysis involves forming a dye by a chemical reaction with the dissolved pollutant, then passing light through the solution and measuring its "absorbance." To calibrate such measurements, the laboratory measures known standard solutions and uses regression to relate absorbance to pollutant concentration. This is usually done every day. Here is one series of data on the absorbance for different levels of nitrates. Nitrates are measured in milligrams per liter of water.[15] NITRATES

Nitrates	Absorbance	Nitrates	Absorbance
50	7.0	800	93.0
50	7.5	1200	138.0
100	12.8	1600	183.0
200	24.0	2000	230.0
400	47.0	2000	226.0

(a) Chemical theory says that these data should lie on a straight line. If the correlation is not at least 0.997, something went wrong and the calibration procedure is repeated. Plot the data and find the correlation. Must the calibration be done again?

(b) What is the equation of the least-squares line for predicting absorbance from concentration? If the lab analyzed a specimen with 500 milligrams of nitrates per liter, what do you expect the absorbance to be? Based on your plot and the correlation, do you expect your predicted absorbance to be very accurate?

2.76 Prices paid to producers and additional land for production. Continue your analysis of the effect of coffee prices on deforestation, begun in Exercise 2.42 (page 98). The slope of the regression line for predicting percent of deforestation from coffee price is an important measure of how serious the problem is. COFFEE

(a) What does the slope tell us? Specifically, the slope for these data is $b = 0.0543$. What does this number say about the relationship between coffee price and deforestation?

(b) What does the estimated intercept's value of -0.976 seem to imply about deforestation rates and coffee prices?

2.77 Influence on correlation. The *Correlation and Regression* applet at www.whfreeman.com/psbe allows you to create a scatterplot and to move points by dragging with the mouse. Click to create a group of 10 points in the lower-left corner of the scatterplot with a strong straight-line pattern (correlation about 0.9).

(a) Add 1 point at the upper right that is in line with the first 10. How does the correlation change?

(b) Drag this last point down until it is opposite the group of 10 points. How small can you make the correlation? Can you make the correlation negative? You see that a single outlier can greatly strengthen or weaken a correlation. Always plot your data to check for outlying points.

2.78 Influence in regression. As in the previous exercise, create a group of 10 points in the lower-left corner of the scatterplot with a strong straight-line pattern (correlation at least 0.9). Click the "Show least-squares line" box to display the regression line. (a) Add 1 point at the upper right that is far from the other 10 points but exactly on the regression line. Why does this outlier have no effect on the line even though it changes the correlation? (b) Now drag this last point down until it is opposite the group of 10 points. You see that one end of the least-squares line chases this single point, while the other end remains near the middle of the original group of 10. What about the last point makes it so influential?

2.79 Employee absenteeism and raises. Data on number of days of work missed and annual salary increase for a company's employees show that in general employees who missed more days of work during the year received smaller raises than those who missed fewer days. Number of days missed explained 64% of the variation in salary increases. What is the numerical value of the correlation between number of days missed and salary increase?

2.80 Always plot your data! Table 2.1 presents four sets of data prepared by the statistician Frank Anscombe to illustrate the dangers of calculating without first plotting the data.[16] ANSCOMBEDATA
(a) Without making scatterplots, find the correlation and the least-squares regression line for all four data sets. What do you notice? Use the regression line to predict y for x = 10.
(b) Make a scatterplot for each of the data sets and add the regression line to each plot.
(c) In which of the four cases would you be willing to use the regression line to describe the dependence of y on x? Explain your answer in each case.

2.81 Best countries for business. Figure 2.7 (page 85) is a scatterplot of the gross domestic product per capita versus the unemployment rate for 99 countries. BESTCOUNTRIES
(a) Plot the data and add the least-squares regression line to the plot.
(b) Is it appropriate to use this least-squares regression line to describe the relationship shown in your plot? Explain your answer.
(c) Plot the residuals versus unemployment rate. Interpret the plot and explain how it helps you to understand this data set.

2.82 Best countries for business with logs. Refer to the previous exercise. Figure 2.8 (page 86) is a scatterplot of the log gross domestic product per capita versus the log unemployment rate for 99 countries. BESTCOUNTRIES
(a) Plot the data and add the least-squares regression line to the plot.
(b) Is it appropriate to use this least-squares regression line to describe the relationship shown in your plot? Explain your answer.
(c) Plot the residuals versus unemployment rate. Interpret the plot and explain how it helps you to understand this data set.

2.83 Delete data for countries with low unemployment rates. Refer to the previous exercise. Delete the countries with log unemployment rates lower than 1.6. This corresponds to an unemployment rate of about 5%. Answer the questions given for the previous exercise and explain the effect of deleting the countries with low unemployment rates. BESTCOUNTRIES

2.84 Move your business here. City officials use a variety of tactics to encourage businesses to open offices in their city. One characteristic of a city that is viewed as desirable is open public space within the city limits. The New York City Open Accessible Space Information System Cooperative (OASIS) is an

TABLE 2.1 Four data sets for exploring correlation and regression

Data Set A

x	10	8	13	9	11	14	6	4	12	7	5
y	8.04	6.95	7.58	8.81	8.33	9.96	7.24	4.26	10.84	4.82	5.68

Data Set B

x	10	8	13	9	11	14	6	4	12	7	5
y	9.14	8.14	8.74	8.77	9.26	8.10	6.13	3.10	9.13	7.26	4.74

Data Set C

x	10	8	13	9	11	14	6	4	12	7	5
y	7.46	6.77	12.74	7.11	7.81	8.84	6.08	5.39	8.15	6.42	5.73

Data Set D

x	8	8	8	8	8	8	8	8	8	8	19
y	6.58	5.76	7.71	8.84	8.47	7.04	5.25	5.56	7.91	6.89	12.50

organization of public- and private-sector representatives that has developed an information system designed to enhance the stewardship of open space.[17] Below are data from the OASIS Web site for 12 large U.S. cities. The variables are population in thousands and total park or open space within city limits in acres. OASIS

City	Population	Open space
Baltimore	651	5,091
Boston	589	4,865
Chicago	2,896	11,645
Long Beach	462	2,887
Los Angeles	3,695	29,801
Miami	362	1,329
Minneapolis	383	5,694
New York	8,008	49,854
Oakland	399	3,712
Philadelphia	1,518	10,685
San Francisco	777	5,916
Washington, D.C.	572	7,504

(a) Make a scatterplot of the data using population as the explanatory variable and open space as the response variable.

(b) Is it reasonable to fit a straight line to these data? Explain your answer.

(c) Find the least-squares regression line. Report the equation of the line and draw the line on your scatterplot.

(d) What proportion of the variation in open space is explained by population?

2.85 Prepare the report card. Refer to the previous exercise. One way to compare cities with respect to the amount of open space that they have is to use the residuals from the regression analysis that you performed in the previous exercise. Cities with positive residuals are doing better than predicted, while those with negative residuals are doing worse. Find the residual for each city and make a table with the city name and the residual, ordered from best to worst by the size of the residual. OASIS

2.86 Is New York an outlier? Refer to Exercises 2.84 and 2.85. Write a short paragraph about the data point corresponding to New York City. Is this point an outlier? Is it influential? Compare the analysis results with and without this observation. OASIS

2.87 Open space per person. Refer to Exercises 2.84, 2.85, and 2.86. Open space in acres per person is an alternative way to report open space. Divide open space by population to compute the value of this variable for each city. Using this new variable as the response variable and population as the explanatory variable, answer the questions given in Exercise 2.84. How do your new results compare with those that you found in that exercise? OASIS

2.88 A different report card. Refer to Exercise 2.85. Prepare a report card based on the analysis of open space per person that you performed in Exercise 2.87. Write a short paragraph comparing this report card with the one that you prepared in Exercise 2.85. Which do you prefer? Give reasons for your answer. OASIS

2.4 Cautions about Correlation and Regression

Correlation and regression are powerful tools for describing the relationship between two variables. When you use these tools, you must be aware of their limitations, beginning with the fact that **correlation and regression describe only linear relationships.** Also remember that **the correlation r and the least-squares regression line are not resistant.** One influential observation or incorrectly entered data point can greatly change these measures. Always plot your data before interpreting regression or correlation. Here are some other cautions to keep in mind when you apply correlation and regression or read accounts of their use.

Beware extrapolation

Associations for variables can be trusted only for the range of values for which data have been collected. Even a very strong relationship may not hold outside the data's range.

TARGET

EXAMPLE 2.24   Predicting the Number of Target Stores in 2008

Here are data on the number of Target stores in operation at the end of each year in the early 1990s and in 2008:[18]

Year (x)	1990	1991	1992	1993	2008
Stores (y)	420	463	506	554	1682

A plot of these data is given in Figure 2.25. The data for 1990 through 1993 lie almost exactly on a straight line, which was calculated using only the 1990s data. The equation of this line is $y = -88,136 + 44.5x$ and $r^2 = 0.9992$. We know that 99.92% of the variation in stores is explained by year for these years. The equation predicts 1220 stores for 2008, but the actual number of stores is much higher, 1682. The prediction is very poor because the very strong linear trend evident in the 1990 to 1993 data did not continue to the year 2008.

FIGURE 2.25 Plot of the number of Target stores versus year with the least-squares regression line calculated using data from 1990, 1991, 1992, and 1993, for Example 2.24. The poor fit to the number of stores in 2008 illustrates the dangers of extrapolation.

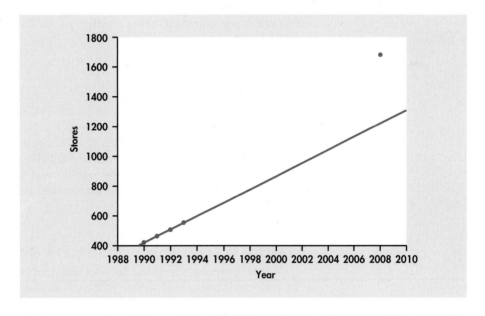

Predictions made far beyond the range for which data have been collected can't be trusted. Few relationships are linear for *all* values of x. It is risky to stray far from the range of x-values that actually appear in your data.

Extrapolation

Extrapolation is the use of a regression line for prediction far outside the range of values of the explanatory variable x that you used to obtain the line. Such predictions are often not accurate.

In general, extrapolation involves using a mathematical relationship beyond the range of the data that were used to estimate the relationship. The scenario described in the previous example is typical: we try to use a least-squares relationship to make predictions for values of the explanatory variable that are much larger than the values in the data that we have. We can encounter the same difficulty when we attempt predictions for values of the explanatory variable that are much smaller than the values in the data that we have.

Careful judgment is needed when making predictions. If the prediction is for values that are within the range of the data that you have or are not too far above or below, then your prediction can be reasonably accurate. Beyond that, you are in danger of making an inaccurate prediction.

Beware correlations based on averaged data

Many regression and correlation studies work with averages or other measures that combine information from many individuals. You should note this carefully and resist the temptation to apply the results of such studies to individuals. **Correlations based on averages are usually too high when applied to individuals.** This is another reminder that it is important to note exactly what variables were measured in a statistical study.

Beware the lurking variable

Correlation and regression describe the relationship between two variables. Often the relationship between two variables is strongly influenced by other variables. We try to measure potentially influential variables. We can then use more advanced statistical methods to examine all the relationships revealed by our data. Sometimes, however, the relationship between two variables is influenced by other variables that we did not measure or even think about. Variables lurking in the background, measured or not, often help explain statistical associations.

> **Lurking Variable**
>
> A **lurking variable** is a variable that is not among the explanatory or response variables in a study and yet may influence the interpretation of relationships among those variables.

A lurking variable can falsely suggest a strong relationship between x and y, or it can hide a relationship that is really there. Here is an example of a negative correlation that is due to a lurking variable.

EXAMPLE 2.25   Gas and Electricity Bills

A single-family household receives bills for gas and electricity each month. The 12 observations for a recent year are plotted with the least-squares regression line in Figure 2.26. We have arbitrarily chosen to put the electricity bill on the x axis and the gas bill on the y axis. There is a clear negative association. Does this mean that a high electricity bill causes the gas bill to be low and vice versa?

To understand the association in this example, we need to know a little more about the two variables. In this household, heating is done by gas and cooling is done by electricity. Therefore, in the winter months the gas bill will be relatively high and the electricity bill will be relatively low. The pattern is reversed in the summer months. The association that we see in this example is due to a lurking variable: time of year.

FIGURE 2.26 Scatterplot with least-squares regression line for predicting monthly charges for gas using monthly charges for electricity for a household, for Example 2.25.

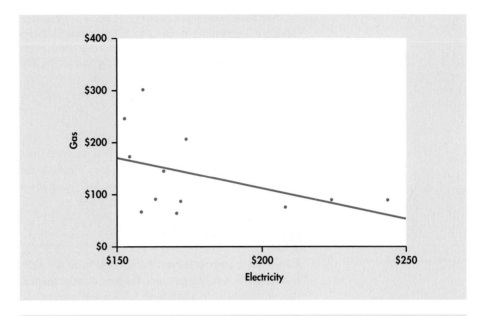

2.89 Education and income. There is a strong positive correlation between years of education and income for economists employed by business firms. In particular, economists with doctorates earn more than economists with only a bachelor's degree. There is also a strong positive correlation between years of education and income for economists employed by colleges and universities. But when all economists are considered, there is a *negative* correlation between education and income. The explanation for this is that business pays high salaries and employs mostly economists with bachelor's degrees, while colleges pay lower salaries and employ mostly economists with doctorates. Sketch a scatterplot with two groups of cases (business and academic) illustrating how a strong positive correlation within each group and a negative overall correlation can occur together.

Association is not causation

When we study the relationship between two variables, we often hope to show that changes in the explanatory variable *cause* changes in the response variable. But a strong association between two variables is not enough to draw conclusions about cause and effect. Sometimes an observed association really does reflect cause and effect. Natural-gas consumption in a household that uses natural gas for heating will be higher in colder months because cold weather requires burning more gas to stay warm. In other cases, an association is explained by lurking variables, and the conclusion that x causes y is either wrong or not proved. Here is an example.

EXAMPLE 2.26 Does Television Extend Life?

Measure the number of television sets per person x and the average life expectancy y for the world's nations. There is a high positive correlation: nations with many TV sets have higher life expectancies.

The basic meaning of causation is that by changing x we can bring about a change in y. Could we lengthen the lives of people in Rwanda by shipping them TV sets? No. Rich nations have more TV sets than poor nations. Rich nations also have longer life expectancies because they offer better nutrition, clean water, and better health care. There is no cause-and-effect tie between TV sets and length of life.

Correlations such as that in Example 2.26 are sometimes called "nonsense correlations." The correlation is real. What is nonsense is the conclusion that changing one of the variables causes changes in the other. A lurking variable—such as national wealth in Example 2.26—that influences both x and y can create a high correlation even though there is no direct connection between x and y.

APPLY YOUR KNOWLEDGE

2.90 How's your self-esteem? People who do well tend to feel good about themselves. Perhaps helping people feel good about themselves will help them do better in their jobs and in life. For a time, raising self-esteem became a goal in many schools and companies. Can you think of explanations for the association between high self-esteem and good performance other than "Self-esteem causes better work"?

2.91 Are big hospitals bad for you? A study shows that there is a positive correlation between the size of a hospital (measured by its number of beds x) and the median number of days y that patients remain in the hospital. Does this mean that you can shorten a hospital stay by choosing a small hospital? Why?

2.92 Do firefighters make fires worse? Someone says, "There is a strong positive correlation between the number of firefighters at a fire and the amount of damage the fire does. So sending lots of firefighters just causes more damage." Explain why this reasoning is wrong.

These and other examples lead us to the most important caution about correlation, regression, and statistical association between variables in general.

> **Association Does Not Imply Causation**
>
> An association between an explanatory variable x and a response variable y, even if it is very strong, is not by itself good evidence that changes in x actually cause changes in y.

experiment   The best way to get good evidence that x causes y is to do an **experiment** in which we change x and keep lurking variables under control. We will discuss experiments in Chapter 3. When experiments cannot be done, finding the explanation for an observed association is often difficult and controversial. Many of the sharpest disputes in which statistics plays a role involve questions of causation that cannot be settled by experiment. Does gun control reduce violent crime? Does using cell phones cause brain tumors? Has increased free trade widened the gap between the incomes of more-educated and less-educated American workers? All of these questions have become public issues. All concern associations among variables. And all have this in common: they try to pinpoint cause and effect in a setting involving complex relations among many interacting variables.

BEYOND THE BASICS: Data Mining

Chapters 1 and 2 of this book are devoted to the important aspect of statistics called *exploratory data analysis* (EDA). We use graphs and numerical summaries to examine data, searching for patterns and paying attention to striking deviations from the patterns we find. In discussing regression, we advanced to using the pattern we find (in this case, a linear pattern) for prediction.

Suppose now that we have a truly enormous data base, such as all purchases recorded by the cash register scanners of our retail chain during the past week. Surely this trove of data contains patterns that might guide business decisions. If we could see clearly the types of activewear preferred in large California cities and compare the preferences of small Midwest cities—right now, not at the end of the season—we might improve profits in both parts of the country by matching stock with demand. This sounds much like EDA, and indeed it is. There is, however, a saying in computer science that a big enough difference of scale amounts to a difference of kind. Exploring really large data bases in the hope of finding useful patterns is called **data mining.** Here are some distinctive features of data mining:

data mining

- When you have 100 gigabytes of data, even straightforward calculations and graphics become impossibly time-consuming. So efficient algorithms are very important.

- The structure of the data base and the process of storing the data, perhaps by unifying data scattered across many departments of a large corporation, require careful thought. The fashionable term is *data warehousing.*

- Data mining requires automated tools that work based on only vague queries by the user. The process is too complex to do step-by-step as we have done in EDA.

All of these features point to the need for sophisticated computer science as a basis for data mining. Indeed, data mining is often thought of as a part of computer science. Yet many statistical ideas and tools—mostly tools for dealing with multidimensional data, not the sort of thing that appears in a first statistics course—are very helpful. Like many modern developments, data mining crosses the boundaries of traditional fields of study. You can learn more about data mining in Chapter 14.

Do remember that the perils we encounter with blind use of correlation and regression are yet more perilous in data mining, where the fog of an immense data base prevents clear vision. Extrapolation, ignoring lurking variables, and confusing association with causation are traps for the unwary data miner.

SECTION 2.4 Summary

- Correlation and regression must be **interpreted with caution. Plot the data** to be sure the relationship is roughly linear and to detect outliers and influential observations.

- Avoid **extrapolation,** the use of a regression line for prediction for values of the explanatory variable far outside the range of the data from which the line was calculated.

- Remember that **correlations based on averages** are usually too high when applied to individuals.

- **Lurking variables** that you did not measure may explain the relations between the variables you did measure. Correlation and regression can be misleading if you ignore important lurking variables.

- Most of all, be careful not to conclude that there is a cause-and-effect relationship between two variables just because they are strongly associated. **High correlation does not imply causation.** The best evidence that an association is due to causation comes from an **experiment** in which the explanatory variable is directly changed and other influences on the response are controlled.

SECTION 2.4 Exercises

For Exercise 2.89, see page 119; and for 2.90 to 2.92, see page 120.

2.93 What's wrong? Each of the following statements contains an error. Describe each error and explain why the statement is wrong.
(a) If the residuals are all positive, this implies that there is a positive relationship between the response variable and the explanatory variable.
(b) A negative relationship can never be due to causation.
(c) A lurking variable is always a response variable.

2.94 What's wrong? Each of the following statements contains an error. Describe each error and explain why the statement is wrong.
(a) High correlation implies causation.
(b) An outlier will always have a small residual.
(c) If we have data at values of x equal to 1, 2, 3, 4, and 5, and we try to predict the value of y using a least-squares regression line, we are extrapolating.

2.95 Use of the Internet and a long life. Exercise 2.22 (page 90) asks the question "Will you live longer if you use the Internet?" Figure 2.13 (page 92) is a scatterplot of life expectancy in years versus Internet use for 181 countries. The scatterplot shows a positive association between these two variables. Do you think that this plot indicates that Internet use causes people to live longer? Give another possible explanation for why these two variables are positively associated. INTERNETANDLIFE

2.96 Older workers and income. The effect of a lurking variable can be surprising when individuals are divided into groups. Explain how, as a nation's population grows older, mean income can go down for workers in each age group but still go up for all workers.

2.97 Marital status and income. Data show that married, divorced, and widowed men earn quite a bit more than men the same age who have never been married. This does not mean that a man can raise his income by getting married, because men who have never been married are different from married men in many ways other than marital status. Suggest several lurking variables that might help explain the association between marital status and income.

2.98 Prices paid to producers and additional land for production. Continue your work on coffee prices and deforestation in Exercise 2.42 (page 98) and Exercise 2.76 (page 114). COFFEE
(a) If the world coffee price settled at 60 cents per pound, what percent of the national park forest do you predict would be cleared each year, on the average? Show your work.
(b) Do part (a) assuming the world coffee price is $1.20 per pound (120 cents per pound).
(c) Do you have any reason to trust your prediction in part (a) more than your prediction in part (b)? Explain your response.

2.99 Does your product have an undesirable side effect? People who use artificial sweeteners in place of sugar tend to be heavier than people who use sugar. Does this mean that artificial sweeteners cause weight gain? Give a more plausible explanation for this association.

2.100 Does your product help nursing-home residents? A group of college students believes that herbal tea has remarkable powers. To test this belief, they make weekly visits to a local nursing home, where they visit with the residents and serve them herbal tea. The nursing-home staff reports that after several months many of the residents are healthier and more cheerful. We should commend the students for their good deeds but doubt that herbal tea helped the residents. Identify the explanatory and response variables in this informal study. Then explain what lurking variables account for the observed association.

2.101 Education and income. There is a strong positive correlation between years of schooling completed x and lifetime earnings y for American men. One possible reason for this association is causation: more education leads to higher-paying jobs. But lurking variables may explain some of the correlation. Suggest some lurking variables that would explain why men with more education earn more.

2.102 Do power lines cause cancer? It has been suggested that electromagnetic fields of the kind present near power lines can cause leukemia in children. Experiments with children and power lines are not ethical. Careful studies have found no association between exposure to electromagnetic fields and childhood leukemia.[19] Suggest several lurking variables that you would want information about in order to investigate the claim that living near power lines is associated with cancer.

2.5 Relations in Categorical Data*

We have concentrated on relationships in which at least the response variable is quantitative. Now we will shift to describing relationships between two or more categorical variables. Some variables—such as gender, race, and occupation—are categorical by nature. Other categorical variables are created by grouping values of a quantitative variable into classes. Published data often appear in grouped form to save space. To analyze categorical data, we use the *counts* or *percents* of cases that fall into various categories.

WINEANDMUSIC

CASE 2.2

Does the Right Music Sell the Product? Market researchers know that background music can influence the mood and the purchasing behavior of customers. One study in a supermarket in Northern Ireland compared three treatments: no music, French accordion music, and Italian string music. Under each condition, the researchers recorded the numbers of bottles of French, Italian, and other wine purchased.[20] Here is the two-way table that summarizes the data:

Counts for wine and music

Wine	Music			Total
	None	French	Italian	
French	30	39	30	99
Italian	11	1	19	31
Other	43	35	35	113
Total	84	75	84	243

two-way table

row and column variables

The data table for Case 2.2 is a **two-way table** because it describes two categorical variables. The type of wine is the **row variable** because each row in the table describes the data for one type of wine. The type of music played is the **column variable** because each column describes the data for one type of music. The entries in the table are the counts of bottles of wine of the particular type sold while the given type of music was playing. The two variables in this example, wine and music, are both categorical variables.

This two-way table is a 3×3 table, to which we have added the marginal totals obtained by summing across rows and columns. For example, the first-row total is $30 + 39 + 30 = 99$. The grand total, the number of bottles of wine in the study, can be computed by summing the row totals, $99 + 31 + 113 = 243$, or the column totals, $84 + 75 + 84 = 243$. It is a good idea to do both as a check on your arithmetic.

Marginal distributions

How can we best grasp the information contained in the wine and music table? First, *look at the distribution of each variable separately*. The distribution of a categorical variable says how often each outcome occurred. The "Total" column at the right margin of the table contains the totals for each of the rows. These are called **marginal row totals.** They give the numbers of bottles of wine sold by the type of wine: 99 bottles of French wine, 31 bottles of Italian wine, and 113 bottles of other types of wine. Similarly, the **marginal column totals** are given in the "Total" row at the bottom margin of the table.

marginal row totals

marginal column totals

---

*This material is important in statistics, but it is needed later in this book only for Chapter 9. You may omit it if you do not plan to read Chapter 9, or you may delay reading it until you reach Chapter 9.

These are the numbers of bottles of wine that were sold while different types of music were being played: 84 bottles when no music was playing, 75 bottles when French music was playing, and 84 bottles when Italian music was playing.

marginal distribution

Percents are often more informative than counts. We can calculate the distribution of wine type in percents by dividing each row total by the table total. This distribution is called the **marginal distribution** of wine type.

> **Marginal Distributions**
>
> To find the marginal distribution for the row variable in a two-way table, divide each row total by the total number of entries in the table. Similarly, to find the marginal distribution for the column variable in a two-way table, divide each column total by the total number of entries in the table.

Although the usual definition of a distribution is in terms of proportions, we often multiply these by 100 to convert them to percents. You can describe a distribution either way as long as you clearly indicate which format you are using.

EXAMPLE 2.27 Calculating a Marginal Distribution

CASE 2.2

 WINEANDMUSIC

Let's find the marginal distribution for the types of wine sold. The counts that we need for these calculations are in the margin at the right of the table:

Wine	Total
French	99
Italian	31
Other	113
Total	243

The percent of bottles of French wine sold is

$$\frac{\text{bottles of French wine sold}}{\text{total sold}} = \frac{99}{243} = 0.4074 = 40.74\%$$

Similar calculations for Italian wine and other wine give the following distribution in percents:

Wine:	French	Italian	Other
Percent:	40.74	12.76	46.50

The total should be 100% because each bottle of wine sold is classified into exactly one of these three categories. In this case the total is exactly 100%. Small deviations from 100% can occur due to roundoff error.

As usual, we prefer to display numerical summaries using a graph. Figure 2.27 is a bar graph of the distribution (in counts) of wine type sold.

In a two-way table, we have two marginal distributions, one for each of the variables that defines the table.

FIGURE 2.27 Marginal distribution of type of wine sold, for Example 2.27.

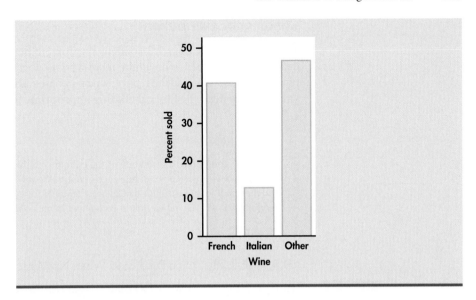

WINEANDMUSIC

APPLY YOUR KNOWLEDGE

CASE 2.2 **2.103 Marginal distribution for type of music.** Find the marginal distribution for the type of music. Display the distribution using a graph.

In working with two-way tables, you must calculate lots of percents. Here's a tip to help you decide what fraction gives the percent you want. Ask, "What group represents the total that I want a percent of?" The count for that group is the denominator of the fraction that leads to the percent. In Example 2.27, we wanted percents "of bottles of the different types of wine sold," so the table total is the denominator.

APPLY YOUR KNOWLEDGE

2.104 Construct a two-way table. Construct a 2 × 3 table. Add the marginal totals and find the two marginal distributions.

FIELDSOFSTUDY

2.105 Fields of study for college students. The following table gives the number of students (in thousands) graduating from college with degrees in several fields of study for seven countries:[21]

Field of study	Canada	France	Germany	Italy	Japan	U.K.	U.S.
Social sciences, business, law	64	153	66	125	259	152	878
Science, mathematics, engineering	35	111	66	80	136	128	355
Arts and humanities	27	74	33	42	123	105	397
Education	20	45	18	16	39	14	167
Other	30	289	35	58	97	76	272

(a) Calculate the marginal totals and add them to the table.
(b) Find the marginal distribution of country and give a graphical display of the distribution.
(c) Do the same for the marginal distribution of field of study.

Conditional distributions

The 3×3 table for Case 2.2 contains much more information than the two marginal distributions. We need to do a little more work to describe the relationship between the type of music playing and the type of wine purchased. **Relationships among categorical variables are described by calculating appropriate percents from the counts given.**

> **Conditional Distributions**
>
> To find the conditional distribution of the column variable for a particular value of the row variable in a two-way table, divide each count in the row by the row total. Similarly, to find the conditional distribution of the row variable for a particular value of the column variable in a two-way table, divide each count in the column by the column total.

EXAMPLE 2.28 Wine Purchased When No Music Was Playing

CASE 2.2

WINEANDMUSIC

What types of wine were purchased when no music was playing? To answer this question we find the marginal distribution of wine type for the value of music equal to none. The counts we need are in the first column of our table:

	Music
Wine	None
French	30
Italian	11
Other	43
Total	84

What percent of French wine was sold when no music was playing? To answer this question we divide the number of bottles of French wine sold when no music was playing by the total number of bottles of wine sold when no music was playing:

$$\frac{30}{84} = 0.3571 = 35.71\%$$

In the same way, we calculate the percents for Italian and other types of wine. Here are the results:

Wine type:	French	Italian	Other
Percent when no music is playing:	35.7	13.1	51.2

Other wine was the most popular choice when no music was playing, but French wine has a reasonably large share. Notice that these percents sum to 100%. There is no roundoff error here. The distribution is displayed in Figure 2.28.

WINEANDMUSIC

APPLY YOUR KNOWLEDGE

CASE 2.2 **2.106 Conditional distribution when French music was playing.**

(a) Write down the column of counts that you need to compute the conditional distribution of the type of wine sold when French music was playing.

(b) Compute this conditional distribution.

FIGURE 2.28 Conditional distribution of types of wine sold when no music was playing, for Example 2.28.

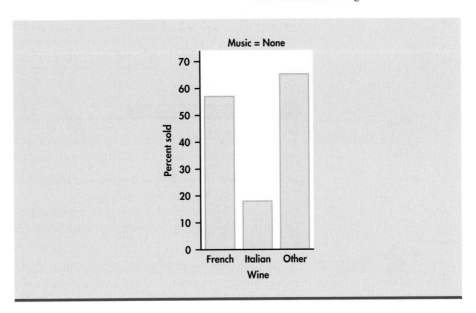

(c) Display this distribution graphically.

(d) Compare this distribution with the one in Example 2.28. Was there an increase in sales of French wine when French music was playing rather than no music?

WINEANDMUSIC

CASE 2.2 **2.107 Conditional distribution when Italian music was playing.**

(a) Write down the column of counts that you need to compute the conditional distribution of the type of wine sold when Italian music was playing.

(b) Compute this conditional distribution.

(c) Display this distribution graphically.

(d) Compare this distribution with the one in Example 2.28. Was there an increase in sales of Italian wine when Italian music was playing rather than no music?

WINEANDMUSIC

CASE 2.2 **2.108 Compare the conditional distributions.** In Example 2.28 we found the distribution of sales by wine type when no music was playing. In Exercise 2.106 you found the distribution when French music was playing, and in Exercise 2.107 you found the distribution when Italian music was playing. Examine these three conditional distributions carefully and write a paragraph summarizing the relationship between sales of different types of wine and the music played.

For Case 2.2 we examined the relationship between sales of different types of wine and the music that was played by studying the three conditional distributions of type of wine sold, one for each music condition. For these computations we used the counts from the 3 × 3 table, one column at a time. We could also have computed conditional distributions using the counts for each row. The result would be the three conditional distributions of the type of music played for each of the three wine types. For this example, we think that conditioning on the type of music played gives us the most useful data summary. Comparing conditional distributions can be particularly useful when the column variable is an explanatory variable.

The choice of which conditional distribution to use depends on the nature of the data and the questions that you want to ask. Sometimes you will prefer to condition on the column variable, and sometimes you will prefer to condition on the row variable.

Occasionally, both sets of conditional distributions will be useful. Statistical software will calculate all of these quantities. You need to select the parts of the output that are needed for your particular questions. Don't let computer software make this choice for you.

FIELDSOFSTUDY

APPLY YOUR KNOWLEDGE

2.109 Fields of study by country for college students. In Exercise 2.105 you examined data on fields of study for graduating college students from seven countries.

(a) Find the seven conditional distributions giving the distribution of graduates in the different fields of study for each country.
(b) Display the conditional distributions graphically.
(c) Write a paragraph summarizing the relationship between field of study and country.

FIELDSOFSTUDY

2.110 Countries by fields of study for college students. Refer to the previous exercise. Answer the same questions for the conditional distribution of country for each field of study.

FIELDSOFSTUDY

2.111 Compare the two analytical approaches. In the previous two exercises you examined the relationship between country and field of study in two different ways.

(a) Compare these two approaches.
(b) Which do you prefer? Give a reason for your answer.
(c) What kinds of questions are most easily answered by each of the two approaches? Explain your answer.

Simpson's paradox

As is the case with quantitative variables, the effects of lurking variables can change or even reverse relationships between two categorical variables. Here is an example that demonstrates the surprises that can await the unsuspecting user of data.

EXAMPLE 2.29   Which Customer Service Representative Is Better?

CUSTSERVICEREP

A customer service center has a goal of resolving customer questions in 10 minutes or less. Here are the records for two representatives:

	Representative	
Goal met	Alexis	Peyton
Yes	172	118
No	28	82
Total	200	200

Alexis has met the goal 172 times out of 200, a success rate of 86%. For Peyton, the success rate is 118 out of 200, or 59%. Alexis clearly has the better success rate.

Let's look at the data in a little more detail. The data summarized come from two different weeks in the year.

CUSTSERVICEREP

EXAMPLE 2.30   Let's Look at the Data More Carefully

Here are the counts broken down by week:

Goal met	Week 1		Week 2	
	Alexis	Peyton	Alexis	Peyton
Yes	162	19	10	99
No	18	1	10	81
Total	180	20	20	180

For Week 1, Alexis met the goal 90% of the time (162/180), while Peyton met the goal 95% of the time (19/20). Peyton had the better performance in Week 1. What about Week 2? Here Alexis met the goal 50% of the time (10/20), while the success rate for Peyton was 55% (99/180). Peyton again had the better performance. How does this analysis compare with the analysis that combined the counts for the two weeks? That analysis clearly showed that Alexis had the better performance, 86% versus 59%.

These results can be explained by a lurking variable related to week. The first week was during a period when the product had been in use for several months. Most of the calls to the customer service center concerned problems that had been encountered before. The representatives were trained to answer these questions and usually had no trouble in meeting the goal of resolving the problems quickly. On the other hand, the second week occurred shortly after the release of a new version of the product. Most of the calls during this week concerned new problems that the representatives had not yet encountered. Many more of these questions took longer than the 10-minute goal to resolve.

Look at the total in the bottom row of the detailed table. During the first week, when calls were easy to resolve, Alexis handled 180 calls and Peyton handled 20. The situation was exactly the opposite during the second week, when calls were difficult to resolve. There were 20 calls for Alexis and 180 for Peyton.

The original two-way table, which did not take account of week, was misleading. This example illustrates *Simpson's paradox*.

> **Simpson's Paradox**
>
> An association or comparison that holds for all of several groups can reverse direction when the data are combined to form a single group. This reversal is called **Simpson's paradox.**

The lurking variables in Simpson's paradox are categorical. That is, they break the individuals into groups, as when calls are classified by week. Simpson's paradox is just an extreme form of the fact that observed associations can be misleading when there are lurking variables.

APPLY YOUR KNOWLEDGE

HOSPITALS

2.112 Which hospital is safer? Insurance companies and consumers are interested in the performance of hospitals. The government releases data about patient outcomes in hospitals that can be useful in making informed health care decisions. Here is a two-way table of data on the survival of patients after surgery in two hospitals. All patients undergoing surgery in a recent time period are included. "Survived" means that the patient lived at least 6 weeks following surgery.

	Hospital A	Hospital B
Died	63	16
Survived	2037	784
Total	2100	800

What percent of Hospital A patients died? What percent of Hospital B patients died? These are the numbers one might see reported in the media.

 HOSPITALS

2.113 Patients in "poor" or "good" condition. Not all surgery cases are equally serious, however. Patients are classified as being in either "poor" or "good" condition before surgery. Here are the data broken down by patient condition. Check that the entries in the original two-way table are just the sums of the "poor" and "good" entries in this pair of tables.

Good Condition		
	Hospital A	Hospital B
Died	6	8
Survived	594	592
Total	600	600

Poor Condition		
	Hospital A	Hospital B
Died	57	8
Survived	1443	192
Total	1500	200

(a) Find the percent of Hospital A patients who died who were classified as "poor" before surgery. Do the same for Hospital B. In which hospital do "poor" patients fare better?

(b) Repeat part (a) for patients classified as "good" before surgery.

(c) What is your recommendation to someone facing surgery and choosing between these two hospitals?

(d) How can Hospital A do better in both groups, yet do worse overall? Look at the data and carefully explain how this can happen.

three-way table   The data in Example 2.30 can be given in a **three-way table** that reports counts for each combination of three categorical variables: week, representative, and whether or not the goal was met. In Example 2.30, we constructed two two-way tables for representative by goal, one for each week. The original table, the one we showed in Example 2.29, can be obtained by adding the corresponding counts for the two tables in Example 2.30. This *aggregation*   process is called **aggregating** the data. When we aggregated data in Example 2.29, we ignored the variable week, which then became a lurking variable. *Conclusions that seem obvious when we look only at aggregated data can become quite different when the data are examined in more detail.*

SECTION 2.5 Summary

• A **two-way table** of counts organizes data about two categorical variables. Values of the **row variable** label the rows that run across the table, and values of the **column variable** label the columns that run down the table. Two-way tables are often used to summarize large amounts of information by grouping outcomes into categories.

- The **row totals** and **column totals** in a two-way table give the **marginal distributions** of the two individual variables. It is clearer to present these distributions as percents of the table total. Marginal distributions tell us nothing about the relationship between the variables.

- To find the **conditional distribution** of the row variable for one specific value of the column variable, look only at that one column in the table. Find each entry in the column as a percent of the column total.

- There is a conditional distribution of the row variable for each column in the table. Comparing these conditional distributions is one way to describe the association between the row and the column variables. It is particularly useful when the column variable is the explanatory variable.

- **Bar graphs** are a flexible means of presenting categorical data. There is no single best way to describe an association between two categorical variables.

- A comparison between two variables that holds for each individual value of a third variable can be changed or even reversed when the data for all values of the third variable are combined. This is **Simpson's paradox.** Simpson's paradox is an example of the effect of lurking variables on an observed association.

SECTION 2.5 Exercises

For Exercises 2.103 to 2.105, see page 125; for 2.106 to 2.108, see pages 126–127; for 2.109 to 2.111, see page 128; and for 2.112 and 2.113, see pages 129–130.

2.114 Remote deposit capture. The Federal Reserve has called remote deposit capture (RDC) "the most important development the (U.S) banking industry has seen in years." This service allows users to scan checks and to transmit the scanned images to a bank for posting.[22] In its annual survey of community banks, the American Bankers Association asked banks whether or not they offered this service.[23] Here are the results classified by the asset size (in millions of dollars) of the bank: RDCSIZE

	Offer RDC	
Asset size	Yes	No
Under $100	63	309
$101 to $200	59	132
$201 or more	112	85

Summarize the results of this survey question numerically and graphically. Write a short paragraph explaining the relationship between the size of a bank, measured by assets, and whether or not RDC is offered.

2.115 How does RDC vary across the country? The survey described in the previous exercise also classified community banks by region. Here is the 6 × 2 table of counts:[24] RDCREGION

	Offer RDC	
Region	Yes	No
Northeast	28	38
Southeast	57	61
Central	53	84
Midwest	63	181
Southwest	27	51
West	61	76

Summarize the results of this survey question numerically and graphically. Write a short paragraph explaining the relationship between the size of a bank, measured by assets, and whether or not remote deposit capture is offered.

2.116 Exercise and adequate sleep. A survey of 656 boys and girls who were 13 to 18 years old asked about adequate sleep and other health-related behaviors. The recommended amount of sleep is six to eight hours per night.[25] In the survey 54% of the respondents reported that they got less than this amount of sleep on school nights. The researchers also developed an exercise scale, which was used to classify the students as above or below the median in how much they exercised. Here is the 2 × 2 table of counts with students classified as getting or not getting adequate sleep and by the exercise variable: SLEEP

	Exercise	
Enough sleep	**High**	**Low**
Yes	151	115
No	148	242

(a) Find the distribution of adequate sleep for the high exercisers.

(b) Do the same for the low exercisers.

(c) Summarize the relationship between adequate sleep and exercise using the results of parts (a) and (b).

2.117 Adequate sleep and exercise. Refer to the previous exercise. SLEEP

(a) Find the distribution of exercise for those who get adequate sleep.

(b) Do the same for those who do not get adequate sleep.

(c) Write a short summary of the relationship between adequate sleep and exercise using the results of parts (a) and (b).

(d) Compare this summary with the summary that you obtained in part (c) of the previous exercise. Which do you prefer? Give a reason for your answer.

2.118 Full-time and part-time college students. The Census Bureau provides estimates of numbers of people in the United States classified in various ways.[26] Let's look at college students. The following table gives us data to examine the relation between age and full-time or part-time status. The numbers in the table are expressed as thousands of U.S. college students. USCOLLEGESTUDENTS

	Status	
Age	**Full-time**	**Part-time**
15–19	3388	389
20–24	5238	1164
25–34	1703	1699
35 and over	762	2045

(a) Find the distribution of age for full-time students.

(b) Do the same for the part-time students.

(c) Use the summaries in parts (a) and (b) to describe the relationship between full- or part-time status and age. Write a brief summary of your conclusions.

2.119 Condition on age. Refer to the previous exercise. USCOLLEGESTUDENTS

(a) For each age group compute the percent of students who are full-time and the percent of students who are part-time.

(b) Make a graphical display of the results that you found in part (a).

(c) In a short paragraph, describe the relationship between age and full- or part-time status using your numerical and graphical summaries.

(d) Explain why you need only the percents of students who are full-time for your summary in part (b).

(e) Compare this way of summarizing the relationship between these two variables with what you presented in part (c) of the previous exercise.

2.120 Lying to a teacher. One of the questions in a survey of high school students asked about lying to teachers.[27] The table below gives the numbers of students who said that they lied to a teacher at least once during the past year, classified by gender. LYING

	Gender	
Lied at least once	**Male**	**Female**
Yes	3,228	10,295
No	9,659	4,620

(a) Add the marginal totals to the table.

(b) Calculate appropriate percents to describe the results of this question.

(c) Summarize your findings in a short paragraph.

2.121 Trust and honesty in the workplace. The students surveyed in the study described in the previous exercise were also asked whether they thought trust and honesty were essential in business and the workplace. Here are the counts classified by gender: TRUST

	Gender	
Trust and honesty are essential	**Male**	**Female**
Agree	11,724	14,169
Disagree	1,163	746

Answer the questions given in the previous exercise for this survey question.

2.122 Class size and course level. College courses taught at lower levels often have larger class sizes. The table below gives the number of classes classified by course level and class size.[28] For example, there were 227 first-year-level courses with between 1 and 9 students. CLASSSIZE

Course level	Class size						
	1–9	**10–19**	**20–29**	**30–39**	**40–49**	**50–99**	**100 or more**
1	227	783	1169	392	46	95	150
2	309	420	503	295	106	116	129
3	262	425	296	120	85	164	33
4	227	287	160	92	63	71	13

(a) Fill in the marginal totals in the table.

(b) Find the marginal distribution for the variable course level.

(c) Do the same for the variable class size.

(d) For each course level, find the conditional distribution of class size.

(e) Summarize your findings in a short paragraph.

2.123 Hiring practices. A company has been accused of age discrimination in hiring for operator positions. Lawyers for both sides look at data on applicants for the past 3 years. They compare hiring rates for applicants younger than 40 years and those 40 years or older. HIRING

Age	Hired	Not hired
Younger than 40	79	1158
40 or older	1	165

(a) Find the two conditional distributions of hired/not hired, one for applicants who are less than 40 years old and one for applicants who are not less than 40 years old.

(b) Based on your calculations, make a graph to show the differences in distribution for the two age categories.

(c) Describe the company's hiring record in words. Does the company appear to discriminate on the basis of age?

(d) What lurking variables might be involved here?

2.124 Nonresponse in a survey of companies. A business school conducted a survey of companies in its state. They mailed a questionnaire to 200 small companies, 200 medium-sized companies, and 200 large companies. The rate of nonresponse is important in deciding how reliable survey results are. Here are the data on response to this survey: NONRESPONSE

	Small	Medium	Large
Response	125	81	40
No response	75	119	160
Total	200	200	200

(a) What was the overall percent of nonresponse?

(b) Describe how nonresponse is related to the size of the business. (Use percents to make your statements precise.)

(c) Draw a bar graph to compare the nonresponse percents for the three size categories.

2.125 Demographics and new products. Companies planning to introduce a new product to the market must define the "target" for the product. Who do we hope to attract with our new product? Age and gender are two of the most important demographic variables. The following two-way table describes the age and marital status of American women in 2007.[29] The table entries are in thousands of women. AGEGENDER

	Marital Status			
Age (years)	Never married	Married	Widowed	Divorced
18 to 24	11,362	2,411	17	185
25 to 39	8,337	19,312	172	2,507
40 to 64	4,810	34,247	2,334	8,538
≥65	753	9,158	8,685	1,978

(a) Find the sum of the entries for each column.

(b) Find the joint distribution, the marginal distributions, and the conditional distributions.

(c) Write a short description of the relationship between marital status and age for women.

2.126 Demographics, continued. AGEGENDER

(a) Using the data in the previous exercise, compare the conditional distributions of marital status for women aged 18 to 24 and women aged 40 to 64. Briefly describe the most important differences between the two groups of women, and back up your description with percents.

(b) Your company is planning a magazine aimed at women who have never been married. Find the conditional distribution of age among never-married women and display it in a bar graph. What age group or groups should your magazine aim to attract?

2.127 Demographics and new products—men. Refer to Exercise 2.125. Here are the corresponding counts for men: AGEGENDER

	Marital Status			
Age (years)	Never married	Married	Widowed	Divorced
18 to 24	12,960	1,343	10	96
25 to 39	11,498	17,035	55	1,813
40 to 64	5,781	34,585	610	6,539
≥ 65	617	11,627	2,018	1,162

Answer the questions from Exercise 2.125 for these counts.

2.128 Discrimination? Wabash Tech has two professional schools, business and law. Here are two-way tables of applicants to both schools, categorized by gender and admission decision. (Although these data are made up, similar situations occur in reality.) DISCRIMINATION

Business	Admit	Deny
Male	480	120
Female	180	20

Law	Admit	Deny
Male	10	90
Female	100	200

(a) Make a two-way table of gender by admission decision for the two professional schools together by summing entries in these tables.

(b) From the two-way table, calculate the percent of male applicants who are admitted and the percent of female applicants who are admitted. Wabash admits a higher percent of male applicants.

(c) Now compute separately the percents of male and female applicants admitted by the business school and by the law school. Each school admits a higher percent of female applicants.

(d) This is Simpson's paradox: both schools admit a higher percent of the women who apply, but overall Wabash admits a lower percent of female applicants than of male applicants. Explain carefully, as if speaking to a skeptical reporter, how it can happen that Wabash appears to favor males when each school individually favors females.

2.129 Obesity and health. Recent studies have shown that earlier reports underestimated the health risks associated with being overweight. The error was due to overlooking lurking variables. In particular, smoking tends both to reduce weight and to lead to earlier death. Illustrate Simpson's paradox by a simplified version of this situation. That is, make up tables of overweight (yes or no) by early death (yes or no) by smoker (yes or no) such that

- Overweight smokers and overweight nonsmokers both tend to die earlier than those not overweight.

- But when smokers and nonsmokers are combined into a two-way table of overweight by early death, persons who are not overweight tend to die earlier.

2.130 Find the table. Here are the row and column totals for a two-way table with two rows and two columns:

a	b	50
c	d	50
60	40	100

Find *two different* sets of counts a, b, c, and d for the body of the table that give these same totals. This shows that the relationship between two variables cannot be obtained from the two individual distributions of the variables.

STATISTICS IN SUMMARY

Chapter 1 dealt with data analysis for a single variable. In this chapter, we have studied analysis of data for two or more variables. The proper analysis depends on whether the variables are categorical or quantitative and on whether one is an explanatory variable and the other a response variable.

When you have a categorical explanatory variable and a quantitative response variable, use the tools of Chapter 1 to compare the distributions of the response variable for the different categories of the explanatory variable. Make side-by-side boxplots, stemplots, or histograms and compare medians or means. If both variables are categorical, there is no satisfactory graph (though bar graphs can help). We describe the relationship numerically by comparing percents. The optional Section 2.5 explains how to do this.

Most of this chapter concentrates on relations between two quantitative variables. The Statistics in Summary figure on the next page organizes the main ideas in a way that stresses that our tactics are the same as when we faced single-variable data in Chapter 1. Here is a review list of the most important skills you should have acquired from your study of this chapter.

A. Data

1. Recognize whether each variable is quantitative or categorical.

2. Identify the explanatory and response variables in situations where one variable explains or influences another.

B. Scatterplots

1. Make a scatterplot to display the relationship between two quantitative variables. Place the explanatory variable (if any) on the horizontal scale of the plot.

2. Add a categorical variable to a scatterplot by using a different plotting symbol or color.

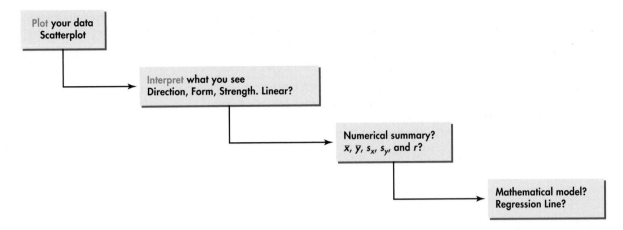

3. Describe the form, direction, and strength of the overall pattern of a scatterplot. In particular, recognize positive or negative association and linear (straight-line) patterns. Recognize outliers in a scatterplot.

C. Correlation

1. Using a calculator or software, find the correlation r between two quantitative variables.

2. Know the basic properties of correlation: r measures the strength and direction of only linear relationships; $-1 \leq r \leq 1$ always; $r = \pm 1$ only for perfect straight-line relations; r moves away from 0 toward ± 1 as the linear relation gets stronger.

D. Straight Lines

1. Explain what the slope b_1 and the intercept b_0 mean in the equation $y = b_0 + b_1 x$ of a straight line.

2. Draw a graph of the straight line when you are given its equation.

E. Regression

1. Using a calculator or software, find the least-squares regression line of a response variable y on an explanatory variable x from data.

2. Find the slope and intercept of the least-squares regression line from the means and standard deviations of x and y and their correlation.

3. Use the regression line to predict y for a given x. Recognize extrapolation and be aware of its dangers.

4. Use r^2 to describe how much of the variation in one variable can be accounted for by a straight-line relationship with another variable.

5. Recognize outliers and potentially influential observations from a scatterplot with the regression line drawn on it.

6. Calculate the residuals and plot them against the explanatory variable x or against other variables. Recognize unusual patterns.

F. Limitations of Correlation and Regression

1. Understand that both r and the least-squares regression line can be strongly influenced by a few extreme observations.

2. Recognize that a correlation based on averages of several observations is usually stronger than the correlation for individual observations.

3. Recognize possible lurking variables that may explain the observed association between two variables x and y.

4. Understand that even a strong correlation does not mean that there is a cause-and-effect relationship between x and y.

G. Categorical Data (Optional)

1. From a two-way table of counts, find the marginal distributions of both variables by obtaining the row sums and column sums.

2. Express any distribution in percents by dividing the category counts by their total.

3. Describe the relationship between two categorical variables by computing and comparing percents. Often this involves comparing the conditional distributions of one variable for the different categories of the other variable.

4. Recognize Simpson's paradox and be able to explain it.

CHAPTER 2 Review Exercises

2.131 Dwelling permits and sales for 21 European countries. The Organization for Economic Co-operation and Development (OECD) collects data on Main Economic Indicators (MEIs) for many countries. Each variable is recorded as an index with the year 2000 serving as a base year. This means that the variable for each year is reported as a ratio of the value for the year divided by the value for 2000. Use of indices in this way makes it easier to compare values for different countries. Table 2.2 gives the values of three MEIs for 21 countries.[30] MEIS

TABLE 2.2 Dwelling permits, sales, and production for 21 European countries			
Country	Dwelling permits	Sales	Production
Australia	116	137	109
Belgium	125	105	112
Canada	224	122	101
Czech Republic	178	134	162
Denmark	121	126	109
Finland	105	136	125
France	145	121	104
Germany	54	100	119
Greece	117	136	102
Hungary	109	140	155
Ireland	92	123	144
Japan	86	99	109
Korea	158	110	156
Luxembourg	145	161	118
Netherlands	160	107	109
New Zealand	127	139	112
Norway	125	136	94
Poland	163	139	159
Portugal	53	112	105
Spain	122	123	108
Sweden	180	142	116

(a) Make a scatterplot with sales as the response variable and permits issued for new dwellings as the explanatory variable. Describe the relationship. Are there any outliers or influential observations?

(b) Find the least-squares regression line and add it to your plot.

(c) What is the predicted value of sales for a country that has an index of 160 for dwelling permits?

(d) The Netherlands has an index of 160 for dwelling permits. Find the residual for this country.

(e) What percent of the variation in sales is explained by dwelling permits?

2.132 Dwelling permits and production. Refer to the previous exercise. MEIS

(a) Make a scatterplot with production as the response variable and permits issued for new dwellings as the explanatory variable. Describe the relationship. Are there any outliers or influential observations?

(b) Find the least-squares regression line and add it to your plot.

(c) What is the predicted value of production for a country that has an index of 160 for dwelling permits?

(d) The Netherlands has an index of 160 for dwelling permits. Find the residual for this country.

(e) What percent of the variation in production is explained by dwelling permits? How does this value compare with the value you found in the previous exercise for the percent of variation in sales that is explained by building permits?

2.133 Sales and production. Refer to the previous two exercises. MEIS

(a) Make a scatterplot with sales as the response variable and production as the explanatory variable. Describe the relationship. Are there any outliers or influential observations?

(b) Find the least-squares regression line and add it to your plot.

(c) What is the predicted value of sales for a country that has an index of 125 for production?

(d) Finland has an index of 125 for production. Find the residual for this country.

(e) What percent of the variation in sales is explained by production? How does this value compare with the percents of variation that you calculated in the two previous exercises?

2.134 Salaries and raises. For this exercise we consider a hypothetical employee who starts working in Year 1 at a salary of $50,000. Each year her salary increases by approximately 5%. By Year 20, she is earning $126,000. The table below gives her salary for each year (in thousands of dollars): RAISES

Year	Salary	Year	Salary	Year	Salary	Year	Salary
1	50	6	63	11	81	16	104
2	53	7	67	12	85	17	109
3	56	8	70	13	90	18	114
4	58	9	74	14	93	19	120
5	61	10	78	15	99	20	126

(a) Figure 2.29 is a scatterplot of salary versus year with the least-squares regression line. Describe the relationship between salary and year for this person.

(b) The value of r^2 for these data is 0.9832. What percent of the variation in salary is explained by year? Would you say that this is an indication of a strong linear relationship? Explain your answer.

FIGURE 2.29 Plot of salary versus year, with the least-squares regression line, for an individual who receives approximately a 5% raise each year for 20 years, for Exercise 2.134.

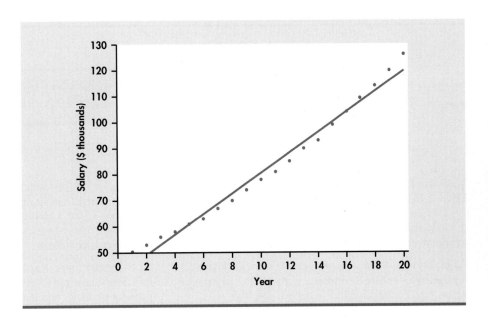

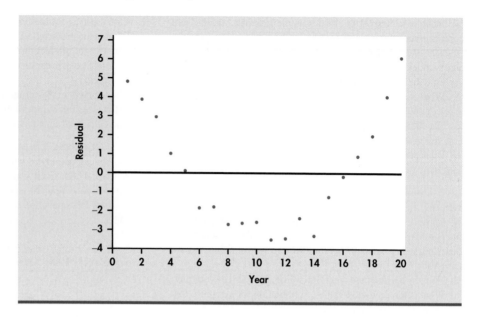

FIGURE 2.30 Plot of residuals versus year for an individual who receives approximately a 5% raise each year for 20 years, for Exercise 2.135.

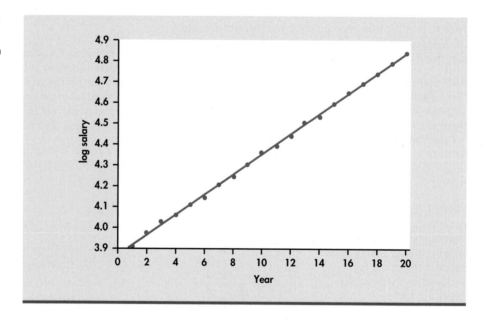

FIGURE 2.31 Plot of log salary versus year, with the least-squares regression line, for an individual who receives approximately a 5% raise each year for 20 years, for Exercise 2.136.

2.135 Look at the residuals. Refer to the previous exercise. Figure 2.30 is a plot of the residuals versus year. 🐾 RAISES
(a) Interpret the residual plot.
(b) Explain how this plot highlights the deviations from the least-squares regression line that you can see in Figure 2.29.

2.136 Try logs. Refer to the previous two exercises. Figure 2.31 is a scatterplot with the least-squares regression line for log salary versus year. For this model, $r^2 = 0.9995$. 🐾 RAISES

(a) Compare this plot with Figure 2.29. Write a short summary of the similarities and the differences.
(b) Figure 2.32 is a plot of the residuals for the model using year to predict log salary. Compare this plot with Figure 2.30 and summarize your findings.

2.137 Predict some salaries. The individual whose salary we have been studying in Exercises 2.134 to 2.136 wants to do some financial planning. Specifically, she would like to predict her

FIGURE 2.32 Plot of residuals, based on log salary, versus year for an individual who receives approximately a 5% raise each year for 20 years, for Exercise 2.136.

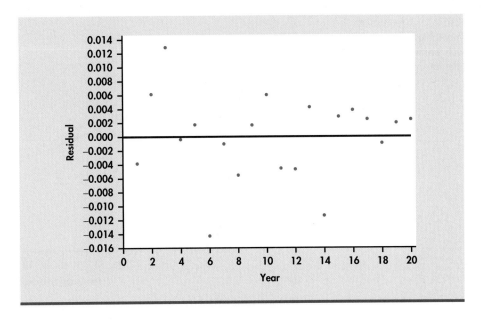

salary 5 years into the future, that is, for Year 25. She is willing to assume that her employment situation will be stable for the next 5 years and that it will be similar to the last 20 years.
RAISES
(a) Use the least-squares regression equation constructed to predict salary from year to predict her salary for Year 25.
(b) Use the least-squares regression equation constructed to predict log salary from year to predict her salary for Year 25. Note that you will need to convert the predicted log salary back to the predicted salary. Many calculators have a function that will perform this operation.
(c) Which prediction do you prefer? Explain your answer.
(d) Someone looking at the numerical summaries and not the plots for these analyses says that because both models have very high values of r^2, they should perform equally well in doing this prediction. Write a response to this comment.
(e) Write a short paragraph about the value of graphical summaries and the problems of extrapolation using what you have learned from studying these salary data.

2.138 Faculty salaries. Data on the salaries of a sample of professors in a mathematics department at a large Midwest university are given below. The salaries are for the academic years 2007–2008 and 2008–2009. FACULTYSALARIES
(a) Construct a scatterplot with the 2008–2009 salaries on the vertical axis and the 2007–2008 salaries on the horizontal axis.
(b) Comment on the form, direction, and strength of the relationship in your scatterplot.
(c) What proportion of the variation in 2008–2009 salaries is explained by 2007–2008 salaries?

2007–2008 salary ($)	2008–2009 salary ($)	2007–2008 salary ($)	2008–2009 salary ($)
141,800	142,900	133,650	136,350
109,800	113,600	129,160	132,485
106,000	110,500	71,972	76,072
95,700	99,800	72,000	76,000
109,000	111,180	79,500	82,700
108,790	111,240	138,850	141,830
100,500	105,100	119,506	122,906
146,000	150,080	112,100	115,200

2.139 Find the line and examine the residuals. Refer to the previous exercise. FACULTYSALARIES
(a) Find the least-squares regression line for predicting 2008–2009 salaries from 2007–2008 salaries.
(b) Analyze the residuals, paying attention to any outliers or influential observations. Write a summary of your findings.

2.140 Bigger raises for those earning less. Refer to the previous two exercises. The 2007–2008 salaries do an excellent job of predicting the 2008–2009 salaries. Is there anything more that we can learn from these data? In this department there is a tradition of giving higher-than-average percent raises to those whose salaries are lower. Let's see if we can find evidence to support this idea in the data. FACULTYSALARIES
(a) Compute the percent raise for each faculty member. Take the difference between the 2008–2009 salary and the 2007–2008 salary, divide by the 2007–2008 salary, and then multiply by 100. Make a scatterplot with the raise as the response variable

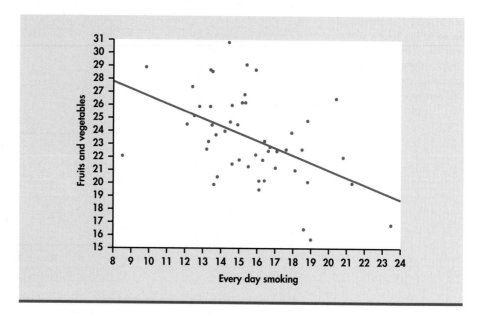

and the 2007–2008 salary as the explanatory variable. Describe the relationship that you see in your plot.

(b) Find the least-squares regression line and add it to your plot.

(c) Analyze the residuals. Are there any outliers or influential cases? Make a graphical display and include it in a short summary of what you conclude.

(d) Is there evidence in the data to support the idea that greater percent raises are given to those with lower salaries? Summarize your findings and include numerical and graphical summaries to support your conclusion.

2.141 Marketing your college. Colleges compete for students, and many students do careful research when choosing a college. One source of information is the rankings compiled by *U.S. News & World Report.* One of the factors used to evaluate undergraduate programs is the proportion of incoming students who graduate. This quantity, called the graduation rate, can be predicted by other variables such as the SAT or ACT scores and the high school records of the incoming students. One of the components in *U.S. News & World Report* rankings is the difference between the actual graduation rate and the rate predicted by a regression equation.[31] In this chapter, we call this quantity the residual. Explain why the residual is a better measure to evaluate college graduation rates than the raw graduation rate.

2.142 Know your customers. Fruits and vegetables are marketed with a heavy emphasis on their health benefits. What are the profiles of those who eat the recommended quantities and those who do not? The Centers for Disease Prevention and Control (CDC) Behavior Risk Factor Surveillance System (BRFSS) collects data related to health conditions and risk behaviors.[32] Aggregated data by state are in the BRFSS data set. Figure 2.33

is a plot of two of the BRFSS variables. Fruits and Vegetables is the percent of adults in the state who report eating at least five servings of fruits and vegetables per day; Smoking is the percent who smoke every day. 🔵 **BRFSS**

(a) Describe the relationship between Fruits and Vegetables and Smoking. Explain why you might expect this type of association.

(b) Find the correlation between the two variables.

(c) For Utah, 22.1% eat at least five servings of fruits and vegetables per day and 8.5% smoke every day. Find Utah on the plot and describe its position relative to the other states.

(d) For California, the percents are 28.9% for Fruits and Vegetables and 9.8% for Smoking. Find California on the plot and describe its position relative to the other states.

(e) Pick your favorite state. What states do you think would have similar values for Smoking and Fruits and Vegetables. Give reasons for your choices. Use Table 2.3 to determine if your ideas are supported by the data. Summarize your results.

2.143 The education level of your customers. Refer to the previous exercise. The BRFSS data set contains a variable called EdCollege, the proportion of adults who have completed college. 🔵 **BRFSS**

(a) Plot the data with Fruits and Vegetables on the *x* axis and EdCollege on the *y* axis. Describe the overall pattern of the data.

(b) Add the least-squares regression line to your plot. Does the line give a summary of the overall pattern? Explain your answer.

(c) Pick out a few states and use their position in the graph to write a short summary of how they compare with other states.

(d) Can you conclude that earning a college degree will cause you to eat five servings of fruits and vegetables per day? Explain your answer.

TABLE 2.3 Fruit and vegetable consumption and smoking

State	Fruits & Vegetables (percent)	Smoking (percent)	State	Fruits & Vegetables (percent)	Smoking (percent)
Alabama	20.1	18.8	Montana	24.7	14.5
Alaska	24.8	18.8	Nebraska	20.2	16.1
Arizona	23.7	13.7	Nevada	22.5	16.6
Arkansas	21.0	18.1	New Hampshire	29.1	15.4
California	28.9	9.8	New Jersey	25.9	12.8
Colorado	24.5	13.5	New Mexico	21.5	14.6
Connecticut	27.4	12.4	New York	26.0	14.6
Delaware	21.3	15.5	North Carolina	22.5	17.1
Florida	26.2	15.2	North Dakota	21.8	15.0
Georgia	23.2	16.4	Ohio	22.6	17.6
Hawaii	24.5	12.1	Oklahoma	15.7	19.0
Idaho	23.2	13.3	Oregon	25.9	13.4
Illinois	24.0	14.2	Pennsylvania	23.9	17.9
Indiana	22.0	20.8	Rhode Island	26.8	15.3
Iowa	19.5	16.1	South Carolina	21.2	17.0
Kansas	19.9	13.6	South Dakota	20.5	13.8
Kentucky	16.8	23.5	Tennessee	26.5	20.4
Louisiana	20.2	16.4	Texas	22.6	13.2
Maine	28.7	15.9	Utah	22.1	8.5
Maryland	28.7	13.4	Vermont	30.8	14.4
Massachusetts	28.6	13.5	Virginia	26.2	15.3
Michigan	22.8	16.7	Washington	25.2	12.5
Minnesota	24.5	14.9	West Virginia	20.0	21.3
Mississippi	16.5	18.6	Wisconsin	22.2	15.9
Missouri	22.6	18.5	Wyoming	21.8	16.3

2.144 Planning for a new product. The editor of a statistics text would like to plan for the next edition. A key variable is the number of pages that will be in the final version. Text files are prepared by the authors using a word processor called LaTeX, and separate files contain figures and tables. For the previous edition of the text, the number of pages in the LaTeX files can easily be determined, as well as the number of pages in the final version of the text. Here are the data: **TEXTPAGES**

	Chapter												
	1	2	3	4	5	6	7	8	9	10	11	12	13
LaTeX pages	77	73	59	80	45	66	81	45	47	43	31	46	26
Text pages	99	89	61	82	47	68	87	45	53	50	36	52	19

(a) Plot the data and describe the overall pattern.
(b) Find the equation of the least-squares regression line and add the line to your plot.

(c) Find the predicted number of pages for the next edition if the number of LaTeX pages for a chapter is 62.
(d) Write a short report for the editor explaining to her how you constructed the regression equation and how she could use it to estimate the number of pages in the next edition of the text.

2.145 Points scored in women's basketball games. Use the Internet to find the scores for the past season's women's basketball team at a college of your choice. Is there a relationship between the points scored by your chosen team and the points scored by their opponents? Summarize the data and write a report on your findings.

2.146 Look at the data for men. Refer to the previous exercise. Analyze the data for the men's team from the same college and compare your results with those for the women.

2.147 Endangered animals and habitat. Endangered animal species often live in isolated patches of habitat. If the population size in a patch varies a lot (due to weather, for example), the species is more likely to disappear from that patch in a bad year. Here is a general question: Is there less variation in population size when a patch of habitat has more diverse vegetation? If so, maintaining habitat diversity can help protect endangered species.

A researcher measured the variation over time in the population of a cricket species in 45 habitat patches. He also measured the diversity of each patch.[33] He reported his results by giving the least-squares equation

$$\text{population variation} = 84.4 - 0.13 \times \text{diversity}$$

along with the fact that $r^2 = 0.34$. Do these results support the idea that more diversity goes with less variation in population size? Is the relationship very strong or only moderately strong?

2.148 Simpson's paradox and regression. Simpson's paradox occurs when a relationship between variables within groups of observations reverses when all of the data are combined. The phenomenon is usually discussed in terms of categorical variables, but it also occurs in other settings. Here is an example: SIMPSONREG

y	x	Group	y	x	Group
10.1	1	1	18.3	6	2
8.9	2	1	17.1	7	2
8.0	3	1	16.2	8	2
6.9	4	1	15.1	9	2
6.1	5	1	14.3	10	2

(a) Make a scatterplot of the data for Group 1. Find the least-squares regression line and add it to your plot. Describe the relationship between y and x for Group 1.
(b) Do the same for Group 2.
(c) Make a scatterplot using all 10 observations. Find the least-squares line and add it to your plot.
(d) Make a plot with all of the data using different symbols for the two groups. Include the three regression lines on the plot. Write a paragraph about Simpson's paradox for regression using this graphical display to illustrate your description.

2.149 Circular saws. The following table gives the weight (in pounds) and amps for 19 circular saws. Saws with higher amp ratings tend to also be heavier than saws with lower amp ratings. We can quantify this fact using regression. CIRCULARSAWS

Weight	Amps	Weight	Amps	Weight	Amps
11	15	9	10	11	13
12	15	11	15	13	14
11	15	12	15	10	12
11	15	12	14	11	12
12	15	10	10	11	12
11	15	12	13	10	12
13	15	11			

(a) We will use amps as the explanatory variable and weight as the response variable. Give a reason for this choice.

(b) Make a scatterplot of the data. What do you notice about the weight and amp values?
(c) Report the equation of the least-squares regression line along with the value of r^2.
(d) Interpret the value of the estimated slope.
(e) How much of an increase in amps would you expect to correspond to a one-pound increase in the weight of a saw, on average, when comparing two saws?
(f) Create a residual plot for the model in part (b). Does the model indicate curvature in the data?

2.150 Circular saws. The table in the previous exercise gives the weight (in pounds) and amps for 19 circular saws. The data contain only five different amp ratings among the 19 saws. CIRCULARSAWS
(a) Calculate the correlation between the weights and the amps of the 19 saws.
(b) Calculate the average weight of the saws for each of the five amp ratings.
(c) Calculate the correlation between the average weights and the amps. Is the correlation between average weights and amps greater than, less than, or equal to the correlation between individual weights and amps?

2.151 What correlation doesn't say. Investment reports now often include correlations. Following a table of correlations among mutual funds, a report adds: "Two funds can have perfect correlation, yet different levels of risk. For example, Fund A and Fund B may be perfectly correlated, yet Fund A moves 20% whenever Fund B moves 10%." Write a brief explanation, for someone who knows no statistics, of how this can happen. Include a sketch to illustrate your explanation.

2.152 A computer game. A multimedia statistics learning system includes a test of skill in using the computer's mouse. The software displays a circle at a random location on the computer screen. The subject clicks in the circle with the mouse as quickly as possible. A new circle appears as soon as the subject clicks the old one. Table 2.4 gives data for one subject's trials, 20 with each hand. Distance is the distance from the cursor location to the center of the new circle, in units whose actual size depends on the size of the screen. Time is the time required to click in the new circle, in milliseconds.[34] COMPUTERGAME
(a) We suspect that time depends on distance. Make a scatterplot of time against distance, using separate symbols for each hand.
(b) Describe the pattern. How can you tell that the subject is right-handed?
(c) Find the regression line of time on distance separately for each hand. Draw these lines on your plot. Which regression does a better job of predicting time from distance? Give numerical measures that describe the success of the two regressions.
(d) It is possible that the subject got better in later trials due to learning. It is also possible that he got worse due to fatigue. Plot the residuals from each regression against the time order of the trials (down the columns in Table 2.4). Is either of these systematic effects of time visible in the data?

TABLE 2.4 Reaction times in a computer game

Time	Distance	Hand	Time	Distance	Hand
115	190.70	right	240	190.70	left
96	138.52	right	190	138.52	left
110	165.08	right	170	165.08	left
100	126.19	right	125	126.19	left
111	163.19	right	315	163.19	left
101	305.66	right	240	305.66	left
111	176.15	right	141	176.15	left
106	162.78	right	210	162.78	left
96	147.87	right	200	147.87	left
96	271.46	right	401	271.46	left
95	40.25	right	320	40.25	left
96	24.76	right	113	24.76	left
96	104.80	right	176	104.80	left
106	136.80	right	211	136.80	left
100	308.60	right	238	308.60	left
113	279.80	right	316	279.80	left
123	125.51	right	176	125.51	left
111	329.80	right	173	329.80	left
95	51.66	right	210	51.66	left
108	201.95	right	170	201.95	left

2.153 Wood products. A wood product manufacturer is interested in replacing solid-wood building material by less-expensive products made from wood flakes.[35] The company collected the following data to examine the relationship between the length (in inches) and the strength (in pounds per square inch) of beams made from wood flakes: 🔵 **WOOD**

Length:	5	6	7	8	9	10	11	12	13	14
Strength:	446	371	334	296	249	254	244	246	239	234

(a) Make a scatterplot that shows how the length of a beam affects its strength.

(b) Describe the overall pattern of the plot. Are there any outliers?

(c) Fit a least-squares line to the entire set of data. Graph the line on your scatterplot. Does a straight line adequately describe these data?

(d) The scatterplot suggests that the relation between length and strength can be described by *two* straight lines, one for lengths of 5 to 9 inches and another for lengths of 9 to 14 inches. Fit least-squares lines to these two subsets of the data, and draw the lines on your plot. Do they describe the data adequately? What question would you now ask the wood experts?

The following exercises concern material in the optional Section 2.5.

2.154 Aspirin and heart attacks. Does taking aspirin regularly help prevent heart attacks? "Nearly five decades of research now link aspirin to the prevention of stroke and heart attacks." So says the Bayer Aspirin Web site, www.bayeraspirin.com. The most important evidence for this claim comes from the Physicians' Health Study. The subjects were 22,071 healthy male doctors at least 40 years old. Half the subjects, chosen at random, took aspirin every other day. The other half took a placebo, a dummy pill that looked and tasted like aspirin. Here are the results.[36] (The row for "None of these" is left out of the two-way table.)

	Aspirin group	Placebo group
Fatal heart attacks	10	26
Other heart attacks	129	213
Strokes	119	98
Total	11,037	11,034

What do the data show about the association between taking aspirin and heart attacks and stroke? Use percents to make your statements precise. Do you think the study provides evidence that aspirin actually reduces heart attacks (cause and effect)? 🔵 **ASPIRIN**

2.155 More smokers live at least 20 more years! You can see the headlines "More smokers than nonsmokers live at least 20 more years after being contacted for study!" A medical study contacted randomly chosen people in a district in England. Here are data on the 1314 women contacted who were either current smokers or who had never smoked. The tables classify these women by their smoking status and age at the time of the survey and whether they were still alive 20 years later.[37] 🔵 **SMOKERS**

	Age 18 to 44		Age 45 to 64		Age 65+	
	Smoker	**Not**	**Smoker**	**Not**	**Smoker**	**Not**
Dead	19	13	78	52	42	165
Alive	269	327	167	147	7	28

(a) From these data make a two-way table of smoking (yes or no) by dead or alive. What percent of the smokers stayed alive for 20 years? What percent of the nonsmokers survived? It seems surprising that a higher percent of smokers stayed alive.

(b) The age of the women at the time of the study is a lurking variable. Show that within each of the three age groups in the data, a higher percent of nonsmokers remained alive 20 years later. This is another example of Simpson's paradox.

(c) The study authors give this explanation: "Few of the older women (over 65 at the original survey) were smokers, but many of them had died by the time of follow-up." Compare the percent of smokers in the three age groups to verify the explanation.

2.156 Recycled product quality. Recycling is supposed to save resources. Some people think recycled products are lower in quality than other products, a fact that makes recycling less practical. People who actually use a recycled product may have different opinions from those who don't use it. Here are data on attitudes toward coffee filters made of recycled paper among people who do and don't buy these filters:[38] 🔵 **RECYCLE**

	Think the quality of the recycled product is:		
	Higher	**The same**	**Lower**
Buyers	20	7	9
Nonbuyers	29	25	43

(a) Find the marginal distribution of opinion about quality. Assuming that these people represent all users of coffee filters, what does this distribution tell us?

(b) How do the opinions of buyers and nonbuyers differ? Use conditional distributions as a basis for your answer. Can you conclude that using recycled filters causes more favorable opinions? If so, giving away samples might increase sales.

CHAPTER 2 Case Study Exercises

CASE STUDY EXERCISE 1: Beef consumption. Read the article titled "Factors Affecting U.S. Beef Consumption" available at `ers.usda.gov/publications/ldp/Oct05/ldpm13502`. Write a summary of the report, paying particular attention to the numerical and graphical summaries used by the authors.

CASE STUDY EXERCISE 2: Predicting college grades. The data set CSDATA contains information about all 224 students who entered a large university in a single year and who planned to major in computer science. We are interested in predicting GPA (grade point average) after three semesters of college from information available before the student enters college. To do this effectively, we must use several explanatory variables together. This is *multiple regression,* the topic of Chapter 11. In this exercise, you will look at individual explanatory variables. 🔵 **CSDA**

A. What is the correlation of each of the explanatory variables with GPA? Explain why knowing the correlations tells us which variables will best predict GPA in a regression with just one explanatory variable. What are the two best predictor variables? Does your finding seem reasonable for computer science majors?

B. Make scatterplots, with the least-squares line added, for GPA versus each of the two best explanatory variables. How well do each of these variables predict GPA? Do the scatterplots

contain unusual observations? In what way is each of these observations unusual?

CASE STUDY EXERCISE 3: Predicting coffee exports. The data set COFFEEEXPORTS contains information on the coffee production and exports for 45 coffee-producing countries. The units are thousands of 60-kilogram bags. The variables are production in 2008 and 2009 and exports of coffee in 2008 and 2009.[39] 🔵 **COFFEEEXPORTS**

A. Use 2008 production to predict 2008 exports. What is the correlation between 2008 production and 2008 exports? Make a scatterplot with the least-squares line added for predicting 2008 exports using 2008 production. What does the value of r^2 tell you? Identify any unusual values by country name.

B. Use 2009 production to predict 2009 exports. What is the correlation between 2009 production and 2009 exports? Make a scatterplot with the least-squares line added for predicting 2009 exports using 2009 production. What does the value of r^2 tell you? Identify any unusual values by country name.

C. Use 2008 production to predict 2009 exports. What is the correlation between 2008 production and 2009 exports? Make a scatterplot with the least-squares line added for predicting 2009 exports using 2008 production. What does the value of r^2 tell you? Identify any unusual values by country name.

D. Use transformations and delete outliers as you see fit. Write a summary of all your analyses.

CHAPTER 2   Appendix

Using Minitab and Excel for Examining Relationships

Scatterplots
Minitab:

Graph ➤ Scatterplot

Select "Simple" for the type of scatterplot, then click **OK.** Click-in the data column associated with the y (response) variable into row 1 of the **Y variables** box, and click-in the data column associated with the x (explanatory) variable into row 1 of the **X variables** box. Click **OK.**

Excel:

To create a scatterplot in Excel, the data for the two variables should be placed in two adjacent columns, with the column associated with the y variable being in the column to the right of the column associated with the x variable. Click and drag the mouse to highlight the cells of the two columns of data. With the cell range highlighted, click the **Insert** tab and then click **Scatter** in the **Charts** group. You want to now choose the scatterplot option with no line connections. If the gridlines are not desired, you can click on them and delete them by hitting the delete key or by right-clicking and selecting **Delete.** The layout of the scatterplot can also be manipulated by choosing among a variety of options offered within the **Charts Layouts** group found under the **Design** tab.

Correlation
Minitab:

Stat ➤ Basic Statistics ➤ Correlation

Click-in the y variable data column and the x variable data column into the **Variables** box and then click **OK.**

Excel:

Select "Correlation" in the **Data Analysis** menu box and click **OK.** Enter the cell range of the data on the two variables (placed in adjacent columns) into the **Input Range** box. Click **OK.**

Least-Squares Regression
Minitab:

Stat ➤ Regression ➤ Regression

Click-in the response variable data column into the **Response** box, and click-in the explanatory variable data column into the **Predictors** box. If you want the residual values, then click the **Storage** button and place a checkmark next to the **Residuals** option and then click **OK** to close the pop-up box. Select the **Fits** option if you want the predicted values. A residual plot similar to Figure 2.22 can be created by clicking the **Graphs** button and then clicking-in the explanatory variable data column into the **Residuals versus the variables** box. When you have closed out the option pop-up boxes, click **OK** to obtain the regression output. If you wish to produce a scatterplot superimposed with the least-squares regression line, do the following pull-down sequence:

Stat ➤ Regression ➤ Fitted Line Plot

Click-in the response variable data column into the **Response (Y)** box, and click-in the explanatory variable data column into the **Predictor (X)** box and then click **OK.**

Excel:

Select "Regression" in the **Data Analysis** menu box and click **OK.** Enter the cell range of the response data into the **Input Y Range** box, and enter the cell range of the explanatory data into the **Input X Range** box. If you want the residual values, then place a checkmark next to the **Residuals** option. By placing a checkmark next to the **Residual Plots** option, a residual plot similar to Figure 2.22 can be produced. When you are finished with your option choices, click **OK** to obtain the regression output. If you wish to produce a scatterplot superimposed with the least-squares regression line, then do not select the **Line Fit Plots.** Instead, create a scatterplot as described earlier in this Appendix and then click the **Design** tab and select the **Layout 9** option found within the **Charts Layouts** group.

Producing Data

Introduction

In Chapters 1 and 2 we learned some basic tools of *data analysis.* We used graphs and numbers to describe data. When we do **exploratory data analysis,** we rely heavily on plotting the data. We look for patterns that suggest interesting conclusions or questions for further study. However, *exploratory analysis alone can rarely provide convincing evidence for its conclusions, because striking patterns we find in data can arise from many sources.*

Anecdotal data

It is tempting to simply draw conclusions from our own experience, making no use of more broadly representative data. A magazine article about Pilates says that men need this form of exercise even more than women. The article describes the benefits that two men received from taking Pilates classes. A newspaper ad states that a particular brand of windows is "considered to be the best" and says that "now is the best time to replace your windows and doors." These types of stories, or *anecdotes,* sometimes provide quantitative data. However, this type of data does not give us a sound basis for drawing conclusions.

> ### Anecdotal Evidence
> **Anecdotal evidence** is based on haphazardly selected individual cases, which often come to our attention because they are striking in some way. These cases need not be representative of any larger group of cases.

3.1 Is this good market research? You and your friends are big fans of "Waverly Place," a Disney Channel show about a family with three children who are training to be wizards. To what extent do you think you can generalize your preference for this show to all students at your college?

3.2 Describe a business anecdote. Find an example from some recent experience where anecdotal evidence is used to make a business decision that is not justified. Describe the example and explain why the evidence should not be used in this way.

3.3 Preference for a brand. Ashley is a hard-core runner. She and all her friends prefer Powerade Ion$^{4©}$ to Gatorade$^©$. Explain why Ashley's experience is not good evidence that most young people prefer Powerade to Gatorade.

3.4 Reliability of a product. A friend has driven a Toyota Camry for more than 200,000 miles with only the usual service maintenance expenses. Explain why not all Camry owners can expect this kind of performance.

Available data

available data

Occasionally, data are collected for a particular purpose but can also serve as the basis for drawing sound conclusions about other research questions. We use the term **available data** for this type of data.

> **Available Data**
>
> **Available data** are data that were produced in the past for some other purpose but that may help answer a present question.

The library and the Internet can be good sources of available data. Because producing new data is expensive, we all use available data whenever possible. Here are two examples:

EXAMPLE 3.1 International Manufacturing Productivity

If you visit the U.S. Bureau of Labor Statistics Web site, `www.bls.gov`, you can find many interesting sets of data and statistical summaries. One recent study compared the economies of 17 countries. The study showed that from 2007 to 2008, the Republic of Korea and the United States each had the highest productivity increases (tied at 1.2%), while Singapore had the largest decline (-6.6%).

EXAMPLE 3.2 Can Our Workforce Compete in a Global Economy?

In preparation to compete in the global economy, students need to improve their mathematics and science skills. At the Web site of the National Center for Education Statistics, `nces.ed.gov/nationsreportcard`, you will find full details about the math skills of schoolchildren in the latest National Assessment of Educational Progress (Figure 3.1). Mathematics scores have slowly but steadily increased since 1990. All racial/ethnic groups, both men and women, and students in most states are getting better in math.

FIGURE 3.1 The Web sites of government statistical offices are prime sources of data. Here is the home page of the National Assessment of Educational Progress, for Example 3.2.

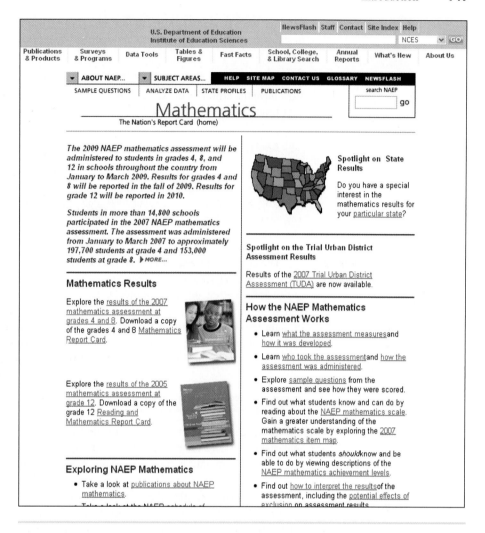

Many nations have a single national statistical office, such as Statistics Canada (www.statcan.ca) and Mexico's INEGI (inegi.gob.mx). More than 70 different U.S. agencies collect data. You can reach most of them through the government's FedStats site (fedstats.gov).

APPLY YOUR KNOWLEDGE

3.5 Find some available business data. Visit the Internet and find an example of available data that is related to business and that is interesting to you. Explain how the data were collected and what questions the study was designed to answer.

In many cases, however, the clearest answers to present questions often require that data be produced to answer those specific questions. A survey of college athletes is designed to estimate the percent who gamble. Do restaurant patrons give higher tips when their server repeats their order carefully? The validity of our conclusions from the analysis of data collected to address these issues rests on a foundation of carefully collected data. In this chapter, we will develop the skills needed to produce trustworthy data and to judge the quality of data produced by others. The techniques for producing data that we will study require no formulas, but they are among the most important ideas in statistics. Statistical designs for producing data rely on either *sampling* or *experiments*.

Sample surveys and experiments

How have the attitudes of Americans, on issues ranging from abortion to work, changed over time? **Sample surveys** are the usual tool for answering questions like these.

EXAMPLE 3.3   Confidence in Banks and Companies

One of the most important sample surveys is the General Social Survey (GSS) conducted by the NORC, a national organization for research and computing affiliated with the University of Chicago.[1] The GSS interviews about 3000 adult residents of the United States every second year. The survey includes questions about how much confidence people have in banks and companies.

---

sample
population

The GSS selects a **sample** of adults to represent the larger **population** of all English-speaking adults living in the United States. The idea of *sampling* is to study a part in order to gain information about the whole. Data are often produced by sampling a population of people or things. Opinion polls, for example, report the views of the entire country based on interviews with a sample of about 1000 people. Government reports on employment and unemployment are produced from a monthly sample of about 60,000 households. The quality of manufactured items is monitored by inspecting small samples each hour or each shift.

APPLY YOUR KNOWLEDGE

3.6 Find a sample survey that relates to business. Use the Internet or some printed material to find an example of a sample survey that relates to business and interests you. Describe the population, how the sample was collected, and some of the conclusions.

census

In all of our examples, the expense of examining every item in the population makes sampling a practical necessity. Timeliness is another reason for preferring a sample to a **census,** which is an attempt to contact every individual in the entire population. We want information on current unemployment and public opinion next week, not next year. Moreover, a carefully conducted sample is often more accurate than a census. Accountants, for example, sample a firm's inventory to verify the accuracy of the records. Attempting to count every last item in the warehouse would be not only expensive but inaccurate. Bored people do not count carefully.

If conclusions based on a sample are to be valid for the entire population, a sound design for selecting the sample is required. Sampling designs are the topic of Section 3.1.

A sample survey collects information about a population by selecting and measuring a sample from the population. The goal is a picture of the population, disturbed as little as possible by the act of gathering information. Sample surveys are one kind of *observational study.*

> **Observation versus Experiment**
>
> In an **observational study** we observe individuals and measure variables of interest but do not attempt to influence the responses.
>
> In an **experiment** we deliberately impose some treatment on individuals and we observe their responses.

3.7 H1N1 vaccines. A report issued by the Centers for Disease Control and Prevention stated that among 120 adults who received an injection of a monovalent H1N1 influenza A vaccine, 116, or 97%, had an effective response by three weeks after the vaccination. They also reported that the rates of adverse events such as headaches were not significantly different from a control group.[2] Is this an observational study or an experiment? Why? What are the explanatory and response variables?

3.8 Violent acts on prime-time TV. A typical hour of prime-time television shows three to five violent acts. Linking family interviews and police records shows a clear association between time spent watching TV as a child and later aggressive behavior.[3] Explain why this is an observational study rather than an experiment. What are the explanatory and response variables?

An observational study, even one based on a statistical sample, is a poor way to determine what will happen if we change something. The best way to see the effects of a *intervention* change is to do an **intervention**—where we actually impose the change. When our goal is to understand cause and effect, experiments are the only source of fully convincing data.

EXAMPLE 3.4   A $56 Billion Market

Child care is a $56 billion business, and there are many opportunities for entrepreneurs to be successful in this market.[4] However, possible negative consequences of using child care are a concern of many parents. A study of child care enrolled 1364 infants and planned to follow them through their sixth year in school. Twelve years later, the researchers published an article finding that "the more time children spent in child care from birth to age four-and-a-half, the more adults tended to rate them, both at age four-and-a-half and at kindergarten, as less likely to get along with others, as more assertive, as disobedient, and as aggressive."[5]

What can we conclude from this study? If parents choose to use child care, are they more likely to see these undesirable behaviors in their children?

EXAMPLE 3.5   Is There a Cause-and-Effect Relationship?

Example 3.4 describes an observational study. Parents made all child care decisions and the study did not attempt to influence them. A summary of the study stated, "The study authors noted that their study was not designed to prove a cause and effect relationship. That is, the study cannot prove whether spending more time in child care causes children to have more problem behaviors."[6] Perhaps employed parents who use child care are under stress and the children react to their parents' stress. Perhaps single parents are more likely to use child care. Perhaps parents are more likely to place in child care children who already have behavior problems.

We can imagine an experiment that would remove these difficulties. From a large group of young children, choose some to be placed in child care and others to remain at home. This is an experiment because the treatment (child care or not) is imposed on the children. Of course, this particular experiment is neither practical nor ethical.

In Examples 3.4 and 3.5, we say that the effect of child care on behavior is *confounded* **confounded** with (mixed up with) other characteristics of families who use child care.

Observational studies that examine the effect of a single variable on an outcome can be misleading when the effects of the explanatory variable are confounded with those of other variables. Because experiments allow us to isolate the effects of specific variables, we generally prefer them. Here is an example.

EXAMPLE 3.6   A Program to Help Consumers

An experiment was designed to examine the effect of a 30-minute instructional session in a food stamp office on food purchases and the dietary behavior of low-income women.[7] A group of women were randomly assigned to either the instructional session or no instruction. Two months later, data were collected on several measures of their behavior.

We begin the discussion of statistical designs for data collection in Section 3.1 with the principles underlying the design of samples.

APPLY YOUR KNOWLEDGE

3.9 Gender and consumer choices. Men and women differ in their choices for many product categories. Are there gender differences in preferences for health insurance plans as well? A market researcher interviews a large sample of consumers, both men and women. She asks each consumer which of two health plans he or she prefers. Is this study an experiment? Why or why not? What are the explanatory and response variables?

3.10 Teaching economics. An educational software company wants to compare the effectiveness of its computer animation for teaching about supply, demand, and market clearing with that of a textbook presentation. The company tests the economic knowledge of each of a group of first-year college students, then divides them into two groups. One group uses the animation, and the other studies the text. The company retests all the students and compares the increase in economic understanding in the two groups. Is this an experiment? Why or why not? What are the explanatory and response variables?

3.11 Does job training work? A state institutes a job-training program for manufacturing workers who lose their jobs. After five years, the state reviews how well the program works. Critics claim that because the state's unemployment rate for manufacturing workers was 6% when the program began and 10% five years later, the program is ineffective. Explain why higher unemployment does not necessarily mean that the training program failed. In particular, identify some lurking variables (see page 118 in Chapter 2) whose effect on unemployment may be confounded with the effect of the training program.

statistical inference   Statistical techniques for producing data are the foundation for **statistical inference,** which answers specific questions with a known degree of confidence. In Section 3.3, we discuss some basic ideas related to inference.

Should an experiment or sample survey that could possibly provide interesting and important information always be performed? How can we safeguard the privacy of subjects in a sample survey? What constitutes the mistreatment of people or animals who *ethics*   are studied in an experiment? These are questions of **ethics.** In Section 3.4, we address ethical issues related to the design of studies and the analysis of data.

3.1 Designing Samples

A political scientist wants to know what percent of college-age adults consider themselves conservatives. An automaker hires a market research firm to learn what percent of adults aged 18 to 35 recall seeing television advertisements for a new sport utility vehicle. Government economists inquire about average household income. In all these cases, we want to gather information about a large group of individuals. We will not, as in an experiment, impose a treatment in order to observe the response. Also, time, cost, and inconvenience forbid contacting every individual. In such cases, we gather information about only part of the group—a *sample*—in order to draw conclusions about the whole. *sample survey* **Sample surveys** are an important kind of observational study.

> **Population and Sample**
>
> The entire group of individuals that we want information about is called the **population.**
>   A **sample** is a part of the population that we actually examine in order to gather information.

Notice that "population" is defined in terms of our desire for knowledge. If we wish to draw conclusions about all U.S. college students, that group is our population even if only local students are available for questioning. The sample is the part from which we *sample design* draw conclusions about the whole. The **design** of a sample survey refers to the method used to choose the sample from the population.

EXAMPLE 3.7   Can We Compete Globally?

A lack of reading skills has been cited as one factor that limits our ability to compete in the global economy.[8] Various efforts have been made to improve this situation. One of these is the Reading Recovery (RR) program. RR has specially trained teachers work one-on-one with at-risk first-grade students to help them learn to read. A study was designed to examine the relationship between the RR teachers' beliefs about their ability to motivate students and the progress of the students whom they teach.[9] The National Data Evaluation Center (NDEC) Web site (`www.ndec.us`) says that there are 13,823 RR teachers. The researchers send a questionnaire to a random sample of 200 of these. The population consists of all 13,823 RR teachers, and the sample is the 200 that were randomly selected.

Unfortunately, our idealized framework of population and sample does not exactly correspond to the situations that we face in many cases. In Example 3.7, the list of teachers was prepared at a particular time in the past. It is very likely that some of the teachers on the list are no longer working as RR teachers today. New teachers have been trained in RR methods and are not on the list. A list of items to be sampled is often called *sampling frame* a **sampling frame.** For our example, we view this list as the population. We may have out-of-date addresses for some who are still working as RR teachers, and some teachers may choose not to respond to our survey questions.

In reporting the results of a sample survey it is important to include all details regarding the procedures used. The proportion of the original sample who actually provide *response rate* usable data is called the **response rate** and should be reported for all surveys. If only 150 of the teachers who were sent questionnaires provided usable data, the response rate would be 150/200, or 75%. Follow-up mailings or phone calls to those who do not initially respond can help increase the response rate.

3.12 Job satisfaction. A research team wanted to examine the relationship between employee participation in decision making and job satisfaction in a company. They are planning to randomly select 300 employees from a list of 2500 employees in the company. The Job Descriptive Index (JDI) will be used to measure job satisfaction, and the Conway Adaptation of the Alutto-Belasco Decisional Participation Scale will be used to measure decision participation. Describe the population and the sample for this study. Can you determine the response rate?

3.13 Taxes and forestland usage. A study was designed to assess the impact of taxes on forestland usage in part of the Upper Wabash River Watershed in Indiana.[10] A survey was sent to 772 forest owners from this region and 348 were returned. Consider the population, the sample, and the response rate for this study. Describe these based on the information given and indicate any additional information that you would need to give a complete answer.

Poor sample designs can produce misleading conclusions. Here is an example.

EXAMPLE 3.8   Sampling Product in a Steel Mill

A mill produces large coils of thin steel for use in manufacturing home appliances. The quality engineer wants to submit a sample of 5-centimeter squares to detailed laboratory examination. She asks a technician to cut a sample of 10 such squares. Wanting to provide "good" pieces of steel, the technician carefully avoids the visible defects in the coil material when cutting the sample. The laboratory results are wonderful, but the customers complain about the material they are receiving.

Online opinion polls are particularly vulnerable to bias because the sample who respond are not representative of the population at large.

In Example 3.8, the sample was selected in a manner that guaranteed that it would not be representative of the entire population. This sampling scheme displays *bias,* or systematic error, in favoring some parts of the population over others. Online polls use *voluntary response samples,* a particularly common form of biased sample.

> **Voluntary Response Sample**
>
> A **voluntary response sample** consists of people who choose themselves by responding to a general appeal. Voluntary response samples are biased because people with strong opinions, especially negative opinions, are most likely to respond.

The remedy for bias in choosing a sample is to allow impersonal chance to do the choosing, so that there is neither favoritism by the sampler nor voluntary response. Random selection of a sample eliminates bias by giving all individuals an equal chance to be chosen, just as randomization eliminates bias in assigning experimental subjects.

convenience sampling   Voluntary response is one common type of bad sample design. Another is **convenience sampling,** which chooses the individuals easiest to reach. Here is an example of convenience sampling.

EXAMPLE 3.9   Interviewing Customers at the Mall

Manufacturers and advertising agencies often use interviews at shopping malls to gather information about the habits of consumers and the effectiveness of ads. A sample of mall customers is fast and cheap. But people contacted at shopping malls are not representative of the entire U.S. population. They are richer, for example, and more likely to be teenagers or retired. Moreover, mall interviewers tend to select neat, safe-looking individuals from the stream of customers. Decisions based on mall interviews may not reflect the preferences of all consumers.

Both voluntary response samples and convenience samples produce samples that are almost guaranteed not to represent the entire population. These sampling methods display *bias*, or systematic error, in favoring some parts of the population over others.

> Bias
>
> The design of a study is **biased** if it systematically favors certain outcomes.

APPLY YOUR KNOWLEDGE

3.14 Sampling women in the workforce. A sociologist wants to know the opinions of employed adult women about government funding for day care. She obtains a list of the 520 members of a local business and professional women's club and mails a questionnaire to 100 of these women selected at random. Only 48 questionnaires are returned. What is the population in this study? What is the sample from whom information is actually obtained? What is the rate (percent) of nonresponse?

3.15 What is the population? For each of the following sampling situations, identify the population as exactly as possible. That is, say what kind of individuals the population consists of and say exactly which individuals fall in the population. If the information given is not sufficient, complete the description of the population in a reasonable way.

(a) Each week, the Gallup Poll questions a sample of about 1500 adult U.S. residents to determine national opinion on a wide variety of issues.

(b) The 2000 census tried to gather basic information from every household in the United States. Also, a "long form" requesting much additional information was sent to a sample of about 17% of households.

(c) A machinery manufacturer purchases voltage regulators from a supplier. There are reports that variation in the output voltage of the regulators is affecting the performance of the finished products. To assess the quality of the supplier's production, the manufacturer sends a sample of 5 regulators from the last shipment to a laboratory for study.

3.16 Market segmentation and movie ratings. You wonder if that new "blockbuster" movie is really any good. Some of your friends like the movie, but you decide to check the Internet Movie Database (`imdb.com`) to see others' ratings. You find that 2497 people chose to rate this movie, with an average rating of only 3.7 out of 10. You are surprised that most of your friends liked the movie, while many people gave low ratings to the movie online. Are you convinced that a majority of those who saw the movie would give it a low rating? What type of sample are your friends? What type of sample are the raters on the Internet Movie Database? Discuss this example in terms of market segmentation (see, for example, `businessplans.org/Segment.html`.)

Simple random samples

The simplest sampling design amounts to placing names in a hat (the population) and drawing out a handful (the sample). This is *simple random sampling.*

Simple Random Sample

A **simple random sample (SRS)** of size *n* consists of *n* individuals from the population chosen in such a way that every set of *n* individuals has an equal chance to be the sample actually selected.

Each treatment group in a completely randomized experimental design is an SRS drawn from the available experimental units. We select an SRS by labeling all the individuals in the population and using software or a table of random digits to select a sample of the desired size, just as in experimental randomization. Notice that an SRS not only gives each individual an equal chance to be chosen (thus avoiding bias in the choice) but gives every possible sample an equal chance to be chosen. There are other random sampling designs that give each individual, but not each sample, an equal chance. One such design, systematic random sampling, is described in Exercise 3.36.

Thinking about random digits helps you to understand randomization even if you will use software in practice. Table B at the back of the book and on the back endpaper is a table of random digits.

Random Digits

A **table of random digits** is a list of the digits 0, 1, 2, 3, 4, 5, 6, 7, 8, 9 that has the following properties:

1. The digit in any position in the list has the same chance of being any one of 0, 1, 2, 3, 4, 5, 6, 7, 8, 9.

2. The digits in different positions are independent in the sense that the value of one has no influence on the value of any other.

You can think of Table B as the result of asking an assistant (or a computer) to mix the digits 0 to 9 in a hat, draw one, then replace the digit drawn, mix again, draw a second digit, and so on. The assistant's mixing and drawing saves us the work of mixing and drawing when we need to randomize. Table B begins with the digits 19223950340575628713. To make the table easier to read, the digits appear in groups of five and in numbered rows. The groups and rows have no meaning—the table is just a long list of digits having the properties 1 and 2 described above.

Our goal is to use random digits to select random samples. We need the following facts about random digits, which are consequences of the basic properties 1 and 2:

- Any *pair* of random digits has the same chance of being any of the 100 possible pairs: 00, 01, 02, ... , 98, 99.

- Any *triple* of random digits has the same chance of being any of the 1000 possible triples: 000, 001, 002, ... , 998, 999.

- ... and so on for groups of four or more random digits.

SPRINGBREAK

EXAMPLE 3.10   Do They Like Students as Customers?

A campus newspaper plans a major article on spring break destinations. The authors intend to call a few randomly chosen resorts at each destination to ask about their attitudes toward groups of students as guests. Here are the resorts listed in one city. The first step is to label the members of this population as shown.

01	Aloha Kai	08	Captiva	15	Palm Tree	22	Sea Shell
02	Anchor Down	09	Casa del Mar	16	Radisson	23	Silver Beach
03	Banana Bay	10	Coconuts	17	Ramada	24	Sunset Beach
04	Banyan Tree	11	Diplomat	18	Sandpiper	25	Tradewinds
05	Beach Castle	12	Holiday Inn	19	Sea Castle	26	Tropical Breeze
06	Best Western	13	Lime Tree	20	Sea Club	27	Tropical Shores
07	Cabana	14	Outrigger	21	Sea Grape	28	Veranda

Now enter Table B, and read two-digit groups until you have chosen three resorts. If you enter at line 185, Banana Bay (03), Palm Tree (15), and Cabana (07) will be called.

APPLET

Most statistical software will select an SRS for you, eliminating the need for Table B. The *Simple Random Sample* applet on the text CD and Web site is a convenient way to automate this task.

Excel and other software can do the job. There are four steps:

1. Create a data set with all the elements of the population in the first column.

2. Assign a random number to each element of the population; put these in the second column.

3. Sort the data set by the random number column.

4. The simple random sample is obtained by taking elements in order from the sorted list until the desired sample size is reached.

We illustrate the procedure with a simplified version of Example 3.10.

EXAMPLE 3.11   Select a Random Sample

Suppose that the population from Example 3.10 is only the first 10 resorts of the display given there:

Aloha Kai	Captiva	Palm Tree	Sea Shell
Anchor Down	Casa del Mar	Radisson	Silver Beach

Note that we do not need numerical labels to identify the cases in the population. Suppose that we want to select a simple random sample of 3 resorts from this population. Figure 3.2(a) gives the spreadsheet with the population labels. The random numbers generated by the RAND() function are given in the second column in Figure 3.2(b). The sorted data set is given in Figure 3.2(c). We have added a third column to the spreadsheet to indicate which resorts were selected for our random sample. They are Captiva, Radisson, and Silver Beach.

(a)   (b)   (c)

FIGURE 3.2 Selection of a simple random sample of resorts using Excel, for Example 3.11. (a) Labels. (b) Random numbers. (c) Randomly sorted labels.

APPLY YOUR KNOWLEDGE

3.17 Ringtones for cell phones. You decide to change the ringtones for your cell phone by choosing 2 from a list of the 10 most popular ringtones.[11] Here is the list:

Changes	Cash Flow	No Se Vivir Sin Ti	I'm Me
Cyclone	Adios Amor Te Vas	Pink Panther	Y Llegaste Tu
Lollipop	Super Mario Brothers Theme		

Select your 2 ringtones using a simple random sample.

3.18 Listen to three songs. The walk to your statistics class takes about 10 minutes, about the amount of time needed to listen to three songs on your iPod. You decide to take a simple random sample of songs from the top 10 songs listed on the Billboard Hot 100.[12] Here is the list:

Single Ladies	Just Dance	Live Your Life	Heartless
Love Story	Hot N Cold	Womanizer	Love Lockdown
Whatever You Like	If I Were a Boy		

Select the three songs for your iPod using a simple random sample.

Stratified samples

The general framework for designs that use chance to choose a sample is a *probability sample*.

> **Probability Sample**
>
> A **probability sample** is a sample chosen by chance. We must know what samples are possible and what chance, or probability, each possible sample has.

Some probability sampling designs (such as an SRS) give each member of the population an *equal* chance to be selected. This may not be true in more elaborate sampling designs. In every case, however, the use of chance to select the sample is the essential principle of statistical sampling.

Designs for sampling from large populations spread out over a wide area are usually more complex than an SRS. For example, it is common to sample important groups within the population separately, then combine these samples. This is the idea of a *stratified sample.*

> **Stratified Random Sample**
>
> To select a **stratified random sample,** first divide the population into groups of similar individuals, called **strata.** Then choose a separate SRS in each stratum and combine these SRSs to form the full sample.

Choose the strata based on facts known before the sample is taken. For example, a population of election districts might be divided into urban, suburban, and rural strata. A stratified design can produce more exact information than an SRS of the same size by taking advantage of the fact that individuals in the same stratum are similar to one another. Think of the extreme case in which all individuals in each stratum are identical: just one individual from each stratum is then enough to completely describe the population.

EXAMPLE 3.12 Fraud against Insurance Companies

A dentist is suspected of defrauding insurance companies by describing some dental procedures incorrectly on claim forms and overcharging for them. An investigation begins by examining a sample of his bills for the past three years. Because there are five suspicious types of procedures, the investigators take a stratified sample. That is, they randomly select bills for each of the five types of procedures separately.

Multistage samples

Another common means of restricting random selection is to choose the sample in stages. This is common practice for national samples of households or people. For example, data on employment and unemployment are gathered by the government's Current Population Survey, which conducts interviews in about 60,000 households each month. The cost of sending interviewers to the widely scattered households in an SRS would be too high. Moreover, the government wants data broken down by states and large cities. *multistage sample* The Current Population Survey therefore uses a **multistage sampling design.** The final sample consists of clusters of nearby households that an interviewer can easily visit. Most opinion polls and other national samples are also multistage, though interviewing in most national samples today is done by telephone rather than in person, eliminating the economic need for clustering. The Current Population Survey sampling design is roughly as follows:[13]

Stage 1. Divide the United States into 2007 geographical areas called Primary Sampling Units, or PSUs. PSUs do not cross state lines. Select a sample of 754 PSUs. This sample includes the 428 PSUs with the largest populations and a stratified sample of 326 of the others.

Stage 2. Divide each PSU selected into smaller areas called "blocks." Stratify the blocks using ethnic and other information and take a stratified sample of the blocks in each PSU.

Stage 3. Sort the housing units in each block into clusters of four nearby units. Interview the households in a probability sample of these clusters.

Analysis of data from sampling designs more complex than an SRS takes us beyond basic statistics. But the SRS is the building block of more elaborate designs, and analysis of other designs differs more in complexity of detail than in fundamental concepts.

APPLY YOUR KNOWLEDGE

WORKSHOP

3.19 Who goes to the market research workshop? A small advertising firm has 30 junior associates and 10 senior associates. The junior associates are

Abel	Fisher	Huber	Miranda	Reinmann
Chen	Ghosh	Jimenez	Moskowitz	Santos
Cordoba	Griswold	Jones	Neyman	Shaw
David	Hein	Kim	O'Brien	Thompson
Deming	Hernandez	Klotz	Pearl	Utts
Elashoff	Holland	Lorenz	Potter	Varga

The senior associates are

Andrews	Fernandez	Kim	Moore	West
Besicovitch	Gupta	Lightman	Vicario	Yang

The firm will send 4 junior associates and 2 senior associates to a workshop on current trends in market research. It decides to choose those who will go by random selection. Use Table B to choose a stratified random sample of 4 junior associates and 2 senior associates. Start at line 121 to choose your sample.

3.20 Sampling by accountants. Accountants use stratified samples during audits to verify a company's records of such things as accounts receivable. The stratification is based on the dollar amount of the item and often includes 100% sampling of the largest items. One company reports 5000 accounts receivable. Of these, 100 are in amounts over $50,000; 500 are in amounts between $1000 and $50,000; and the remaining 4400 are in amounts under $1000. Using these groups as strata, you decide to verify all of the largest accounts and to sample 5% of the midsize accounts and 1% of the small accounts. How would you label the two strata from which you will sample? Use Table B, starting at line 115, to select *only the first 5* accounts from each of these strata.

Cautions about sample surveys

Random selection eliminates bias in the choice of a sample from a list of the population. Sample surveys of large human populations, however, require much more than a good sampling design. To begin, we need an accurate and complete list of the population. Because such a list is rarely available, most samples suffer from some degree of *undercoverage*. A sample survey of households, for example, will miss not only homeless people but prison inmates and students in dormitories. An opinion poll conducted by telephone will miss the 6% of American households without residential phones. The results of national sample surveys therefore have some bias if the people not covered—who most often are poor people—differ from the rest of the population.

A more serious source of bias in most sample surveys is *nonresponse*, which occurs when a selected individual cannot be contacted or refuses to cooperate. Nonresponse to sample surveys often reaches 50% or more, even with careful planning and several callbacks. Because nonresponse is higher in urban areas, most sample surveys substitute other people in the same area to avoid favoring rural areas in the final sample. If the people contacted differ from those who are rarely at home or who refuse to answer questions, some bias remains.

> **Undercoverage and Nonresponse**
>
> **Undercoverage** occurs when some groups in the population are left out of the process of choosing the sample.
>
> **Nonresponse** occurs when an individual chosen for the sample cannot be contacted or does not cooperate.

EXAMPLE 3.13   Nonresponse in the Current Population Survey

How bad is nonresponse? The Current Population Survey (CPS) has the lowest nonresponse rate of any poll we know: only about 4% of the households in the CPS sample refuse to take part, and another 3% or 4% can't be contacted. People are more likely to respond to a government survey such as the CPS, and the CPS contacts its sample in person before doing later interviews by phone.

The General Social Survey (Figure 3.3) is the nation's most important social science research survey. The GSS also contacts its sample in person, and it is run by a university. Despite these advantages, its most recent survey had a 30% rate of nonresponse.[14]

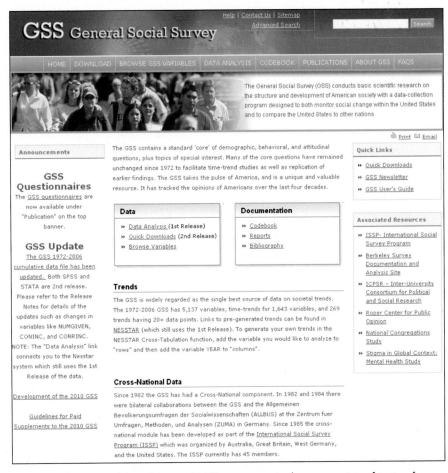

FIGURE 3.3 The General Social Survey (GSS) assesses attitudes on a variety of topics, for Example 3.13.

What about polls done by the media and by market research and opinion-polling firms? We don't know their rates of nonresponse, because they won't say. That itself is a bad sign.

Most sample surveys, and almost all opinion polls, are now carried out by telephone. This and other details of the interview method can affect the results. When presented with several options for a reply, such as completely agree, mostly agree, mostly disagree, and completely disagree, people tend to be a little more likely to respond to the first one or two options presented.

response bias   The behavior of the respondent or of the interviewer can cause **response bias** in sample results. Respondents may lie, especially if asked about illegal or unpopular behavior. The race or gender of the interviewer can influence responses to questions about race relations or attitudes toward feminism. Answers to questions that ask respondents to recall past events are often inaccurate because of faulty memory. For example, many people "telescope" events in the past, bringing them forward in memory to more recent time periods. "Have you visited a dentist in the last 6 months?" will often elicit a "Yes" from someone who last visited a dentist 8 months ago.[15]

EXAMPLE 3.14   Overreporting of Voter Behavior

"One of the most frequently observed survey measurement errors is the overreporting of voting behavior."[16] People know they should vote, so those who didn't vote tend to save face by saying that they did. Here are the data from a typical sample of 663 people after an election:

	What they said:	
What they did:	I voted	I didn't
Voted	358	13
Didn't vote	120	172

You can see that 478 people (72%) said that they voted, but only 371 people (56%) actually did vote.

wording of questions   The **wording of questions** is the most important influence on the answers given to a sample survey. Confusing or leading questions can introduce strong bias, and even minor changes in wording can change a survey's outcome. Here are some examples.

EXAMPLE 3.15   The Form of the Question Is Important

In response to the question "Are you heterosexual, homosexual, or bisexual?" in a social science research survey, one woman answered, "It's just me and my husband, so bisexual." The issue is serious, even if the example seems silly: reporting about sexual behavior is difficult because people understand and misunderstand sexual terms in many ways.

How do Americans feel about government help for the poor? Only 13% think we are spending too much on "assistance to the poor," but 44% think we are spending too much on "welfare." How do the Scots feel about the movement to become independent from England? Well, 51% would vote for "independence for Scotland," but only 34% support "an independent Scotland separate from the United Kingdom." It seems that "assistance to the poor" and "independence" are nice, hopeful words. "Welfare" and "separate" are negative words.[17]

3.21 Random digit dialing. The list of individuals from which a sample is actually selected is called the sampling frame. Ideally, the frame should include every individual in the population, but in practice this is often difficult. A frame that leaves out part of the population is a common source of undercoverage.

(a) Suppose that a sample of households in a community is selected at random from the telephone directory. What households are omitted from this frame? What types of people do you think are likely to live in these households? These people will probably be underrepresented in the sample.

(b) It is usual in telephone surveys to use random digit dialing equipment that selects the last four digits of a telephone number at random after being given the exchange (the first three digits). Which of the households that you mentioned in your answer to (a) will be included in the sampling frame by random digit dialing?

3.22 Ring-no-answer. A common form of nonresponse in telephone surveys is "ring-no-answer." That is, a call is made to an active number, but no one answers. Some of these numbers probably have caller ID where people can choose not to answer calls from people that they do not know. The Italian National Statistical Institute looked at nonresponse to a government survey of households in Italy during two periods, January 1 to Easter and July 1 to August 31. All calls were made between 7 and 10 P.M., but 21.4% gave "ring-no-answer" in one period versus 41.5% "ring-no-answer" in the other period.[18] Which period do you think had the higher rate of no answers? Why? Explain why a high rate of nonresponse makes sample results less reliable.

The statistical design of sample surveys is a science, but this science is only part of the art of sampling. Because of nonresponse, response bias, and the difficulty of posing clear and neutral questions, you should hesitate to fully trust reports about complicated issues based on surveys of large human populations. *Insist on knowing the exact questions asked, the rate of nonresponse, and the date and method of the survey before you trust a poll result.*

BEYOND THE BASICS: Capture-Recapture Sampling

Pacific salmon return to reproduce in the river where they were hatched three or four years earlier. How many salmon made it back this year? The answer will help determine quotas for commercial fishing on the west coast of Canada and the United States. Biologists estimate the size of animal populations with a special kind of repeated sampling, called *capture-recapture sampling*. More recently, capture-recapture methods have been used on human populations as well.

EXAMPLE 3.16   Sampling for a Major Industry in British Columbia

The old method of counting returning salmon involved placing a "counting fence" in a stream and counting all the fish caught by the fence. This is expensive and difficult. For example, fences are often damaged by high water. Sampling using small nets is more practical.[19]

During this year's spawning run in the Chase River in British Columbia, Canada, you net 200 coho salmon, tag the fish, and release them. Later in the week, your nets capture 120 coho salmon in the river, of which 12 have tags.

The proportion of your second sample that have tags should estimate the proportion in the entire population of returning salmon that are tagged. So if N is the unknown number of coho salmon in the Chase River this year, we should have approximately

$$\text{proportion tagged in sample} = \text{proportion tagged in population}$$
$$\frac{12}{120} = \frac{200}{N}$$

Solve for N to estimate that the total number of salmon in this year's spawning run in the Chase River is approximately

$$N = 200 \times \frac{120}{12} = 2000$$

The capture-recapture idea extends the use of a sample proportion to estimate a population proportion. The idea works well if both samples are SRSs from the population and the population remains unchanged between samples. In practice, complications arise. For example, some tagged fish might be caught by bears or otherwise die between the first and second samples. Variations on capture-recapture samples are widely used in wildlife studies and are now finding other applications. One way to estimate the census undercount in a district is to consider the census as "capturing and marking" the households that respond. Census workers then visit the district, take an SRS of households, and see how many of those counted by the census show up in the sample. Capture-recapture estimates the total count of households in the district. As with estimating wildlife populations, there are many practical pitfalls. Our final word is as before: the real world is less orderly than statistics textbooks imply.

SECTION 3.1 Summary

- A sample survey selects a **sample** from the **population** of all individuals about which we desire information. We base conclusions about the population on data about the sample.

- The **design** of a sample refers to the method used to select the sample from the population. **Probability sampling designs** use impersonal chance to select a sample.

- The basic probability sample is a **simple random sample (SRS).** An SRS gives every possible sample of a given size the same chance to be chosen.

- Choose an SRS by labeling the members of the population and using a **table of random digits** to select the sample. Software can automate this process.

- To choose a **stratified random sample,** divide the population into **strata,** groups of individuals that are similar in some way that is important to the response. Then choose a separate SRS from each stratum and combine them to form the full sample.

- **Multistage samples** select successively smaller groups within the population in stages, resulting in a sample consisting of clusters of individuals. Each stage may employ an SRS, a stratified sample, or another type of sample.

- Failure to use probability sampling often results in **bias,** or systematic errors in the way the sample represents the population. **Voluntary response** samples, in which the respondents choose themselves, are particularly prone to large bias.

- In human populations, even probability samples can suffer from bias due to **undercoverage** or **nonresponse,** from **response bias** due to the behavior of the interviewer or the respondent, or from misleading results due to **poorly worded questions.**

SECTION 3.1 Exercises

For Exercises 3.1 to 3.4, see page 148; for 3.5, see page 149; for 3.6, see page 150; for 3.7 and 3.8, see page 151; for 3.9 to 3.11, see page 152; for 3.12 and 3.13, see page 154; for 3.14 to 3.16, see page 155; for 3.17 and 3.18, see page 158; for 3.19 and 3.20, see page 160; and for 3.21 and 3.22, see page 163.

3.23 Make it an experiment! In the following observational studies, describe changes that could be made to the data collection process that would result in an experiment rather than an observational study. Also, offer suggestions about unseen biases or lurking variables that may be present in the studies as they are described here.
(a) A friend of yours likes to play Texas hold'em. Every time that he tells you about his playing, he says that he won.
(b) In an introductory statistics class you notice that the students who sit in the first two rows of seats had a higher score on the first exam than the other students in the class.

3.24 What's wrong? Explain what is wrong in each of the following scenarios.
(a) The population consists of all individuals selected in a simple random sample.
(b) In a poll of an SRS of residents in a local community, respondents are asked to indicate the level of their concern about the dangers of dihydrogen monoxide, a substance that is a major component of acid rain and in its gaseous state can cause severe burns. (*Hint:* Ask a friend who is majoring in chemistry about this substance or search the Internet for information about it.)
(c) Students in a class are asked to raise their hands if they have cheated on an exam one or more times within the past year.

3.25 What's wrong? Explain what is wrong with each of the following random selection procedures and explain how you would do the randomization correctly.
(a) To determine the reading level of an introductory statistics text, you evaluate all of the written material in the third chapter.
(b) You want to sample student opinions about a proposed change in procedures for changing majors. You hand out questionnaires to 100 students as they arrive for class at 7:30 A.M.
(c) A population of subjects is put in alphabetical order, and a simple random sample of size 10 is taken by selecting the first 10 subjects in the list.

3.26 Importance of students as customers. A committee on community relations in a college town plans to survey local businesses about the importance of students as customers. From telephone book listings, the committee chooses 150 businesses at random. Of these, 73 return the questionnaire mailed by the committee. What is the population for this sample survey? What is the sample? What is the rate (percent) of nonresponse?

3.27 Popularity of news personalities can affect market share. A Gallup Poll conducted telephone interviews with 1001 U.S. adults aged 18. One of the questions asked whether the respondents had a favorable or an unfavorable opinion of 17 news personalities. Diane Sawyer received the highest rating, with 80% of the respondents giving her a favorable rating.[20]
(a) What is the population for this sample survey? What was the sample size?
(b) The report on the survey states that 8% of the respondents either never heard of Sawyer or had no opinion about her. When they included only those who provided an opinion, Sawyer's approval percent rose to 88% and she was still at the top of the list. Charles Gibson, on the other hand, was ranked eighth on the original list, with a 55% favorable rating. When only those providing an opinion were counted, his rank rose to second, with 87% approving. Discuss the advantages and disadvantages of the two different ways of reporting the approval percent. State which one you prefer and why.

3.28 Identify the populations. For each of the following sampling situations, identify the population as exactly as possible. That is, say what kind of individuals the population consists of and say exactly which individuals fall in the population. If the information given is not complete, complete the description of the population in a reasonable way.
(a) A college has changed its core curriculum and wants to obtain detailed feedback information from the students during each of the first 12 weeks of the coming semester. Each week, a random sample of 5 students will be selected to be interviewed.
(b) The American Community Survey (ACS) will replace the census "long form" starting with the 2010 census. The main part of the ACS contacts 250,000 addresses by mail each month, with follow-up by phone and in person if there is no response. Each household answers questions about their housing, economic, and social status.
(c) An opinion poll contacts 1161 adults and asks them, "Which political party do you think has better ideas for leading the country in the twenty-first century?"

3.29 Interview potential customers. You have been hired by a company that is planning to build a new apartment complex for students in a college town. They want you to collect information about preferences of potential customers for their complex. Most of the college students who live in apartments live in one of 33 complexes. You decide to select 5 apartment complexes at random for in-depth interviews with residents. Select a simple random sample of 5 of the following apartment complexes. If you use Table B, start at line 137. RESIDENTS

Ashley Oaks	Country View	Mayfair Village
Bay Pointe	Country Villa	Nobb Hill
Beau Jardin	Crestview	Pemberly Courts
Bluffs	Del-Lynn	Peppermill
Brandon Place	Fairington	Pheasant Run
Briarwood	Fairway Knolls	Richfield
Brownstone	Fowler	Sagamore Ridge
Burberry	Franklin Park	Salem Courthouse
Cambridge	Georgetown	Village Manor
Chauncey Village	Greenacres	Waterford Court
Country Squire	Lahr House	Williamsburg

3.30 Using GIS to identify mint field conditions. A Geographic Information System (GIS) is to be used to distinguish different conditions in mint fields. Ground observations will be used to classify regions of each field as either healthy mint, diseased mint, or weed-infested mint. The GIS divides mint-growing areas into regions called pixels. An experimental area contains 200 pixels. For a random sample of 25 pixels, ground measurements will be made to determine the status of the mint, and these observations will be compared with informa-tion obtained by the GIS. Select the random sample. If you use Table B, start at line 112 and choose only the first 5 pixels in the sample.

3.31 Select a simple random sample. After you have labeled the individuals in a population, the *Simple Random Sample* applet automates the task of choosing an SRS. Use the applet to choose the sample in the previous exercise.

3.32 Select a simple random sample. There are approximately 380 active telephone area codes covering Canada, the United States, and some Caribbean areas. (More are created regularly.) You want to choose an SRS of 25 of these area codes for a study of available telephone numbers. Label the codes 001 to 380 and use the *Simple Random Sample* applet to choose your sample. (If you use Table B, start at line 130 and choose only the first 5 codes in the sample.)

3.33 Census tracts. The Census Bureau divides the entire country into "census tracts" that contain about 4000 people. Each tract is in turn divided into small "blocks," which in urban areas are bounded by local streets. An SRS of blocks from a census tract is often the next-to-last stage in a multistage sample. Figure 3.4 shows part of census tract 8051.12, in Cook County, Illinois, west of Chicago. The 44 blocks in this tract are divided into three "block groups." Group 1 contains 6 blocks numbered 1000 to 1005; Group 2 (outlined in Figure 3.4) contains 12 blocks numbered 2000 to 2011; Group 3 contains 26 blocks numbered 3000

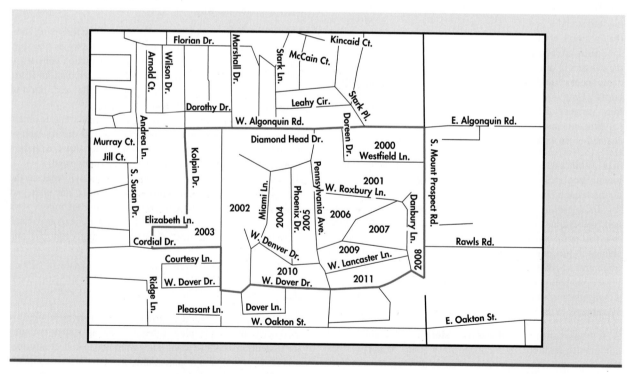

FIGURE 3.4 Census blocks in Cook County, Illinois. The outlined area is a block group. (From factfinder.census.gov.)

to 3025. Use Table B, beginning at line 115, to choose an SRS of 5 of the 44 blocks in this census tract. Explain carefully how you labeled the blocks.

3.34 Repeated use of Table B. In using Table B repeatedly to choose samples, you should not always begin at the same place, such as line 101. Why not?

3.35 A stratified sample. Exercise 3.33 asks you to choose an SRS of blocks from the census tract pictured in Figure 3.4. You might instead choose a stratified sample of one block from the 6 blocks in Group 1, two from the 12 blocks in Group 2, and three from the 26 blocks in Group 3. Choose such a sample, explaining carefully how you labeled blocks and used Table B.

3.36 Systematic random samples. *Systematic random samples* are often used to choose a sample of apartments in a large building or dwelling units in a block at the last stage of a multi-stage sample. An example will illustrate the idea of a systematic sample. Suppose that we must choose 4 addresses out of 100. Because 100/4 = 25, we can think of the list as four lists of 25 addresses. Choose 1 of the first 25 at random, using Table B. The sample contains this address and the addresses 25, 50, and 75 places down the list from it. If 13 is chosen, for example, then the systematic random sample consists of the addresses numbered 13, 38, 63, and 88.

(a) A study of dating among college students wanted a sample of 200 of the 9000 single male students on campus. The sample consisted of every 45th name from a list of the 9000 students. Explain why the survey chooses every 45th name.

(b) Use Table B at line 135 to choose the starting point for this systematic sample.

3.37 Systematic random samples versus simple random samples. The previous exercise introduces systematic random samples. Explain carefully why a systematic random sample *does* give every individual the same chance to be chosen but is *not* a simple random sample.

3.38 Random digit telephone dialing for market research. A market research firm in California uses random digit dialing to choose telephone numbers at random. Numbers are selected separately within each California area code. The size of the sample in each area code is proportional to the population living there.

(a) What is the name for this kind of sampling design?

(b) California area codes, in rough order from north to south, are

530	707	916	209	415	925	510	650	408	831	805	559	760
661	818	213	626	323	562	709	310	949	909	858	619

Another California survey does not call numbers in all area codes but starts with an SRS of 8 area codes. Choose such an SRS. If you use Table B, start at line 128. **AREACODES**

3.39 Stratified samples of forest areas in the Amazon basin. Stratified samples are widely used to study large areas of forest. Based on satellite images, a forest area in the Amazon basin is divided into 14 types. Foresters studied the four most commercially valuable types: alluvial climax forests of quality levels 1, 2, and 3, and mature secondary forest. They divided the area of each type into large parcels, chose parcels of each type at random, and counted tree species in a 20- by 25-meter rectangle randomly placed within each parcel selected. Here is some detail:

Forest type	Total parcels	Sample size
Climax 1	36	4
Climax 2	72	7
Climax 3	31	3
Secondary	42	4

Choose the stratified sample of 18 parcels. Be sure to explain how you assigned labels to parcels. If you use Table B, start at line 130.

3.40 Select employees for an awards committee. A department has 30 hourly workers and 10 salaried workers. The hourly workers are

Abel	Fisher	Huber	Moran	Reinmann
Carson	Golomb	Jimenez	Moskowitz	Santos
Chen	Griswold	Jones	Neyman	Shaw
David	Hein	Kiefer	O'Brien	Thompson
Deming	Hernandez	Klotz	Pearl	Utts
Elashoff	Holland	Liu	Potter	Vlasic

and the salaried workers are

Andrews	Fernandez	Kim	Moore	Rabinowitz
Besicovitch	Gupta	Lightman	Phillips	Yang

The committee will have 6 hourly workers and 2 salaried workers. Random selection will be used to select the committee members. Select a stratified random sample of 6 hourly workers and 2 salaried workers. **COMMITTEEMEMBERS**

3.41 Survey questions. Comment on each of the following as a potential sample survey question. Is the question clear? Is it slanted toward a desired response?

(a) "Some cell phone users have developed brain cancer. Should all cell phones come with a warning label explaining the danger of using cell phones?"

(b) "Do you agree that a national system of health insurance should be favored because it would provide health insurance for everyone and would reduce administrative costs?"

(c) "In view of escalating environmental degradation and incipient resource depletion, would you favor economic incentives for recycling of resource-intensive consumer goods?"

3.42 When do you ask? When observations are taken over time, it is important to check for patterns that may be important for the interpretation of the data. In Section 1.1 we learned to use a time plot for this purpose. Describe and discuss a sample survey question where you would expect to have variation over time for the following situations:

(a) Data are taken at each hour of the day from 8 A.M. to 6 P.M.

(b) Date are taken on each of the 7 days of the week.

(c) Data are taken during each of the 12 months of the year.

3.43 How many children are in your family? A teacher asks her class, "How many children are there in your family, including yourself?" The mean response is about 3 children. According to the Bureau of the Census, in 2008, families that have children average 1.86 children. Why is a sample like this biased toward higher outcomes?

3.44 Bad survey questions. Write your own examples of bad sample survey questions.

(a) Write a biased question designed to get one answer rather than another.

(b) Write a question that is confusing, so that it is hard to answer.

3.45 Economic attitudes of Spaniards. Spain's Centro de Investigaciones Sociológicos carried out a sample survey on the economic attitudes of Spaniards.[21] Of the 2496 adults interviewed, 72% agreed that "employees with higher performance must get higher pay." On the other hand, 71% agreed that "everything a society produces should be distributed among its members as equally as possible and there should be no major differences." Use these conflicting results as an example in a short explanation of why opinion polls often fail to reveal public attitudes clearly.

3.2 Designing Experiments

A study is an experiment when we actually do something to people, animals, or objects in order to observe the response. Here is the basic vocabulary of experiments.

> **Experimental Units, Subjects, Treatment**
>
> The individuals on which the experiment is done are the **experimental units.** When the units are human beings, they are called **subjects.** A specific experimental condition applied to the units is called a **treatment.**

factors

level of a factor

Because the purpose of an experiment is to reveal the response of one variable to changes in other variables, the distinction between explanatory and response variables is important. The explanatory variables in an experiment are often called **factors.** Many experiments study the joint effects of several factors. In such an experiment, each treatment is formed by combining a specific value (often called a **level**) of each of the factors.

EXAMPLE 3.17 Is the Cost Justified?

The increased costs for teacher salaries and facilities associated with smaller class sizes can be substantial. Are smaller classes really better? We might do an observational study that compares students who happened to be in smaller and larger classes in their early school years. Small classes are expensive, so they are more common in schools that serve richer communities. Students in small classes tend to also have other advantages: their schools have more resources, their parents are better educated, and so on. Confounding makes it impossible to isolate the effects of small classes.

The Tennessee STAR program was an experiment on the effects of class size. It has been called "one of the most important educational investigations ever carried out." The *subjects* were 6385 students who were beginning kindergarten. Each student was assigned to one of three *treatments:* regular class (22 to 25 students) with one teacher, regular class with a teacher and a full-time teacher's aide, and small class (13 to 17 students). These treatments are levels of a single *factor,* the type of class. The students stayed in the same type of class for four years, then all returned to regular classes. In later years, students from the small classes had higher scores on standard tests, were less likely to fail a grade, had better high school grades, and so on. The benefits of small classes were greatest for minority students.[22]

Example 3.17 illustrates the big advantage of experiments over observational studies. **In principle, experiments can give good evidence for causation.** In an experiment, we study the specific factors we are interested in, while controlling the effects of lurking variables. All the students in the Tennessee STAR program followed the usual curriculum at their schools. Because students were assigned to different class types within their schools, school resources and family backgrounds were not confounded with class type. The only systematic difference was the type of class. When students from the small classes did better than those in the other two types, we can be confident that class size made the difference.

EXAMPLE 3.18   Effects of TV Advertising

What are the effects of repeated exposure to an advertising message? The answer may depend both on the length of the ad and on how often it is repeated. An experiment investigates this question using undergraduate students as subjects. All subjects view a 40-minute television program that includes ads for a digital camera. Some subjects see a 30-second commercial; others, a 90-second version. The same commercial is repeated either 1, 3, or 5 times during the program. After viewing, all of the subjects answer questions about their recall of the ad, their attitude toward the camera, and their intention to purchase it. These are the response variables.[23]

This experiment has two factors: length of the commercial, with 2 levels; and repetitions, with 3 levels. All possible combinations of the 2×3 factor levels form 6 treatment combinations. Figure 3.5 shows the layout of these treatments.

FIGURE 3.5 The treatments in the experimental design of Example 3.18. Combinations of levels of the two factors form six treatments.

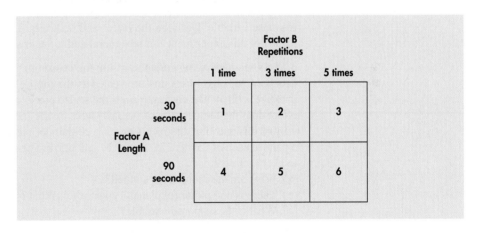

Experimentation allows us to study the effects of the specific treatments we are interested in. Moreover, we can control the environment of the subjects to hold constant the factors that are of no interest to us, such as the specific product advertised in Example 3.18. In one sense, the ideal case is a laboratory experiment in which we control all lurking variables and so see only the effect of the treatments on the response. On the other hand, the effects of being in an artificial environment such as a laboratory may also affect the outcomes. *The balance between control and realism is an important consideration in the design of experiments.*

Another advantage of experiments is that we can study the combined effects of several factors simultaneously. The interaction of several factors can produce effects that could not be predicted from looking at the effect of each factor alone. Perhaps longer commercials increase interest in a product, and more commercials also increase interest, but if we both make a commercial longer and show it more often, viewers get annoyed and their interest in the product drops. The two-factor experiment in Example 3.18 will help us find out.

APPLY YOUR KNOWLEDGE

3.46 Radiation and storage time for food products. Storing food for long periods of time is a major challenge for those planning for human space travel beyond the moon. One problem is that exposure to radiation decreases the length of time that food can be stored. One experiment examined the effects of nine different levels of radiation on a particular type of fat, or lipid.[24] The amount of oxidation of the lipid is the measure of the extent of the damage due to the radiation. Three samples are exposed to each radiation level. Give the experimental units, the treatments, and the response variable. Describe the factor and its levels. There are many different types of lipids. To what extent do you think the results of this experiment can be generalized to other lipids?

3.47 Can they use the Web? A course in computer graphics technology requires students to learn multiview drawing concepts. This topic is traditionally taught using supplementary material printed on paper. The instructor of the course believes that a Web-based interactive drawing program will be more effective in increasing the drawing skills of the students.[25] The 50 students who are enrolled in the course will be randomly assigned to either the paper-based instruction or the Web-based instruction. A standardized drawing test will be given before and after the instruction. Explain why this study is an experiment, and give the experimental units, the treatments, and the response variable. Describe the factor and its levels. To what extent do you think the results of this experiment can be generalized to other settings?

3.48 Is the packaging convenient for the customer? A manufacturer of food products uses package liners that are sealed at the top by applying heated jaws after the package is filled. The customer peels the sealed pieces apart to open the package. What effect does the temperature of the jaws have on the force needed to peel the liner? To answer this question, engineers prepare 20 pairs of pieces of package liner. They seal 5 pairs at each of 250°F, 275°F, 300°F, and 325°F. Then they measure the force needed to peel each seal.
(a) What are the individuals studied?
(b) There is one factor (explanatory variable). What is it, and what are its levels?
(c) What is the response variable?

Comparative experiments

Experiments in the laboratory often have a simple design: impose the treatment and see what happens. We can outline that design like this:

$$\text{Subjects} \longrightarrow \text{Treatment} \longrightarrow \text{Response}$$

In the laboratory, we try to avoid confounding by rigorously controlling the environment of the experiment so that nothing except the experimental treatment influences the response. Once we get out of the laboratory, however, there are almost always lurking variables waiting to confound us. When our subjects are people or animals rather than electrons or chemical compounds, confounding can happen even in the controlled environment of a laboratory or medical clinic. Here is an example that helps explain why careful experimental design is a key issue for pharmaceutical companies and other makers of medical products.

EXAMPLE 3.19   Gastric Freezing to Treat Ulcers

"Gastric freezing" is a clever treatment for ulcers. The patient swallows a deflated balloon with tubes attached, then a refrigerated liquid is pumped through the balloon for an hour. The idea is that cooling the stomach will reduce its production of acid and so relieve ulcers. An experiment reported in the *Journal of the American Medical Association* showed that gastric freezing did reduce acid production and relieve ulcer pain. The treatment was widely used for several years. The design of the experiment was

Subjects ⟶ **Gastric freezing** ⟶ **Observe pain relief**

placebo effect

This experiment is poorly designed. The patients' response may be due to the **placebo effect.** A placebo is a dummy treatment. Many patients respond favorably to *any* treatment, even a placebo, presumably because of trust in the doctor and expectations of a cure. This response to a dummy treatment is the placebo effect.

A later experiment divided ulcer patients into two groups. One group was treated by gastric freezing as before. The other group received a placebo treatment in which the liquid in the balloon was at body temperature rather than freezing. The results: 34% of the 82 patients in the treatment group improved, but so did 38% of the 78 patients in the placebo group. This and other properly designed experiments showed that gastric freezing was no better than a placebo, and its use was abandoned.[26]

The first gastric-freezing experiment gave misleading results because the effects of the explanatory variable were confounded with the placebo effect. We can defeat confounding by *comparing* two groups of patients, as in the second gastric-freezing experiment. The placebo effect and other lurking variables now operate on both groups. The only difference between the groups is the actual effect of gastric freezing. The group

control group

of patients who received a sham treatment is called a **control group,** because it enables us to control the effects of outside variables on the outcome. Control is the first basic principle of statistical design of experiments. Comparison of several treatments in the same environment is the simplest form of control.

Uncontrolled experiments in medicine and the behavioral sciences can be dominated by such influences as the details of the experimental arrangement, the selection of subjects, and the placebo effect. The result is often *bias.*

> **Bias**
>
> The design of a study is **biased** if it systematically favors certain outcomes.

An uncontrolled study of a new medical therapy, for example, is biased in favor of finding the treatment effective because of the placebo effect. It should not surprise you to learn that uncontrolled studies in medicine give new therapies a much higher success rate than proper comparative experiments do. Well-designed experiments usually compare several treatments.

3.49 Does using statistical software improve exam scores? An instructor in an elementary statistics course wants to know if using a new statistical software package will improve students' final-exam scores. He asks for volunteers and about half of the class agrees to work with the new software. He compares the final-exam scores of the students who used the new software with the scores of those who did not. Discuss possible sources of bias in this study.

Randomized comparative experiments

experiment design

The **design of an experiment** first describes the response variables, the factors (explanatory variables), and the layout of the treatments, with *comparison* as the leading principle. The second aspect of design is the rule used to assign the subjects to the treatments. Comparison of the effects of several treatments is valid only when all treatments are applied to similar groups of subjects. If one corn variety is planted on more fertile ground, or if one cancer drug is given to less seriously ill patients, comparisons among treatments are biased. How can we assign individuals to treatments in a way that is fair to all the treatments?

randomization

randomized comparative experiment

Our answer is the same as in sampling: let impersonal chance make the assignment. The use of chance to divide subjects into groups is called **randomization.** Groups formed by randomization don't depend on any characteristic of the subjects or on the judgment of the experimenter. An experiment that uses both comparison and randomization is a **randomized comparative experiment.** Here is an example.

EXAMPLE 3.20   Testing a Breakfast Food

A food company assesses the nutritional quality of a new "instant breakfast" product by feeding it to newly weaned male white rats. The response variable is a rat's weight gain over a 28-day period. A control group of rats eats a standard diet but otherwise receives exactly the same treatment as the experimental group.

This experiment has one factor (the diet) with two levels. The researchers use 30 rats for the experiment and so must divide them into two groups of 15. To do this in an unbiased fashion, put the cage numbers of the 30 rats in a hat, mix them up, and draw 15. These rats form the experimental group and the remaining 15 make up the control group. *Each group is an SRS of the available rats.* Figure 3.6 outlines the design of this experiment.

FIGURE 3.6 Outline of a randomized comparative experiment, for Example 3.20.

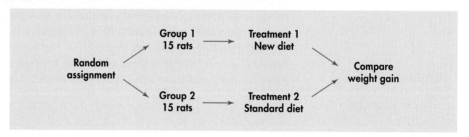

We can use software or the table of random digits to randomize. Label the rats 01 to 30. Enter Table B at (say) line 130. Run your finger along this line (and continue to lines 131 and 132 as needed) until 15 rats are chosen. They are the rats labeled

05 16 17 20 19 04 25 29 18 07 13 02 23 27 21

These rats form the experimental group; the remaining 15 are the control group.

3.50 Diagram the food storage experiment. Refer to Exercise 3.46 (page 170). Draw a diagram similar to Figure 3.5 that describes the food for space travel experiment.

3.51 Diagram the Web use. Refer to Exercise 3.47 (page 170). Draw a diagram similar to Figure 3.5 that describes the computer graphics drawing experiment.

Completely randomized designs

The design in Figure 3.6 combines comparison and randomization to arrive at the simplest statistical design for an experiment. This "flowchart" outline presents all the essentials: randomization, the sizes of the groups and which treatment they receive, and the response variable. There are, as we will see later, statistical reasons for generally using treatment groups that are about equal in size. We call designs like that in Figure 3.6 *completely randomized*.

> **Completely Randomized Design**
>
> In a **completely randomized** experimental design, all the subjects are allocated at random among all the treatments.

Completely randomized designs can compare any number of treatments. Here is an example that compares three treatments.

EXAMPLE 3.21   Utility Companies and Energy Conservation

Many utility companies have introduced programs to encourage energy conservation among their customers. An electric company considers placing electronic meters in households to show what the cost would be if the electricity use at that moment continued for a month. Will these meters reduce electricity use? Would cheaper methods work almost as well? The company decides to design an experiment.

One cheaper approach is to give customers a chart and information about monitoring their electricity use. The experiment compares these two approaches (meter, chart) and also a control. The control group of customers receives information about energy conservation but no help in monitoring electricity use. The response variable is total electricity used in a year. The company finds 60 single-family residences in the same city willing to participate, so it assigns 20 residences at random to each of the 3 treatments. Figure 3.7 outlines the design.

To carry out the random assignment, label the 60 households 01 to 60. Enter Table B (or use software) to select an SRS of 20 to receive the meters. Continue in Table B, selecting 20 more to receive charts. The remaining 20 form the control group.

FIGURE 3.7 Outline of a completely randomized design comparing three treatments, for Example 3.21.

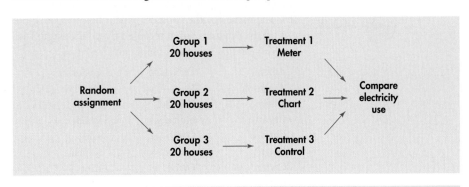

Examples 3.20 and 3.21 describe completely randomized designs that compare levels of a single factor. In Example 3.20, the factor is the diet fed to the rats. In Example 3.21, it is the method used to encourage energy conservation. Completely randomized designs can have more than one factor. The advertising experiment of Example 3.18 has two factors: the length and the number of repetitions of a television commercial. Their combinations form the six treatments outlined in Figure 3.5 (page 169). A completely randomized design assigns subjects at random to these six treatments. Once the layout of treatments is set, the randomization needed for a completely randomized design is tedious but straightforward.

APPLY YOUR KNOWLEDGE

3.52 Gastric freezing. Example 3.19 describes an experiment that helped end the use of gastric freezing to treat ulcers. The subjects were 160 ulcer patients.
(a) Make a diagram to outline the design of this experiment, using the information in Example 3.19. (Show the size of the groups, the treatment each group receives, and the response variable. Figures 3.6 and 3.7 are models to follow.)
(b) The 82 patients in the gastric-freezing group are an SRS of the 160 subjects. Label the subjects and use Table B, starting at line 121, to choose the first 5 members of the gastric-freezing group.

3.53 Sealing food packages. Use a diagram to describe a completely randomized experimental design for the package liner experiment of Exercise 3.48. (Show the size of the groups, the treatment each group receives, and the response variable. Figures 3.6 and 3.7 are models to follow.) Use software or Table B, starting at line 140, to do the randomization required by your design.

3.54 Does child care help recruit employees? Will providing child care for employees make a company more attractive to women, even those who are unmarried? You are designing an experiment to answer this question. You prepare recruiting material for two fictitious companies, both in similar businesses in the same location. Company A's brochure does not mention child care. There are two versions of Company B's material, identical except that one describes the company's on-site child care facility. Your subjects are 40 unmarried women who are college seniors seeking employment. Each subject will read recruiting material for both companies and choose the one she would prefer to work for. You will give each version of Company B's brochure to half the women. You expect that a higher percent of those who read the description that includes child care will choose Company B.
(a) Outline an appropriate design for the experiment.
(b) The names of the subjects appear below. Use Table B, beginning at line 112, to do the randomization required by your design. List the subjects who will read the version that mentions child care.

 CHILDCARE

Abrams	Danielson	Gutierrez	Lippman	Rosen
Adamson	Durr	Howard	Martinez	Sugiwara
Afifi	Edwards	Hwang	McNeill	Thompson
Brown	Fluharty	Iselin	Morse	Travers
Cansico	Garcia	Janle	Ng	Turing
Chen	Gerson	Kaplan	Quinones	Ullmann
Cortez	Green	Kim	Rivera	Williams
Curzakis	Gupta	Lattimore	Roberts	Wong

The logic of randomized comparative experiments

Randomized comparative experiments are designed to give good evidence that differences in the treatments actually *cause* the differences we see in the response. The logic is as follows:

- Random assignment of subjects forms groups that should be similar in all respects before the treatments are applied.

- Comparative design ensures that influences other than the experimental treatments operate equally on all groups.

- Therefore, differences in average response must be due either to the treatments or to the play of chance in the random assignment of subjects to the treatments.

That "either-or" deserves more thought. In Example 3.20, we cannot say that *any* difference in the average weight gains of rats fed the two diets must be caused by a difference between the diets. There would be some difference even if both groups received the same diet because the natural variability among rats means that some grow faster than others. Chance assigns the faster-growing rats to one group or the other, and this creates a chance difference between the groups. We would not trust an experiment with just one rat in each group, for example. The results would depend on which group got lucky and received the faster-growing rat. If we assign many rats to each diet, however, the effects of chance will average out and there will be little difference in the average weight gains in the two groups unless the diets themselves cause a difference. "Use enough subjects to reduce chance variation" is the third big idea of statistical design of experiments.

> **Principles of Experimental Design**
>
> 1. **Control** the effects of lurking variables on the response, most simply by comparing two or more treatments.
> 2. **Randomize**—use impersonal chance to assign subjects to treatments.
> 3. **Replicate** each treatment on enough subjects to reduce chance variation in the results.

EXAMPLE 3.22   Cell Phones and Driving

Comstock/Getty Images

Does talking on a hands-free cell phone distract drivers? Undergraduate students "drove" in a high-fidelity driving simulator equipped with a hands-free cell phone. The car ahead brakes: how quickly does the subject respond? Twenty students (the control group) simply drove. Another 20 (the experimental group) talked on the cell phone while driving. The simulator gave the same driving conditions to both groups.[27]

This experimental design has good control because the only difference in the conditions for the two groups is the use of the cell phone. Students are randomized to the two groups, so we satisfy the second principle. Based on past experience with the simulators, the length of the drive and the number of subjects were judged to provide sufficient information to make the comparison. (We will learn more about choosing sample sizes for experiments in Chapter 6.)

We hope to see a difference in the responses so large that it is unlikely to happen just because of chance variation. We can use the laws of probability, which give a mathematical description of chance behavior, to learn if the treatment effects are larger than we would expect to see if only chance were operating. If they are, we call them *statistically significant*.

> **Statistical Significance**
>
> An observed effect so large that it would rarely occur by chance is called **statistically significant.**

If we observe statistically significant differences among the groups in a comparative randomized experiment, we have good evidence that the treatments actually caused these differences. You will often see the phrase "statistically significant" in reports of investigations in many fields of study. The great advantage of randomized comparative experiments is that they can produce data that give good evidence for a cause-and-effect relationship between the explanatory and response variables. We know that in general a strong association does not imply causation. A statistically significant association in data from a well-designed experiment does imply causation.

APPLY YOUR KNOWLEDGE

3.55 Utility companies. Example 3.21 describes an experiment to learn whether providing households with electronic meters or charts will reduce their electricity consumption. An executive of the utility company objects to including a control group. He says, "It would be simpler to just compare electricity use last year (before the meter or chart was provided) with consumption in the same period this year. If households use less electricity this year, the meter or chart must be working." Explain clearly why this design is inferior to that in Example 3.21.

3.56 Exercise and heart attacks. Does regular exercise reduce the risk of a heart attack? Here are two ways to study this question. Explain clearly why the second design will produce more trustworthy data.

1. A researcher finds 2000 men over 40 who exercise regularly and have not had heart attacks. She matches each with a similar man who does not exercise regularly, and she follows both groups for 5 years.

2. Another researcher finds 4000 men over 40 who have not had heart attacks and are willing to participate in a study. She assigns 2000 of the men to a regular program of supervised exercise. The other 2000 continue their usual habits. The researcher follows both groups for 5 years.

3.57 Statistical significance. The financial aid office of a university asks a sample of students about their employment and earnings. The report says that "for academic year earnings, a significant difference was found between the sexes, with men earning more on the average. No significant difference was found between the earnings of black and white students." Explain the meaning of "a significant difference" and "no significant difference" in plain language.

How to randomize

The idea of randomization is to assign subjects to treatments by drawing names from a hat. In practice, experimenters use software to carry out randomization. Most statistical software will choose 20 out of a list of 40 at random, for example. The list might contain the names of 40 human subjects. The 20 chosen form one group, and the 20 that remain form the second group. The *Simple Random Sample* applet on the text CD and Web site makes it particularly easy to choose treatment groups at random.

You can randomize without software by using a *table of random digits*.

EXAMPLE 3.23   Randomize the Students

In the cell phone experiment of Example 3.22, we must divide 40 students at random into two groups of 20 students each.

Step 1: Label. Give each student a numerical label, using as few digits as possible. Two digits are needed to label 40 students, so we use labels

$$01, \ 02, \ 03, \ \dots, \ 39, \ 40$$

It is also correct to use labels 00 to 39 or some other choice of 40 two-digit labels.

Step 2: Table. Start anywhere in Table B and read two-digit groups. Suppose we begin at line 130, which is

$$69051 \ \ 64817 \ \ 87174 \ \ 09517 \ \ 84534 \ \ 06489 \ \ 87201 \ \ 97245$$

The first 10 two-digit groups in this line are

$$69 \ \ 05 \ \ 16 \ \ 48 \ \ 17 \ \ 87 \ \ 17 \ \ 40 \ \ 95 \ \ 17$$

Each of these two-digit groups is a label. The labels 00 and 41 to 99 are not used in this example, so we ignore them. The first 20 labels between 01 and 40 that we encounter in the table choose students for the experimental group. Of the first 10 labels in line 130, we ignore four because they are too high (over 40). The others are 05, 16, 17, 17, 40, and 17. The students labeled 05, 16, 17, and 40 go into the experimental group. Ignore the second and third 17s because that student is already in the group. Run your finger across line 130 (and continue to the following lines) until you have chosen 20 students. They are the students labeled

$$05, \ 16, \ 17, \ 40, \ 20, \ 19, \ 32, \ 04, \ 25, \ 29, \ 37, \ 39, \ 31, \ 18, \ 07, \ 13, \ 33, \ 02, \ 36, \ 23$$

You should check at least the first few of these. These students form the experimental group. The remaining 20 are the control group.

As Example 3.23 illustrates, randomization requires two steps: assign labels to the experimental units and then use Table B to select labels at random. Be sure that all labels are the same length so that all have the same chance to be chosen. Use the shortest possible labels—one digit for 9 or fewer individuals, two digits for 10 to 100 individuals, and so on. Don't try to scramble the labels as you assign them. Table B will do the required randomizing, so assign labels in any convenient manner, such as in alphabetical order for human subjects. You can read digits from Table B in any order—along a row, down a column, and so on—because the table has no order. As an easy standard practice, we recommend reading along rows.

It is easy to use statistical software or Excel to randomize. Here are the steps:

Step 1: Label. Assigning labels to the experimental units is similar to the procedure we just described except that we are not restricted to using numerical labels. Any system where each experimental unit has a unique label identifier will work.

Step 2: Use the computer. Once we have the labels, we create a data file with the labels and generate a random number for each label. Finally, we sort the entire data set based on the random numbers. Groups are formed by selecting units in order from the sorted list.

This process is essentially the same as writing the labels on a deck of cards, shuffling the cards, and dealing them out one at a time.

EXAMPLE 3.24   Using Software for the Randomization

Let's do a randomization similar to the one we did in Example 3.23, but this time using Excel. Here we will use 10 experimental units. We will assign 5 to the treatment group and 5 to the control group. We first create a data set with the numbers 1 to 10 in the first column. See Figure 3.8(a). Then we use RAND() to generate 10 random numbers in the second column. See Figure 3.8(b). Finally, we sort the data set based on the numbers in the second column. See Figure 3.8(c). The first 5 labels (8, 5, 9, 4, and 6) are assigned to the experimental group. The remaining 5 labels (3, 10, 7, 2, and 1) correspond to the control group.

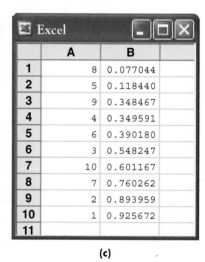

	A	B
1	1	
2	2	
3	3	
4	4	
5	5	
6	6	
7	7	
8	8	
9	9	
10	10	
11		

(a)

	A	B
1	1	0.925672
2	2	0.893959
3	3	0.548247
4	4	0.349591
5	5	0.118440
6	6	0.390180
7	7	0.760262
8	8	0.077044
9	9	0.348467
10	10	0.601167
11		

(b)

	A	B
1	8	0.077044
2	5	0.118440
3	9	0.348467
4	4	0.349591
5	6	0.390180
6	3	0.548247
7	10	0.601167
8	7	0.760262
9	2	0.893959
10	1	0.925672
11		

(c)

FIGURE 3.8 Randomization of 10 experimental units using Excel, for Example 3.24. (a) Labels. (b) Random numbers. (c) Randomly sorted labels.

Completely randomized designs can compare any number of treatments. The treatments can be formed by levels of a single factor or by more than one factor. Here is an example with two factors:

EXAMPLE 3.25   Randomization for the TV Commercial Experiment

Figure 3.5 (page 169) displays six treatments formed by the two factors in an experiment on response to a TV commercial. Suppose that we have 150 students who are willing to serve as subjects. We must assign 25 students at random to each group. Figure 3.9 outlines the completely randomized design.

To carry out the random assignment, label the 150 students 001 to 150. (Three digits are needed to label 150 subjects.) Enter Table B and read three-digit groups until you have selected 25 students to receive Treatment 1 (a 30-second ad shown once). If you start at line 140, the first few labels for Treatment 1 subjects are 129, 048, and 003.

Continue in Table B to select 25 more students to receive Treatment 2 (a 30-second ad shown 3 times). Then select another 25 for Treatment 3 and so on until you have assigned 125 of the 150 students to Treatments 1 through 5. The 25 students who remain get Treatment 6. The randomization is straightforward but very tedious to do by hand. We recommend the *Simple Random Sample* applet. Exercise 3.69 shows how to use the applet to do the randomization for this example.

FIGURE 3.9 Outline of a completely randomized design for comparing six treatments, for Example 3.25.

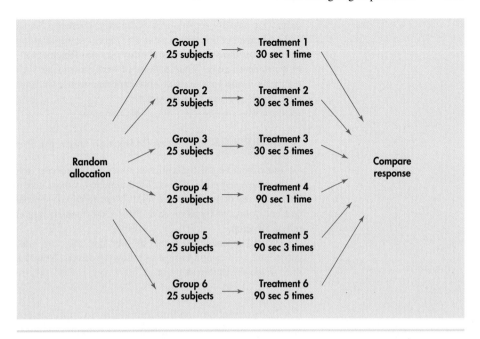

3.58 Do the randomization. Use computer software to carry out the randomization in Example 3.25.

Cautions about experimentation

The logic of a randomized comparative experiment depends on our ability to treat all the subjects identically in every way except for the actual treatments being compared. Good experiments therefore require careful attention to details. For example, the subjects in both groups of the second gastric-freezing experiment (Example 3.19, page 171) all got the same medical attention over the several years of the study. The researchers paid attention to such details as ensuring that the tube in the mouth of each subject was cold, whether or not the fluid in the balloon was refrigerated. Moreover, the study was *double-blind*   **double-blind**—neither the subjects themselves nor the medical personnel who worked with them knew which treatment any subject had received. The double-blind method avoids unconscious bias by, for example, a doctor who doesn't think that "just a placebo" can benefit a patient.

Many—perhaps most—experiments have some weaknesses in detail. The environment of an experiment can influence the outcomes in unexpected ways. Although experiments are the gold standard for evidence of cause and effect, really convincing evidence usually requires that a number of studies in different places with different details produce *lack of realism*   similar results. The most serious potential weakness of experiments is **lack of realism.** The subjects or treatments or setting of an experiment may not realistically duplicate the conditions we really want to study. Here are two examples.

EXAMPLE 3.26   Layoffs and Feeling Bad

How do layoffs at a workplace affect the workers who remain on the job? Psychologists asked student subjects to proofread text for extra course credit, then "let go" some of the workers (who were actually accomplices of the experimenters). Some subjects were told that those let go had performed poorly (Treatment 1). Others were told that not all could be kept and that it was just luck that they were kept and others let go (Treatment 2). We can't be sure that the reactions of the students are the same as those of workers who survive a layoff in which other workers lose their jobs. Many behavioral science experiments use student subjects in a campus setting. Do the conclusions apply to the real world?

EXAMPLE 3.27 Does the Regulation Make the Product Safer?

Do those high center brake lights, required on all cars sold in the United States since 1986, really reduce rear-end collisions? Randomized comparative experiments with fleets of rental and business cars, done before the lights were required, showed that the third brake light reduced rear-end collisions by as much as 50%. Unfortunately, requiring the third light in all cars led to only a 5% drop.

What happened? Most cars did not have the extra brake light when the experiments were carried out, so it caught the eye of following drivers. Now that almost all cars have the third light, they no longer capture attention.

Lack of realism can limit our ability to apply the conclusions of an experiment to the settings of greatest interest. Most experimenters want to generalize their conclusions to some setting wider than that of the actual experiment. Statistical analysis of the original experiment cannot tell us how far the results will generalize. Nonetheless, the randomized comparative experiment, because of its ability to give convincing evidence for causation, is one of the most important ideas in statistics.

APPLY YOUR KNOWLEDGE

3.59 Managers and stress. Some companies employ consultants to train their managers in meditation in the hope that this practice will relieve stress and make the managers more effective on the job. An experiment that claimed to show that meditation reduces anxiety proceeded as follows. The experimenter interviewed the subjects and rated their level of anxiety. Then the subjects were randomly assigned to two groups. The experimenter taught one group how to meditate and they meditated daily for a month. The other group was simply told to relax more. At the end of the month, the experimenter interviewed all the subjects again and rated their anxiety level. The meditation group now had less anxiety. Psychologists said that the results were suspect because the ratings were not blind. Explain what this means and how lack of blindness could bias the reported results.

3.60 Frustration and teamwork. A psychologist wants to study the effects of failure and frustration on the relationships among members of a work team. She forms a team of students, brings them to the psychology laboratory, and has them play a game that requires teamwork. The game is rigged so that they lose regularly. The psychologist observes the students through a one-way window and notes the changes in their behavior during an evening of game playing. Why is it doubtful that the findings of this study tell us much about the effect of working for months developing a new product that never works right and is finally abandoned by your company?

Matched pairs designs

Completely randomized designs are the simplest statistical designs for experiments. They illustrate clearly the principles of control, randomization, and replication of treatments on a number of subjects. However, completely randomized designs are often inferior to more elaborate statistical designs. In particular, matching the subjects in various ways can produce more precise results than simple randomization.

matched pairs design

One common design that combines matching with randomization is the **matched pairs design.** A matched pairs design compares just two treatments. Choose pairs of subjects that are as closely matched as possible. Assign one of the treatments to each subject in a pair by tossing a coin or reading odd and even digits from Table B. Sometimes each "pair" in a matched pairs design consists of just one subject, who gets both treatments one after the other. Each subject serves as his or her own control. The *order* of the treatments can influence the subject's response, so we randomize the order for each subject, again by a coin toss.

EXAMPLE 3.28   Matched Pairs for the Cell Phone Experiment

Example 3.22 describes an experiment on the effects of talking on a cell phone while driving. The experiment compared two treatments, driving in a simulator and driving in a simulator while talking on a hands-free cell phone. The response variable is the time the driver takes to apply the brake when the car in front brakes suddenly. In Example 3.22, 40 student subjects were assigned at random, 20 students to each treatment. Subjects differ in driving skill and reaction times. The completely randomized design relies on chance to create two similar groups of subjects.

In fact, the experimenters used a matched pairs design in which all subjects drove both with and without using the cell phone. They compared each individual's reaction times with and without the phone. If all subjects drove first with the phone and then without it, the effect of talking on the cell phone would be confounded with the fact that this is the first run in the simulator. The proper procedure requires that all subjects first be trained in using the simulator, that the *order* in which a subject drives with and without the phone be random, and that the two drives be on separate days to reduce the chance that the results of the second treatment will be affected by the first treatment.

The completely randomized design uses chance to decide which 20 subjects will drive with the cell phone. The other 20 drive without it. The matched pairs design uses chance to decide which 20 subjects will drive first with and then without the cell phone. The other 20 drive first without and then with the phone.

Block designs

Matched pairs designs apply the principles of comparison of treatments, randomization, and replication. However, the randomization is not complete—we do not randomly assign all the subjects at once to the two treatments. Instead, we only randomize within each matched pair. This allows matching to reduce the effect of variation among the subjects. Matched pairs are an example of *block designs*.

> **Block Design**
>
> A **block** is a group of subjects that are known before the experiment to be similar in some way expected to affect the response to the treatments. In a **block design,** the random assignment of individuals to treatments is carried out separately within each block.

A block design combines the idea of creating equivalent treatment groups by matching with the principle of forming treatment groups at random. Blocks in experiments are similar to starta for sampling. We have two names because the idea of grouping similar units before randomizing arose separately in experiments and in sampling. Blocks are another form of *control*. They control the effects of some outside variables by bringing those variables into the experiment to form the blocks. Here is a typical example of a block design.

EXAMPLE 3.29   Men, Women, and Advertising

Women and men respond differently to advertising. An experiment to compare the effectiveness of three television commercials for the same product will want to look separately at the reactions of men and women, as well as assess the overall response to the ads.

A completely randomized design considers all subjects, both men and women, as a single pool. The randomization assigns subjects to three treatment groups without regard to their gender. This ignores the differences between men and women. A better design considers women and men separately. Randomly assign the women to three groups, one to view each commercial. Then separately assign the men at random to three groups. Figure 3.10 outlines this improved design.

FIGURE 3.10 Outline of a block design, for Example 3.29.

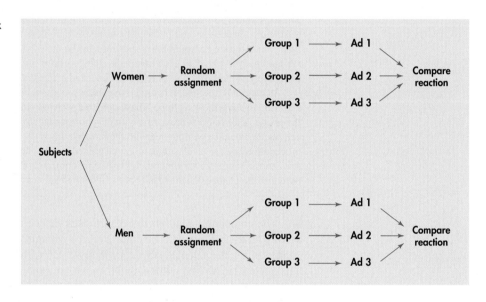

A block is a group of subjects formed before an experiment starts. We reserve the word "treatment" for a condition that we impose on the subjects. We don't speak of 6 treatments in Example 3.29 even though we can compare the responses of 6 groups of subjects formed by the 2 blocks (men, women) and the 3 commercials. Block designs are similar to stratified samples. Blocks and strata both group similar individuals together. We use two different names only because the idea developed separately for sampling and experiments. The advantages of block designs are the same as the advantages of stratified samples. Blocks allow us to draw separate conclusions about each block—for example, about men and women in the advertising study in Example 3.29. Blocking also allows more precise overall conclusions because the systematic differences between men and women can be removed when we study the overall effects of the three commercials. The idea of blocking is an important additional principle of statistical design of experiments. A wise experimenter will form blocks based on the most important unavoidable sources of variability among the experimental subjects. Randomization will then average out the effects of the remaining variation and allow an unbiased comparison of the treatments.

Like the design of samples, the design of complex experiments is a job for experts. Now that we have seen a bit of what is involved, we will usually just act as if most experiments were completely randomized.

APPLY YOUR KNOWLEDGE

3.61 Does charting help investors? Some investment advisers believe that charts of past trends in the prices of securities can help predict future prices. Most economists disagree. In an experiment to examine the effects of using charts, business students trade (hypothetically) a foreign currency at computer screens. There are 20 student subjects available, named for convenience A, B, C, ..., T. Their goal is to make as much money as possible, and the best performances are rewarded with small prizes. The student traders have the price history of the foreign currency in dollars in their computers. They may or may not also have software that highlights trends. Describe two designs for this experiment, a completely randomized design and a matched pairs design in which each student serves as his or her own control. In both cases, carry out the randomization required by the design.

SECTION 3.2 Summary

- In an experiment, we impose one or more **treatments** on the **experimental units** or **subjects.** Each treatment is a combination of levels of the explanatory variables, which we call **factors.**

- The **design** of an experiment describes the choice of treatments and the manner in which the subjects are assigned to the treatments.

- The basic principles of statistical design of experiments are **control, randomization,** and **replication.**

- The simplest form of control is **comparison.** Experiments should compare two or more treatments in order to avoid **confounding** of the effect of a treatment with other influences, such as lurking variables.

- **Randomization** uses chance to assign subjects to the treatments. Randomization creates treatment groups that are similar (except for chance variation) before the treatments are applied. Randomization and comparison together prevent **bias,** or systematic favoritism, in experiments.

- You can carry out randomization by giving numerical labels to the subjects and using a **table of random digits** to choose treatment groups.

- **Replication** of each treatment on many subjects reduces the role of chance variation and makes the experiment more sensitive to differences among the treatments.

- Good experiments require attention to detail as well as good statistical design. Many behavioral and medical experiments are **double-blind. Lack of realism** in an experiment can prevent us from generalizing its results.

- In addition to comparison, a second form of control is to restrict randomization by forming **blocks** of subjects that are similar in some way that is important to the response. Randomization is then carried out separately within each block.

- **Matched pairs** are a common form of blocking for comparing just two treatments. In some matched pairs designs, each subject receives both treatments in a random order. In others, the subjects are matched in pairs as closely as possible, and one subject in each pair receives each treatment.

SECTION 3.2 Exercises

For Exercises 3.46 to 3.48, see page 170; for 3.49, see page 172; for 3.50 and 3.51, see page 173; for 3.52 to 3.54, see page 174; for 3.55 to 3.57, see page 176; for 3.58, see page 179; for 3.59 and 3.60, see page 180; and for 3.61, see page 183.

3.62 What is needed? Explain what is deficient in each of the following proposed experiments and explain how you would improve the experiment.
(a) Two forms of a lab exercise are to be compared. There are 10 rows in the classroom. Students who sit in the first 5 rows of the class are given the first form, and students who sit in the last 5 rows are given the second form.
(b) The effectiveness of a leadership program for high school students is evaluated by examining the change in scores on a standardized test of leadership skills.
(c) An innovative method for teaching introductory biology courses is examined by using the traditional method in the fall zoology course and the new method in the spring botany course.

3.63 What is wrong? Explain what is wrong with each of the following randomization procedures and describe how you would do the randomization correctly.
(a) A list of 50 subjects is entered into a computer file and then sorted by last name. The subjects are assigned to five treatments by taking the first 10 subjects for Treatment 1, the next 10 subjects for Treatment 2, and so forth.
(b) Eight subjects are to be assigned to two treatments, four to each. For each subject, a coin is tossed. If the coin comes up heads, the subject is assigned to the first treatment; if the coin comes up tails, the subject is assigned to the second treatment.
(c) An experiment will assign 80 rats to four different treatment conditions. The rats arrive from the supplier in batches of 20 and the treatment lasts two weeks. The first batch of 20 rats is randomly assigned to one of the four treatments, and data for these rats are collected. After a one-week break, another batch of 20 rats arrives and is assigned to one of the three remaining treatments. The process continues until the last batch of rats is given the treatment that has not been assigned to the three previous batches.

3.64 Evaluate a new method for training new employees. A new method for training new employees is to be evaluated by randomly assigning new employees to either the current training program or the new method. A questionnaire will be used to evaluate the satisfaction of the new employees with the training. Explain how this experiment should be done in a double-blind fashion.

3.65 Can you change attitudes of workers about teamwork? You will conduct an experiment designed to change attitudes of workers about teamwork. Discuss some variables that you might use if you were to use a block design for this experiment.

3.66 An experiment for a new product. Compost tea is rich in microorganisms that help plants grow. It is made by soaking compost in water.[28] Design a comparative experiment that will provide evidence about whether or not compost tea works for a particular type of plant that interests you. Be sure to provide all details regarding your experiment, including the response variable or variables that you will measure. Assuming that the experiment shows positive results, write a short description about how you would use the results in a marketing campaign for compost tea.

3.67 Marketing your training materials. Water quality of streams and lakes is an issue of concern to the public. Although trained professionals typically are used to take reliable measurements, many volunteer groups are gathering and distributing information based on data that they collect.[29] You are part of a team to train volunteers to collect accurate water quality data. Design an experiment to evaluate the effectiveness of the training. Write a summary of your proposed design to present to your team. Be sure to include all the details that they will need to evaluate your proposal. How would you use the results of the experiment to market your training materials?

3.68 Randomly assign the subjects. You can use the *Simple Random Sample* applet to choose a treatment group at random once you have labeled the subjects. Example 3.23 (page 177) uses Table B to choose 20 students from a group of 40 for the treatment group in a study of the effect of cell phones on driving. Use the applet to choose the 20 students for the experimental group. Which students did you choose? The remaining 20 students make up the control group.

3.69 Randomly assign the subjects. The *Simple Random Sample* applet allows you to randomly assign experimental units to more than two groups without difficulty. Example 3.25 (page 178) describes a randomized comparative experiment in which 150 students are randomly assigned to six groups of 25.
(a) Use the applet to randomly choose 25 out of 150 students to form the first group. Which students are in this group?
(b) The "Population hopper" now contains the 125 students that were not chosen, in scrambled order. Click "Sample" again to choose 25 of these remaining students to make up the second group. Which students were chosen?
(c) Click "Sample" three more times to choose the third, fourth, and fifth groups. Don't take the time to write down these groups. Check that there are only 25 students remaining in the "Population hopper." These subjects get Treatment 6. Which students are they?

3.70 Random digits. Table B is a table of random digits. Which of the following statements are true of a table of random digits, and which are false? Explain your answers.
(a) There are exactly four 0s in each row of 40 digits.
(b) Each pair of digits has chance 1/100 of being 00.

(c) The digits 0000 can never appear as a group, because this pattern is not random.

3.71 Drug use and willingness to work. How does smoking marijuana affect willingness to work? Canadian researchers persuaded young adult men who used marijuana to live for 98 days in a "planned environment." The men earned money by weaving belts. They used their earnings to pay for meals and other consumption and could keep any money left over. One group smoked two potent marijuana cigarettes every evening. The other group smoked two weak marijuana cigarettes. All subjects could buy more cigarettes but were given strong or weak cigarettes, depending on their group. Did the weak and strong groups differ in work output and earnings?[30] 🔘 MARIJUANA
(a) Outline the design of this experiment.
(b) Here are the names of the 20 subjects. Use software or Table B at line 131 to carry out the randomization your design requires.

Abate	Dubois	Gutierrez	Lucero	Rosen
Afifi	Engel	Hwang	McNeill	Thompson
Brown	Fluharty	Iselin	Morse	Travers
Chen	Gerson	Kaplan	Quinones	Ullmann

3.72 Price cuts on athletic shoes. Stores advertise price reductions to attract customers. What type of price cut is most attractive? Market researchers prepared ads for athletic shoes announcing different levels of discounts (20%, 40%, or 60%). The student subjects who read the ads were also given "inside information" about the fraction of shoes on sale (50% or 100%). Each subject then rated the attractiveness of the sale on a scale of 1 to 7.[31]
(a) There are two factors. Make a sketch like Figure 3.5 (page 169) that displays the treatments formed by all combinations of levels of the factors.
(b) Outline a completely randomized design using 60 student subjects. Use software or Table B at line 111 to choose the subjects for the first treatment.

3.73 I'll have a Mocha Light. Here's the opening of a press release: "Starbucks Corp. on Monday said it would roll out a line of blended coffee drinks intended to tap into the growing popularity of reduced-calorie and reduced-fat menu choices for Americans." You wonder if Starbucks customers like the new "Mocha Frappuccino Light" as well as the regular version of this drink.
(a) Describe a matched pairs design to answer this question. Be sure to include proper blinding of your subjects.
(b) You have 20 regular Starbucks customers on hand. Use Table B at line 141 to do the randomization that your design requires.

3.74 Public housing. A study of the effect of living in public housing on the income and other variables in poverty-level households was carried out as follows. The researchers obtained a list of all applicants for public housing during the previous year. Some applicants had been accepted, while others had been turned down by the housing authority. Both groups were interviewed and compared. Is this study an experiment or an observational study? Why? What are the explanatory and response variables? Why will confounding make it difficult to see the effect of the explanatory variable on the response variables?

3.75 Effects of price promotions. A researcher studying the effect of price promotions on consumers' expectations makes up a history of the store price of a hypothetical brand of laundry detergent for the past year. Students in a marketing course view the price history on a computer. Some students see a steady price, while others see regular promotions that temporarily cut the price. Then the students are asked what price they would expect to pay for the detergent. Is this study an experiment? Why? What are the explanatory and response variables?

3.76 Should you charge a flat rate? You can use your computer to make telephone calls over the Internet. How will the cost affect the behavior of users of this service? You will offer the service to all 200 rooms in a college dormitory. Some rooms will pay a low flat rate. Others will pay higher rates at peak periods and very low rates off-peak. You are interested in the amount and time of use and in the effect on the congestion of the network. Outline the design of an experiment to study the effect of rate structure.

3.77 Marketing to children. If children are given more choices within a class of products, will they tend to prefer that product to a competing product that offers fewer choices? Marketers want to know. An experiment prepared three sets of beverages. Set 1 contained two milk drinks and two fruit drinks. Set 2 had two fruit drinks and four milk drinks. Set 3 contained four fruit drinks but only two milk drinks. The researchers divided 210 children aged 4 to 12 years into three groups at random. They offered each group one of the sets. As each child chose a beverage to drink from the set presented, the researchers noted whether the choice was a milk drink or a fruit drink.
(a) What are the experimental subjects?
(b) What is the factor and what are its levels? What is the response variable?
(c) Use a diagram to outline a completely randomized design for the study.
(d) Explain how you would assign labels to the subjects. Use Table B at line 125 to choose the first 5 subjects assigned to the first treatment.

3.78 Aspirin and heart attacks. "Nearly five decades of research now link aspirin to the prevention of stroke and heart attacks." So says the Bayer Aspirin Web site, www.bayeraspirin.com. The most important evidence for this claim comes from the Physicians' Health Study, a large medical experiment involving 22,000 male physicians. One group of about 11,000 physicians took an aspirin every second day, while the rest took a placebo. After several years the study found that subjects in the aspirin group had significantly fewer heart attacks than subjects in the placebo group.
(a) Identify the experimental subjects, the factor and its levels, and the response variable in the Physicians' Health Study.

(b) Use a diagram to outline a completely randomized design for the Physicians' Health Study.

(c) What does it mean to say that the aspirin group had "significantly fewer heart attacks"?

3.79 Reducing health care spending. Will people spend less on health care if their health insurance requires them to pay some part of the cost themselves? An experiment on this issue asked if the percent of medical costs that are paid by health insurance has an effect either on the amount of medical care that people use or on their health. The treatments were four insurance plans. Each plan paid all medical costs above a ceiling. Below the ceiling, the plans paid 100%, 75%, 50%, or 0% of costs incurred.

(a) Outline the design of a randomized comparative experiment suitable for this study.

(b) Describe briefly the practical and ethical difficulties that might arise in such an experiment.

3.80 Effects of TV advertising. You decide to use a completely randomized design in the two-factor experiment on response to advertising described in Example 3.18 (page 169). The 36 students named below will serve as subjects. Outline the design. Then use Table B at line 110 to randomly assign the subjects to the 6 treatments. 🌐 **TVADS**

Alomar	Denman	Han	Liang	Padilla	Valasco
Asihiro	Durr	Howard	Maldonado	Plochman	Vaughn
Bennett	Edwards	Hruska	Marsden	Rosen	Wei
Bikalis	Farouk	Imrani	Montoya	Solomon	Wilder
Chao	Fleming	James	O'Brian	Trujillo	Willis
Clemente	George	Kaplan	Ogle	Tullock	Zhang

3.81 Temperature and work performance. An expert on worker performance is interested in the effect of room temperature on the performance of tasks requiring manual dexterity. She chooses temperatures of 20°C (68°F) and 30°C (86°F) as treatments. The response variable is the number of correct insertions, during a 30-minute period, in a peg-and-hole apparatus that requires the use of both hands simultaneously. Each subject is trained on the apparatus and then asked to make as many insertions as possible in 30 minutes of continuous effort.

(a) Outline a completely randomized design to compare dexterity at 20° and 30°. Twenty subjects are available.

(b) Because individuals differ greatly in dexterity, the wide variation in individual scores may hide the systematic effect of temperature unless there are many subjects in each group. Describe in detail the design of a matched pairs experiment in which each subject serves as his or her own control.

3.82 Reaching Mexican Americans. Advertising that hopes to attract Mexican Americans must keep in mind the cultural orientation of these consumers. There are several psychological tests available to measure the extent to which Mexican Americans are oriented toward Mexican/Spanish or Anglo/English culture. Two such tests are the Bicultural Inventory (BI) and the Acculturation Rating Scale for Mexican Americans (ARSMA). To study the relationship between the scores on these two tests, researchers will give both tests to a group of 22 Mexican Americans.

(a) Briefly describe a matched pairs design for this study. In particular, how will you use randomization in your design?

(b) You have an alphabetized list of the subjects (numbered 1 to 22). Carry out the randomization required by your design and report the result.

3.83 Survivor guilt in layoffs. Workers who survive a layoff of other employees at their location may suffer from "survivor guilt." A study of survivor guilt and its effects used as subjects 120 students who were offered an opportunity to earn extra course credit by doing proofreading. Each subject worked in the same cubicle as another student, who was an accomplice of the experimenters. At a break midway through the work, one of three things happened:

> Treatment 1: The accomplice was told to leave; it was explained that this was because she performed poorly.

> Treatment 2: It was explained that unforeseen circumstances meant there was only enough work for one person. By "chance," the accomplice was chosen to be laid off.

> Treatment 3: Both students continued to work after the break.

The subjects' work performance after the break was compared with performance before the break.[32]

(a) Outline the design of this completely randomized experiment. Follow the model of Figure 3.7 (page 173).

(b) If you are using software, choose the subjects for Treatment 1. If not, use Table B at line 123 to choose the first four subjects for Treatment 1.

3.3 Toward Statistical Inference

A poll asked a random sample of 1009 adults about eating dinner in restaurants. Result: 60% reported that they ate dinner out at least once during the past week. That's the truth about the 1009 people in the sample. What is the truth about the almost 230 million American adults who make up the population? Because the sample was chosen at random, it's reasonable to think that these 1009 people represent the entire population pretty well.

So the market researchers turn the *fact* that 60% of the *sample* ate dinner at a restaurant during the past week into an *estimate* that about 60% of *all adults* ate dinner at a restaurant during the past week. That's a basic concept in statistics: use a fact about a sample to estimate the truth about the whole population. We call this **statistical inference** because we infer conclusions about the wider population from data on selected individuals. To think about inference, we must keep straight whether a statistical summary describes a sample or a population. Here is the vocabulary we use.

statistical inference

> **Parameters and Statistics**
>
> A **parameter** is a number that describes the **population.** A parameter is a fixed number, but in practice we do not know its value.
>
>   A **statistic** is a number that describes a **sample.** The value of a statistic is known when we have taken a sample, but it can change from sample to sample. We often use a statistic to estimate an unknown parameter.

CASE 3.1

Changes in Consumer Behavior The economic recession of 2008 led many consumers to reduce their spending on nonessential items. A Gallup Poll taken December 11–14, 2008, asked a random sample of 1008 U.S. adults, aged 18 and over, about eating out in restaurants during the previous week. The results from this survey can be compared with the results from similar surveys taken in 2007 and 2006 to assess the extent to which there has been a change.[33]

The proportion of the sample who said that they ate out in a restaurant last week is

$$\hat{p} = \frac{605}{1009} = 0.60 = 60\%$$

The number $\hat{p} = 0.60$ is a *statistic*. The corresponding *parameter* is the proportion (call it p) of all adult U.S. residents who would have said that they ate out in a restaurant last week if asked the same question. We don't know the value of the parameter p, so we use the statistic $\hat{p}$ as an estimate.

APPLY YOUR KNOWLEDGE

3.84 Sexual harassment of college students. A recent survey of 2036 undergraduate college students aged 18 to 24 reports that 62% of college students say they have encountered some type of sexual harassment while at college.[34] Describe the sample and the population for this setting.

3.85 Consumer preference poll. A Web site asks viewers to choose which of two energy drinks they prefer. Can you apply the ideas about populations and samples that we have just discussed to this poll? Explain why or why not.

Sampling variability, sampling distributions

If Gallup took a second random sample of 1009 adults, the new sample would have different people in it. It is almost certain that there would not be exactly 605 positive responses. That is, the value of the statistic $\hat{p}$ will *vary* from sample to sample. This basic fact is called **sampling variability.** Could it happen that one random sample finds

sampling variability

that 60% said that they ate out in a restaurant last week and a second random sample finds that only 40% say that they ate out? Random samples eliminate *bias* from the act of choosing a sample. However, they can still be inadequate for what we need because of the *variability*. If the variation is too great, the information obtained from our sample may not be very useful.

We are saved by the second great advantage of random samples. The first advantage is that choosing at random eliminates favoritism. That is, random sampling attacks bias. The second advantage is that if we take lots of random samples of the same size from the same population, the variation from sample to sample will follow a predictable pattern. **All of statistical inference is based on one idea: to see how trustworthy a procedure is, ask what would happen if we repeated it many times.**

To understand why sampling variability is not fatal, we therefore ask, "What would happen if we took many samples?" Here's how to answer that question:

- Assume that you have a population with many individuals and take a large number of samples from this population.

- Calculate the sample proportion $\hat{p}$ for each sample.

- Make a histogram of the values of $\hat{p}$.

- Examine the distribution displayed in the histogram for shape, center, and spread, as well as outliers or other deviations.

We can't afford to actually take many samples from a large population such as all adult U.S. residents. But we can imitate taking many samples by using random digits. Using random digits from a table or computer software to imitate chance behavior is called *simulation* **simulation.**

EXAMPLE 3.30 Samples of Size 100

CASE 3.1 Suppose that in fact (unknown to Gallup), exactly 60% of all adults ate out last week. That is, the truth about the population is that $p = 0.6$. What if we select an SRS of size 100 from this population and use the sample proportion $\hat{p}$ to estimate the unknown value of the population proportion p? Using software, we simulated doing this 1000 times.

Figure 3.11 illustrates the process of choosing many samples and finding $\hat{p}$ for each one. In the first sample, 56 of the 100 people say they find shopping frustrating, so $\hat{p} = 56/100 = 0.56$.

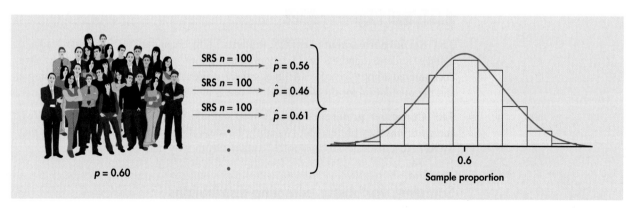

FIGURE 3.11 The results of many SRSs have a regular pattern. Here, we draw 1000 SRSs of size 100 from the same population. The population proportion is $p = 0.60$. The histogram shows the distribution of 1000 sample proportions $\hat{p}$.

Only 46 in the next sample feel this way, so for that sample $\hat{p} = 0.46$. Choose 1000 samples and make a histogram of the 1000 values of $\hat{p}$. That's the graph at the right of Figure 3.11.

What do you think would happen if we used a larger sample size? We can use simulation to get an answer to this question.

EXAMPLE 3.31   Samples of Size 2500

CASE 3.1

Suppose that Gallup used 2500 people for their sample, not just 100. Figure 3.12 shows the process of choosing 1000 SRSs, each of size 2500, from a population in which the true sample proportion is $p = 0.6$. The 1000 values of $\hat{p}$ from these samples form the histogram at the right of the figure. Figures 3.11 and 3.12 are drawn on the same scale. Comparing them shows what happens when we increase the size of our samples from 100 to 2500. These histograms display the *sampling distribution* of the statistic $\hat{p}$ for two sample sizes.

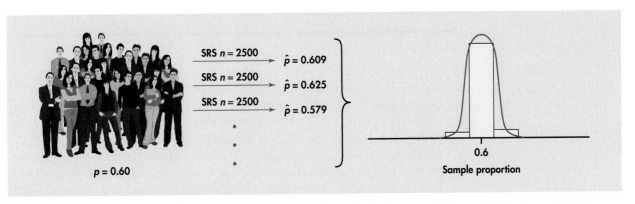

FIGURE 3.12 The distribution of the sample proportions $\hat{p}$ for 1000 SRSs of size 2500 drawn from the same population as in Figure 3.11. The two histograms have the same scale. The statistic from the larger sample is less variable.

Sampling Distribution

The **sampling distribution** of a statistic is the distribution of values taken by the statistic in all possible samples of the same size from the same population.

Strictly speaking, the sampling distribution is the ideal pattern that would emerge if we looked at all possible samples of the same size from our population. A distribution obtained from a fixed number of samples, like the 1000 samples in Figures 3.11 and 3.12, is only an approximation to the sampling distribution. One of the uses of probability theory in statistics is to obtain sampling distributions without simulation. The interpretation of a sampling distribution is the same, however, whether we obtain it by simulation or by the mathematics of probability.

We can use the tools of data analysis to describe any distribution. Let's apply those tools to Figures 3.11 and 3.12.

- **Shape:** The histograms look Normal. Figure 3.13 is a Normal quantile plot of the values of $\hat{p}$ for our samples of size 100. It confirms that the distribution in

FIGURE 3.13 Normal quantile plot of the sample proportions in Figure 3.11. The distribution is close to Normal except for some clustering due to the fact that the sample proportions from a sample of size 100 can take only values that are multiples of 0.01. Because a plot of 1000 points is hard to read, this plot presents only every 10th value.

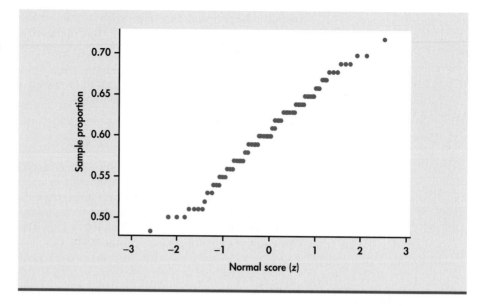

Figure 3.11 is close to Normal. The 1000 values for samples of size 2500 in Figure 3.12 are even closer to Normal. The Normal curves drawn through the histograms describe the overall shape quite well.

- **Center:** In both cases, the values of the sample proportion $\hat{p}$ vary from sample to sample, but the values are centered at 0.6. Recall that $p = 0.6$ is the true population parameter. Some samples have a $\hat{p}$ less than 0.6 and some greater, but there is no tendency to be always low or always high. That is, $\hat{p}$ has no *bias* as an estimator of p. This is true for both large and small samples. (Want the details? The mean of the 1000 values of $\hat{p}$ is 0.598 for samples of size 100 and 0.6002 for samples of size 2500. The median value of $\hat{p}$ is exactly 0.6 for samples of both sizes.)

- **Spread:** The values of $\hat{p}$ from samples of size 2500 are much less spread out than the values from samples of size 100. In fact, the standard deviations are 0.051 for Figure 3.11 and 0.0097, or about 0.01, for Figure 3.12.

Although these results describe just two sets of simulations, they reflect facts that are true whenever we use random sampling.

APPLY YOUR KNOWLEDGE

CASE 3.1 **3.86 Simulation using random digits.** You can use a table of random digits to simulate sampling from a population. Suppose that 60% of the population ate in a restaurant during the past week. That is, the population proportion is $p = 0.6$.

(a) Let each digit in the table stand for one person in this population. Digits 0 to 5 stand for people who ate in a restaurant last week, and 6 to 9 stand for people who did not. Why does looking at one digit from Table B simulate drawing one person at random from a population with $p = 0.6$?

(b) The first 100 entries in Table B contain 63 digits between 0 and 5, so $\hat{p} = 63/100 = 0.63$ for this sample. Why do the first 100 digits simulate drawing 100 people at random?

(c) Simulate a second SRS from this population, using 100 entries in Table B, starting at line 114. What is $\hat{p}$ for this sample? You see that the two sample results are different, and neither is equal to the true population value $p = 0.6$. That's the issue we face in this section.

3.87 Effect of sample size on the sampling distribution. You are planning a study and are considering taking an SRS of either 200 or 400 observations. Explain how the sampling distribution would differ for these two scenarios.

Bias and variability

Our simulations show that a sample of size 2500 will almost always give an estimate $\hat{p}$ that is close to the truth about the population. Figure 3.12 illustrates this fact for just one population, but it is true for any population. Samples of size 100, on the other hand, might give an estimate of 50% or 70% when the truth is 60%.

Thinking about Figures 3.11 and 3.12 helps us restate the idea of bias when we use a statistic like $\hat{p}$ to estimate a parameter like p. It also reminds us that variability matters as much as bias.

> **Bias and Variability**
>
> **Bias** concerns the center of the sampling distribution. A statistic used to estimate a parameter is **unbiased** if the mean of its sampling distribution is equal to the true value of the parameter being estimated.
>
> The **variability of a statistic** is described by the spread of its sampling distribution. This spread is determined by the sampling design and the sample size n. Statistics from larger samples have smaller spreads.

We can think of the true value of the population parameter as the bull's-eye on a target, and of the sample statistic as an arrow fired at the bull's-eye. Bias and variability describe what happens when an archer fires many arrows at the target. *Bias* means that the aim is off, and the arrows land consistently off the bull's-eye in the same direction. The sample values do not center about the population value. Large *variability* means that repeated shots are widely scattered on the target. Repeated samples do not give similar results but differ widely among themselves. Figure 3.14 shows this target illustration of the two types of error.

Notice that small variability (repeated shots are close together) can accompany large bias (the arrows are consistently away from the bull's-eye in one direction). And small bias (the arrows center on the bull's-eye) can accompany large variability (repeated shots are widely scattered). A good sampling scheme, like a good archer, must have both small bias and small variability. Here's how we do this.

> **Managing Bias and Variability**
>
> **To reduce bias,** use random sampling. When we start with a list of the entire population, simple random sampling produces **unbiased estimates**—the values of a statistic computed from an SRS neither consistently overestimate nor consistently underestimate the value of the population parameter.
>
> **To reduce the variability** of a statistic from an SRS, use a larger sample. You can make the variability as small as you want by taking a large enough sample.

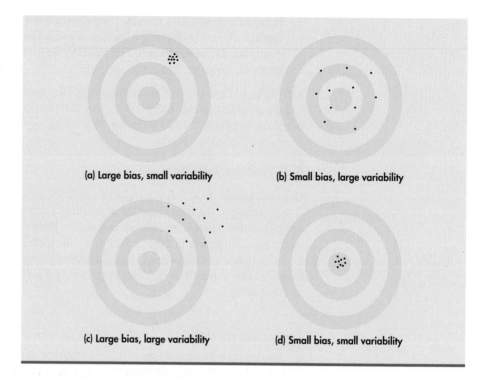

(a) Large bias, small variability   (b) Small bias, large variability

(c) Large bias, large variability   (d) Small bias, small variability

In practice, Gallup takes only one sample. We don't know how close to the truth an estimate from this one sample is, because we don't know what the truth about the population is. But *large random samples almost always give an estimate that is close to the truth.* Looking at the pattern of many samples shows that we can trust the result of one sample. The Current Population Survey's sample of 60,000 households estimates the national unemployment rate very accurately. Of course, only probability samples carry this guarantee. A voluntary response sample is worthless even if a very large number of people respond. Using a probability sampling design and taking care to deal with practical difficulties reduce bias in a sample. The size of the sample then determines how close to the population truth the sample result is likely to fall. Results from a sample survey usually come with a *margin of error* that sets bounds on the size of the likely error. How to do this is part of the detail of statistical inference. We will describe the reasoning in Chapter 6.

APPLY YOUR KNOWLEDGE

3.88 Inuit Survey. Inuit are the original inhabitants of the far north of what is now known as Canada. They live in an area known as Inuit Nunaat, which consists of four regions. These are the Inuvialt Region (in the Northwest Territories), Nunavut, Nunavik (north of the 55th parallel in the province of Quebec), and Nunatsiavut (in northern Labrador). A survey to determine the health status of Inuit collected data from 5000 adults. There are 50,485 Inuit living in the four regions of Inuit Nunaat.[35]
(a) What is the population for this sample survey? What is the sample?
(b) The survey found that 56% of Inuit had seen or talked on the phone with a medical doctor in the past 12 months. Do you think these estimates are close to the truth about the entire population? Why?

3.89 Ask more people. Just before a presidential election, a national opinion-polling firm increases the size of its weekly sample from the usual 1500 people to 4000 people. Why do you think the firm does this?

Sampling from large populations

Gallup's sample of 1009 adults is only about 1 out of every 228,000 adults in the United States. Does it matter whether we sample 1 in 100 individuals in the population or 1 in 228,000?

> **Population Size Doesn't Matter**
>
> The variability of a statistic from a random sample does not depend on the size of the population, as long as the population is at least 100 times larger than the sample.

Why does the size of the population have little influence on the behavior of statistics from random samples? To see why this is plausible, imagine sampling harvested corn by thrusting a scoop into a lot of corn kernels. The scoop doesn't know whether it is surrounded by a bag of corn or by an entire truckload. As long as the corn is well mixed (so that the scoop selects a random sample), the variability of the result depends only on the size of the scoop.

The fact that the variability of sample results is controlled by the size of the sample has important consequences for sampling design. An SRS of size 2500 from the 300 million residents of the United States gives results as precise as an SRS of size 2500 from the 765,000 inhabitants of San Francisco. This is good news for designers of national samples but bad news for those who want accurate information about the citizens of San Francisco. If both use an SRS, both must use the same size sample to obtain equally trustworthy results.

Why randomize?

Why randomize? The act of randomizing guarantees that our data are subject to the laws of probability. The behavior of statistics is described by a sampling distribution. The form of the distribution is known, and in many cases is approximately Normal. Often, the center of the distribution lies at the true parameter value, so the notion that randomization eliminates bias is made more precise. The spread of the distribution describes the variability of the statistic and can be made as small as we wish by choosing a large enough sample. Randomized experiments behave similarly: we can reduce variability by choosing larger groups of subjects for each treatment.

These facts are at the heart of formal statistical inference. Later chapters will have much to say in more technical language about sampling distributions and the way statistical conclusions are based on them. What any user of statistics must understand is that all the technical talk has its basis in a simple question: What would happen if the sample or the experiment were repeated many times? The reasoning applies not only to an SRS but also to the complex sampling designs actually used by opinion polls and other national sample surveys. The same conclusions hold as well for randomized experimental designs. The details vary with the design but the basic facts are true whenever randomization is used to produce data.

Remember that proper statistical design is not the only aspect of a good sample or experiment. The sampling distribution shows only how a statistic varies due to the operation of chance in randomization. It reveals nothing about possible bias due to undercoverage or nonresponse in a sample or to lack of realism in an experiment. The true distance of a statistic from the parameter it is estimating can be much larger than the sampling distribution suggests. What is worse, there is no way to say how large the added error is. The real world is less orderly than statistics textbooks imply.

SECTION 3.3 Summary

- A number that describes a population is a **parameter.** A number that can be computed from the data is a **statistic.** The purpose of sampling or experimentation is usually to use statistics to make statements about unknown parameters.

- A statistic from a probability sample or randomized experiment has a **sampling distribution** that describes how the statistic varies in repeated data production. The sampling distribution answers the question "What would happen if we repeated the sample or experiment many times?" Formal statistical inference is based on the sampling distributions of statistics.

- A statistic as an estimator of a parameter may suffer from **bias** or from high **variability.** Bias means that the center of the sampling distribution is not equal to the true value of the parameter. The variability of the statistic is described by the spread of its sampling distribution.

- Properly chosen statistics from randomized data production designs have no bias resulting from the way the sample is selected or the way the subjects are assigned to treatments. We can reduce the variability of the statistic by increasing the size of the sample or the size of the experimental groups.

SECTION 3.3 Exercises

For Exercises 3.84 and 3.85, see page 187; for 3.86 and 3.87, see pages 190–191; and for 3.88 and 3.89, see pages 192–193.

3.90 What's wrong? State what is wrong in each of the following scenarios.
(a) A sampling distribution describes the distribution of some characteristic in a population.
(b) A statistic will have a large amount of bias whenever it has high variability.
(c) The variability of a statistic based on a small sample from a population will be the same as the variability of a large sample from the same population.

3.91 Describe the population and the sample. For each of the following situations, describe the population and the sample.
(a) A survey of 17,096 students in U.S. four-year colleges reported that 19.4% were binge drinkers.
(b) In a study of work stress, 100 restaurant workers were asked about the impact of work stress on their personal lives.
(c) A tract of forest has 584 longleaf pine trees. The diameters of 40 of these trees were measured.

3.92 Gallup Canada polls. Gallup Canada bases its polls of Canadian public opinion on telephone samples of about 1000

adults, the same sample size as Gallup uses in the United States. Canada's population is about one-ninth as large as that of the United States, so the percent of adults that Gallup interviews in Canada is nine times as large as in the United States. Does this mean that the variability of estimates from a Gallup Canada poll is smaller? Explain your answer.

3.93 Real estate ownership. An agency of the federal government plans to take an SRS of residents in each state to estimate the proportion of owners of real estate in each state's population. The populations of the states range from less than 530,000 people in Wyoming to about 37 million in California.
(a) Will the variability of the sample proportion vary from state to state if an SRS of size 2000 is taken in each state? Explain your answer.
(b) Will the variability of the sample proportion change from state to state if an SRS of 1/10 of 1% (0.001) of the state's population is taken in each state? Explain your answer.

3.94 Simulate some coin tosses. The *Probability* applet simulates tossing a coin, with the advantage that you can choose the true long-term proportion, or probability, of a head. Example 3.30 discusses sampling from a population in which proportion $p = 0.6$ (the parameter) ate in a restaurant during the past week. Tossing a coin with probability $p = 0.6$ of a head simulates this situation: each head is a person who ate at a restaurant during the past week, and each tail is a person who did not. Set the "Probability of heads" in the applet to 0.6 and the number of tosses to 25. This simulates an SRS of size 25 from this population. By alternating between "Toss" and "Reset" you can take many samples quickly.

(a) Take 50 samples, recording the number of heads in each sample. Make a histogram of the 50 sample proportions (count of heads divided by 25). You are constructing the sampling distribution of this statistic.

(b) Another population contains only 20% who ate at a restaurant in the past week. Take 50 samples of size 25 from this population, record the number in each sample who ate at a restaurant in the past week, and make a histogram of the 50 sample proportions. How do the centers of your two histograms reflect the differing truths about the two populations?

3.95 Use statistical software for simulations. Statistical software can speed simulations. We are interested in the sampling distribution of the proportion $\hat{p}$ of people who ate at a restaurant during the past week in an SRS from a population in which proportion p ate at a restaurant during the past week. Here, p is a parameter and $\hat{p}$ is a statistic used to estimate p. We will see in Chapter 5 that "binomial" is the key word to look for in the software menus.

(a) Set $n = 50$ and $p = 0.6$ and generate 100 binomial observations. These are the counts for 100 SRSs of size 50 when 60% of the population ate at a restaurant in the past week. Save these counts and divide them by 50 to get values of $\hat{p}$ from 100 SRSs. Make a stemplot of the 100 values of $\hat{p}$.

(b) Repeat this process with $p = 0.3$, representing a population in which only 30% of people ate at a restaurant during the past week. Compare your two stemplots. How does changing the parameter p affect the center and spread of the sampling distribution?

(c) Now generate 100 binomial observations with $n = 200$ and $p = 0.6$. This simulates 100 SRSs, each of size 200. Obtain the 100 sample proportions $\hat{p}$ and make a stemplot. Compare this with your stemplot from (a). How does changing the sample size n affect the center and spread of the sampling distribution?

3.96 Use Table B for a simulation. We can construct a sampling distribution by hand in the case of a very small sample from a very small population. The population contains 10 students. Here are their scores on an exam:

Student:	0	1	2	3	4	5	6	7	8	9
Score:	82	62	80	58	72	73	65	66	74	62

The parameter of interest is the mean score, which is 69.4. The sample is an SRS of $n = 4$ students drawn from this population. The students are labeled 0 to 9 so that a single random digit from Table B chooses one student for the sample.

(a) Use Table B to draw an SRS of size 4 from this population. Write the four scores in your sample and calculate the mean $\bar{x}$ of the sample scores. This statistic is an estimate of the population parameter.

(b) Repeat this process 9 more times. Make a histogram of the 10 values of $\bar{x}$. You are constructing the sampling distribution of $\bar{x}$. Is the center of your histogram close to 69.4? (Ten repetitions give only a crude approximation to the sampling distribution. If possible, pool your work with that of other students—using different parts of Table B—to obtain several hundred repetitions and make a histogram of the values of $\bar{x}$. This histogram is a better approximation to the sampling distribution.)

3.97 Illustrate the idea of a sampling distribution. The *Simple Random Sample* applet can illustrate the idea of a sampling distribution. Form a population labeled 1 to 100. We will choose an SRS of 10 of these numbers. That is, in this exercise, the numbers themselves are the population, not just labels for 100 individuals. The mean of the whole numbers 1 to 100 is 50.5. This is the parameter, the mean of the population.

(a) Use the applet to choose an SRS of size 10. Which 10 numbers were chosen? What is their mean? This is a statistic, the sample mean $\bar{x}$.

(b) Although the population and its mean 50.5 remain fixed, the sample mean changes as we take more samples. Take another SRS of size 10. (Use the "Reset" button to return to the original population before taking the second sample.) What are the 10 numbers in your sample? What is their mean? This is another value of $\bar{x}$.

(c) Take 8 more SRSs from this same population and record their means. You now have 10 values of the sample mean $\bar{x}$ from 10 SRSs of the same size from the same population. Make a histogram of the 10 values and mark the population mean 50.5 on the horizontal axis. Are your 10 sample values roughly centered at the population value? (If you kept going forever, your $\bar{x}$-values would form the sampling distribution of the sample mean; the population mean would indeed be the center of this distribution.)

3.98 Sampling students. A management student is planning to take a survey of student attitudes toward part-time work while attending college. He develops a questionnaire and plans to ask 25 randomly selected students to fill it out. His faculty adviser approves the questionnaire but urges that the sample size be increased to at least 100 students. Why is the larger sample helpful?

3.99 Bias and variability. Figure 3.15 shows histograms of four sampling distributions of statistics intended to estimate the same parameter. Label each distribution relative to the others as high or low bias and as high or low variability.

3.100 Sampling in the states. The Internal Revenue Service plans to examine an SRS of individual federal income tax returns from each state. One variable of interest is the proportion of returns claiming itemized deductions. The total number of individual tax returns in a state varies from 14 million in California to 227,000 in Wyoming.

FIGURE 3.15 Determine which of these sampling distributions displays high or low bias and high or low variability, for Exercise 3.99.

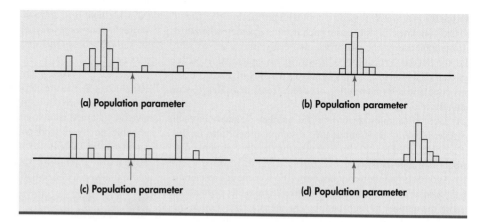

(a) Population parameter

(b) Population parameter

(c) Population parameter

(d) Population parameter

(a) Will the variability of the sample proportion vary from state to state if an SRS of size 2000 is taken in each state? Explain your answer.

(b) Will the variability of the sample proportion change from state to state if an SRS of 1/10 of 1% (0.001) of the state's population is taken in each state? Explain your answer.

3.101 Coin tossing. Coin tossing can illustrate the idea of a sampling distribution. The population is all outcomes (heads or tails) we would get if we tossed a coin forever. The parameter p is the proportion of heads in this population. We suspect that p is close to 0.5. That is, we think the coin will show about one-half heads in the long run. The sample is the outcomes of 20 tosses, and the statistic $\hat{p}$ is the proportion of heads in these 20 tosses.

(a) Toss a coin 20 times and record the value of $\hat{p}$.

(b) Repeat this sampling process 10 times. Make a stemplot of the 10 values of $\hat{p}$. Is the center of this distribution close to 0.5? (Ten repetitions give only a crude approximation to the sampling distribution. If possible, pool your work with that of other students to obtain at least 100 repetitions and make a histogram of the values of $\hat{p}$.)

3.102 Sampling invoices. We will illustrate the idea of a sampling distribution in the case of a very small sample from a very small population. The population contains 10 past-due invoices. Here are the number of days each invoice is past due:

Invoice:	0	1	2	3	4	5	6	7	8	9
Days past due:	8	12	10	5	7	3	15	9	7	6

The parameter of interest is the mean of this population, which is 8.2 days. The sample is an SRS of $n = 4$ invoices drawn from the population. Because the invoices are labeled 0 to 9, a single random digit from Table B chooses one invoice for the sample.

(a) Use Table B to draw an SRS of size 4 from this population. Write the past-due values for the days in your sample and calculate their mean $\bar{x}$. This statistic is an estimate of the population parameter.

(b) Repeat this process 10 times. Make a histogram of the 10 values of $\bar{x}$. You are constructing the sampling distribution of $\bar{x}$. Is the center of your histogram close to 8.2? (Ten repetitions give only a crude approximation to the sampling distribution. If possible, pool your work with that of other students—using different parts of Table B—to obtain at least 100 repetitions. A histogram of these values of $\bar{x}$ is a better approximation to the sampling distribution.)

3.103 A sampling applet experiment. The *Simple Random Sample* applet can animate the idea of a sampling distribution. Form a population labeled 1 to 100. We will choose an SRS of 10 of these numbers. That is, in this exercise the numbers themselves are the population, not just labels for 100 individuals. The proportion of the whole numbers 1 to 100 that are equal to or less than 60 is $p = 0.6$. This is the population proportion.

(a) Use the applet to choose an SRS of size 25. Which 25 numbers were chosen? Count the numbers ≤ 60 in your sample and divide this count by 25. This is the sample proportion $\hat{p}$.

(b) Although the population and the parameter $p = 0.6$ remain fixed, the sample proportion changes as we take more samples. Take another SRS of size 25. (Use the "Reset" button to return to the original population before taking the second sample.) What are the 25 numbers in your sample? What proportion of these is ≤ 60? This is another value of $\hat{p}$.

(c) Take 8 more SRSs from this same population and record their sample proportions. You now have 10 values of the statistic $\hat{p}$ from 10 SRSs of the same size from the same population. Make a histogram of the 10 values and mark the population parameter $p = 0.6$ on the horizontal axis. Are your 10 sample values roughly centered at the population value p? (If you kept going forever, your $\hat{p}$-values would form the sampling distribution of the sample proportion; the population proportion p would indeed be the center of this distribution.)

CASE 3.1 **3.104 A sampling experiment.** Figures 3.11 and 3.12 show how the sample proportion $\hat{p}$ behaves when we take many samples from a population in which the population

FIGURE 3.16 A population of 100 individuals for Exercise 3.104. Some individuals (white circles) ate in a restaurant last week, and the others did not.

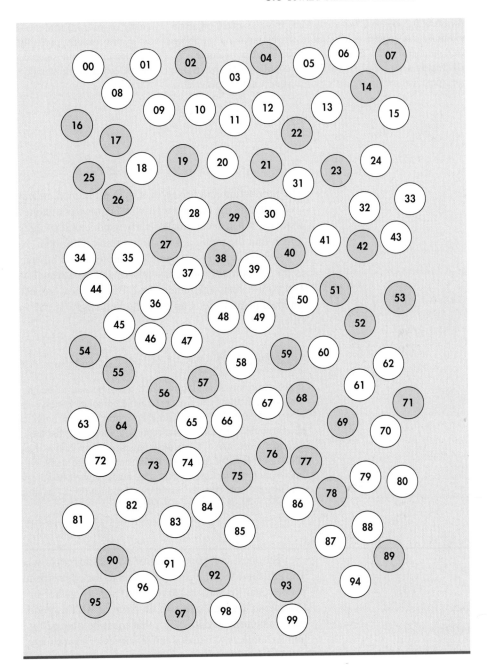

proportion is $p = 0.6$. You can follow the steps in this process on a small scale.

Figure 3.16 is a small population. Each circle represents an adult. The white circles are people who ate in a restaurant last week, and the colored circles are people who did not. You can check that 60 of the 100 circles are white, so in this population the proportion who ate in a restaurant last week is $p = 60/100 = 0.6$.

(a) The circles are labeled 00, 01, ..., 99. Use line 111 of Table B to draw an SRS of size 5. What is the proportion $\hat{p}$ of the people in your sample who ate in a restaurant last week?

(b) Take 9 more SRSs of size 5 (10 in all), using lines 122 to 130 of Table B, a different line for each sample. You now have 10 values of the sample proportion $\hat{p}$.

(c) Because your samples have only 5 people, the only values $\hat{p}$ can take are 0/5, 1/5, 2/5, 3/5, 4/5, and 5/5. That is, $\hat{p}$ is

always 0, 0.2, 0.4, 0.6, 0.8, or 1. Mark these numbers on a line and make a histogram of your 10 results by putting a bar above each number to show how many samples had that outcome.
(d) Taking samples of size 5 from a population of size 100 is not a practical setting, but let's look at your results anyway. How many of your 10 samples estimated the population proportion $p = 0.6$ exactly correctly? Is the true value 0.6 roughly in the center of your sample values? Explain why 0.6 would be in the center of the sample values if you took a large number of samples.

3.4 Commentary: Data Ethics*

The production and use of data, like all human endeavors, raise ethical questions. We won't discuss the telemarketer who begins a telephone sales pitch with "I'm conducting a survey." Such deception is clearly unethical. It enrages legitimate survey organizations, which find the public less willing to talk with them. Neither will we discuss those few researchers who, in the pursuit of professional advancement, publish fake data. There is no ethical question here—faking data to advance your career is just wrong. It will end your career when uncovered. But just how honest must researchers be about real, unfaked data? Here is an example that suggests the answer is "More honest than they often are."

EXAMPLE 3.32   Provide All the Critical Information

Papers reporting scientific research are supposed to be short, with no extra baggage. Brevity, however, can allow researchers to avoid complete honesty about their data. Did they choose their subjects in a biased way? Did they report data on only some of their subjects? Did they try several statistical analyses and report only the ones that looked best? The statistician John Bailar screened more than 4000 medical papers in more than a decade as consultant to the *New England Journal of Medicine.* He says, "When it came to the statistical review, it was often clear that critical information was lacking, and the gaps nearly always had the practical effect of making the authors' conclusions look stronger than they should have."[36] The situation is no doubt worse in fields that screen published work less carefully.

The most complex issues of data ethics arise when we collect data from people. The ethical difficulties are more severe for experiments that impose some treatment on people than for sample surveys that simply gather information. Trials of new medical treatments, for example, can do harm as well as good to their subjects. Here are some basic standards of data ethics that must be obeyed by any study that gathers data from human subjects, whether sample survey or experiment.

> **Basic Data Ethics**
>
> The organization that carries out the study must have an **institutional review board** that reviews all planned studies in advance in order to protect the subjects from possible harm.
>
> All individuals who are subjects in a study must give their **informed consent** before data are collected.
>
> All individual data must be kept **confidential.** Only statistical summaries for groups of subjects may be made public.

---

*This short essay concerns a very important topic, but the material is not needed to read the rest of the book.

The law requires that studies carried out or funded by the federal government obey these principles.[37] But neither the law nor the consensus of experts is completely clear about the details of their application.

Institutional review boards

The purpose of an institutional review board is not to decide whether a proposed study will produce valuable information or whether it is statistically sound. The board's purpose is, in the words of one university's board, "to protect the rights and welfare of human subjects (including patients) recruited to participate in research activities." The board reviews the plan of the study and can require changes. It reviews the consent form to ensure that subjects are informed about the nature of the study and about any potential risks. Once research begins, the board monitors the study's progress at least once a year.

The most pressing issue concerning institutional review boards is whether their workload has become so large that their effectiveness in protecting subjects drops. When the government temporarily stopped human-subject research at Duke University Medical Center in 1999 due to inadequate protection of subjects, more than 2000 studies were going on. That's a lot of review work. There are shorter review procedures for projects that involve only minimal risks to subjects, such as most sample surveys. When a board is overloaded, there is a temptation to put more proposals in the minimal-risk category to speed the work.

APPLY YOUR KNOWLEDGE

The exercises in this section on ethics are designed to help you think about the issues that we are discussing and to formulate some opinions. In general, there are no wrong or right answers but you need to give reasons for your answers.

3.105 Do these proposals involve minimal risk? You are a member of your college's institutional review board. You must decide whether several research proposals qualify for lighter review because they involve only minimal risk to subjects. Federal regulations say that "minimal risk" means the risks are no greater than "those ordinarily encountered in daily life or during the performance of routine physical or psychological examinations or tests." That's vague. Which of these do you think qualifies as "minimal risk"?
(a) Draw a drop of blood by pricking a finger in order to measure blood sugar.
(b) Draw blood from the arm for a full set of blood tests.
(c) Insert a tube that remains in the arm, so that blood can be drawn regularly.

3.106 Who should be on an institutional review board? Government regulations require that institutional review boards consist of at least five people, including at least one scientist, one nonscientist, and one person from outside the institution. Most boards are larger, but many contain just one outsider.
(a) Why should review boards contain people who are not scientists?
(b) Do you think that one outside member is enough? How would you choose that member? (For example, would you prefer a medical doctor? A member of the clergy? An activist for patients' rights?)

Informed consent

Both words in the phrase "informed consent" are important, and both can be controversial. Subjects must be *informed* in advance about the nature of a study and any risk of harm it may bring. In the case of a sample survey, physical harm is not possible. The subjects should be told what kinds of questions the survey will ask and about how much of their time it will take. Experimenters must tell subjects the nature and purpose of the study and outline possible risks. Subjects must then *consent* in writing.

EXAMPLE 3.33 Who Can Give Informed Consent?

Are there some subjects who can't give informed consent? It was once common, for example, to test new vaccines on prison inmates who gave their consent in return for good-behavior credit. Now we worry that prisoners are not really free to refuse, and the law forbids almost all medical research in prisons.

Children can't give fully informed consent, so the usual procedure is to ask their parents. A study of new ways to teach reading is about to start at a local elementary school, so the study team sends consent forms home to parents. Many parents don't return the forms. Can their children take part in the study because the parents did not say "No," or should we allow only children whose parents returned the form and said "Yes"?

What about research into new medical treatments for people with mental disorders? What about studies of new ways to help emergency room patients who may be unconscious? In most cases, there is not time to get the consent of the family. Does the principle of informed consent bar realistic trials of new treatments for unconscious patients?

These are questions without clear answers. Reasonable people differ strongly on all of them. There is nothing simple about informed consent.[38]

The difficulties of informed consent do not vanish even for capable subjects. Some researchers, especially in medical trials, regard consent as a barrier to getting patients to participate in research. They may not explain all possible risks; they may not point out that there are other therapies that might be better than those being studied; they may be too optimistic in talking with patients even when the consent form has all the right details. On the other hand, mentioning every possible risk leads to very long consent forms that really are barriers. "They are like rental car contracts," one lawyer said. Some subjects don't read forms that run five or six printed pages. Others are frightened by the large number of possible (but unlikely) disasters that might happen and so refuse to participate. Of course, unlikely disasters sometimes happen. When they do, lawsuits follow and the consent forms become yet longer and more detailed.

Confidentiality

Ethical problems do not disappear once a study has been cleared by the review board, has obtained consent from its subjects, and has actually collected data about the subjects. It is important to protect the subjects' privacy by keeping all data about individuals confidential. The report of an opinion poll may say what percent of the 1200 respondents felt that legal immigration should be reduced. It may not report what *you* said about this or any other issue.

anonymity   Confidentiality is not the same as **anonymity.** Anonymity means that subjects are anonymous—their names are not known even to the director of the study. Anonymity is rare in statistical studies. Even where it is possible (mainly in surveys conducted by mail), anonymity prevents any follow-up to improve nonresponse or inform subjects of results.

Any breach of confidentiality is a serious violation of data ethics. The best practice is to separate the identity of the subjects from the rest of the data at once. Sample surveys, for example, use the identification only to check on who did or did not respond. In an era of advanced technology, however, it is no longer enough to be sure that each individual set of data protects people's privacy. The government, for example, maintains a vast amount of information about citizens in many separate data bases—census responses, tax returns, Social Security information, data from surveys such as the Current Population Survey, and so on. Many of these data bases can be searched by computers for statistical studies. A clever computer search of several data bases might be able, by combining information, to identify you and learn a great deal about you even if your name and other identification have been removed from the data available for search. A colleague from Germany once remarked that "female full professor of statistics with PhD from the United States" was enough to identify her among all the 83 million residents of Germany. Privacy and confidentiality of data are hot issues among statisticians in the computer age.

> **EXAMPLE 3.34**   Data Collected by the Government

Citizens are required to give information to the government. Think of tax returns and Social Security contributions. The government needs these data for administrative purposes—to see if we paid the right amount of tax and how large a Social Security benefit we are owed when we retire. Some people feel that individuals should be able to forbid any other use of their data, even with all identification removed. This would prevent using government records to study, say, the ages, incomes, and household sizes of Social Security recipients. Such a study could well be vital to debates on reforming Social Security.

> **APPLY YOUR KNOWLEDGE**
>
> **3.107 How can we obtain informed consent?** A researcher suspects that traditional religious beliefs tend to be associated with an authoritarian personality. She prepares a questionnaire that measures authoritarian tendencies and also asks many religious questions. Write a description of the purpose of this research to be read by subjects in order to obtain their informed consent. You must balance the conflicting goals of not deceiving the subjects as to what the questionnaire will tell about them and of not biasing the sample by scaring off religious people.
>
> **3.108 Should we allow this personal information to be collected?** In which of the circumstances below would you allow collecting personal information without the subjects' consent?
>
> (a) A government agency takes a random sample of income tax returns to obtain information on the average income of people in different occupations. Only the incomes and occupations are recorded from the returns, not the names.
>
> (b) A social psychologist attends public meetings of a religious group to study the behavior patterns of members.
>
> (c) A social psychologist pretends to be converted to membership in a religious group and attends private meetings to study the behavior patterns of members.

Clinical trials

Clinical trials are experiments that study the effectiveness of medical treatments on actual patients. Medical treatments can harm as well as heal, so clinical trials spotlight the ethical problems of experiments with human subjects. Here are the starting points for a discussion:

- Randomized comparative experiments are the only way to see the true effects of new treatments. Without them, risky treatments that are no more effective than placebos will become common.

- Clinical trials produce great benefits, but most of these benefits go to future patients. The trials also pose risks, and these risks are borne by the subjects of the trial. So we must balance future benefits against present risks.

- Both medical ethics and international human rights standards say that "the interests of the subject must always prevail over the interests of science and society."

The quoted words are from the 1964 Helsinki Declaration of the World Medical Association, the most respected international standard. The most outrageous examples of unethical experiments are those that ignore the interests of the subjects.

EXAMPLE 3.35   The Tuskegee Study

In the 1930s, syphilis was common among black men in the rural South, a group that had almost no access to medical care. The Public Health Service Tuskegee study recruited 399 poor black sharecroppers with syphilis and 201 others without the disease in order to observe how syphilis progressed when no treatment was given. Beginning in 1943, penicillin became available to treat syphilis. The study subjects were not treated. In fact, the Public Health Service prevented any treatment until word leaked out and forced an end to the study in the 1970s.

The Tuskegee study is an extreme example of investigators following their own interests and ignoring the well-being of their subjects. A 1996 review said, "It has come to symbolize racism in medicine, ethical misconduct in human research, paternalism by physicians, and government abuse of vulnerable people." In 1997, President Clinton formally apologized to the surviving participants in a White House ceremony.[39]

Because "the interests of the subject must always prevail," medical treatments can be tested in clinical trials only when there is reason to hope that they will help the patients who are subjects in the trials. Future benefits aren't enough to justify experiments with human subjects. Of course, if there is already strong evidence that a treatment works and is safe, it is unethical *not* to give it. Here are the words of Dr. Charles Hennekens of the Harvard Medical School, who directed the large clinical trial that showed that aspirin reduces the risk of heart attacks:

> *There's a delicate balance between when to do or not do a randomized trial. On the one hand, there must be sufficient belief in the agent's potential to justify exposing half the subjects to it. On the other hand, there must be sufficient doubt about its efficacy to justify withholding it from the other half of subjects who might be assigned to placebos.*[40]

Why is it ethical to give a control group of patients a placebo? Well, we know that placebos often work. Moreover, placebos have no harmful side effects. So in the state of balanced doubt described by Dr. Hennekens, the placebo group may be getting a better treatment than the drug group. If we *knew* which treatment was better, we would give it to everyone. When we don't know, it is ethical to try both and compare them.

The idea of using a control or a placebo is a fundamental principle to be considered in designing experiments. In many situations deciding what to use as an appropriate control requires some careful thought. The choice of the control can have a substantial impact on the conclusions drawn from an experiment. Here is an example.

EXAMPLE 3.36   Was the Claim Misleading?

The manufacturer of a breakfast cereal designed for children claims that eating this cereal has been clinically shown to improve attentiveness by nearly 20%. The study used two groups of children who were tested before and after breakfast. One group received the cereal for breakfast, while breakfast for the control group was water. The results of tests taken three hours after breakfast were used to make the claim.

The Federal Trade Commission investigated the marketing of this product. They charged that the claim was false and violated federal law. The charges were settled, and the company agreed to not use misleading claims in their advertising.[41]

It is not sufficient to obtain appropriate controls. The data from all groups must be collected and analyzed in the same way. Here is an example of this type of flawed design.

EXAMPLE 3.37   The Product Doesn't Work!

Two scientists published a paper claiming to have developed a very exciting new method to detect ovarian cancer using blood samples. The potential market for such a procedure is substantial, and there is no specific screening test currently available. When other scientists were unable to reproduce the results in different labs, the original work was examined more carefully. The original study used blood samples from women with ovarian cancer and from healthy controls. The blood samples were all analyzed using a mass spectrometer. The control samples were analyzed on one day and the cancer samples were analyzed on the next day. This design was flawed because it could not control for changes over time in the measuring instrument.[42]

APPLY YOUR KNOWLEDGE

3.109 Is this study ethical? Researchers on aging proposed to investigate the effect of supplemental health services on the quality of life of older people. Eligible patients on the rolls of a large medical clinic were to be randomly assigned to treatment and control groups. The treatment group would be offered hearing aids, dentures, transportation, and other services not available without charge to the control group. The review board felt that providing these services to some but not other persons in the same institution raised ethical questions. Do you agree?

3.110 Should the treatments be given to everyone? Effective drugs for treating AIDS are very expensive, so most African nations cannot afford to give them to large numbers of people. Yet AIDS is more common in parts of Africa than anywhere else. Several clinical trials being conducted in Africa are looking at ways to prevent pregnant mothers infected with HIV from passing the infection to their unborn children, a major source of HIV infections in Africa. Some people say these trials are unethical because they do not give effective AIDS drugs to their subjects, as would be required in rich nations. Others reply that the trials are looking for treatments that can work in the real world in Africa and that they promise benefits at least to the children of their subjects. What do you think?

Behavioral and social science experiments

When we move from medicine to the behavioral and social sciences, the direct risks to experimental subjects are less acute, but so are the possible benefits to the subjects. Consider, for example, the experiments conducted by psychologists in their study of human behavior.

EXAMPLE 3.38   Personal Space

Psychologists observe that people have a "personal space" and are uneasy if others come too close to them. We don't like strangers to sit at our table in a coffee shop if other tables are available, and we see people move apart in elevators if there is room to do so. Americans tend to require more personal space than people in most other cultures. Can violations of personal space have physical, as well as emotional, effects?

Investigators set up shop in a men's public restroom. They blocked off urinals to force men walking in to use either a urinal next to an experimenter (treatment group) or a urinal separated from the experimenter (control group). Another experimenter, using a periscope from a toilet stall, measured how long the subject took to start urinating and how long he continued.[43]

This personal space experiment illustrates the difficulties facing those who plan and review behavioral studies.

- There is no risk of harm to the subjects, although they would certainly object to being watched through a periscope. What should we protect subjects from when physical harm is unlikely? Possible emotional harm? Undignified situations? Invasion of privacy?

- What about informed consent? The subjects did not even know they were participating in an experiment. Many behavioral experiments rely on hiding the true purpose of the study. The subjects would change their behavior if told in advance what the investigators were looking for. Subjects are asked to consent on the basis of vague information. They receive full information only after the experiment.

The "Ethical Principles" of the American Psychological Association require consent unless a study merely observes behavior in a public place. They allow deception only when it is necessary to the study, does not hide information that might influence a subject's willingness to participate, and is explained to subjects as soon as possible. The personal space study (from the 1970s) does not meet current ethical standards.

We see that the basic requirement for informed consent is understood differently in medicine and psychology. Here is an example of another setting with yet another interpretation of what is ethical. The subjects get no information and give no consent. They don't even know that an experiment may be sending them to jail for the night.

EXAMPLE 3.39   Reducing Domestic Violence

How should police respond to domestic-violence calls? In the past, the usual practice was to remove the offender and order him to stay out of the household overnight. Police were reluctant to make arrests because the victims rarely pressed charges. Women's groups argued that arresting offenders would help prevent future violence even if no charges were filed. Is there evidence that arrest will reduce future offenses? That's a question that experiments have tried to answer.

A typical domestic-violence experiment compares two treatments: arrest the suspect and hold him overnight, or warn the suspect and release him. When police officers reach the scene of a domestic-violence call, they calm the participants and investigate. Weapons or death threats require an arrest. If the facts permit an arrest but do not require it, an officer radios headquarters for instructions. The person on duty opens the next envelope in a file prepared in advance by a statistician. The envelopes contain the treatments in random order. The police either arrest the suspect or warn and release him, depending on the contents of the envelope. The researchers then watch police records and visit the victim to see if the domestic violence reoccurs.

Such experiments show that arresting domestic-violence suspects does reduce their future violent behavior.[44] As a result of this evidence, arrest has become the common police response to domestic violence.

The domestic-violence experiments shed light on an important issue of public policy. Because there is no informed consent, the ethical rules that govern clinical trials and most social science studies would forbid these experiments. They were cleared by review boards because, in the words of one domestic-violence researcher, "These people became subjects by committing acts that allow the police to arrest them. You don't need consent to arrest someone."

SECTION 3.4 Summary

- The purpose of an **institutional review board** is to protect the rights and welfare of the human subjects in a study. Institutional review boards review **informed consent** forms that subjects will sign before participating in a study.

- Information about individuals in a study must be kept **confidential,** but statistical summaries of groups of individuals may be made public.

- **Clinical trials** are experiments that study the effectiveness of medical treatments on actual patients.

- Some studies in the **behavioral** and **social sciences** are observational while others are designed experiments.

SECTION 3.4 Exercises

For Exercises 3.105 and 3.106, see page 199; for 3.107 and 3.108, see page 201; and for 3.109 and 3.110, see page 203.

Most of these exercises pose issues for discussion. There are no right or wrong answers, but there are more and less thoughtful answers.

3.111 What is wrong? Explain what is wrong in each of the following scenarios.

(a) Clinical trials are always ethical as long as they randomly assign patients to the treatments.

(b) The job of an institutional review board is complete when they decide to allow a study to be conducted.

(c) A treatment that has no risk of physical harm to subjects is always ethical.

3.112 How should the samples have been analyzed? Refer to the ovarian cancer diagnostic test study in Example 3.37 (page 203). Describe how you would process the samples through the mass spectrometer.

3.113 The Vytorin controversy. Vytorin is a combination pill designed to lower cholesterol. It consists of a relatively inexpensive and widely used drug, Zocor, and a newer drug called Zetia. Early study results suggested that Vytorin was no more effective than Zetia. Critics claimed that the makers of the drugs tried to change the response variable for the study, and two congressional panels investigated why there was a two-year delay in the release of the results. Use the Web to search for more information about this controversy, and write a report of what you find. Include an evaluation in the framework of ethical use of experiments and data. A good place to start your search would be to look for the phrase "Vytorin's Shortcomings."

3.114 Facebook and academic performance. *First Monday* is a peer-reviewed journal on the Internet. They recently published two articles concerning Facebook and academic performance. Visit their Web site, firstmonday.org, and look at the first three articles in Volume 14, Number 5–4, May 2009. Identify the key controversial issues that involve the use of statistics in these articles, and write a report summarizing the facts as you see them. Be sure to include your opinions regarding ethical issues related to this work.

3.115 Anonymity and confidentiality in mail surveys. Some common practices may appear to offer anonymity while actually delivering only confidentiality. Market researchers often use mail surveys that do not ask the respondent's identity but contain hidden codes on the questionnaire that identify the respondent. A false claim of anonymity is clearly unethical. If only confidentiality is promised, is it also unethical to say nothing about the identifying code, perhaps causing respondents to believe their replies are anonymous?

3.116 Studying your blood. Long ago, doctors drew a blood specimen from you as part of treating minor anemia. Unknown to you, the sample was stored. Now researchers plan to use stored samples from you and many other people to look for genetic factors that may influence anemia. It is no longer possible to ask your consent. Modern technology can read your entire genetic makeup from the blood sample.
(a) Do you think it violates the principle of informed consent to use your blood sample if your name is on it but you were not told that it might be saved and studied later?
(b) Suppose that your identity is not attached. The blood sample is known only to come from (say) "a 20-year-old white female being treated for anemia." Is it now OK to use the sample for research?
(c) Perhaps we should use biological materials such as blood samples only from patients who have agreed to allow the material to be stored for later use in research. It isn't possible to say in advance what kind of research, so this falls short of the usual standard for informed consent. Is it nonetheless acceptable, given complete confidentiality and the fact that using the sample can't physically harm the patient?

3.117 Anonymous? Confidential? One of the most important nongovernment surveys in the United States is the National Opinion Research Center's General Social Survey. The GSS regularly monitors public opinion on a wide variety of political and social issues. Interviews are conducted in person in the subject's home. Are a subject's responses to GSS questions anonymous, confidential, or both? Explain your answer.

3.118 Anonymous? Confidential? Texas A&M, like many universities, offers free screening for HIV, the virus that causes AIDS. The announcement says, "Persons who sign up for the HIV Screening will be assigned a number so that they do not have to give their name." They can learn the results of the test by telephone, still without giving their name. Does this practice offer *anonymity* or just *confidentiality?*

3.119 Political polls. The presidential election campaign is in full swing, and the candidates have hired polling organizations to take sample surveys to find out what the voters think about the issues. What information should the pollsters be required to give out?

(a) What does the standard of informed consent require the pollsters to tell potential respondents?
(b) The standards accepted by polling organizations also require giving respondents the name and address of the organization that carries out the poll. Why do you think this is required?
(c) The polling organization usually has a professional name such as "Samples Incorporated," so respondents don't know that the poll is being paid for by a political party or candidate. Would revealing the sponsor to respondents bias the poll? Should the sponsor always be announced whenever poll results are made public?

3.120 Making poll results public. Some people think that the law should require that all political poll results be made public. Otherwise, the possessors of poll results can use the information to their own advantage. They can act on the information, release only selected parts of it, or time the release for best effect. A candidate's organization replies that they are paying for the poll in order to gain information for their own use, not to amuse the public. Do you favor requiring complete disclosure of political poll results? What about other private surveys, such as market research surveys of consumer tastes?

3.121 Student subjects. Students taking Psychology 001 are required to serve as experimental subjects. Students in Psychology 002 are not required to serve, but they are given extra credit if they do so. Students in Psychology 003 are required either to sign up as subjects or to write a term paper. Serving as an experimental subject may be educational, but current ethical standards frown on using "dependent subjects" such as prisoners or charity medical patients. Students are certainly somewhat dependent on their teachers. Do you object to any of these course policies? If so, which ones, and why?

3.122 How many have HIV? Researchers from Yale, working with medical teams in Tanzania, wanted to know how common infection with HIV, the virus that causes AIDS, is among pregnant women in that African country. To do this, they planned to test blood samples drawn from pregnant women.

Yale's institutional review board insisted that the researchers get the informed consent of each woman and tell her the results of the test. This is the usual procedure in developed nations. The Tanzanian government did not want to tell the women why blood was drawn or tell them the test results. The government feared panic if many people turned out to have an incurable disease for which the country's medical system could not provide care. The study was canceled. Do you think that Yale was right to apply its usual standards for protecting subjects?

3.123 AIDS trials in Africa. One of the most important goals of AIDS research is to find a vaccine that will protect against HIV infection. Because AIDS is so common in parts of Africa, that is the easiest place to test a vaccine. It is likely, however, that a vaccine would be so expensive that it could not (at least at first) be widely used in Africa. Is it ethical to test in Africa if the benefits go mainly to rich countries? The treatment group of subjects would get the vaccine, and the placebo group would later be given the vaccine if it proved effective. So the actual subjects would benefit—it is the future benefits that would go elsewhere. What do you think?

3.124 Asking teens about sex. The Centers for Disease Control and Prevention, in a survey of teenagers, asked the subjects if they were sexually active. Those who said "Yes" were then asked, "How old were you when you had sexual intercourse for the first time?" Should consent of parents be required to ask minors about sex, drugs, and other such issues, or is consent of the minors themselves enough? Give reasons for your opinion.

3.125 Deceiving subjects. Students sign up to be subjects in a psychology experiment. When they arrive, they are told that interviews are running late and are taken to a waiting room. The experimenters then stage a theft of a valuable object left in the waiting room. Some subjects are alone with the thief, and others are in pairs—these are the treatments being compared. Will the subject report the theft?

The students had agreed to take part in an unspecified study, and the true nature of the experiment is explained to them afterward. Do you think this study is ethically OK?

3.126 Deceiving subjects. A psychologist conducts the following experiment: she measures the attitude of subjects toward cheating, then has them play a game rigged so that winning without cheating is impossible. The computer that organizes the game also records—unknown to the subjects—whether or not they cheat. Then attitude toward cheating is retested.

Subjects who cheat tend to change their attitudes to find cheating more acceptable. Those who resist the temptation to cheat tend to condemn cheating more strongly on the second test of attitude. These results confirm the psychologist's theory.

This experiment tempts subjects to cheat. The subjects are led to believe that they can cheat secretly when in fact they are observed. Is this experiment ethically objectionable? Explain your position.

STATISTICS IN SUMMARY

Designs for producing data are essential parts of statistics in practice. The Statistics in Summary figure below displays the big ideas visually. Random sampling and randomized comparative experiments are perhaps the most important statistical inventions of the twentieth century. Both were slow to gain acceptance, and you will still see many voluntary response samples and uncontrolled experiments. This chapter has explained good techniques for producing data and has also explained why bad techniques often produce worthless data.

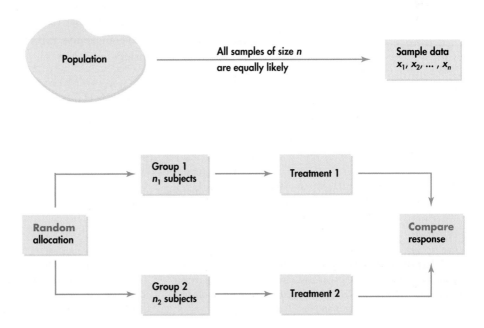

The deliberate use of chance in producing data is a central idea in statistics. It allows use of the laws of probability to analyze data, as we will see in the following chapters. Here is a review list of the most important skills you should have acquired from your study of this chapter.

A. Sampling

1. Identify the population in a sampling situation.

2. Recognize bias due to voluntary response samples and other inferior sampling methods.

3. Use software or Table B of random digits to select a simple random sample (SRS) from a population.

4. Recognize the presence of undercoverage and nonresponse as sources of error in a sample survey. Recognize the effect of the wording of questions on the responses.

5. Use random digits to select a stratified random sample from a population when the strata are identified.

B. Experiments

1. Recognize whether a study is an observational study or an experiment.

2. Recognize bias due to confounding of explanatory variables with lurking variables in either an observational study or an experiment.

3. Identify the factors (explanatory variables), treatments, response variables, and experimental units in an experiment.

4. Outline the design of a completely randomized experiment using a diagram like those in Figures 3.6 and 3.7. The diagram in a specific case should show the sizes of the groups, the specific treatments, and the response variable.

5. Use Table B of random digits to carry out the random assignment of subjects to groups in a completely randomized experiment.

6. Recognize the placebo effect. Recognize when the double-blind technique should be used. Be aware of weaknesses in an experiment, especially lack of realism that makes it hard to generalize conclusions.

7. Explain why a randomized comparative experiment can give good evidence for cause-and-effect relationships.

C. Toward Inference

1. Identify parameters and statistics in a sample or experiment.

2. Recognize the fact of sampling variability: a statistic will take different values when you repeat a sample or experiment.

3. Interpret a sampling distribution as describing the values taken by a statistic in all possible repetitions of a sample or experiment under the same conditions.

4. Understand the effect of sample size on the variability of sample statistics.

CHAPTER 3   Review Exercises

3.127 Experiments and surveys for business. Write a short report describing the differences and similarities between experiments and surveys that would be used in business. Include a discussion of the advantages and disadvantages of each.

3.128 Online behavioral advertising (optional). The Federal Trade Commission Staff Report "Self-Regulatory Principles for Online Behavioral Advertising" defines behavioral advertising as "the tracking of a consumer's online activities over time—including the searches the consumer has conducted, the web pages visited and the content viewed—in order to deliver advertising targeted to the individual consumer's interests." The report suggests four governing concepts for their proposals:

1. Transparency and control: when companies collect information from consumers for advertising, they should tell the consumers about how the data will be collected and consumers should be given a choice about whether to allow the data to be collected.

2. Security and data retention: data should be kept secure and should be retained only as long as needed.

3. Privacy: before data are used in a way that differs from how companies originally said they would use the information, companies should obtain consent from consumers.

4. Sensitive data: consent should be obtained before using any sensitive data.[45]

Write a report discussing your opinions concerning online behavioral advertising and the four governing concepts. Pay particular attention to issues related to the ethical collection and use of statistical data.

3.129 Confidentiality at NORC (optional). The National Opinion Research Center conducts a large number of surveys and has established procedures for protecting the confidentiality of their survey participants. For their Survey of Consumer Finances, they provide a pledge to participants regarding confidentiality. This pledge is available at norc.org/projects/scf/ Confidentiality.html. Review the pledge and summarize its key parts. Do you think that the pledge adequately addresses issues related to the ethical collection and use of data? Explain your answer.

3.130 What's wrong? Explain what is wrong in each of the following statements. Give reasons for your answers.

(a) A simple random sample was used to assign a group of 30 subjects to three treatments.

(b) It is better to use a table of random numbers to select a simple sample than it is to use a computer.

(c) Matched pairs designs and block designs are complicated and should be avoided if possible.

3.131 Price promotions and consumer expectations. A researcher studying the effect of price promotions on consumer expectations makes up two different histories of the store price of a hypothetical brand of laundry detergent for the past year. Students in a marketing course view one or the other price history on a computer. Some students see a steady price, while others see regular promotions that temporarily cut the price. Then the students are asked what price they would expect to pay for the detergent. Is this study an experiment? Why? What are the explanatory and response variables?

3.132 What type of study? What is the best way to answer each of the questions below: an experiment, a sample survey, or an observational study that is not a sample survey? Explain your choices.

(a) Are people generally satisfied with the service they receive from a customer call center?

(b) Do new employees learn basic facts about your company better in a workshop or using an online set of materials?

(c) How long do your customers have to wait to resolve a problem with a new purchase?

3.133 Choose the type of study. Give an example of a question about your customers, their behavior, or their opinions that would best be answered by

(a) a sample survey.

(b) an observational study that is not a sample survey.

(c) an experiment.

3.134 Compare the fries. Do consumers prefer the fries from Burger King or from McDonald's? Design a blind test in which neither source of the fries is identified. Describe briefly the design of a matched pairs experiment to investigate this question. How will you use randomization?

3.135 Coupons and customer expectations. A researcher studying the effect of coupons on consumers' expectations makes up two different series of ads for a hypothetical brand of cola for the past year. Students in a family science course view one or the other sequence of ads on a computer. Some students see a sequence of ads with no coupon offered on the cola, while others see regular coupon offerings that effectively lower the price of the cola temporarily. Next, the students are asked what price they would expect to pay for the cola.

(a) Is this study an experiment? Why?

(b) What are the explanatory and response variables?

3.136 Can you remember how many? An opinion poll calls 1800 randomly chosen residential telephone numbers, then asks to speak with an adult member of the household. The interviewer asks, "How many movies have you watched in a movie theater in the past 12 months?"

(a) What population do you think the poll has in mind?

(b) In all, 1231 people respond. What is the rate (percent) of nonresponse?

(c) What source of response error is likely for the question asked?

(d) Write a variation on this question that would be likely to lessen the associated response error.

3.137 Marketing a dietary supplement. Your company produces a dietary supplement that contains a significant amount of calcium as one of its ingredients. The company would like to be able to market this fact successfully to one of the target groups for the supplement: men with high blood pressure. To this end, you must design an experiment to demonstrate that added calcium in the diet reduces blood pressure. You have available 40 men with high blood pressure who are willing to serve as subjects.
CALCIUMBP

(a) Outline an appropriate design for the experiment, taking the placebo effect into account.

(b) The names of the subjects appear below. Do the randomization required by your design, and list the subjects to whom you will give the drug. (If you use Table B, enter the table at line 131.)

Alomar	Denman	Han	Liang	Rosen
Asihiro	Durr	Howard	Maldonado	Solomon
Bennett	Edwards	Hruska	Marsden	Tompkins
Bikalis	Farouk	Imrani	Moore	Townsend
Chen	Fratianna	James	O'Brian	Tullock
Clemente	George	Kaplan	Ogle	Underwood
Cranston	Green	Krushchev	Plochman	Willis
Curtis	Guillen	Lawless	Rodriguez	Zhang

3.138 A hot fund. A large mutual funds group assigns a young securities analyst to manage its small biotechnology stock fund. The fund's share value increases an impressive 43% during the first year under the new manager. Explain why this performance does not necessarily establish the manager's ability.

3.139 Employee meditation. You see a news report of an experiment that claims to show that a meditation technique increased job satisfaction of employees. The experimenter interviewed the employees and assessed their levels of job satisfaction. The subjects then learned how to meditate and did so regularly for a month. The experimenter reinterviewed them at the end of the month and assessed their job satisfaction levels again.
(a) There was no control group in this experiment. Why is this a blunder? What lurking variables might be confounded with the effect of meditation?
(b) The experimenter who diagnosed the effect of the treatment knew that the subjects had been meditating. Explain how this knowledge could bias the experimental conclusions.
(c) Briefly discuss a proper experimental design, with controls and blind diagnosis, to assess the effect of meditation on job satisfaction.

3.140 Executives and exercise. A study of the relationship between physical fitness and leadership uses as subjects middle-aged executives who have volunteered for an exercise program. The executives are divided into a low-fitness group and a high-fitness group on the basis of a physical examination. All subjects then take a psychological test designed to measure leadership, and the results for the two groups are compared. Is this an observational study or an experiment? Explain your answer.

3.141 Does the new product taste better? Before a new variety of frozen muffins is put on the market, it is subjected to extensive taste testing. People are asked to taste the new muffin and a competing brand and to say which they prefer. (Both muffins are unidentified in the test.) Is this an observational study or an experiment? Why?

3.142 Questions about attitudes. Write two questions about an attitude that concerns you for use in a sample survey. Make the first question so that it is biased in one direction and make the second question biased in the opposite direction. Explain why your questions are biased and then write a third question that has little or no bias.

3.143 Will regulation make the product safer? Canada requires that cars be equipped with "daytime running lights," head-

lights that automatically come on at a low level when the car is started. Some manufacturers are now equipping cars sold in the United States with running lights. Will running lights reduce accidents by making cars more visible?
(a) Briefly discuss the design of an experiment to help answer this question. In particular, what response variables will you examine?
(b) Example 3.27 (page 180) discusses center brake lights. What cautions do you draw from that example that apply to an experiment on the effects of running lights?

3.144 Learning about markets. Your economics professor wonders if playing market games online will help students understand how markets set prices. You suggest an experiment: have some students use the online games, while others discuss markets in recitation sections. The course has two lectures, at 8:30 A.M. and 2:30 P.M. There are 10 recitation sections attached to each lecture. The students are already assigned to recitations. For practical reasons, all students in each recitation must follow the same program.
(a) The professor says, "Let's just have the 8:30 group do online work in recitation and the 2:30 group do discussion." Why is this a bad idea?
(b) Outline the design of an experiment with the 20 recitation sections as individuals. Carry out your randomization and include in your outline the recitation numbers assigned to each treatment.

3.145 How much do students earn? A university's financial aid office wants to know how much it can expect students to earn from summer employment. This information will be used to set the level of financial aid. The population contains 3478 students who have completed at least one year of study but have not yet graduated. The university will send a questionnaire to an SRS of 100 of these students, drawn from an alphabetized list.
(a) Describe how you will label the students in order to select the sample.
(b) Use Table B, beginning at line 125, to select the first 6 students in the sample.

3.146 Attitudes toward collective bargaining. A labor organization wants to study the attitudes of college faculty members toward collective bargaining. These attitudes appear to be different depending on the type of college. The American Association of University Professors classifies colleges as follows:
Class I. Offer doctorate degrees and award at least 15 per year.
Class IIA. Award degrees above the bachelor's but are not in Class I.
Class IIB. Award no degrees beyond the bachelor's.
Class III. Two-year colleges.
Discuss the design of a sample of faculty from colleges in your state, with total sample size about 200.

3.147 Student attitudes concerning labor practices. You want to investigate the attitudes of students at your school about the labor practices of factories that make college-brand apparel. You have a grant that will pay the costs of contacting about 500 students.

(a) Specify the exact population for your study. For example, will you include part-time students?

(b) Describe your sample design. Will you use a stratified sample?

(c) Briefly discuss the practical difficulties that you anticipate. For example, how will you contact the students in your sample?

3.148 Treating drunk drivers. Once a person has been convicted of drunk driving, one purpose of court-mandated treatment or punishment is to prevent future offenses of the same kind. Suggest three different treatments that a court might require. Then outline the design of an experiment to compare their effectiveness. Be sure to specify the response variables you will measure.

3.149 Stocks go down on Monday. Puzzling but true: stocks tend to go down on Mondays, both in the United States and in overseas markets. There is no convincing explanation for this fact. A recent study looked at this "Monday effect" in more detail, using data on the daily returns of stocks on several U.S. exchanges over a 30-year period. Here are some of the findings:

> To summarize, our results indicate that the well-known Monday effect is caused largely by the Mondays of the last two weeks of the month. The mean Monday return of the first three weeks of the month is, in general, not significantly different from zero and is generally significantly higher than the mean Monday return of the last two weeks. Our finding seems to make it more difficult to explain the Monday effect.[46]

A friend thinks that "significantly" in this article has its plain English meaning, roughly "I think this is important." Explain in simple language what "significantly higher" and "not significantly different from zero" actually tell us here.

3.150 The product should not be discolored. Few people want to eat discolored french fries. Potatoes are kept refrigerated before being cut for french fries to prevent spoiling and preserve flavor. But immediate processing of cold potatoes causes discoloring due to complex chemical reactions. The potatoes must therefore be brought to room temperature before processing. Fast-food chains and other sellers of french fries must understand potato behavior. Design an experiment in which tasters will rate the color and flavor of french fries prepared from several groups of potatoes. The potatoes will be freshly harvested or stored for a month at room temperature or stored for a month refrigerated. They will then be sliced and cooked either immediately or after an hour at room temperature.

(a) What are the factors and their levels, the treatments, and the response variables?

(b) Describe and outline the design of this experiment.

(c) It is efficient to have each taster rate fries from all treatments. How will you use randomization in presenting fries to the tasters?

3.151 Quality of service. Statistical studies can often help service providers assess the quality of their service. The United States Postal Service is one such provider of services. We wonder if the number of days a letter takes to reach another city is affected by the time of day it is mailed and whether or not the zip code is used. Describe briefly the design of a two-factor experiment to investigate this question. Be sure to specify the treatments exactly and to tell how you will handle lurking variables such as the day of the week on which the letter is mailed.

3.152 Mac versus PC. Many people hold very strong opinions about the superiority of the computer that they use. Design an experiment to compare customer satisfaction with the Mac versus the PC. Consider whether or not you will include individuals who routinely use both types of computers and whether or not your will block on the type of computer currently being used. Write a summary of your design including your reasons for the choices you make. Be sure to include the question or questions that you will use to measure customer satisfaction.

3.153 Design your own experiment. The previous two exercises illustrate the use of statistically designed experiments to answer questions of interest to consumers as well as to businesses. Select a question of interest to you that an experiment might answer and briefly discuss the design of an appropriate experiment.

3.154 Randomization for testing a breakfast food. To demonstrate how randomization reduces confounding, return to the breakfast food testing experiment described in Example 3.20 (page 172). Label the 30 rats 01 to 30. Suppose that, unknown to the experimenter, the 10 rats labeled 01 to 10 have a genetic defect that will cause them to grow more slowly than normal rats. If the experimenter simply puts rats 01 to 15 in the experimental group and rats 16 to 30 in the control group, this lurking variable will bias the experiment against the new food product.

Use Table B to assign 15 rats at random to the experimental group as in Example 3.22. Record how many of the 10 rats with genetic defects are placed in the experimental group and how many are in the control group. Repeat the randomization using different lines in Table B until you have done five random assignments. What is the mean number of genetically defective rats in experimental and control groups in your five repetitions?

Simulating samples. *Statistical software offers a shortcut to simulate the results of, say, 100 SRSs of size n drawn from a population in which proportion p would say "Yes" to a certain question. We will see in Chapter 5 that "binomial" is the key word to look for in the software menus. Set the sample size n and the sample proportion p. The software will generate the number of "Yes" answers in an SRS. You can divide by n to get the sample proportion $\hat{p}$. Use simulation to complete Exercises 3.155 and 3.156.*

3.155 Changing the population. Draw 100 samples of size $n = 60$ from populations with $p = 0.1$, $p = 0.3$, and $p = 0.5$. Make a stemplot of the 100 values of $\hat{p}$ obtained in each simulation. Compare your three stemplots. Do they show about the same variability? How does changing the parameter p affect the sampling distribution? If your software permits, make a Normal quantile plot of the $\hat{p}$-values from the population with $p = 0.5$. Is the sampling distribution approximately Normal?

3.156 Changing the sample size. Draw 100 samples of each of the sizes $n = 60$, $n = 200$, and $n = 800$ from a population with $p = 0.6$. Make a histogram of the $\hat{p}$-values for each simulation, using the same horizontal and vertical scales so that the three graphs can be compared easily. How does increasing the size of an SRS affect the sampling distribution of $\hat{p}$?

3.157 Two ways to ask sensitive questions. Sample survey questions are usually read from a computer screen. In a Computer Aided Personal Interview (CAPI), the interviewer reads the questions and enters the responses. In a Computer Aided Self

Interview (CASI), the interviewer stands aside and the respondent reads the questions and enters responses. One method almost always shows a higher percent of subjects admitting use of illegal drugs. Which method? Explain why.

3.158 Your institutional review board (optional). Your college or university has an institutional review board that screens all studies that use human subjects. Get a copy of the document that describes this board (you can probably find it online).
(a) According to this document, what are the duties of the board?
(b) How are members of the board chosen? How many members are not scientists? How many members are not employees of the college? Do these members have some special expertise, or are they simply members of the "general public"?

CHAPTER 3   Case Study Exercises

CASE STUDY EXERCISE 1. Consumer preferences in car colors. The most popular colors for cars sold in 2008 are white (including White Pearl) and black (including Black Effect). A survey shows that 20% of luxury cars made in North America in 2008 were white and 17% were black.[47] The preferences of some groups of consumers may of course differ from national patterns. Undertake a study to determine the most popular colors among the vehicles driven by students at your school. You might collect data by questioning a sample of students or by looking at cars in student parking areas. Explain carefully how you attempted to get data that are close to an SRS of student cars, including how you used random selection. Then report your findings on student preferences in motor vehicle colors.

CASE STUDY EXERCISE 2. Calls to a customer service center. Example 1.13 (page 13) describes a data set that gives the lengths of 31,492 calls made to the customer service center of a small bank in a month.

A. **The population.** Describe the distribution of the call lengths in the population. Include a histogram and the five-number summary.

B. **Samples and sampling distributions.** Choose an SRS of 500 calls from this population. Make a histogram of the 500 call lengths in the sample and find the five-number summary.

Briefly compare the shape, center, and spread of the call length distributions in the sample and in the population. Then repeat the process of choosing an SRS of size 500 four more times (five in all). Does it seem reasonable to you from this small trial that an SRS of 500 calls will usually produce a sample whose shape is generally representative of the population?

C. **Statistical estimation.** Do the medians and quartiles of the samples provide reasonable estimates of the population median and quartiles? Explain why we expect that the minimum and maximum of a sample will *not* satisfactorily estimate the population minimum and maximum. Now examine estimation of mean call length in more detail. Use your software to choose 50 SRSs of size 500 from this population. Find the mean length for each sample and save these 50 sample means in a separate file. Make a histogram of the distribution of the 50 sample means. How do the shape, center, and spread of this distribution compare with the distribution of individual call lengths from part A? Does it appear that the sample mean from an SRS of size 500 is usually a reasonable estimator of the population mean? (The sampling distribution of the sample mean for samples of size 500 from this population is the distribution of the means of all possible samples. Your 50 samples give a rough idea of the nature of the sampling distribution.)

CHAPTER 3 Appendix

Using Minitab and Excel for Simple Random Sampling

Simple Random Sample

Minitab:

Calc ➤ Random Data ➤ Sample from Columns

With this option, Minitab is basically picking from a hat. You can have a list of population elements in a worksheet column. For example, the 28 spring break destinations from Example 3.10 can be placed in a column of Minitab. Alternatively, as explained in Section 3.1, you can assign a numerical label to each population element and place these numerical labels in a worksheet column. Input the value of n (sample size) in the **Number of rows to sample** box. Click-in the column that has the list of population elements of the numerical labels into the **From columns** box and then type in the column name in the **Store Samples in** box to indicate where you want to place the simple random sample. Click **OK** to find the random sample in the worksheet.

If the randomly generated sample is a column of numerical values and you wish to sort them in ascending order, then you can do the following pull-down sequence:

Data ➤ Sort

Click-in the column which has the random numbers into the **Sort column(s)** box and then click-in the same column into the **By column** box. To put the sorted numbers into the current worksheet, select the **Column(s) of current worksheet** option and then type in the column name in the box below to indicate where you want to place the sorted numbers. Click **OK.**

Excel:

To conduct random sampling in Excel, you need to assign a numerical label in the form of an integer to each population element. For example, as suggested with Example 3.10, we would assign the numbers $1, \ldots, 28$ to the 28 population elements. These numerical labels should be put in the second column of the spreadsheet. Now select "Sampling" in the **Data Analysis** menu box and click **OK.** Enter the cell range of the numerical labels into the **Input Range** box. Choose the **Random** option and input the number of random selections you wish Excel to make. Repeated selections are possible, so it is advisable that you generate several more random numbers than the sample size. Click **OK** and find the random numbers outputted to a new worksheet. If you wish to sort the random numbers, you should first highlight the cells to be sorted and then click the **Sort & Filter** option, found under the **Home** tab. You can now pick the option to have the numbers sorted from smallest to largest.

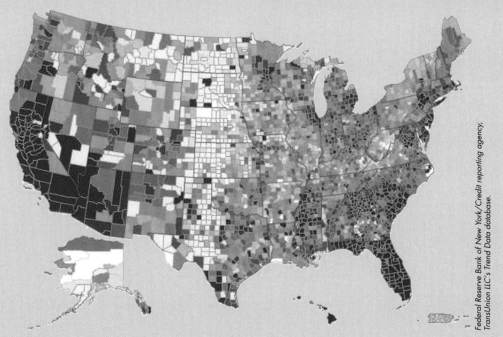

Federal Reserve Bank of New York/Credit reporting agency, TransUnion LLC's Trend Data database.

This Federal Reserve Bank of New York figure shows the ratio of mortgage borrowers currently 90 days or more past due, in the fourth quarter of 2009. What can we infer from this simple map about the extent and severity of the housing problem at the end of 2009? This part of the book examines inference from data.

PART II

Probability and Inference

Probability and Sampling Distributions

Statistical inference rests on random sampling. When we use chance, the laws of probability answer the question: "What would happen if we did this many times?" A simple example that shows the essence of probability is tossing a coin, as Example 4.1 reveals.

Introduction

The reasoning of statistical inference rests on asking, "How often would this method give a correct answer if I used it very many times?" Inference is most secure when we produce data by random sampling or randomized comparative experiments. The reason is that when we use chance to choose respondents or assign subjects, the laws of probability answer the question "What would happen if we did this many times?" The purpose of this chapter is to see what the laws of probability tell us, but without going into the mathematics of probability theory.

4.1 Randomness

What is the mean income of households in the United States? The Bureau of Labor Statistics contacted a random sample of 57,000 households in March 2007 for the Current Population Survey (CPS). The mean income of the 57,000 households for the year 2007 was $\bar{x} = \$67,609$.[1] That $67,609 is a *statistic* that describes the CPS sample households. We use it to estimate an unknown *parameter,* the mean income of all 117 million American households. We know that $\bar{x}$ would take many different values if the Bureau of Labor Statistics had taken several samples in March 2007. Fortunately, we also know that this sampling variability follows a regular pattern that can tell us how accurate the sample result is likely to be. That pattern obeys the laws of probability. Our starting point in understanding probability is the phenomenon we observed in the simulations of Section 3.3: **chance behavior is unpredictable in the short run but has a regular and predictable pattern in the long run.**

The idea of probability

Toss a coin, or choose an SRS. The result can't be predicted in advance, because the result will vary when you toss the coin or choose the sample repeatedly. But there is still a regular pattern in the results, a pattern that emerges clearly only after many repetitions. This remarkable fact is the basis for the idea of probability.

EXAMPLE 4.1 Coin Tossing

When you toss a coin, there are only two possible outcomes, heads or tails. Figure 4.1 shows the results of tossing a coin 5000 times twice. For each number of tosses from 1 to 5000, we have plotted the proportion of those tosses that gave a head. Trial A (solid line) begins tail, head, tail, tail. You can see that the proportion of heads for Trial A starts at 0 on the first toss, rises to 0.5 when the second toss gives a head, then falls to 0.33 and 0.25 as we get two more tails. Trial B, on the other hand, starts with five straight heads, so the proportion of heads is 1 until the sixth toss.

The proportion of tosses that produce heads is quite variable at first. Trial A starts low and Trial B starts high. As we make more and more tosses, however, the proportion of heads for both trials gets close to 0.5 and stays there. If we made yet a third trial at tossing the coin a great many times, the proportion of heads would again settle down to 0.5 in the long run. We say that 0.5 is the *probability* of a head. The probability 0.5 appears as a horizontal line on the graph.

FIGURE 4.1 The proportion of tosses of a coin that give a head changes as we make more tosses. Eventually, however, the proportion approaches 0.5, the probability of a head. This figure shows the results of two trials of 5000 tosses each.

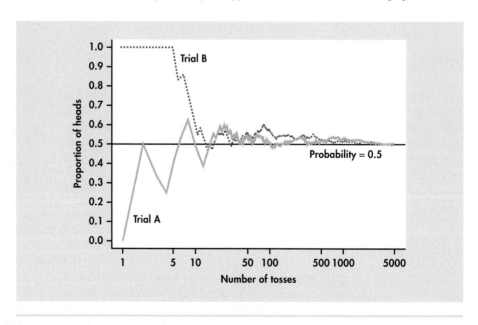

The *Probability* applet available on the Web site for this book, www.whfreeman.com/psbe3e, animates Figure 4.1. It allows you to choose the probability of a head and simulate any number of tosses of a coin with that probability. Experience shows that the proportion of heads gradually settles down close to the probability. Equally important, it also shows that the proportion in a small or moderate number of tosses can be far from the probability. Probability describes *only* what happens in the long run.

"Random" in statistics is not a synonym for "haphazard" but a description of a kind of order that emerges only in the long run. We often encounter the unpredictable side of randomness in our everyday experience, but we rarely see enough repetitions of the same random phenomenon to observe the long-term regularity that probability describes. You can see that regularity emerging in Figure 4.1. In the very long run, the proportion of tosses that give a head is 0.5. This is the intuitive idea of probability. Probability 0.5 means "occurs half the time in a very large number of trials."

We might suspect that a coin has probability 0.5 of coming up heads just because the coin has two sides. As Exercises 4.1 and 4.2 illustrate, such suspicions are not always correct. The idea of probability is *empirical*. That is, it is based on observation rather than theorizing. Probability describes what happens in very many trials, and we must actually observe many trials to pin down a probability. In the case of tossing a coin, some diligent people have in fact made thousands of tosses.

EXAMPLE 4.2   Some Coin Tossers

The French naturalist Count Buffon (1707–1788) tossed a coin 4040 times. Result: 2048 heads, or proportion $2048/4040 = 0.5069$ for heads.

Around 1900, the English statistician Karl Pearson heroically tossed a coin 24,000 times. Result: 12,012 heads, a proportion of 0.5005.

While imprisoned by the Germans during World War II, the South African mathematician John Kerrich tossed a coin 10,000 times. Result: 5067 heads, a proportion of 0.5067.

Randomness and Probability

We call a phenomenon **random** if individual outcomes are uncertain but there is nonetheless a regular distribution of outcomes in a large number of repetitions.

The **probability** of any outcome of a random phenomenon is the proportion of times the outcome would occur in a very long series of repetitions.

Why talk about coin tossing?

Disputes are often settled by a coin toss. NFL games start out with a coin toss. It is intuitive even to young children that a coin toss is an unpredictable phenomenon with a 50% chance that it will result in heads and a 50% chance that it will result in tails. As such, the simple coin toss experiment serves as a common starting point to help in the understanding of randomness and probability. More generally, thinking about coin tossing helps us understand the workings of many real-world problems. For example, many economists and finance researchers suggest that the up-and-down price movements of the stock market over time can be compared with the random outcomes of tossing a coin. The probability concepts underlying repeated coin tossing also fit many other business problems: respondents answer "Yes" or "No" to a customer survey question; consumers buy or don't buy a product after viewing an advertisement; an inspector finds good or defective parts. Of course, as we will learn, we must allow the probability of a "head" to be something other than one-half in these settings.

APPLY YOUR KNOWLEDGE

4.1 Nickels spinning. Hold a nickel upright on its edge under your forefinger on a hard surface, then snap it with your other forefinger so that it spins for some time before falling. Based on 50 spins, estimate the probability of heads.

4.2 Nickels falling over. You may feel that it is obvious that the probability of a head in a coin experiment is about 1/2 because the coin has two faces. Such opinions are not always correct. The previous exercise asked you to spin a nickel rather than toss it—that changes the probability of a head. Now try another variation. Stand a nickel on edge on a hard, flat surface. Pound the surface with your hand so that the nickel falls over. What is the probability that it falls with heads upward? Make at least 50 trials to estimate the probability of a head.

Thinking about randomness

Randomness surrounds us. The outcomes of coin tossing, outdoor temperatures, blood pressure readings, commuting times to work or school, the times between customers arriving at an ATM machine, monthly demand for a product or service, and the prices of a share of a company's stock at the close of the market are all random. So is the outcome of a random sample or a randomized experiment. Probability theory is the branch of mathematics that describes random behavior. Of course, we can never observe a probability exactly. We could always continue tossing the coin, for example. Mathematical probability is an idealization based on imagining what would happen in an indefinitely long series of trials.

The best way to understand randomness is to observe random behavior—not only the long-run regularity but the unpredictable results of short runs. You can do this with physical devices, as in Exercises 4.1 and 4.2, but computer simulations (imitations) of random behavior allow faster exploration. Exercises 4.11 to 4.15 suggest some simulations of random behavior. As you explore randomness, remember:

independence
- You must have a long series of **independent** trials. That is, the outcome of one trial must not influence the outcome of any other. Imagine a crooked gambling house where the operator of a roulette wheel can stop it where she chooses—she can prevent the proportion of "red" from settling down to a fixed number. These trials are not independent.

- The idea of probability is empirical. Computer simulations start with given probabilities and imitate random behavior, but we can estimate a real-world probability only by actually observing many trials.

- Nonetheless, computer simulations are very useful because we need long runs of trials. In situations such as coin tossing, the proportion of an outcome often requires several hundred trials to settle down to the probability of that outcome. The kinds of physical techniques suggested in the exercises are too slow for this. Short runs give only rough estimates of a probability.

Getting a fair coin toss

Use your thumb to flick a coin into the air. Is the probability of getting a head 0.50? The answer: it depends. It appears that the initial conditions of the coin and Newton's laws of mechanics govern the outcomes. In fact, a coin-tossing machine can be adjusted so a coin starting heads up will land heads up 100% of the time. Unlike machines, human tossing is far less predictable, but there is a slight bias in outcome depending on the initial position of the coin. In particular, based on the laws of physics and experimental data, researchers have concluded that a coin starting heads up has a probability of 0.51 of ending up as a head.[2] The same is true for a coin starting tails up. Take away the certainty of the initial starting position (for example, pull a coin out of your pocket and take it straight into a toss or shake a coin in cupped hands prior to tossing) then these very same researchers conclude that the classical assumptions of independence of outcomes with probability 0.50 are pretty solid.

SECTION 4.1 Summary

- A **random phenomenon** has outcomes that we cannot predict but that nonetheless have a regular distribution in very many repetitions.

- The **probability** of an event is the proportion of times the event occurs in many repeated trials of a random phenomenon.

SECTION 4.1 Exercises

For Exercises 4.1 and 4.2, see page 220.

4.3 Financial fraud. In December 2006, the Experian-Gallup Personal Credit Index survey found that 17% of fraud victims knew the perpetrator as a friend or acquaintance. Financial fraud includes crimes like unauthorized credit card charges, withdrawal of money from a savings or checking account, and opening an account in someone else's name. Because 17% is approximately 1-in-6, a single fair die can be used to study the probability that a fraud victim knew the perpetrator. Each roll of the die will represent the random selection of a single fraud victim. If the die lands with the number 6 facing up, record this as a fraud victim who knew the perpetrator. If the die lands with a 1, 2, 3, 4, or 5 facing up, record this as a fraud victim who did not know the perpetrator.

(a) Roll the die 6 times. How many of these 6 simulated fraud victims knew their perpetrators? If your 6 rolls did not result in exactly 1 fraud victim who knew the perpetrator, do you have reason to doubt the fairness of the die?

(b) Roll the die a total of 30 times. How many of these 30 simulated fraud victims knew their perpetrators?

(c) If possible, combine your 30 rolls with the rolls of fellow students. How many fraud victims are represented by the combined rolls? How many knew their perpetrators? Compare the percent who knew their perpetrators in your combined rolls with 1-in-6 (16.67%).

4.4 Credit monitoring. In a recent study of consumers, 25% reported purchasing a credit-monitoring product that alerts them to any activity on their credit report. To study this situation with a physical device providing the randomness as opposed to a computer simulation, we will need to be a bit more clever than flipping a single coin or rolling a single die. When flipping 2 fair coins, there are 4 possible results: HH, HT, TH, TT. Each of these possibilities has a 25% chance of happening. Each flip of a pair of fair coins will represent the random selection of a single consumer. If our coin flip results in HH (both coins land "heads up"), record this as a consumer who purchased a credit-monitoring product.

(a) Flip two fair coins 8 times. How many of the 8 flips resulted in a consumer who purchased a credit-monitoring product?

(b) Steve comments: "You should always get two HH flips out of 8 coin flips like those described in part (a)." Respond to Steve's statement.

(c) Suggest another physical mechanism for studying this situation.

4.5 Random digits. As discussed in Chapter 3, generation of random numbers is one approach for obtaining a simple random sample (SRS). If we were to look at the random generation of digits, the mechanism should give each digit probability 0.1. Consider the digit "0" in particular.

(a) The table of random digits (Table B) was produced by a random mechanism that gives each digit probability 0.1 of being a 0. What proportion of the first 200 digits in the table are 0s? This proportion is an estimate, based on 200 repetitions, of the true probability, which in this case is known to be 0.1.

(b) Now try software:

- *Excel users:* Input the digits 0, 1, 2, 3, 4, 5, 6, 7, 8, 9 in column A. Now choose "Sampling" from the **Data Analysis** menu box (refer to the Appendix of Chapter 1 for installing the Data Analysis ToolPak). Enter the cell range of the digits in the **Input Range** box. Choose the "Random" option and input 10,000 in the **Number of Samples** box. Finally, select an output range in the worksheet.

- *Minitab users:* Do the following pull-down sequence: Calc → Random Data → Integer. Enter "10000" in the **Number of rows of data to generate** box, type "c1" in the **Store in column(s)** box, enter "0" in the **Minimum value** box, and enter "9" in the **Maximum** box. Click **OK** to find 10,000 realizations of random digits outputted in the worksheet.

Whether you used Excel or Minitab, sort the 10,000 random digits from smallest to largest to make your counting easier. What proportion of the 10,000 digits in the column are 0s? Is this proportion close to 0.1?

4.6 How many tosses to get a head? When we toss a penny, experience shows that the probability (long-term proportion) of a head is close to 1/2. Suppose now that we toss the penny repeatedly until we get a head. What is the probability that the first head comes up in an odd number of tosses (1, 3, 5, and so on)? To find out, repeat this experiment 50 times, and keep a record of the number of tosses needed to get a head on each of your 50 trials.

(a) From your experiment, estimate the probability of a head on the first toss. What value should we expect this probability to have?

(b) Use your results to estimate the probability that the first head appears on an odd-numbered toss.

4.7 Tossing a thumbtack. Toss a thumbtack on a hard surface 100 times. How many times did it land with the point up? What is the approximate probability of landing point up?

4.8 Three of a kind. You read in a book on poker that the probability of being dealt three of a kind in a five-card poker hand is 1/50. Explain in simple language what this means.

4.9 Thinking about probability statements. Probability is a measure of how likely an event is to occur. Match one of the probabilities that follow with each statement of likelihood given. (The probability is usually a more exact measure of likelihood than is the verbal statement.)

$$0 \quad 0.01 \quad 0.3 \quad 0.6 \quad 0.99 \quad 1$$

(a) This event is impossible. It can never occur.

(b) This event is certain. It will occur on every trial.

(c) This event is very unlikely, but it will occur once in a while in a long sequence of trials.

(d) This event will occur more often than not.

4.10 What probability doesn't say. The idea of probability is that the *proportion* of heads in many tosses of a balanced coin eventually gets close to 0.5. But does the actual *count* of heads get close to one-half the number of tosses? Let's find out. Set the "Probability of heads" in the *Probability* applet to 0.5 and the number of tosses to 40. You can extend the number of tosses by clicking "Toss" again to get 40 more. Don't click "Reset" during this exercise.

(a) After 40 tosses, what is the proportion of heads? What is the count of heads? What is the difference between the count of heads and 20 (one-half the number of tosses)?

(b) Keep going to 120 tosses. Again record the proportion and count of heads and the difference between the count and 60 (half the number of tosses).

(c) Keep going. Stop at 240 tosses and again at 480 tosses to record the same facts. Although it may take a long time, the laws of probability say that the proportion of heads will always get close to 0.5 and also that the difference between the count of heads and half the number of tosses will approach 0.

4.11 Simulating consumer behavior. About half of the customers entering a local computer store will purchase a new computer before leaving the store. Use the *Probability* applet or your statistical software to simulate 100 customers independently entering the store with each having a probability of 0.5 of purchasing a new computer on this visit to the store. (In most software, the key phrase to look for is "Bernoulli trials." This is the technical term for independent trials with Yes/No outcomes. Our outcomes here are "buy" and "not buy.")

(a) What percent of the 100 simulated customers bought a new computer?

(b) Examine the sequence of "buys" and "not buys." How long was the longest run of "buys"? Of "not buys"? (Sequences of random outcomes often show runs longer than our intuition thinks likely.)

4.12 Simulating an opinion poll. A recent opinion poll showed that about 59% of the American public approve of labor unions.[3] Suppose that this is exactly true. Choosing a person at random then has probability 0.59 of getting one who approves of labor unions. Use the *Probability* applet or your statistical software to simulate choosing many people independently. (In most software, the key phrase to look for is "Bernoulli trials." This is the technical term for independent trials with Yes/No outcomes. Our outcomes here are "favorable" or not.)

(a) Simulate drawing 20 people, then 80 people, then 320 people. What proportion approve of labor unions in each case? We expect (but because of chance variation we can't be sure) that the proportion will be closer to 0.59 in larger runs.

(b) Simulate drawing 20 people 10 times and record the percents in each trial who approve of labor unions. Then simulate drawing 320 people 10 times and again record the 10 percents. Which set of 10 results is less variable? We expect the results of larger trials to be more predictable (less variable) than the results of smaller trials. That is "long-run regularity" showing itself.

4.13 More efficient simulation. Continue the exploration begun in Exercise 4.12. Software allows you to simulate many independent Yes/No trials more quickly if all you want to save is the count of Yes outcomes. The key word "Binomial" simulates n independent Bernoulli trials, each with probability p of a Yes, and records just the count of Yes outcomes.

(a) Simulate 100 draws of 20 people from the population in Exercise 4.12. Record the number who approve of labor unions on each draw. What is the approximate probability that out of 20 people drawn at random at least 14 approve of labor unions?

(b) Convert the "approval" counts into percents of the 20 people in each trial. Make a histogram of these 100 percents. Describe the shape, center, and spread of this distribution.

(c) Now simulate drawing 320 people. Do this 100 times and record the percent who approve on each of the 100 draws. Make a histogram of the percents and describe the shape, center, and spread of the distribution.

(d) In what ways are the distributions in parts (b) and (c) alike? In what ways do they differ? (Because regularity emerges in the long run, we expect the results of drawing 320 individuals many times to be less variable than the results of drawing 20 individuals many times.)

4.14 Financial fraud. Continue with the Experian-Gallup Personal Credit Index survey described in Exercise 4.3. In Exercise 4.3, a fair die was used to simulate the selection of a fraud victim. The die provided a 1-in-6 (or 16.66667%) chance of a fraud victim's knowing the perpetrator of the fraud. The survey result was actually reported as 17% though. Use your statistical software to complete this exercise. (In most software, the key phrase to look for is "Bernoulli trials.")

(a) Set the probability for Bernoulli trials to 1/6 (or 0.166667 if your software doesn't allow you to enter a fraction). Generate 300 Bernoulli trials. How many of the 300 trials resulted in a "success" (likely represented by a number 1 in your software)? What percent of the trials does this represent? (*Note:* The word

"success" is simply a label to describe the event of interest, which in this case is the event that a randomly selected fraud victim knew the perpetrator.)

(b) Now generate 30,000 Bernoulli trials. How many of the 30,000 trials resulted in a "success"? What percent of the trials does this represent?

(c) Redo parts (a) and (b) with the Bernoulli probability set to 0.17.

(d) In which simulation (300 trials or 30,000 trials) did the difference between 1/6 and 17% make the most difference in the results? How is this consistent with the definition of probability given in this section?

4.15 Customer returns. Shoppers return about 6% of everything they buy. Use your statistical software to complete this exercise.

(a) Set the probability for Bernoulli trials to 0.06. Generate 100 Bernoulli trials. How many of the 100 trials resulted in a "success"? What percent of the trials does this represent?

(b) Write a clear explanation of what the simulation in part (a) actually represents in terms of "Shoppers return about 6% of everything they buy." Your explanation provides the context for the statistical simulation of part (a).

4.2 Probability Models

CASE 4.1

Uncovering Fraud by Digital Analysis "Digital analysis" is one of the big new tools among auditors and investigators looking for fraud. Faked numbers in tax returns, payment records, invoices, expense account claims, and many other settings often display patterns that aren't present in legitimate records. Some patterns, like too many round numbers, are obvious and easily avoided by a clever crook. Others are more subtle. It is a striking fact that the first digits of numbers in legitimate records often follow a distribution known as *Benford's law.* Here it is:

First digit:	1	2	3	4	5	6	7	8	9
Proportion:	0.301	0.176	0.125	0.097	0.079	0.067	0.058	0.051	0.046

These proportions are in fact *probabilities.* That is, they are the proportions specified by Benford's law if we examine a very large number of similar records. Set a computer to work comparing invoices from your company's vendors with this distribution. Aha—here is a vendor whose invoices show a very different pattern. All the invoices were approved by the same manager. Confronted, he admits that he was raising the amounts of the invoices and buying from his wife's company at the inflated amounts. Digital analysis has uncovered another case of fraud.

Of course, not all sets of data follow Benford's law. Numbers that are assigned, such as Social Security numbers, do not. Nor do data with a fixed maximum, such as deductible contributions to individual retirement accounts (IRAs). Nor of course do random numbers. But a surprising variety of data from natural science, social affairs, and business obeys this distribution.[4]

Gamblers have known for centuries that the fall of coins, cards, and dice displays clear patterns in the long run. Benford's law says that first digits also follow a clear pattern. The idea of probability rests on the observed fact that the average result of many thousands of chance outcomes can be known with near certainty. How can we give a mathematical description of long-run regularity?

To see how to proceed, think first about a very simple random phenomenon, tossing a coin once. When we toss a coin, we cannot know the outcome in advance. What do we know? We are willing to say that the outcome will be either heads or tails. We believe that each of these outcomes has probability 1/2. This description of coin tossing has two parts:

- a list of possible outcomes
- a probability for each outcome

This description is the basis for all probability models. Here is the vocabulary we use.

Probability Models

The **sample space** of a random phenomenon is the set of all possible outcomes. S is used to denote sample space.

An **event** is an outcome or a set of outcomes of a random phenomenon. That is, an event is a subset of the sample space.

A **probability model** is a mathematical description of a random phenomenon. The model has two parts: a sample space S and a way of assigning probabilities to events.

The sample space S can be very simple or very complex. When we toss a coin once, there are only two outcomes, heads and tails. The sample space is $S = \{H, T\}$. If we draw a random sample of 57,000 U.S. households, as the Current Population Survey does, the sample space contains all possible choices of 57,000 of the 117 million households in the country. This S is extremely large. Each member of S is a possible sample, which explains the term *sample space.*

As we proceed to talk about probability, it should be noted that each outcome in the sample space does not necessarily have the same chance of occurring. For Case 4.1, the sample space is composed of the first digits, that is, $S = \{1, 2, 3, 4, 5, 6, 7, 8, 9\}$. However, according to Benford's law the probabilities of these outcomes are not all the same. In the special case where the outcomes have the same probability of occurring, we refer to these outcomes as being *equally likely.* For example, if you toss a fair coin, the outcomes of heads and tails are equally likely.

EXAMPLE 4.3   Rolling Dice

Rolling two dice is a common way to lose money in casinos. There are 36 possible outcomes when we roll two dice and record the up-faces in order (first die, second die). Figure 4.2 displays these outcomes. They make up the sample space S. "Roll a 5" is an event, call it A, that contains four of these 36 outcomes:

$$A = \left\{ \boxed{\cdot}\,\boxed{\vdots} \quad \boxed{\because}\,\boxed{\therefore} \quad \boxed{\therefore}\,\boxed{\because} \quad \boxed{\vdots}\,\boxed{\cdot} \right\}$$

In craps and other games, all that matters is the *sum* of the spots on the up-faces. Let's change the random outcomes we are interested in: roll two dice and count the spots on the up-faces. Now there are only 11 possible outcomes, from a sum of 2 for rolling a double one through 3, 4, 5, and

FIGURE 4.2 The 36 possible outcomes in rolling two dice.

on up to 12 for rolling a double six. The sample space is now

$$S = \{2, 3, 4, 5, 6, 7, 8, 9, 10, 11, 12\}$$

Comparing this S with Figure 4.2 reminds us that we can change S by changing the detailed description of the random phenomenon we are describing.

APPLY YOUR KNOWLEDGE

4.16 Describing sample spaces. In each of the following situations, describe a sample space S for the random phenomenon. In some cases, you have some freedom in your choice of S.

(a) A new business is started. After two years, it is either still in business or it has closed.

(b) A rust prevention treatment is applied to a new car. The response variable is the length of time before rust begins to develop on the vehicle.

(c) A student enrolls in a statistics course and at the end of the semester receives a letter grade.

(d) A quality inspector examines four portable CD players and rates each as either "acceptable" or "unacceptable." You record the sequence of ratings.

(e) A quality inspector examines four portable CD players and rates each as either "acceptable" or "unacceptable." You record the number of units rated "acceptable."

4.17 Describing sample spaces. In each of the following situations, describe a sample space S for the random phenomenon. In some cases you have some freedom in specifying S, especially in setting the largest and the smallest value in S.

(a) Choose a student in your class at random. Ask how much time that student spent studying during the past 24 hours.

(b) The Physicians' Health Study asked 11,000 physicians to take an aspirin every other day and observed how many of them had a heart attack in a 5-year period.

(c) In a test of a new package design, you drop a carton of a dozen eggs from a height of 1 foot and count the number of broken eggs.

(d) Choose a Fortune 500 company at random. Look up how much the CEO of the company makes annually.

(e) The iTunes store records the number of purchased songs in a given 24-hour period.

Probability rules

The true probability of any outcome—say, "roll a 5 when we toss two dice"—can be found only by actually tossing two dice many times, and then only approximately. How then can we describe probability mathematically? Rather than try to give "correct" probabilities, we start by laying down facts that must be true for any assignment of probabilities. These facts follow from the idea of probability as "the long-run proportion of repetitions in which an event occurs."

1. **Any probability is a number between 0 and 1.** Any proportion is a number between 0 and 1, so any probability is also a number between 0 and 1. An event with probability 0 never occurs, and an event with probability 1 occurs on every trial. An event with probability 0.5 occurs in half the trials in the long run.

2. **All possible outcomes together must have probability 1.** Because some outcome must occur on every trial, the sum of the probabilities for all possible outcomes must be exactly 1.

3. **The probability that an event does not occur is 1 minus the probability that the event does occur.** If an event occurs in (say) 70% of all trials, it fails to occur in the other 30%. The probability that an event occurs and the probability that it does not occur always add to 100%, or 1.

4. **If two events have no outcomes in common, the probability that one or the other occurs is the sum of their individual probabilities.** If one event occurs in 40% of all trials, a different event occurs in 25% of all trials, and the two can never occur together, then one or the other occurs on 65% of all trials because $40\% + 25\% = 65\%$.

We can use mathematical notation to state Facts 1 to 4 more concisely. Capital letters near the beginning of the alphabet denote events. If A is any event, we write its probability as $P(A)$. Here are our probability facts in formal language. As you apply these rules, remember that they are just another form of intuitively true facts about long-run proportions.

Probability Rules

Rule 1. The probability $P(A)$ of any event A satisfies $0 \leq P(A) \leq 1$.

Rule 2. If S is the sample space in a probability model, then $P(S) = 1$.

Rule 3. The **complement** of any event A is the event that A does not occur, written as A^c. The **complement rule** states that

$$P(A^c) = 1 - P(A)$$

Rule 4. Two events A and B are **disjoint** if they have no outcomes in common and so can never occur simultaneously. If A and B are disjoint,

$$P(A \text{ or } B) = P(A) + P(B)$$

This is the **addition rule for disjoint events.**

EXAMPLE 4.4 Benford's Law

CASE 4.1

The probabilities assigned to first digits by Benford's law are all between 0 and 1. They add to 1 because the first digits listed make up the sample space S. (Note that a first digit can't be 0.)

The probability that a first digit is anything other than 1 is, by Rule 3,

$$P(\text{not a } 1) = 1 - P(\text{first digit is } 1)$$
$$= 1 - 0.301 = 0.699$$

That is, if 30.1% of first digits are 1s, the other 69.9% are not 1s.

"First digit is a 1" and "first digit is a 2" are disjoint events. So the addition rule (Rule 4) says

$$P(1 \text{ or } 2) = P(1) + P(2)$$
$$= 0.301 + 0.176 = 0.477$$

About 48% of first digits in data governed by Benford's law are 1s or 2s. Fraudulent records generally have many fewer 1s and 2s.

EXAMPLE 4.5   Probabilities for Rolling Dice

Figure 4.2 displays the 36 possible outcomes of rolling two dice. What probabilities shall we assign to these outcomes?

Casino dice are carefully made. Their spots are not hollowed out, which would give the faces different weights, but are filled with white plastic of the same density as the colored plastic of the body. For casino dice it is reasonable to assign the same probability to each of the 36 outcomes in Figure 4.2. Because all 36 outcomes together must have probability 1 (Rule 2), each outcome must have probability 1/36.

What is the probability of rolling a 5? Because the event "roll a 5" contains the four outcomes displayed in Example 4.3, the addition rule (Rule 4) says that its probability is

$$P(\text{roll a } 5) = P\left(\boxed{\cdot}\ \boxed{::}\right) + P\left(\boxed{\cdot\cdot}\ \boxed{\cdot\cdot}\right) + P\left(\boxed{\cdot\cdot}\ \boxed{\cdot\cdot}\right) + P\left(\boxed{::}\ \boxed{\cdot}\right)$$
$$= \frac{1}{36} + \frac{1}{36} + \frac{1}{36} + \frac{1}{36}$$
$$= \frac{4}{36} = 0.111$$

What about the probability of rolling a 7? In Figure 4.2 you will find six outcomes for which the sum of the spots is 7. The probability is 6/36, or about 0.167.

Example 4.5 considers a situation where all the outcomes are equally likely. In the special case of a sample space made up of equally likely outcomes, the probability of an event A is the ratio of the number of outcomes in A to the number of outcomes in the sample space. Referring to Example 4.5, there are 4 outcomes associated with the event "roll a 5," and there are 36 outcomes in the sample space. Accordingly, the probability is the ratio of 4 to 36, or 0.111. If it cannot be assumed that outcomes are equally likely, then this approach of obtaining probabilities would not be correct. Since most random phenomena do not have equally likely outcomes, it is very important to be familiar with the general method of adding the probabilities of the individual outcomes as done in Example 4.5 and will be done in the next example.

APPLY YOUR KNOWLEDGE

4.18 Moving up. An economist studying economic class mobility finds that the probability that the son of a lower-class father remains in the lower class is 0.46. What is the probability that the son moves to one of the higher classes?

4.19 Causes of death. Government data on job-related deaths assign a single occupation for each such death that occurs in the United States. The data on occupational deaths in 2007 show that the probability is 0.105 that a randomly chosen death was agriculture related and 0.072 that it was manufacturing related. What is the probability that a death was either agriculture related or manufacturing related? What is the probability that the death was related to some other occupation?

4.20 Grading Canadian health care. Annually, the Canadian Medical Association uses the marketing research firm Ipsos Canada to measure public opinion with respect to the Canadian health care system. Between June 10 and June 12 of 2008, Ipsos Canada interviewed a random sample of 1002 adults.[5] The people in the sample were asked to grade the overall quality of health care services as an A, B, C, or F, where an A is the highest grade and an F is a failing grade. Here are the results:

Outcome	Probability
A	0.23
B	0.43
C	?
F	0.07

These proportions are probabilities for the random phenomenon of choosing an adult at random and asking the person's opinion on the Canadian health care system.
(a) What must be the probability that the person chosen gives a grade of C? Why?
(b) If a "positive" grade is defined as A or B, what is this probability?

Assigning probabilities: finite number of outcomes

Examples 4.4 and 4.5 illustrate one way to assign probabilities to events: assign a probability to every individual outcome, then add these probabilities to find the probability of any event. If such an assignment is to satisfy the rules of probability, the probabilities of all the individual outcomes must sum to exactly 1.

> **Probabilities in a Finite Sample Space**
>
> Assign a probability to each individual outcome. These probabilities must be numbers between 0 and 1 and must have sum 1.
>   The probability of any event is the sum of the probabilities of the outcomes making up the event.

EXAMPLE 4.6   Random Digits versus Benford's Law

CASE 4.1   You might think that first digits in financial records are distributed "at random" among the digits 1 to 9. The 9 possible outcomes would then be equally likely. The sample space for a single digit is

$$S = \{1, 2, 3, 4, 5, 6, 7, 8, 9\}$$

Call a randomly chosen first digit X for short. The probability model for X is completely described by this table:

First digit X:	1	2	3	4	5	6	7	8	9
Probability:	1/9	1/9	1/9	1/9	1/9	1/9	1/9	1/9	1/9

The probability that a first digit is equal to or greater than 6 is

$$P(X \geq 6) = P(X = 6) + P(X = 7) + P(X = 8) + P(X = 9)$$
$$= \frac{1}{9} + \frac{1}{9} + \frac{1}{9} + \frac{1}{9} = \frac{4}{9} = 0.444$$

Note that this is not the same as the probability that a random digit is strictly greater than 6, $P(X > 6)$. The outcome $X = 6$ is included in the event $\{X \geq 6\}$ and is omitted from $\{X > 6\}$.

Compare this with the probability of the same event among first digits from financial records that obey Benford's law. A first digit V chosen from such records has the distribution

First digit V:	1	2	3	4	5	6	7	8	9
Probability:	0.301	0.176	0.125	0.097	0.079	0.067	0.058	0.051	0.046

The probability that this digit is equal to or greater than 6 is

$$P(V \geq 6) = 0.067 + 0.058 + 0.051 + 0.046 = 0.222$$

Benford's law allows easy detection of phony financial records based on randomly generated numbers. These records tend to have too few 1s and 2s as first digits and too many first digits of 6 or greater.

---

probability histogram

We can use histograms to display probability distributions as well as distributions of data. Figure 4.3 displays **probability histograms** that compare the probability model for random digits with the model given by Benford's law. The height of each bar shows the probability of the outcome at its base. Because the heights are probabilities, they add to 1.

FIGURE 4.3 Probability histograms for (a) random digits 1 to 9 and (b) Benford's law. The height of each bar shows the probability assigned to a single outcome.

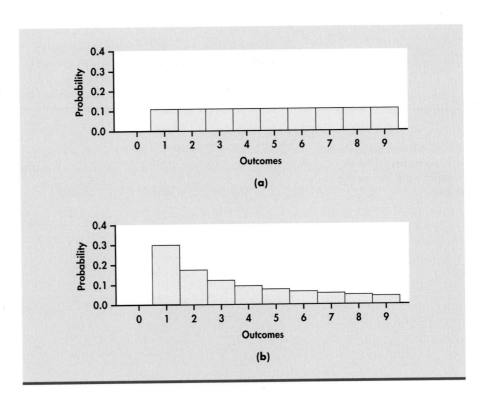

As usual, all the bars in a histogram have the same width. So the areas also display the assignment of probability to outcomes. Think of these histograms as idealized pictures of the results of very many trials.

4.21 Rolling a die. Figure 4.4 displays several assignments of probabilities to the six faces of a die. We can learn which assignment is actually *accurate* for a particular die only by rolling the die many times. However, some of the assignments are not *legitimate* assignments of probability. That is, they do not obey the rules. Which are legitimate and which are not? In the case of the illegitimate models, explain what is wrong.

4.22 U.S. power plant energy consumption. Draw a U.S. power plant at random and record the primary source of energy that the power plant relies on to produce electricity and/or heat for the public. "At random" means that we give every such power plant the same chance to be the one we choose. That is, we choose an SRS of size 1. The probability of any primary source of energy is just the proportion of all power plants that use that source—if we drew many such power plants, this is the proportion we would get. Here is the probability model:[6]

Primary source	Probability
Petroleum	0.02
Natural gas	0.17
Coal	0.51
Renewable energy	0.09
Nuclear power	0.21

(a) Show that this is a legitimate probability model.
(b) What is the probability that a randomly chosen U.S. power plant uses renewable energy as its primary source of energy?
(c) What is the probability that a randomly chosen U.S. power plant uses fossil-based energy, that is, petroleum, natural gas, or coal?

FIGURE 4.4 Four assignments of probabilities for the six faces of a die, for Exercise 4.21.

		Probability		
Outcome	Model 1	Model 2	Model 3	Model 4
⚀	1/7	1/3	1/3	1
⚁	1/7	1/6	1/6	1
⚂	1/7	1/6	1/6	2
⚃	1/7	0	1/6	1
⚄	1/7	1/6	1/6	1
⚅	1/7	1/6	1/6	2

4.23 Job satisfaction. We can use the results of a 2008 poll on job satisfaction to give a probability model for the job satisfaction rating of a randomly chosen employed (full-time or part-time) American.[7] Here is the model:

Rating:	Completely satisfied	Somewhat satisfied	Somewhat dissatisfied	Completely dissatisfied	No opinion
Probability:	?	0.42	0.07	0.02	0.01

(a) What is the probability of a randomly selected employed American being completely satisfied with his or her job? Why?

(b) What is the probability that a randomly selected employed American will be dissatisfied with his or her job?

Assigning probabilities: intervals of outcomes

A software random number generator is designed to produce a number between 0 and 1 chosen at random. It is only a slight idealization to consider *any* number in this range as a possible outcome. The sample space is then

$$S = \{\text{all numbers between 0 and 1}\}$$

Call the outcome of the random number generator Y for short. How can we assign probabilities to such events as $\{0.3 \leq Y \leq 0.7\}$? As in the case of selecting a random digit, we would like all possible outcomes to be equally likely. But we cannot assign probabilities to each individual value of Y and then add them, because over any continuous interval there are infinitely many possible decimal values.

We use a new way of assigning probabilities directly to events—as *areas under a density curve.* Any density curve has area exactly 1 underneath it, corresponding to total probability 1. We first met density curves as models for data in Chapter 1 (page 42).

EXAMPLE 4.7   Random Numbers

uniform density curve

The random number generator will spread its output uniformly across the entire interval from 0 to 1 as we allow it to generate a long sequence of numbers. The results of many trials are represented by the **uniform density curve,** as shown in Figure 4.5. This density curve has height 1 over the interval from 0 to 1. The area under the curve is 1, and the probability of any event is the area under the curve and above the event in question.

As Figure 4.5(a) illustrates, the probability that the random number generator produces a number between 0.3 and 0.7 is

$$P(0.3 \leq Y \leq 0.7) = 0.4$$

because the area under the density curve and above the interval from 0.3 to 0.7 is 0.4. The height of the curve is 1 and the area of a rectangle is the product of height and length, so the probability of any interval of outcomes is just the length of the interval.

Similarly,

$$P(Y \leq 0.5) = 0.5$$
$$P(Y > 0.8) = 0.2$$
$$P(Y \leq 0.5 \text{ or } Y > 0.8) = 0.7$$

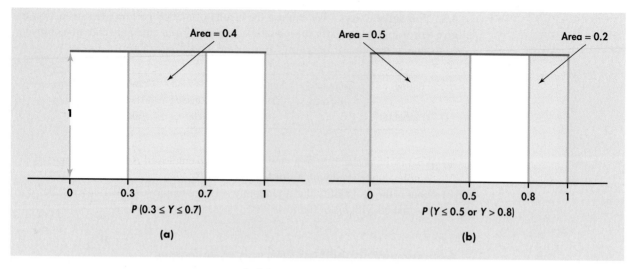

FIGURE 4.5 Probability as area under a density curve, for Example 4.7. These uniform density curves spread probability evenly between 0 and 1.

Notice that the last event consists of two nonoverlapping intervals, so the total area above the event is found by adding two areas, as illustrated by Figure 4.5(b). This assignment of probabilities obeys all our rules for probability.

Compare the density curves in Figure 4.5 with the probability histogram in Figure 4.3(a). Both describe uniform distributions. In 4.3(a), there are only 9 possible outcomes, the whole numbers 1, 2, ..., 9. In Figure 4.5, the outcome can be any decimal value between 0 and 1. In both figures, probability is given by area. Although the mathematics required to deal with density curves is more advanced, the most important ideas of probability apply to both kinds of probability models.

APPLY YOUR KNOWLEDGE

4.24 Random numbers. Let Y be a random number between 0 and 1 produced by the idealized uniform random number generator described in Example 4.7 and Figure 4.5. Find the following probabilities:
(a) $P(0 \leq Y \leq 0.4)$
(b) $P(0.4 \leq Y \leq 1)$
(c) $P(0.3 \leq Y \leq 0.5)$
(d) $P(0.3 < Y < 0.5)$

4.25 Adding random numbers. Generate two random numbers between 0 and 1 and take T to be their sum. The sum T can take any value between 0 and 2. The density curve of T is the triangle shown in Figure 4.6.
(a) Verify by geometry that the area under this curve is 1.
(b) What is the probability that T is less than 1? (Sketch the density curve, shade the area that represents the probability, then find that area. Do this for (c) also.)
(c) What is $P(T < 0.5)$?

Normal probability models

Any density curve can be used to assign probabilities. The density curves that are most familiar to us are the Normal curves introduced in Section 1.3. So **Normal distributions**

FIGURE 4.6 The density curve for the sum of two random numbers, for Exercise 4.25. This density curve spreads probability between 0 and 2.

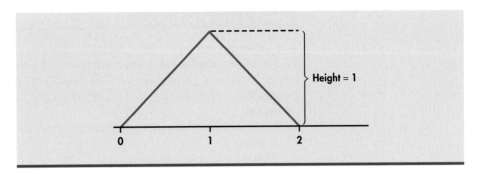

are probability models. There is a close connection between a Normal distribution as an idealized description for data and a Normal probability model. If we look at the weights of all 9-ounce bags of a particular brand of potato chip, we find that they closely follow the Normal distribution with mean $\mu = 9.12$ ounces and standard deviation $\sigma = 0.15$ ounces, $N(9.12, 0.15)$. That is a distribution for a large set of data. Now choose one 9-ounce bag at random. Call its weight W. If we repeat the random choice very many times, the distribution of values of W is the same Normal distribution.

EXAMPLE 4.8   The Weight of Bags of Potato Chips

What is the probability that a randomly chosen 9-ounce bag has weight between 9.33 and 9.45 ounces?

The weight W of the bag we choose has the $N(9.12, 0.15)$ distribution. Software can be used to find the probability directly, or we can standardize and use Table A, the table of standard Normal probabilities. We will reserve capital Z for a standard Normal variable, $N(0, 1)$.

$$P(9.33 \leq W \leq 9.45) = P\left(\frac{9.33 - 9.12}{0.15} \leq \frac{W - 9.12}{0.15} \leq \frac{9.45 - 9.12}{0.15}\right)$$
$$= P(1.4 \leq Z \leq 2.2)$$
$$= 0.9861 - 0.9192 = 0.0669$$

Figure 4.7 shows the areas under the standard Normal curve. The calculation is the same as those we did in Chapter 1. Only the language of probability is new.

FIGURE 4.7 The probability in Example 4.8 as an area under the standard Normal curve.

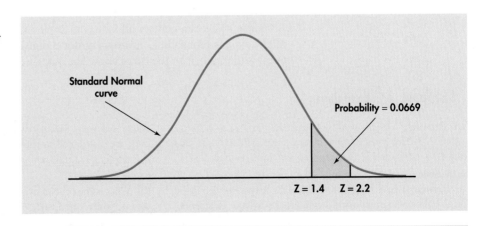

APPLY YOUR KNOWLEDGE

4.26 Miles per gallon for 2009 vehicles. The Normal distribution with mean $\mu = 17.3$ miles per gallon and standard deviation $\sigma = 4.2$ miles per gallon is an approximate model for the city gas mileage of 2009 model year vehicles. Figure 1.17 (page 41) pictures this distribution. Let X be the city miles per gallon of one 2009 vehicle chosen at random.
(a) Write the event "the vehicle chosen has a city miles per gallon of 30 or higher" in terms of X.
(b) Find the probability of this event.

4.27 SAT Math Reasoning scores. Scores on the SAT Math Reasoning test for 2008 high school graduates are approximately Normal with mean 515 points (out of a possible 800 points) and standard deviation 116 points. Let Y stand for the score of a randomly chosen student. Express each of the following events in terms of Y and use the 68–95–99.7 rule to give the approximate probability.
(a) The student has a score above 515.
(b) The student's score is above 631.

SECTION 4.2 Summary

- A **probability model** for a random phenomenon consists of a sample space S and an assignment of probabilities P.

- The **sample space** S is the set of all possible outcomes of the random phenomenon. Sets of outcomes are called **events**. P assigns a number $P(A)$ to an event A as its probability.

- Any assignment of probability must obey the rules that state the basic properties of probability:

  1. $0 \leq P(A) \leq 1$ for any event A.
  2. $P(S) = 1$.
  3. For any event A, $P(A \text{ does not occur}) = 1 - P(A)$.
  4. **Addition rule:** Events A and B are **disjoint** if they have no outcomes in common. If A and B are disjoint, then $P(A \text{ or } B) = P(A) + P(B)$.

- When a sample space S contains finitely many possible outcomes, a probability model assigns each of these outcomes a probability between 0 and 1 such that the sum of all the probabilities is exactly 1. The probability of any event is the sum of the probabilities of all the outcomes that make up the event.

- A sample space can contain all values in some interval of numbers. A probability model assigns probabilities as **areas under a density curve.** The probability of any event is the area under the curve above the outcomes that make up the event.

SECTION 4.2 Exercises

For Exercises 4.16 and 4.17, see page 225; for 4.18 to 4.20, see pages 227–228; for 4.21 to 4.23, see pages 230–231; for 4.24 and 4.25, see page 232; and for 4.26 and 4.27, see page 234.

4.28 Second Life around the world. Second Life (www.secondlife.com) is a free social network service based on an online 3D virtual world in which users, called "Residents," create avatars to interact with each other. Active Second Life users are distributed worldwide:[8]

(a) What probability should replace "?" in the distribution found on the next page?
(b) What is the probability that a randomly chosen Second Life user is North American?

Country:	U.S.	France	Germany	U.K.	Netherlands	Spain	Brazil	Canada	Other
Probability:	0.312	0.127	0.105	0.081	0.066	0.038	0.038	0.033	?

4.29 Confidence in institutions. A Gallup Poll (June 9 to 12, 2008) interviewed a random sample of 822 adults (18 years or older). The people in the sample were asked about their level of confidence in a variety of institutions in the United States. Here are the results for small and big businesses:

	Great deal	Quite a lot	Some	Very little	None	No opinion
Small business	0.28	0.32	0.31	0.07	0.00	0.02
Big business	0.07	0.13	0.43	0.32	0.03	0.02

(a) What is the probability that a randomly chosen person has either no opinion, no confidence, or very little confidence in small businesses? Find the similar probability for big businesses.

(b) Using your answer from part (a), determine the probability that a randomly chosen person has *at least* some confidence in small businesses. Again based on part (a), find the similar probability for big businesses.

4.30 Demographics—language. Canada has two official languages, English and French. Choose a Canadian at random and ask, "What is your mother tongue?" Here is the distribution of responses, combining many separate languages from the broad Asian/Pacific region:[9]

Language:	English	French	Asian/Pacific	Other
Probability:	0.59	0.23	0.07	?

(a) What probability should replace "?" in the distribution?
(b) What is the probability that a Canadian's mother tongue is not English?

4.31 Online health information. Based on a random sample of 1010 adults (18 years or older), a Harris Poll (July 8 to 13, 2008) estimates that 150 million U.S. adults have gone online for health information. Such individuals have been labeled as "cyberchondriacs." Cyberchondriacs in the sample were asked about the success of their online search for information about health topics. The distribution of responses is shown at the bottom of the page.
(a) Show that this is a legitimate probability distribution.
(b) What is the probability that a randomly chosen cyberchondriac feels that his or her search for health information was somewhat or very successful?

4.32 Spill or spell? Spell-checking software catches "nonword errors" that result in a string of letters that is not a word, as when "the" is typed as "teh." When undergraduates are asked to write a 250-word essay (without spell-checking), the number X of nonword errors has the following distribution:

Value of X:	0	1	2	3	4
Probability:	0.1	0.2	0.3	0.3	0.1

(a) Write the event "at least one nonword error" in terms of X. What is the probability of this event?
(b) Describe the event $X \leq 2$ in words. What is its probability? What is the probability that $X < 2$?

4.33 Modes of transportation. Governments (local and national) find it important to gather data on modes of transportation for commercial and workplace movement. Such information is useful for policy making as it pertains to infrastructure (like roads and railways), urban development, energy use, and pollution. Based on 2006 Canadian and U.S. government data, here are the distributions of the primary means of transportation to work for employees working outside the home:[10]

	Automobile (self or pool)	Public transportation	Bicycle or motorcycle	Walk	Other
Canada	?	0.110	0.013	0.064	0.013
U.S.	?	0.045	0.007	0.030	0.016

(a) What is the probability that a randomly chosen Canadian employee who works outside the home uses an automobile? What is the probability that a randomly chosen U.S. employee who works outside the home uses an automobile?
(b) Transportation systems primarily based on the automobile are regarded as unsustainable because of the excessive energy consumption and the effects on the health of populations. The Canadian government includes public transit, walking, and cycles as "sustainable" modes of transportation. For both countries, determine the probability that a randomly chosen employee who works outside the home uses sustainable transportation. How do you assess the relative status of sustainable transportation for these two countries?

4.34 World Internet usage. Approximately 23.6% of the world's population uses the Internet (as of December 2008).[11]

	Very successful	Somewhat successful	Neither successful nor unsuccessful	Somewhat unsuccessful	Very unsuccessful	Decline to answer
Probability	0.41	0.48	0.03	0.02	0.02	0.04

Furthermore, a randomly chosen Internet user has the following probabilities of being from the given region of the world:

Region:	Asia	Europe	North America	Latin America/Caribbean
Probability:	0.411	0.247	0.156	0.110

(a) What is the probability of a randomly chosen Internet user not being from one of the four regions explicitly listed in this table?

(b) What is the probability of a randomly chosen Internet user living in either Asia or Europe?

(c) What is the probability of a randomly chosen Internet user not living in North America?

4.35 Great Britain household size. In government data, a household consists of all occupants of a dwelling unit. Here is the distribution of household size in Great Britain according to the 2005 General Household Survey:

Number of persons:	1	2	3	4	5	6 or more
Probability:	0.31	0.35	0.15	0.13	0.04	0.02

Choose a Great Britain household at random and let the random variable Y be the number of persons living in the household.

(a) Express "more than one person lives in this household" in terms of Y. What is the probability of this event?

(b) What is $P(2 < Y \leq 4)$?

(c) What is $P(Y \neq 2)$?

4.36 Stock market movements. You watch the price of the Dow Jones Industrial Index for four days. Give a sample space for each of the following random phenomena.

(a) You record the sequence of up-days and down-days.

(b) You record the number of up-days.

4.37 Land in Iowa. Choose an acre of land in Iowa at random. The probability is 0.92 that it is farmland and 0.01 that it is forest.

(a) What is the probability that the acre chosen is not farmland?

(b) What is the probability that it is either farmland or forest?

(c) What is the probability that a randomly chosen acre in Iowa is something other than farmland or forest?

4.38 Car colors. Choose a new car or light truck at random and note its color. Here are the probabilities of the most popular colors for cars purchased in China in 2006:[12]

Color:	Silver	Black	Blue	White	Red	Yellow/Gold
Probability:	0.24	0.19	0.17	0.16	0.09	0.07

What is the probability that the car you choose has any color other than the six listed? What is the probability that a randomly chosen car is either silver or black? In North America, the probability of a new car being gray is 0.13. What can you say about the probability of a new car in China being gray?

4.39 Colors of M&M's. The colors of candies such as M&M's are carefully chosen to match consumer preferences. The color of an M&M drawn at random from a bag has a probability distribution determined by the proportions of colors among all M&M's of that type.

(a) Here is the distribution for plain M&M's:

Color:	Blue	Orange	Green	Brown	Yellow	Red
Probability:	0.24	0.20	0.16	0.14	0.14	?

What must be the probability of drawing a red candy?

(b) What is the probability that a plain M&M is any of orange, green, or yellow?

4.40 Almond M&M's. Exercise 4.39 gives the probabilities that an M&M candy is each of blue, orange, green, brown, yellow, and red. If "Almond" M&M's are equally likely to be any of these colors, what is the probability of any one color?

4.41 Legitimate probabilities? In each of the following situations, state whether or not the given assignment of probabilities to individual outcomes is legitimate, that is, satisfies the rules of probability. If not, give specific reasons for your answer.

(a) When a coin is spun, $P(H) = 0.55$ and $P(T) = 0.45$.

(b) When two coins are tossed, $P(HH) = 0.4$, $P(HT) = 0.4$, $P(TH) = 0.4$, and $P(TT) = 0.4$.

(c) Plain M&M's have not always had the mixture of colors given in Exercise 4.39. In the past there were no red candies and no blue candies. Tan had probability 0.10, and the other four colors had the same probabilities that are given in Exercise 4.39.

4.42 Who goes to Paris? Abby, Deborah, Sam, Tonya, and Roberto work in a firm's public relations office. Their employer must choose two of them to attend a conference in Paris. To avoid unfairness, the choice will be made by drawing two names from a hat. (This is an SRS of size 2.)

(a) Write down all possible choices of two of the five names. This is the sample space.

(b) The random drawing makes all choices equally likely. What is the probability of each choice?

(c) What is the probability that Tonya is chosen?

(d) What is the probability that neither of the two men (Sam and Roberto) is chosen?

4.43 How big are farms? Choose an American farm at random and measure its size in acres. The probabilities that the farm chosen falls in several acreage categories are shown below.

Acres:	<10	10–49	50–99	100–179	180–499	500–999	1000–1999	≥2000
Probability:	0.09	0.20	0.15	0.16	0.22	0.09	0.05	0.04

Let A be the event that the farm is less than 50 acres in size, and let B be the event that it is 500 acres or more.
(a) Find $P(A)$ and $P(B)$.
(b) Describe the event "A does not occur" in words and find its probability by Rule 3.
(c) Describe the event "A or B" in words and find its probability by the addition rule.

4.44 Roulette. A roulette wheel has 38 slots, numbered 0, 00, and 1 to 36. The slots 0 and 00 are colored green, 18 of the others are red, and 18 are black. The dealer spins the wheel and at the same time rolls a small ball along the wheel in the opposite direction. The wheel is carefully balanced so that the ball is equally likely to land in any slot when the wheel slows. Gamblers can bet on various combinations of numbers and colors.
(a) What is the probability of any one of the 38 possible outcomes? Explain your answer.
(b) If you bet on "red," you win if the ball lands in a red slot. What is the probability of winning?
(c) The slot numbers are laid out on a board on which gamblers place their bets. One column of numbers on the board contains all multiples of 3, that is, 3, 6, 9, ..., 36. You place a "column bet" that wins if any of these numbers comes up. What is your probability of winning?

4.45 Consumer preference. Suppose that half of all computer users prefer the new version of your company's best-selling software, and half prefer the old version. If you randomly choose three consumers and record their preferences, there are 8 possible outcomes. For example, NNO means the first two consumers prefer the new (N) version and the third consumer prefers the old version (O). All 8 arrangements are equally likely.
(a) Write down all 8 arrangements of preferences of three consumers. What is the probability of any one of these arrangements?
(b) Let X be the number of consumers who prefer the new version of the product. What is the probability that $X = 2$?
(c) Starting from your work in (a), list the values X can take and the probability for each value.

4.46 Moving up. A study of class mobility in England looked at the economic class reached by the sons of lower-class fathers. Economic classes are numbered from 1 (low) to 5 (high). Take X to be the class of a randomly chosen son of a father in Class 1. The study found that the probability model for X is

Son's class X:	1	2	3	4	5
Probability:	0.48	0.38	0.08	0.05	0.01

(a) Check that this distribution satisfies the two requirements for a legitimate assignment of probabilities to individual outcomes.

(b) What is $P(X \leq 3)$? (Be careful: the event "$X \leq 3$" includes the value 3.)
(c) What is $P(X < 3)$?
(d) Write the event "a son of a lower-class father reaches one of the two highest classes" in terms of values of X. What is the probability of this event?

4.47 How large are households? Choose an American household at random and let X be the number of persons living in the household. If we ignore the few households with more than seven inhabitants, the probability model for X is as follows:

Household size X:	1	2	3	4	5	6	7
Probability:	0.25	0.32	0.17	0.15	0.07	0.03	0.01

(a) Verify that this is a legitimate probability distribution.
(b) What is $P(X \geq 5)$?
(c) What is $P(X > 5)$?
(d) What is $P(2 < X \leq 4)$?
(e) What is $P(X \neq 1)$?
(f) Write the event that a randomly chosen household contains more than two persons in terms of X. What is the probability of this event?

4.48 Random numbers. Many random number generators allow users to specify the range of the random numbers to be produced. Suppose that you specify that the random number Y can take any value between 0 and 2. Then the density curve of the outcomes has constant height between 0 and 2, and height 0 elsewhere.
(a) What is the height of the density curve between 0 and 2? Draw a graph of the density curve.
(b) Use your graph from (a) and the fact that probability is area under the curve to find $P(Y \leq 1)$.
(c) Find $P(0.5 < Y < 1.3)$.
(d) Find $P(Y \geq 0.8)$.

4.49 SAT Math Reasoning Scores. Exercise 4.27 indicates that SAT Math Reasoning test scores are approximately Normal with mean 515 points and standard deviation 116 points.
(a) What is the probability that a randomly selected student has a score greater than 550?
(b) What is the probability that a randomly selected student has a score between 600 and 700?

4.50 Miles per gallon for 2009 vehicles. Exercise 4.26 indicates that city gas mileage for 2009 model year vehicles is approximately Normal with mean 17.3 miles per gallon and standard deviation 4.2 miles per gallon.
(a) What is the probability that a randomly selected 2009 vehicle has city mpg greater than 20?
(b) What is the probability that a randomly selected 2009 vehicle has city mpg between 18 and 25?

4.3 Random Variables

Not all sample spaces are made up of numbers. When we toss a coin four times, we can record the outcome as a string of heads and tails, such as HTTH. In statistics, however, we are most often interested in numerical outcomes such as the count of heads in the four tosses. It is convenient to use a shorthand notation: Let X be the number of heads. If our outcome is HTTH, then $X = 2$. If the next outcome is TTTH, the value of X changes to $X = 1$. The possible values of X are 0, 1, 2, 3, and 4. Tossing a coin four times will give X one of these possible values. We call X a *random variable* because prior to tossing the coin we are uncertain of what its value will be. Examples 4.6 and 4.7 used this shorthand notation. In Example 4.8, we let W stand for the weight of a randomly chosen 9-ounce bag of potato chips. We know that W would take a different value if we took another random sample. Because its value changes from one sample to another, W is also a random variable.

> **Random Variable**
>
> A **random variable** is a variable whose value is a numerical outcome of a random phenomenon.

We usually denote random variables by capital letters near the end of the alphabet, such as X or Y. However, when dealing with random variables associated with statistics of a random sample, such as the sample mean $\overline{x}$, we will keep the familiar lowercase notation.[13] As we progress from general rules of probability toward statistical inference, we will concentrate on random variables. When a random variable X describes a random phenomenon, the sample space S just lists the possible values of the random variable. We usually do not mention S separately.

We will meet two types of random variables, corresponding to two ways of assigning probability: *discrete* and *continuous*. Compare the random variable X in Example 4.6 (page 228) and the random variable Y in Example 4.7 (page 231). Both have uniform distributions that distribute probability evenly across the possible outcomes, but one is discrete and the other is continuous.

The possible values of X in Example 4.6 are the whole numbers $\{1, 2, 3, 4, 5, 6, 7, 8, 9\}$. These nine values are separated on the number line by other numbers that are not possible values of X:

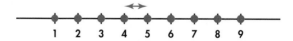

discrete random variable X is a **discrete random variable.** "Discrete" describes the nature of the sample space for X. In contrast, the sample space for Y in Example 4.7 is {all numbers between 0 and 1}. These possible values are not separated on the number line. As you move continuously from the value 0 to the value 1, you do not leave the sample space for Y:

continuous random variable Y is a **continuous random variable.** Once again, "continuous" describes the nature of the sample space.

APPLY YOUR KNOWLEDGE

4.51 Discrete or continuous? For each exercise listed below, decide whether the random variable described is discrete or continuous and give a brief explanation for your response.
(a) Exercise 4.16(b), page 225
(b) Exercise 4.16(e), page 225
(c) Exercise 4.26, page 234
(d) Exercise 4.47, page 237

4.52 Discrete or continuous? For each exercise listed below, decide whether the random variable described is discrete or continuous and sketch the sample space on a number line.
(a) Exercise 4.17(a), page 225
(b) Exercise 4.17(c), page 225
(c) Exercise 4.17(e), page 225

Probability distributions

The starting point for studying any random variable is its probability distribution, which is just the probability model for the outcomes.

> **Probability Distribution**
>
> The **probability distribution** of a random variable X tells us what values X can take and how to assign probabilities to those values.

Because of the difference in the nature of sample spaces for discrete and continuous random variables, we describe probability distributions for the two types of random variables separately.

Discrete random variables

A discrete random variable like the uniform random digit X in Example 4.6 (page 228) has a finite number of possible values.[14] The probability distribution of X therefore simply lists the values and their probabilities.

> **Discrete Probability Distributions**
>
> The **probability distribution** of a discrete random variable X lists the possible values of X and their probabilities:
>
> | Value of X: | x_1 | x_2 | x_3 | $\cdots$ | x_k |
> |---|---|---|---|---|---|
> | Probability: | p_1 | p_2 | p_3 | $\cdots$ | p_k |
>
> The probabilities p_i must satisfy two requirements:
>
> **1.** Every probability p_i is a number between 0 and 1, $0 \le p_i \le 1$.
> **2.** The sum of the probabilities is exactly 1, $p_1 + p_2 + \cdots + p_k = 1$.
>
> To find the probability of any event, add the probabilities p_i of the individual values x_i that make up the event.

We can use a probability histogram to display a discrete distribution. Figure 4.3 (page 229) compares the distributions of random digits and first digits that obey Benford's law.

EXAMPLE 4.9   Hard-Drive Sizes

Buyers of a laptop computer model may choose to purchase either a 120-, 160-, 250-, or 320-gigabyte (GB) internal hard drive. Choose customers from the last 60 days at random to ask what size internal hard drive they purchased for their computer. To "choose at random" means to give every customer of the last 60 days the same chance to be chosen. The size (in gigabytes) of the internal hard drive purchased by a randomly selected customer is a random variable X.

The value of X changes when we repeatedly choose customers at random, but it is always one of 120, 160, 250, or 320 GB. Here is the distribution of X:

Hard-drive size X:	120	160	250	320
Probability:	0.50	0.25	0.15	0.10

The probability histogram in Figure 4.8 pictures this distribution.

The probability that a randomly selected customer purchased at least a 250-GB hard drive is

$$P(\text{size is 250 or 320}) = P(X = 250) + P(X = 320)$$
$$= 0.15 + 0.10 = 0.25$$

FIGURE 4.8 Probability histogram for the distribution of hard-drive sizes in Example 4.9.

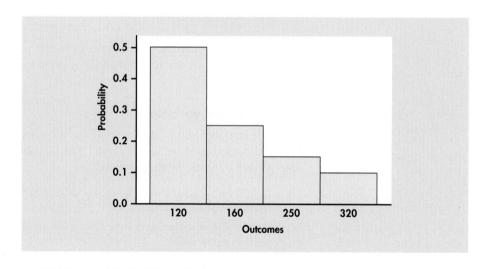

EXAMPLE 4.10   Four Coin Tosses

Toss a balanced coin four times; the discrete random variable X counts the number of heads. How shall we find the probability distribution of X?

The outcome of four tosses is a sequence of heads and tails such as HTTH. There are 16 possible outcomes in all. Figure 4.9 lists these outcomes along with the value of X for each outcome. A reasonable probability model says that the 16 outcomes are all equally likely; that is, each has probability 1/16. (This model is justified both by empirical observation and by a theoretical derivation that appears in the Chapter 5.)

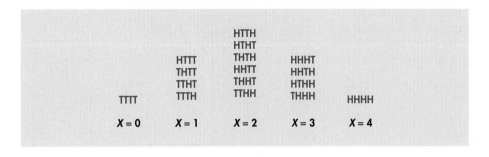

The number of heads X has possible values 0, 1, 2, 3, and 4. These values are *not* equally likely. As Figure 4.9 shows, $X = 0$ can occur in only one way, when the outcome is TTTT. So $P(X = 0) = 1/16$. The outcome $X = 2$, on the other hand, can occur in six different ways, so that

$$P(X = 2) = \frac{\text{count of ways } X = 2 \text{ can occur}}{16}$$

$$= \frac{6}{16}$$

We can find the probability of each value of X from Figure 4.9 in the same way. Here is the result:

$$P(X = 0) = \frac{1}{16} = 0.0625$$

$$P(X = 1) = \frac{4}{16} = 0.25$$

$$P(X = 2) = \frac{6}{16} = 0.375$$

$$P(X = 3) = \frac{4}{16} = 0.25$$

$$P(X = 4) = \frac{1}{16} = 0.0625$$

These probabilities have sum 1, so this is a legitimate probability distribution. In table form, the distribution is

Number of heads X:	0	1	2	3	4
Probability:	0.0625	0.25	0.375	0.25	0.0625

Figure 4.10 is a probability histogram for this distribution. The probability distribution is exactly symmetric. It is an idealization of the distribution of the proportions for numbers of heads after many tosses of four coins, which would be nearly symmetric but is unlikely to be exactly symmetric.

Any event involving the number of heads observed can be expressed in terms of X, and its probability can be found from the distribution of X. For example, the probability of tossing at least two heads is

$$P(X \geq 2) = 0.375 + 0.25 + 0.0625 = 0.6875$$

The probability of at least one head is most simply found by

$$P(X \geq 1) = 1 - P(X = 0)$$
$$= 1 - 0.0625 = 0.9375$$

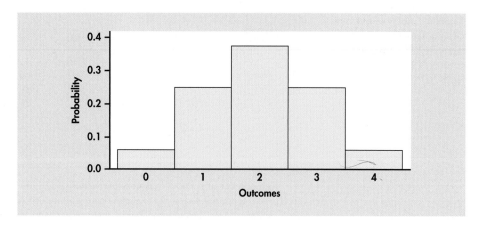

Continuous random variables

A continuous random variable like the uniform random number Y in Example 4.7 (page 231) or the Normally distributed package weight W in Example 4.8 (page 233) has an infinite number of possible values. Continuous probability distributions therefore assign probabilities directly to events as areas under a density curve.

> **Continuous Probability Distributions**
>
> The **probability distribution** of a continuous random variable X is described by a density curve. The probability of any event is the area under the density curve and above the values of X that make up the event.

EXAMPLE 4.11   Tread Life

The actual tread life X of a 40,000-mile automobile tire has a Normal probability distribution with $\mu = 50{,}000$ miles and $\sigma = 5500$ miles. We say X has an $N(50{,}000, 5500)$ distribution. From a manufacturer's perspective, it would be useful to know the percent of tires that fail to meet the guaranteed wear life of 40,000 miles. This implies we want to find the probability that a randomly selected tire has a tread life less than 40,000 miles:

$$
\begin{aligned}
P(X < 40{,}000) &= P\left(\frac{X - 50{,}000}{5500} < \frac{40{,}000 - 50{,}000}{5500} \right) \\
&= P(Z < -1.82) \\
&= 0.0344
\end{aligned}
$$

The $N(50{,}000, 5500)$ distribution is shown in Figure 4.11 with the area $P(X < 40{,}000)$ shaded. The manufacturer should expect to incur warranty costs for about 3.4% of its tires.

The probability distribution for a continuous random variable assigns probabilities to intervals of outcomes rather than to individual outcomes. In fact, **all continuous probability distributions assign probability 0 to every individual outcome.** Only intervals of values have positive probability. To see that this makes sense, return to the uniform random number generator of Example 4.7.

FIGURE 4.11 The Normal distribution with $\mu = 50{,}000$ and $\sigma = 5500$. The shaded area is $P(X < 40{,}000)$, calculated in Example 4.11.

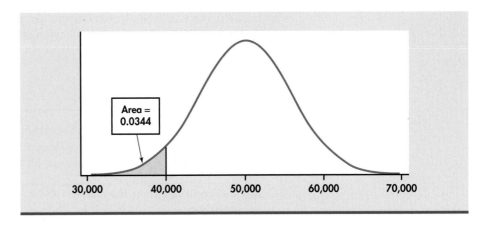

EXAMPLE 4.12   Can a Random Number Be Exactly 0.8?

What is $P(Y = 0.8)$, the probability that a random number generator produces *exactly* 0.8? Figure 4.12 shows the uniform density curve for outcomes between 0 and 1. Probabilities are areas under this curve. The probability of any interval of outcomes is equal to its length. This implies that because the point 0.8 has no length, its probability is 0.

FIGURE 4.12 The density curve of the uniform distribution of numbers chosen at random between 0 and 1, for Example 4.12. This model assigns probability 0 to the event that the number generated has any one specific value, such as 0.8.

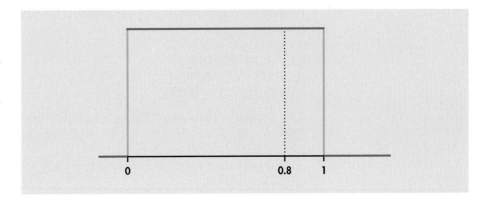

Although this fact may seem odd, it makes intuitive, as well as mathematical, sense. The random number generator produces a number between 0.79 and 0.81 with probability 0.02. An outcome between 0.799 and 0.801 has probability 0.002. A result between 0.799999 and 0.800001 has probability 0.000002. You see that as we home in on 0.8 the probability gets closer to 0. To be consistent, the probability of an outcome *exactly* equal to 0.8 must be 0.

Because there is a zero probability exactly at $Y = 0.8$, the two events $\{Y > 0.8\}$ and $\{Y \geq 0.8\}$ have the same probability. We can ignore the distinction between $>$ and $\geq$ when finding probabilities for continuous (but not discrete) random variables.

Example 4.12 reminds us that continuous probability distributions are idealizations. We don't measure the weight of potato chip bags or the tread life of tires to an unlimited number of decimal places. If we measure tread life to the nearest mile, we could use a discrete distribution with a very small probability for each specific tread life. It is much more convenient to use a continuous Normal distribution. There is an element of "not exactly right but good enough to use in practice" in most choices of a probability model.

We began this chapter with a general discussion of the idea of probability and the properties of probability models. Two specific types of probability models are distributions of discrete and continuous random variables. Our study of statistics will employ only these two types of probability models.

APPLY YOUR KNOWLEDGE

4.53 How many cars? Choose an American household at random and let the random variable X be the number of cars (including SUVs and light trucks) they own. Here is the probability model if we ignore the few households that own more than 5 cars:

Number of cars X:	0	1	2	3	4	5
Probability:	0.09	0.36	0.35	0.13	0.05	0.02

(a) Verify that this is a legitimate discrete distribution. Display the distribution in a probability histogram.
(b) Say in words what the event $\{X \geq 1\}$ is. Find $P(X \geq 1)$.
(c) Your company builds houses with two-car garages. What percent of households have more cars than the garage can hold?

4.54 Uniform probabilities. Let Y be a random number between 0 and 1 produced by the idealized uniform random number generator with density curve pictured in Figure 4.12. Find the following probabilities:
(a) $P(0 \leq Y \leq 0.4)$
(b) $P(0.4 \leq Y \leq 1)$
(c) $P(0.3 \leq Y \leq 0.5)$
(d) $P(0.3 < Y < 0.5)$
(e) $P(0.226 \leq Y \leq 0.713)$

4.55 Normal probabilities. Example 4.11 gives the Normal distribution $N(50,000, 5500)$ for the tread life X of a type of tire (in miles). Calculate the following probabilities:
(a) The probability that a tire lasts more than 50,000 miles.
(b) $P(X > 60,000)$
(c) $P(X \geq 60,000)$

The mean of a random variable

Probability is the mathematical language that describes the long-run regular behavior of random phenomena. The probability distribution of a random variable is an idealized distribution of the proportions of outcomes in very many observations. The probability histograms and density curves that picture probability distributions resemble our earlier pictures of distributions of data. In describing data, we moved from graphs to numerical measures such as means and standard deviations. Now we will make the same move to expand our descriptions of the distributions of random variables. We can speak of the mean winnings in a game of chance or the standard deviation of the randomly varying number of calls a travel agency receives in an hour. We will learn more about how to compute these descriptive measures and about the laws they obey.

The mean $\bar{x}$ of a set of observations is their ordinary average. The mean of a random variable X is also an average of the possible values of X, but with an essential change to take into account the fact that not all outcomes need be equally likely. An example will show what we must do.

EXAMPLE 4.13 Pick 3

Most states and Canadian provinces have government-sponsored lotteries. Here is a simple lottery wager, from the Tri-State Pick 3 game that New Hampshire shares with Maine and Vermont. You choose a three-digit number; the state chooses a three-digit winning number at random and pays you $500 if your number is chosen. Because there are 1000 three-digit numbers, you have probability 1/1000 of winning. Taking X to be the amount your ticket pays you, the probability distribution of X is

Payoff X:	$0	$500
Probability:	0.999	0.001

What is your average payoff from many tickets? The ordinary average of the two possible outcomes $0 and $500 is $250, but that makes no sense as the average because $500 is much less likely than $0. In the long run you receive $500 once in every 1000 tickets and $0 on the remaining 999 of 1000 tickets. The long-run average payoff is

$$\$500\frac{1}{1000} + \$0\frac{999}{1000} = \$0.50$$

or fifty cents. That number is the mean of the random variable X. (Tickets cost $1, so in the long run the state keeps half the money you wager.)

If you play Tri-State Pick 3 several times, we would as usual call the mean of the actual amounts your tickets pay $\bar{x}$. The mean in Example 4.13 is a different quantity—it is the long-run average payoff you expect if you play a very large number of times. Just as probabilities are an idealized description of long-run proportions, the mean of a probability distribution describes the long-run average outcome. The common symbol for the **mean of a probability distribution** is μ, the Greek letter mu. We used μ in Chapter 1 for the mean of a Normal distribution, so this is not a new notation. We will often be interested in several random variables, each having a different probability distribution with a different mean. To remind ourselves that we are talking about the mean of X, we can write μ_X rather than simply μ. In Example 4.13, $\mu_X = \$0.50$. Notice that, as often happens, the mean is not a possible value of X. You will often find the mean of a random variable X called the **expected value** of X. This term can be misleading, for we don't necessarily expect one observation on X to be close to its mean.

mean μ

expected value

The mean of any discrete random variable is found just as in Example 4.13. It is an average of the possible outcomes, but a weighted average in which each outcome is weighted by its probability. Because the probabilities add to 1, we have total weight 1 to distribute among the outcomes. An outcome that occurs half the time has probability one-half and gets one-half the weight in calculating the mean. Here is the general definition.

> **Mean of a Discrete Random Variable**
>
> Suppose that X is a discrete random variable whose distribution is
>
> | Value of X: | x_1 | x_2 | x_3 | $\cdots$ | x_k |
> |----------------|-------|-------|-------|----------|-------|
> | Probability: | p_1 | p_2 | p_3 | $\cdots$ | p_k |
>
> To find the **mean** of X, multiply each possible value by its probability, then add all the products:
>
> $$\mu_X = x_1 p_1 + x_2 p_2 + \cdots + x_k p_k$$
> $$= \sum x_i p_i$$

EXAMPLE 4.14 First Digits

If first digits in a set of records appear "at random," the nine possible digits 1 to 9 all have the same probability. The probability distribution of the first digit X is then

First digit X:	1	2	3	4	5	6	7	8	9
Probability:	1/9	1/9	1/9	1/9	1/9	1/9	1/9	1/9	1/9

The mean of this distribution is

$$\mu_X = 1\left(\frac{1}{9}\right) + 2\left(\frac{1}{9}\right) + 3\left(\frac{1}{9}\right) + 4\left(\frac{1}{9}\right) + 5\left(\frac{1}{9}\right) + 6\left(\frac{1}{9}\right) + 7\left(\frac{1}{9}\right) + 8\left(\frac{1}{9}\right) + 9\left(\frac{1}{9}\right)$$

$$= 45 \times \frac{1}{9} = 5$$

If, on the other hand, the records obey Benford's law, the distribution of the first digit V is

First digit V:	1	2	3	4	5	6	7	8	9
Probability:	0.301	0.176	0.125	0.097	0.079	0.067	0.058	0.051	0.046

The mean of V is

$$\mu_V = (1)(0.301) + (2)(0.176) + (3)(0.125) + (4)(0.097) + (5)(0.079)$$
$$+ (6)(0.067) + (7)(0.058) + (8)(0.051) + (9)(0.046)$$
$$= 3.441$$

The means reflect the greater probability of smaller first digits under Benford's law.

Figure 4.13 locates the means of X and V on the two probability histograms. Because the discrete uniform distribution of Figure 4.13(a) is symmetric, the mean lies at the center of symmetry. We can't locate the mean of the right-skewed distribution of Figure 4.13(b) by eye—calculation is needed.

FIGURE 4.13 Locating the mean of a discrete random variable on the probability histogram for (a) digits between 1 and 9 chosen at random; (b) digits between 1 and 9 chosen from records that obey Benford's law.

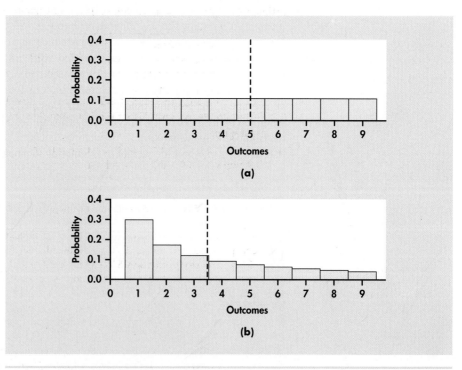

What about continuous random variables? The probability distribution of a continuous random variable X is described by a density curve. Chapter 1 showed how to find the mean of the distribution: it is the point at which the area under the density curve would balance if it were made out of solid material. The mean lies at the center of symmetric density curves such as the Normal curves. Exact calculation of the mean of a distribution with a skewed density curve requires advanced mathematics.[15] The idea that the mean is the balance point of the distribution applies to discrete random variables as well, but in the discrete case we have a formula that gives us the numerical value of μ.

APPLY YOUR KNOWLEDGE

4.56 Household size. Choose an American household at random and let X be the number of persons living in the household. If we ignore the few households with more than seven inhabitants, the probability model for X is as follows:

Household size X:	1	2	3	4	5	6	7
Probability:	0.25	0.32	0.17	0.15	0.07	0.03	0.01

Find the mean μ_X. Sketch a probability histogram of the distribution of X and mark the mean on your sketch. The Census Bureau reports that the mean size of American households is 2.65 people. Why is your result slightly less than this?

4.57 Hard-drive sizes. Example 4.9 (page 240) gives the distribution of customer choices of hard-drive size for a laptop computer model. Find the mean μ of this probability distribution. Explain in simple language what μ tells us about customer choices. Also explain why knowing μ is not very helpful to the computer maker.

4.58 Customer orders. Each week your business receives orders from customers who wish to have air-conditioning units installed in their homes. From past data, you estimate the distribution of the number of units ordered in a week X to be

Units ordered X:	0	1	2	3	4	5
Probability:	0.05	0.15	0.27	0.33	0.13	0.07

(a) Sketch a probability histogram for the distribution of X.
(b) Calculate the mean of X and mark it on the horizontal axis of your probability histogram.
(c) If you hire workers sufficient to handle the mean demand, what is the probability that you will be unable to handle all of a week's installation orders?

Rules for means

Portfolio Analysis The *rate of return* of an investment over a time period is the percent change in the price during the time period, plus any income received. For example, Apple Computer's stock price was $169.53 at the beginning of August 2008 and $113.66 at the end of that month. Apple pays no dividends, so Apple's rate of return for that time period was

$$\frac{\text{change in price}}{\text{starting price}} = \frac{113.66 - 169.53}{169.63} = -0.33, \text{ or } -33\%$$

Investors want high positive returns, but they also want safety. Over the course of a nearly 20-year period (1988 to 2008), Apple's monthly returns have swung to as low as -57%

and to as high as $+45\%$. The variability of returns, called *volatility* in finance, is a measure of the risk of an investment. A highly volatile stock, which may go either up or down a lot, is more risky than a Treasury bill, whose return is very predictable.

A *portfolio* is a collection of investments held by an individual or an institution. *Portfolio analysis* begins by studying how the risk and return of a portfolio are determined by the risk and return of the individual investments it contains. That's where statistics comes in: the return on an investment over some period of time is a random variable. We are interested in the *mean* return and we measure volatility by the *standard deviation* of returns. Let's say that Sadie is interested in building a computer technology–based portfolio. Her intuition is that mixing a stock based on a Microsoft-based PC and an Apple stock will hedge the risk of one stock versus the other. Sadie decides to place 60% of her funds in Dell Inc. and 40% in Apple Inc. If X is the monthly return on Dell and Y the monthly return on Apple, the portfolio rate of return is

$$R = 0.6X + 0.4Y$$

How can we find the mean and standard deviation of the portfolio return R starting from information about X and Y? We must now develop the machinery to do this.

Think first not about investments but about making refrigerators. Dimples and paint sags are two kinds of flaws in the painted finish of refrigerators. Not all refrigerators have the same number of dimples: many have none, some have one, some two, and so on. The inspectors report finding an average of 0.7 dimples and 1.4 sags per refrigerator. How many total imperfections of both kinds (on the average) are on a refrigerator? That's easy: if the average number of dimples is 0.7 and the average number of sags is 1.4, then counting both gives an average of $0.7 + 1.4 = 2.1$ flaws.

In more formal language, the number of dimples on a refrigerator is a random variable X that varies as we inspect one refrigerator after another. We know only that the mean number of dimples is $\mu_X = 0.7$. The number of paint sags is a second random variable Y having mean $\mu_Y = 1.4$. (As usual, the subscripts keep straight which variable we are talking about.) The total number of both dimples and sags is another random variable, the sum $X + Y$. Its mean μ_{X+Y} is the average number of dimples and sags together. It is just the sum of the individual means μ_X and μ_Y. That's an important rule for how means of random variables behave.

Here's another rule. A large lot of plastic coffee-can lids has mean diameter 4.2 inches. What is the mean in centimeters? There are 2.54 centimeters in an inch, so the diameter in centimeters of any lid is 2.54 times its diameter in inches. If we multiply every observation by 2.54, we also multiply their average by 2.54. The mean in centimeters must be 2.54×4.2, or about 10.7 centimeters. More formally, the diameter in inches of a lid chosen at random from the lot is a random variable X with mean μ_X. The diameter in centimeters is $2.54X$, and this new random variable has mean $2.54\mu_X$. Here are the rules we need.

Rules for Means

Rule 1. If X is a random variable and a and b are fixed numbers, then

$$\mu_{a+bX} = a + b\mu_X$$

Rule 2. If X and Y are random variables, then

$$\mu_{X+Y} = \mu_X + \mu_Y$$

This is the **addition rule for means.**

EXAMPLE 4.15 Portfolio Analysis

PORTFOLIO

The past behavior of the two securities in Sadie's portfolio is pictured in Figure 4.14, which plots the monthly returns for Dell stocks against Apple stocks from January 1990 to August 2008. We can see that the returns on the two stocks are correlated. This fact will be used later for gaining a complete assessment of the expected performance of the portfolio. For now, we can calculate mean returns from the 224 data points shown on the plot:[16]

$$X = \text{monthly return for Dell} \qquad \mu_X = 2.02\%$$
$$Y = \text{monthly return for Apple} \qquad \mu_Y = 1.75\%$$

Sadie invests 60% in Dell and 40% in Apple. We can find the mean return on her portfolio by combining Rules 1 and 2:

$$R = 0.6X + 0.4Y$$
$$\mu_R = 0.6\mu_X + 0.4\mu_Y$$
$$= (0.6)(2.02) + (0.4)(1.75) = 1.91\%$$

This calculation uses historical data on returns. Next month may of course be very different. It is usual in finance to use the term *expected return* in place of mean return. Remember our warning that we can't generally expect a future value to be close to the mean.

FIGURE 4.14 Monthly returns on Dell stock versus returns on Apple stock (January 1990 to August 2008) for Example 4.15.

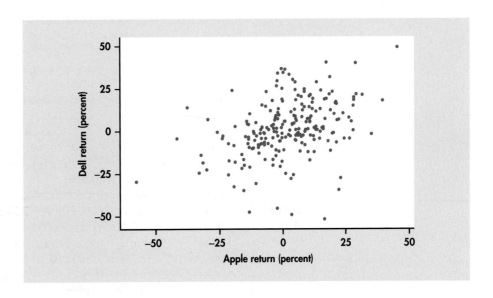

EXAMPLE 4.16 Gain Communications

Gain Communications sells aircraft communications units to both the military and the civilian markets. Next year's sales depend on market conditions that cannot be predicted exactly. Gain follows the modern practice of using probability estimates of sales. The military division estimates its sales as follows:

Units sold:	1000	3000	5000	10,000
Probability:	0.1	0.3	0.4	0.2

These are *personal probabilities* that express the informed opinion of Gain's executives. The corresponding sales estimates for the civilian division are

Units sold:	300	500	750
Probability:	0.4	0.5	0.1

Take X to be the number of military units sold and Y the number of civilian units sold. From the probability distributions we compute that

$$\mu_X = (1000)(0.1) + (3000)(0.3) + (5000)(0.4) + (10,000)(0.2)$$
$$= 100 + 900 + 2000 + 2000 = 5000 \text{ units}$$
$$\mu_Y = (300)(0.4) + (500)(0.5) + (750)(0.1)$$
$$= 120 + 250 + 75 = 445 \text{ units}$$

Gain makes a profit of \$2000 on each military unit sold and \$3500 on each civilian unit. Next year's profit from military sales will be $2000X$, \$2000 times the number X of units sold. By Rule 1 for means, the mean military profit is

$$\mu_{2000X} = 2000\mu_X = (2000)(5000) = \$10,000,000$$

Similarly, the civilian profit is $3500Y$, and the mean profit from civilian sales is

$$\mu_{3500Y} = 3500\mu_Y = (3500)(445) = \$1,557,500$$

The total profit is the sum of the military and civilian profit:

$$Z = 2000X + 3500Y$$

Rule 2 for means says that the mean of this sum of two variables is the sum of the two individual means:

$$\mu_Z = \mu_{2000X} + \mu_{3500Y}$$
$$= \$10,000,000 + \$1,557,500$$
$$= \$11,557,500$$

This mean is the company's best estimate of next year's profit, combining the probability estimates of the two divisions. We can do this calculation more quickly by combining Rules 1 and 2:

$$\mu_Z = \mu_{2000X+3500Y}$$
$$= 2000\mu_X + 3500\mu_Y$$
$$= (2000)(5000) + (3500)(445) = \$11,557,500$$

APPLY YOUR KNOWLEDGE

4.59 Managing new-product development process. Managers often have to oversee a series of related activities directed to a desired goal or output. As a new-product development manager, you are responsible for two sequential steps of the product development process, namely, the development of product specifications followed by the design of the manufacturing process. Let X be the number of weeks required to complete the development of product specifications, and let Y be the number of weeks required to complete the design of the manufacturing process. Based on experience, you estimate the following probability distribution for the first step:

Weeks (X):	1	2	3
Probability:	0.3	0.5	0.2

For the second step, your estimated distribution is

Weeks (Y):	1	2	3	4	5
Probability:	0.1	0.15	0.4	0.30	0.05

(a) Calculate μ_X and μ_Y.
(b) The cost per week for the activity of developing product specifications is $8000, while the cost per week for the activity of designing the manufacturing process is $30,000. Calculate the mean cost for each step.
(c) Calculate the mean completion time and mean cost for the two steps combined.

CASE 4.2 **4.60 Mutual funds.** The addition rule for means extends to sums of any number of random variables. Let's look at a portfolio containing three mutual funds. The monthly returns on Fidelity Magellan Fund, Fidelity Energy Fund, and Fidelity Japan Fund for the 60 months ending in August 2008 had approximately these means:

$$W = \text{Magellan monthly return} \qquad \mu_W = 0.28\%$$

$$X = \text{Energy monthly return} \qquad \mu_X = 1.54\%$$

$$Y = \text{Japan monthly return} \qquad \mu_Y = 0.47\%$$

What is the mean monthly return for a portfolio consisting of 50% Magellan, 30% Energy, and 20% Japan?

The variance of a random variable

The mean portfolio return in Example 4.15 is less than the mean return for Dell alone. What does Sadie gain from including Apple and giving up some average return? We will soon see that Apple's stock is slightly less volatile than Dell's stock. By combining the different stocks into a portfolio, Sadie hopes that her portfolio returns will vary less from month to month than would be the case if she invested all her funds in Dell. The mean is a measure of the center of a distribution. Even the most basic numerical description requires in addition a measure of the spread or variability of the distribution.

The variance and the standard deviation are the measures of spread that accompany the choice of the mean to measure center. Just as for the mean, we need a distinct symbol to distinguish the variance of a random variable from the variance s^2 of a data set. We write the variance of a random variable X as σ_X^2. The definition of the variance σ_X^2 of a random variable is similar to the definition of the sample variance s^2 given in Chapter 1. That is, the variance is an average of the squared deviation $(X - \mu_X)^2$ of the variable X from its mean μ_X. As for the mean, the average we use is a weighted average in which each outcome is weighted by its probability to take account of outcomes that are not equally likely. Calculating this weighted average is straightforward for discrete random variables but requires advanced mathematics in the continuous case. Here is the definition.

Variance of a Discrete Random Variable

Suppose that X is a discrete random variable whose distribution is

Value of X:	x_1	x_2	x_3	$\cdots$	x_k
Probability:	p_1	p_2	p_3	$\cdots$	p_k

and that μ_X is the mean of X. The **variance** of X

$$\sigma_X^2 = (x_1 - \mu_X)^2 p_1 + (x_2 - \mu_X)^2 p_2 + \cdots + (x_k - \mu_X)^2 p_k$$
$$= \sum (x_i - \mu_X)^2 p_i$$

The **standard deviation** σ_X of X is the square root of the variance.

EXAMPLE 4.17 Gain Communications

In Example 4.16 we saw that the number X of communications units that the Gain Communications military division hopes to sell has distribution

Units sold:	1000	3000	5000	10,000
Probability:	0.1	0.3	0.4	0.2

We can find the mean and variance of X by arranging the calculation in the form of a table. Both μ_X and σ_X^2 are sums of columns in this table.

x_i	p_i	$x_i p_i$	$(x_i - \mu_X)^2 p_i$
1,000	0.1	100	$(1{,}000 - 5{,}000)^2 (0.1) = 1{,}600{,}000$
3,000	0.3	900	$(3{,}000 - 5{,}000)^2 (0.3) = 1{,}200{,}000$
5,000	0.4	2,000	$(5{,}000 - 5{,}000)^2 (0.4) = \quad\quad 0$
10,000	0.2	2,000	$(10{,}000 - 5{,}000)^2 (0.2) = 5{,}000{,}000$
		$\mu_X = 5{,}000$	$\sigma_X^2 = 7{,}800{,}000$

We see that $\sigma_X^2 = 7{,}800{,}000$. The standard deviation of X is $\sigma_X = \sqrt{7{,}800{,}000} = 2792.8$. The standard deviation is a measure of how variable the number of units sold is. As in the case of distributions for data, the standard deviation of a probability distribution is easiest to understand for Normal distributions.

APPLY YOUR KNOWLEDGE

4.61 Civilian sales. Example 4.16 also gives the distribution of Gain's civilian sales Y. Find the variance σ_Y^2 and the standard deviation σ_Y.

4.62 Hard-drive sizes. Example 4.9 (page 240) gives the distribution of hard-drive sizes for sales of a particular model computer. You found the mean hard-drive size in Exercise 4.57 (page 247). Find the standard deviation of the distribution of hard-drive sizes.

4.63 Managing new-product development process. Exercise 4.59 (page 250) gives the distribution of time to complete two steps in the new-product development process. (a) Calculate the variance and the standard deviation of the number of weeks to complete the development of product specifications. (b) Calculate σ_Y^2 and σ_Y for the design of the manufacturing-process step.

Rules for variances

What are the facts for variances that parallel Rules 1 and 2 for means? The mean of a sum of random variables is always the sum of their means, but this addition rule is not always true for variances. To understand why, take X to be the percent of a family's after-tax income that is spent and Y the percent that is saved. When X increases, Y decreases by the same amount. Though X and Y may vary widely from year to year, their sum $X + Y$ is always 100% and does not vary at all. It is the association between the variables X and Y that prevents their variances from adding. If random variables are *independent,* this kind of association between their values is ruled out and their variances do add. Two random variables X and Y are **independent** if knowing that any event involving X alone did or did not occur tells us nothing about the occurrence of any event involving Y alone. Probability models often assume independence when the random variables describe outcomes that appear unrelated to each other. You should ask in each instance whether the assumption of independence seems reasonable.

independence

When random variables are not independent, the variance of their sum depends on the **correlation** between them as well as on their individual variances. In Chapter 2, we met the correlation r between two observed variables measured on the same individuals. We defined (page 93) the correlation r as an average of the products of the standardized x and y observations. The correlation between two random variables is defined in the same way, once again using a weighted average with probabilities as weights. We won't give the details—it is enough to know that the correlation between two random variables has the same basic properties as the correlation r calculated from data. We use ρ, the Greek letter rho, for the correlation between two random variables. The correlation ρ is a number between -1 and 1 that measures the direction and strength of the linear relationship between two variables. **The correlation between two independent random variables is zero.**

correlation

Returning to family finances, if X is the percent of a family's after-tax income that is spent and Y the percent that is saved, then $Y = 100 - X$. This is a perfect linear relationship with a negative slope, so the correlation between X and Y is $\rho = -1$. With the correlation at hand, we can state the rules for manipulating variances.

Rules for Variances

Rule 1. If X is a random variable and a and b are fixed numbers, then

$$\sigma^2_{a+bX} = b^2 \sigma^2_X$$

Rule 2. If X and Y are independent random variables, then

$$\sigma^2_{X+Y} = \sigma^2_X + \sigma^2_Y$$
$$\sigma^2_{X-Y} = \sigma^2_X + \sigma^2_Y$$

This is the **addition rule for variances of independent random variables.**

Rule 3. If X and Y have correlation ρ, then

$$\sigma^2_{X+Y} = \sigma^2_X + \sigma^2_Y + 2\rho\sigma_X\sigma_Y$$
$$\sigma^2_{X-Y} = \sigma^2_X + \sigma^2_Y - 2\rho\sigma_X\sigma_Y$$

This is the **general addition rule for variances of random variables.**

Notice that because a variance is the average of *squared* deviations from the mean, multiplying X by a constant b multiplies σ^2_X by the *square* of the constant. Adding a constant a to a random variable changes its mean but does not change its variability.

The variance of $X + a$ is therefore the same as the variance of X. Because the square of -1 is 1, the addition rule says that the variance of a difference of independent random variables is the *sum* of the variances. For independent random variables, the difference $X - Y$ is more variable than either X or Y alone because variations in both X and Y contribute to variation in their difference.

As with data, we prefer the standard deviation to the variance as a measure of the variability of a random variable. Rule 2 for variances implies that standard deviations of independent random variables do *not* add. To combine standard deviations, use the rules for variances rather than trying to remember separate rules for standard deviations. For example, the standard deviations of $2X$ and $-2X$ are both equal to $2\sigma_X$ because this is the square root of the variance $4\sigma_X^2$.

EXAMPLE 4.18 Pick 3

The payoff X of a \$1 ticket in the Tri-State Pick 3 game is \$500 with probability 1/1000 and 0 the rest of the time. Here is the combined calculation of mean and variance:

x_i	p_i	$x_i p_i$	$(x_i - \mu_X)^2 p_i$
0	0.999	0	$(0 - 0.5)^2(0.999) =$ 0.24975
500	0.001	0.5	$(500 - 0.5)^2(0.001) =$ 249.50025
		$\mu_X = 0.5$	$\sigma_X^2 =$ 249.75

The standard deviation is $\sigma_X = \sqrt{249.75} = \$15.80$. Games of chance usually have large standard deviations because large variability makes gambling exciting.

If you buy a Pick 3 ticket, your net winnings are $W = X - 1$ because the dollar you pay for the ticket must be subtracted from the payoff. By the rules for means, the mean amount you win is

$$\mu_W = \mu_X - 1 = -\$0.50$$

That is, you lose an average of 50 cents on a ticket. The rules for variances remind us that the variance and standard deviation of the winnings $W = X - 1$ are the same as those of the payoff X. Subtracting a fixed number changes the mean but not the variance.

Suppose now that you buy a \$1 ticket on each of two different days. The payoffs X and Y on the two tickets are independent because separate drawings are held each day. Your total payoff $X + Y$ has mean

$$\mu_{X+Y} = \mu_X + \mu_Y = \$0.50 + \$0.50 = \$1.00$$

Because X and Y are independent, the variance of $X + Y$ is

$$\sigma_{X+Y}^2 = \sigma_X^2 + \sigma_Y^2 = 249.75 + 249.75 = 499.5$$

The standard deviation of the total payoff is

$$\sigma_{X+Y} = \sqrt{499.5} = \$22.35$$

This is not the same as the sum of the individual standard deviations, which is \$15.80 + \$15.80 = \$31.60. Variances of independent random variables add; standard deviations do not.

EXAMPLE 4.19 Risk Pooling in a Supply Chain

To remain competitive, companies worldwide are increasingly recognizing the need to effectively manage their supply chains. Simply stated, a supply chain encompasses all processes involving the movement of goods and information from the raw materials stage, supply, production, and distribution to the ultimate customer. For many companies managing their supply chains, there is a critical need to use statistical methods to understand the fluctuations in customer demand with the goal of holding an appropriate level of inventory.

Let us consider a simple but realistic supply chain scenario.[17] ElectroWorks is a company that manufactures and distributes electronic parts to various regions in the United States. To serve the Chicago-Milwaukee region, the company has a warehouse in Milwaukee and another in Chicago. Because the company produces thousands of parts, it is considering an alternative strategy of locating a single warehouse between the two markets, say in Kenosha, Wisconsin; that will serve all customer orders. Delivery time, referred to as *lead time,* from manufacturing to warehouse(s) and ultimately to customers is unaffected by the new strategy.

To illustrate the implications of the centralized warehouse, let us focus on one specific part; SurgeArrester. The lead time for this part from manufacturing to warehouses is one week. Based on historical data, the lead time demands for the part in each of the markets are Normally distributed with

$$X = \text{Milwaukee warehouse} \qquad \mu_X = 415 \text{ units} \qquad \sigma_X = 48 \text{ units}$$
$$Y = \text{Chicago warehouse} \qquad \mu_Y = 2689 \text{ units} \qquad \sigma_Y = 272 \text{ units}$$

If the company were to centralize, what would be the mean and standard deviation of the total lead time demand $X + Y$?

The mean overall lead time demand is

$$\mu_{X+Y} = \mu_X + \mu_Y = 415 + 2689 = 3104$$

The variance and standard deviation of the total *cannot be computed* from the information given. Not surprisingly, demands in the two markets are not independent, because of the proximity of the regions. Therefore, Rule 2 does not apply, and we need to know ρ, the correlation between X and Y, to apply Rule 3.

Historically, the correlation between Milwaukee demand and Chicago demand is about $\rho = 0.52$. To find the variance of the overall demand, we use Rule 3:

$$\sigma_{X+Y}^2 = \sigma_X^2 + \sigma_Y^2 + 2\rho\sigma_X\sigma_Y$$
$$= (48)^2 + (272)^2 + (2)(0.52)(48)(272)$$
$$= 89{,}866.24$$

The variance of the sum $X + Y$ is greater than the sum of the variances $\sigma_X^2 + \sigma_Y^2$ because of the positive correlation between the two markets. We find the standard deviation from the variance,

$$\sigma_{X+Y} = \sqrt{89{,}866.24} = 299.78$$

Notice that even though the variance of the sum is greater than the sum of the variances, the standard deviation of the sum is less than the sum of the standard deviations. Here lies the potential benefit of a centralized warehouse. To protect against stockouts, ElectroWorks maintains safety stock for a given product at each warehouse. Safety stock is extra stock in hand over and above the mean demand. For example, if ElectroWorks has a policy of holding two standard deviations of safety stock, then the amount of safety stock (rounded to the nearest integer) at warehouses would be

Location	Safety stock
Milwaukee warehouse	2(48) = 96 units
Chicago warehouse	2(272) = 544 units
Centralized warehouse	2(299.78) = 600 units

The combined safety stock for the Milwaukee and Chicago warehouses is 640 units, which is 40 more units required than if distribution was operated out of a centralized warehouse. Now imagine the implication for safety stock when you take into consideration not just one part but *thousands* of parts that need to be stored. This example illustrates the important supply chain concept known as risk pooling. Many companies such as Walmart and e-commerce retailer Amazon take advantage of the benefits of risk pooling as illustrated by this example.

EXAMPLE 4.20   Portfolio Analysis

CASE 4.2

Now we can complete our initial analysis of Sadie's portfolio. Based on monthly returns between 1990 and 2008, we have

$$X = \text{monthly return for Dell} \qquad \mu_X = 2.02\% \quad \sigma_X = 16.59\%$$
$$Y = \text{monthly return for Apple} \qquad \mu_Y = 1.75\% \quad \sigma_Y = 15.09\%$$
$$\text{Correlation between } X \text{ and } Y: \qquad \rho = 0.379$$

We see that Dell has a higher mean return than Apple but is also more volatile. This is not an uncommon relationship in finance: a more volatile investment often offers a higher return, compensating investors for its greater risk. For the return R on Sadie's portfolio of 60% for Dell and 40% for Apple,

$$R = 0.6X + 0.4Y$$
$$\mu_R = 0.6\mu_X + 0.4\mu_Y$$
$$= (0.6 \times 2.02) + (0.4 \times 1.75) = 1.91\%$$

To find the variance of the portfolio return, combine Rules 1 and 3:

$$\sigma_R^2 = \sigma_{0.6X}^2 + \sigma_{0.4Y}^2 + 2\rho\sigma_{0.6X}\sigma_{0.4Y}$$
$$= (0.6)^2\sigma_X^2 + (0.4)^2\sigma_Y^2 + 2\rho(0.6 \times \sigma_X)(0.4 \times \sigma_Y)$$
$$= (0.6)^2(16.59)^2 + (0.4)^2(15.09)^2 + (2)(0.379)(0.6 \times 16.59)(0.4 \times 15.09)$$
$$= 181.06$$
$$\sigma_R = \sqrt{181.06} = 13.46\%$$

The portfolio has a smaller mean return than investing in all Dell stocks would, but it is also less volatile. As a proportion of the Dell values, the reduction in standard deviation is greater than the reduction in mean return. That's why Sadie put some funds into Apple stock.

Example 4.20 illustrates the first step in modern finance, using the mean and standard deviation to describe the behavior of a portfolio. The next step is to seek best-possible portfolios based on a given set of securities by finding the combination with the highest return for a specified level of risk. The investor says how much risk (measured as standard deviation of returns) she is willing to bear; the best portfolio has the highest mean return for this standard deviation. Our examples also suggest the characteristic weakness of portfolio analysis: risk and return in the future may be very different from the σ and μ we get from past returns.

APPLY YOUR KNOWLEDGE

4.64 Comparing sales. Tamara and Derek are sales associates in a large electronics and appliance store. Their store tracks each associate's daily sales in dollars. Tamara's sales total X varies from day to day with mean and standard deviation

$$\mu_X = \$1100 \quad \text{and} \quad \sigma_X = \$100$$

Derek's sales total Y also varies, with

$$\mu_Y = \$1000 \quad \text{and} \quad \sigma_Y = \$80$$

Because the store is large and Tamara and Derek work in different departments, we might assume that their daily sales totals vary independently of each other. What are the mean and standard deviation of the difference $X - Y$ between Tamara's daily sales and Derek's daily sales? Tamara sells more on the average. Do you think she sells more every day? Why?

4.65 Comparing sales. It is unlikely that the daily sales of Tamara and Derek in the previous problem are independent. They will both sell more during the Christmas season, for example. Suppose that the correlation between their sales is $\rho = 0.4$. What are now the mean and standard deviation of the difference $X - Y$? Can you explain conceptually why positive correlation between two variables reduces the variability of the difference between them?

4.66 Managing new-product development process. Exercise 4.59 (page 250) gives the distributions of X, the number of weeks to complete the development of product specifications, and Y, the number of weeks to complete the design of the manufacturing process. You did some useful variance calculations in Exercise 4.63 (page 252). The cost per week for developing product specifications is $8000, while the cost per week for designing the manufacturing process is $30,000.

(a) Calculate the standard deviation of the cost for each of the two activities using Rule 1 for variances.

(b) Assuming the activity times are independent, calculate the standard deviation for the total cost of both activities combined.

(c) Assuming $\rho = 0.8$, calculate the standard deviation for the total cost of both activities combined.

(d) Assuming $\rho = 0$, calculate the standard deviation for the total cost of both activities combined. How does this compare with your result in part (b)? In part (c)?

(e) Assuming $\rho = -0.8$, calculate the standard deviation for the total cost of both activities combined. How does this compare with your result in part (b)? In part (c)? In part (d)?

SECTION 4.3 Summary

- A **random variable** is a variable taking numerical values determined by the outcome of a random phenomenon.

- The **probability distribution** of a random variable X tells us what the possible values of X are and how probabilities are assigned to those values.

- A random variable X and its distribution can be **discrete** or **continuous.**

- A **discrete random variable** has finitely many possible values. The probability distribution assigns each of these values a probability between 0 and 1 such that the sum of all the probabilities is exactly 1. The probability of any event is the sum of the probabilities of all the values that make up the event.

- A **continuous random variable** takes all values in some interval of numbers. A **density curve** describes the probability distribution of a continuous random variable. The probability of any event is the area under the curve above the values that make up the event. **Normal distributions** are one type of continuous probability distribution.

- You can picture a probability distribution by drawing a **probability histogram** in the discrete case or by graphing the density curve in the continuous case.

- The probability distribution of a random variable X, like a distribution of data, has a **mean** μ_X and a **standard deviation** σ_X.

- The **mean** μ is the balance point of the probability histogram or density curve. If X is discrete with possible values x_i having probabilities p_i, the mean is the average of the values of X, each weighted by its probability:

$$\mu_X = x_1 p_1 + x_2 p_2 + \cdots + x_k p_k$$

- The **variance** σ_X^2 is the average squared deviation of the values of the variable from their mean. For a discrete random variable,

$$\sigma_X^2 = (x_1 - \mu)^2 p_1 + (x_2 - \mu)^2 p_2 + \cdots + (x_k - \mu)^2 p_k$$

- The **standard deviation** σ_X is the square root of the variance. The standard deviation measures the variability of the distribution about the mean. It is easiest to interpret for Normal distributions.

- The mean and variance of a continuous random variable can be computed from the density curve, but to do so requires more advanced mathematics.

- The means and variances of random variables obey the following rules. If a and b are fixed numbers, then

$$\mu_{a+bX} = a + b\mu_X$$
$$\sigma_{a+bX}^2 = b^2 \sigma_X^2$$

If X and Y are any two random variables having correlation ρ, then

$$\mu_{X+Y} = \mu_X + \mu_Y$$
$$\sigma_{X+Y}^2 = \sigma_X^2 + \sigma_Y^2 + 2\rho\sigma_X\sigma_Y$$
$$\sigma_{X-Y}^2 = \sigma_X^2 + \sigma_Y^2 - 2\rho\sigma_X\sigma_Y$$

If X and Y are **independent,** then $\rho = 0$. In this case,

$$\sigma_{X+Y}^2 = \sigma_X^2 + \sigma_Y^2$$
$$\sigma_{X-Y}^2 = \sigma_X^2 + \sigma_Y^2$$

SECTION 4.3 Exercises

For Exercises 4.51 and 4.52, see page 239; for 4.53 to 4.55, see page 244; for 4.56 to 4.58, see page 247; for 4.59 and 4.60, see pages 250–251; for 4.61 to 4.63, see page 252; and for 4.64 to 4.66, see pages 256–257.

4.67 How many transactions? A small investment firm is beginning to reach its limit for the number of clients it can handle without expanding operations significantly. Before making this decision, the firm collects data from its own records to obtain a better understanding of how much business it really is handling. Each client's file is examined, and the number of transactions requested by the client in the past 10 business days is recorded. The data are summarized in a table.

	Number of Transactions					
	0	1	2	3	4	5
Number of clients	56	122	180	140	132	90
Proportion of clients						

Fifty-six of the firm's clients requested no transactions in the past 10 business days, 122 of the firm's clients requested only 1 transaction in the past 10 business days, etc.

(a) Assuming all clients of the firm are included in the table, how many clients does the firm service?

(b) Calculate the proportion of clients who have requested each number of transactions to complete the last row of the table. Round each proportion to 2 decimal places (for example, 8% is the proportion 0.08).

(c) What is the total of the (rounded) proportions? Are the first and third rows of this table a proper probability distribution? Explain your response.

4.68 How many transactions? Continue with the data of the previous exercise. You will need to have filled in the last row of the table to complete this exercise.

(a) Make a probability histogram for the number of transactions per client. The vertical scale should measure the proportion of clients requesting each number of transactions.

(b) Calculate the mean number of transactions per client for this firm. Mark the mean on the probability histogram you made in part (a).

(c) Calculate the variance and the standard deviation of the number of transactions per client for this firm. What are the units of measure for the variance? What are the units of measure for the standard deviation?

4.69 Online betting exchange. Betfair is an online betting exchange. Individuals can make bets and take bets at whatever odds someone is willing to offer. You might think of Betfair as a mixture of eBay and the stock exchange applied to sports betting. The company is a U.K. company and does not knowingly allow any U.S. customers to register with the site. According to Betfair, the company processes "more than 300 bets a second." Assume the number of bets per second at Betfair has a Normal distribution with a mean of 320 and a standard deviation of 58.

(a) Calculate the probability that Betfair processes more than 500 bets in a particular second.

(b) Calculate the probability that Betfair processes fewer than 250 bets in a particular second.

(c) In the top 10% of its busiest times, Betfair processes at least how many bets per second?

4.70 Exporting wood to Japan. Timber is a common raw material for the manufacture of many products such as building materials for homes, furniture, and paper products. The United States exports hundreds of thousands of metric tons of wood chips each year to Japan. The amount exported varies from year to year with an approximately Normal distribution with a mean of 278 thousand metric tons (278,000 metric tons) and a standard deviation of 52 thousand metric tons.

(a) What is the probability of U.S. wood chip exports to Japan exceeding 400 thousand metric tons in a given year?

(b) What is the probability of U.S. wood chip exports to Japan being between 250 and 350 thousand metric tons?

4.71 Insurance policy value. Consider an insurance policy that will pay the holder of the policy the market price of her house if the house is completely destroyed by fire. The market price of a particular house is set at $220,000. The probability of this house being completely destroyed by fire during the next year is determined by the insurance company to be 0.3%, and the cost of the policy is set at $1280 for the next year. Let X represent the value of the insurance policy to the holder of the policy at the end of the next year.

(a) If the house is completely destroyed by fire in the next year, the value of the policy will be $220,000 - 1280 = 218,720$ dollars. If the house is not completely destroyed by fire in the next year, what will be the value of the policy?

(b) Write down the probability distribution of X in the form of a table.

(c) Make a probability histogram of the probability distribution of X.

4.72 Insurance policy value. Continue analyzing the insurance policy described in the previous exercise.

(a) Calculate the mean value of the insurance policy.

(b) To the policyholder, what does the mean value of the policy represent? Under what conditions will this number become meaningful to the policyholder?

(c) To the insurance company, what does the mean value of the policy represent? Under what conditions will this number become meaningful to the company? Does the insurance company have any control over the mean value of the policy?

4.73 How many rooms? Furniture makers and others are interested in how many rooms housing units have, because more rooms can generate more sales. Here are the distributions of the number of rooms for owner-occupied units and renter-occupied units in the United States:[18]

	Rooms				
	1	2	3	4	5
Owned	0.0001	0.0006	0.0139	0.0911	0.2262
Rented	0.0109	0.0278	0.2254	0.3333	0.2302

	Rooms				
	6	7	8	9	10
Owned	0.2601	0.1746	0.1067	0.0519	0.0747
Rented	0.1062	0.0367	0.0131	0.0047	0.0118

(a) Make probability histograms of these two distributions, using the same scales. What are the most important differences between the distributions for owner-occupied and rented housing units?

(b) Find the mean number of rooms for both types of housing unit. How do the means reflect the differences you found in (a)?

4.74 Households and families. In government data, a household consists of all occupants of a dwelling unit, while a family consists of two or more persons who live together and are related by blood or marriage. Here are the distributions of household size and of family size in the United States:

	Number of Persons						
	1	2	3	4	5	6	7
Household probability	0.25	0.32	0.17	0.15	0.07	0.03	0.01
Family probability	0	0.42	0.23	0.21	0.09	0.03	0.02

You have considered the distribution of household size in Exercises 4.47 and 4.56. Compare the two distributions using probability histograms, means, and standard deviations. Write a brief comparison, using your calculations to back up your statements.

4.75 How many rooms? Which of the two distributions for room counts in Exercise 4.73 appears more spread out in the probability histograms? Why? Find the standard deviation for each distribution. The standard deviation provides a numerical measure of spread.

4.76 Tossing four coins. The distribution of the count X of heads in four tosses of a balanced coin was found in Example 4.10 to be

Number of heads x_i:	0	1	2	3	4
Probability p_i:	0.0625	0.25	0.375	0.25	0.0625

Find the mean μ_x for this distribution. Then find the mean number of heads for a single coin toss and show that your two results are related by the addition rule for means.

4.77 Gain Communications. Example 4.16 provides the probability distribution for military sales X and the mean of X. From this section, we learned that the mean of a probability distribution describes the long-run average outcome. In this exercise, you will explore this concept using technology.

• *Excel users:* Input the values 1000, 3000, 5000, and 10000 in the first four rows of column A. Now input the corresponding probabilities of 0.1, 0.3, 0.4, and 0.2 in the first four rows of column B. Now choose "Random Number Generation" from the **Data Analysis** menu box (refer to the Appendix of Chapter 1 for installing the Data Analysis ToolPak). Enter "1" in the **Number of Variables** box, enter "20000" in the **Number of Random Numbers** box, choose "Discrete" for the **Distribution** option, enter the cell range of the X-values and their probabilities ($A\$1:\$B\$4$) in the **Value and Probability Input Range** box, and finally select Row 1 of any empty column for the **Output Range**. Click **OK** to find 20,000 realizations of X outputted in the worksheet. Using Excel's AVERAGE( ) function, find the average of the 20,000 X-values.

• *Minitab users:* Input the values 1000, 3000, 5000, and 10000 in the first four rows of column 1 (c1). Now input the corresponding probabilities of 0.1, 0.3, 0.4, and 0.2 in the first four rows of column 2 (c2). Do the following pull-down sequence: Calc → Random Data → Discrete. Enter "20000" in the **Number of rows of data to generate** box, type "c3" in the **Store in column(s)** box, click in "c1" in the **Values in** box, and click in "c2" in the **Probabilities in** box. Click **OK** to find 20,000 realizations of X outputted in the worksheet. Find the average of the 20,000 X-values.

Whether you used Excel or Minitab, how does the average value of the 20,000 X-values compare with the mean reported in Example 4.16?

4.78 Gain Communications. Refer to the previous exercise and perform the same Excel or Minitab experiment on the civilian sales distribution shown in Example 4.16.

4.79 Pick 3 once more. The Tri-State Pick 3 lottery game offers a choice of several bets. You choose a three-digit number. The lottery commission announces the winning three-digit number, chosen at random, at the end of each day. The "box" pays $83.33 if the number you choose has the same digits as the winning number, in any order. Find the expected payoff for a $1 bet on the box. (Assume that you chose a number having three different digits.)

4.80 Tossing four coins. The distribution of outcomes for tossing four coins given in Exercise 4.76 assumes that the tosses are independent of each other. Find the variance for four tosses. Then find the variance of the number of heads (0 or 1) on a single toss and show that your two results are related by the addition rule for variances of independent random variables.

4.81 Independent random variables? For each of the following situations, would you expect the random variables X and Y to be independent? Explain your answers.
(a) X is the rainfall (in inches) on November 6 of this year and Y is the rainfall at the same location on November 6 of next year.
(b) X is the amount of rainfall today and Y is the rainfall at the same location tomorrow.
(c) X is today's rainfall at the Orlando, Florida, airport, and Y is today's rainfall at Disney World just outside Orlando.

4.82 Independent random variables? In which of the following games of chance would you be willing to assume independence of X and Y in making a probability model? Explain your answer in each case.
(a) In blackjack, you are dealt two cards and examine the total points X on the cards (face cards count 10 points). You can choose to be dealt another card and compete based on the total points Y on all three cards.
(b) In craps, the betting is based on successive rolls of two dice. X is the sum of the faces on the first roll, and Y the sum of the faces on the next roll.

4.83 Time-and-motion studies. A time-and-motion study measures the time required for an assembly-line worker to perform a repetitive task. The data show that the time required to bring a part from a bin to its position on an automobile chassis varies from car to car with mean 11 seconds and standard deviation 2 seconds. The time required to attach the part to the chassis varies with mean 20 seconds and standard deviation 4 seconds.
(a) What is the mean time required for the entire operation of positioning and attaching the part?
(b) If the variation in the worker's performance is reduced by better training, the standard deviations will decrease. Will this decrease change the mean you found in (a) if the mean times for the two steps remain as before?
(c) The study finds that the times required for the two steps are independent. A part that takes a long time to position, for example, does not take more or less time to attach than other parts. How would your answer in (a) change if the two variables were dependent, with correlation 0.8? With correlation 0.3?

4.84 A chemical production process. Laboratory data show that the time required to complete two chemical reactions in a production process varies. The first reaction has a mean time of 40 minutes and a standard deviation of 2 minutes; the second has a mean time of 25 minutes and a standard deviation of 1 minute.

The two reactions are run in sequence during production. There is a fixed period of 5 minutes between them as the product of the first reaction is pumped into the vessel where the second reaction will take place. What is the mean time required for the entire process?

4.85 Time-and-motion studies. Find the standard deviation of the time required for the two-step assembly operation studied in Exercise 4.83, assuming that the study shows the two times to be independent. Redo the calculation assuming that the two times are dependent, with correlation 0.3. Can you explain in nontechnical language why positive correlation increases the variability of the total time?

4.86 A chemical production process. The times for the two reactions in the chemical production process described in Exercise 4.84 are independent. Find the standard deviation of the time required to complete the process.

4.87 Get the mean you want. Here is a simple way to create a random variable X that has mean μ and standard deviation σ: X takes only the two values $\mu - \sigma$ and $\mu + \sigma$, each with probability 0.5. Use the definition of the mean and variance for discrete random variables to show that X does have mean μ and standard deviation σ.

4.88 Combining measurements. You have two scales for measuring weight. Both scales give answers that vary a bit in repeated weighings of the same item. If the true weight of an item is 2 grams (g), the first scale produces readings X that have mean 2.000 g and standard deviation 0.002 g. The second scale's readings Y have mean 2.001 g and standard deviation 0.001 g.
(a) What are the mean and standard deviation of the difference $Y - X$ between the readings? (The readings X and Y are independent.)
(b) You measure once with each scale and average the readings. Your result is $Z = (X + Y)/2$. What are μ_Z and σ_Z? Is the average Z more or less variable than the reading Y of the less variable scale?

4.89 Gain Communications. Examples 4.16 and 4.17 concern a probabilistic projection of sales and profits by an electronics firm, Gain Communications. The mean and variance of military sales X appear in Example 4.17 (page 252). You found the mean and variance of civilian sales Y in Exercise 4.61 (page 252).
(a) Because the military budget and the civilian economy are not closely linked, Gain is willing to assume that its military and civilian sales vary independently. What is the standard deviation of Gain's total sales $X + Y$?
(b) Find the standard deviation of the estimated profit, $Z = 2000X + 3500Y$.

4.90 Gain Communications, continued. Redo Exercise 4.89 assuming correlation 0.01 between civilian and military sales. Redo the exercise assuming correlation 0.99. Comment on the effect of small and large correlations on the uncertainty of Gain's sales projections.

4.91 Study habits. The academic motivation and study habits of female students as a group are better than those of males. The Survey of Study Habits and Attitudes (SSHA) is a psychological test that measures these factors. The distribution of SSHA scores among the women at a college has mean 120 and standard deviation 28, and the distribution of scores among men students has mean 105 and standard deviation 35. You select a single male student and a single female student at random and give them the SSHA test.
(a) Explain why it is reasonable to assume that the scores of the two students are independent.
(b) What are the mean and standard deviation of the difference (female minus male) between their scores?
(c) From the information given, can you find the probability that the woman chosen scores higher than the man? If so, find this probability. If not, explain why you cannot.

4.92 SAT combined score. Exercise 4.27 (page 234) reported that scores on the SAT Math Reasoning test are approximately Normal with mean 515 points and standard deviation 116 points. The scores on the Verbal Reasoning test are also approximately Normal but with mean 502 and standard deviation 112. The correlation between Math Reasoning and Verbal Reasoning scores is about $\rho = 0.70$. Define the random variable X as Math Reasoning score and the random variable Y as Verbal Reasoning score.
(a) Determine the mean and standard deviation for the combined score $(X + Y)$.
(b) Given that the Math and Verbal scores are approximately Normal, the combined score will also be approximately Normal. Determine the probability that a randomly chosen student will have a combined score of at least 1250.

4.93 Vegas versus actual NFL point spreads. Based on all 224 NFL games in 2007, the mean Vegas point spread (= Underdog − Favorite) is −5.82 and the standard deviation is 3.94. The actual point spreads have mean −6.69 and standard deviation 12.25. The correlation between Vegas and actual point spreads is 0.28. Define the random variable Y as the Vegas point spread and the random variable X as the actual point spread.
(a) What are the mean and standard deviation for the difference $Y - X$ between point spreads?
(b) It turns out that the differences between point spreads are well approximated by the Normal distribution. Suppose that if a bettor takes the Vegas point spread and bets on the underdog, the bet is regarded as "even" money. Convince yourself that if the difference as defined in part (a) is positive, then the bettor loses the bet or, equivalently, Vegas wins the bet. Given the numbers of part (a), determine the probability that the difference is positive for any given game.

CASE 4.2 *Portfolio analysis. Here are the means, standard deviations, and correlations for the monthly returns from three Fidelity mutual funds for the 60 months ending in August 2008. Because there are three random variables, there are three correlations. We use subscripts to show which pair of random variables a correlation refers to.*

W = monthly return on Magellan Fund $\mu_W = 0.28\%$ $\sigma_W = 3.98\%$

X = monthly return on Energy Fund $\mu_X = 1.54\%$ $\sigma_X = 7.06\%$

Y = monthly return on Japan Fund $\mu_Y = 0.47\%$ $\sigma_Y = 5.20\%$

Correlations

$\rho_{WX} = 0.65$ $\rho_{WY} = 0.52$ $\rho_{XY} = 0.40$

Exercises 4.94 to 4.96 make use of these historical data.

CASE 4.2 **4.94 Diversification.** Many advisers recommend using roughly 20% foreign stocks to diversify portfolios of U.S. stocks. Michael owns Fidelity Magellan Fund, which concentrates on stocks of large American companies. He decides to move to a portfolio of 80% Magellan and 20% Fidelity Japan Fund. Show that (based on historical data) this portfolio has both a *higher* mean return and *less* volatility than Magellan alone. This illustrates the beneficial effects of diversifying among investments.

CASE 4.2 **4.95 More on diversification.** Diversification works better when the investments in a portfolio have small correlations. To demonstrate this, suppose that returns on Magellan Fund and Japan Fund had the means and standard deviations we have given but were uncorrelated ($\rho_{WY} = 0$). Show that the standard deviation of a portfolio that combines 80% Magellan with 20% Japan is then smaller than your result from the previous exercise. What happens to the mean return if the correlation is 0?

CASE 4.2 **4.96 Larger portfolios (optional).** Portfolios often contain more than two investments. The rules for means and variances continue to apply, though the arithmetic gets messier. A portfolio containing proportions a of Magellan Fund, b of Energy Fund, and c of Japan Fund has return $R = aW + bX + cY$. Because a, b, and c are the proportions invested in the three funds, $a + b + c = 1$. The mean and variance of the portfolio return R are

$$\mu_R = a\mu_W + b\mu_X + c\mu_Y$$
$$\sigma_R^2 = a^2\sigma_W^2 + b^2\sigma_X^2 + c^2\sigma_Y^2 + 2ab\rho_{WX}\sigma_W\sigma_X$$
$$+ 2ac\rho_{WY}\sigma_W\sigma_Y + 2bc\rho_{XY}\sigma_X\sigma_Y$$

Having seen the advantages of diversification, Michael decides to invest his funds 60% in Magellan, 20% in Energy, and 20% in Japan. What are the (historical) mean and standard deviation of the monthly returns for this portfolio?

4.97 Perfectly correlated variables (optional). We know that variances add if the random variables involved are uncorrelated ($\rho = 0$), but not otherwise. The opposite extreme is perfect positive correlation ($\rho = 1$). Show by using the general addition rule for variances that in this case the standard deviations add. That is, $\sigma_{X+Y} = \sigma_X + \sigma_Y$ if $\rho_{XY} = 1$.

4.98 A hot stock. You purchase a hot stock for $1000. The stock either gains 30% or loses 25% each day, each with probability 0.5. Its returns on consecutive days are independent of each other. This implies that all four possible combinations of gains and losses in two days are equally likely, each having probability 0.25. You plan to sell the stock after two days.
(a) What are the possible values of the stock after two days, and what is the probability for each value? What is the probability that the stock is worth more after two days than the $1000 you paid for it?
(b) What is the mean value of the stock after two days? You see that these two criteria give different answers to the question "Should I invest?"

4.99 Making glassware. In a process for manufacturing glassware, glass stems are sealed by heating them in a flame. The temperature of the flame varies a bit. Here is the distribution of the temperature X measured in degrees Celsius:

Temperature:	540°	545°	550°	555°	560°
Probability:	0.1	0.25	0.3	0.25	0.1

(a) Find the mean temperature μ_X and the standard deviation σ_X.
(b) The target temperature is 550°C. Use the rules for means and variances to find the mean and standard deviation of the number of degrees off target, $X - 550$.
(c) A manager asks for results in degrees Fahrenheit. The conversion of X into degrees Fahrenheit is given by

$$Y = \frac{9}{5}X + 32$$

What are the mean μ_Y and standard deviation σ_Y of the temperature of the flame in the Fahrenheit scale?

4.100 Irrational decision making? The psychologist Amos Tversky did many studies of our perception of chance behavior. In its obituary of Tversky (June 6, 1996), the *New York Times* cited the following example.
(a) Tversky asked subjects to choose between two public health programs that affect 600 people. One has probability 1/2 of saving all 600 and probability 1/2 that all 600 will die. The other is guaranteed to save exactly 400 of the 600 people. Find the mean number of people saved by the first program.
(b) Tversky then offered a different choice. One program has probability 1/2 of saving all 600 and probability 1/2 of losing all 600, while the other will definitely lose exactly 200 lives. What is the difference between this choice and that in (a)?
(c) Given option (a), most people choose the second program. Given option (b), most people choose the first program. Do people appear to use means in making their decisions? Why do you think their choices differed in the two cases?

4.4 The Sampling Distribution of a Sample Mean

Section 3.3 (page 186) motivated our study of probability by looking at *sampling variability* and *sampling distributions*. A statistic from a random sample will take different values if we take more samples from the same population. That is, sample statistics are random variables. Even though the values of a statistic vary randomly from sample to sample, they do have a regular pattern, the sampling distribution, over many samples. This is the distribution of the random variable. We can now use the language of probability to examine one very important sampling distribution, that of a sample mean $\bar{x}$. Some new and important facts about probability emerge from this examination. Here is the example we will follow to introduce the role of probability in statistical inference.

EXAMPLE 4.21   Does this wine smell bad?

Sulfur compounds such as dimethyl sulfide (DMS) are sometimes present in wine. DMS causes "off-odors" in wine, which from the winemakers' perspective is an important product quality attribute to monitor and control. Too high a DMS level can result in loss of sales. On the other hand, there is a cost for lowering DMS levels. It makes good business sense for winemakers to study the odor threshold, the lowest concentration of DMS that the human nose can detect. Different people have different thresholds, so we start by asking about the mean threshold μ in the population of all adults. The number μ is a *parameter* that describes this population.

To estimate μ, we present tasters with both natural wine and the same wine spiked with DMS at different concentrations to find the lowest concentration at which they can identify the spiked wine. Here are the odor thresholds (measured in micrograms of DMS per liter of wine) for 10 randomly chosen subjects:

$$28 \quad 40 \quad 28 \quad 33 \quad 20 \quad 31 \quad 29 \quad 27 \quad 17 \quad 21$$

The mean threshold for these subjects is $\bar{x} = 27.4$. This sample mean is a *statistic* that we use to estimate the parameter μ, but it is probably not exactly equal to μ. Moreover, we know that a different 10 subjects would give us a different $\bar{x}$.

A parameter, such as the mean odor threshold μ of all adults, is in practice a fixed but unknown number. A statistic, such as the mean threshold $\bar{x}$ of a random sample of 10 adults, is a random variable. It seems reasonable to use $\bar{x}$ to estimate μ. An SRS should fairly represent the population, so the mean $\bar{x}$ of the sample should be somewhere near the mean μ of the population. Of course, we don't expect $\bar{x}$ to be exactly equal to μ, and we realize that, if we choose another SRS, the luck of the draw will probably produce a different $\bar{x}$.

Statistical estimation and the law of large numbers

If $\bar{x}$ is rarely exactly right and varies from sample to sample, why is it nonetheless a reasonable estimate of the population mean μ? Here is one answer: if we keep on taking larger and larger samples, the statistic $\bar{x}$ is *guaranteed* to get closer and closer to the parameter μ. We have the comfort of knowing that if we can afford to keep on measuring more subjects, eventually we will estimate the mean odor threshold of all adults very accurately. This remarkable fact is called the *law of large numbers*. It is remarkable because it holds for *any* population, not just for some special class such as Normal distributions.

> **Law of Large Numbers**
>
> Draw independent observations at random from any population with finite mean μ. As the number of observations drawn increases, the mean $\bar{x}$ of the observed values gets closer and closer to the mean μ of the population.

The mean μ of a random variable is the average value of the variable in two senses. By its definition, μ is the average of the possible values, weighted by their probability of occurring. The law of large numbers says that μ is also the long-run average of many independent observations on the variable. The law of large numbers can be proved mathematically starting from the basic laws of probability. The behavior of $\bar{x}$ is similar to the idea of probability. In the long run, the proportion of outcomes taking any value gets close to the probability of that value, and the average outcome gets close to the population mean. Figure 4.1 (page 218) shows how proportions approach the probability in one example. Here is an example of how sample means approach the population mean.

EXAMPLE 4.22   The Law of Large Numbers in Action

In fact, the distribution of DMS odor thresholds among all adults has mean 25. The mean $\mu = 25$ is the true value of the parameter we seek to estimate. Typically, the value of a parameter like μ is unknown. We use an example in which μ is known to illustrate facts about the general behavior of $\bar{x}$ that hold even when μ is unknown. Figure 4.15 shows how the sample mean $\bar{x}$ of an SRS drawn from this population changes as we add more subjects to our sample.

The first subject in Example 4.21 had threshold 28, so the line in Figure 4.15 starts there. The mean for the first two subjects is

$$\bar{x} = \frac{28 + 40}{2} = 34$$

This is the second point on the graph. At first, the graph shows that the mean of the sample changes as we take more observations. Eventually, however, the mean of the observations gets close to the population mean $\mu = 25$ and settles down at that value.

FIGURE 4.15 The law of large numbers in action: as we take more observations, the sample mean $\bar{x}$ always approaches the mean (μ) of the population.

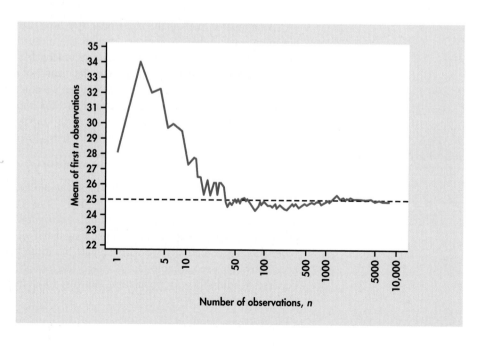

If we started over, again choosing people at random from the population, we would get a different path from left to right in Figure 4.15. The law of large numbers says that whatever path we get will always settle down at 25 as we draw more and more people.

As in the case of probability, an animated version of Figure 4.15 makes the idea of the mean as long-run average outcome clearer. The *Law of Large Numbers* applet at the text Web site, www.whfreeman.com/psbe3e, gives you control of an animation.

The law of large numbers is the foundation of such business enterprises as gambling casinos and insurance companies. The winnings (or losses) of a gambler on a few plays are uncertain—that's why gambling is exciting. In Figure 4.15, the mean of even 100 observations is not yet very close to μ. It is only *in the long run* that the mean outcome is predictable. The house plays tens of thousands of times. So the house, unlike individual gamblers, can count on the long-run regularity described by the law of large numbers. The average winnings of the house on tens of thousands of plays will be very close to the mean of the distribution of winnings. Needless to say, this mean guarantees the house a profit. That's why gambling can be a business.

APPLY YOUR KNOWLEDGE

4.101 Comparing computers.   Pfeiffer Consulting, a technology consulting group, designed benchmark tests to compare the speed with which computer models complete a variety of tasks. Pfeiffer announced that the mean completion time system startup was **56.6** seconds for 2.0-GHz Power Mac G5 models and **30.1** seconds for 3.2-GHz Mac Pro 8-core models.[19] Do you think the bold numbers are parameters or statistics? Explain your reasoning carefully.

4.102 Successive averages.   Figure 4.15 shows how the mean of n observations behaves as we keep adding more observations to those already in hand. The first 10 observations are given in Example 4.21. Demonstrate that you grasp the idea of Figure 4.15: find the mean of the first one, then two, three, four, and five of these observations and plot the successive means against n. Verify that your plot agrees with the first part of the plot in Figure 4.15.

4.103 Playing the numbers.   The numbers racket is a well-entrenched illegal gambling operation in most large cities. One version works as follows. You choose one of the 1000 three-digit numbers 000 to 999 and pay your local numbers runner a dollar to enter your bet. Each day, one three-digit number is chosen at random and pays off $600. The mean payoff for the population of thousands of bets is $\mu = 60$ cents. Joe makes one bet every day for many years. Explain what the law of large numbers says about Joe's results as he keeps on betting.

Thinking about the law of large numbers

The law of large numbers says broadly that the average results of many independent observations are stable and predictable. Casinos are not the only businesses that base forecasts on this fact. A grocery store deciding how many gallons of milk to stock and a fast-food restaurant deciding how many beef patties to prepare can predict demand even though their many customers make independent decisions. The law of large numbers says that these many individual decisions will produce a stable result. It is worth the effort to think a bit more closely about so important a fact.

The "law of small numbers"

Both the rules of probability and the law of large numbers describe the regular behavior of chance phenomena *in the long run*. Psychologists have discovered that the popular understanding of randomness is quite different from the true laws of chance.[20] Most people believe in an incorrect "law of small numbers." That is, we expect even short sequences of random events to show the kind of average behavior that in fact appears only in the long run.

Try this experiment: Write down a sequence of heads and tails that you think imitates 10 tosses of a balanced coin. How long was the longest string (called a *run*) of consecutive heads or consecutive tails in your tosses? Most people will write a sequence with no runs of more than two consecutive heads or tails. Longer runs don't seem "random" to us. In fact, the probability of a run of three or more consecutive heads or tails in 10 tosses is greater than 0.8, and the probability of *both* a run of three or more heads and a run of three or more tails is almost 0.2.[21] This and other probability calculations suggest that a short sequence of coin tosses will often not seem random to most people. The runs of consecutive heads or consecutive tails that appear in real coin tossing (and that are predicted by the mathematics of probability) surprise us. Because we don't expect to see long runs, we may conclude that the coin tosses are not independent or that some influence is disturbing the random behavior of the coin.

Belief in the law of small numbers influences behavior. If a basketball player makes several consecutive shots, both the fans and his teammates believe that he has a "hot hand" and is more likely to make the next shot. This is doubtful. Careful study suggests that runs of baskets made or missed are no more frequent in basketball than would be expected if each shot were independent of the player's previous shots. Players perform consistently, not in streaks. (Of course, some players make a higher percent of their shots in the long run than others.) Our perception of hot or cold streaks simply shows that we don't perceive random behavior very well.[22]

Gamblers often follow the hot-hand theory, betting that a run will continue. At other times, however, they draw the opposite conclusion when confronted with a run of outcomes. If a coin gives 10 straight heads, some gamblers feel that it must now produce some extra tails to get back to the average of half heads and half tails. Not so. If the next 10,000 tosses give about 50% tails, those 10 straight heads will be swamped by the later thousands of heads and tails. No compensation is needed to get back to the average in the long run. Remember that it is *only* in the long run that the regularity described by probability and the law of large numbers takes over.

Our inability to accurately distinguish random behavior from systematic influences points out once more the need for statistical inference to supplement exploratory analysis of data. Probability calculations can help verify that what we see in the data is more than a random pattern.

How large is a large number?

The law of large numbers says that the actual mean outcome of many trials gets close to the distribution mean μ as more trials are made. It doesn't say how many trials are needed to guarantee a mean outcome close to μ. That depends on the *variability* of the random outcomes. The more variable the outcomes, the more trials are needed to ensure that the mean outcome $\bar{x}$ is close to the distribution mean μ.

BEYOND THE BASICS: More Laws of Large Numbers

The law of large numbers is one of the central facts about probability. It helps us understand the mean μ of a random variable. It explains why gambling casinos and insurance companies make money. It assures us that statistical estimation will be accurate if we can afford enough observations. The basic law of large numbers applies to independent observations that all have the same distribution. Mathematicians have extended the law to many more general settings. Here are two of these.

Is there a winning system for gambling?

Serious gamblers often follow a system of betting in which the amount bet on each play depends on the outcome of previous plays. You might, for example, double your bet on each spin of the roulette wheel until you win—or, of course, until your fortune is exhausted. Such a system tries to take advantage of the fact that you have a memory even though the roulette wheel does not. Can you beat the odds with a system based on the outcomes of past plays? No. Mathematicians have established a stronger version of the law of large numbers that says that if you do not have an infinite fortune to gamble with, your long-run average winnings μ remain the same as long as successive trials of the game (such as spins of the roulette wheel) are independent.

What if observations are not independent?

You are in charge of a process that manufactures video screens for computer monitors. Your equipment measures the tension on the metal mesh that lies behind each screen and is critical to its image quality. You want to estimate the mean tension μ for the process by the average $\overline{x}$ of the measurements. Alas, the tension measurements are not independent. If the tension on one screen is a bit too high, the tension on the next is more likely to also be high. Many real-world processes are like this—the process stays stable in the long run, but observations made close together are likely to be both above or both below the long-run mean. Again the mathematicians come to the rescue: as long as the dependence dies out fast enough as we take measurements farther and farther apart in time, the law of large numbers still holds.

APPLY YOUR KNOWLEDGE

4.104 Help this man.

(a) A gambler knows that red and black are equally likely to occur on each spin of a roulette wheel. He observes five consecutive reds and bets heavily on red at the next spin. Asked why, he says that "red is hot" and that the run of reds is likely to continue. Explain to the gambler what is wrong with this reasoning.

(b) After hearing you explain why red and black remain equally probable after five reds on the roulette wheel, the gambler moves to a poker game. He is dealt five straight red cards. He remembers what you said and assumes that the next card dealt in the same hand is equally likely to be red or black. Is the gambler right or wrong? Why?

4.105 The "law of averages." The baseball Hall of Famer Tony Gwynn got a hit about 34% of the time over his 20-year career. After he failed to hit safely in six straight at-bats, the TV commentator said, "Tony is due for a hit by the law of averages." Is that right? Why?

4.106 The law of large numbers. Figure 4.2 (page 225) shows the 36 possible outcomes of rolling two dice and counting the spots on the up-faces. These 36 outcomes are equally likely. You can calculate that the mean for the sum of the two up-faces is $\mu = 7$. This is the population mean μ for the idealized population that contains the results of rolling two dice forever. The law of large numbers says that the average $\overline{x}$ from a finite number of rolls gets closer and closer to 7 as we do more and more rolls.

(a) Go to the *Law of Large Numbers* applet. Click "More dice" once to get two dice. Click "Show mean" to see the mean 7 on the graph. Leaving the number of rolls at 1, click "Roll dice" three times. Note the count of spots for each roll (what were they?) and the average for the three rolls. You see that the graph displays at each point the average number of spots for all rolls up to the last one. Now you understand the display.

(b) Set the number of rolls to 100 and click "Roll dice." The applet rolls the two dice 100 times. The graph shows how the average count of spots changes as we make more rolls. That is, the graph shows $\bar{x}$ as we continue to roll the dice. Make a rough sketch of the final graph.

(c) Repeat your work from (b). Click "Reset" to start over, then roll two dice 100 times. Make a sketch of the final graph of the mean $\bar{x}$ against the number of rolls. Your two graphs will often look very different. What they have in common is that the average eventually gets close to the population mean $\mu = 7$. The law of large numbers says that this will *always* happen if you keep on rolling the dice.

Sampling distributions

The law of large numbers assures us that if we measure enough subjects, the statistic $\bar{x}$ will eventually get very close to the unknown parameter μ. But our study in Example 4.21 had just 10 subjects. What can we say about $\bar{x}$ from 10 subjects as an estimate of μ? We ask: "What would happen if we took many samples of 10 subjects from this population?" We learned in Section 3.3 how to answer this question:

- Take a large number of samples of size 10 from the same population.
- Calculate the sample mean $\bar{x}$ for each sample.
- Make a histogram of the values of $\bar{x}$. This histogram shows how $\bar{x}$ varies in many samples.

In Section 3.3 (page 186) we used computer simulation to take many samples in a different setting. The histogram of values of the statistic approximates the *sampling distribution* that we would see if we kept on sampling forever. The idea of a sampling distribution is the foundation of statistical inference. One reason for studying probability is that the laws of probability can tell us about sampling distributions without the need to actually choose or simulate a large number of samples.

The mean and the standard deviation of $\bar{x}$

The first important fact about the sampling distribution of $\bar{x}$ follows from using the rules for means and variances, though we won't do the algebra. Here is the result.[23]

> **Mean and Standard Deviation of a Sample Mean**
>
> Suppose that $\bar{x}$ is the mean of an SRS of size n drawn from a large population with mean μ and standard deviation σ. Then the **mean** of the sampling distribution of $\bar{x}$ is μ and its **standard deviation** is $\sigma/\sqrt{n}$.

Both the mean and the standard deviation of the sampling distribution of $\bar{x}$ have important implications for statistical inference.

- The mean of the statistic $\bar{x}$ is always the same as the mean μ of the population. The sampling distribution of $\bar{x}$ is centered at μ. In repeated sampling, $\bar{x}$ will sometimes fall above the true value of the parameter μ and sometimes below, but there is no systematic tendency to overestimate or underestimate the parameter. This makes the idea of lack of bias in the sense of "no favoritism" more precise. Because the mean of $\bar{x}$ is equal to μ, we say that the statistic $\bar{x}$ is an **unbiased estimator** of the parameter μ.

unbiased estimator

• An unbiased estimator is "correct on the average" in many samples. How close the estimator falls to the parameter in most samples is determined by the spread of the sampling distribution. If individual observations have standard deviation σ, then sample means $\bar{x}$ from samples of size n have standard deviation $\sigma/\sqrt{n}$. **Averages are less variable than individual observations.**

Not only is the standard deviation of the distribution of $\bar{x}$ smaller than the standard deviation of individual observations, but it gets smaller as we take larger samples. **The results of large samples are less variable than the results of small samples.** If n is large, the standard deviation of $\bar{x}$ is small and almost all samples will give values of $\bar{x}$ that lie very close to the true parameter μ. That is, the sample mean from a large sample can be trusted to estimate the population mean accurately. Notice, however, that the standard deviation of the sampling distribution gets smaller only at the rate $\sqrt{n}$. To cut the standard deviation of $\bar{x}$ in half, we must take four times as many observations, not just twice as many.

APPLY YOUR KNOWLEDGE

4.107 Generating a sampling distribution. Let's illustrate the idea of a sampling distribution in the case of a very small sample from a very small population. The population is the sizes of 10 medium-sized businesses where size is measured in terms of the number of employees. For convenience, the 10 companies have been labeled with the integers 0 to 9.

Company:	0	1	2	3	4	5	6	7	8	9
Size:	82	62	80	58	72	73	65	66	74	62

The parameter of interest is the mean size μ in this population. The sample is an SRS of size $n = 4$ drawn from the population. Software can be used to generate an SRS. Alternatively, because the companies are labeled 0 to 9, a single random digit from Table B chooses one company for the sample.

(a) Find the mean of the 10 sizes in the population. This is the population mean μ.

(b) Input the 10 sizes in Minitab or Excel. Refer to the Appendix of Chapter 3 to see how to use each of these software programs to generate an SRS. Alternatively, you can use Table B to draw an SRS of size 4 from this population. Draw an SRS using software or Table B and write the four sizes in your sample and calculate the mean $\bar{x}$ of the sample sizes. This statistic is an estimate of μ.

(c) Repeat this process 10 times using software or using different parts of Table B. Make a histogram of the 10 values of $\bar{x}$. You are constructing the sampling distribution of $\bar{x}$. Is the center of your histogram close to μ?

4.108 Measurements on the production line. Sodium content (in milligrams) is measured for bags of potato chips sampled from a production line. The standard deviation of the sodium content measurements is $\sigma = 10$ mg. The sodium content is measured 3 times and the mean $\bar{x}$ of the 3 measurements is recorded.

(a) What is the standard deviation of the mean result? (That is, if you kept on making 3 measurements and averaging them, what would be the standard deviation of all the $\bar{x}$'s?)

(b) How many times must we repeat the measurement to reduce the standard deviation of $\bar{x}$ to 5? Explain to someone who knows no statistics the advantage of reporting the average of several measurements rather than the result of a single measurement.

4.109 Measuring blood cholesterol. A study of the health of teenagers plans to measure the blood cholesterol level of an SRS of youth of ages 13 to 16 years. The researchers will report the mean $\bar{x}$ from their sample as an estimate of the mean cholesterol level μ in this population.

(a) Explain to someone who knows no statistics what it means to say that $\bar{x}$ is an "unbiased" estimator of μ.

(b) The sample result $\bar{x}$ is an unbiased estimator of the population mean μ no matter what size SRS the study chooses. Explain to someone who knows no statistics why a large sample gives more trustworthy results than a small sample.

The central limit theorem

We have described the center and spread of the sampling distribution of a sample mean $\bar{x}$, but not its shape. The shape of the distribution of $\bar{x}$ depends on the shape of the population distribution. Here is one important case: if the population distribution is Normal, then so is the distribution of the sample mean.

> **Sampling Distribution of a Sample Mean**
>
> If a population has the $N(\mu, \sigma)$ distribution, then the sample mean $\bar{x}$ of n independent observations has the $N(\mu, \sigma/\sqrt{n})$ distribution.

EXAMPLE 4.23 Estimating Odor Thresholds

Adults differ in the smallest amount of DMS they can detect in wine. Extensive studies have found that the DMS odor threshold of adults follows roughly a Normal distribution with mean $\mu = 25$ micrograms per liter (μg/l) and standard deviation $\sigma = 7$ μg/l. Because the population distribution is Normal, the sampling distribution of $\bar{x}$ is also Normal. Figure 4.16 displays the

FIGURE 4.16 The distribution of single observations compared with the distribution of the means $\bar{x}$ of 10 observations, for Example 4.23. Averages are less variable than individual observations.

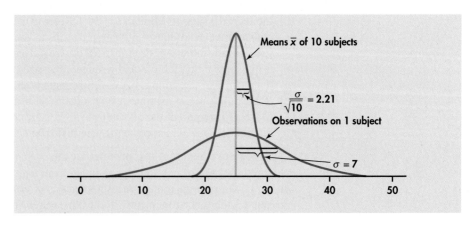

Normal curve for odor thresholds in the adult population and also the Normal curve for the average threshold in random samples of size 10.

Both distributions have the same mean. But means $\bar{x}$ from samples of 10 adults vary less than do measurements on individual adults. The standard deviation of $\bar{x}$ is

$$\frac{\sigma}{\sqrt{n}} = \frac{7}{\sqrt{10}} = 2.21 \ \mu g/l$$

What happens when the population distribution is not Normal? As the sample size increases, the distribution of $\bar{x}$ changes shape: it looks less like that of the population and more like a Normal distribution. When the sample is large enough, the distribution of $\bar{x}$ is very close to Normal. This is true no matter what shape the population distribution has, as long as the population has a finite standard deviation σ. This famous fact of probability theory is called the *central limit theorem*. It is much more useful than the fact that the distribution of $\bar{x}$ is exactly Normal if the population is exactly Normal.

> **Central Limit Theorem**
>
> Draw an SRS of size n from any population with mean μ and finite standard deviation σ. When n is large, the sampling distribution of the sample mean $\bar{x}$ is approximately Normal:
>
> $$\bar{x} \text{ is approximately } N\left(\mu, \frac{\sigma}{\sqrt{n}}\right)$$

More general versions of the central limit theorem say that the distribution of a sum or average of many small random quantities is close to Normal. This is true even if the quantities are not independent (as long as they are not too highly correlated) and even if they have different distributions (as long as no one random quantity is so large that it dominates the others). The central limit theorem suggests why the Normal distributions are common models for observed data. Any variable that is a sum of many small influences will have approximately a Normal distribution.

How large a sample size n is needed for $\bar{x}$ to be close to Normal depends on the population distribution. More observations are required if the shape of the population distribution is far from Normal. Here are two examples in which the population is far from Normal.

EXAMPLE 4.24   The Central Limit Theorem in Action

 UNITED

In 2009, there were 73,423 departures for United Airlines from its largest hub airport, O'Hare Airport in Chicago. Figure 4.17(a) is a histogram of the departure delay times (in minutes) for the entire population of United flights from O'Hare.[24] A negative departure delay represents a flight that left earlier than its scheduled departure time. The distribution is clearly very different from the Normal distribution. It is strongly skewed to the right and very spread out. The right tail of the distribution is even longer than what appears in the figure because there are too few high delay times for the histogram bars to be visible on this scale. In fact, we cut off the departure delay time scale at 350 minutes to save space.

The mean μ of the population is 11.793 min. Suppose we take an SRS of 100 flights. The mean delay time in this sample is $\bar{x} = 8.09$ min. That's less than the mean of the population. Take another SRS of size 100. The mean for this sample is $\bar{x} = 17.37$ min. That's higher than the mean of the population. If we take more samples of size 100, we will get different values of $\bar{x}$. To find the sampling distribution of $\bar{x}$, we take many random samples of size 100 and calculate $\bar{x}$ for each sample. Figure 4.17(b) is a histogram of the mean departure delay times for 1000 samples, each of size 100. Notice something remarkable. Even though the distribution of the individual delay times

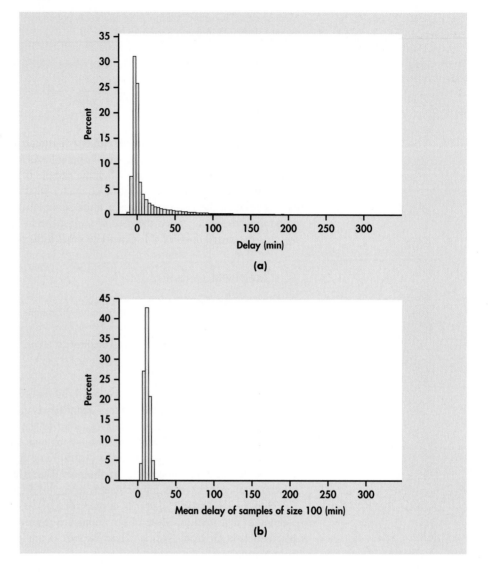

is strongly skewed and very spread out, the distribution of the sample means is quite symmetric and much less spread out.

Figure 4.18(a) is the histogram of the sample means on a scale that more clearly shows its shape. We can see that the distribution of sample means is close to the Normal distribution. The Normal quantile plot of Figure 4.18(b) further confirms the compatibility of the distribution of sample means with the Normal distribution. Furthermore, as expected, the histogram in Figure 4.18(a) appears to be essentially centered on the population mean μ value. Specifically, the mean of the 1000 sample means is 11.831, which is nearly equal to the μ-value of 11.793. If more and more random samples are taken, the mean of the sample mean values will eventually equal μ exactly.

Example 4.24 illustrates three important points of this section that are worth repeating. First, sample means are less variable than individual observations. Second, the distribution of the sample means is centered at μ. Third, sample means are more Normal than individual observations. These facts contribute to the popularity of sample means in statistical inference.

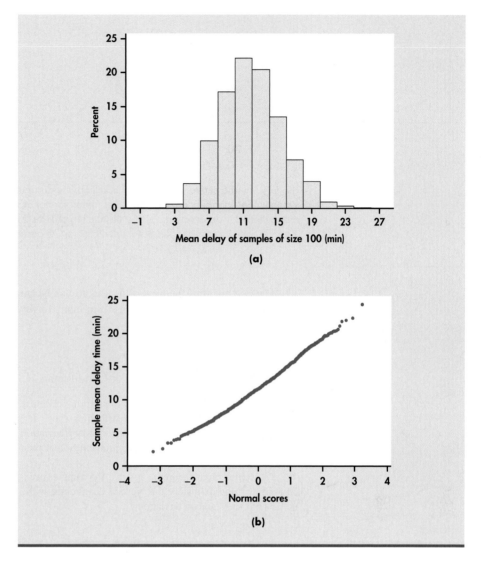

EXAMPLE 4.25   The Central Limit Theorem in Action

Figure 4.19 shows how the central limit theorem works for a very non-Normal population. Figure 4.19(a) displays the density curve of a single observation, that is, of the population. The distribution is strongly right-skewed, and the most probable outcomes are near 0. The mean μ of this distribution is 1, and its standard deviation σ is also 1. This particular distribution is called an *exponential distribution*. Exponential distributions are used as models for the lifetime in service of electronic components and for the time required to serve a customer or repair a machine.

Figures 4.19(b), (c), and (d) are the density curves of the sample means of 2, 10, and 25 observations from this population. As n increases, the shape becomes more Normal. The mean remains at $\mu = 1$, and the standard deviation decreases, taking the value $1/\sqrt{n}$. The density curve for the sample mean of 10 observations is still somewhat skewed to the right but already resembles a Normal curve having $\mu = 1$ and $\sigma = 1/\sqrt{10} = 0.32$. The density curve for $n = 25$ is yet more Normal. The contrast between the shapes of the population distribution and of the distribution of the mean of 10 or 25 observations is striking.

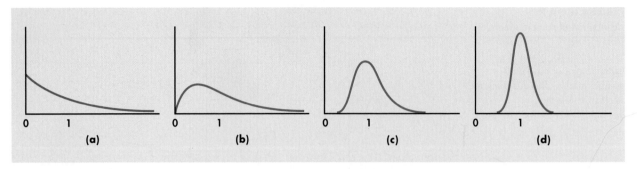

FIGURE 4.19 The central limit theorem in action: the distribution of sample means $\bar{x}$ from a strongly non-Normal population becomes more Normal as the sample size increases. (a) The distribution of 1 observation. (b) The distribution of $\bar{x}$ for 2 observations. (c) The distribution of $\bar{x}$ for 10 observations. (d) The distribution of $\bar{x}$ for 25 observations.

The central limit theorem allows us to use Normal probability calculations to answer questions about sample means from many observations even when the population distribution is not Normal.

EXAMPLE 4.26 Maintaining Air Conditioners

The time X that a technician requires to perform preventive maintenance on an air-conditioning unit is governed by the exponential distribution whose density curve appears in Figure 4.19(a). The mean time is $\mu = 1$ hour and the standard deviation is $\sigma = 1$ hour. Your company operates 70 of these units. What is the probability that their average maintenance time exceeds 50 minutes?

The central limit theorem says that the sample mean time $\bar{x}$ (in hours) spent working on 70 units has approximately the Normal distribution with mean equal to the population mean $\mu = 1$ hour and standard deviation

$$\frac{\sigma}{\sqrt{70}} = \frac{1}{\sqrt{70}} = 0.12 \text{ hour}$$

The distribution of $\bar{x}$ is therefore approximately $N(1, 0.12)$. This Normal curve is the solid curve in Figure 4.20.

FIGURE 4.20 The exact distribution (dashed) and the Normal approximation from the central limit theorem (solid) for the average time needed to maintain an air conditioner, for Example 4.26.

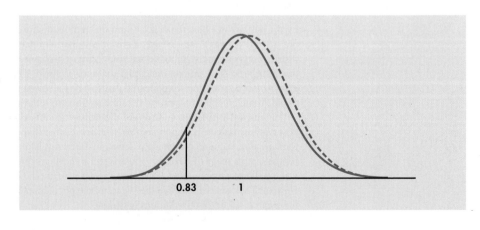

Because 50 minutes is 50/60 of an hour, or 0.83 hour, the probability we want is $P(\bar{x} > 0.83)$. A Normal distribution calculation gives this probability as 0.9222. This is the area to the right of 0.83 under the solid Normal curve in Figure 4.20.

Using more mathematics, we could start with the exponential distribution and find the actual density curve of $\bar{x}$ for 70 observations. This is the dashed curve in Figure 4.20. You can see that the solid Normal curve is a good approximation. The exactly correct probability is the area under the dashed density curve. It is 0.9294. The central limit theorem Normal approximation is off by only about 0.007.

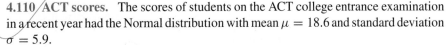

APPLY YOUR KNOWLEDGE

4.110 ACT scores. The scores of students on the ACT college entrance examination in a recent year had the Normal distribution with mean $\mu = 18.6$ and standard deviation $\sigma = 5.9$.

(a) What is the probability that a single student randomly chosen from all those taking the test scores 21 or higher?

(b) Now take an SRS of 50 students who took the test. What are the mean and standard deviation of the sample mean score $\bar{x}$ of these 50 students?

(c) What is the probability that the mean score $\bar{x}$ of these students is 21 or higher?

4.111 Flaws in carpets. The number of flaws per square yard in a type of carpet material varies with mean 1.6 flaws per square yard and standard deviation 1.2 flaws per square yard. The population distribution cannot be Normal, because a count takes only whole-number values. An inspector samples 200 square yards of the material, records the number of flaws found in each square yard, and calculates $\bar{x}$, the mean number of flaws per square yard inspected. Use the central limit theorem to find the approximate probability that the mean number of flaws exceeds 2 per square yard.

4.112 Returns on stocks. The distribution of annual returns on common stocks is roughly symmetric, but extreme observations are somewhat more frequent than in Normal distributions. Because the distribution is not strongly non-Normal, the mean return over a number of years is close to Normal. Historically, annual returns on common stocks (not adjusted for inflation) have varied with mean about 13% and standard deviation about 17%. Andrew plans to retire in 45 years. He assumes (this is dubious) that the past pattern of variation will continue. What is the probability that the mean annual return on common stocks over the next 45 years will exceed 15%? What is the probability that the mean return will be less than 7%?

SECTION 4.4 Summary

- When we want information about the **population mean** μ for some variable, we often take an SRS and use the **sample mean** $\bar{x}$ to estimate the unknown parameter μ.

- The **law of large numbers** states that the actually observed mean outcome $\bar{x}$ must approach the mean μ of the population as the number of observations increases.

- The **sampling distribution** of $\bar{x}$ describes how the statistic $\bar{x}$ varies in all possible samples of the same size from the same population.

- The **mean** of the sampling distribution is μ, so that $\bar{x}$ is an **unbiased estimator** of μ.

- The **standard deviation** of the sampling distribution of $\bar{x}$ is $\sigma/\sqrt{n}$ for an SRS of size n if the population has standard deviation σ. That is, averages are less variable than individual observations.

- If the population has a Normal distribution, so does $\overline{x}$.
- The **central limit theorem** states that for large n the sampling distribution of $\overline{x}$ is approximately Normal for any population with finite standard deviation σ. That is, averages are more Normal than individual observations. We can use the $N(\mu, \sigma/\sqrt{n})$ distribution to calculate approximate probabilities for events involving $\overline{x}$.

SECTION 4.4 Exercises

For Exercises 4.101 to 4.103, see page 265; for 4.104 to 4.106, see pages 267–268; for 4.107 to 4.109, see pages 269–270; and for 4.110 to 4.112, see page 275.

4.113 Legal music downloads. According to Apple, Inc. press reports, the iTunes Music Store sold **500,000,000** songs between July 17, 2005, and February 24, 2006. In an analysis of over 2 million credit and debit card transactions, Forrester Research (Forrester.com) calculated that "iTunes households" make an average of **5.6** iTunes transactions per year. Is each of the bold numbers a parameter or a statistic?

4.114 What Americans buy. In a story on consumer spending, *Time* magazine reports that Americans buy **34** Porsche 911s and **88,163** Apple iPods each day. Is each of the bold numbers a parameter or a statistic?

4.115 Business employees. There are more than 5 million businesses in the United States. The mean number of employees in these businesses is about **19.** A university selects a random sample of 100 businesses in North Dakota and finds that they average about **14** employees. Is each of the bold numbers a parameter or a statistic?

4.116 Ages of Canadian farmers. Every five years, Statistics Canada conducts a Census of Agriculture. The target of the census is all farms selling one or more agricultural products. In 2006, Statistics Canada reported that the average age of a farmer on Prince Edward Island was **51.4** years old, while the average age of a farmer in Nova Scotia was **53.2** years old. Is each of the bold numbers a parameter or a statistic?

4.117 Dust in coal mines. A laboratory weighs filters from a coal mine to measure the amount of dust in the mine atmosphere. Repeated measurements of the weight of dust on the same filter vary Normally with standard deviation $\sigma = 0.08$ milligram (mg) because the weighing is not perfectly precise. The dust on a particular filter actually weighs 123 mg. Repeated weighings will then have the Normal distribution with mean 123 mg and standard deviation 0.08 mg.
(a) The laboratory reports the mean of 3 weighings. What is the distribution of this mean?
(b) What is the probability that the laboratory reports a weight of 124 mg or higher for this filter?

4.118 Making auto parts. An automatic grinding machine in an auto parts plant prepares axles with a target diameter $\mu = 40.125$ millimeters (mm). The machine has some

variability, so the standard deviation of the diameters is $\sigma = 0.002$ mm. A sample of 4 axles is inspected each hour for process control purposes, and records are kept of the sample mean diameter. What will be the mean and standard deviation of the numbers recorded?

4.119 Safe flying weight. In response to the increasing weight of airline passengers, the Federal Aviation Administration told airlines to assume that passengers average 190 pounds in the summer, including clothing and carry-on baggage. But passengers vary: the FAA gave a mean but not a standard deviation. A reasonable standard deviation is 35 pounds. Weights are not Normally distributed, especially when the population includes both men and women, but they are not very non-Normal. A commuter plane carries 19 passengers. What is the approximate probability that the total weight of the passengers exceeds 4000 pounds? (*Hint:* To apply the central limit theorem, restate the problem in terms of the mean weight.)

4.120 Supplier delivery times. Supplier on-time delivery performance is critical to enabling the buyer's organization to meet its customer service commitments. Therefore, monitoring supplier delivery times is critical. Based on a great deal of historical data, a manufacturer of personal computers finds for one of its just-in-time suppliers that the delivery times are random and well approximated by the Normal distribution with mean 48.2 minutes and standard deviation 13.4 minutes.
(a) What is the probability that a particular delivery will exceed one hour?
(b) Based on part (a), what is the probability that a particular delivery arrives in less than one hour?
(c) What is the probability that the mean time of 5 deliveries will exceed one hour?

4.121 The cost of Internet access. The amount that households pay service providers for access to the Internet varies quite a bit, but the mean monthly fee is $28 and the standard deviation is $10. The distribution is not Normal: many households pay about $10 for limited dial-up access or about $25 for unlimited dial-up access, but many pay more for broadband connections. A sample survey asks an SRS of 500 households with Internet access how much they pay. What is the probability that the average fee paid by the sample households exceeds $29?

4.122 The business of insurance. The idea of insurance is that we all face risks that are unlikely but carry high cost. Think of a fire destroying your home. So we form a group to share the risk:

we all pay a small amount, and the insurance policy pays a large amount to those few of us whose homes burn down. An insurance company looks at the records for millions of homeowners and sees that the mean loss from fire in a year is $\mu = \$250$ per house and that the standard deviation of the loss is $\sigma = \$1000$. (The distribution of losses is extremely right-skewed: most people have $0 loss, but a few have large losses.) The company plans to sell fire insurance for $250 plus enough to cover its costs and profit.

(a) Explain clearly why it would be unwise to sell only 12 policies. Then explain why selling many thousands of such policies is a safe business.

(b) If the company sells 10,000 policies, what is the approximate probability that the average loss in a year will be greater than $275?

4.123 Manufacturing defects. Newly manufactured automobile radiators may have small leaks. Most have no leaks, but some have 1, 2, or more. The number of leaks in radiators made by one supplier has mean 0.15 and standard deviation 0.4. The distribution of number of leaks cannot be Normal because only whole-number counts are possible. The supplier ships 400 radiators per day to an auto assembly plant. Take $\bar{x}$ to be the mean number of leaks in these 400 radiators. Over several years of daily shipments, what range of values will contain the middle 95% of the many $\bar{x}$'s?

4.124 Auto accidents. The number of accidents per week at a hazardous intersection varies with mean 2.2 and standard deviation 1.4. This distribution takes only whole-number values, so it is certainly not Normal.

(a) Let $\bar{x}$ be the mean number of accidents per week at the intersection during a year (52 weeks). What is the approximate distribution of $\bar{x}$ according to the central limit theorem?

(b) What is the approximate probability that $\bar{x}$ is less than 2?

(c) What is the approximate probability that there are fewer than 100 accidents at the intersection in a year? (*Hint:* Restate this event in terms of $\bar{x}$.)

4.125 Budgeting for expenses. Weekly postage expenses for your company have a mean of $312 and a standard deviation of $58. Your company has allowed $400 for postage per week in its budget.

(a) What is the approximate probability that the average weekly postage expense for the past 52 weeks will exceed the budgeted amount of $400?

(b) What information would you need to determine the probability that postage for one particular week will exceed $400? Why wasn't this information required for (a)?

STATISTICS IN SUMMARY

This chapter lays the foundations for the study of statistical inference. Statistical inference uses data to draw conclusions about the population or process from which the data come. What is special about inference is that the conclusions include a statement, in the language of probability, about how reliable they are. The statement gives a probability that answers the question "What would happen if I used this method very many times?" The probabilities we need come from the sampling distributions of sample statistics.

Sampling distributions are the key to understanding statistical inference. The Statistics in Summary figure below summarizes the facts about the sampling distribution of $\bar{x}$ in a way that reminds us of the big idea of a sampling distribution. Keep taking random samples of size n from a population with mean μ. Find the sample mean $\bar{x}$ for each sample. Collect all the $\bar{x}$'s and display their distribution. That's the sampling distribution of $\bar{x}$. Keep this figure in mind as you go forward.

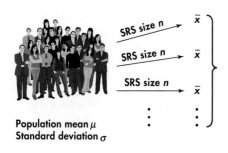

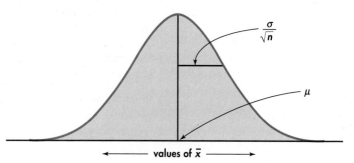

To think more effectively about sampling distributions, we use the language of probability. Probability, the mathematics that describes randomness, is important in many areas of study. Here, we concentrate on informal probability as the conceptual foundation for statistical inference. Because random samples and randomized comparative experiments use chance, their results vary according to the laws of probability. Here is a review list of the most important skills you should have acquired from your study of this chapter.

A. Probability

1. Recognize that some phenomena are random. Probability describes the long-run regularity of random phenomena.

2. Understand that the probability of an event is the proportion of times the event occurs in very many repetitions of a random phenomenon. Use the idea of probability as long-run proportion to think about probability.

3. Know the rules that are obeyed by any legitimate assignment of probabilities to events. Use these rules to find the probabilities of events that are formed from other events.

4. Know what properties an assignment of probabilities to a finite number of outcomes must satisfy. Find probabilities of events in a finite sample space by adding the probabilities of their outcomes.

5. Know what a density curve is. Find probabilities of events as areas under a density curve.

B. Random Variables

1. Use the language of random variables to give compact descriptions of events and their probabilities, such as $P(X \geq 2.5)$.

2. Verify that the probability distribution of a discrete random variable is legitimate. Use the distribution to find probabilities of events involving the random variable.

3. Use the density curve of a continuous random variable to find probabilities of events involving the random variable.

4. Calculate the mean and standard deviation of a discrete random variable from its probability distribution.

5. Know and use the rules for means and variances of sums and differences of random variables, both for independent random variables and correlated random variables.

C. The Sampling Distribution of a Sample Mean

1. Interpret the sampling distribution of a statistic as describing the probabilities of its possible values.

2. Recognize when a problem involves the mean $\bar{x}$ of a sample. Understand that $\bar{x}$ estimates the mean μ of the population from which the sample is drawn.

3. Use the law of large numbers to describe the behavior of $\bar{x}$ as the size of the sample increases.

4. Find the mean and standard deviation of a sample mean $\bar{x}$ from an SRS of size n when the mean μ and standard deviation σ of the population are known.

5. Understand that $\bar{x}$ is an unbiased estimator of μ and that the variability of $\bar{x}$ about its mean μ gets smaller as the sample size increases.

6. Understand that $\bar{x}$ has approximately a Normal distribution when the sample is large (central limit theorem). Use this Normal distribution to calculate probabilities that concern $\bar{x}$.

CHAPTER 4 Review Exercises

4.126 Insuring an asset. Consider an insurance policy that will pay the holder of the policy the market price of her house if the house is completely destroyed by fire. The market price of a particular house is set at $220,000. The probability of this house being completely destroyed by fire during the next year is determined by the insurance company to be 0.3%, and the cost of the policy is set at $1280 for the next year. Let X represent the value of the insurance policy to the holder of the policy at the end of the next year, and let Y represent the value of the house at the end of the next year. Assume the house will be worth $220,000 unless it is completely destroyed by fire.
(a) If the house is completely destroyed by fire in the next year, the value of the policy will be $220,000 - 1280 = 218,720$ dollars. If the house is not completely destroyed by fire in the next year, what will be the value of the policy?
(b) Write down the probability distribution of X in the form of a table.
(c) Write down the probability distribution of Y in the form of a table.

4.127 Insuring an asset. Continue analyzing the insurance scenario described in the previous exercise.
(a) Calculate the mean and standard deviation of X, the value of the insurance policy at the end of the next year.
(b) Calculate the mean and standard deviation of Y, the value of the house at the end of the next year.
(c) What do you notice about the risk (as measured by the standard deviation) of owning the house compared to the risk of holding the insurance policy?

4.128 Insuring an asset. Continue analyzing the insurance scenario described in the previous two exercises. Consider the value of holding the insurance policy *and* owning the house, $X + Y$. Let $W = X + Y$ in this exercise.
(a) Write down the probability distribution of W in the form of a table. What do you notice about the possible values of W? What does this mean for the owner of the house who also is the holder of the insurance policy?
(b) Calculate the mean value of W.
(c) Calculate the standard deviation of W. What does this imply about the risk associated with owning the house *and* holding the insurance policy on the house?
(d) X and Y are perfectly correlated. What is the value of ρ in this case?

4.129 Weights of bags of potato chips. In Example 4.8 (page 233), the weight of a single 9-ounce bag of potato chips has an $N(9.12, 0.15)$ distribution. Samuel works on the production line that manufactures these bags of chips. He weighs 3 randomly chosen bags and records their mean.
(a) What is the standard deviation of Samuel's mean result?
(b) How many bags must Samuel weigh to reduce the standard deviation of his mean to 0.05 ounces?

(c) Explain to someone who knows no statistics the advantage of reporting the average of several weights rather than just a single weight.

4.130 Weights of bags of potato chips. In Example 4.8 (page 233), the probability of a single randomly chosen 9-ounce bag weighing between 9.33 and 9.45 ounces was calculated to be 0.0669. In the setting of the previous exercise, Samuel regularly samples 3 bags of chips and calculates their average weight.
(a) What is the probability of Samuel's 3-bag average falling between 9.33 and 9.45 ounces?
(b) What is the probability of a 6-bag average falling between 9.33 and 9.45 ounces?

4.131 Weights of bags of potato chips. In Example 4.8 (page 233), the weight of a single 9-ounce bag of potato chips has an $N(9.12, 0.15)$ distribution. Consider taking a sample of bags of chips and calculating the average weight of the sample.
(a) What is the probability of a 3-bag average falling between 9.08 and 9.16 ounces?
(b) What is the probability of a 30-bag average falling between 9.08 and 9.16 ounces?
(c) What is the probability of a 150-bag average falling between 9.08 and 9.16 ounces?
(d) If the weights were known to have a mean of 9.12 ounces and a standard deviation of 0.15 ounces but the distribution of the weights was unknown, what would you be able to say about the probabilities you calculated in parts (a), (b), and (c)? Would any of them still be reasonably accurate?

4.132 Customer backgrounds. A company that offers courses to prepare would-be MBA students for the GMAT examination has the following information about its customers: 20% are currently undergraduate students in business; 15% are undergraduate students in other fields of study; 60% are college graduates who are currently employed; and 5% are college graduates who are not employed.
(a) This is a legitimate assignment of probabilities to customer backgrounds. Why?
(b) What percent of customers are currently undergraduates?

4.133 Who gets promoted? Exactly one of Brown, Chavez, and Williams will be promoted to partner in the law firm that employs them all. Brown thinks that she has probability 0.25 of winning the promotion and that Williams has probability 0.2. What probability does Brown assign to the outcome that Chavez is the one promoted?

4.134 Predicting the ACC champion. Las Vegas Zeke, when asked to predict the Atlantic Coast Conference basketball champion, follows the modern practice of giving probabilistic predictions. He says, "North Carolina's probability of winning is twice Duke's. North Carolina State and Virginia each have probability 0.1 of winning, but Duke's probability is three times that. Nobody else has a chance." Has Zeke given a legitimate assignment of probabilities to the eight teams in the conference? Explain your answer.

4.135 Global warming in Canadian minds. Given Canada's northern location and the shrinking of the ice caps (both Canadian and Arctic), global-warming issues are reported daily in the media. In a nationwide random sample of 1000 Canadians (September 29, 2008), the public opinion research firm Compas asked the people in the sample a variety of questions related to global warming.[25] One question concerned their general opinion on global warming. Here are the results:

Outcome	Probability
Is taking place and caused by humans	0.62
Is taking place and is natural	0.15
Is taking place but unsure why	0.12
Debatable if it is taking place	0.08
Is not taking place	?
No opinion	0.02

(a) What is the probability that a randomly chosen Canadian does not believe global warming is taking place?

(b) If you take 1 minus the probability found in part (a), is that the probability that a randomly chosen Canadian is sure that global warming is taking place? If not, explain and determine this probability.

4.136 Age profile of social networkers. Millions of people regularly use social networking services on the Internet to communicate, sharing interests and activities. In a 2008 comprehensive study by the Web search firm RapLeaf, the age distribution of MySpace, Facebook, and LinkedIn users were as follows:[26]

	Age Group						
	14–17	18–24	25–34	35–44	45–54	55–64	65+
MySpace	0.198	0.464	0.241	0.063	0.024	0.008	0.003
Facebook	0.268	0.387	0.221	0.077	0.032	0.011	0.004
LinkedIn	0.011	0.107	0.510	0.247	0.089	0.031	0.006

(a) Verify for each service that there is a legitimate assignment of probabilities to the age groups.

(b) What is the probability that a randomly chosen MySpace user is in the age group of 18 to 24 years old? Find the probabilities of randomly chosen Facebook and LinkedIn users being in the same age group.

(c) For each of the services, what is the probability that a randomly chosen user is an adult, defined as 18 years old or older? Explain two different ways the probability of being an adult can be determined.

(d) Compare the probabilities across the three services. For what age groups are the probabilities similar and for what age groups

are the probabilities markedly different? Provide an explanation for the differences.

4.137 Classifying occupations. Choose an American worker at random and assign his or her occupation to one of the following classes. These classes are used in government employment data.

A   Managerial and professional
B   Technical, sales, administrative support
C   Service occupations
D   Precision production, craft, and repair
E   Operators, fabricators, and laborers
F   Farming, forestry, and fishing

The table below gives the probabilities that a randomly chosen worker falls into each of 12 sex-by-occupation classes:

	Class					
	A	B	C	D	E	F
Male	0.14	0.11	0.06	0.11	0.12	0.03
Female	0.09	0.20	0.08	0.01	0.04	0.01

(a) Verify that this is a legitimate assignment of probabilities to these outcomes.

(b) What is the probability that the worker is female?

(c) What is the probability that the worker is not engaged in farming, forestry, or fishing?

(d) Classes D and E include most mechanical and factory jobs. What is the probability that the worker holds a job in one of these classes?

(e) What is the probability that the worker does not hold a job in Classes D or E?

4.138 Rolling dice. Figure 4.2 (page 225) shows the possible outcomes for rolling two dice. If the dice are carefully made, each of these 36 outcomes has probability 1/36. The outcome of interest to a gambler is the sum of the spots on the two up-faces. Call this random variable X. Example 4.5 shows that $P(X = 5) = 4/36$.

(a) Give the complete probability distribution of X in the form of a table of the possible values and their probabilities. Draw a probability histogram to display the distribution.

(b) One bet available in craps wins if a 7 or an 11 comes up on the next roll of two dice. What is the probability of rolling a 7 or an 11 on the next roll?

(c) Several bets in craps lose if a 7 is rolled. If any outcome other than 7 occurs, these bets either win or continue to the next roll. What is the probability that anything other than a 7 is rolled?

4.139 Nonstandard dice. You have two balanced, six-sided dice. One is a standard die, with faces having 1, 2, 3, 4, 5, and 6 spots. The other die has three faces with 0 spots and three faces with 6 spots. Find the probability distribution for the total number of spots Y on the up-faces when you roll these two dice.

4.140 An IQ test. The Wechsler Adult Intelligence Scale (WAIS) is a common "IQ test" for adults. The distribution of

WAIS scores for persons over 16 years of age is approximately Normal with mean 100 and standard deviation 15.

(a) What is the probability that a randomly chosen individual has a WAIS score of 105 or higher?

(b) What are the mean and standard deviation of the average WAIS score $\bar{x}$ for an SRS of 60 people?

(c) What is the probability that the average WAIS score of an SRS of 60 people is 105 or higher?

(d) Would your answers to any of (a), (b), or (c) be affected if the distribution of WAIS scores in the adult population were distinctly non-Normal?

4.141 Weights of eggs. The weight of the eggs produced by a certain breed of hen is Normally distributed with mean 65 grams (g) and standard deviation 5 g. Think of cartons of such eggs as SRSs of size 12 from the population of all eggs. What is the probability that the weight of a carton falls between 750 and 825 g?

4.142 How many people in a car? A study of rush-hour traffic in San Francisco counts the number of people in each car entering a freeway at a suburban interchange. Suppose that this count has mean 1.5 and standard deviation 0.75 in the population of all cars that enter at this interchange during rush hours.

(a) Could the exact distribution of the count be Normal? Why or why not?

(b) Traffic engineers estimate that the capacity of the interchange is 700 cars per hour. According to the central limit theorem, what is the approximate distribution of the mean number of persons $\bar{x}$ in 700 randomly selected cars at this interchange?

(c) What is the probability that 700 cars will carry more than 1075 people? (*Hint:* Restate this event in terms of the mean number of people $\bar{x}$ per car.)

4.143 A grade distribution. North Carolina State University has posted the grade distributions for its courses online. You can find that the distribution of grades in a large section of Accounting 210 in past semester to be:

Grade:	A	B	C	D	F
Probability:	0.18	0.32	0.34	0.09	0.07

(a) Verify that this is a legitimate assignment of probabilities to grades.

(b) Using the common scale A= 4, B= 3, C= 2, D= 1, F= 0, what is the mean grade in Accounting 210?

4.144 Risk pooling in a supply chain. Example 4.19 (page 254) compares a decentralized versus a centralized inventory system as it ultimately relates to the amount of safety stock (extra inventory over and above mean demand) held in the system. Suppose that the CEO of ElectroWorks requires a 99% customer service level. This means that the probability of satisfying customer demand during the lead time is 0.99. Assume that lead time demands for the Milwaukee warehouse, Chicago warehouse, and centralized warehouse are Normally distributed with the means and standard deviations found in the example.

(a) For a 99% service level, how much safety stock of the part SurgeArrester does the Milwaukee warehouse need to hold? Round your answer to the nearest integer.

(b) For a 99% service level, how much safety stock of the part SurgeArrester does the Chicago warehouse need to hold? Round your answer to the nearest integer.

(c) For a 99% service level, how much safety stock of the part SurgeArrester does the centralized warehouse need to hold? Round your answer to the nearest integer. How many more units of the part need to be held in the decentralized system than in the centralized system?

4.145 Risk pooling in a supply chain. Example 4.19 (page 254) shows that the standard deviation of demand on the centralized warehouse is less than the sum of standard deviations of the demands on each of the decentralized warehouses. The smaller the standard deviation of demand on the centralized warehouse, the less safety stock needs to be held. For given standard deviations of demand on two decentralized warehouses, what is the ideal value of ρ that will result in the least safety stock needed at a centralized warehouse? (*Hint:* Refer to Rule 3 of the rules for variances.)

4.146 Sadie's portfolio. Examples 4.15 (page 249) and 4.20 (page 256) provide the means and standard deviations of monthly returns for Dell, Apple, and Sadie's portfolio (60% Dell and 40% Apple). As noted in Exercise 4.112 (page 275), stock returns follow a distribution that is slightly non-Normal. It is pretty safe to assume that the mean monthly return $\bar{x}$ based on 12 months is Normally distributed. Since monthly returns can be shown to behave as a purely random process over time, we can view a sample of 12 consecutive months as an SRS of size 12.

(a) For each of the investment options, fill in the table with the probabilities for the mean monthly return:

	Dell only	Apple only	Portfolio
P(mean return less than −5%)	−	−	−
P(positive mean return)	−	−	−
P(mean return greater than +5%)	−	−	−

(b) An investor willing to take risks in the hope of big gains accept the possibility of a big loss. Based on the probabilities you found in part (a), which investment is most attractive to a risky investor? Which investment is most attractive to an investor who wishes to reduce the risk of big losses and is content with a greater chance of a moderate return?

4.147 Sadie's portfolio. Continuing the previous exercise, for each of the three investment strategies, 90% of the time the mean monthly return $\bar{x}$ based on 12 months will be greater than what value?

4.148 Simulating a mean. One consequence of the law of large numbers is that once we have a probability distribution for a random variable, we can find its mean by simulating many outcomes and averaging them. The law of large numbers says that if we take enough outcomes, their average value is sure to approach the mean of the distribution.

I have a little bet to offer you. Toss a coin 10 times. If there is no run of three or more straight heads or tails in the 10 outcomes, I'll pay you $2. If there is a run of three or more, you pay me just $1. Surely you will want to take advantage of me and play this game?

Simulate enough plays of this game (the outcomes are +$2 if you win and −$1 if you lose) to estimate the mean outcome. Is it to your advantage to play?

Insurance. The business of selling insurance is based on probability and the law of large numbers. Consumers (including businesses) buy insurance because we all face risks that are unlikely but carry high cost—think of a fire destroying your home. So we form a group to share the risk: we all pay a small amount, and the insurance policy pays a large amount to those few of us whose homes burn down. The insurance company sells many policies, so it can rely on the law of large numbers. Exercises 4.149 to 4.154 explore aspects of insurance.

4.149 Fire insurance. An insurance company looks at the records for millions of homeowners and sees that the mean loss from fire in a year is $\mu = \$400$ per person. (Most of us have no loss, but a few lose their homes. The $400 is the average loss.) The company plans to sell fire insurance for $400 plus enough to cover its costs and profit. Explain clearly why it would be stupid to sell only 10 policies. Then explain why selling thousands of such policies is a safe business.

4.150 More about fire insurance. In fact, the insurance company sees that in the entire population of homeowners, the mean loss from fire is $\mu = \$400$ and the standard deviation of the loss is $\sigma = \$300$. The distribution of losses is strongly right-skewed: many policies have $0 loss, but a few have large losses. If the company sells 10,000 policies, what is the approximate probability that the average loss will be greater than $410?

4.151 Life insurance. A life insurance company sells a term insurance policy to a 21-year-old male that pays $100,000 if the insured dies within the next 5 years. The probability that a randomly chosen male will die each year can be found in mortality tables. The company collects a premium of $250 each year as payment for the insurance. The amount X that the company earns on this policy is $250 per year, less the $100,000 that it must pay if the insured dies. The distribution of X is shown below. Fill in the missing probability in the table and calculate the mean earnings μ_X.

4.152 More about life insurance. It would be quite risky for you to insure the life of a 21-year-old friend under the terms of Exercise 4.151. There is a high probability that your friend would live and you would gain $1250 in premiums. But if he were to die, you would lose almost $100,000. Explain carefully why selling insurance is not risky for an insurance company that insures many thousands of 21-year-old men.

4.153 The risk of selling insurance. We have seen that the risk of an investment is often measured by the standard deviation of the return on the investment. The more variable the return is (the larger σ is), the riskier the investment. We can measure the great risk of insuring one person's life in Exercise 4.151 by computing the standard deviation of the income X that the insurer will receive. Find σ_X, using the distribution and mean you found in Exercise 4.151.

4.154 The risk of selling insurance, continued. The risk of insuring one person's life is reduced if we insure many people. Use the result of the previous exercise and the rules for means and variances to answer the following questions.
(a) Suppose that we insure two 21-year-old males, and that their ages at death are independent. If X and Y are the insurer's income from the two insurance policies, the insurer's average income on the two policies is

$$Z = \frac{X + Y}{2} = 0.5X + 0.5Y$$

Find the mean and standard deviation of Z. You see that the mean income is the same as for a single policy but the standard deviation is less.
(b) If four 21-year-old men are insured, the insurer's average income is

$$Z = \frac{1}{4}(X_1 + X_2 + X_3 + X_4)$$

where X_i is the income from insuring one man. The X_i are independent and each has the same distribution as before. Find the mean and standard deviation of Z. Compare your results with the results of (a). We see that averaging over many insured individuals reduces risk.

4.155 Finite sample spaces. Choose one employee of your company at random and let X be the number of years that person has been employed by the company (rounded to the nearest year).
(a) Give a reasonable sample space S for the possible values of X.
(b) Is X a discrete random variable or a continuous random variable? Why?

	Age at Death (years)					
	21	22	23	24	25	≥ 26
Earnings X	−$99,750	−$99,500	−$99,250	−$99,000	−$98,750	$1250
Probability	0.00183	0.00186	0.00189	0.00191	0.00193	?

(c) How many possible values of X did you include in your definition of S?

4.156 Infinite sample spaces, Part I (optional). Buy one share of Apple Computer (AAPL) stock, and let Y be the number of days until the closing market value of your share is double what you initially paid for the share.
(a) Give the sample space S for the possible values of Y.
(b) Is Y a discrete random variable or a continuous random variable? Why?
(c) How many possible values of Y did you include in your definition of S? (This exercise illustrates that "discrete" and "finite" are not the same thing, though we have made use of only finite discrete sample spaces.)

4.157 Infinite sample spaces, Part II (optional). Randomly choose one 30-milliliter (ml) ink cartridge from the lot of cartridges produced at your plant in the last hour. Let W be the actual amount of ink in the cartridge. The cartridges are supposed to contain 30 ml of ink but are designed to hold up to a maximum of 35 ml of ink.
(a) Give a reasonable sample space S for the possible values of W.
(b) Is W a discrete random variable or a continuous random variable? Why?
(c) How many possible values of W did you include in your definition of S?

CHAPTER 4 Case Study Exercises

CASE STUDY EXERCISE 1: Benford's law. Case 4.1 (page 223) concerns Benford's law, which gives the distribution of first digits in many sets of data from business and science. Locate at least two long tables whose entries could plausibly begin with any digit 1 through 9. You may choose data tables, such as populations of many cities or the number of shares traded on the New York Stock Exchange on many days, or mathematical tables such as logarithms or square roots. We hope it's clear that you can't use the table of random digits. Let's require that your examples each contain at least 300 numbers. Tally the first digits of all entries in each table. Report the distributions (in percents) and compare them with each other, with Benford's law, and with the "equally likely" distribution.

CASE STUDY EXERCISE 2: State lotteries. Most American states and Canadian provinces operate lotteries. State lotteries combine probability with economics and politics. States sometimes add opportunities for gambling in hard times to increase their take from operating the games or from taxing games run by others. Investigate the presence of legal gambling in your state. Include probability aspects, such as the probability of winning and the expected payoff on different games. Also look at economic aspects of gambling: how much does the state earn? What percent of the state budget is this? Has the state's take from gambling changed in recent years? Probability and economics are related, because a state lottery keeps only what it doesn't pay out in winnings to gamblers. What percent of the money bet in the lottery does the state pay out in prizes? How does this compare with casino games?

Probability Theory

Introduction

The mathematics of probability can provide models to describe the flow of traffic through a highway system, a telephone interchange, or a computer processor; the product preferences of consumers; the spread of epidemics or computer viruses; and the rate of return on risky investments. Although we are interested in probability because of its usefulness in statistics, the mathematics of chance is important in many fields of study. This chapter presents a bit more of the theory of probability.

5.1 General Probability Rules

Our study of probability in Chapter 4 concentrated on sampling distributions. Now we return to the general laws of probability. With more probability at our command, we can model more complex random phenomena. We have already met and used four rules.

The mathematics of probability can help solve business problems. What if a manufacturer uses two suppliers for an identical part and some parts are defective? Probability rules can tell us whether defects likely are caused by one of the two suppliers, as Example 5.2 reveals.

Rules of Probability

Rule 1. $0 \leq P(A) \leq 1$ for any event A

Rule 2. $P(S) = 1$

Rule 3. Complement rule: For any event A,

$$P(A^c) = 1 - P(A)$$

Rule 4. Addition rule: If A and B are **disjoint** events, then

$$P(A \text{ or } B) = P(A) + P(B)$$

The complement rule takes its name from the fact that the set of all outcomes that are *not* in an event A is often called the **complement** of A. As a convenient short notation, we will write the complement of A as A^c. For example, if A is the event that a randomly chosen corporate CEO is female, then A^c is the event that the CEO is male.

complement

Independence and the multiplication rule

Rule 4, the addition rule for disjoint events, describes the probability that *one or the other* of two events A and B occurs when A and B cannot occur together. Now we will describe the probability that *both* events A and B occur, again only in a special situation.

You may find it helpful to draw a picture to display relations among several events. A picture like Figure 5.1 that shows the sample space S as a rectangular area and events as areas within S is called a **Venn diagram.** The events A and B in Figure 5.1 are disjoint because they do not overlap. The Venn diagram in Figure 5.2 illustrates two events that are not disjoint. The event $\{A$ and $B\}$ appears as the overlapping area that is common to both A and B.

Venn diagram

Suppose that you toss a balanced coin twice. You are counting heads, so two events of interest are

$$A = \text{first toss is a head}$$
$$B = \text{second toss is a head}$$

The events A and B are not disjoint. They occur together whenever both tosses give heads. We want to find the probability of the event $\{A$ and $B\}$ that *both* tosses are heads.

The coin tossing of Buffon, Pearson, and Kerrich described at the beginning of Chapter 4 convinces us to assign probability 1/2 to a head when we toss a coin. So,

$$P(A) = 0.5$$
$$P(B) = 0.5$$

FIGURE 5.1 Venn diagram showing disjoint events A and B.

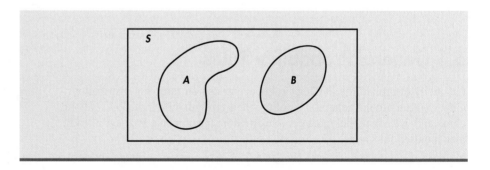

FIGURE 5.2 Venn diagram showing events A and B that are not disjoint. The event $\{A$ and $B\}$ consists of outcomes common to A and B.

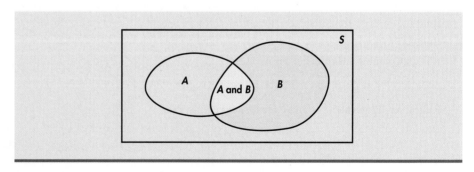

What is $P(A \text{ and } B)$? Our common sense says that it is 1/4. The first coin will give a head half the time and then the second will give a head on half of those trials, so both coins will give heads on $1/2 \times 1/2 = 1/4$ of all trials in the long run. This reasoning assumes that the second coin still has probability 1/2 of a head after the first has given a head. This is true—we can verify it by tossing two coins many times and observing the proportion of heads on the second toss after the first toss has produced a head. We say that the events "head on the first toss" and "head on the second toss" are **independent.** Independence means that the outcome of the first toss cannot influence the outcome of the second toss.

independence

EXAMPLE 5.1   Independent or Not?

Because a coin has no memory and most coin tossers cannot influence the fall of the coin, it is safe to assume that successive coin tosses are independent. For a balanced coin this means that, after we see the outcome of the first toss, we still assign probability 1/2 to heads on the second toss.

On the other hand, the colors of successive cards dealt from the same deck are not independent. A standard 52-card deck contains 26 red and 26 black cards. For the first card dealt from a shuffled deck, the probability of a red card is $26/52 = 0.50$ (equally likely outcomes). Once we see that the first card is red, we know that there are only 25 reds among the remaining 51 cards. The probability that the second card is red is therefore only $25/51 = 0.49$. Knowing the outcome of the first deal changes the probabilities for the second.

If two individuals request a credit report from a particular credit agency (for example, Equifax, TransUnion, or Experian), it is reasonable to assume that the two credit scores are independent because the credit history of one customer does not influence the credit history of the other customer. But if one individual were to request a credit report in two successive years, then the two credit scores would likely not be independent. Knowing the initial credit score might lead the individual to try to improve his credit record which would influence the subsequent credit score.

> **Multiplication Rule for Independent Events**
>
> Two events A and B are **independent** if knowing that one occurs does not change the probability that the other occurs. If A and B are independent,
>
> $$P(A \text{ and } B) = P(A)P(B)$$
>
> If the above equality does not hold true, then A and B are not independent events.

EXAMPLE 5.2   Determining Independence Using the Multiplication Rule

Consider a manufacturer that uses two suppliers for supplying an identical part that enters the production line. Sixty percent of the parts come from one supplier, while the remaining 40% come from the other supplier. Internal quality audits find that there is a 1% chance that a randomly chosen part from the production line is defective. External supplier audits reveal that two parts per thousand are defective from Supplier 1. Are the events of a part coming from a particular supplier, say Supplier 1, and a part being defective independent?

Define the two events as follows:

$$S1 = \text{A randomly chosen part comes from Supplier 1.}$$
$$D = \text{A randomly chosen part is defective.}$$

We have $P(S1) = 0.60$ and $P(D) = 0.01$. The product of these probabilities is

$$P(S1)P(D) = (0.60)(0.01) = 0.006$$

However, supplier audits of Supplier 1 indicate that $P(S1 \text{ and } D) = 0.002$. Given that $P(S1 \text{ and } D) \neq P(S1)P(D)$, we conclude that the supplier and defective part events are not independent.

Independence versus disjoint: avoid the confusion

The multiplication rule $P(A \text{ and } B) = P(A)P(B)$ holds if A and B are *independent* but not otherwise. The addition rule $P(A \text{ or } B) = P(A) + P(B)$ holds if A and B are *disjoint* but not otherwise. Resist the temptation to use these simple rules when the circumstances that justify them are not present. You must also be certain not to confuse disjointness and independence. If A and B are disjoint, then the fact that A occurs tells us that B cannot occur—look again at Figure 5.1. So disjoint events are not independent. Unlike disjointness, we cannot picture independence in a Venn diagram, because it involves the probabilities of the events rather than just the outcomes that make up the events.

APPLY YOUR KNOWLEDGE

5.1 High school rank. Select a first-year college student at random and ask what his or her academic rank was in high school. Here are the probabilities, based on proportions from a large sample survey of first-year students:

Rank:	Top 20%	Second 20%	Third 20%	Fourth 20%	Lowest 20%
Probability:	0.41	0.23	0.29	0.06	0.01

(a) Choose two first-year college students at random. Why is it reasonable to assume that their high school ranks are independent?
(b) What is the probability that both were in the top 20% of their high school classes?
(c) What is the probability that the first was in the top 20% and the second was in the lowest 20%?

5.2 College-educated part-time workers? For people aged 25 years or older, government data show that 34% of employed people have at least 4 years of college and that 20% of employed people work part-time. Can you conclude that because $(0.34)(0.20) = 0.068$ about 6.8% of employed people aged 25 years or older are college-educated part-time workers? Explain your answer.

Applying the multiplication rule

If two events A and B are independent, the event that A does not occur is also independent of B, and so on. Suppose, for example, that 75% of all registered voters in a suburban district are Republicans. If an opinion poll interviews two voters chosen independently, the probability that the first is a Republican and the second is not a Republican is $(0.75)(0.25) = 0.1875$. The multiplication rule also extends to collections of more than two events, provided that all are independent. Independence of events A, B, and C means that no information about any one or any two can change the probability of the remaining events. Independence is often assumed in setting up a probability model when the events we are describing seem to have no connection. We can then use the multiplication rule freely, as in the example on the next page.

EXAMPLE 5.3   Undersea Cables

The first successful transatlantic telegraph cable was laid in 1866. The first telephone cable across the Atlantic did not appear until 1956—the barrier was designing "repeaters," amplifiers needed to boost the signal, that could operate for years on the sea bottom. This first cable had 52 repeaters. The last copper cable, laid in 1983 and retired in 1994, had 662 repeaters. The first fiber-optic cable was laid in 1988 and has 109 repeaters. There are now more than 400,000 miles of undersea cable, with more being laid every year to handle the flood of Internet traffic.

Repeaters in undersea cables must be very reliable. To see why, suppose that each repeater has probability 0.999 of functioning without failure for 25 years. Repeaters fail independently of each other. (This assumption means that there are no "common causes" such as earthquakes that would affect several repeaters at once.) Denote by A_i the event that the ith repeater operates successfully for 25 years.

The probability that 2 repeaters both last 25 years is

$$P(A_1 \text{ and } A_2) = P(A_1)P(A_2)$$
$$= 0.999 \times 0.999 = 0.998$$

For a cable with 10 repeaters the probability of no failures in 25 years is

$$P(A_1 \text{ and } A_2 \text{ and } \ldots \text{ and } A_{10}) = P(A_1)P(A_2)\cdots P(A_{10})$$
$$= 0.999 \times 0.999 \times \cdots \times 0.999$$
$$= 0.999^{10} = 0.990$$

Cables with 2 or 10 repeaters would be quite reliable. Unfortunately, the last copper transatlantic cable had 662 repeaters. The probability that all 662 work for 25 years is

$$P(A_1 \text{ and } A_2 \text{ and } \ldots \text{ and } A_{662}) = 0.999^{662} = 0.516$$

This cable will fail to reach its 25-year design life about half the time if each repeater is 99.9% reliable over that period. The multiplication rule for probabilities shows that repeaters must be much more than 99.9% reliable.

By combining the rules we have learned, we can compute probabilities for rather complex events. Here is an example.

EXAMPLE 5.4   False Positives in Job Drug Testing

Job applicants in both the public and the private sector are often finding that preemployment drug testing is a requirement. The Society for Human Resource Management found that 84% of employers require drug testing of new job applicants and that 39% of employers randomly test hired employees.[1] From an applicant's or employee's perspective, one primary concern with drug testing is a "false positive" result, that is, an indication of drug use when the individual has indeed not used drugs. If a job applicant tests positive, some companies allow the applicant to pay for a retest. For existing employees, a positive result is sometimes followed up with a more sophisticated and expensive test. Beyond cost considerations, there are issues of defamation, wrongful discharge, and emotional distress.

The enzyme multiplied immunoassay technique, or EMIT, applied to urine samples is one of the most common tests for illegal drugs because it is fast and inexpensive. Applied to people who are free of illegal drugs, EMIT has been reported to have false positive rates ranging from 0.2% to 2.5%. If 150 employees are tested and all 150 are free of illegal drugs, what is the probability that at least 1 false positive will occur, assuming a 0.2% false positive rate?

It is reasonable to assume as part of the probability model that the test results for different individuals are independent. The probability that the test is positive for a single person is 0.2% or 0.002, so the probability of a negative result is $1 - 0.002 = 0.998$ by the complement rule. The probability of at least 1 false positive among the 150 people tested is therefore

$$P(\text{at least 1 positive}) = 1 - P(\text{no positives})$$
$$= 1 - P(150 \text{ negatives})$$
$$= 1 - 0.998^{150}$$
$$= 1 - 0.741 = 0.259$$

The probability is greater than 1/4 that at least 1 of the 150 people will test positive for illegal drugs even though no one has taken such drugs.

APPLY YOUR KNOWLEDGE

5.3 Misleading résumés.   For more than two decades, Jude Werra, president of an executive recruiting firm, has tracked executive résumés to determine the rate of misrepresenting education credentials and/or employment information. On a biannual basis, Werra reports a now nationally recognized statistic known as the "Liars Index." In 2008, Werra reported that 15.9% of executive job applicants lied on their résumés.[2]

(a) Suppose 5 résumés are randomly selected from an executive job applicant pool. What is the probability that all of the résumés are truthful?

(b) What is the probability that at least 1 of 5 randomly selected résumés has a misrepresentation?

5.4 Failing to detect drug use.   In Example 5.4, we considered how drug tests can indicate illegal drug use when no illegal drugs were actually used. Consider now another type of false test result. Suppose an employee is suspected of having used an illegal drug and is given two tests that operate independently of each other. Test A has probability 0.9 of being positive if the illegal drug has been used. Test B has probability 0.8 of being positive if the illegal drug has been used. What is the probability that *neither* test is positive if the illegal drug has been used?

5.5 Bright lights?   A string of holiday lights contains 20 lights. The lights are wired in series, so that if any light fails the whole string will go dark. Each light has probability 0.02 of failing during a 3-year period. The lights fail independently of each other. What is the probability that the string of lights will remain bright for 3 years?

The general addition rule

We know that if A and B are disjoint events, then $P(A \text{ or } B) = P(A) + P(B)$. This addition rule extends to more than two events that are disjoint in the sense that no two have any outcomes in common. The Venn diagram in Figure 5.3 shows three disjoint events A, B, and C. The probability that one of these events occurs is $P(A) + P(B) + P(C)$.

FIGURE 5.3 The addition rule for disjoint events:
$P(A \text{ or } B \text{ or } C) =$
$P(A) + P(B) + P(C)$ when events A, B, and C are disjoint.

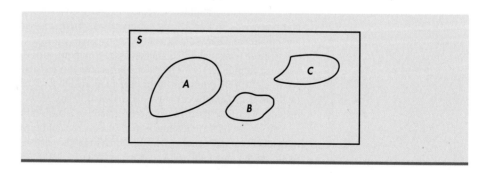

FIGURE 5.4 The general addition rule: $P(A \text{ or } B) = P(A) + P(B) - P(A \text{ and } B)$ for any events A and B.

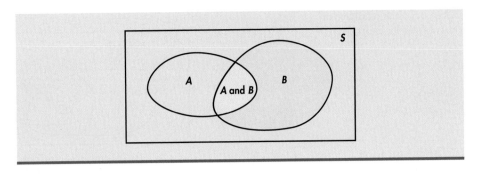

If events A and B are *not* disjoint, they can occur simultaneously. The probability that one or the other occurs is then *less* than the sum of their probabilities. As Figure 5.4 suggests, the outcomes common to both are counted twice when we add probabilities, so we must subtract this probability once. Here is the addition rule for any two events, disjoint or not.

> **General Addition Rule for Any Two Events**
>
> For any two events A and B,
>
> $$P(A \text{ or } B) = P(A) + P(B) - P(A \text{ and } B)$$

If A and B are disjoint, the event $\{A \text{ and } B\}$ that both occur contains no outcomes and therefore has probability 0. So the general addition rule includes Rule 4, the addition rule for disjoint events.

EXAMPLE 5.5   Making Partner

Deborah and Matthew are anxiously awaiting word on whether they have been made partners of their law firm. Deborah guesses that her probability of making partner is 0.7 and that Matthew's is 0.5. (These are personal probabilities reflecting Deborah's assessment of chance.) This assignment of probabilities does not give us enough information to compute the probability that at least one of the two is promoted. In particular, adding the individual probabilities of promotion gives the impossible result 1.2. If Deborah also guesses that the probability that *both* she and Matthew are made partners is 0.3, then by the general addition rule

$$P(\text{at least one is promoted}) = 0.7 + 0.5 - 0.3 = 0.9$$

The probability that *neither* is promoted is then 0.1 by the complement rule.

Venn diagrams are a great help in finding probabilities because you can just think of adding and subtracting areas. Figure 5.5 shows some events and their probabilities for Example 5.5. What is the probability that Deborah is promoted and Matthew is not? The Venn diagram shows that this is the probability that Deborah is promoted minus the probability that both are promoted, $0.7 - 0.3 = 0.4$. Similarly, the probability that Matthew is promoted and Deborah is not is $0.5 - 0.3 = 0.2$. The four probabilities that appear in the figure add to 1 because they refer to four disjoint events that make up the entire sample space.

FIGURE 5.5 Venn diagram and probabilities for Example 5.5.

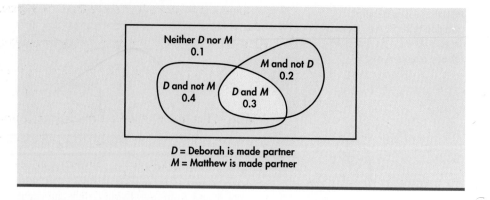

APPLY YOUR KNOWLEDGE

5.6 Prosperity and education. Call a household prosperous if its income exceeds $100,000. Call the household educated if the householder completed college. Select an American household at random, and let A be the event that the selected household is prosperous and B the event that it is educated. According to the Current Population Survey, $P(A) = 0.180$, $P(B) = 0.290$, and the probability that a household is both prosperous and educated is $P(A \text{ and } B) = 0.105$.

(a) Draw a Venn diagram that shows the relation between the events A and B. What is the probability $P(A \text{ or } B)$ that the household selected is either prosperous or educated?

(b) In your diagram, shade the event that the household is educated but not prosperous. What is the probability of this event?

5.7 Caffeine in the diet. Common sources of caffeine are coffee, tea, and cola drinks. Suppose that

    55% of adults drink coffee

    25% of adults drink tea

    45% of adults drink cola

and also that

    15% drink both coffee and tea

    5% drink all three beverages

    25% drink both coffee and cola

    5% drink only tea

Draw a Venn diagram marked with this information. Use it along with the addition rules to answer the following questions.

(a) What percent of adults drink only cola?

(b) What percent drink none of these beverages?

SECTION 5.1 Summary

- Events A and B are **disjoint** if they have no outcomes in common. Events A and B are **independent** if knowing that one event occurs does not change the probability we would assign to the other event.

- Any assignment of probability obeys these more general rules in addition to those stated in Chapter 4:

Addition rule: If events A, B, C, ... are all **disjoint** in pairs, then

$$P(\text{at least one of these events occurs}) = P(A \text{ or } B \text{ or } C \text{ or } \ldots)$$
$$= P(A) + P(B) + P(C) + \cdots$$

Multiplication rule: If events A and B are **independent,** then

$$P(A \text{ and } B) = P(A)P(B)$$

General addition rule: For any two events A and B,

$$P(A \text{ or } B) = P(A) + P(B) - P(A \text{ and } B)$$

SECTION 5.1 Exercises

For Exercises 5.1 and 5.2, see page 288; for 5.3 to 5.5, see page 290; and for 5.6 and 5.7, see page 292.

5.8 Using Internet sources. Internet sites often vanish or move, so that references to them can't be followed. In fact, 13% of Internet sites referenced in major scientific journals are lost within two years after publication.
(a) If a paper contains seven Internet references, what is the probability that all seven are still good two years later?
(b) What specific assumptions did you make in order to calculate this probability?

5.9 Demographics in an SRS. The Census Bureau reports that 27% of California residents are foreign-born. Suppose that you choose three Californians at random, so that each has probability 0.27 of being foreign-born and the three are independent of each other. Let W be the number of foreign-born people you chose.
(a) What are the possible values of W?
(b) Look at your three people in order. There are eight possible arrangements of foreign (F) and domestic (D) birth. For example, FFD means the first two are foreign-born and the third is not. What is the probability of each arrangement? (Use the multiplication rule.)
(c) What is the value of W for each arrangement in (b)? What is the probability of each possible value of W? (This is the distribution of a Yes/No response for an SRS of size 3. In principle, the same idea works for an SRS of any size.)

5.10 I'll switch! In a 2007 buying-intention survey, Goldman Sachs found that 71% of the respondents indicated interest in buying an Apple mobile phone. Of the respondents interested in buying an Apple mobile phone, 15% indicated that they would switch cellular carriers to get an Apple mobile phone. Consider randomly selecting one survey respondent. Let A be the event that the respondent is interested in buying an Apple mobile phone, and let B be the event that the respondent would switch carriers to get an Apple mobile phone.
(a) Draw a Venn diagram that shows the relation between the events A and B.

(b) Apply the general addition rule to calculate $P(A \text{ or } B)$.
(c) Calculate $P(A \text{ and } B)$ and calculate $P(A)P(B)$. Are the two calculations equal? Are A and B independent?

5.11 Everyone gets audited. Wallen Accounting Services specializes in tax preparation for individual tax returns. Data collected from past records reveals that 9% of the returns prepared by Wallen have been selected for audit by the Internal Revenue Service. Today, Wallen has six new customers. Assume the chances of these six customers being audited are independent.
(a) What is the probability that all six new customers will be selected for audit?
(b) What is the probability that none of the six new customers will be selected for audit?
(c) What is the probability that exactly one of the six new customers will be selected for audit?

5.12 Hiring strategy. A chief executive officer (CEO) has resources to hire one vice-president or three managers. He believes that he has probability 0.6 of successfully recruiting the vice-president candidate and probability 0.8 of successfully recruiting each of the manager candidates. The three candidates for manager will make their decisions independently of each other. The CEO must successfully recruit either the vice-president or all three managers to consider his hiring strategy a success. Which strategy should he choose?

5.13 Playing the lottery. An instant lottery game gives you probability 0.02 of winning on any one play. Plays are independent of each other. If you play 5 times, what is the probability that you win at least once?

5.14 Nonconforming chips. Automobiles use semiconductor chips for engine and emission control, repair diagnosis, and other purposes. An auto manufacturer buys chips from a supplier. The supplier sends a shipment of which 5% fail to conform to performance specifications. Each chip chosen from this shipment has probability 0.05 of being nonconforming, and each automobile uses 12 chips selected independently. What is the probability that all 12 chips in a car will work properly?

5.15 A random walk on Wall Street? The "random walk" theory of securities prices holds that price movements in disjoint time periods are independent of each other. Suppose that we record only whether the price is up or down each year, and that the probability that our portfolio rises in price in any one year is 0.65. (This probability is approximately correct for a portfolio containing equal dollar amounts of all common stocks listed on the New York Stock Exchange.)

(a) What is the probability that our portfolio goes up for three consecutive years?

(b) If you know that the portfolio has risen in price 2 years in a row, what probability do you assign to the event that it will go down next year?

(c) What is the probability that the portfolio's value moves in the same direction in both of the next 2 years?

5.16 Getting into an MBA program. Ramon has applied to MBA programs at both Harvard and Stanford. He thinks the probability that Harvard will admit him is 0.4, the probability that Stanford will admit him is 0.5, and the probability that both will admit him is 0.2.

(a) Make a Venn diagram with the probabilities given marked.

(b) What is the probability that neither university admits Ramon?

(c) What is the probability that he gets into Stanford but not Harvard?

5.17 Will we get the jobs? Consolidated Builders has bid on two large construction projects. The company president believes that the probability of winning the first contract (event A) is 0.6, that the probability of winning the second (event B) is 0.5, and that the probability of winning both jobs (event $\{A \text{ and } B\}$) is 0.3. What is the probability of the event $\{A \text{ or } B\}$ that Consolidated will win at least one of the jobs?

5.18 Tastes in music. Musical styles other than rock and pop are becoming more popular. A survey of college students finds that 40% like country music, 30% like gospel music, and 10% like both.

(a) Make a Venn diagram with these results.

(b) What percent of college students like country but not gospel?

(c) What percent like neither country nor gospel?

5.19 Independent? In the setting of Exercise 5.17, are events A and B independent? Do a calculation that proves your answer.

5.20 Customer satisfaction. An airline company conducts a customer satisfaction survey of its passengers to investigate differences between its mileage reward members and nonmembers. Here are the results for 500 randomly selected passengers:

Very satisfied	Reward Member	
	Yes	No
Yes	203	129
No	87	81

(a) What is the estimated probability that a passenger is a mileage reward member?

(b) What is the estimated probability that a passenger is very satisfied with the airline?

(c) What is the estimated probability that a passenger is a mileage reward member and very satisfied?

(d) Assuming independence between satisfaction and member status, use the probabilities from parts (a) and (b) to determine the probability that a passenger is a mileage reward member and very satisfied.

(e) Are satisfaction and member status independent? Explain why or why not.

(f) What is the probability that a passenger is not a reward member or not very satisfied?

5.21 Is that independence? While preparing for the business statistics midterm, your study partner comments that if $P(A) > 0$, $P(B) > 0$, and $P(A \text{ or } B) = P(A) + P(B)$, then the events A and B are independent. Is your study partner on mark or confused? Explain.

5.22 Age effects in medical care. The type of medical care a patient receives may vary with the age of the patient. A large study of women who had a breast lump investigated whether or not each woman received a mammogram and a biopsy when the lump was discovered. Here are some probabilities estimated by the study. The entries in the table are the probabilities that *both* of two events occur; for example, 0.321 is the probability that a patient is under 65 years of age *and* the tests were done. The four probabilities in the table have sum 1 because the table lists all possible outcomes.

	Tests Done?	
Age	Yes	No
Under 65	0.321	0.124
65 or over	0.365	0.190

(a) What is the probability that a patient in this study is under 65? That a patient is 65 or over?

(b) What is the probability that the tests were done for a patient? That they were not done?

(c) Are the events $A = \{$the patient was 65 or older$\}$ and $B = \{$the tests were done$\}$ independent? Were the tests omitted on older patients more or less frequently than would be the case if testing were independent of age?

5.23 Playing the odds? A writer on casino games says that the odds against throwing an 11 in the dice game craps are 17 to 1. He then says that the odds against three 11s in a row are $17 \times 17 \times 17$ to 1, or 4913 to 1.[3]

(a) What is the probability that the sum of the up-faces is 11 when you throw two balanced dice? (See Figure 4.2 on page 225.)

What is the probability of three 11s in three independent throws?

(b) If an event A has probability P, the odds against A are

$$\text{odds against } A = \frac{1 - P}{P}$$

Gamblers often speak of odds rather than probabilities. The odds against an event that has probability 1/3 are 2 to 1, for example. Find the odds against throwing an 11 and the odds against throwing three straight 11s. Which of the writer's statements are correct?

5.2 Conditional Probability

In Section 2.5 we met the idea of a *conditional distribution,* the distribution of a variable, given that a condition is satisfied. Now we will introduce the probability language for this idea.

EXAMPLE 5.6 Employment Status

Each month the Bureau of Labor Statistics (BLS) announces a variety of statistics on employment status in the United States. Employment statistics are important gauges of the economy as a whole. To understand the reported statistics, we need to understand how the government defines "labor force." The labor force includes all people who are either currently employed or who are jobless but are looking for jobs and available for work. The latter group is viewed as unemployed. People who have no job and are not actively looking for one are not considered to be in the labor force. There are a variety of reasons for people not to be in the labor force, including being retired, going to school, having certain disabilities, or being too discouraged to look for a job.

Averaged over the year 2007, the following table contains counts (in thousands) of persons aged 25 and older, classified by education attained and employment status:

Education	Employed	Unemployed	Not in labor force	Total
Did not finish high school	11,521	886	14,226	26,633
High school degree	36,857	1,682	22,834	61,373
Some college	34,612	1,275	13,944	49,831
Bachelor's degree or higher	43,182	892	12,546	56,620
Total	126,172	4,735	63,550	194,457

Randomly choose a person aged 25 or older. What is the probability that the person is employed? Because "choose at random" gives all 194,457,000 such persons the same chance, the probability is just the proportion that are employed. In thousands,

$$P(\text{employed}) = \frac{126,172}{194,457} = 0.6488$$

Now we are told that the person chosen has only a high school degree. The probability that the person is employed, *given the information that the person has only a high school degree,* is

$$P(\text{employed} \mid \text{HS degree}) = \frac{36,857}{61,373} = 0.6005$$

conditional probability   This is a **conditional probability.**

The conditional probability 0.6005 in Example 5.6 gives the probability of one event (the person chosen is employed) under the condition that we know another event (the person has only a high school degree). You can read the bar | as "given the information that." We found the conditional probability by applying common sense to the two-way table.

We want to turn this common sense into something more general. To do this, we reason as follows. To find the proportion of people aged 25 or older who have only a high school degree *and* are employed, first find the proportion who have only a high school degree in the group of interest (aged 25 or older). Then out of the population of people who have only a high school degree find the proportion who are employed. Multiply the two proportions. The actual proportions from Example 5.6 are

$$P(\text{HS degree } and \text{ employed}) = P(\text{HS degree}) \times P(\text{employed} \mid \text{HS degree})$$
$$= \left(\frac{61{,}373}{194{,}457} \right)(0.6005) = 0.1895$$

You can check that this is right: the probability that a randomly chosen person from this group who has only a high school degree and who is employed is

$$P(\text{HS degree and employed}) = \frac{36{,}857}{194{,}457} = 0.1895$$

Try to think your way through this in words before looking at the formal notation. We have just discovered the general multiplication rule of probability.

> **General Multiplication Rule for Any Two Events**
>
> The probability that both of two events A and B happen together can be found by
>
> $$P(A \text{ and } B) = P(A)P(B \mid A)$$
>
> Here $P(B \mid A)$ is the conditional probability that B occurs, given the information that A occurs.

In words, this rule says that for both of two events to occur, first one must occur and then, given that the first event has occurred, the second must occur.

EXAMPLE 5.7   Focus Group Probabilities

A focus group of 10 consumers has been selected to view a new TV commercial. After the viewing, 2 members of the focus group will be randomly selected and asked to answer detailed questions about the commercial. The group contains 4 men and 6 women. What is the probability that the 2 chosen to answer questions will both be women?

To find the probability of randomly selecting 2 women, first calculate

$$P(\text{first person is female}) = \frac{6}{10}$$
$$P(\text{second person is female} \mid \text{first person is female}) = \frac{5}{9}$$

Both probabilities are found by counting group members. The probability that the first person selected is a female is 6/10 because 6 of the 10 group members are female. If the first person is a female, that leaves 5 females among the 9 remaining people. So the *conditional* probability of another female is 5/9. The multiplication rule now says that

$$P(\text{both people are female}) = \frac{6}{10} \times \frac{5}{9} = \frac{1}{3} = 0.3333$$

One-third of the time, randomly picking 2 people from a group of 4 males and 6 females will result in a pair of females.

Remember that events A and B play different roles in the conditional probability $P(B \mid A)$. Event A represents the information we are given, and B is the event whose probability we are computing.

EXAMPLE 5.8   Internet Users

About 20% of all Web surfers use Macintosh computers. About 90% of all Macintosh users surf the Web. If you know someone who uses a Macintosh computer, then the probability that that person surfs the Web is

$$P(\text{surfs the Web} \mid \text{Macintosh user}) = 0.90$$

The 20% is a different conditional probability that does not apply when you are considering someone who you know uses a Macintosh computer.

The general multiplication rule also extends to the probability that all of several events occur. The key is to condition each event on the occurrence of *all* of the preceding events. For example, for three events A, B, and C,

$$P(A \text{ and } B \text{ and } C) = P(A)P(B \mid A)P(C \mid A \text{ and } B)$$

EXAMPLE 5.9   Career in Big Business: NFL

Worldwide, the sports industry has become synonymous with big business. It has been estimated by the United Nations that sports account for nearly 3% of global economic activity.[4] The most profitable sport in the world is professional football under the management of the National Football League (NFL).[5] With multimillion-dollar signing contracts, the economic appeal of pursuing a career as a professional sports athlete is unquestionably strong. But what are the realities? Only 5.7% of high school football players go on to play at the college level. Of these, only 1.8% will play in the NFL.[6] About 40% of the NFL players have a career of more than 3 years. Define these events for the sport of football:

$$A = \{\text{competes in college}\}$$
$$B = \{\text{competes in the NFL}\}$$
$$C = \{\text{has an NFL career longer than 3 years}\}$$

What is the probability that a high school football player competes in college and then goes on to have an NFL career of more than 3 years? We know that

$$P(A) = 0.057$$
$$P(B \mid A) = 0.018$$
$$P(C \mid A \text{ and } B) = 0.4$$

The probability we want is therefore

$$P(A \text{ and } B \text{ and } C) = P(A)P(B \mid A)P(C \mid A \text{ and } B)$$
$$= 0.057 \times 0.018 \times 0.40 = 0.00041$$

Only about 4 of every 10,000 high school football players can expect to compete in college and have an NFL career of more than 3 years. High school football players would be wise to concentrate on studies rather than on unrealistic hopes of fortune from pro football. Hopes of fortune may be further tempered by the statistical fact that NFL players with careers longer than 3 years have an average life expectancy that is 20 years less than that of the typical U.S. male.

5.24 Reward members.   Refer to Exercise 5.20 to find that 290 passengers in the sample are reward members. Suppose two members are randomly selected from the 290 members. Find the probability that both members are very satisfied with the airline.

5.25 Woman managers.   Choose an employed person at random. Let A be the event that the person chosen is a woman, and B the event that the person holds a managerial or professional job. Government data tell us that $P(A) = 0.46$ and the probability of managerial and professional jobs among women is $P(B \mid A) = 0.32$. Find the probability that a randomly chosen employed person is a woman holding a managerial or professional position.

5.26 Buying from Japan.   Functional Robotics Corporation buys electrical controllers from a Japanese supplier. The company's treasurer thinks that there is probability 0.4 that the dollar will fall in value against the Japanese yen in the next month. The treasurer also believes that *if* the dollar falls there is probability 0.8 that the supplier will demand renegotiation of the contract. What probability has the treasurer assigned to the event that the dollar falls and the supplier demands renegotiation?

5.27 Employment status.   Use the two-way table in Example 5.6 to find these conditional probabilities.
(a) $P(\text{employed} \mid \text{some college})$
(b) $P(\text{employed} \mid \text{bachelor's degree or higher})$
(c) $P(\text{high school degree} \mid \text{employed})$
(d) $P(\text{unemployed} \mid \text{bachelor's degree or higher})$
(e) $P(\text{bachelor's degree or higher} \mid \text{unemployed})$

Conditional probability and independence

If we know $P(A)$ and $P(A \text{ and } B)$, we can rearrange the multiplication rule to produce a *definition* of the conditional probability $P(B \mid A)$ in terms of unconditional probabilities.

> **Definition of Conditional Probability**
>
> When $P(A) > 0$, the **conditional probability** of B, given A, is
>
> $$P(B \mid A) = \frac{P(A \text{ and } B)}{P(A)}$$

The conditional probability $P(B \mid A)$ makes no sense if the event A can never occur, so we require that $P(A) > 0$ whenever we talk about $P(B \mid A)$. The definition of conditional probability reminds us that in principle all probabilities, including conditional probabilities, can be found from the assignment of probabilities to events that describe a random phenomenon. More often, as in Examples 5.6 and 5.7, conditional probabilities are part of the information given to us in a probability model, and the multiplication rule is used to compute $P(A \text{ and } B)$.

The conditional probability $P(B \mid A)$ is generally not equal to the unconditional probability $P(B)$. That is because the occurrence of event A generally gives us some additional information about whether or not event B occurs. If knowing that A occurs gives no additional information about B, then A and B are independent events. The precise definition of independence is expressed in terms of conditional probability.

> **Independent Events**
>
> Two events A and B that both have positive probability are **independent** if
>
> $$P(B \mid A) = P(B)$$

This definition makes precise the informal description of independence given in Section 5.1. We now see that the multiplication rule for independent events, $P(A \text{ and } B) = P(A)P(B)$, is a special case of the general multiplication rule, $P(A \text{ and } B) = P(A)P(B \mid A)$, just as the addition rule for disjoint events is a special case of the general addition rule. We will rarely use the definition of independence, because most often independence is part of the information given to us in a probability model.

APPLY YOUR KNOWLEDGE

5.28 College degrees. Here are the counts (in thousands) of earned degrees in the United States in the 2005–2006 academic year, classified by level and by the gender of the degree recipient:[7]

	Bachelor's	Master's	Professional	Doctorate	Total
Female	855	356	44	27	1282
Male	631	238	44	29	942
Total	1486	594	88	56	2224

(a) If you choose a degree recipient at random, what is the probability that the person you choose is a woman?

(b) What is the conditional probability that you choose a woman, given that the person chosen received a professional degree?

(c) Are the events "choose a woman" and "choose a professional degree recipient" independent? How do you know?

5.29 Prosperity and education. Call a household prosperous if its income exceeds $100,000. Call the household educated if the householder completed college. Select an American householder at random, and let A be the event that the selected household is prosperous and B the event that the householder is educated. According to the 2005 U.S. Census Bureau estimates, $P(A) = 0.180$, $P(B) = 0.290$, and the probability that a household is both prosperous and educated is $P(A \text{ and } B) = 0.105$.

(a) Find the conditional probability that a household is educated, given that it is prosperous.

(b) Find the conditional probability that a household is prosperous, given that it is educated.

(c) Are events A and B independent? How do you know?

Tree diagrams and Bayes's rule

Probability problems often require us to combine several of the basic rules into a more elaborate calculation. Here is an example that illustrates how to solve problems that have several stages.

EXAMPLE 5.10   How Many Go to MLB?

In Example 5.9, we investigated the likelihood of a high school football player going on to play collegiately and then have an NFL career of more than 3 years. The sports of football and basketball are unique in that players are prohibited from going straight into professional ranks from high school. Baseball, however, has no such restriction.

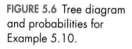

For baseball, 6.1% of high school players go on to play at the college level. Of these, 9.4% will play in Major League Baseball (MLB). Borrowing the notation of Example 5.9, the probability of a high school player ultimately playing professionally is $P(B)$. To find $P(B)$, use the **tree diagram** in Figure 5.6 to organize your thinking.

Each segment in the tree is one stage of the problem. Each complete branch shows a path that a player can take. The probability written on each segment is the conditional probability that a player follows that segment given that he has reached the point from which it branches. Starting at the left, high school baseball players either do or do not compete in college. We know that the probability of competing in college is $P(A) = 0.061$, so the probability of not competing is $P(A^c) = 0.939$. These probabilities mark the leftmost branches in the tree.

Conditional on competing in college, the probability of playing in MLB is $P(B \mid A) = 0.094$. So the conditional probability of *not* playing in MLB is

$$P(B^c \mid A) = 1 - P(B \mid A) = 1 - 0.094 = 0.906$$

These conditional probabilities mark the paths branching out from A in Figure 5.6.

The lower half of the tree diagram describes players who do not compete in college (A^c). For baseball, in years past, the majority of destined professional players did not take the route through college. However, nowadays, it is relatively unusual for players to go straight from high school to MLB. Studies have shown that the conditional probability that a high school athlete reaches MLB, given that he does not compete in college, is $P(B \mid A^c) = 0.002$.[8] We can now mark the two paths branching from A^c in Figure 5.6.

FIGURE 5.6 Tree diagram and probabilities for Example 5.10.

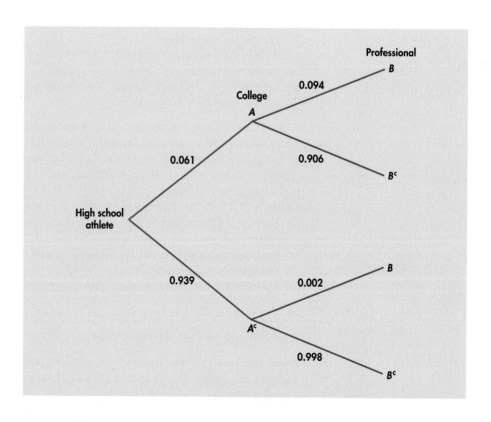

There are two disjoint paths to B (MLB play). By the addition rule, $P(B)$ is the sum of their probabilities. The probability of reaching B through college (top half of the tree) is

$$P(B \text{ and } A) = P(A)P(B \mid A)$$
$$= 0.061 \times 0.094 = 0.00573$$

The probability of reaching B without college is

$$P(B \text{ and } A^c) = P(A^c)P(B \mid A^c)$$
$$= 0.939 \times 0.002 = 0.00188$$

The final result is

$$P(B) = 0.00573 + 0.00188 = 0.00761$$

About 8 high school baseball players out of 1000 will play professionally. Even though this probability is quite small, it is comparatively much greater than the chances of making it to the professional ranks in basketball and football.

Tree diagrams combine the addition and multiplication rules. The multiplication rule says that the probability of reaching the end of any complete branch is the product of the probabilities written on its segments. The probability of any outcome, such as the event B that a high school baseball player plays in MLB, is then found by adding the probabilities of all branches that are part of that event.

There is another kind of probability question that we might ask in the context of studies of athletes. Our earlier calculations look forward toward professional sports as the final stage of an athlete's career. Now let's concentrate on professional athletes and look back at their earlier careers.

EXAMPLE 5.11 Professional Athletes' Pasts

What proportion of professional athletes competed in college? In the notation of Examples 5.9 and 5.10, this is the conditional probability $P(A \mid B)$. In the case of baseball players, we start from the definition of conditional probability:

$$P(A \mid B) = \frac{P(A \text{ and } B)}{P(B)}$$
$$= \frac{0.00573}{0.00761} = 0.753$$

About 75% of MLB players competed in college.

We know the probabilities $P(A)$ and $P(A^c)$ that a high school baseball player does and does not compete in college. We also know the conditional probabilities $P(B \mid A)$ and $P(B \mid A^c)$ that a player from each group reaches MLB. Example 5.10 shows how to use this information to calculate $P(B)$. The method can be summarized in a single expression that adds the probabilities of the two paths to B in the tree diagram:

$$P(B) = P(A)P(B \mid A) + P(A^c)P(B \mid A^c)$$

In Example 5.11 we calculated the "reverse" conditional probability $P(A \mid B)$. The denominator 0.00761 in that example came from the expression just above. Put in this general notation, we have another probability law.

Bayes's Rule

If A and B are any events whose probabilities are not 0 or 1,

$$P(A \mid B) = \frac{P(B \mid A)P(A)}{P(B \mid A)P(A) + P(B \mid A^c)P(A^c)}$$

Bayes's rule is named after Thomas Bayes, who wrestled with arguing from outcomes like B back to antecedents like A in a book published in 1763. It is far better to think your way through problems like Examples 5.10 and 5.11 rather than memorize these formal expressions.

APPLY YOUR KNOWLEDGE

5.30 Where to manufacture? Zipdrive, Inc., has developed a new high-capacity external drive to be used primarily by notebook and laptop users. The demand for the new product is uncertain but can be described as "high" or "low" in any one year. After 4 years, the product is expected to be obsolete. Management must decide whether to build a plant or to contract with a factory in Hong Kong to manufacture the new drive. Building a plant will be profitable if demand remains high but could lead to a loss if demand drops in future years.

After careful study of the market and of all relevant costs, Zipdrive's planning office provides the following information. Let A be the event that the first year's demand is high, and B be the event that the following 3 years' demand is high. The planning office's best estimate of the probabilities is

$$P(A) = 0.9$$
$$P(B \mid A) = 0.36$$
$$P(B \mid A^c) = 0$$

The probability that building a plant is more profitable than contracting the production to Hong Kong is 0.95 if demand is high all 4 years, 0.3 if demand is high only in the first year, and 0.1 if demand is low all 4 years.

Draw a tree diagram that organizes this information. The tree will have three stages: first year's demand, next 3 years' demand, and whether building or contracting is more profitable. Which decision has the higher probability of being more profitable? (When probability analysis is used for investment decisions like this, firms usually compare the mean profits rather than the probability of a profit. We ignore this complication.)

5.31 PDA screens. A manufacturer of Personal Digital Assistants (PDAs) purchases screens from two different suppliers. The company receives 55% of its screens from Screensource and the remaining screens from Brightscreens. The quality of the screens varies between the suppliers: Screensource supplies 1% unsatisfactory screens while 4% of the screens from Brightscreens are unsatisfactory. Given that a randomly chosen screen is unsatisfactory, what is the probability it came from Brightscreens? (*Hint:* In the notation of this section, take A to be the event that the screen came from Brightscreens, and let B be the event that a randomly chosen screen is unsatisfactory.)

SECTION 5.2 Summary

- The **conditional probability** $P(B \mid A)$ of an event B, given an event A, is defined by

$$P(B \mid A) = \frac{P(A \text{ and } B)}{P(A)}$$

when $P(A) > 0$. In practice, we most often find conditional probabilities from directly available information rather than from the definition.

- Any assignment of probability obeys the **general multiplication rule** $P(A \text{ and } B) = P(A)P(B \mid A)$. This rule is often used along with **tree diagrams** in calculating probabilities in settings with several stages.

- A and B are **independent** when $P(B \mid A) = P(B)$. The multiplication rule then becomes $P(A \text{ and } B) = P(A)P(B)$.

- When $P(A)$, $P(B \mid A)$, and $P(B \mid A^c)$ are known, **Bayes's rule** can be used to calculate $P(A \mid B)$ as follows:

$$P(A \mid B) = \frac{P(B \mid A)P(A)}{P(B \mid A)P(A) + P(B \mid A^c)P(A^c)}$$

SECTION 5.2 Exercises

For Exercises 5.24 to 5.27, see page 298; for 5.28 and 5.29, see page 299; and for 5.30 and 5.31, see page 302.

5.32 Self-service gas pumps. At a self-service gas station, 40% of the customers pump regular gas, 35% pump midgrade, and 25% pump premium gas. Of those who pump regular, 30% pay at least $30. Of those who pump midgrade, 50% pay at least $30. And of those who pump premium, 60% pay at least $30. What is the probability that the next customer pays at least $30? (Draw a tree diagram to organize the information given.)

5.33 Self-service gas pumps. In the setting of the previous exercise, what percent of customers who pay at least $30 pump premium gas? (Write this as a conditional probability and use your result from the previous exercise.)

5.34 Loan officer decision. A loan officer is considering a loan request from a customer of the bank. Based on data collected from the bank's records over many years, there is an 8% chance that a customer who has overdrawn an account will default on the loan. However, there is only a 0.6% chance that a customer who has never overdrawn an account will default on the loan. Based on the customer's credit history, the loan officer believes there is a 40% chance that this customer will overdraw his account. Let D be the event that the customer defaults on the loan, and let O be the event that the customer overdraws his account.
(a) Express the three probabilities given in the problem in the notation of probability and conditional probability.
(b) What is the probability that the customer will default on the loan?

5.35 Loan officer decision. Considering the information provided in the previous exercise, calculate $P(O|D)$. Show your work. Also, express this probability in words in the context of the loan officer's decision. If new information about the customer becomes available before the loan officer makes her decision and if this information indicates that there is only a 25% chance that this customer will overdraw his account, rather than a 40% chance, how does this change $P(O|D)$?

5.36 Income tax returns. In 2006, the Internal Revenue Service received 138,394,754 individual tax returns. Of these, 16,153,307 reported an adjusted gross income of at least $100,000, and 354,093 reported at least $1 million.
(a) What is the probability that a randomly chosen individual tax return reports an income of at least $100,000? At least $1 million?
(b) If you know that the return chosen shows an income of $100,000 or more, what is the conditional probability that the income is at least $1 million?

5.37 Tastes in music. Musical styles other than rock and pop are becoming more popular. A survey of college students finds that 40% like country music, 30% like gospel music, and 10% like both.
(a) What is the conditional probability that a student likes gospel music if we know that he or she likes country music?
(b) What is the conditional probability that a student who does not like country music likes gospel music? (A Venn diagram may help you.)

5.38 High school football players. Using the information in Example 5.9, determine the proportion of high school football players expected to play professionally in the NFL.

5.39 High school baseball players. It is estimated that 56% of MLB players have careers of 3 or more years.[9] Using the information in Example 5.10, determine the proportion of high school players expected to play 3 or more years in MLB.

5.40 College degrees. Exercise 5.28 (page 299) gives the counts (in thousands) of earned degrees in the United States in a recent year. Use these data to answer the following questions.
(a) What is the probability that a randomly chosen degree recipient is a man?
(b) What is the conditional probability that the person chosen received a bachelor's degree, given that he is a man?
(c) Use the multiplication rule to find the probability of choosing a male bachelor's degree recipient. Check your result by finding this probability directly from the table of counts.
(d) What is the probability that a randomly chosen degree recipient is a male or has earned a bachelor's degree?

5.41 A little geometry. Choose a point at random in the square with sides $0 \le x \le 1$ and $0 \le y \le 1$. This means that the probability that the point falls in any region within the square is equal to the area of that region. Let X be the x coordinate and Y the y coordinate of the point chosen. Find the conditional probability $P(Y < 1/2 \mid Y > X)$. (*Hint:* Draw a diagram of the square and the events $Y < 1/2$ and $Y > X$.)

5.42 Classifying occupations. Exercise 4.137 (page 280) gives the probability distribution of the gender and occupation of a randomly chosen American worker. Use this distribution to answer the following questions.
(a) Given that the worker chosen holds a managerial (Class A) job, what is the conditional probability that the worker is female?
(b) Classes D and E include most mechanical and factory jobs. What is the conditional probability that a worker is female, given that a worker holds a job in one of these classes?
(c) Are gender and job type independent? How do you know?

5.43 Employment status. As noted in Example 5.6 (page 295), in the language of government statistics, you are "in the labor force" if you are available for work and either working or actively seeking work. The unemployment rate is the proportion of the labor force (not of the entire population) who are unemployed. Based on the table given in Example 5.6, find the unemployment rate for people with each level of education. How does the unemployment rate change with education? Explain carefully why your results show that level of education and being employed are not independent.

5.44 Employment status, continued.
(a) What is the probability that a randomly chosen person 25 years of age or older is in the labor force?
(b) If you know that the person chosen is a college graduate, what is the conditional probability that he or she is in the labor force?

(c) Are the events "in the labor force" and "college graduate" independent? How do you know?

5.45 Preparing for the GMAT. A company that offers courses to prepare would-be MBA students for the GMAT examination finds that 40% of its customers are currently undergraduate students and 60% are college graduates. After completing the course, 50% of the undergraduates and 70% of the graduates achieve scores of at least 600 on the GMAT. Use a tree diagram to organize this information.
(a) What percent of customers are undergraduates *and* score at least 600? What percent of customers are graduates *and* score at least 600?
(b) What percent of all customers score at least 600 on the GMAT?

5.46 Telemarketing. A telemarketing company calls telephone numbers chosen at random. It finds that 70% of calls are not completed (the party does not answer or refuses to talk), that 20% result in talking to a woman, and that 10% result in talking to a man. After that point, 30% of the women and 20% of the men actually buy something. What percent of calls result in a sale? (Draw a tree diagram.)

5.47 Success on the GMAT. In the setting of Exercise 5.45, what percent of the customers who score at least 600 on the GMAT are undergraduates? (Write this as a conditional probability.)

5.48 Sales to women. In the setting of Exercise 5.46, what percent of sales are made to women? (Write this as a conditional probability.)

5.49 Credit card defaults. The credit manager for a local department store discovers that 88% of all the store's credit card holders who defaulted on their payments were late (by a week or more) with two or more of their monthly payments before failing to pay entirely (defaulting). This prompts the manager to suggest that future credit be denied to any customer who is late with two monthly payments. Further study shows that 3% of all credit customers default on their payments and 40% of those who have not defaulted have had at least two late monthly payments in the past.
(a) What is the probability that a customer who has two or more late payments will default?
(b) Under the credit manager's policy, in a group of 100 customers who have their future credit denied, how many would we expect *not* to default on their payments?
(c) Does the credit manager's policy seem reasonable? Explain your response.

5.50 Successful bids. Consolidated Builders has bid on two large construction projects. The company president believes that the probability of winning the first contract (event A) is 0.6, that the probability of winning the second (event B) is 0.5, and that the probability of winning both jobs (event $\{A \text{ and } B\}$) is 0.3. What is the probability of the event $\{A \text{ or } B\}$ that Consolidated will win at least one of the jobs?

5.51 Independence? In the setting of the previous exercise, are events A and B independent? Do a calculation that proves your answer.

5.52 Successful bids, continued. Draw a Venn diagram that illustrates the relation between events A and B in Exercise 5.50. Write each of the following events in terms of A, B, A^c, and B^c. Indicate the events on your diagram and use the information in Exercise 5.50 to calculate the probability of each.

(a) Consolidated wins both jobs.

(b) Consolidated wins the first job but not the second.

(c) Consolidated does not win the first job but does win the second.

(d) Consolidated does not win either job.

5.53 Inspecting final products. Final products are sometimes selected to go through a complete inspection before leaving the production facility. Suppose that 8% of all products made at a particular facility fail to conform to specifications. Furthermore, 55% of all nonconforming items are selected for complete inspection while 20% of all conforming items are selected for complete inspection. Given that a randomly chosen item has gone through a complete inspection, what is the probability the item is nonconforming?

5.3 The Binomial Distributions

A company's human resources manager asks 100 employees if job stress is affecting their personal lives. How many will say "Yes"? A company develops 30 new products that are subject to market testing. How many products will be successful in market testing? A store sells 60 computers with extended 3-year warranties. How many will not need repair within 3 years? In all these situations, we want a probability model for a *count* of successful outcomes.

The binomial setting

The distribution of a count depends on how the data are produced. Here is a common situation.

> **The Binomial Setting**
>
> 1. There are a fixed number n of observations.
> 2. The n observations are all **independent.** That is, knowing the result of one observation tells you nothing about the other observations.
> 3. Each observation falls into one of just two categories, which for convenience we call "success" and "failure."
> 4. The probability of a success, call it p, is the same for each observation.

Think of tossing a coin n times as an example of the binomial setting. Each toss gives either heads or tails. Knowing the outcome of one toss doesn't tell us anything about other tosses, so the n tosses are independent. If we call heads a success, then p is the probability of a head and remains the same as long as we toss the same coin. The number of heads we count is a random variable X. The distribution of X is called a *binomial distribution.*

> **Binomial Distribution**
>
> The distribution of the count X of successes in the binomial setting is the **binomial distribution** with parameters n and p. The parameter n is the number of observations, and p is the probability of a success on any one observation. The possible values of X are the whole numbers from 0 to n.

The binomial distributions are an important class of probability distributions. Pay attention to the binomial setting, because not all counts have binomial distributions.

EXAMPLE 5.12 Determining Consumer Preferences

Market research to determine the product preferences of consumers is an increasingly important area in the intersection of business and statistics. With some companies competing in markets with little product discrimination, determining what features consumers most prefer is critical to the success of a product. Market research is interested in finding the probability of a "typical" consumer purchasing a product with a particular combination of features.

Suppose that your product is actually preferred over competitors' products by 25% of all consumers. If X is the count of the number of consumers who prefer your product in a group of 5 consumers, then X has a binomial distribution with $n = 5$ and $p = 0.25$, provided the 5 consumers make choices independently. Given their importance, the methods for collecting independent consumer data are part of most marketing curricula found in business schools. These methods are routinely employed by marketing research consulting firms, such as the Nielsen Company (www.nielsen.com) and Burke, Inc. (www.burke.com), as well as by in-house marketing research departments found in most major firms.

EXAMPLE 5.13 Dealing Cards

Deal 10 cards from a shuffled deck and count the number X of red cards. There are 10 observations, and each gives either a red or a black card. A "success" is a red card. But the observations are *not* independent. If the first card is black, the second is more likely to be red because there are more red cards than black cards left in the deck. The count X does *not* have a binomial distribution.

CASE 5.1

Inspecting a Supplier's Products A manufacturing firm purchases components for its products from suppliers. Good practice calls for suppliers to manage their production processes to ensure good quality. You can find some discussion of statistical methods for managing and improving quality in Chapter 12. There have, however, been quality lapses in the switches supplied by a regular vendor. While working with the supplier to improve its processes, the manufacturing firm temporarily institutes an *acceptance sampling* plan to assess the quality of shipments of switches. If a random sample from a shipment contains too many switches that don't conform to specifications, the firm will not accept the shipment.

An engineer at the firm chooses an SRS of 10 switches from a shipment of 10,000 switches. Suppose that (unknown to the engineer) 10% of the switches in the shipment are nonconforming. The engineer counts the number X of nonconforming switches in the sample.

This is not quite a binomial setting. Just as removing 1 card in Example 5.13 changed the makeup of the deck, removing 1 switch changes the proportion of nonconforming switches remaining in the shipment. If there are initially 1000 nonconforming switches, the proportion remaining is $1000/9999 = 0.10001$ if the first switch drawn is OK and $999/9999 = 0.09991$ if the first switch fails inspection. That is, the state of the second switch chosen is not independent of the first. But removing 1 switch from a shipment of 10,000 changes the makeup of the remaining 9999 switches very little. In practice, the distribution of X is very close to the binomial distribution with $n = 10$ and $p = 0.1$.

Case 5.1 shows how we can use the binomial distributions in the statistical setting of selecting an SRS. When the population is much larger than the sample, a count of successes in an SRS of size n has approximately the binomial distribution with n equal to the sample size and p equal to the proportion of successes in the population.

APPLY YOUR KNOWLEDGE

In each of Exercises 5.54 to 5.56, X is a count. Does X have a binomial distribution? Give your reasons in each case.

5.54 Deliveries. A courier service audits the records of 100 next–business-morning (before 10 A.M.) deliveries; X is the number of on-time deliveries among them.

5.55 Customer satisfaction calls. The service department of an automobile dealership follows up each service encounter with a customer satisfaction survey by means of a phone call. On a given day, let X be the number of customers a service representative has to call until a customer is willing to participate in the survey.

5.56 Teaching office software. A company uses a computer-based system to teach clerical employees new office software. After a lesson, the computer presents 10 exercises. The student solves each exercise and enters the answer. The computer gives additional instruction between exercises if the answer is wrong. The count X is the number of exercises that the student gets right.

Binomial probabilities*

We can find a formula for the probability that a binomial random variable takes any value by adding probabilities for the different ways of getting exactly that many successes in n observations. An example will guide us toward the formula we want.

EXAMPLE 5.14 Determining Consumer Preferences

Each consumer has probability 0.25 of preferring your product over competitors' products. If we question 5 consumers, what is the probability that exactly 2 of them prefer your product?

The count of consumers preferring your product is a binomial random variable X with $n = 5$ tries and probability $p = 0.25$ of a success on each try. We want $P(X = 2)$.

Because the method doesn't depend on the specific example, let's use "S" for success and "F" for failure for short. Do the work in two steps.

Step 1. Find the probability that a specific 2 of the 5 tries, say the first and the third, give successes. This is the outcome SFSFF. Because tries are independent, the multiplication rule for independent events applies. The probability we want is

$$P(\text{SFSFF}) = P(S)P(F)P(S)P(F)P(F)$$
$$= (0.25)(0.75)(0.25)(0.75)(0.75)$$
$$= (0.25)^2(0.75)^3$$

Step 2. Observe that the probability of *any one* arrangement of 2 S's and 3 F's has this same probability. This is true because we multiply together 0.25 twice and 0.75 three times whenever we have 2 S's and 3 F's. The probability that $X = 2$ is the probability of getting 2 S's and 3 F's in any arrangement whatsoever. Here are all the possible arrangements:

SSFFF   SFSFF   SFFSF   SFFFS   FSSFF
FSFSF   FSFFS   FFSSF   FFSFS   FFFSS

*The derivation and use of the exact formula for binomial probabilities are optional.

There are 10 of them, all with the same probability. The overall probability of 2 successes is therefore

$$P(X = 2) = 10(0.25)^2(0.75)^3 = 0.2637$$

Approximately 26% of the time, samples of 5 independent consumers will produce exactly 2 who prefer your product over competitors' products.

The pattern of this calculation works for any binomial probability. To use it, we must count the number of arrangements of k successes in n observations. We use the following fact to do the counting without actually listing all the arrangements.

Binomial Coefficient

The number of ways of arranging k successes among n observations is given by the **binomial coefficient**

$$\binom{n}{k} = \frac{n!}{k!\,(n-k)!}$$

for $k = 0, 1, 2, \ldots, n$.

---

factorial

The formula for binomial coefficients uses the **factorial** notation. For any positive whole number n, its factorial $n!$ is

$$n! = n \times (n-1) \times (n-2) \times \cdots \times 3 \times 2 \times 1$$

Also, $0! = 1$ by definition.

The larger of the two factorials in the denominator of a binomial coefficient will cancel much of the $n!$ in the numerator. For example, the binomial coefficient we need for Example 5.14 is

$$
\begin{aligned}
\binom{5}{2} &= \frac{5!}{2!\,3!} \\
&= \frac{(5)(4)(3)(2)(1)}{(2)(1) \times (3)(2)(1)} \\
&= \frac{(5)(4)}{(2)(1)} = \frac{20}{2} = 10
\end{aligned}
$$

Careful with notation interpretation

The notation $\binom{n}{k}$ is *not* related to the fraction $\frac{n}{k}$. A helpful way to remember its meaning is to read it as "binomial coefficient n choose k." Binomial coefficients have many uses in mathematics, but we are interested in them only as an aid to finding binomial probabilities. The binomial coefficient $\binom{n}{k}$ counts the number of different ways in which k successes can be arranged among n observations. The binomial probability $P(X = k)$ is this count multiplied by the probability of any specific arrangement of the k successes. Here is the result we seek.

> **Binomial Probability**
>
> If X has the binomial distribution with n observations and probability p of success on each observation, the possible values of X are $0, 1, 2, \ldots, n$. If k is any one of these values,
>
> $$P(X = k) = \binom{n}{k} p^k (1 - p)^{n-k}$$

EXAMPLE 5.15   Inspecting Switches

CASE 5.1

The number X of switches that fail inspection in Case 5.1 closely follows the binomial distribution with $n = 10$ and $p = 0.1$.

The probability that no more than 1 switch fails is

$$
\begin{aligned}
P(X \le 1) &= P(X = 1) + P(X = 0) \\
&= \binom{10}{1} (0.1)^1 (0.9)^9 + \binom{10}{0} (0.1)^0 (0.9)^{10} \\
&= \frac{10!}{1!\,9!} (0.1)(0.3874) + \frac{10!}{0!\,10!} (1)(0.3487) \\
&= (10)(0.1)(0.3874) + (1)(1)(0.3487) \\
&= 0.3874 + 0.3487 = 0.7361
\end{aligned}
$$

This calculation uses the facts that $0! = 1$ and that $a^0 = 1$ for any number a other than 0. We see that about 74% of all samples will contain no more than 1 bad switch. In fact, 35% of the samples will contain no bad switches. A sample of size 10 cannot be trusted to alert the engineer to the presence of unacceptable items in the shipment. Calculations such as this are used to design acceptance sampling schemes.

APPLY YOUR KNOWLEDGE

5.57 Misleading résumés. In Exercise 5.3 (page 290), we found that 15.9% of executive job applicants lied on their résumés. Suppose an executive job hunter randomly selects 5 résumés from an executive job applicant pool. Let X be the number of misleading résumés found in the sample. So X has the binomial distribution with $n = 5$ and $p = 0.159$.
(a) What are the possible values of X?
(b) Find the probability of each value of X. Draw a probability histogram to display this distribution. (Because probabilities are long-run proportions, a histogram with the probabilities as the heights of the bars shows what the distribution of X would be in very many repetitions.)

5.58 Hispanic representation. A factory employs several thousand workers, of whom 30% are Hispanic. If the 15 members of the union executive committee were chosen from the workers at random, the number of Hispanics on the committee would have the binomial distribution with $n = 15$ and $p = 0.3$.
(a) What is the probability that exactly 3 members of the committee are Hispanic?
(b) What is the probability that 3 or fewer members of the committee are Hispanic?

5.59 Do our athletes graduate? A university claims that 80% of its basketball players get degrees. An investigation examines the fate of all 20 players who entered the program over a period of several years that ended six years ago. Of these players,

11 graduated and the remaining 9 are no longer in school. If the university's claim is true, the number of players who graduate among the 20 should have the binomial distribution with $n = 20$ and $p = 0.8$. What is the probability that exactly 11 out of 20 players graduate? 0.0222

Finding binomial probabilities: tables

The formula given on page 309 for binomial probabilities is practical for hand calculations when n is small. However, in practice, you will rarely have to use this formula for calculations. Some calculators and most statistical software packages calculate binomial probabilities. If you do not have suitable computing facilities, you can look up the probabilities for some values of n and p in Table C in the back of this book. The entries in the table are the probabilities $P(X = k)$ of individual outcomes for a binomial random variable X.

EXAMPLE 5.16   Inspecting Switches

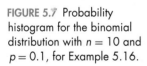

The quality engineer in Case 5.1 inspects an SRS of 10 switches from a large shipment of which 10% fail to conform to specifications. What is the probability that no more than 1 of the 10 switches in the sample fails inspection?

The count X of nonconforming switches in the sample has approximately the binomial distribution with $n = 10$ and $p = 0.1$. Figure 5.7 is a probability histogram for this distribution. The distribution is strongly skewed. Although X can take any whole-number value from 0 to 10, the probabilities of values larger than 5 are so small that they do not appear in the histogram.

We want to calculate

$$P(X \leq 1) = P(X = 1) + P(X = 0)$$

when X is binomial with $n = 10$ and $p = 0.1$. Your software may do this—look for the key word "Binomial." To use Table C for this calculation, look opposite $n = 10$ and under $p = 0.10$. This part of the table appears at the left. The entry opposite each k is $P(X = k)$. We find

$$P(X \leq 1) = P(X = 1) + P(X = 0)$$
$$= 0.3874 + 0.3487 = 0.7361$$

About 74% of all samples will contain no more than 1 bad switch. This matches our calculation in Example 5.15.

FIGURE 5.7 Probability histogram for the binomial distribution with $n = 10$ and $p = 0.1$, for Example 5.16.

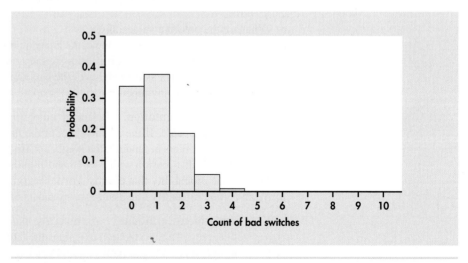

Count of bad switches

n	k	p 0.10
10	0	0.3487
	1	0.3874
	2	0.1937
	3	0.0574
	4	0.0112
	5	0.0015
	6	0.0001
	7	0.0000
	8	0.0000
	9	0.0000
	10	0.0000

The excerpt from Table C in the margin contains the full binomial distribution for $n = 10$ and $p = 0.1$. The probabilities are rounded to four decimal places. Outcomes larger than 6 do not have probability exactly 0, but their probabilities are so small that the rounded values are 0.0000. Check that the sum of the probabilities given is 1, as it should be.

The values of p that appear in Table C are all 0.5 or smaller. When the probability of a success is greater than 0.5, restate the problem in terms of the number of failures. The probability of a failure is less than 0.5 when the probability of a success exceeds 0.5. When using the table, always stop to ask whether you must count successes or failures.

EXAMPLE 5.17   Free Throws

Corinne is a basketball player who makes 75% of her free throws over the course of a season. In a key game, Corinne shoots 12 free throws and misses 5 of them. The fans think that she failed because she was nervous. Is it unusual for Corinne to perform this poorly?

To answer this question, assume that free throws are independent with probability 0.75 of a success on each shot. (Studies of long sequences of basketball shots—free throws and long-distance shots—have found no evidence that they are dependent, so this is a reasonable assumption.)[10] Because the probability of making a free throw is greater than 0.5, we count misses in order to use Table C. The probability of a miss is $1 - 0.75$, or 0.25. The number X of misses in 12 attempts has the binomial distribution with $n = 12$ and $p = 0.25$.

We want the probability of missing 5 or more. This is

$$P(X \geq 5) = P(X = 5) + P(X = 6) + \cdots + P(X = 12)$$
$$= 0.1032 + 0.0401 + \cdots + 0.0000 = 0.1576$$

Corinne will miss 5 or more out of 12 free throws about 16% of the time, or roughly one of every six games. While below her average level, her performance in this game was well within the range of the usual chance variation in her shooting.

APPLY YOUR KNOWLEDGE

5.60 Restaurant survey.   You operate a restaurant. You read that a sample survey by the National Restaurant Association shows that 40% of adults are committed to eating nutritious food when eating away from home. To help plan your menu, you decide to conduct a sample survey in your own area. You will use random digit dialing to contact an SRS of 20 households by telephone.

(a) If the national result holds in your area, it is reasonable to use the binomial distribution with $n = 20$ and $p = 0.4$ to describe the count X of respondents who seek nutritious food when eating out. Explain why.

(b) Ten of the 20 respondents say they are concerned about nutrition. Is this reason to believe that the percent in your area is higher than the national 40%? To answer this question, use software or Table C to find the probability that X is 10 or larger if $p = 0.4$ is true. If this probability is very small, that is reason to think that p is actually greater than 0.4 in your area.

Binomial mean and standard deviation

If a count X has the binomial distribution based on n observations with probability p of success, what is its mean μ? That is, in very many repetitions of the binomial setting, what will be the average count of successes? We can guess the answer. If a basketball player makes 75% of her free throws, the mean number made in 12 tries should be 75% of 12, or 9. In general, the mean of a binomial distribution should be $\mu = np$. Here are the facts.

Binomial Mean and Standard Deviation

If a count X has the binomial distribution with number of observations n and probability of success p, the **mean** and **standard deviation** of X are

$$\mu = np$$
$$\sigma = \sqrt{np(1-p)}$$

Remember that these short formulas are good only for binomial distributions. They can't be used for other distributions.

EXAMPLE 5.18   Inspecting Switches

CASE 5.1

Continuing Case 5.1, the count X of bad switches is binomial with $n = 10$ and $p = 0.1$. The mean and standard deviation of this binomial distribution are

$$\mu = np$$
$$= (10)(0.1) = 1$$
$$\sigma = \sqrt{np(1-p)}$$
$$= \sqrt{(10)(0.1)(0.9)} = \sqrt{0.9} = 0.9487$$

In Figure 5.8, we have added the mean to the probability histogram of the distribution.

FIGURE 5.8 Probability histogram for the binomial distribution with $n = 10$ and $p = 0.1$ and $\mu = 1$ marked, for Example 5.18.

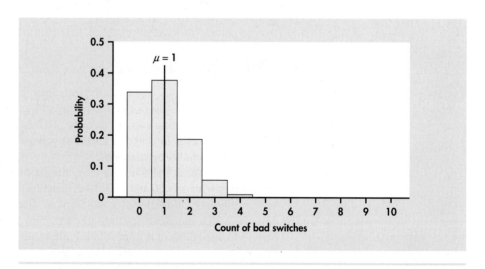

APPLY YOUR KNOWLEDGE

5.61 Restaurant survey.   As in Exercise 5.60, you ask an SRS of 20 adults from your restaurant's target area if they are concerned about nutrition when eating away from home. If the national proportion $p = 0.4$ holds in your area, what will be the mean number of "Yes" responses? What is the standard deviation of the count of "Yes" answers?

5.62 Hispanic representation.

(a) What is the mean number of Hispanics on randomly chosen committees of 15 workers in Exercise 5.58?

(b) What is the standard deviation σ of the count X of Hispanic members?

(c) Suppose that 10% of the factory workers were Hispanic. Then $p = 0.1$. What is σ in this case? What is σ if $p = 0.01$? What does your work show about the behavior of the standard deviation of a binomial distribution as the probability of a success gets closer to 0?

5.63 Do our athletes graduate?

(a) Find the mean number of graduates out of 20 players in the setting of Exercise 5.59 if the university's claim is true.

(b) Find the standard deviation σ of the count X.

(c) Suppose that the 20 players came from a population of which $p = 0.9$ graduated. What is the standard deviation σ of the count of graduates? If $p = 0.99$, what is σ? What does your work show about the behavior of the standard deviation of a binomial distribution as the probability p of success gets closer to 1?

The Normal approximation to binomial distributions

The binomial probability formula and tables are practical only when the number of trials n is small. Even software and statistical calculators are unable to handle calculations for very large n. Figure 5.9 shows the binomial distribution for different values of p and n. From these graphs, we see that, for a given p, the shape of the binomial distribution becomes more symmetrical as n gets larger. In particular, *as the number of trials n gets larger, the binomial distribution gets closer to a Normal distribution.* We can also see from Figure 5.9 that, for a given n, the binomial distribution is more symmetrical as p approaches 0.5. The upshot is that the accuracy of Normal approximation depends on the values of both n and p. In the next example, we use Normal probability calculations to approximate hard-to-calculate binomial probabilities because n is so large.

EXAMPLE 5.19 Life Satisfaction in Economically Hard Times

In 2008, the U.S. economy revealed its strain with skyrocketing energy costs, a depressed housing market, a weakening U.S. dollar, and an unprecedented financial market meltdown. But, amazingly, surveys showed that people were no less satisfied with the lives they were leading in comparison to prior years. A survey asked a nationwide random sample of 1010 adults if "On the whole, are you very satisfied, fairly satisfied, not very satisfied or not at all satisfied with the life you lead?"[11] The population that the poll wants to draw conclusions about is *all* U.S. residents aged 18 and over. Suppose that in fact 60% of *all* adult U.S. residents would respond "very satisfied" if asked the same question. What is the probability that 610 or more of the sample are very satisfied with their lives?

Because there are almost 230 million adults, we can take the responses of 1010 randomly chosen adults to be independent. The number in our sample who feel very satisfied with life is a random variable X having the binomial distribution with $n = 1010$ and $p = 0.6$. To find the probability that at least 610 of the people in the sample are very satisfied with their lives, we must add the binomial probabilities of all outcomes from $X = 610$ to $X = 1010$. This isn't practical. Here are three ways to do this problem:

1. Statistical software (including Excel) can do the calculation. The result is

$$P(X \geq 610) = 0.4119$$

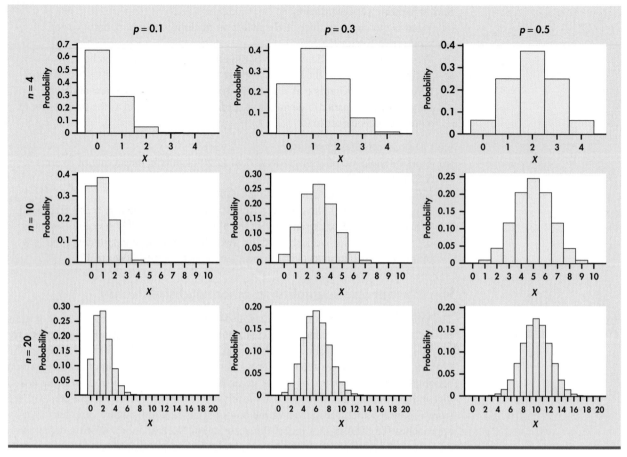

FIGURE 5.9 The shape of the binomial distribution for different values of *n* and *p*.

2. We can simulate a large number of repetitions of the sample. Figure 5.10 displays a histogram of the counts X from 10,000 samples of size 1010 when the truth about the population is $p = 0.6$. Because 4096 of these 10,000 samples have X at least 610, the probability estimated from this particular simulation is

$$P(X \geq 610) = \frac{4096}{10,000} = 0.4096$$

3. Both of the previous methods require software. Instead, look at the Normal curve in Figure 5.10. This is the density curve of the Normal distribution with the same mean and standard deviation as the binomial variable X:

$$\mu = np = (1010)(0.6) = 606$$
$$\sigma = \sqrt{np(1-p)} = \sqrt{(1010)(0.6)(0.4)} = 15.57$$

As the figure shows, this Normal distribution approximates the binomial distribution quite well. So we can do a Normal calculation.

FIGURE 5.10 Histogram of 10,000 binomial counts ($n = 1010$, $p = 0.60$) and the Normal density curve that approximates the binomial distribution.

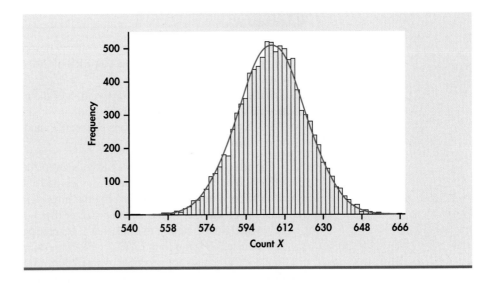

EXAMPLE 5.20   Normal Calculation of a Binomial Probability

Act as though the count X had the $N(606, 15.57)$ distribution. Here is the probability we want, using Table A:

$$P(X \geq 610) = P\left(\frac{X - 606}{15.57} \geq \frac{610 - 606}{15.57}\right)$$
$$= P(Z \geq 0.26)$$
$$= 1 - 0.6026 = 0.3974$$

The Normal approximation 0.3974 differs from the software result 0.4119 by only about 0.015.

Normal Approximation for Binomial Distributions

Suppose that a count X has the binomial distribution with n trials and success probability p. When n is large, the distribution of X is approximately Normal, $N(np, \sqrt{np(1 - p)})$.

As a rule of thumb, we will use the Normal approximation when n and p satisfy $np \geq 10$ and $n(1 - p) \geq 10$.

The Normal approximation is easy to remember because it says that X is Normal with its binomial mean and standard deviation. The accuracy of the Normal approximation improves as the sample size n increases. It is most accurate for any fixed n when p is close to $1/2$, and least accurate when p is near 0 or 1. In an end-of-section exercise (Exercise 5.82), you will be asked to slightly improve the Normal approximation described here by using a method known as **continuity correction.** Continuity correction is basically a small adjustment that takes into account the fact that the approximating distribution (Normal) is continuous while the binomial distribution is discrete. Whether or not you use the Normal approximation (with or without continuity correction) should depend on how accurate your calculations need to be. For most statistical purposes great accuracy is not required. Our "rule of thumb" for use of the Normal approximation reflects this judgment.

continuity correction

APPLY YOUR KNOWLEDGE

5.64 Restaurant survey.   Return to the survey described in Exercise 5.60. You plan to use random digit dialing to contact an SRS of 200 households by telephone rather than just 20.

(a) What are the mean and standard deviation of the number of nutrition-conscious people in your sample if $p = 0.4$ is true?

(b) What is the probability that X lies between 75 and 85? (Use the Normal approximation.)

5.65 The effect of sample size.   The SRS of size 200 described in the previous exercise finds that 100 of the 200 respondents are concerned about nutrition. We wonder if this is reason to conclude that the percent in your area is higher than the national 40%.

(a) Find the probability that X is 100 or larger if $p = 0.4$ is true. If this probability is very small, that is reason to think that p is actually greater than 0.4 in your area.

(b) In Exercise 5.60, you found $P(X \geq 10)$ for a sample of size 20. In (a), you have found $P(X \geq 100)$ for a sample of size 200 from the same population. Both of these probabilities answer the question "How likely is a sample with at least 50% successes when the population has 40% successes?" What does comparing these probabilities suggest about the importance of sample size?

SECTION 5.3 Summary

- A count X of successes has a **binomial distribution** in the **binomial setting:** the number of observations n is fixed in advance; the observations are independent of each other; each observation results in a success or a failure; and each observation has the same probability p of a success.

- The binomial distribution with n observations and probability p of success gives a good approximation to the sampling distribution of the count of successes in an SRS of size n from a large population containing proportion p of successes.

- If X has the binomial distribution with parameters n and p, the possible values of X are the whole numbers $0, 1, 2, \ldots, n$. The **binomial probability** that X takes any value is

$$P(X = k) = \binom{n}{k} p^k (1 - p)^{n-k}$$

Binomial probabilities are most easily found by software. This formula is practical for calculations when n is small. Table C contains binomial probabilities for some values of n and p. For large n, you can use the Normal approximation.

- The **binomial coefficient**

$$\binom{n}{k} = \frac{n!}{k! \, (n - k)!}$$

counts the number of ways k successes can be arranged among n observations. Here the **factorial $n!$** is

$$n! = n \times (n - 1) \times (n - 2) \times \cdots \times 3 \times 2 \times 1$$

for positive whole numbers n, and $0! = 1$.

- The **mean** and **standard deviation** of a binomial count X are

$$\mu = np$$
$$\sigma = \sqrt{np(1-p)}$$

- The **Normal approximation** to the binomial distribution says that if X is a count having the binomial distribution with parameters n and p, then when n is large, X is approximately $N(np, \sqrt{np(1-p)}\,)$. We will use this approximation when $np \geq 10$ and $n(1-p) \geq 10$.

SECTION 5.3 Exercises

For Exercises 5.54 to 5.56, see page 307; for 5.57 to 5.59, see pages 309–310; for 5.60, see page 311; for 5.61 to 5.63, see pages 312–313; and for 5.64 and 5.65, see page 316.

All of the binomial probability calculations required in these exercises can be done by using Table C or the Normal approximation. Your instructor may request that you use the binomial probability formula or software.

5.66 Web site traffic. What kinds of Web sites do males aged 18 to 34 visit most often? Pornographic sites take first place,[12] but about 50% of male Internet users in this age group visit an auction site such as eBay at least once a month. Interview a random sample of 12 male Internet users aged 18 to 34.
(a) What is the distribution of the number who have visited an online auction site in the past month?
(b) What is the probability that exactly 8 of the 12 have visited an auction site in the past month? If you have software, also find the probability that at least 8 of the 12 have visited an auction site in the past month.

5.67 Web site traffic. Suppose that 50% of male Internet users aged 18 to 34 have visited an auction site at least once in the past month.
(a) If you interview 12 at random, what is the mean of the count X who have visited an auction site? What is the mean of the proportion $\hat{p}$ in your sample who have visited an auction site?
(b) Repeat the calculations in (a) for samples of size 120 and 1200. What happens to the mean count of successes as the sample size increases? What happens to the mean proportion of successes?

5.68 Carpooling stats. Although cities encourage carpooling to reduce traffic congestion, most vehicles carry only one person. For example, 70% of vehicles on the roads in the Minneapolis–St. Paul metropolitan area are occupied by just the driver.
(a) If you choose 10 vehicles at random, what is the probability that more than half (that is, 6 or more) carry just one person?
(b) If you choose 100 vehicles at random, what is the probability that more than half (that is, 51 or more) carry just one person?

5.69 Internet postings. Suppose (as is roughly true) that 20% of all Internet users have posted photos online. A sample survey interviews an SRS of 1555 Internet users.
(a) What is the actual distribution of the number X in the sample who have posted photos online?
(b) What is the probability that 300 or fewer of the people in the sample have posted photos online? (Use software or a suitable approximation.)

5.70 Binomial setting? In each situation below, is it reasonable to use a binomial distribution for the random variable X? Give reasons for your answer in each case.
(a) An auto manufacturer chooses one car from each hour's production for a detailed quality inspection. One variable recorded is the count X of finish defects (dimples, ripples, etc.) in the car's paint.
(b) Joe buys a ticket in his state's "Pick 3" lottery game every week; X is the number of times in a year that he wins a prize.

5.71 Binomial setting? In each of the following cases, decide whether or not a binomial distribution is an appropriate model, and give your reasons.
(a) A firm uses a computer-based training module to prepare 20 machinists to use new numerically controlled lathes. The module contains a test at the end of the course; X is the number who perform satisfactorily on the test.
(b) The list of potential product testers for a new product contains 100 persons chosen at random from the adult residents of a large city. Each person on the list is asked whether he or she would participate in the study if given the chance; X is the number who say "Yes."

5.72 Random digits. Each entry in a table of random digits like Table B has probability 0.1 of being a 0, and digits are independent of each other.
(a) What is the probability that a group of five digits from the table will contain at least one 0?
(b) What is the mean number of 0s in lines 40 digits long?

5.73 Unmarried women. Among employed women, 25% have never been married. Select 10 employed women at random.

(a) The number in your sample who have never been married has a binomial distribution. What are n and p?

(b) What is the probability that exactly 2 of the 10 women in your sample have never been married?

(c) What is the probability that 2 or fewer have never been married?

(d) What is the mean number of women in such samples who have never been married? What is the standard deviation?

5.74 Generic brand soda. In a taste test of a generic soda versus a brand name soda, 25% of tasters can distinguish between the colas. Twenty tasters are asked to take the taste test and guess which cup contains the brand name soda. The tests are done independently in separate locations, so that the tasters do not interact with each other during the test.

(a) The count of correct guesses in 20 taste tests has a binomial distribution. What are n and p?

(b) What is the mean number of correct guesses in many repetitions?

(c) What is the probability of exactly 5 correct guesses?

5.75 Random stock prices. A believer in the "random walk" theory of stock markets thinks that an index of stock prices has probability 0.65 of increasing in any year. Moreover, the change in the index in any given year is not influenced by whether it rose or fell in earlier years. Let X be the number of years among the next 5 years in which the index rises.

(a) X has a binomial distribution. What are n and p?

(b) What are the possible values that X can take?

(c) Find the probability of each value of X. Draw a probability histogram for the distribution of X.

(d) What are the mean and standard deviation of this distribution? Mark the location of the mean on your histogram.

5.76 Lie detectors. A federal report finds that lie detector tests given to truthful persons have probability about 0.2 of suggesting that the person is deceptive.[13]

(a) A company asks 12 job applicants about thefts from previous employers, using a lie detector to assess their truthfulness. Suppose that all 12 answer truthfully. What is the probability that the lie detector says all 12 are truthful? What is the probability that the lie detector says at least 1 is deceptive?

(b) What is the mean number among 12 truthful persons who will be classified as deceptive? What is the standard deviation of this number?

(c) What is the probability that the number classified as deceptive is less than the mean?

5.77 Multiple-choice tests. Here is a simple probability model for multiple-choice tests. Suppose that each student has probability p of correctly answering a question chosen at random from a universe of possible questions. (A strong student has a higher p than a weak student.) Answers to different questions are independent. Jodi is a good student for whom $p = 0.75$.

(a) Use the Normal approximation to find the probability that Jodi scores 70% or lower on a 100-question test.

(b) If the test contains 250 questions, what is the probability that Jodi will score 70% or lower?

5.78 Ichiro Suzuki's hits. In 2004, Ichiro Suzuki of the Seattle Mariners had 262 hits, a new Major League record breaking an 84-year-old mark held by George Sisler of the St. Louis Browns, who had 257 hits. "I just hope people realize the monumental effort it took to surpass this record, which has stood so long," said Seattle hitting coach Paul Molitor, a Hall of Famer.[14] Was this feat as surprising as suggested in this quotation and as many people thought? In the three seasons before 2004, Suzuki hit 0.328. He went to bat 704 times in 2004. Assume that Suzuki's hit count in 704 times at bat has approximately the binomial distribution with $n = 704$ and $p = 0.328$.

(a) What is the mean number of hits he will hit in 704 times at bat?

(b) What is the probability of 262 or more hits? Use the Normal approximation.

(c) In part (b), you determined the approximate probability that Suzuki would perform this feat given his intrinsic batting ability. Historically, an average Major League Baseball player hits around 0.265. What is the likelihood that an average MLB player would get 262 or more hits? Looking at this probability and the probability found in part (b), do these probabilities confirm or refute the feat as being a "monumental effort"?

(d) Compare your answer in (b) with the actual probability of 0.007457 found using software.

	A	B	C
1	1-BINOMDIST(261,704,0.328,1) =	0.007457	
2			

5.79 Planning a survey. You are planning a sample survey of small businesses in your area. You will choose an SRS of businesses listed in the telephone book's Yellow Pages. Experience shows that only about half the businesses you contact will respond.

(a) If you contact 150 businesses, it is reasonable to use the binomial distribution with $n = 150$ and $p = 0.5$ for the number X who respond. Explain why.

(b) What is the expected number (the mean) who will respond?

(c) What is the probability that 70 or fewer will respond? (Use the Normal approximation.)

(d) How large a sample must you take to increase the mean number of respondents to 100?

5.80 Are we shipping on time? Your mail-order company advertises that it ships 90% of its orders within three working days. You select an SRS of 100 of the 5000 orders received in the past week for an audit. The audit reveals that 86 of these orders were shipped on time.

(a) If the company really ships 90% of its orders on time, what is the probability that 86 or fewer in an SRS of 100 orders are shipped on time?

(b) A critic says, "Aha! You claim 90%, but in your sample the on-time percent is only 86%. So the 90% claim is wrong." Explain in simple language why your probability calculation in (a) shows that the result of the sample does not refute the 90% claim.

5.81 Checking for survey errors. One way of checking the effect of undercoverage, nonresponse, and other sources of error in a sample survey is to compare the sample with known facts about the population. About 12% of American adults are black. The number X of blacks in a random sample of 1500 adults should therefore vary with the binomial ($n = 1500$, $p = 0.12$) distribution.
(a) What are the mean and standard deviation of X?
(b) Use the Normal approximation to find the probability that the sample will contain 170 or fewer blacks. Be sure to check that you can safely use the approximation.

5.82 Continuity correction. We have found that when the sample size n is large enough, the Normal distribution well approximates the binomial distribution. There is, however, a fine-tuning that can be done to make the approximation even better. The fine-tuning, known as "continuity correction," stems from the fact that the Normal distribution takes on all real numbers (is continuous) while the binomial distribution takes on only integers (is discrete). The correction is accomplished by treating an integer value for a binomial random variable as an interval given by the integer value ± 0.5.
(a) Refer to Example 5.20 (page 315), in which we used the Normal distribution to find that $P(X \geq 610)$ is approximately equal to 0.3974. Now use the continuity correction to improve the approximation. Since the integer value of 610 would be viewed as the interval from 609.5 to 610.5, we will use the Normal distribution to find $P(X \geq 609.5)$. Find this probability and compare it with both the approximate probability of 0.3974 found in Example 5.20 and the exact probability found on page 313.
(b) Use the Normal approximation with continuity correction to find the probability that 580 or fewer of the sample are very satisfied with their lives.
(c) Without continuity correction, what will the Normal approximation give as the probability that exactly 600 of the sample are very satisfied with their lives? Now find the same probability using continuity correction.

5.83 Ichiro Suzuki's hits. Using the continuity correction discussed in Exercise 5.82, redo part (b) of Exercise 5.78. How does this value compare with the original approximation and the exact answer of part (d)?

5.4 The Poisson Distributions

Not all counts have binomial distributions. It is common to meet counts that are open-ended, that is, that do not have the fixed number of observations required by the binomial model. Count the number of finish defects in the sheet metal of a car or a refrigerator: the count could be 0, 1, 2, 3, and so on indefinitely. A bank counts the number of automatic teller machine (ATM) customers arriving at a particular ATM between 2:00 P.M. and 4:00 P.M. A railyard counts the number of work injuries that happen in a month. All of these count examples share common characteristics.

The Poisson setting

The Poisson distribution is another distribution for count random variables. Count the number of events (call them "successes") that occur in some fixed unit of measure such as an area of sheet metal, a period of time, or a length of cable. The Poisson distribution is appropriate in the following situation.

> **The Poisson Setting**
>
> 1. The number of successes that occur in any unit of measure is **independent** of the number of successes that occur in any nonoverlapping unit of measure.
> 2. The probability that a success will occur in a unit of measure is the same for all units of equal size and is proportional to the size of the unit.
> 3. The probability that 2 or more successes will occur in a unit approaches 0 as the size of the unit becomes smaller.

For binomial distributions, the important quantities were n, the fixed number of observations, and p, the probability of success on any given observation. The quantity important in specifying Poisson distributions is the mean number of successes occurring per unit of measure.

Poisson Distribution

The distribution of the count X of successes in the Poisson setting is the **Poisson distribution** with **mean** μ. The parameter μ is the mean number of successes per unit of measure. The possible values of X are the whole numbers 0, 1, 2, 3, ... If k is any whole number 0 or greater, then*

$$P(X = k) = \frac{e^{-\mu}\mu^k}{k!}$$

The **standard deviation** of the distribution is $\sqrt{\mu}$.

EXAMPLE 5.21   Flaws in Carpets

A carpet manufacturer knows that the number of flaws per square yard in a type of carpet material varies with an average of 1.6 flaws per square yard. The count X of flaws per square yard can be modeled by the Poisson distribution with $\mu = 1.6$. The unit of measure is a square yard of carpet material. What is the probability of no more than 2 defects in a randomly chosen square yard of this material?

We will calculate $P(X \le 2)$ in two ways:

1. Software can do the calculation:

	A	B	C
1	Poisson(0,1.6,0)=	0.2019	
2	Poisson(1,1.6,0)=	0.3230	
3	Poisson(2,1.6,0)=	0.2584	
4	SUM(B1:B3)=	0.7834	
5			

2. We can use the Poisson probability formula:

$$P(X \le 2) = P(X = 0) + P(X = 1) + P(X = 2)$$
$$= \frac{e^{-1.6}(1.6)^0}{0!} + \frac{e^{-1.6}(1.6)^1}{1!} + \frac{e^{-1.6}(1.6)^2}{2!}$$
$$= 0.2019 + 0.3230 + 0.2584 = 0.7833$$

The software answer and the hand calculation differ by 0.0001 due to roundoff error in the hand calculation. The software calculates the individual probabilities to many significant digits even though it displays only four significant digits.

---

cumulative probability   Recall that Table A gives **cumulative probabilities** of the form $P(X \le k)$ for the standard Normal distribution. Most software will calculate cumulative probabilities for other distributions, including the Poisson family. Cumulative probability calculations make solving many problems less tedious.

---

* The quantity e in the Poisson probability formula is a mathematical constant, $e = 2.71828$ to six significant digits. Many calculators have an e^x function.

Mark Dyball/Alamy

EXAMPLE 5.22 Counting ATM Customers

Suppose the number of persons using an ATM in any given hour can be modeled by a Poisson distribution with $\mu = 5.5$. What is the probability of more than 8 persons using the machine during the next hour? Calculating this probability requires two steps:

1. Write $P(X > 8)$ as an expression involving a cumulative probability:

$$P(X > 8) = 1 - P(X \leq 8)$$

2. Calculate $P(X \leq 8)$ and subtract the value from 1.

	A	B	C
1	1-Poisson(8,5.5,1)=	0.1056	
2			

In this case, relying on software to get the cumulative probability is quicker and less prone to error than the method of Example 5.21, which here would require determining nine individual probabilities and then summing their values.

The Poisson model

If we add counts of successes in nonoverlapping areas of space or time, we are just counting the successes in a larger area. That count still meets the conditions of the Poisson setting. Put more formally, if X is a Poisson random variable with mean μ_X and Y is a Poisson random variable with mean μ_Y and Y is independent of X, then $X + Y$ is a Poisson random variable with mean $\mu_X + \mu_Y$. This fact is important in using Poisson models. We can combine areas or look at just a portion of an area and still use Poisson distributions for counts of successes.

EXAMPLE 5.23 Paint Finish Flaws

Auto bodies are painted during manufacture by robots programmed to move in such a way that the paint is uniform in thickness and quality. You are testing a newly programmed robot by counting paint sags caused by small areas receiving too much paint. Sags are more common on vertical surfaces. Suppose that counts of sags on the roof follow the Poisson model with mean 0.7 sags per square yard and that counts on the side panels of the auto body follow the Poisson model with mean 1.4 sags per square yard. Counts in nonoverlapping areas are independent. Then

- The number of sags in 2 square yards of roof is a Poisson random variable with mean $0.7 + 0.7 = 1.4$.
- The total roof area of the auto body is 4.8 square yards. The number of paint sags on a roof is a Poisson random variable with mean $4.8 \times 0.7 = 3.36$.
- A square foot is 1/9 square yard. The number of paint sags in a square foot of roof is a Poisson random variable with mean $1/9 \times 0.7 = 0.078$.
- If we examine 1 square yard of roof and 1 square yard of side panel, the number of sags is a Poisson random variable with mean $0.7 + 1.4 = 2.1$.

APPLY YOUR KNOWLEDGE

5.84 Industrial accidents. A large manufacturing plant has averaged 7 "reportable accidents" per month. Suppose that accident counts over time follow a Poisson distribution with mean 7 per month.

(a) What is the probability of exactly 7 accidents in a month?

(b) What is the probability of 7 or fewer accidents in a month?

5.85 A safety initiative. This year, a "safety culture change" initiative attempts to reduce the number of accidents at the plant described in the previous exercise. There are 66 reportable accidents during the year. Suppose that the Poisson distribution of the previous exercise continues to apply.

(a) What is the distribution of the number of reportable accidents in a year?

(b) What is the probability of 66 or fewer accidents in a year? (Use software.) The probability is small, which is evidence that the initiative did reduce the accident rate.

BEYOND THE BASICS: More Distribution Approximations

In Section 5.3, we observed that the Normal distribution could be used to calculate binomial probabilities when n, the number of trials, is large. When n is large, the binomial probability histogram has the familiar mound shape of the Normal density curve (page 315). This fact allows us to use Normal probabilities and to avoid tedious hand calculations or the need to use software to calculate binomial probabilities.

Using the Normal distribution to approximate the binomial distribution is just one example of using one distribution to approximate another to make probability calculations more convenient. With the distributions we have studied, two more approximations are common:

k	Poisson	Normal
190	0.2529	0.2397
200	0.5188	0.5000
210	0.7727	0.7603

- **Normal approximation to the Poisson.** Some software is unable to calculate Poisson probabilities when μ is large.* What can we do if our software cannot handle Poisson distributions with large means (for example, a wireless company is monitoring the number of dropped calls in a metro area per day)? Fortunately, when μ is large, Poisson probabilities can be approximated using the Normal distribution with mean μ and standard deviation $\sqrt{\mu}$. The table in the left margin compares $P(X \leq k)$ for a Poisson random variable X with $\mu = 200$ with approximations using the $N(200, 14.142)$ distribution.

  The Normal approximation is adequate for many practical purposes, but we recommend statistical software that can give exact Poisson probabilities.

k	Binomial	Poisson
0	0.3677	0.3679
1	0.7358	0.7358
2	0.9198	0.9197
3	0.9811	0.9810
4	0.9964	0.9963
5	0.9994	0.9994
6	0.9999	0.9999
7	1.0000	1.0000

- **Poisson approximation to the binomial.** We recommend using the Normal approximation to a binomial distribution only when n and p satisfy $np \geq 10$ and $n(1 - p) \geq 10$. In cases where p is so small that $np < 10$, using the Poisson distribution with $\mu = np$ to calculate binomial probabilities yields more accurate results. The table in the left margin compares $P(X \leq k)$ for a binomial distribution with $n = 1000$ and $p = 0.001$ with Poisson probabilities calculated using $\mu = np = (1000)(0.001) = 1$.

  The Poisson approximation gives very accurate probability calculations for the binomial distribution with $n = 1000$ and $p = 0.001$.

---

*Minitab is unable to calculate Poisson probabilities when $\mu > 709$, and earlier versions of Excel would return an error for considerably smaller values for μ.

Even statistical software has its limits, and some binomial and Poisson probability calculations can exceed those limits. In many cases, however, one of the approximations we have discussed will make calculations possible.

SECTION 5.4 Summary

- A count X of successes has a **Poisson distribution** in the **Poisson setting:** the number of successes in any unit of measure is independent of the number of successes in any other nonoverlapping unit; the probability of a success in a unit of measure is the same for all units of equal size and is proportional to the size of the unit; the probability of 2 or more successes in a unit approaches 0 as the size of the unit becomes smaller.

- If X has the Poisson distribution with mean μ, then the standard deviation of X is $\sqrt{\mu}$, and the possible values of X are all the whole numbers 0, 1, 2, 3, and so on. The **Poisson probability** that X takes any one of these values is

$$P(X = k) = \frac{e^{-\mu}\mu^k}{k!} \qquad k = 0, 1, 2, 3, \ldots$$

- Poisson probabilities are most easily found by software. The formula above is practical when only a small number of probabilities is needed and k is not large.

- Sums of independent Poisson random variables also have the Poisson distribution. In a Poisson model with mean μ per unit of space or time, the count of successes in a units is a Poisson random variable with mean $a\mu$.

SECTION 5.4 Exercises

For Exercises 5.84 and 5.85, see page 322.

Use software to calculate the Poisson probabilities in the following exercises.

5.86 How many calls? Calls to the customer service department of a cable TV provider are made randomly and independently at a rate of 11 per minute. The company has a staff of 20 customer service specialists who handle all the calls. Assume that none of the specialists are on a call at this moment and that a Poisson model is appropriate for the number of incoming calls per minute.
(a) What is the probability of the customer service department receiving more than 20 calls in the next minute?
(b) What is the probability of the customer service department receiving exactly 20 calls in the next minute?
(c) What is the probability of the customer service department receiving fewer than 11 calls in the next minute?

5.87 Hospitalization claims. A national home improvement and hardware chain provides health insurance to its full-time employees. On the basis of past experience with companies similar to this one, the insurance carrier for this chain of stores expects 6 hospitalizations per month, on average. Employee turnover within the chain of stores is approximately 32%; however, the overall size and general demographic characteristics of the workforce remain fairly consistent over time. This stability within the workforce makes the Poisson model reasonable.
(a) What assumption must be made about hospitalizations for different employees?
(b) What is the probability of 2 or fewer hospitalizations in the next month?
(c) What is the probability of more than 2 hospitalizations in the next month?
(d) What is the probability of at least 8 hospitalizations in the next month?

5.88 Bad checks. Duck Worth Wearing, a children's consignment shop, accepts cash, check, and credit card for payment. About 30% of all transactions are paid by check. Duck Worth Wearing anticipates only 3 bad checks per month, on average. Assume that the number of bad checks per month can be modeled as a Poisson random variable.
(a) What is the probability of no bad checks in the next month?
(b) What is the probability of one or more bad checks in the next month?
(c) What is the probability of more than 7 bad checks in the next month?

5.89 Switchboard errors. At a large software company, calls coming to the general contact phone number are answered by a switchboard operator. The operator's job is to manually forward calls to the desired party within the company. The switchboard operator's past performance indicates that he forwards calls incorrectly at a rate of 2.5 per week. Assuming a Poisson model is appropriate, answer the following questions.

(a) What is the probability of no call-forwarding errors in the next week?

(b) What is the probability of 4 or more call-forwarding errors in the next week?

5.90 Too much email? According to email logs, one employee at your company receives an average of 110 emails per week. Suppose the count of emails received can be adequately modeled as a Poisson random variable.

(a) What is the probability of this employee receiving exactly 110 emails in a given week?

(b) What is the probability of receiving 100 or fewer emails in a given week?

(c) What is the probability of receiving more than 125 emails in a given week?

(d) What is the probability of receiving 125 or more emails in a given week? (Be careful: this is not the same event as in part (c).)

5.91 Traffic model. The number of vehicles passing a particular mile marker during 15-minute units of time can be modeled as a Poisson random variable. Counting devices show that the average number of vehicles passing the mile marker every 15 minutes is 48.7.

(a) What is the probability of 50 or more vehicles passing the marker during a 15-minute time period?

(b) What is the standard deviation of the number of vehicles passing the marker in a 15-minute time period? A 30-minute time period?

(c) What is the probability of 100 or more vehicles passing the marker during a 30-minute time period?

5.92 Too much email? According to email logs, one employee at your company receives an average of 110 emails per week. Suppose the count of emails received can be adequately modeled as a Poisson random variable.

(a) What is the distribution of the number of emails in a two-week period?

(b) What is the probability of receiving 200 or fewer emails in a two-week period?

5.93 Work-related deaths. Based on 2007 government data, work-related deaths in the United States have a mean of 15 per day. Suppose the count of work-related deaths per day follows an approximate Poisson distribution.

(a) What is the standard deviation for daily work-related deaths?

(b) What is the probability of 10 or fewer work-related deaths in one day?

(c) What is the probability of more than 30 work-related deaths in one day?

5.94 Flaws in carpets. Flaws in carpet material follow the Poisson model with mean 1.6 flaws per square yard. An inspector examines 100 randomly selected square yard specimens of the material, records the number of flaws found in each specimen, and calculates $\overline{x}$, the average number of flaws per square yard inspected.

(a) The total number of flaws $100\overline{x}$ is a Poisson random variable. What is its mean?

(b) What is the probability that the total number of flaws $100\overline{x}$ exceeds 110?

(c) We can use the central limit theorem (page 270) to calculate the same probability as in part (b) by realizing that $P(100\overline{x} > 110) = P(\overline{x} > 110/100)$. What is the probability found using the central limit theorem?

(d) Compare your answers to (b) and (c). How close are the two answers? Which one is more accurate and why?

5.95 Calling tech support. The number of calls received between 8 A.M. and 9 A.M. by a software developer's technical support line has a Poisson distribution with a mean of 14.

(a) What is the probability of at least 5 calls between 8 A.M. and 9 A.M.?

(b) What is the probability of at least 5 calls between 8:15 A.M. and 8:45 A.M.?

(c) What is the probability of at least 5 calls between 8:15 A.M. and 8:30 A.M.?

5.96 Web site hits. A "hit" for a Web site is a request for a file from the Web site's server computer. Some popular Web sites have thousands of hits per minute. One popular Web site boasts an average of 6500 hits per minute between the hours of 9 A.M. and 6 P.M. Some weaker software packages will have trouble calculating Poisson probabilities with such a large value of μ.

(a) Try calculating the probability of 6400 or more hits during the minute beginning at 10:05 A.M. using the software that you have available. Did you get an answer? If not, how did the software respond?

(b) Now, use the central limit theorem to calculate the probability of 6400 or more hits during the minute beginning at 10:05 A.M. To do this, think of the number of hits in this minute as the sum of the number of hits for each of the 60 seconds in this minute. We can express $P(\text{sum of hits for each of the 60 seconds} \geq 6400)$ as $P(\overline{x} \geq 6400/60)$ where $\overline{x}$ is the average number of hits per second for the 60 seconds in the minute of interest.

5.97 Credit card manufacturing. Large sheets of plastic are cut into smaller pieces to be pressed into credit cards. One manufacturer uses sheets of plastic known to have approximately 2.3 defects per square yard. The number of defects can be modeled as a Poisson random variable X.

(a) What is the standard deviation of the number of defects per square yard?

(b) What is the probability of an inspector finding more than 5 defects in a randomly chosen square yard?

(c) Using trial and error with your software, find the largest value k such that $P(X > k) \geq 0.15$.

5.98 Initial public offerings. The number of companies making their initial public offering of stock (IPO) can be modeled by a Poisson distribution with a mean of 15 per month.
(a) What is the probability of fewer than 3 IPOs in a month?
(b) What is the probability of fewer than 15 IPOs in a month?
(c) What is the probability of fewer than 30 IPOs in a two-month period?

(d) What is the probability of fewer than 180 IPOs in a year?

5.99 Calls to 911 system. The City of Chicago reports that on an average day 15,000 emergency calls come into the city's 911 system.[15] Assuming a Poisson distribution with this reported mean rate, use the Normal distribution to find the range in which we would expect 99.7% of the calls to fall.

STATISTICS IN SUMMARY

This chapter concerns some further facts about probability that are useful in modeling but are not needed in our study of statistics. Section 5.1 discusses general rules that all probability models must obey, including the important multiplication rule for independent events. There are many specific probability models for specific situations. When events are not independent, we need the idea of conditional probability. That is the topic of Section 5.2. At this point, we finally reach the fully general form of the basic rules of probability. Section 5.3 uses the multiplication rule to obtain one of the most important probability models, the binomial distribution for counts. Remember that not all counts have a binomial distribution, just as not all measured variables have a Normal distribution. In Section 5.4 we considered the Poisson distribution, an alternative model for counts. Here is a review list of the most important skills you should have acquired from your study of this chapter.

A. Probability Rules

1. Use Venn diagrams to picture relationships among several events.
2. Use the general addition rule to find probabilities that involve overlapping events.
3. Understand the idea of independence. Judge when it is reasonable to assume independence as part of a probability model.
4. Use the multiplication rule for independent events to find the probability that all of several independent events occur.
5. Use the multiplication rule for independent events in combination with other probability rules to find the probabilities of complex events.

B. Conditional Probability

1. Understand the idea of conditional probability. Identify the two events required from a verbal description of conditional probability. Find conditional probabilities for individuals chosen at random from a two-way table of counts of outcomes.
2. Use the general multiplication rule to find $P(A \text{ and } B)$ from $P(A)$ and the conditional probability $P(B \mid A)$.
3. Use a tree diagram to organize several-stage probability models.
4. Use Bayes's rule to calculate conditional probabilities when given other "reverse" conditional probabilities.

C. Binomial Distributions

1. Recognize the binomial setting: we have a fixed number n of independent success-failure trials with the same probability p of success on each trial.

2. Recognize and use the binomial distribution of the count of successes in a binomial setting.

3. (Optional) Use the binomial probability formula to find probabilities of events involving the count X of successes in a binomial setting for small values of n.

4. Use binomial tables to find binomial probabilities.

5. Find the mean and standard deviation of a binomial count X.

6. Recognize when you can use the Normal approximation to a binomial distribution. Use the Normal approximation to calculate probabilities that concern a binomial count X.

D. Poisson Distributions

1. Recognize the Poisson setting: we are counting the number of successes in a fixed unit of measure (time, area, volume, or length).

2. Given a Poisson model with stated mean count per unit, find the Poisson distribution for the count in a multiple or a fractional number of units.

3. Use software to calculate Poisson probabilities.

4. Find the mean and standard deviation of a Poisson count X.

5. Use the central limit theorem to approximate Poisson probabilities when μ is too large for your software by dividing the basic unit of measure into many smaller units of measure and viewing the Poisson random variable as the sum of many independent Poisson random variables.

CHAPTER 5   Review Exercises

5.100 Motor vehicle sales. Motor vehicles sold to individuals are classified as either cars or light trucks (including SUVs) and as either domestic or imported. From January to October 2008, 48% of vehicles sold were light trucks, 75% were domestic, and 40% were domestic light trucks.

(a) Draw a Venn diagram that represents the given percents.

(b) Calculate the probability that a randomly selected vehicle is a car.

(c) Calculate the probability that a randomly selected vehicle is an imported car.

5.101 Using a mailing list. The mailing list of an agency that markets scuba-diving trips to the Florida Keys contains 70% males and 30% females. The agency calls 30 people chosen at random from its list.

(a) What is the probability that 20 of the 30 are men? (Use the binomial probability formula.)

(b) What is the probability that the first woman is reached on the fourth call? (That is, the first 4 calls give MMMF.)

(c) List any assumptions necessary for using the binomial distribution in this context.

5.102 Motor vehicle sales. Using the information in Exercise 5.100, answer these questions.

(a) Given that a vehicle is imported, what is the conditional probability that it is a car?

(b) Are the events "vehicle is a car" and "vehicle is imported" independent? Justify your answer.

5.103 Human spell-checker. Spelling errors in a text can be either nonword errors or word errors. Nonword errors make up 25% of all errors. A human proofreader will catch 90% of nonword errors and 70% of word errors. What percent of all errors will the proofreader catch? (Draw a tree diagram to organize the information given.)

5.104 Playing the slots. Slot machines are now video games, with winning determined by electronic random number generators. In the old days, slot machines worked like this: you pull the lever to spin three wheels; each wheel has 20 symbols, all equally likely to show when the wheel stops spinning; the three wheels are independent of each other. Suppose that the middle wheel has 9 bells among its 20 symbols, and the left and right wheels have 1 bell each.

(a) You win the jackpot if all three wheels show bells. What is the probability of winning the jackpot?

(b) What is the probability that the wheels stop with exactly 2 bells showing?

5.105 Medical risks. You have torn a tendon and are facing surgery to repair it. The surgeon explains the risks to you: infection occurs in 3% of such operations, the repair fails in 14%, and both infection and failure occur together in 1%. What percent of these operations succeed and are free from infection?

5.106 Income and savings. A sample survey chooses a sample of households and measures their annual income and their savings. Some events of interest are

A = the household chosen has income at least $100,000

C = the household chosen has at least $50,000 in savings

Based on this sample survey, we estimate that $P(A) = 0.13$ and $P(C) = 0.25$.

(a) We want to find the probability that a household has either income at least $100,000 *or* savings at least $50,000. Explain why we do not have enough information to find this probability. What additional information is needed?

(b) We want to find the probability that a household has income at least $100,000 *and* savings at least $50,000. Explain why we do not have enough information to find this probability. What additional information is needed?

5.107 Applying for student financial aid. To determine eligibility for federal student financial aid, students (undergraduate and graduate) must fill out a form known as "The Free Application for Federal Student Aid" (FAFSA). Many students are unaware of the FAFSA, while other students learn about it from a variety of sources. The U.S. Department of Education hopes that more students become aware of the FAFSA because it seems likely that the more students who apply for aid the more students will attend and complete college. To understand what sources most influence students to complete the FAFSA, the Center for Community College Student Engagement conducted an extensive survey of community college students in which students were asked how they originally learned about the FAFSA and whether they filled out the form.[16] The results are given below:

Source	Completed FAFSA	Didn't complete FAFSA
Family	57,624	28,675
High school staff	49,941	31,693
College staff	48,020	21,129
Friend	32,654	16,601
Didn't learn from others	5,762	51,313

(a) What is the probability that a randomly chosen student completed the FAFSA?

(b) What is the probability that a randomly chosen student first learned about the FAFSA from a high school staff member?

(c) What is the probability that a randomly chosen student first learned about the FAFSA from a friend and did not complete the application?

(d) If you know that the student first learned about the FAFSA from a family member, what is the probability that the student completed the FAFSA?

5.108 Canadian children and family structure. It is common for governments to gather data on family structure in the population. Information on family structure distribution is important because of the known associations between family structure and child well-being, behavior, and general success in life. Additionally, information on family structure in the population has business implications for such things as day care for employees and location of work (for example, from home or at organization). Here are data from 2006 on Canadian families with at least one child:[17]

Number of children	Married couple	Common-law couple	Mother only	Father only
One child at home	1,267,625	291,255	682,025	188,790
Two children at home	1,497,755	234,755	327,660	72,655
Three or more children at home	678,405	92,140	122,605	20,320

(a) What is the probability that a randomly chosen family with at least one child is a single-parent family?

(b) If you know that the family chosen is a married couple, what is the conditional probability that the family has two children?

(c) Are the events "mother-only family" and "one child at home" independent? How do you know?

5.109 Testing for HIV. Enzyme immunoassay (EIA) tests are used to screen blood specimens for the presence of antibodies to HIV, the virus that causes AIDS. Antibodies indicate the presence of the virus. The test is quite accurate but is not always correct. Here are approximate probabilities of positive and negative EIA outcomes when the blood tested does and does not actually contain antibodies to HIV:[18]

	Test Result	
	+	−
Antibodies present	0.9985	0.0015
Antibodies absent	0.006	0.994

Suppose that 1% of a large population carries antibodies to HIV in their blood.

(a) Draw a tree diagram for selecting a person from this population (outcomes: antibodies present or absent) and for testing his or her blood (outcomes: EIA positive or negative).

(b) What is the probability that the EIA is positive for a randomly chosen person from this population?

(c) What is the probability that a person has the antibody, given that the EIA test is positive?

(*Comment:* This exercise illustrates a fact that is important when considering proposals for widespread testing for HIV, illegal drugs, or agents of biological warfare: if the condition being tested is uncommon in the population, many positives will be false positives.)

5.110 Testing for HIV, continued. The previous exercise gives data on the results of EIA tests for the presence of antibodies to HIV. Repeat part (c) of that exercise for two different populations:

(a) Blood donors are prescreened for HIV risk factors, so perhaps only 0.1% (0.001) of this population carries HIV antibodies.

(b) Clients of a drug rehab clinic are a high-risk group, so perhaps 10% of this population carries HIV antibodies.

(c) What general lesson do your calculations illustrate?

5.111 The geometric distributions. In some manufacturing environments, all manufactured items are inspected, and when the first defect is encountered, the operation is halted for investigation of possible causes. Suppose that the rate of defective items is 1% and that the quality of each item is independent of the quality of all other items. We are interested in how long we must wait to get the first defective item.

(a) The probability of the first inspected item being defective is 0.01. What is the probability that the first item is not defective and the second is defective?

(b) What is the probability that the first two items are not defective and the third item is defective? This is the probability that the first defective item is the third item inspected.

(c) Now you see the pattern. What is the probability that the first defective item is the fourth item inspected? The fifth item? Give the general result: what is the probability that the first defective item is the kth item inspected?

(*Comment:* The distribution of the number of trials to the first success is called a *geometric distribution*. In this exercise you have found geometric distribution probabilities when the probability of a success on each trial is $p = 0.01$. The same idea works for any p.)

5.112 Teenage drivers. An insurance company has the following information about drivers aged 16 to 18 years: 20% are involved in accidents each year; 10% in this age group are A students; among those involved in an accident, 5% are A students.

(a) Let A be the event that a young driver is an A student and C the event that a young driver is involved in an accident this year. State the information given in terms of probabilities and conditional probabilities for the events A and C.

(b) What is the probability that a randomly chosen young driver is an A student and is involved in an accident?

5.113 Race and ethnicity. Annually, the U.S. Census Bureau provides estimates of a variety of demographic characteristics of the U.S. population. In terms of categorizing people by race, the bureau divides the population into those of "Hispanic" origin and those not. Within these two origin categories, a person can be associated with one of several races. For 2007, the bureau gives the following probability estimates:

	Hispanic	Not Hispanic
Asian	0.001	0.043
Black	0.006	0.123
White	0.139	0.660
Other	0.005	0.023

(a) What is the probability that a randomly chosen person is white?

(b) You know that the person chosen is of Hispanic origin. What is the conditional probability that this person is white?

5.114 More on teenage drivers. Use your work from Exercise 5.112 to find the percent of A students who are involved in accidents. (Start by expressing this as a conditional probability.)

5.115 More on race and ethnicity. Use the information in Exercise 5.113 to answer these questions.

(a) What is the probability that a randomly chosen American is Hispanic?

(b) You know that the person chosen is black. What is the conditional probability that this person is Hispanic?

5.116 Screening job applicants. A company retains a psychologist to assess whether job applicants are suited for assembly-line work. The psychologist classifies applicants as A (well suited), B (marginal), or C (not suited). The company is concerned about event D: an employee leaves the company within a year of being hired. Data on all people hired in the past five years give these probabilities:

$$P(A) = 0.4 \qquad P(B) = 0.3 \qquad P(C) = 0.3$$
$$P(A \text{ and } D) = 0.1 \quad P(B \text{ and } D) = 0.1 \quad P(C \text{ and } D) = 0.2$$

Sketch a Venn diagram of the events A, B, C, and D and mark on your diagram the probabilities of all combinations of psychological assessment and leaving (or not) within a year. What is $P(D)$, the probability that an employee leaves within a year?

5.117 Who buys iMacs? When the iMac computer was first introduced by Apple Computer, it quickly became one of the company's best-selling products. The iMac was particularly aimed at first-time computer buyers. Approximately 5 months after the introduction of the iMac, Apple reported that 32% of iMac buyers were first-time computer buyers. At this same time, approximately 5% of all computer sales were of iMacs. Of buyers who did not purchase an iMac, approximately 40% were first-time computer buyers. Among first-time computer buyers during this time, what percent bought iMacs?

5.118 Stealing software. Employees sometimes install on their home computers software that was purchased by their employer for use on their work computers. For most commercial software packages, this is illegal. Suppose that 5% of all employees at a large corporation have illegally installed corporate software on their home computers knowing the act is illegal and an additional 2% have installed corporate software on their home computers not realizing that this is illegal. Of the 5% aware that the home installation is illegal, 80% will deny that they knew the act was illegal if confronted by a "software auditor." If an employee who has illegally installed software at home is confronted and denies knowing it was an illegal act, what is the probability that the employee knew the home installation was illegal?

5.119 Leaking gas tanks. Leakage from underground gasoline tanks at service stations can damage the environment. It is estimated that 25% of these tanks leak. You examine 15 tanks chosen at random, independently of each other.
(a) What is the mean number of leaking tanks in such samples of 15?
(b) What is the probability that 10 or more of the 15 tanks leak?
(c) Now you do a larger study, examining a random sample of 1000 tanks nationally. What is the probability that at least 275 of these tanks are leaking?

5.120 Environmental credits. An opinion poll asks an SRS of 500 adults whether they favor tax credits for companies that demonstrate a commitment to preserving the environment. Suppose that in fact 45% of the population favor this idea. What is the probability that more than half of the sample are in favor?

5.121 Computer training. Macintosh users make up about 5% of all computer users. A computer training school that wants to attract Macintosh users mails an advertising flyer to 25,000 computer users.
(a) If the mailing list can be considered a random sample of the population, what is the mean number of Macintosh users who will receive the flyer?
(b) What is the probability that at least 1245 Macintosh users will receive the flyer?

5.122 Is this coin balanced? While he was a prisoner of the Germans during World War II, John Kerrich tossed a coin 10,000 times. He got 5067 heads. Take Kerrich's tosses to be an SRS from the population of all possible tosses of his coin. If the coin is perfectly balanced, $p = 0.5$. Is there reason to think that Kerrich's coin gave too many heads to be balanced? To answer this question, find the probability that a balanced coin would give 5067 or more heads in 10,000 tosses. What do you conclude?

5.123 Who is driving? A sociology professor asks her class to observe cars having a man and a woman in the front seat and record which of the two is the driver.
(a) Explain why it is reasonable to use the binomial distribution for the number of male drivers in n cars if all observations are made in the same location at the same time of day.
(b) Explain why the binomial model may not apply if half the observations are made outside a church on Sunday morning and half are made on campus after a dance.
(c) The professor requires students to observe 10 cars during business hours in a retail district close to campus. Past observations have shown that the man is driving about 85% of cars in this location. What is the probability that the man is driving 8 or fewer of the 10 cars?

(d) The class has 10 students, who will observe 100 cars in all. What is the probability that the man is driving 80 or fewer of these?

5.124 Binomial distribution? Suppose a manufacturing colleague tells you that 1% of items produced in first shift are defective while 1.5% in second shift are defective and 2% in third shift are defective. He notes that the number of items produced is approximately the same from shift to shift, which implies an average defective rate of 1.5%. He further states that since the items produced are independent of each other, the binomial distribution with $p = 0.015$ will represent the number of defective items in an SRS of items taken in any given day. What is your reaction?

5.125 Poisson distribution? Suppose you find in your spam folder an average of 2 spam emails every 10 minutes. Furthermore, you find that the rate of spam mail from midnight to 6 A.M. is twice the rate during other parts of the day. Explain whether or not the Poisson distribution is an appropriate model for the spam process.

5.126 Life satisfaction. Refer to Example 5.19 (page 313). We learned that the probability that at least 610 out of 1010 randomly selected adults would be "very satisfied" with their lives is 0.4119.
(a) In Excel, fill the first column of the spreadsheet with the numbers $610, \ldots, 1010$. In the second column, put the binomial probabilities $P(X = k)$ for $k = 610, \ldots, 1010$. (Refer to this chapter's Appendix for guidance on Excel's binomial built-in function.) Now, add up all the probabilities to confirm that $P(X \geq 610)$ is equal to 0.4119.
(b) In part (a), you needed to find 401 probabilities and then add them all up. There is, however, an easier option. Excel has a built-in function to compute $P(X \leq k)$ (refer again to the Appendix). Use this built-in function to find the desired probability of 0.4119 (Hint: Be careful what you choose k to be).
(c) Do parts (a) and (b) using statistical software available to you.

5.127 Six Sigma. Six Sigma is a quality improvement strategy that strives to identify and remove the causes of defects. Processes that operate with six-sigma quality produce defects at a level of 3.4 defects per million. Suppose 10,000 independent items are produced from a six-sigma process. What is the probability that there will be at least one defect produced?

5.128 Airline overbooking. Airlines regularly overbook flights to compensate for no-show passengers. In doing so, airlines are balancing the risk of having to compensate bumped passengers against lost revenue associated with empty seats. In a *USA Today* analysis of airline statistics, it was found that the average no-show rate was 12%.[19] Assuming this no-show rate, what is the probability that no passenger will be bumped if an airline books 215 passengers on a 200-seat plane (use software)?

5.129 Inventory control. OfficeShop experiences a one-week order time to restock its HP printer cartridges. During this reorder time, also known as *lead time,* OfficeShop wants to ensure a high level of customer service by not running out of cartridges. Suppose the average lead time demand for a particular HP cartridge is 15 cartridges. OfficeShop makes a restocking order when there are 18 cartridges on the shelf. Assuming the Poisson distribution models the lead time demand process, what is the probability that OfficeShop will be short of cartridges during the lead time?

5.130 More about inventory control. Refer to the previous exercise. In practice, the amount of inventory held on the shelf during the lead time is known as the *reorder point.* Firms use the term *service level* to indicate the percentage of the time that the amount of inventory is sufficient to meet demand during the reorder period. Use software and the Poisson distribution to determine the reorder points so that the service level is minimally

(a) 90%.

(b) 95%.

(c) 99%.

CHAPTER 5   Case Study Exercises

CASE STUDY EXERCISE 1: The Pentium FDIV bug. Intel's Pentium processors power a majority of the world's computers. In 1994, a bug was discovered in the Pentium chip. The bug, which came to be known as the Pentium FDIV bug, caused some floating-point division operations to give incorrect values. Intel claimed that the probability of an error was only 1 in 9 billion. At that rate, said the chip firm, a spreadsheet user who performs 1000 floating-point divisions per day will encounter an incorrect division only about once in 25,000 years. The probability of one or more errors due to the FDIV bug in 365 working days is only about 0.00004.

An IBM research group disagreed. Based on the results of probability simulations, IBM concluded that a more reasonable estimate of the probability of a floating-point division error was 1 in 100 million, 90 times the estimate given by Intel. What is more, simulations of typical financial calculations done by spreadsheet users found that an average user would perform nearly 4.2 million divisions per day. The probability of one or more errors due to the FDIV bug in 365 days, said IBM, is not 0.00004 but 0.9999998.

At first, Intel claimed that the bug would cause errors so rarely that it would not replace the faulty chips that had already been sold—as many as 2 million chips. However, faced with results like IBM's and growing consumer concern, Intel agreed to replace the chips for anyone who wanted a replacement. The Pentium FDIV bug incident is reported to have cost Intel approximately 475 million dollars.[20]

The probability of one or more errors in 365 days is calculated using the multiplication rule as illustrated in Example 5.3 (page 289) and Example 5.4 (page 289). The formula can be expressed as

$$P(\text{one or more errors in 365 days}) = 1 - (1 - a)^{365 \times b}$$

where a is the P(error for a single division) and b is the assumed number of divisions per day for a typical user.

A. Intel's estimates. Using Intel's estimates for a and b, calculate the probability of one or more errors in 365 days. You will need to use statistical or mathematical software to do this calculation. Verify the probability claimed by Intel.

B. IBM's estimates. Using IBM's estimates for a and b, calculate the probability of one or more errors in 365 days. You will need to use statistical or mathematical software to do this calculation. Verify the probability claimed by IBM.

CASE STUDY EXERCISE 2: More on the Pentium FDIV bug.

A. More with IBM's estimates. Using IBM's estimates for a and b, calculate the probability of one or more errors in 24 days. You will need to use statistical or mathematical software to do this calculation. In an IBM report dated December 1994, the authors state that a typical user could make a mistake every 24 days. Do you agree with this statement? Explain your reasoning and supporting calculations.

B. Even more with IBM's estimates. Using IBM's estimates for a and b, calculate the probability of one or more errors in a single day (round your answer to 5 decimal places). For 100,000 typical users, how many errors would you expect in a single day? In the IBM report dated December 1994, the authors state that 100,000 Pentium users could expect 4000 errors to occur each day. Do you agree with this statement? Explain your reasoning and supporting calculations.

CHAPTER 5 Appendix

Using Minitab and Excel for Probability Theory

Binomial Distribution

Minitab:

Calc ➤ Probability Distributions ➤ Binomial

Choose the **Probability** option if you wish to find the probability that the binomial random variable takes on a specific value k; that is, $P(X = k)$. Choose the **Cumulative probability** option if you wish to find the probability that the binomial random variable takes on a value less than or equal to a specific value k; that is, $P(X \leq k)$. Input the value of n (number of observations) in the **Number of trials** box, and input the value of p (probability of success) in the **Event probability** box. You can then select the **Input constant** option. In the box next to this option, enter the specified value of k. Click **OK** to find the results reported in the Session window.

Excel:

Click an empty cell in the spreadsheet and then proceed to the **Statistical** function menu as described in the overview section of Chapter 1's Appendix. Scroll down the list of functions and double-click on the **BINOMDIST** function choice. In the **Number_s** box, input the specified value of k (the number of successes). In the **Trials** box, input the value of n (number of observations). In the **Probability_s** box, input the value of p (probability of success). If you wish to have $P(X = k)$, then input "0" or "FALSE" in the **Cumulative** box. If you wish to have $P(X \leq k)$, then input "1" or "TRUE" in the **Cumulative**

box. Click **OK** to find the results reported in the Session window.

Poisson Distribution

Minitab:

Calc ➤ Probability Distributions ➤ Poisson

Choose the **Probability** option if you wish to find the probability that the Poisson random variable takes on a specific value k; that is, $P(X = k)$. Choose the **Cumulative probability** option if you wish to find the probability that the Poisson random variable takes on a value less than or equal to a specific value k; that is, $P(X \leq k)$. Input the value of μ (mean number of successes per unit of measure) in the **Mean** box, and input the value of p (probability of success) in the **Event probability** box. You can then select the **Input constant** option. In the box next to this option, enter the specified value of k. Click **OK** to find the results reported in the Session window.

Excel:

Click an empty cell in the spreadsheet and then proceed to the **Statistical** function menu as described in the overview section of Chapter 1's Appendix. Scroll down the list of functions and double-click on the **POISSON** function choice. In the **X** box, input the specified value of k (the number of successes). In the **Mean** box, input the value of μ (mean number of successes per unit of measure). If you wish to have $P(X = k)$, then input "0" "FALSE" in the **Cumulative** box. If you wish to have $P(X \leq k)$, then input "1" or "TRUE" in the **Cumulative** box. Click **OK** to find the results reported in the Session window.

CHAPTER 6

Introduction to Inference

Statistical inference draws conclusions from data that take into account the natural variability in the data. Are differences in response to television commercials significant or not? Example 6.1 explores this.

Introduction

The purpose of statistical inference is to draw conclusions from data, conclusions that take into account the natural variability in the data. To do this, formal inference relies on probability to describe chance variation. We can then correct our "eyeball" judgment by calculation. Here is an example.

EXAMPLE 6.1 Compare TV Commercials

Suppose we show a new TV commercial and the present commercial to 20 consumers each. Twelve of those who see the new ad declare an interest in buying the product, versus only 8 who watch the current version. Is the new commercial more effective?

Probability calculations tell us that a difference this large or larger between the results in the two groups would occur about one time in five simply because of chance variation. Since this probability is not particularly small, it is reasonable to conclude that the difference is due to chance rather than a real difference between the two commercials.

Here is an example with a different conclusion.

EXAMPLE 6.2 Pattern in Trading Volumes

Investors and stock analysts are continually seeking ways to understand price movements in stocks. One factor many have considered in the prediction of price movements is trading volume, that is, the quantity of shares that change owners. In studying the relationship between trading volumes and price movements, an understanding of the trading-volume process becomes a source of interest. Are trading volumes random from trading period to trading period? Or is there some pattern in the trading volumes over

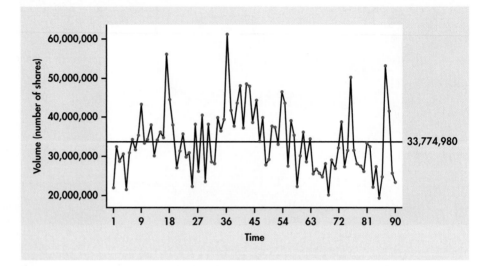

time? Figure 6.1 is a time plot of daily trading volumes for AT&T stock from January 2 to May 12, 2009, with the mean trading volume indicated. Do the volumes appear to be randomly distributed around the mean trading volume over time? More specifically, could you simulate the trading-volume process with the process of flipping a fair coin, with heads being an observation below the mean and tails being an observation above the mean? Or does it appear that the volumes move in patterns around the mean level? In particular, are there "strings" of observations either all below or all above the mean? As you will learn in Chapter 13, one approach to the analysis of these data, known as runs test analysis, indicates that observing a sequencing of observations as patterned as or more patterned than that shown in Figure 6.1 would occur only 2.1% of the time if, in fact, trading volumes are truly random over time.

Because this chance is fairly small, we are inclined to believe that the trading volumes are *not* random over time. Our probability calculation helps us to distinguish between patterns that are consistent and patterns that are inconsistent with a random process scenario.

In this chapter we introduce the two most prominent types of formal statistical inference. Section 6.1 considers *confidence intervals* for estimating the value of a population parameter. Section 6.2 presents *tests of significance,* which assess the evidence for a claim. Both types of inference are based on the sampling distributions of statistics. That is, both report probabilities that state *what would happen if we used the inference method many times.* This kind of probability statement is characteristic of standard statistical inference.

Formal inference procedures are based on sampling distributions. This means that they require a probability model that describes how the data are generated.

EXAMPLE 6.3 A Probability Model

Consider sampling from a Normal population. The mean is μ and the standard deviation is σ. A picture of this model is given in Figure 6.2.

In this chapter we are concerned with general ideas about inference. We illustrate these ideas in the setting of inference for means. The sampling distribution for the mean is the sampling distribution that is relevant for the inference.

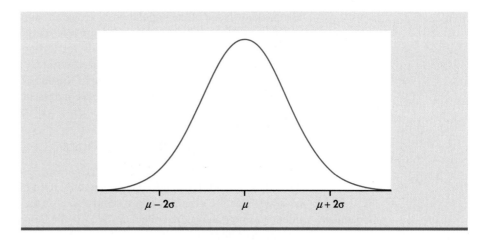

EXAMPLE 6.4   The Sampling Distribution

Suppose we take a random sample of size 5 from the Normal population in Example 6.3. From Section 4.4 we know that the mean of this sampling distribution is μ and the standard deviation is $\sigma/\sqrt{n}$, or $\sigma/\sqrt{5}$ for our sample size of $n = 5$. A picture of this sampling distribution is given in Figure 6.3.

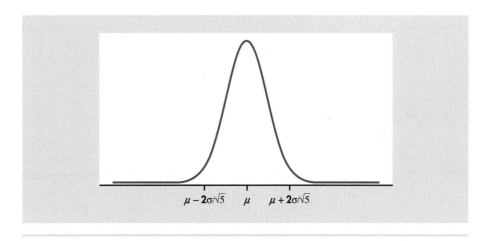

The sampling procedures and experimental designs that we studied in Chapter 3 lead to probability models that generate sampling distributions that are the basis for most of the inference procedures that we discuss in this text. For some business applications, however, we encounter situations that differ from these idealized settings. Here, we often argue that our inference is based on assuming an underlying process that generates the data rather than some random mechanism associated with its collection.

Because we are concentrating on the reasoning of statistical inference in this chapter, we temporarily make a simplifying assumption: that we know the standard deviation σ of the population. In applications based on larger sample sizes, this assumption is not unrealistic since the greater amount of information provides us with a reasonable sense of the value of σ. In other applications, we may have prior knowledge of the population (for example, past results from an ongoing manufacturing process) that allows us to assume knowledge of the population σ. However, in practice, we will more commonly encounter settings that challenge us to learn by means of exploration of the data in hand. Later chapters present inference methods for use in such situations. More elaborate statistical techniques are available for many different applications. Informed use of all these methods requires an understanding of the underlying reasoning. A computer will do the arithmetic, but *you* must still exercise judgment based on understanding.

6.1 Estimating with Confidence

population

One way to characterize a collection of businesses is to determine the average of some measure of size. Total assets is one commonly used measure. If the collection of businesses is large, we generally take a sample and use the information gathered to make an inference about the entire collection. We use the term **population** to refer to the entire collection of interest.

EXAMPLE 6.5 Health Insurance

The Kaiser Family Foundation (Kaiser) and the Health Research & Educational Trust (HRET) conduct an annual national survey of firms on a variety of health coverage issues, including premiums and employee contributions.[1] The 2008 survey included 2832 randomly selected public and private firms with three or more employees. Firms are categorized as "large" if they have 200 or more employees. In the sample, nearly 100% of large firms offer health benefits.

The survey considers different health plans such as health maintenance organizations (HMOs), preferred-provider organizations (PPOs), point-of-service plans (POS), and high-deductible health plans with a savings option (HDHP/SO). Let us consider the mean monthly cost of the premium for single-coverage HMO plans for workers in large firms. For this situation, we find 453 firms in the sample. For these firms, the mean monthly premium is $\bar{x} = \$405.02$. What can we say about μ, the mean monthly single-coverage premium per worker for all large firms in the United States?

The sample mean $\bar{x}$ is the natural estimator of the unknown population mean μ. We know that $\bar{x}$ is an unbiased estimator of μ. More important, the law of large numbers says that the sample mean must approach the population mean as the size of the sample grows. The value $\bar{x} = 405.02$ therefore appears to be a reasonable estimate of the mean monthly premium μ for all large firms. But how reliable is this estimate? A second sample would surely not give 405.02 again. Unbiasedness says only that there is no systematic tendency to underestimate or overestimate the truth. Could we plausibly get a sample mean of 500 or 300 on repeated samples? An estimate without an indication of its variability is of limited value.

Statistical confidence

Just as unbiasedness of an estimator concerns the center of its sampling distribution, questions about variation are answered by looking at the spread. The central limit theorem tells us that, if the entire population of large firms has mean μ and standard deviation σ, then in repeated samples of size 453 the sample mean $\bar{x}$ approximately follows the $N(\mu, \sigma/\sqrt{453})$ distribution. Suppose that the true standard deviation σ is equal to the sample standard deviation $s = 112.08$. Although this will not be exactly true, it will give quite accurate results for samples as large as 453. In the next chapter, we will learn how to proceed when σ is not known. But for now, we are more interested in statistical

FIGURE 6.4 In 95% of all samples, $\bar{x}$ lies within ± 10.54 of μ. So μ also lies within ± 10.54 of $\bar{x}$ in those samples.

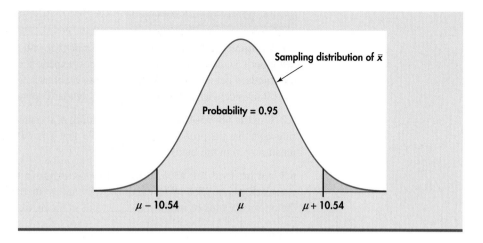

reasoning than in such details of our methods. In repeated sampling the sample mean $\bar{x}$ is approximately Normal, centered at the unknown population mean μ, with standard deviation

$$\sigma_{\bar{x}} = \frac{\$112.08}{\sqrt{453}} = \$5.27$$

Now we can talk about estimating μ. Consider this line of thought, which is illustrated by Figure 6.4:

- The 68–95–99.7 rule says that the probability is about 0.95 that $\bar{x}$ is within 10.54 (two standard deviations of $\bar{x}$) of the population mean monthly premium μ.
- To say that $\bar{x}$ lies within 10.54 of μ is the same as saying that μ is within 10.54 of $\bar{x}$.
- So 95% of all samples will capture the true μ in the interval from $\bar{x} - 10.54$ to $\bar{x} + 10.54$.

We have simply restated a fact about the sampling distribution of $\bar{x}$. *The language of statistical inference uses this fact about what would happen in the long run to express our confidence in the results of any one sample.* Our sample gave $\bar{x} = 405.02$. We say that we are *95% confident* that the unknown mean monthly premium for all large firms lies between

$$\bar{x} - 10.54 = 405.02 - 10.54 = \$394.48$$

and

$$\bar{x} + 10.54 = 405.02 + 10.54 = \$415.56$$

Be sure you understand the grounds for our confidence. There are only two possibilities:

1. The interval between $394.48 and $415.56 contains the true μ.
2. Our SRS was one of the few samples for which $\bar{x}$ is not within 10.54 of the true μ. Only 5% of all samples give such inaccurate results.

We cannot know whether our sample is one of the 95% for which the interval $\bar{x} \pm 10.54$ catches μ or one of the unlucky 5%. The statement that we are 95% confident that the unknown μ lies between $394.48 and $415.56 is shorthand for saying, "We arrived at these numbers by a process that gives correct results 95% of the time."

APPLY YOUR KNOWLEDGE

6.1 Company invoices. The mean amount μ for all the invoices for your company last month is not known. Based on your past experience, you are willing to assume that the standard deviation of invoice amounts is about \$220. If you take a random sample of 100 invoices, what is the value of the standard deviation for $\overline{x}$?

6.2 Use the 68–95–99.7 rule. In the setting of the previous exercise, the 68–95–99.7 rule says that the probability is about 0.95 that $\overline{x}$ is within _____ of the population mean μ. Fill in the blank.

6.3 An interval for 95% of the sample means. In the setting of the previous two exercises, about 95% of all samples will capture the true mean of all the invoices in the interval $\overline{x}$ plus or minus _____. Fill in the blank.

Confidence intervals

The interval of numbers between the values $\overline{x} \pm 10.54$ is called a *95% confidence interval* for μ. Like most confidence intervals we will meet, this one has the form

$$\text{estimate} \pm \text{margin of error}$$

margin of error

The estimate ($\overline{x}$ in this case) is our guess for the value of the unknown parameter. The **margin of error** of \$10.54 shows how precise we believe our guess is, based on the variability of the estimate. This is a 95% confidence interval because it catches the unknown μ in 95% of all possible samples.

> **Confidence Interval**
>
> A **level C confidence interval** for a parameter has two parts:
>
> - An interval calculated from the data, usually of the form
>
> $$\text{estimate} \pm \text{margin of error}$$
>
> - A **confidence level C,** which gives the probability that the interval will capture the true parameter value in repeated samples.

Figure 6.5 illustrates the behavior of 95% confidence intervals in repeated sampling. The center of each interval is at $\overline{x}$ and therefore varies from sample to sample. The sampling distribution of $\overline{x}$ appears at the top of the figure. The 95% confidence intervals $\overline{x} \pm 10.54$ from 25 SRSs appear below. The center $\overline{x}$ of each interval is marked by a dot. The arrows on either side of the dot span the confidence interval. All except one of the 25 intervals cover the true value of μ. In a very large number of samples, 95% of the confidence intervals would contain μ. You can choose the confidence level. Common practice is to choose 95%, but 90% and 99% are also popular.

The *Confidence Interval* applet at `www.whfreeman.com/psbe3e` animates Figure 6.5 and allows you to choose among several levels of confidence. This interactive applet is an excellent way to grasp the idea of a confidence interval.

APPLY YOUR KNOWLEDGE

6.4 80% confidence intervals. The idea of an 80% confidence interval is that the interval captures the true parameter value in 80% of all samples. That's not high enough

FIGURE 6.5 Twenty-five samples from the same population gave these 95% confidence intervals. In the long run, 95% of all samples give an interval that covers μ.

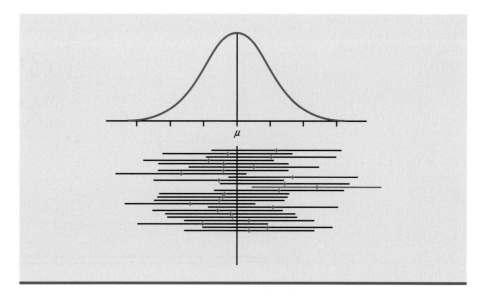

confidence for practical use, but 80% hits and 20% misses make it easy to see how a confidence interval behaves in repeated samples from the same population.

(a) Set the confidence level in the *Confidence Interval* applet to 80%. Click "Sample" to choose an SRS and calculate the confidence interval. Do this 10 times to simulate 10 SRSs with their 10 confidence intervals. How many of the 10 intervals captured the true mean μ? How many missed?

(b) You see that we can't predict whether the next sample will hit or miss. The confidence level, however, tells us what percent will hit in the long run. Reset the applet and click "Sample 50" to get the confidence intervals from 50 SRSs. How many hit? Keep clicking "Sample 50" and record the percent of hits among 100, 200, 300, 400, 500, 600, 700, 800, and 1000 SRSs. Even 1000 samples is not truly "the long run," but we expect the percent of hits in 1000 samples to be fairly close to the confidence level, 80%.

Confidence interval for a population mean

We use the sampling distribution of the sample mean $\bar{x}$ to construct a level C confidence interval for the mean μ of a population. We assume that the data are an SRS of size n. The sampling distribution is exactly $N(\mu, \sigma/\sqrt{n})$ when the population has the $N(\mu, \sigma)$ distribution. The central limit theorem says that this same sampling distribution is approximately correct for large samples whenever the population mean and standard deviation are μ and σ.

Our construction of a 95% confidence interval for the mean premium of large firms began by noting that any Normal distribution has probability about 0.95 OF TAKING A VALUE within ± 2 standard deviations of its mean. To construct a level C confidence interval, we first catch the central C area under a Normal curve. Because all Normal distributions are the same in the standard scale, we can obtain everything we need from the standard Normal curve. Figure 6.6 shows the relationship between the central area C and the points z^* that mark off this area. Values of z^* for many choices of C appear in the row labeled z^* in Table D at the back of the book. Here are the most important entries from that part of the table:

z^*:	1.645	1.960	2.576
C:	90%	95%	99%

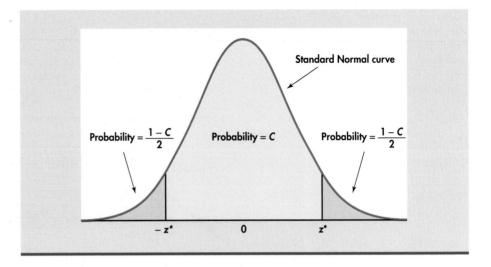

critical value   Values z^* that mark off specified areas are called **critical values** of the standard Normal distribution. Notice that for $C = 95\%$ the table gives $z^* = 1.960$. This is slightly more precise than the value $z^* = 2$ based on the 68–95–99.7 rule.

Any Normal curve has probability *C OF TAKING A VALUE* between the point z^* standard deviations below the mean and the point z^* above the mean, as Figure 6.6 reminds us. The sample mean $\overline{x}$ has the Normal distribution with mean μ and standard deviation $\sigma/\sqrt{n}$. So there is probability C that $\overline{x}$ lies between

$$\mu - z^* \frac{\sigma}{\sqrt{n}} \quad \text{and} \quad \mu + z^* \frac{\sigma}{\sqrt{n}}$$

This is exactly the same as saying that the unknown population mean μ lies between

$$\overline{x} - z^* \frac{\sigma}{\sqrt{n}} \quad \text{and} \quad \overline{x} + z^* \frac{\sigma}{\sqrt{n}}$$

So, the probability that the interval

$$\overline{x} \pm z^* \sigma/\sqrt{n}$$

contains μ is C. This is our confidence interval. The estimate of the unknown μ is $\overline{x}$, and the margin of error is $z^*\sigma/\sqrt{n}$.

Confidence Interval for a Population Mean

Choose an SRS of size n from a population having unknown mean μ and known standard deviation σ. A **level C confidence interval** for μ is

$$\overline{x} \pm z^* \frac{\sigma}{\sqrt{n}}$$

Here z^* is the **critical value** with area C between $-z^*$ and z^* under the standard Normal curve. The quantity

$$z^*\sigma/\sqrt{n}$$

is the **margin of error.** The interval is exact when the population distribution is Normal and is approximately correct when n is large in other cases.

Bankruptcy Attorney Fees In 2005, Congress passed the Bankruptcy Abuse Prevention and Consumer Protection Act. This act made significant changes to the administration of bankruptcy relief. The Government Accountability Office (GAO) was commissioned to study the effects of the bankruptcy reform law on consumers. Attorney fees in consumer bankruptcy cases were studied by the GAO and reported in their 2008 Bankruptcy Reform report.[2] In the period February to March 2007, there were 71,106 Chapter 7 consumer bankruptcy cases.*

To estimate legal fees for Chapter 7 consumer bankruptcy cases, the GAO conducted a nationwide random sample of Chapter 7 bankruptcy filings. Business bankruptcy cases and cases not involving attorneys were excluded. In the end, the GAO had $n = 292$ consumer cases with attorney involvement. The mean attorney fee in the sample is $\bar{x} = \$1078$ and the standard deviation is $s = \$592$.

The sample mean of $1078 provides us with a single-value estimate of the population mean. However, by itself, this number gives us no information to assess just how close we might be to the population mean. A confidence interval tells us a much more complete story. Let us turn to its construction for this Case Study.

EXAMPLE 6.6   Confidence Interval for Bankruptcy Attorney Fees

CASE 6.1

The study of Case 6.1 is associated with a fairly large sample size. This allows us to comfortably use s as the population σ here. Let us now find a 95% confidence interval for the mean attorney fee in Chapter 7 consumer bankruptcy cases.

For 95% confidence, we see from Table D that $z^* = 1.96$. The margin of error is

$$z^* \frac{\sigma}{\sqrt{n}} = 1.96 \frac{592}{\sqrt{292}} = 67.9$$

A 95% confidence interval for μ is therefore

$$\bar{x} \pm z^* \frac{\sigma}{\sqrt{n}} = 1078 \pm 67.9$$
$$= (1010.1, 1145.9)$$

We are 95% confident that the mean attorney fee is between $1010.1 and $1145.9.

In this example we used an estimate for σ based on a large sample. Sometimes we may know σ quite accurately from past experience with similar data. When we have few data, past experience may be better than estimation from our sample. In general, however, we can't act as if we know the population standard deviation σ. The next chapter presents methods that don't assume that we know σ.

APPLY YOUR KNOWLEDGE

6.5 Health insurance. For the 453 large firms in the Kaiser health coverage survey (Example 6.5), the mean monthly cost of the premium for single-coverage HMO plans per worker was $405.02. The sample standard deviation of premium costs was $112.08. Assume that the sample standard deviation can be used in place of the population

---

*Chapter 7 is the most common type of bankruptcy applied in the United States. Under Chapter 7, nonexempt assets of an individual or business are liquidated and distributed to creditors in accordance with the Bankruptcy Code.

standard deviation. We have already computed an approximate 95% confidence interval for the mean cost. Construct more precisely the 95% confidence interval for μ, the mean premium cost for all large firms.

6.6 Is the margin of error larger or smaller? In the setting of the previous exercise, would the margin of error for 99% confidence be larger or smaller? Verify your answer by performing the calculations.

How confidence intervals behave

The margin of error $z^*\sigma/\sqrt{n}$ for estimating the mean of a Normal population illustrates several important properties that are shared by all confidence intervals in common use. In particular, a higher confidence level increases z^* and so increases the margin of error for intervals based on the same data. We would like high confidence and a small margin of error, but improving one degrades the other. If a confidence interval's margin of error is too large, there are only three ways to reduce it:

- Use a lower level of confidence (smaller C, hence smaller z^*).
- Reduce σ.
- Increase the sample size (larger n).

The common choices of confidence level are 99%, 95%, and 90%. The critical values z^* for these levels are 2.576, 1.960, and 1.645, decreasing as the confidence level drops. Figure 6.6 makes it clear why z^* is smaller for lower confidence (smaller C). If n and σ are unchanged, settling for lower confidence will reduce the margin of error.

The standard deviation σ measures variation in the population. Think of the variation among individuals in the population as noise that obscures the average value μ. It is harder to pin down the mean μ of a highly variable population. This is why the margin of error of a confidence interval increases with σ. Sometimes we can reduce σ by carefully controlling the measurement process or by restricting our attention to only part of a large population.

Increasing the sample size n also reduces the margin of error. Suppose we want to cut the margin of error in half. The square root in the formula implies that we must have four times as many observations, not just twice as many.

EXAMPLE 6.7   How Confidence Intervals Behave

CASE 6.1

Suppose there were only 100 attorney-involved Chapter 7 bankruptcy cases in the GAO study, and that $\bar{x}$ and σ are unchanged from Example 6.6. The margin of error increases from 67.9 to

$$z^*\frac{\sigma}{\sqrt{n}} = 1.96\frac{592}{\sqrt{100}} = 116$$

A 95% confidence interval for μ is therefore

$$\bar{x} \pm z^*\frac{\sigma}{\sqrt{n}} = 1078 \pm 116$$
$$= (962, 1194)$$

Figure 6.7 illustrates how the margin of error increases for the smaller sample.

Suppose next that we demand 99% confidence for the mean attorney fees in Example 6.6, rather than 95%. Table D tells us that for 99% confidence, $z^* = 2.576$. The margin of error

FIGURE 6.7 95% confidence intervals for $n = 100$ and $n = 292$, for Example 6.7.

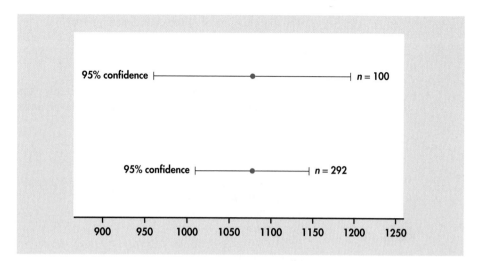

FIGURE 6.8 95% and 99% confidence intervals with $n = 292$ in both cases, for Example 6.7.

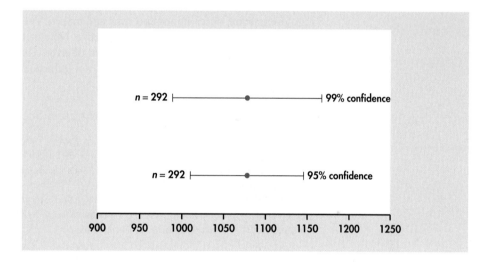

increases from 67.9 to

$$z^* \frac{\sigma}{\sqrt{n}} = 2.576 \frac{592}{\sqrt{292}}$$
$$= 89.2$$

The 99% confidence interval is therefore

$$\bar{x} \pm \text{margin of error} = 1078 \pm 89.2$$
$$= (988.8, 1167.2)$$

Requiring 99% confidence rather than 95% confidence has increased the margin of error from 67.9 to 89.2. Figure 6.8 compares the two intervals.

Choosing the sample size

A wise user of statistics never plans data collection without at the same time planning the inference. You can arrange to have both high confidence and a small margin of error.

The margin of error of the confidence interval $\bar{x} \pm z^*\sigma/\sqrt{n}$ for a Normal mean is

$$z^* \frac{\sigma}{\sqrt{n}}$$

To obtain a desired margin of error m, just set this expression equal to m, substitute the critical value z^* for your desired confidence level, and solve for the sample size n. Here is the result.

Sample Size for Desired Margin of Error

The confidence interval for a population mean will have a specified margin of error m when the sample size is

$$n = \left(\frac{z^*\sigma}{m}\right)^2$$

In practice, observations cost time and money. The sample size you calculate from this formula may turn out to be too expensive. Note that it is the *sample* size and not the population size that determines the margin of error. The size of the *population* (as long as it is much larger than the sample) does not influence the margin of error.

EXAMPLE 6.8   How Many Bankruptcy Cases Should Be Sampled?

CASE 6.1

In Example 6.6, we found that the margin of error was 67.9 for estimating the mean attorney fees in Chapter 7 consumer bankruptcy cases with $n = 292$ and 95% confidence. If the GAO is willing to settle for a margin of error of \$100 when it does a subsequent study, how many cases should the GAO sample?

For 95% confidence, Table D gives $z^* = 1.960$. Assume that the standard deviation in a future study will be approximately the same as the $\sigma = 592$ that the GAO obtained in its most recent study. For a margin of error of \$100 we have

$$n = \left(\frac{z^*\sigma}{m}\right)^2 = \left(\frac{1.96 \times 592}{100}\right)^2 = 134.6$$

Because 134 observations will give a slightly wider interval than desired and 135 observations will give a slightly narrower interval, we recommend that the GAO have 135 randomly selected cases in the study. With this choice, the estimate of mean attorney fees will be within \$100 of the true value with 95% confidence.

Always round *up* to the next higher whole number when using the formula for the sample size n. In practice we often calculate the margins of error corresponding to a range of values of n. We then decide what margin of error we can afford.

APPLY YOUR KNOWLEDGE

6.7 Starting salaries. You are planning a survey of starting salaries for recent business major graduates from your college. From a pilot study you estimate that the standard deviation is about \$9000. What sample size do you need to have a margin of error equal to \$500 with 95% confidence?

6.8 Will the sample size needed be larger or smaller? Suppose that in the setting of the previous exercise you are willing to settle for a margin of error of $1000. Will the required sample size be larger or smaller? Verify your answer by performing the calculations.

Some cautions

We have already seen that small margins of error and high confidence can require large numbers of observations. You should also be keenly aware that *any formula for inference is correct only in specific circumstances*. If the government required statistical procedures to carry warning labels like those on drugs, most inference methods would have long labels indeed. Our formula $\bar{x} \pm z^* \sigma/\sqrt{n}$ for estimating a Normal mean comes with the following list of warnings for the user:

- The data must be an SRS from the population. We are completely safe if we actually did a randomization and drew an SRS. The GAO study that lies behind Examples 6.6 to 6.8 is based on a random sample of Chapter 7 bankruptcies. We are not in great danger if the data can plausibly be thought of as independent observations from a population.

- The formula is not correct for probability sampling designs more complex than an SRS. Correct methods for other designs are available. We will not discuss confidence intervals based on multistage or stratified samples. If you plan such samples, be sure that you (or your statistical consultant) know how to carry out the inference you desire.

- There is no correct method for inference from data haphazardly collected with bias of unknown size. Fancy formulas cannot rescue badly produced data.

- Because $\bar{x}$ is not resistant, outliers can have a large effect on the confidence interval. You should search for outliers and try to correct them or justify their removal before computing the interval. If the outliers cannot be removed, ask your statistical consultant about procedures that are not sensitive to outliers.

- If the sample size is small and the population is not Normal, the true confidence level will be different from the value C used in computing the interval. Examine your data carefully for skewness and other signs of non-Normality. The interval relies only on the distribution of $\bar{x}$, which even for quite small sample sizes is much closer to Normal than that of the individual observations. When $n \geq 15$, the confidence level is not greatly disturbed by non-Normal populations unless extreme outliers or quite strong skewness are present. We will discuss this issue in more detail in the next chapter.

- You must know the standard deviation σ of the population. In practice, this requirement is often unrealistic. Therefore, the interval $\bar{x} \pm z^* \sigma/\sqrt{n}$ is of limited application in statistical practice. We will learn in the next chapter what to do when σ is unknown. If, however, the sample is large, the sample standard deviation s will be close to the unknown σ. The formula $\bar{x} \pm z^* s/\sqrt{n}$ is then an approximate confidence interval for μ.

The most important caution concerning confidence intervals is a consequence of the first of these warnings. *The margin of error in a confidence interval covers only random sampling errors.* The margin of error is obtained from the sampling distribution and indicates how much error can be expected because of chance variation in randomized data production. Practical difficulties such as undercoverage and nonresponse in a sample survey cause additional errors. These errors can be larger than the random sampling error, particularly when the sample size is large. Remember this unpleasant fact when reading the results of an opinion poll. The practical conduct of a survey influences the trustworthiness of its results in ways that are not included in the announced margin of error.

Every inference procedure that we will meet has its own list of warnings. Because many of the warnings are similar to those above, we will not print the full warning label each time. Using the mathematics of probability, it is easy to state conditions under which a method of inference is exactly correct. These conditions are *never* fully met in practice. For example, no population is exactly Normal. Deciding when a statistical procedure should be used in practice often requires judgment assisted by exploratory analysis of the data. Mathematical facts are therefore only a part of statistics. The difference between statistics and mathematics can be stated thus: mathematical theorems are true; statistical methods are often effective when used with skill.

Finally, you should understand what statistical confidence does not say. We are 95% confident that the mean monthly HMO premium in Example 6.5 lies between $394.48 and $415.56. This says that these numbers were calculated by a method that gives correct results in 95% of all possible samples. It does *not* say that the probability is 95% that the true mean falls between $394.48 and $415.56. No randomness remains after we draw a particular sample and get from it a particular interval. The true mean either is or is not between $394.48 and $415.56. Probability in its interpretation as long-term relative frequency makes no sense in this situation. The probability calculations of standard statistical inference describe how often the *method* gives correct answers.

APPLY YOUR KNOWLEDGE

6.9 Confidence interval mistakes and misunderstandings. Consider the following scenario. Suppose 100 randomly selected customers were asked to rate a particular service they recently had on a 0 to 10 scale. The sample mean ($\overline{x}$) was found to be 7.3. Assume that the population standard deviation is known to be $\sigma = 0.8$.

(a) A customer relations manager computes the 95% confidence interval for the mean satisfaction as $7.3 \pm 1.96(0.8)$. A common mistake was made here. What is it?

(b) The manager fixes the mistake in part (a). After constructing the correct interval, the manager states, "I am 95% confident that the sample mean falls between 7.1432 and 7.4568." What misinterpretation is the manager making?

(c) The manager states, "95% of customers in the population rate our service between 7.1432 and 7.4568." What misinterpretation is the manager making?

(d) The manager states, "Because our sample size is so large, the population of customer ratings is Normal." In what way is the manager confused?

BEYOND THE BASICS: The Bootstrap

Confidence intervals are based on sampling distributions. The interval $\overline{x} \pm z^*\sigma/\sqrt{n}$ follows from the fact that the sampling distribution of $\overline{x}$ is $N(\mu, \sigma/\sqrt{n})$ when the data are an SRS from a $N(\mu, \sigma)$ population. The central limit theorem says that this sampling distribution is also approximately right for large samples from non-Normal populations.

bootstrap

What if the population is clearly non-Normal and we have only a small sample? Then we do not know what the sampling distribution of $\overline{x}$ looks like. The **bootstrap** is a procedure for approximating sampling distributions when theory cannot tell us their shape.[3]

resample

The basic idea is to act as if our sample were the population. We take many samples from it. Each of these is called a **resample.** For each resample, we calculate the mean $\overline{x}$.

We get different results from different resamples because we sample *with replacement*. That is, an observation in the original sample can appear more than once in a resample.

The HMO premium data described in Example 6.5 are somewhat skewed toward high values. The mean $405.02 is larger than the median, $390.20. There are two firms with HMO premiums of $1114.71 and $1185 that could be considered outliers. Because we have no reason to suspect that these firms are qualitatively different from the other firms in the sample, we are reluctant to discard them. Rather, we will reexamine our inference with a method that is not sensitive to the Normality assumption.

If we had 1000 SRSs from the population of firms, the distribution of the 1000 values of $\overline{x}$ would be close to the sampling distribution of $\overline{x}$. We have only one sample. The bootstrap idea says: take 1000 resamples of size 453 (with replacement) from our one sample of 453 firms. For each resample, compute the mean. Treat the distribution of $\overline{x}$'s from the 1000 resamples as if it were the sampling distribution. Because the sample is like the population, taking resamples from the sample is similar to taking samples from the population.

The bootstrap is practical only when you can use a computer to take 1000 or more resamples quickly. It is an example of how fast and cheap computing has changed the way we do statistics.

EXAMPLE 6.9   Use the Bootstrap to Check the Interval

In the discussion of Example 6.5 we found a 95% confidence interval for the mean HMO premium using the traditional statistical procedure based on the middle 95% of a Normal distribution. The interval was (394.48, 415.56). Let's check to see if we made a serious error. The middle 95% of the 1000 $\overline{x}$'s from 1000 resamples runs from 394.92 to 415.74. Try it again with 1000 new resamples: we get the interval (394.76, 415.52). The two bootstrap estimates are very close to each other and also very close to the standard interval that assumes Normality. The standard interval is reasonably accurate for these data.

If the bootstrap method applied to this problem gave very different results, we would need to reconsider the analysis. Obtaining similar results from different methods gives us confidence that the answers we seek are coming from the data, not from the particular method that we used.

SECTION 6.1 Summary

- The purpose of a **confidence interval** is to estimate an unknown parameter with an indication of how accurate the estimate is and of how confident we are that the result is correct.

- Any confidence interval has two parts: an interval computed from the data and a confidence level. The interval often has the form

$$\text{estimate} \pm \text{margin of error}$$

- The **confidence level** states the probability that the method will give a correct answer. That is, if you use 95% confidence intervals often, in the long run 95% of your intervals will contain the true parameter value. When you apply the method once, you do not know if your interval gave a correct value (this happens 95% of the time) or not (this happens 5% of the time).

- A level C confidence interval for the mean μ of a Normal population with known standard deviation σ, based on an SRS of size n, is given by

$$\bar{x} \pm z^* \frac{\sigma}{\sqrt{n}}$$

Here z^* is a **critical value** obtained from the bottom row in Table D. The probability is C that a standard Normal random variable takes a value between $-z^*$ and z^*.

- The **margin of error** in the confidence interval for μ is

$$z^* \frac{\sigma}{\sqrt{n}}$$

- Other things being equal, the margin of error of a confidence interval decreases as

  the confidence level C decreases,

  the population standard deviation σ decreases, or

  the sample size n increases.

- The sample size required to obtain a confidence interval of specified margin of error m for a Normal mean is

$$n = \left(\frac{z^* \sigma}{m} \right)^2$$

where z^* is the critical value for the desired level of confidence.

- A specific confidence interval formula is correct only under specific conditions. The most important conditions concern the method used to produce the data. Other factors such as the form of the population distribution may also be important.

SECTION 6.1 Exercises

For Exercises 6.1 to 6.3, see page 338; for 6.4, see pages 338–339; for 6.5 and 6.6, see pages 341–342; for 6.7 and 6.8, see pages 344–345; and for 6.9, see page 346.

6.10 Margin of error and the confidence interval. A study based on a sample of size 25 reported a mean of 76 with a margin of error of 12 for 95% confidence.
(a) Give the 95% confidence interval.
(b) If you wanted 99% confidence for the same study, would your margin of error be greater than, equal to, or less than 12? Explain your answer.

6.11 Change the sample size. Suppose that the sample mean is 50 and the standard deviation is assumed to be 5. Make a diagram similar to Figure 6.7 (page 343) that illustrates the effect of sample size on the width of a 95% interval. Use the following sample sizes: 10, 20, 40, and 100. Summarize what the diagram shows.

6.12 Change the confidence. A study with 25 observations gave a mean of 70. Assume that the standard deviation is 15. Make a diagram similar to Figure 6.8 (page 343) that illustrates the effect of the confidence level on the width of the interval. Use 80%, 90%, 95%, and 99%. Summarize what the diagram shows.

6.13 Populations sampled and margins of error. Consider the following two scenarios. (A) Take a simple random sample of 100 sophomore students at your college or university. (B) Take a simple random sample 100 sophomore students in your major at your college or university. For each of these samples you will record the amount spent on textbooks used for classes during the fall semester. Which sample should have the smaller margin of error for 95% confidence? Explain your answer.

6.14 Reporting margins of error. A *Wall Street Journal* article ("Home Construction at Record Slow Pace," January 23, 2009) reported Commerce Department estimates of changes in the construction industry:

Construction of new single- and multifamily homes sank 15.5% in December from the previous month to a seasonally adjusted annual rate of 550,000 units—the slowest pace since monthly records began in 1959.

If we turn to the original Commerce Department report (released on January 22, 2009), we read:

Privately-owned housing starts in December were at a seasonally adjusted annual rate of 550,000. This is 15.5 percent (9.3%) below the revised November estimate of 651,000.

(a) The 9.3% figure is the margin of error based on a 90% level of confidence. Given that fact, what is the 90% confidence interval for the percent change in housing starts from November to December? How do you feel about the media's report of only 15.5% in light of the confidence interval?

(b) The Commerce Department report also states:

Privately-owned housing completions in December were at a seasonally adjusted annual rate of 1,015,000. This is 5.2 percent (11.9%) below the revised November estimate of 1,071,000.

What is the 90% confidence interval for the percent change in housing completions from November to December?

(c) As a followup to part (b), explain why a credible media report should state: "The Commerce Department has no evidence that privately-owned housing completions rose or fell in December from the previous month" rather than "Privately-owned housing completions sank 5.2% in December from the previous month."

6.15 In the extremes. As suggested in our discussions, 90%, 95%, and 99% are probably the most common confidence levels chosen in practice.

(a) In general, what would be a 100% confidence interval for the mean μ? Explain why such an interval is of no practical use.

(b) What would be a 0% confidence interval? Explain why it makes sense that the resulting interval provides you with 0% confidence.

6.16 Fuel efficiency. Computers in some vehicles calculate various quantities related to performance. One of these is the fuel efficiency, or gas mileage, usually expressed as miles per gallon (mpg). One of the authors of this book conducted an experiment with his 2006 Toyota Highlander Hybrid by randomly recording mpg readings shown on the vehicle computer while the car was set to 60 miles per hour by cruise control. Here are the mpg values from the experiment:

37.2	21.0	17.4	24.9	27.0	36.9	38.8	35.3	32.3	23.9
19.0	26.1	25.8	41.4	34.4	32.5	25.3	26.5	28.2	22.1

Suppose that the standard deviation of the population of mpg readings of this vehicle is known to be $\sigma = 6.5$ mpg. **MILEAGE**

(a) What is $\sigma_{\bar{x}}$, the standard deviation of $\bar{x}$?

(b) Based on a 95% confidence level, what is the margin of error for the mean estimate?

(c) Given the margin of error computed in part (b), give a 95% confidence interval for μ, the mean highway mpg for this vehicle. The vehicle sticker information for the 2006 Toyota Highlander Hybrid states a highway average of 27 mpg. Are the results of this experiment consistent with the vehicle sticker?

6.17 Wanting another confidence interval. Suppose you were told that the 90% confidence interval for the mean μ based on some known σ is (329.87, 356.46). You, however, want a 95% confidence interval. With only the information provided here determine the 95% confidence interval.

6.18 Confidence intervals for informal sector. The International Labor Organization (an agency of the United Nations) coined the term "informal sector" to refer to economic activities that are not regulated by labor or taxation laws and are not included in gross domestic product estimates. This sector is a major component underlying the economies of developing countries. In January 2009, the National Statistics Office of the Republic of the Philippines (www.census.gov.ph) released a report on various estimates related to the informal sector in the Philippines. The report estimated the number of male and female workers in the informal sector:

	Estimate	Standard error	Lower	Upper
Male	6,894,875	85,594	6,727,040	7,062,711
Female	3,559,392	53,974	?	?

(a) The "Lower" and "Upper" headers signify lower and upper confidence interval limits. As will be noted in Chapter 7, the term "standard error" stands for the estimated standard deviation of the statistic used to obtain the estimate. In this exercise, assume the standard error to be equal to the standard deviation of the statistic used. The limits were constructed based on the standard Normal distribution in a manner similar to that presented in Section 6.1. Using the numbers for males, determine the confidence level used to construct the confidence interval limits.

(b) Using the same level of confidence determined in part (a), determine the lower and upper limits for the female population.

6.19 Survey response and margin of error. Suppose that a business conducts a marketing survey. As is often done, the survey is conducted by telephone. As it turns out, the business was only able to illicit responses from less than 10% of the randomly chosen customers. The low response rate is attributable to many factors, including caller ID screening. Undaunted, the marketing manager was pleased with the sample results because the margin of error was quite small, and thus the manager felt that the business had a good sense of the customers' perceptions on various issues. Do you think the small margin of error is a good measure of the accuracy of the survey's results? Explain.

6.20 Convert to metric. In Exercise 6.16 you found an estimate with a margin of error for the average fuel efficiency expressed in miles per gallon (mpg). Convert your estimate and margin of error to kilometers per liter (kpl). To change mpg to kpl, multiply by 0.4251.

6.21 What is the cost? In Exercise 6.16 you found an estimate with a margin of error for the fuel efficiency expressed in miles per gallon. Suppose that fuel costs $3.10 per gallon. Find the estimate and margin of error for fuel efficiency in terms of miles per dollar. To convert miles per gallon to miles per dollar, divide miles per gallon by the cost in dollars per gallon.

6.22 Apartment rental rates. You want to rent an unfurnished one-bedroom apartment for next semester. The mean monthly rent for a random sample of 10 apartments advertised in the local newspaper is $640. Assume that the standard deviation is $90.

Find a 95% confidence interval for the mean monthly rent for unfurnished one-bedroom apartments available for rent in this community.

6.23 Study habits. A questionnaire about study habits was given to a random sample of students taking a large introductory statistics class. The sample of 35 students reported that they spent an average of 115 minutes per week studying statistics. Assume that the standard deviation is 40 minutes.
(a) Give a 95% confidence interval for the mean time spent studying statistics by students in this class.
(b) Is it true that 95% of the students in the class have weekly study times that lie in the interval you found in part (a)? Explain your answer.

6.24 Clothing for runners. Your company sells exercise clothing and equipment on the Internet. To design the clothing, you collect data on the physical characteristics of your different types of customers. Here are the weights (in kilograms) for a sample of 24 male runners. Assume that these runners can be viewed as a random sample of your potential male customers. Suppose also that the standard deviation of the population is known to be $\sigma = 4.5$ kg. 🔵 CLOTHING

68.7	61.8	63.2	53.1	62.3	59.7	55.4	58.9	60.9	69.2	63.7	67.8
65.6	65.5	56.0	57.8	66.0	62.9	53.6	65.0	55.8	60.4	69.3	62.1

(a) What is $\sigma_{\bar{x}}$, the standard deviation of $\bar{x}$?
(b) Give a 95% confidence interval for μ, the mean of the population from which the sample is drawn.
(c) Will the interval contain the weights of approximately 95% of similar runners? Explain your answer.

6.25 Pounds versus kilograms. Suppose that the weights of the runners in Exercise 6.24 were recorded in pounds rather than kilograms. Use your answers to Exercise 6.24, and the fact that 1 kilogram equals 2.2 pounds, to answer these questions.
(a) What is the mean weight of these runners?
(b) What is the standard deviation of the mean weight?
(c) Give a 95% confidence interval for the mean weight of the population of runners that these runners represent.

6.26 99% versus 95% confidence interval. Find a 99% confidence interval for the mean weight μ of the population of male runners in Exercise 6.24. Is the 99% confidence interval wider or narrower than the 95% interval found in Exercise 6.24? Explain in plain language why this is true.

6.27 Hotel managers. In a study of the career paths of hotel general managers, questionnaires were sent to an SRS of 160 hotels belonging to major U.S. hotel chains. There were 114 responses. The average time these 114 general managers had spent with their current company was 11.8 years. Give a 99% confidence interval for the mean number of years general managers of major-chain hotels have spent with their current company. (Take it as known that the standard deviation of time with the company for all general managers is 3.2 years.)

6.28 Supermarket shoppers. A marketing consultant observed 40 consecutive shoppers at a supermarket. One variable of interest was how much each shopper spent in the store. Here are the data (in dollars), arranged in increasing order: 🔵 SHOPPERS

5.32	8.88	9.26	10.81	12.69	15.23	15.62	17.00
17.35	18.43	19.50	19.54	20.59	22.22	23.04	24.47
25.13	26.24	26.26	27.65	28.08	28.38	32.03	34.98
37.37	38.64	39.16	41.02	42.97	44.67	45.40	46.69
49.39	52.75	54.80	59.07	60.22	84.36	85.77	94.38

Assume that the standard deviation is $21.00. Find a 95% confidence interval for the mean amount spent by shoppers in similar circumstances.

6.29 Supermarket shoppers. Suppose that you want to perform a study similar to the survey of supermarket shoppers described in Exercise 6.28. If you want the margin of error to be about $5.00, how many shoppers would you need in your sample? (Use $22.00 for the standard deviation in your calculations.)

6.30 Hotel managers. How large a sample of the hotel managers in Exercise 6.27 would be needed to estimate the mean μ within ± 1 year with 99% confidence?

6.31 Like your job? A Gallup Poll asked 1001 adult workers about their job satisfaction. One question was "All in all, which best describes how you feel about your job?" The possible answers were "love job," "like job," "dislike job," and "hate job." Fifty-nine percent of the sample responded that they liked their job. Material provided with the results of the poll noted:

> *Results are based on telephone interviews with 1,001 national adults, aged 18 and older, conducted Aug. 8–11, 2005. For results based on the total sample of national adults, one can say with 95% confidence that the maximum margin of sampling error is 3 percentage points.*

The Gallup Poll uses a complex multistage sample design, but the sample percent has approximately a Normal sampling distribution.
(a) The announced poll result was 59% $\pm$ 3%. Can we be certain that the true population percent falls in this interval?
(b) Explain to someone who knows no statistics what the announced result 59% $\pm$ 3% means.
(c) This confidence interval has the same form we met earlier:

$$\text{estimate} \pm z^* \sigma_{\text{estimate}}$$

What is the standard deviation σ_{estimate} of the estimated percent?
(d) Does the announced margin of error include errors due to practical problems such as undercoverage and nonresponse?

6.32 Like or love your job? Refer to the previous exercise. In the same poll, 32% of the respondents said that they loved their job.

(a) Combine the respondents who like their job with those who love their job. Give the percent who either love their job or like their job.

(b) Assuming that the same margin of error applies to this percent, give the 95% confidence interval.

6.33 Other variables. Refer to the previous two exercises. Gallup polls such as this often present the results for subgroups of respondents such as females and males. Suggest subgroups for this job satisfaction question. Give reasons for your answer.

6.34 More than one confidence interval. As we prepare to take a sample and compute a 95% confidence interval, we know that the probability that the interval we compute will cover the parameter is 0.95. That's the meaning of 95% confidence. If we use several such intervals, however, our confidence that *all* give correct results is less than 95%.

Suppose we are interested in confidence intervals for the median household incomes for three states. We compute a 95% interval for each of the three, based on independent samples in the three states.

(a) What is the probability that all three intervals cover the true median incomes? This probability (expressed as a percent) is our overall confidence level for the three simultaneous statements.

(b) What is the probability that at least two of the three intervals cover the true median incomes?

6.35 An election poll. A newspaper headline describing a poll of registered voters taken two weeks before a recent election reads "Ringel leads with 52%." The accompanying article says that the margin of error is 3% with 95% confidence.

(a) Explain in plain language to someone who knows no statistics what "95% confidence" means.

(b) The poll shows Ringel leading. But the newspaper article says that the election was too close to call. Explain why.

6.36 Manager trainee wages. A newspaper ad for a manager trainee position contained the statement "Our manager trainees have a first-year earnings average of $20,000 to $30,000." Do you think that the ad is describing a confidence interval? Explain your answer.

6.37 Talk show poll. A radio talk show invites listeners to enter a dispute about a proposed pay increase for city council members. "What yearly pay do you think council members should get? Call us with your number." In all, 958 people call. The station calculates the 95% confidence interval for the mean pay μ that all citizens would propose for council members to be $9669 to $9811. Is this result trustworthy? Explain your answer.

6.38 An outlier. Exercise 6.24 gives the weights of a sample of 24 male runners. Suppose that the sample actually contained 25 runners. The extra runner claimed to weigh 92.3 kg.

(a) Compute the 95% confidence interval for the mean with this observation included.

(b) Would you report the interval you calculated in part (a) of this question or the interval you calculated in Exercise 6.24? Explain the reasons for your choice.

6.2 Tests of Significance

Confidence intervals are one of the two most common types of formal statistical inference. They are appropriate when our goal is to estimate a population parameter. The second common type of inference is directed at a different goal: to assess the evidence provided by the data in favor of some claim about the population.

The reasoning of significance tests

A significance test is a formal procedure for comparing observed data with a hypothesis whose truth we want to assess. The hypothesis is a statement about the parameters in a population or model. The results of a test are expressed in terms of a probability that measures how well the data and the hypothesis agree. We use the following Case Study and subsequent examples to illustrate these ideas.

CASE 6.2

Fill the Bottles Perhaps one of the most common applications of hypothesis testing of the mean is the quality control problem of assessing whether or not the underlying population mean is on "target." Consider the case of Bestea Bottlers. One of Bestea's most popular products is the 500 milliliters (ml) bottle of raspberry-flavored green tea. There is some variation from bottle to bottle because the filling machinery is not perfectly precise. Bestea has two concerns: whether there is a problem of underfilling (customers are then being shortchanged, which is a form of false advertising), or whether there is a problem of overfilling (unnecessary cost to the bottler).

Notice that in Case 6.2, there is an intimate understanding of what is important to be discovered. In particular, is the population mean too high or too low relative to a desired level? With an understanding of what role the data play in the discovery process, we are able to formulate appropriate hypotheses. If the bottler were concerned only about the possible underfilling of bottles, then the hypotheses of interest would change. Let us proceed with the question of whether the bottling process is either underfilling or overfilling bottles.

EXAMPLE 6.10   Are the Bottles Being Filled as Advertised?

CASE 6.2

BOTTLEFILL

The filling process is not new to Bestea. Data on past production shows that the distribution of the contents is Normal with standard deviation $\sigma = 2$ ml. To assess the state of the bottling process, an inspector measures the contents of 10 randomly selected bottles. The results are

502.9   499.8   503.2   502.8   500.9   503.9   498.2   502.5   503.8   501.4

For this sample of 10 observations, the mean content ($\bar{x}$) is 501.94 ml. Is a sample mean of 501.94 ml convincing evidence that the mean fill of all bottles produced by the current process differs from the advertised 500 ml?

If we lack proper statistical thinking, this is a juncture to knee-jerk in one of two ways. One knee-jerk response is to say, "The mean of the bottles sampled is not 500 ml so the process is not filling the bottles at a mean level of 500 ml." Another knee-jerk response is to say, "The difference of 1.94 ml is small relative to the 500 ml baseline so there is nothing unusual going on here." What these responses fail to do is consider the underlying variability of the population, which ultimately implies a failure to consider the sampling variability of the mean statistic.

To proceed, we step back and suppose that the truth about the population is $\mu = 500$ ml. This is our hypothesis. We now ask, "*What is the probability of observing a sample mean at least as far from 500 ml as 1.94 ml?*" Taking into account sampling variability, the answer is 0.002. (You will learn how to find this probability in Example 6.15.) Because this probability is so small, we see that the sample mean $\bar{x} = 501.94$ is incompatible with a population mean of $\mu = 500$. With this evidence, we are led to the conclusion that the bottling process does not have a mean of $\mu = 500$ ml. This suggests the need for corrective action.

What are the key steps in Example 6.10? We want to know if the sample gives good evidence that the mean fill of all bottles produced from the current process differs from the advertised 500 ml. Suppose that the mean fill is on target, that is, that $\mu = 500$. Compare the sample outcome $\bar{x} = 501.94$ with the on-target population mean $\mu = 500$. The comparison takes the form of a probability: if $\mu = 500$, then a sample of size 10 will have a sample mean at least as far from 500 as 1.94 with probability equal to 0.002.

The fact that the calculated probability is so small leads us to believe that the mean fill amount is *not* in fact 500. Here's the logic. We need to envision two possibilities:

1. μ is equal to 500 and we have observed something very unusual given this reality, or

2. μ is some value other than 500 that makes the observed data more probable.

We calculated a probability assuming that the first possibility is true, that is, $\mu = 500$. That probability guides our final choice. If the probability is very small, observing our data result is highly unlikely if the mean is indeed equal to 500. This leads us to doubt that $\mu = 500$ and to accept a more plausible explanation for the data result: namely, we conclude that the mean is not in fact 500. It should be emphasized that to "conclude" does not mean we know the truth or that we are right. There is always a chance that *our conclusion is wrong*. Always bear in mind that when dealing with data, there are no guarantees. We now turn to an example in which the data suggest a different conclusion.

EXAMPLE 6.11   Is It Right Now?

CASE 6.2

In Example 6.10, sample evidence suggested that the mean fill amount was not at the desired target of 500 ml. In particular, it appeared that the process was overfilling the bottles on average. In response, Bestea's production staff made adjustments to the process and collected a sample of 10 bottles from the "corrected" process. From this sample, we find $\bar{x} = 499.47$ ml. (We assume that the standard deviation is the same, $\sigma = 2$ ml.) In this case, the sample mean is less than 500 ml—to be exact, 0.53 ml less than 500 ml. Did the production staff overreact and adjust the mean level too low? We need to ask a similar question as in Example 6.10. In particular, what is the probability that the mean of a sample of size $n = 10$ from a Normal population with mean $\mu = 500$ and standard deviation $\sigma = 2$ is as far away or farther away from 500 as 0.53? The answer is 0.402. A sample result this far from 500 would happen just by chance in 40.2% of samples from a population having a true mean of 500. An outcome that could so easily happen just by chance is not convincing evidence that the population mean differs from 500. Bestea can feel confident that the corrections made to the filling process fixed the problem.

The probabilities of Examples 6.10 and 6.11 measure the compatibility of the outcomes $\bar{x} = 501.94$ and $\bar{x} = 499.47$ with the hypothesis that $\mu = 500$. Figure 6.9 compares the two results graphically. The Normal curve is the sampling distribution of $\bar{x}$ when $\mu = 500$. You can see that we are not particularly surprised to observe $\bar{x} = 499.47$, but $\bar{x} = 501.94$ is an extreme outcome. Herein lies the core reasoning of statistical tests: *a sample outcome that would be extreme if a hypothesis were true is evidence that the hypothesis may not be true.* Now we proceed to the formal details of testing.

Stating hypotheses

In Example 6.10, we asked whether the fill data are plausible if, in fact, the true mean fill amount for all bottles (μ) is 500. That is, we ask if the data provide evidence *against* the claim that the population mean is 500. The first step in a test of significance is to state a claim that we will try to find evidence against.

> **Null Hypothesis H_0**
>
> The statement being tested in a test of significance is called the **null hypothesis.** The test of significance is designed to assess the strength of the evidence against the null hypothesis. Usually the null hypothesis is a statement of "no effect" or "no difference." We abbreviate "null hypothesis" as H_0.

FIGURE 6.9 The mean fill amount for a sample of 10 bottles will have this sampling distribution if the mean for all bottles is $\mu = 500$. A sample mean $\bar{x} = 501.94$ is so far out on the curve that it would rarely happen just by chance.

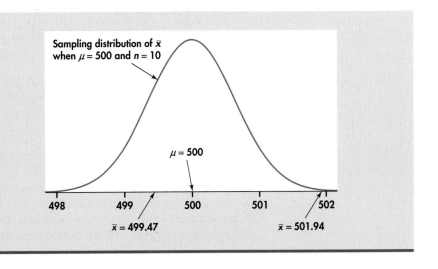

A null hypothesis is a statement about a population, expressed in terms of some parameter or parameters. The null hypothesis for Examples 6.10 and 6.11 is

$$H_0: \mu = 500$$

alternative hypothesis

It is convenient also to give a name to the statement that we hope or suspect is true instead of H_0. This is called the **alternative hypothesis** and is abbreviated as H_a. In Examples 6.10 and 6.11, the alternative hypothesis states that the mean fill amount is not 500. We write this as

$$H_a: \mu \neq 500$$

Hypotheses always refer to some population or model, not to a particular outcome. For this reason, we always state H_0 and H_a in terms of population parameters.

Because H_a expresses the effect that we hope to find evidence *for*, we often begin with H_a and then set up H_0 as the statement that the hoped-for effect is not present. Stating H_a is not always straightforward. It is not always clear, in particular, whether H_a

one-sided and two-sided

should be **one-sided** or **two-sided.**

The alternative $H_a: \mu \neq 500$ in the bottle-filling examples is two-sided. In both examples, we simply asked if mean fill amount is off target. The process can be off target in that it fills too much or too little on average, so we include both possibilities in the alternative hypothesis. Here is a setting in which a one-sided alternative is appropriate.

EXAMPLE 6.12   Have We Reduced Processing Time?

Your company hopes to reduce the mean time μ required to process customer orders. At present, this mean is 3.8 days. You study the process and eliminate some unnecessary steps. Did you succeed in decreasing the average processing time? You hope to show that the mean is now less than 3.8 days, so the alternative hypothesis is one-sided, $H_a: \mu < 3.8$. The null hypothesis is as usual the "no-change" value, $H_0: \mu = 3.8$ days.

The alternative hypothesis should express the hopes or suspicions we had in mind when we decided to collect the data. It is cheating to first look at the data and then frame H_a to fit what the data show. If you do not have a specific direction firmly in mind in advance, use a two-sided alternative. In fact, some users of statistics argue that we should *always* work with the two-sided alternative.

The choice of the hypotheses in Example 6.12 as

$$H_0: \mu = 3.8$$
$$H_a: \mu < 3.8$$

deserves a final comment. We do not expect that elimination of steps in order processing would actually increase the processing time. However, we can allow for an increase by including this case in the null hypothesis. Then we would write

$$H_0: \mu \geq 3.8$$
$$H_a: \mu < 3.8$$

This statement is logically satisfying because the hypotheses account for all possible values of μ. However, only the parameter value in H_0 that is closest to H_a influences the form of the test in all common significance-testing situations. Think of it this way: if the data lead us away from $\mu = 3.8$ to believing that $\mu < 3.8$, then the data would certainly lead us away from believing that $\mu > 3.8$ since this involves values of μ that are in the opposite direction to which the data are pointing. We will therefore take H_0 to be the simpler statement that the parameter has a specific value, in this case H_0: $\mu = 3.8$.

APPLY YOUR KNOWLEDGE

6.39 Customer feedback. Feedback from your customers shows that many think it takes too long to fill out the online order form for your products. You redesign the form and plan a survey of customers to determine whether or not they think that the new form is actually an improvement. Sampled customers will respond using a five-point scale: -2 if the new form takes much less time than the old form; -1 if the new form takes a little less time; 0 if the new form takes about the same time; $+1$ if the new form takes a little more time; and $+2$ if the new form takes much more time. The mean response from the sample is $\bar{x}$, and the mean response for all of your customers is μ. State null and alternative hypotheses that provide a framework for examining whether or not the new form is an improvement.

6.40 DEXA scanners. A dual-energy X-ray absorptiometry (DEXA) scanner is used to measure bone mineral density for people who may be at risk for osteoporosis. Customers want assurance that your company's latest-model DEXA scanner is accurate. You therefore supply an object called a "phantom" that has known mineral density $\mu = 1.4$ grams per square centimeter. A user scans the phantom 10 times and compares the sample mean reading $\bar{x}$ with the theoretical mean μ using a significance test. State the null and alternative hypotheses for this test.

Test statistics

We will learn the form of significance tests in a number of common situations. Here are some principles that apply to most tests and that help in understanding the form of tests:

- The test is based on a statistic that estimates the parameter appearing in the hypotheses. Usually this is the same estimate we would use in a confidence interval for the parameter. When H_0 is true, we expect the estimate to take a value near the parameter value specified by H_0.

- Values of the estimate far from the parameter value specified by H_0 give evidence against H_0. The alternative hypothesis determines which directions count against H_0.

EXAMPLE 6.13   Bottle Fill Amount: The Hypotheses

CASE 6.2

For Example 6.10, the hypotheses are stated in terms of the mean fill amount for all bottles:

$$H_0: \mu = 500$$
$$H_a: \mu \neq 500$$

The estimate of μ is the sample mean $\bar{x}$. Because H_a is two-sided, values of $\bar{x}$ far from 500 on either the low or the high side count as evidence against the null hypothesis.

test statistic   A **test statistic** measures compatibility between the null hypothesis and the data. Many test statistics can be thought of as a distance between a sample estimate of a parameter and the value of the parameter specified by the null hypothesis.

EXAMPLE 6.14   Bottle Fill Amount: The Test Statistic

CASE 6.2

For Example 6.10, the null hypothesis is $H_0: \mu = 500$, and a sample gave $\bar{x} = 501.94$. The test statistic for this problem is the standardized version of $\bar{x}$:

$$z = \frac{\bar{x} - \mu}{\sigma/\sqrt{n}}$$

This statistic is the distance between the sample mean and the hypothesized population mean in the standard scale of z-scores. In this example,

$$z = \frac{501.94 - 500}{2/\sqrt{10}} = 3.07$$

Even without a formal probability calculation, we know that standard score $z = 3.07$ lies far out on a Normal curve, which, in turn, suggests incompatibility of the observed sample result with the null hypothesis.

As stated in Example 6.10, past production shows that the fill amounts of the individual bottles follow the Normal distribution. As a result, we know that $\bar{x}$ follows exactly the Normal distribution. In applications where the distribution of the population is unknown, we can lean on the central limit theorem to ensure that $\bar{x}$ follows at least approximately the Normal distribution for sufficiently large sample sizes. So if the null hypothesis $H_0: \mu = 500$ is true, the test statistic z of Example 6.14 is a random variable that has the standard Normal distribution $N(0, 1)$. To move from the test statistic z to a probability, we must do Normal probability calculations. Review Sections 1.3 and 4.4 if you need to refresh your skills.

P-values

A test of significance assesses the evidence against the null hypothesis and provides a numerical summary of this evidence in terms of a probability. The idea is that "surprising" outcomes are evidence against H_0. A surprising outcome is one that is far from what we would expect if H_0 were true. In many cases, including our bottle fill example, the standard scale of z-scores is one way to measure "far from what we would expect." In Example 6.14, the standardized distance of the sample outcome $\bar{x}$ from the hypothesized population mean $\mu = 500$ was $z = 3.07$. This suggests that the sample is not compatible with the null hypothesis $H_0: \mu = 500$. In fact, the Supreme Court of the United States has said that "two or three standard deviations" ($z = 2$ or 3) is its criterion for rejecting H_0, and this is the criterion used in most applications involving the law.

Because not all test statistics produce z-scores, we translate the values of test statistics into a common language, the language of probability. A test of significance finds the probability of getting an outcome *as extreme or more extreme than the actually observed outcome.* "Extreme" means "far from what we would expect if H_0 were true." The direction or directions that count as "far from what we would expect" are determined by the alternative hypothesis H_a.

> **P-Value**
>
> The probability, computed assuming that H_0 is true, that the test statistic would take a value as extreme or more extreme than that actually observed is called the **P-value** of the test. The smaller the P-value, the stronger the evidence against H_0 provided by the data.

To calculate the P-value we must use the sampling distribution of the test statistic. In this chapter, we need only the standard Normal distribution for the test statistic z.

EXAMPLE 6.15   Bottle Fill Amount: The P-Value

CASE 6.2

In Example 6.10 the observations are an SRS of size $n = 10$ from a population of bottles with $\sigma = 2$. The observed average fill amount is $\bar{x} = 501.94$. In Example 6.14, we found that the test statistic for testing H_0: $\mu = 500$ versus H_a: $\mu \neq 500$ is

$$z = \frac{501.94 - 500}{2/\sqrt{10}} = 3.07$$

If the null hypothesis is true, we expect z to take a value not too far from 0. Because the alternative is two-sided, values of z far from 0 *in either direction* count as evidence against H_0 and in favor of H_a. So the P-value is

$$P(z \leq -3.07) + P(z \geq 3.07)$$

If H_0 is true, then z has the standard Normal $N(0, 1)$ distribution. Figure 6.10 shows the P-value as a very small area under the standard Normal curve.

From Table A, we find

$$P(z \leq -3.07) = 0.0011$$

Similarly, from Table A or using symmetry of the Normal distribution,

$$P(z \geq 3.07) = 1 - 0.9989 = 0.0011$$

Since $P(z \leq -3.07)$ equals $P(z \geq 3.07)$, we see that the sum of these probabilities is simply twice either of the probabilities. So this means

$$
\begin{aligned}
P &= P(z \leq -3.07) + P(z \geq 3.07) \\
&= 2P(Z \leq -3.07) \\
&= 2P(Z \geq 3.07) \\
&= 0.0022
\end{aligned}
$$

In Example 6.10 we rounded this value to 0.002.

FIGURE 6.10 The P-value for Example 6.15. The two-sided P-value is the probability (when H_0 is true) that $\bar{x}$ takes a value at least as far from 0 as the actually observed value.

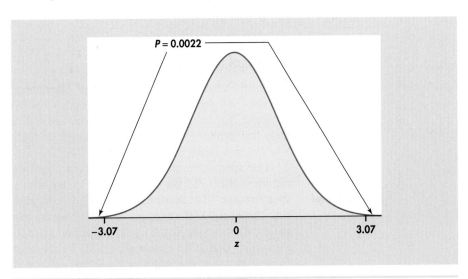

APPLY YOUR KNOWLEDGE

6.41 Spending on housing. The Census Bureau reports that households spend an average of 31% of their total spending on housing. A homebuilders association in Cleveland wonders if the national finding applies in their area. They interview a sample of 40 households in the Cleveland metropolitan area to learn what percent of their spending goes toward housing. Take μ to be the mean percent of spending devoted to housing among all Cleveland households. We want to test the hypotheses

$$H_0: \mu = 31\%$$
$$H_a: \mu \neq 31\%$$

The population standard deviation is $\sigma = 9.6\%$.
(a) The study finds $\bar{x} = 28.6\%$ for the 40 households in the sample. What is the value of the test statistic z? Sketch a standard Normal curve and mark z on the axis. Shade the area under the curve that represents the P-value.
(b) Calculate the P-value. Are you convinced that Cleveland differs from the national average?

6.42 State null and alternative hypotheses. In the setting of the previous exercise, suppose that the Cleveland homebuilders were convinced, before interviewing their sample, that residents of Cleveland spend less than the national average on housing. Do the interviews support their conviction? State null and alternative hypotheses. Find the P-value, using the interview results given in the previous problem. Why do the same data give different P-values in these two problems?

6.43 Why is this wrong? The homebuilders wonder if the national finding applies in the Cleveland area. They have no idea whether Cleveland residents spend more or less than the national average. Because their interviews find that $\bar{x} = 28.6\%$, less than the national 31%, their analyst tests

$$H_0: \mu = 31\%$$
$$H_a: \mu < 31\%$$

Explain why this is incorrect.

Statistical significance

Statistical software automates the task of calculating the test statistic and its P-value. You must still decide which test is appropriate and whether to use a one-sided or two-sided test. You must also decide what conclusion the computer's numbers support. We know that smaller P-values indicate stronger evidence against the null hypothesis. But how strong is strong enough?

significance level   One approach is to announce in advance how much evidence against H_0 we will require to reject H_0. We compare the P-value with a level that says "this evidence is strong enough." The decisive level is called the **significance level.** It is denoted by α, the Greek letter alpha. If we choose $\alpha = 0.05$, we are requiring that the data give evidence against H_0 so strong that it would happen no more than 5% of the time (1 time in 20) when H_0 is true. If we choose $\alpha = 0.01$, we are insisting on stronger evidence against H_0, evidence so strong that it would appear only 1% of the time (1 time in 100) if H_0 is in fact true.

Statistical Significance

If the *P*-value is as small or smaller than α, we say that the data are **statistically significant at level α.**

"Significant" in the statistical sense does not mean "important." The original meaning of the word is "signifying something." In statistics, the term is used to indicate only that the evidence against the null hypothesis reached the standard set by α. Significance at level 0.01 is often expressed by the statement "The results were significant ($P < 0.01$)." Here *P* stands for the *P*-value. The *P*-value is more informative than a statement of significance because we can then assess significance at any level we choose. For example, a result with $P = 0.03$ is significant at the $\alpha = 0.05$ level but is not significant at the $\alpha = 0.01$ level.

You need not actually find the *P*-value to assess significance at a fixed level α. You need only compare the observed statistic *z* with a *critical value* that marks off area α in one or both tails of the standard Normal curve.

EXAMPLE 6.16   Significant?

CASE 6.2   In Example 6.14, the test statistic took the value $z = 3.07$. Is this significant at the $\alpha = 0.01$ level? Given the magnitude of the test statistic value, we are pretty certain about the answer to this question. Let us still walk through the thinking process. Values in the extreme 0.01 area of the standard Normal curve are significant at the 0.01 level. Because the alternative is two-sided, this area is divided equally between the two tails of the curve. So *z* is significant if it lies in the extreme area 0.005 at either end. Look at the z^* row at the bottom of Table D. The extreme 0.005 in the right tail starts at $z^* = 2.576$. That is, the *z*-values that are significant at the $\alpha = 0.01$ level are $z > 2.576$ and $z < -2.576$. Figure 6.11 shows why. The sample outcome $z = 3.07$ is thus statistically significant ($P < 0.01$).

FIGURE 6.11 The *P*-value for Example 6.16. The two-sided *P*-value is the probability (when H_0 is true) that $\bar{x}$ takes a value at least as far from 0 as the actually observed value.

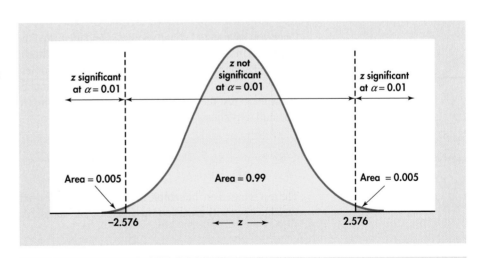

APPLY YOUR KNOWLEDGE

6.44 How to show that you are rich. Every society has its own marks of wealth and prestige. In ancient China, it appears that owning pigs was such a mark. Evidence comes from examining burial sites. If the skulls of sacrificed pigs tend to appear along with expensive ornaments, that suggests that the pigs, like the ornaments, signal the wealth and prestige of the person buried. A study of burials from around 3500 B.C. concluded that "there are striking differences in grave goods between burials with pig skulls and burials without them.... A test indicates that the two samples of total artifacts are significantly different at the 0.01 level."[4] Explain clearly why "significantly different at the 0.01 level" gives good reason to think that there really is a systematic difference between burials that contain pig skulls and those that lack them.

6.45 Significance. You are testing H_0: $\mu = 0$ against H_a: $\mu \neq 0$ based on an SRS of 30 observations from a Normal population. What values of the z statistic are statistically significant at the $\alpha = 0.005$ level?

6.46 Significance. You are testing H_0: $\mu = 0$ against H_a: $\mu > 0$ based on an SRS of 30 observations from a Normal population. What values of the z statistic are statistically significant at the $\alpha = 0.005$ level?

6.47 The Supreme Court speaks. Court cases in such areas as employment discrimination often involve statistical evidence. The Supreme Court has said that z-scores beyond $z^* = 2$ or 3 are generally convincing statistical evidence. For a two-sided test, what significance level corresponds to $z^* = 2$? To $z^* = 3$?

Tests for a population mean

There are four steps in carrying out a significance test:

1. State the hypotheses.
2. Calculate the test statistic.
3. Find the P-value.
4. State your conclusion in the context of your specific setting.

Once you have stated your hypotheses and identified the proper test, you or your software can perform Steps 2 and 3. We now summarize Steps 2 and 3 for conducting a test for the population mean as we have done in our examples.

We have an SRS of size n drawn from a Normal population with unknown mean μ. We want to test the hypothesis that μ has a specified value. Call the specified value μ_0. The null hypothesis is

$$H_0: \mu = \mu_0$$

The test is based on the sample mean $\bar{x}$. Because Normal calculations require standardized variables, we will use as our test statistic the standardized sample mean

$$z = \frac{\bar{x} - \mu_0}{\sigma/\sqrt{n}}$$

one-sample z statistic This **one-sample z statistic** has the standard Normal distribution when H_0 is true. The P-value of the test is the probability that z takes a value at least as extreme as the value for our sample. What counts as extreme is determined by the alternative hypothesis H_a. Here is a summary.

z Test for a Population Mean

To test the hypothesis H_0: $\mu = \mu_0$ based on an SRS of size n from a population with unknown mean μ and known standard deviation σ, compute the **one-sample z statistic**

$$z = \frac{\bar{x} - \mu_0}{\sigma/\sqrt{n}}$$

In terms of a variable Z having the standard Normal distribution, the P-value for a test of H_0 against

H_a: $\mu > \mu_0$ is $P(Z \geq z)$

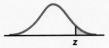

H_a: $\mu < \mu_0$ is $P(Z \leq z)$

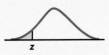

H_a: $\mu \neq \mu_0$ is $2P(Z \geq |z|)$

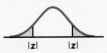

These P-values are exact if the population distribution is Normal and are approximately correct for large n in other cases.

EXAMPLE 6.17   Blood Pressures of Executives

The medical director of a large company is concerned about the effects of stress on the company's younger executives. According to the National Center for Health Statistics, the mean systolic blood pressure for males 35 to 44 years of age is 128 and the standard deviation in this population is 15. The medical director examines the records of 72 executives in this age group and finds that their mean systolic blood pressure is $\bar{x} = 129.93$. Is this evidence that the mean blood pressure for all the company's young male executives is higher than the national average? As usual in this chapter, we make the unrealistic assumption that the population standard deviation is known, in this case that executives have the same $\sigma = 15$ as the general population.

Step 1: Hypotheses. The hypotheses about the unknown mean μ of the executive population are

$$H_0: \mu = 128$$
$$H_a: \mu > 128$$

Step 2: Test statistic. The z test requires that the 72 executives in the sample are an SRS from the population of the company's young male executives. We must ask how the data were produced. If records are available only for executives with recent medical problems, for example, the data are of little value for our purpose. It turns out that all executives are given a free annual medical exam and that the medical director selected 72 exam results at random. The one-sample z statistic is

$$z = \frac{\bar{x} - \mu_0}{\sigma/\sqrt{n}} = \frac{129.93 - 128}{15/\sqrt{72}}$$
$$= 1.09$$

Step 3: P-value. Draw a picture to help find the P-value. Figure 6.12 shows that the P-value is the probability that a standard Normal variable Z takes a value of 1.09 or greater. From Table A

FIGURE 6.12 The *P*-value for
the one-sided test in
Example 6.17.

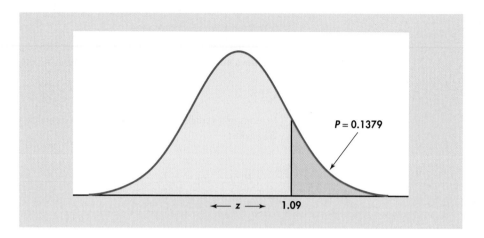

we find that this probability is

$$P = P(Z \geq 1.09) = 1 - 0.8621 = 0.1379$$

Step 4: Conclusion. About 14% of the time an SRS of size 72 from the general male population would have a mean blood pressure as high as that of the executive sample. The observed $\overline{x} = 129.93$ is not significantly higher than the national average at any of the usual significance levels.

The data in Example 6.17 do *not* establish that the mean blood pressure μ for this company's middle-aged male executives is 128. We sought evidence that μ was higher than 128 and failed to find convincing evidence. That is all we can say. No doubt the mean blood pressure of the entire executive population is not exactly equal to 128. A large enough sample would give evidence of the difference, even if it is very small. Tests of significance assess the evidence *against* H_0. If the evidence is strong, we can confidently reject H_0 in favor of the alternative. Failing to find evidence against H_0 means only that the data are consistent with H_0, not that we have clear evidence that H_0 is true.

EXAMPLE 6.18   A Company-Wide Health Promotion

The company medical director institutes a health promotion campaign to encourage employees to exercise more and eat a healthier diet. One measure of the effectiveness of such a program is a drop in blood pressure. Choose a random sample of 50 employees, and compare their blood pressures from annual physical examinations given before the campaign and again a year later. The mean change in blood pressure for these $n = 50$ employees is $\overline{x} = -6$. We take the population standard deviation to be $\sigma = 20$.

Step 1: Hypotheses. We want to know if the health campaign reduced average blood pressure. Taking μ to be the mean change in blood pressure for all employees,

$$H_0: \mu = 0$$
$$H_a: \mu < 0$$

Step 2: Test statistic. The one-sample z test is appropriate:

$$z = \frac{\overline{x} - \mu_0}{\sigma/\sqrt{n}} = \frac{-6 - 0}{20/\sqrt{50}}$$
$$= -2.12$$

FIGURE 6.13 The *P*-value
for the one-sided test in
Example 6.18.

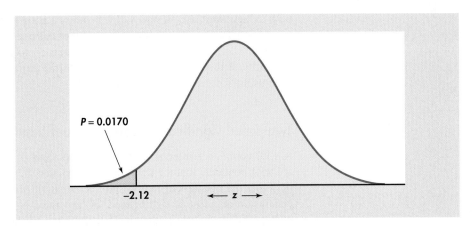

P = 0.0170

−2.12   ← z →

Step 3: *P*-value. Because H_a is one-sided on the low side, large negative values of z count against H_0. See Figure 6.13. From Table A, we find that the *P*-value is

$$P = P(Z \leq -2.12) = 0.0170$$

Step 4: Conclusion. A mean change in blood pressure of −6 or better would occur only 17 times in 1000 samples if the campaign had no effect on the blood pressures of the employees. This is convincing evidence that the mean blood pressure in the population of all employees has decreased.

Our conclusion in Example 6.18 is cautious. We would like to conclude that the health campaign *caused* the drop in mean blood pressure. But the data are not protected from confounding. If, for example, the local television station runs a series on the risk of heart attacks and the value of better diet and exercise, many employees may be moved to reform their health habits even without encouragement from the company. Only a randomized comparative experiment protects against such lurking variables. The medical director preferred to launch a company-wide campaign that appealed to all employees. This is no doubt a sound medical decision, but the absence of a control group weakens the statistical conclusion.

APPLY YOUR KNOWLEDGE

6.48 Testing a random number generator. Statistical software has a "random number generator" that is supposed to produce numbers uniformly distributed between 0 to 1. If this is true, the numbers generated come from a population with $\mu = 0.5$. A command to generate 100 random numbers gives outcomes with mean $\bar{x} = 0.522$ and $s = 0.316$. Because the sample is reasonably large, take the population standard deviation also to be $\sigma = 0.316$. Do we have evidence that the mean of all numbers produced by this software is not 0.5?

6.49 The effects of different alternative hypotheses. A test of the null hypothesis $H_0: \mu = \mu_0$ gives test statistic $z = 1.9$.
(a) What is the *P*-value if the alternative is $H_a: \mu > \mu_0$?
(b) What is the *P*-value if the alternative is $H_a: \mu < \mu_0$?
(c) What is the *P*-value if the alternative is $H_a: \mu \neq \mu_0$?

6.50 A new supplier. A new supplier offers a good price on a catalyst used in your production process. You compare the purity of this catalyst with that from your current supplier. The P-value for a test of "no difference" is 0.27. Can you be confident that the purity of the new product is the same as the purity of the product that you have been using? Discuss.

Two-sided significance tests and confidence intervals

A 95% confidence interval captures the true value of μ in 95% of all samples. If we are 95% confident that the true μ lies in our interval, we are also confident that values of μ that fall outside our interval are incompatible with the data. That sounds like the conclusion of a test of significance. In fact, there is an intimate connection between 95% confidence intervals and significance at the 5% level. The same connection holds between 99% confidence intervals and significance at the 1% level, and so on.

> **Confidence Intervals and Two-Sided Tests**
>
> A level α two-sided significance test rejects a hypothesis H_0: $\mu = \mu_0$ exactly when the value μ_0 falls outside a level $1 - \alpha$ confidence interval for μ.

EXAMPLE 6.19 Performance of IPO Firms

The decision to go public is clearly one of the most significant decisions to be made by a privately owned company. The first sale of stock to the public by a private company is referred to as an initial public offering (IPO). The natural question to ask is if the decision to go public is financially beneficial to the firm? In a study of a sample of 3964 IPO firms from a thirty-year period, it was found that the mean change in firm profitability, measured as return on equity (ROE),* was −4.29% over a three-year span starting from the initial offering.[5] The sample standard deviation s was 15.56. Clearly, the very large sample size allows us to safely use the sample standard deviation s as if it were the population standard deviation σ. In addition, the very large sample size guarantees that the sampling distribution of $\bar{x}$ is indistinguishable from the Normal distribution.

The researchers of this study asked whether the −4.29% mean change in ROE in the *sample* provides evidence that there is a nonzero mean change in ROE in the whole population of IPO firms?** This calls for a test of the hypotheses

$$H_0: \mu = 0$$
$$H_a: \mu \neq 0$$

We will carry out the test twice, first with the usual significance test and then with a 99% confidence interval.

---

*Return on equity is the amount of net income returned as a percent of shareholder equity (assets minus liabilities). This ratio measures a company's efficiency at generating profits from every dollar of shareholder equity.

**Though the sample of 3964 IPO firms is large, it represents only about half of IPO firms during the time frame of the study in the United States and a much smaller proportion worldwide. More generally, the researchers implicitly view the population on an even much grander scale in terms of the underlying process generating IPO firms over time, which includes outcomes prior to the study and outcomes yet to be observed.

First, the test. The mean of the sample is $\bar{x} = -4.29$. The one-sample z test statistic is therefore

$$z = \frac{\bar{x} - \mu_0}{\sigma/\sqrt{n}} = \frac{-4.29 - 0}{15.56/\sqrt{3964}} = -17.36$$

We do not need to find the exact P-value to assess significance at the $\alpha = 0.01$ level. Look in Table D under tail area 0.005 because the alternative is two-sided. The z-values that are significant at the 1% level are $z > 2.576$ and $z < -2.576$. Our observed $z = -17.36$ is clearly significant ($P < 0.01$).

To compute a 99% confidence interval for the mean change in ROE, find in Table D the critical value for 99% confidence. It is $z^* = 2.576$, the same critical value that marked off significant z's in our test. The confidence interval is

$$\bar{x} \pm z^* \frac{\sigma}{\sqrt{n}} = -4.29 \pm 2.576 \frac{15.56}{\sqrt{3964}}$$

$$= -4.29 \pm 0.6366$$

$$= (-4.927, -3.653)$$

The hypothesized value $\mu_0 = 0$ falls well outside this confidence interval. We are therefore 99% confident that μ is *not* equal to 0, so we can reject

$$H_0: \mu = 0$$

at the 1% significance level. On the other hand, we cannot reject values of μ that fall in the range -4.927 to -3.653. For example, we cannot reject

$$H_0: \mu = -3.9$$

at the 1% level in favor of the two-sided alternative $H_a: \mu \neq -3.9$. Figure 6.14 illustrates both cases.

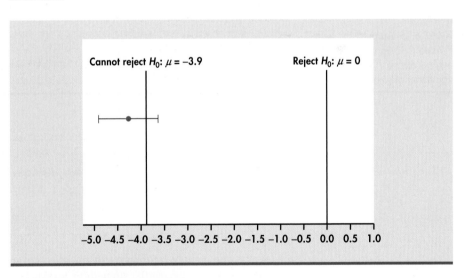

FIGURE 6.14 Values of μ falling outside a 99% confidence interval can be rejected at the 1% level. Values falling inside the interval cannot be rejected.

APPLY YOUR KNOWLEDGE

6.51 Does the confidence interval include μ_0? The P-value for a two-sided test of the null hypothesis $H_0: \mu = 15$ is 0.07.
(a) Does the 95% confidence interval include the value 15? Why?
(b) Does the 90% confidence interval include the value 15? Why?

6.52 Can you reject the null hypothesis? A 95% confidence interval for a population mean is (16, 32).

(a) Can you reject the null hypothesis that $\mu = 34$ at the 5% significance level? Why?

(b) Can you reject the null hypothesis that $\mu = 17$ at the 5% significance level? Why?

P-values versus fixed α

For Example 6.19, we concluded that the test statistic $z = -17.36$ is significant at the 1% level. This isn't the whole story. The observed z is far beyond the critical value for $\alpha = 0.01$, and the evidence against H_0 is far stronger than 1% significance suggests. The *P*-value $P = 1.657 \times 10^{-67}$ (from software) gives a sense of how strong the evidence is. In the IPO case, the evidence is astronomically strong against the hypothesis $\mu = 0$. *In general, the P-value is the smallest level α at which the data are significant.* Knowing the *P*-value allows us to assess significance at any level.

EXAMPLE 6.20 How Significant?

In Example 6.18, we tested the hypotheses

$$H_0: \mu = 0$$

$$H_a: \mu < 0$$

to evaluate the effectiveness of a health promotion program on blood pressure. The test had the *P*-value $P = 0.0170$. This result is not significant at the $\alpha = 0.01$ level, because $0.0170 \geq 0.01$. However, it is significant at the $\alpha = 0.05$ level, because the *P*-value is less than 0.05. See Figure 6.15.

FIGURE 6.15 An outcome with *P*-value *P* is significant at all levels α above *P* and is not significant at smaller levels of α.

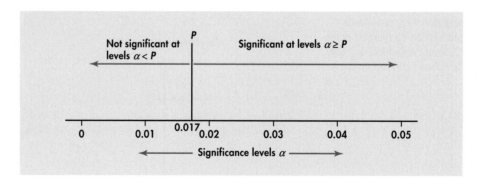

A *P*-value is more informative than a reject-or-not finding at a fixed significance level in that it summarizes the strength or weakness of the evidence against H_0. But assessing significance at a fixed level α is easier, because no probability calculation is required. You need only look up a critical value in a table. Because the practice of statistics almost always employs software that calculates *P*-values automatically, tables of critical values are becoming outdated. Notwithstanding, we include the usual tables of critical values (such as Table D) at the end of the book for learning purposes and as a reference when computing is not available.

APPLY YOUR KNOWLEDGE

6.53 Do you reject the null hypothesis? The P-value for a significance test is 0.049.
(a) Do you reject the null hypothesis at level $\alpha = 0.05$?
(b) Do you reject the null hypothesis at level $\alpha = 0.01$?
(c) Explain your answers.

6.54 Suppose the P-value is a little higher? The P-value for a significance test is 0.051.
(a) Do you reject the null hypothesis at level $\alpha = 0.05$?
(b) Do you reject the null hypothesis at level $\alpha = 0.01$?
(c) Explain your answers.

SECTION 6.2 Summary

- A **test of significance** assesses the evidence provided by data against a **null hypothesis** H_0 and in favor of an **alternative hypothesis** H_a. It provides a method for ruling out chance as an explanation for data that deviate from what we expect under H_0.

- The hypotheses are stated in terms of population parameters. Usually, H_0 is a statement that no effect is present, and H_a says that a parameter differs from its null value in a specific direction (**one-sided alternative**) or in either direction (**two-sided alternative**).

- The test is based on a **test statistic.** The **P-value** is the probability, computed assuming that H_0 is true, that the test statistic will take a value at least as extreme as that actually observed. Small P-values indicate strong evidence against H_0. Calculating P-values requires knowledge of the sampling distribution of the test statistic when H_0 is true.

- If the P-value is as small or smaller than a specified value α, the data are **statistically significant** at significance level α.

- Significance tests for the hypothesis H_0: $\mu = \mu_0$ concerning the unknown mean μ of a population are based on the z **statistic:**

$$z = \frac{\bar{x} - \mu_0}{\sigma/\sqrt{n}}$$

- The z test assumes an SRS of size n, known population standard deviation σ, and either a Normal population or a large sample. P-values are computed from the Normal distribution (Table A). Fixed α tests use the table of **standard Normal critical values** (z^* row in Table D).

SECTION 6.2 Exercises

For Exercises 6.39 and 6.40, see page 355; for 6.41 to 6.43, see page 358; for 6.44 to 6.47, see pages 359–360; for 6.48 to 6.50, see pages 363–364; for 6.51 and 6.52, see pages 365–366; and for 6.53 and 6.54, see page 367.

6.55 What's wrong? Here are several situations where there is an incorrect application of the ideas presented in this section. Write a short paragraph explaining what is wrong in each situation and why it is wrong.

(a) A fleet manager for a trucking company wants to test the null hypothesis that the price of diesel fuel will not increase next month.
(b) A random sample of size 30 is taken from a population that is assumed to have a standard deviation of 18. The standard deviation of the sample mean is 18/30.
(c) A researcher tests the following null hypothesis: H_0: $\bar{x} = 20$.

6.56 What's wrong? Here are several situations where there is an incorrect application of the ideas presented in this section. Write a short paragraph explaining what is wrong in each situation and why it is wrong.

(a) A change is made that should improve student satisfaction with the way grades are processed at your college. The null hypothesis, that there is an improvement, is tested versus the alternative, that there is no change.

(b) A significance test rejected the null hypothesis that the sample mean is 35.

(c) A report on a study says that the results are statistically significant and the P-value is 0.85.

(d) The z statistic had a value of 0.023, and the null hypothesis was rejected at the 5% level because $0.023 < 0.05$.

6.57 Compare student credit card debt for different regions. Nellie Mae, a subsidiary of Sallie Mae, is a national student loan company providing loans to undergraduate and graduate students. Nellie Mae conducts statistical studies to estimate various types of debt in the student population. One such study is the Graduate Student Credit Card Study, which attempts to estimate the amount of credit card debt held by different subgroups of graduate students.[6] There were 1149 graduate students included in the study. The study found that the 207 borrowers who attended graduate school in the West region of the United States had a mean credit card debt of $9835, while those who attended a graduate school in the South region of the United States had a mean credit card debt of $8894. The difference of $941 is not a trivial amount, but we know that these numbers are estimates of the true means. If we took a different sample, we would get different estimates. Can we conclude from these data that graduate students in the West region have a different credit card debt level than graduate students in the South region? We answer this question by computing the probability of obtaining a difference as large or larger than the observed $941 assuming that, in fact, there is no difference in the true means. This probability is 0.213. What do you conclude? Illustrate the probability result with a sketch, and write a short paragraph explaining your answer.

6.58 Compare student credit card debt for different age groups. The credit card study in the previous exercise examined the amount of debt held by different age groups. It was found that the mean credit card debt for graduate students in the United States aged 22 to 29 years was $6479, while the debt for students aged 30 to 59 years was $12,593. The difference is $6114. Can we conclude that older graduate students have a different debt level than younger students? To determine if the data provide evidence for a difference, we calculate the probability of observing a difference as large or larger than the observed $6114 assuming that there is no difference in the true means. The calculated probability is 4.4×10^{-27}. What do you conclude? Illustrate the probability result with a sketch, and write a short paragraph explaining your answer.

6.59 Hypotheses. Each of the following situations requires a significance test about a population mean μ. State the appropriate null hypothesis H_0 and alternative hypothesis H_a in each case.

(a) Larry's car averages 31 miles per gallon on the highway. He now switches to a new motor oil that is advertised as increasing gas mileage. After driving 2500 highway miles with the new oil, he wants to determine if his gas mileage actually has increased.

(b) The diameter of a spindle in a small motor is supposed to be 4 millimeters. If the spindle is either too small or too large, the motor will not perform properly. The manufacturer measures the diameter in a sample of motors to determine whether the mean diameter has moved away from the target.

(c) The mean area of the several thousand apartments in a new development is advertised to be 1400 square feet. A tenant group thinks that the apartments are smaller than advertised. They hire an engineer to measure a sample of apartments to test their suspicion.

6.60 Hypotheses. In each of the following situations, a significance test for a population mean μ is called for. State the null hypothesis H_0 and the alternative hypothesis H_a in each case.

(a) A university gives credit in French language courses to students who pass a placement test. The language department wants to know if students who get credit in this way differ in their understanding of spoken French from students who actually take the French courses. Experience has shown that the mean score of students in the courses on a standard listening test is 26. The language department gives the same listening test to a sample of 35 students who passed the credit examination to see if their performance is different.

(b) Experiments on learning in animals sometimes measure how long it takes a mouse to find its way through a maze. The mean time is 22 seconds for one particular maze. A researcher thinks that a loud noise will cause the mice to complete the maze faster. She measures how long each of 12 mice takes with a noise as stimulus.

(c) The examinations in a large accounting class are scaled after grading so that the mean score is 75. A self-confident teaching assistant thinks that his students have a higher mean score than the class as a whole. His students this semester can be considered a sample from the population of all students he might teach, so he compares their mean score with 75.

6.61 Hypotheses. In each of the following situations, state an appropriate null hypothesis H_0 and alternative hypothesis H_a. Be sure to identify the parameters that you use to state the hypotheses. (We have not yet learned how to test these hypotheses.)

(a) A sociologist asks a large sample of high school students which academic subject they like best. She suspects that a higher percent of males than of females will name economics as their favorite subject.

(b) An education researcher randomly divides sixth-grade students into two groups for physical education class. He teaches both groups basketball skills, using the same methods of instruction in both classes. He encourages Group A with compliments and other positive behavior but acts cool and neutral toward Group B. He hopes to show that positive teacher attitudes result in a higher mean score on a test of basketball skills than do neutral attitudes.

(c) An economist believes that among employed young adults there is a positive correlation between income and the percent of disposable income that is saved. To test this, she gathers income and savings data from a sample of employed persons in her city aged 25 to 34.

6.62 Hypotheses. Translate each of the following research questions into appropriate H_0 and H_a.
(a) Census Bureau data show that the mean household income in the area served by a shopping mall is $62,500 per year. A market research firm questions shoppers at the mall to find out whether the mean household income of mall shoppers is higher than that of the general population.
(b) Last year, your company's service technicians took an average of 2.6 hours to respond to trouble calls from business customers who had purchased service contracts. Do this year's data show a different average response time?

6.63 Responding to customer complaints. Maritz is a leading marketing research firm (`www.maritz.com`) that conducts marketing analysis for various industries worldwide. In a 2004 study, Maritz conducted a survey of new-vehicle buyers. The focus of the study was on customer complaint resolution and customer loyalty. In the study, the following conclusions were made:

> Positive dealership resolution of customers' vehicle complaints seemed to have little impact on intent to repurchase the same make (this difference was not statistically significant). However, when dealerships resolved vehicle complaints poorly (which occurred more than twice as often as resolving vehicle complaints satisfactorily) intent to repurchase the same make dropped considerably compared to people who did not complain and compared to people who complained but received good resolution. These differences were statistically significant at the p < .001 level.

Explain these conclusions, especially the *P*-value, as if you were speaking to a manager of a car dealership who knows no statistics.

6.64 Handshakes and getting a job. When building business relationships or forming new friendships, first impressions are regarded as crucial. One of those first-impression opportunities is the handshake. In the sales arena, it is conventional wisdom that a firm handshake makes for a good first impression with one's clientele. Until recently, the importance of handshakes has been anecdotal in the business environment. A research experiment was conducted to examine how a job applicant's handshake quality influences hiring recommendations.[7] Here is an excerpt from the study:

> Hypothesis 1 predicted a relationship between a firm handshake and interview ratings and was supported (r = 0.29, p < 0.05).

How would you explain this statement to someone who knows no statistics? Include an explanation of both the description given by *r* and the statistical significance.

6.65 Exercise and statistics exams. A study examined whether exercise affects how students perform on their final exam in statistics. The *P*-value was given as 0.68.
(a) State null and alternative hypotheses that could be used for this study. (Note that there is more than one correct answer.)
(b) Do you reject the null hypothesis? State your conclusion in plain language.
(c) What other facts about the study would you like to know for a proper interpretation of the results?

6.66 Financial aid. The financial aid office of a university asks a sample of students about their employment and earnings. The report says that "for academic year earnings, a significant difference ($P = 0.038$) was found between the sexes, with men earning more on the average. No difference ($P = 0.476$) was found between the earnings of black and white students."[8] Explain both of these conclusions, for the effects of sex and of race on mean earnings, in language understandable to someone who knows no statistics.

6.67 Who is the author? Statistics can help decide the authorship of literary works. Sonnets by a certain Elizabethan poet are known to contain an average of $\mu = 6.9$ new words (words not used in the poet's other works). The standard deviation of the number of new words is $\sigma = 2.7$. Now a manuscript with 5 new sonnets has come to light, and scholars are debating whether it is the poet's work. The new sonnets contain an average of $\overline{x} = 11.2$ words not used in the poet's known works. We expect poems by another author to contain more new words, so to see if we have evidence that the new sonnets are not by our poet we test

$$H_0: \mu = 6.9$$
$$H_a: \mu > 6.9$$

Give the *z* test statistic and its *P*-value. What do you conclude about the authorship of the new poems?

6.68 Study habits. The Survey of Study Habits and Attitudes (SSHA) is a psychological test that measures the motivation, attitude toward school, and study habits of students. Scores range from 0 to 200. The mean score for U.S. college students is about 115, and the standard deviation is about 30. A teacher who suspects that older students have better attitudes toward school gives the SSHA to 25 students who are at least 30 years of age. Their mean score is $\overline{x} = 133.2$.
(a) Assuming that $\sigma = 30$ for the population of older students, carry out a test of

$$H_0: \mu = 115$$
$$H_a: \mu > 115$$

Report the *P*-value of your test, draw a sketch illustrating the *P*-value, and state your conclusion clearly.
(b) Your test in (a) required two important assumptions in addition to the assumption that the value of σ is known. What are they? Which of these assumptions is most important to the validity of your conclusion in (a)?

6.69 Corn yield. The mean yield of corn in the United States is about 135 bushels per acre. A survey of 50 farmers this year gives a sample mean yield of $\bar{x} = 138.4$ bushels per acre. We want to know whether this is good evidence that the national mean this year is not 135 bushels per acre. Assume that the farmers surveyed are an SRS from the population of all commercial corn growers and that the standard deviation of the yield in this population is $\sigma = 10$ bushels per acre. Report the value of the test statistic z, give a sketch illustrating the P-value and report the P-value for the test of

$$H_0: \mu = 135$$
$$H_a: \mu \neq 135$$

Are you convinced that the population mean is not 135 bushels per acre? Is your conclusion correct if the distribution of corn yields is somewhat non-Normal? Why?

6.70 Cigarettes. According to data from the Tobacco Institute Testing Laboratory, Camel Lights King Size cigarettes contain an average of 1.4 milligrams of nicotine. An advocacy group commissions an independent test to see if the mean nicotine content is higher than the industry laboratory claims.
(a) What are H_0 and H_a?
(b) Suppose that the test statistic is $z = 2.36$. Is this result significant at the 5% level?
(c) Is the result significant at the 1% level?

6.71 Academic probation and TV watching. There are other z statistics that we have not yet met. We can use Table D to assess the significance of any z statistic. A study compares the habits of students who are on academic probation with students whose grades are satisfactory. One variable measured is the hours spent watching television last week. The null hypothesis is "no difference" between the means for the two populations. The alternative hypothesis is two-sided. The value of the test statistic is $z = -1.27$.
(a) Is the result significant at the 5% level?
(b) Is the result significant at the 1% level?

6.72 Why is it significant at the 5% level? Explain in plain language why a significance test that is significant at the 1% level must always be significant at the 5% level.

6.73 Use Table D. You have performed a two-sided test of significance and obtained a value of $z = 3.1$. Use Table D to find the approximate P-value for this test.

6.74 Find a P-value. You have performed a one-sided test of significance and obtained a value of $z = 0.35$. Use Table D to find the approximate P-value for this test.

6.75 What values will lead to rejection of H_0? You will perform a significance test of $H_0: \mu = 0$ versus $H_a: \mu > 0$.
(a) What values of z would lead you to reject H_0 at the 5% level?
(b) If the alternative hypothesis was $H_a: \mu \neq 0$, what values of z would lead you to reject H_0 at the 5% level?
(c) Explain why your answers to parts (a) and (b) are different.

6.76 Check the P-value. Between what critical values from Table D does the P-value for the outcome $z = -1.27$ in Exercise 6.71 lie? Calculate the P-value using Table A, and verify that it lies between the values you found from Table D.

6.77 Radon. Radon is a colorless, odorless gas that is naturally released by rocks and soils and may concentrate in tightly closed houses. Because radon is slightly radioactive, there is some concern that it may be a health hazard. Radon detectors are sold to homeowners worried about this risk, but the detectors may be inaccurate. University researchers placed 12 detectors in a chamber where they were exposed to 105 picocuries per liter (pCi/l) of radon over 3 days.[9] Here are the readings given by the detectors:

RADON

91.9	97.8	111.4	122.3	105.4	95.0
103.8	99.6	96.6	119.3	104.8	101.7

Assume (unrealistically) that you know that the standard deviation of readings for all detectors of this type is $\sigma = 9$.
(a) Give a 95% confidence interval for the mean reading μ for this type of detector.
(b) Is there significant evidence at the 5% level that the mean reading differs from the true value 105? State hypotheses and base a test on your confidence interval from (a).

6.78 Clothing for runners. Your company sells exercise clothing and equipment on the Internet. To design the clothing, you collect data on the physical characteristics of your different types of customers. Here are the weights for a sample of 24 male runners. Assume that these runners can be viewed as a random sample of your potential customers. The weights are expressed in kilograms. **CLOTHING**

68.7	61.8	63.2	53.1	62.3	59.7	55.4	58.9	60.9	69.2	63.7	67.8
65.6	65.5	56.0	57.8	66.0	62.9	53.6	65.0	55.8	60.4	69.3	62.1

Exercise 6.24 (page 350) asks you to find a 95% confidence interval for the mean weight of the population of all such runners, assuming that the population standard deviation is $\sigma = 4.5$ kg.
(a) Give the confidence interval from that exercise, or calculate the interval if you did not do the exercise.
(b) Based on this confidence interval, does a test of

$$H_0: \mu = 61.3 \text{ kg}$$
$$H_a: \mu \neq 61.3 \text{ kg}$$

reject H_0 at the 5% significance level?
(c) Would $H_0: \mu = 63$ be rejected at the 5% level if tested against a two-sided alternative?

6.79 Cockroaches. Your company is developing a better means to eliminate cockroaches from buildings. In the process, your research and development team studies the absorption of sugar by these insects.[10] They feed cockroaches a diet containing measured amounts of a particular sugar. After 10 hours, the cockroaches are killed and the concentration of the sugar in various body parts is determined by a chemical analysis. The paper that reports the research states that a 95% confidence interval for the mean amount (in milligrams) of the sugar in the hindguts of the cockroaches is 4.2 ± 2.3.

(a) Does this paper give evidence that the mean amount of sugar in the hindguts under these conditions is not equal to 7 mg? State H_0 and H_a and base a test on the confidence interval.

(b) Would the hypothesis that $\mu = 5$ mg be rejected at the 5% level in favor of a two-sided alternative?

6.80 Market pioneers. Market pioneers, companies that are among the first to develop a new product or service, tend to have higher market shares than latecomers to the market. What accounts for this advantage? Here is an excerpt from the conclusions of a study of a sample of 1209 manufacturers of industrial goods:

Can patent protection explain pioneer share advantages? Only 21% of the pioneers claim a significant benefit from either a product patent or a trade secret. Though their average share is two points higher than that of pioneers without this benefit, the increase is not statistically significant ($z = 1.13$). Thus, at least in mature industrial markets, product patents and trade secrets have little connection to pioneer share advantages.[11]

Find the P-value for the given z. Then explain to someone who knows no statistics what "not statistically significant" in the study's conclusion means. Why does the author conclude that patents and trade secrets don't help, even though they contributed 2 percentage points to average market share?

6.3 Using Significance Tests

Significance tests are widely used in reporting the results of research in many fields of applied science and in industry. New pharmaceutical products require significant evidence of effectiveness and safety. Courts inquire about statistical significance in hearing class action discrimination cases. Marketers want to know whether a new ad campaign significantly outperforms the old one, and medical researchers want to know whether a new therapy performs significantly better. In all these uses, statistical significance is valued because it points to an effect that is unlikely to occur simply by chance.

Carrying out a test of significance is often quite simple, especially if you get a P-value effortlessly from a calculator or computer. Using tests wisely is not so simple. Here are some points to keep in mind when using or interpreting significance tests.

How small a *P* is convincing?

The purpose of a test of significance is to describe the degree of evidence provided by the sample against the null hypothesis. The P-value does this. But how small a P-value is convincing evidence against the null hypothesis? This depends mainly on two circumstances:

- *How plausible is H_0?* If H_0 represents an assumption that the people you must convince have believed for years, strong evidence (small P) will be needed to persuade them.

- *What are the consequences of rejecting H_0?* If rejecting H_0 in favor of H_a means making an expensive changeover from one type of product packaging to another, you need strong evidence that the new packaging will boost sales.

These criteria are a bit subjective. Different people will often insist on different levels of significance. Giving the P-value allows each of us to decide individually if the evidence is sufficiently strong.

Users of statistics have often emphasized standard levels of significance such as 10%, 5%, and 1%. This emphasis reflects the time when tables of critical values rather than computer software dominated statistical practice. The 5% level ($\alpha = 0.05$) is particularly common. **There is no sharp border between "significant" and "insignificant," only increasingly strong evidence as the P-value decreases.** There is no practical distinction between the P-values 0.049 and 0.051. It makes no sense to treat $P \leq 0.05$ as a universal rule for what is significant.

6.81 Is it significant? More than 200,000 people worldwide take the GMAT examination each year as they apply for MBA programs. Their scores vary Normally with mean about $\mu = 525$ and standard deviation about $\sigma = 100$. One hundred students go through a rigorous training program designed to raise their GMAT scores. Test the hypotheses

$$H_0 : \mu = 525$$
$$H_a : \mu > 525$$

in each of the following situations:
(a) The students' average score is $\overline{x} = 541.4$. Is this result significant at the 5% level?
(b) The average score is $\overline{x} = 541.5$. Is this result significant at the 5% level?
The difference between the two outcomes in (a) and (b) is of no importance. Beware attempts to treat $\alpha = 0.05$ as sacred.

Statistical significance and practical significance

When a null hypothesis ("no effect" or "no difference") can be rejected at the usual levels, $\alpha = 0.05$ or $\alpha = 0.01$, there is good evidence that an effect is present. But that effect may be extremely small. When large samples are available, even tiny deviations from the null hypothesis will be significant.

EXAMPLE 6.21 It's Significant. So What?

We are testing the hypothesis of no correlation between two variables. With 1000 observations, an observed correlation of only $r = 0.08$ is significant evidence at the $\alpha = 0.01$ level that the correlation in the population is not zero but positive. The low significance level does not mean there is a strong association, only that there is strong evidence of some association. The true population correlation is probably quite close to the observed sample value, $r = 0.08$. We might well conclude, however, that for practical purposes we can ignore the association between these variables, even though we are confident (at the 1% level) that the correlation is positive.

Remember the wise saying: *statistical significance is not the same as practical significance.* On the other hand, if we fail to reject the null hypothesis, it may be because H_0 is true or because our sample size is insufficient to detect the alternative. Exercise 6.91 (page 375) demonstrates in detail the effect on P of increasing the sample size.

The remedy for attaching too much importance to statistical significance is to pay attention to the actual experimental results as well as to the P-value. Plot your data and examine them carefully. Are there outliers or other deviations from a consistent pattern? A few outlying observations can produce highly significant results if you blindly apply common tests of significance. Outliers can also destroy the significance of otherwise-convincing data. The foolish user of statistics who feeds the data to a computer without exploratory analysis will often be embarrassed. Is the effect that you are seeking visible in your plots? If not, ask yourself how can the effect be of practical importance if it is not large enough to even be seen. Even if the effect is visible, you can still ask yourself if it is large enough to be of practical importance. In either case, remember that what is considered large enough is application dependent. It may be that detection of tiny deviations is of great practical importance. For example, in many of today's manufacturing environments, parts are produced to very exacting tolerances with the minutest of deviations (for example, ten-thousandths of a millimeter) resulting in defective product. It is usually wise to give a confidence interval for the parameter in which you are interested. A confidence interval actually estimates the size of an effect rather than simply asking if it is too large to reasonably occur by chance alone. At which point, understanding and background knowledge of the practical application will guide you to assess whether the estimated effect size is important enough for action. Confidence intervals are not used as often as they should be, while tests of significance are perhaps overused.

APPLY YOUR KNOWLEDGE

6.82 How far do rich parents take us? How much education children get is strongly associated with the wealth and social status of their parents. In social science jargon, this is "socioeconomic status," or SES. But the SES of parents has little influence on whether children who have graduated from college go on to yet more education. One study looked at whether college graduates took the graduate admissions tests for business, law, and other graduate programs. The effects of the parents' SES on taking the LSAT test for law school were "both statistically insignificant and small."
(a) What does "statistically insignificant" mean?
(b) Why is it important that the effects were small in size as well as insignificant?

Statistical inference is not valid for all sets of data

We know that badly designed surveys or experiments often produce useless results. Formal statistical inference cannot correct basic flaws in the design. A statistical test is valid only in certain circumstances, with properly produced data being particularly important. The z test, for example, should bear the same warning label that we attached on page 345 to the z confidence interval. Similar warnings accompany the other tests that we will learn.

Tests of significance and confidence intervals are based on the laws of probability. Randomization in sampling or experimentation ensures that these laws apply. But we must often analyze data that do not arise from randomized samples or experiments. To apply statistical inference to such data, we must have confidence in a probability model for the data. The diameters of successive holes bored in auto engine blocks during production, for example, may behave like independent observations on a Normal distribution. We can check this probability model by examining the data. If the model appears correct, we can apply the methods of this chapter to do inference about the process mean diameter μ. Do ask how the data were produced, and don't be too impressed by P-values on a printout until you are confident that the data deserve a formal analysis.

APPLY YOUR KNOWLEDGE

6.83 Give an example. Give an example of a set of data for which statistical inference is not valid.

Beware of searching for significance

Statistical significance ought to mean that you have found an effect that you were looking for. The reasoning behind statistical significance works well if you decide what effect you are seeking, design a study to search for it, and use a test of significance to weigh the evidence you get. In other settings, significance may have little meaning.

EXAMPLE 6.22   Cell Phones and Brain Cancer

Might the radiation from cell phones be harmful to users? Many studies have found little or no connection between using cell phones and various illnesses. Here is part of a news account of one study:

> A hospital study that compared brain cancer patients and a similar group without brain cancer found no statistically significant association between cell phone use and a group of brain cancers known as gliomas. But when 20 types of glioma were considered separately an association was found between phone use and one rare form. Puzzlingly, however, this risk appeared to decrease rather than increase with greater mobile phone use.[12]

Think for a moment: Suppose that the 20 null hypotheses for these 20 significance tests are all true. Then each test has a 5% chance of being significant at the 5% level. That's what $\alpha = 0.05$ means—results this extreme occur only 5% of the time just by chance when the null hypothesis is true. Because 5% is 1/20, we expect about 1 of 20 tests to give a significant result just by chance. Running 1 test and reaching the $\alpha = 0.05$ level is reasonably good evidence that you have found something; running 20 tests and reaching that level only once is not.

The peril of multiple analyses is increased now that a few simple commands will set software to work performing all manner of complicated tests and operations on your data. We will state it as a law that any large set of data—even several pages of a table of random digits—contains some unusual pattern. Sufficient computer time will discover that pattern, and when you test specifically for the pattern that turned up, the result will be significant. That's much like testing for 20 kinds of cancer and finding one test significant. The proper use of significance levels involves choosing one hypothesis before searching your data and then testing that one hypothesis.

Searching data for suggestive patterns is certainly legitimate. Exploratory data analysis is an important part of statistics. But the reasoning of formal inference does not apply when your search for a striking effect in the data is successful. The remedy is clear. Once you have a hypothesis, design a study to search specifically for the effect you now think is there. If the result of this study is statistically significant, you have real evidence.

APPLY YOUR KNOWLEDGE

6.84 Predicting success of trainees. What distinguishes managerial trainees who eventually become executives from those who, after expensive training, don't succeed and leave the company? We have abundant data on past trainees—data on their personalities and goals, their college preparation and performance, even their family backgrounds and their hobbies. Statistical software makes it easy to perform dozens of significance tests on these dozens of variables to see which ones best predict later success. We find that future executives are significantly more likely than washouts to have an urban or suburban upbringing and an undergraduate degree in a technical field.

Explain clearly why using these "significant" variables to select future trainees is not wise. Then suggest a follow-up study using this year's trainees as subjects that should clarify the importance of the variables identified by the first study.

SECTION 6.3 Summary

- *P*-values are more informative than the reject-or-not result of a fixed level α test. Beware of placing too much weight on traditional values of α, such as $\alpha = 0.05$.

- Very small effects can be highly significant (small P), especially when a test is based on a large sample. A statistically significant effect need not be practically important. Plot the data to display the effect you are seeking, and use confidence intervals to estimate the actual value of parameters.

- On the other hand, lack of significance does not imply that H_0 is true, especially when the test is based on a small sample.

- Significance tests are not always valid. Faulty data collection, outliers in the data, and testing a hypothesis on the same data that suggested the hypothesis can invalidate a test. Many tests run at once will probably produce some significant results by chance alone, even if all the null hypotheses are true.

SECTION 6.3 Exercises

For Exercise 6.81, see page 372; for 6.82, see page 373; for 6.83, see page 373; and for 6.84, see page 374.

6.85 Your role on a team. You are the statistical expert on a team that is planning a study. After you have made a careful presentation of the mechanics of significance testing, one of the team members suggests using $\alpha = 0.50$ for the study because you would be more likely to obtain statistically significant results with this choice. Explain in simple terms why this would not be a good use of statistical methods.

6.86 What do you know? A research report described two results that both achieved statistical significance at the 5% level. The P-value for the first is 0.049; for the second it is 0.00002. Do the P-values add any useful information beyond that conveyed by the statement that both results are statistically significant? Write a short paragraph explaining your views on this question.

6.87 Find some journal articles. Find two journal articles that report results with statistical analyses. For each article, summarize how the results are reported and write a critique of the presentation. Be sure to include details regarding use of significance testing at a particular level of significance, P-values, and confidence intervals.

6.88 Discuss these views. State whether or not you agree with each of the following statements, and provide a short summary of the reasons for your answers.
(a) If the P-value is not less than 0.05, the null hypothesis is proven.
(b) Practical significance is not the same as statistical significance.
(c) With a sufficiently powerful computer, we can perform a statistical analysis for any set of data.
(d) If you find an interesting pattern in a set of data, it is appropriate to then use a significance test to determine if the pattern is due to chance.

6.89 Significance tests. Which of the following questions does a test of significance answer?
(a) Is the sample or experiment properly designed?
(b) Is the observed effect due to chance?
(c) Is the observed effect important?

6.90 Vitamin C and colds. In a study of the suggestion that taking vitamin C will prevent colds, 400 subjects are assigned at random to one of two groups. The experimental group takes a vitamin C tablet daily, while the control group takes a placebo. At the end of the experiment, the researchers calculate the difference between the percents of subjects in the two groups who were free of colds. This difference is statistically significant ($P = 0.03$) in favor of the vitamin C group. Can we conclude that vitamin C has a strong effect in preventing colds? Explain your answer.

6.91 Coaching for the SAT. Every user of statistics should understand the distinction between statistical significance and practical importance. A sufficiently large sample will declare very small effects statistically significant. Let us suppose that SAT Mathematics (SATM) scores in the absence of coaching vary Normally with mean $\mu = 515$ and $\sigma = 100$. Suppose further that coaching may change μ but does not change σ. An increase in the SATM score from 515 to 518 is of no importance in seeking admission to college, but this unimportant change can be statistically very significant. To see this, calculate the P-value for the test of

$$H_0: \mu = 515$$
$$H_a: \mu > 515$$

in each of the following situations:
(a) A coaching service coaches 100 students; their SATM scores average $\bar{x} = 518$.
(b) By the next year, the service has coached 1000 students; their SATM scores average $\bar{x} = 518$.
(c) An advertising campaign brings the number of students coached to 10,000; their average score is still $\bar{x} = 518$.

6.92 Coaching for the SAT. Give a 99% confidence interval for the mean SATM score μ after coaching in each part of the previous exercise. For large samples, the confidence interval says, "Yes, the mean score is higher after coaching, but only by a small amount."

6.93 Student loan poll. A local television station announces a question for a call-in opinion poll on the six o'clock news and then gives the response on the eleven o'clock news. Today's question concerns a proposed increase in funds for student loans. Of the 2372 calls received, 1921 oppose the increase. The station, following standard statistical practice, makes a confidence statement: "81% of the Channel 13 Pulse Poll sample oppose the increase. We can be 95% confident that the proportion of all viewers who oppose the increase is within 1.6% of the sample result." Is the station's conclusion justified? Explain your answer.

6.94 Extrasensory perception. A researcher looking for evidence of extrasensory perception (ESP) tests 500 subjects. Four of these subjects do significantly better ($P < 0.01$) than random guessing.
(a) Is it proper to conclude that these four people have ESP? Explain your answer.
(b) What should the researcher now do to test whether any of these four subjects have ESP?

6.95 More than one test. A P-value based on a single test is misleading if you perform several tests. The *Bonferroni procedure* gives a significance level for several tests together. Level α then means that if *all* the null hypotheses are true, the probability is α that *any* of the tests rejects its null hypothesis.

If you perform 2 tests and want to use the $\alpha = 5\%$ significance level, Bonferroni says to require a P-value of $0.05/2 = 0.025$ to declare either one of the tests significant. In general, if you perform k tests and want protection at level α, use α/k as your cutoff for statistical significance for each test.

You perform 6 tests and obtain individual P-values of 0.476, 0.032, 0.241, 0.008, 0.010, and 0.001. Which of these are statistically significant using the Bonferroni procedure with $\alpha = 0.05$?

6.96 More than one test. Refer to the previous exercise. A researcher has performed 12 tests of significance and wants to apply the Bonferroni procedure with $\alpha = 0.05$. The calculated P-values are 0.041, 0.569, 0.050, 0.416, 0.001, 0.004, 0.256, 0.041, 0.888, 0.010, 0.002, and 0.433. Which of these tests reject their null hypotheses with this procedure?

6.97 Many tests. Long ago, a group of psychologists carried out 77 separate significance tests and found that 2 were significant at the 5% level. Suppose that these tests are independent of each other. (In fact, they were not independent, because all involved the same subjects.) If all of the null hypotheses are true, each test has probability 0.05 of being significant at the 5% level.
(a) What is the distribution of the number X of tests that are significant?
(b) Find the probability that 2 or more of the tests are significant.

6.4 Power and Inference as a Decision

Although we prefer to use P-values rather than the reject-or-not view of the fixed α significance test, the latter view is important for planning statistical studies. We will first explain why, then discuss different views of statistical tests.

The power of a statistical test

In examining the usefulness of a confidence interval, we are concerned with both the level of confidence and the margin of error. The confidence level tells us how reliable the method is in repeated use. The margin of error tells us how sensitive the method is, that is, how closely the interval pins down the parameter being estimated. Fixed level α significance tests are closely related to confidence intervals—in fact, we saw that a two-sided test can be carried out directly from a confidence interval. The significance level, like the confidence level, says how reliable the method is in repeated use. If we use 5% significance tests repeatedly when H_0 is in fact true, we will be wrong (the test will reject H_0) 5% of the time and right (the test will fail to reject H_0) 95% of the time.

High confidence is of little value if the interval is so wide that few values of the parameter are excluded. Similarly, a test with a small level of α is of little value if it almost never rejects H_0 even when the true parameter value is far from the hypothesized value. We must be concerned with the ability of a test to detect that H_0 is false, just as we are concerned with the margin of error of a confidence interval. This ability is measured by the probability that the test will reject H_0 when an alternative is true. The higher this probability is, the more sensitive the test is.

> **Power**
>
> The probability that a fixed level α significance test will reject H_0 when a particular alternative value of the parameter is true is called the **power** of the test.

EXAMPLE 6.23   Exercise and Bones

Can a 6-month exercise program increase the total body bone mineral content (TBBMC) of young women? A team of researchers is planning a study to examine this question. Based on the results of a previous study, they are willing to assume that $\sigma = 2$ for the percent change in TBBMC over the 6-month period. A change in TBBMC of 1% would be considered important, and the researchers would like to have a reasonable chance of detecting a change this large or larger. Are 25 subjects a large enough sample for this project?

We will answer this question by calculating the power of the significance test that will be used to evaluate the data to be collected. The calculation consists of three steps:

1. State H_0, H_a (the particular alternative we want to detect), and the significance level α.

2. Find the values of $\bar{x}$ that will lead us to reject H_0.

3. Calculate the probability of observing these values of $\bar{x}$ when the alternative is true.

Step 1 The null hypothesis is that the exercise program has no effect on TBBMC. In other words, the mean percent change is zero. The alternative is that exercise is beneficial; that is, the mean change is positive. Formally, we have

$$H_0: \mu = 0$$
$$H_a: \mu > 0$$

Even though the alternative hypothesis considers all values of μ greater than 0, we are specifically interested in detecting a μ-value of at least 1. We will perform our power calculations with the *conservative* choice of $\mu = 1$. If in reality the value of μ departs even farther from the null hypothesis value of 0 than 1, then the actual power will be even greater than what we would calculate for $\mu = 1$. A 5% test of significance will be used.

Step 2 The z test rejects H_0 at the $\alpha = 0.05$ level whenever

$$z = \frac{\bar{x} - \mu_0}{\sigma/\sqrt{n}} = \frac{\bar{x} - 0}{2/\sqrt{25}} \geq 1.645$$

Be sure you understand why we use 1.645. Rewrite this in terms of $\bar{x}$:

$$\text{reject } H_0 \text{ when } \bar{x} \geq 1.645 \frac{2}{\sqrt{25}}$$
$$\text{reject } H_0 \text{ when } \bar{x} \geq 0.658$$

Because the significance level is $\alpha = 0.05$, this event has probability 0.05 of occurring *when the population mean μ is 0*.

Step 3 The power for the alternative $\mu = 1\%$ increase in TBBMC is the probability that H_0 will be rejected *when in fact $\mu = 1$*. We calculate this probability by standardizing

FIGURE 6.16 The sampling distributions of $\bar{x}$ when $\mu = 0$ and when $\mu = 1$, with α and the power. Power is the probability that the test rejects H_0 when the alternative is true.

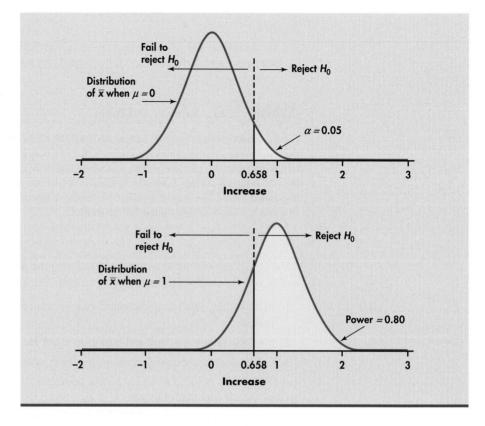

$\bar{x}$, using the value $\mu = 1$, the population standard deviation $\sigma = 2$, and the sample size $n = 25$. The power is

$$P(\bar{x} \geq 0.658 \quad \text{when} \quad \mu = 1) = P\left(\frac{\bar{x} - \mu}{\sigma/\sqrt{n}} \geq \frac{0.658 - 1}{2/\sqrt{25}}\right)$$
$$= P(Z \geq -0.855) = 0.80$$

Figure 6.16 illustrates the power with the sampling distribution of $\bar{x}$ when $\mu = 1$. This significance test rejects the null hypothesis that exercise has no effect on TBBMC 80% of the time if the true effect of exercise is a 1% increase in TBBMC. If the true effect of exercise is greater than a 1% increase, the test will have greater power; it will reject with a higher probability.

High power is desirable. Along with 95% confidence intervals and 5% significance tests, 80% power is becoming a standard. Many U.S. government agencies that provide research funds require that the sample size for the funded studies be sufficient to detect important results 80% of the time using a 5% test of significance.

Increasing the power

Suppose you have performed a power calculation and found that the power is too small. What can you do to increase it? Here are four ways:

- Increase α. A 5% test of significance will have a greater chance of rejecting the null hypothesis than a 1% test because the strength of evidence required for rejection is less.

- Consider a particular alternative that is farther away from μ_0. Values of μ that are in H_a but lie close to the hypothesized value μ_0 are harder to detect (lower power) than values of μ that are far from μ_0.

- Increase the sample size. More data will provide more information about $\bar{x}$ so we have a better chance of distinguishing values of μ.

- Decrease σ. This has the same effect as increasing the sample size: more information about μ. Improving the measurement process and restricting attention to a subpopulation are two common ways to decrease σ. Additionally, the use of better technology (for example, in the manufacturing of items) and improved standard operating procedures (SOPs) can have the desired result of reduced variability in both manufacturing and service applications.

Power calculations are important in planning studies. Using a significance test with low power makes it unlikely that you will find a significant effect even if the truth is far from the null hypothesis. A null hypothesis that is in fact false can become widely believed if repeated attempts to find evidence against it fail because of low power. The following example illustrates this point.

EXAMPLE 6.24   Are Stock Markets Efficient?

The "efficient market hypothesis" for stock prices says that future stock prices show only random variation. No information available now will help us predict stock prices in the future, because the efficient working of the market has already incorporated all available information in the present price. Many studies tested the claim that one or another kind of information is helpful. In these studies, the efficient market hypothesis is H_0, and the claim that prediction is possible is H_a. Almost all studies failed to find good evidence against H_0. As a result, the efficient market theory became quite popular.

An examination of the significance tests employed found that the power was generally low. Failure to reject H_0 when using tests of low power is not evidence that H_0 is true. As one expert said, "The widespread impression that there is strong evidence for market efficiency may be due just to a lack of appreciation of the low power of many statistical tests."[13] More careful studies later showed that the size of a company and measures of value such as the ratio of stock price to earnings do help predict future stock price movements.

Here is another example of a power calculation, this time for a two-sided z test.

EXAMPLE 6.25   Find the Power for a Two-Sided z Test

CASE 6.2   Case 6.2 considered the following competing hypotheses:

$$H_0: \mu = 500$$
$$H_a: \mu \neq 500$$

In Example 6.10, we learned that $\sigma = 2$ ml for the filling process. Suppose that the bottler Bestea wishes to conduct tests of the filling process mean at a 1% level of significance. Assume, as in Example 6.10, that 10 bottles are randomly chosen for inspection. What is the power of this test against the specific alternative $\mu = 498.5$?

From Table D, we find that z-values less than -2.576 or greater than 2.576 would be viewed as significant at the 1% level. In other words, the test rejects H_0 when $|z| \geq 2.576$. The test statistic is

$$z = \frac{\bar{x} - 500}{2/\sqrt{10}}$$

Some arithmetic shows that the test rejects when either of the following is true:

$$z \geq 2.576 \text{ (in other words, } \overline{x} \geq 501.629)$$

$$z \leq -2.576 \text{ (in other words, } \overline{x} \leq 498.371)$$

These are disjoint events, so the power is the sum of their probabilities, *computed assuming that the alternative* $\mu = 498.5$ *is true*. We find that

$$P(\overline{x} \geq 501.629) = P\left(\frac{\overline{x} - \mu}{\sigma/\sqrt{n}} \geq \frac{501.629 - 498.5}{2/\sqrt{10}}\right)$$

$$= P(Z \geq 4.95) \doteq 0$$

$$P(\overline{x} \leq 498.371) = P\left(\frac{\overline{x} - \mu}{\sigma/\sqrt{n}} \leq \frac{498.371 - 498.5}{2/\sqrt{10}}\right)$$

$$= P(Z \leq -0.204) = 0.4192$$

Figure 6.17 illustrates this calculation. Because the power is only about 0.42, we are not strongly confident that the test will reject H_0 when this alternative is true.

FIGURE 6.17 Power for Example 6.25.

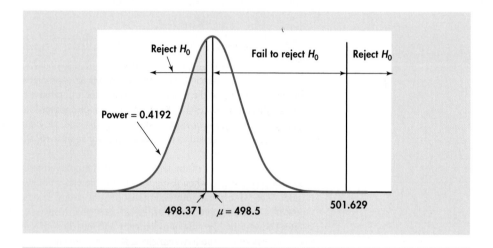

In Example 6.25, the specific alternative of $\mu = 498.5$ was considered. Since we are dealing with a two-sided test, the power of the test would also be 0.4192 for the alternative of $\mu = 501.5$. At the heart of the power calculations is the difference between the null hypothesis value and the alternative of interest. In Example 6.25, we are considering the test's ability to detect a departure of 1.5 ml from the null hypothesis value of 500 ml. Using the same σ, sample size, and significance level of Example 6.25, we would equally find the power to be 0.4192 for detecting any alternative hypothesis that is 1.5 units away from the null hypothesis.

If the power of 0.4192 found in Example 6.25 is unsatisfactory to the bottler, we noted earlier that one option is to increase the sample size. Just how large should the sample be? The following example explores this question.

EXAMPLE 6.26 Choosing Sample Size for a Desired Power?

CASE 6.2

Suppose the bottler Bestea desires a power of 0.9 in the detection of the specific alternative of $\mu = 498.5$. From Example 6.25, we found that a sample size of 10 offers a power of only 0.4192. Manually, we can repeat the calculations found in Example 6.25 for different values of n larger than 10 until we find the smallest sample size giving at least a power of 0.9. Fortunately, most statistical software saves us from such a tedium. Below is Minitab output with inputs of 0.9 for power, 1% for significance level, 2 for σ, and -1.5 ($= 498.5 - 500$) for the departure amount from the null hypothesis:

```
Power and Sample Size

1-Sample Z Test

Testing mean = null (versus not = null)
Calculating power for mean = null + difference
Alpha = 0.01  Assumed standard deviation = 2

              Sample  Target
Difference    Size    Power   Actual Power
      -1.5     27      0.9      0.906797
```

From the output, we learn that a sample size of at least 27 is needed to have a power of at least 0.9. If we used a sample size of 26, the actual power would be less than the target power of 0.9.

In addition to the output shown in Example 6.26, statistical software will often show graphically the power of the test for several possible departures from the null hypothesis. Such a graph is known as a **power curve.** Figure 6.18 shows the power curves for the bottling application for $n = 10$ and $n = 27$. The power values of 0.4192 and 0.906797 found in Examples 6.25 and 6.26 are labeled on the curves. The curves are intuitive in that we find that the power of the test for a given sample size becomes stronger the larger the departure (low or high) is from the null hypothesis.

power curve

Inference as decision

We have presented tests of significance as methods for assessing the strength of evidence against the null hypothesis. This assessment is made by the P-value, which is a probability computed under the assumption that H_0 is true. The alternative hypothesis (the statement

FIGURE 6.18 Power curves for two departures from the null hypothesis mean for a population with $\sigma = 2$ for a testing procedure using a 1% significance level and $n = 10$ and 27. The power values found in Examples 6.25 and 6.26 are labeled on the curves.

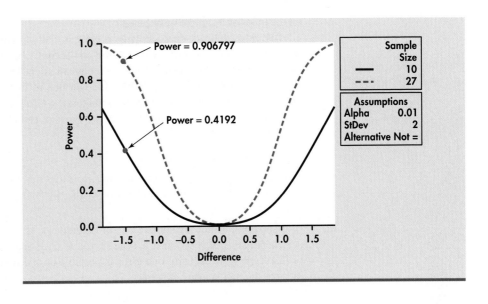

we seek evidence for) enters the test only to show us what outcomes count against the null hypothesis. Most users of statistics in practice think of tests in this way.

But signs of another way of thinking were present in our discussion of significance tests with fixed level α. A level of significance α chosen in advance points to the outcome of the test as a *decision*. If the P-value is less than α, we reject H_0 in favor of H_a. Otherwise, we fail to reject H_0. There is a big distinction between measuring the strength of evidence and making a decision. Many statisticians feel that making decisions is too grand a goal for statistics alone. Decision makers should take account of the results of statistical studies, but decisions should be based on many additional factors that are hard to reduce to numbers.

Yet there are circumstances in which a statistical test leads directly to a decision. The quality control application of Case 6.2 is one circumstance. In that application, the bottler needs to decide whether or not to make adjustments to the filling process based on a sample outcome. Consider another example. A producer of ball bearings and the consumer of the ball bearings agree that each shipment of bearings shall meet certain quality standards. When a shipment arrives, the consumer inspects a random sample of bearings from the thousands of bearings found in the shipment. On the basis of the sample outcome, the consumer either accepts or rejects the shipment. Some statisticians argue that if "decision" is given a broad meaning, almost all problems of statistical inference can be posed as problems of making decisions in the presence of uncertainty. We will not venture further into the arguments over how we ought to think about inference. We do want to show how a different concept—inference as decision—changes the reasoning used in tests of significance.

Two types of error

Tests of significance concentrate on H_0, the null hypothesis. If a decision is called for, however, there is no reason to single out H_0. There are simply two hypotheses, and we must accept one and reject the other. It is convenient to call the two hypotheses H_0 and H_a, but H_0 no longer has the special status (the statement we try to find evidence against) that it had in tests of significance. In the ball bearing problem, we must decide between

H_0: the shipment of bearings meets standards

H_a: the shipment does not meet standards

on the basis of a sample of bearings.

We hope that our decision will be correct, but sometimes it will be wrong. There are two types of incorrect decisions. We can accept a bad shipment of bearings, or we can reject a good shipment. Accepting a bad shipment leads to a variety of costs to the consumer (for example, machine breakdown due to faulty bearings or injury to end-product users such as skateboarders or bikers), while rejecting a good shipment hurts the producer. To help distinguish these two types of error, we give them specific names.

Type I and Type II Errors

If we reject H_0 (accept H_a) when in fact H_0 is true, this is a **Type I error.**

If we accept H_0 (reject H_a) when in fact H_a is true, this is a **Type II error.**

The possibilities are summed up in Figure 6.19. If H_0 is true, our decision either is correct (if we accept H_0) or is a Type I error. If H_a is true, our decision either is correct or is a Type II error. Only one error is possible at one time. Figure 6.20 applies these ideas to the ball bearing example.

FIGURE 6.19 The two types of error in testing hypotheses.

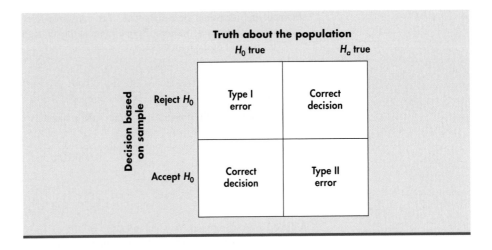

FIGURE 6.20 The two types of error for the sampling of ball bearings application.

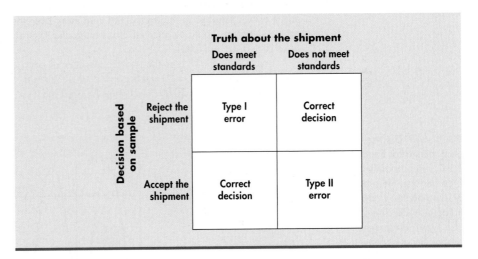

Error probabilities

We can assess any rule for making decisions in terms of the probabilities of the two types of error. This is in keeping with the idea that statistical inference is based on probability. We cannot (short of inspecting the whole shipment) guarantee that good shipments of bearings will never be rejected and bad shipments will never be accepted. But by random sampling and the laws of probability, we can say what the probabilities of both kinds of error are.

Significance tests with fixed level α give a rule for making decisions because the test either rejects H_0 or fails to reject it. If we adopt the decision-making way of thought, failing to reject H_0 means deciding to act as if H_0 is true. We can then describe the performance of a test by the probabilities of Type I and Type II errors.

EXAMPLE 6.27   Diameters of Bearings

The diameter of a particular precision ball bearing has a target value of 20 millimeters (mm) with tolerance limits of ± 0.001 mm around the target. Suppose that the bearing diameters vary Normally with standard deviation of sixty-five hundred-thousandths of a millimeter, that is, $\sigma = 0.00065$ mm. When a shipment of the bearings arrives, the consumer takes an SRS of 5 bearings from the shipment and measures their diameters. The consumer rejects the bearings if the sample mean diameter is significantly different from 20 mm at the 5% significance level.

This is a test of the hypotheses

$$H_0: \mu = 20$$

$$H_a: \mu \neq 20$$

To carry out the test, the consumer computes the z statistic:

$$z = \frac{\bar{x} - 20}{0.00065/\sqrt{5}}$$

and rejects H_0 if

$$z < -1.96 \quad \text{or} \quad z > 1.96$$

A Type I error is to reject H_0 when in fact $\mu = 20$.

What about Type II errors? Because there are many values of μ in H_a, we will concentrate on one value. Based on the tolerance limits, the producer agrees that if there is evidence that the mean of ball bearings in the lot is 0.001 mm away from the desired mean of 20 mm, then the whole shipment should be rejected. So, a particular Type II error is to accept H_0 when in fact $\mu = 20 + 0.001 = 20.001$.

Figure 6.21 shows how the two probabilities of error are obtained from the two sampling distributions of $\bar{x}$, for $\mu = 20$ and for $\mu = 20.001$. When $\mu = 20$, H_0 is true and to reject H_0 is a Type I error. When $\mu = 20.001$, accepting H_0 is a Type II error. We will now calculate these error probabilities.

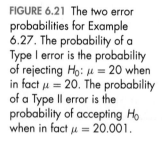

FIGURE 6.21 The two error probabilities for Example 6.27. The probability of a Type I error is the probability of rejecting H_0: $\mu = 20$ when in fact $\mu = 20$. The probability of a Type II error is the probability of accepting H_0 when in fact $\mu = 20.001$.

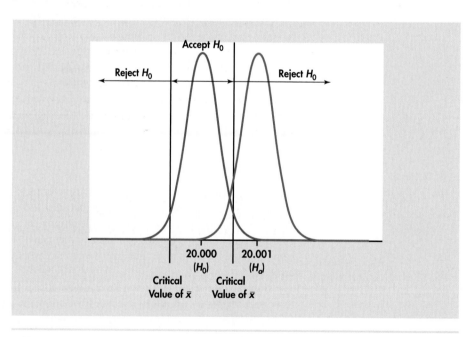

The probability of a Type I error is the probability of rejecting H_0 when it is really true. In Example 6.27, this is the probability that $|z| \geq 1.96$ when $\mu = 20$. But this is exactly the significance level of the test. The critical value 1.96 was chosen to make this probability 0.05, so we do not have to compute it again. The definition of "significant at level 0.05" is that sample outcomes this extreme will occur with probability 0.05 when H_0 is true.

> **Significance and Type I Error**
>
> The significance level α of any fixed level test is the probability of a Type I error. That is, α is the probability that the test will reject the null hypothesis H_0 when H_0 is in fact true.

The probability of a Type II error for the particular alternative $\mu = 20.001$ in Example 6.27 is the probability that the test will fail to reject H_0 when μ has this alternative value. The *power* of the test for the alternative $\mu = 20.001$ is just the probability that the test *does* reject H_0 when H_a is true. By following the method of Example 6.25, we can calculate that the power is about 0.93. Therefore, the probability of a Type II error is equal to $1 - 0.93$, or 0.07. It would also be the case that the probability of a Type II error is 0.07 if the value of the alternative μ is 19.999, that is, 0.001 less than the null hypothesis mean of 20.

> **Power and Type II Error**
>
> The power of a fixed level test for a particular alternative is 1 minus the probability of a Type II error for that alternative.

The two types of error and their probabilities give another interpretation of the significance level and power of a test. The distinction between tests of significance and tests as rules for deciding between two hypotheses lies, not in the calculations, but in the reasoning that motivates the calculations. In a test of significance we focus on a single hypothesis (H_0) and a single probability (the P-value). The goal is to measure the strength of the sample evidence against H_0. Calculations of power are done to check the sensitivity of the test. If we cannot reject H_0, we conclude only that there is not sufficient evidence against H_0, not that H_0 is actually true. If the same inference problem is thought of as a decision problem, we focus on two hypotheses and give a rule for deciding between them based on the sample evidence. We therefore must focus equally on two probabilities, the probabilities of the two types of error. We must choose one or the other hypothesis and cannot abstain on grounds of insufficient evidence.

The common practice of testing hypotheses

Such a clear distinction between the two ways of thinking is helpful for understanding. In practice, the two approaches often merge. We continued to call one of the hypotheses in a decision problem H_0. The common practice of *testing hypotheses* mixes the reasoning of significance tests and decision rules as follows:

1. State H_0 and H_a just as in a test of significance.
2. Think of the problem as a decision problem, so that the probabilities of Type I and Type II errors are relevant.
3. Because of Step 1, Type I errors are more serious. So, choose an α (significance level) and consider only tests with probability of Type I error no greater than α.
4. Among these tests, select one that makes the probability of a Type II error as small as possible (that is, power as large as possible). If this probability is too large, you will have to take a larger sample to reduce the chance of an error.

Testing hypotheses may seem to be a hybrid approach. It was, historically, the effective beginning of decision-oriented ideas in statistics. An impressive mathematical theory of hypothesis testing was developed between 1928 and 1938 by Jerzy Neyman and Egon Pearson. The decision-making approach came later (1940s). Because decision theory in its pure form leaves you with two error probabilities and no simple rule on how to balance them, it has been used less often than either tests of significance or tests of hypotheses.

SECTION 6.4 Summary

- The **power** of a significance test measures its ability to detect an alternative hypothesis. The power for a specific alternative is calculated as the probability that the test will reject H_0 when that alternative is true. This calculation requires knowledge of the sampling distribution of the test statistic under the alternative hypothesis. Increasing the size of the sample increases the power when the significance level remains fixed.

- In the case of testing H_0 versus H_a, decision analysis chooses a decision rule on the basis of the probabilities of two types of error. A **Type I error** occurs if H_0 is rejected when it is in fact true. A **Type II error** occurs if H_0 is accepted when in fact H_a is true.

- In a fixed level α significance test, the significance level α is the probability of a Type I error, and the power for a specific alternative is 1 minus the probability of a Type II error for that alternative.

SECTION 6.4 Exercises

6.98 Make a recommendation. Your manager has asked you to review a research proposal that includes a section on sample size justification. A careful reading of this section indicates that the power is 20% for detecting an effect that you would consider important. Write a short report for your manager explaining what this means, and make a recommendation on whether or not this study should be run.

6.99 Explain power and sample size. Two studies are identical in all respects except for the sample sizes. Consider the power versus a particular sample size. Will the study with the larger sample size have more power or less power than the one with the smaller sample size? Explain your answer in terms that could be understood by someone with very little knowledge of statistics.

6.100 Power versus a different alternative. The power for a two-sided test of the null hypothesis $\mu = 0$ versus the alternative $\mu = 10$ is 0.82. What is the power versus the alternative $\mu = -10$? Draw a picture and use this to explain your answer.

6.101 Power versus a different alternative. A one-sided test of the null hypothesis $\mu = 50$ versus the alternative $\mu = 70$ has power equal to 0.5. Will the power for the alternative $\mu = 80$ be higher or lower than 0.5? Draw a picture and use this to explain your answer.

6.102 Profitability of IPO firms. In Example 6.19 (page 364), we ask if the mean of the profitability of IPO firms has changed during the initial three-year period since becoming public. If μ is the mean percent change for the population of IPO firms, the hypotheses are $H_0: \mu = 0$ and $H_a: \mu \neq 0$. The data come from

a sample of 3964 firms, and we assume that the population standard deviation is $\sigma = 15.56$. The test rejects H_0 at the 1% level of significance when $z \geq 2.576$ or $z < -2.576$, where

$$z = \frac{\bar{x} - 0}{15.56/\sqrt{3964}}$$

What is the power of the test for detecting a 1% change in mean profitability?

6.103 Sample size determination. Example 6.23 (page 377) discusses a test of $H_0 : \mu = 0$ against $H_0 : \mu > 0$ where μ is the mean percent change in total body bone mineral content of young women. The population standard deviation is given to be $\sigma = 2$, and the testing was performed at a 5% significance level. Use software for the following:

(a) What sample size is minimally required to give at least 0.9 power? What is the actual power for this sample size?

(b) Produce a power curve for a sample size of 25. How does the power curve differ in look from the power curves of Figure 6.18?

(c) Looking at the graph, for what alternative mean values is the detection of the alternative almost certain?

6.104 Mail-order catalog sales. You want to see if a redesign of the cover of a mail-order catalog will increase sales. A very large number of customers will receive the original catalog, and a random sample of customers will receive the one with the new cover. For planning purposes, you are willing to assume that the mean sales for the new catalog will be approximately Normal with $\sigma = 50$ dollars and that the mean sales for the original catalog will be $\mu = 25$ dollars. You decide to use a sample size of $n = 900$. You wish to test

$$H_0: \mu = 25$$
$$H_a: \mu > 25$$

You decide to reject H_0 if $\bar{x} > 26$ and to accept H_0 otherwise.
(a) Find the probability of a Type I error, that is, the probability that your test rejects H_0 when in fact $\mu = 25$ dollars.
(b) Find the probability of a Type II error when $\mu = 28$ dollars. This is the probability that your test accepts H_0 when in fact $\mu = 28$.
(c) Find the probability of a Type II error when $\mu = 30$.
(d) The distribution of sales is not Normal, because many customers buy nothing. Why is it nonetheless reasonable in this circumstance to assume that the mean will be approximately Normal?

6.105 Decreasing Type II error probability. What are four distinct ways to decrease the probability of a Type II error? Explain why these ways will decrease the error probability.

6.106 Profitability of IPO firms. Use the result of Exercise 6.102 to give the probabilities of Type I and Type II errors for the test discussed there.

6.107 Decreasing population standard deviation. Improved measurement systems, better technology, and changes to standard operating procedures are among various strategies to reduce population variability in manufacturing and service applications. Suppose variation reduction strategies are implemented and reduce the population standard deviation by 50%; that is, it is half of its original value.
(a) If n is the sample size used for hypothesis testing under the original standard deviation, what sample size in terms of n is now required to maintain some specified power or Type II error probability?
(b) If the new sample size were used, what might you be concerned about? (*Hint:* Think about the sampling distribution.)

6.108 Decide. You must decide which of two discrete distributions a random variable X has. We will call the distributions p_0 and p_1. Here are the probabilities that the distributions assign to the values x of X:

x:	0	1	2	3	4	5	6
p_0:	0.1	0.1	0.1	0.1	0.2	0.1	0.3
p_1:	0.2	0.1	0.1	0.2	0.2	0.1	0.1

You have a single observation on X and wish to test

$$H_0: p_0 \text{ is correct}$$
$$H_a: p_1 \text{ is correct}$$

One possible decision procedure is to accept H_0 if $X = 4$ or $X = 6$ and reject H_0 otherwise.
(a) Find the probability of a Type I error, that is, the probability that you reject H_0 when p_0 is the correct distribution.
(b) Find the probability of a Type II error.

6.109 A Web-based business. You are in charge of marketing for a Web site that offers automated medical diagnoses. The program will scan the results of routine medical tests (pulse rate, blood pressure, urinalysis, etc.) and either clear the patient or refer the case to a doctor. You are marketing the program for use as part of a preventive-medicine system to screen many thousands of persons who do not have specific medical complaints. The program makes a decision about each patient.
(a) What are the two hypotheses and the two types of error that the program can make? Describe the two types of error in terms of "false positive" and "false negative" test results.
(b) The program can be adjusted to decrease one error probability at the cost of an increase in the other error probability. Which error probability would you choose to make smaller, and why? (This is a matter of judgment. There is no single correct answer.)

6.110 Inspecting shipments. In Example 6.27 (page 383), we considered the situation where a sample from a shipment of many parts serves as a basis for rejecting or accepting the shipment. Suppose that a sampling test has probability 0.05 of rejecting a good shipment of manufactured parts and probability 0.08 of accepting a bad shipment. The consumer of the manufactured parts may imagine that such a sampling procedure guarantees that most accepted shipments are good. Alas, it is not so. Suppose that 90% of all shipments delivered by the producer are bad.
(a) Draw a tree diagram for delivering a shipment (the branches are "bad" and "good") and then inspecting it (the branches at this stage are "accept" and "reject").
(b) Write the appropriate probabilities on the branches, and find the probability that a shipment is accepted.
(c) Use the definition of conditional probability or Bayes's formula (page 302) to find the probability that a shipment is bad, given that it has been accepted. This is the proportion of bad shipments among the shipments that the sampling procedure accepts.

STATISTICS IN SUMMARY

Statistical inference draws conclusions about a population on the basis of sample data and uses probability to indicate how reliable the conclusions are. A confidence interval estimates an unknown parameter. A significance test shows how strong the evidence is for some claim about a parameter.

The probabilities in both confidence intervals and tests tell us what would happen if we used the same procedure for the interval construction or test computation very many times. A confidence level is the probability that the method for constructing a confidence interval actually produces an interval that contains the unknown parameter. A 95% confidence interval gives a correct result 95% of the time when we use it repeatedly. A *P*-value is the probability that the sample would produce a result at least as extreme as the observed result if the null hypothesis really were true. That is, a *P*-value tells us how surprising the observed outcome is. Very surprising outcomes (small *P*-values) are good evidence that the null hypothesis is not true.

These ideas are the foundation for the rest of this book. We will have much to say about many statistical methods and their use in practice. In every case, the basic reasoning of confidence intervals and significance tests remains the same. Here is a review list of the most important skills you should have acquired from your study of this chapter.

A. Confidence Intervals

1. State in nontechnical language what is meant by "95% confidence" or other statements of confidence in statistical reports.

2. Construct a confidence interval for the mean μ of a Normal population with known standard deviation σ by calculating $\overline{x} \pm z^*\sigma/\sqrt{n}$.

3. Recognize when you can safely interpret a confidence interval and when the data collection design or a small sample from a skewed population makes it inaccurate. Understand that the margin of error does not include the effects of undercoverage, nonresponse, or other practical difficulties.

4. Understand how the margin of error of a confidence interval changes with the sample size and the level of confidence C.

5. Find the sample size required to obtain a confidence interval of specified margin of error m when the confidence level and other information are given.

B. Significance Tests

1. State the null and alternative hypotheses in a testing situation when the parameter in question is a population mean μ.

2. Explain in nontechnical language the meaning of the *P*-value when you are given the numerical value of *P* for a test.

3. Calculate the one-sample z statistic and the *P*-value for both one-sided and two-sided tests about the mean μ of a Normal population.

4. Assess statistical significance at standard levels α, either by comparing *P* to α or by comparing z to standard Normal critical values z^*.

5. Recognize that significance testing does not measure the size or importance of an effect.

6. Recognize when you can use the z test and when the data collection design or a small sample from a skewed population makes it inappropriate.

C. Power and Errors

1. Understand that the power of a significance test measures its ability to detect an alternative hypothesis.

2. Calculate the power of a test of the mean against a two-sided or one-sided alternative.

3. Recognize that increasing the size of the sample increases the power of a significance test.

4. Understand that in the testing of hypotheses, there is a risk of committing one of two errors: rejecting the null hypothesis when it is in fact true (Type I) or accepting the null hypothesis when in fact the alternative is true (Type II).

5. Know the relationship between power and the probability of a Type II error.

CHAPTER 6 Review Exercises

6.111 Full-time employment and age. A study reported average months of full-time employment for individuals aged 18 to 26.[14] Here are the means:

Age:	18	19	20	21	22	23	24	25	26
Months employed:	2.9	4.2	5.0	5.3	6.4	7.4	8.5	8.9	9.3

Assume that the standard deviation for each of these means is 4.5 months and that each sample size is 750.
(a) Calculate the 95% confidence interval for each mean.
(b) Plot the means versus age. Draw a vertical line through the first mean extending up to the upper confidence limit and down to the lower limit. At each end of the line, draw a short dash. Do the same for each of the other means.
(c) Write a summary of what the data show. In circumstances such as this, it is common practice not to make any adjustments for the fact that several confidence intervals are being reported. Be sure to include comments about this in your summary.

6.112 Workers' perceptions about safety. The Safety Climate Index (SCI) measures workers' perceptions about the safety of their work environment. A study of safe work practices of industrial workers reported mean SCI scores for workers classified by workplace size.[15] Here are the means:

Workplace size:	Fewer than 50 workers	50 to 200 workers	More than 200 workers
Mean SCI:	67.23	70.37	74.83

Assume that the standard deviation is 19 and the sample sizes are all 180. (We will discuss ways to compare three means such as these in Chapter 12.)
(a) Calculate the 95% confidence interval for each mean.
(b) Plot the means versus workplace size. Draw a vertical line through the first mean extending up to the upper confidence limit and down to the lower limit. At each end of the line, draw a short dash. Do the same for each of the other means.
(c) One way to adjust for the fact that we are reporting three confidence intervals is a procedure that uses a larger value of z^* in the calculation of the margin of error. For this problem one recommendation would be to use $z^* = 2.40$. Repeat parts (a) and (b) making this adjustment.
(d) Summarize your results. Be sure to include comments on the effects of the adjustment on your results.

6.113 Generate some confidence intervals. For this exercise you will use the *Confidence Interval* applet. Set the confidence level at 95% and click the "Sample" button 10 times to simulate 10 confidence intervals. Record the percent hit. Simulate another 10 intervals by clicking another 10 times (do not click the "Reset" button). Record the percent hit for your 20 intervals. Repeat the process of simulating 10 additional intervals and recording the results until you have a total of 200 intervals. Plot your results and write a summary of what you have found.

6.114 Change the confidence level. Refer to the previous exercise. Do the simulations and report the results for 90% confidence.

6.115 Change the confidence level. Refer to Example 6.19 (page 364) and construct a 95% confidence interval for the mean change in profitability for the population of IPO firms.

6.116 Inheriting control and company performance. "The only reason I was on the payroll is because I was the son of the boss," testified John Tyson, chief executive officer (CEO) of Tyson Food, Inc., in a legal proceeding.[16] Promotion of one's kin in the business world is quite common. Showing favoritism based on relationship rather than an objective evaluation of a person's ability can be a source of controversy. A study of 335 firms investigated the effects of nepotism on corporate performance.[17] Each of the firms in the study underwent management transition in that there was a departing CEO who was replaced by a new CEO. All the firms were characterized as having concentrated ownership or founding-family involvement. The sample of 335 firms was split into two groups: (1) the incoming CEO was related by blood or marriage to the departing CEO, to the founder, or to the largest shareholder of the corporation, and (2) the incoming CEO had no relationship to the departing CEO. An incoming CEO classified in Group 1 is referred to as a "family" CEO. Here is an excerpt from the study:

> When firms are classified by incoming CEOs' family links, I find that firms that promote family CEOs undergo average declines in unadjusted OROA [operating return on assets] of 1.88 percentage points, significant at the one-percent level. In contrast, the difference in profitability of firms that appoint non-family CEOs is +0.21 percentage points, and is not different from zero at conventional levels. The resulting difference-in-differences is −2.09 percentage points, significant at the one-percent level.

How would you explain the three conclusions found in the excerpt to someone who knows no statistics? Include an explanation of the statistical significance for each of the conclusions.

6.117 Wine. Many food products contain small quantities of substances that would give an undesirable taste or smell if they were present in large amounts. An example is the "off-odors" caused by sulfur compounds in wine. Oenologists (wine experts) have determined the odor threshold, the lowest concentration of a compound that the human nose can detect. For example, the odor threshold for dimethyl sulfide (DMS) is given in the oenology literature as 25 micrograms per liter of wine (μg/l). Untrained noses may be less sensitive, however. Here are the DMS odor thresholds for 10 beginning students of oenology:

31 31 43 36 23 34 32 30 20 24

Assume (this is not realistic) that the standard deviation of the odor threshold for untrained noses is known to be $\sigma = 7$ μg/l.
(a) Make a stemplot to verify that the distribution is roughly symmetric with no outliers. (A Normal quantile plot confirms that there are no systematic departures from Normality.)
(b) Give a 95% confidence interval for the mean DMS odor threshold among all beginning oenology students.
(c) Are you convinced that the mean odor threshold for beginning students is higher than the published threshold, 25 μg/l? Carry out a significance test to justify your answer.

6.118 Annual household income. A government report gives a 90% confidence interval for the 2007 median annual household income as $50,233 \pm $230. This result was calculated by advanced methods from the Current Population Survey, a multistage random sample of about 50,000 households.
(a) Would a 95% confidence interval be wider or narrower? Explain your answer.
(b) Would the null hypothesis that the 2007 median household income was $50,000 be rejected at the 10% significance level in favor of the two-sided alternative?

6.119 Annual household income. Refer to the previous exercise. Give a 90% confidence interval for the 2007 median *weekly* household income. Use 52.14 ($= 365/7$) as the number of weeks in a year.

6.120 Too much cellulose to be profitable? Excess cellulose in alfalfa reduces the "relative feed value" of the product that will be fed to dairy cows. If the cellulose content is too high, the price will be lower and the producer will have less profit. An agronomist examines the cellulose content of one type of alfalfa hay. Suppose that the cellulose content in the population has standard deviation $\sigma = 8$ milligrams per gram (mg/g). A sample of 15 cuttings has mean cellulose content $\bar{x} = 145$ mg/g.
(a) Give a 90% confidence interval for the mean cellulose content in the population.
(b) A previous study claimed that the mean cellulose content was $\mu = 140$ mg/g, but the agronomist believes that the mean is higher than that figure. State H_0 and H_a and carry out a significance test to see if the new data support this belief.

(c) The statistical procedures used in (a) and (b) are valid when several assumptions are met. What are these assumptions?

6.121 Where do you buy? Consumers can purchase nonprescription medications at food stores, mass merchandise stores such as Kmart and Walmart, or pharmacies. About 45% of consumers make such purchases at pharmacies. What accounts for the popularity of pharmacies, which often charge higher prices?

A study examined consumers' perceptions of overall performance of the three types of store using a long questionnaire that asked about such things as "neat and attractive store," "knowledgeable staff," and "assistance in choosing among various types of nonprescription medication." A performance score was based on 27 such questions. The subjects were 201 people chosen at random from the Indianapolis telephone directory. Here are the means and standard deviations of the performance scores for the sample:[18]

Store type	$\bar{x}$	s
Food stores	18.67	24.95
Mass merchandisers	32.38	33.37
Pharmacies	48.60	35.62

We do not know the population standard deviations, but a sample standard deviation s from so large a sample is usually close to σ. Use s in place of the unknown σ in this exercise.
(a) What population do you think the authors of the study want to draw conclusions about? What population are you certain they can draw conclusions about?
(b) Give 95% confidence intervals for the mean performance for each type of store.
(c) Based on these confidence intervals, are you convinced that consumers think that pharmacies offer higher performance than the other types of stores? In Chapter 12 we will study a statistical method for comparing the means of several groups.

6.122 CEO pay. A study of the pay of corporate chief executive officers (CEOs) examined the increase in cash compensation of the CEOs of 104 companies, adjusted for inflation, in a recent year. The mean increase in real compensation was $\bar{x} = 6.9\%$, and the standard deviation of the increases was $s = 55\%$. Is this good evidence that the mean real compensation μ of all CEOs increased that year? The hypotheses are

$$H_0: \mu = 0 \quad \text{(no increase)}$$

$$H_a: \mu > 0 \quad \text{(an increase)}$$

Because the sample size is large, the sample s is close to the population σ, so take $\sigma = 55\%$.
(a) Sketch the Normal curve for the sampling distribution of $\bar{x}$ when H_0 is true. Shade the area that represents the P-value for the observed outcome $\bar{x} = 6.9\%$.
(b) Calculate the P-value.
(c) Is the result significant at the $\alpha = 0.05$ level? Do you think the study gives strong evidence that the mean compensation of all CEOs went up?

6.123 Large samples. Statisticians prefer large samples. Describe briefly the effect of increasing the size of a sample (or the number of subjects in an experiment) on each of the following.
(a) The width of a level C confidence interval.
(b) The P-value of a test, when H_0 is false and all facts about the population remain unchanged as n increases.
(c) The power of a fixed level α test, when α, the alternative hypothesis, and all facts about the population remain unchanged.

6.124 Roulette. A roulette wheel has 18 red slots among its 38 slots. You observe many spins and record the number of times that red occurs. Now you want to use these data to test whether the probability of a red has the value that is correct for a fair roulette wheel. State the hypotheses H_0 and H_a that you will test. (We will describe the test for this situation in Chapter 8.)

6.125 Significant. When asked to explain the meaning of "statistically significant at the $\alpha = 0.05$ level," a student says, "This means there is only probability 0.05 that the null hypothesis is true." Is this a correct explanation of statistical significance? Explain your answer.

6.126 Significant. Another student, when asked why statistical significance appears so often in research reports, says, "Because saying that results are significant tells us that they cannot easily be explained by chance variation alone." Do you think that this statement is essentially correct? Explain your answer.

6.127 Welfare reform. A study compares two groups of mothers with young children who were on welfare two years ago. One group attended a voluntary training program offered free of charge at a local vocational school and advertised in the local news media. The other group did not choose to attend the training program. The study finds a significant difference ($P < 0.01$) between the proportions of the mothers in the two groups who are still on welfare. The difference is not only significant but quite large. The report says that with 95% confidence the percent of the nonattending group still on welfare is 21% ± 4% higher than that of the group who attended the program. You are on the staff of a member of Congress who is interested in the plight of welfare mothers and who asks you about the report.
(a) Explain briefly and in nontechnical language what "a significant difference ($P < 0.01$)" means.
(b) Explain clearly and briefly what "95% confidence" means.
(c) Is this study good evidence that requiring job training of all welfare mothers would greatly reduce the percent who remain on welfare for several years?

6.128 Simulation. Use a computer to generate $n = 5$ observations from the Normal distribution $N(20, 5)$. Find the 95% confidence interval for μ. Repeat this process 100 times and then count the number of times that the confidence interval includes the value $\mu = 20$. Explain your results.

6.129 Simulation. Use a computer to generate $n = 5$ observations from the Normal distribution $N(20, 5)$. Test $H_0: \mu = 20$ versus $H_a: \mu \neq 20$ at the $\alpha = 0.05$ significance level. Repeat this process 100 times and then count the number of times that you reject H_0. Explain your results.

6.130 Simulation. Use the same procedure for generating data as in the previous exercise. Now test the null hypothesis that $\mu = 22.5$. Explain your results.

CASE 6.1 **6.131 Simulation.** Figure 6.5 (page 339) demonstrates the behavior of a confidence interval in repeated sampling by showing the results of 25 samples from the same population. Now you will do a similar demonstration, though in an artificial setting. Suppose that the attorney fees for Chapter 7 consumer bankruptcies for the population of Chapter 7 consumer bankruptcies follow the uniform distribution between $250 and $1750. Then the mean attorney fees is $\mu = 1000$ and the standard deviation is $\sigma = 433$.
(a) Simulate the drawing of 25 SRSs of size $n = 100$ from this population.
(b) For calculating the confidence intervals, you are willing to assume that the sample means are approximately Normal. Explain why this is a reasonable assumption.
(c) The 50% confidence interval for the population mean μ has the form $\bar{x} \pm m$. What is the margin of error m? (Use 433 for the standard deviation.)
(d) Use your software to calculate the 50% confidence interval for μ for each of your 25 samples. Verify the computer's calculations by checking the interval given for the first sample against your result in (b). Use the $\bar{x}$ reported by the software.
(e) How many of the 25 confidence intervals contain the true mean $\mu = 1000$? If you repeated the simulation, would you expect exactly the same number of intervals to contain μ? In a very large number of samples, what percent of the confidence intervals would contain μ?

CASE 6.1 **6.132 Simulation.** In the previous exercise you simulated the attorney fees of 25 SRSs of 100 Chapter 7 consumer bankruptcy cases. Now use these samples to demonstrate the behavior of a significance test. We know that the population standard deviation is $\sigma = 433$, and we are willing to assume that the sample means are approximately Normal.
(a) Use your software to carry out a test of

$$H_0: \mu = 1000$$
$$H_a: \mu \neq 1000$$

for each of the 25 samples.
(b) Verify the computer's calculations by using Table A to find the P-value of the test for the first of your samples. Use the $\bar{x}$ reported by your software.
(c) How many of your 25 tests reject the null hypothesis at the $\alpha = 0.05$ significance level? (That is, how many have P-value 0.05 or smaller?)
(d) Because the simulation was done with $\mu = 1000$, samples that lead to rejecting H_0 produce the wrong conclusion. In a very large number of samples, what percent would falsely reject the null hypothesis?

CHAPTER 6   Case Study Exercises

CASE STUDY EXERCISE 1: Older customers in restaurants. Persons aged 55 and over represented 21.3% of the U.S. population in the year 2000. This group is expected to increase to 30.5% by 2025. In terms of actual numbers of people, the increase is from 58.6 million to 101.4 million. Restauranteurs have found this market to be important and would like to make their businesses attractive to older customers. One study used a questionnaire to collect data from people aged 50 and over.[19] For one part of the analysis, individuals were classified into two age groups: 50–64 and 65–79. There were 267 people in the first group and 263 in the second. One set of items concerned ambience, menu design, and service. A series of statements were rated on a 1 to 5 scale with 1 representing "strongly disagree" and 5 representing "strongly agree." In some cases the wording of questions has been shortened in the table below. Here are the means:

Item	50–64	65–79
Ambience:		
Most restaurants are too dark	2.75	2.93
Most restaurants are too noisy	3.33	3.43
Background music is often too loud	3.27	3.55
Restaurants are too smoky	3.17	3.12
Tables are too small	3.00	3.19
Tables are too close together	3.79	3.81
Menu design:		
Print size is not large enough	3.68	3.77
Glare makes menus difficult to read	2.81	3.01
Colors of menus make them difficult to read	2.53	2.72
Service:		
It is difficult to hear the service staff	2.65	3.00
I would rather be served than serve myself	4.23	4.14
I would rather pay the server than a cashier	3.88	3.48
Service is too slow	3.13	3.10

First examine the means of the people who are 50 to 64. Order the statements according to the means and describe the results. Then do the same for the older group. For each statement compute the z statistic and the associated P-value for the comparison between the two groups. For these calculations you can assume that the denominator in the test statistic is 0.08, so z is simply the difference in the means divided by 0.08. Note that you are performing 13 significance tests in this exercise. Keep this in mind when you interpret your results. Write a report summarizing your work.

CASE STUDY EXERCISE 2: Accessibility concerns. Refer to the previous question. The questionnaire also asked about accessibility both inside the restaurant and outside. Analyze the following data and write a report.

Item	50–64	65–79
Accessibility and comfort inside:		
Most chairs are too small	2.49	2.56
Bench seats are usually too narrow	3.03	3.25
Salad bars and buffets are difficult to reach	3.04	3.09
Serving myself from salad bars and buffets is difficult	2.58	2.75
Floors around salad bars and buffets are often slippery	2.84	3.01
Aisles are too narrow	3.04	3.20
Bathroom stalls are too narrow	2.82	3.10
Outside accessibility:		
Doors are too heavy	2.51	3.01
Curbs near entrance are difficult	2.54	3.07
Parking spaces are too narrow	2.83	3.16
Distance from parking lot is too far	2.33	2.64
Parking lots are too dark at night	2.84	3.26

CHAPTER 6 Appendix

Using Minitab and Excel for Inference for the Mean (σ Known)

Confidence Intervals

Minitab:

Stat ➤ Basic Statistics ➤ 1-Sample Z

If the sample data are in the worksheet, then choose the **Samples in columns** option and then click the column containing the data into the box below. With this option, you can construct confidences intervals for more than one data set simultaneously. Alternatively, if you know the value of the sample mean, you can choose the **Summarized data** option and input the sample mean and sample size in the boxes found below. Since this routine assumes that the population standard deviation (σ) is known, you need to input its value in the **Standard deviation** box. In default mode, Minitab will construct a 95% confidence interval.

If you wish to change the level of confidence, click the **Options** button and input your desired confidence level in the **Confidence level** box. Click **OK** to close the pop-up box and then click **OK** to find the confidence interval, along with some basic statistics which are reported in the Session window.

Excel:

Excel does not have a routine for reporting the lower and upper limits of a confidence interval for the mean. However, you can use Excel to compute the margin of error, which can then be subtracted and added to the sample mean to get the confidence interval. To compute the margin of error, click an empty cell in the spreadsheet and then proceed to the **Statistical Functions** menu as described in the overview section of Chapter 1's Appendix. Scroll down the list of functions and click on the **CONFIDENCE** function choice. In the **Alpha** box, input the value of α (for example, 0.05) to obtain a $(1 - \alpha)$ confidence interval. In the **Standard-dev box,** input the known population standard deviation (σ). In the **Size** box, input the sample size and then click **OK.**

Tests of Significance

Minitab:

Stat ➤ Basic Statistics ➤ 1-Sample Z

This is the same routine described in this Appendix for obtaining the confidence interval for the mean. If you wish to conduct a hypothesis test, select the **Perform hypothesis test** option and input the null hypothesis mean value (μ_0) in the box below. In default mode, Minitab will assume a two-sided alternative. If you wish to have a one-sided alternative, click the **Options** button and choose either the "Less than" or "Greater than" alternative hypothesis choice. Click **OK** to close the pop-up box and then click **OK** to find the test statistic and corresponding P-value, which are reported in the Session window.

Excel:

To conduct a test of the mean with known population standard deviation, Excel provides a statistical function that will report the P-value for a one-sided, greater-than alternative hypothesis. To obtain this P-value, click an empty cell in the spreadsheet and then proceed to the **Statistical Functions** menu as described in the overview section of Chapter 1's Appendix. Scroll down the list of functions and click on the **ZTEST** function choice. In the **Array** box, input the cell range of the data. In the **X** box, input the null hypothesis mean value (μ_0). In the **Sigma** box, input the known population standard deviation (σ). Click **OK** to find the P-value, which is reported in the selected cell. If your alternative hypothesis is a less-than alternative, subtract the reported P-value from 1 to obtain the appropriate P-value. For a two-sided alternative, if the reported P-value is less than 0.5, you should double it; otherwise, you should double $(1 - \text{reported } P\text{-value})$.

Power

Minitab:

Stat ➤ Power and Sample Size ➤ 1-Sample Z

After invoking this routine, you can specify values for any two of the following three dialog boxes: **Sample sizes, Differences,** or **Power values.** Minitab will then compute the item not specified. For example, if you wish to compute the power of a test, input the sample size and the difference between the mean value of the particular alternative of interest (μ) and the null hypothesis mean value (μ_0), that is, $\mu - \mu_0$. In the case of Example 6.25 (page 379), you would input "10" in the **Sample sizes** box and "-1.5" ($= 498.5 - 500$) in the **Differences** box. This routine can also be used to determine what minimal sample size is required to have a desired power in relation to a particular alternative. In addition to providing values for two of the three dialog boxes, you must input the known population standard deviation (σ) in the **Standard deviation** box. In default mode, Minitab will assume a two-sided alternative. If you wish to have a one-sided alternative, click the **Options** button and choose either the "Less than" or the "Greater than" alternative hypothesis choice. In addition, if you wish to change the value of α from the default value of 0.05, then input your desired significance level in the **Significance level** box. Click **OK** to close the pop-up box and then click **OK** to find the results, which are reported in the Session window. Additionally, Minitab will produce a "Power Curve," which is simply a graph of the different power values against a range of difference values.

Excel:

Automated power calculations are not available in standard Excel or the WHFStat Add-In for Excel. To compute power in Excel, follow the general steps for computing power described in this chapter and then use Excel's **NORMSDIST** function to compute probabilities from the Normal distribution (refer to Chapter 1's Appendix for details on using the **NORMSDIST** function).

CHAPTER 7

Inference for Distributions

In 2007, monthly personal consumption expenditures for cellular telephone service were larger than for landline service for the first time. Are landlines becoming a thing of the past? Case 7.1 pursues this.

Introduction

We began our study of data analysis in Chapter 1 by learning graphical and numerical tools for describing the distribution of a single variable and for comparing several distributions. Our study of the practice of statistical inference begins in the same way, with inference about a single distribution and comparison of two distributions. Comparing more than two distributions requires more elaborate methods, which are presented in Chapters 14 and 15.

Two important aspects of any distribution are its center and spread. If the distribution is Normal, we describe its center by the mean μ and its spread by the standard deviation σ. In this chapter, we will meet confidence intervals and significance tests for inference about a population mean μ and the difference between population means $\mu_1 - \mu_2$. The previous chapter emphasized the reasoning of significance tests and confidence intervals; now we emphasize statistical practice, so we no longer assume that population standard deviations are known. The resulting t procedures for inference about means are among the most commonly used statistical methods.

7.1 Inference for the Mean of a Population

Both confidence intervals and tests of significance for the mean μ of a Normal population are based on the sample mean $\bar{x}$, which estimates the unknown μ. The sampling distribution of $\bar{x}$ depends on σ. This fact causes no difficulty when σ is known. However, when σ is unknown, we must estimate σ even though we are primarily interested in μ. We will use the sample standard deviation s to estimate the population standard deviation σ.

t distributions

Suppose that we have a simple random sample (SRS) of size n from a Normally distributed population with mean μ and standard deviation σ. The sample mean $\overline{x}$ then has the Normal distribution with mean μ and standard deviation $\sigma/\sqrt{n}$. When σ is not known, we estimate it with the sample standard deviation s, and then we estimate the standard deviation of $\overline{x}$ by $s/\sqrt{n}$. This quantity is called the *standard error* of the sample mean $\overline{x}$ and we denote it by SE.

Standard Error

When the standard deviation of a statistic is estimated from the data, the result is called the **standard error** of the statistic. The standard error of the sample mean is

$$\text{SE} = \frac{s}{\sqrt{n}}$$

The term "standard error" is sometimes used for the actual standard deviation of a statistic, $\sigma/\sqrt{n}$ in the case of $\overline{x}$. The estimated value $s/\sqrt{n}$ is then called the "estimated standard error." In this book we will use the term "standard error" only when the standard deviation of a statistic is estimated from the data. The term has this meaning in the output of many statistical computer packages and in reports of research in many fields that apply statistical methods.

The standardized sample mean, or one-sample z statistic,

$$z = \frac{\overline{x} - \mu}{\sigma/\sqrt{n}}$$

is the basis of the z procedures for inference about μ when σ is known. This statistic has the standard Normal distribution $N(0, 1)$. When we substitute the standard error $s/\sqrt{n}$ for the standard deviation $\sigma/\sqrt{n}$ of $\overline{x}$, the statistic does *not* have a Normal distribution. It has a distribution that is new to us, called a t *distribution*.

The *t* Distributions

Suppose that an SRS of size n is drawn from an $N(\mu, \sigma)$ population. Then the **one-sample *t* statistic**

$$t = \frac{\overline{x} - \mu}{s/\sqrt{n}}$$

has the *t* **distribution** with $n - 1$ **degrees of freedom.**

---

degrees of freedom

A particular t distribution is specified by its **degrees of freedom.** The degrees of freedom for this t statistic come from the sample standard deviation s in the denominator of t. We saw in Chapter 1 (page 34) that s has $n - 1$ degrees of freedom. This means there is a different t distribution for each sample size. There are other t statistics with different degrees of freedom, some of which we will meet later in this chapter.

We use $t(k)$ to stand for the t distribution with k degrees of freedom.* The density curves of the $t(k)$ distributions are similar in shape to the standard Normal curve. That is, they are symmetric about 0 and are bell-shaped. Figure 7.1 compares the density curves

---

*The t distributions were discovered in 1908 by William S. Gosset. Gosset was a statistician employed by the Guinness brewing company, which prohibited its employees from publishing their discoveries that were brewing related. In this case, the company let him publish under the pen name "Student" using an example that did not involve brewing. The t distribution is often called "Student's t" in his honor.

FIGURE 7.1 Density curves for the standard Normal and $t(5)$ distributions. Both are symmetric with center 0. The t distribution has more probability in the tails than the standard Normal distribution.

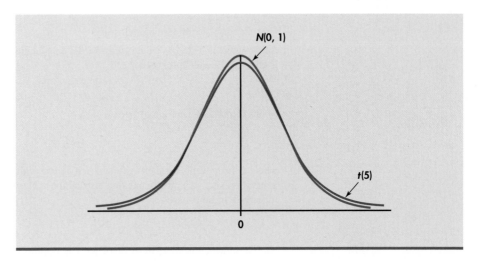

of the standard Normal distribution and the t distribution with 5 degrees of freedom. The similarity in shape is apparent, as is the fact that the t distribution has more probability in the tails and less in the center. This greater spread is due to the extra variability caused by substituting the random variable s for the fixed parameter σ. As the degrees of freedom k increase, the $t(k)$ density curve gets closer to the $N(0, 1)$ curve. This reflects the fact that s approaches σ as the sample size increases.

Table D in the back of the book gives critical values t^* for the t distributions. For convenience, we have labeled the table entries both by the value of p needed for significance tests and by the confidence level C (in percent) required for confidence intervals. The standard Normal critical values in the bottom row of entries are labeled z^*. As in the case of the Normal table (Table A), computer software often makes Table D unnecessary.

APPLY YOUR KNOWLEDGE

7.1 Apartment rents. Your local newspaper contains a large number of advertisements for unfurnished one-bedroom apartments. You choose 16 at random and calculate that their mean monthly rent is $508 and that the standard deviation of their rents is $78.

$$\frac{78}{\sqrt{16}} = \frac{78}{4}$$

(a) What is the standard error of the mean?
(b) What are the degrees of freedom for a one-sample t statistic?

7.2 Increasing the sample size. Refer to the previous exercise. Suppose that instead of an SRS of 16, you had sampled 25 advertisements. Would the standard error of the mean be larger or smaller in this case? Verify your answer by doing the calculation.

The one-sample t confidence interval

With the t distributions to help us, we can analyze samples from Normal populations with unknown σ. The one-sample t procedures are similar in both reasoning and computational detail to the z procedures of Chapter 6. By replacing the standard deviation $\sigma/\sqrt{n}$ of $\bar{x}$ by its standard error $s/\sqrt{n}$, the z statistic becomes the t statistic. Thus, we must now employ P-values or critical values from t in place of the corresponding Normal values (z). Here are the details for the confidence interval.

One-Sample t Confidence Interval

Suppose that an SRS of size n is drawn from a population having unknown mean μ. A level C **confidence interval** for μ is

$$\bar{x} \pm m$$

In this formula, the margin of error is

$$m = t^*\text{SE} = t^* \frac{s}{\sqrt{n}}$$

where t^* is the value for the $t(n-1)$ density curve with area C between $-t^*$ and t^*. This interval is exact when the population distribution is Normal and is approximately correct for large n in other cases.

In Chapter 6, the margin of error for the population mean was $z^*\sigma/\sqrt{n}$. Here, we replace σ by its estimate s and z^* by t^*. So the margin of error for the population mean when we use the data to estimate σ is $m = t^*s/\sqrt{n}$.

CASE 7.1

Monthly Household Telephone Expenditures *Trends in Telephone Service* is published by the Industry Analysis and Technology Division of the Federal Communications Commission (FCC). The annual publication contains summary information from a variety of published reports and is designed to answer frequently asked questions about the telephone industry. One section of the report summarizes monthly household expenditures and household personal consumption expenditures for telephone services. The year 2007 marked the first time monthly personal consumption expenditures for cellular service were larger than personal consumption expenditures for landline service.[1] You're interested in how much, on average, a household in your community spends on landline service.

EXAMPLE 7.1   Estimating Monthly Expenditures for Landline Telephone Service

CASE 7.1

TELEPHONE

The following data are the monthly dollar amounts for landline telephone service for a random sample of 8 households in your community:

$$43 \quad 47 \quad 51 \quad 36 \quad 50 \quad 42 \quad 38 \quad 41$$

We want to find a 95% confidence interval for μ, the average monthly expenditure for landline telephone service.

The sample mean is

$$\bar{x} = \frac{43 + 47 + \cdots + 41}{8} = 43.5$$

and the standard deviation is

$$s = \sqrt{\frac{(43-43.5)^2 + (47-43.5)^2 + \cdots + (41-43.5)^2}{8-1}} = 5.42$$

with degrees of freedom $n - 1 = 7$. The standard error of $\bar{x}$ is

$$\text{SE} = \frac{s}{\sqrt{n}} = \frac{5.42}{\sqrt{8}} = 1.92$$

df = 7		
t^*	1.895	2.365
C	90%	95%

From Table D we find $t^* = 2.365$. The margin of error is

$$m = 2.365\text{SE} = (2.365)(1.92) = 4.5$$

The 95% confidence interval is

$$\bar{x} \pm m = 43.5 \pm 4.5$$
$$= (39.0,\ 48.0)$$

We are 95% confident that the average monthly expenditure for landline telephone service is between \$39 and \$48.

In this example we have given the actual interval (39.0, 48.0) as our answer. Sometimes, we prefer to report the mean and margin of error: the average expenditure is \$43.5 with a margin of error of \$4.5.

The use of the t confidence interval in Example 7.1 rests on conditions that appear reasonable here. First, we assume that our random sample is an SRS from the community of households. Second, we assume that the distribution of monthly expenditures is Normal. With only 8 observations, this assumption cannot be effectively checked. In fact, since some households may use only cellular telephone service, this distribution could be quite irregular. With these data, however, when we make a stemplot, we can see clearly that there are no outliers:

5	01
4	1237
3	56

APPLY YOUR KNOWLEDGE

7.3 More on apartment rents. Refer to Exercise 7.1 (page 397). Construct a 95% confidence interval for the mean monthly rent of all advertised one-bedroom apartments.

7.4 90% versus 95% confidence interval. If you chose 90%, rather than 95%, confidence in the previous exercise, would your margin of error be larger or smaller? Explain your answer.

The one-sample t test

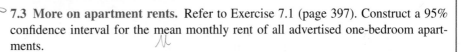

In tests as in confidence intervals, we allow for unknown σ by using the standard error and replacing z by t. This changes the distribution we use to find the P-value, but we still carry out the four steps required to do a significance test (page 360). Here are the details.

> **One-Sample t Test**
>
> Suppose that an SRS of size n is drawn from a population having unknown mean μ. To test the hypothesis $H_0: \mu = \mu_0$ based on an SRS of size n, compute the one-sample t statistic
>
> $$t = \frac{\bar{x} - \mu_0}{s/\sqrt{n}}$$

In terms of a random variable T having the $t(n-1)$ distribution, the P-value for a test of H_0 against

$H_a: \mu > \mu_0$ is $P(T \geq t)$

$H_a: \mu < \mu_0$ is $P(T \leq t)$

$H_a: \mu \neq \mu_0$ is $2P(T \geq |t|)$

These P-values are exact if the population distribution is Normal and are approximately correct for large n in other cases.

EXAMPLE 7.2   Do Average Monthly Landline Expenditures in Your Community Differ from the U.S. Average?

CASE 7.1

TELEPHONE

Suppose that the overall U.S. average monthly expenditure for landline service is $49. We can test a null hypothesis that the average monthly landline expenditure in your community is equal to this amount. First, we state our hypotheses:

$$H_0: \mu = 49$$
$$H_a: \mu \neq 49$$

Next, we calculate our test statistic. Recall that $n = 8, \overline{x} = 43.5,$ and $s = 5.42$. The t test statistic is

$$t = \frac{\overline{x} - \mu_0}{s/\sqrt{n}} = \frac{43.5 - 49}{5.42/\sqrt{8}} = -2.87$$

df = 7	
p	0.02	0.01
t^*	2.517	2.998

Because the degrees of freedom are $n - 1 = 7$, this t statistic has the $t(7)$ distribution. Figure 7.2 shows that the P-value is $2P(T \geq 2.87)$, where T has the $t(7)$ distribution. From Table D we see that $P(T \geq 2.517) = 0.02$ and $P(T \geq 2.998) = 0.01$. Therefore, we conclude that the P-value is between $2 \times 0.01 = 0.02$ and $2 \times 0.02 = 0.04$. Software gives the exact value as $P = 0.0241$. Clearly, these data are incompatible with an average of $\mu = \$49$.

FIGURE 7.2 Sketch of the P-value calculation for Example 7.2.

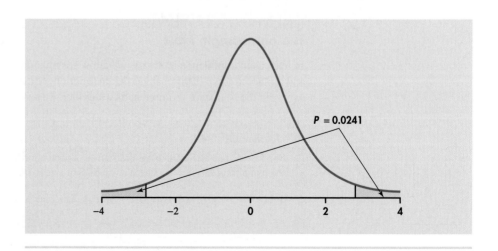

$P = 0.0241$

In this example we tested the null hypothesis $\mu = \$49$ against the two-sided alternative $\mu \neq \$49$. This is the alternative to use when there is no prior suspicion that the mean is larger or smaller. If we had suspected that the average was smaller, we would have used a one-sided test. *It is wrong to examine the data first and then decide to do a one-sided test in the direction indicated by the data.* If in doubt, always use a two-sided test.

EXAMPLE 7.3   One-sided Test for Monthly Landline Expenditures

CASE 7.1

TELEPHONE

To test whether the average monthly landline expenditure is less than the U.S. average (perhaps expected because your community's cost of living is below the average) our hypotheses are

$$H_0: \mu = 49$$
$$H_a: \mu < 49$$

The t test statistic does not change: $t = -2.87$. As Figure 7.3 illustrates, however, the P-value is now $P(T \leq -2.87)$, half of the value in Example 7.2. From Table D we can determine that $0.01 < P < 0.02$; software gives the exact value as $P = 0.0120$. We conclude that the average monthly expenditure for landline service in your community is less than the U.S. average.

FIGURE 7.3 Sketch of the P-value calculation for Example 7.3.

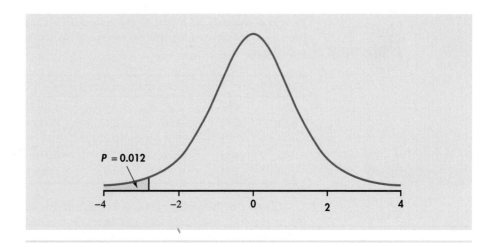

$P = 0.012$

APPLY YOUR KNOWLEDGE

7.5 Apartment rents. Refer to Exercise 7.1 (page 397). Do these data give good reason to believe that the average rent for all advertised one-bedroom apartments is greater than $550 per month? Make sure to state the hypotheses, find the t statistic and its P-value, and state your conclusion using the 5% significance level.

7.6 Significant? A test of a null hypothesis versus a two-sided alternative gives $t = 2.18$.
(a) The sample size is 16. Is the test result significant at the 5% level? Explain how you obtained your answer.
(b) The sample size is 10. Is the test result significant at the 5% level?
(c) Sketch the two t distributions to illustrate your answers.

7.7 Have sales changed? You will have complete sales information for last month in a week, but right now you have data from a random sample of 40 stores. The mean change in sales in the sample is -3.2%, and the standard deviation of the changes is 12%. Are average sales for all stores different from last month?

(a) State appropriate null and alternative hypotheses. Explain how you decided between the one- and two-sided alternatives.

(b) Find the t statistic and its P-value. State your conclusion using the $\alpha = 0.05$ significance level.

(c) If the test gives strong evidence against the null hypothesis, would you conclude that monthly sales are different in every one of your stores? Explain your answer.

Using software

For small data sets such as the one in Example 7.1 it is easy to perform the computations for confidence intervals and significance tests with an ordinary calculator. For larger data sets, however, software or a statistical calculator eases our work.

EXAMPLE 7.4   Diversify or Be Sued

 DIVERSIFY

An investor with a stock portfolio worth several hundred thousand dollars sued his broker and brokerage firm because lack of diversification in his portfolio led to poor performance. The conflict was settled by an arbitration panel that gave "substantial damages" to the investor.[2] Table 7.1 gives the rates of return for the 39 months that the account was managed by the broker. The arbitration panel compared these returns with the average of the Standard & Poor's 500-stock index for the same period.

TABLE 7.1   Monthly rates of return on a portfolio (percent)

-8.36	1.63	-2.27	-2.93	-2.70	-2.93	-9.14	-2.64
6.82	-2.35	-3.58	6.13	7.00	-15.25	-8.66	-1.03
-9.16	-1.25	-1.22	-10.27	-5.11	-0.80	-1.44	1.28
-0.65	4.34	12.22	-7.21	-0.09	7.34	5.04	-7.24
-2.14	-1.01	-1.41	12.03	-2.56	4.33	2.35	

Consider the 39 monthly returns as a random sample from the population of monthly returns that the brokerage would generate if it managed the account forever. Are these returns compatible with a population mean of $\mu = 0.95\%$, the S&P 500 average? Our hypotheses are

$$H_0: \mu = 0.95$$
$$H_a: \mu \neq 0.95$$

Minitab and SPSS outputs appear in Figure 7.4. Output from other software will look similar.

Here is one way to report the conclusion: the mean monthly return on investment for this client's account was $\bar{x} = -1.1\%$. This differs significantly from the performance of the S&P 500 for the same period ($t = -2.14$, df = 38, $P = 0.039$).

The hypothesis test in Example 7.4 leads us to conclude that the mean return on the client's account differs from that of the stock index. Now let's assess the return on the client's account with a confidence interval.

EXAMPLE 7.5   Estimating Mean Monthly Return

The mean monthly return on the client's portfolio was $\bar{x} = -1.1\%$ and the standard deviation was $s = 5.99\%$. Figure 7.4 gives the Minitab and SPSS outputs and Figure 7.5 gives the Excel and

FIGURE 7.4 Minitab and SPSS outputs for Examples 7.4 and 7.5.

Minitab

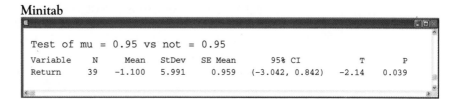

```
Test of mu = 0.95 vs not = 0.95

Variable   N    Mean   StDev   SE Mean      95% CI         T      P
Return    39  -1.100   5.991    0.959  (-3.042, 0.842)  -2.14  0.039
```

SPSS

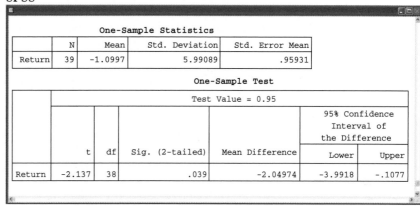

One-Sample Statistics

	N	Mean	Std. Deviation	Std. Error Mean
Return	39	-1.0997	5.99089	.95931

One-Sample Test

	Test Value = 0.95					
					95% Confidence Interval of the Difference	
	t	df	Sig. (2-tailed)	Mean Difference	Lower	Upper
Return	-2.137	38	.039	-2.04974	-3.9918	-.1077

FIGURE 7.5 Excel and JMP outputs for Example 7.5.

Excel

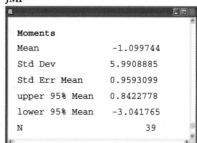

	A	B
1	Return	
2	Mean	-1.09974359
3	Standard Error	0.95930991
4	Median	-1.41
5	Mode	-2.93
6	Standard Deviation	5.990888471
7	Sample Variance	35.89074467
8	Kurtosis	0.226609438
9	Skewness	0.158971057
10	Range	27.47
11	Minimum	-15.25
12	Maximum	12.22
13	Sum	-42.89
14	Count	39
15	Confidence Level(95.0%)	1.942021368

JMP

```
Moments
Mean             -1.099744
Std Dev           5.9908885
Std Err Mean      0.9593099
upper 95% Mean    0.8422778
lower 95% Mean   -3.041765
N                        39
```

DIVERSIFY

JMP outputs for a 95% confidence interval for the population mean μ. Note that Excel gives the margin of error next to the label "Confidence Level (95.0%)" rather than the actual confidence interval. We see that the 95% confidence interval is $(-3.04,\ 0.84)$, or (from Excel) -1.0997 ± 1.9420.

Because the S&P 500 return, 0.95%, falls outside this interval, we know that μ differs significantly from 0.95% at the $\alpha = 0.05$ level. Example 7.4 gave the actual P-value as $P = 0.039$.

The confidence interval suggests that the broker's management of this account had a long-term mean somewhere between a loss of 3.04% and a gain of 0.84% per month. We are interested not in the actual mean but in the difference between the broker's process and the diversified S&P 500 index.

EXAMPLE 7.6   Estimating Difference from a Standard

DIVERSIFY

Following the analysis accepted by the arbitration panel, we are considering the S&P 500 monthly average return as a constant standard. (It is easy to envision scenarios where we would want to treat this type of quantity as random.) The difference between the mean of the investor's account and the S&P 500 is $\bar{x} - \mu = -1.10 - 0.95 = -2.05\%$. In Example 7.5 we found that the 95% confidence interval for the investor's account was $(-3.04,\ 0.84)$. To obtain the corresponding interval for the difference, subtract 0.95 from each of the endpoints. The resulting interval is $(-3.04 - 0.95,\ 0.84 - 0.95)$, or $(-3.99,\ -0.11)$. We conclude with 95% confidence that the underperformance was between -3.99% and -0.11%. This estimate helps to set the compensation owed to the investor.

APPLY YOUR KNOWLEDGE

CASE 7.1 **7.8 Using software to compute a confidence interval.** In Example 7.1 (page 398) we calculated the 95% confidence interval for the average household expenditure for landline telephone service in your community. Use software to compute this interval and verify that you obtain the same interval.

CASE 7.1 **7.9 Using software to perform a significance test.** In Example 7.2 (page 400) we tested whether the average household expenditure for landline telephone service in your community was different from the U.S. average. Use software to perform this test and verify that you obtain the same P-value.

Matched pairs *t* procedures

The telephone expenditure problem of Case 7.1 concerns only a single population. We know that comparative studies are usually preferred to single-sample investigations because of the protection they offer against confounding. For that reason, inference about a parameter of a single distribution is less common than comparative inference.

matched pairs   One common comparative design, however, makes use of single-sample procedures. In a **matched pairs** study, subjects are matched in pairs and the outcomes are compared within each matched pair. For example, an experiment to compare two advertising campaigns might use pairs of subjects that are the same age, sex, and income level. The experimenter could toss a coin to assign the two campaigns to the two subjects in each pair. The idea is that matched subjects are more similar than unmatched subjects, so comparing outcomes within each pair is more efficient (page 181). Matched pairs are also common when randomization is not possible. One situation calling for matched pairs is before-and-after observations on the same subjects, as illustrated in the next example.

EXAMPLE 7.7   The Effects of Language Instruction

A company contracts with a language institute to provide individualized instruction in foreign languages for its executives who will be posted overseas. Is the instruction effective?

Last year, 20 executives studied French. All had some knowledge of French, so they were given the Modern Language Association's listening test of understanding of spoken French before the instruction began. After several weeks of immersion in French, the executives took the listening test again. (The actual French spoken in the two tests was different, so that simply taking the first test should not improve the score on the second test.)

FRENCH

Table 7.2 gives the pretest and posttest scores. The maximum possible score on the test is 36.[3] To analyze these data, subtract the pretest score from the posttest score to obtain the improvement for each executive. These 20 differences form a single sample. They appear in the "Gain" columns in Table 7.2. The first executive, for example, improved from 32 to 34, so the gain is $34 - 32 = 2$.

TABLE 7.2   French listening scores for executives

Executive	Pretest	Posttest	Gain	Executive	Pretest	Posttest	Gain
1	32	34	2	11	30	36	6
2	31	31	0	12	20	26	6
3	29	35	6	13	24	27	3
4	10	16	6	14	24	24	0
5	30	33	3	15	31	32	1
6	33	36	3	16	30	31	1
7	22	24	2	17	15	15	0
8	25	28	3	18	32	34	2
9	32	26	−6	19	23	26	3
10	20	26	6	20	23	26	3

To assess whether the institute significantly improved the executives' comprehension of spoken French, we test

$$H_0: \mu = 0$$

$$H_a: \mu > 0$$

Here μ is the mean improvement that would be achieved if the entire population of executives received similar instruction. The null hypothesis says that no improvement occurs, and H_a says that posttest scores are higher on the average.

The 20 differences have

$$\bar{x} = 2.5 \text{ and } s = 2.893$$

The one-sample t statistic is therefore

$$t = \frac{\bar{x} - 0}{s/\sqrt{n}} = \frac{2.5}{2.893/\sqrt{20}}$$

$$= 3.86$$

The P-value is found from the $t(19)$ distribution. Remember that the degrees of freedom are 1 less than the sample size.

	df = 19	
p	0.001	0.0005
t^*	3.579	3.883

Table D shows that 3.86 lies between the upper 0.001 and 0.0005 critical values of the $t(19)$ distribution. The P-value therefore lies between these values. Software gives the value $P = 0.00053$. The improvement in listening scores is very unlikely to be due to chance alone. We have strong evidence that the posttest scores are systematically higher. When reporting results it is usual to omit the details of routine statistical procedures; our test would be reported in the form: "The improvement in scores was significant ($t = 3.86$, df = 19, $P = 0.00053$)."

The significance test shows that the mean score has improved. A critic could argue that the test does not show that the instruction *caused* the improvement—for example, the executives might simply be less nervous when taking the test a second time. A more elaborate study with a control group who receive no instruction would be more convincing, but the company is unlikely to assign executives to "no instruction." Also note that the subjects all knew that they would receive overseas assignments. Our conclusions apply only to the population of such executives, not to the larger population of all the company's managers.

A statistically significant but very small improvement in language ability would not justify the expense of the individualized instruction. A confidence interval allows us to estimate the *amount* of the improvement.

EXAMPLE 7.8   The Effects of Language Instruction

Let's compute a 90% confidence interval for the mean improvement in the entire population. The standard error is

$$\text{SE} = \frac{s}{\sqrt{n}} = \frac{2.893}{\sqrt{20}} = 0.6469$$

The margin of error is

$$m = t^*\text{SE} = (1.729)(0.6469) = 1.12$$

where the critical value $t^* = 1.729$ comes from Table D. The confidence interval is

$$\overline{x} \pm m = 2.5 \pm 1.12$$
$$= (1.38, \ 3.62)$$

The estimated average improvement is 2.5 points, with margin of error 1.12 for 90% confidence. Though statistically significant, the effect of the instruction was rather small.

A look at the data discloses one reason for the small average improvement. Several of the executives had pretest scores close to the maximum of 36. They could not improve their scores very much even if their mastery of French increased substantially. This is a weakness in the listening test that is the measuring instrument in this study.

Here are some key points to remember concerning matched pairs:

- A matched pairs analysis is needed when subjects or experimental units are matched in pairs or there are two measurements or observations on each individual or experimental unit and our question of interest concerns the difference between the two measurements. Frequently, the observations are "before" and "after" measures.

- The analysis is based on the *difference* between the two measures for each individual or pair of experimental units.

- Use the one-sample confidence interval and significance-testing procedures that we learned in this section.

APPLY YOUR KNOWLEDGE

7.10 Comparison of two energy drinks. Consider the following study to compare two popular energy drinks. Each drink was rated on a 0 to 100 scale, with 100 being the highest rating. For each subject, a coin was tossed to see which drink would be tried first. The drinks were served in identical Styrofoam cups.

Subject:	1	2	3	4	5
Drink A:	48	83	67	76	73
Drink B:	55	85	61	71	70

Is there a difference in preference? State appropriate hypotheses and carry out a matched pairs t test using $\alpha = 0.05$.

7.11 95% confidence interval for the difference in energy drinks. For the companies producing these drinks, the real question is how much difference there is between the two preferences. Use the data above to give a 95% confidence interval for the difference in preference between Drink A and Drink B.

Robustness of the *t* procedures

The matched pairs t procedures use one-sample t confidence intervals and significance tests for differences. They are therefore based on an assumption that the *population of differences* has a Normal distribution. Look carefully at the gain scores in Table 7.2. One executive actually lost 6 points between the pretest and the posttest. This one subject lowered the sample mean from 2.95 for the other 19 subjects to 2.5 for all 20. A Normal quantile plot (Figure 7.6) displays this outlier as well as granularity due to the fact that only whole-number scores are possible. The data do not appear to be Normal. Does this non-Normality suggest that we should not use the t procedures for these data?

   The results of one-sample t procedures are exactly correct only when the population is Normal. Real populations are never exactly Normal. All inference procedures are based on some conditions, such as Normality: procedures that are not strongly affected by violations of a condition are called *robust*.

> **Robust Procedures**
>
> A statistical inference procedure is called **robust** if the probability calculations required are insensitive to violations of the conditions that usually justify the procedure.

FIGURE 7.6 Normal quantile plot for the change in French listening score.

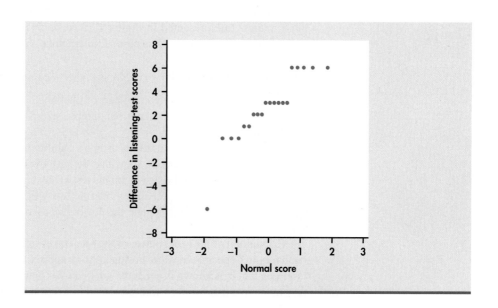

The condition that the population be Normal rules out outliers, so the presence of outliers shows that this condition is not fulfilled. The t procedures are not robust against outliers, because $\bar{x}$ and s are not resistant to outliers.

Fortunately, the t procedures are quite robust against non-Normality of the population except when outliers or strong skewness are present, particularly when the sample size is large. The t procedures rely only on the Normality of the sample mean $\bar{x}$. This condition is satisfied when the population is Normal, but the central limit theorem tells us that a mean $\bar{x}$ from a large sample follows a Normal distribution closely even when individual observations are not Normally distributed.

Before using the t procedures for small samples, make a Normal quantile plot, stemplot, or boxplot to check for skewness and outliers. For most purposes, the one-sample t procedures can be safely used when $n \geq 15$ unless an outlier or clearly marked skewness is present. Except in the case of small samples, the condition that the data are an SRS from the population of interest is more crucial than the condition that the population distribution is Normal. Here are practical guidelines for inference on a single mean:[4]

- *Sample size less than 15:* Use t procedures if the data are close to Normal. If the data are clearly non-Normal or if outliers are present, do not use t.
- *Sample size at least 15:* The t procedures can be used except in the presence of outliers or strong skewness.
- *Large samples:* The t procedures can be used even for clearly skewed distributions when the sample is large, roughly $n \geq 40$.

The French instruction data are a borderline case: there are 20 observations, with one clear outlier. The t procedures give approximately correct results, but we would be happier if a good reason, such as illness on the posttest date, justified removing the outlier.

APPLY YOUR KNOWLEDGE

7.12 Significance test for the average time to start a business? Consider the sample of time data presented in Figure 1.30 (page 57). Would you feel comfortable applying the t procedures in this case? Explain your answer.

7.13 Significance test for the average T-bill interest rate? Consider data on the T-bill interest rate presented in Figure 1.29 (page 56). Would you feel comfortable applying the t procedures in this case? Explain your answer.

The power of the t test

Power calculations are an important part of planning a study. One piece of information that they provide is whether or not our sample size is sufficiently large to answer our research questions.

The power of a statistical test measures its ability to detect deviations from the null hypothesis. Usually, we hope to show that the null hypothesis is false, so high power is important. The power of the one-sample t test against a specific alternative value of the population mean μ is the probability that the test will reject the null hypothesis when this alternative is true. To calculate the power, we assume a fixed level of significance, usually $\alpha = 0.05$.

Calculation of the exact power of the t test takes into account the estimation of σ by s and requires special software. Fortunately, an approximate calculation that is based on assuming that σ is known is generally adequate for planning a study. This calculation is very much like that for the z test, presented in Section 6.4 (page 376).

1. Write the event, in terms of $\bar{x}$, that the test rejects H_0.

2. Find the probability of this event when the population mean has the alternative value.

Here is an example.

EXAMPLE 7.9 Is the Sample Size Large Enough?

The company in Example 7.7 is planning to evaluate the next language instruction session. They decide that an improvement of 2 points or more is large enough to be useful in practice. Can they rely on a sample of 20 executives to detect an improvement of this size?

They wish to compute the power of the t test for

$$H_0: \mu = 0$$
$$H_a: \mu > 0$$

against the alternative $\mu = 2$ when $n = 20$. This gives us most of the information we need to compute the power. The other important piece is a rough guess of the size of σ. In planning a large study, a pilot study is often run for this and other purposes. In this case, listening-score improvements in past language instruction programs for similar executives have had sample standard deviations of about 3. We will therefore take both $\sigma = 3$ and $s = 3$ in our approximate calculation.

Step 1. *The t test with 20 observations rejects H_0 at the 5% significance level if the t statistic*

$$t = \frac{\bar{x} - 0}{s/\sqrt{20}}$$

exceeds the upper 5% point of $t(19)$, which is 1.729. Taking $s = 3$, the event that the test rejects H_0 is therefore

$$t = \frac{\bar{x}}{3/\sqrt{20}} \geq 1.729$$
$$\bar{x} \geq 1.729\frac{3}{\sqrt{20}}$$
$$\bar{x} \geq 1.160$$

Step 2. *The power is the probability that $\bar{x} \geq 1.160$ when $\mu = 2$. Taking $\sigma = 3$, we find this probability by standardizing $\bar{x}$:*

$$P(\bar{x} \geq 1.160 \text{ when } \mu = 2) = P\left(\frac{\bar{x} - 2}{3/\sqrt{20}} \geq \frac{1.160 - 2}{3/\sqrt{20}}\right)$$
$$= P(Z \geq -1.252)$$
$$= 1 - 0.1056 = 0.8944$$

A true difference of 2 points in the population mean scores will produce significance at the 5% level in 89% of all possible samples. The company can be reasonably confident of detecting a difference this large.

APPLY YOUR KNOWLEDGE

7.14 Power for other values of μ. If you repeat the calculation in Example 7.9 for values of μ that are larger than 2, would you expect the power to be higher or lower than 0.8944? Why?

7.15 Another power calculation. Verify your answer to the previous exercise by doing the calculation for the alternative $\mu = 2.4$.

Inference for non-Normal populations

We have not discussed how to do inference about the mean of a clearly non-Normal distribution based on a small sample. If you face this problem, you should consult an expert. Three general strategies are available:

- In some cases a distribution other than a Normal distribution describes the data well. There are many non-Normal models for data, and inference procedures for these models are available.

- Because skewness is the chief barrier to the use of t procedures on data without outliers, you can attempt to transform skewed data so that the distribution is symmetric and as close to Normal as possible. Confidence levels and P-values from the t procedures applied to the transformed data will be quite accurate for even moderate sample sizes. Methods are generally available for transforming the results back to the original scale.

distribution-free procedures

nonparametric procedures

- The third strategy is to use a **distribution-free** inference procedure. Such procedures do not assume that the population distribution has any specific form, such as Normal. Distribution-free procedures are often called **nonparametric procedures.** The *bootstrap* (page 346) is a modern computer-intensive nonparametric procedure that is especially useful for confidence intervals. Chapter 16 discusses traditional nonparametric procedures, especially significance tests.

Each of these strategies quickly carries us beyond the basic practice of statistics. We emphasize procedures based on Normal distributions because they are the most common in practice, because their robustness makes them widely useful, and (most important) because we are first of all concerned with understanding the principles of inference.

Distribution-free significance tests do not require that the data follow any specific type of distribution such as Normal. This gain in generality isn't free: if the data really are close to Normal, distribution-free tests have less power than t tests. They also don't quite answer the same question. The t tests concern the population *mean.* Distribution-free tests ask about the population *median,* as is natural for distributions that may be skewed.

The sign test

sign test

The simplest distribution-free test, and one of the most useful, is the **sign test.** The test gets its name from the fact that we look only at the signs of the differences, not their actual values. The following example illustrates this test.

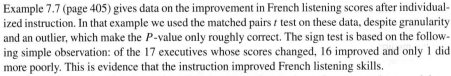

EXAMPLE 7.10 The Effects of Language Instruction

FRENCH

Example 7.7 (page 405) gives data on the improvement in French listening scores after individualized instruction. In that example we used the matched pairs t test on these data, despite granularity and an outlier, which make the P-value only roughly correct. The sign test is based on the following simple observation: of the 17 executives whose scores changed, 16 improved and only 1 did more poorly. This is evidence that the instruction improved French listening skills.

To perform a significance test based on the count of executives whose scores improved, let p be the probability that a randomly chosen executive would improve if she attended the institute. The null hypothesis of "no effect" says that the posttest is just a repeat of the pretest with no change in ability, so an executive is equally likely to do better on either test. We therefore want to test

$$H_0: p = 1/2$$
$$H_a: p > 1/2$$

The 17 executives whose scores changed are 17 independent trials, so the number who improve has the binomial distribution $B(17, 1/2)$ if H_0 is true. The P-value for the observed count 16 is therefore $P(X \geq 16)$, where X has the $B(17, 1/2)$ distribution. You can compute this probability with software or from the binomial probability formula (page 309):

$$P(X \geq 16) = P(X = 16) + P(X = 17)$$

$$= \binom{17}{16} \left(\frac{1}{2}\right)^{16} \left(\frac{1}{2}\right)^{1} + \binom{17}{17} \left(\frac{1}{2}\right)^{17} \left(\frac{1}{2}\right)^{0}$$

$$= (17) \left(\frac{1}{2}\right)^{17} + \left(\frac{1}{2}\right)^{17}$$

$$= 0.00014$$

As in Example 7.7, there is very strong evidence that participation in the institute has improved performance on the listening test.

There are several varieties of sign test, all based on counts and the binomial distribution. The sign test for matched pairs (Example 7.10) is the most useful. The null hypothesis of "no effect" is then always H_0: $p = 1/2$. The alternative can be one-sided in either direction or two-sided, depending on the type of change we are considering.

Sign Test for Matched Pairs

Ignore pairs with difference 0; the number of trials n is the count of the remaining pairs. The test statistic is the count X of pairs with a positive difference. P-values for X are based on the binomial $B(n, 1/2)$ distribution.

The matched pairs t test in Example 7.7 (page 405) tested the hypothesis that the mean of the distribution of differences (score after the instruction minus score before) is 0. The sign test in Example 7.10 is in fact testing the hypothesis that the *median* of the differences is 0. If p is the probability that a difference is positive, then $p = 1/2$ when the median is 0. This is true because the median of the distribution is the point with probability $1/2$ lying to its right. As Figure 7.7 illustrates, $p > 1/2$ when the median is greater than 0, again because the probability to the right of the median is always $1/2$. The sign test of H_0: $p = 1/2$ against H_a: $p > 1/2$ is a test of

$$H_0: \text{population median} = 0$$

$$H_a: \text{population median} > 0$$

The sign test in Example 7.10 makes no use of the actual scores—it just counts how many executives improved. The executives whose scores did not change were ignored

FIGURE 7.7 Why the sign test tests the median difference: when the median is greater than 0, the probability p of a positive difference is greater than 1/2, and vice versa.

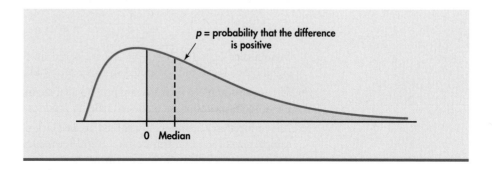

altogether. Because the sign test uses so little of the available information, it is much less powerful than the t test when the population is close to Normal. Chapter 16 describes other distribution-free tests that are more powerful than the sign test.

APPLY YOUR KNOWLEDGE

7.16 Sign test for the energy drink comparison. Exercise 7.10 (page 406) gives data on the appeal of two popular energy drinks. Is there evidence that the medians are different? State the hypotheses, carry out the sign test, and report your conclusion.

SECTION 7.1 Summary

- Significance tests and confidence intervals for the mean μ of a Normal population are based on the sample mean $\bar{x}$ of an SRS. Because of the central limit theorem, the resulting procedures are approximately correct for other population distributions when the sample is large.

- The **standard error** of the sample mean is

$$SE = \frac{s}{\sqrt{n}}$$

- The standardized sample mean, or **one-sample z statistic,**

$$z = \frac{\bar{x} - \mu}{\sigma/\sqrt{n}}$$

has the $N(0, 1)$ distribution. If the standard deviation $\sigma/\sqrt{n}$ of $\bar{x}$ is replaced by the **standard error** $SE = s/\sqrt{n}$, the **one-sample t statistic**

$$t = \frac{\bar{x} - \mu}{s/\sqrt{n}}$$

has the **t distribution** with $n - 1$ degrees of freedom.

- There is a t distribution for every positive **degrees of freedom** k. All are symmetric distributions similar in shape to Normal distributions. The $t(k)$ distribution approaches the $N(0, 1)$ distribution as k increases.

- The **margin of error** for level C confidence is

$$m = t^* SE = t^* \frac{s}{\sqrt{n}}$$

where t^* is the value for the $t(n - 1)$ density curve with area C between $-t^*$ and t^*.

- A level C **confidence interval for the mean** μ of a Normal population is

$$\bar{x} \pm m$$

- Significance tests for $H_0: \mu = \mu_0$ are based on the t statistic. P-values or fixed significance levels are computed from the $t(n - 1)$ distribution.

- These one-sample procedures are used to analyze **matched pairs** data by first taking the differences within the matched pairs to produce a single sample.

- The t procedures are relatively **robust** against lack of Normality, especially for larger sample sizes. The t procedures are useful for non-Normal data when $n \geq 15$ unless the data show outliers or strong skewness.

- The **power** of the t test is calculated like that of the z test, using an approximate value for both σ and s.

- The **sign test** is a **distribution-free test** because it uses probability calculations that are correct for a wide range of population distributions.

- The sign test for "no treatment effect" in matched pairs counts the number of positive differences. The P-value is computed from the $B(n, 1/2)$ distribution, where n is the number of non-0 differences. The sign test is less powerful than the t test in cases where use of the t test is justified.

SECTION 7.1 Exercises

For Exercises 7.1 and 7.2, see page 397; for 7.3 and 7.4, see page 399; for 7.5 to 7.7, see page 401; for 7.8 and 7.9, see page 404; for 7.10 and 7.11, see pages 406–407; for 7.12 and 7.13, see page 408; for 7.14 and 7.15, see page 409; and for 7.16, see page 412.

7.17 Finding the critical value t^*. What critical value t^* from Table D should be used to calculate the margin of error for a confidence interval for the mean of the population in each of the following situations?
(a) A 95% confidence interval based on $n = 13$ observations.
(b) A 95% confidence interval from an SRS of 26 observations.
(c) A 99% confidence interval from a sample of size 26.
(d) These cases illustrate how the size of the margin of error depends on the confidence level and on the sample size. Summarize the relationships illustrated.

7.18 Finding critical t^*-values. What critical value t^* from Table D should be used to calculate the margin of error for a confidence interval for the mean of the population in each of the following situations?
(a) A 90% confidence interval based on $n = 9$ observations.
(b) A 90% confidence interval from an SRS of 36 observations.
(c) A 95% confidence interval from a sample of size 36.
(d) These cases illustrate how the size of the margin of error depends on the confidence level and on the sample size. Summarize the relationships illustrated.

7.19 A one-sample t test. The one-sample t statistic for testing

$$H_0: \mu = 10$$
$$H_a: \mu > 10$$

from a sample of $n = 18$ observations has the value $t = 1.83$.
(a) What are the degrees of freedom for this statistic?
(b) Give the two critical values t^* from Table D that bracket t.
(c) What are the right-tail probabilities p for these two entries?
(d) Between what two values does the P-value of the test fall?
(e) Is the value $t = 1.83$ significant at the 5% level? Is it significant at the 1% level?
(f) If you have software available, find the exact P-value.

7.20 Another one-sample t test. The one-sample t statistic for testing

$$H_0: \mu = 60$$
$$H_a: \mu \neq 60$$

from a sample of $n = 28$ observations has the value $t = 0.75$.
(a) What are the degrees of freedom for t?
(b) Locate the two critical values t^* from Table D that bracket t. What are the right-tail probabilities p for these two values?
(c) How would you report the P-value for this test?
(d) Is the value $t = 0.75$ statistically significant at the 5% level? At the 1% level?
(e) If you have software available, find the exact P-value.

7.21 A final one-sample t test. The one-sample t statistic for testing

$$H_0: \mu = 20$$
$$H_a: \mu < 20$$

based on $n = 150$ observations has the value $t = -3.83$.
(a) What are the degrees of freedom for this statistic?
(b) How would you report the P-value based on Table D?
(c) If you have software available, find the exact P-value.

7.22 Critical values. What critical value t^* would you use to construct
(a) a 90% confidence interval when $n = 15$?
(b) a 95% confidence interval when $n = 26$?
(c) a 99% confidence interval when $n = 150$?

7.23 Fuel efficiency. Computers in some vehicles calculate various quantities related to performance. One of these is the fuel efficiency, or gas mileage, usually expressed as miles per gallon (mpg). One of the authors of this book conducted an experiment with his Toyota Highlander Hybrid by randomly recording mpg readings shown on the vehicle computer while the car was set to 60 miles per hour by cruise control. Here are the mpg values from the experiment: **MILEAGE**

37.2 21.0 17.4 24.9 27.0 36.9 38.8 35.3 32.3 23.9
19.0 26.1 25.8 41.4 34.4 32.5 25.3 26.5 28.2 22.1

Give a 95% confidence interval for μ, the mean mpg for this vehicle cruising at 60 miles per hour.

7.24 Testing the sticker information. Refer to the previous exercise. The vehicle sticker information for this Toyota Highlander Hybrid states a highway average of 27 mpg. Are the results of this experiment consistent with the vehicle sticker? Perform a significance test using the 0.05 significance level. Be sure to specify the hypotheses, your test statistic, the P-value, and your conclusion.

7.25 Bankruptcies in Canada. From 1980 to 1992, the number of company bankruptcies in Canada more than doubled. Dramatic improvements, however, have occurred since that time. By 2005, the number of bankruptcies had returned to where it was 25 years ago. One survey conducted by Statistics Canada, reported that the average number of bankruptcies per 1000 businesses from 1999 to 2005 was 8.98 with a standard error of 1.25.[5] Construct a 95% confidence interval for the mean number of bankruptcies per 1000 businesses between 1999 and 2005.

7.26 Health insurance costs. The Consumer Expenditure Survey provides information on the buying habits of U.S. consumers.[6] In 2007, the average amount a household spent on health insurance was reported to be $1545. Assuming a sample standard deviation of $750 and sample size of $n = 200$, calculate a 90% confidence interval for the mean amount spent on health insurance.

7.27 Counts of seeds in one-pound scoops. A leading agricultural company must maintain strict control over the size, weight, and number of seeds they package for sale to customers. An SRS of 81 one-pound scoops of seeds was collected as part of a Six Sigma quality improvement effort within the company. The number of seeds in each scoop is shown below. SEEDCOUNT

1471	1489	1475	1547	1497	1490	1889	1881	1877
1448	1503	1492	1553	1557	1504	1666	1717	1670
1703	1649	1649	1323	1311	1315	1469	1428	1471
1626	1658	1662	1517	1517	1519	1529	1549	1539
1858	1843	1857	1547	1470	1453	1412	1398	1398
1698	1692	1688	1435	1421	1428	1712	1722	1721
1426	1433	1422	1562	1583	1581	1720	1721	1743
1441	1434	1444	1500	1509	1521	1575	1548	1529
1735	1759	1745	1483	1464	1481	1900	1930	1953

(a) Create a histogram, boxplot, and a Normal quantile plot of these counts.

(b) Write a careful description of the distribution. Make sure to note any outliers and comment on the skewness or Normality of the data.

(c) Based on your observations in part (b), is it appropriate to analyze these data using the t procedures? Briefly explain your response.

7.28 How many seeds on average? Refer to the previous exercise.

(a) Find the mean, the standard deviation, and the standard error of the mean for this sample.

(b) If you were to calculate the margin of error for the average number of seeds at 90% and 95% confidence, which would be smaller? Briefly explain your reasoning without doing the calculations.

(c) Calculate the 90% and 95% confidence intervals for the mean number of seeds in a one-pound scoop.

(d) Compare the widths of these two intervals. Does this comparison support your answer to part (b)? Explain.

7.29 Significance test for the average number of seeds. Refer to the previous two exercises.

(a) Do these data provide evidence that the average number of seeds in a one-pound scoop is greater than 1550? Using a significance level of 5%, state your hypotheses, the P-value, and your conclusion.

(b) Do these data provide evidence that the average number of seeds in a one-pound scoop is greater than 1560? Using a significance level of 5%, state your hypotheses, the P-value, and your conclusion.

(c) Explain the relationship between your conclusions to parts (a) and (b) and the 90% confidence interval calculated in the previous exercise.

7.30 ReRuns R Fun. The largest nonprofit childrenís consignment sale in America is called ReRuns R Fun.[7] Consignors to the sale keep 35% of the profits generated by the sale of their items, and the rest goes to a variety of charities. A random sample of $n = 4$ consignor gross sales records (in dollars) for the Fall sale in 2008 is shown below:

1511.00	1639.50	1276.50	1136.50

(a) With so few observations, we cannot expect a histogram to reveal the shape of the population very well. Create a Normal quantile plot and comment on the apparent Normality or non-Normality of these data.

(b) Calculate the mean, standard deviation, and standard error of the mean for these data.

(c) Calculate a 95% confidence interval for the average gross sales of consignors at the Fall 2008 sale. Write a one-sentence interpretation of the confidence interval.

7.31 Plant capacity. A leading company chemically treats its product before packaging. The company monitors the weight of product per hour that each machine treats. An SRS of 90 hours of production data for a particular machine is collected. The measured variable is in pounds. CAPACITY

(a) Describe the distribution of pounds treated using graphical methods. Is it appropriate to analyze these data using t distribution methods? Explain.

(b) Calculate the mean, standard deviation, standard error, and margin of error for 90% confidence.

(c) Report the 90% confidence interval for the mean pounds treated per hour by this particular machine.

(d) Test whether these data provide evidence that the mean pounds of product treated in one hour is greater than 33,000.

Use a significance level of 5% and state your hypotheses, the P-value, and your conclusion.

7.32 How large is the pollution problem? Pollution of water resources is a serious problem that can require substantial efforts to improve. To determine the financial resources required, an accurate assessment of the extent of the problem is needed. The index of biotic integrity (IBI) is a measure of the water quality in streams. Here are IBI measurements for a sample of streams in the Ozark Highland ecoregion of Arkansas that were collected as part of a study:[8] IBI

47	61	39	59	72	76	85	89	74	89
33	32	46	80	80	53	78	43	88	84
62	55	29	29	54	78	71	55	71	58
33	81	59	71	75	64	41	82	60	83
84	82	82	86	79	67	56	85	91	

(a) Make a stemplot of the data. Also, make a Normal quantile plot if your software permits. Are there outliers present? Is the distribution skewed? If so, in what way? Based on your answers, do you think that t procedures should be used to analyze these data? Explain your answer.
(b) Give a 95% confidence interval for the mean IBI for streams in the Ozark Highland ecoregion of Arkansas that were sampled in this study.

7.33 Classifying poor-quality streams. Refer to the previous exercise. Streams with IBI scores of less than 20 are classified as "very poor," while those with scores greater than 20 but less than 40 are classified as "poor."
(a) Count the number of streams in the sample that are classified in these two categories, and use this result to compute an estimate of the proportion of streams that have either very poor or poor IBI scores.
(b) Can you use the confidence interval that you found in Exercise 7.32 to determine a confidence interval for the proportion of Ozark Highland ecoregion streams that have very poor or poor IBI scores? Explain your answer.

7.34 Is all the output helpful? When the data for Exercise 7.32 were fed to a statistical package, the output included a t statistic under the label "Test for Location: Mu0=0." Explain why this test is meaningless for the IBI water quality data.

7.35 Supermarket shoppers. A marketing consultant observed 40 consecutive shoppers at a supermarket. One variable of interest was how much each shopper spent in the store. Here are the data (in dollars), arranged in increasing order: SHOPPERS

5.32	8.88	9.26	10.81	12.69	15.23	15.62	17.00
17.35	18.43	19.50	19.54	20.59	22.22	23.04	24.47
25.13	26.24	26.26	27.65	28.08	28.38	32.03	34.98
37.37	38.64	39.16	41.02	42.97	44.67	45.40	46.69
49.39	52.75	54.80	59.07	60.22	84.36	85.77	94.38

(a) Display the data using a stemplot. Make a Normal quantile plot if your software allows. The data are clearly non-Normal.

In what way? Because $n = 40$, the t procedures remain quite accurate.
(b) Calculate the mean, the standard deviation, and the standard error of the mean.
(c) Find a 95% t confidence interval for the mean spending for all shoppers at this store.

7.36 The influence of big shoppers. Eliminate the three largest observations and redo parts (a), (b), and (c) of the previous exercise. Do these observations have a large influence on the results?

7.37 Corn prices. The U.S. Department of Agriculture (USDA) uses sample surveys to obtain important economic estimates. One USDA pilot study estimated the price received by farmers for corn sold in January from a sample of 20 farms. The mean price was reported as $3.64 per bushel with a standard error of $0.0835 per bushel. Give a 95% confidence interval for the mean price received by farmers for corn sold in January.[9]

7.38 Health care costs. The cost of health care is the subject of many studies that use statistical methods. One such study estimated that the average length of service for home health care among people aged 65 and over who use this type of service is 304 days with a standard error of 25.2 days. Assuming a large sample, calculate a 90% confidence interval for the mean length of service for all users of home health care.[10]

7.39 A customer satisfaction survey. Many organizations do surveys to determine the satisfaction of their customers. Attitudes toward various aspects of campus life were the subject of one such study conducted at Purdue University. Each question was rated on a 1 to 5 scale, with 5 being the highest rating. The mean response of 2535 first-year students to "Feeling welcomed at Purdue" was 3.75, and the standard deviation of the responses was 1.26. Assuming that the respondents are an SRS, give a 99% confidence interval for the mean of all first-year students.

7.40 Credit card fees. A bank wonders whether omitting the annual credit card fee for customers who charge at least $4000 in a year would increase the amount charged on its credit card. The bank makes this offer to an SRS of 100 of its existing credit card customers. It then compares how much these customers charge this year with the amount that they charged last year. The mean is $585, and the standard deviation is $928.
(a) Is there significant evidence at the 1% level that the mean amount charged increases under the no-fee offer? State H_0 and H_a and carry out a t test.
(b) Give a 95% confidence interval for the mean amount of the increase.
(c) The distributions of the amount charged are skewed to the right, but outliers are prevented by the credit limit that the bank enforces on each card. Use of the t procedures is justified in this case even though the population distribution is not Normal. Explain why.
(d) A critic points out that the customers would probably have charged more this year than last even without the new offer because the economy is more prosperous and interest rates are

lower. Briefly describe the design of an experiment to study the effect of the no-fee offer that would avoid this criticism.

7.41 Truth in advertising. A leading agricultural company advertises that their bags of seeds contain at least a certain number of seeds. An SRS of bags from one warehouse is chosen, and these bags are opened and the number of seeds in each bag is counted. The actual counts are compared with the advertised target number of seeds. The difference is calculated as actual minus target, so that positive differences indicate bags that meet the advertising claim, and negative differences indicate bags that fall below the advertising claim. **SEEDTARGET**

(a) Using the data file, create a histogram, a boxplot, and a Normal quantile plot of these data. What percent of these bags fall below the advertised target?

(b) From inspecting the plots in part (a), comment on skewness, Normality, and the presence of outliers. Do any of your observations shed doubt on the appropriateness of using t procedures to analyze these data? Briefly explain your response.

(c) To assess its claim, the company accounts for the inherent variability in seed count by testing that the mean difference is at least 500 seeds. Using an alpha of 0.05, perform this test making sure to specify the hypotheses and the P-value.

(d) The data include at least one negative outlier. If you were to exclude this outlier from the analysis, would your P-value decrease or increase? Explain your response and support your explanation with a sketch.

(e) Redo the test with this low outlier excluded and verify your answer to part (d).

7.42 Design of controls. The design of controls and instruments has a large effect on how easily people can use them. A student project investigated this effect by asking 25 right-handed students to turn a knob (with their right hands) that moved an indicator by screw action. There were two identical instruments, one with a right-hand thread (the knob turns clockwise) and the other with a left-hand thread (the knob turns counterclockwise). The table below gives the times required (in seconds) to move the indicator a fixed distance:[11] **CONTROLS**

Subject	Right thread	Left thread	Subject	Right thread	Left thread
1	113	137	14	107	87
2	105	105	15	118	166
3	130	133	16	103	146
4	101	108	17	111	123
5	138	115	18	104	135
6	118	170	19	111	112
7	87	103	20	89	93
8	116	145	21	78	76
9	75	78	22	100	116
10	96	107	23	89	78
11	122	84	24	85	101
12	103	148	25	88	123
13	116	147			

(a) Each of the 25 students used both instruments. Discuss briefly how the experiment should be arranged and how randomization should be used.

(b) The project hoped to show that right-handed people find right-hand threads easier to use. State the appropriate H_0 and H_a about the mean time required to complete the task.

(c) Carry out a test of your hypotheses. Give the P-value and report your conclusions.

7.43 Is the difference important? Give a 90% confidence interval for the mean time advantage of right-hand over left-hand threads in the setting of the previous exercise. Do you think that the time saved would be of practical importance if the task were performed many times—for example, by an assembly-line worker? To help answer this question, find the mean time for right-hand threads as a percent of the mean time for left-hand threads.

7.44 Potential insurance fraud? Insurance adjusters are concerned about the high estimates they are receiving from Jocko's Garage. To see if the estimates are unreasonably high, each of 10 damaged cars was taken to Jocko's and to a "trusted" garage and the estimates recorded. Here are the results: **FRAUD**

Car:	1	2	3	4	5
Jocko's:	500	1550	1250	1300	750
Other:	430	1500	1300	1280	780
Car:	6	7	8	9	10
Jocko's:	1000	1210	1300	800	2500
Other:	870	1120	1140	650	2290

Test the null hypothesis that there is no difference between the two garages. Be sure to specify the null and alternative hypotheses, the test statistic with degrees of freedom, and the P-value. What do you conclude?

7.45 Fuel efficiency comparison t test. One of the authors of this book records the mpg of his car each time he fills the tank. He does this by dividing the miles driven since the last fill-up by the amount of gallons at fill-up. He wants to determine if these calculations differ from what his car's computer estimates. **FUELCOMP**

Fill-up:	1	2	3	4	5	6	7	8	9	10
Computer:	41.5	50.7	36.6	37.3	34.2	45.0	48.0	43.2	47.7	42.2
Driver:	36.5	44.2	37.2	35.6	30.5	40.5	40.0	41.0	42.8	39.2
Fill-up:	11	12	13	14	15	16	17	18	19	20
Computer:	43.2	44.6	48.4	46.4	46.8	39.2	37.3	43.5	44.3	43.3
Driver:	38.8	44.5	45.4	45.3	45.7	34.2	35.2	39.8	44.9	47.5

(a) State the appropriate H_0 and H_a.

(b) Carry out the test. Give the P-value, and then interpret the result.

7.46 Executives learn Spanish. The table below gives the pretest and posttest scores on the MLA listening test in Spanish for 20 executives who received intensive training in Spanish.[12] The setting is identical to the one described in Example 7.7 (page 405). SPANISH

Subject	Pretest	Posttest	Subject	Pretest	Posttest
1	30	29	11	30	32
2	28	30	12	29	28
3	31	32	13	31	34
4	26	30	14	29	32
5	20	16	15	34	32
6	30	25	16	20	27
7	34	31	17	26	28
8	15	18	18	25	29
9	28	33	19	31	32
10	20	25	20	29	32

(a) We hope to show that the training improves listening skills. State an appropriate H_0 and H_a. Describe in words the parameters that appear in your hypotheses.

(b) Make a graphical check for outliers or strong skewness in the data that you will use in your statistical test, and report your conclusions on the validity of the test.

(c) Carry out a test. Can you reject H_0 at the 5% significance level? At the 1% significance level?

(d) Give a 90% confidence interval for the mean increase in listening score due to the intensive training.

7.47 A field trial. An agricultural field trial compares the yield of two varieties of tomatoes for commercial use. The researchers divide in half each of 10 small plots of land in different locations and plant each tomato variety on one half of each plot. After harvest, they compare the yields in pounds per plant at each location. The 10 differences (Variety A − Variety B) give the following statistics: $\bar{x} = 0.55$ and $s = 1.23$. Is there convincing evidence that Variety A has the higher mean yield? State H_0 and H_a, and give a P-value to answer this question.

7.48 Which design? The following situations all require inference about a mean or means. Identify each as (1) a single sample, (2) matched pairs, or (3) two independent samples. The procedures of this section apply to the first two situations. We will learn procedures for the third in the next section. Explain your answers.

(a) Your customers are college students. You want to compare students who live in the dorms and those who live elsewhere with regard to their interest in a new product that you are developing.

(b) Your customers are college students. You are interested in comparing which of two new product labels is more appealing.

(c) Your customers are college students. You are interested in assessing their interest in a new product.

7.49 Which design? The following situations all require inference about a mean or means. Identify each as (1) a single sample,

(2) matched pairs, or (3) two independent samples. The procedures of this section apply to the first two situations. We will learn procedures for the third in the next section. Explain your answers.

(a) You want to estimate the average age of your customers.

(b) You survey an SRS of your customers every year. One of the questions on the survey asks respondents to rate their satisfaction on a seven-point scale with 1 indicating "very dissatisfied" and 7 indicating "very satisfied." You want to see if the mean customer satisfaction has improved since last year.

7.50 Areas of continents. While browsing the Internet, you find a site that gives the land area (in square kilometers) for Africa, Antarctica, Asia, Australia, Europe, North America, and South America. You feed the data into your statistical software package and the output produced includes a confidence interval. Is this a valid use of the procedure? Explain your answer.

The following exercises concern the material in the sections on the power of the t test and on non-Normal populations.

7.51 Credit card fees. The bank in Exercise 7.40 tested a new idea on a sample of 100 customers. Suppose that the bank wanted to be quite certain of detecting a mean increase of $\mu = \$300$ in the credit card amount charged, at the $\alpha = 0.01$ significance level. Perhaps a sample of only $n = 60$ customers would accomplish this. Find the approximate power of the test with $n = 60$ for the alternative $\mu = \$300$ as follows:

(a) What is the t critical value for the one-sided test with $\alpha = 0.01$ and $n = 60$?

(b) Write the criterion for rejecting H_0: $\mu = 0$ in terms of the t statistic. Then take $s = 928$ and state the rejection criterion in terms of $\bar{x}$.

(c) Assume that $\mu = 300$ (the given alternative) and that $\sigma = 928$. The approximate power is the probability of the event you found in part (b), calculated under these assumptions. Find the power. Would you recommend that the bank do a test on 60 customers, or should more customers be included?

7.52 A field trial. The tomato experts who carried out the field trial described in Exercise 7.47 suspect that the relative lack of significance is due to low power. They would like to be able to detect a mean difference in yields of 0.6 pound per plant at the 0.05 significance level. Based on the previous study, use 1.23 as an estimate of both the population σ and the value of s in future samples.

(a) What is the power of the test from Exercise 7.47 with $n = 10$ for the alternative $\mu = 0.6$?

(b) If the sample size is increased to $n = 25$ plots of land, what will be the power for the same alternative?

7.53 Design of controls. Apply the sign test to the data in Exercise 7.42 to assess whether the subjects can complete a task with a right-hand thread significantly faster than with a left-hand thread.

(a) State the hypotheses two ways, in terms of a population median and in terms of the probability of completing the task faster with a right-hand thread.

(b) Carry out the sign test. Find the approximate P-value using the Normal approximation to the binomial distributions, and report your conclusion.

7.54 Learning Spanish. Use the sign test to assess whether the intensive language training of Exercise 7.46 improves Spanish listening skills. State the hypotheses, give the P-value using the binomial table (Table C), and report your conclusion

7.55 Sign test for potential insurance fraud. The differences in the repair estimates in Exercise 7.44 can also be analyzed

using a sign test. Set up the appropriate null and alternative hypotheses, carry out the test, and summarize the results. How do these results compare with those that you obtained in Exercise 7.44?

7.56 Sign test for fuel efficiency comparison. Use the sign test to assess whether the computer calculates a higher mpg than the driver in Exercise 7.45. State the hypotheses, give the P-value using the binomial table (Table C), and report your conclusion.

7.2 Comparing Two Means

How do small businesses that fail differ from those that succeed? Business school researchers compare two samples of new firms: one sample of failed businesses and one of firms that are still going after two years. This study *compares two random samples,* one from each of two different populations. Which of two incentive packages will lead to higher use of a bank's credit cards? The bank designs a *randomized comparative experiment to compare two treatments.* Credit card customers are assigned at random to receive one or the other incentive offer. After six months, the bank compares the amounts charged. *Two-sample problems* such as these are among the most common situations encountered in statistical practice.

Two-Sample Problems

- The goal of inference is to compare the responses in two groups.

- Each group is considered to be a sample from a distinct population.

- The responses in each group are independent of those in the other group.

You must carefully distinguish two-sample problems from the matched pairs designs studied earlier. In two-sample problems, there is no matching of the units in the two samples, and the two samples may be of different sizes. Inference procedures for two-sample data differ from those for matched pairs.

We can present two-sample data graphically with a back-to-back stemplot (for small samples) or with side-by-side boxplots (for larger samples). Now we will apply the ideas of formal inference in this setting. When both population distributions are symmetric, and especially when they are at least approximately Normal, a comparison of the mean responses in the two populations is most often the goal of inference.

We have two independent samples, from two distinct populations (such as failed businesses and successful businesses). We measure the same variable (such as initial capital) in both samples. We will call the variable x_1 in the first population and x_2 in the second because the variable may have different distributions in the two populations. Here is the notation that we will use to describe the two populations:

Population	Variable	Mean	Standard deviation
1	x_1	μ_1	σ_1
2	x_2	μ_2	σ_2

We want to compare the two population means, either by giving a confidence interval for $\mu_1 - \mu_2$ or by testing the hypothesis of no difference, H_0: $\mu_1 = \mu_2$. We base inference on two independent SRSs, one from each population. Here is the notation that describes the samples:

Population	Sample size	Sample mean	Sample standard deviation
1	n_1	$\overline{x}_1$	s_1
2	n_2	$\overline{x}_2$	s_2

Throughout this section, the subscripts 1 and 2 show the population to which a parameter or a sample statistic refers.

The two-sample z statistic

The natural estimator of the difference $\mu_1 - \mu_2$ is the difference between the sample means, $\overline{x}_1 - \overline{x}_2$. If we are to base inference on this statistic, we must know its sampling distribution. Our knowledge of probability is equal to the task. First, the mean of the difference $\overline{x}_1 - \overline{x}_2$ is the difference of the means $\mu_1 - \mu_2$. This follows from the addition rule for means (page 248) and the fact that the mean of any $\overline{x}$ is the same as the mean of the population. Because the samples are independent, their sample means $\overline{x}_1$ and $\overline{x}_2$ are independent random variables. The addition rule for variances (page 253) says that the variance of the difference $\overline{x}_1 - \overline{x}_2$ is the sum of their variances, which is

$$\frac{\sigma_1^2}{n_1} + \frac{\sigma_2^2}{n_2}$$

We now know the mean and variance of the distribution of $\overline{x}_1 - \overline{x}_2$ in terms of the parameters of the two populations. If the two population distributions are both Normal, then the distribution of $\overline{x}_1 - \overline{x}_2$ is also Normal. This is true because each sample mean alone is Normally distributed and because a difference of independent Normal random variables is also Normal.

Since any Normal random variable has the $N(0, 1)$ distribution when standardized, we have arrived at a new z statistic.

Two-Sample z Statistic

Suppose that $\overline{x}_1$ is the mean of an SRS of size n_1 drawn from an $N(\mu_1, \sigma_1)$ population and that $\overline{x}_2$ is the mean of an independent SRS of size n_2 drawn from an $N(\mu_2, \sigma_2)$ population. Then the **two-sample z statistic**

$$z = \frac{(\overline{x}_1 - \overline{x}_2) - (\mu_1 - \mu_2)}{\sqrt{\dfrac{\sigma_1^2}{n_1} + \dfrac{\sigma_2^2}{n_2}}}$$

has the standard Normal $N(0, 1)$ sampling distribution.

In the unlikely event that both population standard deviations are known, the two-sample z statistic is the basis for inference about $\mu_1 - \mu_2$. Exact z procedures are seldom used, however, because σ_1 and σ_2 are rarely known. In Chapter 6, we discussed the one-sample z procedures to introduce the ideas of inference. Now we pass immediately to the more useful t procedures.

The two-sample *t* procedures

In practice, the two population standard deviations σ_1 and σ_2 are not known. We estimate them by the sample standard deviations s_1 and s_2 from our two samples. Following the pattern of the one-sample case, we substitute the standard errors for the standard deviations in the two-sample z statistic. The result is the *two-sample t statistic:*

$$t = \frac{(\overline{x}_1 - \overline{x}_2) - (\mu_1 - \mu_2)}{\sqrt{\dfrac{s_1^2}{n_1} + \dfrac{s_2^2}{n_2}}}$$

Unfortunately, this statistic does *not* have a t distribution. A t distribution replaces an $N(0, 1)$ distribution only when a single standard deviation (σ) in a z statistic is replaced by an estimate (s). In this case, we replaced two standard deviations (σ_1 and σ_2) by their estimates (s_1 and s_2), which does not produce a statistic having a t distribution.

Nonetheless, we can approximate the distribution of the two-sample t statistic by using the $t(k)$ distribution with an **approximation for the degrees of freedom k**. We use these approximations to find approximate values of t^* for confidence intervals and to find approximate P-values for significance tests. There are two procedures used in practice:

$k = n - 1$

Satterthwaite approximation

1. Use an approximation known as the **Satterthwaite approximation** to calculate a value of k from the data. In general, this k will not be an integer.

2. Use degrees of freedom k equal to the smaller of $n_1 - 1$ and $n_2 - 1$.

Most statistical software uses the Satterthwaite approximation for two-sample problems unless the user requests another method. Use of this approximation without software is a bit complicated; we will give the details later in this section. If you are not using software, we recommend the second approximation. This approximation is appealing because it is conservative.[13] That is, margins of error for confidence intervals are a bit larger than they need to be, so the true confidence level is larger than C. For significance testing, the true P-values are a bit smaller than those we obtain from the approximation; for tests at a fixed significance level, we are a little less likely to reject H_0 when it is true. In practice, the choice of approximation almost never makes a difference in our practical conclusion.

The two-sample *t* significance test

We now apply the basic ideas about t procedures to the problem of comparing two means when the standard deviations are unknown. We start with significance tests.

> **Two-Sample *t* Significance Test**
>
> Draw an SRS of size n_1 from a Normal population with unknown mean μ_1 and an independent SRS of size n_2 from another Normal population with unknown mean μ_2. To test the hypothesis H_0: $\mu_1 = \mu_2$, compute the **two-sample *t* statistic**
>
> $$t = \frac{(\overline{x}_1 - \overline{x}_2) - (\mu_1 - \mu_2)}{\sqrt{\dfrac{s_1^2}{n_1} + \dfrac{s_2^2}{n_2}}}$$
>
> and use P-values or critical values for the $t(k)$ distribution, where the degrees of freedom k are either approximated by software or are the smaller of $n_1 - 1$ and $n_2 - 1$.

EXAMPLE 7.11 Wine Labels with Animals?

Traditional brand research argues that successful logos are ones that are highly relevant to the product they represent. For example, pictures of grapes or a vineyard would be considered highly appropriate for a table wine. However, a market research firm recently reported that nearly 20% of all table wine brands introduced in the last three years feature an animal on the label. Since animals have little to do with the product, why are marketers using this tactic?

A recent study sheds some light on this tactic's success.[14] The researchers found that consumers have an easier time processing visual information when they are "primed" – in other words, when they've thought about the image earlier in an unrelated context. To demonstrate this, one of the researchers' experiments compared product preference for a group of consumers who were "primed" and a group of consumers who were not primed (actually, the study involved three groups but we will focus on just these two here). A bottle of MagicCoat pet shampoo, with a picture of a collie on the label, was the product, and participants indicated their attitude toward this product on a seven-point scale (from 1-dislike very much to 7-like very much). Prior to giving this score, however, participants were asked to do a "word find" where four of the words were common across groups (pet, grooming, bottle, label) and four were either related to the image (dog, collie, puppy, woof) or conflicted with the image (cat, feline, kitten, meow). Since their journal article reported only summary information, we generated the following data to approximate their results. We assume that there are 22 subjects in the "primed" group and 20 subjects in the "nonprimed" group.

BRANDPREFERENCE

A back-to-back stemplot of the results suggests a slight skewness in the responses, perhaps more so in the primed group, but has no obvious outliers. Because the responses are integers, the distributions are certainly non-Normal, but given the sample sizes and lack of serious departures from Normality, we will press on using the t procedures.

Nonprimed		Primed
xx	1	
xx	2	xx,
xxxxxxxxxxx	3	xxx
xxx	4	xxxxxxxxxx
x	5	xxxxxxx
	6	
	7	

In general, the scores of the primed group appear to be higher than those of the nonprimed group. The summary statistics are

Group	n	$\bar{x}$	s
Primed	22	4.00	0.93
Nonprimed	20	2.95	0.94

(handwritten annotations: "sample size" above n, "sample mean" above $\bar{x}$)

Because the researchers hope to show that the primed group (Group 1) views the product more favorably than the nonprimed group (Group 2), the hypotheses are

$$H_0: \mu_1 = \mu_2$$
$$H_a: \mu_1 > \mu_2$$

The two-sample t test statistic is

$$t = \frac{\bar{x}_1 - \bar{x}_2}{\sqrt{\dfrac{s_1^2}{n_1} + \dfrac{s_2^2}{n_2}}}$$

$$= \frac{4.00 - 2.95}{\sqrt{\dfrac{0.93^2}{22} + \dfrac{0.94^2}{20}}} = 3.63$$

The P-value for the one-sided test is $P(T \geq 3.63)$. Software gives the approximate P-value as 0.0004 and uses 39.5 as the degrees of freedom. For the second approximation, the degrees of freedom k are equal to the smaller of

$$n_1 - 1 = 22 - 1 = 21 \quad \text{and} \quad n_2 - 1 = 20 - 1 = 19$$

df = 19		
p	0.001	0.0005
t^*	3.579	3.883

Comparing $t = 3.63$ with the entries in Table D for 19 degrees of freedom, we see that P lies between 0.0005 and 0.001. The data strongly suggest that consumers who were primed view the dog shampoo more favorably ($t = 3.63$, df $= 19$, $P < 0.001$).

Some software gives P-values only for the two-sided alternative. In this case, divide the reported value by 2 to get the one-sided P-value. However, be sure to check that the means differ in the direction specified by the alternative hypothesis.

APPLY YOUR KNOWLEDGE

7.57 Selling on the Internet. You want to compare the daily sales for two different designs of Web pages for your Internet business. You assign the next 90 days to either Design A or Design B, 45 days to each.
(a) Would you use a one-sided or two-sided significance test for this problem? Explain your choice.
(b) If you use Table D to find the critical value, what are the degrees of freedom using the second approximation?
(c) If you perform the significance test using $\alpha = 0.05$, how large (positive or negative) must the t statistic be to reject the null hypothesis that the two designs result in the same average daily sales?

7.58 Comparison of two Web designs. Consider the previous problem. If the t statistic for comparing the daily sales were 1.93, what P-value would you report? What would you conclude using $\alpha = 0.05$?

7.59 Changing the experiment. Suppose you know that online sales differ across days of the week. To compare two Web page designs, you choose two successive weeks in the middle of a month. You flip a coin to assign one Monday to the first design and the other Monday to the second. You repeat this for each of the seven days of the week. You now have 7 sales amounts for each design. It is incorrect to use the two-sample t test to see if the mean sales differ for the two designs. Carefully explain why.

The two-sample t confidence interval

The same ideas that we used for the two-sample t significance tests also apply to *two-sample t confidence intervals*. We can use either software or the conservative approach with Table D to approximate the value of t^*.

Two-Sample t Confidence Interval

Draw an SRS of size n_1 from a Normal population with unknown mean μ_1 and an independent SRS of size n_2 from another Normal population with unknown mean μ_2. The **confidence interval for $\mu_1 - \mu_2$** given by

$$(\bar{x}_1 - \bar{x}_2) \pm t^* \sqrt{\frac{s_1^2}{n_1} + \frac{s_2^2}{n_2}}$$

has confidence level at least C no matter what the population standard deviations may be. The margin of error is

$t_{\alpha/2,\,n-1}$

$$t^* \sqrt{\frac{s_1^2}{n_1} + \frac{s_2^2}{n_2}}$$

Here, t^* is the value for the $t(k)$ density curve with area C between $-t^*$ and t^*. The value of the degrees of freedom k is approximated by software or we use the smaller of $n_1 - 1$ and $n_2 - 1$.

To complete the analysis of the preference scores we examined in Example 7.11, we need to describe the size of the treatment effect. We do this with a confidence interval for the difference between the group means.

EXAMPLE 7.12   How Large an Increase in Preference?

BRANDPREFERENCE

We will find a 95% confidence interval for the mean increase. The interval is

$$(\bar{x}_1 - \bar{x}_2) \pm t^* \sqrt{\frac{s_1^2}{n_1} + \frac{s_2^2}{n_2}} = (4.00 - 2.95) \pm t^* \sqrt{\frac{0.93^2}{22} + \frac{0.94^2}{20}}$$

$$= 1.05 \pm (t^* \times 0.29)$$

$t_{0.05/2,\,n-1}$

$t_{.025,\,39.5} = 2.022$

Using software, the degrees of freedom are 39.5 and $t^* = 2.022$. This approximation gives

$$1.05 \pm (2.022 \times 0.29) = 1.05 \pm 0.58 = (0.47,\ 1.63)$$

The conservative approach uses the $t(19)$ distribution. Table D gives $t^* = 2.093$. With this approximation we have

$$1.05 \pm (2.093 \times 0.29) = 1.05 \pm 0.60 = (0.45,\ 1.65)$$

The conservative approach does give a larger interval than the more accurate approximation used by software. However, the difference is very small.

We estimate the mean increase in preference when primed to be about 1 point, but with a margin of error of slightly more than half a point. Although we have good evidence of an increase, the data do not allow a very precise estimate of the size of the average increase.

This study is typical of many situations in which an experiment is carried out using a sample that is not an SRS from the population of interest. In this case, the participants were undergraduate students who were native speakers of English. In addition, each participant was paid $2 to complete the experiment. Although it is unlikely that this sample would not be representative of the population of U.S. consumers, there is a risk in extending these conclusions beyond the studied population. The degree of risk depends on the research area and should be considered when designing the study.

APPLY YOUR KNOWLEDGE

7.60 New computer monitors? The purchasing department has suggested that all new computer monitors for your company should be a new style. You want data to be assured that employees will like these new monitors. You designate the next 20 employees needing a new computer to participate in an experiment in which 10 will be randomly assigned to receive the standard monitor and the remainder will receive the new monitor. After a month of use, these employees will express their satisfaction with their new monitors by responding to the statement "I like my new monitor" on a scale from 1 to 5, where 1 represents "strongly disagree," 2 is "disagree," 3 is "neutral," 4 is "agree," and 5 stands for "strongly agree."

(a) The employees with the new monitors have an average satisfaction score of 4.1 with standard deviation 0.8. The employees with the standard monitors have an average of 2.9 with standard deviation 1.6. Give a 95% confidence interval for the difference in the mean satisfaction scores for all employees.

(b) Would you reject the null hypothesis that the mean satisfaction for the two types of monitors is the same versus the two-sided alternative at significance level 0.05? Use your confidence interval to answer this question. Explain why you do not need to calculate the test statistic.

7.61 Why randomize? A coworker suggested that you give the new monitors to the next 10 employees who need new computers and the standard monitor to the following 10. Explain why your randomized design in the previous exercise is better.

Robustness of the two-sample procedures

The two-sample t procedures are more robust than the one-sample t methods. When the sizes of the two samples are equal and the distributions of the two populations being compared have similar shapes, probability values from the t table are quite accurate for a broad range of distributions when the sample sizes are as small as $n_1 = n_2 = 5$.[15] When the two population distributions have different shapes, larger samples are needed. The guidelines given on page 408 for the use of one-sample t procedures can be adapted to two-sample procedures by replacing "sample size" with the "sum of the sample sizes" $n_1 + n_2$. These guidelines are rather conservative, especially when the two samples are of equal size. In planning a two-sample study, you should usually choose equal sample sizes. The two-sample t procedures are most robust against non-Normality in this case, and the conservative probability values are most accurate.

EXAMPLE 7.13   Wheat Prices

The U.S. Department of Agriculture (USDA) uses sample surveys to produce important economic estimates.[16] One pilot study estimated wheat prices in July and in September using independent samples of wheat producers in the two months. Here are the summary statistics, in dollars per bushel:

Month	n	$\overline{x}$	s
September	45	$7.43	$0.24
July	90	$7.16	$0.27

The September prices are higher on the average. But we have data from only a limited number of producers each month. Can we conclude that national average prices in July and September are not the same? Or are these differences merely what we would expect to see due to random variation?

Because we did not specify a direction for the difference before looking at the data, we choose a two-sided alternative. The hypotheses are

$$H_0: \mu_1 = \mu_2$$
$$H_a: \mu_1 \neq \mu_2$$

Because the samples are moderately large, we can confidently use the t procedures even though we lack the detailed data and so cannot verify the Normality condition.

The two-sample t statistic is

$$t = \frac{\bar{x}_1 - \bar{x}_2}{\sqrt{\dfrac{s_1^2}{n_1} + \dfrac{s_2^2}{n_2}}} = \frac{7.43 - 7.16}{\sqrt{\dfrac{0.24^2}{45} + \dfrac{0.27^2}{90}}} = 5.91$$

The conservative approach finds the P-value by comparing 5.91 to critical values for the $t(44)$ distribution because the smaller sample has 45 observations. We must double the table tail area p because the alternative is two-sided.

Table D does not have entries for 44 degrees of freedom. When this happens, we use the next smaller degrees of freedom. Our calculated value of t is larger than the $p = 0.0005$ entry in the table. Doubling 0.0005, we conclude that the P-value is less than 0.001. The data give conclusive evidence that the mean wheat prices were higher in September than they were in July ($t = 5.91$, df = 44, $P < 0.001$).

df = 40	
p	0.0005
t^*	3.551

In this example the exact P-value is very small because $t = 5.91$ says that the observed mean is almost 6 standard deviations above the hypothesized mean. The difference in mean prices is not only highly significant but large enough (27 cents per bushel) to be important to producers.

In this and other examples, we can choose which population to label 1 and which to label 2. After inspecting the data, we chose September as Population 1 because this choice makes the t statistic a positive number. This avoids any possible confusion from reporting a negative value for t. Choosing the population labels is *not* the same as choosing a one-sided alternative after looking at the data. Choosing hypotheses after seeing a result in the data is a violation of sound statistical practice.

Inference for small samples

Small samples require special care. We do not have enough observations to examine the distribution shapes, and only extreme outliers stand out. The power of significance tests tends to be low, and the margins of error of confidence intervals tend to be large. Despite these difficulties, we can often draw important conclusions from studies with small sample sizes. If the size of an effect is as large as it was in the wheat price example, it should still be evident even if the n's are small.

EXAMPLE 7.14   More about Wheat Prices

 WHEAT

In the setting of Example 7.13, a quick survey collects prices from only 5 producers each month. The data are

Month	Price of wheat ($/bushel)				
September	$7.3325	$7.5575	$7.3125	$7.3600	$7.5550
July	$6.8250	$7.3025	$7.0275	$7.0825	$7.3000

The prices are reported to the nearest quarter of a cent. First, examine the distributions with a back-to-back stemplot after rounding each price to the nearest cent.

September		July
	6.8	3
	6.9	
	7.0	38
	7.1	
	7.2	
631	7.3	00
	7.4	
66	7.5	

The pattern is reasonably clear. Although there is variation among prices within each month, the top 5 prices are all from September and the 5 lowest prices are from July.

A significance test can confirm that the difference between months is too large to easily arise just by chance. We test

$$H_0: \mu_1 = \mu_2$$
$$H_a: \mu_1 \neq \mu_2$$

The price is higher in September ($t = 3.00$, df $= 6.6$, $P = 0.0214$). The difference in sample means is 31.6 cents.

Figure 7.8 gives outputs for this analysis from several software systems. Although the formats differ, the basic information is the same. All report the sample sizes, the sample means and standard deviations (or variances), the t statistic, and its P-value. All agree that the P-value is very small, though some give more detail than others. Software often labels the groups in alphabetical order. In this example, July is then the first population and $t = -3.00$, the negative of our result. Always check the means first and report the statistic (you may need to change the sign) in an appropriate way. Be sure to also mention the size of the effect you observed, such as "The mean price for September was 31.6 cents higher than in July."

SAS and SPSS report the results of *two t* procedures: a special procedure that assumes that the two population variances are equal and the general two-sample procedure that we have just studied. We don't recommend the "equal-variances" procedures, but we describe them later, in the section on pooled two-sample t procedures.

Satterthwaite approximation for the degrees of freedom

We noted earlier that the two-sample t statistic does not have an exact t distribution. Moreover, the exact distribution changes as the unknown population standard deviations σ_1 and σ_2 change. However, the distribution can be approximated by a t distribution with degrees of freedom given by

$$df = \frac{\left(\dfrac{s_1^2}{n_1} + \dfrac{s_2^2}{n_2}\right)^2}{\dfrac{1}{n_1 - 1}\left(\dfrac{s_1^2}{n_1}\right)^2 + \dfrac{1}{n_2 - 1}\left(\dfrac{s_2^2}{n_2}\right)^2}$$

Most statistical software uses this approximation. It was discovered by a statistician named Satterthwaite, and his name is frequently used to identify the approximation on output. The Satterthwaite approximation is quite accurate when both sample sizes n_1 and n_2 are 5 or larger.

EXAMPLE 7.15   Outputs for Wheat Prices

Outputs for the wheat prices example with samples of size 5 appear in Figure 7.8. SAS, SPSS, and JMP report the degrees of freedom for the Satterthwaite approximation as 6.6 or 6.603. Excel rounds to the nearest integer and Minitab rounds down. All of these df's are larger than the conservative df $= 4$ that we would use in the absence of software. In this example, the t statistic is so large that all choices lead to the same practical conclusion.

Excel

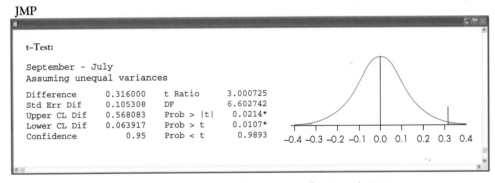

	A	B	C
1		July	September
2	Mean	7.1075	7.4235
3	Variance	0.040478125	0.014970625
4	Observations	5	5
5	Hypothesized Mean Difference	0	
6	df	7	
7	t Stat	-3.00072506	
8	P(T<=t) one-tail	0.009960842	
9	t Critical one-tail	1.894578604	
10	P(T<=t) two-tail	0.019921684	
11	t Critical two-tail	2.364624251	

t-Test: Two-Sample Assuming Unequal Variances

Minitab

Two-Sample T-Test and CI: Price, Month

```
Two-sample T for Price

Month        N      Mean     StDev    SE Mean
July         5      7.107    0.201     0.090
September    5      7.423    0.122     0.055

Difference = mu (July) - mu (September)
Estimate for difference: -0.316
95% CI for difference: (-0.574, -0.058)
T-Test of difference = 0 (vs not =): T-Value = -3.00 P-Value = 0.024 DF = 6
```

JMP

```
t-Test:

September - July
Assuming unequal variances

Difference    0.316000    t Ratio      3.000725
Std Err Dif   0.105308    DF           6.602742
Upper CL Dif  0.568083    Prob > |t|   0.0214*
Lower CL Dif  0.063917    Prob > t     0.0107*
Confidence        0.95    Prob < t     0.9893
```

-0.4 -0.3 -0.2 -0.1 0.0 0.1 0.2 0.3 0.4

FIGURE 7.8 Excel, Minitab, JMP, SAS, and SPSS outputs for Example 7.14.

The number df given by the Satterthwaite approximation is always at least as large as the smaller of $n_1 - 1$ and $n_2 - 1$. On the other hand, df is never larger than the sum $n_1 + n_2 - 2$ of the two individual degrees of freedom. The number of degrees of freedom is generally not a whole number. There is a t distribution with any positive degrees of freedom, even though Table D contains entries only for whole-number degrees of freedom. Because of this and the need to calculate df, we do not recommend using the approximation unless a computer is doing the arithmetic. With software, however, the more accurate procedures are painless.

SAS

```
                    The TTEST Procedure
                        Statistics


                       Lower CL        Upper CL  Lower CL        Upper CL
Variable   Month    N    Mean    Mean    Mean   Std Dev  Std Dev  Std Dev  Std Err

Price      July     5   6.8577  7.1075  7.3573  0.1205   0.2012   0.5781   0.09
Price      September 5  7.2716  7.4235  7.5754  0.0733   0.1224   0.3516   0.0547
Price      Diff (1-2)      -0.559  -0.316  -0.073  0.1125   0.1665   0.319    0.1053

                             T-Tests
          Variable   Method          Variances   DF   t Value   Pr > |t|
          Price      Pooled          Equal        8    -3.00     0.0171
          Price      Satterthwaite   Unequal     6.6   -3.00     0.0214
```

SPSS

T-Test

Group Statistics

	Month 1	N	Mean	Std. Deviation	Std. Error Mean
Price	July	5	7.107500	.2011918	.0899757
	Sept	5	7.423500	.1223545	.0547186

Independent Samples Test

		Levene's Test for Equality of Variances		t-test for Equality of Means						
									95% Confidence Interval of the Difference	
		F	Sig.	t	df	Sig. (2-tailed)	Mean Difference	Std. Error Difference	Lower	Upper
Price	Equal variances assumed	1.052	.335	-3.001	8	.017	-.3160000	.1053079	-.5588404	-.0731596
	Equal variances not assumed			-3.001	6.603	.021	-.3160000	.1053079	-.5680831	-.0639169

FIGURE 7.8 *(Continued)*

APPLY YOUR KNOWLEDGE

7.62 Calculating the Satterthwaite df. The SAS output in Figure 7.8 gives df $= 6.6$. Use the sample standard deviations in that output and the Satterthwaite formula to verify this value.

7.63 Compare the outputs. Figure 7.8 gives the outputs from five software systems for the two-sample t procedures. Make tables to compare the outputs, giving the numerical values reported by the software.

(a) How do they report the means?

(b) How do they report variability for each of the groups?

(c) How do they report the test statistic, degrees of freedom, and the P-value?

(d) Do they give a confidence interval for the mean difference?

(e) Write a short summary comparing the five outputs from the viewpoint of a user of statistics. Which do you like the best? Which do you like the least?

7.64 Can you do better? Design your own output for a two-sample t analysis and illustrate it with the results given in Figure 7.8. Pay particular attention to the use of labels and how numbers are rounded.

The pooled two-sample t procedures

There is one situation in which a t statistic for comparing two means has exactly a t distribution. Suppose that the two Normal population distributions have the *same* standard deviation. In this case we need substitute only a single standard error in a z statistic, and the resulting t statistic has a t distribution. We will develop the z statistic first, as usual, and from it the t statistic.

Call the common—but still unknown—standard deviation of both populations σ. Both sample variances s_1^2 and s_2^2 estimate σ^2. The best way to combine these two estimates is to average them with weights equal to their degrees of freedom. This gives more weight to the information from the larger sample. The resulting estimator of σ^2 is

$$s_p^2 = \frac{(n_1 - 1)s_1^2 + (n_2 - 1)s_2^2}{n_1 + n_2 - 2}$$

pooled estimator of σ^2   This is called the **pooled estimator of σ^2** because it combines the information in both samples.

When both populations have variance σ^2, the addition rule for variances says that $\bar{x}_1 - \bar{x}_2$ has variance equal to the *sum* of the individual variances, which is

$$\frac{\sigma^2}{n_1} + \frac{\sigma^2}{n_2} = \sigma^2 \left(\frac{1}{n_1} + \frac{1}{n_2} \right)$$

The standardized difference of means in this equal-variance case is therefore

$$z = \frac{(\bar{x}_1 - \bar{x}_2) - (\mu_1 - \mu_2)}{\sigma\sqrt{\dfrac{1}{n_1} + \dfrac{1}{n_2}}}$$

This is a special two-sample z statistic for the case in which the populations have the same σ. Replacing the unknown σ by the estimate s_p gives a t statistic. The degrees of freedom are $n_1 + n_2 - 2$, the sum of the degrees of freedom of the two sample variances. This statistic is the basis of the pooled two-sample t inference procedures.

Pooled Two-Sample *t* Procedures

Draw an SRS of size n_1 from a Normal population with unknown mean μ_1 and an independent SRS of size n_2 from another Normal population with unknown mean μ_2. Suppose that the two populations have the same unknown standard deviation. A level C confidence interval for $\mu_1 - \mu_2$ is

$$(\bar{x}_1 - \bar{x}_2) \pm t^* s_p \sqrt{\frac{1}{n_1} + \frac{1}{n_2}}$$

Here t^* is the value for the $t(n_1 + n_2 - 2)$ density curve with area C between $-t^*$ and t^*.

To test the hypothesis H_0: $\mu_1 = \mu_2$, compute the **pooled two-sample *t* statistic**

$$t = \frac{\bar{x}_1 - \bar{x}_2}{s_p \sqrt{\dfrac{1}{n_1} + \dfrac{1}{n_2}}}$$

and use P-values from the $t(n_1 + n_2 - 2)$ distribution.

CASE 7.2

COMPANIES

Healthy Companies versus Failed Companies In what ways are companies that fail different from those that continue to do business? To answer this question, one study compared various characteristics of 68 healthy and 33 failed firms.[17] One of the variables was the ratio of current assets to current liabilities. Roughly speaking, this is the amount that the firm is worth divided by what it owes. The data appear in Table 7.3.

TABLE 7.3   Ratio of current assets to current liabilities

Healthy firms						Failed firms		
1.50	0.10	1.76	1.14	1.84	2.21	0.82	0.89	1.31
2.08	1.43	0.68	3.15	1.24	2.03	0.05	0.83	0.90
2.23	2.50	2.02	1.44	1.39	1.64	1.68	0.99	0.62
0.89	0.23	1.20	2.16	1.80	1.87	0.91	0.52	1.45
1.91	1.67	1.87	1.21	2.05	1.06	1.16	1.32	1.17
0.93	2.17	2.61	3.05	1.52	1.93	0.42	0.48	0.93
1.95	2.61	1.11	0.95	0.96	2.25	0.88	1.10	0.23
2.73	1.56	2.73	0.90	2.12	1.42	1.11	0.19	0.13
1.62	1.76	2.22	2.80	1.85	0.96	2.03	0.51	1.12
1.71	1.02	2.50	1.55	1.69	1.64	0.92	0.26	1.15
1.03	1.80	0.67	2.44	2.30	2.21	0.13	0.88	0.09
1.96	1.81							

Chris Hondros/Getty Images

As usual, we first examine the data. Histograms for the two groups of firms are given in Figure 7.9. Normal curves with mean and standard deviation equal to the sample values are superimposed on the histograms. The distribution for the healthy firms looks more Normal than the distribution for the failed firms. It appears that there may actually be two subgroups of data for the failed firms. This is seen more clearly in the back-to-back stemplot given in Figure 7.10. However, there are no outliers or strong departures from Normality that will prevent us from using the *t* procedures for these data. Let's compare the mean current assets to current liabilities ratio for the two groups of firms using a significance test.

FIGURE 7.9 Histograms for assets to liabilities ratios, for Example 7.16.

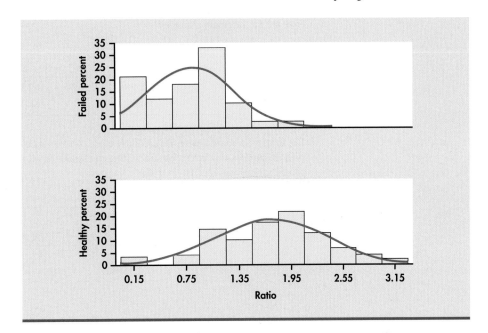

FIGURE 7.10 Back-to-back stemplot for assets to liabilities ratios, for Example 7.16.

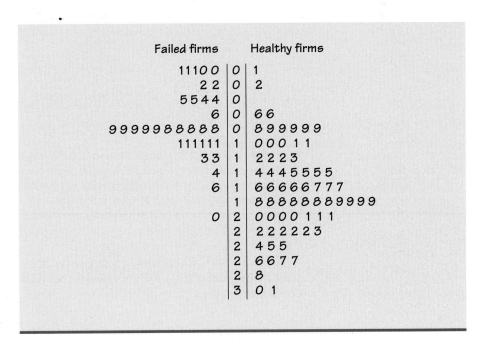

EXAMPLE 7.16 Do Mean Asset/Liability Ratios Differ?

CASE 7.2

COMPANIES

Take Group 1 to be the firms that were healthy and Group 2 to be those that failed. The question of interest is whether or not the mean ratio of current assets to current liabilities is different for the two groups. We therefore test

$$H_0: \mu_1 = \mu_2$$
$$H_a: \mu_1 \neq \mu_2$$

Here are the summary statistics:

Group	Firms	n	$\bar{x}$	s
1	Healthy	68	1.7256	0.6393
2	Failed	33	0.8236	0.4811

The sample standard deviations are fairly close. A difference this large is not particularly unusual even in samples this large. We are willing to assume equal population standard deviations. The pooled sample variance is

$$s_p^2 = \frac{(n_1 - 1)s_1^2 + (n_2 - 1)s_2^2}{n_1 + n_2 - 2}$$

$$= \frac{(67)(0.6393)^2 + (32)(0.4811)^2}{68 + 33 - 2}$$

$$= 0.35141$$

so that

$$s_p = \sqrt{0.35141} = 0.5928$$

The pooled two-sample t statistic is

$$t = \frac{\bar{x}_1 - \bar{x}_2}{s_p\sqrt{\dfrac{1}{n_1} + \dfrac{1}{n_2}}}$$

$$= \frac{1.7256 - 0.8236}{0.5928\sqrt{\dfrac{1}{68} + \dfrac{1}{33}}} = 7.17$$

The P-value is $P(T \geq 7.17)$, where T has the $t(99)$ distribution.

In Table D we have entries for 80 and 100 degrees of freedom. We will use the entries for 100. Our calculated value of t is larger than the $p = 0.0005$ entry in the table. Doubling 0.0005, we conclude that the two-sided P-value is less than 0.001. Statistical software gives a similar result—there is no practical need to report the exact P-value when it is smaller than 0.001.

df = 100	
p	0.0005
t^*	3.300

Of course, a P-value is rarely a complete summary of a statistical analysis. To make a judgment regarding the size of the difference between the two groups of firms, we need a confidence interval.

EXAMPLE 7.17 How Different Are Mean Asset/Liability Ratios?

CASE 7.2

COMPANIES

The difference in mean current assets to current liabilities ratios for healthy versus failed firms is

$$\bar{x}_1 - \bar{x}_2 = 1.7256 - 0.8236 = 0.902$$

For a 95% margin of error we will use the critical value $t^* = 1.984$ from the $t(100)$ distribution. The margin of error is

$$t^* s_p \sqrt{\frac{1}{n_1} + \frac{1}{n_2}} = (1.984)(0.5928)\sqrt{\frac{1}{68} + \frac{1}{33}}$$

$$= 0.25$$

We report that the successful firms have current assets to current liabilities ratios that average 0.90 higher than failed firms, with margin of error 0.25 for 95% confidence. Alternatively, we are 95% confident that the difference is between 0.65 and 1.15.

The pooled two-sample t procedures are anchored in statistical theory and have long been the standard version of the two-sample t in textbooks. But they require the condition that the two unknown population standard deviations are equal. As we shall see in Section 7.3, this condition is hard to verify.

The pooled t procedures are therefore a bit risky. They are reasonably robust against both non-Normality and unequal standard deviations when the sample sizes are nearly the same. When the samples are quite different in size, the pooled t procedures become sensitive to unequal standard deviations and should be used with caution unless the samples are large. Unequal standard deviations are quite common. In particular, it is common for the spread of data to increase when the center moves up, as in the data given in Case 7.2. We recommend regular use of the unpooled t procedures, particularly when software automates the Satterthwaite approximation.

APPLY YOUR KNOWLEDGE

7.65 Wheat prices revisited. Example 7.13 (page 424) gives summary statistics for prices received by wheat producers in September and July. The two sample standard deviations are very similar, so we may be willing to assume equal population standard deviations. Calculate the pooled t test statistic and its degrees of freedom from the summary statistics. Use Table D to assess significance. How do your results compare with the unpooled analysis in the example?

7.66 Using software. Figure 7.8 (pages 427 and 428) gives the outputs from five software systems for comparing prices received by wheat producers in July and September for small samples of 5 producers in each month. Some of the software reports both pooled and unpooled analyses. Which outputs give the pooled results? What are the pooled t and its P-value?

7.67 Pooled equals unpooled? The software outputs in Figure 7.8 give the *same value* for the pooled and unpooled t statistics. Do some simple algebra to show that this is always true when the two sample sizes n_1 and n_2 are the same. In other cases, the two t statistics usually differ.

SECTION 7.2 Summary

- Significance tests and confidence intervals for the difference of the means μ_1 and μ_2 of two Normal populations are based on the difference $\bar{x}_1 - \bar{x}_2$ of the sample means from two independent SRSs. Because of the central limit theorem, the resulting procedures are approximately correct for other population distributions when the sample sizes are large.

- When independent SRSs of sizes n_1 and n_2 are drawn from two Normal populations with parameters μ_1, σ_1 and μ_2, σ_2 the **two-sample z statistic**

$$z = \frac{(\bar{x}_1 - \bar{x}_2) - (\mu_1 - \mu_2)}{\sqrt{\dfrac{\sigma_1^2}{n_1} + \dfrac{\sigma_2^2}{n_2}}}$$

has the $N(0, 1)$ distribution.

- The **two-sample t statistic**

$$t = \frac{(\bar{x}_1 - \bar{x}_2) - (\mu_1 - \mu_2)}{\sqrt{\dfrac{s_1^2}{n_1} + \dfrac{s_2^2}{n_2}}}$$

does *not* have a t distribution. However, software can give accurate P-values and critical values using the **Satterthwaite approximation.**

- **Conservative inference procedures** for comparing μ_1 and μ_2 use the two-sample t statistic and the $t(k)$ distribution with degrees of freedom k equal to the smaller of $n_1 - 1$ and $n_2 - 1$. Use this method unless you are using software.

- An approximate level C **confidence interval** for $\mu_1 - \mu_2$ is given by

$$(\bar{x}_1 - \bar{x}_2) \pm t^* \sqrt{\frac{s_1^2}{n_1} + \frac{s_2^2}{n_2}}$$

Here, t^* is the value for the $t(k)$ density curve with area C between $-t^*$ and t^*, where k either is found by the Satterthwaite approximation or is the smaller of $n_1 - 1$ and $n_2 - 1$. The **margin of error** is

$$t^* \sqrt{\frac{s_1^2}{n_1} + \frac{s_2^2}{n_2}}$$

- Significance tests for H_0: $\mu_1 = \mu_2$ are based on the **two-sample t statistic**

$$t = \frac{\bar{x}_1 - \bar{x}_2}{\sqrt{\dfrac{s_1^2}{n_1} + \dfrac{s_2^2}{n_2}}}$$

The P-value is approximated using the $t(k)$ distribution, where k either is found by the Satterthwaite approximation or is the smaller of $n_1 - 1$ and $n_2 - 1$.

- The guidelines for practical use of two-sample t procedures are similar to those for one-sample t procedures. Equal sample sizes are recommended.

- If we can assume that the two populations have equal variances, **pooled two-sample t procedures** can be used. These are based on the **pooled estimator**

$$s_p^2 = \frac{(n_1 - 1)s_1^2 + (n_2 - 1)s_2^2}{n_1 + n_2 - 2}$$

of the unknown common variance and the $t(n_1 + n_2 - 2)$ distribution.

SECTION 7.2 Exercises

For Exercises 7.57 to 7.59, see page 422; for 7.60 and 7.61, see pages 423–424; for 7.62 to 7.64, see pages 428–429; and for 7.65 to 7.67, see page 433.

In exercises that call for two-sample t procedures, you may use either of the two approximations for the degrees of freedom that we have discussed: the value given by your software or the smaller of $n_1 - 1$ and $n_2 - 1$. Be sure to state clearly which approximation you have used.

7.68 What's wrong? In each of the following situations explain what is wrong and why.

(a) A researcher wants to test H_0: $\bar{x}_1 = \bar{x}_2$ versus the two-sided alternative H_a: $\bar{x}_1 \neq \bar{x}_2$.

(b) A study recorded the credit card IQ scores of 100 college freshmen. The scores of the 48 males in the study were compared with the scores of all 100 freshmen using the two-sample methods of this section.

(c) A two-sample t statistic gave a P-value of 0.97. From this we can reject the null hypothesis with 95% confidence.

(d) A researcher is interested in testing the one-sided alternative $H_a: \mu_1 < \mu_2$. The significance test for $\mu_1 - \mu_2$ gave $t = 2.41$. Since the P-value for the two-sided alternative is 0.024, he concluded that his P-value was 0.012.

7.69 Understanding concepts. For each of the following, answer the question and give a short explanation of your reasoning.

(a) A 95% confidence interval for the difference between two means is reported as (0.6, 1.3). What can you conclude about the results of a significance test of the null hypothesis that the population means are equal versus the two-sided alternative?

(b) Will larger samples generally give a larger or smaller margin of error for the difference between two sample means?

7.70 Determining significance. For each of the following, answer the question and give a short explanation of your reasoning.

(a) A significance test for comparing two means gave $t = -2.16$ with 11 degrees of freedom. Can you reject the null hypothesis that the μ's are equal versus the two-sided alternative at the 5% significance level?

(b) Answer part (a) for the one-sided alternative that the difference in means is negative.

(c) Answer part (a) for the one-sided alternative that the difference in means is positive.

7.71 Effect of the confidence level. Assume $\overline{x}_1 = 120$, $\overline{x}_2 = 110$, $s_1 = 12$, $s_2 = 11$, $n_1 = 50$, and $n_2 = 50$. Find a 95% confidence interval for the difference in the corresponding values of μ. Does this interval include more or fewer values than a 99% confidence interval? Explain your answer.

7.72 Sadness and spending. The "misery is not miserly" phenomenon refers to a sad person's spending judgment going haywire. In a recent study, 31 young adults were given $10 and randomly assigned to either a sad or a neutral group. The participants in the sad group watched a video about the death of a boy's mentor (from *The Champ*), and those in the neutral group watched a video on the Great Barrier Reef. After the video, each participant was offered the chance to trade $0.50 increments of the $10 for an insulated water bottle.[18] Here are the data: 🌀 **SADNESS**

Group	Purchase price ($)
Neutral	0.00 2.00 0.00 1.00 0.50 0.00 0.50
	2.00 1.00 0.00 0.00 0.00 0.00 1.00
Sad	3.00 4.00 0.50 1.00 2.50 2.00 1.50 0.00 1.00
	1.50 1.50 2.50 4.00 3.00 3.50 1.00 3.50

(a) Examine each group's prices graphically. Is use of the t procedures appropriate for these data? Carefully explain your answer.

(b) Make a table with the sample size, mean, and standard deviation for each of the two groups.

(c) State appropriate null and alternative hypotheses for comparing these two groups.

(d) Perform the significance test at the $\alpha = 0.05$ level, making sure to report the test statistic, degrees of freedom, and P-value. What is your conclusion?

(e) Construct a 95% confidence interval for the mean difference in purchase price between the two groups.

7.73 Comparison of blood lipid levels in males and females. Coronary heart disease (CHD) begins in young adulthood and is the fifth leading cause of death among adults aged 20 to 24 years. However, studies of serum cholesterol levels among college students are scarce. One study at a southern university investigated the lipid levels in a cohort of sedentary university students.[19] A total of 108 students volunteered for the study and met the eligibility criteria. The following table summarizes the blood lipid levels, in milligrams per deciliter (mg/dl), of the participants broken down by gender:

	Females ($n = 71$)		Males ($n = 37$)	
	$\overline{x}$	s	$\overline{x}$	s
Total cholesterol	173.70	34.79	171.81	33.24
LDL	96.38	29.78	109.44	31.05
HDL	61.62	13.75	46.47	7.94

(a) Is it appropriate to use the two-sample t procedures that we studied in this section to analyze these data for gender differences? Give reasons for your answer.

(b) Describe appropriate null and alternative hypotheses for comparing male and female total cholesterol levels.

(c) Carry out the significance test using $\alpha = 0.05$. Report the test statistic with the degrees of freedom and the P-value. Write a short summary of your conclusion.

(d) Find a 95% confidence interval for the difference between the two means. Compare the information given by the interval with the information given by the significance test.

(e) The participants in this study were all taking an introductory health class. To what extent do you think the results can be generalized to other populations?

7.74 More on blood lipid levels. Refer to the previous exercise. LDL is also known as "bad" cholesterol. Suppose the researchers wanted to test the hypothesis that LDL levels are higher in sedentary males than in sedentary females. Describe appropriate null and alternative hypotheses, and carry out the significance test using $\alpha = 0.05$. Report the test statistic with the degrees of freedom and the P-value. Write a short summary of your conclusion.

7.75 Counts of seeds in one-pound scoops. Refer to Exercise 7.27 (page 414). As part of the Six Sigma quality improvement effort, the company wants to compare scoops of seeds from two different packaging plants. An SRS of 50 one-pound scoops of seeds was collected from Plant 1746 and an SRS of 19 one-pound scoops of seeds was collected from Plant 1748. The number of seeds in each scoop were recorded. 🌀 **SEEDCOUNT2**

(a) Using this data set, create a histogram, boxplot, and Normal quantile plot of the seed counts from Plant 1746. Do the same for

Plant 1748. Are the distributions reasonably Normal? Summarize the distributions in words.

(b) Are the t procedures appropriate given your observations in part (a)? Explain your answer.

(c) Compare the mean number of seeds per one-pound scoop for these two manufacturing plants using a 99% confidence interval.

(d) Test the equality of the means using a two-sided alternative and a significance level of 1%. Make sure to specify the test statistic, degrees of freedom, and P-value.

(e) Write a brief summary of your t procedures assuming your audience is the company CEO and the two plant managers.

7.76 More on counts of seeds. Refer to the previous exercise.
(a) When would a one-sided alternative hypothesis be appropriate in this setting? Explain.
(b) What alternative hypothesis would we be testing if we halved the P-value from the previous exercise?

7.77 Drive-thru speaker clarity. *QSR Magazine.com* surveyed 689 adults on their drive-thru window experiences at quick-service restaurants.[20] One question was "Thinking about your most recent drive-thru experience, please rate how satisfied you were with the clarity of communication through the speaker." Responses ranged from "Very Dissatisfied (1)" to "Very Satisfied (5)." The table below breaks down the responses according to gender. 🔊 SPEAKERCLARITY

Gender	Rating 1	2	3	4	5
Female	5	44	48	183	188
Male	5	29	30	91	66

(a) Report the means and standard deviations of the rating for the male and female participants separately.

(b) Comment on the appropriateness of t procedures for these data.

(c) Test whether males and females are, on average, equally satisfied with speaker clarity. Use a two-sided alternative hypothesis and a significance level of 5%.

(d) Construct a 95% confidence interval for the difference in average satisfaction.

(e) Given the coarseness of the rating, the owner of the Sir Beef-a-lot chain only considers a difference in the means of at least 0.25 units meaningful. Based on your results in parts (c) and (d), what would you tell this owner?

7.78 A multimedia program designed to change behavior. A multimedia program designed to improve dietary behavior among low-income women was evaluated by comparing women who were randomly assigned to intervention and control groups. The intervention was a 30-minute session in a computer kiosk in the food stamp office.[21] One of the outcomes was the score on a knowledge test taken about 2 months after the program. Here is a summary of the data:

Group	n	$\bar{x}$	s
Intervention	165	5.08	1.15
Control	212	4.33	1.16

(a) The test had six multiple-choice items that were scored as correct or incorrect, so the total score was an integer between 0 and 6. Do you think that these data are Normally distributed? Explain why or why not.

(b) Is it appropriate to use the two-sample t procedures that we studied in this section to analyze these data? Give reasons for your answer.

(c) Describe appropriate null and alternative hypotheses for evaluating the intervention. Some people would prefer a two-sided alternative in this situation, while others would use a one-sided significance test. Give reasons for each point of view.

(d) Carry out the significance test. Report the test statistic with the degrees of freedom and the P-value. Write a short summary of your conclusion.

(e) Find a 95% confidence interval for the difference between the two means. Compare the information given by the interval with the information given by the significance test.

(f) The women in this study were all residents of Durham, North Carolina. To what extent do you think the results can be generalized to other populations?

7.79 Self-control and food. Self-efficacy is a general concept that measures how well we think we can control different situations. In the study described in the previous exercise, the participants were asked, "How sure are you that you can eat foods low in fat over the next month?" The response was measured on a five-point scale with 1 corresponding to "not sure at all" and 5 corresponding to "very sure." Here is a summary of the self-efficacy scores obtained about 2 months after the program:

Group	n	$\bar{x}$	s
Intervention	165	4.10	1.19
Control	212	3.67	1.12

Analyze the data using the questions in the previous exercise as a guide.

7.80 Dust exposure at work. Exposure to dust at work can lead to lung disease later in life. One study measured the workplace exposure of tunnel construction workers.[22] Part of the study compared 115 drill and blast workers with 220 outdoor concrete workers. Total dust exposure was measured in milligram years per cubic meter (mg.y/m^3). The mean exposure for the drill and blast workers was 18.0 mg.y/m^3 with a standard deviation of 7.8 mg.y/m^3. For the outdoor concrete workers, the corresponding values were 6.5 and 3.4 mg.y/m^3.

(a) The sample included all workers for a tunnel construction company who received medical examinations as part of routine health checkups. Discuss the extent to which you think these results apply to other similar types of workers.

(b) Use a 95% confidence interval to describe the difference in the exposures. Write a sentence that gives the interval and provides the meaning of 95% confidence.

(c) Test the null hypothesis that the exposures for these two types of workers are the same. Justify your choice of a one-sided or two-sided alternative. Report the test statistic, the degrees of freedom, and the P-value. Give a short summary of your conclusion.

(d) The authors of the article describing these results note that the distributions are somewhat skewed. Do you think that this fact makes your analysis invalid? Give reasons for your answer.

7.81 Not all dust is the same. Not all dust particles that are in the air around us cause problems for our lungs. Some particles are too large and stick to other areas of our body before they can get to our lungs. Others are so small that we can breathe them in and out and they will not deposit in our lungs. The researchers in the study described in the previous exercise also measured respirable dust. This is dust that deposits in our lungs when we breathe it. For the drill and blast workers, the mean exposure to respirable dust was 6.3 mg.y/m^3 with a standard deviation of 2.8 mg.y/m^3. The corresponding values for the outdoor concrete workers were 1.4 and 0.7 mg.y/m^3. Analyze these data using the questions in the previous exercise as a guide.

7.82 Change in portion size. A recent study of food portion sizes reported that over a 17-year period, the average size of a soft drink consumed by Americans aged 2 years and older increased from 13.1 ounces (oz) to 19.9 oz. The authors state that the difference is statistically significant with $P < 0.01$.[23] Explain what additional information you would need to compute a confidence interval for the increase and outline the procedure that you would use for the computations. Do you think that a confidence interval would provide useful additional information? Explain why or why not.

7.83 Compare two groups of consumers. The results in the previous exercise were based on two national surveys with a very large number of individuals. Here is a study that also looked at beverage consumption but with much smaller samples. One part of this study compared 20 children who were 7 to 10 years old with 5 who were 11 to 13.[24] The younger children consumed an average of 8.2 oz of sweetened drinks per day, while the older ones averaged 14.5 oz. The standard deviations were 10.7 oz and 8.2 oz respectively.

(a) Do you think that it is reasonable to assume that these data are Normally distributed? Explain why or why not. (*Hint:* Think about the 68–95–99.7 rule.)

(b) Using the methods in this section, test the null hypothesis that the two groups of children consume equal amounts of sweetened drinks versus the two-sided alternative. Report all details of the significance-testing procedure with your conclusion.

(c) Give a 95% confidence interval for the difference in means.

(d) Do you think that the analyses performed in parts (b) and (c) are appropriate for these data? Explain why or why not.

(e) The children in this study were all participants in an intervention study at the Cornell Summer Day Camp at Cornell University. To what extent do you think that these results apply to other groups of children?

CASE 7.2 **7.84 Healthy companies versus failed companies.** Examples 7.16 and 7.17 (pages 431–432) compare healthy and failed companies under the special assumption that the two populations of firms have the same standard deviation. In practice, we prefer not to make this assumption, so let's analyze the data without making this assumption. We expect healthy firms to have a higher ratio of assets to liabilities on the average. Do the data give good evidence in favor of this expectation? By how much on the average does the ratio for healthy firms exceed that for failed firms (use 99% confidence)?

7.85 Effect of storage on a product. Does bread lose vitamins when it is stored? Researchers prepared four loaves of bread with flour fortified with a known amount of vitamins. After baking, they measured the vitamin C content of two loaves. The other two loaves were baked at the same time but were stored for three days and then measured for vitamin C content. The units are milligrams of vitamin C per hundred grams of flour (mg/100 g). Here are the data:[25] **BREADVITC**

Immediately after baking:	47.62	49.79
Three days after baking:	21.25	22.34

(a) Does bread lose vitamin C when it is stored for three days? Use a two-sample t test to answer this question. (State the hypotheses; give the test statistic, degrees of freedom, and P-value; and state your conclusion in nontechnical language.)

(b) Give a 90% confidence interval for the amount of vitamin C lost.

7.86 Study design matters! The researchers in the previous exercise might have measured only two loaves of bread, first immediately after baking and again after three days. Suppose that the data given had come from this study design. (The values are for the first loaf and second loaf, from left to right.)

(a) Explain carefully why your analysis from Exercise 7.85 is *not correct* now, even though the numbers are the same.

(b) Redo the analysis for the design based on measuring the same loaves twice.

7.87 Another ingredient. The researchers of Exercise 7.85 also measured the amount of vitamin E (in mg/100 g of flour) in the same four loaves. Here are the data: **BREADVITE**

Immediately after baking:	94.6	96.0
Three days after baking:	97.4	94.3

(a) Does bread lose vitamin E in short-term storage? State hypotheses for a statistical test; give the test statistic, degrees of freedom, and P-value; and state your conclusion.

(b) Give a 90% confidence interval for the amount of vitamin E lost.

7.88 Are the samples too small? Exercises 7.85 and 7.87 are based on samples of just two loaves of bread. Some people claim that significance tests with very small samples never lead to rejection of the null hypothesis. Discuss this claim using the results of these two exercises.

7.89 Fitness and ego. Employers sometimes seem to prefer executives who appear physically fit, despite the legal troubles that may result. Employers may also favor certain personality characteristics. Fitness and personality are related. In one study, middle-aged college faculty who had volunteered for a fitness program were divided into low-fitness and high-fitness groups based on a physical examination. The subjects then took the Cattell Sixteen Personality Factor Questionnaire.[26] Here are the data for the "ego strength" personality factor: 🅖 **EGO**

Low fitness			High fitness		
4.99	5.53	3.12	6.68	5.93	5.71
4.24	4.12	3.77	6.42	7.08	6.20
4.74	5.10	5.09	7.32	6.37	6.04
4.93	4.47	5.40	6.38	6.53	6.51
4.16	5.30		6.16	6.68	

(a) Is the difference in mean ego strength significant at the 5% level? At the 1% level? Be sure to state H_0 and H_a.

(b) Can you generalize these results to the population of all middle-aged men? Give reasons for your answer.

(c) Can you conclude that increasing fitness *causes* an increase in ego strength? Give reasons for your answer.

7.90 Study design matters! In the previous exercise you analyzed data on the ego strength of high-fitness and low-fitness participants in a campus fitness program. Suppose that instead you had data on the ego strengths of the *same* men before and after six months in the program. You wonder if the program has affected their ego scores. Explain carefully how the statistical procedures you would use would differ from those you applied in Exercise 7.89.

7.91 Sales of small appliances. A market research firm supplies manufacturers with estimates of the retail sales of their products from samples of retail stores. Marketing managers are prone to look at the estimate and ignore sampling error. Suppose that an SRS of 70 stores this month shows mean sales of 53 units of a small appliance, with standard deviation 12 units. During the same month last year, an SRS of 58 stores gave mean sales of 50 units, with standard deviation 10 units. An increase from 50 to 53 is a rise of 6%. The marketing manager is happy, because sales are up 6%.

(a) Use the two-sample t procedure to give a 95% confidence interval for the difference in mean number of units sold at all retail stores.

(b) Explain in language that the manager can understand why he cannot be confident that sales rose by 6%, and that in fact sales may even have dropped.

7.92 Compare two marketing strategies. A bank compares two proposals to increase the amount that its credit card cus-

tomers charge on their cards. (The bank earns a percentage of the amount charged, paid by the stores that accept the card.) Proposal A offers to eliminate the annual fee for customers who charge $3600 or more during the year. Proposal B offers a small percent of the total amount charged as a cash rebate at the end of the year. The bank offers each proposal to an SRS of 150 of its existing credit card customers. At the end of the year, the total amount charged by each customer is recorded. Here are the summary statistics:

Group	n	$\bar{x}$	s
A	150	$3385	$468
B	150	$3124	$411

(a) Do the data show a significant difference between the mean amounts charged by customers offered the two plans? Give the null and alternative hypotheses, and calculate the two-sample t statistic. Obtain the P-value (either approximately from Table D or more accurately from software). State your practical conclusions.

(b) The distributions of amounts charged are skewed to the right, but outliers are prevented by the limits that the bank imposes on credit balances. Do you think that skewness threatens the validity of the test that you used in part (a)? Explain your answer.

7.93 Study habits. The Survey of Study Habits and Attitudes (SSHA) is a psychological test designed to measure the motivation, study habits, and attitudes toward learning of college students. These factors, along with ability, are important in explaining success in school. Scores on the SSHA range from 0 to 200. A selective private college gives the SSHA to an SRS of both male and female first-year students. The data for the women are as follows: 🅢 **STUDYHABITS**

```
156  109  137  115  152  140  154  178  111
123  126  126  137  165  165  129  200  150
```

Here are the scores of the men:

```
118  140  114   91  180  115  126   92  169  139
121  132   75   88  113  151   70  115  187  114
```

(a) Examine each sample graphically, with special attention to outliers and skewness. Is use of a t procedure acceptable for these data?

(b) Most studies have found that the mean SSHA score for men is lower than the mean score in a comparable group of women. Test this supposition here. That is, state hypotheses, carry out the test and obtain a P-value, and give your conclusions.

(c) Give a 90% confidence interval for the mean difference between the SSHA scores of male and female first-year students at this college.

7.94 Summer earnings of college students. College financial aid offices expect students to use summer earnings to help pay for college. But how large are these earnings? One college studied this question by asking a sample of students how much they earned.[27] Omitting students who were not employed, 1324 responses were received. Here are the data in summary form:

Group	n	$\bar{x}$	s
Males	670	$3295.38	$2349.56
Females	654	$2580.86	$1815.55

(a) Use the two-sample t procedures to give a 90% confidence interval for the difference between the mean summer earnings of male and female students.

(b) The distribution of earnings is strongly skewed to the right. Nevertheless, use of t procedures is justified. Why?

(c) Once the sample size was decided, the sample was chosen by taking every kth name from an alphabetical list of undergraduates. Is it reasonable to consider the samples as SRSs chosen from the male and female undergraduate populations?

(d) What other information about the study would you request before accepting the results as describing all undergraduates?

The following exercises concern material on the Satterthwaite approximation and the pooled two-sample t procedures.

7.95 Satterthwaite approximation. Example 7.11 (page 421) reports an analysis comparing product preference for a group of primed and a group of nonprimed consumers. Starting from the computer's results for $\bar{x}_i$ and s_i, verify that the Satterthwaite approximation for the degrees of freedom is 39.5.

7.96 Pooled procedures. Exercise 7.60 (page 423) compares a new type of computer monitor with a standard design. Reanalyze the data using the pooled procedure. Does the conclusion depend on the choice of method? The standard deviations are quite different for these data, so we do not recommend use of the pooled procedures in this case.

7.97 The advantage of pooling. For the analysis of wheat prices in Example 7.14 (page 425), there are only five observations per month. When sample sizes are small, we have very little information to make a judgment about whether the population standard deviations are equal. The potential gain from pooling is large when the sample sizes are very small. Assume that we will perform a two-sided test using the 5% significance level.

(a) Find the critical value for the unpooled t test statistic that does not assume equal variances. Use the minimum of $n_1 - 1$ and $n_2 - 1$ for the degrees of freedom.

(b) Find the critical value for the pooled t test statistic.

(c) How does comparing these critical values show an advantage of the pooled test?

7.98 The advantage of pooling. Suppose that in the setting of the previous exercise you are interested in 95% confidence intervals for the difference rather than significance testing. Find the widths of the intervals for the two procedures (assuming or not assuming equal standard deviations). How do they compare?

7.3 Optional Topics in Comparing Distributions

In this section we discuss two topics that are related to the procedures we have learned for inference about population means. First, it is natural to ask if we can do inference about population spread as well. The answer is yes, but there are many cautions. We also show how to find the power for the two-sample t test. This is a bit technical, but necessary if you plan to design studies.

Inference for population spread

The two most basic descriptive features of a distribution are its center and spread. In a Normal population, we measure center and spread by the mean and the standard deviation. We have described procedures for inference about population means for Normal populations and found that these procedures are often useful for non-Normal populations as well. It is natural to turn next to inference about the standard deviations of Normal populations. Our recommendation here is short and clear: Don't do it without expert advice.

There are indeed inference procedures appropriate for the standard deviations of Normal populations. We will describe the most common such procedure, the F test for comparing the spread of two Normal populations. Unlike the t procedures for means, the F test and other procedures for standard deviations are extremely sensitive to non-Normal distributions.[28] This lack of robustness does not improve in large samples. It is difficult in practice to tell whether a significant F-value is evidence of unequal population spreads or simply evidence that the populations are not Normal.

The deeper difficulty that underlies the very poor robustness of Normal population procedures for inference about spread already appeared in our work on describing data. The standard deviation is a natural measure of spread for Normal distributions but not for distributions in general. In fact, because skewed distributions have unequally spread tails, no single numerical measure is adequate to describe the spread of a skewed distribution. Thus, the standard deviation is not always a useful parameter, and even when it is, the results of inference about it are not trustworthy. Consequently, we do not recommend use of inference about population standard deviations in basic statistical practice.[29]

It was once common to test equality of standard deviations as a preliminary to performing the pooled two-sample t test for equality of two population means. It is better practice to check the distributions graphically, with special attention to skewness and outliers, and to use the more general two-sample t that does not require equal standard deviations.

Chapters 14 and 15 discuss procedures called "analysis of variance" for comparing several means. These procedures are extensions of the pooled t test and require that the populations have a common standard deviation. Fortunately, like the t test, analysis of variance comparisons of means are quite robust. (Analysis of variance uses F statistics, but these are not the same as the F statistic for comparing two population standard deviations.) Formal tests for the equality of two standard deviations are not very robust and will often give misleading results. In the words of one distinguished statistician, "To make a preliminary test on variances is rather like putting to sea in a rowing boat to find out whether conditions are sufficiently calm for an ocean liner to leave port!"[30]

The *F* test for equality of spread

Because of the limited usefulness of procedures for inference about the standard deviations of Normal distributions, we will present only one such procedure. Suppose that we have independent SRSs from two Normal populations, a sample of size n_1 from $N(\mu_1, \sigma_1)$ and a sample of size n_2 from $N(\mu_2, \sigma_2)$. The population means and standard deviations are all unknown. The hypothesis of equal spread,

$$H_0: \sigma_1 = \sigma_2$$

$$H_a: \sigma_1 \neq \sigma_2$$

is tested by a simple statistic, the ratio of the two sample variances.

The F Statistic and F Distributions

When s_1^2 and s_2^2 are sample variances from independent SRSs of sizes n_1 and n_2 drawn from Normal populations, the **F statistic**

$$F = \frac{s_1^2}{s_2^2}$$

has the **F distribution** with $n_1 - 1$ and $n_2 - 1$ degrees of freedom when $H_0: \sigma_1 = \sigma_2$ is true.

F distributions

The **F distributions** are a family of distributions with two parameters, the degrees of freedom of the sample variances in the numerator and denominator of the F statistic.*

---

*The F distributions are another of R. A. Fisher's contributions to statistics and are called F in his honor. Fisher introduced F statistics for comparing several means. We will meet these useful statistics in Chapters 14 and 15.

FIGURE 7.11 The density curve for the $F(9, 10)$ distribution. The F distributions are skewed to the right.

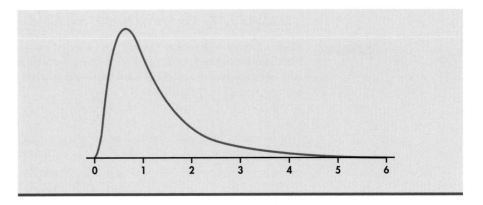

The numerator degrees of freedom are always mentioned first. Interchanging the degrees of freedom changes the distribution, so the order is important. Our brief notation will be $F(j, k)$ for the F distribution with j degrees of freedom in the numerator and k in the denominator. The F distributions are not symmetric but are right-skewed. The density curve in Figure 7.11 illustrates the shape. Because sample variances cannot be negative, the F statistic takes only positive values and the F distribution has no probability below 0. The peak of the F density curve is near 1; values far from 1 in either direction provide evidence against the hypothesis of equal standard deviations.

Tables of F critical values are awkward, because a separate table is needed for every pair of degrees of freedom j and k. Table E in the back of the book gives upper p critical values of the F distributions for $p = 0.10, 0.05, 0.025, 0.01,$ and 0.001. For example, these critical values for the $F(9, 10)$ distribution shown in Figure 7.11 are

p	0.10	0.05	0.025	0.01	0.001
F^*	2.35	3.02	3.78	4.94	8.96

The skewness of the F distributions causes additional complications. In the symmetric Normal and t distributions, the point with probability 0.05 below it is just the negative of the point with probability 0.05 above it. This is not true for F distributions. We therefore require either tables of both the upper and lower tails or some way of getting by without lower-tail critical values. Statistical software that eliminates the need for tables is plainly very convenient. If you do not use statistical software, arrange the F test as follows:

1. Take the test statistic to be

$$F = \frac{\text{larger } s^2}{\text{smaller } s^2}$$

This amounts to naming the populations so that s_1^2 is the larger of the observed sample variances. The resulting F is always 1 or greater.

2. Compare the value of F with critical values from Table E. Then *double* the significance levels from the table to obtain the significance level for the two-sided F test.

The idea is that we calculate the probability in the upper tail and double it to obtain the probability of all ratios on either side of 1 that are at least as improbable as that observed. Remember that the order of the degrees of freedom is important in using Table E.

EXAMPLE 7.18   Comparing Healthy and Failed Firms

CASE 7.2

Case 7.2 (page 430) recounts a study that compared the mean ratios of current assets to current liabilities for successful and failed firms. There we used the pooled two-sample t procedures. Because these procedures require equal population standard deviations, it is tempting to first test

$$H_0: \sigma_1 = \sigma_2$$
$$H_a: \sigma_1 \neq \sigma_2$$

The 68 healthy firms had a standard deviation of $s = 0.6393$, while the 33 failed firms had a standard deviation of $s = 0.4811$. The F test statistic is therefore

$$F = \frac{\text{larger } s^2}{\text{smaller } s^2} = \frac{0.6393^2}{0.4811^2} = 1.77$$

We compare the calculated value $F = 1.77$ with critical points for the $F(67, 32)$ distribution. One set of entries in Table E that is close to this distribution is the $F(60, 30)$ distribution. Our observed value $F = 1.77$ falls between the critical values $F^* = 1.74$ and $F^* = 1.94$, corresponding to tail areas 0.05 and 0.025. The two-sided P-value therefore lies between 0.10 and 0.05. The standard deviations are significantly different at the 10% level but not at the 5% level.

In most significance tests, we insist on quite strong evidence before we abandon a null hypothesis. That's because we are trying to convince ourselves and others that the effect described by the alternative hypothesis really is present in the population. In this example, however, the null hypothesis $H_0: \sigma_1 = \sigma_2$ is a requirement for using the pooled t to compare two means. Evidence much weaker than the usual significance levels of 5% or 1% should lead you to question H_0 and abandon the pooled t for the more general t that we have recommended.

Statistical output often includes the results of this test with the output for the two-sample t test. Here is the output from SAS:

```
                  Equality of Variances
Variable   Method    Num DF   Den DF   F Value   Pr > F
ratio      Folded F    67       32       1.77    0.0792
```

We see that the exact P-value, *if* the populations are Normal, is 0.0792. When we examined the histograms and the stemplot (Figures 7.09 and 7.10) for these data, we noted that the data for the failed firms did not look particularly Normal, but we were not concerned about that fact because we were interested in comparing the means. For the question of equality of variances, we are not in the same situation. To the extent that our populations are not Normal, we are less confident in our conclusion that the standard deviations are not significantly different.

APPLY YOUR KNOWLEDGE

7.99 The F statistic. The F statistic $F = s_1^2/s_2^2$ is calculated from samples of size $n_1 = 10$ and $n_2 = 16$. (Remember that n_1 is the numerator sample size.)
(a) What is the upper 5% critical value for this F?
(b) In a test of equality of standard deviations against the two-sided alternative, this statistic has the value $F = 2.65$. Is this value significant at the 10% level? Is it significant at the 5% level?

7.100 Statistical significance? The F statistic for equality of standard deviations based on samples of sizes $n_1 = 15$ and $n_2 = 7$ takes the value $F = 5.12$.
(a) Is this significant evidence of unequal population standard deviations at the 5% level?
(b) Use Table E to give an upper and a lower bound for the P-value.

The power of the two-sample *t* test

The two-sample t test is one of the most used statistical procedures. Unfortunately, because of inadequate planning, users frequently fail to find evidence for the effects that they believe to be present. Power calculations should be part of the planning of any statistical study.

In Section 7.1, we learned how to approximate the power of the one-sample t test. The basic idea is the same for the two-sample case, but we give the exact method rather than an approximation. The exact power calculation involves a new distribution, the **noncentral t distribution.** This calculation is not practical by hand but is easy with software that calculates probabilities for this new distribution.

noncentral t distribution

We consider only the common case where the null hypothesis is "no difference," $\mu_1 - \mu_2 = 0$. We illustrate the calculation for the pooled two-sample t test. A simple modification is needed when we do not pool. The unknown parameters in the pooled t setting are μ_1, μ_2, and a single common standard deviation σ. To find the power for the pooled two-sample t test, follow these steps.

Step 1 Specify these quantities:
(a) an alternative value for $\mu_1 - \mu_2$ that you consider important to detect;
(b) the sample sizes, n_1 and n_2;
(c) a fixed significance level α, often $\alpha = 0.05$; and
(d) an estimate of the standard deviation σ from a pilot study or previous studies under similar conditions.

Step 2 Find the degrees of freedom df $= n_1 + n_2 - 2$ and the value of t^* that will lead to rejecting H_0 at your chosen level α.

noncentrality parameter

Step 3 Calculate the **noncentrality parameter**

$$\delta = \frac{|\mu_1 - \mu_2|}{\sigma \sqrt{\dfrac{1}{n_1} + \dfrac{1}{n_2}}}$$

Step 4 The power is the probability that a noncentral t random variable with degrees of freedom df and noncentrality parameter δ will be less than t^*. Use software to calculate this probability. In SAS, the command is $1 - \text{PROBT}(\text{tstar, df, delta})$. If you do not have software that can perform this calculation, you can approximate the power as the probability that a standard Normal random variable is greater than $t^* - \delta$, that is, $P(Z > t^* - \delta)$. Use Table A or software for standard Normal probabilities.

Note that the denominator in the noncentrality parameter,

$$\sigma \sqrt{\frac{1}{n_1} + \frac{1}{n_2}}$$

is our guess at the standard error for the difference in the sample means. Therefore, if we wanted to assess a possible study in terms of the margin of error for the estimated difference, we would examine t^* times this quantity.

If we do not assume that the standard deviations are equal, we need to guess both standard deviations and then combine these to get an estimate of the standard error:

$$\sqrt{\frac{\sigma_1^2}{n_1} + \frac{\sigma_2^2}{n_2}}$$

This guess is then used in the denominator of the noncentrality parameter. Use the conservative value, the smaller of $n_1 - 1$ and $n_2 - 1$, for the degrees of freedom.

EXAMPLE 7.19 Healthy versus Failed Firms

CASE 7.2

In Case 7.2 we compared the ratio of current assets to current liabilities for 68 successful and 33 failed firms. Using the pooled two-sample procedure, the difference was statistically significant ($t = 7.71$, df $= 99$, $P < 0.001$). Now that this study is several years old, we are planning a similar study to determine if the finding continues to hold.

Should our new sample have similar numbers of firms? Or could we save resources by using smaller samples and still be able to declare that the successful and failed firms are different? To answer this question, we do a power calculation.

Step 1 We want to be able to detect a difference in the means that is about the same as the value that we observed in our previous study. So in our calculations we will use $\mu_1 - \mu_2 = 1.15$. We are willing to assume that the standard deviations will be about the same as in the earlier study, so we take the standard deviation for each of the two groups of firms to be the pooled value from our previous study, $\sigma = 0.5928$.

We need only two pieces of additional information: a significance level α and the sample sizes n_1 and n_2. For the first, we will choose the standard value $\alpha = 0.05$. For the sample sizes we want to try several different values. Let's start with $n_1 = 15$ and $n_2 = 15$.

Step 2 The degrees of freedom are $n_1 + n_2 - 2 = 28$. The critical value is $t^* = 2.048$, the value from Table D for a two-sided $\alpha = 0.05$ significance test based on 28 degrees of freedom.

Step 3 The noncentrality parameter is

$$\delta = \frac{1.15}{0.5928\sqrt{\dfrac{1}{15} + \dfrac{1}{15}}} = \frac{1.15}{0.2165} = 5.31$$

Step 4 Software gives the power as 0.99922, over 99.9%. The Normal approximation is very accurate:

$$P(Z > t^* - \delta) = P(Z > -3.262) = 0.99945$$

If we repeat the calculation with $n_1 = 10$ and $n_2 = 10$, we still have power greater than 0.98. A difference of means as large as that found in the earlier study can be detected with quite small samples.

A different, and perhaps more important, issue is the margin of error for the estimated difference. For $n_1 = 15$ and $n_2 = 15$, it is

$$t^* \times 0.5928\sqrt{\frac{1}{15} + \frac{1}{15}} = 2.048 \times 5.31 = 0.44$$

Repeating the calculation for $n_1 = 10$ and $n_2 = 10$ gives 0.56. If we want to reduce the margin of error for the difference in the means of asset/liability ratios, we will need larger sample sizes.

APPLY YOUR KNOWLEDGE

7.101 Power and $\mu_1 - \mu_2$. If you repeat the calculation in Example 7.19 for other values of $\mu_1 - \mu_2$ that are smaller than 1.15, would you expect the power to increase or decrease? Explain.

7.102 Power and the standard deviation. If the true population standard deviation were 0.7 instead of the 0.5928 hypothesized in Example 7.19, would the power increase or decrease? Explain.

SECTION 7.3 Summary

- Inference procedures for comparing the standard deviations of two Normal populations are based on the **F statistic,** which is the ratio of sample variances:

$$F = \frac{s_1^2}{s_2^2}$$

- When an SRS of size n_1 is drawn from the x_1 population and an independent SRS of size n_2 is drawn from the x_2 population, the F statistic has the **F distribution** $F(n_1 - 1, n_2 - 1)$ if the two population standard deviations σ_1 and σ_2 are in fact equal.

- The **F test for equality of standard deviations** tests $H_0: \sigma_1 = \sigma_2$ versus $H_a: \sigma_1 \neq \sigma_2$ using the statistic

$$F = \frac{\text{larger } s^2}{\text{smaller } s^2}$$

and doubles the upper-tail probability to obtain the P-value.

- The t procedures are quite **robust** when the distributions are not Normal. The F tests and other procedures for inference about the spread of one or more Normal distributions are not robust. They are so strongly affected by non-Normality that we do not recommend them for regular use.

- The **power** of the two-sample t test is found by first finding the critical value for the significance test, the degrees of freedom, and the **noncentrality parameter** for the alternative of interest. These are used to calculate the power from a **noncentral t distribution.** A Normal approximation works quite well. Calculating margins of error for various study designs and conditions is an alternative procedure for evaluating designs.

SECTION 7.3 Exercises

For Exercises 7.99 and 7.100, see page 442; and for 7.101 and 7.102, see page 444.
In all exercises calling for use of the F test, assume that both population distributions are very close to Normal. The actual data are not always sufficiently Normal to justify use of the F test.

7.103 Comparison of standard deviations. Here are some summary statistics from two independent samples from Normal distributions:

Sample	n	s^2
1	10	3.2
2	16	11.5

You want to test the null hypothesis that the two population standard deviations are equal versus the two-sided alternative at the 5% significance level.
(a) Calculate the test statistic.
(b) Find the appropriate value from Table E that you need to perform the significance test.
(c) What do you conclude?

7.104 Revisiting the cholesterol comparison. Compare the standard deviations of total cholesterol in Exercise 7.73 (page 435). Give the test statistic, the degrees of freedom, and the P-value. Write a short summary of your analysis, including comments on the assumptions for the test.

7.105 Vitamin C loss in storage. Exercise 7.85 (page 437) presents data on the loss of vitamin C when bread is stored. Two loaves were measured immediately after baking and another two loaves were measured after three days of storage. These are very small sample sizes.
(a) Use Table E to find the value that the ratio of variances would have to exceed for us to reject (at the 5% level) the null hypothesis that the standard deviations are equal. What does this suggest about the power of the test?
(b) Perform the test and state your conclusion.

7.106 Wheat prices. Example 7.14 (page 425) describes a USDA survey used to estimate wheat prices in July and September. Calculate the two sample standard deviations and perform the test for equality of standard deviations and summarize your conclusion.

7.107 Study habits of men and women. Return to the SSHA data in Exercise 7.93 (page 438). SSHA scores are generally less variable among women than among men. We want to know whether this is true for this college.

(a) State H_0 and H_a. Note that H_a is one-sided.

(b) Because Table E contains only upper critical values for F, a one-sided test requires the numerator s^2 in the F statistic to belong to the group that H_a claims to have the larger σ. Calculate this F.

(c) Compare F to the entries in Table E (no doubling of p) to obtain the P-value. Be sure the degrees of freedom are in the proper order. What do you conclude about the variation in SSHA scores?

7.108 Comparison of packaging plants: power. Exercise 7.75 (page 435) summarizes data on the number of seeds in one-pound scoops from two different packaging plants. Suppose that you are designing a new study for their next improvement effort. Based on information from the company, you want to identify a difference in these plants of 150 seeds. For planning purposes assume that you will have 20 scoops from each plant and that the common standard deviation is 190 seeds, a guess that is roughly the pooled sample standard deviation. If you use a pooled two-sample t test

with significance level 0.05, what is the power of the test for this design?

7.109 Power, continued. Repeat the power calculation in the previous exercise, for 25, 30, 35, and 40 scoops from each plant. Summarize your power study. A graph of the power against sample size will help.

7.110 Margins of error. For each of the sample sizes considered in the previous two exercises, estimate the margin of error for the 95% confidence interval for the difference in seed counts. Display these results with a graph or a sketch.

7.111 Ego strength: power. You want to compare the ego strengths of MBA students who plan to seek work at consulting firms and those who favor manufacturing firms. Based on the data from Exercise 7.89 (page 438), you will use $\sigma = 0.7$ for planning purposes. The pooled two-sample t test with $\alpha = 0.01$ will be used to make the comparison. You judge a difference of 0.5 points to be of interest.

(a) Find the power for the design with 20 MBA students in each group.

(b) The power in part (a) is not acceptable. Redo the calculations for 30 students in each group and $\alpha = 0.05$.

STATISTICS IN SUMMARY

This chapter presents t tests and confidence intervals for inference about the mean of a single population and for comparing the means of two populations. The one-sample t procedures do inference about one mean, and the two-sample t procedures compare two means. Matched pairs studies use one-sample procedures because you first create a single sample by taking the differences in the responses within each pair. These t procedures are among the most common methods of statistical inference.

The t procedures require that the data be random samples and that the distribution of the population or populations be Normal. One reason for the wide use of t procedures is that they are not very strongly affected by lack of Normality. If you can't regard your data as a random sample, however, the results of inference may be of little value.

The chapter exercises are important in this and later chapters. You must now recognize problem settings and decide which of the methods presented in the chapter fits. In this chapter, you must recognize one-sample studies, matched pairs studies, and two-sample studies. Here is a review list of the most important skills you should have acquired from your study of this chapter.

A. Recognition

1. Recognize when a problem requires inference about a mean or comparing two means.

2. Recognize from the design of a study whether one-sample, matched pairs, or two-sample procedures are needed.

B. One-Sample t Procedures

1. Recognize when the t procedures are appropriate in practice, in particular that they are quite robust against lack of Normality but are influenced by outliers.

2. Also recognize when the design of the study, outliers, or a small sample from a skewed distribution make the t procedures risky.

3. Use t to obtain a confidence interval at a stated level of confidence for the mean μ of a population.

4. Carry out a t test for the hypothesis that a population mean μ has a specified value against either a one-sided or a two-sided alternative. Use Table D of t distribution critical values to approximate the P-value or carry out a fixed α test.

5. Recognize matched pairs data and use the t procedures to obtain confidence intervals and to perform tests of significance for such data.

C. Two-Sample t Procedures

1. Recognize when the two-sample t procedures are appropriate in practice.

2. Give a confidence interval for the difference between two means. Use the two-sample t statistic with conservative degrees of freedom if you do not have statistical software. Use software if you have it.

3. Test the hypothesis that two populations have equal means against either a one-sided or a two-sided alternative. Use the two-sample t test with conservative degrees of freedom if you do not have statistical software. Use software if you have it.

4. Know that procedures for comparing the standard deviations of two Normal populations are available, but that these procedures are risky because they are not at all robust against non-Normal distributions.

CHAPTER 7 Review Exercises

7.112 LSAT scores. The scores of four roommates on the Law School Admission Test are

$$156 \quad 122 \quad 140 \quad 135$$

Find the mean, the standard deviation, and the standard error of the mean. Is it appropriate to calculate a confidence interval based on these data? Explain why or why not.

7.113 t is robust. A manufacturer of small appliances employs a market research firm to estimate retail sales of its products. Here are last month's sales of electric can openers from an SRS of 50 stores in the Midwest sales region: 🔲 RETAIL

19	19	16	19	25	26	24	63	22	16
13	26	34	10	48	16	20	14	13	24
34	14	25	16	26	25	25	26	11	79
17	25	18	15	13	35	17	15	21	12
19	20	32	19	24	19	17	41	24	27

(a) Make a stemplot of the data. The distribution is skewed to the right and has several high outliers. The *bootstrap* (page 346) is a modern computer-intensive tool for getting accurate confidence intervals without the Normality condition. Three bootstrap simulations, each with 10,000 repetitions, give these 95% confidence intervals for mean sales in the entire region: (20.42, 27.26), (20.40, 27.18), and (20.48, 27.28).

(b) Find the 95% t confidence interval for the mean. It is essentially the same as the bootstrap intervals. The lesson is that for sample sizes as large as $n = 50$, t procedures are very robust.

7.114 Number of critical food violations. The results of a major city's restaurant inspections are available through its online newspaper.[31] Critical food violations are those that put patrons at risk of getting sick and must be immediately corrected by the restaurant. An SRS of $n = 200$ inspections from the more than 4400 inspections since January 2008 had $\bar{x} = 1.20$ violations and $s = 1.81$ violations.

(a) Test the hypothesis, using $\alpha = 0.05$, that the average number of critical violations is less than 1.5. State the two hypotheses, the test statistic, and P-value.

(b) Construct a 95% confidence interval for the average number of critical violations and summarize your result.

(c) Which of the two summaries (significance test versus confidence interval) do you find more helpful in this case? Explain your answer.

(d) These data are integers ranging from 0 to 9. The data are also skewed to the right, with 70% of the values either a 0 or 1. Given this information, do you feel use of the t procedures is appropriate? Explain your answer.

7.115 Interpreting software output. You use statistical software to perform a significance test of the null hypothesis that two means are equal. The software reports P-values for the two-sided alternative. Your alternative is that the first mean is less than the second mean.

(a) The software reports $t = -1.88$ with a P-value of 0.08. Would you reject H_0 with $\alpha = 0.05$? Explain your answer.

(b) The software reports $t = 1.88$ with a P-value of 0.08. Would you reject H_0 with $\alpha = 0.05$? Explain your answer.

7.116 The wine makes the meal? In a recent study, 39 diners were given a free glass of Cabernet Sauvignon to accompany a French meal.[32] Although the wine was identical, half the bottle labels claimed the wine was from California and the other half claimed it was from North Dakota. The table below summarizes the grams of entrée and wine consumed during the meal.

	Wine label	n	Mean	St. dev.
Entrée	California	24	499.8	87.2
	North Dakota	15	439.0	89.2
Wine	California	24	100.8	23.3
	North Dakota	15	110.4	9.0

Did the patrons who thought the wine was from California consume more? Analyze the data and write a report summarizing your work. Be sure to include details regarding the statistical methods you used, your assumptions, and your conclusions.

7.117 Study design information. In the previous study, diners were seated alone or in groups of two, three, four, and, in one case, nine (for a total of $n = 16$ tables). Also, each table, not each patron, was randomly assigned a particular wine label. Does this information alter how you might do the analysis in the previous exercise? Explain your answer.

7.118 Does dress affect competence and intelligence ratings? Researchers performed a study to examine whether or not women are perceived as less competent and less intelligent when they dress in a sexy manner versus a business-like manner. Competence was rated from 1 (not at all) to 7 (extremely), and a 1 to 5 scale was used for intelligence. Under each condition, 17 subjects provided data. Here are summary statistics:[33]

Rating	Sexy		Business-like	
	Mean	SD	Mean	SD
Competence	4.13	0.99	5.42	0.85
Intelligence	2.91	0.74	3.50	0.71

Analyze the two variables and write a report summarizing your work. Be sure to include details regarding the statistical methods you used, your assumptions, and your conclusions.

7.119 Perceived quality of high- and low-performing restaurants. A study classified 394 quick-service restaurants (QSR) into high-performing and low-performing groups based on their total sales. Each restaurant was rated on a collection of perceived measures of quality by a large number of diners using a 1 to 7 scale. In this study we view the diners as a measuring instrument, and our major interest is in comparing the 170 high-sales restaurants with the 224 low-sales restaurants. Here are summary statistics:[34]

Perceived quality	High Sales		Low Sales	
	Mean	SD	Mean	SD
Food served in promised time	4.60	1.35	4.15	1.38
Quickly corrects mistakes	3.59	1.25	3.43	1.19
Well-dressed staff	4.25	1.31	3.86	1.38
Visually attractive menu	4.54	1.45	4.33	1.35
Ordered food served accurately	4.22	1.26	3.85	1.13
Well-trained personnel	4.01	1.27	3.95	1.18
Clean dining area and restrooms	4.05	1.26	3.97	1.24
Employees adjust to needs	3.94	1.19	3.80	1.24
Employees know menu	4.17	1.19	3.88	1.15
Convenient operating hours	4.83	1.32	4.59	1.19

Use this information to compare the two groups of restaurants and write a report summarizing your work. Be sure to include details regarding the statistical methods you used, your assumptions, and your conclusions.

7.120 Evaluate the dress study. Refer to Exercise 7.118. Participants in the study viewed a videotape of a woman described as a 28-year-old senior manager for a Chicago advertising firm who had been working for this firm for 7 years. The same woman was used for each of the two conditions, but she wore different clothing each time. For the business-like condition, the woman wore little makeup, black slacks, a turtleneck, a business jacket, and flat shoes. For the sexy condition, the same woman wore a tight knee-length skirt, a low-cut shirt with a cardigan over it, high-heeled shoes, and more makeup, and her hair was tousled. The subjects who evaluated the videotape were male and female undergraduate students who were predominantly Caucasian, from middle- to upper-class backgrounds, and between the ages of 18 and 24. The content of the videotape was identical in both conditions. The woman described her general background, life in college, and hobbies.

(a) Write a critique of this study with particular emphasis on its limitations and how you would take these into account when drawing conclusions based on the study.

(b) Propose an alternative study that would address a similar question. Be sure to provide details about how your study would be run.

7.121 Evaluate the restaurant study. Refer to Exercise 7.119. Subjects who provided the evaluation data were shoppers who entered a mall in Seoul, South Korea, between the hours of 1:00 and 6:00 P.M., on Monday, Wednesday, and Saturday during two weeks. A total of 950 questionnaires were distributed, and 394 provided data that could be used in the study. Seven QSRs were studied: McDonald's, Burger King, Hardee's, Jakob's, KFC, Lotteria, and Popeye's. Jakob's and Lotteria are two major domestic QSR brands. To account for size differences, total sales were divided by the number of units in operation during the survey period to determine sales per unit. This was the variable used to classify the QSRs into the two groups. Age, education, number of visits to QSRs per month, and average amount spent per visit for the sample were compared with other surveys of QSR customers. No statistically significant differences in these variables were found.

(a) Write a critique of this study with particular emphasis on the limitations and how you would take these into account when drawing conclusions based on the study.

(b) Propose an alternative study that would address a similar question. Be sure to provide details about how your study would be run.

7.122 Air in poultry-processing plants. The air in poultry-processing plants often contains fungus spores. If the ventilation is inadequate, this can affect the health of the workers. The problem is most serious during the summer. To measure the presence of spores, air samples are pumped to an agar plate and "colony-forming units (CFUs)" are counted after an incubation period. Here are data from two locations in a plant that processes 37,000 turkeys per day, taken on four days in the summer. The units are CFUs per cubic meter of air.[35] 🌀 AIRPOULTRY

	Day 1	Day 2	Day 3	Day 4
Kill room	3175	2526	1763	1090
Processing	529	141	362	224

(a) Explain carefully why these are matched pairs data.

(b) The spore count is clearly higher in the kill room. Give sample means and a 90% confidence interval to estimate how much higher. Be sure to state your conclusion in plain English.

7.123 Personalities of hotel managers. Successful hotel managers must have personality characteristics often thought of as feminine (such as "compassionate") as well as those often thought of as masculine (such as "forceful"). The Bem Sex-Role Inventory (BSRI) is a personality test that gives separate ratings for female and male stereotypes, both on a scale of 1 to 7. Here are summary statistics for a sample of 148 male general mangers of three-star and four-star hotels.[36] The data come from a comprehensive mailing to these hotels. The response rate was 48%, which is good for mail surveys of this kind. Although nonresponse remains an issue, users of statistics usually act as if they have an SRS when the response rate is "good enough."

	Masculinity score	Femininity score
Mean	$\bar{x} = 5.91$	$\bar{x} = 5.29$
Standard deviation	$s = 0.57$	$s = 0.75$

The mean BSRI masculinity score for the general male population is $\mu = 4.88$. Is there evidence that hotel managers on the average score higher in masculinity than the general male population?

7.124 Another personality trait of hotel managers. Continue your study from the previous exercise. The mean BSRI femininity score in the general male population is $\mu = 5.19$. (It does seem odd that the mean femininity score is higher than the mean masculinity score, but such is the world of personality tests. The

two scales are separate.) Is there evidence that hotel managers on the average score higher in femininity than the general male population?

7.125 Alcohol content of wine. The alcohol content of wine depends on the grape variety, the way in which the wine is produced from the grapes, the weather, and other influences. Here are data on the percent of alcohol in wine produced from the same grape variety in the same year by 48 winemakers in the same region of Italy:[37] 🌀 ALCOHOLCONTENT

12.86	12.88	12.81	12.70	12.51
12.60	12.25	12.53	13.49	12.84
12.93	13.36	13.52	13.62	12.25
13.16	13.88	12.87	13.32	13.08
13.50	12.79	13.11	13.23	12.58
13.17	13.84	12.45	14.34	13.48
12.36	13.69	12.85	12.96	13.78
13.73	13.45	12.82	13.58	13.40
12.20	12.77	14.16	13.71	13.40
13.27	13.17	14.13		

(a) Make a stemplot of the data. The distribution is a bit irregular, but there is no reason to avoid use of t procedures for $n = 48$.

(b) Give a 95% confidence interval for the mean alcohol content of wine of this type.

7.126 Air in poultry-processing plants: summer versus winter. The air in poultry-processing plants often contains fungus spores. If the ventilation is inadequate, this can affect the health of the workers. The problem is most serious during the summer and least serious during the winter. To measure the presence of spores, air samples are pumped to an agar plate and "colony-forming units (CFUs)" are counted after an incubation period. Here are data from the "kill room" of a plant that processes 37,000 turkeys per day, taken on four separate days in the summer and in the winter. The units are CFUs per cubic meter of air.[38] 🌀 AIRSEASON

Summer:	3175	2526	1763	1090
Winter:	384	104	251	97

The counts are clearly much higher in the summer. Give a 90% confidence interval to estimate how much higher the mean count is during the summer. Be sure to write a conclusion in plain English.

7.127 The manufacture of dyed clothing fabrics. Different fabrics respond differently when dyed. This matters to clothing manufacturers, who want the color of the fabric to be just right. Fabrics made of cotton and of ramie are dyed with the same "procion blue" die applied in the same way. A colorimeter is used to measure the lightness of the color on a scale in which black is 0 and white is 100. Here are the data for 8 pieces of each fabric:[39] 🌀 DYECOLOR

Cotton	48.82	48.88	48.98	49.04
	48.68	49.34	48.75	49.12
Ramie	41.72	41.83	42.05	41.44
	41.27	42.27	41.12	41.49

Which fabric is darker when dyed in this way? Write an answer to this question that includes summary statistics and a test of significance.

7.128 Durable press and breaking strength. "Durable press" cotton fabrics are treated to improve their recovery from wrinkles after washing. Unfortunately, the treatment also reduces the strength of the fabric. A study compared the breaking strength of fabric treated by two commercial durable press processes. Five specimens of the same fabric were assigned at random to each process. Here are the data, in pounds of pull needed to tear the fabric:[40] BREAKSTRENGTH

| Permafresh 55: | 29.9 | 30.7 | 30.0 | 29.5 | 27.6 |
| Hylite LF: | 28.8 | 23.9 | 27.0 | 22.1 | 24.2 |

Is there good evidence that the two processes result in different mean breaking strengths?

7.129 Find a confidence interval. Continue your work from the previous exercise. A fabric manufacturer wants to know how large a strength advantage fabrics treated by the Permafresh method have over fabrics treated by the Hylite process. Give a 95% confidence interval for the difference in mean breaking strengths.

7.130 Recovery from wrinkles. Of course, the reason for durable press treatment is to reduce wrinkling. "Wrinkle recovery angle" measures how well a fabric recovers from wrinkles. Higher is better. Here are data on the wrinkle recovery angle (in degrees) for the same fabric specimens discussed in the previous two exercises: WRINKLES

| Permafresh 55: | 136 | 135 | 132 | 137 | 134 |
| Hylite LF: | 143 | 141 | 146 | 141 | 145 |

Which process has better wrinkle resistance? Is the difference statistically significant?

7.131 Insulation failures. A manufacturer of electric motors tests insulation at a high temperature (250°C) and records the number of hours until the insulation fails. The data for 5 specimens are[41] INSULATION

300 324 372 372 444

The small sample size makes judgment difficult, but engineering experience suggests that the logarithm of the failure time will have a Normal distribution. Take the logarithms of the 5 observations, and use t procedures to give a 90% confidence interval for the mean of the log failure time for insulation of this type.

7.132 Brain training. The assessment of computerized brain-training programs is a rapidly growing area of research. Researchers are now focusing on who this training benefits most, what brain functions are most susceptible to improvement, and

which products are most effective. A recent study looked at 487 community-dwelling adults aged 65 and older, each randomly assigned to one of two training groups. In one group, the participants used a computerized program 1 hour per day. In the other, DVD-based educational programs were shown and quizzes were administered after each video. The training period lasted 8 weeks. The response was the improvement in a composite score obtained from an auditory memory/attention survey given before and after the 8 weeks.[42] The results are summarized below.

Group	n	$\bar{x}$	s
Computer program	242	3.9	8.28
DVD program	245	1.8	8.33

(a) Given that other studies have shown a benefit of computerized brain training, state the null and alternative hypotheses.

(b) Report the test statistic, its degrees of freedom, and the P-value. What is your conclusion using significance level $\alpha = 0.05$?

(c) Can you conclude that this computerized brain training always improves a person's auditory memory/perception better than the DVD program? If not, explain why.

7.133 Competitive prices? A retailer entered into an exclusive agreement with a supplier who guaranteed to provide all products at competitive prices. The retailer eventually began to purchase supplies from other vendors who offered better prices. The original supplier filed a legal action claiming violation of the agreement. In defense, the retailer had an audit performed on a random sample of invoices. For each audited invoice, all purchases made from other suppliers were examined and the prices were compared with those offered by the original supplier. For each invoice, the percent of purchases for which the alternate supplier offered a lower price than the original supplier was recorded. Here are the data:[43] COMPETITIVE

100	0	0	100	33	45	100	34	78
100	77	33	100	69	100	89	100	100
100	100	100	100	100	100	100		

Report the average of the percents with a 95% margin of error. Do the sample invoices suggest that the original supplier's prices are not competitive on the average?

CASE 7.1 **7.134 Sign test for telephone expenditures.** Example 7.1 (page 398) gives data on the monthly expenditures for landline telephone service for 8 households. Is there evidence that the median monthly expenditure is less than $49? State the hypotheses, carry out the sign test, and report your conclusion. TELEPHONE

7.135 Testing job applicants. The one-hole test is used to test the manipulative skill of job applicants. This test requires subjects to grasp a pin, move it to a hole, insert it, and return for another pin. The score on the test is the number of pins inserted in a fixed time interval. One study compared male college students with experienced female industrial workers. Here are the data for the first minute of the test:[44]

Group	n	$\bar{x}$	s
Students	750	35.12	4.31
Workers	412	37.32	3.83

(a) We expect that the experienced workers will outperform the students, at least during the first minute, before learning occurs. State the hypotheses for a statistical test of this expectation and perform the test. Give a P-value and state your conclusions.

(b) The distribution of scores is slightly skewed to the left. Explain why the procedure you used in part (a) is nonetheless acceptable.

(c) One purpose of the study was to develop performance norms for job applicants. Based on the data above, what is the range that covers the middle 95% of experienced workers? (Be careful! This is not the same as a 95% confidence interval for the mean score of experienced workers.)

(d) The five-number summary of the distribution of scores among the workers is

$$23 \quad 33.5 \quad 37 \quad 40.5 \quad 46$$

for the first minute and

$$32 \quad 39 \quad 44 \quad 49 \quad 59$$

for the fifteenth minute of the test. Display these summaries graphically, and describe briefly the differences between the distributions of scores in the first and fifteenth minutes.

7.136 Occupation and diet. Do various occupational groups differ in their diets? A British study of this question compared 98 drivers and 83 conductors of London double-decker buses. The conductors' jobs require more physical activity. The article reporting the study gives the data as "mean daily consumption ($\pm$ se)."[45] Some of the study results appear below:

	Drivers	Conductors
Total calories	2821 ± 44	2844 ± 48
Alcohol (grams)	0.24 ± 0.06	0.39 ± 0.11

(a) What does "se" stand for? Give $\bar{x}$ and s for each of the four sets of measurements.

(b) Is there significant evidence at the 5% level that conductors consume more calories per day than do drivers? Use a t test to give a P-value, and then assess significance.

(c) How significant is the observed difference in mean alcohol consumption? Use a t test to obtain the P-value.

(d) Use of the pooled two-sample t test is justified in part (b). Explain why. Also find the P-value for the pooled t statistic and compare with your result in part (b).

(e) The report says that the distributions of alcohol consumption among the individuals studied are "grossly skew." Do you think that this skewness prevents the use of the two-sample t test for equality of means? Explain your answer.

7.137 Occupation and diet, continued. Use the data in the previous exercise to give two confidence intervals:

(a) A 90% confidence interval for the mean daily alcohol consumption of London double-decker bus conductors.

(b) An 80% confidence interval for the difference in mean daily alcohol consumption between drivers and conductors.

(c) Write a one-sentence description of each of the confidence intervals calculated above.

7.138 Ego strengths of MBA graduates: power. In Exercise 7.111 (page 446) you found the power for a study designed to compare the "ego strengths" of two groups of MBA students. Now you must design a study to compare MBA graduates who reached partner in a large consulting firm with those who joined the firm but failed to become partners.

Assume the same value of $\sigma = 0.7$ and use $\alpha = 0.05$. You are planning to have 20 subjects in each group. Calculate the power of the pooled two-sample t test that compares the mean ego strengths of these two groups of MBA graduates for several values of the true difference. Include values that have a very small chance of being detected and some that are virtually certain to be seen in your sample. Plot the power versus the true difference and write a short summary of what you have found.

7.139 t approaches z. As the degrees of freedom increase, the t distributions get closer and closer to the $N(0, 1)$, or z, distribution. One way to see this is to look at how the critical value t^* for a 95% confidence interval changes with the degrees of freedom. Make a plot with degrees of freedom from 2 to 100 on the x axis and t^* on the y axis. Draw a horizontal line on the plot corresponding to the value of $z^* = 1.96$. Summarize the main features of the plot.

7.140 Sample size and margin of error. The margin of error for the one-sample t confidence interval depends on the confidence level, the standard deviation, and the sample size. Fix the confidence level at 95% and the standard error at $s = 1$ to examine the effect of the sample size n. Find the margin of error for sample sizes of 5 to 100 by 5s. That is, let $n = 5, 10, 15, \ldots, 100$. Plot the margins of error versus the sample size and summarize the relationship.

CHAPTER 7 Case Study Exercises

CASE STUDY EXERCISE 1: Architectural firms. The data file ARCHITECT gives characteristics of 21 large Indianapolis-area architectural firms. The variables are NAME, the name of the firm; TOTBILL02, total billings in 2002; ARCHBILL02, architectural billings in 2002; ARCHBILL01, architectural billings in 2001; N_ARCH, the number of architects in the firm; N_ENG, the number of engineers in the firm; N_STAFF, the number of staff members in the firm; and YR_ESTAB, the year that the firm was established. Table 7.4 show the data for the first six firms.

TABLE 7.4   Characteristics of six large Indianapolis-area architectural firms

NAME	TOTBILL02	ARCHBILL02	ARCHBILL01	N_ARCH	N_ENG	N_STAFF	YR_ESTAB
BSA	29.5	24.8	23.7	39	36	240	1975
CSO	12.1	9.0	11.1	17	1	66	1961
American	18.1	4.9	5.1	9	35	168	1966
Schmidt	10.5	10.5	9.4	17	5	80	1976
Browning	12.2	12.2	10.4	22	0	70	1967
OdleMcG	5.1	2.4	4.2	6	2	47	1916

Make a table giving the mean, the standard deviation, the 95% confidence interval, and the five-number summary for the variables 2002 architectural billings, 2001 architectural billings, the number of architects employed, the number of engineers employed, and the number of staff members. Classify each firm as "old" or "new" based on whether or not they began doing business in the area before 1970. Compare the means of the old and new firms for the variables 2002 architectural billings, 2001 architectural billings, the number of architects employed, the number of engineers employed, and the number of staff members. Be sure to state whether or not you used the pooled procedures and why. Give your results with graphical and numerical summaries. Then write a short paragraph explaining any differences that you have found.

CASE STUDY EXERCISE 2: Three or four bedrooms. How much more would you expect to pay for a home that has four bedrooms than for a home that has three? Here are the selling prices (in dollars) that the owners of the homes in West Lafayette, Indiana are asking[46]:   **3VS4BEDROOM**

Four-bedroom homes

121,900	139,900	157,000	159,900	176,900
224,900	235,000	245,000	294,000	

Three-bedroom homes

65,500	79,900	79,900	79,900	82,900
87,900	94,000	97,500	105,000	111,900

116,900	117,900	119,900	122,900	124,000
125,000	126,900	127,900	127,900	127,900
132,900	145,000	145,500	157,500	194,000
205,900	259,900	265,000		

Plot the selling prices for the two sets of homes and describe the two distributions. Test the null hypothesis that the mean asking prices for the two sets of homes are equal versus the two-sided alternative. Give the test statistic with degrees of freedom, the P-value, and your conclusion. Would you consider using a one-sided alternative for this analysis? Explain why or why not. Give a 95% confidence interval for the difference in mean selling prices. These data are not SRSs from a population. Give a justification for use of the two-sample t procedures in this case.

Go to the Web site www.realtor.com and select two geographical areas of interest to you. You will compare the prices of similar types of homes in these two areas. State clearly how you define the areas and the type of homes. For example, you can use city names or zip codes to define the area and you can select single-family homes or condominiums. We view these homes as representative of the asking prices of homes for these areas at the time of your search. If the search gives a large number of homes, select a random sample. Be sure to explain exactly how you do this. Use the methods you have learned in this chapter to compare the selling prices. Be sure to include a graphical summary.

CHAPTER 7   Appendix

Using Minitab and Excel for Inference for Distributions

Confidence Interval for the Mean (σ Unknown)

Minitab:

Stat ➤ Basic Statistics ➤ 1-Sample t

If the data are in the worksheet, choose the **Samples in columns** option and click the column containing the data into the box below. With this option, you can construct confidence intervals for more than one data set simultaneously. Alternatively, if you know the values of the sample mean and sample standard deviation, you can choose the **Summarized data** option and then input the sample size, sample mean, and sample standard deviation in the boxes found below. In default mode, Minitab will construct a 95% confidence interval. If you wish to change the level of confidence, click the **Options** button and input your desired confidence level in the **Confidence level** box. Click **OK** to close the pop-up box and then click **OK** to find the confidence interval along with some basic statistics, which are reported in the Session window.

Excel:

Excel does not have a routine for reporting the lower and upper limits of a confidence interval for the mean. However, you can use Excel to compute the margin of error, which can then be subtracted from and added to the sample mean to get the confidence interval. To find the margin of error, select "Descriptive Statistics" in the **Data Analysis** menu box and click **OK.** Enter the cell range of the data into the **Input Range** box. Place a checkmark next to the **Confidence Level for Mean** option. If you wish to change the level of confidence, input your desired confidence level in the adjacent box. Since you will need the sample mean value to ultimately construct the confidence interval, place a checkmark next to the **Summary statistics** option. Click **OK** and you will find the margin of error reported in a spreadsheet next to a cell labeled "Confidence Level."

Test for the Population Mean (σ Unknown)

Minitab:

Stat ➤ Basic Statistics ➤ 1-Sample t

This is the same routine described in this Appendix for obtaining the confidence interval for the mean. If you wish to conduct a hypothesis test, select the **Perform hypothesis test** option and input the null hypothesis mean value (μ_0) in the box below. In default mode, Minitab will assume a two-sided alternative. If you wish to have a one-sided alternative, click the **Options** button and choose either the "less than" or the "greater than" alternative hypothesis choice. Click **OK** to close the pop-up box and then click **OK** to find the test statistic and corresponding P-value, which are reported in the Session window.

Excel:

Interestingly, even though Excel provides a statistical function for conducting a test of the mean with known population standard deviation, Excel does not provide a dedicated statistical function or routine for testing the mean when the population standard deviation is unknown. The one-sample t test is available in the WHFStat Add-In for Excel.

Even though there is no dedicated routine in standard Excel, there is a way to utilize the software. First, compute the value of the one-sample t statistic. The values of the sample mean and sample standard deviation needed for this computation can be obtained from the "Descriptive Statistics" option found in the **Data Analysis** menu box. At this point, we can use Excel to compute areas under the t distribution density curve to obtain the P-value for the observed one-sample t statistic. To do so, click an empty cell in the spreadsheet and then proceed to the **Statistical** function menu as described in the overview section of Chapter 1's appendix. Scroll down the list of functions and click on the **TDIST** function choice. This function will find the area to the right of only a positive t-value. Enter the magnitude of the observed t statistic in the **X** box. In the **Deg freedom** box, enter the appropriate degrees of freedom. In the **Tails** box, enter "1" if you have a one-sided alternative and "2" if you have a two-sided alternative. Click **OK** to find a probability reported in the selected cell. If the observed t statistic is negative and you have a greater-than alternative, then subtract the reported probability from 1 to obtain the appropriate P-value. If the observed t statistic is positive and you have a less-than alternative, then subtract the reported probability from 1 to obtain the appropriate P-value. In all other cases, the reported probability is the appropriate P-value.

Matched Pairs t Test

Minitab:

Stat ➤ Basic Statistics ➤ Paired t

If the data are in the worksheet, then choose the **Samples in columns** option and click the two columns containing the data into the boxes below. Alternatively, if you know the values of the sample mean and sample standard deviation, you can choose the **Summarized data (differences)** option and input the sample size, sample mean of the differences, and sample standard deviation of the differences in the boxes found below. In default mode, Minitab will construct a 95% confidence interval for the mean of the differences. If you wish to change the level of confidence, click the **Options** button and input your desired confidence level in the **Confidence level** box. In addition to reporting a confidence interval, this routine will conduct a test for the mean difference. In default mode, Minitab will assume a two-sided alternative. If you wish to have a one-sided alternative, click the **Options** button and choose either the "less than" or the "greater than" alternative hypothesis choice. It is important to realize that your choice of an appropriate one-sided alternative hypothesis should take into account the fact that Minitab obtains the differences by subtracting the second sample from the first sample. If you wish to test against a mean difference other than 0, input the null hypothesis value in **Test mean** box. Click **OK** to close the pop-up box and then click **OK** to find the confidence interval along with the test statistic and corresponding P-value, which are reported in the Session window.

Excel:

Select "t-Test: Paired Two Sample for Means" in the **Data Analysis** menu box and click **OK.** Enter the cell range of the data for the two samples into the **Variable 1 Range** and **Variable 2 Range** boxes. Excel obtains the differences by subtracting the second sample from the first sample. If you wish to test against a mean difference other than 0, input the null hypothesis value in the **Hypothesized Mean Difference** box. Click **OK** to find descriptive statistics on the samples reported along with a t statistic and both a one-sided P-value and a two-sided P-value. The one-sided P-value is computed as the probability that the t statistic is more negative if the observed statistic is negative or as the probability that the t statistic is more positive if the observed statistic is positive. No regard to the alternative hypothesis is given. If you are testing a greater-than alternative hypothesis and the observed t statistic is negative, then subtract the reported one-sided P-value from 1 to obtain the appropriate P-value. Similarly, if you are testing a less-than alternative hypothesis and the observed t statistic is positive, then subtract the reported one-sided P-value from 1 to obtain the appropriate P-value. In all other cases, the reported probability is the appropriate P-value.

Power (One-Sample *t*)

Minitab:

Stat ➤ Power and Sample Size ➤ 1-Sample *t*

This routine is nearly identical to the routine for finding the power of a one-sample z test discussed in Chapter 6's Appendix. The only difference is that, in the **Standard deviation** box, you need to enter an estimate of the population standard deviation (σ). For a matched pairs study, enter the estimated standard deviation of the differences between the pairs.

Excel:

Automated power calculations are not available in standard Excel or the WHFStat Add-In for Excel. As done in Example 7.9, the Normal distribution can be used to compute the approximate power. Accordingly, you can use Excel's **NORMSDIST** function to compute probabilities from the Normal distribution (refer to Chapter 1's Appendix for details on using the **NORMSDIST** function).

Sign Test

Minitab:

Stat ➤ Nonparametrics ➤ 1-Sample Sign

The sign test can be used to test against any general median value. In this chapter, we focused on testing the differences of matched pairs against a median value of 0. For such an application, click the column containing the differences into the **Variables** box. Select the **Test median** option and keep the default test median value at 0. Choose the appropriate alternative hypothesis from the **Alternative** menu box. Click **OK** to find the P-value for the test reported in the Session window.

Excel:

The sign test is not available in standard Excel or the WHFStat Add-In for Excel.

Comparing Two Means

Minitab:

Stat ➤ Basic Statistics ➤ 2-Sample *t*

There are three ways to enter the data with this routine. If the data for the two samples are in two separate columns, choose the **Samples in different columns** option and then click the columns containing the data into the two boxes below. You can also have the data for the two samples all in one column with a group label in a second column. If the data are stored in this manner, choose the **Samples in one column** option and then click the data column into the **Samples** box and click the column of labels into the **Subscripts** box. As a third option, if you know the values of the sample means and sample standard deviations, you can choose the **Summarized data** option and input the sample sizes, sample means, and sample standard deviations in the boxes found below. The default for this routine is not to assume equal population variances. If you wish to assume equal population variances, select the **Assume equal variances** option.

In default mode, Minitab will construct a 95% confidence interval for the difference between the two population means. If you wish to change the level of confidence, click the **Options** button and input your desired confidence level in the **Confidence level** box. In addition to reporting a confidence interval, this routine will conduct a test for the difference between the means. In default mode, Minitab will assume a two-sided alternative. If you wish to have a one-sided alternative, then click the **Options** button and choose either the "less than" or the "greater than" alternative hypothesis choice. Your choice of the appropriate one-sided alternative hypothesis should take into account what you specified as the first and second samples. If you wish to test against a difference between means other than 0, input the null hypothesis value in the **Test difference** box. Click **OK** to close the pop-up box and then click **OK** to find the confidence interval along with the test statistic and corresponding P-value, which are reported in the Session window.

Excel:

If you do not want to assume equal population variances, select "t-Test: Two-Sample Assuming Unequal Variances" in the **Data Analysis** menu box and click **OK.** If you wish to assume equal population variances, select "t-Test: Two-Sample Assuming Equal Variances" in the **Data Analysis** menu box and click **OK.** Enter the cell range of the data for the two samples into the **Variable 1 Range** and **Variable 2 Range** boxes. If you wish to test against a difference between means other than 0, input the null hypothesis value in **Hypothesized Mean Difference** box. Click **OK** to find descriptive statistics on the samples reported along with a t statistic and both a one-sided P-value and a two-sided P-value. The one-sided P-value is computed as the probability that the t statistic is more negative if the observed statistic is negative or as the probability that the t statistic is more positive if the observed statistic is positive. No regard to the alternative hypothesis is given. If you are testing a greater-than alternative hypothesis and the observed t statistic is negative, then subtract the reported one-sided P-value from 1 to obtain the appropriate P-value. Similarly, if you are testing a less-than alternative hypothesis and the observed t statistic is positive, then subtract the reported one-sided P-value from 1 to obtain the appropriate P-value. In all other cases, the reported probability is the appropriate P-value.

F Test For Equality of Spread

Minitab:

Stat ➤ Basic Statistics ➤ 2 Variances

There are three ways to enter the data with this routine. If the data for the two samples are in two separate columns, choose the **Samples in different columns** option and then click the columns containing the data into the two boxes below. You can also have the data for the two samples all in one column with a group label in a second column. If the data are stored in this manner, choose the **Samples in one column** option and then click the data column into the **Samples** box and click the column of labels into the **Subscripts** box. As a third option, if you know the values

of the sample variances, then you can choose the **Summarized data** option and input the sample sizes and sample variances in the boxes found below. Click **OK** to find the F test statistic and corresponding P-value, which are reported in the Session window. Minitab will also produce a plot showing the confidence intervals for the population standard deviations along with a boxplot for each of the samples. Finally, you will notice that Minitab also reports a Levene's test statistic along with a corresponding P-value. The Levene's test is an alternative to the F test when we cannot safely assume Normality for the two populations.

Excel:

Select "F Test Two-Sample for Variances" in the **Data Analysis** menu box and click **OK.** Enter the cell range of the data for the two samples into the **Variable 1 Range** and **Variable 2 Range** boxes. If you wish to test against a difference between means other than 0, input the null hypothesis value in the **Hypothesized Mean Difference** box. Click **OK** to find descriptive statistics on the samples reported along with the F statistic and corresponding P-value.

Power (Two-Sample t)

Minitab:

Stat ➤ Power and Sample Size ➤ 2-Sample t

This routine is essentially identical to the routine for finding the power of a one-sample t test discussed earlier in this Appendix. This routine assumes that the two sample sizes (n_1 and n_2) are equal. Thus, you need to enter only one sample size in the **Sample sizes** box. Furthermore, as done in Example 7.19, Minitab assumes that the standard deviations of the two populations are equal. The estimated standard deviation should be entered in the **Standard deviation** box.

Excel:

Automated power calculations are not available in standard Excel or the WHFStat Add-In for Excel. As done in Example 7.19, the Normal distribution can be used to compute the approximate power. Accordingly, you can use Excel's **NORMSDIST** function to compute probabilities from the Normal distribution (refer to Chapter 1's Appendix for details on using the **NORMSDIST** function).

RICHARD KEPPELSMITH/GETTY

Inference for Proportions

The U.S. video game market is approximately $8.2 billion and growing. Case 8.1 discusses a PEW survey that has been used to collect data on gamers.

Introduction

We frequently collect data on *categorical variables,* such as whether or not a person is a full-time college student or a part-time college student, the brand name of a cell phone, or the country where a college student studies abroad. When we record categorical variables, our data consist of *counts* or of *percents* obtained from counts.

The parameters we want to do inference about in these settings are *population proportions*. Just as in the case of inference about population means, we may be concerned with a single population or with comparing two populations. Inference about one or two proportions is very similar to inference about means, which we discussed in Chapter 7. In particular, inference for both means and proportions is based on sampling distributions that are approximately Normal.

We begin in Section 8.1 with inference about a single population proportion. Section 8.2 concerns methods for comparing two proportions.

CHAPTER OUTLINE

8.1 Inference for a Single Proportion

8.2 Comparing Two Proportions

8.1 Inference for a Single Proportion

CASE 8.1

Adults and Video Games A PriceWaterhouseCooper report estimates that the U.S. video game market was approximately $8.6 billion in 2007 and is expected to increase at an annual rate of 6.3% through 2012.[1] Who plays video games? A PEW survey, conducted by Princeton Survey Research International, reports that over half of American adults aged 18 and over play video games.[2] The PEW survey used a nationally representative sample of 2054 adults. Of the total, 1063 adults said that they played video games.

For problems involving a single proportion, we will use n for the sample size and X for the count of the outcome of interest. Often we will use the terms "success" and "failure" for the two possible outcomes. When we do this, X is the number of successes.

> **EXAMPLE 8.1**   Data for the Video Game Case

CASE 8.1

The *count* of people who responded "Yes" to the question about whether or not they played video games in the sample of Case 8.1 is $X = 1063$. The sample size is $n = 2054$.

population proportion
sample proportion

We would like to know the proportion of video game players in the adult U.S. population. This **population proportion** is the *parameter* of interest. The *statistic* used to estimate this unknown parameter is the **sample proportion.** The sample proportion is $\hat{p} = X/n$.

> **EXAMPLE 8.2**   Estimating the Proportion of Adults Who Play Video Games

CASE 8.1

The sample proportion $\hat{p}$ in Case 8.1 is a discrete random variable that can take the values 0, 1/2054, 2/2054, ..., 2053/2054, or 1. For our particular sample, we have

$$\hat{p} = \frac{1063}{2054} = 0.52$$

In many cases, a probability model for $\hat{p}$ can be based on the binomial distributions for counts, discussed in Chapter 5. If the sample size n is very small, we can base tests and confidence intervals for p on the discrete distribution of $\hat{p}$. We will focus on situations where the sample size is sufficiently large that we can approximate the distribution of $\hat{p}$ by a Normal distribution.

Sampling Distribution of a Sample Proportion

Choose an SRS of size n from a large population that contains population proportion p of "successes." Let X be the count of successes in the sample and let $\hat{p}$ be the **sample proportion** of successes,

$$\hat{p} = \frac{X}{n}$$

Then:

- For large sample sizes, the distribution of $\hat{p}$ is **approximately Normal.**
- The **mean** of the distribution of $\hat{p}$ is p.
- The **standard deviation** of $\hat{p}$ is

$$\sqrt{\frac{p(1-p)}{n}}$$

Figure 8.1 summarizes these facts in a form that recalls the idea of sampling distributions. Our inference procedures are based on this Normal approximation. These procedures are similar to those for inference about the mean of a Normal distribution. We will see, however, that there are a few extra details involved, caused by the added difficulty in approximating the discrete distribution of $\hat{p}$ by a continuous Normal distribution.

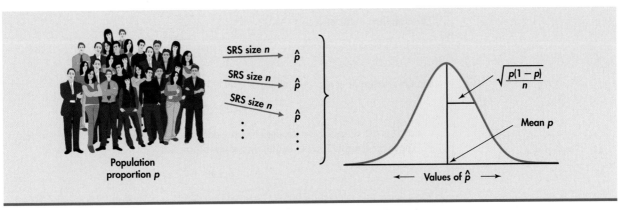

FIGURE 8.1 Draw a large SRS from a population in which the proportion p are successes. The sampling distribution of the sample proportion $\hat{p}$ of successes has approximately a Normal distribution.

APPLY YOUR KNOWLEDGE

8.1 Bank acquisitions. The American Bankers Association Community Bank Competitiveness Survey for 2008 had responses from 760 community banks. Of these, 283 reported that they expected to acquire another bank within five years.[3]
(a) What is the sample size n for this survey?
(b) What is the count X? Describe the count in a short sentence.
(c) Find the sample proportion $\hat{p}$.

CASE 8.1 **8.2 How often do they play?** In the PEW survey described in Case 8.1, those who played video games were asked how often they played. In this subpopulation, 223 adults said that they played every day or almost every day.
(a) What is the sample size n for the subpopulation of U.S. adults who play video games? (*Hint:* Look at Case 8.1.)
(b) What is the count X of those who said that they played every day or almost every day?
(c) Find the sample proportion $\hat{p}$.

Large-sample confidence interval for a single proportion

The sample proportion $\hat{p} = X/n$ is the natural estimator of the population proportion p. Notice that $\sqrt{p(1-p)/n}$, the standard deviation of $\hat{p}$, depends upon the unknown parameter p. In our calculations, we estimate it by replacing the population parameter p with the sample estimate $\hat{p}$. Therefore, our estimated standard error is $SE_{\hat{p}} = \sqrt{\hat{p}(1-\hat{p})/n}$. If the sample size is large, the distribution of $\hat{p}$ will be approximately Normal with mean p and standard deviation $SE_{\hat{p}}$. It follows that $\hat{p}$ will be within two standard deviations $(2SE_{\hat{p}})$ of the unknown parameter p about 95% of the time. This is how we use the Normal approximation to construct the large-sample confidence interval for p. Here are the details.

> z **Confidence Interval for a Population Proportion**
>
> Choose an SRS of size n from a large population with unknown proportion p of successes. The **sample proportion** is
>
> $$\hat{p} = \frac{X}{n}$$

The **standard error of** $\hat{p}$ is

$$SE_{\hat{p}} = \sqrt{\frac{\hat{p}(1-\hat{p})}{n}}$$

and the **margin of error** for confidence level C is

$$m = z^*SE_{\hat{p}}$$

where z^* is the value for the standard Normal density curve with area C between $-z^*$ and z^*. The **large-sample level** C **confidence interval** for p is

$$\hat{p} \pm m$$

You can use this interval for 90% ($z^* = 1.645$), 95% ($z^* = 1.960$), or 99% ($z^* = 2.576$) confidence when the number of successes and the number of failures are both at least 15.

EXAMPLE 8.3   Confidence Interval for the Proportion of Adults Who Play Video Games

CASE 8.1

The sample survey in Case 8.1 found that 1063 of a sample of 2054 adults reported that they played video games. So, the sample size is $n = 2054$ and the count is $X = 1063$. The sample proportion of adults who play video games is

$$\hat{p} = \frac{X}{n} = \frac{1063}{2054} = 0.51753$$

The standard error is

$$SE_{\hat{p}} = \sqrt{\frac{\hat{p}(1-\hat{p})}{n}} = \sqrt{\frac{0.5175(1-0.5175)}{2054}} = 0.011026$$

The z critical value for 95% confidence is $z^* = 1.96$, so the margin of error is

$$m = 1.96SE_{\hat{p}} = (1.96)(0.011026) = 0.021610$$

The confidence interval is

$$\hat{p} \pm m = 0.52 \pm 0.02$$

We are 95% confident that between 50% and 54% of adults play video games.

In performing these calculations we have kept a large number of digits for our intermediate calculations. However, when reporting the results we prefer to use rounded values. For example, "52% with a margin of error of 2%." In this way we focus attention on what is important. There is no additional information to be gained by reporting 0.51753 with a margin of error of 0.021610.

Remember that the margin of error in any confidence interval includes only random sampling error. If people do not respond honestly to the questions asked, for example, your estimate is likely to miss by more than the margin of error.

Because the calculations for statistical inference for a single proportion are relatively straightforward, we often do them with a calculator or in a spreadsheet. Figure 8.2 gives output from Minitab and SAS for the data in Case 8.1. As usual, the output reports more digits than are useful. When you use software, be sure to think about how many digits are meaningful for your purposes. Do not clutter your report with information that is not meaningful. SAS gives the standard error next to the label ASE, which stands for

FIGURE 8.2 Minitab and SAS outputs for the confidence interval in Example 8.3.

Minitab

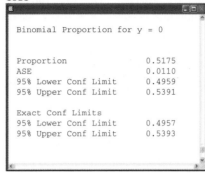

```
Test and CI for One Proportion

Sample     X      N    Sample p       95% CI
1        1063   2054   0.517527   (0.495917, 0.539137)

Using the normal approximation.
```

SAS

```
Binomial Proportion for y = 0


Proportion                  0.5175
ASE                         0.0110
95% Lower Conf Limit        0.4959
95% Upper Conf Limit        0.5391

Exact Conf Limits
95% Lower Conf Limit        0.4957
95% Upper Conf Limit        0.5393
```

asymptotic standard error. The SAS output also includes an alternative interval based on an "exact" method.

APPLY YOUR KNOWLEDGE

8.3 Bank acquisitions. Refer to Exercise 8.1 (page 459).
(a) Find $SE_{\hat{p}}$, the standard error of $\hat{p}$.
(b) Give the 95% confidence interval for p in the form of estimate plus or minus the margin of error.
(c) Give the confidence interval as an interval of percents.

CASE 8.1 **8.4 How often do they play?** Refer to Exercise 8.2 (page 459).
(a) Find $SE_{\hat{p}}$, the standard error of $\hat{p}$.
(b) Give the 95% confidence interval for p in the form of estimate plus or minus the margin of error.
(c) Give the confidence interval as an interval of percents.

Plus four confidence interval for a single proportion*

Suppose we have a sample where the count is $X = 0$. Then, because $\hat{p} = 0$, the standard error and the margin of error based on this estimate will both be 0. The confidence interval for any confidence level would be the single point 0. Confidence intervals based on the large-sample Normal approximation do not make sense in this situation.

Both computer studies and careful mathematics show that we can do better by moving the sample proportion $\hat{p}$ slightly away from 0 and 1.[4] There are several ways to do this. Here is a simple adjustment that works very well in practice.

The adjustment is based on the following idea: act as if we have 4 additional observations, 2 of which are successes and 2 of which are failures. The new sample size is

---

*The material on the plus four confidence interval is optional and can be omitted without loss of continuity.

Wilson estimate

$n + 4$ and the count of successes is $X + 2$. Because this estimate was first suggested by Edwin Bidwell Wilson in 1927 (though rarely used in practice until recently), we call it the **Wilson estimate.**

To compute a confidence interval based on the Wilson estimate, first replace the value of X by $X + 2$ and the value of n by $n + 4$. Then use these values in the formulas for the z confidence interval.

In Example 8.1, we had $X = 1063$ and $n = 2054$. To apply the plus four approach we use the z procedure with $X = 1065$ and $n = 2058$. You can use this interval when the sample size is at least $n = 10$ and the confidence level is 90%, 95%, or 99%.

APPLY YOUR KNOWLEDGE

8.5 Use plus-four for adults who play video games. Refer to Example 8.3 (page 460). Compute the plus four 95% confidence interval and compare this interval with the one given in that example.

8.6 New-product sales. Yesterday, your top salesperson called on 6 customers and obtained orders for your new product from all 6. Suppose that it is reasonable to view these 6 customers as a random sample of all of her customers.
(a) Give the plus four estimate of the proportion of her customers who would buy the new product. Notice that we don't estimate that all customers will buy, even though all 6 in the sample did.
(b) Give the margin of error for 95% confidence. (You may see that the upper endpoint of the confidence interval is greater than 1. In that case, take the upper endpoint to be 1.)
(c) Do the results apply to all of your sales force? Explain why or why not.

8.7 Construct an example. Make up an example where the large-sample method and the plus four method give very different intervals. Do not use a case where either $\hat{p} = 0$ or $\hat{p} = 1$.

Significance test for a single proportion

We know that the sample proportion $\hat{p} = X/n$ is approximately Normal, with mean $\mu_{\hat{p}} = p$ and standard deviation $\sigma_{\hat{p}} = \sqrt{p(1-p)/n}$. To construct confidence intervals, we need to use an estimate of the standard deviation based on the data because the standard deviation depends upon the unknown parameter p. When performing a significance test, however, the null hypothesis specifies a value for p, which we will call p_0. When we calculate P-values, we act as if the hypothesized p were actually true. When we test $H_0: p = p_0$, we substitute p_0 for p in the expression for $\sigma_{\hat{p}}$ and then standardize $\hat{p}$. Here are the details.

---

z Significance Test for a Population Proportion

Choose an SRS of size n from a large population with unknown proportion p of successes. To test the hypothesis $H_0: p = p_0$, compute the z **statistic**

$$z = \frac{\hat{p} - p_0}{\sqrt{\dfrac{p_0(1 - p_0)}{n}}}$$

In terms of a standard Normal random variable Z, the approximate P-value for a test of H_0 against

H_a: $p > p_0$ is $P(Z \geq z)$

H_a: $p < p_0$ is $P(Z \leq z)$

H_a: $p \neq p_0$ is $2P(Z \geq |z|)$

Use this test when the expected number of successes np_0 and the expected number of failures $n(1 - p_0)$ are both at least 10.

We call this z test a "large-sample test" because it is based on a Normal approximation to the sampling distribution of $\hat{p}$ that becomes more accurate as the sample size increases. For small samples, or if the population is less than 10 times as large as the sample, consult an expert for other procedures.

EXAMPLE 8.4   Comparing Two Sun Block Lotions

Your company produces a sun block lotion designed to protect the skin from both UVA and UVB exposure to the sun. You hire a company to compare your product with the product sold by your major competitor. The testing company exposes skin on the backs of a sample of 20 people to UVA and UVB rays and measures the protection provided by each product. For 13 of the subjects, your product provided better protection, while for the other 7 subjects, your competitor's product provided better protection. Do you have evidence to support a commercial claiming that your product provides superior UVA and UVB protection? For the data we have $n = 20$ subjects and $X = 13$ successes. To answer the claim question, we test

$$H_0: p = 0.5$$
$$H_a: p \neq 0.5$$

The expected numbers of successes (your product provides better protection) and failures (your competitor's product provides better protection) are $20 \times 0.5 = 10$ and $20 \times 0.5 = 10$. Both are at least 10, so we can use the z test. The sample proportion is

$$\hat{p} = \frac{X}{n} = \frac{13}{20} = 0.65$$

The test statistic is

$$z = \frac{\hat{p} - p_0}{\sqrt{\dfrac{p_0(1 - p_0)}{n}}} = \frac{0.65 - 0.5}{\sqrt{\dfrac{(0.5)(0.5)}{20}}} = 1.34$$

From Table A we find $P(Z \geq 1.34) = 0.9099$, so the probability in the upper tail is $1 - 0.9099 = 0.0901$. The P-value is the area in both tails, $P = 2 \times 0.0901 = 0.1802$. Minitab and SAS outputs for the analysis appear in Figure 8.3. We conclude that the sun block testing data are compatible with the hypothesis of no difference between your product and your competitor's ($\hat{p} = 0.65$, $z = 1.34$, $P = 0.18$). The data do not provide you with a basis to support your advertising claim.

FIGURE 8.3 Minitab and SAS outputs for the significance test in Example 8.4.

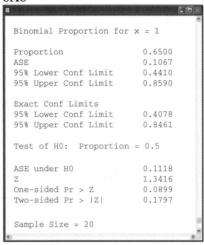

Minitab

```
Test and CI for One Proportion

Test of p=0.5 vs p not=0.5

Sample     X     N    Sample p        95% CI          Z-Value   P-Value
1          13    20   0.650000  (0.440963, 0.859037)   1.34     0.180
Using the normal approximation.
```

SAS

```
Binomial Proportion for x = 1

Proportion               0.6500
ASE                      0.1067
95% Lower Conf Limit     0.4410
95% Upper Conf Limit     0.8590

Exact Conf Limits
95% Lower Conf Limit     0.4078
95% Upper Conf Limit     0.8461

Test of H0:  Proportion = 0.5

ASE under H0             0.1118
Z                        1.3416
One-sided Pr > Z         0.0899
Two-sided Pr > |Z|       0.1797

Sample Size = 20
```

Note that we used a two-sided hypothesis test when we compared the two sun block lotions in Example 8.4. In settings like this, we must start with the view that either product could be better if we want to prove a claim of superiority. Thinking or hoping that your product is superior cannot be used to justify a one-sided test.

APPLY YOUR KNOWLEDGE

8.8 Draw a picture. Draw a picture of a standard Normal curve and shade the tail areas to illustrate the calculation of the P-value for Example 8.4.

8.9 What does the confidence interval tell us? Inspect the outputs in Figure 8.3 and report the confidence interval for the percent of people who would get better sun protection from your product than from your competitor's. Be sure to convert from proportions to percents and to round appropriately. Interpret the confidence interval and compare this way of analyzing data with the significance test.

8.10 The effect of X. In Example 8.4, suppose that your product provided better UVA and UVB protection for 15 of the 20 subjects. Perform the significance test and summarize the results.

8.11 The effect of n. In Example 8.4, consider what would have happened if you had paid for twice as many subjects to be tested. Assume that the results would be the same as what you obtained for 20 subjects; that is 65% had better UVA and UVB protection with your product. Perform the significance test and summarize the results.

In Example 8.4, we treated an outcome as a success whenever your product provided better sun protection. Would we get the same results if we defined success as an outcome where your competitor's product was superior? In this setting the null hypothesis is still H_0: $p = 0.5$. You will find that the z test statistic is unchanged except for its sign and that the P-value remains the same.

APPLY YOUR KNOWLEDGE

8.12 Yes or no? In Example 8.4 we performed a significance test to compare your product with your competitor's. Success was defined as the outcome where your product provided better protection. Now, take the viewpoint of your competitor and define success as the outcome where your competitor's product provides better protection. In other words, n remains the same (20) but X is now 7.
(a) Perform the two-sided significance test and report the results. How do these compare with what we found in Example 8.4?
(b) Find the 95% confidence interval for this setting and compare it with the interval calculated where success is defined as the outcome when your product provides better protection.

Choosing a sample size

In Chapter 6, we showed how to choose the sample size n to obtain a confidence interval with specified margin of error m for a Normal mean. Because we are using a Normal approximation for inference about a population proportion, sample size selection proceeds in much the same way.

Recall that the margin of error for the large-sample confidence interval for a population proportion is

$$m = z^*\mathrm{SE}_{\hat{p}} = z^*\sqrt{\frac{\hat{p}(1 - \hat{p})}{n}}$$

Choosing a confidence level C fixes the critical value z^*. The margin of error also depends on the value of $\hat{p}$ and the sample size n. Because we don't know the value of $\hat{p}$ until we gather the data, we must guess a value to use in the calculations. We will call the guessed value p^*. Here are two ways to get p^*:

• Use the sample estimate from a pilot study or from similar studies done earlier.
• Use $p^* = 0.5$. Because the margin of error is largest when $\hat{p} = 0.5$, this choice gives a sample size that is somewhat larger than we really need for the confidence level we choose. It is a safe choice no matter what the data later show.

Once we have chosen p^* and the margin of error m that we want, we can find the n we need to achieve this margin of error. Here is the result.

> **Sample Size for Desired Margin of Error**
>
> The level C confidence interval for a proportion p will have a margin of error approximately equal to a specified value m when the sample size is
>
> $$n = \left(\frac{z^*}{m}\right)^2 p^*(1 - p^*)$$

Here z^* is the critical value for confidence C, and p^* is a guessed value for the proportion of successes in the future sample.

The margin of error will be less than or equal to m if p^* is chosen to be 0.5. The sample size required is then given by

$$n = \left(\frac{z^*}{2m}\right)^2$$

The value of n obtained by this method is not particularly sensitive to the choice of p^* as long as p^* is not too far from 0.5. However, if your actual sample turns out to have $\hat{p}$ smaller than about 0.3 or larger than about 0.7, the sample size based on $p^* = 0.5$ may be much larger than needed.

EXAMPLE 8.5   Planning a Sample of Customers

Your company has received complaints about its customer support service. You intend to hire a consulting company to carry out a sample survey of customers. Before contacting the consultant, you want some idea of the sample size you will have to pay for. One critical question is the degree of satisfaction with your customer service, measured on a five-point scale. You want to estimate the proportion p of your customers who are satisfied (that is, who choose either "satisfied" or "very satisfied," the two highest levels on the five-point scale).

You want to estimate p with 95% confidence and a margin of error less than or equal to 3%, or 0.03. For planning purposes, you are willing to use $p^* = 0.5$. To find the sample size required,

$$n = \left(\frac{z^*}{2m}\right)^2 = \left[\frac{1.96}{(2)(0.03)}\right]^2 = 1067.1$$

Round up to get $n = 1068$. (Always round up. Rounding down would give a margin of error slightly greater than 0.03.)

Similarly, for a 2.5% margin of error we have (after rounding up)

$$n = \left[\frac{1.96}{(2)(0.025)}\right]^2 = 1537$$

and for a 2% margin of error,

$$n = \left[\frac{1.96}{(2)(0.02)}\right]^2 = 2401$$

News reports frequently describe the results of surveys with sample sizes between 1000 and 1500 and a margin of error of about 3%. These surveys generally use sampling procedures more complicated than simple random sampling, so the calculation of confidence intervals is more involved than what we have studied in this section. The calculations in Example 8.5 nonetheless show in principle how such surveys are planned.

In practice, many factors influence the choice of a sample size. Case 8.2 illustrates one set of factors.

CASE 8.2 **Marketing Christmas Trees** An association of Christmas tree growers in Indiana sponsored a sample survey of Indiana households to help improve the marketing of Christmas trees.[5] The researchers decided to use a telephone survey and estimated that each telephone interview would take about 2 minutes. Nine trained students in agribusiness marketing were to make the phone calls between 1:00 P.M. and 8:00 P.M. on a Sunday. After discussing problems related to people not being at home or being unwilling to answer the questions, the survey team proposed a sample size of 500. Several of the questions asked demographic information about the household. The key questions of interest had responses of "Yes" or "No," for example, "Did you have a Christmas tree last year?" The primary purpose of the survey was to estimate various sample proportions for Indiana households. An important issue in designing the survey was therefore whether the proposed sample size of $n = 500$ would be adequate to provide the sponsors of the survey with the information they required.

To address this question, we calculate the margins of error of 95% confidence intervals for various values of $\hat{p}$.

EXAMPLE 8.6   Margins of Error

CASE 8.2   In the Christmas tree market survey, the margin of error of a 95% confidence interval for any value of $\hat{p}$ and $n = 500$ is

$$m = z^* SE_{\hat{p}}$$

$$= 1.96\sqrt{\frac{\hat{p}(1 - \hat{p})}{500}}$$

The results for various values of $\hat{p}$ are

$\hat{p}$	m	$\hat{p}$	m
0.05	0.019	0.60	0.043
0.10	0.026	0.70	0.040
0.20	0.035	0.80	0.035
0.30	0.040	0.90	0.026
0.40	0.043	0.95	0.019
0.50	0.044		

The survey team judged these margins of error to be acceptable, and they used a sample size of 500 in their survey.

The table in Example 8.6 illustrates two points. First, the margins of error for $\hat{p} = 0.05$ and $\hat{p} = 0.95$ are the same. The margins of error will always be the same for $\hat{p}$ and $1 - \hat{p}$. This is a direct consequence of the form of the confidence interval. Second, the margin of error varies only between 0.040 and 0.044 as $\hat{p}$ varies from 0.3 to 0.7, and the margin of error is greatest when $\hat{p} = 0.5$, as we claimed earlier. It is true in general that the margin of error will vary relatively little for values of $\hat{p}$ between 0.3 and 0.7. Therefore, when planning a study, it is not necessary to have a very precise guess for p. If $p^* = 0.5$ is used and the observed $\hat{p}$ is between 0.3 and 0.7, the actual interval will be a little shorter than needed, but the difference will be quite small.

APPLY YOUR KNOWLEDGE

8.13 Is there interest in a new product? One of your employees has suggested that your company develop a new product. You decide to take a random sample of your customers and ask whether or not there is interest in the new product. The response is on a 1 to 5 scale, with 1 indicating "definitely would not purchase"; 2, "probably would not purchase"; 3, "not sure"; 4, "probably would purchase"; and 5, "definitely would purchase." For an initial analysis, you will record the responses 1, 2, and 3 as "No" and 4 and 5 as "Yes." What sample size would you use if you wanted the 95% margin of error to be 0.1 or less?

8.14 More information is needed. Refer to the previous exercise. Suppose that, after reviewing the results of the previous survey, you proceeded with preliminary development of the product. Now you are at the stage where you need to decide whether or not to make a major investment to produce and market the product. You will use another random sample of your customers, but now you want the margin of error to be smaller. What sample size would you use if you wanted the 95% margin of error to be 0.05 or less?

SECTION 8.1 Summary

- Inference about a population proportion is based on an SRS of size n. When n is large, the distribution of the **sample proportion** $\hat{p} = X/n$ is approximately Normal with mean p and standard deviation $\sqrt{p(1-p)/n}$.

- The **standard error of $\hat{p}$** is

$$SE_{\hat{p}} = \sqrt{\frac{\hat{p}(1-\hat{p})}{n}}$$

- The z **margin of error** for confidence level C is

$$m = z^* SE_{\hat{p}}$$

where z^* is the value for the standard Normal density curve with area C between $-z^*$ and z^*.

- The z **large-sample level C confidence interval** for p is

$$\hat{p} \pm m$$

We recommend using this method when the number of successes and the number of failures are both at least 15.

- The **plus four estimate of a population proportion** is obtained by adding two successes and two failures to the sample and then using the z procedure. We recommend using this method when the sample size is at least 10 and the confidence level is 90%, 95%, or 99%.

- The **sample size** required to obtain a confidence interval of approximate margin of error m for a proportion is found from

$$n = \left(\frac{z^*}{m}\right)^2 p^*(1-p^*)$$

where p^* is a guessed value for the proportion, and z^* is the standard Normal critical value for the desired level of confidence. To ensure that the margin of error of the interval is less than or equal to m no matter what $\hat{p}$ may be, use

$$n = \left(\frac{z^*}{2m}\right)^2$$

- Tests of H_0: $p = p_0$ are based on the z **statistic**

$$z = \frac{\hat{p} - p_0}{\sqrt{\dfrac{p_0(1-p_0)}{n}}}$$

with P-values calculated from the $N(0, 1)$ distribution. Use this test when the expected number of successes np_0 and the expected number of failures $n(1-p_0)$ are both at least 10.

SECTION 8.1 Exercises

For Exercises 8.1 and 8.2, see page 459; for 8.3 and 8.4, see page 461; for 8.5 to 8.7, see page 462; for 8.8 to 8.11, see page 464; for 8.12, see page 465; and for 8.13 and 8.14, see pages 467–468.

8.15 What's wrong? Explain what is wrong with each of the following.
(a) You can use a significance test to evaluate the hypothesis H_0: $\hat{p} = 0.6$ versus the two-sided alternative.
(b) The large-sample significance test for a population proportion is based on a t statistic.
(c) A large-sample 95% confidence interval for an unknown proportion p is $\hat{p}$ plus or minus its standard error.

8.16 What's wrong? Explain what is wrong with each of the following.
(a) The margin of error for a confidence interval used for an opinion poll takes into account that fact that people who did not answer the poll questions may have had different responses from those who did answer the questions.
(b) If the P-value for a significance test is 0.35, we can conclude that the null hypothesis has a 35% chance of being true.
(c) A student project used a confidence interval to describe the results in a final report. The confidence level was 110%.

8.17 Draw some pictures. Consider the binomial setting with $n = 50$ and $p = 0.4$.
(a) The sample proportion $\hat{p}$ will have a distribution that is approximately Normal. Give the mean and the standard deviation of this Normal distribution.
(b) Draw a sketch of this Normal distribution. Mark the location of the mean.
(c) Find a value p^* for which the probability is 95% that $\hat{p}$ will be between $\pm p^*$. Mark these two values on your sketch.

8.18 Country food and Inuits. Country food includes seal, caribou, whale, duck, fish, and berries and is an important part of the diet of the aboriginal people called Inuits who inhabit Inuit Nunaat, the northern region of what is now called Canada. A survey of Inuits in Inuit Nunaat reported that 3274 out of 5000 respondents said that at least half of the meat and fish that they eat is country food.[6] Find the sample proportion and a 95% confidence interval for the population proportion of Inuits who eat meat and fish that are at least half country food.

8.19 Most desirable mates. A poll of 5000 residents in Brazil, Canada, China, France, Malaysia, South Africa, and the United States asked about what profession they would prefer their marriage partner to have. The choice receiving the highest percent, 805 of the responses, was doctors, nurses, and other health care professionals.[7]
(a) Find the sample proportion and a 95% confidence interval for the proportion of people who would prefer a doctor, nurse, or other health care professional as a marriage partner.
(b) Convert the estimate and the confidence interval to percents.

8.20 Guitar Hero and Rock Band. An electronic survey of 7061 reported that 67% of players of Guitar Hero and Rock Band who do not currently play a musical instrument said that they are likely to begin playing a real musical instrument in the next two years.[8] The reports describing the survey do not give the number of respondents who do not currently play a musical instrument.
(a) Explain why it is important to know the number of respondents who do not currently play a musical instrument.
(b) Assume that half of the respondents do not currently play a musical instrument. Find the count of players who said that they are likely to begin playing a real musical instrument in the next two years.
(c) Give a 99% confidence interval for the population proportion who would say that they are likely to begin playing a real musical instrument in the next two years.
(d) The survey collected data from two separate consumer panels. There were 3300 respondents from the LightSpeed consumer panel and the others were from Guitar Center's proprietary consumer panel. Comment on the sampling procedure used for this survey and how it would influence your interpretation of the findings.

8.21 Guitar Hero and Rock Band. Refer to the previous exercise.
(a) How would the result that you reported in part (c) of the previous exercise change if only 25% of the respondents said that they did not currently play a musical instrument?
(b) Do the same calculations if the percent was 75%.
(c) The main conclusion of the survey that appeared in many news stories was that 67% of players of Guitar Hero and Rock Band who do not currently play a musical instrument said that they are likely to begin playing a real musical instrument in the next two years. What can you conclude about the effect of the three scenarios—part (b) in the previous exercise and parts (a) and (b) in this exercise—on the margin of error for the main result?

8.22 Gambling and college athletics. Gambling is an issue of great concern to those involved in intercollegiate athletics. Because of this, the National Collegiate Athletic Association (NCAA) surveyed student-athletes concerning their gambling-related behaviors.[9] There were 5594 Division I male athletes in the survey. Of these, 3547 reported participation in some gambling behavior. This includes playing cards, betting on games of skill, buying lottery tickets, betting on sports, and similar activities.
(a) Find the sample proportion and the large-sample margin of error for 95% confidence. Explain in simple terms the meaning of the 95%.
(b) Because of the way that the study was designed to protect the anonymity of the student-athletes who responded, it was not possible to calculate the number of students who were asked to respond but did not. Does this fact affect the way that you interpret the results? Write a short paragraph explaining your answer.

8.23 Women athletes and gambling. In the study described in the previous exercise, 1447 out of a total of 3469 female student-athletes reported participation in some gambling activity.
(a) Use the large-sample methods to find an estimate of the true proportion with a 95% confidence interval.
(b) The margin of error for this sample is not the same as the margin of error calculated for the previous exercise. Explain why.

8.24 Students doing community service. In a sample of 159,949 first-year college students, the National Survey of Student Engagement reported that 39% participated in community service or volunteer work.[10]
(a) Find the margin of error for 99% confidence.
(b) Here are some facts from the report that summarizes the survey. The students were from 617 four-year colleges and universities. The response rate was 36%. Institutions paid a participation fee of between $1800 and $7800 based on the size of their undergraduate enrollment. Discuss these facts as possible sources of error in this study. How do you think these errors would compare with the error that you calculated in part (a)?

8.25 Plans to study abroad. The survey described in the previous exercise also asked about items related to academics. In response to one of these questions, 42% of first-year students reported that they plan to study abroad.
(a) Based on the information available, what is the value of the count of students who plan to study abroad?
(b) Give a 99% confidence interval for the population proportion of first-year college students who plan to study abroad.

8.26 Dogs or rats to find cocaine (optional). Dogs are big and expensive. Rats are small and cheap. Can rats be trained to replace dogs in sniffing out illegal drugs? One study trained six male albino Sprague-Dawley rats to rear up on their hind legs in response to the smell of cocaine.[11] After training, each rat was tested 80 times. In the test a rat was presented with a large number of cups, one of which smelled like cocaine. A success was recorded if the rat correctly identified the cup containing cocaine by rearing up in front of it. The numbers of successes for the six rats were 80, 80, 73, 80, 74, and 80. You want to estimate the success rate in the future for each of the six rats. Compare the large-sample estimates with the plus four estimates for this problem and make a recommendation concerning which is better. Write a short summary giving reasons for your recommendation.

8.27 Long sermons. The National Congregations Study collected data in a one-hour interview with a key informant—that is, a minister, priest, rabbi, or other staff person or leader.[12] One question concerned the length of the typical sermon. For 390 out of 1191 congregations, the typical sermon lasted more than 30 minutes.
(a) Use the large-sample inference procedures to estimate the true proportion for this question with a 95% confidence interval.
(b) (Optional) Compute the interval using the plus four method. Compare these results with those from part (a) and summarize what the comparison tells you about the two methods.

(c) There were 1236 congregations surveyed in this study. Calculate the nonresponse rate for this question. Does this influence how you interpret the results? Write a short discussion of this issue.
(d) The respondents to this question were not asked to use a stopwatch to record the lengths of a random sample of sermons at their congregations. They responded based on their impressions of the sermons. Do you think that ministers, priests, rabbis, or other staff persons or leaders might perceive sermon lengths differently from the people listening to the sermons? Discuss how your ideas influence your interpretation of the results of this study.

8.28 Are the congregations conservative? The study described in the previous exercise also asked each respondent to classify his or her congregation according to theological orientation. For this question, 707 out of 1191 congregations were classified as "more conservative." Using the questions in the previous exercise as a guide, analyze and interpret these data. Compare your answers to parts (c) and (d) in the two exercises and discuss reasons why you think the answers should be similar or different.

8.29 Student credit cards. In a survey of 1430 undergraduate students, 1087 reported that they had one or more credit cards.[13] Give a 95% confidence interval for the proportion of all college students who have at least one credit card.

8.30 How many credit cards? The survey described in the previous exercise reported that 43% of undergraduates had four or more credit cards. Give a 95% confidence interval for the proportion of all college students who have four or more credit cards.

8.31 How would the confidence interval change? Refer to Exercise 8.25. Would a 95% confidence interval be wider or narrower than the one that you found in that exercise? Verify your results by computing the interval.

8.32 How would the confidence interval change? Refer to Exercise 8.23. Would a 90% confidence interval be wider or narrower than the one that you found in that exercise? Verify your results by computing the interval.

8.33 College students and diets. For a study of unhealthy eating behaviors, 267 college women aged 18 to 25 years were surveyed.[14] Of these, 69% reported that they had been on a diet sometime during the past year. Give a 95% confidence interval for the true proportion of college women aged 18 to 25 years in this population who dieted last year.

8.34 High school students and diets. In the study described in the previous exercise, the researchers also surveyed 266 high school students who were 18 years old. In this sample 58.3% reported that they had dieted sometime in the past year. Give a 95% confidence interval for the true proportion of 18-year-old high school students in this population who were on a diet sometime during the past year.

8.35 Marketing pet care products to older adults. You have been asked to investigate the possibility of a marketing campaign to promote your company's pet care products to older adults. Your report will include information about your potential market. In a study of the relationship between pet ownership and physical activity in older adults, 594 subjects reported that they owned a pet, while 1939 reported that they did not.[15] Give a 95% confidence interval for the proportion of older adults in this population who are pet owners.

CASE 8.2 **8.36 Christmas tree marketing.** One question in the Christmas tree market survey described in Case 8.2 was "Did you have a Christmas tree last year?" Of the 500 respondents, 421 answered "Yes."
(a) What proportion of the sampled households responded "Yes"?
(b) Give the standard error for your estimate in part (a).
(c) Find a 95% confidence interval for the proportion of Indiana households that had a Christmas tree last year.

8.37 Shipping the orders on time. As part of a quality improvement program, your mail-order company is studying the process of filling customer orders. According to company standards, an order is shipped on time if it is sent within 2 working days of the time it is received. You select an SRS of 150 of the 5000 orders received in the past month for an audit. The audit reveals that 124 of these orders were shipped on time. Find a 95% confidence interval for the true proportion of the month's orders that were shipped on time.

8.38 Power companies and trimming trees. Large trees growing near power lines can cause power failures during storms when their branches fall on the lines. Power companies spend a great deal of time and money trimming and removing trees to prevent this problem. Researchers are developing hormone and chemical treatments that will stunt or slow tree growth. If the treatment is too severe, however, the tree will die. In one series of laboratory experiments on 216 sycamore trees, 41 trees died. Give a 90% confidence interval for the proportion of sycamore trees that would be expected to die from this particular treatment.

8.39 Financial goals of college students. In recent years over 70% of first-year college students responding to a national survey have identified "being well-off financially" as an important personal goal. A state university finds that 141 of an SRS of 200 of its first-year students say that this goal is important. Give a 95% confidence interval for the proportion of all first-year students at the university who would identify being well-off as an important personal goal.

8.40 Can we use the z test? In each of the following cases, is the sample large enough to permit safe use of the z test? (The population is very large.)
(a) $n = 12$ and H_0: $p = 0.6$.
(b) $n = 100$ and H_0: $p = 0.4$.
(c) $n = 1000$ and H_0: $p = 0.98$.
(d) $n = 500$ and H_0: $p = 0.3$.

CASE 8.2 **8.41 Checking the demographics of a sample.** Of the 500 households that responded to the Christmas tree market-

ing survey, 38% were from rural areas (including small towns), and the other 62% were from urban areas (including suburbs). According to the census, 36% of Indiana households are in rural areas, and the remaining 64% are in urban areas. Let p be the proportion of rural respondents. Set up hypotheses about p_0 and perform a test of significance to examine how well the sample represents the state in regard to rural versus urban residence. Summarize your results.

8.42 More on demographics. In the previous exercise we arbitrarily chose to state the hypotheses in terms of the proportion of rural respondents. We could as easily have used the proportion of *urban* respondents.
(a) Write hypotheses in terms of the proportion of urban residents to examine how well the sample represents the state in regard to rural versus urban residence.
(b) Perform the test of significance and summarize the results.
(c) Compare your results with the results of the previous exercise. Summarize and generalize your conclusion.

8.43 Vouchers for schools? A national opinion poll found that 42% of all American adults agree that parents should be given vouchers good for education at any public or private school of their choice. The result was based on a small sample. How large an SRS is required to obtain a margin of error of ±0.035 (that is, ±3.5%) in a 95% confidence interval? (Use the previous poll's result to obtain the guessed value p^*.)

CASE 8.2 **8.44 Profile of the survey respondents.** Of the 500 respondents in the Christmas tree market survey of Case 8.2, 44% had no children at home and 56% had at least one child at home. The corresponding census figures are 48% with no children and 52% with at least one child. Test the null hypothesis that the telephone survey technique has a probability of selecting a household with no children that is equal to the value obtained by the census. Give the z statistic and the P-value. What do you conclude?

8.45 Mathematician tosses coin 10,000 times! The South African mathematician John Kerrich, while a prisoner of war during World War II, tossed a coin 10,000 times and obtained 5067 heads.
(a) Is this significant evidence at the 5% level that the probability that Kerrich's coin comes up heads is not 0.5?
(b) Give a 95% confidence interval to see what probabilities of heads are roughly consistent with Kerrich's result.

8.46 Instant versus fresh-brewed coffee. A matched pairs experiment compares the taste of instant coffee with fresh-brewed coffee. Each subject tastes two unmarked cups of coffee, one of each type, in random order and states which he or she prefers. Of the 60 subjects who participate in the study, 25 prefer the instant coffee and the other 35 prefer fresh-brewed. Take p to be the proportion of the population that prefers fresh-brewed coffee.
(a) Test the claim that a majority of people prefer the taste of fresh-brewed coffee. Report the z statistic and its P-value. Is

your result significant at the 5% level? What is your practical conclusion?

(b) Find a 90% confidence interval for p.

8.47 High-income households on a mailing list. Land's Beginning sells merchandise through the mail. It is considering buying a list of addresses from a magazine. The magazine claims that at least 25% of its subscribers have high incomes (that is, household income in excess of $100,000). Land's Beginning would like to estimate the proportion of high-income people on the list. Verifying income is difficult, but another company offers this service. Land's Beginning will pay to verify the incomes of an SRS of people on the magazine's list. They would like the margin of error of the 95% confidence interval for the proportion to be 0.05 or less. Use the guessed value $p^* = 0.25$ to find the required sample size.

8.48 Change the specs. Refer to the previous exercise. For each of the following variations on the design specifications, state whether the required sample size will be higher, lower, or the same as that found above.

(a) Use a 99% confidence interval.

(b) Change the allowable margin of error to 0.01.

(c) Use a planning value of $p^* = 0.15$.

(d) Use a different company to do the income verification.

8.49 Start a student nightclub? A student organization wants to start a nightclub for students under the age of 21. To assess support for this proposal, the organization will select an SRS of students and ask each respondent if he or she would patronize this type of establishment. About 75% of the student body are expected to respond favorably. What sample size is required to obtain a 95% confidence interval with an approximate margin of error of 0.06? Suppose that 50% of the sample responds favorably. Calculate the margin of error of the 95% confidence interval.

8.50 Are the customers dissatisfied? A cell phone manufacturer would like to know what proportion of its customers are dissatisfied with the service received from their local distributor. The customer relations department will survey a random sample of customers and compute a 99% confidence interval for the proportion that are dissatisfied. From past studies, they believe that this proportion will be about 0.1. Find the sample size needed if the margin of error of the confidence interval is to be about 0.02. Suppose 18% of the sample say that they are dissatisfied. What is the margin of error of the 99% confidence interval?

8.51 Increase student fees? You have been asked to survey students at a large college to determine the proportion that favor an increase in student fees to support an expansion of the student newspaper. Each student will be asked whether he or she is in favor of the proposed increase. Using records provided by the registrar you can select a random sample of students from the college. After careful consideration of your resources, you decide that it is reasonable to conduct a study with a sample of 150 students. Construct a table of the margins of error for 95% confidence when $\hat{p}$ takes the values 0.1, 0.2, 0.3, 0.4, 0.5, 0.6, 0.7, 0.8, and 0.9.

8.52 Justify the cost of the survey. A former editor of the student newspaper agrees to underwrite the study in the previous exercise because she believes the results will demonstrate that most students support an increase in fees. She is willing to provide funds for a sample of size 500. Write a short summary for your benefactor of why the increased sample size will provide better results.

8.2 Comparing Two Proportions

Because comparative studies are so common, we often want to compare the proportions of two groups (such as men and women) that have some characteristic. We call the two groups being compared Population 1 and Population 2, and the two population proportions of "successes" p_1 and p_2. The data consist of two independent SRSs. The sample sizes are n_1 for Population 1 and n_2 for Population 2. The proportion of successes in each sample estimates the corresponding population proportion. Here is the notation we will use in this section:

Population	Population proportion	Sample size	Count of successes	Sample proportion
1	p_1	n_1	X_1	$\hat{p}_1 = X_1/n_1$
2	p_2	n_2	X_2	$\hat{p}_2 = X_2/n_2$

To compare the two unknown population proportions, start with the observed difference between the two sample proportions,

$$D = \hat{p}_1 - \hat{p}_2$$

FIGURE 8.4 The sampling distribution of the difference between two sample proportions is approximately Normal. The mean and standard deviation are found from the two population proportions of successes, p_1 and p_2.

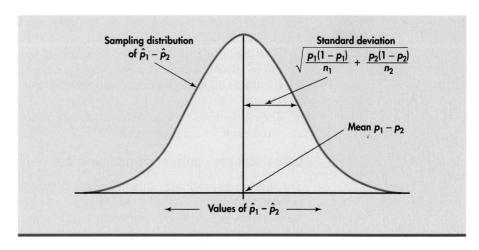

When both sample sizes are sufficiently large, the sampling distribution of the difference D is approximately Normal. What are the mean and the standard deviation of D? Each of the two $\hat{p}$'s has the mean and standard deviation given in the box on page 458. Because the two samples are independent, the two $\hat{p}$'s are also independent. We can apply the rules for means and variances of sums of random variables. Here is the result, which is summarized in Figure 8.4.

Sampling Distribution of $\hat{p}_1 - \hat{p}_2$

 Choose independent SRSs of sizes n_1 and n_2 from two populations with proportions p_1 and p_2 of successes. Let $D = \hat{p}_1 - \hat{p}_2$ be the difference between the two sample proportions of successes. Then

- As both sample sizes increase, the sampling distribution of D becomes **approximately Normal.**

- The **mean** of the sampling distribution is $p_1 - p_2$.

- The **standard deviation** of the sampling distribution is

$$\sigma_D = \sqrt{\frac{p_1(1 - p_1)}{n_1} + \frac{p_2(1 - p_2)}{n_2}}$$

APPLY YOUR KNOWLEDGE

8.53 Rules for means and variances. Suppose $p_1 = 0.4$, $n_1 = 25$, $p_2 = 0.5$, $n_2 = 30$. Find the mean and the standard deviation of the sampling distribution of $p_1 - p_2$.

8.54 Effect of the sample sizes. Suppose $p_1 = 0.4$, $n_1 = 100$, $p_2 = 0.5$, $n_2 = 120$.
(a) Find the mean and the standard deviation of the sampling distribution of $p_1 - p_2$.
(b) The sample sizes here are four times as large as those in the previous exercise, while the population proportions are the same. Compare the results for this exercise with those that you found in the previous exercise. What is the effect of multiplying the sample sizes by 4?

8.55 Rules for means and variances. It is quite easy to verify the mean and standard deviation of the difference D.

(a) What are the means and standard deviations of the two sample proportions $\hat{p}_1$ and $\hat{p}_2$? (Look at the box on page 460 if you need to review this.)

(b) Use the addition rule for means of random variables: what is the mean of $D = \hat{p}_1 - \hat{p}_2$?

(c) The two samples are independent. Use the addition rule for variances of random variables: what is the variance of D?

Large-sample confidence intervals for a difference in proportions

The large-sample estimate of the difference in two proportions $p_1 - p_2$ is the corresponding difference in sample proportions $\hat{p}_1 - \hat{p}_2$. To obtain a confidence interval for the difference, we once again replace the unknown parameters in the standard deviation by estimates to obtain an estimated standard deviation, or standard error. Here is the confidence interval we want.

---

z Confidence Interval for Comparing Two Proportions

Choose an SRS of size n_1 from a large population having proportion p_1 of successes and an independent SRS of size n_2 from another population having proportion p_2 of successes.

The large-sample estimate of the difference in proportions is

$$D = \hat{p}_1 - \hat{p}_2 = \frac{X_1}{n_1} - \frac{X_2}{n_2}$$

The **standard error of the difference** is

$$SE_D = \sqrt{\frac{\hat{p}_1(1 - \hat{p}_1)}{n_1} + \frac{\hat{p}_2(1 - \hat{p}_2)}{n_2}}$$

and the **margin of error for confidence level C** is

$$m = z^* SE_D$$

where z^* is the value for the standard Normal density curve with area C between $-z^*$ and z^*. The **large-sample level C confidence interval** for $p_1 - p_2$ is

$$(\hat{p}_1 - \hat{p}_2) \pm m$$

Use this method when the number of successes and the number of failures in each of the samples are at least 10.

Paul Galipeau

CASE 8.3

"No Sweat" Garment Labels Following complaints about the working conditions in some apparel factories both in the United States and abroad, a joint government and industry commission recommended that companies that monitor and enforce proper standards be allowed to display a "No Sweat" label on their products. Does the presence of these labels influence consumer behavior?

A survey of U.S. residents aged 18 or older asked a series of questions about how likely they would be to purchase a garment under various conditions. For some conditions, it was stated that the garment had a "No Sweat" label; for others, there was no mention of such a label. On the basis of the responses, each person was classified as a "label user" or a "label nonuser."[16] About 16.5% of those surveyed were label users. One purpose of the study was to describe the demographic characteristics of users and nonusers.

Here is a summary of the data. We let X denote the number of label users.

Population	n	X	$\hat{p} = X/n$
1 (women)	296	63	0.213
2 (men)	251	27	0.108

The study in Case 8.3 suggested that there is a gender difference in the proportion of label users. Let's explore this possibility using a confidence interval.

EXAMPLE 8.7 Gender Differences in Label Use

CASE 8.3

First, we find the estimate of the difference:

$$D = \hat{p}_1 - \hat{p}_2 = \frac{X_1}{n_1} - \frac{X_2}{n_2} = 0.213 - 0.108 = 0.105$$

Next, we calculate the standard error:

$$SE_D = \sqrt{\frac{0.213(1 - 0.213)}{63} + \frac{0.108(1 - 0.108)}{27}} = 0.0308$$

For 95% confidence, we use $z^* = 1.96$, so the margin of error is

$$m = z^*SE_D = (1.96)(0.0308) = 0.060$$

The large-sample 95% confidence interval is

$$D \pm m = 0.105 \pm 0.060 = (0.04,\ 0.16)$$

With 95% confidence we can say that the difference in the proportions is between 0.04 and 0.16. Alternatively, we can report that the gender difference is about 10% in favor of women, with a 95% margin of error of 6%.

Minitab and SAS output for Example 8.7 appear in Figure 8.5. Other statistical packages provide output that is similar.

In surveys such as this, men and women are typically not sampled separately. The respondents to a single sample are divided after the fact into men and women. The sample sizes are then random and reflect the characteristics of the population sampled. Two-sample significance tests and confidence intervals are still approximately correct in this situation, even though the two sample sizes were not fixed in advance.

FIGURE 8.5 Minitab and SAS outputs for Example 8.7.

Minitab

```
Sample                  X    N      Sample p
1                       63   296    0.212838
2                       27   251    0.107570

Difference = p (1) - p (2)
Estimate for difference: 0.105268
95% CI for difference: (0.0449066, 0.165630)
```

SAS

	Risk	ASE	(Asymptotic) 95% Confidence Limits	
Row 1	0.2128	0.0238	0.1662	0.2595
Row 2	0.1076	0.0196	0.0692	0.1459
Total	0.1645	0.0159	0.1335	0.1956
Difference	0.1053	0.0308	0.0449	0.1656

In Example 8.7 we chose women to be the first population. Had we chosen men as the first population, the estimate of the difference would be negative (-0.104). Because it is easier to discuss positive numbers, we generally choose the first population to be the one with the higher proportion. The choice doesn't affect the substance of the analysis.

APPLY YOUR KNOWLEDGE

8.56 Lying and online dating profiles. JupiterResearch estimates that the U.S. online dating market will reach $932 million by 2011 and that the European online dating sites will double revenues from 243 million euros in 2006 to 549 million euros in 2011.[17] When trying to start a new relationship, people want to make a favorable impression. Sometimes they will even stretch the truth a bit when disclosing information about themselves. A study of deception in online dating studied the accuracy of the information given in their online dating profiles by 80 online daters.[18] The study found that 22 of 40 men lied about their height, while 17 of 40 women were deceptive in this way. A difference between the person's actual height and that reported in the online dating profile was classified as a lie if it was greater than 0.5 inches.
(a) Find the sample proportion of men who lied about their height. Do the same for the women.
(b) Give the estimate of the difference between the proportion of men who lie about their height and the proportion of women who lie about their height.
(c) Find the standard error for the estimated difference.
(d) Give the 95% confidence interval for the difference.

8.57 Lying about weight. The study described in the previous exercise also described results for lying about weight. They reported that 24 men and 23 women lied about their weight. Answer parts (a) through (d) from the previous exercise for these data.

The previous two exercises ask questions about lying about weight and lying about height. Suppose we wanted to look at the men only and compare the lying rates for height and weight. Can we do this using the methods that we just studied? Stop for a moment to review the material in the box on page 474, paying particular attention to the assumptions that are needed for this method to be valid. The assumptions state that we have independent samples from the two populations. In our examples, however, we are using data from the same people to examine lying about height and lying about weight. The z confidence interval for comparing two proportions that we have been studying is not valid for this situation. Be sure to check your assumptions before applying any statistical inference procedure.

Plus four confidence intervals for a difference in proportions*

Just as in the case of estimating a single proportion, a small modification of the sample proportions greatly improves the accuracy of confidence intervals.[19] As before, we first add 2 successes and 2 failures to the actual data, dividing them equally between the two samples. That is, *add 1 success and 1 failure to each sample*. Note that each sample size is increased by 2. We then perform the calculations for the z procedure with the modified data. As in the case of a single sample, we use the term **Wilson estimates** for

Wilson estimates

---

*The material on the plus four confidence interval for a difference in proportions is optional and can be omitted without loss of continuity.

the estimates produced in this way. We recommend using this method when both sample sizes are at least 5 and the confidence level is 90%, 95%, or 99%.

In Example 8.7, we had $X_1 = 63$, $n_1 = 296$, $X_2 = 27$, and $n_2 = 251$. For the plus four procedure, we would use $X_1 = 64$, $n_1 = 298$, $X_2 = 28$, and $n_2 = 253$.

APPLY YOUR KNOWLEDGE

8.58 Gender and labels using plus four. Refer to Example 8.7 (page 475), where we computed a 95% confidence interval for the difference in the proportions of men and women who were likely to use "No Sweat" labels when deciding to purchase clothing. Redo the computations using the plus four method and compare your results with those obtained in Example 8.7.

8.59 Gender and labels using plus four. Refer to the previous exercise and to Example 8.7. Suppose that the sample sizes were smaller but that the proportions remained approximately the same. Specifically, assume that 6 out of 30 women were label users and 3 out of 25 men were label users. Compute the plus four interval for 95% confidence. Then, compute the corresponding z interval and compare the results.

8.60 Lying about age. Refer to Exercises 8.56 and 8.57, where you analyzed data about lying about height and weight in online dating profiles. The study also reported that 10 men and 5 women lied about their age.
(a) The z confidence interval for comparing two proportions should not be used for these data. Why?
(b) Compute the plus four confidence interval for the difference in proportions.

Significance tests

Although we prefer to compare two proportions by giving a confidence interval for the difference between the two population proportions, it is sometimes useful to test the null hypothesis that the two population proportions are the same.

We standardize $D = \hat{p}_1 - \hat{p}_2$ by subtracting its mean $p_1 - p_2$ and then dividing by its standard deviation

$$\sigma_D = \sqrt{\frac{p_1(1 - p_1)}{n_1} + \frac{p_2(1 - p_2)}{n_2}}$$

If n_1 and n_2 are large, the standardized difference is approximately $N(0, 1)$. To get a confidence interval, we used sample estimates in place of the unknown population proportions p_1 and p_2 in the expression for σ_D. Although this approach would lead to a valid significance test, we follow the more common practice of replacing the unknown σ_D with an estimate that takes into account the null hypothesis that $p_1 = p_2$. If these two proportions are equal, we can view all of the data as coming from a single population. Let p denote the common value of p_1 and p_2. The standard deviation of $D = \hat{p}_1 - \hat{p}_2$ is then

$$\sigma_{Dp} = \sqrt{\frac{p(1 - p)}{n_1} + \frac{p(1 - p)}{n_2}}$$

$$= \sqrt{p(1 - p)\left(\frac{1}{n_1} + \frac{1}{n_2}\right)}$$

The subscript on σ_{Dp} reminds us that this is the standard deviation under the special condition that the two populations share a common proportion p of successes.

We estimate the common value of p by the overall proportion of successes in the two samples:

$$\hat{p} = \frac{\text{number of successes in both samples}}{\text{number of observations in both samples}} = \frac{X_1 + X_2}{n_1 + n_2}$$

pooled estimate of p

This estimate of p is called the **pooled estimate** because it combines, or pools, the information from both samples.

To estimate the standard deviation of D, substitute $\hat{p}$ for p in the expression for σ_{Dp}. The result is a standard error for D under the condition that the null hypothesis H_0: $p_1 = p_2$ is true. The test statistic uses this standard error to standardize the difference between the two sample proportions.

Significance Tests for Comparing Two Proportions

Choose an SRS of size n_1 from a large population having proportion p_1 of successes and an independent SRS of size n_2 from another population having proportion p_2 of successes. To test the hypothesis

$$H_0:\ p_1 = p_2$$

compute the z **statistic**

$$z = \frac{\hat{p}_1 - \hat{p}_2}{\mathrm{SE}_{Dp}}$$

where the **pooled standard error** is

$$\mathrm{SE}_{Dp} = \sqrt{\hat{p}(1 - \hat{p})\left(\frac{1}{n_1} + \frac{1}{n_2}\right)}$$

based on the **pooled estimate** of the common proportion of successes

$$\hat{p} = \frac{X_1 + X_2}{n_1 + n_2}$$

In terms of a standard Normal random variable Z, the P-value for a test of H_0 against

H_a: $p_1 > p_2$ is $P(Z \geq z)$

H_a: $p_1 < p_2$ is $P(Z \leq z)$

H_a: $p_1 \neq p_2$ is $2P(Z \geq |z|)$

Use this test when the number of successes and the number of failures in each of the samples are at least 5.

EXAMPLE 8.8   Men, Women, and Garment Labels

CASE 8.3

Example 8.7 (page 475) presents survey data on whether consumers are "label users" who pay attention to label details when buying a garment. Are men and women equally likely to be label users? Here is the data summary:

Population	n	X	$\hat{p} = X/n$
1 (women)	296	63	0.213
2 (men)	251	27	0.108

The sample proportions are certainly quite different, but we need a significance test to verify that the difference is too large to easily result from the play of chance in choosing the sample. Formally, we compare the proportions of label users in the two populations (women and men) by testing the hypotheses

$$H_0:\ p_1 = p_2$$
$$H_a:\ p_1 \neq p_2$$

The pooled estimate of the common value of p is

$$\hat{p} = \frac{63 + 27}{296 + 251} = \frac{90}{547} = 0.1645$$

This is just the proportion of label users in the entire sample.

First, we compute the standard error

$$SE_{Dp} = \sqrt{(0.1645)(0.8355)\left(\frac{1}{296} + \frac{1}{251}\right)} = 0.03181$$

and then we use this in the calculation of the test statistic

$$z = \frac{\hat{p}_1 - \hat{p}_2}{SE_{Dp}} = \frac{0.213 - 0.108}{0.03181} = 3.30$$

The difference in the sample proportions is more than 3 standard deviations away from zero. The P-value is $2P(Z \geq 3.30)$. From Table A we have $P = 2 \times 0.0005 = 0.0010$. Software gives $P = 0.0009$. We report: 21% of women are label users versus only 11% of men; the difference is statistically significant ($z = 3.30$, $P < 0.001$).

Figure 8.6 gives the Minitab and SAS outputs for Example 8.8. Carefully examine the output to find all the important pieces that you would need to report the results of the analysis and to draw a conclusion.

Many market researchers would expect the proportion of label users to be higher among women than among men. That is, we might choose the one-sided alternative $H_a: p_1 > p_2$. The P-value would be half of the value obtained for the two-sided test. Because the z statistic is so large, this distinction is of no practical importance.

APPLY YOUR KNOWLEDGE

8.61 Do men lie more often about their height than women? Refer to Exercise 8.56 (page 476) about lying and online dating profiles.
(a) State appropriate null and alternative hypotheses for this setting. Give a justification for your choice.
(b) Use the data given in Exercise 8.56 to perform a two-sided significance test. Give the test statistic and the P-value.
(c) Summarize the results of your significance test.

8.62 What about weight? Refer to Exercise 8.57 (page 476) for the data on lying about weight. Answer the questions given in the previous exercise for weight.

FIGURE 8.6 Minitab and SAS outputs for Example 8.8.

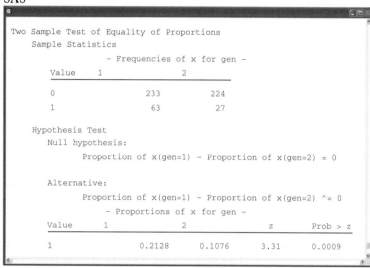

Minitab

```
Sample                          X    N      Sample p
1                               63   296    0.212838
2                               27   251    0.107570

Difference = p(1)- p(2)
Estimate for difference: 0.105268
Test for difference = 0(vs not = 0): z = 3.31 P-Value = 0.001
```

SAS

```
Two Sample Test of Equality of Proportions
     Sample Statistics
                      - Frequencies of x for gen -
          Value    1              2

          0                  233            224
          1                  63             27

  Hypothesis Test
    Null hypothesis:
          Proportion of x(gen=1) - Proportion of x(gen=2) = 0

    Alternative:
          Proportion of x(gen=1) - Proportion of x(gen=2) ^= 0
                   - Proportions of x for gen -
          Value    1              2              z        Prob > z

          1              0.2128        0.1076      3.31      0.0009
```

BEYOND THE BASICS: Relative Risk

relative risk

In Example 8.7 (page 475) we compared the proportions of women and men who are "label users" when they shop for clothing by giving a confidence interval for the *difference* of proportions. Alternatively, we might choose to make this comparison by giving the *ratio* of the two proportions. This ratio is often called the **relative risk** (RR). A relative risk of 1 means that the proportions $\hat{p}_1$ and $\hat{p}_2$ are equal. Confidence intervals for relative risk apply the principles that we have studied, but the details are somewhat complicated. Fortunately, we can leave the details to software and concentrate on interpreting and communicating the results.

EXAMPLE 8.9   Relative Risk for Use of Labels

CASE 8.3

The following table summarizes the data on the proportions of men and women who use labels when buying clothing:

Population	n	X	$\hat{p} = X/n$
1 (women)	296	63	0.2128
2 (men)	251	27	0.1076

The relative risk for this sample is

$$\text{RR} = \frac{\hat{p}_1}{\hat{p}_2} = \frac{0.2128}{0.1076} = 1.98$$

Confidence intervals for the relative risk in the entire population of shoppers are based on this sample relative risk. Software (for example, PROC FREQ with the MEASURES option in SAS) gives a 95% confidence interval as 1.30 to 3.01. Our summary: Women are about twice as likely as men to use labels; the 95% confidence interval is (1.30, 3.01).

In Example 8.9 the confidence interval is clearly not symmetric about the estimate: that is, 1.98 is not the midpoint of 1.30 and 3.01. This is true in general for confidence intervals for relative risk.

Relative risk—that is, comparing proportions by a ratio rather than by a difference—is particularly useful when the proportions are small. This is often the case in epidemiology and medical statistics. Here is a typical epidemiological example.

EXAMPLE 8.10   Smoking and Colorectal Cancer

Colorectal cancer is fourth in the list of types of cancers that lead to death. Many studies have examined the relationship between cigarette smoking and colorectal cancer but the results have been inconsistent. Twenty-six studies gave relative risk estimates for people who had ever smoked relative to those who had never smoked. A recent study combined the results of these studies to obtain a summary measure of relative risk.[20] The smokers are Population 1 and the nonsmokers are Population 2. The report of the study stated that the relative risk was 1.18 with a 95% confidence interval of 1.11 to 1.25. Since the confidence interval does not include the value of 1, which would correspond to equal risks in the two populations, we conclude that there is a higher risk of colorectal cancer for cigarette smokers. The estimated increase in risk is 18% with a 95% confidence interval of 11% to 25%.

SECTION 8.2 Summary

- The **estimate of the difference in two population proportions** is

$$D = \hat{p}_1 - \hat{p}_2$$

where

$$\hat{p}_1 = \frac{X_1}{n_1} \quad \text{and} \quad \hat{p}_2 = \frac{X_2}{n_2}$$

The **standard error of the difference** is

$$\text{SE}_D = \sqrt{\frac{\hat{p}_1(1 - \hat{p}_1)}{n_1} + \frac{\hat{p}_2(1 - \hat{p}_2)}{n_2}}$$

and the **margin of error for confidence level C** is

$$m = z^*\text{SE}_D$$

where z^* is the value for the standard Normal density curve with area C between $-z^*$ and z^*.

- The z **large-sample level C** confidence interval for the difference in two proportions $p_1 - p_2$ is

$$(\hat{p}_1 - \hat{p}_2) \pm m$$

We recommend using this method when the number of successes and the number of failures in both samples are at least 10.

- The **plus four confidence interval for comparing two proportions** is obtained by adding one success and one failure to each sample and then using the z procedure. We recommend using this method when both sample sizes are at least 5 and the confidence level is 90%, 95%, or 99%.

- Significance tests of H_0: $p_1 = p_2$ use the z **statistic**

$$z = \frac{\hat{p}_1 - \hat{p}_2}{SE_{Dp}}$$

with P-values from the $N(0, 1)$ distribution. In this statistic,

$$SE_{Dp} = \sqrt{\hat{p}(1 - \hat{p})\left(\frac{1}{n_1} + \frac{1}{n_2}\right)}$$

where $\hat{p}$ is the **pooled estimate** of the common value of p_1 and p_2,

$$\hat{p} = \frac{X_1 + X_2}{n_1 + n_2}$$

We recommend using this test when the number of successes and the number of failures in each of the samples are at least 5.

- **Relative risk** is the ratio of two sample proportions:

$$RR = \frac{\hat{p}_1}{\hat{p}_2}$$

Confidence intervals for relative risk are an alternative to confidence intervals for the difference when we want to compare two proportions.

SECTION 8.2 Exercises

For Exercises 8.53 to 8.55, see pages 473–474; for 8.56 and 8.57, see page 476; for 8.58 to 8.60, see page 477; and for 8.61 and 8.62, see page 479.

8.63 Podcast downloading. The Podcast Alley Web site recently reported that they have 53,501 podcasts available for downloading, with 3,447,545 episodes.[21] A Pew survey of Internet users described the results of two surveys about podcast downloading. The first was conducted between February and April 2006 and surveyed 2822 Internet users. They found that 198 of these said that they had downloaded a podcast to listen to it or view it later at least once. In a more recent survey, conducted in May 2008, there were 1553 Internet users. Of this total, 295 said that they had downloaded a podcast to listen to it or view it later.[22]

(a) Refer to the table that appears at the beginning of this section (page 472). Fill in the numerical values of all quantities that are known.

(b) Find the estimate of the difference between the proportion of Internet users who had downloaded podcasts as of February to April 2006 and the proportion as of May 2008.

(c) Is the large-sample confidence interval for the difference in two proportions appropriate to use in this setting? Explain your answer.

(d) Find the 95% confidence interval for the difference.

(e) Convert your estimated difference and confidence interval to percents.

(f) One of the surveys was conducted between February and April, whereas the other was conducted in May. Do you think that this difference should have any effect on the interpretation of the results? Be sure to explain your answer.

8.64 Significance test for podcast downloading. Refer to the previous exercise. Test the null hypothesis that the two proportions are equal. Report the test statistic with the P-value and summarize your conclusion.

8.65 Are more Internet users downloading podcasts? Refer to the previous two exercises. The ratio of the proportion in the 2008 sample to the proportion in the 2006 sample is about 2.7.
(a) Can you conclude that 2.7 times as many people are downloading podcasts? Explain why or why not.
(b) Can you conclude from the data available that there has been an increase from 2006 to 2008 in the number of people who download podcasts? If your answer is no, explain what additional data you would need or what additional assumptions you would have to make to be able to draw this conclusion.

8.66 Adult gamers versus teen gamers. A Pew Internet Project Data Memo presented data comparing adult gamers with teen gamers with respect to the devices on which they play. The data are from two surveys. The adult survey had 1063 gamers, and the teen survey had 1064 gamers. The memo reports that 54% of adult gamers played on game consoles (Xbox, PlayStation, Wii, etc.), and 89% of teen gamers played on game consoles.[23]
(a) Refer to the table that appears at the beginning of this section (page 472). Fill in the numerical values of all quantities that are known.
(b) Find the estimate of the difference between the proportion of teen gamers who played on game consoles and the proportion of adults who played on these devices.
(c) Is the large-sample confidence interval for the difference in two proportions appropriate to use in this setting? Explain your answer.
(d) Find the 95% confidence interval for the difference.
(e) Convert your estimated difference and confidence interval to percents.
(f) The adult survey was conducted between October and December 2008, whereas the teen survey was conducted between November 2007 and February 2008. Do you think that this difference should have any effect on the interpretation of the results? Be sure to explain your answer.

8.67 Significance test for gaming on consoles. Refer to the previous exercise. Test the null hypothesis that the two proportions are equal. Report the test statistic with the P-value and summarize your conclusion.

8.68 Gamers on computers. The report described in Exercise 8.66 also presented data from the same surveys for gaming on computers (desktops or laptops). These devices were used by 73% of adult gamers and by 76% of teen gamers. Answer the questions given in Exercise 8.66 for gaming on computers.

8.69 Significance test for gaming on computers. Refer to the previous exercise. Test the null hypothesis that the two proportions are equal. Report the test statistic with the P-value and summarize your conclusion.

8.70 Can we compare gaming on consoles with gaming on computers? Refer to the previous four exercises. Do you think that you can use the large-sample confidence intervals for a

difference in proportions to compare teens' use of computers with teens' use of consoles? Write a short paragraph giving the reason for your answer. (*Hint:* Look carefully in the box giving the assumptions needed for this procedure.)

8.71 Draw a picture. Suppose that there are two binomial populations. For the first, the true proportion of successes is 0.4; for the second, it is 0.5. Consider taking independent samples from these populations, 50 from the first and 60 from the second.
(a) Find the mean and the standard deviation of the distribution of $\hat{p}_1 - \hat{p}_2$.
(b) This distribution is approximately Normal. Sketch this Normal distribution and mark the location of the mean.
(c) Find a value d for which the probability is 0.95 that the difference in sample proportions is within $\pm d$. Mark these values on your sketch.

8.72 What's wrong? For each of the following, explain what is wrong and why.
(a) A z statistic is used to test the null hypothesis that $\hat{p}_1 = \hat{p}_2$.
(b) If two sample proportions are equal, then the sample counts are equal.
(c) A 95% confidence interval for the difference in two proportions includes errors due to nonresponse.

8.73 College student summer employment. Suppose (as is roughly true) that 85% of college men and 83% of college women were employed last summer. A sample survey interviews SRSs of 400 college men and 400 college women. The two samples are of course independent.
(a) What is the approximate distribution of the proportion $\hat{p}_F$ of women who worked last summer? What is the approximate distribution of the proportion $\hat{p}_M$ of men who worked?
(b) The survey wants to compare men and women. What is the approximate distribution of the difference in the proportions who worked, $\hat{p}_M - \hat{p}_F$?

8.74 A corporate liability trial. A major court case on liability for contamination of groundwater took place in the town of Woburn, Massachusetts. A town well in Woburn was contaminated by industrial chemicals. During the period that residents drank water from this well, there were 16 birth defects among 414 births. In years when the contaminated well was shut off and water was supplied from other wells, there were 3 birth defects among 228 births. The plaintiffs suing the firms responsible for the contamination claimed that these data show that the rate of birth defects was higher when the contaminated well was in use.[24] How statistically significant is the evidence? Be sure to state what assumptions your analysis requires and to what extent these assumptions seem reasonable in this case.

CASE 8.2 **8.75 Natural versus artificial Christmas trees.** In the Christmas tree survey introduced in Case 8.2 (page 466), respondents who had a tree during the holiday season were asked whether the tree was natural or artificial. Respondents were also asked if they lived in an urban area or in a rural area. Of the 421 households displaying a Christmas tree, 160 lived in rural areas and 261 were urban residents. The tree growers want to know if there is a difference in preference for natural trees versus artificial trees between urban and rural households. Here are the data:

Population	n	X(natural)
1 (rural)	160	64
2 (urban)	261	89

$P_1 \neq P_2$

(a) Give the null and alternative hypotheses that are appropriate for this problem assuming that we have no prior information suggesting that one population would have a higher preference than the other.

(b) Test the null hypothesis. Give the test statistic and the P-value, and summarize the results.

(c) Give a 90% confidence interval for the difference in proportions.

8.76 **Summer employment of college students.** A university financial aid office polled an SRS of undergraduate students to study their summer employment. Not all students were employed the previous summer. Here are the results for men and women:

	Men	Women
Employed	712	623
Not employed	68	92
Total	780	715

(a) Is there evidence that the proportion of male students employed during the summer differs from the proportion of female students who were employed? State H_0 and H_a, compute the test statistic, and give its P-value.

(b) Give a 95% confidence interval for the difference between the proportions of male and female students who were employed during the summer. Does the difference seem practically important to you?

8.77 **Gender bias in textbooks.** To what extent do textbooks on syntax (analysis of sentence structure) display gender bias? A study of this question sampled sentences from 10 texts.[25] One part of the study examined the use of the words "girl," "boy," "man," and "woman." Call the first two words *juvenile* and the last two *adult*. Is the proportion of female references that are juvenile ("girl" rather than "woman") equal to the proportion

of male references that are juvenile ("boy" rather than "man")? Here are data from one of the texts:

Gender	n	X (juvenile)
Female	60	48
Male	132	52

(a) Find the sample proportions of juvenile references for females and for males.

(b) Give a 95% confidence interval for the difference and briefly summarize what the data show.

8.78 **Is the gender bias statistically significant?** The previous exercise addresses a question about gender bias with a confidence interval. Set up the problem as a significance test. Carry out the test and summarize the results.

8.79 **Effect of the sample size.** Return to the study of undergraduate student summer employment described in Exercise 8.76. Similar results from a smaller number of students may not have the same statistical significance. Specifically, suppose that 71 of 78 men surveyed were employed and 62 of 71 women surveyed were employed. The sample proportions are essentially the same as in the earlier exercise.

(a) Compute the z statistic for these data and report the P-value. What do you conclude?

(b) Compare the results of this significance test with your results in Exercise 8.76. What do you observe about the effect of the sample size on the results of these significance tests?

8.80 **Relative risk for gamers.** Refer to the Pew data about gaming on game consoles (Xbox, PlayStation, Wii, etc.) by adults and teens in Exercises 8.66 and 8.67 (page 483). Now, compare the adults with the teens using the relative risk approach.

(a) Find the proportion of adult gamers who use game consoles. Do the same for the teen gamers.

(b) Find the relative risk using the teen proportion in the numerator.

(c) Repeat the computation of the relative risk using percents in place of proportions. Compare this calculation with the one that you performed in part (b) and explain what you have learned.

(d) Do you expect the 95% confidence interval for the relative risk to include the value 1? Explain why or why not.

(e) Find the 95% confidence interval if you have access to software that can do this calculation.

STATISTICS IN SUMMARY

Inference about population proportions is based on sample proportions. We rely on the fact that a sample proportion has a distribution that is close to Normal unless the sample is small. All the z procedures in this chapter work well when the samples are large enough. You must check this before using them. Here is a review list of the most important skills you should have acquired from your study of this chapter.

A. Recognition

1. Recognize from the design of a study whether one-sample or two-sample procedures are needed.

2. Recognize what parameter or parameters an inference problem concerns. In particular, distinguish between settings that require inference about a proportion and comparing two proportions.

3. Calculate from sample counts the sample proportion or proportions.

B. Inference about One Proportion

1. Use the z procedure to give a confidence interval for a population proportion p.

2. Use the z statistic to carry out a test of significance for the hypothesis H_0: $p = p_0$ about a population proportion p against either a one-sided or a two-sided alternative.

3. Check that you can safely use these z procedures in a particular setting.

C. Comparing Two Proportions

1. Use the two-sample z procedure to give a confidence interval for the difference $p_1 - p_2$ between proportions in two populations based on independent samples from the populations.

2. Use a z statistic to test the hypothesis H_0: $p_1 = p_2$ that proportions in two distinct populations are equal.

3. Check that you can safely use these z procedures in a particular setting.

Statistical inference always draws conclusions about one or more parameters of a population. When you think about doing inference, ask first what the population is and what parameter you are interested in. The t procedures of Chapter 7 allow us to give confidence intervals and carry out tests about population means. We use the z procedures of this chapter for inference about population proportions.

CHAPTER 8 Review Exercises

8.81 Changes in credit card usage by undergraduates. In Exercise 8.30 (page 470) we looked at data from a survey of 1430 undergraduate students and their credit card use. These students were surveyed 2004. In the sample, 43% said that they had four or more credit cards. A similar study performed in 2000 by the same organization reported that 32% of the sample said that they had four or more credit cards.[26] Assume that the sample sizes for the two studies are the same. Find a 95% confidence interval for the change from 2000 to 2004 in the percent of undergraduates who report having four or more credit cards.

8.82 Do the significance test for the change. Refer to the previous exercise. Perform the significance test for comparing the two proportions. Report your test statistic, the P-value, and summarize your conclusion.

8.83 We did not know the sample size. Refer to the previous two exercises. We did not report the sample size for the 2000 study, but it is reasonable to assume that it is fairly close to the sample size for the 2004 study.
(a) Suppose that the sample size for the 2000 study was only 1000. Redo the confidence interval and significance testing calculations for this scenario.

(b) Suppose that the sample size for the 2000 study was 2000. Redo the confidence interval and significance testing calculations for this scenario.
(c) Compare your results for parts (a) and (b) of this exercise with the results that you found in the previous two exercises. Write a short paragraph about the effects of assuming a value for the sample size on your conclusions.

8.84 Student employment during the school year. A study of 1430 undergraduate students reported that 994 work 10 or more hours a week during the school year. Give a 95% confidence interval for the proportion of all undergraduate students who work 10 or more hours a week during the school year.

8.85 Examine the effect of the sample size. Refer to the previous exercise. Assume a variety of different scenarios where the sample size changes but the proportion in the sample who work 10 or more hours a week during the school year remains the same. Write a short report summarizing your results and conclusions. Be sure to include numerical and graphical summaries of what you have found.

8.86 Video game genres. U.S. computer and video game software sales were $9.5 billion in 2007.[27] A survey of 1102 teens collected data about their video game use. The table below lists the most popular game genres.[28]

Genre	Examples	Percent who play
Racing	NASCAR, Mario Kart, Burnout	74
Puzzle	Bejeweled, Tetris, Solitaire	72
Sports	Madden, FIFA, Tony Hawk	68
Action	Grand Theft Auto, Devil May Cry, Ratchet and Clank	67
Adventure	Legend of Zelda, Tomb Raider	66
Rhythm	Guitar Hero, Dance Dance Revolution, Lumines	61

Give a 95% confidence interval for the proportion who play games in each of these six genres.

8.87 Too many errors. Refer to the previous exercise. The chance that each of the six intervals that you calculated includes the true proportion for that genre is approximately 95%. In other words, the chance that you make an error and your interval misses the true value is approximately 5%.

(a) Explain why the chance that at least one of your intervals does not contain the true value of the parameter is greater than 5%.

(b) One way to deal with this problem is to adjust the confidence level for each interval so that the overall probability of at least one miss is 5%. One simple way to do this is to use a **Bonferroni procedure.** Here is the basic idea: You have an error budget of 5% and you choose to spend it equally on six intervals. Each interval has a budget of $0.05/6 = 0.0083$. So each confidence interval should have a 0.83% chance of missing the true value. In other words, the confidence level for each interval should be $1 - 0.0083 = 0.9917$. Use Table A to find the value of z for a large-sample confidence interval for a single proportion corresponding to 99.17% confidence.

(c) Calculate the six confidence intervals using the Bonferroni procedure.

8.88 Wireless only. Are customers giving up their landlines and relying on wireless for all their phone needs? Surveys have collected data to answer this question.[29] In December 2003, 4.2% of households were wireless only. Assume that this survey is based on sampling 15,000 households.

(a) Convert the percent to a proportion. Then use the proportion and the sample size to find the count of households who were wireless only.

(b) Find a 95% confidence interval for the proportion of households that were wireless only in December 2003.

8.89 Change in wireless only. Refer to the previous exercise. The percent increased to 16.4% in December 2007. Assume the same sample size for this sample.

(a) Find the proportion and the count for this sample.

(b) Compute the 95% confidence interval for the proportion.

(c) Convert the estimate and confidence interval in terms of proportions to an estimate and confidence interval in terms of percents.

(d) Find the estimate of the difference in the proportions

of households that were wireless only in December 2007 and the households that were wireless only in December 2003.

(e) Give the margin of error for 95% confidence for the difference in proportions.

8.90 Analyze the change in terms of relative risk. Refer to the previous two exercises.

(a) Summarize the change data in terms of relative risk. The term "relative risk" is a poor description of the ratio that you are using for this exercise. Give a better term for this ratio.

(b) Analyze the data in terms of relative risk and write a summary of your results.

(c) Compare your results in part (b) with your findings in terms of a difference in proportions from the previous exercise.

(d) Which approach do you prefer? Give reasons for your answer.

8.91 Gambling and student-athletes. Gambling behaviors of Division I intercollegiate male student-athletes were analyzed in Exercise 8.22 (page 469). Similar data for women were given in Exercise 8.23. Compare the males and females with a significance test and give an estimate of the difference in proportions of student-athletes who participate in any gambling activity with a 95% margin of error. We noted in Exercise 8.22 that we do not have any information available to assess nonresponse. Consider the possibility that the response rates differ by gender and by whether or not the person participates in any gambling activity. Write a short summary of how these differences might affect inference on these issues.

8.92 Effects of reducing air pollution. A study that evaluated the effects of a reduction in exposure to traffic-related air pollutants compared respiratory symptoms of 283 residents of an area with congested streets with 165 residents in a similar area where the congestion was removed because a bypass was constructed. The symptoms of the residents of both areas were evaluated at baseline and again a year after the bypass was completed.[30] For the residents of the congested streets, 17 reported that their symptoms of wheezing improved between baseline and one year later, while 35 of the residents of the bypass streets reported improvement.

(a) Find the two sample proportions.

(b) Report the difference in the proportions and the standard error of the difference.

(c) What are the appropriate null and alternative hypotheses for examining the question of interest? Be sure to explain your choice of the alternative hypothesis.

(d) Find the test statistic. Construct a sketch of the distribution of the test statistic under the assumption that the null hypothesis is true. Find the P-value and use your sketch to explain its meaning.

(e) Is no evidence of an effect the same as evidence that there is no effect? Use a 95% confidence interval to answer this question. Summarize your ideas in a way that could be understood by someone who has very little experience with statistics.

(f) The study was done in the United Kingdom. To what extent do you think that the results can be generalized to other circumstances?

8.93 Downloading music from the Internet. The following quote is from a survey of Internet users.[31] The sample size for the survey was 1371. Since 18% of those surveyed said they download music, the sample size for this subsample is 247.

Among current music downloaders, 38% say they are downloading less because of the RIAA suits. . . . About a third of current music downloaders say they use peer-to-peer networks. . . . 24% of them say they swap files using email and instant messaging; 20% download files from music-related Web sites like those run by music magazines or musician homepages. And while online music services like iTunes are far from trumping the popularity of file-sharing networks, 17% of current music downloaders say they are using these paid services. Overall, 7% of Internet users say they have bought music at these new services at one time or another, including 3% who currently use paid services.

(a) For each percent quoted, give the 95% margin of error. You should express these in percents, as given in the quote.
(b) Rewrite the paragraph in a shorter form but include the margins of error.
(c) Pick either side (A) or side (B) and give arguments in favor of the view that you select. (A) The margins of error should be included because they are necessary for the reader to properly interpret the results. (B) The margins of error interfere with the flow of the important ideas. It would be better to just report one margin of error and say that all of the others are no greater than this number. If you choose view (B), be sure to give the value of the margin of error that you report.

8.94 Other effects of reducing air pollution. In Exercise 8.92 the effects of a reduction in air pollution on wheezing was examined by comparing the one-year change in symptoms in a group of residents who lived on congested streets with a group who lived in an area that had been congested but from which the congestion was removed when a bypass was built. The effect of the reduction in air pollution was assessed by comparing the proportions of residents in the two groups who reported that their wheezing symptoms improved. Here are some additional data from the same study:

Symptom	Bypass n	Bypass Improved	Congested n	Congested Improved
Number of wheezing attacks	282	45	163	21
Wheezing disturbs sleep	282	45	164	12
Wheezing limits speech	282	12	164	4
Wheezing affects activities	281	26	165	13
Winter cough	261	15	156	14
Winter phlegm	253	12	144	10
Consulted doctor	247	29	140	18

The table gives the number of subjects in each group and the number reporting improvement. So, for example, the proportion who reported improvement in the number of wheezing attacks was 21/163 in the congested group.
(a) The reported sample sizes vary from symptom to symptom. Give possible reasons for this and discuss the possible impact on the results.
(b) Calculate the difference in the proportions for each symptom. Make a table of symptoms ordered from highest to lowest based on these differences. Include the estimates of the differences and the 95% confidence intervals in the table. Summarize your conclusions.
(c) Can you justify a one-sided alternative in this situation? Give reasons for your answer.
(d) Perform a significance test to compare the two groups for each of the symptoms. Summarize the results.
(e) Reanalyze the data using only the data from the bypass group. Give confidence intervals for the proportions that reported improved symptoms. Compare the conclusions that someone might make from these results with those you presented in part (b). Use your analyses of the data in this exercise to discuss the importance of a control group in studies such as this.

8.95 The parrot effect: how to increase your tips. An experiment examined the relationship between tips and server behavior in a restaurant.[32] In one condition, the server repeated the customer's order word for word, while in the other condition, the orders were not repeated. Tips were received in 47 of the 60 trials under the repeat condition and in 31 of the 60 trials under the no-repeat condition.
(a) Find the sample proportions and compute a 95% confidence interval for the difference in population proportions.
(b) Use a significance test to compare the two conditions. Summarize the results.
(c) The study was performed in a restaurant in the Netherlands. Two waitresses performed the tasks. How do these facts relate to the type of conclusions that can be drawn from this study? Do you think that the parrot effect would apply in other countries?
(d) Design a study to test the parrot effect in a setting that is familiar to you. Be sure to include complete details about how the study will be conducted and how you will analyze the results.

8.96 Brand loyalty and the Chicago Cubs. According to literature on brand loyalty, consumers who are loyal to a brand are likely to consistently select the same product. This type of consistency may come from a positive childhood association. To examine brand loyalty among fans of the Chicago Cubs, 371 Cubs fans among patrons of a restaurant located in Wrigleyville were surveyed before a game at Wrigley Field, the Cubs home field.[33] The respondents were classified as "die-hard fans" or "less loyal fans." Of the 134 die-hard fans, 90.3% reported that they had watched or listened to Cubs games when they were children. Among the 237 less loyal fans, 67.9% said that they had watched or listened as children.

(a) Find the numbers of die-hard Cubs fans who watched or listened to games when they were children. Do the same for the less loyal fans.

(b) Use a significance test to compare the die-hard fans with the less loyal fans with respect to their childhood experiences of the team.

(c) Express the results with a 95% confidence interval for the difference in proportions.

8.97 Brand loyalty in action. The study mentioned in the previous exercise found that two-thirds of the die-hard fans attended Cubs games at least once a month, but only 20% of the less loyal fans attended this often. Analyze these data using a significance test and a confidence interval. Write a short summary of your findings.

8.98 Frequent lottery players. A study of state lotteries included a random digit dialing (RDD) survey conducted by the National Opinion Research Center (NORC). The survey asked 2406 adults about their lottery spending.[34] A total of 248 individuals were classified as "heavy" players. Of these, 152 were male. The study notes that 48.5% of U.S. adults are male. Use a significance test to compare the proportion of males among heavy lottery players with the proportion of males in the U.S. adult population and write a short summary of your results. For this analysis, assume that the 248 heavy lottery players are a random sample of all heavy lottery players and that the margin of error for the 48.5% estimate of the percent of males in the U.S. adult population is so small that it can be neglected.

8.99 Use a confidence interval. Use a confidence interval to give an alternative analysis for the previous exercise.

8.100 Time to repair golf clubs. The Ping Company makes custom-built golf clubs and competes in the $4 billion golf equipment industry. To improve its business processes, Ping decided to seek ISO 9001 certification.[35] As part of this process, a study of the time it took to repair golf clubs that were sent to the company by mail determined that 16% of orders were sent back to the customers in 5 days or less. Ping examined the processing of repair orders and made changes. Following the changes, 90% of orders were completed within 5 days. Assume that each of the estimated percents is based on a random sample of 200 orders.

(a) How many orders were completed in 5 days or less before the changes? Give a 95% confidence interval for the proportion of orders completed in this time.

(b) Do the same for orders after the changes.

(c) Give a 95% confidence interval for the improvement. Express this both for a difference in proportions and for a difference in percents.

8.101 Does the new process give a better product? Eleven percent of the products produced by an industrial process over the past several months fail to conform to the specifications. The company modifies the process in an attempt to reduce the rate of nonconformities. In a trial run, the modified process produces 16 nonconforming items out of a total of 300 produced. Do these results demonstrate that the modification is effective? Support

your conclusion with a clear statement of your assumptions and the results of your statistical calculations.

8.102 How much is the improvement? In the setting of the previous exercise, give a 95% confidence interval for the proportion of nonconforming items for the modified process. Then, taking $p_0 = 0.11$ to be the old proportion and p the proportion for the modified process, give a 95% confidence interval for $p - p_0$.

8.103 Choosing sample sizes. For a single proportion the margin of error of a confidence interval is largest for any given sample size n and confidence level C when $\hat{p} = 0.5$. This led us to use $p^* = 0.5$ for planning purposes. A similar result is true for the two-sample problem. The margin of error of the confidence interval for the difference between two proportions is largest when $\hat{p}_1 = \hat{p}_2 = 0.5$. Use these conservative values in the following calculations, and assume that the sample sizes n_1 and n_2 have the common value n. Calculate the margins of error of the 95% confidence intervals for the difference in two proportions for the following choices of n: 10, 25, 50, 100, 150, 200, 400, and 500. Present the results in a table and with a graph. Summarize your conclusions.

8.104 Choosing sample sizes, continued. As the previous exercise noted, using the guessed value 0.5 for both $\hat{p}_1$ and $\hat{p}_2$ gives a conservative margin of error in confidence intervals for the difference between two population proportions. You are planning a survey and will calculate a 95% confidence interval for the difference in two proportions when the data are collected. You would like the margin of error of the interval to be less than or equal to 0.04. You will use the same sample size n for both populations.

(a) How large a value of n is needed?

(b) Give a general formula for n in terms of the desired margin of error m and the critical value z^*.

8.105 Unequal sample sizes. You are planning a survey in which a 95% confidence interval for the difference between two proportions will present the results. You will use the conservative guessed value 0.5 for $\hat{p}_1$ and $\hat{p}_2$ in your planning. You would like the margin of error of the confidence interval to be less than or equal to 0.15. It is very difficult to sample from the first population, so that it will be impossible for you to obtain more than 25 observations from this population. Taking $n_1 = 25$, can you find a value of n_2 that will guarantee the desired margin of error? If so, report the value; if not, explain why not.

8.106 Students change their majors. In a random sample of 950 students from a large public university, it was found that 444 of the students changed majors during their college years.

(a) Give a 99% confidence interval for the proportion of students at this university who change majors.

(b) Express your results from (a) in terms of the *percent* of students who change majors.

(c) University officials are more interested in the *number* of students who change majors than in the proportion. The university has 30,000 undergraduate students. Convert your confidence interval in (a) to a confidence interval for the number of students who change majors during their college years.

8.107 Statistics and the law. *Casteneda v. Partida* is an important court case in which statistical methods were used as part of a legal argument. When reviewing this case, the Supreme Court used the phrase "two or three standard deviations" as a criterion for statistical significance. This Supreme Court review has served as the basis for many subsequent applications of statistical methods in legal settings. (The two or three standard deviations referred to by the Court are values of the z statistic and correspond to P-values of approximately 0.05 and 0.0026.) In *Casteneda* the plaintiffs alleged that the method for selecting juries in a county in Texas was biased against Mexican Americans.[36] For the period of time at issue, there were 181,535 persons eligible for jury duty, of whom 143,611 were Mexican Americans. Of the 870 people selected for jury duty, 339 were Mexican Americans. (a) What proportion of eligible jurors were Mexican Americans? Let this value be p_0.

(b) Let p be the probability that a randomly selected juror is a Mexican American. The null hypothesis to be tested is $H_0: p = p_0$. Find the value of $\hat{p}$ for this problem, compute the z statistic, and find the P-value. What do you conclude? (A finding of statistical significance in this circumstance does not constitute proof of discrimination. It can be used, however, to establish a prima facie case. The burden of proof then shifts to the defense.)

(c) We can reformulate this exercise as a two-sample problem. Here we wish to compare the proportion of Mexican Americans among those selected as jurors with the proportion of Mexican Americans among those not selected as jurors. Let p_1 be the probability that a randomly selected juror is a Mexican American, and let p_2 be the probability that a randomly selected nonjuror is a Mexican American. Find the z statistic and its P-value. How do your answers compare with your results in (b)?

CHAPTER 8 Case Study Exercises

CASE STUDY EXERCISE 1: Gender bias in textbooks. Exercise 8.77 (page 484) reports a study of gender bias in 10 syntax textbooks. Here are the counts of "girl," "woman," "boy," and "man" for all the texts. The data in Exercise 8.77 are for text number 6.

	\multicolumn{10}{c}{Text Number}									
	1	2	3	4	5	6	7	8	9	10
Girl	2	5	25	11	2	48	38	5	48	13
Woman	3	2	31	65	1	12	2	13	24	5
Boy	7	18	14	19	12	52	70	6	128	32
Man	27	45	51	138	31	80	2	27	48	95

Analyze the data and write a report summarizing your conclusions. The researchers who conducted the study note that the authors of texts 8, 9, and 10 are women, while the other seven were written by men. Do you see any pattern that suggests that the gender of the author is associated with the results?

CASE STUDY EXERCISE 2: Sample size, P-value, and the margin of error. In this Case Study we examine the effects of the sample size on the significance test and the confidence interval for comparing two proportions. For each calculation, suppose that $\hat{p}_1 = 0.75$ and $\hat{p}_2 = 0.5$, and take n to be the common value of n_1 and n_2. Use the z statistic to test $H_0: p_1 = p_2$ versus the alternative $H_a: p_1 \neq p_2$. Compute the statistic and the associated P-value for the following values of n: 12, 20, 40, 80, 100, 200, and 500. Summarize the results in a table and make a plot. Explain what you observe about the effect of the sample size on statistical significance when the sample proportions $\hat{p}_1$ and $\hat{p}_2$ are unchanged.

Now we will do similar calculations for the confidence interval. Here, we suppose that $\hat{p}_1 = 0.75$ and $\hat{p}_2 = 0.5$. Compute the margin of error for the 95% confidence interval for the difference in the two proportions for $n = 12, 20, 40, 80, 100, 200,$ and 500. Summarize and explain your results.

CHAPTER 8 Appendix

Using Minitab and Excel for Inference for Proportions

Confidence Interval for a Single Proportion

Minitab:

Stat ► Basic Statistics ► 1 Proportion

If the data are in the worksheet, then choose the **Samples in columns** option and click the column containing the data into the box below. With this option, you can construct confidence intervals for more than one data set simultaneously. With respect to the nature of the data, data entries can be any two distinct values (numeric or text) where one value represents "success" and the other represents "failure." Alternatively, if you know the number of successes, you can choose the **Summarized data** option and then input the number of successes in the **Number of events** box and input the sample size in the **Number of trials** box. Now it is important that you click the **Options** button and select the **Use test and interval based on normal distribution** option. You can also input your desired confidence level in the **Confidence level** box if other than 95%. Click **OK** to close the pop-up box and then click **OK** to find the sample proportion and confidence interval reported in the Session window.

Excel:

Confidence intervals for the proportion are not available in standard Excel but they are available in the WHFStat Add-In for Excel.

Test for a Single Proportion

Minitab:

$$Stat \blacktriangleright Basic\ Statistics \blacktriangleright 1\ Proportion$$

This is the same routine described in this Appendix for obtaining the confidence interval for the proportion. If you wish to conduct a hypothesis test, select the **Perform hypothesis test** option and input the null hypothesis mean value (p_0) in the box below. Now click the **Options** button and select the **Use test and interval based on normal distribution** option. With this pop-up box, you can also select your alternative hypothesis from the **Alternative** menu box. Click **OK** to close the pop-up box and then click **OK** to find the test statistic and corresponding P-value reported in the Session window.

Excel:

Testing for the proportion is not available in standard Excel but it is available in the WHFStat Add-In for Excel.

Confidence Interval for Comparing Two Proportions

Minitab:

$$Stat \blacktriangleright Basic\ Statistics \blacktriangleright 2\ Proportions$$

If the data for the two samples are in two separate columns, choose the **Samples in different columns** option and then click the columns containing the data into the two boxes below. As with the single proportion, data entries can be any two distinct values (numeric or text) where one value represents "success" and the other represents "failure." Minitab will allow you to store the data for the two samples all in one column with no requirement in terms of placement order of the data. When this is done, a second column in the worksheet is required in which there are two distinct labels (numerical or text) indicating for a given row whether the corresponding data observation comes from the first or second sample. If the data are stored in this manner, then choose the **Samples in one column** option and click the data column into the **Samples** box and click the column of labels into the **Subscripts** box. As a third option, if you know the number of successes in each of the samples, you can choose the **Summarized data** option and input the number of successes in the **Number of events** box and input the sample size in the **Number of trials** box for each of the samples. If you wish to change the level of confidence from the default value of 95%, click the **Options** button and input your desired confidence level in the **Confidence level** box. Click **OK** to close the pop-up box and then click **OK** to find the sample proportions and confidence interval for the difference between population proportions reported in the Session window.

Excel:

Confidence intervals for comparing two proportions are not available in standard Excel but they are available in the WHFStat Add-In for Excel.

Test for Comparing Two Proportions

Minitab:

$$Stat \blacktriangleright Basic\ Statistics \blacktriangleright 2\ Proportions$$

This is the same routine described in this Appendix for obtaining the confidence interval for the difference between two proportions. To have the test based on a pooled standard error as described in the chapter, click the **Options** button and select the **Use pooled estimate of p for test** option. With this pop-up box, you can also select your alternative hypothesis from the **Alternative** menu box. Click **OK** to close the pop-up box and then click **OK** to find the test statistic and corresponding P-value reported in the Session window.

Excel:

Testing for two proportions is not available in standard Excel but it is available in the WHFStat Add-In for Excel.

À John Pyle/Cal Sport Media

PART III

Topics in Inference

We use samples to infer characteristics of populations. In business applications, the sample is often not a simple random sample from a population and we must decide how well it represents a population of interest. What population do you think a sample of Ohio State football fans represents?

Inference for Two-Way Tables

9.1 Analysis of Two-Way Tables

When we compared two proportions in Section 8.2, we started by summarizing the raw data by giving the number of observations in each group (n) and how many of these were classified as "successes" (X).

EXAMPLE 9.1   No Sweat Labels

Case 8.3 (page 474) concerned the response of men and women to "No Sweat" labels on a garment. These labels supposedly guarantee fair working conditions for the workers who made the garment. Here is the data summary. The table gives the number n of subjects in each gender. The count X is the number who were classified as No Sweat label users.

Population	n	X
1 (women)	296	63
2 (men)	251	27
Total	547	90

To compare men and women, we calculated sample proportions from these counts.

Two-way tables

In this chapter we start with a different summary of the same data. Rather than recording just the count of label users, we record counts for all outcomes in a **two-way table.**

Garment labels sometimes show that the workers who produced the garments worked under fair conditions. Do these labels influence purchase decisions? Are the influences different for men and women? Examples 9.1 and 9.2 provide data that can be used to answer these questions.

CHAPTER OUTLINE

two-way table

EXAMPLE 9.2 No Sweat Labels

Here is the two-way table classifying customers by gender and whether or not they are label users:

Label user	Gender		
	Women	Men	Total
Yes	63	27	90
No	233	224	457
Total	296	251	547

Check that this table simply rearranges the information in Example 9.1. Because we are interested in how gender influences label use, we view gender as an explanatory variable and label use as a response variable. This is why we put gender in the columns (like the x axis in a regression) and label use in the rows (like the y axis in a regression). Be sure that you understand how this table is obtained from the table in Example 9.1. Most errors in the use of categorical-data methods come from a misunderstanding of how these tables are constructed.

We call this particular two-way table a 2×2 table because there are two rows (Yes and No for label use) and two columns (Women and Men). The advantage of two-way tables is that they can present data for variables having more than two categories by simply increasing the number of rows or columns. Suppose, for example, that we recorded the influence of No Sweat labels on a consumer using a 5-point scale rather than simply "Yes" and "No." The response variable would then have 5 levels, so our table would be 5×2, with 5 rows and 2 columns.

In this section we advance from describing data to inference in the setting of two-way tables. Our data are counts of observations, classified according to two categorical variables. The question of interest is whether there is a relation between the row variable and the column variable. For example, is there a relation between gender and label use? In Example 8.8 (page 478) we found that there was a statistically significant difference in the proportions of women and men label users: 21.3% for women versus 10.8% for men. We now think about this problem from a slightly different point of view: is there a relationship between gender and the use of No Sweat labels? The 2×2 table is the starting point for examining relationships in two-way tables.

We introduce inference for two-way tables with data that form a 2×2 table. The methodology applies to tables in general and we will discuss a 3×3 table (two categorical variables, each with three categories) in the next section.

CASE 9.1

FRANCHISES

Exclusive Territories and the Success of New Franchise Chains Many popular businesses are franchises—think of McDonald's. The owner of a local franchise benefits from the brand recognition, national advertising, and detailed guidelines provided by the franchise chain. In return, he or she pays fees to the franchise firm and agrees to follow its policies. The relationship between the local entrepreneur and the franchise firm is spelled out in a detailed contract.

One clause that the contract may or may not contain is the entrepreneur's right to an exclusive territory. This means that the new outlet will be the only representative of the franchise in a specified territory and will not have to compete with other outlets of the same chain. How does the presence of an exclusive-territory clause in the contract relate to the survival of the business? A study designed to address this question collected data from a sample of 170 new franchise firms.[1]

To start our analysis of the relationship between exclusive territories and success we will organize the data in a two-way table. The following example gives the details.

EXAMPLE 9.3   The Two-Way Table

Two categorical variables were measured for each firm. First, the firm was classified as successful or not based on whether or not it was still franchising as of a certain date. Second, the contract of each firm was classified according to whether or not there was an exclusive-territory clause. Here is the 2×2 table with the marginal totals:

Observed numbers of firms

Success	Exclusive Territory		Total
	Yes	No	
Yes	108	15	123
No	34	13	47
Total	142	28	170

The entries in the two-way table in Example 9.3 are the observed, or sample, counts of the numbers of firms in each category. For example, there were 108 successful firms with an exclusive-territory contract and 15 successful firms without one. The table includes the marginal totals, calculated by summing over the rows or columns. The grand total, 170, is the sum of the row totals and is also the sum of the column totals. It is the total number of firms in the study.

The rows and columns of a two-way table represent values of two categorical variables. These are called "Success" and "Exclusive Territory" in Example 9.3. We are interested in the influence of exclusive territories on success, so Exclusive Territory is the explanatory variable (the column variable) and Success is the response (row) variable. Each combination of values for these two variables defines a **cell.** A two-way table with r rows and c columns contains $r \times c$ cells. The 2×2 table in Example 9.3 has 4 cells. In this study, we have data on two variables for a single sample of 170 firms. The same table might also have arisen from two separate samples, one from successful firms and the other from unsuccessful firms. Fortunately, the same inference applies in both cases. When we studied relationships between quantitative variables in Chapter 2, we noted that not all relationships involve an explanatory variable and a response variable. The same is true for categorical variables that we study here. Two-way tables can be used to display the relationship between any two categorical variables.

cell

APPLY YOUR KNOWLEDGE

9.1 Lying and online dating profiles.   In Exercise 8.56 (page 476) we analyzed data from a study where it was reported that 22 of 40 men lied about their height in their online profile for a dating Web site. For women, the numbers were 17 out of 40.
(a) For these data do you want to consider one of these categorical variables as an explanatory variable and the other as a response variable? Give a reason for your answer.
(b) Display these data using an $r \times c$ table. What are the values of r and c? Which variable do you choose to be the column variable and which is the row variable?
(c) How many cells will that table have?
(d) Give the table with the marginal totals.

9.2 A reduction in force.   A human resources manager wants to assess the impact of a planned reduction in force (RIF) on employees over the age of 40. (Various laws state that discrimination against this group is illegal.) The company has 800 employees over 40 and 575 who are 40 years of age or less. The current plan for the RIF will terminate 120 employees: 85 who are over 40, and 35 who are 40 or less. Display these data in a two-way table. (Be careful. Remember that each employee should be counted in exactly one cell.)

Describing relations in two-way tables

Section 2.5 (page 123) discusses methods for describing relationships in two-way tables. You should review that material now if you have not already studied it.

Analysis of two-way tables in practice uses statistical software to carry out the considerable arithmetic required. We will use output from some typical software packages for the data of Case 9.1 to describe inference for two-way tables. In Section 9.2 we will give more detail, so that you can do the work with a calculator if software is not available.

joint distribution
conditional distribution

To describe relations between categorical variables, we compute and compare percents. The count in each cell can be viewed as a percent of the grand total, of the row total, or of the column total. In the first case, we are describing the **joint distribution** of the two variables; in the other two cases, we are examining the **conditional distributions.** We learned many of the ideas related to conditional distributions when we studied conditional probability in Section 5.2 (page 295). When analyzing data, you should use the context of the problem to decide which percents are most appropriate. Software usually prints out all three, but not all are of interest in a specific problem.

EXAMPLE 9.4   Software Output

CASE 9.1

Figure 9.1 shows the output from Minitab, SPSS, and SAS for the data of Case 9.1. We named the variables SUCCESS and EXCL. The two-way table appears in the outputs in expanded form. Each cell contains four entries. They appear in different orders or with different labels, but all three outputs contain the same information. The frequency or count is the first entry in all three outputs. The row and column totals appear in the margins, just as in Example 9.3. The cell count as a percent of the row total is variously labeled as "% of Row" or "% within SUCCESS" or "Row Pct." For the first cell, this is 108/123, or 87.8%. The cell count as a percent of the column total is also given. You can request yet another entry, cell count divided by the total number of observations (the joint distribution). This is often not very useful and tends to clutter up the output.

In Case 9.1 we are interested in comparing the two types of firms, those with exclusive territories and those without exclusive territories. The columns of the table contain the information about each firm. Therefore, we examine the column percents to compare the success of the two types of firms. Here they are, rounded from the output for clarity:

Column percents for firms

Success	Exclusive Territory	
	Yes	No
Yes	76%	54%
No	24%	46%
Total	100%	100%

FIGURE 9.1 Minitab, SPSS, and SAS output, for Example 9.4.

Minitab

```
Tabulated statistics: success, excl

Rows:  success  Columns:   excl

             1         2       All

1          108        15       123
         87.80     12.20    100.00
         76.06     53.57     72.35
        102.74     20.26    123.00

2           34        13        47
         72.34     27.66    100.00
         23.94     46.43     27.65
         39.26      7.74     47.00

All        142        28       170
         83.53     16.47    100.00
        100.00    100.00    100.00
        142.00     28.00    170.00


Cell Contents:      Count
                    % of Row
                    % of Column
                    Expected count


Pearson Chi-Square = 5.911, DF = 1, P-Value = 0.015
```

SPSS

```
SUCCESS *EXCL Crosstabulation

                                    EXCL                  Total

                                  1         2

SUCCESS   1Count                108        15        123
            Expected Count     102.7      20.3      123.0
            % within SUCCESS   87.8%     12.2%     100.0%
            % within EXCL      76.1%     53.6%      72.4%
          2Count                 34        13         47
            Expected Count      39.3       7.7       47.0
            % within SUCCESS   72.3%     27.7%     100.0%
            % within EXCL      23.9%     46.4%      27.6%
  Total   Count                 142        28        170
            Expected Count     142.0      28.0      170.0
            % within SUCCESS   83.5%     16.5%     100.0%
            % within EXCL     100.0%    100.0%     100.0%

Chi-Square Tests

                     Value    df    Asymp. Sig. (2-sided)

Pearson Chi-Square   5.911     1                    .015
 N of Valid Cases      170
```

FIGURE 9.1 (*Continued*)

SAS

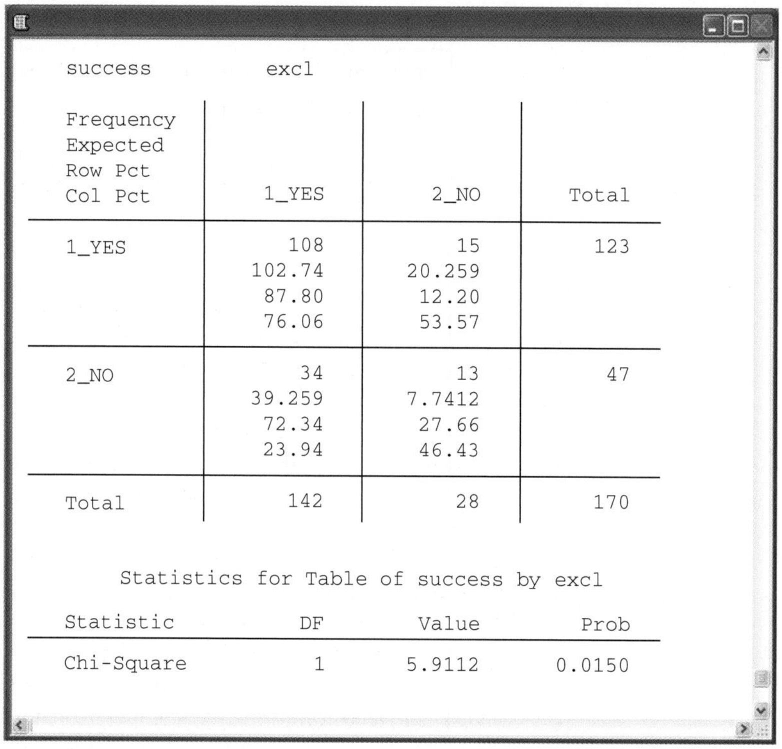

```
success          excl

Frequency
Expected
Row Pct
Col Pct           1_YES         2_NO        Total

1_YES              108            15          123
                102.74        20.259
                 87.80         12.20
                 76.06         53.57

2_NO                34            13           47
                39.259        7.7412
                 72.34         27.66
                 23.94         46.43

Total              142            28          170


        Statistics for Table of success by excl

Statistic          DF         Value          Prob

Chi-Square          1         5.9112        0.0150
```

The "Total" row reminds us that 100% of each type of firm have been classified as successful or not. (The sums sometimes differ slightly from 100% because of roundoff error.) The bar graph in Figure 9.2 compares the percents. The data reveal a clear relationship: 76% of firms with exclusive territories are successful, as opposed to only 54% of firms that did not have an exclusive-territory contract.

FIGURE 9.2 Bar graph of the percents of successful firms, for Example 9.4.

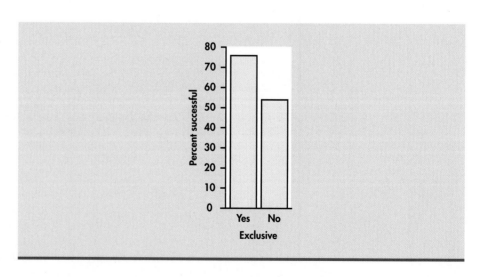

APPLY YOUR KNOWLEDGE

9.3 Reading software output. Look at Figure 9.1. What percent of successful firms had an exclusive-territory contract? What percent of unsuccessful firms had an exclusive-territory contract?

9.4 Reading software output. Look at Figure 9.1. What percent of firms are successful? What percent of firms had an exclusive-territory contract?

The hypothesis: no association

The difference between the percents of successes among the two types of firms is quite large. A statistical test will tell us whether or not these differences can be plausibly attributed to chance. Specifically, if there is no association between success and having an exclusive territory, how likely is it that a sample would show differences as large or larger than those displayed in Figure 9.2?

The null hypothesis H_0 of interest in a two-way table is: there is *no association* between the row variable and the column variable. For Case 9.1, this null hypothesis says that success and having an exclusive territory are not related. The alternative hypothesis H_a is that there is an association between these two variables. The alternative H_a does not specify any particular direction for the association. For $r \times c$ tables in general, the alternative includes many different possibilities. Because it includes all the many kinds of association that are possible, we cannot describe H_a as either one-sided or two-sided.

In our example, the hypothesis H_0 that there is no association between success and having an exclusive territory is equivalent to the statement that the distributions of the success variable are the same among firms with and without exclusive territories. For $r \times c$ tables like that in Example 9.3, there are c distributions for the row variable, one for each population. The null hypothesis then says that the c distributions of the row variable are identical. The alternative hypothesis is that the distributions are not all the same.

Expected cell counts

expected cell counts

To test the null hypothesis in $r \times c$ tables, we compare the observed cell counts with **expected cell counts** calculated under the assumption that the null hypothesis is true. Our test statistic is a numerical measure of the distance between the observed and expected cell counts.

EXAMPLE 9.5   Expected Counts from Software

CASE 9.1

The expected counts for the successful-firms example appear in the computer outputs shown in Figure 9.1. For example, the expected count for the first cell is 102.74.

How is this expected count obtained? Look at the percents in the right margin of the table in Figure 9.1. We see that 72.35% of all firms are successful. If the null hypothesis of no relation between success and exclusive territories is true, we expect this overall percent to apply to both firms with and firms without exclusive territories. In particular, we expect 72.35% of the firms with exclusive territories to be successful. Since there are 142 firms with exclusive territories, the expected count is 72.35% of 142, or 102.74. The other expected counts are calculated in the same way.

The reasoning of Example 9.5 leads to a simple formula for calculating expected cell counts. To compute the expected count for successful firms with exclusive territories, we multiplied the proportion of successful firms (123/170) by the number of firms with exclusive territories (142). From Figure 9.1 we see that the numbers 123 and 142 are the row and column totals for the cell of interest and that 170 is n, the total number of observations for the table. The expected cell count is therefore the product of the row and column totals divided by the table total.

> **Expected Cell Counts**
>
> The **expected count** in any cell of a two-way table when the null hypothesis of no association is true is
>
> $$\text{expected count} = \frac{\text{row total} \times \text{column total}}{n}$$

APPLY YOUR KNOWLEDGE

9.5 Expected counts. We want to calculate the expected count of unsuccessful firms that have an exclusive territory. From Figure 9.1, how many firms have exclusive territories? What percent of all firms are unsuccessful? Explain in words why, if there is no association between success and exclusive territories, the expected count we want is the product of these two numbers. Verify that the formula gives the same answer.

9.6 An alternative view. Refer to Figure 9.1. Verify that you can obtain the expected count for the first cell by multiplying the number of successful firms by the percent of firms that have exclusive territories. Explain your calculations in words.

The chi-square test

To test the H_0 that there is no association between the row and column classifications, we use a statistic that compares the entire set of observed counts with the set of expected counts. First, take the difference between each observed count and its corresponding expected count, and then square these values so that they are all 0 or positive. A large difference means less if it comes from a cell that we think will have a large count, so divide each squared difference by the expected count, a kind of standardization. Finally, sum over all cells. The result is called the *chi-square statistic* X^2.*

> **Chi-Square Statistic**
>
> The **chi-square statistic** is a measure of how much the observed cell counts in a two-way table diverge from the expected cell counts. The recipe for the statistic is
>
> $$X^2 = \sum \frac{(\text{observed count} - \text{expected count})^2}{\text{expected count}}$$
>
> where "observed" represents an observed sample count, "expected" represents the expected count for the same cell, and the sum is over all $r \times c$ cells in the table.

If the expected counts and the observed counts are very different, a large value of X^2 will result. Therefore, large values of X^2 provide evidence against the null hypothesis.

---

*The chi-square statistic was invented by the English statistician Karl Pearson (1857–1936) in 1900, for purposes slightly different from ours. It is the oldest inference procedure still used in its original form. With the work of Pearson and his contemporaries at the beginning of the twentieth century, statistics first emerges as a separate discipline.

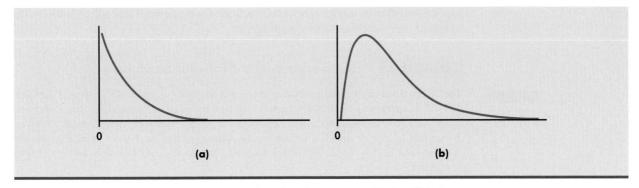

FIGURE 9.3 (a) The $\chi^2(2)$ density curve. (b) The $\chi^2(4)$ density curve.

chi-square distribution

To obtain a P-value for the test, we need the sampling distribution of X^2 under the assumption that H_0 (no association between the row and column variables) is true. We once again use an approximation, related to the Normal approximations that we employed in Chapter 8. The result is a new distribution, the **chi-square distribution,** which we denote by χ^2 (χ is the lowercase form of the Greek letter chi).

Like the t distributions, the χ^2 distributions form a family described by a single parameter, the degrees of freedom. We use $\chi^2(\text{df})$ to indicate a particular member of this family. Figure 9.3 displays the density curves of the $\chi^2(2)$ and $\chi^2(4)$ distributions. As the figure suggests, χ^2 distributions take only positive values and are skewed to the right. Table F in the back of the book gives upper critical values for the χ^2 distributions.

> ### Chi-Square Test for Two-Way Tables
>
> The null hypothesis H_0 is that there is no association between the row and column variables in a two-way table. The alternative is that these variables are related.
>
> If H_0 is true, the chi-square statistic X^2 has approximately a χ^2 distribution with $(r - 1)(c - 1)$ degrees of freedom.
>
> The P-value for the chi-square test is
>
> $$P(\chi^2 \geq X^2)$$
>
>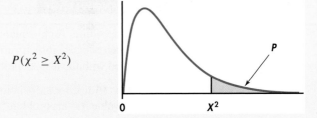
>
> where χ^2 is a random variable having the $\chi^2(\text{df})$ distribution with df $= (r - 1)(c - 1)$. If the P-value is sufficiently small, we reject the null hypothesis of no association. In this case, we say that the data provide evidence for us to conclude that there is an association.

The chi-square test always uses the upper tail of the χ^2 distribution, because any deviation from the null hypothesis makes the statistic larger. The approximation of the distribution of X^2 by χ^2 becomes more accurate as the cell counts increase. Moreover, it is more accurate for tables larger than 2×2 tables. For tables larger than 2×2, we will use this approximation whenever the average of the expected counts is 5 or more and the smallest expected count is 1 or more. For 2×2 tables, we require that all four expected cell counts be 5 or more.[2] When the data are not suitable for the chi-square approximation to be useful, other exact methods are available. These are provided in the output of many statistical software programs.

> ### EXAMPLE 9.6   Exclusive Territories and Franchise Firm Success
>
> **CASE 9.1**
>
> The results of the chi-square significance test for the successful-firms example appear in the lower parts of the computer outputs in Figure 9.1, labeled Chi-Square or Pearson Chi-Square. Because all the expected cell counts are moderately large, the χ^2 distribution provides accurate P-values. We see that $X^2 = 5.91$ and df $= 1$. Examine the outputs and find the P-value in each output. The rounded value is $P = 0.015$. As a check we verify that the degrees of freedom are correct for a 2×2 table:
>
> $$\text{df} = (r - 1)(c - 1) = (2 - 1)(2 - 1) = 1$$
>
> The chi-square test confirms that the data contain clear evidence against the null hypothesis that there is no relationship between success and exclusive territories. Under H_0, the chance of obtaining a value of X^2 greater than or equal to the calculated value of 5.91 is small—less than 15 times in 1000. The test does not tell us what kind of relationship is present. It is up to us to see that the data show that firms with exclusive territories are more likely to be successful. You should always accompany a chi-square test by percents such as those in the table on page 494 and Figure 9.2 and by a description of the nature of the relationship.

The observational study of Case 9.1 cannot tell us whether having exclusive territories is a *cause* of success in a franchise firm. The association may be explained by confounding of contract terms with other variables such as type of business. Our data don't allow us to investigate possible confounding variables. A randomized comparative experiment that assigns firms to have or not have exclusive territories would settle the issue of causation. As is often the case, however, an experiment isn't practical. The authors of this study supplemented the data we used with interviews with founders of franchises and a theoretical framework for the results. Still, they limit their conclusion to stating that the firms with exclusive territories are more likely to survive.

APPLY YOUR KNOWLEDGE

9.7 Degrees of freedom. A chi-square significance test is performed to examine the association between two categorical variables in a 6×4 table. What are the degrees of freedom associated with the test statistic?

9.8 The P-value. A test for association gives $X^2 = 23.21$ with df $= 6$. How would you report the P-value for this problem? Use Table F in the back of the book. Illustrate your solution with a sketch.

The chi-square test and the z test

We began this chapter by converting a "compare two proportions" setting (Example 9.1) into a 2×2 table. We now have two ways to test the hypothesis of equality of two population proportions: the chi-square test and the two-sample z test from Section 8.2. In fact, *these tests always give exactly the same result,* because the chi-square statistic is equal to the square of the z statistic, and $\chi^2(1)$ critical values are equal to the squares of the corresponding $N(0, 1)$ critical values. Exercise 9.9 asks you to verify this for Example 9.1. The advantage of the z test is that we can test either one-sided or two-sided alternatives and add confidence intervals to the significance test. The chi-square test always tests the two-sided alternative for a 2×2 table. The advantage of the chi-square test is that it is much more general: we can compare more than two population proportions or, more generally yet, ask about relations in two-way tables of any size.

APPLY YOUR KNOWLEDGE

9.9 No Sweat labels.   Sample proportions from Example 9.1 and the two-way table in Example 9.2 (page 494) report the same information in different ways. We saw in Chapter 8 (page 479) that the z statistic for the hypothesis of equal population proportions is $z = 3.30$ with $P < 0.001$.

(a) Find the chi-square statistic X^2 for this two-way table and verify that it is equal (up to roundoff error) to z^2.

(b) Verify that the 0.001 critical value for chi-square with df = 1 (Table F) is the square of the 0.0005 critical value for the standard Normal distribution (Table D). The 0.0005 critical value corresponds to a P-value of 0.001 for the two-sided z test.

(c) Explain carefully why the two hypotheses

$$H_0: p_1 \neq p_2 \qquad (z \text{ test})$$

$$H_0: \text{no relation between gender and label use} \qquad (X^2 \text{ test})$$

say the same thing about the population.

BEYOND THE BASICS: Meta-analysis

meta-analysis

Policymakers wanting to make decisions based on research are sometimes faced with the problem of summarizing the results of many studies. These studies may show effects of different magnitudes, some highly significant and some not significant. What *overall conclusion* can we draw? **Meta-analysis** is a collection of statistical techniques designed to combine information from different but similar studies. Each individual study must be examined with care to ensure that its design and data quality are adequate. The basic idea is to compute a measure of the effect of interest for each study. These are then combined, usually by taking some sort of weighted average, to produce a summary measure for all of the studies. Of course, a confidence interval for the summary is included in the results. Here is an example.

EXAMPLE 9.7   Vitamin A Saves Lives of Young Children

Vitamin A is often given to young children in developing countries to prevent night blindness. It was observed that children receiving vitamin A appear to have reduced death rates. To investigate the possible relationship between vitamin A supplementation and death, a large field trial with over 25,000 children was undertaken in Aceh Province of Indonesia. About half of the children were given large doses of vitamin A, and the other half were controls. The researchers reported a 34% reduction in mortality (deaths) for the treated children who were 1 to 6 years old compared with the controls. Several additional studies were then undertaken. Most of the results confirmed the association: treatment of young children in developing countries with vitamin A reduces the death rate; but the size of the effect varied quite a bit.

How can we use the results of these studies to guide policy decisions? To address this question, a meta-analysis was performed on data from eight studies.[3] Although the designs varied, each study provided a two-way table of counts. Here is the table for the study conducted in Aceh Province. A total of $n = 25,200$ children were enrolled in the study. Approximately half received vitamin A supplements. One year after the start of the study, the number of children who had died was determined.

	Vitamin A	Control
Dead	101	130
Alive	12,890	12,079
Total	12,991	12,209

relative risk

The summary measure chosen was the **relative risk:** the ratio formed by dividing the proportion of children who died in the vitamin A group by the proportion of children who died in the control group. For Aceh, the proportion who died in the vitamin A group was

$$\frac{101}{12,991} = 0.00777$$

or 7.7 per thousand. For the control group, the proportion who died was

$$\frac{130}{12,209} = 0.01065$$

or 10.6 per thousand. The relative risk is therefore

$$\frac{0.00777}{0.01065} = 0.73$$

Relative risk less than 1 means that the vitamin A group has the lower mortality rate.

The relative risks for the eight studies were

$$0.73 \quad 0.50 \quad 0.94 \quad 0.71 \quad 0.70 \quad 1.04 \quad 0.74 \quad 0.80$$

A meta-analysis combined these eight results to produce a relative risk estimate of 0.77 with a 95% confidence interval of (0.68, 0.88). That is, vitamin A supplementation reduced the mortality rate to 77% of its value in an untreated group. The confidence interval does not include 1, so we can reject the null hypothesis of no effect (a relative risk of 1). The researchers examined many variations of this meta-analysis, such as using different weights and leaving out one study at a time. These variations had little effect on the final estimate.

After these findings were published, large-scale programs to distribute high-potency vitamin A supplements were started. These programs have saved hundreds of thousands of lives since the meta-analysis was conducted and the arguments and uncertainties were resolved.

SECTION 9.1 Summary

- The **null hypothesis** for $r \times c$ tables of count data is that there is no relationship between the row variable and the column variable.
- **Expected cell counts** under the null hypothesis are computed using the formula

$$\text{expected count} = \frac{\text{row total} \times \text{column total}}{n}$$

- The null hypothesis is tested by the **chi-square statistic,** which compares the observed counts with the expected counts:

$$X^2 = \sum \frac{(\text{observed} - \text{expected})^2}{\text{expected}}$$

- Under the null hypothesis, X^2 has approximately the **chi-square distribution** with $(r-1)(c-1)$ degrees of freedom. The P-value for the test is

$$P(\chi^2 \geq X^2)$$

where χ^2 is a random variable having the $\chi^2(\text{df})$ distribution with $\text{df} = (r-1)(c-1)$.

- The chi-square approximation is adequate for practical use when the average expected cell count is 5 or greater and all individual expected counts are 1 or greater, except in the case of 2×2 tables. All four expected counts in a 2×2 table should be 5 or greater.

The section we just completed assumed that you have access to software or a statistical calculator. If you do, you can now work the exercises that appear at the end of the chapter. If not, you should read the part of the optional Section 9.2 that illustrates the details of chi-square calculations before attempting the exercises.

9.2 Formulas and Models for Two-Way Tables*

The calculations required to analyze a two-way table are straightforward but tedious. In practice, we recommend using software, but it is possible to do the work with a calculator, and some insight can be gained by examining the details. Here is an outline of the steps required.

Computations for Two-Way Tables

1. Calculate descriptive statistics that convey the important information in the table. Usually these will be column or row percents.
2. Find the expected counts and use these to compute the X^2 statistic.
3. Use chi-square critical values from Table F to find the approximate P-value.
4. Draw a conclusion about the association between the row and column variables.

The following example illustrates these steps.

CASE 9.2

 MUSIC

Background Music and Consumer Behavior Market researchers know that background music can influence the mood and purchasing behavior of customers. One study in a supermarket in Northern Ireland compared three treatments: no music, French accordion music, and Italian string music. Under each condition, the researchers recorded the numbers of bottles of French, Italian, and other wine purchased.[4] Here is the two-way table that summarizes the data:

	Counts for wine and music			
	Music			
Wine	None	French	Italian	Total
French	30	39	30	99
Italian	11	1	19	31
Other	43	35	35	113
Total	84	75	84	243

This is a 3×3 table, to which we have added the marginal totals obtained by summing across rows and columns. For example, the first-row total is $30 + 39 + 30 = 99$. The grand total, the number of bottles of wine in the study, can be computed by summing the row totals, $99 + 31 + 113 = 243$, or the column totals, $84 + 75 + 84 = 243$. It is a good idea to do both as a check on your arithmetic.

---

*The analysis of two-way tables is based on computations that are a bit messy and on statistical models that require a fair amount of notation to describe. This section gives the details. By studying this material you will deepen your understanding of the methods described in this chapter, but this section is optional.

Conditional distributions

First, we summarize the observed relation between the music being played and the type of wine purchased. The researchers expected that music would influence sales, so music type is the explanatory variable and the type of wine purchased is the response variable. In general, the clearest way to describe this kind of relationship is to compare the conditional distributions of the response variable for each value of the explanatory variable. So we will compare the column percents that give the conditional distribution of purchases for each type of music played.

EXAMPLE 9.8   Wine Sales, Given Music Type

CASE 9.2

When no music was played, there were 84 bottles of wine sold. Of these, 30 were French wine. Therefore, the column proportion for this cell is

$$\frac{30}{84} = 0.357$$

That is, 35.7% of the wine sold was French when no music was played. Similarly, 11 bottles of Italian wine were sold under this condition, and this is 13.1% of the sales.

$$\frac{11}{84} = 0.131$$

In all, we calculate nine percents. Here are the results:

Column percents for wine and music

Wine	Music			
	None	French	Italian	Total
French	35.7	52.0	35.7	40.7
Italian	13.1	1.3	22.6	12.8
Other	51.9	46.7	41.7	46.5
Total	100.0	100.0	100.0	100.0

In addition to the conditional distributions of types of wine sold for each kind of music being played, the table also gives the marginal distribution of the types of wine sold. These percents appear in the rightmost column, labeled "Total."

The sum of the percents in each column should be 100, except for possible small roundoff errors. It is good practice to calculate each percent separately and then sum each column as a check. In this way we can find arithmetic errors that would not be uncovered if, for example, we calculated the column percent for the "Other" row by subtracting the sum of the percents for "French" and "Italian" from 100.

Figure 9.4 compares the distributions of types of wine sold for each of the three music conditions. There appears to be an association between the music played and the type of wine that customers buy. Sales of Italian wine are very low when French music is playing but are higher when Italian music or no music is playing. French wine is popular in this market, selling well under all music conditions but notably better when French music is playing.

Another way to look at these data is to examine the row percents. These fix a type of wine and compare its sales when different types of music are playing. Figure 9.5 displays these results. We see that more French wine is sold when French music is playing.

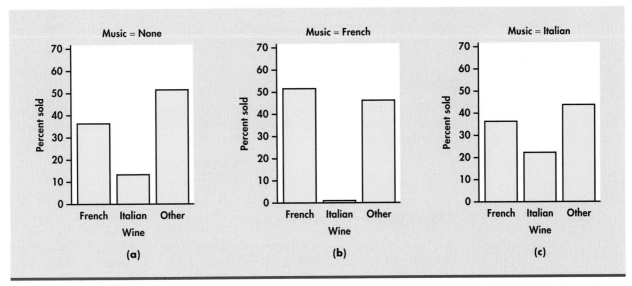

FIGURE 9.4 Comparison of the percents of different types of wine sold for different music conditions, for Example 9.8.

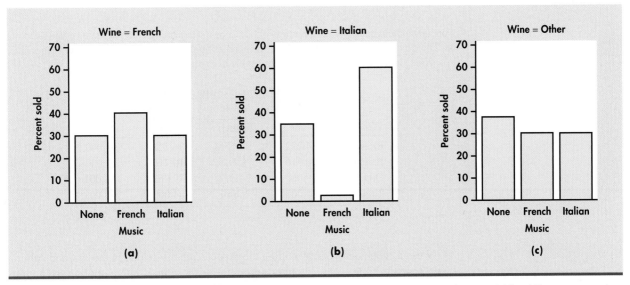

FIGURE 9.5 Comparison of the percents of different types of wine sold for different music conditions, for Example 9.8.

Similarly for Italian wine. The negative effect of French music on sales of Italian wine is dramatic.

We observe a clear relationship between music type and wine sales for the 243 bottles sold during the study. The chi-square test assesses whether this observed association is statistically significant, that is, too strong to occur often just by chance. The test only confirms that there is some relationship. The percents we have compared describe the nature of the relationship. What is more, the chi-square test does not in itself tell us what population our conclusion describes. If the study was done in one market on a Saturday, the results may apply only to Saturday shoppers at this market. The researchers may invoke their understanding of consumer behavior to argue that their findings apply more generally, but that is beyond the scope of the statistical analysis.

Expected cell counts

The null hypothesis is that there is no relationship between music and wine sales. The alternative is that these two variables are related. We know that under this hypothesis the expected cell counts are

$$\text{expected count} = \frac{\text{row total} \times \text{column total}}{n}$$

EXAMPLE 9.9   Expected Cell Counts

CASE 9.2   What is the expected count in the upper-left cell in the table for Case 9.2, bottles of French wine sold when no music is playing, under the null hypothesis that music and wine sales are independent?

The column total, the number of bottles of wine sold when no music is playing, is 84. The row total shows that 99 bottles of French wine were sold during the study. The total sales were 243. The expected cell count is therefore

$$\frac{(84)(99)}{243} = 34.222$$

Note that although any count of bottles sold must be a whole number, an expected count need not be. The expected count is the mean over many repetitions of the study, assuming no relationship.

Nine similar calculations produce this table of expected counts:

Expected counts for wine and music

Wine	Music None	French	Italian	Total
French	34.222	30.556	34.222	99.000
Italian	10.716	9.568	10.716	31.000
Other	39.062	34.877	39.062	113.001
Total	84.000	75.001	84.000	243.001

We can check our work by adding the expected counts to obtain the row and column totals, as in the table. These should be the same as those in the table of observed counts, except for small roundoff errors, such as 113.001 rather than 113 for the total number of bottles of other wine sold.

The X^2 statistic and its P-value

The expected counts are all large, so we proceed with the chi-square test. We compare the table of observed counts with the table of expected counts using the X^2 statistic.[5] We must calculate the term for each cell, then sum over all nine cells. For French wine with no music, the observed count is 30 bottles and the expected count is 34.222. The contribution to the X^2 statistic for this cell is therefore

$$\frac{(30 - 34.222)^2}{34.222} = 0.5209$$

The X^2 statistic is the sum of nine such terms:

$$X^2 = \sum \frac{(\text{observed} - \text{expected})^2}{\text{expected}}$$

$$= \frac{(30 - 34.222)^2}{34.222} + \frac{(39 - 30.556)^2}{30.556} + \frac{(30 - 34.222)^2}{34.222}$$

$$+ \frac{(11 - 10.716)^2}{10.716} + \frac{(1 - 9.568)^2}{9.568} + \frac{(19 - 10.716)^2}{10.716}$$

$$+ \frac{(43 - 39.062)^2}{39.062} + \frac{(35 - 34.877)^2}{34.877} + \frac{(35 - 39.062)^2}{39.062}$$

$$= 0.5209 + 2.3337 + 0.5209 + 0.0075 + 7.6724 + 6.4038$$

$$+ 0.3971 + 0.0004 + 0.4223$$

$$= 18.28$$

Because there are $r = 3$ types of wine and $c = 3$ music conditions, the degrees of freedom for this statistic are

$$\text{df} = (r - 1)(c - 1) = (3 - 1)(3 - 1) = 4$$

Under the null hypothesis that music and wine sales are independent, the test statistic X^2 has a $\chi^2(4)$ distribution. To obtain the P-value, look at the df = 4 row in Table F.

	df = 4	
p	0.0025	0.001
χ^2	16.42	18.47

The calculated value $X^2 = 18.28$ lies between the critical points for probabilities 0.0025 and 0.001. The P-value is therefore between 0.0025 and 0.001. Because the expected cell counts are all large, the P-value from Table F will be quite accurate. There is strong evidence ($X^2 = 18.28$, df = 4, $P < 0.0025$) that the type of music being played has an effect on wine sales.

The size and nature of the relationship between music and wine sales are described by row and column percents. These are displayed in Figures 9.4 and 9.5. Here is another way to look at the data: we see that just two of the nine terms that make up the chi-square sum contribute about 14 of the total $X^2 = 18.28$. Comparing the observed and expected counts in these two cells, we see that sales of Italian wine are much below expectation when French music is playing and much above expectation when Italian music is playing. We are led to a specific conclusion: sales of Italian wine are strongly affected by Italian and French music. Figure 9.5(b) displays this effect.

Models for two-way tables

The chi-square test for the presence of a relationship between the two directions in a two-way table is valid for data produced from several different study designs. The precise statement of the null hypothesis "no relationship" in terms of population parameters is different for different designs.

The entries in a two-way table count the individuals studied—franchise firms in Case 9.1, bottles of wine sold in Case 9.2. Each individual counts in only one cell of the table. The two cases differ in design. The data in Case 9.1 concern a single sample of franchise firms, each classified in two ways: whether or not they were successful (rows) and whether or not they had exclusive-territory contracts (columns). In Case 9.2, the market researchers changed the background music and took three samples of wine sales in three distinct environments. Each column in the table represents one of these samples, and each row a type of wine.

The first model: comparing several populations

Case 9.2 (wine sales in three environments) is an example of *separate and independent random samples* from each of c populations. The c columns of the two-way table represent the populations. There is a single categorical response variable, wine type. The r rows of the table correspond to the values of the response variable.

We know that the z test for comparing the two proportions of successes and the chi-square test for the 2×2 table are equivalent. The $r \times c$ table allows us to compare more than two populations or more than two categories of response, or both. In this setting, the null hypothesis "no relationship between column variable and row variable" becomes

H_0: The distribution of the response variable is the same in all c populations.

Because the response variable is categorical, its distribution just consists of the probabilities of its r values. The null hypothesis says that these probabilities (or population proportions) are the same in all c populations.

EXAMPLE 9.10   Music and Wine Sales

CASE 9.2

In the market research study of Case 9.2, we compare three populations:

Population 1: bottles of wine sold when no music is playing

Population 2: bottles of wine sold when French music is playing

Population 3: bottles of wine sold when Italian music is playing

We have three samples, of sizes 84, 75, and 84, a separate sample from each population. The null hypothesis for the chi-square test is

H_0: The proportions of each wine type sold are the same in all three populations.

The parameters of the model are the proportions of the three types of wine that would be sold in each of the three environments. There are three proportions (for French wine, Italian wine, and other wine) for each environment.

> **Model for Comparing Several Populations Using Two-Way Tables**
>
> Select independent SRSs from each of c populations, of sizes $n_1, n_2, \ldots, n_c$. Classify each individual in a sample according to a categorical response variable with r possible values. There are c different probability distributions, one for each population.
>
> The null hypothesis is that the distributions of the response variable are the same in all c populations. The alternative hypothesis says that these c distributions are not all the same.

The second model: testing independence

A second model for which our analysis of $r \times c$ tables is valid is illustrated by the exclusive-territory study of Case 9.1. There, a *single* sample from a *single* population was classified according to two categorical variables.

EXAMPLE 9.11 Exclusive Territories and Franchise Success

CASE 9.1 The single population studied is

Population: recently established franchise firms

The researchers had a sample of 170 such firms. They measured two categorical variables for each firm:

Column variable: Exclusive territory? (Yes/No)

Row variable: Firm successful? (Yes/No)

The null hypothesis for the chi-square test is

H_0: The row variable and the column variable are independent.

The parameters of the model are the probabilities for each of the four possible combinations of values of the row and column variables. If the null hypothesis is true, the multiplication rule for independent events (page 287) says that these can be found as the products of outcome probabilities for each variable alone.

> **Model for Examining Independence in Two-Way Tables**
>
> Select an SRS of size n from a population. Measure two categorical variables for each individual.
>   The null hypothesis is that the row and column variables are independent. The alternative hypothesis is that the row and column variables are dependent.

Concluding remarks

You can distinguish between the two models by examining the design of the study. In the independence model, there is a single sample. The column totals and row totals are random variables. The total sample size n is set by the researcher; the column and row sums are known only after the data are collected. For the comparison-of-populations model, on the other hand, there is a sample from each of two or more populations. The column sums are the sample sizes selected at the design phase of the research. The null hypothesis in both models says that there is no relationship between the column variable and the row variable. The precise statement of the hypothesis differs, depending on the sampling design. Fortunately, *the test of the hypothesis of "no relationship" is the same for both models:* it is the chi-square test. There are yet other statistical models for two-way tables that justify the chi-square test of the null hypothesis "no relation," made precise in ways suitable for these models. Statistical methods related to the chi-square test also allow the analysis of three-way and higher-way tables of count data. You can find a discussion of these topics in texts on analysis of categorical data.[6]

Both models require that the data can be viewed as random samples from well-defined populations. When you use the chi-square test (and other inference procedures), you are acting as if the data come from random samples or randomized experiments. In practice, there are often some weaknesses in study design. Case 9.2, in counting bottles of wine sold, ignores the fact that some customers buy one bottle and others buy several bottles. If a single customer buys many bottles during the study period, the data may not be a random sample of bottles sold because they give too much weight to this one customer's taste. The franchising study in Case 9.1 took place at a specific time and therefore under business conditions that no doubt differ from later conditions. Fortunately, our results in these two examples are sufficiently clear that we are still comfortable with our conclusions.

SECTION 9.2 Summary

- To analyze a two-way table, first **compute percents or proportions** that describe the relationship between the row and column variables. Then calculate **expected counts, the chi-square statistic,** and the **P-value.**

- Two different models for generating $r \times c$ tables lead to the chi-square test. In the first model, independent SRSs are drawn from each of c populations, and each observation is classified according to a categorical variable with r possible values. The null hypothesis is that the distributions of the row categorical variable are the same for all c populations. In the second model, a single SRS is drawn from a population, and observations are classified according to two categorical variables having r and c possible values. In this model, H_0 states that the row and column variables are independent.

STATISTICS IN SUMMARY

The chi-square test is one of the most common statistical procedures because two-way tables of counts are frequently constructed in practice. The central idea is always to ask if there is some relationship between the column categories and the row categories. The chi-square test assesses the statistical significance of the column-row relationship but does not by itself tell us the nature or size of the relationship. Here is a review list of the most important skills you should have acquired from your study of this chapter.

A. Two-Way Tables

1. Arrange data on two categorical variables measured on the same group of individuals in a two-way table of counts for the possible outcomes.

2. Use percents to describe the relationship between two categorical variables, starting from the counts in a two-way table.

B. Interpreting Chi-Square Tests

1. Locate expected cell counts, the X^2 statistic, and its P-value in output from your software or calculator.

2. Explain what null hypothesis the X^2 statistic tests in a specific two-way table.

3. If the test is significant, use percents or comparison of expected and observed counts to see what deviations from the null hypothesis are most important.

C. Doing Chi-Square Tests (Optional)

1. Calculate the expected count for any cell from the observed counts in a two-way table.

2. Calculate the component of the X^2 statistic for any cell, as well as the overall statistic. Make a quick assessment of the significance of the statistic by comparing the observed and expected counts.

3. Give the degrees of freedom of a X^2 statistic.

4. Use the X^2 critical values in Table F to approximate the P-value of a chi-square test.

CHAPTER 9 Review Exercises

For Exercises 9.1 and 9.2, see pages 495–496; for 9.3 and 9.4, see page 499; for 9.5 and 9.6, see page 500; for 9.7 and 9.8, see page 502; and for 9.9, see page 503.

9.10 Remote deposit capture. The Federal Reserve has called remote deposit capture (RDC) "the most important development the (U.S) banking industry has seen in years." This service allows users to scan checks and to transmit the scanned images to a bank for posting.[7] In its annual survey of community banks, the American Bankers Association asked banks whether or not they offered this service.[8] Here are the results classified by the asset size (in millions of dollars) of the bank: **RCDSIZE**

	Offer RDC	
Asset size	Yes	No
Under $100	63	309
$101 to $200	59	132
$201 or more	112	85

(a) Summarize the results of this survey question numerically and graphically. (In Exercise 2.114 (page 131), you were asked to do this.)

(b) Test the null hypothesis that there is no association between the size of a bank, measured by assets, and whether or not it offers RDC. Report the test statistic, the *P*-value, and your conclusion.

9.11 How does RDC vary across the country? The survey described in the previous exercise also classified community banks by region.[9] Here is the 6 × 2 table of counts: **RCDREGION**

	Offer RDC	
Region	Yes	No
Northeast	28	38
Southeast	57	61
Central	53	84
Midwest	63	181
Southwest	27	51
West	61	76

(a) Summarize the results of this survey question numerically and graphically. (In Exercise 2.115 (page 131), you were asked to do this.)

(b) Test the null hypothesis that there is no association between region and whether or not a community bank offers RDC. Report the test statistic with the degrees of freedom.

(c) Report the *P*-value and make a sketch similar to the one on page 501 to illustrate the calculation.

(d) Write a summary of your analysis and conclusion. Be sure to include numerical and graphical summaries.

9.12 Exercise and adequate sleep. A survey of 656 boys and girls who were 13 to 18 years old asked about adequate sleep and other health-related behaviors. The recommended amount of sleep is six to eight hours per night.[10] In the survey 54% of the respondents reported that they got less than this amount of sleep on school nights. An exercise scale was developed and was used to classify the students as above or below the median in this domain. Here is the 2 × 2 table of counts with students classified as getting or not getting adequate sleep and by the exercise variable: **SLEEP**

	Exercise	
Enough sleep	High	Low
Yes	151	115
No	148	242

Note that you answered parts (a) through (c) of this exercise if you completed Exercise 2.116 (page 131).

(a) Find the distribution of adequate sleep for the high exercisers.

(b) Do the same for the low exercisers.

(c) Summarize the relationship between adequate sleep and exercise using the results of parts (a) and (b).

(d) Perform the significance test for examining the relationship between exercise and getting enough sleep. Give the test statistic and the *P*-value (with a sketch similar to the one on page 501) and summarize your conclusion. Be sure to include numerical and graphical summaries.

9.13 Lying to a teacher. One of the questions in a survey of high school students asked about lying to teachers.[11] The table below gives the numbers of students who said that they lied to a teacher at least once during the past year, classified by gender. **LYING**

	Gender	
Lied at least once	Male	Female
Yes	3,228	10,295
No	9,659	4,620

Note that you answered parts (a) through (c) of this exercise if you completed Exercise 2.120 (page 132).

(a) Add the marginal totals to the table.

(b) Calculate appropriate percents to describe the results of this question.

(c) Summarize your findings in a short paragraph.

(d) Test the null hypothesis that there is no association between gender and lying to teachers. Give the test statistic and the *P*-value (with a sketch similar to the one on page 501) and summarize your conclusion. Be sure to include numerical and graphical summaries.

9.14 Trust and honesty in the workplace. The students surveyed in the study described in the previous exercise were also asked whether they thought trust and honesty were essential in business and the workplace. Here are the counts classified by gender: 🔵 TRUST

| | Gender | |
Trust and honesty are essential	Male	Female
Agree	11,724	14,169
Disagree	1,163	746

Note that you answered parts (a) through (c) of this exercise if you completed Exercise 2.121 (page 132). Answer questions the given in the previous exercise for this survey question.

9.15 Hiring practices. A company has been accused of age discrimination in hiring for operator positions. Lawyers for both sides look at data on applicants for the past 3 years. They compare hiring rates for applicants younger than 40 years and those 40 years or older. 🔵 HIRING

Age	Hired	Not hired
Younger than 40	79	1158
40 or older	1	165

Note that you answered parts (a) through (d) of this exercise if you completed Exercise 2.123 (page 133).
(a) Find the two conditional distributions of hired/not hired, one for applicants who are less than 40 years old and one for applicants who are not less than 40 years old.
(b) Based on your calculations, make a graph to show the differences in distribution for the two age categories.
(c) Describe the company's hiring record in words. Does the company appear to discriminate on the basis of age?
(d) What lurking variables might be involved here?
(e) Use a significance test to determine whether or not the data indicate that there is a relationship between age and whether or not an applicant is hired.

9.16 Nonresponse in a survey. A business school conducted a survey of companies in its state. They mailed a questionnaire to 200 small companies, 200 medium-sized companies, and 200 large companies. The rate of nonresponse is important in deciding how reliable survey results are. Here are the data on response to this survey: 🔵 NONRESPONSE

	Small	Medium	Large
Response	125	81	40
No response	75	119	160
Total	200	200	200

Note that you answered parts (a) through (c) of this exercise if you completed Exercise 2.124 (page 133).
(a) What was the overall percent of nonresponse?
(b) Describe how nonresponse is related to the size of the business. (Use percents to make your statements precise.)

(c) Draw a bar graph to compare the nonresponse percents for the three size categories.
(d) State and test an appropriate null hypothesis for these data.

9.17 Discrimination? Wabash Tech has two professional schools, business and law. Here are two-way tables of applicants to both schools, categorized by gender and admission decision. (Although these data are made up, similar situations occur in reality.) 🔵 DISCRIMINATION

| | Business | | | Law | |
	Admit	Deny		Admit	Deny
Male	480	120	Male	10	90
Female	180	20	Female	100	200

Note that you answered parts (a) through (d) of this exercise if you completed Exercise 2.128 (page 133).
(a) Make a two-way table of gender by admission decision for the two professional schools together by summing entries in these tables.
(b) From the two-way table, calculate the percent of male applicants who are admitted and the percent of female applicants who are admitted. Wabash admits a higher percent of male applicants.
(c) Now compute separately the percents of male and female applicants admitted by the business school and by the law school. Each school admits a higher percent of female applicants.
(d) This is Simpson's paradox: both schools admit a higher percent of the women who apply, but overall Wabash admits a lower percent of female applicants than of male applicants. Explain carefully, as if speaking to a skeptical reporter, how it can happen that Wabash appears to favor males when each school individually favors females.
(e) Use the data summary that you prepared in part (a) to test the null hypothesis that there is no relationship between gender and whether or not an applicant is admitted to a professional school at Wabash Tech.
(f) Test the same null hypothesis using the business school data only. Do the same for the law school data and then compare the results.

9.18 Obesity and health. Recent studies have shown that earlier reports underestimated the health risks associated with being overweight. The error was due to overlooking lurking variables. In particular, smoking tends both to reduce weight and to lead to earlier death. Note that you answered part (a) of this exercise if you completed Exercise 2.129 (page 134).
(a) Illustrate Simpson's paradox by a simplified version of this situation. That is, make up tables of overweight (yes or no) by early death (yes or no) by smoker (yes or no) such that

- Overweight smokers and overweight nonsmokers both tend to die earlier than those not overweight.
- But when smokers and nonsmokers are combined into a two-way table of overweight by early death, persons who are not overweight tend to die earlier.

(b) Perform significance tests for the combined data set and for the smokers and nonsmokers separately. If all P-values are not less than 0.05, redo your tables so that all results are statistically significant at this level.

9.19 Plot the test statistic and the P-values. Here is a 2×2 two-way table of counts. The two categorical variables are U and V, and the possible values for each of these variables are 0 and 1. Notice that the second row depends upon a quantity that we call a. For this exercise you will examine how the test statistic and its corresponding P-value depend upon this quantity. Notice that the row sums are both 100.

		V
U	0	1
0	50	50
1	$50 + a$	$50 - a$

(a) Consider setting a equal to zero. Find the percent of zeros for the variable V when $U = 0$. Do the same for the case where $U = 1$. With this choice of a, the data match the null hypothesis as closely as possible. Explain why.
(b) Consider the tables where the values of a are equal to 0, 5, 10, 15, 20, and 25. For each of these scenarios, find the percent of zeros for V when $U = 1$. Notice that this percent does not vary with a for $U = 0$.
(c) Compute the test statistic and P-value for testing the null hypothesis that there is no association between the row and column variables for each of the values of a given in part (b).
(d) Plot the values of the X^2 test statistic versus the percent of zeros for V when $U = 1$. Do the same for the P-values. Summarize what you have learned from this exercise in a short paragraph.

9.20 Plot the test statistic and the P-values. Here is a 2×2 two-way table of counts. The two categorical variables are U and V, and the possible values for each of these variables are 0 and 1.
🔵 COUNTS

		U
V	0	1
0	5	5
1	7	3

(a) Find the percent of zeros for V when $U = 0$. Do the same for the case where $U = 1$. Find the value of the test statistic and its P-value.
(b) Now multiply all of the counts in the table by 2. Verify that the percent of zeros for V when $U = 0$ and the percent of zeros for the V when $U = 1$ do not change. Find the value of the test statistic and its P-value for this table.
(c) Answer part (b) for tables where all counts are multiplied by 4, 6, and 8. Summarize all your results graphically, and write a short paragraph describing what you have learned from this exercise.

9.21 Trends in broadband use over time. The Pew Internet and American Life Project collects data about the impact of the Internet on various aspects of American life.[12] One set of surveys has tracked the use of broadband in homes over a period of several years.[13] Here are some data on the percent of homes that access the Internet using broadband:

Date of survey:	April 2001	April 2004	March 2007	April 2008
Homes with broadband:	5%	24%	48%	55%

Assume a sample size of 2250 for each survey.
(a) Display the data in a two-way table of counts.
(b) Test the null hypothesis that the proportion of homes that access the Internet using broadband has not changed over this period of time. Report your test statistic with degrees of freedom and the P-value. What do you conclude?
(c) Now analyze the data from March 2007 and April 2008 only. Summarize your results and compare them with what you found in part (b).

9.22 What about dial-up? Refer to the previous exercise. The same surveys provided data on access to the Internet using dial-up. Here are the data:

Date of survey:	April 2001	April 2004	March 2007	April 2008
Homes with dial-up:	41%	30%	16%	12%

(a) to (c) Answer the questions given in the previous exercise for these data.
(d) Write a short report summarizing the changes in broadband access that have occurred over this period of time using your analysis from this exercise and the previous one. Include a graph with information about both broadband and dial-up access over time.

9.23 How robust are the conclusions? Refer to Exercise 9.21 on the use of broadband to access the Internet. In that exercise, the percents were read from a graph, and we assumed that the sample size was 2250 for all the surveys. Investigate the robustness of your conclusions in Exercise 9.21 against the use of 2250 as the sample size for all surveys and to roundoff and slight errors in reading the graph. Assume that the actual sample sizes ranged from 2200 to 2600. Assume also that the percents reported are all accurate to within $\pm 2\%$. In other words, if the reported percent is 16%, then we can assume that the actual survey percent is between 14% and 18%. Reanalyze the data using at least five scenarios that vary the percents and the sample sizes within the assumed ranges. Summarize your results in a report, paying particular attention to the consequences for your conclusions in Exercise 9.21.

9.24 Switching from dial-up to broadband. A Pew survey asked dial-up users about their interest in switching to broadband.[14] Here is the summary given in the report:

Date of survey:	October 2002	February 2004	December 2005	May 2008
Interested in switch:	38%	40%	39%	36%

(a) The percents reported in the table were computed by dividing the number of respondents who expressed an interest in switching from dial-up to broadband by the number of respondents who said that they used dial-up. Use the data given in Exercise 9.22 to estimate the number of dial-up users in each of these surveys. Note that the dates do not match. Explain carefully how you obtain your estimates.

(b) Using the results from part (a), construct the 2 × 4 table of counts.

(c) Analyze the table and write a summary of what you have found.

(d) The Pew report states:

The roughly 60% of dial-up users consistently saying that they are not interested in broadband, in the face of the shrinking pool of dial-up users, suggests that the preferences of dial-up users change over time. That is, assuming that many of those interested in getting broadband switched over to it from December 2005 to May 2008, some of those who said that they were not interested in broadband replenished the supply of "yes, interested in broadband" responses in order to maintain the 40–60 ratio of those interested in broadband versus not interested.

Approximately what percent of dial-up users who said they were not interested in broadband in December 2005 would need to change their mind and become interested in broadband for this statement to be correct. Explain how you determined your answer.

9.25 Health care fraud. Most errors in billing insurance providers for health care services involve honest mistakes by patients, physicians, or others involved in the health care system. However, fraud is a serious problem. The National Health Care Anti-fraud Association estimates that approximately $68 billion is lost to health care fraud each year.[15] When fraud is suspected, an audit of randomly selected billings is often conducted. The selected claims are then reviewed by experts and each claim is classified as allowed or not allowed. The distributions of the amounts of claims are frequently highly skewed, with a large number of small claims and small number of large claims. Since simple random sampling would likely be overwhelmed by small claims and would tend to miss the large claims, stratification is often used. See the section on stratified sampling in Chapter 3 (page 158). Here are data from an audit that used three strata

based on the sizes of the claims (small, medium, and large):[16]

BILLINGERRORS

Stratum	Sampled claims	Number not allowed
Small	57	6
Medium	17	5
Large	5	1

(a) Construct the 3 × 2 table of counts for these data and include the marginal totals.

(b) Find the percent of claims that were not allowed in each of the three strata.

(c) State an appropriate null hypothesis to be tested for these data.

(d) Perform the significance test and report your test statistic with degrees of freedom and the P-value. State your conclusion.

9.26 Population estimates. Refer to the previous exercise. One reason to do an audit such as this is to estimate the number of claims that would not be allowed if all claims in a population were examined by experts. We have estimates of the proportions of such claims from each stratum based on our sample. With our simple random sampling of claims from each stratum, we have unbiased estimates of the corresponding population proportion for each stratum. Therefore, if we take the sample proportions and multiply by the population sizes, we would have the estimates that we need. Here are the population sizes for the three strata:

Stratum	Claims in strata
Small	3342
Medium	246
Large	58

(a) For each stratum, estimate the total number of claims that would not be allowed if all claims in the stratum had been audited.

(b) (Optional) Give margins of error for your estimates. *Hint:* You first need to find standard errors for your sample estimates; see Chapter 8, page 460. Then you need to use the rules for variances given in Chapter 4 (page 253) to find the standard errors for the population estimates. Finally, you need to multiply by z^* to determine the margins of error.

9.27 DFW rates. Colleges and universities want the students that they admit to graduate. Studies indicate that a key issue in retaining students is their performance in so-called gateway courses.[17] These are courses that serve as prerequisites for other key courses that are essential for student success. One measure of student performance in these courses is the DFW rate, the percent of students who receive grades of D, F, or W (withdraw). A major project was undertaken to improve the DFW rate in a gateway course at a large midwestern university. The course was revised to make it more relevant to the majors of the students taking the course, a small group of excellent teachers taught the course, technology (including clickers and online homework) was introduced, and there was an increase in student support outside the classroom. The table below gives data on the DFW rates for the course for three years.[18] In Year 1, the traditional course was given; in Year 2, a few changes were introduced; and in Year 3, the course was substantially revised.

Year	DFW rate	Number of students taking course
Year 1	42.3%	2408
Year 2	24.9%	2325
Year 3	19.9%	2126

Do you think that the changes in this gateway course had an impact on the DFW rate? Write a report giving your answer to this question. Support your answer with an analysis of the data.

9.28 Class attendance and DFW rates. One study that looked at DFW rates surveyed 719 students who were enrolled in one or more of seven gateway courses in business, mathematics, and science.[19] If a student was enrolled in more than one course, then a single course was randomly selected for analysis. In the survey, students were asked how often they attended the gateway class. Here are the data:

Class attendance	ABC percent	DFW percent
Less than 50%	2%	5%
51% to 74%	8%	14%
75% to 94%	25%	30%
95% or more	66%	51%
Total	100%	100%

In this table students are classified as earning an A, B, or C in the gateway course or earning a D or an F or withdrawing from the course. Notice that the data are given in terms of the marginal distributions of class attendance for each (ABC or DFW) group. In the survey, there were 539 students in the ABC group and 180 students in the DFW group.
(a) Use the data given to construct the 4×2 table of counts and add the marginal totals to the table. Sum the row totals and the marginal totals separately to verify that you have the correct total sample size, 719.
(b) Test the null hypothesis that there is no association between class attendance and a DFW grade. Give the test statistic with degrees of freedom and the P-value.
(c) Summarize your conclusion in a short paragraph.
(d) Can you conclude that the data indicate that not going to class causes a student to get a bad grade? Can you conclude that the data are consistent with this scenario?

9.29 When do Canadian students enter private career colleges? A survey of Canadian students who enrolled in private career colleges was conducted to understand student participation in the private postsecondary educational system.[20] In one part of the survey, students were asked about their field of study and when they entered college. Here are the results:

Field of study	Number of students	Time of Entry	
		Right after high school	Later
Trades	942	34%	66%
Design	584	47%	53%
Health	5085	40%	60%
Media/IT	3148	31%	69%
Service	1350	36%	64%
Other	2255	52%	48%

In this table, the second column gives the number of students in each field of study. The next two columns give the marginal distribution of time of entry for each field of study.
(a) Use the data provided to make the 6×2 table of counts for this problem.
(b) Analyze the data.
(c) Write a summary of your conclusions. Be sure to include the results of your significance testing as well as a graphical summary.

9.30 Government loans for Canadian students in private career colleges. Refer to the previous exercise. The survey also asked about how these college students paid for their education. A major source of funding was government loans. Here are the survey percents of Canadian private students who used government loans classified by field of study:

Field of study	Number of students	Percent using government loans
Trades	942	45%
Design	599	53%
Health	5234	55%
Media/IT	3238	55%
Service	1378	60%
Other	2300	47%

(a) Construct the 6×2 table of counts for this exercise.
(b) Test the null hypothesis that the percent of students using government loans to finance their education does not vary with field of study. Be sure to provide all the details of your significance test.
(c) Summarize your analysis and conclusions. Be sure to include a graphical summary.

9.31 Other funding for Canadian students in private career colleges? Refer to the previous exercise. Another major source of funding was parents, family, or spouse. The table below gives the survey percents of Canadian private students who relied on these sources classified by field of study:

Field of study	Number of students	Percent using parents/family/spouse
Trades	942	20%
Design	599	37%
Health	5234	26%
Media/IT	3238	16%
Service	1378	18%
Other	2300	41%

Answer the questions in the previous exercise for these data.

9.32 What's wrong? Explain what is wrong with each of the following:

(a) A chi-square test was used to test the null hypothesis that there is an association between two categorical variables.

(b) The P-value for a chi-square significance test was 1.05.

(c) Expected cell counts are computed under the assumption that the alternative hypothesis is true.

9.33 Construct a table. Construct a 3×3 table of counts where there is no apparent association between the row and column variables.

9.34 Find the P-value. For each of the following situations give the degrees of freedom and an appropriate bound on the P-value (give the exact value if you have software available) for the X^2 statistic for testing the null hypothesis of no association between the row and column variables.

(a) A 4×3 table with $X^2 = 20.41$.

(b) A 4×4 table with $X^2 = 20.41$.

(c) A 3×4 table with $X^2 = 20.41$.

(d) A 5×6 table with $X^2 = 20.41$.

9.35 Class size and departments. A university classifies its classes as either "small" (fewer than 40 students) or "large." A dean sees that 62% of Department A's classes are small, while Department B has only 40% small classes. She wonders if she should cut Department A's budget and insist on larger classes. Department A responds to the dean by pointing out that classes for third- and fourth-year students tend to be smaller than classes for first- and second-year students. The three-way table below gives the counts of classes by department, size, and student audience. Write a short report for the dean that summarizes these data. Start by computing the percents of small classes in the two departments, and include other numerical and graphical comparisons as needed. (Do not perform any statistical significance tests.) Here are the numbers of classes to be analyzed:

	Department A			Department B		
Year	Large	Small	Total	Large	Small	Total
First	2	0	2	18	2	20
Second	9	1	10	40	10	50
Third	5	15	20	4	16	20
Fourth	4	16	20	2	14	16

9.36 Sexual imagery in magazine ads. In what ways do advertisers in magazines use sexual imagery to appeal to youth? One study classified each of 1509 full-page or larger ads as "not sexual" or "sexual," according to the amount and style of the dress of the male or female model in the ad. The ads were also classified according to the target readership of the magazine.[21] **IMAGERY** Here is the two-way table of counts:

	Magazine Readership			
Model dress	Women	Men	General interest	Total
Not sexual	351	514	248	1113
Sexual	225	105	66	396
Total	576	619	314	1509

(a) Summarize the data numerically and graphically.

(b) Perform the significance test that compares the model dress for the three categories of magazine readership. Summarize the results of your test and give your conclusion.

(c) All of the ads were taken from the March, July, and November issues of six magazines in one year. Discuss this fact from the viewpoint of the validity of the significance test and the interpretation of the results.

9.37 Intended readership of ads with sexual imagery. The ads in the study described in the previous exercise were also classified according to the age group of the intended readership. Here is a summary of the data:

	Magazine Readership Age Group	
Model dress	Young adult	Mature adult
Not sexual (percent)	72.3%	76.1%
Sexual (percent)	27.7%	23.9%
Number of ads	1006	503

Using parts (a) and (b) in the previous exercise as a guide, analyze these data and write a report summarizing your work.

9.38 Identity theft and college students. A study of identity theft looked at how well consumers protect themselves from this increasingly prevalent crime. The behaviors of 61 college students were compared with the behaviors of 59 nonstudents. One of the questions was "When asked to create a password, I have used either my mother's maiden name, or my pet's name, or my birth date, or the last four digits of my social security number, or a series of consecutive numbers."[22] For the students, 22 agreed with this statement, while 30 of the nonstudents agreed.

(a) Display the data in a two-way table and perform the chi-square test. Summarize the results.

(b) Reanalyze the data using the methods for comparing two proportions that we studied in the previous chapter. Compare the results and verify that the chi-square statistic is the square of the z statistic.

(c) The students in this study were junior and senior college students from two sections of a course in Internet marketing at a large northeastern university. The nonstudents were a group of individuals who were recruited to attend commercial focus groups on the West Coast conducted by a lifestyle marketing organization. Discuss how the method of selecting the subjects in this study relates to the conclusions that can be drawn from it.

9.39 Student athletes and gambling. A survey of student athletes that asked questions about gambling behavior classified students according to the National Collegiate Athletic Association (NCAA) division.[23] For male student athletes, the percent who reported wagering on collegiate sports are given here along with the numbers of respondents in each division:

	Division		
	I	II	III
Percent	17.2%	21.0%	24.4%
Number	5619	2957	4089

(a) Use a significance test to compare the percents for the three NCAA divisions. Give details and a short summary of your conclusion.

(b) The percents in the table above are given in the NCAA report, but the numbers of male student athletes in each division who responded to the survey question are estimated based on other information in the report. To what extent do you think this has an effect on the results? (*Hint:* Rerun your analysis a few times, with slightly different numbers of students but the same percents.)

(c) Some student athletes may be reluctant to provide this kind of information, even in a survey where there is no possibility that they can be identified. Discuss how this fact may affect your conclusions.

(d) The chi-square test for this set of data assumes that the responses of the student athletes are independent. However, some of the students are at the same school and even on the same team. Discuss how you think this might affect the results.

9.40 Air pollution from a steel mill. One possible effect of air pollution is genetic damage. A study designed to examine this problem exposed one group of mice to air near a steel mill and another group to air in a rural area and compared the numbers of mutations in each group.[24] Here are the data for a mutation at the *Hm-2* gene locus:

	Location	
Mutation	Steel mill air	Rural air
Yes	30	23
No		
Total	96	150

(a) Fill in the missing entries in the table.

(b) Summarize the data numerically and graphically.

(c) Is there evidence to conclude that the location is related to the occurrence of mutations? Perform the significance test and summarize the results.

9.41 Hummingbirds in Santa Lucia. *E. jugularis* is a type of hummingbird that lives in the forest preserves of the Caribbean island of Santa Lucia. The males and the females of this species have bills that are shaped somewhat differently. Researchers who study these birds thought that the bill shape might be related to the shape of the flowers that they visit for food. The researchers observed 49 females and 21 males. Of the females, 20 visited the flowers of *H. bihai,* while none of the males visited these flowers.[25] Display the data in a two-way table and perform the chi-square test. Summarize the results and give a brief statement of your conclusion. Your two-way table has a count of zero in one cell. Does this invalidate your significance test? Explain why or why not.

9.42 Internet references in prominent journals. The World Wide Web has led to an enormous increase in the amount of information that is easily available to anyone with Internet access. References to Internet pages are becoming quite common in the scientific literature. One study examined Internet references in articles in three prominent journals: the *New England Journal of Medicine* (NEJM), the *Journal of the American Medical Association* (JAMA), and *Science.*[26] In one part of the study, Internet references were classified according to the top-level domain. Here are the data: **INTERNETREFERENCES**

	Journal		
Top-level domain	NEJM	JAMA	Science
.gov	41	103	111
.org	37	46	162
.com	6	17	14
.edu	4	8	47
Other	9	15	52

Analyze the data. Include numerical and graphical summaries as well as a significance test. Summarize your results and conclusions.

9.43 Students explain statistical data. The National Survey of Student Engagement conducts surveys to study various aspects of undergraduate education.[27] In a recent survey, students were asked if they needed to explain the meaning of numerical or statistical data in a written assignment. For first-year students, 43% responded positively, while for seniors, the percent was 50%. A total of 184,457 first-year students and 194,858 seniors from 722 U.S. four-year colleges and universities responded to the survey.

(a) Construct the two-way table of counts.

(b) State an appropriate null hypothesis that can be tested with these data.

(c) Perform the significance test and summarize the results. What do you conclude?

(d) The sample sizes here are very large, so even relatively small effects will be detected through a significance test. Do you think that the difference in percents is important and/or interesting? Explain your answer.

9.44 A reduction in force. In economic downturns or to improve their competitiveness, corporations may undertake a "reduction in force" (RIF), in which substantial numbers of employees are laid off. Federal and state laws require that employees be treated equally regardless of their age. In particular, employees over the age of 40 years are a "protected class." Many allegations of discrimination focus on comparing employees over 40 with their younger coworkers. Here are the data for a recent RIF: RIF1

	Over 40	
Released	No	Yes
Yes	7	41
No	504	765

(a) Complete this two-way table by adding marginal and table totals. What percent of each employee age group (over 40 or not) were laid off? Does there appear to be a relationship between age and being laid off?

(b) Perform the chi-square test. Give the test statistic, the degrees of freedom, the P-value, and your conclusion.

9.45 Performance appraisal. A major issue that arises in RIFs like that in the previous exercise is the extent to which employees in various groups are similar. If, for example, employees over 40 receive generally lower performance ratings than younger workers, that might explain why more older employees were laid off. We have data on the last performance appraisal. The possible values are "partially meets expectations," "fully meets expectations," "usually exceeds expectations," and "continually exceeds expectations." Because there were very few employees who partially met expectations, we combine the first two categories. Here are the data: RIF2

	Over 40	
Performance appraisal	No	Yes
Partially or fully meets expectations	82	230
Usually exceeds expectations	353	496
Continually exceeds expectations	61	32

Note that the total number of employees in this table is less than the number in the previous exercise because some employees do not have a performance appraisal. Analyze the data. Do the older employees appear to have lower performance evaluations?

9.46 Jury selection. Exercise 8.107 (page 489) concerns *Casteneda v. Partida,* the case in which the Supreme Court decision used the phrase "two or three standard deviations" as a criterion for statistical significance. There were 181,535 persons eligible for jury duty, of whom 143,611 were Mexican Americans. Of the 870 people selected for jury duty, 339 were Mexican Americans. We are interested in finding out if there is an association between being a Mexican American and being selected as a juror. Formulate this problem using a two-way table of counts. Construct the 2×2 table using the variables "Mexican American or not" and "juror or not." Find the X^2 statistic and its P-value. Square the z statistic that you obtained in Exercise 8.107 and verify that the result is equal to the X^2 statistic.

9.47 Which model? This exercise concerns the optional material in Section 9.2 on models for two-way tables. Look at Exercises 9.10, 9.12, 9.16, 9.28, and 9.30. For each exercise, state whether you are comparing several populations based on separate samples from each population (the first model for two-way tables) or testing independence between two categorical variables based on a single sample (the second model).

9.48 Computations for exercise and adequate sleep. Refer to the 2×2 table of data for exercise and adequate sleep in Exercise 9.12 (page 513).

(a) Compute the expected count for each cell in the table.

(b) Compute the X^2 test statistic.

(c) What are the degrees of freedom for this statistic?

(d) Sketch the appropriate χ^2 distribution for this statistic and mark the values from Table F that bracket the computed value of the test statistic. What is the P-value that you would report if you did not use software and relied solely on Table F for your work?

9.49 Computations for RDC and bank size. Refer to the 3×2 table of data for bank asset size and remote deposit capture offering in Exercise 9.10 (page 513).

(a) Compute the expected count for each cell in the table.

(b) Compute the X^2 test statistic.

(c) What are the degrees of freedom for this statistic?

(d) Sketch the appropriate χ^2 distribution for this statistic and mark the values from Table F that bracket the computed value of the test statistic. What is the P-value that you would report if you did not use software and relied solely on Table F for your work?

9.50 *Titanic!* In 1912 the luxury liner *Titanic,* on its first voyage, struck an iceberg and sank. Some passengers got off the ship in lifeboats, but many died. Think of the *Titanic* disaster as an experiment in how the people of that time behaved when faced with death in a situation where only some can escape. The passengers are a sample from the population of their peers. Here is information about who lived and who died, by gender and economic status. (The data leave out a few passengers whose economic status is unknown.)[28] TITANIC

Men		
Status	Died	Survived
Highest	111	61
Middle	150	22
Lowest	419	85
Total	680	168

Women		
Status	Died	Survived
Highest	6	126
Middle	13	90
Lowest	107	101
Total	126	317

(a) Compare the percents of men and of women who died. Is there strong evidence that a higher proportion of men die in such situations? Why do you think this happened?

(b) Look only at the women. Describe how the three economic classes differ in the percent of women who died. Are these differences statistically significant?

(c) Now look only at the men and answer the same questions.

CHAPTER 9 Case Study Exercises

CASE STUDY Exercise 1: Pew survey of broadband internet use. The Pew Home Broadband Survey of 2008 is described at pewresearch.org/pubs/888 and the responses of the 2251 adults aged 18 and older are given at pewinternet.org/Reports/2008/Home-Broadband-2008 .aspx. Study the results and pick a few of the questions that interest you. Analyze the data using the methods that you learned in this chapter and in Chapter 2, Section 2.5, to prepare a report on broadband use. Be sure to include graphical and numerical summaries as well as the results of any significance tests that you decide are appropriate. Include an Executive Summary at the beginning of your report describing your major findings.

CASE STUDY Exercise 2: Canadian students in private career colleges. Exercises 9.29 to 9.31 (pages 517 and 518) gave data about time of entry into college and sources of funding for Canadian students in private career colleges. Below is some additional information about the students in this survey. Analyze the data and write a report summarizing your findings. For some of the tables you might want to analyze the data with and without those who did not respond. Describe each field of study according to how it is similar to or different from other fields of study. Be sure to include graphical and numerical summaries.

CANADACOLLEGES

Employment status

Field of study	Number of students	Percent employed
Trades	942	25%
Design	599	45%
Health	5234	42%
Media/IT	3238	32%
Service	1378	38%
Other	2300	38%

Salary for those employed full-time

Field of study	Number of students	Salary Range (thousands)				
		$19 or less	$20 to $29	$30 to $39	$40 or more	No response
Trades	68	45%	17%	6%	25%	7%
Design	21	64%	21%	3%	7%	6%
Health	318	48%	24%	12%	11%	6%
Media/IT	282	47%	27%	10%	12%	4%
Service	103	49%	22%	15%	10%	5%
Other	106	47%	24%	15%	9%	4%

Preprogram employment status

Field of study	Number of students	Status				
		Employed	Not working	In school	Other	No response
Trades	942	43%	23%	11%	5%	18%
Design	599	49%	14%	18%	6%	12%
Health	5234	49%	13%	12%	9%	17%
Media/IT	3238	39%	24%	11%	11%	15%
Service	1378	46%	10%	8%	11%	19%
Other	2300	50%	21%	9%	6%	12%

Satisfaction with course content

Field of study	Number of students	Percent very satisfied or satisfied
Trades	942	83%
Design	599	86%
Health	5234	85%
Media/IT	3238	83%
Service	1378	87%
Other	2300	90%

Satisfaction with job market preparation

Field of study	Number of students	Percent very satisfied or satisfied
Trades	942	76%
Design	599	75%
Health	5234	77%
Media/IT	3238	74%
Service	1378	79%
Other	2300	76%

CHAPTER 9 Appendix

Using Minitab and Excel for Analysis of Two-Way Tables

Analysis of Two-Way Tables

Minitab:

Stat ➤ Tables ➤ Cross Tabulation and Chi-Square

With this pull-down sequence, you can have Minitab tabulate the occurrences of one variable against another variable to ultimately create a two-way table. To create a two-way table with r rows and c columns, you should have two columns of data in the worksheet. One column must have r distinct labels (numeric or text), and the other column must have c distinct labels (numeric or text). Any given row will then reflect the simultaneous occurrence of one of the r categories with one of the c categories. For example, if you open the Minitab file for Case 9.1 (FRANCHISES), you will find two columns named "success" and "excl" in which there are two distinct values, namely, "1" and "2," which represent the categories of "Yes" and "No," respectively. In the first row, the data entry for success is "2" and the data entry for "excl" is "2"; this row then represents 1 of the 13 unsuccessful firms that did not have an exclusive-territory contract; see Case 9.1 (page 494). In general, to implement this Minitab routine, click the column with the r categories into the **For rows** box and click the column with the c categories into the **For columns** box. In addition to reporting counts, you can request percents (row, column, and total) to be reported by making selections in the **Display** option section. If the counts have been pretabulated for each combination of categories, click the column having these counts in the optional **Frequencies are in** dialog box. To conduct a chi-square analysis, click the **Chi-Square** button and then select the **Chi-Square analysis** option and, if you wish, the **Expected cell counts** option. Click **OK** to close the pop-up box and then click **OK** to find the two-way table along with the chi-square analysis reported in the Session window. It should be noted that Minitab will report two chi-square statistics (each with a P-value). The Pearson chi-square statistic is the test statistic discussed in this chapter.

If the cell counts for the two-way table have been pretabulated and the counts are laid out as a two-way table in the worksheet, then consider the following pull-down sequence:

Stat ➤ Tables ➤ Chi-Square Test

(Two-Way Table in Worksheet)

In general, the worksheet should have c columns and each column should have r rows. The entries in the worksheet cells are the two-way table cell counts. For example, if you open the Minitab file for Case Study Exercise 2 (*MUSIC*), you will find data in three rows of three columns. In the first row of column one (c1), the data entry is "30," which represents the 30 bottles of French wine purchased when no background music was played; see Case 9.2 (page 505). In general, to implement this routine, click all the data columns into the **Columns containing the table** box and then click **OK** to find the two-way table along with the chi-square analysis reported in the Session window.

Excel:

Excel provides a statistical function that will perform a chi-square analysis. This function requires as inputs the two-way table cell counts and the expected counts. Excel does not automatically calculate the expected counts. As a first step, we need to create the two-way table. You should have two adjacent columns of data in the spreadsheet, where one column has r distinct labels (numeric or text) and the other column has c distinct labels (numeric or text). For each of these columns, there should be a column name in the top row of each column. Now click any cell in the data range. Then click the **Insert** tab, click **PivotTable** in the **Tables** group, and finally click **PivotTable**. The cell range that includes the two column names and the data will be inputted into the **Table/Range** box. Click **OK** and Excel will produce a **PivotTable Field List** box. You will find the column name(s) that you highlighted listed as field(s). Select the field(s) presented to you by clicking a checkmark next to the name(s). Now, click and hold the field name containing the r distinct labels and drag the field from the field section to the **Row Labels** section. Similarly, click and hold the field name containing the c distinct labels and drag the field from the field section to the **Column Labels** section. At this stage, drag one of the fields (not both) into the **Σ Values** section. To have counts reported, click the field name found in the **Σ Values** section, select the **Value Field Setting** option, click the **Summarize by** tab, select "Count" from the menu, and then click **OK** to find the two-way table in the spreadsheet. If the cell counts for the two-way table have been pretabulated, simply put the cell counts directly into the spreadsheet to construct the two-way table. It is then useful to sum the counts to obtain row and column totals, which can be placed in the margins of the table.

With the two-way table created, you now need to create a two-way table of expected counts in a separate area of the spreadsheet. This is done by inserting in a given cell the formula of row total times column total divided by the total sample size. To calculate the chi-square statistic, click an empty cell in the spreadsheet, and then proceed to the **Statistical** function menu as described in the overview section of Chapter 1's Appendix. Scroll down the list of functions and click on the **CHITEST** function choice. In the **Actual_range** box, input the cell range of the observed cell counts. In the **Expected_range** box, input the cell range of the expected cell counts. Click **OK** to find the P-value for the chi-square test reported in the selected cell.

CHAPTER **10**

Inference for Regression

Is job stress associated with the extent to which a person feels in control of events? Feeling less in control is associated with more job stress, as Case 10.1 shows.

Introduction

One of the most common uses of statistical methods is to predict a response based on one or several explanatory variables. Prediction is most straightforward when there is a straight-line relationship between a quantitative response variable and a single quantitative explanatory variable. This is **simple linear regression,** the topic of this chapter. Here are some regression problems we will consider:

- Is feeling more in control of one's destiny associated with lower job stress in a sample of accountants? If so, how can we generalize the result from this sample to the population of accountants?

- Among North American universities, is there a positive correlation between the student binge-drinking rate and the average price for a bottle of beer at establishments within a two-mile radius of the campus?

- How is a company's reputation (a subjective measure) related to objective performance measures such as profitability?

As we saw in Chapter 2, when a scatterplot shows a linear relationship between a quantitative explanatory variable x and a quantitative response variable y, we can use the least-squares line fitted to the data to predict y for a given value of x. Now we want to do tests and confidence intervals in this setting.

To do this, we will think of the least-squares line, $b_0 + b_1 x$, as an estimate of a regression line for the population, just as the sample mean $\overline{x}$ is an estimate of

population regression line

the population mean μ. We write the **population regression line** as $\beta_0 + \beta_1 x$. The numbers β_0 and β_1 are *parameters* that describe the population. The numbers b_0 and b_1 are *statistics* calculated from a sample. The intercept b_0 estimates the intercept β_0 of the population line, and the slope b_1 estimates the slope β_1.

We can give confidence intervals and use significance tests for inference about the slope β_1 and the intercept β_0. Because regression lines are often used for prediction, we also consider inference about either the mean response or an individual future observation on y for a given value of the explanatory variable x. Finally, we discuss statistical inference about the correlation between two variables x and y.

10.1 Inference about the Regression Model

Simple linear regression studies the relationship between a response variable y and an explanatory variable x. We expect that different values of x will produce different mean responses. We encountered a similar but simpler situation in Chapter 7 when we discussed methods for comparing two population means. Figure 10.1 illustrates a statistical model for comparing the items per hour entered by two groups of financial clerks using new data entry software. Group 2 received some training in the software while Group 1 did not. Entries per hour is the response variable. The treatment (training or not) is the explanatory variable. The model has two important parts:

- The mean entries per hour may be different in the two populations. These means are μ_1 and μ_2 in Figure 10.1.
- Individual subjects' entries per hour vary within each population according to a Normal distribution. The two Normal curves in Figure 10.1 describe the individual responses. These Normal distributions have the same standard deviation.

Statistical model for simple linear regression

subpopulation

Now imagine giving different lengths x of training to different groups of subjects. We can think of these groups as belonging to **subpopulations,** one for each possible value of x. Each subpopulation consists of all individuals in the population having the same value of x. If we gave $x = 15$ hours of training to some subjects, $x = 30$ hours of training to some others, and $x = 60$ hours of training to some others, these three groups of subjects would be considered samples from the corresponding three subpopulations.

FIGURE 10.1 The statistical model for comparing the responses to two treatments. The mean response depends on the treatment.

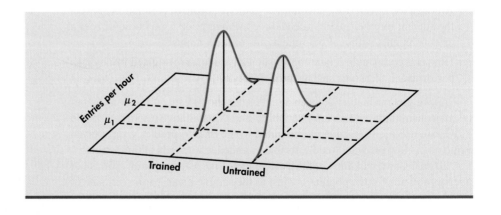

FIGURE 10.2 The statistical model for linear regression. The mean response is a straight-line function of the explanatory variable.

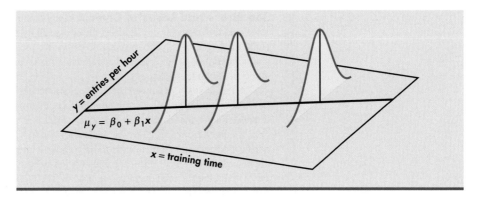

The statistical model for simple linear regression also assumes that for each value of x the response variable y is Normally distributed with a mean that depends on x. We use μ_y to represent these means. This model also has two important parts:

- The mean entries per hour μ_y changes as the number of hours of training x changes. The means all lie on a straight line. That is, $\mu_y = \beta_0 + \beta_1 x$.

- Individual entries per hour y for subjects with the same amount of training x vary according to a Normal distribution. These Normal distributions all have the same standard deviation.

Compare Figure 10.1 for two groups with Figure 10.2, which pictures the regression model in which the explanatory variable x can take many values. Rather than just two means μ_1 and μ_2, we are interested in how the many means μ_y change as x changes. Although the means μ_y might follow any sort of pattern as x changes, in simple linear regression we assume that all these means lie on a line when plotted against x. The equation of the line is $\mu_y = \beta_0 + \beta_1 x$, with intercept β_0 and slope β_1. This is the population regression line; it describes how the mean response changes with x. The line in Figure 10.2 is the population regression line. The three Normal curves show how responses y will vary for three different values of x. Each curve is centered at the mean response, that is, at the μ_y given by the population regression line for a specific x. All three curves have the same spread, measured by their common standard deviation σ.

From data analysis to inference

The data for a regression problem are observed values of x and y. The model takes each x to be a fixed known quantity, like the hours of training a worker has received.[1] The response y for a given x is a Normal random variable. The model describes the mean and standard deviation of this random variable.

We will use Case 10.1 to explain the fundamentals of simple linear regression. Because regression calculations in practice are always done by software, we will rely on computer output for the arithmetic for Case 10.1. Later in the chapter, we will show formulas for doing the calculations. These are useful in understanding analysis of variance (see Section 10.3) and multiple regression (see Chapter 11).

LOC

Job Stress and Locus of Control Many factors, such as the type of job, education level, and job experience, can affect the stress felt by workers on the job. Locus of control (LOC) is a term in psychology that describes the extent to which a person believes he or she is in control of the events that influence his or her life. Is feeling "more in control" associated with less job stress?

A recent study examined the relationship between LOC and several work-related behavioral measures among certified public accountants in Taiwan.[2] LOC was assessed using a questionnaire that asked respondents to select one of two options for each of 23 items. Scores ranged from 0 to 23. Individuals with low LOC believe that their own behavior and attributes determine their rewards in life. Those with high LOC believe that these rewards are beyond their control. Each accountant's job stress was assessed using the averaged score on 22 items, each scored on a five-point scale. The higher the score, the higher the perceived job stress. We will consider a random sample of 100 accountants. Figure 10.3 is a scatterplot of the data. The explanatory variable x is the respondent's LOC. The response variable y is the job stress score STRESS.

FIGURE 10.3 Scatterplot, with regression line, of perceived job stress versus locus of control for a sample of Taiwanese accountants.

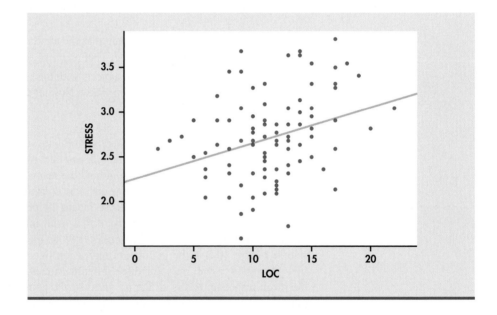

Let's briefly review some of the ideas from Chapter 2 regarding least-squares regression. We start with a plot of the data, as in Figure 10.3, and verify that the relationship is approximately linear with no outliers. We then apply the least-squares method to find the least-squares regression line.

This idea is illustrated in Figure 10.4, which includes only the data points for the accountants with the three highest values of LOC and part of the least-squares line. The middle point corresponds to Subject 31, who has an LOC (x) equal to 20. The observed value of job stress (y) is 2.818. For an LOC equal to 20, the value of job stress on the least-squares line is 3.053. This is the predicted value of job stress ($\hat{y}$) for an individual who has an LOC of 20. The difference between the observed (y) and predicted ($\hat{y}$) values of job stress is the residual ($y - \hat{y}$). The least-squares regression line is the line that minimizes the sum of the squares of the residuals.

FIGURE 10.4 The least-squares idea. For each observation, find the vertical distance between each point and a regression line. The least-squares regression line makes the sum of the squares of these distances as small as possible.

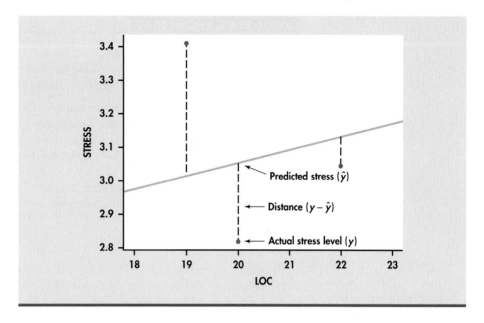

EXAMPLE 10.1   Prediction of Job Stress from LOC

CASE 10.1

The line in Figures 10.3 and 10.4 is the least-squares regression line for predicting job stress from LOC. The equation of this line is

$$\text{predicted STRESS} = 2.2555 + 0.0399 \times \text{LOC}$$

In Section 2.2 we discussed the correlation as a measure of association between two quantitative variables. In Section 2.3 we learned to interpret the square of the correlation as the fraction of the variation in y that is explained by x in a simple linear regression.

EXAMPLE 10.2   Correlation between STRESS and LOC

CASE 10.1

For Case 10.1, the correlation between LOC and STRESS is $r = 0.3123$. Because the squared correlation $r^2 = 0.0975$, the change in STRESS along the regression line as LOC increases explains only about 10% of the variation. The remaining 90% is due to other differences among these accountants. For example, they do not all have the same education and job experience.

We can use the least-squares regression equation to find predicted STRESS corresponding to a given value of LOC. The difference between the observed and predicted values is the residual.

EXAMPLE 10.3   Predicted Values and Residuals for STRESS

CASE 10.1

Suppose an accountant has an LOC of 16. We predict that this person will have a job stress score of

$$2.2555 + (0.0399)(16) = 2.894$$

If the actual job stress score is 1.909, then the residual would be

$$y - \hat{y} = 1.909 - 2.894 = -0.985$$

10.1 Predict the level of job stress.   In Case 10.1, the third accountant had an LOC of 3 and a stress level of 2.68. Using the least-squares regression equation in Example 10.1, find the predicted stress level and the residual for this individual.

10.2 Understanding a linear regression model.   Consider a linear regression model with $\mu_y = 100.35 - 5.8x$ and standard deviation $\sigma = 3.2$.
(a) What is the slope of the population regression line?
(b) Explain clearly what this slope says about the change in the mean of y for a unit change in x.
(c) What is the subpopulation mean when $x = 12$?
(d) Between what two values would approximately 95% of the observed responses y fall when $x = 12$?

Having reviewed the basics of least-squares regression, we are now ready to proceed with a discussion of inference for regression. Here's what is new in this chapter:

- We regard the 100 accountants for whom we have data as a simple random sample from the population of all certified public accountants in Taiwan.

- We use the regression line calculated from this sample as a basis for inference about the population. For example, we want to give not simply a prediction but a prediction with a margin of error and a level of confidence for the job stress score of any accountant in Taiwan.

The population regression line $\mu_y = \beta_0 + \beta_1 x$ describes the relationship between mean stress score μ_y and the locus of control measure x in the population. The slope β_1 is the mean increase in stress score for each unit increase in LOC. The intercept β_0 is also meaningful in this example. It is the mean stress score when an accountant has an extreme internal locus of control. By "extreme internal" we mean an individual who believes that his or her destiny is controlled completely by his or her own actions. Luck or other external circumstances play no role.

We cannot observe the population regression line, because the observed responses y vary about their means. In Figure 10.3 we see the least-squares regression line that describes the overall pattern of the data, along with the scatter of individual points about this line. The statistical model for regression makes the same distinction. The population regression line describes only the on-the-average relationship. Think of the model in the form

$$\text{DATA} = \text{FIT} + \text{RESIDUAL}$$

The FIT part of the model consists of the subpopulation means, given by the expression $\beta_0 + \beta_1 x$. The RESIDUAL part represents deviations of the data from the line of population means. The model assumes that these deviations are Normally distributed with standard deviation σ. We use ϵ (the lowercase Greek letter epsilon) to stand for the RESIDUAL part of the statistical model. A response y is the sum of its mean and a chance deviation ϵ from the mean. The deviations ϵ represent "noise," variation in y due to other causes that prevent the observed (x, y)-values from forming a perfectly straight line on the scatterplot.

> **Simple Linear Regression Model**
>
> The data are n observations on an explanatory variable x and a response variable y,
>
> $$(x_1, y_1), \quad (x_2, y_2), \quad \ldots, \quad (x_n, y_n)$$
>
> The **statistical model for simple linear regression** states that the observed response y_i when the explanatory variable takes the value x_i is
>
> $$y_i = \beta_0 + \beta_1 x_i + \epsilon_i$$
>
> Here $\mu_y = \beta_0 + \beta_1 x_i$ is the mean response when $x = x_i$. The deviations ϵ_i are independent and Normally distributed with mean 0 and standard deviation σ.
>
> The parameters of the model are the intercept β_0 and slope β_1 of the population regression line and the variability σ of the response y about this line.

The simple linear regression model can be justified in a wide variety of circumstances. Sometimes we observe the values of two variables and we formulate a model with one of these as the response variable and the other as the explanatory variable. This was the setting for Case 10.1, where the response variable was perceived job stress and the explanatory variable was locus of control. In other settings the values of the explanatory variable are chosen by the persons designing the study. The scenario illustrated by Figure 10.2 is an example of this setting. Here the explanatory variable is training time, which is set at a few carefully selected values. The response variable is the number of entries per hour.

For the simple linear regression model to be valid, one essential assumption is that the relationship between the response variable and the explanatory variable is approximately linear. This is the FIT part of the model. Another essential assumption concerns the RESIDUAL part of the model. The assumption states that the residuals are approximately Normal with mean zero and standard deviation σ. If the data are collected through some sort of random sampling, this assumption is often easy to justify. This is the case in the two scenarios described above, in which both variables are observed in a random sample from a population or the response variable is measured at predetermined values of the explanatory variable. In many other settings, particularly in business applications, we analyze all of the data available and there is no random sampling. Here we often justify the use of inference for simple linear regression by viewing the data as coming from some sort of process. The line gives a good description of the relationship, the fit, and we model the deviations from the fit, the residuals, as coming from a Normal probability distribution.

EXAMPLE 10.4 Retail Sales and Floor Space

It is customary in retail operations to assess the performance of stores partly in terms of their annual sales relative to their floor area (square feet). We might expect sales to increase linearly as stores get larger, with, of course, individual variation among stores of the same size. The regression model for a population of stores says that

$$\text{SALES} = \beta_0 + \beta_1 \times \text{AREA} + \epsilon$$

The slope β_1 is, as usual, a rate of change: it is the expected increase in annual sales associated with each additional square foot of floor space. The intercept β_0 is needed to describe the line but has no statistical importance because no stores have area close to zero. Floor space does not completely determine sales. The ϵ term in the model accounts for differences among individual stores with the same floor space. A store's location, for example, could be important but is not included in the FIT part of the model.

10.3 Domestic and foreign stock markets. Returns on common stocks in the United States and overseas appear to be growing more closely correlated as economies become more interdependent. Suppose that the following population regression line connects the total annual returns (in percent) on two indexes of stock prices:

$$\text{MEAN OVERSEAS RETURN} = 0.3 + 0.85 \times \text{U.S. RETURN}$$

(a) What is β_0 in this line? What does this number say about overseas returns when the U.S. market is flat (0% return)?

(b) What is β_1 in this line? What does this number say about the relationship between U.S. and overseas returns?

(c) We know that overseas returns will vary in years that have the same return on U.S. common stocks. Write the regression model based on the population regression line given above. What part of this model allows overseas returns to vary when U.S. returns remain the same?

10.4 Fixed and variable costs. In some mass production settings there is a linear relationship between the number x of units of a product in a production run and the total cost y of making these x units.

(a) Write a population regression model to describe this relationship.

(b) Which parameter in your model is the fixed cost (for example, the cost of setting up the production line) that does not change as x increases?

(c) Which parameter in your model shows how total cost changes as more units are produced? Do you expect this number to be greater than 0 or less than 0?

(d) Actual data from several production runs will not fit a straight line exactly. What term in your model allows variation among runs of the same size x?

Estimating the regression parameters

The method of least squares presented in Chapter 2 fits the least-squares line to summarize a relationship between the observed values of an explanatory variable and a response variable. Now we want to use this line as a basis for inference about a population from which our observations are a sample. We can do this when the statistical model for regression holds. In that setting, the slope b_1 and intercept b_0 of the least-squares line

$$\hat{y} = b_0 + b_1 x$$

estimate the slope β_1 and the intercept β_0 of the population regression line.

Recalling the formulas from Chapter 2, the slope of the least-squares line is

$$b_1 = r \frac{s_y}{s_x}$$

and the intercept is

$$b_0 = \bar{y} - b_1 \bar{x}$$

Here, r is the correlation between the observed values of y and x, s_y is the standard deviation of the sample of y's, and s_x is the standard deviation of the sample of x's.

The remaining parameter to be estimated is σ, which measures the variation of y about the population regression line. More precisely, σ is the standard deviation of the Normal distribution of the deviations ϵ_i in the regression model. It should come as no surprise that we use the deviations of the observed responses from the fitted line to estimate σ.

residuals

    Recall that the vertical deviations of the points in a scatterplot from the fitted regression line are the **residuals.** We will use e_i for the residual of the ith observation:

$$e_i = \text{observed response} - \text{predicted response}$$
$$= y_i - \hat{y}_i$$
$$= y_i - b_0 - b_1 x_i$$

The residuals e_i are the sample quantities that correspond to the model deviations ϵ_i. The e_i sum to 0, and the ϵ_i come from a population with mean 0.

    To estimate σ, we work first with the variance and take the square root to obtain the standard deviation. For simple linear regression the estimate of σ^2 is the average squared residual

$$s^2 = \frac{1}{n-2} \sum e_i^2$$
$$= \frac{1}{n-2} \sum (y_i - \hat{y}_i)^2$$

We average by dividing the sum by $n - 2$ in order to make s^2 an unbiased estimator of σ^2. Recall that the sample variance of n observations uses the divisor $n - 1$ because the n deviations $x_i - \bar{x}$ are not n separate quantities but rather must sum to 0. Similarly, the residuals e_i are not n separate quantities. When any $n - 2$ residuals are known, we can

degrees of freedom

find the other two. We say that the residuals and s^2 have $n - 2$ **degrees of freedom.** To estimate σ, use

$$s = \sqrt{s^2}$$

We call s the *regression standard error.*

Estimating the Regression Parameters

    In the simple linear regression setting, we use the slope b_1 and intercept b_0 of the **least-squares regression line** to estimate the slope β_1 and intercept β_0 of the population regression line.

    The standard deviation σ in the model is estimated by the **regression standard error**

$$s = \sqrt{\frac{1}{n-2} \sum (y_i - \hat{y}_i)^2}$$

    In practice, we use software to calculate b_1, b_0, and s from data on x and y. Here are the results for the job stress example of Case 10.1.

EXAMPLE 10.5   Job Stress and Locus of Control

CASE 10.1

Figure 10.5 displays the Excel output for the regression of perceived job stress (STRESS) on locus of control (LOC) for our sample of 100 accountants in Taiwan. In this output, we find the correlation $r = 0.3123$ and the squared correlation that we used in Example 10.2, along with the intercept and slope of the least-squares line. The regression standard error s is labeled simply "Standard Error."

    The three parameter estimates are

$$b_0 = 2.2554967 \quad b_1 = 0.03990857 \quad s = 0.451276599$$

After rounding, the fitted regression line is

$$\hat{y} = 2.2555 + 0.0399x$$

Excel

	A	B	C	D	E	F	G
	F14 ▼	fx					
1	SUMMARY OUTPUT						
2							
3	*Regression Statistics*						
4	Multiple R	0.31227647					
5	R Square	0.097516594					
6	Adjusted R Square	0.088307579					
7	Standard Error	0.451276599					
8	Observations	100					
9							
10	ANOVA						
11		*df*	*SS*	*MS*	*F*	*Significance F*	
12	Regression	1	2.156507639	2.156508	10.58925418	0.001561535	
13	Residual	98	19.95775576	0.203651			
14	Total	99	22.1142634				
15							
16		*Coefficients*	*Standard Error*	*t Stat*	*P-value*	*Lower 95%*	*Upper 95%*
17	Intercept	2.2554967	0.146912731	15.35263	8.02318E-28	1.963953174	2.547040226
18	LOC	0.03990857	0.012264038	3.254113	0.001561535	0.015570987	0.064246153
19							

FIGURE 10.5 Excel output for the regression of job stress on locus of control, for Example 10.5.

As usual, we ignore the parts of the output that we do not yet need. We will return to the output for additional information later.

Figure 10.6 shows the regression output from two other software packages. Although the formats differ, you can easily find the results you need. Once you know what to look for, you can understand statistical output from almost any software.

APPLY YOUR KNOWLEDGE

10.5 Research and development spending. The National Science Foundation collects data on the research and development spending by universities and colleges in the United States.[3] Here are the data for the years 2003 to 2007:

Year:	2003	2004	2005	2006	2007
Spending (billions of dollars):	40.1	43.3	45.8	47.7	49.4

(a) Make a scatterplot that shows the increase in research and development spending over time. Does the pattern suggest that the spending is increasing linearly over time?

(b) Find the equation of the least-squares regression line for predicting spending from year. Add this line to your scatterplot.

(c) For each of the five years find the residual. Use these residuals to calculate the standard error s. (Do these calculations with a calculator.)

Minitab

```
The regression equation is
STRESS = 2.26 + 0.0399 LOC

Predictor          Coef    SE Coef        T        P
Constant         2.2555     0.1469    15.35    0.000
LOC             0.03991    0.01226     3.25    0.002


S = 0.451277     R-Sq = 9.8%        R-Sq(adj) = 8.8%

Analysis of Variance

Source             DF         SS        MS        F        P
Regression          1     2.1565    2.1565    10.59    0.002
Residual Error     98    19.9578    0.2037
Total              99    22.1143
```

SPSS

Model Summary

Model	R	R Square	Adjusted R Square	Std. Error of the Estimate
1	.312[a]	.098	.088	.45127660

a. Predictors: (Constant), LOC

ANOVA[b]

Model		Sum of Squares	df	Mean Square	F	Sig.
1	Regression	2.157	1	2.157	10.589	.002[a]
	Residual	19.958	98	.204		
	Total	22.114	99			

a. Predictors: (Constant), LOC
b. Dependent Variable: STRESS

Coefficients[a]

Model		Unstandardized Coefficients		Standardized Coefficients		
		B	Std. Error	Beta	t	Sig.
1	(Constant)	2.255	.147		15.353	.000
	LOC	.040	.012	.312	3.254	.002

a. Dependent Variable: STRESS

FIGURE 10.6 Minitab and SPSS outputs for the regression of job stress on locus of control. The data are the same as in Figure 10.5.

(d) Write the regression model for this setting. What are your estimates of the unknown parameters in this model?
(e) Use your least-squares equation to predict research and development spending for the year 2001. The actual spending for that year was $32.8 billion. Add this point to your plot and comment on why your equation performed so poorly.

(*Comment:* These are *time series data.* Simple regression is often a good fit to time series data over a limited span of time. See Chapter 13 for methods designed specifically for use with time series.)

Conditions for regression inference

You can fit a least-squares line to any set of explanatory-response data when both variables are quantitative. The simple linear regression model, which is the basis for inference, imposes several conditions on this fit. *We should always verify these conditions before proceeding to inference.* There is no point in trying to do statistical inference if the data do not, at least approximately, meet the conditions that are the foundation for the inference.

The conditions concern the population, but we can observe only our sample. Thus, in doing inference, we act as if **the sample is an SRS from the population.** For the study described in Case 10.1, the researchers sent out 620 surveys to various accounting firms in Taiwan. With a return rate of only 33.7%, response bias may be a concern, but we will treat this sample as an SRS.

The next condition is that **there is a linear relationship in the population,** described by the population regression line. We can't observe the population line, so we check this condition by asking if the sample data show a roughly linear pattern in a scatterplot. We also check for any outliers or influential observations (page 111) that could impact the least-squares fit. The model also says that **the standard deviation of the responses about the population line is the same for all values of the explanatory variable.** In practice, the spread of observations above and below the least-squares line should be roughly uniform as x varies.

Plotting the residuals against the explanatory variable is another helpful and frequently used visual aid to check these conditions, because a residual plot magnifies patterns. The residual plot in Figure 10.7 for the data of Case 10.1 looks satisfactory. There is no curved pattern or data points that seem out of the ordinary, and whatever narrowing of the spread there may be for small and large values of LOC can be explained by the fact that there are fewer observations there.

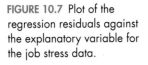

FIGURE 10.7 Plot of the regression residuals against the explanatory variable for the job stress data.

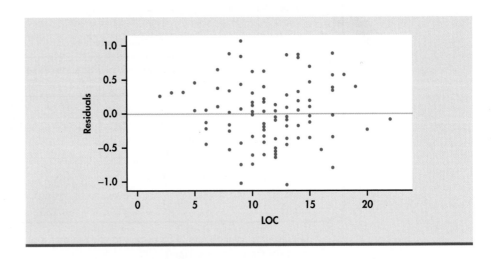

FIGURE 10.8 Normal quantile plot of the regression residuals for the job stress data.

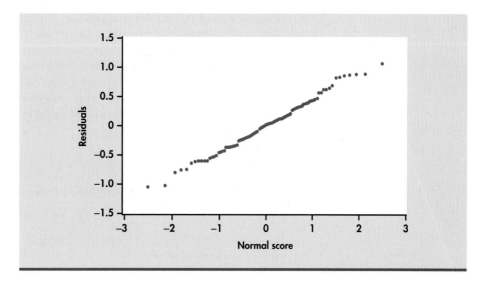

The final condition is that **the response varies Normally about the population regression line.** In that case, we expect the residuals e_i to also be Normally distributed.[4] A Normal quantile plot of the residuals (Figure 10.8) shows no serious deviations from a Normal distribution. The data give no reason to doubt the simple linear regression model, so we proceed to inference.

Confidence intervals and significance tests

Chapter 7 presented confidence intervals and significance tests for means and differences in means. In each case, inference rested on the standard errors of estimates and on t distributions. Inference for the slope and intercept in linear regression is similar in principle, although the recipes are more complicated. All the confidence intervals, for example, have the form

$$\text{estimate} \pm t^* \text{SE}_{\text{estimate}}$$

where t^* is a critical value of a t distribution.

Confidence intervals and tests for the slope and intercept are based on the sampling distributions of the estimates b_1 and b_0. Here are the facts:

- When the simple linear regression model is true, each of b_0 and b_1 has a **Normal distribution.**
- The **mean** of b_0 is β_0 and the mean of b_1 is β_1. That is, the intercept and slope of the fitted line are unbiased estimators of the intercept and slope of the population regression line.
- The **standard deviations** of b_0 and b_1 are multiples of the model standard deviation σ. (We will give details later.)

Normality of b_0 and b_1 is a consequence of Normality of the individual deviations ϵ_i in the regression model. Fortunately, a general form of the central limit theorem tells us that the distributions of b_0 and b_1 will be approximately Normal when we have a large sample, even if the ϵ_i are not. **Regression inference is robust against moderate lack of Normality.** On the other hand, outliers and influential observations can invalidate the results of inference for regression.

Because b_0 and b_1 have Normal sampling distributions, standardizing these estimates gives standard Normal z statistics. The standard deviations of these estimates are multiples of σ, the model parameter that describes the variability about the population regression line. Because we do not know σ, we estimate it by s, the variability of the data about the least-squares line. When we do this, we get t distributions with degrees of freedom $n - 2$, the degrees of freedom of s. We give formulas for the standard errors SE_{b_1} and SE_{b_0} later. For now we will concentrate on the basic ideas and let software do the calculations.

Inference for Regression Slope

A **level C confidence interval** for the slope β_1 of the population regression line is

$$b_1 \pm t^* SE_{b_1}$$

In this expression t^* is the value for the $t(n-2)$ density curve with area C between $-t^*$ and t^*. The **margin of error** is $t^* SE_{b_1}$.

To test the hypothesis $H_0\colon \beta_1 = 0$, compute the t **statistic**

$$t = \frac{b_1}{SE_{b_1}}$$

The **degrees of freedom** are $n - 2$. In terms of a random variable T having the $t(n-2)$ distribution, the P-value for a test of H_0 against

$H_a\colon \beta_1 > 0$ is $P(T \geq t)$

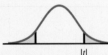

$H_a\colon \beta_1 < 0$ is $P(T \leq t)$

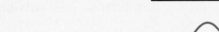

$H_a\colon \beta_1 \neq 0$ is $2P(T \geq |t|)$

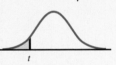

Confidence intervals and significance tests for the intercept β_0 are exactly the same, replacing b_1 and SE_{b_1} by b_0 and its standard error SE_{b_0}. Although computer outputs often include a test of $H_0\colon \beta_0 = 0$, this information usually has little practical value. From the equation for the population regression line, $\mu_y = \beta_0 + \beta_1 x$, we see that β_0 is the mean response corresponding to $x = 0$. In many practical situations, this subpopulation does not exist or is not interesting.

On the other hand, the test of $H_0\colon \beta_1 = 0$ is quite useful. When we substitute $\beta_1 = 0$ in the model, the x term drops out and we are left with

$$\mu_y = \beta_0$$

This model says that the mean of y does not vary with x. In other words, all the y's come from a single population with mean β_0, which we would estimate by $\overline{y}$. The hypothesis $H_0\colon \beta_1 = 0$ therefore says that there is no straight-line relationship between y and x and that linear regression of y on x is of no value for predicting y. The t test for this hypothesis asks whether the data provide sufficient evidence to conclude that the slope is not zero.

EXAMPLE 10.6   Does Job Stress Increase with LOC?

CASE 10.1

The Excel regression output in Figure 10.5 for the job stress problem contains the information needed for inference about the regression coefficients. You can see that the slope of the least-squares line is $b_1 = 0.0399$ and that the standard error of this statistic is $SE_{b_1} = 0.012264$.

The t statistic and P-value for the test of H_0: $\beta_1 = 0$ against the two-sided alternative H_a: $\beta_1 \neq 0$ appear in the columns labeled "t Stat" and "P-value." The t statistic for the significance of the regression is

$$t = \frac{b_1}{SE_{b_1}} = \frac{0.0399}{0.012264} = 3.25$$

and the P-value for the two-sided alternative is 0.0016. If we expected beforehand that job stress rises with locus of control, our alternative hypothesis would be one-sided, H_a: $\beta_1 > 0$. The P-value for this H_a is one-half the two-sided value given by Excel; that is, $P = 0.0008$. In both cases there is strong evidence that the mean stress level increases as LOC increases.

A 95% confidence interval for the slope β_1 of the regression line in the population of all accountants in Taiwan is

$$b_1 \pm t^* SE_{b_1} = 0.0399 \pm (1.990)(0.012264)$$
$$= 0.0399 \pm 0.02441$$
$$= 0.0155 \text{ to } 0.0643$$

This interval contains only positive values, suggesting an increase in mean stress level for a unit increase in LOC. However, the increase in mean stress level is quite small relative to the range of stress scores.

The t distributions for this problem have $n - 2 = 98$ degrees of freedom. Table D has no entry for 98 degrees of freedom, so we used the table entry $t^* = 1.990$ for 80 degrees of freedom. As a result, our confidence interval agrees only approximately with the more accurate software result. Note that using the next *lower* degrees of freedom in Table D makes our interval a bit wider than we actually need for 95% confidence. Use this conservative approach when you don't know t^* for the exact degrees of freedom.

APPLY YOUR KNOWLEDGE

INFLATION

Treasury bills and inflation. *When inflation is high, lenders require higher interest rates to make up for the loss of purchasing power of their money while it is loaned out. Table 10.1 displays the return of six-month Treasury bills (annualized) and the rate of inflation as measured by the change in the government's Consumer Price Index in the same year.[5] An inflation rate of 5% means that the same set of goods and services costs 5% more. The data cover 51 years, from 1958 to 2008. Figure 10.9 is a scatterplot of these data. Figure 10.10 shows Excel regression output for predicting T-bill return from inflation rate. Exercises 10.6 to 10.8 ask you to use this information.*

10.6 Look at the data. Give a brief description of the form, direction, and strength of the relationship between the inflation rate and the return on Treasury bills. What is the equation of the least-squares regression line for predicting T-bill return?

10.7 Is there a relationship? What are the slope b_1 of the fitted line and its standard error? Use these numbers to calculate by hand the t statistic for testing the hypothesis that there is no straight-line relationship between inflation rate and T-bill return against the alternative that the return on T-bills increases as the rate of inflation increases. (State hypotheses, give both t and its degrees of freedom, and use Table D to approximate

TABLE 10.1 Return on Treasury bills and rate of inflation

Year	T-bill percent	Inflation percent	Year	T-bill percent	Inflation percent	Year	T-bill percent	Inflation percent
1958	3.01	1.76	1975	6.10	6.94	1992	3.54	2.90
1959	3.81	1.73	1976	5.26	4.86	1993	3.12	2.75
1960	3.20	1.36	1977	5.52	6.70	1994	4.64	2.67
1961	2.59	0.67	1978	7.58	9.02	1995	5.56	2.54
1962	2.90	1.33	1979	10.04	13.20	1996	5.08	3.32
1963	3.26	1.64	1980	11.32	12.50	1997	5.18	1.70
1964	3.68	0.97	1981	13.81	8.92	1998	4.83	1.61
1965	4.05	1.92	1982	11.06	3.83	1999	4.75	2.68
1966	5.06	3.46	1983	8.74	3.79	2000	5.90	3.39
1967	4.61	3.04	1984	9.78	3.95	2001	3.34	1.55
1968	5.47	4.72	1985	7.65	3.80	2002	1.68	2.38
1969	6.86	6.20	1986	6.02	1.10	2003	1.05	1.88
1970	6.51	5.57	1987	6.03	4.43	2004	1.58	3.26
1971	4.52	3.27	1988	6.91	4.42	2005	3.39	3.42
1972	4.47	3.41	1989	8.03	4.65	2006	4.81	2.54
1973	7.20	8.71	1990	7.46	6.11	2007	4.44	4.08
1974	7.95	12.34	1991	5.44	3.06	2008	1.62	0.09

FIGURE 10.9 Scatterplot of the percent return on Treasury bills against the rate of inflation the same year, for Exercises 10.6 to 10.8.

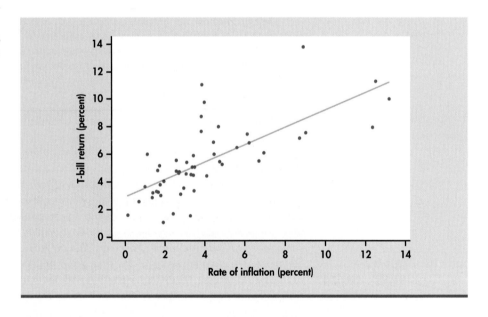

the *P*-value. Then compare your results with those given by Excel. Excel's *P*-value 5.55E-09 is shorthand for 0.00000000555. We would report this as "< 0.0001.")

10.8 Estimating the slope. Using Excel's values for b_1 and its standard error, find a 95% confidence interval for the slope β_1 of the population regression line. Compare your result with Excel's 95% confidence interval. What does the confidence interval tell you about the change in the T-bill return rate for a change in the inflation rate?

Excel

	A	B	C	D	E	F	G
1	SUMMARY OUTPUT						
2							
3	*Regression Statistics*						
4	Multiple R	0.7096					
5	R Square	0.5036					
6	Adjusted R Square	0.4934					
7	Standard Error	1.8761					
8	Observations	51					
9							
10	**ANOVA**						
11	Regression	*df*	*SS*	*MS*	*F*	*Significance F*	
12	Residual	1	174.956	174.956	49.705	5.5476E-09	
13	Total	49	172.476	3.520			
14		50	347.431				
15							
16			*Standard*				
17		*Coefficients*	*Error*	*t Stat*	*P-value*	*Lower 95%*	*Upper 95%*
18	Intercept	2.9645	0.4452	6.6591	2.24E-08	2.0698	3.8591
19	INFLATION	0.6269	0.0889	7.0501	5.55E-09	0.4482	0.8055

FIGURE 10.10 Excel output for the regression of the percent return on Treasury bills against the rate of inflation the same year, for Exercises 10.6 to 10.8.

The word "regression"

To "regress" means to go backward. Why are statistical methods for predicting a response from an explanatory variable called "regression"? Sir Francis Galton (1822–1911) was the first to apply regression to biological and psychological data. He looked at examples such as the heights of children versus the heights of their parents. He found that the taller-than-average parents tended to have children who were also taller than average, but not as tall as their parents. Galton called this fact "regression toward mediocrity," and the name came to be applied to the statistical method. Galton also invented the correlation coefficient r and named it "correlation."

Why are the children of tall parents shorter on the average than their parents? The parents are tall in part because of their genes. But they are also tall in part by chance. Looking at tall parents selects those in whom chance produced height. Their children inherit their genes, but not their good luck. As a group, the children are taller than average (genes) but their heights vary by chance about the average, some upward and some downward. The children, unlike the parents, were not selected because they were tall and thus, on average, are shorter. A similar argument can be used to describe why children of short parents tend to be taller than their parents.

Here's another example. Students who score at the top on the first exam in a course are likely to do less well on the second exam. Does this show that they stopped studying? No—they scored high in part because they knew the material but also in part because they were lucky. On the second exam they may still know the material but be less lucky. As a group, they will still do better than average but not as well as they did on the first exam. The students at the bottom on the first exam will tend to move up on the second exam, for the same reason.

regression fallacy   The **regression fallacy** is the assertion that regression toward the mean shows that there is some systematic effect at work: students with top scores now work less hard; or managers of last year's best-performing mutual funds lose their touch this year; or heights get less variable with each passing generation as tall parents have shorter children and short parents have taller children. The Nobel economist Milton Friedman says, "I suspect that the regression fallacy is the most common fallacy in the statistical analysis of economic data."[6] Beware.

APPLY YOUR KNOWLEDGE

10.9 Hot funds?   Explain carefully to a naive investor why the mutual funds that had the highest returns this year will as a group probably do less well relative to other funds next year.

10.10 Mediocrity triumphant?   In the early 1930s a man named Horace Secrist wrote a book titled *The Triumph of Mediocrity in Business*. Secrist found that businesses that did unusually well or unusually poorly in one year tended to be nearer the average in profitability at a later year. Why is it a fallacy to say that this fact demonstrates an overall movement toward "mediocrity"?

Inference about correlation

The correlation between perceived job stress and locus of control for the 100 accountants is $r = 0.3123$. This value appears in the Excel output in Figure 10.5 (page 534), where it is labeled "Multiple R."[7] We might expect a positive correlation between these two measures in the population of all accountants in Taiwan. Is the sample result convincing evidence that this is true?

population correlation ρ   This question concerns a new population parameter, the **population correlation.** This is the correlation between locus of control and job stress when we measure these variables for every member of the population. We will call the population correlation ρ, the Greek letter rho. To assess the evidence that $\rho > 0$ in the accountant population, we must test the hypotheses

$$H_0: \rho = 0$$
$$H_a: \rho > 0$$

It is natural to base the test on the sample correlation $r = 0.3123$. Table G in the back of the book shows the one-sided critical values of r. To use software for the test, we exploit the close link between correlation and the regression slope. The population correlation ρ is zero, positive, or negative exactly when the slope β_1 of the population regression line is zero, positive, or negative. In fact, the t statistic for testing $H_0: \beta_1 = 0$ also tests $H_0: \rho = 0$. What is more, this t statistic can be written in terms of the sample correlation r.

Test for Zero Population Correlation

To test the hypothesis $H_0: \rho = 0$ that the population correlation is 0, compare the sample correlation r with critical values in Table G or use the t statistic for regression slope.

The t statistic for the slope can be calculated from the sample correlation r:

$$t = \frac{r\sqrt{n-2}}{\sqrt{1-r^2}}$$

This t statistic has $n - 2$ degrees of freedom.

EXAMPLE 10.7   Stress and Locus of Control

CASE 10.1

The sample correlation between job stress and locus of control is $r = 0.3123$ from a sample of size $n = 100$. We can use Table G to test

$$H_0: \rho = 0$$

$$H_a: \rho > 0$$

The table has no entry for $n = 100$. If we follow the conservative practice of using the next smaller entry, for $n = 80$, we find that the P-value for $r = 0.3123$ lies between 0.001 and 0.0025.

We can get a more accurate result from the Excel output in Figure 10.5 (page 534). In the "LOC" line we see that $t = 3.254$ with two-sided P-value 0.00156. That is, $P = 0.00078$ for our one-sided alternative.

Finally, we can calculate t directly from r as follows:

$$
\begin{aligned}
t &= \frac{r\sqrt{n-2}}{\sqrt{1-r^2}} \\
&= \frac{0.3123\sqrt{100-2}}{\sqrt{1-(0.3123)^2}} \\
&= \frac{3.0916}{0.9500} = 3.254
\end{aligned}
$$

If we are not using software, we can compare $t = 3.254$ with critical values from the t table (Table D) with $n - 2 = 98$ degrees of freedom.

The alternative formula for the test statistic is convenient because it uses only the sample correlation r and the sample size n. Remember that correlation, unlike regression, does not require the distinction between explanatory and response variables. For variables x and y, there are two regressions (y on x and x on y) but just one correlation. Both regressions produce the same t statistic.

The distinction between the regression setting and correlation is important only for understanding the conditions under which the test for 0 population correlation makes sense. In the regression model, we take the values of the explanatory variable x as given. The values of the response y are Normal random variables, with means that are a straight-line function of x. In the model for testing correlation, we think of the setting where we obtain a random sample from a population and measure both x and y. Both are assumed to be Normal random variables. In fact, they are taken to be **jointly Normal.** This implies that the conditional distribution of y for each possible value of x is Normal, just as in the regression model.

jointly Normal

APPLY YOUR KNOWLEDGE

10.11 T-bills and inflation.   We expect the interest rates on Treasury bills to rise when the rate of inflation rises and fall when inflation falls. That is, we expect a positive correlation between the return on T-bills and the inflation rate.

(a) Find the sample correlation r for the 51 years in Table 10.1 in the Excel output in Figure 10.10. Use Table G to get an approximate P-value. What do you conclude?

(b) From r, calculate the t statistic for testing correlation. What are its degrees of freedom? Use Table D to give an approximate P-value. Compare your result with the P-value from part (a).

(c) Verify that your t for correlation calculated in part (b) has the same value as the t for slope in the Excel output.

DATA FILE
LOC

CASE 10.1 **10.12 Two regressions.** We have regressed perceived stress level of accountants on their locus of control, with the results appearing in Figures 10.5 and 10.6 (pages 534 and 535). Use software to regress locus of control on stress level for the same data.

(a) What is the equation of the least-squares line for predicting locus of control from stress level? It is a different line than the regression line from Figure 10.3 (page 528).

(b) Verify that the two lines cross at the mean values of the two variables. That is, substitute the mean locus of control into the line from Figure 10.5 and show that the predicted stress level equals the mean of the stress levels of the 100 subjects. Then substitute the mean stress level into your new line and show that the predicted locus of control equals the mean locus of control for the accountants.

(c) Verify that the two regressions give the same value of the t statistic for testing the hypothesis of zero population slope. You could use either regression to test the hypothesis of zero population correlation.

SECTION 10.1 Summary

- **Least-squares regression** fits a straight line to data in order to predict a response variable y from an explanatory variable x. Inference about regression requires additional conditions.

- The **regression model** says that there is a **population regression line** $\mu_y = \beta_0 + \beta_1 x$ that describes how the mean response in an entire population varies as x changes. The observed response y for any x has a Normal distribution with mean given by the population regression line and with the same standard deviation σ for any value of x. The parameters of the regression model are the intercept β_0, the slope β_1, and the standard deviation σ.

- The slope b_0 and intercept b_1 of the least-squares line estimate the slope β_0 and intercept β_1 of the population regression line. To estimate σ, use the **regression standard error s**.

- The standard error s has $n - 2$ **degrees of freedom.** Inference about β_0 and β_1 uses t distributions with $n - 2$ degrees of freedom.

- **Confidence intervals for the slope** of the population regression line have the form $b_1 \pm t^* SE_{b_1}$. In practice, use software to find the slope b_1 of the least-squares line and its standard error SE_{b_1}.

- To test the hypothesis that the population slope is zero, use the **t statistic** $t = b_1/SE_{b_1}$, also given by software. This null hypothesis says that straight-line dependence on x has no value for predicting y.

- The t test for zero population slope also tests the null hypothesis that the **population correlation** is zero. This t statistic can be expressed in terms of the sample correlation, $t = r\sqrt{n - 2}/\sqrt{1 - r^2}$.

SECTION 10.1 Exercises

For Exercises 10.1 and 10.2, see page 530; for 10.3 and 10.4, see page 532; for 10.5, see page 534; for 10.6 to 10.8, see pages 539–540; for 10.9 and 10.10, see page 542; and for 10.11 and 10.12, see pages 543–544.

10.13 Assessment value versus sales price. Real estate is typically reassessed annually for property tax purposes. This

assessed value, however, is not necessarily the same as the fair market value of the property. Table 10.2 summarizes an SRS of 30 properties recently sold in a midwestern city.[8] Both variables are measured in thousands of dollars. 🅓 **HOMESALES**

(a) Inspect the data. How many have a selling price greater than the assessed value? Do you think this trend would be true for

TABLE 10.2 Sales price and assessed value (in thousands of $) of 30 homes in a midwestern city

Property	Sales price	Assessed value	Property	Sales price	Assessed value	Property	Sales price	Assessed value
1	167.9	152.7	11	230.0	225.4	21	283.0	303.9
2	168.0	163.8	12	230.0	170.4	22	269.0	233.7
3	155.0	167.6	13	222.5	200.4	23	255.0	233.6
4	158.5	127.3	14	225.5	209.6	24	285.0	234.2
5	159.9	155.7	15	220.0	205.2	25	146.0	145.1
6	162.0	169.0	16	216.0	220.9	26	128.0	108.3
7	165.0	187.1	17	215.0	194.9	27	126.5	136.2
8	174.5	153.6	18	228.0	231.4	28	129.9	113.3
9	175.0	167.1	19	209.0	224.2	29	150.0	121.4
10	159.0	148.9	20	267.0	235.1	30	195.0	184.0

the larger population of all homes recently sold? Explain your answer.

(b) Make a scatterplot with assessed value on the horizontal axis. Briefly describe the relationship between assessed value and selling price.

(c) Report the least-squares regression line for predicting selling price from assessed value.

(d) Obtain the residuals and plot them versus assessed value. Is there anything unusual to report? If so, explain.

(e) Do the residuals appear to be approximately Normal? Explain your answer.

(f) Based on your answers to parts (b), (d), and (e), do the assumptions for the linear regression analysis appear reasonable? Explain your answer.

10.14 Assessment value versus sales price, continued. Refer to the previous exercise. ⬤ HOMESALES

(a) Calculate the predicted selling prices for homes currently assessed at $155,000, $220,000, and $285,000.

(b) Suppose these houses sold for $142,900, $224,000, and $286,000 respectively. Calculate the residual for each of these sales.

(c) Construct a 95% confidence interval for both the slope and the intercept.

(d) Using the result from part (c), compare the estimated regression line with $y = x$, which says that, on average, the selling price is equal to the assessed value. Is there evidence that this model is not reasonable? In other words, is the selling price typically larger or smaller than the assessed value? Explain your answer.

10.15 Public university tuition: 2000 versus 2008. Table 10.3 shows the in-state undergraduate tuition and required fees for 33 public universities in 2000 and 2008.[9] ⬤ TUITION

(a) Plot the data with the 2000 tuition on the x axis and describe the relationship. Are there any outliers or unusual values? Does a linear relationship between the tuition in 2000 and 2008 seem reasonable?

TABLE 10.3 In-state tuition and fees (in dollars) for 33 public universities

School	2000	2008	School	2000	2008	School	2000	2008
Penn. State	7018	13,706	Virginia	4335	9300	Iowa State	3132	5524
Pittsburgh	7002	13,642	Indiana	4405	8231	Oregon	3819	6435
Michigan	6926	11,738	Texas A&M	3374	7844	Iowa	3204	6544
Rutgers	6333	11,540	Texas	3575	8532	Washington	3761	6802
Illinois	4994	12,106	Cal.-Irvine	3970	8046	Nebraska	3450	6584
Minnesota	4877	10,634	Cal.-San Diego	3848	8062	Kansas	2725	7042
Mich. State	5432	10,690	Cal.-Berkeley	4047	7656	Colorado	3188	7278
Ohio State	4383	8679	UCLA	3698	8310	North Carolina	2768	5397
Maryland	5136	8005	Purdue	3872	7750	Arizona	2348	5542
Cal.-Davis	4072	9497	Wisconsin	3791	7569	Florida	2256	3256
Missouri	4726	7386	Buffalo	4715	6385	Georgia Tech.	3308	6040

(b) Run the simple linear regression and give the least-squares regression line.

(c) Obtain the residuals and plot them versus the 2000 tuition amount. Is there anything unusual in the plot?

(d) Do the residuals appear to be approximately Normal? Explain.

(e) Give the null and alternative hypotheses for examining the relationship between 2000 and 2008 tuition amounts.

(f) Write down the test statistic and P-value for the hypotheses stated in part (e). State your conclusions.

10.16 More on public university tuition. Refer to the previous exercise.

(a) Construct a 95% confidence interval for the slope. What does this interval tell you about the annual percent increase in tuition between 2000 and 2008?

(b) What percent of the variability in 2008 tuition is explained by a linear regression model using the 2000 tuition?

(c) The tuition at BusStat U was $5800 in 2000. What is the predicted tuition in 2008?

(d) The tuition at Moneypit U was $8700 in 2000. What is the predicted tuition in 2008?

(e) Discuss the appropriateness of using the fitted equation to predict tuition for each of these universities.

10.17 Incentive pay and job performance. Tying compensation to performance is typically an efficient method of motivating employees. In the National Football League (NFL), incentive bonuses now account for roughly 25% of player compensation.[10] Does tying a player's salary to performance bonuses result in better individual or team success on the field? *usatoday.com* contains payroll data for most current NFL players. *cbssports.com* contains a player rating system that uses game statistics.[11] Focusing on linebackers, let's look at the relationship between a player's overall 2008 rating and the percent of his 2008 salary from incentive payments. ⬤ **LINEBACKERS**

(a) Use numerical and graphical methods to describe the two variables and summarize your results.

(b) Neither variable is Normally distributed. Does that necessarily pose a problem for performing linear regression? Explain.

(c) Construct a scatterplot of the data using the percent as the predictor variable and describe the relationship. Are there any outliers or unusual values? Does a linear relationship between the percent of salary from incentive payments and player rating seem reasonable? Is it a very strong relationship? Explain.

(d) Run the simple linear regression and give the least-squares regression line.

(e) Obtain the residuals and assess whether the assumptions for the linear regression analysis are reasonable. Include all plots and numerical summaries that you used to make this assessment.

10.18 Incentive pay, continued. Refer to the previous exercise. ⬤ **LINEBACKERS**

(a) Now run the simple linear regression for the variables square root of rating and percent of salary from incentive payments.

(b) Obtain the residuals and assess whether the assumptions for the linear regression analysis are reasonable. Include all

plots and numerical summaries that you used to make this assessment.

(c) Construct a 95% confidence interval for the square root increase in rating given a 1% increase in the percent of salary from incentive payments.

(d) Consider the values 0%, 20%, 40%, 60% and 80% salary from incentives. Compute the predicted rating for this model and for the one in Exercise 10.17. For the model in this exercise, you will need to square the predicted value to get back to the original units.

(e) Plot the predicted values versus the percents and connect those values from the same model. For which regions of percent do the predicted values from the two models vary the most?

(f) Based on your comparison of the regression models (both predicted values and residuals), which model do you prefer? Explain.

10.19 Predicting water quality. Pollution of water resources is a serious problem that can require substantial efforts to improve. To determine the financial resources required, the extent of the problem needs to be assessed. The index of biotic integrity (IBI) is a measure of the water quality in streams that is fairly expensive to obtain. Land use measures, on the other hand, are relatively inexpensive to obtain. IBI and land use measures for a collection of streams in the Ozark Highland ecoregion of Arkansas were collected as part of a study.[12] Table 10.4 gives the data for IBI and the area of the watershed in square kilometers for streams in the original sample with area less than or equal to 70 km^2. ⬤ **IBI**

TABLE 10.4 Watershed area (square kilometers) and index of biotic integrity

Area	IBI	Area	IBI	Area	IBI	Area	IBI	Area	IBI
21	47	29	61	31	39	32	59	34	72
34	76	49	85	52	89	2	74	70	89
6	33	28	46	21	32	59	80	69	80
47	78	8	53	8	43	58	88	54	84
10	62	57	55	18	29	19	29	39	54
49	78	9	71	5	55	14	58	9	71
23	33	31	59	18	81	16	71	21	75
32	64	10	41	26	82	9	60	54	84
12	83	21	82	27	82	23	86	26	79
16	67	26	56	26	85	28	91		

(a) Use numerical and graphical methods to describe the variable IBI. Do the same for area. Summarize your results.

(b) Plot the data with area on the x axis and describe the relationship. Are there any outliers or unusual patterns?

(c) Give the statistical model for simple linear regression for this problem.

(d) State the null and alternative hypotheses for examining the relationship between IBI and area.

(e) Run the simple linear regression and summarize the results.

(f) Obtain the residuals and plot them versus area. Is there anything unusual in the plot?

(g) Do the residuals appear to be approximately Normal? Give reasons for your answer.

(h) Do the assumptions for the analysis of these data using the model you gave in part (c) appear to be reasonable? Explain your answer.

10.20 Predicting water quality using a different explanatory variable. The researchers who conducted the study described in the previous exercise also recorded the percent of the watershed area that was forest for each of the streams. The data are given in Table 10.5. Analyze these data using the questions in the previous exercise as a guide. IBI

TABLE 10.5 Percent forest and index of biotic integrity

Forest	IBI	Forest	IBI	Forest	IBI	Forest	IBI	Forest	IBI
0	47	0	61	0	39	0	59	0	72
0	76	3	85	3	89	7	74	8	89
9	33	10	46	10	32	11	80	14	80
17	78	17	53	18	43	21	88	22	84
25	62	31	55	32	29	33	29	33	54
33	78	39	71	41	55	43	58	43	71
47	33	49	59	49	81	52	71	52	75
59	64	63	41	68	82	75	60	79	84
79	83	80	82	86	82	89	86	90	79
95	67	95	56	100	85	100	91		

10.21 Comparing the analyses. In Exercises 10.19 and 10.20 you used two different explanatory variables to predict IBI. Summarize the two analyses and compare the results. If you had to choose between the two, which one would you prefer? Give reasons for your answer.

10.22 Using the correlation to describe the results. Refer to Exercise 10.19. Find the correlation and test the null hypothesis that the corresponding population correlation is zero versus the two-sided alternative. Is the correlation a good numerical measure to describe these data? Compare the results of your significance test with those in Exercise 10.19 and explain the relationship.

10.23 Using the other explanatory variable. Refer to Exercise 10.20. Using the questions in the previous exercise as a guide, use the correlation to analyze the relationship between IBI and percent of watershed area that was forest.

10.24 An outlier can destroy statistical significance. Consider the data in Table 10.4 and the relationship between IBI and watershed area. The relationship between these two variables is statistically significant. Demonstrate the potential effect of an outlier on statistical significance by changing the IBI value of one of the observations in such a way that the relationship between the two variables is no longer statistically significant. The relationship between these two variables is positive. Can you change the data in such a way that the apparent relationship is more positive, that is, by raising the value of IBI for an observation with a large watershed area? Write a short summary of what you have learned from this exercise.

Age and income. *How do the incomes of working-age people change with age? Because many older women have been out of the labor force for much of their lives, we look only at men between the ages of 25 and 65. Because education strongly influences income, we look only at men who have a bachelor's degree but no higher degree. The data file for the following exercises contains the age and income of a random sample of 5712 such men. Figure 10.11 is a scatterplot of these data. Figure 10.12 displays*

FIGURE 10.11 Scatterplot of income against age for a random sample of 5712 men aged 25 to 65, for Exercises 10.25 to 10.27.

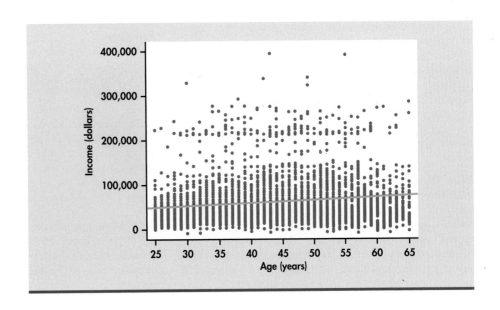

Excel

	A	B	C	D	E	F	G
1	SUMMARY OUTPUT						
2							
3	*Regression Statistics*						
4	Multiple R	0.18775					
5	R Square	0.03525					
6	Adjusted R Square	0.03508					
7	Standard Error	47620.29					
8	Observations	5712					
9							
10	ANOVA						
11		*df*	*SS*	*MS*	*F*	*Significance F*	
12	Regression	1	4.7310E+11	4.73E+11	208.6271	1.7913E-46	
13	Residual	5710	1.2949E+13	2.27E+09			
14	Total	5711	1.3422E+13				
15							
16			*Standard*				
17		*Coefficients*	*Error*	*t Stat*	*P-value*	*Lower 95%*	*Upper 95%*
18	Intercept	24874.3745	2637.4198	9.4313	5.75E-21	19704.031	30044.718
19	AGE	892.1135	61.7639	14.4439	1.79E-46	771.033	1013.194

FIGURE 10.12 Excel output for the regression of income on age, for Exercises 10.25 to 10.27.

the Excel output for regressing income on age. The line in the scatterplot is the least-squares regression line. Exercises 10.25 to 10.27 ask you to interpret this information. **INCOMEAGE**

10.25 Looking at age and income. The scatterplot in Figure 10.11 has a distinctive form.
(a) Age is recorded as of the last birthday. How does this explain the vertical stacks in the scatterplot?
(b) Give some reasons why older men in this population might earn more than younger men. Give some reasons why younger men might earn more than older men. What do the data show about the relationship between age and income in the sample? Is the relationship very strong?
(c) What is the equation of the least-squares line for predicting income from age? What specifically does the slope of this line tell us?

10.26 Income increases with age. We see that older men do on the average earn more than younger men, but the increase is not very rapid. (Note that the regression line describes many men of different ages—data on the same men over time might show a different pattern.)
(a) We know even without looking at the Excel output that there is highly significant evidence that the slope of the population regression line is greater than 0. Why do we know this?

(b) Excel gives a 95% confidence interval for the slope of the population regression line. What is this interval?
(c) Give a 99% confidence interval for the slope of the population regression line.

10.27 Was inference justified? You see from Figure 10.11 that the incomes of men at each age are (as expected) not Normal but right-skewed.
(a) How is this apparent on the plot?
(b) Nonetheless, your confidence interval in the previous exercise will be quite accurate even though it is based on Normal distributions. Why?

10.28 Stress level for an extreme internal locus of control. The intercept β_0 of the population regression line of job stress among accountants on locus of control measures the job stress felt by an accountant who feels his or her fate is completely in his or her own hands. Use the Excel output in Figure 10.5 (page 534) to give a 90% confidence interval for β_0.

10.29 T-bills and inflation. Exercises 10.6 to 10.8 interpret the part of the Excel output in Figure 10.10 (page 541) that concerns the slope, the rate at which T-bill returns increase as the rate of inflation increases. Use this output to answer questions about the intercept.

(a) The intercept β_0 in the regression model is meaningful in this example. Explain what β_0 represents. Why should we expect β_0 to be greater than 0?
(b) What values does Excel give for the estimated intercept b_0 and its standard error SE_{b_0}?
(c) Is there good evidence that β_0 is greater than 0?
(d) Write the formula for a 95% confidence interval for β_0. Verify that hand calculation (using the Excel values for b_0 and SE_{b_0}) agrees approximately with the output in Figure 10.10.

10.30 Is the correlation significant? A study reports correlation $r = -0.5$ based on a sample of size $n = 20$. Another study reports the same correlation based on a sample of size $n = 10$. For each, use Table G to test the null hypothesis that the population correlation $\rho = 0$ against the one-sided alternative $\rho < 0$. Are the results significant at the 5% level? Explain why the conclusions of the two studies differ.

10.31 Correlation between binge drinking and the average price of beer. A recent study looked at 118 colleges to investigate the association between the binge-drinking rate and the average price for a bottle of beer at establishments within a two-mile radius of campus.[13] A correlation of -0.36 was found. Explain this correlation.

10.32 Is this relationship significant? Refer to the previous exercise. Test the null hypothesis that the correlation between the binge-drinking rate and the average price for a bottle of beer within a two-mile radius of campus is zero.

10.33 Size and selling price of houses. Table 10.6 describes a random sample of 30 houses sold in a Midwest city during a recent year.[14] We will examine the relationship between size and price. HOUSESIZE

TABLE 10.6 Selling price and size of homes

Price ($1000)	Size (sq. ft.)	Price ($1000)	Size (sq. ft.)	Price ($1000)	Size (sq. ft.)
268	1897	142	1329	83	1378
131	1157	107	1040	125	1668
112	1024	110	951	60	1248
112	935	187	1628	85	1229
122	1236	94	816	117	1308
128	1248	99	1060	57	892
158	1620	78	800	110	1981
135	1124	56	492	127	1098
146	1248	70	792	119	1858
126	1139	54	980	172	2010

(a) Plot the selling price versus the number of square feet. Describe the pattern. Does r^2 suggest that size is quite helpful for predicting selling price?

(b) Do a linear regression analysis. Give the least-squares line and the results of the significance test for the slope. What does your test tell you about the relationship between size and selling price?

10.34 Stocks and bonds. How is the flow of investors' money into stock mutual funds related to the flow of money into bond mutual funds? Table 10.7 shows the net new money flowing into stock and bond mutual funds in the years 1985 to 2008, in billions of dollars.[15] "Net" means that funds flowing out are subtracted from those flowing in. If more money leaves than arrives, the net flow will be negative. To eliminate the effect of inflation, all dollar amounts are in "real dollars" with constant buying power equal to that of a dollar in the year 2000. MONEYFLOW

TABLE 10.7 Net new money (billions of $) flowing into stock and bond mutual funds

Year	Stocks	Bonds	Year	Stocks	Bonds	Year	Stocks	Bonds
1985	12.8	100.8	1993	151.3	84.6	2001	31.1	85.0
1986	34.6	161.8	1994	133.6	−72.0	2002	−25.8	134.4
1987	28.8	10.6	1995	140.1	−6.8	2003	143.0	29.4
1988	−23.3	−5.8	1996	238.2	3.3	2004	161.8	−9.4
1989	8.3	−1.4	1997	243.5	30.0	2005	119.7	27.7
1990	17.1	9.2	1998	165.9	79.2	2006	134.9	51.9
1991	50.6	74.6	1999	194.3	−6.2	2007	76.8	90.1
1992	97.0	87.1	2000	309.0	−48.0	2008	−187.4	24.9

(a) Make a scatterplot with cash flow into stock funds as the explanatory variable. Find the least-squares line for predicting net bond investments from net stock investments. What do the data suggest?
(b) Is there statistically significant evidence that there is some straight-line relationship between the flows of cash into bond funds and stock funds? (State hypotheses, give a test statistic and its P-value, and state your conclusion.)
(c) Remove the data for 2008 and refit the remaining years. Is there now statistically significant evidence of a straight-line relationship?
(d) How would you report these results in a paper? In other words, how would you handle the change in statistical significance caused by this one observation?

10.35 Do larger houses have higher prices? We expect that there is a positive correlation between the sizes of houses in the same market and their selling prices. HOUSESIZE
(a) Use the data in Table 10.6 to test this hypothesis. (State hypotheses, find the sample correlation r and the t statistic based on it, and give an approximate P-value and your conclusion.)
(b) To what extent do you think that these results would apply to other cities in the United States?

10.36 Are inflows into stocks and bonds correlated? Is the correlation between net flow of money into stock mutual funds and into bond mutual funds significantly different from 0? Use the regression analysis you did in Exercise 10.34 to answer this question with no additional calculations.

10.37 Influence? Your scatterplot in Exercise 10.33 shows one house whose selling price is quite high for its size. Rerun the analysis without this outlier. Does this one house influence r^2, the location of the least-squares line, or the t statistic for the slope in a way that would change your conclusions?

10.38 Beer and blood alcohol. How well does the number of beers a student drinks predict his or her blood alcohol content (BAC)? Sixteen student volunteers at Ohio State University drank a randomly assigned number of 12-ounce cans of beer. Thirty minutes later, a police officer measured their BAC.[16] ⬤ **BAC**

Student	Beers	BAC	Student	Beers	BAC
1	5	0.10	9	3	0.02
2	2	0.03	10	5	0.05
3	9	0.19	11	4	0.07
4	8	0.12	12	6	0.10
5	3	0.04	13	5	0.085
6	7	0.095	14	7	0.09
7	3	0.07	15	1	0.01
8	5	0.06	16	4	0.05

The students were equally divided between men and women and differed in weight and usual drinking habits. Because of this variation, many students don't believe that number of drinks predicts BAC well.

(a) Make a scatterplot of the data. Find the equation of the least-squares regression line for predicting BAC from number of beers and add this line to your plot. What is r^2 for these data? Briefly summarize what your data analysis shows.

(b) Is there significant evidence that drinking more beers increases BAC on the average in the population of all students? State hypotheses, give a test statistic and P-value, and state your conclusion.

10.39 Computer memory. The capacity (bits) of the largest DRAM (dynamic random access memory) chips commonly available at retail has increased as follows:[17] ⬤ **DRAM**

Year	Bits	Year	Bits
1971	1,024	1993	16,384,000
1980	64,000	1999	256,000,000
1987	1,024,000	2000	512,000,000

(a) Make a scatterplot of the data. Growth is much faster than linear.
(b) Plot the logarithm of DRAM capacity against year. These points are close to a straight line.
(c) Regress the logarithm of DRAM capacity on year. Give a 90% confidence interval for the slope of the population regression line.

10.40 Influence? Your scatterplot in Exercise 10.38 shows one unusual point: Student 3, who drank 9 beers.
(a) Does Student 3 have the largest residual from the fitted line? (You can use the scatterplot to see this.) Is this observation extreme in the x direction, so that it may be influential?
(b) Do the regression again, omitting Student 3. Add the new regression line to your scatterplot. Does removing this observation greatly change predicted BAC? Does r^2 change greatly? Does the P-value of your test change greatly? What do you conclude: did your work in the previous problem depend heavily on this one student?

10.2 Using the Regression Line

One of the most common reasons to fit a line to data is to predict the response to a particular value of the explanatory variable. The method is simple: just substitute the value of x into the equation of the line. The least-squares line for predicting perceived job stress in accountants from their locus of control (Case 10.1) is

$$\hat{y} = 2.2555 + 0.0399x$$

For an LOC score of 16, our least-squares regression equation gives

$$\hat{y} = 2.2555 + (0.0399)(16) = 2.894$$

There are two different uses of this prediction. First, we can estimate the *mean* level of stress in the subpopulation of accountants with an LOC of 16. Second, we can predict the job stress of *one individual accountant* with an LOC of 16. For each use, the actual prediction is the same, $\hat{y} = 2.894$. But the margin of error is different for the two cases. Individual workers with an LOC of 16 don't all feel the same level of stress. Thus, we need a larger margin of error when predicting an individual's level than when estimating the mean stress level of all accountants who have an LOC of 16.

Write the given value of the explanatory variable x as x^*. In the example, $x^* = 16$. The distinction between predicting a single outcome and estimating the mean of all outcomes when $x = x^*$ determines what margin of error is correct. To emphasize the distinction, we use different terms for the two intervals.

- To estimate the *mean* response, we use a *confidence interval*. This is an ordinary confidence interval for the parameter

$$\mu_y = \beta_0 + \beta_1 x^*$$

The regression model says that μ_y is the mean of responses y when x has the value x^*. It is a fixed number whose value we don't know.

prediction interval   • To estimate an *individual* response y, we use a **prediction interval.** A prediction interval estimates a single random response y rather than a parameter like μ_y. The response y is not a fixed number. If we took more observations with $x = x^*$, we would get different responses.

Fortunately, the meaning of a prediction interval is very much like the meaning of a confidence interval. A 95% prediction interval, like a 95% confidence interval, is right 95% of the time in repeated use. "Repeated use" now means that we take an observation on y for each of the n values of x in the original data, and then take one more observation y with $x = x^*$. Form the prediction interval from the n observations, then see if it covers the one more y. The interval will cover this one more y in 95% of all repetitions.

Each interval has the usual form

$$\hat{y} \pm t^* \text{SE}$$

where $t^* \text{SE}$ is the margin of error. The main distinction is that since it is harder to predict for an individual than for the mean of a population of individuals, the margin of error for the prediction interval is wider than the margin of error for the confidence interval. Formulas for computing these quantities are given in Section 10.3. For now, we rely on software to do the arithmetic.

Confidence and Prediction Intervals for Regression Response

A level C **confidence interval for the mean response** μ_y when x takes the value x^* is

$$\hat{y} \pm t^* \text{SE}_{\hat{\mu}}$$

Here $\text{SE}_{\hat{\mu}}$ is the standard error for estimating a mean response.

A level C **prediction interval for a single observation** on y when x takes the value x^* is

$$\hat{y} \pm t^* \text{SE}_{\hat{y}}$$

The standard error $\text{SE}_{\hat{y}}$ for estimating an individual response is larger than the standard error $\text{SE}_{\hat{\mu}}$ for a mean response to the same x^*.

In both cases, t^* is the value for the $t(n-2)$ density curve with area C between $-t^*$ and t^*.

Predicting an individual response is an exception to the general fact that regression inference is robust against lack of Normality. The prediction interval relies on Normality of individual observations, not just on the approximate Normality of statistics like the slope b_1 and intercept b_0 of the least-squares line. In practice, this means that you should regard prediction intervals as rough approximations.

EXAMPLE 10.8   Predicting Job Stress from Locus of Control

CASE 10.1

Seung Jin Lim has worked as an accountant for several years and has LOC = 16. We don't know his level of job stress, but we can use the data on other accountants to predict his level.

Statistical software usually allows prediction of the response for each x-value in the data and also for new values of x. Here is the output from the prediction option in the Minitab regression command for $x^* = 16$ when we ask for 95% intervals:

```
Predicted Values

  Fit      StDev Fit     95.0% CI            95.0% PI
2.8940       0.0722    (2.7507, 3.0374)   (1.9871, 3.8010)
```

The "Fit" entry gives the predicted stress level, 2.894. This agrees with our hand calculation. Minitab gives both 95% intervals. You must choose which one you want. We are predicting a single response, so the prediction interval "95.0% PI" is the right choice. We are 95% confident that Seung Jin's stress level lies between 1.9871 and 3.8010. This is a wide range because the data are widely scattered about the least-squares line. The 95% confidence interval for the mean stress level of all accountants with LOC = 16, given as "95.0% CI," is much narrower.

Note that Minitab reports only one of the two standard errors. It is the standard error for estimating the mean response, $\text{SE}_{\hat{\mu}} = 0.0722$. A graph will help us to understand the difference between the two types of intervals.

EXAMPLE 10.9   Comparing the Two Intervals

CASE 10.1

Figure 10.13 displays the data, the least-squares line, and both intervals. The confidence interval for the mean is solid. The prediction interval for Seung Jin's individual stress level is dashed. You can see that the prediction interval is much wider and that it matches the vertical spread of accountants' stress levels about the regression line.

FIGURE 10.13 Confidence interval for mean stress level (solid) and prediction interval for individual stress level (dashed) for an accountant with an LOC of 16. Both intervals are centered at the predicted value from the least-squares line, which is $\hat{y} = 2.894$ for $x^* = 16$.

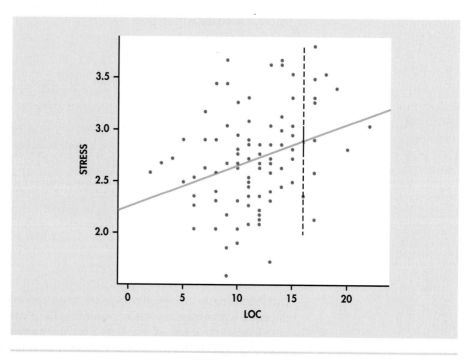

Some software packages will graph the intervals for all values of the explanatory variable within the range of the data. With this type of display, it is easy to see the difference between the two types of intervals.

EXAMPLE 10.10 Graphing the Confidence Intervals

CASE 10.1

The confidence intervals for the job stress data are graphed in Figure 10.14. For each value of LOC, we see the predicted value on the solid line and the confidence limits on the dashed curves.

FIGURE 10.14 95% confidence intervals for mean response for the job stress data.

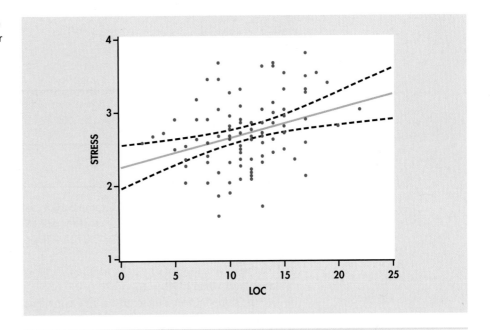

Notice that the intervals get wider as the values of locus of control move away from the mean of this variable. This phenomenon reflects the fact that we have less information for estimating means that correspond to extreme values of the explanatory variable.

EXAMPLE 10.11 Graphing the Prediction Intervals

CASE 10.1

The prediction intervals for the job stress data are graphed in Figure 10.15. As with the confidence intervals, we see the predicted values on the solid line and the prediction limits on the dashed curves.

It is much easier to see the curvature of the confidence limits in Figure 10.14 than the curvature of the prediction limits in Figure 10.15. One reason for this is that the prediction intervals in Figure 10.15 are dominated by the accountant-to-accountant variation. Notice that because prediction intervals are concerned with individual predictions, they contain a very large proportion of the data. On the other hand, the confidence intervals are designed to contain mean values and are not concerned with individual points.

FIGURE 10.15 95% prediction
intervals for individual
response for the job stress
data.

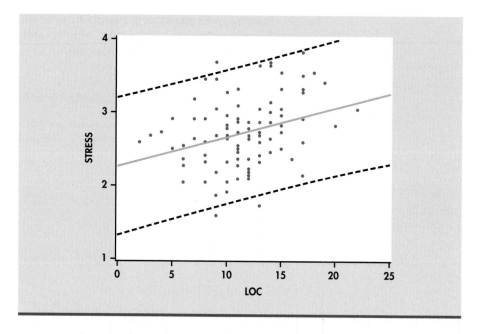

10.41 Predicting stress level.   In Example 10.8, software predicts the mean stress level of accountants with an LOC of 16 to be $\hat{y} = 2.894$. We also see that the standard error of this estimated mean is $SE_{\hat{\mu}} = 0.0722$. These results come from data on 100 accountants.

(a) Use these facts to verify by hand Minitab's 95% confidence interval for the mean stress level when LOC $= 16$.

(b) Use the same information to give a 90% confidence interval for the mean stress level.

10.42 Predicting the return on Treasury bills.   Table 10.1 (page 540) gives data on the rate of inflation and the percent return on Treasury bills for 51 years. Figures 10.9 and 10.10 analyze these data. You think that next year's inflation rate will be 3.7%. Figure 10.16 displays part of the Minitab regression output, including predicted values for $x^* = 3.7$. The basic output agrees with the Excel results in Figure 10.10.

FIGURE 10.16 Minitab output
for the regression of the
percent return on Treasury bills
against the rate of inflation the
same year, for Exercise 10.42.
The output includes predictions
of the T-bill return when the
inflation rate is 3.7%.

Minitab

```
The regression equation is
TBILL = 2.96 + 0.627 INFLATION

Predictor      Coef   SE Coef      T      P
Constant     2.9645    0.4452   6.66  0.000
INFLATION   0.62685   0.08891   7.05  0.000

S = 1.87614    R-Sq = 50.4%    R-Sq(adj) = 49.3%

Predicted Values for New Observations

New Obs    Fit   SE Fit      95% CI           95% PI
   1      5.284   0.264  (4.752, 5.815)   (1.476, 9.091)
```

(a) Verify the predicted value $\hat{y} = 5.284$ from the equation of the least-squares line.

(b) What is your 95% interval for predicting next year's return on Treasury bills?

BEYOND THE BASICS: Nonlinear Regression

The simple linear regression model assumes that the relationship between the response variable and the explanatory variable can be summarized with a straight line. When the relationship is not linear, we can sometimes transform one or both of the variables so that the relationship becomes linear. Exercise 10.39 (page 550) is an example in which the relationship of log y with x is linear. In other circumstances, we use models that directly express a curved relationship using parameters that are not just intercepts and slopes.

nonlinear models

These are **nonlinear models.**

Here is a typical example of a model that involves parameters β_0 and β_1 in a nonlinear way:

$$y_i = \beta_0 x_i^{\beta_1} + \epsilon_i$$

This nonlinear model still has the form

$$\text{DATA} = \text{FIT} + \text{RESIDUAL}$$

The FIT term describes how the mean response μ_y depends on x. Figure 10.17 shows the form of the mean response for several values of β_1 when $\beta_0 = 1$. Choosing $\beta_1 = 1$ produces a straight line, but other values of β_1 result in a variety of curved relationships.

We cannot write simple formulas for the estimates of the parameters β_0 and β_1, but software can calculate both estimates and approximate standard errors for the estimates. If the deviations ϵ_i follow a Normal distribution, we can do inference both on the model parameters and for prediction. The details become more complex, but the ideas remain the same as those we have studied.

FIGURE 10.17 The nonlinear model $\mu_{\hat{y}} = \beta_0 x^{\beta_1}$ includes these and other relationships between the explanatory variable x and the mean response.

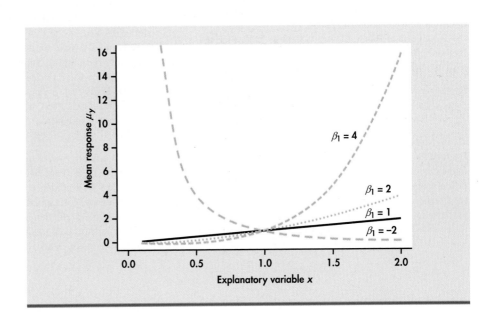

SECTION 10.2 Summary

- The **estimated mean response** for the subpopulation corresponding to the value x^* of the explanatory variable is found by substituting $x = x^*$ in the equation of the least-squares regression line:

$$\text{estimated mean response} = \hat{y} = b_0 + b_1 x^*$$

- The **predicted value of the response** y for a single observation from the subpopulation corresponding to the value x^* of the explanatory variable is found in exactly the same way:

$$\text{predicted individual response} = \hat{y} = b_0 + b_1 x^*$$

- **Confidence intervals for the mean response** μ_y when x has the value x^* have the form

$$\hat{y} \pm t^* \text{SE}_{\hat{\mu}}$$

- **Prediction intervals** for an individual response y have a similar form with a larger standard error:

$$\hat{y} \pm t^* \text{SE}_{\hat{y}}$$

In both cases, t^* is the value for the $t(n-2)$ density curve with area C between $-t^*$ and t^*. Software often gives these intervals. The standard error $\text{SE}_{\hat{y}}$ for an individual response is larger than the standard error $\text{SE}_{\hat{\mu}}$ for a mean response because it must account for the variation of individual responses around their mean.

SECTION 10.2 Exercises

For Exercises 10.41 and 10.42, see page 554.

10.43 More on public university tuition. Refer to Exercise 10.15 (page 545). 🌐 TUITION
(a) Construct a 95% confidence interval for the percent increase in tuition between 2000 and 2008.
(b) The tuition at BusStat U was $5800 in 2000. Find the 95% prediction interval for its tuition in 2008.
(c) The tuition at Moneypit U was $8700 in 2000. Find the 95% prediction interval for its tuition in 2008.
(d) Compare the widths of these two intervals. Which is wider and why?

10.44 More on assessment value versus sales price. Refer to Exercises 10.13 and 10.14 (pages 544 and 545). Suppose we're interested in determining whether the population regression line differs from $y = x$. We'll look at this three ways. 🌐 HOMESALES
(a) Construct a 95% prediction interval of sales price for each property in the data set. If the model $y = x$ is reasonable, then the assessed value used to predict the sales price should be in the interval. Is this true for all cases?

(b) The model $y = x$ means $\beta_0 = 0$ and $\beta_1 = 1$. Test these two hypotheses. Is there enough evidence to reject either of these two hypotheses?
(c) Recall that not rejecting H_0 does not imply H_0 is true. A test of "equivalence" would be a more appropriate method to assess similarity. The key to this approach is determining a "zone of indifference," which represents true differences that aren't scientifically meaningful. Suppose that for the slope a difference within ±0.05% is considered not different. Construct a 90% confidence interval for the slope and see if it falls entirely within the interval (0.95, 1.05). If it does, we would conclude that the slope is not different from 1. What is your conclusion using this method?

10.45 Predicting water quality from watershed area. Refer to Exercise 10.19 (page 546). 🌐 IBI
(a) Find a 95% confidence interval for mean response (index of biotic integrity) corresponding to an area of 54 km^2.
(b) Find a 95% prediction interval for a future response.
(c) Write a short paragraph interpreting the meaning of the intervals in terms of Ozark Highland streams.

(d) Do you think that these results can be applied to other streams in Arkansas or in other states? Explain why or why not.

10.46 Predicting water quality from the percent of forested area. Refer to Exercise 10.20 (page 547). **IBI**
(a) Find a 95% confidence interval for mean response (index of biotic integrity) corresponding to 22% forest.
(b) Find a 95% prediction interval for a future response.
(c) Write a short paragraph interpreting the meaning of the intervals in terms of Ozark Highland streams.
(d) Do you think that these results can be applied to other streams in Arkansas or in other states? Explain why or why not.

10.47 Compare the estimates. Case 20 in Table 10.4 and Table 10.5 corresponds to the same watershed area. For this case the area is 54 km^2 and the percent forest is 22%. A predicted index of biotic integrity based on area was computed in Exercise 10.45, while one based on percent forest was computed in Exercise 10.46. Compare these two estimates and explain why they differ. Use the idea of a prediction interval to interpret these results.

10.48 Is the price right? Refer to Exercise 10.33 (page 549), where the relationship between the size of a home and its selling price is examined.
(a) Suppose that you have a client who is thinking about purchasing a home in this area that is 1500 square feet in size. The asking price is $140,000. What advice would you give this client?
(b) Answer the same question for a client who is looking at a 1200-square-foot home that is selling for $100,000.

10.49 Predicting income from age. Figures 10.11 and 10.12 (pages 547 and 548) analyze data on the age and income of 5712 men between the ages of 25 and 65. Here is Minitab output predicting the income for ages 30, 40, 50, and 60 years:

```
Predicted Values
  Fit  StDev Fit    95.0% CI          95.0% PI
51638       948  (49780, 53496)  (-41735, 145010)
60559       637  (59311, 61807)  (-32803, 153921)
69480       822  (67870, 71091)  (-23888, 162848)
78401      1307  (75840, 80963)  (-14988, 171790)
```

(a) Use the regression line from Figure 10.11 to verify the "Fit" for age 30 years.
(b) Give a 95% confidence interval for the income of all 30-year-old men.
(c) Rabi is 30 years old. You don't know his income, so give a 95% prediction interval based on his age alone. How useful do you think this interval is?

10.50 Predict what? The two 95% intervals for the income of 30-year-olds given in Exercise 10.49 are very different. Explain briefly to someone who knows no statistics why the second interval is so much wider than the first. Start by looking at 30-year-olds in Figure 10.11 (page 547).

10.51 Predicting income from age, continued. Use the computer outputs in Figure 10.12 (page 548) and Exercise 10.49 to give a 99% confidence interval for the mean income of all 40-year-old men.

10.52 T-bills and inflation. Figure 10.16 (page 554) gives part of a regression analysis of the data in Table 10.1 relating the return on Treasury bills to the rate of inflation. The output includes prediction of the T-bill return when the inflation rate is 3.7%.
(a) Use the output to give a 90% confidence interval for the mean return on T-bills in all years having 3.7% inflation.
(b) You think that next year's inflation rate will be 3.7%. It isn't possible, without complicated arithmetic, to give a 90% prediction interval for next year's T-bill return based on the output displayed. Why not?

10.53 Two confidence intervals. The data used for Exercise 10.49 include 195 men 30 years old. The mean income of these men is $\bar{y} = \$49,880$ and the standard deviation of these 195 incomes is $s_y = \$38,250$.
(a) Use the one-sample t procedure to give a 95% confidence interval for the mean income μ_y of 30-year-old men.
(b) Why is this interval different from the 95% confidence interval for μ_y in the regression output? (*Hint:* What data are used by each method?)

The following exercises require use of software that will calculate the intervals required for predicting mean response and individual response.

10.54 Size and selling price of houses. Table 10.6 (page 549) gives data on the size in square feet of a random sample of houses sold in a Midwest city along with their selling prices. **HOUSESIZE**
(a) Find the mean size $\bar{x}$ of these houses and also their mean selling price $\bar{y}$. Give the equation of the least-squares regression line for predicting price from size, and use it to predict the selling price of a house of mean size. (You knew the answer, right?)
(b) Jasmine and Woodie are selling a house whose size is equal to the mean of our sample. Give an interval that predicts the price they will receive with 95% confidence.

10.55 Beer and blood alcohol. Exercise 10.38 (page 550) gives data from measuring the blood alcohol content (BAC) of students 30 minutes after they drank an assigned number of cans of beer. Steve thinks he can drive legally 30 minutes after he drinks 5 beers. The legal limit is BAC = 0.08. Give a 90% confidence interval for Steve's BAC. Can he be confident he won't be arrested if he drives and is stopped? **BAC**

10.56 Selling a large house. Among the houses for which we have data in Table 10.6, just four have floor areas of 1800 square feet or more. Give a 90% confidence interval for the mean selling price of houses with floor areas of 1800 square feet or more. **HOUSESIZE**

10.3 Some Details of Regression Inference

We have assumed that you will use software to handle regression in practice. If you do, it is much more important to understand what the standard error of the slope SE_{b_1} means than it is to know the formula your software uses to find its numerical value. For that reason, we have not yet given formulas for the standard errors. We have also not explained the block of output from software that is labeled ANOVA or Analysis of Variance. This section remedies both of these omissions.

Standard errors

We will give the formulas for all the standard errors we have met, for two reasons. First, you may want to see how these formulas can be obtained from facts you already know. The second reason is more practical: some software (in particular, spreadsheet programs) does not automate inference for prediction. We will see that the hard work lies in calculating the regression standard error s, which almost any regression software will do for you. With s in hand, the rest is easy, but only if you know more details.

Tests and confidence intervals for the slope of a population regression line start with the slope b_1 of the least-squares line and with its standard error SE_{b_1}. If you are willing to skip some messy algebra, it is easy to see where SE_{b_1} and the similar standard error SE_{b_0} of the intercept come from.

1. The regression model takes the explanatory values x_i to be fixed numbers and the response values y_i to be independent random variables all having the same standard deviation σ.

2. The least-squares slope is $b_1 = rs_y/s_x$. Here is the first bit of messy algebra that we skip: it is possible to write the slope b_1 as a linear function of the responses, $b_1 = \sum a_i y_i$. The coefficients a_i depend on the x_i.

3. Because the a_i are constants, we can find the variance of b_1 by applying the rule for the variance of a sum of independent random variables (page 253). It is just $\sigma^2 \sum a_i^2$. A second piece of messy algebra shows that this simplifies to

$$\sigma_{b_1}^2 = \frac{\sigma^2}{\sum (x_i - \overline{x})^2}$$

The standard deviation σ about the population regression line is of course not known. If we estimate it by the regression standard error s based on the residuals from the least-squares line, we get the standard error of b_1. Here are the results for both slope and intercept.

Standard Errors for Slope and Intercept

The standard error of the slope b_1 of the least-squares regression line is

$$SE_{b_1} = \frac{s}{\sqrt{\sum (x_i - \overline{x})^2}}$$

The standard error of the intercept b_0 is

$$SE_{b_0} = s\sqrt{\frac{1}{n} + \frac{\overline{x}^2}{\sum (x_i - \overline{x})^2}}$$

The critical fact is that both standard errors are multiples of the regression standard error s. In a similar manner, accepting the results of yet more messy algebra, we get the standard errors for the two uses of the regression line that we have studied.

Standard Errors for Two Uses of the Regression Line

The standard error for estimating the mean response when the explanatory variable x takes the value x^* is

$$\text{SE}_{\hat{\mu}} = s\sqrt{\frac{1}{n} + \frac{(x^* - \overline{x})^2}{\sum(x_i - \overline{x})^2}}$$

The standard error for predicting an individual response when $x = x^*$ is

$$\text{SE}_{\hat{y}} = s\sqrt{1 + \frac{1}{n} + \frac{(x^* - \overline{x})^2}{\sum(x_i - \overline{x})^2}}$$

Once again both standard errors are multiples of s. The only difference between the two prediction standard errors is the extra 1 under the square root sign in the standard error for predicting an individual response. This added term reflects the additional variation in individual responses. It means that, as we have said earlier, $\text{SE}_{\hat{y}}$ is always greater than $\text{SE}_{\hat{\mu}}$.

EXAMPLE 10.12   Prediction Intervals from a Spreadsheet

In Example 10.8, we used statistical software to predict the stress level of Seung Jin, who has an LOC = 16. Suppose that we have only the Excel spreadsheet. The prediction interval then requires some additional work.

Step 1. From the Excel output in Figure 10.5 (page 534), we know that $s = 0.4513$. Excel can also find the mean and variance of the LOC scores x for the 100 workers. They are $\overline{x} = 11.40$ and $s_x^2 = 13.6768$.

Step 2. We need the value of $\sum(x_i - \overline{x})^2$. Recalling the definition of the variance, we see that this is just

$$\sum(x_i - \overline{x})^2 = (n-1)s_x^2$$
$$= (99)(13.6768) = 1354.00$$

Step 3. The standard error for predicting Seung Jin's stress level from his LOC, $x^* = 16$, is

$$\text{SE}_{\hat{y}} = s\sqrt{1 + \frac{1}{n} + \frac{(x^* - \overline{x})^2}{\sum(x_i - \overline{x})^2}}$$
$$= 0.4513\sqrt{1 + \frac{1}{100} + \frac{(16 - 11.40)^2}{1354.00}}$$
$$= 0.4513\sqrt{1 + \frac{1}{100} + \frac{21.16}{1354.00}}$$
$$= (0.4513)(1.02563) = 0.4629$$

Step 4. We predict Seung Jin's stress level from the least-squares line (Figure 10.5 again):

$$\hat{y} = 2.2555 + (0.0399)(16) = 2.894$$

This agrees with the "Fit" from software in Example 10.8. The 95% prediction interval requires the 95% critical value for $t(98)$. For hand calculation we use $t^* = 1.990$ from Table D with df = 80. The interval is

$$\hat{y} \pm t^*\text{SE}_{\hat{y}} = 2.894 \pm (1.990)(0.4629)$$

$$= 2.894 \pm 0.9212$$

$$= 1.9728 \text{ to } 3.8152$$

This agrees with the software result in Example 10.8, with a small difference due to roundoff and especially to not having the exact t^*.

The formulas for the standard errors for prediction show us one more thing about prediction. They both contain the term $(x^* - \bar{x})^2$, the squared distance of the value x^* for which we want to do prediction from the mean $\bar{x}$ of the x-values in our data. We see that prediction is most accurate (smallest margin of error) near the mean and grows less accurate as we move away from the mean of the explanatory variable. If you know what values of x you want to do prediction for, try to collect data centered near these values.

APPLY YOUR KNOWLEDGE

10.57 T-bills and inflation. Figure 10.10 (page 541) gives the Excel output for regressing the annual return on Treasury bills on the annual rate of inflation. The data appear in Table 10.1 (page 540). Starting with the regression standard error $s = 1.8761$ from the output and the variance of the inflation rates in Table 10.1 (use your calculator), find the standard error of the regression slope, SE_{b_1}. Check your result against the Excel output.

10.58 Predicting T-bill return. Figure 10.16 (page 554) uses statistical software to predict the return on Treasury bills in a year when the inflation rate is 3.7%. Let's do this without specialized software. Figure 10.10 contains Excel regression output. Use a calculator or software to find the variance s_x^2 of the annual inflation rates in Table 10.1. From this information, find the 95% prediction interval for one year's T-bill return. Check your result against the software output in Figure 10.16.

Analysis of variance for regression

Software output for regression problems, such as those in Figures 10.5, 10.6, and 10.10, reports values under the heading of ANOVA or Analysis of Variance. You can ignore this part of the output for simple linear regression, but it becomes useful in *multiple regression,* where several explanatory variables are used together to predict a response.

analysis of variance   **Analysis of variance** is the term for statistical analyses that break down the variation in data into separate pieces that correspond to different sources of variation. In the regression setting, the observed variation in the responses y_i comes from two sources:

- As the explanatory variable x moves, it pulls the response with it along the regression line. In Figure 10.3, for example, accountants with an LOC of 15 have generally higher stress levels than those accountants with an LOC of 8. The least-squares line drawn on the scatterplot describes this tie between x and y.

- When x is held fixed, y still varies because not all individuals who share a common x have the same response y. There are several accountants with LOC near 8, and their stress level values are scattered above and below the least-squares line.

We discussed these sources of variation in Chapter 2 (page 105), where the main point was that the squared correlation r^2 is the proportion of the total variation in the responses that comes from the first source, the straight-line tie between x and y.

Analysis of variance for regression expresses these two sources of variation in algebraic form so that we can calculate the breakdown of overall variation into two parts. Skipping quite a bit of messy algebra, we just state that this **analysis of variance** *ANOVA equation* **equation** always holds:

$$\text{Total variation in } y = \text{Variation along the line} + \text{Variation about the line}$$
$$\sum(y_i - \overline{y})^2 = \sum(\hat{y}_i - \overline{y})^2 + \sum(y_i - \hat{y}_i)^2$$

Understanding the ANOVA equation requires some thought. The "Total variation" in the responses y_i is expressed by the sum of the squares of the deviations $y_i - \overline{y}$. If all responses were the same, all would equal the mean response $\overline{y}$ and the total variation would be zero. The total variation term is just $n - 1$ times the variance of the responses. The "Variation along the line" term has the same form: it is the variation among the *predicted* responses $\hat{y}_i$. The predicted responses lie on the least-squares line—they show how y moves in response to x. The "Variation about the line" term is the sum of squares of the *residuals* $y_i - \hat{y}_i$. It measures the size of the scatter of the observed responses above and below the line. If all the responses fell exactly on a straight line, the residuals would all be 0 and there would be no variation about the line. The total variation would equal the variation along the line.

EXAMPLE 10.13 ANOVA for Accountants' Job Stress

CASE 10.1   Figure 10.18 repeats Figure 10.5. It is the Excel output for the regression of job stress on locus of control (Case 10.1). The three terms in the analysis of variance equation appear under the "SS"

Excel

	A	B	C	D	E	F	G
1	SUMMARY OUTPUT						
2							
3	*Regression Statistics*						
4	Multiple R	0.31227647					
5	R Square	0.097516594					
6	Adjusted R Square	0.088307579					
7	Standard Error	0.451276599					
8	Observations	100					
9							
10	ANOVA						
11		*df*	*SS*	*MS*	*F*	*Significance F*	
12	Regression	1	2.156507639	2.156508	10.58925418	0.001561535	
13	Residual	98	19.95775576	0.203651			
14	Total	99	22.1142634				
15							
16			*Standard*				
17		*Coefficients*	*Error*	*t Stat*	*P-value*	*Lower 95%*	*Upper 95%*
18	Intercept	2.2554967	0.146912731	15.35263	8.02318E-28	1.963953174	2.547040226
19	LOC	0.03990857	0.012264038	3.254113	0.001561535	0.015570987	0.064246153

FIGURE 10.18 Excel output for the regression of perceived job stress on locus of control, for Example 10.13. We now concentrate on the analysis of variance part of the output.

sum of squares   heading. SS stands for **sum of squares,** reflecting the fact that each of the three terms is a sum of squared quantities. You can read the output as follows:

Total variation in y = Variation along the line + Variation about the line

Total SS	=	Regression SS	+	Residual SS
22.1143	=	2.1565	+	19.9578

The proportion of variation in stress levels explained by regressing LOC is

$$r^2 = \frac{\text{Regression SS}}{\text{Total SS}}$$

$$= \frac{2.1565}{22.1143} = 0.0975$$

This agrees with the "R Square" value in the output. Only about 10% of the variation in stress level is explained by the linear relationship between stress level and LOC. The rest is variation in stress among workers with similar locus of controls.

There is more to the ANOVA table in Figure 10.18. Each sum of squares has a *degrees of freedom*   **degrees of freedom.** The total degrees of freedom are $n - 1 = 99$, the degrees of freedom for the variance of $n = 100$ observations. This matches the total sum of squares, which is the sum of squares that appears in the definition of the variance. We know that the degrees of freedom for the residuals and for t statistics in simple linear regression are $n - 2$. It is therefore no surprise that the degrees of freedom for the residual sum of squares are also $n - 2 = 98$. That leaves just 1 degree of freedom for regression, because degrees of freedom in ANOVA also add:

Total df	=	Regression df	+	Residual df
$n - 1$	=	1	+	$n - 2$

mean square   Dividing a sum of squares by its degrees of freedom gives a **mean square (MS).** The total mean square (not given in the output) is just the variance of the responses y_i. The residual mean square is the square of our old friend the regression standard error:

$$\text{Residual mean square} = \frac{\text{Residual SS}}{\text{Residual df}}$$

$$= \frac{\sum (y_i - \hat{y}_i)^2}{n - 2}$$

$$= s^2$$

You see that the analysis of variance table reports in a different way quantities such as r^2 and s that are needed in regression analysis. It also reports in a different way the test for the overall significance of the regression. If regression on x has no value for predicting y, we expect the slope of the population regression line to be close to zero. That is, the null hypothesis of "no linear relationship" is $H_0: \beta_1 = 0$. To test H_0, we standardize the slope of the least-squares line to get a t statistic. The ANOVA approach starts instead with sums of squares. If regression on x has no value for predicting y, we expect the regression SS to be only a small part of the total SS, most of which will be made up of the residual SS. It turns out that the proper way to standardize this comparison is to use the ratio

$$F = \frac{\text{Regression MS}}{\text{Residual MS}}$$

ANOVA F statistic   This **ANOVA F statistic** appears in the second column from the right in the ANOVA table in Figure 10.18. If H_0 is true, we expect F to be small. For simple linear regression, the ANOVA F statistic always equals the square of the t statistic for testing $H_0: \beta_1 = 0$. That is, the two tests amount to the same thing.

EXAMPLE 10.14 ANOVA for Accountants' Job Stress, Continued

The Excel output in Figure 10.18 contains the values for the analysis of variance equation for sums of squares and also the corresponding degrees of freedom. The residual mean square is

$$\text{Residual MS} = \frac{\text{Residual SS}}{\text{Residual df}}$$

$$= \frac{19.9578}{98} = 0.2037$$

The square root of the residual MS is $\sqrt{0.2037} = 0.4513$. This is the regression standard error s, as claimed. The ANOVA F statistic is

$$F = \frac{\text{Regression MS}}{\text{Residual MS}}$$

$$= \frac{2.1565}{0.20365} = 10.589$$

The square root of F is $\sqrt{10.589} = 3.254$. Sure enough, this is the value of the t statistic for testing the significance of the regression, which also appears in the Excel output. The P-value for F, $P = 0.0016$, is the same as the two-sided P-value for t.

We have now explained almost all the results that appear in a typical regression output such as Figure 10.18. ANOVA shows exactly what r^2 means in regression. Aside from this, ANOVA seems redundant; it repeats in less clear form information that is found elsewhere in the output. This is true in simple linear regression, but ANOVA comes into its own in *multiple regression,* the topic of the next chapter.

APPLY YOUR KNOWLEDGE

T-bills and inflation. *Figure 10.10 (page 541) gives the Excel output for the regression of the rate of return on Treasury bills against the rate of inflation during the same year. Exercises 10.59 to 10.61 use this output.*

10.59 A significant relationship?   The output reports *two* tests of the null hypothesis that regressing on inflation does help to explain the return on T-bills. State the hypotheses carefully, give the two test statistics, show how they are related, and give the common P-value.

10.60 The ANOVA table.   Use the numerical results in the Excel output to verify each of these relationships.
(a) The ANOVA equation for sums of squares.
(b) How to obtain the total degrees of freedom and the residual degrees of freedom from the number of observations.
(c) How to obtain each mean square from a sum of squares and its degrees of freedom.
(d) How to obtain the F statistic from the mean squares.

10.61 ANOVA by-products.

(a) The output gives $r^2 = 0.5036$. How can you obtain this from the ANOVA table?

(b) The output gives the regression standard error as $s = 1.8761$. How can you obtain this from the ANOVA table?

SECTION 10.3 Summary

- The **analysis of variance equation** for simple linear regression expresses the total variation in the responses as the sum of two sources: the linear relationship of y with x and the residual variation in responses for the same x. The equation is expressed in terms of **sums of squares.**

- Each sum of squares has a **degrees of freedom.** A sum of squares divided by its degrees of freedom is a **mean square.** The residual mean square is the square of the regression standard error.

- The **ANOVA table** gives the degrees of freedom, sums of squares, and mean squares for total, regression, and residual variation. The **ANOVA F statistic** is the ratio $F = \text{Regression MS}/\text{Residual MS}$. In simple linear regression, F is the square of the t statistic for the hypothesis that regression on x does not help explain y.

SECTION 10.3 Exercises

For Exercises 10.57 and 10.58, see page 560; and for 10.59 to 10.61, see pages 563–564.

U.S. versus overseas stock returns. *How are returns on common stocks in overseas markets related to returns in U.S. markets? Consider measuring U.S. returns by the annual rate of return on the Standard & Poor's 500 stock index and overseas returns by the annual rate of return on the Morgan Stanley Europe, Australasia, Far East (EAFE) index. Both are recorded in percents. Here is part of the Minitab output for regressing the EAFE returns on the S&P 500 returns for the 20 years 1989 to 2008.*

```
The regression equation is
EAFE = - 2.58 + 0.775 S&P


Analysis of Variance

Source          DF     SS     MS     F
Regression            4560.6
Residual Error
Total           19   8556.0
```

Exercises 10.62 to 10.64 use this output. **EAFE**

10.62 The ANOVA table. Complete the analysis of variance table by filling in the "Residual Error" row and the other missing items in the DF, MS, and F columns.

10.63 s and r^2. What are the values of the regression standard error s and the squared correlation r^2?

10.64 Estimating the slope. The standard deviation of the S&P 500 returns for these years is 19.99%. From this and your work in

the previous exercise, find the standard error for the least-squares slope b_1. Give a 90% confidence interval for the slope β_1 of the population regression line.

Corporate reputation and profitability. *Is a company's reputation (a subjective assessment) related to objective measures of corporate performance such as its profitability? One study of this relationship examined the records of 154 Fortune 500 firms.[18] Corporate reputation was measured on a scale of 1 to 10 by a* Fortune *magazine survey. Profitability was defined as the rate of return on invested capital. Figure 10.19 contains SAS output for the regression of profitability (PROFIT) on reputation score (REPUTAT). The format is very similar to the Excel and Minitab output we have seen, with minor differences in labels. Exercises 10.65 to 10.72 concern this study. You can take it as given that examination of the data shows no serious violations of the conditions required for regression inference.*

10.65 Significance in two senses.

(a) Is there good evidence that reputation helps explain profitability? (State hypotheses, give a test statistic and P-value, and state a conclusion.)

(b) What percent of the variation in profitability among these companies is explained by regression on reputation?

(c) Use your findings in parts (a) and (b) as the basis for a short description of the distinction between statistical significance and practical significance.

10.66 Estimating the slope. Explain clearly what the slope β_1 of the population regression line tells us in this setting. Give a 99% confidence interval for this slope.

FIGURE 10.19 SAS output for the regression of the profitability of 154 companies on their reputation scores, for Exercises 10.65 to 10.72.

```
SAS

Dependent Variable: PROFIT

                  Analysis of Variance

                       Sum of      Mean
  Source      DF      Squares     Square   F Value   Prob>F

  Model        1      0.18957    0.18957   36.492    0.0001
  Error      152      0.78963    0.00519
  C Total    153      0.97920

      Root MSE      0.07208   R-Square     0.1936
      Dep Mean      0.10000   Adj R-sq     0.1883
      C.V.         72.07575

                  Parameter Estimates

                   Parameter    Standard   T for H0:
  Variable DF       Estimate       Error   Parameter=0  Prob> |T|

  INTERCEP 1      -0.147573  0.04139259    -3.565       0.0005
  REPUTAT  1       0.039111  0.00647442     6.041       0.0001
```

10.67 Predicting profitability. An additional calculation shows that the variance of the reputation scores for these 154 firms is $s_x^2 = 0.8101$. SAS labels the regression standard error s as "Root MSE." Starting from these facts, give a 95% confidence interval for the mean profitability (return on investment) for all companies with reputation score $x = 7$.

10.68 Predicting profitability. A company not covered by the *Fortune* survey has reputation score $x = 7$. Will a 95% prediction interval for this company's profitability be wider or narrower than the confidence interval found in the previous exercise? Explain why we should expect this. Then give the 95% prediction interval.

10.69 *F* versus *t*. How do the ANOVA F statistic and its P-value relate to the t statistic for the slope and its P-value? Identify these results on the output and verify their relationship (up to roundoff error).

10.70 The regression standard error. SAS labels the regression standard error s as "Root MSE." How can you obtain s from the ANOVA table? Do this, and verify that your result agrees with Root MSE.

10.71 Squared correlation. SAS gives the squared correlation r^2 as "R-Square." How can you obtain r^2 from the ANOVA table? Do this, and verify that your result agrees with R-Square.

10.72 Correlation. The regression in Figure 10.19 takes reputation as explaining profitability. We could as well take reputation as in part explained by profitability. We would then reverse the roles of the variables, regressing REPUTAT on PROFIT. Both regressions lead to the same conclusions about the correlation between PROFIT and REPUTAT. What is this correlation r? Is there good evidence that it is positive?

STATISTICS IN SUMMARY

The methods of data analysis apply to any set of data. We can make a scatterplot and calculate the correlation and the least-squares regression line whenever we have data on two quantitative variables. Statistical inference makes sense only in more restrictive circumstances. The regression model describes the circumstances in which we can do inference about regression. The regression model includes a new parameter, the standard deviation σ that describes how much variation there is in responses y when x is held fixed. Estimating σ is the key to inference about regression. We use the regression standard error s (roughly, the sample standard deviation of the residuals) to estimate σ. Here is a review list of the most important skills you should have acquired from your study of this chapter.

A. Preliminaries

1. Make a scatterplot to show the relationship between an explanatory and a response variable.

2. Use a calculator or software to find the correlation and the equation of the least-squares regression line.

B. Recognition

1. Recognize the regression setting: a straight-line relationship between an explanatory variable x and a response variable y.

2. Recognize which type of inference you need in a particular regression setting.

3. Inspect the data to recognize situations in which inference isn't safe: a nonlinear relationship, influential observations, strongly skewed residuals in a small sample, or nonconstant variation of the data points about the regression line.

C. Do Inference Using Software Output

1. Explain in any specific regression setting the meaning of the slope β_1 of the population regression line.

2. Understand software output for regression. Find in the output the slope and intercept of the least-squares line, their standard errors, the regression standard error s, and the squared correlation r^2.

3. Carry out tests and calculate confidence intervals for the slope β_1 of the population regression line.

4. Explain the distinction between a confidence interval for the mean response when x has a specific value and a prediction interval for an individual response for the same value of x.

5. If software gives output for prediction, use that output to give either confidence or prediction intervals.

D. Details of Regression

1. Give either type of interval for prediction starting with basic regression output and descriptive statistics for the explanatory variable x.

2. Find the regression standard deviation s and the squared correlation r^2 from an ANOVA table.

3. Use the ANOVA F statistic to test the significance of regression. Find the t statistic for this test from F.

CHAPTER 10 Review Exercises

10.73 What's wrong? For each of the following, explain what is wrong and why.
(a) The slope describes the change in x for a unit change in y.
(b) The population regression line is $y = b_0 + b_1 x$.
(c) A 95% confidence interval for the mean response is the same width regardless of x.

10.74 What's wrong? For each of the following, explain what is wrong and why.
(a) The parameters of the simple linear regression model are b_0, b_1, and s.

(b) To test H_0: $b_1 = 0$, use a t test.
(c) For any value of the explanatory variable x, the confidence interval for the mean response will be wider than the prediction interval for a future observation.

10.75 Are the two fuel efficiency measurements similar? Refer to Exercise 7.45 (page 416). The driver wants to determine if these two mpg calculations are different. 🅐 FUELCOMP

Fill-up:	1	2	3	4	5	6	7	8	9	10
Computer:	41.5	50.7	36.6	37.3	34.2	45.0	48.0	43.2	47.7	42.2
Driver:	36.5	44.2	37.2	35.6	30.5	40.5	40.0	41.0	42.8	39.2

Fill-up:	11	12	13	14	15	16	17	18	19	20
Computer:	43.2	44.6	48.4	46.4	46.8	39.2	37.3	43.5	44.3	43.3
Driver:	38.8	44.5	45.4	45.3	45.7	34.2	35.2	39.8	44.9	47.5

(a) Consider the driver's mpg calculations as the explanatory variable. Plot the data and describe the relationship. Are there any outliers or unusual values? Does a linear relationship seem reasonable?

(b) Run the simple linear regression and give the equation for the least-squares regression line.

(c) Summarize the results. Does it appear that the computer and driver calculations are the same? Explain.

10.76 Yearly number of tornadoes. The Storm Prediction Center of the National Oceanic and Atmospheric Administration maintains a list of tornadoes, floods, and other weather phenomena. Table 10.8 summarizes the annual number of tornadoes in the United States between 1953 and 2008.[19] **TORNADOES**

(a) Make a plot of the total number of tornadoes by year. Is there a linear trend over time? Are there any outliers or unusual patterns? Explain your answer.

(b) Run the simple linear regression and summarize the results, making sure to construct a 95% confidence interval for the average annual increase in the number of tornadoes.

(c) Obtain the residuals and plot them versus year. Is there anything unusual in the plot?

(d) Are the residuals Normal? Justify your answer.

10.77 Plot indicates model assumptions. Construct a plot with data and a regression line that fits the simple linear regression model framework. Then construct another plot that has the same slope and intercept but a much smaller value of the regression standard error s.

10.78 Significance tests and confidence intervals. The significance test for the slope in a simple linear regression gave a value $t = 4.12$ with 50 degrees of freedom. Would the 95% confidence interval for the slope include the value zero? Give a reason for your answer.

10.79 Bias in Medicare payments. Payments to hospitals for pharmaceuticals under Medicare's Outpatient Prospective Payment System (OPPS) are based on estimates of costs. These are determined by multiplying the cost of the pharmaceutical to the hospital by a department-specific average markup. Some evidence suggests, however, that when hospitals set charges for the pharmaceuticals sold through their pharmacies, they do not simply add a fixed percent of the cost to each item to cover the expense of operating the pharmacy. It has been suggested that hospitals apply below-average markups for pharmaceutical products with above-average costs. This is the "charge compression" hypothesis. If it is true, then the system for reimbursing hospitals for pharmaceutical costs is based on biased estimates that underestimate true costs. A study designed to test the charge compression hypothesis used Medicare OPPS data on 139 different drugs.[20] One analysis presented the results of a simple linear regression analysis using (log of) cost to predict (log of) markup.

(a) A negative coefficient for cost in this model would be consistent with support for the charge compression hypothesis. Explain why.

(b) The researchers who conducted the study reported that the constant was 2.885 and the coefficient for (log) cost was −0.295. Give the equation of the fitted regression line.

TABLE 10.8 Annual number of tornadoes in the United States between 1953 and 2008

Year	Number of tornadoes	Year	Number of tornadoes	Year	Number of tornadoes	Year	Number of tornadoes
1953	421	1967	926	1981	783	1995	1235
1954	550	1968	660	1982	1046	1996	1173
1955	593	1969	608	1983	931	1997	1148
1956	504	1970	653	1984	907	1998	1449
1957	856	1971	888	1985	684	1999	1340
1958	564	1972	741	1986	764	2000	1076
1959	604	1973	1102	1987	656	2001	1213
1960	616	1974	947	1988	702	2002	934
1961	697	1975	920	1989	856	2003	1372
1962	657	1976	835	1990	1133	2004	1819
1963	464	1977	852	1991	1132	2005	1194
1964	704	1978	788	1992	1298	2006	1103
1965	906	1979	852	1993	1176	2007	1098
1966	585	1980	866	1994	1082	2008	1691

(c) A t statistic is given for the coefficient of (log) cost. The value is -12.26. Interpret this statistic and what it means in terms of the charge compression hypothesis.

10.80 Correlations for restaurant spending. A study of spending by customers in restaurants examined 1413 transactions.[21] Five of the variables measured were check (total check amount divided by the number of customers at the table), minutes (the amount of time spent dining), table (the size of the table), party (the number of people in the party), and SPM (spending per minute, check divided by minutes). Here is a table of the correlations for these variables:

	Check	Minutes	Table	Party
Minutes	0.251			
Table	0.045	0.083		
Party	-0.306	0.137	0.379	
SPM	0.689	-0.397	-0.031	-0.400

(a) Perform a significance test for each of these correlations and summarize the results.
(b) Discuss the pattern in the correlations, paying particular attention to the sign, the size, and the results of the significance test.
(c) You performed 10 significance tests in this exercise. Suppose that you wanted to do a Bonferroni correction to take into account this fact and to keep the overall false rejection rate at 5%. In that case, you would declare a correlation to be statistically significant only if the P-value is $0.05/10$ or less. Perform the significance tests using the Bonferroni correction. Summarize the results and compare them with the results you obtained in part (a).

10.81 What can we infer? The study described in the previous exercise analyzed data that were collected from a Chevys Fresh-mex Restaurant in suburban Phoenix during busy times at the restaurant on Fridays, Saturdays, and Sundays.
(a) To what extent do you think that the results that you obtained in the previous exercise apply to other restaurants in other cities? Give reasons for your answers.
(b) Design a similar study for a different restaurant. Give details regarding what data you will collect and how you will collect it.

10.82 Brand equity and sales. Brand equity is one of the most important assets of a business. It includes brand loyalty, brand awareness, perceived quality, and brand image. One study examined the relationship between brand equity and sales using simple linear regression analysis.[22] The correlation between brand equity and sales was reported to be 0.757 with a significance level of 0.001.
(a) Explain in simple language the meaning of these results.
(b) The study examined quick-service restaurants in Korea and was based on 394 usable surveys from a total of 950 that were distributed to shoppers at a mall. Write a short narrative commenting on the design of the study and how well you think the results would apply to other settings.

10.83 Hotel sizes and numbers of employees. A human resources study of hotels collected data on the size, measured by number of rooms, and the number of employees for 14 hotels in Canada.[23] Here are the data: HOTELSIZE

Employees	Rooms	Employees	Rooms
1200	1388	275	424
180	348	105	240
350	294	435	601
250	413	585	1590
415	346	560	380
139	353	166	297
121	191	228	108

(a) To what extent can the number of employees be predicted by the size of the hotel? Plot the data and summarize the relationship.
(b) Is this the type of relationship that you would expect to see before examining the data? Explain why or why not.
(c) Calculate the least-squares regression line and add it to the plot.
(d) Give the results of the significance test for the regression slope with your conclusion.
(e) Find a 95% confidence interval for the slope.

10.84 How can we use the results? Refer to the previous exercise.
(a) If one hotel had 100 more rooms than another, how many additional employees would you expect that hotel to have?
(b) Give a 95% confidence interval for your answer in part (a).
(c) The study collected these data from 14 hotels in Toronto. Discuss how well you think the results can be generalized to other hotels in Toronto, to hotels in Canada, and to hotels in other countries.

10.85 Check the outliers. The plot that you generated in Exercise 10.83 has two observations that appear to be outliers.
(a) Identify these points on a plot of the data.
(b) Rerun the analysis with the other 12 hotels and summarize the effect of the two possible outliers on the results that you gave in Exercise 10.83.

10.86 Growth in grocery store size. Here are data giving the median store size (in square feet) by year for grocery stores:[24] GROCERY

Year	Store size	Year	Store size	Year	Store size
1993	33.0	1998	40.5	2003	44.0
1994	35.1	1999	44.8	2004	45.6
1995	37.2	2000	44.6	2005	48.1
1996	38.6	2001	44.0	2006	48.8
1997	39.3	2002	44.0	2007	47.5

(a) Use a simple linear regression and a prediction interval to give an estimate, along with a measure of uncertainty, for the median grocery store size in 2008.

(b) Plot the data with the regression line. Based on what you see, do you think that the answer that you computed in part (a) is a good prediction? Explain why or why not.

10.87 Expenditures of school districts. There are 315 school corporations in the state of Indiana. Various information about these corporations is available on the Internet.[25] Here are data on the numbers of students (NStu), expressed in thousands, and the total yearly expenditures (Exp), expressed in millions of dollars, for a simple random sample of 30 of these school corporations: SCHOOLEXP

NStu	Exp	NStu	Exp	NStu	Exp
9	5.3	3	3.7	45	24.5
32	17.8	37	19.5	18	9.8
62	35.2	68	35.1	17	10.0
14	8.2	8	5.1	16	9.4
90	42.6	17	9.5	17	9.2
14	9.7	5	3.1	8	4.6
16	9.3	34	17.7	11	6.8
5	3.2	16	9.1	24	12.5
16	8.3	23	10.7	4	2.8
32	17.2	102	60.1	11	5.8

Use these data to explain how the number of students in a school corporation can be used to predict the total yearly expenditures. Write a report summarizing your work and conclusions.

10.88 Some unusual observations. Refer to the previous exercise. Two corporations that were not included in the simple random sample are the School System of Hammond and the Northwest Indiana Special Education Cooperative. Hammond is a very large corporation with 137 thousand students and 83 million dollars in expenditures. On the other hand, Northwest Indiana is listed as having 39 students (which would be rounded to 0 thousand students) and expenditures of 13.9 million dollars. We assume that this school corporation provides services to students with special education needs within the state. Reanalyze the data from the previous exercise with each of these corporations added to the data set (31 observations) and with both added (32 observations). Use graphical and numerical summaries to describe how the results and inferences change for these altered data sets.

10.89 Predicting success in a statistics course. Can a pretest on mathematics skills predict success in a statistics course? The 55 students in an introductory statistics class took a pretest at the beginning of the semester. The least-squares regression line for predicting the score y on the final exam from the pretest score x was $\hat{y} = 10.5 + 0.82x$. The standard error of b_1 was 0.38. Test the null hypothesis that there is no linear relationship between the pretest score and the score on the final exam against the two-sided alternative.

CHAPTER 10 Case Study Exercises

CASE STUDY EXERCISE 1: Agricultural productivity. Few sectors of the economy have increased their productivity as rapidly as agriculture. Productivity is defined as output per unit input. "Total factor productivity" (TFP) takes all inputs (labor, capital, fuels, and so on) into account. Table 10.9 gives the total factor productivity of U.S. farms from 1948 to 2006.[26] The entries are index numbers. That is, they give each year's TFP as a percent of the value for 1948. Your assignment is to describe in some detail how U.S. farm productivity has increased. AGPRODUCTIVITY

TABLE 10.9	Total factor productivity of U.S. farms between 1948 and 2006								
Year	TFP	Year	TFP	Year	TFP	Year	TFP	Year	TFP
1948	100.0	1960	111.5	1972	130.8	1984	163.3	1996	206.0
1949	94.4	1961	114.4	1973	134.9	1985	174.5	1997	210.8
1950	92.6	1962	113.8	1974	126.5	1986	172.9	1998	209.1
1951	95.3	1963	116.7	1975	139.1	1987	175.9	1999	211.7
1952	97.2	1964	117.1	1976	137.0	1988	167.3	2000	221.6
1953	98.3	1965	120.7	1977	144.7	1989	181.6	2001	224.5
1954	101.1	1966	118.9	1978	138.0	1990	188.2	2002	220.4
1955	101.2	1967	122.8	1979	142.8	1991	188.2	2003	227.4
1956	101.5	1968	123.5	1980	138.1	1992	202.1	2004	242.2
1957	100.6	1969	124.3	1981	153.9	1993	190.3	2005	240.0
1958	105.7	1970	123.1	1982	158.6	1994	205.0	2006	238.4
1959	106.9	1971	131.8	1983	139.2	1995	190.4		

(a) Plot TFP against year. It appears that around 1980 the rate of increase in TFP changed. How is this apparent from the plot? What was the nature of the change?

(b) Regress TFP on year using only the data for the years 1948 to 1980. Add the least-squares line to your scatterplot. The line makes the finding in part (a) clearer.

(c) Give a 95% confidence interval for the annual rate of change in TFP during the period 1948 to 1980.

(d) Regress TFP on year for the years 1981 to 2006. Add this line to your plot. Give a 95% confidence interval for the annual rate of improvement in TFP during these years.

(e) Write a brief report on trends in U.S. farm productivity since 1948, making use of your analysis in parts (a) to (d).

CASE STUDY Exercise 2: Philip Morris versus the market.
Table 10.10 shows the monthly returns on the common stock of Philip Morris (MO), as the company was then named, and the returns on the Standard & Poor's 500 stock index for the same months. Return is measured in percent. The data are for 83 consecutive months from mid-1990 to mid-1997, a period chosen to avoid the stock market bubble of the late 1990s. (The time order runs down the columns, but we will ignore trends over time.)

We expect a stock to move in concert with the market to some extent, but the strength of the relationship varies greatly among different stocks. We will examine the relationship between MO and market returns in detail.

(a) Regress the MO returns on the S&P returns. Make a scatterplot and draw the least-squares line on the plot. Explain carefully what the slope and intercept of this line mean, in terms understandable to an investor. Also give a measure of the strength of the relationship and explain its meaning to an investor.

(b) Suppose an investor is willing to take these data as a sample of future returns. Explain why we cannot expect future returns (the population) to have exactly the same slope as the least-squares line. Then give a 90% confidence interval for the slope of the population regression line.

(c) Find the residuals from your regression. (Most software will do this for you.) Are any of the residuals outliers by the $1.5 \times IQR$ criterion (page 39)? If so, circle the corresponding points on your scatterplot. Is it likely that these points strongly influence the regression? Why? Verify your answer by redoing the regression without the outlier(s) and adding the new line to your plot for easy comparison with the original line.

(d) If your software allows, make a Normal quantile plot of the residuals. Aside from any outliers detected in part (c) is the distribution approximately Normal?

(e) An investor believes that a market rally is imminent and that the S&P 500 will rise 7.5% in the next month. Give a 90% confidence interval for the return on MO next month.

TABLE 10.10 Monthly returns (percents) on Philip Morris (MO) and the Standard & Poor's 500 stock index (S&P)

Month	MO	S&P	Month	MO	S&P	Month	MO	S&P	Month	MO	S&P
1	−5.7	−9.0	22	−0.5	0.5	43	−7.1	−2.7	64	4.2	4.4
2	1.2	−5.5	23	−4.5	−2.1	44	−8.4	−5.0	65	4.0	0.7
3	4.1	−0.4	24	8.7	4.0	45	7.7	2.0	66	2.8	3.4
4	3.2	6.4	25	2.7	−2.1	46	−9.6	1.6	67	6.7	0.9
5	7.3	0.5	26	4.1	0.6	47	6.0	−2.9	68	−10.4	0.5
6	7.5	6.5	27	−10.3	0.3	48	6.8	3.8	69	2.7	1.5
7	18.6	7.1	28	4.8	3.4	49	10.9	4.1	70	10.3	2.5
8	3.7	1.7	29	−2.3	0.6	50	1.6	−2.9	71	5.7	0.0
9	−1.8	0.9	30	−3.1	1.5	51	0.2	2.2	72	0.6	−4.4
10	2.4	4.3	31	−10.2	1.4	52	−2.4	−3.7	73	−14.2	2.1
11	−6.5	−5.0	32	−3.7	1.5	53	−2.4	0.0	74	1.3	5.2
12	6.7	5.1	33	−26.6	−1.8	54	3.9	4.0	75	2.9	2.8
13	9.4	2.3	34	7.2	2.7	55	1.7	3.9	76	11.8	7.6
14	−2.0	−2.1	35	−2.9	−0.3	56	9.0	2.5	77	10.6	−3.1
15	−2.8	1.3	36	−2.3	0.1	57	3.6	3.4	78	5.2	6.2
16	−3.4	−4	37	3.5	3.8	58	7.6	4.0	79	13.8	0.8
17	19.2	9.5	38	−4.6	−1.3	59	3.2	1.9	80	−14.7	−4.5
18	−4.8	−0.2	39	17.2	2.1	60	−3.7	3.3	81	3.5	6.0
19	0.5	1.2	40	4.2	−1.0	61	4.2	0.3	82	11.7	6.1
20	−0.6	−2.5	41	0.5	0.2	62	13.2	3.8	83	1.3	5.8
21	2.8	3.5	42	8.3	4.4	63	0.9	0.0			

CHAPTER 10 Appendix

Using Minitab and Excel for Inference for Regression

Regression was introduced in Chapter 2 and you can refer to its Appendix for guidance on creating scatterplots, computing correlation, getting simple regression output, and obtaining residual plots. In this Appendix, we will provide more details on the software in relation to the topic of inference.

Simple Regression – Confidence and Prediction Limits
Minitab:

$$\text{Stat} \blacktriangleright \text{Regression} \blacktriangleright \text{Regression}$$

Minitab does not compute confidence intervals for the regression coefficients (β_0 and β_1). However, for a given value of the explanatory variable, Minitab can compute confidence intervals for the mean and prediction intervals. Once you have clicked-in the response and explanatory data columns, click the **Options** button. Now input the value of the explanatory variable (x^*) in the **Prediction intervals for new observations** box. If you wish, you can change the confidence level from the default value of 95 as seen in the **Confidence level** box. Click **OK** to close the pop-up box and then click **OK** to get the regression output along with the confidence and prediction intervals, which are reported at the bottom of the output.

Excel:

Excel does not compute confidence intervals for the mean or prediction intervals based on an estimated regression model. These intervals are available in the WHFStat Add-In for Excel. Interestingly, unlike Minitab, Excel can report confidence intervals for the regression coefficients (β_0 and β_1). To obtain these confidence intervals, place a checkmark next to the **Confidence Level** option. If you wish, you can change the confidence level from the default value of 95 as seen in the box to the right of the **Confidence Level** option.

Correlation – Inference
Minitab:

$$\text{Stat} \blacktriangleright \text{Basic Statistics} \blacktriangleright \text{Correlation}$$

Obtaining correlation output was described in the Appendix of Chapter 2. In the correlation output, a P-value is reported only for the two-sided test of H_0: $\rho = 0$ versus H_0: $\rho \neq 0$.

Excel:

Testing of the correlation parameter is not available in standard Excel, but one-sided and two-sided tests of the correlation parameter are available in the WHFStat Add-In for Excel.

DANIEL BARRY/BLOOMBERG VIA GETTY IMAGES

Multiple Regression

Can opening weekend revenue successfully predict total box office revenue for a new movie? What if we include the number of theaters where the movie was shown? Case 11.2 examines what best predicts a new movie's success.

CHAPTER OUTLINE

11.1 Data Analysis for Multiple Regression

11.2 Inference for Multiple Regression

11.3 Multiple Regression Model Building

Introduction

In Chapters 2 and 10 we studied the relationship between one explanatory variable and one response variable. In this chapter we look at situations where several explanatory variables work together to explain the response variable. We do this by building on the descriptive tools we learned in Chapter 2— scatterplots, least-squares regression, and correlation—and on the basics of regression inference from Chapter 10. Many of these ideas carry directly over. For example, we will continue to use scatterplots and correlation for pairs of variables. However, the presence of several explanatory variables, which may assist or substitute for each other in predicting the response, leads to several new ideas. We start by exploring some of these ideas with an example.

EXAMPLE 11.1 A Space Model

Allocation of space or other resource within a business organization is often done using quantitative methods. Characteristics for a subunit of the organization are determined, and then a mathematical formula is used to decide the required needs.

A university has used this approach to determine office space needs in square feet (ft^2) for each department.[1] The formula allocates 210 ft^2 for the department head (HEAD), 160 ft^2 for each faculty member (FAC), 160 ft^2 for each manager (MGR), 150 ft^2 for each administrator and lecturer (LECT), 65 ft^2 for each postdoctorate and graduate assistant (GRAD), and 120 ft^2 for each clerical and service worker (CLSV).

The Chemistry Department in this university has 1 department head, 45.25 faculty, 15.50 managers, 41.52 lecturers, 411.88 graduate assistants, and 25.24 clerical and service workers. Note that fractions of people are possible in these calculations because individuals may have appointments in more than one department. For example, a person with an even split between two departments would be counted as 0.50 in each.

EXAMPLE 11.2   Office Space Needs for the Chemistry Department

Let's calculate the office space needs for the Chemistry Department based on these personnel numbers. We start with 210 ft^2 for the department head. We have 45.25 faculty, each needing 160 ft^2. Therefore, the total office space needed for faculty is 45.25 × 160 ft^2, which is 7240 ft^2. We do the same type of calculation for each personnel category and then sum the results.

Here are the calculations in a table:

Category	Number of employees	Ft^2 per employee	Number times ft^2 per employee
HEAD	1.00	210	210.0
FAC	45.25	160	7,240.0
MGR	15.50	160	2,480.0
LECT	41.52	150	6,228.0
GRAD	411.88	65	26,772.2
CLSV	25.24	120	3028.8
Total			45,959.0

The calculations that we just performed use a set of predictors, HEAD, FAC, MGR, LECT, GRAD, and CLSV, to find the office space needs for the Chemistry Department. Given values of these predictors for any other department in the university, we can perform the same calculations to find the office space needs. We organized our calculations for the Chemistry Department in the table given above. Another way to organize calculations of this type is to give a formula.

EXAMPLE 11.3   The Office Space Needs Formula

Let's assume that each department has exactly one head. So the first term in our equation will be the space need for this position, 210 ft^2. To this we add the space needs for the faculty, 160 ft^2 for each, or 160FAC. Similarly, we add the number of square feet for each category of personnel times the number of employees in the category. The result is the office space needs predicted by the space model. Here is the formula:

$$\text{PredSpace} = 210 + 160\text{FAC} + 160\text{MGR} + 150\text{LECT} + 65\text{GRAD} + 120\text{CLSV}$$

The formula combines information from the predictor variables and computes the office space needs for any department. This prediction generally will not match the actual space being used by a department. The difference between the value predicted by the model and the actual space being used is of interest to the people who assign space to departments.

EXAMPLE 11.4 Com...

The Chemistry Department cur...dicted Space with Actual Space
predicts a space need of 45,959.0...

50,075 ft² of space. On the other hand, the model

Residual ...nce between these two quantities is a residual:

$$= ...$$
$$= 411...$$

PredSpace

According to the university space needs model, ...
more office space than it needs.

Because of this, the university director of space ...
some of this excess space to a department that has actual... has about 4116 ft²
predicts. Of course, the Chemistry Department does not th...
negotiations with the space management office, they will ex...
space that they currently have and that their needs are not fully...

...iving
...

APPLY YOUR KNOWLEDGE

11.1 Check the formula. The table that appears before Example 11.3...
predicted office space needed by the Chemistry Department is 45,959.0 ft...
the formula given in Example 11.3 gives the same predicted value.

11.2 Needs of the Department of Mathematics. The Department of Mathematics
has 1 department head, 57.5 faculty, 2 managers, 49.75 administrators and lecturers,
198.74 graduate assistants, and 10.64 clerical and service workers.
(a) Find the office needs for the Mathematics Department that are predicted by the
model.
(b) The actual office space for this department is 27,326 ft². Find the residual and
explain what it means in a few sentences.

These space allocation examples illustrate two key ideas that we need for multiple
regression. First, we have several explanatory variables that are combined in a prediction equation. Second, residuals are the differences between the actual values and the
predicted values. We will now illustrate the techniques of multiple regression, including
some new ideas, through a series of case studies. In all examples, we will use software
to do the calculations.

DATA FILE   DJU

CASE 11.1

Assets, Sales, and Profits Table 11.1 shows some characteristics of the 15 prominent
utility companies that make up the Dow Jones Utility Average (DJU).[2] Included are the
stock symbol, company name, assets at the end of 2008, and sales and profits for the
year 2008 (all in billions of dollars).[3] How are profits related to sales and assets? In this
case, profits represents the response variable, and sales and assets are two explanatory
variables.

11.1 Data Analysis for Multiple Regression

As with any statistical analysis, we begin our multiple regression analysis with a careful
examination of the data.

CHAPTER 11 Multiple Reg~

TABLE 11.~y

...s Utility Average: assets, sales, and profits

		Assets ($ billions)	Sales ($ billions)	Profits ($ billions)
	...S Corporation	34.806	16.070	1.234
	American Electric Power	45.155	14.440	1.380
S~	CenterPoint Energy, Inc.	19.676	10.725	0.391
	Consolidated Edison, Inc.	33.498	13.429	1.077
	Dominion Resources, Inc.	42.053	16.290	1.834
DUK	Duke Energy Corporation	53.077	13.207	2.195
EIX	Edison International	44.615	14.112	1.215
EXC	Exelon Corporation	47.817	19.065	2.867
FE	FirstEnergy	33.521	13.627	1.342
FPL	FPL Group Corporation	44.821	16.680	1.754
NI	NiSource	20.032	8.309	−0.502
PCG	PG&E	40.860	14.628	1.338
PEG	Public Service Enterprise Group	29.049	13.322	1.192
SO	Southern Company, Inc.	48.347	17.110	1.492
WMB	Williams Companies, Inc.	26.006	11.256	0.694

Data for multiple regression

The data for a simple linear regression problem consist of observations on an explanatory variable x and a response variable y. We use n for the number of cases, or individuals. The major difference in the data for multiple regression is that we have more than one explanatory variable.

EXAMPLE 11.5   Data for Assets, Sales, and Profits

CASE 11.1   In Case 11.1, the cases are the 15 companies. Each observation consists of a value for a response variable (profits) and values for the two explanatory variables (assets and sales).

In general, we have data on n cases and we use p for the number of explanatory variables. Data are often entered into spreadsheets and computer regression programs in a format where each row describes an individual and each column corresponds to a different variable.

EXAMPLE 11.6   Spreadsheet Data for Assets, Sales, and Profits

CASE 11.1   In Case 11.1, there are 15 companies; assets and sales are the explanatory variables. Therefore, $n = 15$ and $p = 2$. Figure 11.1 shows the part of an Excel spreadsheet with the first 10 cases.

APPLY YOUR KNOWLEDGE

BANKS

11.3 Deposits, assets, and number of banks. Table 11.2 gives data for insured commercial banks, by state or other area.[4] The cases are the 50 states, the District of Columbia, and Puerto Rico. Bank assets and deposits are given in billions of dollars. We are interested in describing how assets are explained by deposits and the number of banks.

FIGURE 11.1 First 10 cases of data in spreadsheet for Example 11.6.

Microsoft Excel – DJU.xls

	A	B	C	D	E	F
1	ID	Symbol	Firm	Assets	Sales	Profits
2	1	AES	AES Corporation	34.806	16.070	1.234
3	2	AEP	American Electric Power	45.155	14.440	1.380
4	3	CNP	CenterPoint Energy, Inc.	19.676	10.725	0.391
5	4	ED	Consolidated Edison, Inc.	33.498	13.429	1.077
6	5	D	Dominion Resources, Inc.	42.053	16.290	1.834
7	6	DUK	Duke Energy Corporation	53.077	13.207	2.195
8	7	EIX	Edison International	44.615	14.112	1.215
9	8	EXC	Exelon Corporation	47.817	19.065	2.867
10	9	FE	FirstEnergy	33.521	13.627	1.342
11	10	FPL	FPL Group Corporation	44.821	16.680	1.754

TABLE 11.2   Insured commercial banks by state or other area

State or area	Number	Assets ($ billions)	Deposits ($ billions)	State or area	Number	Assets ($ billions)	Deposits ($ billions)
Alabama	148	266.2	183.0	Montana	74	18.7	14.2
Alaska	5	31.0	9.6	Nebraska	231	40.4	32.4
Arizona	55	15.7	12.2	Nevada	37	1255.6	767.3
Arkansas	137	50.2	39.3	New Hampshire	9	2.4	1.8
California	288	408.2	267.3	New Jersey	68	55.8	40.9
Colorado	142	49.9	40.8	New Mexico	49	17.3	13.8
Connecticut	25	24.8	17.9	New York	129	449.2	329.8
Delaware	24	437.2	190.7	North Carolina	79	2342.8	1545.4
District of Columbia	5	1.4	1.3	North Dakota	94	20.6	12.6
Florida	270	116.6	89.2	Ohio	155	2432.9	1525.1
Georgia	314	293.4	204.9	Oklahoma	248	64.6	48.3
Hawaii	7	30.4	22.7	Oregon	37	34.8	18.3
Idaho	16	6.4	4.7	Pennsylvania	145	257.1	178.3
Illinois	570	332.9	265.3	Rhode Island	7	169.8	76.8
Indiana	111	58.7	43.9	South Carolina	66	48.3	37.5
Iowa	361	56.2	43.9	South Dakota	83	651.5	408.6
Kansas	331	50.8	40.0	Tennessee	181	92.9	63.8
Kentucky	179	48.2	37.7	Texas	594	273.5	205.1
Louisiana	134	48.9	39.0	Utah	63	461.3	268.4
Maine	8	6.2	4.2	Vermont	9	3.5	2.7
Maryland	53	28.1	21.5	Virginia	103	391.5	285.4
Massachusetts	36	190.3	128.7	Washington	81	55.7	43.1
Michigan	143	108.1	81.3	West Virginia	61	24.2	18.0
Minnesota	408	72.3	54.8	Wisconsin	249	136.8	97.3
Mississippi	90	56.3	41.8	Wyoming	36	6.2	5.3
Missouri	321	116.3	91.0	Puerto Rico	10	99.5	64.0

(a) What is the response variable?
(b) What are the explanatory variables?
(c) What is p, the number of explanatory variables?
(d) What is n, the sample size?

11.4 Describing a multiple regression. As part of a study, data from 50 Fortune 500 companies were obtained.[5] Based on these data, the researchers described the relationship between a company's annual profits and the age, attractiveness, and facial maturity level of its CEO.
(a) What is the response variable?
(b) What is n, the number of cases?
(c) What is p, the number of explanatory variables?
(d) What are the explanatory variables?

Preliminary data analysis for multiple regression

Following our principles of data analysis, we look first at each variable separately, then at relationships among the variables. We again combine plots and numerical description.

EXAMPLE 11.7   Describing Assets, Sales, and Profits

CASE 11.1   Figure 11.2 shows descriptive statistics from Minitab for the 15 companies in the DJU. Figure 11.3 presents histograms for each variable. Each distribution appears relatively symmetric (mean and median approximately equal) with no large outliers. The response variable Profits has two observations (one large and one small) that are a bit unusual relative to the others but not to a degree that would raise concern.

FIGURE 11.2 Descriptive statistics for Example 11.7.

```
Descriptive Statistics: Assets, Sales, Profits
Variable    N     Mean  SE Mean   StDev  Minimum      Q1  Median      Q3  Maximum
Assets     15    37.56     2.70   10.44    19.68   29.05   40.86   45.16    53.08
Sales      15   14.151    0.705   2.732    8.309  13.207  14.112  16.290   19.065
Profits    15    1.300    0.199   0.770   -0.502   1.077   1.338   1.754    2.867
```

Later in this chapter, we will describe a statistical model that is the basis for inference in multiple regression. This model does not require Normality for the distributions of the response or explanatory variables. The Normality assumption applies to the distribution of the residuals, as was the case for inference in simple linear regression. We look at the distribution of each variable to be used in a multiple regression to determine if there are any unusual patterns that may be important in building our regression analysis.

APPLY YOUR KNOWLEDGE

11.5 Is there a problem? Refer to Exercise 11.4. The researchers selected the 25 highest and 25 lowest Fortune 500 companies for their study. Suppose this resulted in the response variable, annual profits, having a bimodal distribution. Since this distribution is not Normal, will this necessarily be a problem for inference in multiple regression? Explain your answer.

FIGURE 11.3 Histograms for Example 11.7.

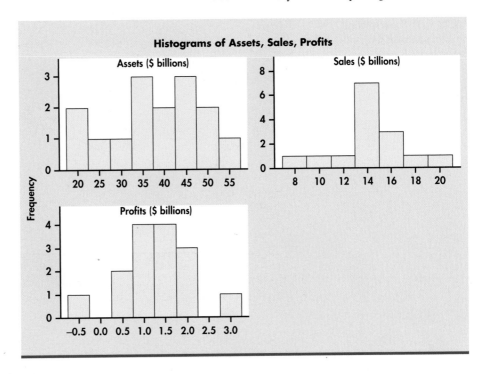

Histograms of Assets, Sales, Profits

11.6 Look at the data. Examine the data for deposits, assets, and number of banks given in Table 11.2. That is, use graphs to display the distribution of each variable. Based on your examination, how would you describe the data? Are there any states or other areas that you consider to be outliers or unusual in any way? Explain your answer.

Now that we know something about the distributions of the individual variables, we look at the relations between pairs of variables.

EXAMPLE 11.8 Assets, Sales, and Profits in Pairs

CASE 11.1

With three variables, we also have three pairs of variables to examine. Figure 11.4 gives the three correlations and Figure 11.5 displays the corresponding scatterplots. We used a scatterplot smoother (page 82) to help us see the overall pattern of each scatterplot.

FIGURE 11.4 Correlations for Example 11.8.

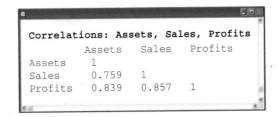

```
Correlations: Assets, Sales, Profits
            Assets   Sales   Profits
Assets      1
Sales       0.759    1
Profits     0.839    0.857   1
```

Both assets and sales have reasonably strong positive correlations with profits. These variables may be useful in explaining profits. Assets and sales are also positively correlated ($r = 0.759$). Because we will use both assets and sales to explain profits, we are somewhat concerned that this correlation is so high. Two highly correlated variables

FIGURE 11.5 Scatterplots of pairs of variables for Example 11.8.

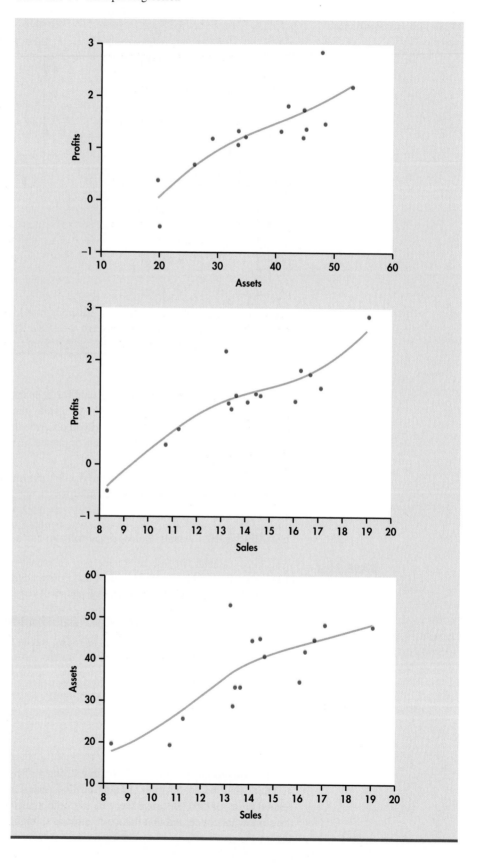

contain about the same information, so that both together may explain profits only a little better than either alone.

The plots are more revealing. The relationship between profits and each of the two explanatory variables appears reasonably linear. So does the relationship between assets and sales. One company, Duke Energy, has far more assets and profits than would be predicted from sales alone. On the other hand, these profits are well predicted using assets alone. This suggests that both variables may be helpful in predicting profits. The portion of profits that is unexplained by one explanatory variable (sales) is explained by the other (assets).

APPLY YOUR KNOWLEDGE

11.7 Examining the pairs of relationships. Examine the relationship between each pair of variables in Table 11.2. That is, compute correlations and construct scatterplots. Based on these summaries, describe these relationships. Are there any states or other areas that you consider unusual in any way? Explain your answer.

11.8 Try logs. The data file for Table 11.2 contains the logarithms of each variable. The logarithm transformation pulls in the long tail of a skewed distribution. It is common to use logarithms to make economic and financial data more symmetric before doing inference. Find the correlations and generate scatterplots for each pair of transformed variables. Interpret the results and then compare with your analysis of the original variables.

Estimating the multiple regression coefficients

Simple linear regression with a response variable y and one explanatory variable x begins by using the least-squares idea (Section 2.3) to fit a straight line $\hat{y} = b_0 + b_1 x$ to data on the two variables. Although we now have p explanatory variables, the principle is the same: we use the least-squares idea to fit a linear function

$$\hat{y} = b_0 + b_1 x_1 + b_2 x_2 + \cdots + b_p x_p$$

to the data.

We use a subscript i to distinguish different cases. For the ith case the predicted response is

$$\hat{y}_i = b_0 + b_1 x_{i1} + b_2 x_{i2} + \cdots + b_p x_{ip}$$

As usual, the residual is the difference between the observed value of the response variable and the value predicted by the model:

$$e = \text{observed response} - \text{predicted response}$$

residual   For the ith case, the **residual** is

$$e_i = y_i - \hat{y}_i$$

method of least squares   The **method of least squares** chooses the b's that make the sum of squares of the residuals as small as possible. In other words, the *least-squares estimates* are the values that minimize the quantity

$$\sum (y_i - \hat{y}_i)^2$$

As in the simple linear regression case, it is possible to give formulas for the least-squares estimates. Because the formulas are complicated and hand calculation is out of the question, we are content to understand the least-squares principle and to let software do the computations.

EXAMPLE 11.9   Predicting Profits from Sales and Assets

CASE 11.1   Our examination of explanatory and response variables separately and then in pairs did not reveal any severely skewed distributions with outliers or potential influential observations. Outputs for the multiple regression analysis from Excel, SPSS, SAS, and Minitab are given in Figure 11.6. Rounding the results to four significant digits gives the least-squares equation

$$\widehat{\text{Profits}} = -2.005 + 0.03285 \times \text{Assets} + 0.1464 \times \text{Sales}$$

In Example 11.9 we rounded off the coefficients to *four significant digits*. Significant digits do not include any zeros before the first nonzero digit. So the four significant digits for the coefficient of Assets are 3285.

APPLY YOUR KNOWLEDGE

11.9 Predicting bank assets. Using the bank data in Table 11.2, do the regression to predict assets using deposits and number of banks. Give the least squares regression equation.

11.10 Regression after transforming. In Exercise 11.8 we considered the logarithm transformation for all variables in Table 11.2. Run the regression using the logarithm-transformed variables and report the least-squares equation. Note that the units differ from those in Exercise 11.9, so the results cannot be directly compared.

Excel

	A	B	C	D	E	F	G
1	SUMMARY OUTPUT						
2							
3	*Regression Statistics*						
4	Multiple R	0.904969435					
5	R Square	0.818969679					
6	Adjusted R Square	0.788797959					
7	Standard Error	0.354000516					
8	Observations	15					
9							
10	ANOVA						
11		*df*	*SS*	*MS*	*F*	*Significance F*	
12	Regression	2	6.80307938	3.40153969	27.1436191	3.51972E-05	
13	Residual	12	1.503796387	0.12531637			
14	Total	14	8.306875767				
15							
16		*Coefficients*	*Standard Error*	*t Stat*	*P-value*	*Lower 95%*	*Upper 95%*
17	Intercept	-2.00538437	0.500886867	-4.00366729	0.00175011	-3.096723096	-0.91404564
18	Assets	0.032851349	0.013907392	2.36215013	0.03591063	0.002549744	0.06315295
19	Sales	0.146407286	0.053165394	2.75380796	0.01747854	0.030569843	0.26224473

FIGURE 11.6 Excel, SPSS, SAS, and Minitab output for Example 11.9.

SPSS

Regression

Model Summary

Model	R	R Square	Adjusted R Square	Std. Error of the Estimate
1	.905[a]	.819	.789	.354001

a. Predictors: (Constant), Sales, Assets

ANOVA[b]

Model		Sum of Squares	df	Mean Square	F	Sig.
1	Regression	6.803	2	3.402	27.144	.000[a]
	Residual	1.504	12	.125		
	Total	8.307	14			

a. Predictors: (Constant), Sales, Assets
b. Dependent Variable: Profits

Coefficients[a]

Model		Unstandardized Coefficients		Standardized Coefficients	t	Sig.
		B	Std. Error	Beta		
1	(Constant)	-2.005	.501		-4.004	.002
	Assets	.033	.014	.445	2.362	.036
	Sales	.146	.053	.519	2.754	.017

a. Dependent Variable: Profits

SAS

```
                    The REG Procedure
                 Dependent Variable: Profits
                   Analysis of Variance

                          Sum of      Mean
Source            DF     Squares     Square    F Value   Pr > F

Model              2     6.80308    3.40154     27.14    <.0001
Error             12     1.50380    0.12532
Corrected Total   14     8.30688

      Root MSE             0.35400     R-Square   0.8190
      Dependent Mean       1.30022     Adj R-Sq   0.7888
      Coeff Var           27.22618

                      Parameter Estimates

                           Parameter    Standard
Variable    Label    DF     Estimate      Error    t Value   Pr > |t|

Intercept   Intercept  1    -2.00538    0.50089     -4.00     0.0018
Assets      Assets     1     0.03285    0.01391      2.36     0.0359
Sales       Sales      1     0.14641    0.05317      2.75     0.0175
```

FIGURE 11.6 (Continued)

Minitab

```
Regression Analysis: Profits versus Assets, Sales
The regression equation is
Profits = -2.01 + 0.0329 Assets + 0.146 Sales

Predictor          Coef      SE Coef          T          P
Constant        -2.0054       0.5009      -4.00      0.002
Assets           0.03285      0.01391      2.36      0.036
Sales            0.14641      0.05317      2.75      0.017

S = 0.354001      R-Sq = 81.9%      R-Sq(adj) = 78.9%

Analysis of Variance

Source            DF           SS          MS         F          P
Regression         2       6.8031      3.4015     27.14      0.000
Residual Error    12       1.5038      0.1253
Total             14       8.3069
```

FIGURE 11.6 (*Continued*)

Regression residuals

The residuals are the errors in predicting the sample responses from the multiple regression equation. Recall that the residuals are the differences between the observed and predicted values of the response variable.

$$e = \text{observed response} - \text{predicted response}$$
$$= y - \hat{y}$$

As with simple linear regression, the residuals sum to zero and the best way to examine them is to use plots.

We first examine the distribution of the residuals using stemplots or histograms. To see if the residuals appear to be approximately Normal, we use a Normal quantile plot.

EXAMPLE 11.10   Distribution of the Residuals

CASE 11.1

Figure 11.7 is a histogram of the residuals. The units are billions of dollars. The distribution appears reasonably symmetric and does not have any outliers. The Normal quantile plot in Figure 11.8

FIGURE 11.7 Histogram of residuals for Example 11.10.

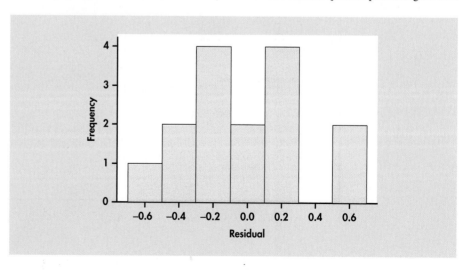

FIGURE 11.8 Normal quantile
plot of residuals for
Example 11.10.

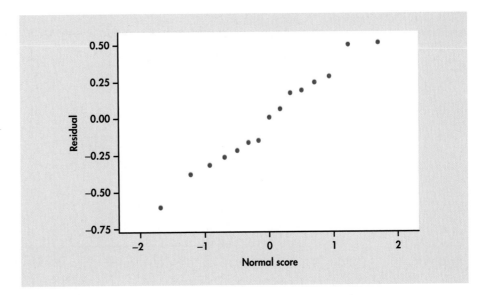

also suggests that the residuals have an approximately Normal distribution. Similar to simple linear regression, inference is robust against moderate lack of Normality, so we're just looking for obvious violations.

Another important aspect of examining the residuals is to plot them against each explanatory variable. Sometimes we can detect unusual patterns when we examine the data in this way.

EXAMPLE 11.11   Residual Plots

CASE 11.1

The residuals are plotted versus assets in Figure 11.9 and versus sales in Figure 11.10. In both cases the residuals appear reasonably randomly scattered above and below zero. It should be noted,

FIGURE 11.9 Plot of residuals
versus assets for
Example 11.11.

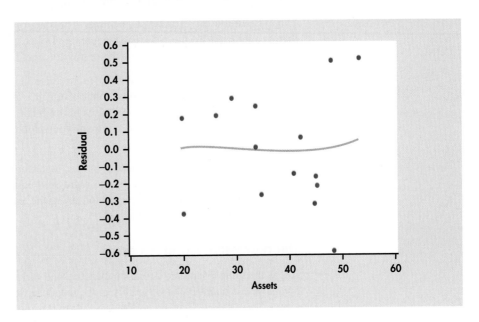

FIGURE 11.10 Plot of residuals versus sales for Example 11.11.

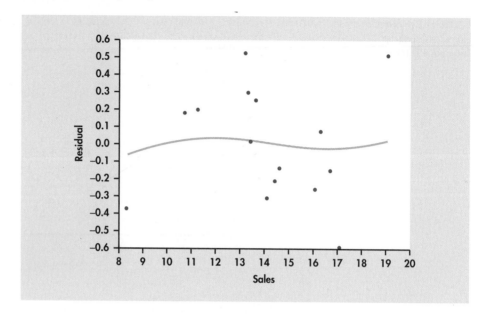

however, that the three largest residuals (in absolute value) are associated with the three largest asset values and two of the three largest sales values. This suggests that the model may not predict as well when the predictors are large, and this is worthy of further analysis.

APPLY YOUR KNOWLEDGE

11.11 Examine the residuals. In Exercise 11.9 (page 582), you ran a multiple regression using the data in Table 11.2. Obtain the residuals from this regression and plot them versus each of the explanatory variables. Also, examine the Normality of the residuals using a histogram or stemplot. If possible, use your software to make a Normal quantile plot. Summarize your conclusions.

11.12 Examine the effect of Delaware. The state of Delaware has far more assets than is predicted by the regression equation. Delete this observation and rerun the multiple regression. Describe how the regression coefficients change.

11.13 Residuals for the log analysis. In Exercise 11.10 (page 582), you carried out multiple regression using the logarithms of all the variables in Table 11.2. Obtain the residuals from this regression and examine them as you did in Exercise 11.11. Summarize your conclusions and compare your plots with the plots for the original variables.

11.14 Examine the effect of Alaska. For the logarithm-transformed data, Alaska has far more assets than is predicted by the regression equation. Delete Alaska from the data set and rerun the multiple regression using the transformed data. Describe how the regression coefficients change.

The regression standard error

Just as the sample standard deviation measures the variability of observations about their mean, we can quantify the variability of the response variable about the predicted values obtained from the multiple regression equation. As in the case of simple linear regression (page 533), we first calculate a variance using the squared residuals:

$$s^2 = \frac{1}{n-p-1} \sum e_i^2$$

$$= \frac{1}{n-p-1} \sum (y_i - \hat{y}_i)^2$$

degrees of freedom   The quantity $n - p - 1$ is the **degrees of freedom** associated with s^2. The number of degrees of freedom equals the sample size n minus $p + 1$, the number of coefficients b_i in the multiple regression model. In the simple linear regression case there is just one explanatory variable, so $p = 1$ and the number of degrees of freedom for s^2 is $n - 2$.

regression standard error   The **regression standard error** s is the square root of the sum of squares of residuals divided by the number of degrees of freedom:

$$s = \sqrt{s^2}$$

APPLY YOUR KNOWLEDGE

CASE 11.1 **11.15 Reading software output.** Regression software usually reports both s^2 and the regression standard error s. For the assets, sales, and profits data of Case 11.1, the approximate values are $s^2 = 0.125$ and $s = 0.354$. Locate s^2 and s in each of the four outputs in Figure 11.6 (pages 582–584). Give the unrounded values from each output. What name does each software give to s?

CASE 11.1 **11.16 Compare the variability.** Figure 11.2 (page 578) gives the standard deviation s_y of the profits of the DJU companies. What is this value? The regression standard error s from Figure 11.6 also measures the variability of profits, this time after taking into account the effect of assets and sales on profits. Explain briefly why we expect s to be smaller than s_y. One way to describe how well multiple regression explains the response variable y is to compare s with s_y.

Case 11.1 uses data on the assets, sales, and profits of the 15 companies included in the Dow Jones Utility Average. These are not an SRS from any population. They are selected by the editors of the *Wall Street Journal* and have changed over time due to corporate acquisitions and mergers.

Data analysis does not require that the cases be a random sample from a larger population. Our analysis of Case 11.1 tells us something about the DJU companies, not about all publicly-traded companies or any other larger group. Inference, as opposed to data analysis, draws conclusions about a population or process from which our data are a sample. Inference is most easily understood when we have an SRS from a clearly defined population. Whether inference from a multiple regression model not based on a random sample is trustworthy is a matter for judgment.

Applications of statistics in business settings frequently involve data that are not random samples. We often justify inference by saying that we are studying an underlying process that generates the data. In salary-discrimination studies, we have data on all employees in a particular group. The salaries of these current employees reflect the process by which the company sets salaries. Multiple regression builds a model of this process, and inference tells us whether gender, for example, has a statistically significant effect in the context of this model.

SECTION 11.1 Summary

- **Data for multiple linear regression** consist of the values of a response variable y and p explanatory variables $x_1, x_2, \ldots, x_p$ for n cases. We write the data and enter them into software in the form

Individual	Variables				
	x_1	x_2	...	x_p	y
1	x_{11}	x_{12}	...	x_{1p}	y_1
2	x_{21}	x_{22}	...	x_{2p}	y_2
$\vdots$					
n	x_{n1}	x_{n2}	...	x_{np}	y_n

- **Data analysis for multiple regression** starts with an examination of the distribution of each of the variables and scatterplots to display the relations between the variables.

- The **multiple regression equation** predicts the response variable by a linear relationship with all the explanatory variables:

$$\hat{y} = b_0 + b_1 x_1 + b_2 x_2 + \cdots + b_p x_p$$

The coefficients b_i in this equation are estimated using the **principle of least squares.**

- The **residuals** for multiple linear regression are

$$e_i = y_i - \hat{y}_i$$

Always examine the **distribution of the residuals** and plot them against the explanatory variables.

- The variability of the responses about the multiple regression equation is measured by the **regression standard error s,** where s is the square root of

$$s^2 = \frac{\sum e_i^2}{n - p - 1}$$

SECTION 11.1 Exercises

For Exercises 11.1 and 11.2, see page 575; for 11.3 and 11.4, see pages 576–578; for 11.5 and 11.6, see pages 578–579; for 11.7 and 11.8, see page 581; for 11.9 and 11.10, see page 582; for 11.11 to 11.14, see page 586, and for 11.15 and 11.16, see page 587.

11.17 Describing a multiple regression. As part of a study, data from 106 Bryant College actuarial graduates were obtained.[6] The researchers were interested in describing how students' overall math grade point averages are explained by SAT Math and Verbal scores, class rank, and Bryant College's mathematics placement score.

(a) What is the response variable?

(b) What is n, the number of cases?

(c) What is p, the number of explanatory variables?

(d) What are the explanatory variables?

11.18 Understanding the fitted regression line. The fitted regression equation for a multiple regression is

$$\hat{y} = 2.5 + 3.8x_1 - 2.3x_2$$

(a) If $x_1 = 4$ and $x_2 = 2$, what is the predicted value of y?

(b) For the answer to part (a) to be valid, is it necessary that the values $x_1 = 4$ and $x_2 = 2$ correspond to a case in the data set? Explain why or why not.

(c) If you hold x_2 at a fixed value, what is the effect of an increase of two units in x_1 on the predicted value of y?

11.19 Predicting the price of circular saws: individual variables. Suppose your construction company needs to buy some circular saws. To help in the purchasing decision, you decide to develop a model to predict the selling price. You decide to obtain price and product characteristic information on 19 circular saws from *Consumer Reports*.[7] The characteristics are weight (pounds), amps, maximum depth of cut (inches), cutting speed, power, ease of use, and construction. The latter four are scored on a 1 to 5 scale. 🔵 SAWS

(a) Make a table giving the mean, median, and standard deviation of each variable.

(b) Use stemplots or histograms to make graphical summaries of each distribution.

(c) Describe these distributions. Are there any unusual observations that may affect a multiple regression? Explain your answer.

11.20 Predicting the price of circular saws: pairs of variables. Refer to the circular saw data described in Exercise 11.19.

(a) Examine the relationship between each pair of variables using correlation and a scatterplot.

(b) Which characteristic is most strongly correlated with price? Is any pair of characteristics strongly correlated?

(c) Summarize the relationships. Are there any unusual or outlying cases?

11.21 Predicting the price of circular saws: multiple regression equation. Refer to the circular saw data described in Exercise 11.19.

(a) Run a multiple regression to predict price using the seven product characteristics. Give the equation for predicted price.

(b) What is the value of the regression standard error s? Verify that this value is the square root of the sum of squares of residuals divided by the degrees of freedom for the residuals.

(c) Obtain the residuals and use graphical summaries to describe the distribution.

11.22 Predicting the price of circular saws. Refer to the previous exercise.

(a) What is the predicted price for the second saw? The characteristics are WEIGHT=12, AMPS=15, DEPTH=2.25, SPEED=5, POWER=5, EASE=4, and CONSTRUCTION=5.

(b) The stated price for this model is $110. Is the predicted price above or below the stated price? Should you consider buying it? Explain your answer.

(c) Explain how you could use the residuals to help determine which saw to buy.

(d) *Consumer Reports* names Saws 2, 6, and 8 as "Best Buys." Based on your regression model, do you agree with this assessment? If not, what other saws would you recommend?

11.23 Data analysis: individual variables. The following table gives data on market share, number of accounts, and assets held by 10 online stock brokerages.[8]

Brokerage	Market share	Accounts	Assets
Charles Schwab	27.5	2500	219.0
E* Trade	12.9	909	21.1
TD Waterhouse	11.6	615	38.8
Datek	10.0	205	5.5
Fidelity	9.3	2300	160.0
Ameritrade	8.4	428	19.5
DLJ Direct	3.6	590	11.2
Discover	2.8	134	5.9
Suretrade	2.2	130	1.3
National Discount Brokers	1.3	125	6.8

Market share is expressed in percents, based on the number of trades per day. The number of accounts is given in thousands, and assets are given in billions of dollars. 🔵 BROKERAGE

(a) Make a table giving the mean, the standard deviation, and the five-number summary for each of these variables.

(b) Use stemplots or histograms to make graphical summaries of the three distributions.

(c) Describe the distributions. Are there any unusual observations?

11.24 Data analysis: pairs of variables. Refer to the previous exercise.

(a) Plot market share versus number of accounts, market share versus assets, and number of accounts versus assets.

(b) Summarize these relationships. Are there any influential observations?

(c) Find the correlation between each pair of variables.

11.25 Multiple regression equation. Refer to the brokerage data in Exercise 11.23. Run a multiple regression to predict market share using number of accounts and assets as explanatory variables.

(a) Give the equation for predicted market share.

(b) What is the value of the regression standard error s?

11.26 Residuals. Refer to the brokerage data in Exercise 11.23. Find the residuals for the multiple regression used to predict market share with number of accounts and assets as explanatory variables.

(a) Give a graphical summary of the distribution of the residuals. Are there any outliers in this distribution?

(b) Plot the residuals versus the number of accounts. Describe the plot and any unusual cases.

(c) Repeat part (b) with assets in place of number of accounts.

Your analyses in Exercises 11.23 to 11.26 point to two firms, Charles Schwab and Fidelity, as unusual in several respects. How influential are these firms? The following four exercises provide answers.

11.27 Rerun Exercise 11.23 without the data for Schwab and Fidelity. Compare your results with what you obtained in that exercise.

11.28 Rerun Exercise 11.24 without the data for Schwab and Fidelity. Compare your results with what you obtained in that exercise.

11.29 Rerun Exercise 11.25 without the data for Schwab and Fidelity. Compare your results with what you obtained in that exercise.

11.30 Rerun Exercise 11.26 without the data for Schwab and Fidelity. Compare your results with what you obtained in that exercise.

11.31 Predicting retail sales. Daily sales at a secondhand shop are recorded over a 25-day period.[9] The daily gross sales and total number of items sold are broken down into items paid by check, cash, and credit card. The owners expect that the daily numbers of cash items, check items, and credit card items sold will accurately predict gross sales. 🔘 RETAIL
(a) Describe the distribution of each of these four variables using both graphical and numerical summaries. Briefly summarize what you find and note any unusual observations.
(b) Use plots and correlations to describe the relationships between each pair of variables. Summarize your results.
(c) Run a multiple regression and give the least-squares equation.

(d) Analyze the residuals from this multiple regression. Are there any patterns of interest?
(e) One of the owners is troubled by the equation because the intercept is not zero (that is, no items sold should result in $0 gross sales). Explain to this owner why this isn't a problem.

11.32 Architectural firm billings. A summary of firms engaged in commercial architecture in the Indianapolis, Indiana, area provides firm characteristics including total annual billing and the number of architects, engineers, and staff employed in the firm.[10] Consider developing a model to predict total billing. 🔘 ARCHITECT
(a) Using numerical and graphical summaries, describe the distribution of total billing and the number of architects, engineers, and staff.
(b) For each of the 6 pairs of variables, use graphical and numerical summaries to describe the relationship.
(c) Carry out a multiple regression. Report the fitted regression equation and the value of the regression standard error s.
(d) Analyze the residuals from the multiple regression. Are there any concerns?
(e) The firm HCO did not report its total billing but employs 3 architects, 1 engineer, and 17 staff members. What is the predicted total billing for this firm?

11.2 Inference for Multiple Regression

To move from using multiple regression for data analysis to inference in the multiple regression setting, we need to make some assumptions about our data. These assumptions are summarized in the form of a statistical model. As with all the models that we have studied, we do not require that the model be exactly correct. We only require that it be approximately true and that the data do not severely violate the assumptions.

Recall that the *simple linear regression model* assumes that the mean of the response variable y depends on the explanatory variable x according to a linear equation

$$\mu_y = \beta_0 + \beta_1 x$$

For any fixed value of x, the response y varies Normally around this mean and has a standard deviation σ that is the same for all values of x.

In the *multiple regression* setting, the response variable y depends on not one but p explanatory variables, denoted by $x_1, x_2, \ldots, x_p$. The mean response is a linear function of the explanatory variables:

$$\mu_y = \beta_0 + \beta_1 x_1 + \beta_2 x_2 + \cdots + \beta_p x_p$$

population regression equation   This expression is the **population regression equation.** We do not observe the mean response, because the observed values of y vary about their means. We can think of subpopulations of responses, each corresponding to a particular set of values for *all* the explanatory variables $x_1, x_2, \ldots, x_p$. In each subpopulation, y varies Normally with a mean given by the population regression equation. The regression model assumes that the standard deviation σ of the responses is the same in all subpopulations.

Multiple linear regression model

To form the multiple regression model, we combine the population regression equation with assumptions about the form of the *variation* of the observations about their mean. We again think of the model in the form

$$\text{DATA} = \text{FIT} + \text{RESIDUAL}$$

The FIT part of the model consists of the subpopulation mean μ_y. The RESIDUAL part represents the variation of the response y around its subpopulation mean. That is, the model is

$$y = \mu_y + \epsilon$$

The symbol ϵ represents the deviation of an individual observation from its subpopulation mean. We assume that these deviations are Normally distributed with mean 0 and an unknown standard deviation σ that does not depend on the values of the x variables.

Multiple Linear Regression Model

The **statistical model for multiple linear regression** is

$$y_i = \beta_0 + \beta_1 x_{i1} + \beta_2 x_{i2} + \cdots + \beta_p x_{ip} + \epsilon_i$$

for $i = 1, 2, \ldots, n$.

The **mean response μ_y** is a linear function of the explanatory variables:

$$\mu_y = \beta_0 + \beta_1 x_1 + \beta_2 x_2 + \cdots + \beta_p x_p$$

The **deviations ϵ_i** are independent and Normally distributed with mean 0 and standard deviation σ. That is, they are an SRS from the $N(0, \sigma)$ distribution.

The parameters of the model are $\beta_0, \beta_1, \beta_2, \ldots, \beta_p$, and σ.

The assumption that the subpopulation means are related to the regression coefficients β by the equation

$$\mu_y = \beta_0 + \beta_1 x_1 + \beta_2 x_2 + \cdots + \beta_p x_p$$

implies that we can estimate all subpopulation means from estimates of the β's. To the extent that this equation is accurate, we have a useful tool for describing how the mean of y varies with the x's.

CASE 11.2

BOXOFFICE

Predicting Movie Revenue Nash Information Services provides information and analytical services to the movie industry, including statistical models for predicting movie revenue. Consider an SRS of 40 movies released over a five-year period.[11] This sample was collected to see if information available soon after the theatrical release can successfully predict total box office revenue. The larger population of interest consists of *all* movies released over this five-year period.

The response variable is a movie's total U.S. box office revenue (USRevenue). Among the explanatory variables are the movie's budget (Budget), opening-weekend revenue (Opening), and how many theaters the movie was in for the opening weekend (Theaters). All dollar amounts are measured in millions of U.S. dollars.

CASE 11.2

EXAMPLE 11.12 A Model for Predicting Movie Revenue

We expect that a movie's budget, opening-weekend revenue, and opening-weekend theater count will help predict total U.S. box office revenue. The multiple regression model then has $p = 3$ explanatory variables: $x_1 = $ Budget, $x_2 = $ Opening, and $x_3 = $ Theaters. Each particular combination of budget, opening-weekend revenue, and opening-weekend theater count defines a particular subpopulation. Our response variable y is the U.S. box office revenue.

The multiple regression model for the subpopulation mean U.S. box office revenue is

$$\mu_{USRevenue} = \beta_0 + \beta_1 Budget + \beta_2 Opening + \beta_3 Theaters$$

For movies with \$35 million budgets that earn \$78 million in 2100 theaters their first weekend, the model gives the subpopulation mean U.S. box office revenue as

$$\mu_{USRevenue} = \beta_0 + \beta_1 \times 35 + \beta_2 \times 78 + \beta_3 \times 2100$$

Estimating the parameters of the model

To estimate mean U.S. box office revenue in Example 11.12, we must estimate the coefficients β_0, β_1, β_2, and β_3. Inference requires that we also estimate the variability of the responses about their mean, represented in the model by the standard deviation σ. In any multiple regression model, the parameters to be estimated from the data are β_0, $\beta_1, \ldots, \beta_p$, and σ.

We estimate these parameters by applying least-squares multiple regression as in Section 11.1. That is, we view the coefficients b_j in the multiple regression equation

$$\hat{y} = b_0 + b_1 x_1 + b_2 x_2 + \cdots + b_p x_p$$

as estimates of the population parameters β_j. The observed variability of the responses about this fitted model is measured by the variance

$$s^2 = \frac{1}{n - p - 1} \sum e_i^2$$

and the regression standard error

$$s = \sqrt{s^2}$$

In the model, the parameters σ^2 and σ measure the variability of the responses about the population regression equation. It is natural to estimate σ^2 by s^2 and σ by s.

Estimating the Regression Parameters

In the multiple linear regression setting, we use the **method of least-squares regression** to estimate the population regression parameters.

The standard deviation σ in the model is estimated by the **regression standard error**

$$s = \sqrt{\frac{1}{n - p - 1} \sum (y_i - \hat{y}_i)^2}$$

Inference about the regression coefficients

Confidence intervals and significance tests for each of the regression coefficients β_j have the same form as in simple linear regression. The standard errors of the b's have more complicated formulas, but all are again multiples of s. Statistical software does the calculations.

Confidence Intervals and Significance Tests for β_j

A **level C confidence interval** for β_j is

$$b_j \pm t^*\mathrm{SE}_{b_j}$$

where SE_{b_j} is the standard error of b_j and t^* is the value for the $t(n - p - 1)$ density curve with area C between $-t^*$ and t^*.

To test the hypothesis $H_0: \beta_j = 0$, compute the **t statistic**

$$t = \frac{b_j}{\mathrm{SE}_{b_j}}$$

In terms of a random variable T having the $t(n - p - 1)$ distribution, the P-value for a test of H_0 against

$H_a: \beta_j > 0$ is $P(T \geq t)$

$H_a: \beta_j < 0$ is $P(T \leq t)$

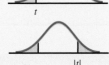

$H_a: \beta_j \neq 0$ is $2P(T \geq |t|)$

EXAMPLE 11.13 Predicting U.S. Box Office Revenue

CASE 11.2

BOXOFFICE

In Example 11.12, there are $p = 3$ explanatory variables, and we have data on $n = 40$ movies. The degrees of freedom for multiple regression are therefore

$$n - p - 1 = 40 - 3 - 1 = 36$$

Statistical software output for this fitted model provides many details of the model's fit and the significance of the independent variables. Figure 11.11 shows multiple regression outputs from Excel, Minitab, and SPSS. You see that the regression equation is

$$\widehat{\mathrm{USRevenue}} = 53.7 - 0.499\mathrm{Budget} + 3.24\mathrm{Opening} - 0.00124\mathrm{Theaters}$$

and that the regression standard error is $s = 41.8471$.

The outputs present the t statistic for each regression coefficient and its two-sided P-value. For example, the t statistic for the coefficient of Opening is 10.93 with a very small P-value. The data give strong evidence against the null hypothesis

$$H_0: \beta_2 = 0$$

that the population coefficient for opening weekend-revenue is zero. We would report this result as $t = 10.93$, df $= 36$, $P < 0.0001$. Two of the three outputs also give the 95% confidence interval for the coefficient β_2. It is (2.64, 3.84). The confidence interval does not include 0, consistent with the fact that the test rejects the null hypothesis at the 5% significance level.

Excel

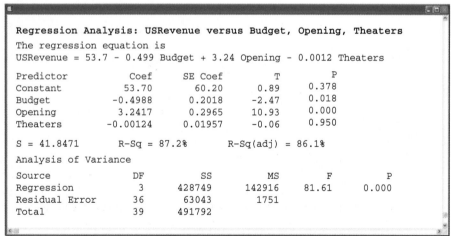

	A	B	C	D	E	F	G
1	SUMMARY OUTPUT						
2							
3	*Regression Statistics*						
4	Multiple R	0.9337079					
5	R Square	0.8718104					
6	Adjusted R Square	0.8611279					
7	Standard Error	41.847129					
8	Observations	40					
9							
10	ANOVA						
11		*df*	*SS*	*MS*	*F*	*Significance F*	
12	Regression	3	428749.0207	142916.3	81.61135	4.00135E-16	
13	Residual	36	63042.55845	1751.182			
14	Total	39	491791.5791				
15							
16		*Coefficients*	*Standard Error*	*t Stat*	*P-value*	*Lower 95%*	*Upper 95%*
17	Intercept	53.701973	60.19527155	0.892129	0.378248	-68.37969545	175.78364
18	Budget	-0.498817	0.201815831	-2.47164	0.018313	-0.908118552	-0.0895156
19	Opening	3.2417285	0.296467651	10.93451	5.42E-13	2.640464253	3.8429927
20	Theaters	-0.00124	0.019571562	-0.06338	0.949813	-0.040933452	0.0384524

Minitab

```
Regression Analysis: USRevenue versus Budget, Opening, Theaters
The regression equation is
USRevenue = 53.7 - 0.499 Budget + 3.24 Opening - 0.0012 Theaters

Predictor         Coef      SE Coef         T         P
Constant         53.70        60.20      0.89     0.378
Budget          -0.4988      0.2018     -2.47     0.018
Opening          3.2417      0.2965     10.93     0.000
Theaters        -0.00124     0.01957    -0.06     0.950

S = 41.8471      R-Sq = 87.2%      R-Sq(adj) = 86.1%
Analysis of Variance

Source            DF         SS           MS         F         P
Regression         3      428749       142916     81.61     0.000
Residual Error    36       63043         1751
Total             39      491792
```

FIGURE 11.11 Multiple regression outputs from Excel, Minitab, and SPSS for Examples 11.13, 11.14, and 11.15.

Be very careful in your interpretation of the *t* tests and confidence intervals for individual regression coefficients. In *simple linear regression,* the model says that $\mu_y = \beta_0 + \beta_1 x$. The null hypothesis $H_0: \beta_1 = 0$ says that regression on x is of no value for predicting the response y, or alternatively, that there is no straight-line relationship between x and y. The corresponding hypothesis for the *multiple regression*

SPSS

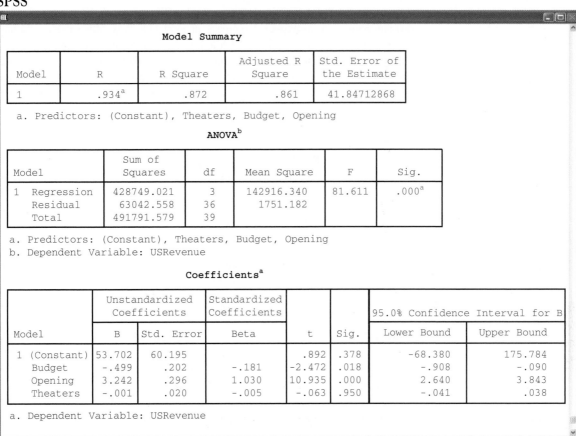

Model Summary

Model	R	R Square	Adjusted R Square	Std. Error of the Estimate
1	.934[a]	.872	.861	41.84712868

a. Predictors: (Constant), Theaters, Budget, Opening

ANOVA[b]

Model		Sum of Squares	df	Mean Square	F	Sig.
1	Regression	428749.021	3	142916.340	81.611	.000[a]
	Residual	63042.558	36	1751.182		
	Total	491791.579	39			

a. Predictors: (Constant), Theaters, Budget, Opening
b. Dependent Variable: USRevenue

Coefficients[a]

Model		Unstandardized Coefficients		Standardized Coefficients	t	Sig.	95.0% Confidence Interval for B	
		B	Std. Error	Beta			Lower Bound	Upper Bound
1	(Constant)	53.702	60.195		.892	.378	-68.380	175.784
	Budget	-.499	.202	-.181	-2.472	.018	-.908	-.090
	Opening	3.242	.296	1.030	10.935	.000	2.640	3.843
	Theaters	-.001	.020	-.005	-.063	.950	-.041	.038

a. Dependent Variable: USRevenue

FIGURE 11.11 (*Continued*)

model $\mu_y = \beta_0 + \beta_1 x_1 + \beta_2 x_2 + \beta_3 x_3$ of Example 11.13 says that x_2 is of no value for predicting y, ***given that x_1 and x_3 are also in the model***. That's a very important difference.

The output in Figure 11.11 shows, for example, that the P-value for opening-weekend theater count is $P = 0.95$. We can conclude that the number of theaters does not help predict U.S. box office revenue, *given that budget and opening-weekend revenue are available to use for prediction*. This does *not* mean that the opening-weekend theater count cannot help predict U.S. box office revenue. In a simple linear regression of U.S. box office revenue on theater count, without budget and opening-weekend revenue present, the regression slope is highly significant (see Exercise 11.34).

The conclusions of inference about any one explanatory variable in multiple regression depend on what other explanatory variables are also in the model. This is a basic principle for understanding multiple regression. The t tests in Example 11.13 show that the opening-weekend theater count does not significantly aid prediction of the U.S. box office revenue *if the budget and opening-weekend revenue are also in the model*. On the other hand, opening-weekend revenue is highly significant *even when the budget and opening-weekend theater count are also in the model*.

The interpretation of a confidence interval for an individual coefficient also depends on the other variables in the model, but in this case only if they remain constant. For example, the 95% confidence interval for Opening implies that, *given the number of theaters and the budget do not change,* a $1 million increase in the opening-weekend revenue results in an expected increase in total U.S. box office revenue somewhere between $2.64 and $3.84 million. While it makes sense for the budget to remain fixed, it may not make sense to keep the number of theaters fixed. The number of theaters and opening-weekend revenue are positively correlated and it may be very unreasonable to assume that opening revenue can increase this much without the number of theaters also increasing.

APPLY YOUR KNOWLEDGE

CASE 11.2 **11.33 Reading software outputs.** Carefully examine the outputs from the three software packages given in Figure 11.11. Make a table giving the estimated regression coefficient for opening-weekend revenue (Opening), its standard error, the *t* statistic with degrees of freedom, and the *P*-value as reported by each of the packages. What do you conclude about this coefficient?

CASE 11.2 **11.34 The model matters.** Carry out the simple linear regression of U.S. box office revenue on opening-weekend theater count, Theaters. What is the *P*-value of the *t* test for the hypothesis that Theaters does not help predict U.S. box office revenue? Explain carefully to someone who knows no statistics why the conclusions about opening-weekend theater count here and in Figure 11.11 differ so greatly.

CASE 11.2 **11.35 A simpler model.** In the multiple regression analysis using all three variables, opening-weekend theater count, Theaters, appears to be the least helpful (given that the other two explanatory variables are in the model). Do a new analysis using only the movie's budget and opening-weekend revenue. Give the estimated regression equation for this analysis and compare it with the analysis using all three variables as predictors. Summarize the inference results for the coefficients.

Inference about prediction

confidence interval for mean response

prediction interval

Inference about the regression coefficients looks much the same in simple and multiple regression, but there are important differences in interpretation. Inference about prediction also looks much the same, and in this case the interpretation is also the same. We may wish to give a **confidence interval for the mean response** for some specific set of values of the explanatory variables. Or we may want a **prediction interval** for an individual response for the same set of specific values.

The distinction between predicting mean and individual response is exactly as in simple regression. The prediction interval is again wider because it must allow for the variation of individual responses about the mean. In most software, the commands for prediction inference are the same for multiple and simple regression. The details of the arithmetic performed by the software are of course more complicated for multiple regression, but this does not affect interpretation of the output.

What about changes in the model, which we saw can greatly influence inference about the regression coefficients? It is often the case that different models give similar predictions. We expect, for example, that the predictions of U.S. box office revenue from budget and opening-weekend revenue will be about the same as predictions based on budget, opening-weekend revenue, and opening-weekend theater count. Because of this, when prediction is the key goal of a multiple regression, it is common to search for a model that predicts well but does not contain unnecessary predictors. Some refer to this as following the KISS principle.[12] In Section 11.3, we discuss some procedures that can be used for this type of search.

APPLY YOUR KNOWLEDGE

CASE 11.2 **11.36 Prediction versus confidence intervals.** For the movie revenue model, would confidence or prediction intervals be used more frequently? Explain your answer.

CASE 11.2 **11.37 Predicting U.S. movie revenue.** *Talladega Nights: The Ballad of Ricky Bobby* was released August 4, 2006. It had a budget of $73.0 million and was shown in 3803 theaters grossing $47.042 million during the first weekend. Use software to construct the following.

(a) A 95% prediction interval based on the model with all three predictors.

(b) A 95% prediction interval based on the model using only opening-weekend revenue and budget.

(c) Compare the two intervals. Do the models give similar predictions?

ANOVA table for multiple regression

The basic ideas of the regression ANOVA table are the same in simple and multiple regression. ANOVA expresses variation in the form of sums of squares. It breaks the total variation into two parts: the sum of squares explained by the regression equation and the sum of squares of the residuals. The ANOVA table has the same form in simple and multiple regression except for the degrees of freedom, which reflect the number p of explanatory variables. Here is the ANOVA table for multiple regression:

Source	Degrees of freedom	Sum of squares	Mean square	F
Regression	$\text{DFR} = p$	$\text{SSR} = \sum(\hat{y}_i - \bar{y})^2$	$\text{MSR} = \text{SSR}/\text{DFR}$	MSR/MSE
Residual	$\text{DFE} = n - p - 1$	$\text{SSE} = \sum(y_i - \hat{y}_i)^2$	$\text{MSE} = \text{SSE}/\text{DFE}$	
Total	$\text{DFT} = n - 1$	$\text{SST} = \sum(y_i - \bar{y})^2$		

The brief notation in the table uses, for example, MSE for the residual mean square. This is common notation; the "E" stands for "error." Of course, no error has been made. "Error" in this context is just a synonym for "residual."

The degrees of freedom and sums of squares add, just as in simple regression:

$$\text{SST} = \text{SSR} + \text{SSE}$$

$$\text{DFT} = \text{DFR} + \text{DFE}$$

The estimate of the variance σ^2 for our model is again given by the MSE in the ANOVA table. That is, $s^2 = \text{MSE}$.

ANOVA F test   The ratio MSR/MSE is again the statistic for the **ANOVA F test.** In simple linear regression the F test from the ANOVA table is equivalent to the two-sided t test of the hypothesis that the slope of the regression line is 0. In the multiple regression setting, the null hypothesis for the F test states that *all* the regression coefficients (with the exception of the intercept) are 0. One way to write this is

$$H_0: \beta_1 = 0 \text{ and } \beta_2 = 0 \text{ and } \cdots \text{ and } \beta_p = 0$$

A shorter way to express this hypothesis is

$$H_0: \beta_1 = \beta_2 = \cdots = \beta_p = 0$$

The alternative hypothesis is

$$H_a: \text{at least one of the } \beta_j \text{ is not } 0$$

The null hypothesis says that none of the explanatory variables helps explain the response, at least when used in the form expressed by the multiple regression equation. The alternative states that at least one of them is linearly related to the response. As in simple linear regression, large values of F give evidence against H_0. When H_0 is true, F has the $F(p, n - p - 1)$ distribution. The degrees of freedom for the F distribution are those associated with the regression and residual terms in the ANOVA table.

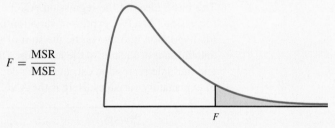

Analysis of Variance F Test

In the multiple regression model, the hypothesis

$$H_0: \beta_1 = \beta_2 = \cdots = \beta_p = 0$$

versus

$$H_a: \text{at least one of these coefficients is not zero}$$

is tested by the analysis of variance F statistic

$$F = \frac{\text{MSR}}{\text{MSE}}$$

The P-value is the probability that a random variable having the $F(p, n - p - 1)$ distribution is greater than or equal to the calculated value of the F statistic.

EXAMPLE 11.14 F Test for Movie Revenue Model

CASE 11.2   Example 11.13 gives the results of multiple regression analysis for predicting U.S. box office revenue. The F statistic is 81.61. The degrees of freedom appear in the ANOVA table. They are 3 and 36. The software packages (see Figure 11.11) report the P-value in different forms: Excel, 4.00135E-16; Minitab, 0.000; and SPSS, .000. We would report the results as follows: a movie's budget, opening-weekend revenue, and opening-weekend theater count contain information that can be used to predict the movie's total U.S. box office revenue ($F = 81.61$, df = 3 and 36, $P < 0.0001$).

A significant F test does not tell us which explanatory variables explain the response. It simply allows us to conclude that at least one of the coefficients is not zero. We may want to refine the model by eliminating some variables that do not appear to be useful (KISS principle). On the other hand, if we fail to reject the null hypothesis, we have found no evidence that *any* of the coefficients are not zero. In this case, there is little point in attempting to refine the model.

APPLY YOUR KNOWLEDGE

CASE 11.2   **11.38 F test for the model without Theaters.** Rerun the multiple regression using the movie's budget and opening weekend revenue to predict U.S. box office revenue. Report the F statistic, the associated degrees of freedom, and the P-value. How do these differ from the corresponding values for the model with the three predictors? What do you conclude?

Squared multiple correlation R^2

For simple linear regression the square r^2 of the sample correlation can be written as the ratio of SSR to SST. We interpret r^2 as the proportion of variation in y explained by linear regression on x. A similar statistic is important in multiple regression.

The Squared Multiple Regression Correlation

The statistic

$$R^2 = \frac{\text{SSR}}{\text{SST}} = \frac{\sum (\hat{y}_i - \bar{y})^2}{\sum (y_i - \bar{y})^2}$$

is the proportion of the variation of the response variable y that is explained by the explanatory variables $x_1, x_2, \ldots, x_p$ in a multiple linear regression.

---

multiple regression correlation coefficient

Often, R^2 is multiplied by 100 and expressed as a percent. The square root of R^2, called the **multiple regression correlation coefficient,** is the correlation between the observations y_i and the predicted values $\hat{y}_i$.

EXAMPLE 11.15 R^2 for Movie Revenue Model

CASE 11.2   Example 11.13 and Figure 11.11 give the results of multiple regression analysis to predict U.S. box office revenue. The value of the R^2 statistic is 0.8718, or 87.18%. Be sure that you can find this statistic in the outputs. We conclude that about 87% of the variation in U.S. box office revenue can be explained by the movies' budgets, opening-weekend revenues, and opening-weekend theater counts.

The F statistic for the multiple regression of U.S. box office revenue on budget, opening-weekend revenue, and opening-weekend theater count is highly significant, $P < 0.0001$. There is strong evidence of a relationship between these three variables and eventual box office revenue. The squared multiple correlation tells us that these variables in this multiple regression model explain about 87% of the variability in box office revenues. The other 13% is represented by the RESIDUAL term in our model and is due to differences among the movies that are not measured by these three variables. These differences may include the movie's rating, time of year the movie was released, and the genre of the movie.

APPLY YOUR KNOWLEDGE

CASE 11.2 **11.39 R^2 for different models.** Use each of the following sets of explanatory variables to predict U.S. box office revenue: (a) Budget, Opening; (b) Budget, Theater; (c) Opening, Theater; (d) Budget; (e) Opening; (f) Theater. Make a table giving the model and the value of R^2 for each. Summarize what you have found.

Inference for a collection of regression coefficients

We have studied two different types of significance tests for multiple regression. The F test examines the hypothesis that the coefficients for *all* the explanatory variables are zero. On the other hand, we used t tests to examine *individual* coefficients. (For simple linear regression with one explanatory variable, these are two different ways to examine the same question.) Often we are interested in an intermediate setting: does a set of explanatory variables contribute to explaining the response, given that another set of explanatory variables is also available? We formulate such questions as follows: start with the multiple regression model that contains all the explanatory variables and test the hypothesis that a set of the coefficients are all zero.

F Test for a Collection of Regression Coefficients

In the multiple regression model with p explanatory variables, the hypothesis

$$H_0:\ q \text{ specific explanatory variables all have zero coefficients}$$

versus the hypothesis

$$H_a:\ \text{at least one of these coefficients is not zero}$$

is tested by an F statistic. The degrees of freedom are q and $n - p - 1$. The P-value is the probability that a random variable having the $F(q,\ n - p - 1)$ distribution is greater than or equal to the calculated value of the F statistic.

Some software allows you to directly state and test hypotheses of this form. Here is a way to find the F statistic by doing two regression runs.

1. Regress y on all p explanatory variables. Read the R^2-value from the output and call it R_1^2.

2. Then regress y on just the $p - q$ variables that remain after you remove the q variables from the model. Again read the R^2-value and call it R_2^2. This will be smaller than R_1^2 because removing variables can only decrease R^2.

3. The test statistic is

$$F = \left(\frac{n - p - 1}{q}\right)\left(\frac{R_1^2 - R_2^2}{1 - R_1^2}\right)$$

with q and $n - p - 1$ degrees of freedom.

EXAMPLE 11.16 Do Budget and Opening-Weekend Theater Count Add Predictive Ability?

CASE 11.2

In the multiple regression analysis using all three predictors, opening-weekend revenue (Opening) appears to be the most helpful (given the other two explanatory variables are in the model). A question we might ask is

> Do these other two variable help predict movie revenue, given that opening-weekend revenue is included?

The same question in another form is

> If we start with a model containing all three predictors, does removing theater count and budget reduce our ability to predict revenue?

The first regression run includes $p = 3$ explanatory variables: Opening, Budget, and Theaters. The R^2 for this model is $R_1^2 = 0.8718$.

Now remove the $q = 2$ variables Budget and Theaters and redo the regression with just Opening as the explanatory variable. For this model we get $R_2^2 = 0.8500$.

The test statistic is

$$F = \left(\frac{n - p - 1}{q}\right)\left(\frac{R_1^2 - R_2^2}{1 - R_1^2}\right)$$

$$= \left(\frac{40 - 3 - 1}{2}\right)\left(\frac{0.8718 - 0.8500}{1 - 0.8718}\right) = 3.061$$

The degrees of freedom are $q = 2$ and $n - p - 1 = 40 - 3 - 1 = 36$.

BOXOFFICE

The hypothesis test in Example 11.16 asks about the coefficients of Budget and Theaters in a model that also contains Opening as an explanatory variable. If we start with a different model, we may get a different answer. For example, we would not be surprised to find that Budget and Theaters help explain movie revenue in a model with only these two predictors. *Individual regression coefficients, their standard errors, and significance tests are meaningful only when interpreted in the context of the other explanatory variables in the model.*

APPLY YOUR KNOWLEDGE

CASE 11.2 **11.40 Are Budget and Theaters useful predictors of USRevenue?** Run the multiple regression to predict movie revenue using all three predictors. Then run the model using only Budget and Theaters.
(a) The R^2 for the second model is 0.4461. Does your work confirm this?
(b) Make a table giving the Budget and Theaters coefficients and their standard errors, t statistics, and P-values for both models. Explain carefully how your assessment of the value of these two predictors of movie revenue depends on whether or not opening-weekend revenue is in the model.

CASE 11.2 **11.41 Is Opening helpful when Budget and Theaters are available?** We saw that Budget and Theaters are not useful in a model that contains the opening-weekend revenue. Now, let's examine the other version of this question. Does Opening help explain USRevenue in a model that contains Budget and Theaters? Run the models with all three predictors and with only Budget and Theaters. Compare the values of R^2. Perform the F test and give its degrees of freedom and P-value. Carefully state a conclusion about the usefulness of the predictor Opening when Budget and Theaters are available. Also compare this F-test and P-value with the t test for the coefficient of Opening in Example 11.13.

SECTION 11.2 Summary

- The statistical model for **multiple linear regression** with response variable y and p explanatory variables $x_1, x_2, \ldots, x_p$ is

$$y_i = \beta_0 + \beta_1 x_{i1} + \beta_2 x_{i2} + \cdots + \beta_p x_{ip} + \epsilon_i$$

where $i = 1, 2, \ldots, n$. The deviations ϵ_i are independent Normal random variables with mean 0 and a common standard deviation σ. The **parameters** of the model are $\beta_0, \beta_1, \beta_2, \ldots, \beta_p$, and σ.

- The β's are estimated by the coefficients $b_0, b_1, b_2, \ldots, b_p$ of the multiple regression equation fitted to the data by **the method of least squares.** The parameter σ is estimated by the **regression standard error**

$$s = \sqrt{\text{MSE}} = \sqrt{\frac{\sum e_i^2}{n - p - 1}}$$

where the e_i are the **residuals,**

$$e_i = y_i - \hat{y}_i$$

- A **level C confidence interval for the regression coefficient** β_j is

$$b_j \pm t^* \text{SE}_{b_j}$$

where t^* is the value for the $t(n - p - 1)$ density curve with area C between $-t^*$ and t^*.

- Tests of the hypothesis H_0: $\beta_j = 0$ are based on the **individual t statistic:**

$$t = \frac{b_j}{\text{SE}_{b_j}}$$

and the $t(n - p - 1)$ distribution.

- The estimate b_j of β_j and the test and confidence interval for β_j are all based on a specific multiple linear regression model. The results of all these procedures change if other explanatory variables are added to or deleted from the model.

- The **ANOVA table** for a multiple linear regression gives the degrees of freedom, sum of squares, and mean squares for the regression and residual sources of variation. The **ANOVA F statistic** is the ratio MSR/MSE and is used to test the null hypothesis

$$H_0: \beta_1 = \beta_2 = \cdots = \beta_p = 0$$

If H_0 is true, this statistic has the $F(p, \ n - p - 1)$ distribution.

- The **squared multiple correlation** is given by the expression

$$R^2 = \frac{\text{SSR}}{\text{SST}}$$

and is interpreted as the proportion of the variability in the response variable y that is explained by the explanatory variables $x_1, x_2, \ldots, x_p$ in the multiple linear regression.

- The null hypothesis that a **collection of q explanatory variables** all have coefficients equal to zero is tested by an **F statistic** with q degrees of freedom in the numerator and $n - p - 1$ degrees of freedom in the denominator. This statistic can be computed from the squared multiple correlations for the model with all the explanatory variables included (R_1^2) and the model with the q variables deleted (R_2^2):

$$F = \left(\frac{n - p - 1}{q} \right) \left(\frac{R_1^2 - R_2^2}{1 - R_1^2} \right)$$

SECTION 11.2 Exercises

For Exercises 11.33 to 11.35, see page 596; for 11.36 and 11.37, see page 597; for 11.38, see page 598; for 11.39, see page 599; and for 11.40 and 11.41, see page 601.

11.42 Confidence interval for a regression coefficient. In each of the following settings, give a 95% confidence interval for the coefficient of x_1.
(a) $n = 28$, $\hat{y} = 10.3 + 14.7x_1 + 17.3x_2$, $\text{SE}_{b_1} = 7.4$.
(b) $n = 53$, $\hat{y} = 10.3 + 14.7x_1 + 17.3x_2$, $\text{SE}_{b_1} = 7.4$.
(c) $n = 28$, $\hat{y} = 10.3 + 14.7x_1 + 17.3x_2 + 2.1x_3$, $\text{SE}_{b_1} = 7.4$.
(d) $n = 53$, $\hat{y} = 10.3 + 14.7x_1 + 17.3x_2 + 2.1x_3$, $\text{SE}_{b_1} = 7.4$.

11.43 Significance test for a regression coefficient. For each of the settings in the previous exercise, test the null hypothesis that the coefficient of x_1 is zero versus the two-sided alternative.

11.44 What's wrong? In each of the following situations, explain what is wrong and why.
(a) One of the assumptions for multiple regression is that the distribution of each explanatory variable is Normal.
(b) The smaller the P-value for the ANOVA F test, the greater the explanatory power of the model.

(c) All explanatory variables that are significantly correlated with the response variable will have a statistically significant regression coefficient in the multiple regression model.

11.45 What's wrong? In each of the following situations, explain what is wrong and why.
(a) The multiple correlation gives the proportion of the variation in the response variable that is explained by the explanatory variables.
(b) In a multiple regression with a sample size of 35 and 4 explanatory variables, the test statistic for the null hypothesis $H_0: b_2 = 0$ is a t statistic that follows the $t(30)$ distribution when the null hypothesis is true.
(c) A small P-value for the ANOVA F test implies that all explanatory variables are statistically different from zero.

11.46 Inference basics. You run a multiple regression with 65 cases and 4 explanatory variables.
(a) What are the degrees of freedom for the F statistic for testing the null hypothesis that all four of the regression coefficients for the explanatory variables are zero?
(b) Software output gives MSE = 42. What is the estimate of the standard deviation σ of the model?
(c) The output gives the estimate of the regression coefficient for the first explanatory variable as 10.3 with a standard error of 3.1. Find a 95% confidence interval for the true value of this coefficient.
(d) Test the null hypothesis that the regression coefficient for the first explanatory variable is zero. Give the test statistic, the degrees of freedom, the P-value, and your conclusion.

11.47 Inference basics. You run a multiple regression with 27 cases and 3 explanatory variables. The ANOVA table includes the sums of squares SSR = 16 and SSE = 120.
(a) Find the F statistic for testing the null hypothesis that the regression coefficients for the 3 explanatory variables are all zero. Carry out the significance test and report the results.
(b) What is the value of R^2 for this model? Explain what this number tells us.

11.48 Effects of incentives on employee motivation in Japan. To improve worker motivation, organizations typically offer employees opportunities for promotion, wage increases, and engagement in challenging work. The Japanese career system is typically lagging in terms of promotion and rapid wage increases. A survey of workers in 75 companies of Toyota Group was performed to investigate the relative strengths of these incentives in Japanese organizations.[13] A questionnaire was distributed and information on a worker's motivation level, job characteristics, recent change in wage, relative wage level, and perceived difficulty and fairness in promotion was collected. The following table summarizes the multiple regression models used to predict a worker's motivation level. The first three explanatory variables represent demographic information, followed by two describing types of wage incentives, two describing types of promotion incentives, and four representing job characteristics.

	White Collar			Blue Collar		
Independent variables	b	t	P	b	t	P
Male	−0.02	0.69	0.490	−0.04	1.14	0.253
Tenure	0.11	3.12	0.002	−0.03	0.83	0.407
Education	0.00	0.09	0.929	0.00	0.03	0.975
Wage level	0.08	2.47	0.014	0.11	3.34	0.001
Wage increase	0.07	2.34	0.020	0.09	2.83	0.005
Promotion difficulty	0.04	1.39	0.165	0.06	1.84	0.066
Promotion fairness	0.23	7.24	0.000	0.26	7.76	0.000
Job range	0.00	0.08	0.939	0.01	0.19	0.852
Discretion	0.17	4.95	0.000	0.09	2.37	0.018
Knowledge and ability	0.09	2.71	0.007	0.05	1.47	0.142
Development opportunity	0.23	7.61	0.000	0.16	4.42	0.000
F		27.17			17.35	
d.f.		11,886			11,773	
P		0.000			0.000	
R^2		0.25			0.20	

(a) The researchers separated the sample into blue- and white-collar workers. Based on the table, how many blue- and white-collar workers were included in the analysis?
(b) Do wage incentives appear to increase worker motivation? For both white- and blue-collar workers? Explain your answer.
(c) Repeat part (b) but now focus on the promotion incentives.
(d) The F statistics for the multiple regressions are highly significant, but the R^2 are relatively low. Explain to a statistical novice how this can occur.

CASE 11.2 **11.49 Effect of potential outliers.** Statistical inference requires us to make some assumptions about our data.
BOXOFFICE
(a) Obtain the residuals for the multiple regression in Example 11.13. Two movies have much larger U.S. box office revenues than predicted. Which are they and how much more revenue did they obtain compared with that predicted?
(b) Remove these two movies and redo the multiple regression. Make a table giving the regression coefficients and their standard errors, t statistics, and P-values.
(c) Compare these results with those presented in Example 11.13. How does the removal of these outlying movies impact the estimated model?
(d) Obtain the residuals from this reduced data set and graphically examine their distribution. Do the residuals appear approximately Normal? Explain your answer.

11.50 Pension benefits. The defined-benefit (DB) pension plan guarantees a certain income to employees when they retire. The defined-contribution (DC) plan specifies that a regular contribution will be made to an investment account, and upon retirement, the employee will receive the income from the investment. The DB has been the traditional plan but in recent years there has been a shift toward DC. One study examined the change in the proportion of DC-covered employees in 40 different industries.[14] The researchers who conducted the study used multiple regression methods to examine the effects of various explanatory variables on the change in the proportion of individuals covered by DC plans over a 20-year period. One set of variables were classified as "supply-side" factors that relate to changes in firm characteristics. One of these focused on quantifying shorter-term employment relationships using the change in the proportion of workers with fewer than 5 years with the company. The researchers called this variable P5. Variables such as age, gender, and region of residence were included in the model to control for these factors. A total of 30 variables were included.

(a) In a model to predict the change in the proportion of DC plans, the regression coefficient for P5 is 0.625 with a standard error of 0.189. Construct the t statistic for testing the null hypothesis that the coefficient for P5 is zero. Carry out the test and summarize the result.

(b) Give a 95% confidence interval for the coefficient for P5. From your calculations in part (a) of this exercise, would you expect the interval to include zero? Explain your answer.

11.51 Decreases in DB plans. Refer to the previous exercise. A model for the change in DB plans was also analyzed. The coefficient for P5 in this model was −1.186 with a standard error of 0.133. Answer the questions in parts (a) and (b) of the previous exercise for this analysis.

11.52 Do the results for change in DC and the results for change in DB agree? Refer to the previous two exercises. Summarize the results and interpret the signs for the coefficient for P5 in the two models. Do the results appear to be consistent? Explain your answer.

11.53 Interpret the "demand-side" explanatory variables. Refer to the previous three exercises. Another set of variables included in the analysis were classified as "demand-side" factors that relate to the changing needs of individuals in the labor force. One of these was the change in the proportion of dual-earner cou-

ples. The coefficient in the DC model was −0.986 with a standard error of 0.118, while the coefficient in the DB model was 0.466 with a standard error of 0.167. Using the previous three exercises as a guide, write a short report summarizing these results.

11.54 Bank auto loans. Banks charge different interest rates for different loans. A random sample of 2229 loans made by banks for the purchase of new automobiles was studied to identify variables that explain the interest rate charged. A multiple regression was run with interest rate as the response variable and 13 explanatory variables.[15]

(a) The F statistic reported is 71.34. State the null and alternative hypotheses for this statistic. Give the degrees of freedom and the P-value for this test. What do you conclude?

(b) The value of R^2 is 0.297. What percent of the variation in interest rates is explained by the 13 explanatory variables?

11.55 Bank auto loans, continued. Table 11.3 gives the coefficients for the fitted model and the individual t statistic for each explanatory variable in the study described in the previous exercise. The t-values are given without the sign, assuming that all tests are two-sided.

(a) State the null and alternative hypotheses tested by an individual t statistic. What are the degrees of freedom for these t statistics? What values of t will lead to rejection of the null hypothesis at the 5% level?

(b) Which of the explanatory variables are significantly different from zero in this model? Explain carefully what you conclude when an individual t statistic is not significant.

(c) The signs of many of the coefficients are what we might expect before looking at the data. For example, the negative coefficient for loan size means that larger loans get a smaller interest rate. This is reasonable. Examine the signs of each of the statistically significant coefficients and give a short explanation of what they tell us.

TABLE 11.3 Regression coefficients and t statistics for Exercise 11.55

Variable	b	t
Intercept	15.47	
Loan size (in dollars)	−0.0015	10.30
Length of loan (in months)	−0.906	4.20
Percent down payment	−0.522	8.35
Cosigner (0 = no, 1 = yes)	−0.009	3.02
Unsecured loan (0 = no, 1 = yes)	0.034	2.19
Total payments (borrower's monthly installment debt)	0.100	1.37
Total income (borrower's total monthly income)	−0.170	2.37
Bad credit report (0 = no, 1 = yes)	0.012	1.99
Young borrower (0 = older than 25, 1 = 25 or younger)	0.027	2.85
Male borrower (0 = female, 1 = male)	−0.001	0.89
Married (0 = no, 1 = yes)	−0.023	1.91
Own home (0 = no, 1 = yes)	−0.011	2.73
Years at current address	−0.124	4.21

11.56 Auto dealer loans. The previous two exercises describe auto loans made directly by a bank. The researchers also looked at 5664 loans made indirectly, that is, through an auto dealer. They again used multiple regression to predict the interest rate using the same set of 13 explanatory variables.
(a) The F statistic reported is 27.97. State the null and alternative hypotheses for this statistic. Give the degrees of freedom and the P-value for this test. What do you conclude?
(b) The value of R^2 is 0.141. What percent of the variation in interest rates is explained by the 13 explanatory variables? Compare this value with the percent explained for direct loans in Exercise 11.54.

11.57 Auto dealer loans, continued. Table 11.4 gives the estimated regression coefficient and individual t statistic for each explanatory variable in the setting of the previous exercise. The t-values are given without the sign, assuming that all tests are two-sided.
(a) What are the degrees of freedom of any individual t statistic for this model? What values of t are significant at the 5% level? Explain carefully what significance tells us about an explanatory variable.
(b) Which of the explanatory variables are significantly different from zero in this model?
(c) The signs of many of these coefficients are what we might expect before looking at the data. For example, the negative coefficient for loan size means that larger loans get a smaller interest rate. This is reasonable. Examine the signs of each of the statistically significant coefficients and give a short explanation of what they tell us.

11.58 Direct versus indirect loans. The previous four exercises describe a study of loans for buying new cars. The authors conclude that banks take higher risks with indirect loans because they do not take into account borrower characteristics when setting the loan rate. Explain how the results of the multiple regressions lead to this conclusion.

11.59 Canada's Small Business Financing Program. The Canada Small Business Financing Program (CSBFP) seeks to increase the availability of loans for establishing and improving small businesses. A survey was performed to better understand the experiences of small businesses when seeking loans and the extent to which they are aware of and satisfied with the CSBFP.[16] A total of 503 survey interviews were completed. To understand the drivers of overall satisfaction with the financing options available, a multiple regression was undertaken. The response variable was the subject's overall satisfaction scored on a 5-point scale, where 1 means "very dissatisfied" and 5 means "very satisfied." The eight predictors were the perceived importance of certain factors when considering or obtaining loans, including the interest rate, service fees, and quality of service from lenders. These were also scored on a 5-point scale, with 1 meaning "not important at all" and 5 meaning "very important."
(a) What are the degrees of freedom for the F statistic of the model that contains all the predictors?
(b) The report states that the complete set of predictors has an R^2 of 0.68. What percent of the variation in the response variable is explained by these eight explanatory variables?
(c) The report also states that the model with just the quality of service received and the possibility of negotiating the terms of financing explained 60% of the variation in the response variable. Test the hypothesis that the other six predictors do not help predict satisfaction when these two predictors are already in the model.

11.60 Compensation and human capital. A study of bank branch manager compensation collected data on the salaries

TABLE 11.4 Regression coefficients and t statistics for Exercise 11.57		
Variable	b	t
Intercept	15.89	
Loan size (in dollars)	−0.0029	17.40
Length of loan (in months)	−1.098	5.63
Percent down payment	−0.308	4.92
Cosigner (0 = no, 1 = yes)	−0.001	1.41
Unsecured loan (0 = no, 1 = yes)	0.028	2.83
Total payments (borrower's monthly installment debt)	−0.513	1.37
Total income (borrower's total monthly income)	0.078	0.75
Bad credit report (0 = no, 1 = yes)	0.039	1.76
Young borrower (0 = older than 25, 1 = 25 or younger)	−0.036	1.33
Male borrower (0 = female, 1 = male)	−0.179	1.03
Married (0 = no, 1 = yes)	−0.043	1.61
Own home (0 = no, 1 = yes)	−0.047	1.59
Years at current address	−0.086	1.73

of 82 managers at branches of a large eastern U.S. bank.[17] Multiple regression models were used to predict how much these branch managers were paid. The researchers examined two sets of explanatory variables. The first set were variables that measured characteristics of the branch and the position of the branch manager. These were number of branch employees, a variable constructed to represent how much competition the branch faced, market share, return on assets, an efficiency ranking, and the rank of the manager. A second set of variables were called human capital variables and measured characteristics of the manager. These were experience in industry, gender, years of schooling, and age. For the multiple regression using all the explanatory variables, the value of R^2 was 0.77. When the human capital variables were deleted, R^2 fell to 0.06. Test the null hypothesis that the coefficients for the human capital variables are all zero in the model that includes all the explanatory variables. Give the test statistic with its degrees of freedom and P-value, and give a short summary of your conclusion in nontechnical language.

11.3 Multiple Regression Model Building

Often we have many explanatory variables, and our goal is to use these to explain the variation in the response variable. A model using just a few of the variables often predicts about as well as the model using all the explanatory variables. We may also find that the reciprocal of a variable is a better choice than the variable itself, or that including the square of an explanatory variable improves prediction. How can we find a good model? That is the **model building** issue. A complete discussion would be quite lengthy, so we will be content to illustrate some of the basic ideas with a Case Study.

model building

HOMES

CASE 11.3

Prices of Homes People wanting to buy a home can find information on the Internet about homes for sale in their community. We will work with online data for homes for sale in Lafayette and West Lafayette, Indiana.[18] The response variable is Price, the asking price of a home. The online data contain the following explanatory variables: (a) SqFt, the number of square feet for the home, (b) BedRooms, the number of bedrooms, (c) Baths, the number of bathrooms, (d) Garage, the number of cars that can fit in the garage, and (e) Zip, the postal zip code for the address. There are 504 homes in the data set.

The analysis starts with a study of the variables involved. Here is a short summary of this work.

Price, as we expect, has a right-skewed distribution. The mean (in thousands of dollars) is $158 and the median is $130. There is one high outlier at $830, which we will delete as unusual in this location. Remember that a skewed distribution for Price does not itself violate the conditions for multiple regression. The model requires that the *residuals* from the fitted regression equation be approximately Normal. We will have to examine how well this condition is satisfied when we build our regression model.

BedRooms ranges from 1 to 5. The Web site uses 5 for all homes with 5 or more bedrooms. The data contain just one home with 1 bedroom. **Baths** includes both full baths (with showers or bathtubs) and half baths (which lack bathing facilities). Typical values are 1, 1.5, 2, and 2.5. **Garage** has values of 0, 1, 2, and 3. The Web site uses the value 3 when 3 or more vehicles can fit in the garage. There are 50 homes that can fit 3 or more vehicles into their garage (or possibly garages). The data set has begun a process of combining some values of these variables, such as 5 or more bedrooms and garages that hold 3 or more vehicles. We will continue this process as we build models for predicting Price.

Zip describes location, traditionally the most important explanatory variable for house prices, but Zip is a quite crude description because a single zip code covers a broad area. All of the postal zip codes in this community have 4790 as the first four digits. The fifth digit is coded as the variable Zip. The possible values are 1, 4, 5, 6, and 9. There is only one home with zip code 47901. We will first look at the houses in each zip code separately.

SqFt, the number of square feet for the home, is a quantitative variable that we expect to strongly influence Price. We will start our analysis by examining the relationship between Price and this explanatory variable. To control for location, we start by examining only the homes in zip code 47904, corresponding to Zip = 4. Most homes for sale in this area are moderately priced.

EXAMPLE 11.17 Price and Square Feet

CASE 11.3   Table 11.5 gives data for the 44 homes for sale in zip code 47904. Preliminary examination of Price reveals that a few homes have prices that are somewhat high relative to the others. Similarly, some values for SqFt are relatively high. Because we do not want our analysis to be overly influenced by these homes, we exclude any home with Price greater than $150,000 and any home with SqFt greater than 1800 ft^2. Seven homes were excluded by these criteria.

Figure 11.12 displays the relationship between SqFt and Price. We have added a "smooth" fit to help us see the pattern. The relationship is approximately linear but curves up somewhat for the higher-priced homes.

Because the relationship is approximately linear and we expect SqFt to be an important explanatory variable, let's start by examining the simple linear regression of Price on SqFt.

TABLE 11.5   Homes for sale in zip code 47904

Id	Price ($ thousands)	SqFt	BedRooms	Baths	Garage	Id	Price ($ thousands)	SqFt	BedRooms	Baths	Garage
01	52,900	932	1	1.0	0	23	75,000	2188	4	1.5	2
02	62,900	760	2	1.0	0	24	76,900	1400	3	1.5	2
03	64,900	900	2	1.0	0	25	81,900	796	2	1.0	2
04	69,900	1504	3	1.0	0	26	84,500	864	2	1.0	2
05	76,900	1030	3	2.0	0	27	84,900	1350	3	1.0	2
06	87,900	1092	3	1.0	0	28	89,600	1504	3	1.0	2
07	94,900	1288	4	2.0	0	29	87,000	1200	2	1.0	2
08	52,000	1370	3	1.0	1	30	89,000	876	2	1.0	2
09	72,500	698	2	1.0	1	31	89,000	1112	3	2.0	2
10	72,900	766	2	1.0	1	32	93,900	1230	3	1.5	2
11	73,900	777	2	1.0	1	33	96,000	1350	3	1.5	2
12	73,900	912	2	1.0	1	34	99,900	1292	3	2.0	2
13	81,500	925	3	1.0	1	35	104,900	1600	3	1.5	2
14	82,900	941	2	1.0	1	36	114,900	1630	3	1.5	2
15	84,900	1108	3	1.5	1	37	124,900	1620	3	2.5	2
16	84,900	1040	2	1.0	1	38	124,900	1923	3	3.0	2
17	89,900	1300	3	2.0	1	39	129,000	2090	3	1.5	2
18	92,800	1026	3	1.0	1	40	173,900	1608	2	2.0	2
19	94,900	1560	3	1.0	1	41	179,900	2250	5	2.5	2
20	114,900	1581	3	1.5	1	42	199,500	1855	2	2.0	2
21	119,900	1576	3	2.5	1	43	80,000	1600	3	1.0	3
22	65,000	853	3	1.0	2	44	129,000	2296	3	2.5	3

FIGURE 11.12 Plot of price versus square feet for Example 11.17.

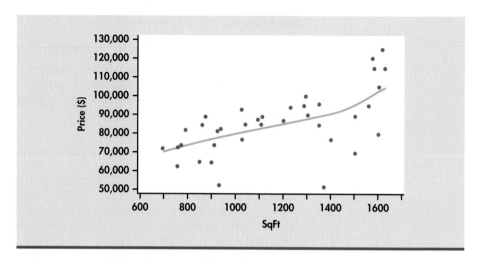

EXAMPLE 11.18 Regression of Price on Square Feet

CASE 11.3

HOMEPRICE

Figure 11.13 gives the regression output. The number of degrees of freedom in the "Corrected Total" line in the ANOVA table is 36. This is correct for the $n = 37$ homes that remain after we excluded 7 of the original 44. The fitted model is

$$\widehat{\text{Price}} = 45{,}298 + 34.32\text{SqFt}$$

The coefficient for SqFt is statistically significant ($t = 4.57, \text{df} = 35, P < 0.0001$). Each additional square foot of area raises selling prices by \$34.32 on the average. From the R^2 we see that 37.3% of the variation in the home prices is explained by a linear relationship with square feet. We hope that multiple regression will allow us to improve on this first attempt to explain selling price.

```
                        Analysis of Variance

                               Sum of          Mean
        Source          DF     Squares        Square    F Value   Pr > F
        Model            1  3780229462     3780229462    20.86   < .0001
        Error           35  6343998647      181257104
        Corrected Total 36 10124228108


        Root MSE            13463    R-Square     0.3734
        Dependent Mean      85524    Adj R-Sq     0.3555
        Coeff Var        15.74193

                        Parameter Estimates

                              Parameter   Standard
        Variable   Label   DF   Estimate     Error   t Value  Pr > |t|

        Intercept  Intercept 1     45298  9082.26322   4.99   < .0001
        SqFt       SqFt      1   34.32362    7.51591    4.57   < .0001
```

FIGURE 11.13 Linear regression output for predicting price using square feet, for Example 11.18.

CASE 11.3 **11.61 Distributions.** Make stemplots or histograms of the prices and of the square feet for the 44 homes in Table 11.5. Do the 7 homes excluded in Example 11.17 appear unusual for this location?

CASE 11.3 **11.62 Plot the residuals.** Obtain the residuals from the simple linear regression in the above example and plot them versus SqFt. Describe the plot. Does it suggest that the relationship might be curved?

CASE 11.3 **11.63 Predicted values.** Use the simple linear regression equation to obtain the predicted price for a home that has 1000 ft^2. Do the same for a home that has 1500 ft^2.

Models for curved relationships

Figure 11.12 suggests that the relationship between SqFt and Price may be slightly curved. One simple kind of curved relationship is a quadratic function. To model a quadratic function with multiple regression, create a new variable that is the square of the explanatory variable and include it in the regression model. There are now $p = 2$ explanatory variables, x and x^2. The model is

$$y = \beta_0 + \beta_1 x + \beta_2 x^2 + \epsilon$$

with the usual conditions on the ϵ_i.

EXAMPLE 11.19 Quadratic Regression of Price on Square Feet

CASE 11.3 To predict price using a quadratic function of square feet, first create a new variable by squaring each value of SqFt. Call this variable SqFt2. Figure 11.14 displays the output for multiple regression

Analysis of Variance

Source	DF	Sum of Squares	Mean Square	F Value	Pr > F
Model	2	3910030335	1955015167	10.70	0.0002
Error	34	6214197773	182770523		
Corrected Total	36	10124228108			

Root MSE	13519	R-Square	0.3862
Dependent Mean	85524	Adj R-Sq	0.3501
Coeff Var	15.80751		

Parameter Estimates

Variable	Label	DF	Parameter Estimate	Standard Error	t Value	Pr > |t|
Intercept	Intercept	1	81273	43653	1.86	0.0713
SqFt	SqFt	1	-30.13753	76.86278	-0.39	0.6974
SqFt2		1	0.02710	0.03216	0.84	0.4053

FIGURE 11.14 Quadratic regression output for predicting price using square feet, for Example 11.19.

of Price on SqFt and SqFt2. The fitted model is

$$\widehat{\text{Price}} = 81{,}273 - 30.14\text{SqFt} + 0.0271\text{SqFt2}$$

This model explains 38.6% of the variation in Price, little more than the 37.3% explained by simple linear regression of Price on SqFt. The coefficient of SqFt2 is not significant ($t = 0.84$, df $= 34$, $P = 0.41$). That is, the squared term does not significantly improve the fit when the SqFt term is present. We conclude that adding SqFt2 to our model is not helpful.

The output in Figure 11.14 is a good example of the need for care in interpreting multiple regression. The individual t tests for *both* SqFt and SqFt2 are not significant. Yet the overall F test for the null hypothesis that both coefficients are zero *is* significant ($F = 10.70$, df $= 2$ and 34, $P < 0.0002$). To resolve this apparent contradiction, remember that a t test assesses the contribution of a single variable, *given that the other variables are present in the model*. Once either SqFt or SqFt2 is present, the other contributes very little. This is a consequence of the fact that these two variables are highly correlated. This

collinearity multicollinearity

phenomenon is called **collinearity** or **multicollinearity.** In extreme cases collinearity can cause numerical instabilities, and the results of the regression calculations can become very imprecise. Collinearity can exist between seemingly unrelated variables and can be hard to detect in models with many explanatory variables. Some statistical software

Variance Inflation Factor VIF

packages will calculate a **Variance Inflation Factor VIF** value for each explanatory variable in a model. VIF values greater than 10 are generally considered an indication that severe collinearity exists among the explanatory variables in a model. Exercise 11.82 (page 623) explores the calculation and use of VIF values. In this particular case, we could dispense with either SqFt or SqFt2, but the F test tells us that we cannot drop both of them. It is natural to keep SqFt and drop its square, SqFt2.

Multiple regression can fit a *polynomial* model of any degree:

$$y = \beta_0 + \beta_1 x + \beta_2 x^2 + \cdots + \beta_k x^k + \epsilon$$

In general, we include all powers up to the highest power in the model. A relationship that curves first up and then down, for example, might be described by a cubic model with explanatory variables x, x^2, and x^3. Other transformations of the explanatory variable, such as the square root and the logarithm, can also be used to model curved relationships.

> **APPLY YOUR KNOWLEDGE**
>
> **CASE 11.3** **11.64 The relationship between SqFt and SqFt2.** Using the data set for Example 11.19, plot SqFt2 versus SqFt. Describe the relationship. We know that it is not linear, but is it approximately linear? What is the correlation between SqFt and SqFt2? The plot and correlation demonstrate that these variables are collinear and explain why neither of them contributes much to a multiple regression once the other is present.
>
> **CASE 11.3** **11.65 Predicted values.** Use the quadratic regression equation in Example 11.19 to predict the price of a home that has 1000 ft^2. Do the same for a home that has 1500 ft^2. Compare these predictions with the ones from an analysis that uses only SqFt as an explanatory variable.

Models with categorical explanatory variables

Although adding the square of SqFt failed to improve our model significantly, Figure 11.12 does suggest that the price rises a bit more steeply for larger homes. Perhaps some of these homes have other desirable characteristics that increase the price. Let's examine another explanatory variable.

EXAMPLE 11.20   Price and the Number of Bedrooms

CASE 11.3

Figure 11.15 gives a plot of Price versus BedRooms. We see that there appears to be a curved relationship. However, all but two of the homes have either two or three bedrooms. One home has one bedroom and another has four. These two cases are why the relationship appears to be curved. To avoid this situation we will group the four-bedroom home with those that have three bedrooms (BedRooms = 3) and the one-bedroom home with the homes that have two bedrooms.

FIGURE 11.15 Plot of price versus the number of bedrooms for Example 11.20.

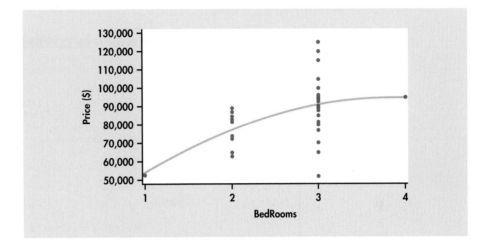

The price of the four-bedroom home is in the middle of the distribution of the prices for the three-bedroom homes. On the other hand, the one-bedroom home has the lowest price of all the homes in the data set. This observation may require special attention later.

"Number of bedrooms" is now a *categorical variable* that places homes in two groups: one/two bedrooms and three/four bedrooms. Software often allows you to simply declare that a variable is categorical. Then the values for the two groups don't matter. We could use the values 2 and 3 for the two groups. If you work directly with the variable, however, it is better to indicate whether or not the home has three or more bedrooms. We will take the "number of bedrooms" categorical variable to be Bed3 = 1 if the home has three or more bedrooms and Bed3 = 0 if it does not. Bed3 is called an *indicator variable*.

Indicator Variables

An indicator variable is a variable with the values 0 and 1. To use a categorical variable that has I possible values in a multiple regression, create $K = I - 1$ indicator variables to use as the explanatory variables. This can be done in many different ways. Here is one common choice:

$$X1 = \begin{cases} 1 & \text{if the categorical variable has the first value} \\ 0 & \text{otherwise} \end{cases}$$

$$X2 = \begin{cases} 1 & \text{if the categorical variable has the second value} \\ 0 & \text{otherwise} \end{cases}$$

$$\vdots$$

$$XK = \begin{cases} 1 & \text{if the categorical variable has the next to last value} \\ 0 & \text{otherwise} \end{cases}$$

We need only $I - 1$ variables to code I different values because the last value is identified by "all $I - 1$ indicator variables are 0."

EXAMPLE 11.21   Price and the Number of Bedrooms

CASE 11.3

Figure 11.16 displays the output for the regression of Price on the indicator variable Bed3. This model explains 19% of the variation in price. This is about one-half of the 37.3% explained by SqFt, but it suggests that Bed3 may be a useful explanatory variable.

```
                        Analysis of Variance

                              Sum of        Mean
    Source          DF       Squares       Square    F Value  Pr > F
    Model            1    1934368525   1934368525       8.27  0.0068
    Error           35    8189859583    233995988
    Corrected Total 36   10124228108


    Root MSE            15297     R-Square      0.1911
    Dependent Mean      85524     Adj R-Sq      0.1680
    Coeff Var        17.88605

                       Parameter Estimates

                            Parameter    Standard
    Variable   Label    DF   Estimate       Error   t Value  Pr > |t|

    Intercept  Intercept 1     75700  4242.60432     17.84   <.0001
    SqFt                 1     15146  5267.78172      2.88    0.0068
```

FIGURE 11.16 Output for predicting price using whether or not there are three or more bedrooms, for Example 11.21.

HOMEPRICE

The fitted equation is

$$\widehat{\text{Price}} = 75{,}700 + 15{,}146\text{Bed3}$$

The coefficient for Bed3 is significantly different from 0 ($t = 2.88$, df $= 35$, $P = 0.0068$). This coefficient is the slope of the least-squares line. That is, it is the increase in the average price when Bed3 increases by 1. The indicator variable Bed3 has only two values, so we can clarify the interpretation.

The predicted price for homes with two or fewer bedrooms (Bed3 $= 0$) is

$$\widehat{\text{Price}} = 75{,}700 + 15{,}146(0) = 75{,}700$$

That is, the intercept 75,700 is the mean price for homes with Bed3 $= 0$. The predicted price for homes with three or more bedrooms (Bed3 $= 1$) is

$$\widehat{\text{Price}} = 75{,}700 + 15{,}146(1) = 90{,}846$$

That is, the slope 15,146 says that homes with three or more bedrooms are priced $15,146 higher on the average than homes with two or fewer bedrooms. When we regress on a single indicator variable, both intercept and slope have simple interpretations.

Example 11.21 shows that regression on one indicator variable essentially models the means of the two groups. A regression model for a categorical variable with I possible values requires $I - 1$ indicator variables and therefore I regression coefficients (including the intercept) to model the I category means. Indicator variables can also be used in models with other variables to allow for different regression lines for different groups. Exercise 11.76 (page 622) explores the use of an indicator variable in a model with another quantitative variable for this purpose.

Here is an example of a categorical explanatory variable with four possible values.

EXAMPLE 11.22 Price and the Number of Bathrooms

CASE 11.3

The homes in our data set have 1, 1.5, 2, or 2.5 bathrooms. Figure 11.17 gives a plot of price versus the number of bathrooms with a "smooth" fit. The relationship does not appear to be linear so we will start by treating the number of bathrooms as a categorical variable. We require three indicator variables for the four values. To use the homes that have 1 bath as the basis for comparisons, we let this correspond to "all indicator variables equal to 0." The indicator variables are

$$\text{Baths15} = \begin{cases} 1 & \text{if the home has 1.5 baths} \\ 0 & \text{otherwise} \end{cases}$$

$$\text{Baths2} = \begin{cases} 1 & \text{if the home has 2 baths} \\ 0 & \text{otherwise} \end{cases}$$

$$\text{Baths25} = \begin{cases} 1 & \text{if the home has 2.5 baths} \\ 0 & \text{otherwise} \end{cases}$$

Multiple regression using these three explanatory variables gives the output in Figure 11.18. The overall model is statistically significant ($F = 12.59, df = 3$ and 33, $P < 0.0001$), and it explains 53.3% of the variation in price. This is somewhat more than the 37.3% explained by square feet. The fitted model is

$$\widehat{\text{Price}} = 77,504 + 20,533\text{Baths15} + 12,616\text{Baths2} + 44,896\text{Baths25}$$

The coefficients of all the indicator variables are statistically significant, indicating that each additional bathroom is associated with higher prices.

FIGURE 11.17 Plot of price versus the number of bathrooms for Example 11.22.

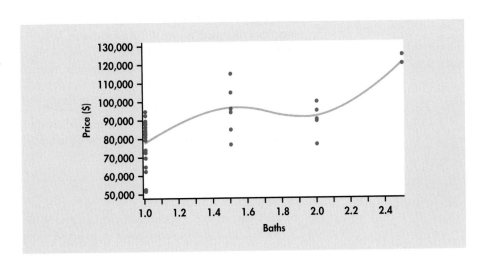

```
                        Analysis of Variance

                               Sum of        Mean
    Source          DF        Squares       Square     F Value  Pr > F
    Model            3     5404093400    1801364467      12.59  <.0001
    Error           33     4720134708     143034385
    Corrected Total 36    10124228108


    Root MSE              11960      R-Square      0.5338
    Dependent Mean        85524      Adj R-Sq      0.4914
    Coeff Var          13.98397

                        Parameter Estimates

                             Parameter     Standard
    Variable   Label    DF    Estimate        Error    t Value  Pr > |t|

    Intercept  Intercept  1       77504   2493.76950      31.08  <.0001
    Baths15                1       20553   5162.59333       3.98  0.0004
    Baths2                 1       12616   5901.33572       2.14  0.0400
    Baths25                1       44896   8816.80661       5.09  <.0001
```

FIGURE 11.18 Output for predicting price using indicator variables for the number of bathrooms, for Example 11.22.

APPLY YOUR KNOWLEDGE

CASE 11.3 **11.66 Find the means.** Using the data set for Example 11.21, find the mean price for the homes that have three or more bedrooms and the mean price for those that do not.

(a) Compare these sample means with the predicted values given in Example 11.21.

(b) What is the difference between the mean price of the homes in the sample that have three or more bedrooms and the mean price of those that do not? Verify that this difference is the coefficient for the indicator variable Bed3 in the regression in Example 11.21.

CASE 11.3 **11.67 Compare the means.** Regression on a single indicator variable compares the mean responses in two groups. It is in fact equivalent to the pooled t test for comparing two means (Chapter 7, page 429). Use the pooled t test to compare the mean price of the homes that have three or more bedrooms with the mean price of those that do not. Verify that the test statistic, degrees of freedom, and P-value agree with the t test for the coefficient of Bed3 in Example 11.21.

CASE 11.3 **11.68 Modeling the means.** Following the pattern in Example 11.21, use the output in Figure 11.18 to write the equations for the predicted mean price for

(a) homes with 1 bathroom.

(b) homes with 1.5 bathrooms.

(c) homes with 2 bathrooms.

(d) homes with 2.5 bathrooms.

(e) How can we interpret the coefficient for one of the indicator variables, say Baths2, in language understandable to house shoppers?

More elaborate models

We now suspect that a model that uses square feet, number of bedrooms, and number of bathrooms may explain price reasonably well. Before examining such a model, we use an insight based on careful data analysis to improve the treatment of number of baths.

Figure 11.17 reminds us that the data describe only two homes with 2.5 bathrooms. Our model with three indicator variables fits a mean for these two observations. The pattern of Figure 11.17 reveals an interesting feature: adding a half bath to either 1 or 2 baths raises the predicted price by a similar, quite substantial amount. If we use this information to construct a model, we can avoid the problem of fitting one parameter to just two houses.

EXAMPLE 11.23   An Alternative Bath Model

CASE 11.3

Starting again with 1-bath homes as the base (all indicator variables 0), let B2 be an indicator variable for an extra full bath and let Bh be an indicator variable for an extra half bath. Thus, a home with 2 baths has $Bh = 0$ and $B2 = 1$. A home with 2.5 baths has $Bh = 1$ and $B2 = 1$. Regressing Price on Bh and B2 gives the output in Figure 11.19.

The overall model is statistically significant ($F = 18.30$, df = 2 and 34, $P < 0.0001$), and it explains 51.8% of the variation in price. This compares favorably with the 53.3% explained by the model with three indicator variables for bathrooms. The fitted model is

$$\widehat{\text{Price}} = 76{,}929 + 15{,}837\text{B2} + 23{,}018\text{Bh}$$

```
                          Analysis of Variance

                                  Sum of          Mean
         Source          DF       Squares        Square     F Value   Pr > F
         Model            2    5248929412     2624464706      18.30    <.0001
         Error           34    4875298697      143391138
         Corrected Total 36   10124228108


         Root MSE             11975     R-Square      0.5185
         Dependent Mean       85524     Adj R-Sq      0.4901
         Coeff Var          14.00140

                          Parameter Estimates

                                 Parameter      Standard
         Variable   Label    DF   Estimate         Error   t Value   Pr > |t|

         Intercept  Intercept 1       76929    2434.86662     31.59   <.0001
         B2                   1       15837    5032.10205      3.15   0.0034
         Bh                   1       23018    4593.65967      5.01   <.0001
```

FIGURE 11.19 Output for predicting price using the alternative coding for bathrooms, for Example 11.23.

That is, an extra full bath adds \$15,837 to the mean price and an extra half bath adds \$23,018. The t statistics show that both regression coefficients are significantly different from zero ($t = 3.15$, df $= 34$, $P = 0.0034$; and $t = 5.01$, df $= 34$, $P < 0.0001$).

So far we have learned that the price of a home is related to the number of square feet, whether or not there are three or more bedrooms, whether or not there is an additional full bathroom, and whether or not there is an additional half bathroom. Let's try a model including all these explanatory variables.

EXAMPLE 11.24   Square Feet, Bedrooms, and Bathrooms

CASE 11.3   Figure 11.20 gives the output for predicting price using SqFt, Bed3, B2, and Bh. The overall model is statistically significant ($F = 11.48$, df $= 4$ and 32, $P < 0.0001$), and it explains 58.9% of the variation in price.

Analysis of Variance

Source		DF	Sum of Squares	Mean Square	F Value	Pr > F
Model		4	5966829910	1491707478	11.48	<.0001
Error		32	4157398198	129918694		
Corrected Total		36	10124228108			

Root MSE	11398	R-Square	0.5894	
Dependent Mean	85524	Adj R-Sq	0.5380	
Coeff Var	13.32742			

Parameter Estimates

Variable	Label	DF	Parameter Estimate	Standard Error	t Value	Pr > |t|
Intercept	Intercept	1	55539	9390.30284	5.91	<.0001
SqFt	SqFt	1	22.86649	10.02864	2.28	0.0294
Bed3		1	-5788.52568	5943.78386	-0.97	0.3374
B2		1	14564	5155.65595	2.82	0.0081
Bh		1	17209	5247.38676	3.28	0.0025

FIGURE 11.20 Output for predicting price using square feet, bedroom, and bathroom information, for Example 11.24.

The individual t for Bed3 is not statistically significant ($t = -0.97$, df $= 32$, $P = 0.3374$). That is, in a model that contains square feet and information about the bathrooms, there is no additional information in the number of bedrooms that is useful for predicting price. This happens because the explanatory variables are related to each other: houses with more bedrooms tend to also have more square feet and more baths.

We therefore redo the regression without Bed3. The output appears in Figure 11.21. The value of R^2 has decreased slightly to 57.7%, but now all the coefficients for the explanatory variables are statistically significant. The fitted regression equation is

$$\widehat{\text{Price}} = 59,268 + 16.78\text{SqFt} + 13,161\text{B2} + 16,859\text{Bh}$$

```
┌─────────────────────────────────────────────────────────────────────┐
│ ▣                                                         ─ □ ⊠       │
│                                                                       │
│                       Analysis of Variance                            │
│                                                                       │
│                            Sum of        Mean                         │
│         Source         DF  Squares       Square    F Value  Pr > F     │
│         Model           3  5843609810  1947869937    15.02  <.0001     │
│         Error          33  4280618298   129715706                      │
│         Corrected Total 36 10124228108                                 │
│                                                                       │
│                                                                       │
│         Root MSE           11389    R-Square     0.5772                │
│         Dependent Mean     85524    Adj R-Sq     0.5388                │
│         Coeff Var       13.31701                                       │
│                                                                       │
│                       Parameter Estimates                             │
│                              Parameter    Standard                     │
│         Variable   Label  DF  Estimate       Error   t Value  Pr > |t| │
│                                                                       │
│         Intercept  Intercept  1    59268  8567.44638    6.92  <.0001   │
│         SqFt       SqFt       1 16.77989     7.83689    2.14  0.0397   │
│         B2                    1    13161  4946.59620    2.66  0.0119   │
│         Bh                    1    16859  5230.97932    3.22  0.0029   │
│                                                                       │
└─────────────────────────────────────────────────────────────────────┘
```

FIGURE 11.21 Output for predicting price using square feet and bathroom information, for Example 11.24.

APPLY YOUR KNOWLEDGE

CASE 11.3 **11.69 What about garages?** We have not yet examined the number of garage spaces as a possible explanatory variable for price. Make a scatterplot of price versus garage spaces. Describe the pattern. Use a "smooth" fit if your software has this capability. Otherwise, find the mean price for each possible value of Garage, plot the means on your scatterplot, and connect the means with lines. Is the relationship approximately linear?

CASE 11.3 **11.70 The home with three garages.** There is only one home with 3 garage spaces. We might either place this house in the Garage = 2 group or remove it as unusual. Either decision leaves Garage with values 0, 1, and 2. Based on your plot in the previous exercise, which choice do you recommend?

Variable selection methods

We have arrived at a reasonably satisfactory model for predicting the asking price of houses. But it is clear that there are many other models we might consider. We have not used the Garage variable, for example. What is more, the explanatory variables can *interact* with each other. This means that the effect of one explanatory variable depends upon the value of another explanatory variable. We account for this situation in a regression model by including **interaction terms.** The simplest way to construct an interaction term is to multiply the two explanatory variables together. Thus, if we wanted to allow the effect of an additional half bath to depend upon whether or not there is an additional full bath, we would create a new explanatory variable by taking the product B2 × Bh. Interaction terms that are the product of an indicator variable and another variable in the model can be used to allow for different slopes for different groups. Exercise 11.78 (page 622) explores the use of such an interaction term in a model.

interaction terms

Considering interactions increases the number of possible explanatory variables. If we start with the 5 explanatory variables SqFt, Bed3, B2, Bh, and Garage, there are 10 interactions between pairs of variables. That is, there are now 15 possible explanatory variables in all. From 15 explanatory variables it is possible to build 32,767 different models for predicting price. We need to automate the process of examining possible models. Modern regression software offers *variable selection methods* that examine, for example, the R^2-values or s-values for all possible multiple regression models. The software then presents us with the models having the highest R^2 for each number of explanatory variables or the smallest s-value overall.

EXAMPLE 11.25   Predicting Asking Price

CASE 11.3   Software tells us that the highest available R^2 increases as we increase the number of explanatory variables as follows:

Variables	R^2
1	0.44
2	0.57
3	0.62
4	0.66
5	0.72
⋮	
13	0.77

Because of collinearity problems, models with 14 or all 15 explanatory variables cannot be used. The highest possible R^2 is 77%, using 13 explanatory variables. There are only 37 houses in the data set. A model with too many explanatory variables will fit the accidental wiggles in the prices of these specific houses and may do a poor job of predicting prices for other houses. This is called *overfitting* **overfitting.**

There is no formula for choosing the "best" multiple regression model. We want to avoid overfitting, and we prefer smaller models to larger models. We want a model that makes intuitive sense. For example, the best single predictor ($R^2 = 0.44$) is SqFtBh, the interaction between square feet and having an extra half bath. In general, we would not use a model that has interaction terms unless the explanatory variables involved in the interaction are also included. The variable selection output does suggest that we include SqFtBh, and therefore SqFt and Bh.

EXAMPLE 11.26   One More Model

CASE 11.3   Figure 11.22 gives the output for the multiple regression of Price on SqFt, Bh, and the interaction between these variables. This model explains 55.7% of the variation in price. The coefficients for SqFt and Bh are significant at the 10% level but not at the 5% level ($t = 1.82$, df $= 33$, $P = 0.0777$; and $t = -1.79$, df $= 33$, $P = 0.0825$). However, the interaction is significantly different from zero ($t = 2.29$, df $= 33$, $P = 0.0284$). Because the interaction is significant, we will keep the terms SqFt and Bh in the model even though their P-values do not quite pass the 0.05 standard. The fitted regression equation is

$$\widehat{\text{Price}} = 63{,}375 + 15.15\text{SqFt} - 58{,}800\text{Bh} + 52.81\text{SqFtBh}$$

```
                          Analysis of Variance

                                  Sum of       Mean
      Source              DF      Squares      Square    F Value  Pr > F
      Model                3   5639442004  1879814001     13.83  <.0001
      Error               33   4484786104   135902609
      Corrected Total     36  10124228108


      Root MSE             11658    R-Square      0.5570
      Dependent Mean       85524    Adj R-Sq      0.5168
      Coeff Var         13.63089

                          Parameter Estimates
                                  Parameter    Standard
      Variable    Label    DF     Estimate       Error   t Value  Pr > |t|

      Intercept   Intercept   1      63375   9263.60218      6.84  <.0001
      SqFt        SqFt        1    15.15440      8.32360      1.82  0.0777
      Bh                      1     -58800        32832     -1.79  0.0825
      SqFtBh                  1    52.81209     23.03886      2.29  0.0284
```

FIGURE 11.22 Output for predicting price using square feet, extra half bathroom, and the interaction, for Example 11.26.

The negative coefficient for Bh seems odd. We expect an extra half bath to increase price. A plot will help us to understand what this model says about the prices of homes.

EXAMPLE 11.27   Interpretation of the Fit

CASE 11.3

Figure 11.23 plots Price against SqFt, using different plot symbols to show the values of the categorical variable Bh. Homes without an extra half bath are plotted as circles, and homes with an extra half bath are plotted as squares. Look carefully: none of the smaller homes have an extra half bath.

FIGURE 11.23 Plot of the data and model for price predicted by square feet, extra half bathroom, and the interaction, for Example 11.27.

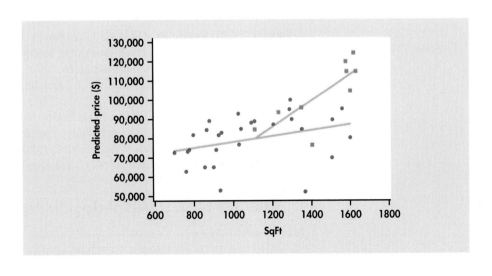

The lines on the plot graph the model from Example 11.26. The presence of the categorical variable Bh and the interaction Bh × SqFt account for the *two* lines. That is, the model expresses the fact that the relationship between price and square feet depends on whether or not there is an additional half bath.

We can determine the equations of the two lines from the fitted regression equation. Homes that lack an extra half bath have Bh = 0. When we set Bh = 0 in the regression equation

$$\widehat{\text{Price}} = 63,375 + 15.15\text{SqFt} - 58,800\text{Bh} + 52.81\text{SqFtBh}$$

we get

$$\widehat{\text{Price}} = 63,375 + 15.15\text{SqFt}$$

For homes that have an extra half bath, Bh = 1 and the regression equation becomes

$$\widehat{\text{Price}} = 63,375 + 15.15\text{SqFt} - 58,800 + 52.81\text{SqFt}$$
$$= (63,375 - 58,800) + (15.15 + 52.81)\text{SqFt}$$
$$= 4575 + 67.96\text{SqFt}$$

In Figure 11.23, we graphed this line starting at the price of the least-expensive home with an extra half bath. These two equations tell us that an additional square foot increases asking price by $15.15 for homes without an extra half bath and by $67.96 for homes that do have an extra half bath.

APPLY YOUR KNOWLEDGE

CASE 11.3 **11.71 Comparing some predicted values.** Consider two homes, both with 1400 ft². Suppose the first has an extra half bath and the second does not. Find the predicted price for each home and then find the difference.

CASE 11.3 **11.72 Suppose the homes are larger?** Consider two additional homes, both with 1600 ft², one with an extra half bath and one without. Find the predicted prices and the difference. How does this difference compare with the difference you obtained in the previous exercise? Explain what you have found.

CASE 11.3 **11.73 How about the smaller homes?** Would it make sense to do the same calculations as in the previous two exercises for homes that have 700 ft²? Explain why or why not.

CASE 11.3 **11.74 Residuals.** Once we have chosen a model, we must examine the residuals for violations of the conditions of the multiple regression model. Examine the residuals from the model in Example 11.26.
(a) Plot the residuals against SqFt. Do the residuals show a random scatter, or is there some systematic pattern?
(b) The residual plot shows three somewhat low residuals, between −$20,000 and −$40,000. Which homes are these? Is there anything unusual about these homes?
(c) Make a Normal quantile plot of the residuals. (Make a histogram if your software does not make Normal quantile plots.) Is the distribution roughly Normal?

BEYOND THE BASICS: Multiple Logistic Regression

Many studies have yes/no or success/failure response variables. A surgery patient lives or dies; a consumer does or does not purchase a product after viewing an advertisement.

Because the response variable in a multiple regression is assumed to have a Normal distribution, this methodology is not suitable for predicting yes/no responses. However, there are models that apply the ideas of regression to response variables with only two possible outcomes.

logistic regression

The most common technique is called **logistic regression.** The starting point is that if each response is 0 or 1 (for failure or success), then the mean response is the probability p of a success. Logistic regression tries to explain p in terms of one or more explanatory variables. Details are even more complicated than those for multiple regression, but the fundamental ideas are very much the same and software handles most details. Here is an example.

EXAMPLE 11.28 Sexual Imagery in Advertisements

Marketers sometimes use sexual imagery in advertisements targeted at teens and young adults. One study designed to examine this issue analyzed how models were dressed in 1509 ads in magazines read by young and mature adults.[19] The clothing of the models in the ads was classified as not sexual or sexual. Logistic regression was used to model the probability that the model's clothing was sexual as a function of four explanatory variables.[20] Here, model clothing with values 1 for sexual and 0 for not sexual is the response variable.

The explanatory variables were x_1, a variable having the value 1 if the median age of the readers of the magazine is 20 to 29, and 0 if the median age of the readers of the magazine is 40 to 49; x_2, the gender of the model, coded as 1 for female and 0 for male; x_3, a code to indicate men's magazines, with values 1 for a men's magazine and 0 otherwise; and x_4, a code to indicate women's magazines, with values 1 for a women's magazine, and 0 otherwise. General-interest magazines are coded as 0 for both x_3 and x_4.

Similar to the F test in multiple regression, there is a chi-square test for multiple logistic regression that tests the null hypothesis that *all* coefficients of the explanatory variables are zero. The value is $X^2 = 168.2$, and the degrees of freedom are the number of explanatory variables, 4 in this case. The P-value is reported as $P = 0.001$. (You can verify that it is less than 0.0005 using Table F.) We conclude that not all the explanatory variables have zero coefficients.

Interpretation of the coefficients is a little more difficult in multiple logistic regression because of the form of the model. For our example, the fitted model is

$$\log\left(\frac{p}{1-p}\right) = -2.32 + 0.50x_1 + 1.31x_2 - 0.05x_3 + 0.45x_4$$

odds

The expression $p/(1-p)$ is the **odds** that the model is sexually dressed. Logistic regression models the "log odds" as a linear combination of the explanatory variables. Positive coefficients are associated with a higher probability that the model is dressed sexually. We see that ads in magazines with younger readers, female models, and women's magazines are more likely to show models dressed sexually.

In place of the t tests for individual coefficients in multiple regression, chi-square tests, each with 1 degree of freedom, are used to test whether individual coefficients are zero. For reader age and model gender, $P < 0.01$, while for the indicator for women's magazines, $P < 0.05$. The indicator for men's magazines is not statistically significant.

SECTION 11.3 Summary

- Start the model-building process by performing **data analysis** for multiple linear regression. Examine the distribution of the variables and the form of relationships among them.

- Note any **categorical explanatory variables** that have very few cases for some values. If it is reasonable, combine these values with other values. If not, delete these cases and examine them separately.

- For **curved relationships,** consider transformations or additional explanatory variables that will account for the curvature. Sometimes adding a **quadratic term** improves the fit.

- Examine the possibility that the effect of one explanatory variable depends upon the value of another explanatory variable. **Interactions** can be used to model this situation.

- Examine software outputs from modern **variable selection methods** to see what models of each subset size have the highest R^2.

SECTION 11.3 Exercises

For Exercises 11.61 to 11.63, see page 609; for 11.64 and 11.65, see page 610; for 11.66 to 11.68, see page 614; for 11.69 and 11.70, see pages 617; and for 11.71 to 11.74, see page 620.

11.75 Quadratic models. Sketch each of the following quadratic equations for values of x between 0 and 5. Then describe the relationship between μ_y and x in your own words.
(a) $\mu_y = 8 + 2x + 2x^2$
(b) $\mu_y = 8 - 2x + 2x^2$
(c) $\mu_y = 8 + 2x - 2x^2$
(d) $\mu_y = 8 - 2x - 2x^2$

11.76 Models with indicator variables. Suppose that x is an indicator variable with the value 0 for Group A and 1 for Group B. The following equations describe relationships between the value of μ_y and membership in Group A or B. For each equation, give the value of the mean response μ_y for Group A and for Group B.
(a) $\mu_y = 8 + 4x$
(b) $\mu_y = 8 + 40x$
(c) $\mu_y = 8 + 400x$

11.77 Differences in means. Verify that the coefficient of x in each part of the previous exercise is equal to the mean for Group B minus the mean for Group A. Do you think that this will be true in general? Explain your answer.

11.78 Models with interactions. Suppose that x_1 is an indicator variable with the value 0 for Group A and 1 for Group B, and x_2 is a quantitative variable. Each of the models below describes a relationship between μ_y and the explanatory variables x_1 and x_2. For each model, substitute the value 0 for x_1 and write the resulting equation for μ_y in terms of x_2 for Group A. Then substitute $x_1 = 1$ to obtain the equation for Group B and sketch the two equations on the same graph. Describe in words the difference in the relationship for the two groups.
(a) $\mu_y = 70 + 30x_1 + 6x_2 + 6x_1x_2$
(b) $\mu_y = 60 - 8x_1 + 7x_2 - 5x_1x_2$
(c) $\mu_y = 130 + 10x_1 - 7x_2 + 14x_1x_2$

11.79 Differences in slopes and intercepts. Refer to the previous exercise. Verify that the coefficient of x_1x_2 is equal to the slope for Group B minus the slope for Group A in each of these cases. Also, verify that the coefficient of x_1 is equal to the intercept for Group B minus the intercept for Group A in each of these cases. Do you think these two results will be true in general? Explain your answer.

11.80 Write the model. For each of the following situations write a model for μ_y of the form

$$\mu_y = \beta_0 + \beta_1 x_1 + \beta_2 x_2 + \cdots + \beta_p x_p$$

where p is the number of explanatory variables. Be sure to give the value of p and, if necessary, explain how each of the x's is coded.
(a) A model where the explanatory variable is a categorical variable with four possible values.
(b) A model where there are three explanatory variables. One of these is categorical with two possible values; another is categorical with four possible values. Include a term that would model an interaction of the first categorical variable and the third (quantitative) explanatory variable.
(c) A quartic regression, where terms up to and including the fourth power of an explanatory variable are included in the model.

11.81 Online stock brokerages. Refer to the online stock brokerage data in Exercise 11.23 (page 589). Plot assets versus the number of accounts. Investigate the possibility that the relationship is curved by running a multiple regression to predict assets using the number of accounts and the square of the number of accounts as explanatory variables. (Note that to use statistical inference for these data, we need to assume that there is some underlying model that generated the data and that it is the properties of this model that are of interest.) ⬤ BROKERAGE
(a) Give the fitted regression equation.
(b) Find a 95% confidence interval for the coefficient of the squared term.

(c) Give the results of the significance test for the coefficient of the squared term. Report the test statistic with its degrees of freedom and P-value, and summarize your conclusion.

(d) Rerun the analysis without the quadratic term. Explain why the coefficient of the number of accounts is not the same as you found for part (a).

CASE 11.2 **11.82 Assessing collinearity in the movie revenue model.** Many software packages will calculate VIF values for each explanatory variable. In this exercise, you will calculate the VIF values using several multiple regressions and then use them to see if there is collinearity among the movie explanatory variables. **BOXOFFICE**

(a) Use statistical software to estimate the multiple regression model for predicting Budget based on Opening and Theaters. Calculate the VIF value for Budget using R^2 from this model and the formula

$$\text{VIF} = \frac{1}{1 - R^2}$$

(b) Use statistical software to estimate the multiple regression model for predicting Opening based on Budget and Theaters. Calculate the VIF value for Opening using R^2 from this model and the formula from part (a).

(c) Use statistical software to estimate the multiple regression model for predicting Theaters based on Budget and Opening. Calculate the VIF value for Theaters using R^2 from this model and the formula from part (a).

(d) Do any of the calculated VIF values indicate severe collinearity among the explanatory variables? Explain your response.

CASE 11.2 **11.83 Predicting movie revenue.** A plot of theater count versus box office revenue suggests that the relationship may be slightly curved. **BOXOFFICE**

(a) Examine this question by running a regression to predict the box office revenue using the theater count and the square of the theater count. Report the relevant test statistic

with its degrees of freedom and P-value, and summarize your conclusion.

(b) Now view this analysis in the framework of testing a hypothesis about a collection of regression coefficients, which you studied in Section 11.2 (page 590). The first model includes theater count and the square of theater count, while the second includes only theater count. Run both regressions and find the value of R^2 for each. Find the F statistic for comparing the models based on the difference in the values of R^2. Carry out the test and report your conclusion.

(c) Verify that the square of the t statistic that you found in the previous exercise for testing the coefficient of the quadratic term is equal to the F statistic that you found for this exercise.

CASE 11.2 **11.84 Predicting movie revenue: model selection.** Refer to the data set on movie revenue in Case 11.2 (page 591). In addition to the movie's budget, opening-weekend revenue, and opening-weekend theater count, the data set also includes a column named Sequel. Sequel is 1 if the corresponding movie is a sequel, and Sequel is 0 if the movie is not a sequel. Assuming opening-weekend revenue (Opening) is in the model, there are 8 possible regression models. For example, one model just includes Opening, another model includes Opening and Theaters, and another model includes Opening, Sequel, and Theaters. Run these 8 regressions and make a table giving the regression coefficients, the value of R^2, and the value of s for each regression. (If an explanatory variable is not included in a particular regression, enter a value 0 for its coefficient in the table.) Mark coefficients that are statistically significant at the 5% level with an asterisk (*). Summarize your results and state which model you prefer. **BOXOFFICE**

CASE 11.2 **11.85 Effect of outliers?** In Exercise 11.49 (page 603), we identified two movies that have much higher revenue than predicted. Remove these two movies and repeat the previous exercise. Does the removal of these two movies change which model you prefer?

STATISTICS IN SUMMARY

Multiple regression uses more than one explanatory variable to explain the variation in a response variable. In particular, multiple regression allows *categorical explanatory variables, polynomial functions* of a quantitative explanatory variable, and *interactions* among explanatory variables. As in the case of simple linear regression, we can fit a multiple regression equation by the *least-squares* method to any set of data. Inference requires more restrictive conditions that are summarized in the *multiple regression model*. To do inference, we must estimate not only the coefficients of the regression equation but also the standard deviation σ that describes how much variation there is in responses y when the x's are held fixed. We use the *regression standard error s* (roughly, the sample standard deviation of the residuals) to estimate σ. Here is a review of the most important skills you should have acquired from your study of this chapter.

A. Preliminaries

1. Write a suitable regression model for your situation. Choose a response variable and promising explanatory variables. Recognize categorical explanatory variables and use indicator functions to include them in the model.

2. Examine the distribution of each explanatory variable and the response variable.

3. Make scatterplots to show the relationship between explanatory variables and the response variable. Decide whether, for example, a quadratic term may be needed.

4. Use software to find by least squares the regression equation that fits your chosen model to the data.

5. Inspect the data to recognize situations in which inference isn't safe: influential observations, strongly skewed residuals in a small sample, or nonconstant variation of the data points about the fitted equation.

B. Inference Using Software Output

1. Explain in any specific regression setting the meaning of the coefficients (β's) of the population regression equation.

2. Understand software output for multiple regression. Find in the output the intercept and the coefficients for the fitted regression equation, their standard errors and individual t statistics, and the regression standard error.

3. Use software output to carry out tests and calculate confidence intervals for the β's. Recognize clearly that conclusions about one explanatory variable depend on what other variables are also present in the model.

4. Use software output to carry out the F test for the significance of the model as a whole. Understand why F can be significant even though all individual t's are not significant.

5. Explain the distinction between a confidence interval for the mean response and a prediction interval for an individual response. If software gives output for prediction, use that output to give either confidence or prediction intervals.

6. Know how to test a null hypothesis concerning the β's for a subset of variables.

CHAPTER 11 Review Exercises

CASE 11.2 **11.86 Alternate movie revenue model.** Refer to the data set on movie revenue in Case 11.2 (page 591). The variables Budget, Opening, and USRevenue all have distributions with long tails. For this problem, let's consider building a model using the logarithm transformation of these variables. **BOXOFFICE**
(a) Run the multiple regression to predict the logarithm of US-Revenue using the logarithm of Budget, the logarithm of Opening, and Theaters and obtain the residuals. There is one outlying movie. Which is it?
(b) Remove this movie and refit the model in part (a). State the regression model and note which coefficients are statistically significant at the 5% level.
(c) Examine the residuals graphically. Does the distribution appear approximately Normal? Explain your answer.
(d) In Exercise 11.37 (page 597), you were asked to predict the revenue of a particular movie. Using the results from this

model, construct a 95% prediction interval for the movie's log USRevenue.

CASE 10.1 **11.87 Job stress.** Recall Case 10.1 (page 528) which looked at the relationship between perceived job stress and a person's locus of control (LOC). In addition to the accountant's locus of control, the person's age and gender were also obtained. **LOCl**
(a) Write the model that you would use for a multiple regression to predict job stress from LOC, age, and gender.
(b) What are the parameters of your model?
(c) Run the multiple regression and give the estimates of the model parameters.
(d) Test the hypothesis that the coefficients for LOC, age, and gender are all zero. Give the test statistic with degrees of freedom and the P-value. What do you conclude?

(e) What is the value of R^2?

(f) Give the results of the hypothesis test for the coefficient for LOC. Include the test statistic, degrees of freedom, and the P-value. Do the same for the other two variables. Summarize your conclusions from these three tests.

11.88 Examine the assumptions. Refer to the previous exercise. We now ask whether the data meet the requirements of the multiple regression model.

(a) Find the residuals and examine their distribution. Summarize what you find.

(b) Plot the residuals versus each of the explanatory variables. Describe the plots. Does your analysis suggest that the model assumptions may not be reasonable for this problem?

11.89 Compare regression coefficients. Again refer to Exercise 11.87. ![LOC1] **LOC1**

(a) In Example 10.5 (page 533), parameter estimates for the model that included just LOC were obtained. Compare those parameter estimates with the ones obtained from the full model that also includes age and gender. Describe any changes.

(b) Consider a 42-year-old male accountant with an LOC of 10. What is his predicted stress level based on the full model? What about for the model that includes only LOC?

(c) Suppose his LOC changed to 11. How much does this change his job stress level for each model?

(d) In Example 10.13 (page 561), we computed r^2 for the model that included only LOC. It was 0.0975. Use r^2 and R^2 to test whether age and gender are helpful predictors, given LOC is already in the model.

11.90 Business to business (B2B) marketing. A group of researchers were interested in determining the likelihood that a business currently purchasing office supplies via a catalog would switch to purchasing from the Web site of the same supplier. To do this, they performed an online survey using the business clients of a large Australian-based stationery provider with both a catalog and a Web-based business.[21] Results from 1809 firms, all currently purchasing via the catalog, were obtained. The following table summarizes the regression model.

Variable	b	t
Staff interpersonal contact with catalog	−0.08	3.34
Trust of supplier	0.11	4.66
Web benefits (access and accuracy)	0.08	3.92
Previous Web purchases	0.18	8.20
Previous Web information search	0.08	3.47
Key catalog benefits (staff, speed, security)	−0.08	3.96
Web benefits (speed and ease of use)	0.36	3.97
Problems with Web ordering and delivery	−0.06	2.65

(a) The F statistic is reported to be 78.15. What degrees of freedom are associated with this statistic?

(b) This F statistic can be expressed in terms of R^2 as

$$F = \left(\frac{n - p - 1}{p} \right) \left(\frac{R^2}{1 - R^2} \right)$$

Use this relationship to determine R^2.

(c) The coefficients listed above are **standardized coefficients**. These are obtained when each variable is standardized (subtract its mean, divide by its standard deviation) prior to fitting the regression model. These coefficients then represent the change in standard deviations of y for a one standard deviation change in x. This typically allows one to determine which independent variables have a greater effect on the dependent variable. Using this idea, what are the top four variables in this analysis?

Exercises 11.91 to 11.94 use the PRICEPROMO data set shown in Table 11.6 on the next page.

11.91 Discount promotions at a supermarket. How does the frequency that a supermarket product is promoted at a discount affect the price that customers expect to pay for the product? Does the percent reduction also affect this expectation? These questions were examined by researchers in a study that used 160 subjects. The treatment conditions corresponded to the number of promotions (1, 3, 5, or 7) that were described during a 10-week period and the percent that the product was discounted (10%, 20%, 30%, and 40%). Ten students were randomly assigned to each of the $4 \times 4 = 16$ treatments.[22]

(a) Plot the expected price versus the number of promotions. Do the same for expected price versus discount. Summarize the results.

(b) These data come from a designed experiment with an equal number of observations for each promotion by discount combination. Find the means and standard deviations for expected price for each of these combinations. Describe any patterns that are evident in these summaries.

(c) Using your summaries from part (b), make a plot of the mean expected price versus the number of promotions for the 10% discount condition. Connect these means with straight lines. On the same plot add the means for the other discount conditions. Summarize the major features of this plot.

11.92 Run the multiple regression. Refer to the previous exercise. Run a multiple regression using promotions and discount to predict expected price. Write a summary of your results.

11.93 Residuals and other models. Refer to the previous exercise. Analyze the residuals from your analysis and investigate the possibility of using quadratic and interaction terms as predictors. Write a report recommending a final model for this problem with a justification for your recommendation.

11.94 Can we generalize the results? The subjects in this experiment were college students at a large Midwest university who were enrolled in an introductory management course. They received the information about the promotions during a 10-week period during their course. Do you think that these facts about

TABLE 11.6 Expected price data

Number of promotions	Percent discount	Expected price ($)									
1	40	4.10	4.50	4.47	4.42	4.56	4.69	4.42	4.17	4.31	4.59
1	30	3.57	3.77	3.90	4.49	4.00	4.66	4.48	4.64	4.31	4.43
1	20	4.94	4.59	4.58	4.48	4.55	4.53	4.59	4.66	4.73	5.24
1	10	5.19	4.88	4.78	4.89	4.69	4.96	5.00	4.93	5.10	4.78
3	40	4.07	4.13	4.25	4.23	4.57	4.33	4.17	4.47	4.60	4.02
3	30	4.20	3.94	4.20	3.88	4.35	3.99	4.01	4.22	3.70	4.48
3	20	4.88	4.80	4.46	4.73	3.96	4.42	4.30	4.68	4.45	4.56
3	10	4.90	5.15	4.68	4.98	4.66	4.46	4.70	4.37	4.69	4.97
5	40	3.89	4.18	3.82	4.09	3.94	4.41	4.14	4.15	4.06	3.90
5	30	3.90	3.77	3.86	4.10	4.10	3.81	3.97	3.67	4.05	3.67
5	20	4.11	4.35	4.17	4.11	4.02	4.41	4.48	3.76	4.66	4.44
5	10	4.31	4.36	4.75	4.62	3.74	4.34	4.52	4.37	4.40	4.52
7	40	3.56	3.91	4.05	3.91	4.11	3.61	3.72	3.69	3.79	3.45
7	30	3.45	4.06	3.35	3.67	3.74	3.80	3.90	4.08	3.52	4.03
7	20	3.89	4.45	3.80	4.15	4.41	3.75	3.98	4.07	4.21	4.23
7	10	4.04	4.22	4.39	3.89	4.26	4.41	4.39	4.52	3.87	4.70

the data would influence how you would interpret and generalize the results? Write a summary of your ideas regarding this issue.

11.95 Predicting water quality. Pollution of water resources is a serious problem that can require substantial efforts to improve. To determine the financial resources required, an accurate assessment of the extent of the problem is necessary. The index of biotic integrity (IBI) is a measure of the water quality in streams that is fairly expensive to determine. Land use measures, on the other hand, are relatively inexpensive to obtain. IBI and land use measures for a collection of streams in the Ozark Highland ecoregion of Arkansas were collected as part of a study.[23] The data set gives the IBI, the area of the watershed in square kilometers (Area) and the percent of the watershed area that is forest (Forest) for streams in the original sample with area less than or equal to 70 km^2. IBI
(a) Use numerical and graphical methods to describe each of the three variables in this data set. Summarize your results.
(b) Plot the data and describe the relationships between each pair of variables. Are there any outliers or unusual patterns?
(c) Give the statistical model for a multiple regression that would use Area and Forest to predict IBI.
(d) State the null and alternative hypotheses for all the significance tests associated with this analysis.
(e) Run the multiple regression and summarize the results.
(f) Obtain the residuals and plot them versus Area and versus Forest. Is there anything unusual in these plots?
(g) Do the residuals appear to be approximately Normal? Give reasons for your answer.

(h) Do the assumptions for the analysis of these data using the model you gave in part (c) appear to be reasonable? Explain your answer.

11.96 More complicated analyses. Refer to the previous exercise. Investigate the possibility that quadratic terms and interactions would improve the estimation of IBI. Write a summary of your results.

11.97 Impact of word of mouth. Word of mouth (WOM) is informal advice passed between consumers that may have a quick and powerful influence on consumer behavior. Word of mouth may be positive (PWOM), encouraging choice of a certain brand, or negative (NWOM), discouraging that choice. A study investigated the impact of WOM on brand purchase probability.[24] Multiple regression was used to assess the effect of six variables on brand choice. These were room for change (PPP), strength of expression of WOM, WOM about main brand, closeness of the communicator, whether advice was sought, and amount of WOM given. The table below summarizes the results for 903 participants who received NWOM.

Variable	b	$s(b)$
PPP	−0.37	0.022
Strength of expression of WOM	−0.22	0.065
WOM about main brand	0.21	0.164
Closeness of communicator	−0.06	0.121
Whether advice was sought	−0.04	0.140
Amount of WOM given	−0.08	0.022

In addition, it is reported that $R^2 = 0.20$.

(a) What percent of the variation in change in brand purchase probability is explained by these explanatory variables?

(b) State which of these variables are statistically significant at the 5% level.

(c) The PPP result implies that the more uncertain someone is about purchasing, the more negative the impact of NWOM. Explain what the "strength of expression of WOM" result implies.

(d) The variable "WOM about main brand" is an indicator variable. It is equal to 1 when the NWOM is about the receiver's main brand and 0 when it is about another brand. Explain the meaning of this result.

11.98 Correlations may not be a good way to screen for multiple regression predictors. We use a constructed data set in this problem to illustrate this point. 📀 DATASETA

(a) Find the correlations between the response variable Y and each of the explanatory variables X_1 and X_2. Plot the data and run the two simple linear regressions to verify that no evidence of a relationship is found by this approach. Some researchers would conclude at this point that there is no point in further exploring the possibility that X_1 and X_2 could be useful in predicting Y.

(b) Analyze the data using X_1 and X_2 in a multiple regression to predict Y. The fit is quite good. Summarize the results of this analysis.

(c) What do you conclude about an analytical strategy that first looks at one candidate predictor at a time and selects from these candidates for a multiple regression based on some threshold level of significance?

11.99 The multiple regression results do not tell the whole story. We use a constructed data set in this problem to illustrate this point. 📀 DATASETB

(a) Run the multiple regression using X_1 and X_2 to predict Y. The F test and the significance tests for the coefficients of the explanatory variables fail to reach the 5% level of significance. Summarize these results.

(b) Now run the two simple linear regressions using each of the explanatory variables in separate analyses. The coefficients of the explanatory variables are statistically significant at the 5% level in each of these analyses. Verify these conclusions with plots and correlations.

(c) What do you conclude about an analytical strategy that looks only at multiple regression results?

11.100 Price-fixing litigation. Multiple regression is sometimes used in litigation. In the case of *Cargill, Inc. v. Hardin,* the prosecution charged that the cash price of wheat was manipulated in violation of the Commodity Exchange Act. In a statistical study conducted for this case, a multiple regression model was constructed to predict the price of wheat using three supply-and-demand explanatory variables.[25] Data for 14 years were used to construct the regression equation, and a prediction for the suspect period was computed from this equation. The value of R^2 was 0.989.

(a) The fitted model gave the predicted value $2.136 with standard error $0.013. Express the prediction as an interval. (The

degrees of freedom were large for this analysis, so use 100 as the df to determine t^*.)

(b) The actual price for the period in question was $2.13. The judge decided that the analysis provided evidence that the price was not artificially depressed, and the opinion was sustained by the court of appeals. Write a short summary of the results of the analysis that relate to the decision and explain why you agree or disagree with it.

11.101 Predicting the total PCB level in fish. Polychlorinated biphenyls (PCBs) are synthetic compounds, called congeners, that are particularly toxic to fetuses and young children. Although PCBs are no longer produced in the United States, they are still found in the environment. Since human exposure to these PCBs is primarily through the consumption of fish, the Environmental Protection Agency (EPA) monitors PCB levels in fish. Unfortunately, there are 209 different congeners and measuring all of them in a fish specimen is an expensive and time-consuming process. You've been asked to see if the total amount of PCBs in a specimen can be estimated with only a few, easily quantifiable congeners.[26] If this can be done, costs can be greatly reduced. Consider the following variables: PCB (the total amount of PCB) and four congeners, PCB52, PCB118, PCB138, and PCB180. 📀 PCB

(a) Using numerical and graphical summaries, describe the distribution of each of these variables.

(b) Using numerical and graphical summaries, describe the relationship between each pair of variables in this set.

(c) Use the four congeners, PCB52, PCB118, PCB138, and PCB180, in a multiple regression to predict PCB.

(d) Examine the residuals. Do they appear to be approximately Normal? When you plot them versus each of the explanatory variables, are any patterns evident?

(e) There may be two outliers, one with a high residual and one with a low residual. Because of safety issues, we are more concerned about underestimating total PCB in a specimen than about overestimating. Give the specimen number for each of the two suspected outliers. Which one corresponds to an overestimate of PCB?

(f) Rerun the analysis with the two suspected outliers deleted, summarize these results, and compare them with those you obtained in part (c).

11.102 Run a new model and compare the results. Run a regression to predict PCB using the variables PCB52, PCB118, and PCB138. Note that this is similar to the analysis that you did in Exercise 11.101 with the change that PCB180 is not included as an explanatory variable.

(a) Summarize the results.

(b) In this analysis, the regression coefficient for PCB118 is not statistically significant. Give the estimate of the coefficient and the associated P-value.

(c) Find the estimate of the coefficient for PCB118 and the associated P-value for the model analyzed in Exercise 11.101.

(d) Using the results in parts (b) and (c), write a short paragraph explaining how the inclusion of other variables in a multiple regression can have an effect on the estimate of a particular coefficient and the results of the associated significance test.

11.103 Try logs. Because distributions of variables such as PCB and the PCB congeners tend to be skewed, researchers frequently analyze the logarithms of the measured variables. Create a data set that has the logs of each of the variables in the PCB data set. Note that zero is a possible value for PCB126; most software packages will eliminate these cases when you request a log transformation.
(a) If you do not do anything about the 16 zero values of PCB126, what does your software do with these cases? Is there an error message of some kind?
(b) If you attempt to run a regression to predict the log of PCB using the log of PCB126 and the log of PCB52, are the cases with the zero values of PCB126 eliminated? Do you think that is a good way to handle this situation?
(c) The smallest nonzero value of PCB126 is 0.0052. One common practice when taking logarithms of measured values is to replace the zeros by one-half of the smallest observed value. Create a logarithm data set using this procedure; that is, replace the 16 zero values of PCB126 by 0.0026 before taking logarithms. Use numerical and graphical summaries to describe the distributions of the log variables.

Exercises 11.104 to 11.110 use the CROP data file which contains the U.S. yield (bushels/acre) of corn and soybeans from 1957 to 2007.[27]

11.104 Corn yield varies over time. Run the simple linear regression using year to predict corn yield.
(a) Summarize the results of your analysis, including the significance test results for the slope and R^2 for this model.
(b) Analyze the residuals with a Normal quantile plot. Is there any indication in the plot that the residuals are not Normal?
(c) Plot the residuals versus soybean yield. Does the plot indicate that soybean yield might be useful in a multiple linear regression with year to predict corn yield? Explain your answer.

11.105 Can soybean yield predict corn yield? Run the simple linear regression using soybean yield to predict corn yield.
(a) Summarize the results of your analysis, including the significance test results for the slope and R^2 for this model.
(b) Analyze the residuals with a Normal quantile plot. Is there any indication in the plot that the residuals are not Normal?
(c) Plot the residuals versus year. Does the plot indicate that year might be useful in a multiple linear regression with soybean yield to predict corn yield? Explain your answer.

11.106 Use both predictors. From the previous two exercises, we conclude that year *and* soybean yield may be useful together in a model for predicting corn yield. Run this multiple regression.
(a) Explain the results of the ANOVA F test. Give the null and alternative hypotheses, the test statistic with degrees of freedom, and the P-value. What do you conclude?
(b) What percent of the variation in corn yield is explained by these two variables? Compare it with the percent explained in the simple linear regression models of the previous two exercises.

(c) Give the fitted model. Why do the coefficients for year and soybean yield differ from those in the previous two exercises?
(d) Summarize the significance test results for the regression coefficients for year and soybean yield.
(e) Give a 95% confidence interval for each of these coefficients.
(f) Plot the residuals versus year and versus soybean yield. What do you conclude?

11.107 Try a quadratic. We need a new variable to model the curved relation that we see between corn yield and year in the residual plot of the last exercise. Let year2 = (year − 1982)^2. (When adding a squared term to a multiple regression model, we sometimes subtract the mean of the variable being squared before squaring. This eliminates the correlation between the linear and quadratic terms in the model and thereby reduces collinearity.)
(a) Run the multiple linear regression using year, year2, and soybean yield to predict corn yield. Give the fitted regression equation.
(b) Give the null and alternative hypotheses for the ANOVA F test. Report the results of this test, giving the test statistic, degrees of freedom, P-value, and conclusion.
(c) What percent of the variation in corn yield is explained by this multiple regression? Compare this with the model in the previous exercise.
(d) Summarize the results of the significance tests for the individual regression coefficients.
(e) Analyze the residuals and summarize your conclusions.

11.108 Compare models. Run the model to predict corn yield using year and the squared term year2 defined in the previous exercise.
(a) Summarize the significance test results.
(b) The coefficient for year2 is not statistically significant in this run, but it was significant in the model analyzed in the previous exercise. Explain how this can happen.
(c) Obtain the fitted values for each year in the data set and use these to sketch the curve on a plot of the data. Plot the least-squares line on this graph for comparison. Do you see much of a difference between these two regression functions? Compute the difference between the two predicted values for each year and plot them versus year. For what years are the differences the largest?

11.109 Do a prediction. Use the simple linear regression model with corn yield as the response variable and year as the explanatory variable to predict the corn yield for the year 2008, and give the 95% prediction interval. Also, use the multiple regression model where year and year2 are both explanatory variables to find another predicted value with the 95% interval. Explain why these two predicted values are so different. Compare the two prediction intervals. Explain why they differ in width even though the predicted values are similar? The actual yield for 2008 was 1539 bushels per acre. How well did your models predict this value?

11.110 Predict the yield for another year. Repeat the previous exercise doing the prediction for 2020. Compare the results of this exercise with the previous one. Also explain why the predicted values are beginning to differ more substantially.

Exercises 11.111 to 11.119 use the CSDATA data file. For this study, the computer science department of a large university was interested in understanding why a large proportion of their first-year students failed to graduate as computer science majors. An examination of records from the registrar indicated that most of the attrition occurred during the first three semesters. Therefore, they decided to study all first-year students entering their program in a particular year and to follow their progress for the first three semesters.

The variables studied included the grade point average after three semesters and a collection of variables that were available when students entered their program. These included scores on standardized tests such as the SAT and high school grades in various subjects. The individuals who conducted the study were also interested in examining differences between men and women in this program. Therefore, sex was included as a variable. A total of 224 students were available for analysis.

11.111 SAT scores and GPA. Use software to make a plot of GPA versus SATM. Do the same for GPA versus SATV. Describe the general patterns. Are there any unusual values?

11.112 High school grades and GPA. Make a plot of GPA versus HSM. Do the same for the other two high school grade variables. Describe the three plots. Are there any outliers or influential points?

11.113 Predict GPA from high school grades. Regress GPA on the three high school grade variables. Calculate and store the residuals from this regression. Plot the residuals versus each of the three predictors and versus the predicted value of GPA. Are there any unusual points or patterns in these four plots?

11.114 Predict GPA from SAT scores. Use the two SAT scores in a multiple regression to predict GPA. Calculate and store the residuals. Plot the residuals versus each of the explanatory variables and versus the predicted GPA. Describe the plots.

11.115 Use the math explanatory variables. It appears that the mathematics explanatory variables are strong predictors of GPA for computer science students. Run a multiple regression using HSM and SATM to predict GPA.
(a) Give the fitted regression equation.
(b) State the H_0 and H_a tested by the ANOVA F statistic, and explain their meaning in plain language. Report the value of the F statistic, its P-value, and your conclusion.
(c) Give 95% confidence intervals for the regression coefficients of HSM and SATM. Do either of these include the point 0?

(d) Report the t statistics and P-values for the tests of the regression coefficients for HSM and SATM. What conclusions do you draw from these tests?
(e) What is the value of s, the estimate of σ?
(f) What percent of the variation in GPA is explained by HSM and SATM in your model?

11.116 Use the verbal explanatory variables. How well do verbal variables predict the performance of computer science students? Perform a multiple regression analysis to predict GPA from HSE and SATV. Summarize the results and compare them with those obtained in the previous exercise. In what ways do the regression results indicate that the mathematics variables are better predictors?

11.117 Analyze the males. The variable Sex has the value 1 for males and 2 for females. Create a data set containing the values for males only. Run a multiple regression analysis for predicting GPA from the three high school grade variables for this group. Interpret the results and state what conclusions can be drawn from this analysis. In what way (if any) do the results for males alone differ from those for all students?

11.118 Analyze the females and compare. Refer to the previous exercise. Perform the analysis using the data for females only. Are there any important differences between female and male students in predicting GPA?

11.119 The effect of gender. In the previous two exercises, you analyzed the males and females using separate multiple regressions with the high school grades as explanatory variables. Here we will run the analyses together. Recode the variable Sex into a new variable Gender that is an indicator variable. (You can do this by setting Gender equal to Sex minus 1.) Construct three interaction variables that model the interaction between gender and each of the high school grade variables. Then run a multiple regression using seven explanatory variables: the three high school grade variables, gender, and the three interaction variables.
(a) Report the fitted equation and the results of the significance tests for the regression coefficients.
(b) Substitute the value 0 for Gender and simplify the fitted model. This is the model for males. Verify that this is the same fitted model that you obtained in Exercise 11.117.
(c) Repeat part (b) to obtain the results for females and verify that this is the same fitted model that you obtained in Exercise 11.118.
(d) Use software or the method for comparing the coefficients of a collection of q explanatory variables using the values of R^2 (page 600) to test the null hypothesis that the three interaction terms are all zero.

CHAPTER 11 Case Study Exercises

CASE STUDY EXERCISE 1: Predict the quality of a product. As cheddar cheese matures, a variety of chemical processes take place. The taste of matured cheese is related to the concentration of several chemicals in the final product. In a study of cheddar cheese from the LaTrobe Valley of Victoria, Australia, samples of cheese were analyzed for their quality. The quality of the cheese was assessed by a panel of tasters whose scores are summarized in a single variable called Taste. This is the response variable. The explanatory variables are three chemicals that are present in the cheese: acetic acid, hydrogen sulfide, and lactic acid. For acetic acid and hydrogen sulfide (natural) log transformations were taken. Thus, the explanatory variables are the transformed concentrations of acetic acid ("Acetic") and hydrogen sulfide ("H2S") and the untransformed concentration of lactic acid ("Lactic").[28] The table below gives the data for the first five samples of cheese.

Case	Taste	Acetic	H2S	Lactic
01	12.3	4.543	3.135	0.86
02	20.9	5.159	5.043	1.53
03	39.0	5.366	5.438	1.57
04	47.9	5.759	7.496	1.81
05	5.6	4.663	3.807	0.99

There are 30 cases in this data set. Analyze these data using multiple regression methods and write a report summarizing your results. Be sure to include appropriate graphical displays. **CHEESE**

CASE STUDY EXERCISE 2: Self-concept and grade point average of seventh-grade students. Data were collected on 78 seventh-grade students in a rural Midwest school to determine the relationship between the students' "self-concept" and their academic performance.[29] The explanatory variables are OBS, a subject identification number; GPA, grade point average; IQ, score on an IQ test; AGE, age in years; SEX, female (1) or male (2); SC, overall score on the Piers-Harris Children's Self-Concept Scale; and C1 to C6, "cluster scores" for specific aspects of self-concept: C1 = behavior, C2 = school status, C3 = physical appearance, C4 = anxiety, C5 = popularity, and C6 = happiness. Here are data for the first five students:

OBS	GPA	IQ	AGE	SEX	SC	C1	C2	C3	C4	C5	C6
001	7.940	111	13	2	67	15	17	13	13	11	9
002	8.292	107	12	2	43	12	12	7	7	6	6
003	4.643	100	13	2	52	11	10	5	8	9	7
004	7.470	107	12	2	66	14	15	11	11	9	9
005	8.882	114	12	1	58	14	15	10	12	11	6

The response variable is grade point average. Use the methods you learned in this chapter to examine models for predicting grade point average from these explanatory variables. Be sure to examine the assumptions needed for the multiple regressions. If there are outliers, be sure to assess their effect on your final results by rerunning the analyses without them. **CONCEPT**

CASE STUDY EXERCISE 3: Is gender related to salary for hourly employees? Executive Order 11246, signed by President Lyndon B. Johnson in 1965, prohibits discrimination in hiring and employment decisions on the basis of race, color, gender, religion, and national origin. It applies to contractors doing business in excess of $10,000 with the U.S. government. The Office of Federal Contract Compliance Programs (OFCCP) is responsible for monitoring compliance with this Executive Order. Setting employees' pay is an important employment decision, and

the OFCCP frequently examines the salary data of government contractors.

Many banks are subject to the OFCCP guidelines and they periodically review their salary data to verify that they are in compliance with the regulations. You will examine records for 1745 hourly workers at one such bank. The data are based on real records but to preserve confidentiality, we simply call the bank National Bank. We have also made minor modifications to the numbers. The variables are ID, an employee identifier; Gender, coded as "F" for female and "M" for male; Race, coded as "Black," "White," and "Other"; Earnings, the employee's annual salary; and Status, with values "PT" for a part time worker and "FT" for a full time worker. Here are the data for the first five employees:

ID	Gender	Race	Earnings	Status
1	F	Black	14,682	FT
2	F	Black	12,655	PT
3	F	Black	12,641	FT
4	F	Black	14,678	PT
5	F	Black	14,714	FT

The response variable is annual salary, and the primary explanatory variable of interest is gender. If there is a statistically significant relationship between salary and gender, the company may face litigation.[30] These employees are called "nonexempt" employees because they are subject to the regulations of the Fair Labor Standards Act of 1938, which control things such as overtime pay. Some of the employees work part-time, and this should be taken into consideration when examining the question of gender differences. Analyze the data and write a report summarizing your conclusions. Be sure to include concise numerical summaries and graphical displays that will explain what you have found. **HOURLY**

CASE STUDY EXERCISE 4: Is gender related to salary for salaried employees? These data are similar to those in Case Study Exercise 3, but here we will look at the salaried employees. These employees are exempt from the Fair Labor Standards Act of 1938. This means, for example, that they do not receive overtime pay. Again, the response variable is annual salary, and the primary explanatory variable of interest is gender. For these employees a major determinant of salary is job level. This variable should be used as an explanatory variable in models where you examine the gender question. Note that the values are integers between 7 and 14. You will need to decide whether to treat job level as a categorical variable. Analyze the data and prepare a report summarizing your findings. **SALARY**

CASE STUDY EXERCISE 5: Prices of homes. Consider the data set used for Case 11.3 (page 606). This data set includes information for several other zip codes. Pick a different zip code and analyze the data. Compare your results with what we found for zip code 47904 in Section 11.3. **HOMES**

CHAPTER 11 Appendix

Using Minitab and Excel for Multiple Regression

Scatterplots–Multiple X's

Minitab:

Graph ➤ Scatterplot

As discussed in the chapter, a preliminary step in the data analysis of multiple variables is to plot the response variable against each of the explanatory variables and, in addition, to plot the explanatory variables against each other. You can do this by creating separate scatterplots for all the pairs of variables (refer to the Appendix of Chapter 2 for guidance on producing individual scatterplots). Alternatively, an option exists to create and present the scatterplots simultaneously. To do this, click the data column for the response variable into row 1 of the **Y variables** box, and click the data column for the first explanatory variable into row 1 of the **X variables** box. Then click the data column for the response variable into row 2 of the **Y variables** box, and click the data column for the second explanatory variable into row 2 of the **X variables** box. Continue this process for as many explanatory variables as you have. Thereafter, you can click all the pairs of explanatory variables into the subsequent rows. To have all these scatterplots presented simultaneously, click the **Multiple Graphs** button and then select the **In separate panels of the same graph** option. Click **OK** to close the pop-up box and click **OK** to produce the plots.

Excel:

Excel does not offer an option of presenting multiple scatterplots simultaneously. You will need to create a scatterplot for each pair of variables (refer to the Appendix of Chapter 2 for guidance on producing individual scatterplots).

Multiple Regression–Estimation, Confidence, and Prediction Limits

Minitab:

Stat ➤ Regression ➤ Regression

The only difference from the steps explained for the simple regression model in the Appendices of Chapters 2 and 10 is

that you need to click more than one explanatory variable into the **Predictors** box. To obtain confidence intervals for means and prediction intervals, enter the values for each of the p explanatory variables in the **Prediction intervals for new observations** box. (*Note:* The values for the explanatory variables should be inputted in the same order that they have been clicked into the **Predictors** box.)

Excel:

The only difference from the steps explained for the simple regression model in the Appendices of Chapters 2 and 10 is that you need to input a cell range of multiple columns for the multiple explanatory variables. In the spreadsheet, the data for the multiple explanatory variables must be placed in adjacent columns ranging over the same rows. Once the data for the explanatory variables have been set up in this fashion, you can then enter the cell range of the explanatory variables into the **Input X Range** box.

Variable Selection

Minitab:

Stat ➤ Regression ➤ Best Subsets

Minitab offers a number of automated variable selection methods to help you find a good subset of explanatory variables without requiring the estimation of all possible subset regressions. As noted in the chapter, one such method is based on examining the R^2-values for all the possible subset regressions and then presenting the models having the highest R^2 for a given number of explanatory variables. In Minitab, this method is known as "Best Subsets Regression" and is invoked by following the above-listed pull-down sequence. In the **Best Subsets Regression** pop-up box click the response variable data column into the **Response** box, and click the explanatory variable data column into the **Free predictors** box. Normally, you would input all the explanatory variables into the **Free predictors** box. However, there is a caveat. Explanatory variables that are fully explained by one or more of the other explanatory variables are regarded as "not free" and should not be inputted into the **Free predictors** box. Such a situation is typically encountered when dealing with indicator variables. As mentioned in Section 11.3, if we have I indicator variables for I categories, then only $I - 1$ indicator variables are needed. So, for best subsets regression, you should input all but one of the indicator variables. In default mode, Minitab reports the top two models based on one explanatory variable, then the top two models based on two explanatory variables, and so on. If you click the **Options** button, you can change the number of models that can be reported in the **Models of each size to print** box. Minitab will allow you to report up to five best mod-els. Click **OK** to close the pop-up box, and click **OK** to produce the best subsets regression output.

Excel:

Variable selection methods are not available in standard Excel, but they are available in the WHFStat Add-In for Excel.

ANDREW DOUGLAS/MASTERFILE

Statistics for Quality: Control and Capability

Introduction

Beginning in the 1980s, the marketplace signals were becoming clear: poor quality in products and services would not be tolerated by customers. Organizations increasingly recognized that what they didn't know about the quality of their products could have devastating results: customers often simply left when encountering poor quality rather than making complaints and hoping that the organization makes changes. To make matters worse, customers would voice their discontent to other customers, resulting in a spiraling negative effect on the organization in question. The twenty-first century's competitive marketplace is pressuring organizations to leave no room for error in the delivery of products and services.

To meet these marketplace challenges, organizations have recognized that a shift to a different paradigm of management thought and action is necessary. The new paradigm calls for developing an organizational system dedicated to customer responsiveness and continuous improvement in the quick development of innovative products and services that at once combine exceptional quality, fast and on-time delivery, and low prices and costs. In the pursuit of developing such an organizational system, there has been an onslaught of recommended management approaches, including Total Quality Management (TQM), Continuous Quality Improvement (CQI), Business Process Engineering (BPR), Business Process Improvement (BPI), and, most recently, Six Sigma (6σ). In addition, the work of numerous individuals has helped shape

Escalating health care costs and growing demand for better patient service has pushed the health care industry to rapidly embrace quality management techniques. Improvements in timeliness and accuracy in lab test results within a hospital is a common focus of health care improvement teams, as Case 12.1 reveals.

CHAPTER OUTLINE

12.1 Statistical Process Control

12.2 Variable Control Charts

12.3 Attribute Control Charts

contemporary quality thinking. These individuals include W. Edwards Deming, Joseph Juran, Armand Feigenbaum, Kaoru Ishikawa, and Genichi Taguchi.[1]

Since no approach or philosophy is one-size-fits-all, organizations are learning to develop their own personalized versions of a quality management system that integrates the aspects of these approaches and philosophies that best suit the challenges of their competitive environments. However, in the end, it is universally accepted that any effective quality management approach must integrate certain basic themes. Four themes are particularly embraced:

- The modern approach to management views work as a process.
- The key to maintaining and improving quality is the systematic use of data in place of intuition or anecdotes.
- It is important to recognize that variation is present in all processes and the goal of an organization should be to understand and respond wisely to variation.
- The tools of process improvement—including the use of statistics and teams—are most effective if the organization culture is supportive and is oriented toward continuously pleasing customers.

The idea of work as a process noted above is fundamental to modern approaches to quality, and even management in general. A process can be simply defined as a collection of activities that are intended to achieve some result. Specific business examples of processes include manufacturing a part to a desired dimension, billing a customer, treating a patient, and delivering products to customers. Manufacturing and service organizations alike have processes. The challenge for organizations is to identify key processes to improve. Key processes are the processes that have significant impact on customers and, more generally, on organizational performance.

To know how a process is performing and whether attempts to improve the process have been successful requires data. Process improvement usually cannot be achieved by armchair reasoning or intuition. To emphasize the importance of data, quality professionals often state, "You can't improve what you can't measure" or "In God we trust; all others, bring data." Examples of process data measures might include

- Average number of days of sales outstanding (Finance/Accounting)
- Time needed to hire new employees (Human Resources)
- Number of on-the-job accidents (Safety)
- Time needed to design a new product or service (Product/Service Design)
- Dimensions of a manufactured part (Manufacturing)
- Time to generate sales invoices (Sales and Marketing)
- Time to ship a product to a customer (Shipping)
- Percent of abandoned calls (Call Center)
- Downtime of a network (Information Technology)
- Waiting times for patients in a hospital clinic (Customer Service)

Our focus has been on processes common within an organization. However, the notion of a process is universal. For instance, we apply the ideas of a process to personal applications such as cooking, playing golf, or controlling one's weight. Or we may consider broader processes such as air pollution levels or crime rates. One of the great contributions of the quality revolution is the recognition that any process can be improved.

Systematic approach to process improvement

Management by intuition, slogans, or exhortation does not provide an environment or strategy conducive to process improvement. One of the key lessons of quality management is that process improvement should be based on an approach that is systematic, scientific, and fact (data) based.

The systematic steps of process improvement involve identifying the key processes to improve, process understanding/description, root cause analysis, assessment of attempted improvement efforts, and implementation of successful improvements. The systematic steps for process improvement are captured in the Plan-Do-Check-Act (PDCA) cycle. This cycle calls for one to *plan* the intended work, which includes identifying the process to improve, describing the current process, and coming up with solutions for improving the process. One must then *do,* which involves the implementation of the solution or change to the process; typically, improvements are first made on a small scale so as not to disrupt the routine activities of the organization. Next, one must *check* to see if the improvement efforts have indeed been successful. Finally, if the process improvement efforts were successful, the organization will *act* by implementing the changes as part of the organization's routine activities. Completion of these general steps represents one PDCA cycle. By continually initiating the PDCA cycle, continuous process improvement is accomplished, as depicted in Figure 12.1.

Advocates of the Six-Sigma approach emphasize that the Six-Sigma improvement model distinguishes itself from other process improvement models in that it calls for projects to be selected only if they are clearly linked to business priorities. This means that not only must projects be linked to customers' needs but they must have a significant financial impact seen in the bottom line. Organizations pursuing process improvement as part of a Six-Sigma effort use a tailored version of the generic PDCA improvement model known as Define-Measure-Analyze-Improve-Control (DMAIC). In the *Define* phase, the goal is to clearly identify an improvement opportunity in measurable terms and establish project goals. In the *Measure* phase, data are gathered to establish the current process performance. In the *Analyze* phase, efforts are made to find the sources (root causes) of less-than-desirable process performance. In many applications, root cause analysis relies on performing appropriately designed experiments and analyzing the resulting data using statistical techniques such as analysis of variance (Chapter 14) and multiple regression (Chapter 11). In the *Improve* phase, solutions are developed and implemented to attack the root causes. Finally, in the *Control* phase, process improvements are institutionalized, and procedures and methods are put into place to hold the process in control so as to

FIGURE 12.1 The PDCA cycle.

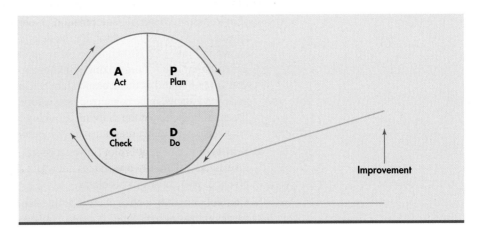

maintain the gains from the improvement efforts. One of the most common statistical tools used in the Control phase is the control chart, which is the focus of this chapter.

Process improvement toolkit

Each of the steps of the PDCA and DMAIC improvement models can potentially make use of a variety of tools. The quality literature is rich with examples of tools useful for process improvement. Indeed, a number of statistical tools that we have already introduced in earlier chapters frequently play a key role in process improvement efforts. Here are some basic tools (statistical and nonstatistical) frequently used for process improvement efforts:

1. **Flowchart.** A flowchart is a picture of the stages of a process. Many organizations have formal standards for making flowcharts. A flowchart can often jump-start the process improvement effort by exposing unexpected complexities (for example, unnecessary loops) or non-value-added activities (for example, waiting points that increase overall cycle time). Figure 12.2(a) is a flowchart showing the steps of an order fulfillment process for an electronic order from a customer.

2. **Run chart.** A run chart is what quality professionals call a time plot (time plots were introduced in Chapter 1). A run chart allows one to observe the performance of a process over time. For example, Motorola's service centers calculate mean response times each month and depict overall performance with a run chart.

3. **Histogram.** Every process is subject to variability. The histogram (Chapter 1) is useful in process improvement efforts because it allows the practitioner to visualize the process behavior in terms of location, variability, and distribution. As we will see in Section 12.2, histograms with superimposed product specification limits can be used to display process "capability."

4. **Pareto chart.** A Pareto chart (Chapter 1) is a bar graph with the bars ordered by height. Pareto charts help focus process improvement efforts on issues of greatest impact ("vital few") as opposed to the less important issues ("trivial many").

5. **Cause-and-effect diagram.** A cause-and-effect diagram is a simple visual tool used by quality improvement teams to show the possible causes of the quality problem under study. Figure 12.2(b) is a cause-and-effect diagram of the process of converting metal billets (ingots) into a forged item.[2] Here the ultimate "effect" is a good forged item. Notice that the main branches (Environment, Material, Equipment, Personnel, Methods) organize the causes and serve as a skeleton for the detailed entries. The main branches shown in Figure 12.2(b) apply to many applications and can serve as a general template for organizing thinking about possible causes. Of course, you are not bound to these branch labels. Once a list of possible causes is generated, they can be organized into natural main groupings that represent the main branches of the diagram. Looking at Figure 12.2(b) you can see why cause-and-effect diagrams are sometimes called fishbone diagrams.

6. **Scatterplot.** The scatterplot (introduced in Chapter 2) can be used to investigate whether two variables are related, which might help in identifying potential root causes of problems.

7. **Control chart.** A control chart is a time-sequenced plot used to study how a process changes over time. A control chart is more than a run chart in that

FIGURE 12.2 Examples of nonstatistical process improvement tools. (a) Flowchart of an ordering process for an electronic order; (b) Cause-and-effect diagram of hypothesized causes related to the making of a good forged item.

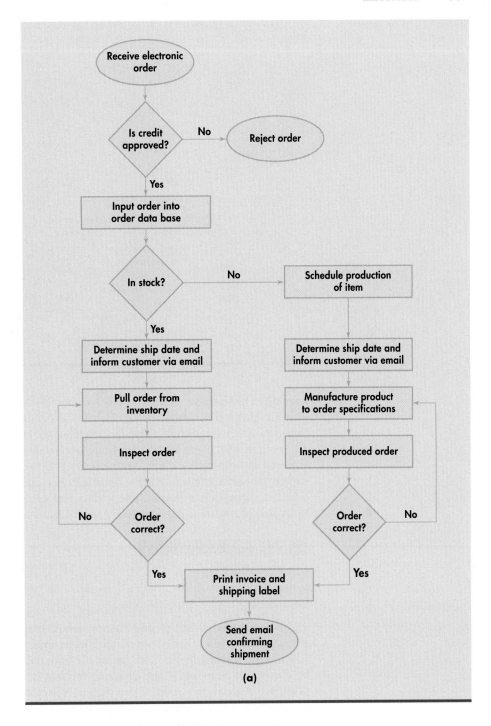

(a)

control limits and a line denoting the average are superimposed on the plot. The control limits help practitioners determine if the process is consistent with past behavior or if there is evidence that the process has changed in some way. This chapter is largely devoted to the control chart technique.

Beyond the application of simple tools, there is an increasing use of more sophisticated statistical tools in the pursuit of quality. For example, the design of a new product as

FIGURE 12.2 (*Continued*)

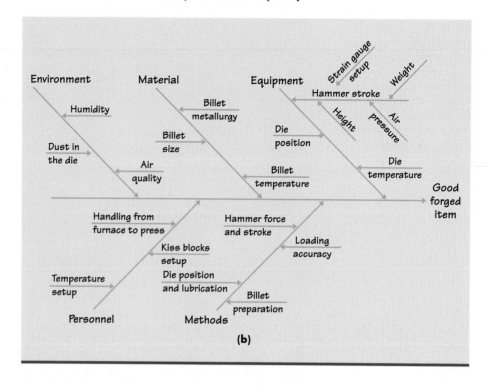

(b)

simple as a multivitamin tablet may involve interviewing samples of consumers to learn what vitamins and minerals they want included and using randomized comparative experiments to design the manufacturing process (Chapter 3). An experiment might discover, for example, what combination of moisture level in the raw vitamin powder and pressure in the tablet-forming press produces the right tablet hardness. In general, well-designed experiments reduce ambiguity about cause and effect and allow practitioners to determine what factors truly affect the quality of products and services. Let us now turn our attention to the area of *statistical process control* and its distinctive tool—the control chart.

APPLY YOUR KNOWLEDGE

12.1 Describe a process. Consider the process of going from curbside to sitting in your assigned airplane seat. Make a flowchart of the process. Do not forget to consider steps that involve Yes/No outcomes.

12.2 Operational definition and measurement. If asked to measure the percent of late departures of an airline, you are faced with an unclear task. Is late departure defined in terms of "leaving the gate" or "taking off from the runway"? What is required is an operational definition of the measurement, that is, an unambiguous definition of what is to be measured so that if you were to collect the data and someone else were to collect the data, both of you would come back with the same measurement values. Provide an example of an operational definition for the following:

(a) Reliable mobile provider.
(b) Clean desk.
(c) Effective teacher.

12.3 Causes of variation. Consider the process of uploading a photo to a Facebook account. Brainstorm at least five possible causes for variation in upload time. Construct a cause-and-effect diagram based on your identified potential causes.

12.1 Statistical Process Control

The goal of statistical process control is to make a process stable over time and then keep it stable unless planned changes are made. You might want, for example, to keep your weight constant over time. A manufacturer of machine parts wants the critical dimensions to be the same for all parts. "Constant over time" and "the same for all" are not realistic requirements. They ignore the fact that *all processes have variation.* Your weight fluctuates from day to day, the critical dimension of a machined part varies a bit from item to item, the time to process a college admission application is not the same for all applications. Variation occurs in even the most precisely made product due to small changes in the raw material, the adjustment of the machine, the behavior of the operator, and even the temperature in the plant. Because variation is always present, we can't expect to hold a variable exactly constant over time. The statistical description of stability over time requires that the *pattern of variation* remain stable, not that there be no variation in the variable measured.

Statistical Control

A process that continues to be described by the same distribution when observed over time is said to be in statistical control, or simply **in control.**

    Control charts are statistical tools that monitor a process and alert us when the process has been changed so that it is now **out of control.** This is a signal to find and respond to the cause of the change.

---

common cause   In the language of statistical quality control, a process that is in control has only common cause variation. **Common cause** variation is the inherent variability of the system, due to many small causes that are always present. Since it is assumed that these many underlying small causes result in small *random* perturbations to which all process outcomes are exposed, their cumulative effect is by definition assumed to be random by nature. Thus, an in-control process is a random process that generates random or independent process outcomes over time.

special cause   When the normal functioning of the process has changed, we say that **special cause** variation is added to the common cause variation. A special cause can be viewed as any factor impinging on the process and resulting in variation not consistent with common cause variation. In contrast to common causes, special causes can often be traced to some clear and identifiable event. Examples might include an operator error, a jammed machine, or a bad batch of raw material. These are classic manufacturing examples where the special cause variation has negative implications on the process. In particular, when dealing with manufacturing processes where the goal is to produce parts as close to targets or specifications as possible, any added variation is undesirable. In such situations, we hope to be able to discover what lies behind special cause variation and eliminate that cause to restore the stable functioning of the process.

    Historically, statistical process control (SPC) methods were devised for the monitoring of manufactured parts with the intention of detecting unwanted special cause variation. However, one of the great contributions of the quality revolution is the recognition that any process, not just classical manufacturing processes, has the potential to be improved. In the business arena, SPC methods are routinely used for monitoring services processes, for example, patient waiting time in a hospital clinic. These same methods, however, can be used to monitor the ratings of a television show, daily stock returns, the level of ozone in the atmosphere, or even your golf scores. With this broader perspective, process change due to a special cause might be viewed favorably: for example, a decrease in waiting times or an increase in monthly customer satisfaction ratings. In such situations, our intention should not be to eliminate the special cause but rather to learn about the special cause and promote its effects.

EXAMPLE 12.1 Common Cause, Special Cause

Imagine yourself doing the same task repeatedly, say folding an advertising flyer, stuffing it into an envelope, and sealing the envelope. The time to complete the task will vary a bit, and it is hard to point to any one reason for the variation. Your completion time shows only common cause variation.

Now the telephone rings. You answer, and though you continue folding and stuffing while talking, your completion time rises beyond the level expected from common causes alone. Answering the telephone adds special cause variation to the common cause variation that is always present. The process has been disturbed and is no longer in its normal and stable state.

If you are paying temporary employees to fold and stuff advertising flyers, you avoid this special cause by not having telephones present and by asking your employees to turn off their cell phones while they are working.

The idea underlying control charts is simple but ingenious.* By setting limits on the natural variability of a process, control charts work by distinguishing the always-present common cause variation in a process from the additional variation that suggests that the process has been changed by a special cause. When a control chart indicates process change, it is a signal to respond, which often entails taking corrective action. On the flip side, when a control chart indicates that there has been no process change, the chart still serves a purpose: it restrains the user from taking unnecessary actions. All too often, time and resources are wasted by misinterpreting common cause variation as special cause variation. When a control chart is not signaling, the best management practice is one of no action.†

A wide variety of control charts are available to quality practitioners. Control charts can be broadly classified based on the type of data collection.

> **Types of Control Charts**
>
> **Variable control charts** are control charts devised for monitoring measurements, such as weights, time, temperature, or dimensions. Variable control charts include charts for monitoring the mean of the process and charts for monitoring the variability of the process.
>
> **Attribute control charts** are control charts for monitoring counting data. Examples of counting data are number (or proportion) of defective items in a production run, number of invoice errors, or number of complaining customers per month. Section 12.3 discusses two of the most common attribute charts: the p chart and the c chart.

APPLY YOUR KNOWLEDGE

12.4 Special causes. Jeannine participates in bicycle road races. She regularly rides 25 kilometers over the same course in training. Her time varies a bit from day to day but is generally stable. Give several examples of special causes that might raise or lower Jeannine's time on a particular day.

---

*Control charts were invented in the 1920s by Walter Shewhart at the Bell Telephone Laboratories. Shewhart's classic book *Economic Control of Quality of Manufactured Product* (Van Nostrand, 1931) organized the application of statistics to improving quality.

†In his classic book *Out of the Crisis* (MIT Center for Advanced Engineering Study, 1986), W. Edwards Deming demonstrates the effects of counterproductive adjustment to an in-control process by means of a physical experiment based on dropping marbles through a funnel onto a tabletop. Participants in the experiment learn that the least scatter on the tabletop is obtained by not moving the funnel, that is, by means of "no action."

12.5 Common causes and special causes. In Exercise 12.1, you described the process of getting on an airplane. What are some sources of common cause variation in this process? What are some special causes that can result in out-of-control variation?

SECTION 12.1 Summary

- Work is organized in **processes,** chains of activities that lead to some result. We use **flowcharts** and **cause-and-effect diagrams** to describe processes. **Pareto charts** and **scatterplots** can be useful in isolating primary root causes for quality problems.

- All processes have variation. Variation due to common causes, called **common cause** variation, reflects the natural variation inherent in every process. A process exhibiting only common cause variation is said to be **in control.** Special cause variation is variation inconsistent with common cause variation. Processes influenced by special cause variation are **out of control.**

- **Control charts** are statistical devices indicating when the process is in control or when it is affected by special cause variation. **Variable control charts** are used for monitoring measurements taken on some continuous scale. **Attribute control charts** are used for monitoring counting data.

SECTION 12.1 Exercises

For Exercises 12.1 to 12.3, see page 638; and for 12.4 and 12.5, see pages 640–641.

12.6 Which type of control chart? For each of the following process outcomes indicate if a variable control chart or an attribute control chart is most applicable:
(a) Number of lost-baggage claims per day.
(b) Time to respond to a field service call.
(c) Thickness (in millimeters) of cold-rolled steel plates.
(d) Percent of late shipments per week.

12.7 Describe a process. Each weekday morning, you must get to work or to your first class on time. Make a flowchart of your daily process for doing this, starting when you wake. Be sure to include the time at which you plan to start each step.

12.8 Common cause, special cause. Each weekday morning, you must get to work or to your first class on time. The time at which you reach work or class varies from day to day, and your planning must allow for this variation. List several common causes of variation in your arrival time. Then list several special causes that might result in unusual variation, such as being late to work or class.

12.9 Pareto charts. Continue the study in the previous two exercises of the process of getting to work or class on time. If you kept good records, you could make a Pareto chart of the reasons (special causes) for late arrivals at work or class. Make a Pareto chart that you think roughly describes your own reasons for lateness. That is, list the reasons from your experience, and chart your estimates of the percent of late arrivals each reason explains.

12.10 Pareto charts. Painting new auto bodies is a multistep process. There is an "electrocoat" that resists corrosion, a primer, a color coat, and a gloss coat. A quality study for one paint shop produced this breakdown of the primary problem type for those autos whose paint did not meet the manufacturer's standards:

Problem	Percent
Electrocoat uneven—redone	4
Poor adherence of color to primer	5
Lack of clarity in color	2
"Orange peel" texture in color	32
"Orange peel" texture in gloss	1
Ripples in color coat	28
Ripples in gloss coat	4
Uneven color thickness	19
Uneven gloss thickness	5
Total	100

Make a Pareto chart. Which stage of the painting process should we look at first?

12.2 Variable Control Charts

This section considers the scenario in which regular samples on measurement data are obtained to monitor process behavior. In the quality area, samples of observations are often referred to as *subgroups*. For each subgroup, pertinent statistics are computed and then charted over time. For example, the subgroup means can be plotted over time to control the overall mean level of the process, while process variability might be controlled by plotting subgroup standard deviations or a more simplistic statistic known as the range statistic.

The effectiveness of a control chart depends on how the subgroups were collected. Three basic issues need to be addressed:

1. **Rational subgrouping.** Walter Shewhart, the founder of statistical process control, conceptualized a basis for forming subgroups. He suggested that subgroups should be chosen in such a way that the individual observations within the subgroups have been measured under similar process conditions. Subgroups formed on this principle are known as *rational subgroups*. The idea is that if the individual observations within the subgroups are as homogeneous as possible, then any special causes disrupting the process will be reflected by greater variability between the subgroups. Thus, when special causes are present, rational subgrouping attempts to maximize the likelihood that subgroup statistics will signal that the process is out of control. In manufacturing settings, the most common way to create rational subgroups is to take individual measurements over a short period of time; often this means measuring consecutive items produced.

2. **Subgroup size.** Subgroup sizes are usually small. Sampling cost is one important consideration. Another is that large samples may span too much time, making it possible for the process to change while the sample is being taken. When sample sizes are large, the subgroups are at risk of not being rational subgroups. Subgroup sizes (n) for variable control charts nearly always range from 1 to 25. For various historical reasons, sample sizes of 4 or 5 are among the most common choices in practice. As we will see, control charts rely on approximate Normality of the subgroup statistics. It is fortunate that for certain statistics, like the mean, the central limit theorem effect will provide approximate Normality for sample sizes as small as 4 or 5.

3. **Sampling frequency.** A final subgroup design issue is the frequency of sampling, that is, the timing between subgroups. Cost factors obviously come to bear. There are not only the costs of sampling (for example, costs of testing and measuring) but also the costs associated with missing significant process changes. If sampling is done infrequently, then there is a risk that the process is out of control between subgroups, resulting in variety of costs ranging from higher rework to customer dissatisfaction for receiving unacceptable product or service. Process stability (or the lack thereof) is another consideration. A process that is erratic needs frequent surveillance but a process that has achieved stability can be less frequently sampled. A common strategy is to start with frequent sampling of the process and then to relax the frequency of sampling as one gains confidence about the stability of the process.

Once the sampling scheme has been designed in terms of establishing rational subgroups, sample size, and sampling frequency, data are collected to form preliminary subgroups. When first applying control charts to a process, the process behavior is not fully understood. By using a control chart to analyze a given set of initial data, we are looking back *retrospectively* on the performance of the process. In this phase, control charts are said to be used *as a judgment*.

If the process is found to be in control, then the control chart can be used *prospectively* to monitor future process performance in real time. In this phase, control charts are said to be used *as an ongoing operation*. If retrospective analysis shows the process to be not in control, then any numerical descriptions of the data used to construct the initial control chart will not serve as reliable guides for monitoring the process into the future. The priority is then to bring the process into control. This may mean uncovering and removing the unwanted effects of special causes. Or it may mean incorporating the favorable effects of the special causes and stabilizing the process at a more favorable position. Later, when the process has been operating in control for some time and you understand its usual behavior, control charts can be used prospectively.

$\bar{x}$ and R charts

We begin with a quantitative variable x that is an important measure of quality. The variable might be the diameter of a part or the time to respond to a customer call. Given this quality characteristic is subject to variation, there is a distribution underlying the process. As discussed earlier, an in-control process is a stable process whose underlying distribution remains the same over time. Associated with this distribution is a **process mean** μ and a **process standard deviation** σ.

To make subgroup control charts, we begin by taking samples of size n from the process at regular intervals. For instance, we might measure 4 or 5 consecutive parts or time the responses to 4 or 5 consecutive customer calls. For each subgroup, we can compute the sample mean $\bar{x}$. From Chapter 4, we learned that $\bar{x}$ is a random variable with mean μ and standard deviation $\sigma/\sqrt{n}$. Furthermore, the central limit theorem tells us that the sampling distribution of the sample mean is approximately Normal.

The 99.7 part of the 68–95–99.7 rule for Normal distributions says that, as long as the process remains in control, 99.7% of the values of $\bar{x}$ will fall between:

$$\text{UCL} = \mu + 3\frac{\sigma}{\sqrt{n}}$$

$$\text{LCL} = \mu - 3\frac{\sigma}{\sqrt{n}}$$

three-sigma control limits

where UCL and LCL stand for "upper control limit" and "lower control limit," respectively. These limits are called *three-sigma control limits* and serve as the basis for most control charts. Along with the control limits, it is standard practice to draw a center line (CL) at the mean.

With three-sigma limits, if the process remains in control and the process mean and standard deviation do not change, we will rarely observe an $\bar{x}$ outside the control limits: only about 3 out of every 1000 sample means, given the assumption of Normality. From that perspective, if we do observe an $\bar{x}$ outside the control limits, it serves as a strong signal that the underlying conditions of the process may have changed.

In practice, μ and σ are typically not known and must be estimated. One obvious estimate for the mean is the average of all the individual observations taken from all the preliminary subgroups. If the sample sizes are all equal, then the average of all the individual observations can be computed as the average of the subgroup means. In particular, if we are basing the construction of the control chart on k subgroups, then the

grand mean

overall average (referred to as the *grand mean*) is computed as

$$\bar{\bar{x}} = \frac{1}{k}(\bar{x}_1 + \bar{x}_2 + \cdots + \bar{x}_k)$$

We now need an estimate of σ. One possibility is to use the subgroup sample standard deviations. However, in practice, it is more common to use a more simplistic estimate of

TABLE 12.1   Control chart constants

Subgroup size n	D_3	D_4	B_3	B_4	A_2	A_3	c_4
2	0.000	3.267	0.000	3.267	1.881	2.659	0.7979
3	0.000	2.574	0.000	2.568	1.023	1.954	0.8862
4	0.000	2.282	0.000	2.266	0.729	1.628	0.9213
5	0.000	2.114	0.000	2.089	0.577	1.427	0.9400
6	0.000	2.004	0.030	1.970	0.483	1.287	0.9515
7	0.076	1.924	0.118	1.882	0.419	1.182	0.9594
8	0.136	1.864	0.185	1.815	0.373	1.099	0.9650
9	0.184	1.816	0.239	1.761	0.337	1.032	0.9693
10	0.223	1.777	0.284	1.716	0.308	0.975	0.9727
11	0.256	1.744	0.321	1.679	0.285	0.927	0.9754
12	0.283	1.717	0.354	1.646	0.266	0.886	0.9776
13	0.307	1.693	0.382	1.618	0.249	0.850	0.9794
14	0.328	1.672	0.406	1.594	0.235	0.817	0.9810
15	0.347	1.653	0.428	1.572	0.223	0.789	0.9823
16	0.363	1.637	0.448	1.552	0.212	0.763	0.9835
17	0.378	1.622	0.466	1.534	0.203	0.739	0.9845
18	0.391	1.609	0.482	1.518	0.194	0.718	0.9854
19	0.404	1.597	0.497	1.503	0.187	0.698	0.9862
20	0.415	1.585	0.510	1.490	0.180	0.680	0.9869
21	0.425	1.575	0.523	1.477	0.173	0.663	0.9876
22	0.435	1.566	0.534	1.466	0.168	0.647	0.9882
23	0.443	1.557	0.545	1.455	0.162	0.633	0.9887
24	0.452	1.548	0.555	1.445	0.157	0.619	0.9892
25	0.459	1.541	0.565	1.435	0.153	0.606	0.9896

range statistic R variation based on the **range statistic R.** The range statistic is simply the sample range, which is the difference between the largest and the smallest observations in a sample. With k preliminary subgroups, the average range $\overline{R}$ can be computed as

$$\overline{R} = \frac{1}{k}(R_1 + R_2 + \cdots + R_k)$$

Based on statistical theory, it can be shown that if $\overline{R}$ is multiplied by a constant (A_2) that is a function of the subgroup size n, we then have a reasonable estimate for $3\sigma/\sqrt{n}$. Table 12.1 provides values for A_2 for various subgroup sizes.

With the mean and variability estimates, the estimated center line and control limits for the mean are given by

$$\text{UCL} = \overline{\overline{x}} + A_2\overline{R}$$

$$\text{CL} = \overline{\overline{x}}$$

$$\text{LCL} = \overline{\overline{x}} - A_2\overline{R}$$

$\overline{x}$ chart The control chart for the mean, called an **$\overline{x}$ chart**, is dedicated to detecting changes in the process level. It is also important to monitor the variability of the process. Because we have the subgroup ranges in hand, it is sensible to develop a control chart for ranges.

R chart Such a chart is known as an **R chart.** The center line and control limits of the R chart are given by

$$\text{UCL} = D_4\overline{R}$$

$$\text{CL} = \overline{R}$$

$$\text{LCL} = D_3\overline{R}$$

The control chart constants D_3 and D_4 are provided in Table 12.1. On the surface, the R chart does not appear to have the same format as the $\overline{x}$-chart in the sense of establishing control limits a certain amount above and below the center line. However, underlying the development of the control chart constants D_3 and D_4 is a three-sigma structure for the range statistic. You will notice that for subgroup sizes of 2 to 6, the D_3 factor is 0, which implies a lower control limit of 0. For small subgroup sizes, it can be shown that the theoretical control limits placed plus/minus three standard deviations of the range statistic above or below the range mean will result in a negative lower control limit. Since the range statistic can never be negative, the lower control limit is accordingly set to 0. Here is a summary of our discussion.

Construction of $\overline{x}$ and R Charts

Take regular samples of size n from a process. The center line and control limits for an $\overline{x}$ **chart** are

$$\text{UCL} = \overline{\overline{x}} + A_2\overline{R}$$

$$\text{CL} = \overline{\overline{x}}$$

$$\text{LCL} = \overline{\overline{x}} - A_2\overline{R}$$

where $\overline{\overline{x}}$ is the average of the subgroup means and $\overline{R}$ is the average of the subgroup ranges. The center line and control limits for an **R chart** are

$$\text{UCL} = D_4\overline{R}$$

$$\text{CL} = \overline{R}$$

$$\text{LCL} = D_3\overline{R}$$

The **control chart constants** A_2, D_3, and D_4 depend on the sample size n.

CASE 12.1

A Health Care Application: Getting Lab Results With escalating health care costs and continual demand for better patient service and patient outcomes, the health care industry is rapidly embracing the use of quality management techniques. Improving the timeliness and accuracy of lab results within a hospital is a common focus of health care improvement teams. Consider the case of a hospital looking at the time from request to receipt of blood tests for the emergency room (ER). One of the most commonly requested tests from the ER is a complete blood count (CBC). A CBC provides doctors with red and white cell counts and blood-clotting measures, all of which can be crucial in an emergency situation. Because of the importance of the turnaround time for CBC requests, hospital management appointed a quality improvement team to study the process. The team selected random samples of 5 CBC requests per shift (day, evening, late night) over the course of 10 days. Thus, the team had $3 \times 10 = 30$ preliminary subgroups. The sampling within shifts associates the individual observations with similar conditions (for example, staffing) and thus abides by the rational subgrouping principle. For each of the CBC requests sampled, the turnaround time (minutes) was recorded. Table 12.2 provides the observation values along with the subgroup means and ranges.

DATA FILE

LABTESTS

TABLE 12.2 Thirty control chart subgroups of lab test turnaround times (in minutes)

Subgroup	Turnaround observations					Subgroup mean	Range
1	39	33	65	50	41	45.6	32
2	46	36	34	53	37	41.2	19
3	37	35	28	37	41	35.6	13
4	50	38	35	60	39	44.4	25
5	29	27	22	43	50	34.2	28
6	32	35	40	27	42	35.2	15
7	42	43	37	44	39	41.0	7
8	40	45	50	43	24	40.4	26
9	34	47	54	39	51	45.0	20
10	43	65	25	45	25	40.6	40
11	35	48	44	45	34	41.2	14
12	55	54	44	36	55	48.8	19
13	29	39	47	42	47	40.8	18
14	41	31	29	37	27	33.0	14
15	41	40	32	33	52	39.6	20
16	32	32	41	47	43	39.0	15
17	24	54	34	53	56	44.2	32
18	36	45	53	31	31	39.2	22
19	48	57	36	31	30	40.4	27
20	38	27	39	35	27	33.2	12
21	53	33	51	50	42	45.8	20
22	53	45	37	44	33	42.4	20
23	27	50	35	29	47	37.6	23
24	39	39	51	49	44	44.4	12
25	33	29	38	68	34	40.4	39
26	34	43	48	49	56	46.0	22
27	43	20	51	49	50	42.6	31
28	33	42	51	58	26	42.0	32
29	25	46	25	43	42	36.2	21
30	30	34	42	36	51	38.6	21

Before we proceed to the construction of control charts for Case 12.1, we must mention a subtle implementation issue with respect to the $\bar{x}$ and R charts. The $\bar{x}$ chart limits rely on the average range $\bar{R}$. If process variability is not stable and is affected by special causes, then $\bar{R}$ is not a reliable estimate of variability, and thus the $\bar{x}$ chart limits are less meaningful. There is a lesson here: *it is difficult to interpret an $\bar{x}$ chart unless the ranges are in control.* When you look at $\bar{x}$ and R charts, always start with the R chart.

EXAMPLE 12.2 Constructing $\bar{x}$ and R Charts

CASE 12.1   We begin the analysis of the turnaround times for the lab-testing process by focusing on the process variability. We use the 30 ranges shown in Table 12.2 to find the average range:

$$\bar{R} = \frac{1}{30}(32 + 19 + \cdots + 21)$$

$$= \frac{659}{30} = 21.967$$

From Table 12.1 (page 644), for subgroup size $n = 5$, the values of D_3 and D_4 are 0 and 2.114, respectively. Accordingly, the center line and control limits for the R chart are

$$\text{UCL} = D_4\overline{R} = 2.114(21.967) = 46.438$$
$$\text{CL} = \overline{R} = 21.967$$
$$\text{LCL} = D_3\overline{R} = 0(21.967) = 0$$

Figure 12.3 shows the R chart for the lab-testing process. The R chart shows no points outside the upper control limit. Furthermore, the ranges plotted over time show no unusual pattern. We can say that from the perspective of process variation, the process is in control.

FIGURE 12.3 R chart for the lab-testing data of Table 12.2.

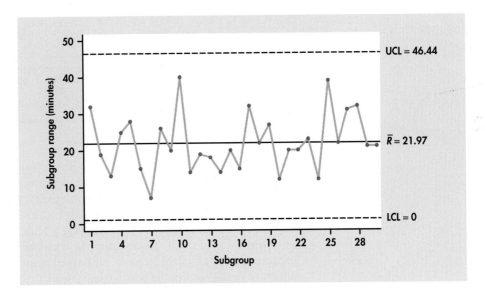

We now construct the $\overline{x}$ chart. Since the R chart exhibited in-control behavior, we can safely use the value of 21.967 computed earlier for $\overline{R}$ in the computation of the $\overline{x}$ chart limits. Referring to Table 12.2, we can find the grand mean:

$$\overline{\overline{x}} = \frac{1}{30}(45.6 + 41.2 + \cdots + 38.6)$$
$$= \frac{1218.6}{30} = 40.62$$

From Table 12.1 (page 644), we find $A_2 = 0.577$. The center line and control limits for the $\overline{x}$ chart are then as follows:

$$\text{UCL} = \overline{\overline{x}} + A_2\overline{R} = 40.62 + 0.577(21.967) = 53.29$$
$$\text{CL} = \overline{\overline{x}} = 40.62$$
$$\text{LCL} = \overline{\overline{x}} - A_2\overline{R} = 40.62 - 0.577(21.967) = 27.95$$

Figure 12.4 shows the $\overline{x}$ chart. The subgroup means of the 30 samples do vary, but all lie within the range of variation marked out by the control limits. We are seeing the common cause variation of a stable process with no indications of special causes.

Example 12.2 shows that the lab-testing process for the ER is stable and in control. Does this mean the process is *acceptable*? This is not a statistical question but rather a managerial question for the ER and the hospital administration. Currently, the mean

FIGURE 12.4 $\bar{x}$ chart for the lab-testing data of Table 12.2.

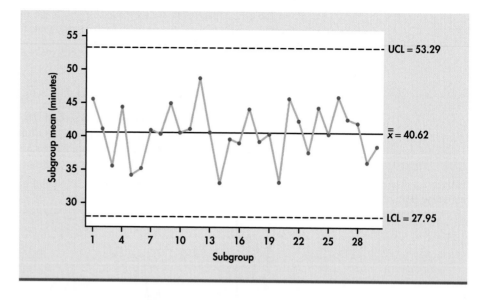

turnaround time is estimated to be 40.62 minutes. The process is stable and in control around that estimated mean. If this process performance is considered acceptable, then the control chart limits for both the $\bar{x}$ chart and the R chart can be projected out into the future to monitor the process so as to maintain performance. If, however, the mean time of around 41 minutes is viewed as unacceptably high, then efforts need to be initiated to find ways to improve the process. Here is where the basic tools of flowcharts, Pareto charts, and cause-and-effect diagrams can be used by quality improvement teams to better understand the lab turnaround process and to search out the underlying causes for delay.

When efforts are made to improve a process, the control chart can play an important role. Control charts can be used to judge whether an attempt to improve a process has resulted in a successful change or not. Checking for successful change is a critical aspect of the "Check" phase of the PDCA cycle and the "Improve" phase of the DMAIC model. Figure 12.5(a) shows a case where the control chart demonstrates a successful attempt to decrease the turnaround time. In this case, control chart limits should be revised, and the new process should be monitored to maintain the gains, as called for in the "Control" phase of the DMAIC model. However, if sample means are as seen in Figure 12.5(b), then the control chart indicates no impact from the attempted process improvement; the organization should seek alternative improvement ideas.

The preliminary samples of Example 12.2 appear to come from an in-control process. Let us now consider control chart implementation issues when there is evidence of special cause effects in the preliminary samples.

CASE 12.2

O-Ring Diameters A manufacturer of synthetic-rubber O-rings must monitor and control their dimensions. O-rings are used in numerous industries, including medical, aerospace, oil refining, automotive, and chemical processing. O-rings are doughnut-shaped gaskets used to seal joints against high pressure from gases or fluids. The two primary dimensions that need to be controlled are the cross-sectional width of the doughnut ring and the inside diameter of the inner circle of the doughnut. Within the O-ring product family, the manufacturer produces an aerospace industry class of O-rings known as AS568A. Table 12.3 gives the observations for 25 preliminary subgroups of size 4 for the inside diameter of AS568A-146 O-rings along with subgroup means and ranges. This O-ring is specified to have an inside diameter of 2.612 inches. The tolerances are set at ±0.02 inch around this specification.

ORING

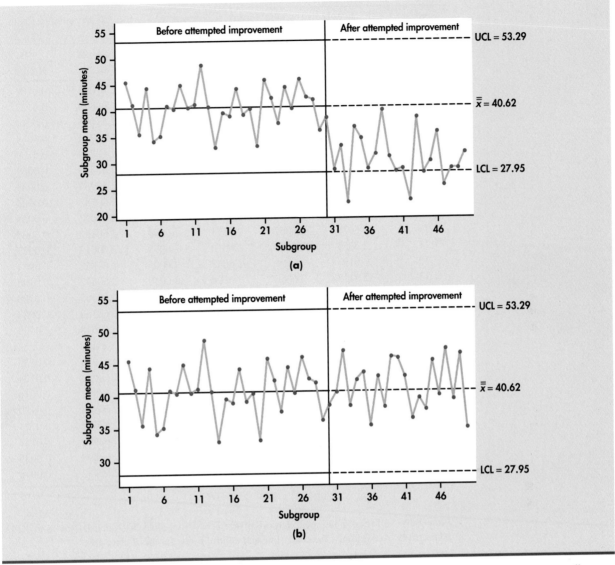

FIGURE 12.5 Using control charts to assess improvement efforts. (a) The improvement effort was successful. (b) The improvement effort was unsuccessful.

EXAMPLE 12.3 $\bar{x}$ and R Charts and Out-of-Control Signals

CASE 12.2 Our initial step is to study the variability of the O-ring process. Using the 25 ranges shown in Table 12.3, we find the average range to be

$$\bar{R} = \frac{1}{25}(0.0108 + 0.0127 + \cdots + 0.0078)$$
$$= \frac{0.2381}{25} = 0.009524$$

TABLE 12.3 Twenty-five control chart subgroups of O-ring measurements (in inches)

Subgroup	O-ring measurements				Sample mean	Range
1	2.6088	2.6120	2.6167	2.6059	2.61085	0.0108
2	2.5993	2.6120	2.6089	2.6046	2.60620	0.0127
3	2.6117	2.6074	2.6118	2.6101	2.61025	0.0044
4	2.6063	2.6055	2.6119	2.6076	2.60783	0.0064
5	2.6139	2.6030	2.6038	2.6097	2.60760	0.0109
6	2.6019	2.6075	2.6086	2.6076	2.60640	0.0067
7	2.6045	2.6005	2.5980	2.5964	2.59985	0.0081
8	2.6114	2.6050	2.6063	2.6086	2.60783	0.0064
9	2.6091	2.6100	2.6146	2.6100	2.61093	0.0055
10	2.6078	2.6067	2.6111	2.6044	2.60750	0.0067
11	2.6055	2.6089	2.6010	2.6093	2.60617	0.0083
12	2.6107	2.6098	2.6043	2.6095	2.60858	0.0064
13	2.6155	2.6050	2.6094	2.6050	2.60872	0.0105
14	2.6068	2.6067	2.6075	2.5975	2.60463	0.0100
15	2.6054	2.6021	2.6103	2.6054	2.60580	0.0082
16	2.6068	2.6084	2.6103	2.6004	2.60648	0.0099
17	2.6061	2.6185	2.5953	2.6075	2.60685	0.0232
18	2.6185	2.6096	2.6077	2.6050	2.61020	0.0135
19	2.6072	2.6067	2.6121	2.6017	2.60693	0.0104
20	2.6091	2.6113	2.6037	2.6092	2.60832	0.0076
21	2.6054	2.6149	2.6114	2.6020	2.60843	0.0129
22	2.6074	2.6092	2.6113	2.5992	2.60677	0.0121
23	2.6034	2.5972	2.6124	2.6070	2.60500	0.0152
24	2.6107	2.6101	2.6079	2.6072	2.60898	0.0035
25	2.6021	2.6073	2.6044	2.5995	2.60332	0.0078

From Table 12.1 (page 644), for subgroups of size $n = 4$, the values of D_3 and D_4 are 0 and 2.282, respectively. Accordingly, the center line and control limits for the R chart are

$$\text{UCL} = D_4\overline{R} = 2.282(0.009524) = 0.021734$$

$$\text{CL} = \overline{R} = 0.009524$$

$$\text{LCL} = D_3\overline{R} = 0(0.009524) = 0$$

Figure 12.6 is the R chart for the O-ring process. Subgroup 17 lies outside the upper control limit. Had we constructed an $\overline{x}$ chart at this time, we would have found that Subgroup 17 would not have signaled out of control. It is not unusual for the R chart to signal out of control while the $\overline{x}$ chart does not, or vice versa. Each chart is looking for different departures. The R chart is looking for changes in variability, and the $\overline{x}$ chart is looking for changes in the process level. It is, of course, possible for both process variation and level to go out of control together, resulting in signals on both charts. At this stage, an explanation should be sought for the out-of-control signal. Suppose that an investigation reveals a machine problem at the time of the out-of-control signal. Since a special cause was discovered, the associated subgroup should be set aside and a new R chart constructed. By deleting Subgroup 17, the revised range estimate based on the remaining 24 subgroups is

$$\overline{R} = \frac{0.2149}{24} = 0.008954$$

FIGURE 12.6 *R* chart for the O-ring data of Table 12.3. Subgroup 17 signals out of control.

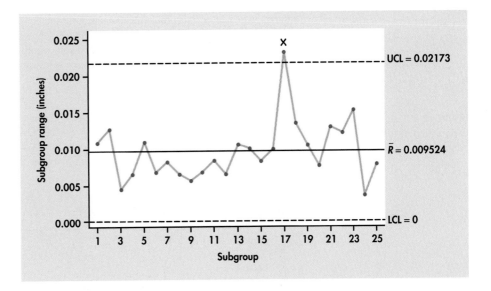

FIGURE 12.7 Updated *R* chart for the O-ring data of Table 12.3 with Subgroup 17 removed.

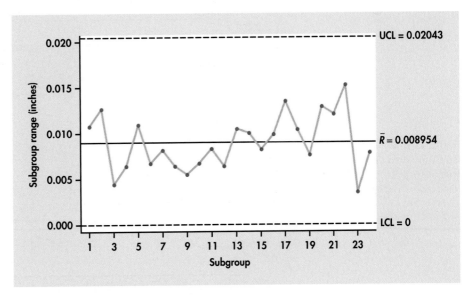

Figure 12.7 shows the updated *R* chart applied to the 24 subgroups. Now, all the subgroup ranges are found to be in control. We can now turn our attention to the construction of the $\bar{x}$ chart limits. For the 24 samples in Table 12.3, the grand mean is

$$\bar{\bar{x}} = \frac{1}{24}(2.61085 + 2.60620 + \cdots + 2.60332)$$

$$= \frac{62.5735}{24} = 2.60723$$

From Table 12.1 (page 644), we find $A_2 = 0.729$. The center line and control limits for the $\bar{x}$ chart are then

$$\text{UCL} = \bar{\bar{x}} + A_2\bar{R} = 2.60723 + 0.729(0.008954) = 2.61375$$

$$\text{CL} = \bar{\bar{x}} = 2.60723$$

$$\text{LCL} = \bar{\bar{x}} - A_2\bar{R} = 2.60723 - 0.729(0.008954) = 2.60070$$

Figure 12.8 shows the $\bar{x}$ chart. The $\bar{x}$ chart shows an out-of-control signal at Subgroup 7. A special cause investigation reveals that the abnormally smaller diameters associated with this subgroup were caused by a problem in the postcuring stage that resulted in too much shrinkage of the rubberized material. With an explanation in hand, Subgroup 7 needs to be discarded and the R chart limits need to be recomputed. Figure 12.9 displays both the $\bar{x}$ chart and the R chart based on the remaining 23 samples. The data for both charts appear well behaved. The two sets of control limits can be used for prospective control.

FIGURE 12.8 $\bar{x}$ chart for the O-ring data of Table 12.3 with Subgroup 17 removed.

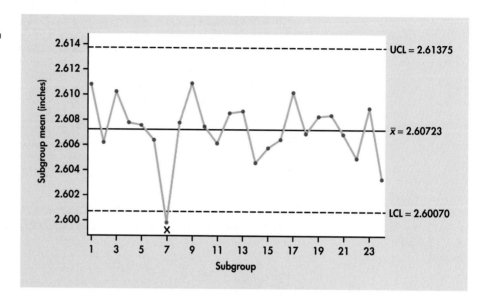

FIGURE 12.9 $\bar{x}$ and R charts for the O-ring data of Table 12.3 with Subgroups 7 and 17 removed.

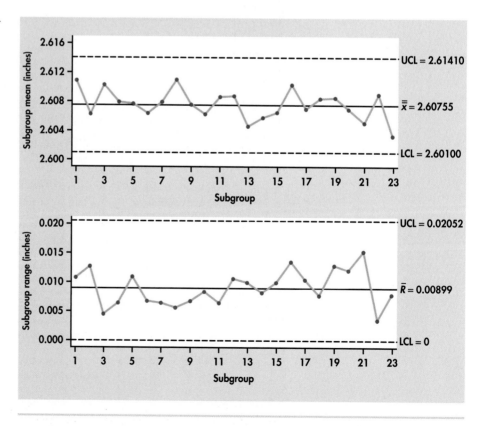

APPLY YOUR KNOWLEDGE

12.11 Interpreting signals. Explain the difference in the interpretation of a point falling beyond the upper control limit of the $\bar{x}$ chart versus a point falling beyond the upper control limit of the R chart.

12.12 Auto thermostats. A maker of auto air conditioners checks a sample of 4 thermostatic controls from each hour's production. The thermostats are set at $75°F$ and then placed in a chamber where the temperature is raised gradually. The temperature at which the thermostat turns on the air conditioner is recorded. The process mean should be $\mu = 75°$. Past experience indicates that the response temperature of properly adjusted thermostats varies with $\sigma = 0.5°$. The mean response temperature $\bar{x}$ for each hour's sample is plotted on an $\bar{x}$ control chart. Calculate the center line and control limits for this chart.

 ORING   *CASE 12.2* **12.13 O-rings.** Show the computations that confirm the limits of the $\bar{x}$ chart and R chart shown in Figure 12.9.

$\bar{x}$ and s charts

In the construction of subgroup control charts, the use of the simplistic range statistic instead of the sample standard deviation statistic is a historical artifact from when calculations were done by hand. Given the availability of computer software to do calculations, the need for computational simplicity is no longer a compelling argument. The fact that the range statistic is still in widespread use is probably due to training issues. It is much easier for corporate trainers to explain and for employees to comprehend the range statistic (largest minus smallest observation) than a statistic based on summing squared deviations, dividing the sum by $n − 1$, and then taking a square root!

The primary advantage of the sample standard deviation is that it uses all the data as opposed to the range statistic, which utilizes only two observation values (largest and smallest). For small subgroup sizes ($n \leq 10$), the range statistic competes well with the sample standard deviation statistic. However, for larger subgroup sizes, it is generally advisable to utilize the more efficient sample standard deviation.*

When using sample standard deviations, the R chart is replaced by an s chart. The s chart is a plot of the subgroup standard deviations with appropriate control limits superimposed. In addition, the construction of the $\bar{x}$ chart is based on the subgroup standard deviation values, not the range values. There is no difference in the calculation of the grand mean ($\bar{\bar{x}}$), but we need to calculate the average sample standard deviation from the k preliminary samples:

$$\bar{s} = \frac{1}{k}(s_1 + s_2 + \cdots + s_k)$$

Here is a summary of how to construct $\bar{x}$ and s charts.

Construction of $\bar{x}$ and s Charts

Take regular samples of size n from a process. The center line and control limits for an $\bar{x}$ **chart** are

$$\text{UCL} = \bar{\bar{x}} + A_3\bar{s}$$
$$\text{CL} = \bar{\bar{x}}$$
$$\text{LCL} = \bar{\bar{x}} - A_3\bar{s}$$

---

*In statistics, the term "efficient" relates to the variance of the sampling distribution of the estimator. In a group of parameter estimators, the estimator with the smallest variation is referred to as an *efficient* estimator.

where $\overline{\overline{x}}$ is the average of the subgroup means and $\overline{s}$ is the average of the subgroup standard deviations. The center line and control limits for an **s chart** are

$$\text{UCL} = B_4\overline{s}$$

$$\text{CL} = \overline{s}$$

$$\text{LCL} = B_3\overline{s}$$

The **control chart constants** A_3, B_3, and B_4 depend on the sample size n and are provided in Table 12.1 (page 644).

LABTESTS

CASE 12.1

EXAMPLE 12.4 $\overline{x}$ and s Charts

We leave the actual specific computations of the $\overline{x}$ and s charts for the lab-testing data to Exercise 12.15. Figure 12.10 shows the $\overline{x}$ and s charts produced by Minitab. Comparing these control charts with the $\overline{x}$ and R charts of Figures 12.3 and 12.4 (Pages 647 and 648), we are left with the same conclusion: the lab-testing process is in control in terms of both level and variability.

FIGURE 12.10 $\overline{x}$ and s charts for the lab-testing data of Table 12.2.

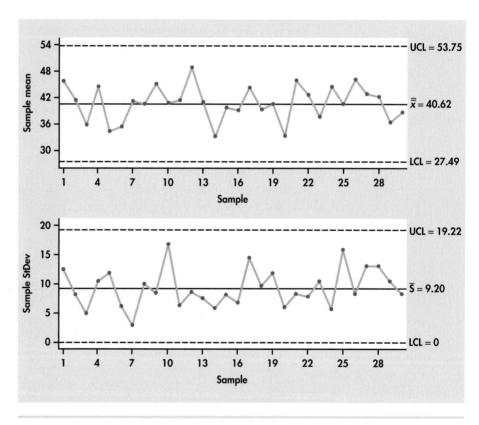

APPLY YOUR KNOWLEDGE

HOSPITALLOSS

12.14 Hospital losses.   Both nonprofit and for-profit hospitals are financially pressed by restrictions on reimbursement by insurers and the government. One hospital looked at their losses broken down by diagnosis. The leading source was joint replacement surgery. Table 12.4 gives data on the losses (in dollars) incurred by this hospital in

TABLE 12.4 Hospital losses for 15 samples of joint replacement patients

Subgroup	Losses (dollars)								Sample mean	Standard deviation
1	6835	5843	6019	6731	6362	5696	7193	6206	6360.63	521.72
2	6452	6764	7083	7352	5239	6911	7479	5549	6603.63	817.12
3	7205	6374	6198	6170	6482	4763	7125	6241	6319.75	749.12
4	6021	6347	7210	6384	6807	5711	7952	6023	6556.88	736.51
5	7000	6495	6893	6127	7417	7044	6159	6091	6653.25	503.72
6	7783	6224	5051	7288	6584	7521	6146	5129	6465.75	1034.26
7	8794	6279	6877	5807	6076	6392	7429	5220	6609.25	1103.96
8	4727	8117	6586	6225	6150	7386	5674	6740	6450.63	1032.96
9	5408	7452	6686	6428	6425	7380	5789	6264	6479.00	704.70
10	5598	7489	6186	5837	6769	5471	5658	6393	6175.13	690.46
11	6559	5855	4928	5897	7532	5663	4746	7879	6132.37	1128.64
12	6824	7320	5331	6204	6027	5987	6033	6177	6237.88	596.56
13	6503	8213	5417	6360	6711	6907	6625	7888	6828.00	879.82
14	5622	6321	6325	6634	5075	6209	4832	6386	5925.50	667.79
15	6269	6756	7653	6065	5835	7337	6615	8181	6838.88	819.46

treating major joint replacement patients.[3] The hospital has taken from its records a random sample of 8 such patients each month for 15 months.

(a) Calculate $\bar{x}$ and s for the first 2 subgroups to verify the table entries.

(b) Make an s control chart using center lines and limits calculated from these past data. There are no points out of control.

(c) Because the s chart is in control, base the $\bar{x}$ chart on all 15 samples. Make this chart. Is it also in control?

    LABTESTS   **CASE 12.1** **12.15 Lab testing.** Show the computations that confirm the limits of the $\bar{x}$ chart and s chart shown in Figure 12.10.

Individuals chart

Up to this point we have concentrated on the application of control charts to statistics based on samples of two or more observations. There are, however, many applications where it is not practical to gather a sample of observations at a given time. For example, in low-volume manufacturing environments, the production rate is often slow, and therefore the time between measurements is too long to allow rational subgroups to be formed. Some processes are just naturally viewed as a series of individual measurements. Data arising once a day, once a week, or once every two weeks do not lend themselves to being grouped into subgroups. Weekly sales or inventory levels are common company performance measurements monitored as individual observations.

A series of individual observations can be viewed as a special case of the $\bar{x}$ chart with *individuals (I) chart* $n = 1$. A control chart for monitoring individual observations is known as an **individuals (I) chart.** Samples of size one each time period do not allow for the computation of the sample range each period. Therefore, the R chart is not implementable for individual observations. To detect changes in variability in a series of individual observations, some practitioners group successive observations and compute the range statistic for each group of two observations. In turn, these ranges are plotted sequentially and monitored *moving-range (MR) chart*   with a control chart known as a **moving-range (MR) chart.** The moving-range chart is offered in most statistical software packages. There is, however, much debate in the SPC community about the effectiveness of the *MR* chart. Many SPC practitioners take the view that the *I* chart is sufficient for monitoring the process for changes in both level and variability. We will focus on the construction of the *I* chart.

For an in-control process with mean μ and standard deviation σ, the sample mean control chart with known parameters and $n = 1$ (page 643) is:

$$\text{UCL} = \mu + 3\frac{\sigma}{\sqrt{1}} = \mu + 3\sigma$$

$$\text{CL} = \mu$$

$$\text{LCL} = \mu - 3\frac{\sigma}{\sqrt{1}} = \mu - 3\sigma$$

If the goal is to maintain the process at a target level mean of μ, then μ is used to establish the control limits. In other cases, we need to estimate μ. For a set of k consecutive observations $x_1, x_2, \ldots, x_k$, the estimate of μ is simply given by the sample mean $\bar{x}$. To estimate σ, we first calculate the sample standard deviation of the k observations. It can be shown, however, that the sample standard deviation is a biased estimate of σ. The corrected estimate is

$$\hat{\sigma} = \frac{s}{c_4}$$

Table 12.1 (page 644) gives values for c_4 for sample sizes ranging from 2 to 25. For series of more than 25 observations, c_4 is well approximated[4] by

$$c_4 \approx \frac{4k - 4}{4k - 3}$$

where k is the number of individual observations. Here is the summary of our discussion.

Construction of an *I* Chart

Take a sample of k consecutive observations from a process. Estimate the process mean μ and the process standard deviation σ from past samples by

$$\hat{\mu} = \bar{x}$$

$$\hat{\sigma} = \frac{s}{c_4}$$

The center line and control limits for an *I* **chart** are

$$\text{UCL} = \hat{\mu} + 3\hat{\sigma}$$

$$\text{CL} = \hat{\mu}$$

$$\text{LCL} = \hat{\mu} - 3\hat{\sigma}$$

The **control chart constant** c_4 depends on the number of observations (k) in the series. For $k \leq 25$, the values of c_4 are provided in Table 12.1 (page 644). For $k > 25$, c_4 can be approximated by the formula $c_4 = 4k - 4/4k - 3$.

EXAMPLE 12.5   Is the NBA's MVP in Control?

Sport enthusiasts use a variety of statistics to prognosticate team and player performance for general entertainment, betting purposes, and fantasy league play. Indeed, there are literally hundreds of thousands of Web sites providing sports statistics on amateur and professional athletes and teams. Let us consider the performance of the professional basketball superstar Kobe Bryant, who won

TABLE 12.5	Points per minute scored by Kobe Bryant for each of the 82 regular-season games (read left to right)					
1.04651	0.57143	0.76744	0.75676	0.75000	0.47368	0.78947
0.51351	0.56250	1.03226	0.62791	0.75676	0.75610	1.00000
0.80000	0.80000	0.66667	0.68966	0.80645	0.73684	0.78947
0.56757	0.88889	0.50000	0.52500	0.48718	0.95122	0.92683
1.06897	0.59459	0.51724	0.72222	0.62500	0.57576	1.00000
1.02778	1.14286	0.76923	0.43590	0.64444	0.88889	0.76744
0.54545	0.82979	1.04545	0.85714	0.15000	0.29730	0.78261
0.78571	0.75610	0.82857	0.76667	0.89130	0.40476	0.80769
0.66667	0.51220	0.76744	1.01961	0.79070	0.51613	0.61905
0.72340	0.94737	0.51064	0.65909	0.65854	0.76667	0.78261
0.57692	0.69231	1.26190	0.55319	0.87805	0.59524	0.80556
0.75556	0.45714	0.67442	0.62500	0.68966		

the NBA's Most Valuable Player (MVP) award in 2008. Basketball players can be tracked on a variety of offensive and defensive measures. Since the number of minutes played from game to game varies, it makes sense to consider measures in terms of a rate, such as points per minute or rebounds per minute.

KOBE

Table 12.5 provides the points per minute (ppm) that Kobe had for each of the 82 consecutive regular-season games in 2007–2008. The sample mean and standard deviation for the 82 observations are

$$\bar{x} = \frac{1.04651 + 0.57143 + \cdots + 0.68966}{82} = 0.726512$$

$$s = \sqrt{\frac{(1.04651 - 0.726512)^2 + \cdots + (0.68966 - 0.726512)^2}{82 - 1}} = 0.194210$$

The control constant c_4 is approximated as

$$c_4 \approx \frac{4(82) - 4}{4(82) - 3} = 0.9969$$

Accordingly, the unbiased estimate for σ is computed to be

$$\hat{\sigma} = \frac{0.194210}{0.9969} = 0.194814$$

The center line and control limits for the I chart are

$$\text{UCL} = \hat{\mu} + 3\hat{\sigma} = 0.726512 + 3(0.194814) = 1.31095$$

$$\text{CL} = \hat{\mu} = 0.726512$$

$$\text{LCL} = \hat{\mu} - 3\hat{\sigma} = 0.726512 - 3(0.194814) = 0.14207$$

Figure 12.11 displays the I chart. For the most part, the control chart shows a process that is stable around the center line. The only points that look suspicious are Observations 47 and 48. Observation 47 almost breaches the lower control limit. Observations 47 and 48 are both beyond two standard deviations from the center line. In fact, there is a common **runs rule** that signals out of control when two out of three consecutive observations are more than two standard deviations from the center line. Investigation of Observation 47 reveals that Kobe dislocated his right pinkie finger in the second quarter of this game (February 5, 2008). Clearly, Observation 48 reflects the continuing adverse effects of the injury. With this information, it is reasonable to consider removing the two observations and reestimating the control limits. Aside from the two

runs rule

noted points, the control chart shows that Kobe's offensive performance is an in-control process. This does not mean that we can predict his future individual game outcomes precisely, but rather it means that the level and average variability of his performance can be predicted with a high degree of confidence.

FIGURE 12.11 Individuals (I) chart for Kobe Bryant's points per minute (ppm) process. Observations 47 and 48 are highlighted as a run of two observations (out of three consecutive observations) that are more than two standard deviations from the mean.

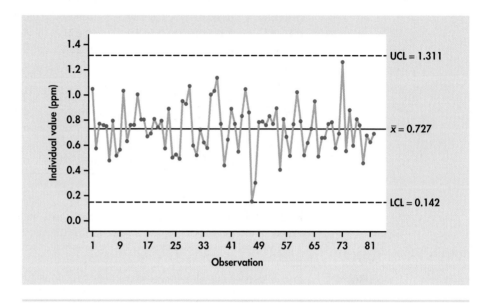

The individuals chart, like other control charts, is a three-sigma control chart with interpretation grounded in the Normal distribution. Unlike the mean chart based on sample sizes greater than one, the individuals chart does not have the advantage of the central limit theorem effect. Checking for Normality of the individual observations then becomes a particularly important step in the implementation of the individuals chart. Figure 12.12 shows the histogram for the points per minute data. We see that the data are compatible with the Normal distribution.

FIGURE 12.12 Histogram of Kobe Bryant's points per minute data.

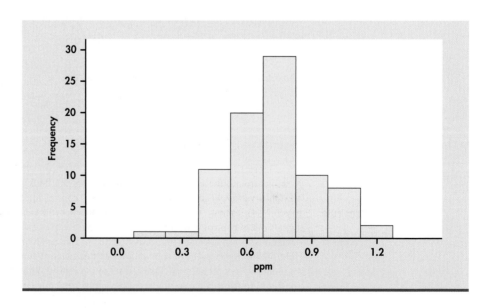

12.16 Kobe Bryant. In Example 12.5, we observed two unusual observations associated with an injury to Kobe Bryant. Remove these two points and reestimate the center line and control limits. Comment on the process relative to the revised limits.

12.17 Personal processes. From your personal life, provide two examples of processes for which you would collect data in the form of individual measurements that ultimately might be monitored by an *I* chart.

12.18 Kobe Bryant in the playoffs. The control charts of Example 12.5 and Exercise 12.16 are based on the regular-season performance. After the conclusion of the regular season, Kobe Bryant's team (LA Lakers) played in the playoffs. Below are his points per minute in the playoffs (read left to right):

0.86486	1.16667	0.61111	0.73810	0.95000	0.89474	0.80952
0.70213	0.63415	0.85000	0.61364	0.61111	0.78947	0.68293
0.90698	0.57143	0.75000	0.80000	0.39535	0.56818	0.51163

Project the control limits from Exercise 12.16 and plot Kobe's playoff numbers. What do you conclude about Kobe's playoff performance?

Don't confuse control with capability!*

A process in control is stable over time. With an in-control process, we can predict the mean and the amount of variation the process will show. Control charts are, so to speak, the voice of the process telling us what state it is in. *There is no guarantee that a process in control produces products of satisfactory quality.* "Satisfactory quality" is measured by comparing the product or service to some standard outside the process, set by technical specifications, customer expectations, or the goals of the organization. These external standards are unrelated to the internal state of the process, which is all that statistical control pays attention to.

> **Capability**
>
> **Capability** refers to the ability of a process to meet or exceed the requirements placed on it.

Capability has nothing to do with control—except for the very important point that, if a process is not in control, it is hard to tell if it is capable or not.

EXAMPLE 12.6   Capability

CASE 12.1

Both the $\bar{x}$ and R charts and the $\bar{x}$ and s charts showed that the lab-testing process is stable and in control. Suppose that the ER stipulates that getting lab results must take no longer than 50 minutes. Figure 12.13 compares the distribution of individual lab test times with an *upper specification limit* (USL) of 50 minutes. We can clearly see that the process is not capable of meeting the specification.

---

*This material is optional and not required for the understanding of control chart construction. However, capability analysis is often conducted in practice in conjunction with control charts and does help eliminate the common misconception that being in control means acceptable quality.

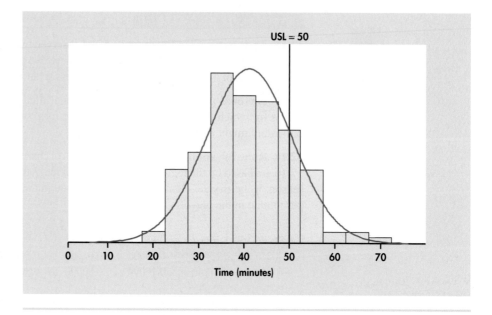

Managers must understand that, *if a process that is in control does not have adequate capability, fundamental changes in the process are needed.* The process is doing as well as it can and displays only the chance variation that is natural to its present state. Slogans to encourage the workers or disciplining the workers for poor performance will not change the state of the process. Better training for workers is a change in the process that may improve capability. New equipment or more uniform material may also help, depending on the findings of a careful investigation.

Figure 12.13 gives us a visual summary of the capability of the process, but managers often like a numerical summary of capability. One measure of capability is simply the *percent of process outcomes that meet the specifications.* When the variable we measure has a Normal distribution, we can use the estimated mean and standard deviation along with the Normal distribution to estimate this percent. When the variable is not Normal, we can use the actual percents of the measurements in the samples that meet the specifications.

EXAMPLE 12.7   Percent Meeting Specifications

CASE 12.1

Figure 12.13 shows that the distribution of individual turnaround times is approximately Normal. We found in Example 12.2 that $\overline{\overline{x}} = 40.62$. We now need an estimate of the standard deviation σ for the process producing the individual measurements.

Now a subtle point arises. We could derive an estimate for σ based on the subgroup variability estimates $\overline{R}$ or $\overline{s}$. However, these estimates are based solely on the within-subgroup variation. That's what we want for control charts, because within-sample variation is likely to be "pure common cause" variation. Even when the process is in control, there is some additional variation from sample to sample, just by chance. So the variation in the process outcomes will be greater than the variation within subgroups. With this in mind, the process standard deviation can be estimated from the single set of all individual measurements. In our discussion of the individuals (I) chart, we noted that s/c_4 is an unbiased estimate of σ, where the correction factor c_4 is based on the sample size n. The standard deviation for all 150 lab test times is

$$s = 9.6033$$

For a sample of size 150, c_4 is approximately 0.9983 (page 656), which gives the unbiased estimate for σ

$$\hat{\sigma} = \frac{9.6033}{0.9983} = 9.6197$$

We can now calculate the percent of lab tests that meet the upper specification:

$$P(\text{lab times} \leq 50) = P\left(Z \leq \frac{50 - 40.62}{9.6197}\right)$$

$$= P(Z \leq 0.9751) \approx 0.8365$$

It is estimated that about 84% of lab tests meet the ER specification of 50 minutes or less turnaround time.

Even though *percent meeting specifications* seems to be a reasonable measure of process capability, there are some situations that can call into question its appropriateness. Figure 12.14 shows why. This figure compares the distributions of the diameter of the same part manufactured by two processes. The target diameter and the specification limits are marked. All the parts produced by Process A meet the specifications, but about 1.5% of those from Process B fail to do so. Nonetheless, Process B is superior to Process A because it is less variable: much more of Process B's output is close to the target. Process A produces many parts close to the lower specification limit (LSL) and the upper specification limit (USL). These parts meet the specifications, but they will fit and perform more poorly than parts with diameters close to the center of the specifications. A distribution like that for Process A might result from inspecting all the parts and discarding those whose diameters fall outside the specifications. That's not an efficient way to achieve quality.

We need a way to measure process capability that pays attention to the variability of the process (smaller is better). The standard deviation does that, but it doesn't measure capability because it takes no account of the specifications that the output must *capability indices* meet. **Capability indices** start with the idea of comparing process variation with the specifications. Process B will beat Process A by such a measure. Capability indices also allow us to measure process improvement—we can continue to drive down variation, and so improve the process, long after 100% of the output meets specifications. Continual

FIGURE 12.14 Two distributions for part diameters. All the parts from Process A meet the specifications, but a higher proportion of parts from Process B have diameters close to target.

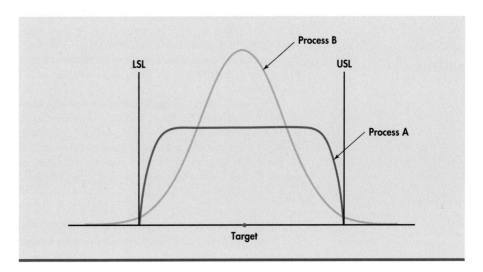

improvement of processes is our goal, not just reaching "satisfactory" performance. The real importance of capability indices is that they give us numerical measures to describe ever-better process quality. Statistical software offers many capability indices, but we will consider only the most basic ones.

Capability Indices for Two-Sided Specifications

Consider a process with lower and upper specification limits (LSL and USL) for some measured characteristic of its output. The process mean for this characteristic is μ and the standard deviation is σ. The **potential capability index C_p** is

$$C_p = \frac{\text{USL} - \text{LSL}}{6\sigma}$$

The **performance capability index C_{pk}** is

$$C_{pk} = \min\left(\frac{\mu - \text{LSL}}{3\sigma}, \frac{\text{USL} - \mu}{3\sigma}\right)$$

Large values of C_p or C_{pk} indicate more capable processes.

Capability indices start from the fact that *Normal distributions are in practice about 6 standard deviations wide.* That's the 99.7 part of the 68–95–99.7 rule. Conceptually, C_p is the specification width as a multiple of the process width 6σ. When $C_p = 1$, the process output will just fit within the specifications if the center is midway between LSL and USL. Larger values of C_p are better—the process output can fit within the specs with room to spare. But a process with high C_p can produce poor-quality product if it is not correctly centered. As we will see with the next example, C_{pk} remedies this deficiency by considering both the center μ and the variability σ of the measurements.

EXAMPLE 12.8 Interpreting Capability Indices

Consider the series of pictures in Figure 12.15. We might think of a process that machines a metal part. Measure a dimension of the part that has LSL and USL as its specification limits. There is of course variation from part to part. The dimensions vary Normally with mean μ and standard deviation σ.

Figure 12.15(a) shows process width equal to the specification width. That is, $C_p = 1$. Almost all the parts will meet specifications *if,* as in this figure, the process mean μ is at the center of the specs. Because the mean is centered, it is 3σ from both LSL and USL, so $C_{pk} = 1$ also. In Figure 12.15(b), the mean has moved down to LSL. Only half the parts will meet the specifications. C_p is unchanged because the process width has not changed. But C_{pk} sees that the center μ is right on the edge of the specifications, $C_{pk} = 0$. The value becomes negative if μ is outside the specifications.

In Figures 12.15(c) and (d) the process σ has been reduced to half the value it had in (a) and (b). The process width 6σ is now half the specification width, so $C_p = 2$. In Figure 12.15(c) the center is just 3 of the new σ's above LSL, so that $C_{pk} = 1$. Figure 12.15(d) shows the same smaller σ accompanied by mean μ correctly centered between LSL and USL. C_{pk} rewards the process for moving the center from 3σ to 6σ away from the nearer limit by increasing from 1 to 2. You see that C_p and C_{pk} are equal if the process is properly centered. If not, C_{pk} is smaller than C_p.

Example 12.8 shows that as a process moves off target C_p remains constant while C_{pk} decreases in value. Off-target processes have the *potential* of having higher capability

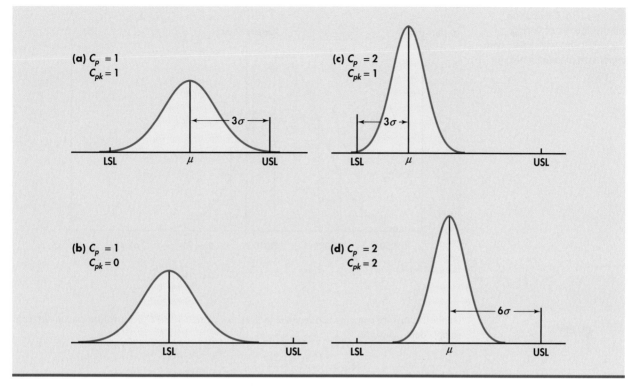

FIGURE 12.15 How capability indices work. (a) Process centered, process width equal to specification width. (b) Process off-center, process width equal to specification width. (c) Process off-center, process width equal to half the specification width. (d) Process centered, process width equal to half the specification width.

if the mean is adjusted to the center of the specs. For these reasons, C_p can be viewed as a measure of process potential while C_{pk} attempts to measure *actual* process performance.

ORING

EXAMPLE 12.9   O-Ring Process Capability

CASE 12.2

At the conclusion of the process study in Example 12.3 (page 649), we found two special causes and eliminated from our data the subgroups on which those causes operated. The remaining 92 individual O-ring measurements have $\overline{\overline{x}} = 2.60755$ and $s = 0.00412328$. As noted in Case 12.2, specification limits for the inside diameter are set at 2.612 ± 0.02 inches, which implies LSL = 2.592 and USL = 2.632. Figure 12.16 shows that the individual measurements are compatible with the Normal distribution.

For a sample of size 92, c_4 is approximately 0.9973 (page 656), which gives the unbiased estimate for σ

$$\hat{\sigma} = \frac{0.00412328}{0.9973} = 0.00413444$$

In addition, the mean estimate $\hat{\mu}$ is simply the grand mean $\overline{\overline{x}}$. These estimates may be quite accurate if we have data on many past samples. Estimates based on only a few observations may, however, be inaccurate because statistics from small samples can have large sampling variability. This important point is often not appreciated when capability indices are used in practice. To

FIGURE 12.16 Comparing the distribution of O-ring measurements with lower and upper specification limits, for Example 12.9.

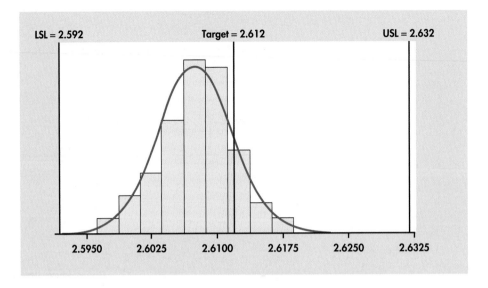

emphasize that we can only estimate the indices, we write $\hat{C}_p$ and $\hat{C}_{pk}$ for values calculated from sample data. They are

$$\hat{C}_p = \frac{\text{USL} - \text{LSL}}{6\hat{\sigma}}$$

$$= \frac{2.632 - 2.592}{(6)(0.00413444)} = \frac{0.04}{0.02480664} = 1.61$$

$$\hat{C}_{pk} = \min\left(\frac{\hat{\mu} - \text{LSL}}{3\hat{\sigma}}, \frac{\text{USL} - \hat{\mu}}{3\hat{\sigma}}\right)$$

$$= \min\left(\frac{2.60755 - 2.592}{(3)(0.00413444)}, \frac{2.632 - 2.60755}{(3)(0.00413444)}\right)$$

$$= \min(1.25, 1.97) = 1.25$$

Both indices are well above 1, which indicates that we have a highly capable process. However, the fact that $\hat{C}_{pk}$ is markedly smaller than $\hat{C}_p$ indicates that the process is off target. Indeed, a close look at Figure 12.16 shows that the center of the distribution is to the left of the center of the specs. If we can adjust the center of the process distribution to the target of 2.612, then $\hat{C}_{pk}$ will increase and will equal $\hat{C}_p$.

Our discussion of capability indices has been focused on capability relative to two specification limits. When there is only one specification limit involved, we can define one-sided indices as follows:

$$C_{pl} = \frac{\mu - \text{LSL}}{3\sigma}$$

$$C_{pu} = \frac{\text{USL} - \mu}{3\sigma}$$

The change in denominator from 6 to 3 standard deviations reflects the focus on one specification rather than two.

EXAMPLE 12.10   Capability of a Lab-Testing Process

CASE 12.1

LABTESTS

In Examples 12.2 and 12.7, we found the estimates for the mean and standard deviation:

$$\hat{\mu} = 40.62$$
$$\hat{\sigma} = 9.6197$$

In this application, the upper specification limit was set to 50 minutes. We estimate the one-sided upper capability index to be

$$\hat{C}_{pu} = \frac{50 - 40.62}{(3)(9.6197)} = 0.33$$

The estimated capability index is considerably less than 1. As can be seen from Figure 12.13 (page 660), this reflects the fact that the process mean is close enough to the upper specification limit to result in a high percent of unacceptable outcomes.

We end our discussion on process capability indices on two cautionary notes. First, their interpretation is based on the assumption that the individual process measurements are Normally distributed. It is hard to interpret indices when the measurements are strongly non-Normal. It is best to apply capability indices only when a Normal quantile plot or histogram shows that the distribution is at least roughly Normal. Second, as we saw with Examples 12.9 and 12.10, process indices need to be estimated from the process data in hand. The implication is that estimated indices are statistics and thus are subject to sampling variation. A supplier under pressure from a large customer to measure C_{pk} often may base calculations on small samples from the process. The resulting estimate $\hat{C}_{pk}$ can differ greatly from the true process C_{pk} in either direction. As a rough rule of thumb, it is best to rely on indices computed from samples of at least 50 measurements.

APPLY YOUR KNOWLEDGE

12.19 Specification limits versus control limits. The manager you report to is confused by LSL and USL versus LCL and UCL. The notations look similar. Carefully explain the conceptual difference between specification limits for individual measurements and control limits for $\bar{x}$.

12.20 C_p **versus** C_{pk}. Sketch Normal curves that represent measurements on products from a process with
(a) $C_p = 3$ and $C_{pk} = 1$.
(b) $C_p = 3$ and $C_{pk} = 2$.
(c) $C_p = 3$ and $C_{pk} = 3$.

ORING

CASE 12.2 **12.21 O-ring capability in terms of percent defective.** Refer to Example 12.9 for the mean and standard deviation estimates for the O-ring application. Using software, estimate the percent of O-rings that do not meet specs. In quality applications, it is common to report defective rates in units of parts per million (ppm). What is the defective rate for the O-ring process in ppm?

SECTION 12.2 Summary

- Standard **three-sigma control charts** plot the values of some statistic for regular samples from the process against the time order of the samples. The **center line** is set at the mean of the plotted statistic. The **control limits** lie three standard deviations of the plotted statistic above and below the center line. A point outside the control limits is an **out-of-control signal.**

- When we measure some quantitative characteristic of the process and gather samples of two or more observations, we use $\bar{x}$ **and R charts** for process control. The R chart monitors variation within individual samples. If the R chart is in control, the $\bar{x}$ chart monitors variation from sample to sample. To interpret the charts, always look first at the R chart. For larger subgroups, the R chart can be replaced by an **s chart.**

- An I **chart** is used for monitoring a process of individual observations. The I chart does not benefit from the central limit theorem effect. As a result, it is important to check if the individual observations follow the Normal distribution before constructing the I chart.

- **Capability indices** measure process variability (C_p) or process center and variability (C_{pk}) against the standard provided by external specifications for the output of the process. Larger values indicate higher capability.

- Interpretation of C_p and C_{pk} requires that measurements on the process output have a roughly Normal distribution. These indices are not meaningful unless the process is in control so that its center and variability are stable.

- Estimates of C_p and C_{pk} can be quite inaccurate when based on small numbers of observations, due to sampling variability. It is generally recommended that capability index estimates be based on at least 50 measurements.

SECTION 12.2 Exercises

For Exercises 12.11 to 12.13, see page 653; for 12.14 and 12.15, see pages 654–655; for 12.16 to 12.18, see page 659; and for 12.19 to 12.21, see page 665.

12.22 Dyeing yarn. The unique colors of the cashmere sweaters your firm makes result from heating undyed yarn in a kettle with a dye liquor. The pH (acidity) of the liquor is critical for regulating dye uptake and hence the final color. There are 5 kettles, all of which receive dye liquor from a common source. Twice each day, the pH of the liquor in each kettle is measured, giving a sample of size 5. The process has been operating in control with $\mu = 4.22$ and $\sigma = 0.127$. Give the center line and control limits for the $\bar{x}$ chart.

12.23 Probability out? An $\bar{x}$ chart plots the means of samples of size 4 against center line CL = 700 and control limits LCL = 687 and UCL = 713. The process has been in control. Now the process is disrupted in a way that changes the mean to $\mu = 693$ and the standard deviation to $\sigma = 12$. What is the probability that the first sample after the disruption gives a point beyond the control limits of the $\bar{x}$ chart?

12.24 Alternative control limits. American and Japanese practice uses three-sigma control charts. That is, the control limits are 3 standard deviations on either side of the mean. When the statistic being plotted has a Normal distribution, the probability of a point outside the limits is about 0.003 (or about 0.0015 in each direction) by the 68–95–99.7 rule. European practice uses control limits placed so that the probability of a point out is 0.001 in each direction. For a Normally distributed statistic, how many standard deviations on either side of the mean do these alternative control limits lie?

12.25 Monitoring packaged products. To control the fill amount of its cereal products, a cereal manufacturer monitors the net weight of the product with $\bar{x}$ and R charts using a subgroup size of $n = 5$. One of its brands, Organic Bran Squares, has a target of 10.6 ounces. Suppose that 20 preliminary subgroups were gathered, and the following summary statistics were found for the 20 subgroups:

$$\sum \bar{x}_i = 211.624 \qquad \sum R_i = 7.44$$

Assume that the process is stable in both variation and level. Compute the control limits for the $\bar{x}$ and R charts.

12.26 Measuring bone density. Loss of bone density is a serious health problem for many people, especially older women. Conventional X-rays often fail to detect loss of bone density until the loss reaches 25% or more. New equipment such as the Lunar bone densitometer is much more sensitive. A health clinic installs one of these machines. The manufacturer supplies a "phantom," an aluminum piece of known density that can be used to keep the machine calibrated. Each morning, the clinic makes two measurements on the phantom before measuring the first patient. Control charts based on these measurements alert the operators if the machine has lost calibration. Table 12.6 contains data for the first 30 days of operation.[5] The units are grams per square centimeter (for technical reasons, area rather than volume is measured).
(a) Calculate $\bar{x}$ and R for the first 2 days to verify the table entries.
(b) Make an R chart and comment on control. If any points are out of control, remove them and recompute the chart limits until all remaining points are in control. (That is, assume that special causes are found and removed.) 🏷 BONE

TABLE 12.6   Daily calibration subgroups for a Lunar bone densitometer (grams per square centimeter)

Subgroup	Measurements		Sample mean	Range
1	1.261	1.260	1.2605	0.001
2	1.261	1.268	1.2645	0.007
3	1.258	1.261	1.2595	0.003
4	1.261	1.262	1.2615	0.001
5	1.259	1.262	1.2605	0.003
6	1.269	1.260	1.2645	0.009
7	1.262	1.263	1.2625	0.001
8	1.264	1.268	1.2660	0.004
9	1.258	1.260	1.2590	0.002
10	1.264	1.265	1.2645	0.001
11	1.264	1.259	1.2615	0.005
12	1.260	1.266	1.2630	0.006
13	1.267	1.266	1.2665	0.001
14	1.264	1.260	1.2620	0.004
15	1.266	1.259	1.2625	0.007
16	1.257	1.266	1.2615	0.009
17	1.257	1.266	1.2615	0.009
18	1.260	1.265	1.2625	0.005
19	1.262	1.266	1.2640	0.004
20	1.265	1.266	1.2655	0.001
21	1.264	1.257	1.2605	0.007
22	1.260	1.257	1.2585	0.003
23	1.255	1.260	1.2575	0.005
24	1.257	1.259	1.2580	0.002
25	1.265	1.260	1.2625	0.005
26	1.261	1.264	1.2625	0.003
27	1.261	1.264	1.2625	0.003
28	1.260	1.262	1.2610	0.002
29	1.260	1.256	1.2580	0.004
30	1.260	1.262	1.2610	0.002

(c) Make an $\bar{x}$ chart using the samples that remain after your work in part (b). What kind of variation will be visible on this chart? Comment on the stability of the machine over these 30 days based on both charts.

12.27 Additional out-of-control signals. A single extreme point outside three-sigma limits represents one possible statistical signal of unusual process behavior. As we saw in the Kobe Bryant application (Example 12.5, page 656), special causes can also give rise to unusual variation *within* control limits. A variety of statistical rules, known as runs rules, have been developed to supplement the three-sigma rule in an effort to more quickly detect special cause variation. A commonly used runs rule for the detection of smaller shifts of gradual process drifts is to signal if nine consecutive points all fall on one side of the center line. We have learned that for an in-control process and the assumption of Normality, the false alarm rate for the three-sigma rule is about 3 in 1000. Assuming Normality of the control chart statistics, what is the false alarm rate for the nine-in-a-row rule if the process is in control?

12.28 Alloy composition—retrospective control. Die casts are used to make molds for molten metal to produce a wide variety of products ranging from kitchen and bathroom fittings to toys, doorknobs, and a variety of auto and electronic components. Die casts themselves are made out of an alloy of metals including zinc, copper, and aluminum. For one particular die cast, the manufacturer must maintain the percent of aluminum between 3.8% and 4.2%. To monitor the percent of aluminum in the casts, three casts are periodically sampled, and their aluminum content is measured. The first 20 rows of Table 12.7 give the data for 20 preliminary subgroups. **ALLOY**

TABLE 12.7   Aluminum percentage measurements

Subgroup	Measurements		
1	3.99	3.90	3.98
2	4.02	3.95	3.95
3	3.99	3.90	3.90
4	3.99	3.94	3.88
5	3.96	3.93	3.91
6	3.94	3.97	3.83
7	3.89	3.95	3.99
8	3.86	3.97	4.02
9	3.98	3.98	3.95
10	3.93	3.88	4.06
11	3.97	3.91	3.92
12	3.86	3.95	3.88
13	3.92	3.97	3.95
14	4.01	3.91	3.91
15	4.00	4.02	3.93
16	4.01	3.97	3.98
17	3.92	3.92	3.95
18	3.96	3.96	3.90
19	4.04	3.93	3.95
20	3.96	3.85	4.03
21	4.06	4.04	3.93
22	3.94	4.02	3.98
23	3.95	4.07	3.99
24	3.90	3.92	3.97
25	3.97	3.96	3.94
26	3.96	4.00	3.91
27	3.90	3.85	3.91
28	3.97	3.87	3.94
29	3.97	3.88	3.83
30	3.82	3.99	3.84
31	3.95	3.87	3.94
32	3.86	3.91	3.98
33	3.94	3.93	3.90
34	3.89	3.90	3.81
35	3.91	3.99	3.83

(a) Make an R chart and comment on control of the process variation.
(b) Using the range estimate, make an $\bar{x}$ chart and comment on the control of the process level.

12.29 Alloy composition—prospective control. Project the $\bar{x}$ and R chart limits found in the previous exercise for prospective control of aluminum content. The last 15 rows of Table 12.7 give data on the next 15 future subgroups. Refer to Exercise 12.27 and apply the nine-in-a-row rule along with the standard three-sigma rule to the new subgroups. Is the process maintaining control? If not, describe the nature of the process change and indicate the subgroups affected.

12.30 Deming speaks. The quality guru W. Edwards Deming (1900–1993) taught (among much else) that
(a) "People work in the system. Management creates the system."
(b) "Putting out fires is not improvement. Finding a point out of control, finding the special cause and removing it, is only putting the process back to where it was in the first place. It is not improvement of the process."
(c) "Eliminate slogans, exhortations and targets for the workforce asking for zero defects and new levels of productivity."

Choose one of Deming's sayings. Explain carefully what facts about improving quality the saying attempts to summarize.

12.31 Accounts receivable. In an attempt to understand the bill-paying behavior of its distributors, a manufacturer samples bills and records the number of days between the issuing of the bill and the receipt of payment. The manufacturer formed subgroups of 10 randomly chosen bills per week over the course of 30 weeks. It found an overall mean $\bar{\bar{x}}$ of 30.6833 days and an average standard deviation $\bar{s}$ of 7.50638 days.
(a) Assume that the process is stable in both variation and level. Compute the control limits for the $\bar{x}$ and s charts.
(b) Here are the means and standard deviations of future subgroups:

Week:	31	32	33	34	35	36
$\bar{x}$:	31.1	29.5	33.0	33.4	33.2	35.8
s:	6.1001	10.5013	8.5114	7.5011	3.7059	4.1846

Week:	37	38	39	40
$\bar{x}$:	37.3	41.5	35.9	36.7
s:	6.5328	8.1548	5.8585	6.7338

Is the accounts receivable process still in control? If not, specify the nature of the process departure.

12.32 Financial market movements. Financial reporters and investors often perceive movements in individual stock prices, mutual funds, and market indices as unusual. Consider the week-to-week changes in the S&P 500 market index from April 6, 2009, to December 28, 2009. 🌑 SP500.
(a) Construct a histogram of market movements. How compatible is the histogram with the Normal distribution?

(b) Determine the mean and standard deviation estimates ($\hat{\mu}$ and $\hat{\sigma}$) that will be used in the construction of an I chart.
(c) Compute the UCL and LCL of the I chart.
(d) Construct the I chart for the market movement series. Discuss the stability of the process.
(e) Explain how an I chart on market movements can help us to better evaluate media reports about the market.

12.33 Control charting your reaction times. Consider the following personal data-generating experiment. Obtain a stopwatch, a capability that many electronic watches offer. Alternatively, you can use one of many Web-based stopwatches easily found with a Google search (make sure to use a site that reports to at least 0.01 seconds). Attempt to start and stop your stopwatch as close as possible to 5 seconds. Record the result to as many decimal places as your stopwatch shows. Repeat the experiment 50 times. Input your results into a statistical software package.
(a) Construct a histogram of your measurements. How compatible is the histogram with the Normal distribution?
(b) Determine the mean and standard deviation estimates ($\hat{\mu}$ and $\hat{\sigma}$) that will be used in the construction of an I chart.
(c) Compute the UCL and LCL of the I chart.
(d) Construct the I chart for your data series. Discuss the stability of your process. Are you in control? Were there any out-of-control signals? If so, provide an explanation for the unusual observation(s).

12.34 Estimating nonconformance rate. Suppose a Normally distributed process is centered on target with the target being halfway between specification limits. If $\hat{C}_p = 0.80$, what is the estimated rate of nonconformance of the process to the specifications?

12.35 Measuring capability (optional). You are in charge of a process that makes metal clips. The critical dimension is the opening of a clip, which has specifications 15 ± 0.5 millimeters (mm). The process is monitored by $\bar{x}$ and s charts based on samples of 5 consecutive clips each hour. Control has recently been excellent. The 200 individual measurements from the past week's 40 samples have

$$\bar{x} = 14.99 \text{ mm} \qquad s = 0.2239 \text{ mm}$$

A Normal quantile plot shows no important deviations from Normality.
(a) What percent of clip openings will meet specifications if the process remains in its current state?
(b) Estimate the capability index C_{pk}.

12.36 Hospital losses again (optional). Table 12.4 (page 655) gives data on a hospital's losses for 120 joint replacement patients, collected as 15 monthly samples of 8 patients each. The process has been in control and losses have a roughly Normal distribution. The sample standard deviation (s) for the individual measurements is 811.53. The hospital decides that suitable specification limit for its loss in treating one such patient is USL = $8000. 🌑 HOSPITALLOSS

(a) Estimate the percent of losses that meet the specification.

(b) Estimate C_{pu}.

12.37 Measuring your personal capability (optional). Refer to Exercise 12.33, in which you collected 50 sequential observations on your ability to measure 5 seconds. Suppose we define acceptable performance as 5 ± 0.15 seconds.

(a) Assume that the Normal distribution is sufficiently adequate to describe your distribution of times. Estimate the percent of stopwatch recordings that will meet specifications if your process remains in its current state.

(b) Estimate your personal C_p.

(c) Estimate your personal C_{pk}.

(d) Are your C_p and C_{pk} close in value? If not, what does that suggest about your stopwatch recording ability?

12.38 Alloy composition process capability (optional). Obtain the individual observations from the 20 preliminary subgroups studied in Exercise 12.28. The acceptable range for the percent of aluminum is 3.8% to 4.2%.

(a) Make a Normal quantile plot of the 60 observations. What do you conclude? (If your software will not make a Normal quantile plot, use a histogram to assess Normality.)

(b) Estimate C_p.

(c) Estimate C_{pk}.

(d) Comparing your results from parts (b) and (c), what would you recommend to improve process capability?

12.39 Six-sigma quality (optional). A process with $C_p \geq 2$ is sometimes said to have "six-sigma quality." Sketch the specification limits and a Normal distribution of individual measurements for such a process when it is properly centered. Explain from your sketch why this is called six-sigma quality.

12.40 More on six-sigma quality (optional). The originators of the "six-sigma quality" standard reasoned as follows. Short-term process variation is described by σ. In the long term, the process mean μ will also vary. Studies show that in most manufacturing processes, $\pm 1.5\sigma$ is adequate to allow for changes in μ. The six-sigma standard is intended to allow the mean μ to be as much as 1.5σ away from the center of the specifications and still meet high standards for percent of output lying outside the specifications.

(a) Sketch the specification limits and a Normal distribution for process output when $C_p = 2$ and the mean is 1.5σ away from the center of the specifications.

(b) What is C_{pk} in this case? Is six-sigma quality as strong a requirement as $C_{pk} \geq 2$?

(c) Because most people don't understand standard deviations, six-sigma quality is usually described as guaranteeing a certain level of parts per million of output that fails to meet specifications. Based on your sketch in part (a), what is the probability of an outcome outside the specification limits when the mean is 1.5σ away from the center? How many parts per million is this? (You will need software or a calculator for Normal probability calculations, because the value you want is beyond the limits of the standard Normal table.)

12.3 Attribute Control Charts

We have considered control charts for just one kind of data: measurements of a quantitative variable in some continuous scale of units. We described the distribution of measurements by its center and spread, and use $\overline{x}$ and R (or $\overline{x}$ and s) charts for process control. In contrast to continuous data, discrete data typically result from counting. Examples of counting are the number (or proportion) of defective parts in a production run, the daily number of patients in a clinic, and the number of invoice errors. In the quality area, discrete data are known as attribute data. We will consider the two most common control charts dedicated to attribute data: namely, the p chart for use when the data are proportions and the c chart for use when the data are counts of events that can occur in some interval of time, area, or volume.

Control charts for sample proportions

We studied the sampling distribution of a sample proportion $\hat{p}$ in Chapter 8. When the binomial distribution underlying the sample proportions is well approximated by the Normal distribution, then the standard three-sigma framework can be applied to the sample proportion data. We ought to call such charts "$\hat{p}$ charts" because they plot sample proportions. Unfortunately, they have always been called p charts in business practice. We will keep the traditional name but also keep our usual notation: p is a *process* proportion and $\hat{p}$ is a *sample* proportion.

Construction of a p Chart

Take regular samples from a process that has been in control. The samples need not all be the same size. Denote the sample size for the ith sample as n_i. Estimate the process proportion p of "successes" by

$$\overline{p} = \frac{\text{total number of successes in past samples}}{\text{total number of opportunities in these samples}}$$

The center line and control limits for the **p chart** are

$$\text{UCL} = \overline{p} + 3\sqrt{\frac{\overline{p}(1 - \overline{p})}{n_i}}$$

$$\text{CL} = \overline{p}$$

$$\text{LCL} = \overline{p} - 3\sqrt{\frac{\overline{p}(1 - \overline{p})}{n_i}}$$

where n_i is the sample size for sample i. If the lower control limit computes to a negative value, then LCL is set to 0 since negative proportions are not possible.

If we have k preliminary samples of the *same* size n, then $\overline{p}$ is just the average of the k sample proportions. In some settings, you may meet samples of unequal size—differing numbers of students enrolled in a month or differing numbers of parts inspected in a shift. The average $\overline{p}$ estimates the process proportion p even when the sample sizes vary. In cases of unequal sample sizes, the width of the control limits will vary from sample to sample, as will be shown in Example 12.12.

CASE 12.3

Reducing Absenteeism Unscheduled absences by clerical and production workers are an important cost in many companies. Reducing the rate of absenteeism is therefore an important goal for a company's human relations department. A rate of absenteeism above 5% is a serious concern. Many companies set 3% absent as a desirable target. You have been asked to improve absenteeism in a production facility where 12% of the workers are now absent on a typical day.

You first do some background study—in greater depth than this very brief summary. Companies try to avoid hiring workers who are likely to miss work often, such as substance abusers. They may have policies that reward good attendance or penalize frequent absences by individual workers. Changing those policies in this facility will have to wait until the union contract is renegotiated. What might you do with the current workers under current policies? Studies of absenteeism of clerical and production workers who do repetitive, routine work under close supervision point to unpleasant work environment and harsh or unfair treatment by supervisors as factors that increase absenteeism. It's now up to you to apply this general knowledge to your specific problem.

First, collect data. Daily absenteeism data are already available. You carry out a sample survey that asks workers about their absences and the reasons for them (responses are anonymous, of course). Workers who are more often absent complain about their supervisor and about the lighting at their workstation. Female workers complain that the restrooms are dirty and unpleasant. You do more data analysis:

- A chart of average absenteeism rate for the past month broken down by supervisor (Figure 12.17) shows important differences among supervisors. Only supervisors B, E, and H meet your goal of 5% or less absenteeism. Workers supervised by I and D have particularly high rates.

FIGURE 12.17 Chart of the average percent of days absent for workers reporting to each of 12 supervisors.

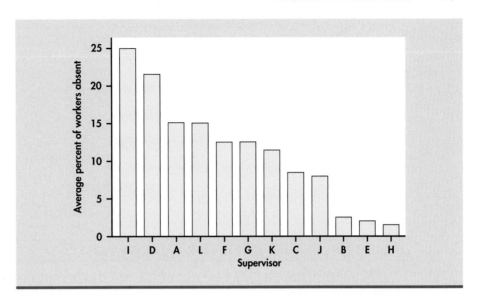

- Further data analysis (not shown) shows that certain workstations have substantially higher rates of absenteeism.

Now you take action. You retrain all the supervisors in human relations skills, using B, E, and H as discussion leaders. In addition, a trainer works individually with supervisors I and D. You ask supervisors to talk with any absent worker when he or she returns to work. Working with the engineering department, you study the workstations with high absenteeism rates and make changes such as better lighting. You refurbish the restrooms (for both genders even though only women complained) and schedule more frequent cleaning.

ABSENTEEISM

CASE 12.3

EXAMPLE 12.11   Absenteeism Rate p Chart

Are your actions effective? You hope to see a reduction in absenteeism. To view progress (or lack of progress), you will keep a p chart of the proportion of absentees. The plant has 987 production workers. For simplicity, you just record the number who are absent from work each day. Only unscheduled absences count, not planned time off such as vacations. Each day you will plot

$$\hat{p} = \frac{\text{number of workers absent}}{987}$$

You first look back at data for the past three months. There were 64 workdays in these months. The total workdays available for the workers was

$$(64)(987) = 63{,}168 \text{ person-days}$$

Absences among all workers totaled 7580 person-days. The average daily proportion absent was therefore

$$\overline{p} = \frac{\text{total days absent}}{\text{total days available for work}}$$
$$= \frac{7580}{63{,}168} = 0.120$$

The daily rate has been in control at this level.

These past data allow you to set up a p chart to monitor future proportions absent:

$$\text{UCL} = \overline{p} + 3\sqrt{\frac{\overline{p}(1 - \overline{p})}{n}} = 0.120 + 3\sqrt{\frac{(0.120)(0.880)}{987}}$$

$$= 0.120 + 0.031 = 0.151$$

$$\text{CL} = \overline{p} = 0.120$$

$$\text{LCL} = \overline{p} - 3\sqrt{\frac{\overline{p}(1 - \overline{p})}{n}} = 0.120 - 3\sqrt{\frac{(0.120)(0.880)}{987}}$$

$$= 0.120 - 0.031 = 0.089$$

Table 12.8 gives the data for the next four weeks. Figure 12.18 is the p chart.

TABLE 12.8	Proportions of workers absent during four weeks									
Day	M	T	W	Th	F	M	T	W	Th	F
Workers absent	129	121	117	109	122	119	103	103	89	105
Proportion $\hat{p}$	0.131	0.123	0.119	0.110	0.124	0.121	0.104	0.104	0.090	0.106
Workers absent	99	92	83	92	92	115	101	106	83	98
Proportion $\hat{p}$	0.100	0.093	0.084	0.093	0.093	0.117	0.102	0.107	0.084	0.099

FIGURE 12.18 Prospective-monitoring p chart for daily proportion of workers absent over a four-week period, for Example 12.11. The lack of control shows an improvement (decrease) in absenteeism. Update the chart to continue monitoring the process.

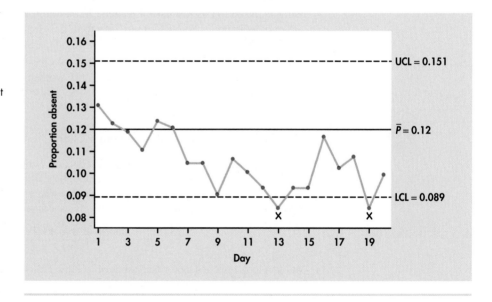

Figure 12.18 shows a clear downward trend in the daily proportion of workers who are absent. Days 13 and 19 lie below LCL, and a run of 9 days below the center line is achieved at Day 15 and continues. (See Exercise 12.27, page 667, for discussion of the "nine-in-a-row" out-of-control signal.) It appears that a special cause (the various actions you took) has reduced the absenteeism rate from around 12% to around 10%. The last two weeks' data suggest that the rate has stabilized at this level. You will update the chart based on the new data. If the rate does not decline further (or even rises again as the effect of your actions wears off), you will consider further changes.

Example 12.11 is a bit oversimplified. The number of workers available did not remain fixed at 987 each day. Hirings, resignations, and planned vacations change the number a bit from day to day. The control limits for a day's $\hat{p}$ depend on n, the number of workers that day. If n varies, the control limits will move in and out from day to day. In this case, n is fairly large, which means that as long as the count of workers remains close to 987, the greater detail provided by variable limits will not likely change your conclusion. We will demonstrate the construction of variable limits in the next example.

A single p chart for all workers is not the only, or even the best, choice in this setting. Because of the important role of supervisors in absenteeism, it would be wise to also keep separate p charts for the workers under each supervisor. These charts may show that you must reassign some supervisors.

EXAMPLE 12.12   Patient Satisfaction p Chart

Nationwide, health care organizations are instituting process improvement methods to improve the quality of health care delivery, including patient outcomes and patient satisfaction. Bellin Health (www.bellin.org) is a leader in the implementation of quality control methods in a health care setting. Located in Green Bay, Wisconsin, Bellin Health serves nearly half a million people in northeastern Wisconsin and in the Upper Peninsula of Michigan. In 2005, Bellin instituted a measurement control system of over 250 quality indicators, which they later expanded to include more than 1200 quality indicators. Most of these quality indicators are monitored by control charts.

BELLIN

Table 12.9 gives the numbers of Bellin ambulatory (outpatient) surgery patients sampled each quarter from the third quarter of 2004 to the third quarter of 2008. Also provided is the number of patients out of each sample who said they would likely recommend Bellin to others for ambulatory surgery.[6] The number of patients who would likely recommend Bellin can be divided by the sample size to give the sample proportion of patients who are likely to recommend Bellin. These proportions are also provided in Table 12.9.

TABLE 12.9	Proportions of ambulatory surgery patients of Bellin Health who are likely to recommend Bellin for ambulatory surgery		
Quarter	Patients likely to recommend	Total number of patients sampled	Proportion
2004 Q3	164	222	0.7387
2004 Q4	239	306	0.7810
2005 Q1	186	245	0.7592
2005 Q2	219	293	0.7474
2005 Q3	219	287	0.7631
2005 Q4	170	216	0.7870
2006 Q1	199	256	0.7773
2006 Q2	189	249	0.7590
2006 Q3	177	245	0.7224
2006 Q4	209	260	0.8038
2007 Q1	227	275	0.8255
2007 Q2	253	322	0.7857
2007 Q3	278	350	0.7943
2007 Q4	247	315	0.7841
2008 Q1	234	285	0.8211
2008 Q2	251	341	0.7361
2008 Q3	319	405	0.7877

The average quarterly proportion of patients likely to recommend Bellin is computed as follows:

$$\overline{p} = \frac{164 + 239 + \cdots + 319}{222 + 306 + \cdots + 405} = \frac{3780}{4872} = 0.7759$$

The upper and lower control limits for each sample are given by

$$\text{UCL} = 0.7759 + 3\sqrt{\frac{(0.7759)(0.2241)}{n_i}} = 0.7759 + \frac{1.2511}{\sqrt{n_i}}$$

$$\text{LCL} = 0.7759 + 3\sqrt{\frac{(0.7759)(0.2241)}{n_i}} = 0.7759 - \frac{1.2511}{\sqrt{n_i}}$$

For the third quarter of 2004, the control limits are

$$\text{UCL} = 0.7759 + \frac{1.2511}{\sqrt{222}} = 0.8599$$

$$\text{LCL} = 0.7759 - \frac{1.2511}{\sqrt{222}} = 0.6919$$

Figure 12.19 displays the p chart for all the proportions. Notice first that the control limits are of varying widths. The sample proportions are behaving as an in-control process around the center line with no out-of-control signals. Even though the stability of the process implies that Bellin is sustaining a fairly high level of satisfaction, management's goal is no doubt to find ways to increase satisfaction to even higher levels and thus cause an upward trend or upward shift in the process.

FIGURE 12.19 The p chart for proportions of ambulatory surgery patients of Bellin Health who are likely to recommend Bellin for ambulatory surgery, for Example 12.12.

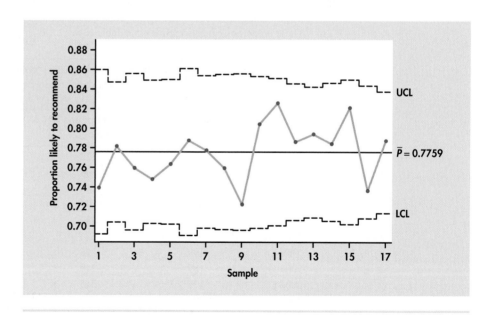

APPLY YOUR KNOWLEDGE

12.41 Unpaid invoices. The controller's office of a corporation is concerned that invoices that remain unpaid after 30 days are damaging relations with vendors. To assess the magnitude of the problem, a manager searches payment records for invoices that arrived in the past 10 months. The average number of invoices is 2875 per month, with relatively little month-to-month variation. Of all these invoices, 960 remained unpaid after 30 days.

(a) What is the total number of opportunities for unpaid invoices? What is $\overline{p}$?

(b) Give the center line and control limits for a p chart on which to plot the future monthly proportions of unpaid invoices.

 ABSENTEEISM

CASE 12.3 **12.42 Setting up a p chart.** After inspecting Figure 12.18 (page 672), you decide to monitor the future absenteeism rates using a center line and control limits calculated from the second two weeks of data recorded in Table 12.8. Find $\overline{p}$ for these 10 days and give the new values of CL, LCL, and UCL.

Control charts for counts per unit of measure

In the discussion of the p chart, there is a limit to the number of occurrences we can count. For example, if 100 parts are inspected, the most defective parts we could find would be 100. In contrast, if we were counting the number of stitch flaws in an area of carpet, then the count could be 0, 1, 2, 3, and so on indefinitely. This latter example represents the Poisson setting discussed in Section 5.4. The Poisson distribution accounts for random occurrence of events within a continuous interval of time, area, or volume. From Section 5.4, we learned that a Poisson distribution with mean μ has a standard deviation of $\sqrt{\mu}$. The Normal distribution can be used to approximate the Poisson distribution. As a rule of thumb, the Normal distribution is an adequate approximation for the Poisson distribution when $\mu \geq 5$. With these facts in mind, we can establish a three-sigma control chart for Poisson count data.

Construction of a c Chart

Suppose that k nonoverlapping units are sampled and $c_1, c_2, \ldots, c_k$ are the observed counts. Estimate the process mean count $\overline{c}$ by

$$\overline{c} = \frac{1}{k}(c_1 + c_2 + \cdots + c_k)$$

The center line and control limits for the **c chart** are

$$\text{UCL} = \overline{c} + 3\sqrt{\overline{c}}$$
$$\text{CL} = \overline{c}$$
$$\text{LCL} = \overline{c} - 3\sqrt{\overline{c}}$$

If the lower control limit computes to a negative value, then LCL is set to 0 since negative counts are not possible.

 SAFETY

EXAMPLE 12.13 Work Safety c Chart

State and federal laws require employers to provide safe working conditions for their workers. Beyond the legal requirements, many companies have implemented process improvement measures, such as six-sigma methods, to improve worker safety. Companies recognize that such efforts have beneficial effects on employee satisfaction and reduce liability and loss of workdays, all of which can improve productivity and corporate profitability. For a manufacturing facility, Table 12.10 provides 24 months of the counts of Occupational Safety and Health Administration (OSHA) reportable injuries.

The average monthly number of injuries is calculated as follows:

$$\overline{c} = \frac{12 + 6 + \cdots + 6}{24} = \frac{161}{24} = 6.7083$$

TABLE 12.10 Counts of OSHA reportable injuries per month for 24 consecutive months

Month:	1	2	3	4	5	6	7	8	9	10	11	12
Injuries:	12	6	6	10	5	2	6	5	5	4	6	6

Month:	13	14	15	16	17	18	19	20	21	22	23	24
Injuries:	16	9	10	5	7	3	5	7	12	4	4	6

The center line and control limits are given by

$$\text{UCL} = \bar{c} + 3\sqrt{\bar{c}} = 6.7083 + 3\sqrt{6.7083} = 14.48$$

$$\text{CL} = \bar{c} = 6.7083$$

$$\text{LCL} = \bar{c} - 3\sqrt{\bar{c}} = 6.7083 - 3\sqrt{6.7083} = -1.06 \rightarrow 0$$

Figure 12.20 shows the sequence plot of the counts along with the above-computed control limits. We can see that the 13th count falls above the upper control limit. This unusually high number of injuries is a signal for management investigation. If a special cause can be found for the out-of-control signal, then the associated observation should be removed from the preliminary data and the control limits should be recomputed. We leave the recomputation of this c chart application to Exercise 12.43.

FIGURE 12.20 The c chart for OSHA reportable injuries, for Example 12.13.

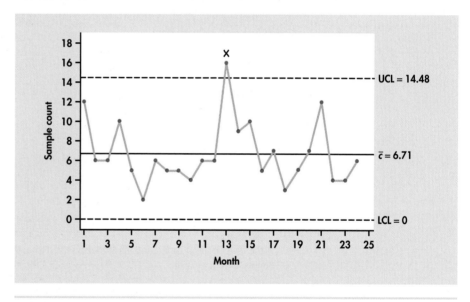

APPLY YOUR KNOWLEDGE

SAFETY

12.43 Worker safety. An investigation of the out-of-control signal seen in Figure 12.20 revealed that not only were there new hires that month but new machinery was installed. The combination of relatively inexperienced employees and unfamiliarity with the new machinery resulted in an unusually high number of injuries. Remove the out-of-control observation from the data count series and recompute the c chart control limits. Comment on control of the remaining counts.

12.44 Positive lower control limit? What values of $\bar{c}$ are associated with a positive lower control limit for the c chart?

SECTION 12.3 Summary

- **Attribute control charts** are dedicated to the monitoring and control of counting type data in the form of proportions or direct counts.

- A *p* **chart** is used to monitor sample proportions $\hat{p}$. However, when monitoring counts over continuous intervals of time, area, or volume when there is no definite limit on the number of counts that can be observed, a *c* **chart** is used.

- The interpretation of *p* and *c* charts is very similar to that of variable control charts. The out-of-control signals are also the same.

SECTION 12.3 Exercises

For Exercises 12.41 and 12.42, see pages 674–675; and for 12.43 and 12.44, see page 676.

12.45 Aircraft rivets. After completion of an aircraft wing assembly, inspectors count the number of missing or deformed rivets. There are hundreds of rivets in each wing, but the total number varies depending on the aircraft type. Recent data for wings with a total of 34,700 rivets show 208 missing or deformed. The next wing contains 1070 rivets. What are the appropriate center line and control limits for plotting the $\hat{p}$ from this wing on a *p* chart?

12.46 Call center. A large nationwide retail chain keeps track of a variety of statistics on its service call center. One of those statistics is the length of time a customer has to wait before talking to a representative. Based on call center research and general experience, the retail chain has determined that it is unacceptable for any customer to be on hold for more than 90 seconds. To monitor the performance of the call center, a random sample of 200 calls per shift (three shifts per day) is obtained. Here are the number of unacceptable calls in each sample for 15 consecutive shifts over the course of one business week: CALLCENTER

Shift:	Shift 1	Shift 2	Shift 3	Shift 1	Shift 2	Shift 3
Unacceptable:	6	17	6	9	16	10
Shift:	Shift 1	Shift 2	Shift 3	Shift 1	Shift 2	Shift 3
Unacceptable:	8	14	5	6	16	6
Shift:	Shift 1	Shift 2	Shift 3			
Unacceptable:	9	14	7			

(a) What is $\overline{p}$ for the call center process?
(b) What are the center line and control limits for a *p* chart for plotting proportions of unacceptable calls?
(c) Label the data points on the *p* chart by the shift. What do you observe that the *p* chart limits failed to pick up?

12.47 School absenteeism. Here are data from an urban school district on the number of eighth-grade students with three or more unexcused absences from school during each month of a school year. Because the total number of eighth-graders changes a bit from month to month, these totals are also given for each month.
SCHOOLABSENT

	Sept.	Oct.	Nov.	Dec.	Jan.	Feb.	Mar.	Apr.	May	June
Students	911	947	939	942	918	920	931	925	902	883
Absent	291	349	364	335	301	322	344	324	303	344

(a) Find $\overline{p}$. Because the number of students varies from month to month, also find $\overline{n}$, the average per month.
(b) Make a *p* chart using control limits based on $\overline{n}$ students each month. Comment on control.
(c) The exact control limits are different each month because the number of students *n* is different each month. This situation is common in using *p* charts. What are the exact limits for October and June, the months with the largest and smallest *n*? Add these limits to your *p* chart, using short lines spanning a single month. Do exact limits affect your conclusions?

12.48 *p* charts and high-quality processes. A manufacturer of consumer electronic equipment makes full use not only of statistical process control but of automated testing equipment that efficiently tests all completed products. Data from the testing equipment show that finished products have only 3.5 defects per million opportunities.
(a) What is $\overline{p}$ for the manufacturing process? If the process turns out 5000 pieces per day, how many defects do you expect to see at this rate? In a typical month of 24 working days, how many defects do you expect to see?
(b) What are the center line and control limits for a *p* chart for plotting daily defect proportions?
(c) Explain why a *p* chart is of no use at such high levels of quality.

12.49 Monitoring lead time demand. Refer to the lead time demand process discussed in Exercise 5.129 (page 329). Assuming the Poisson distribution given in the exercise, what would be the appropriate control chart limits for monitoring lead time demand?

12.50 Purchase order errors. Purchase orders are checked for two primary mistakes: incorrect charge account number and missing required information. Each day, 10 purchase orders are randomly selected, and the number of mistakes in the sample is recorded. Here are the numbers of mistakes observed for 20 consecutive days: ⬤ POERRORS

$$
\begin{array}{cccccccccc}
6 & 4 & 11 & 6 & 3 & 7 & 3 & 10 & 14 & 6 \\
3 & 5 & 6 & 7 & 5 & 7 & 7 & 4 & 3 & 7
\end{array}
$$

(a) What is $\bar{c}$ for the purchase order process? How many mistakes would you expect to see in 50 randomly selected purchase orders?

(b) What are the center line and control limits for a c chart for plotting counts of purchase order mistakes per ten orders? Are there any indications of out-of-control behavior?

(c) Remove any out-of-control observation(s) from the data count series and recompute the c chart control limits. Comment on control of the remaining counts.

12.51 Implications of out-of-control signal. For attribute control charts, explain the difference in implications for a process and in actions to be taken when the plotted statistic falls beyond the upper control limit versus beyond the lower control limit.

STATISTICS IN SUMMARY

Statistical process control uses a combination of graphical and numerical descriptions to decide when a process has become unstable and requires intervention. Control charts combine time plots with limits based on the sampling distributions of the statistic being charted. This chapter presents only the most common control charts, $\bar{x}$ and R charts, $\bar{x}$ and s charts, I charts, p charts, and c charts. Other types are also in common use, but in all cases the details are simple to understand and the ideas and interpretation are similar. This chapter emphasizes some aspects of practical use of control charts and places them in context by also discussing process capability. Here is a review list of the most important skills you should have acquired from your study of this chapter.

A. PROCESSES

1. Describe the process leading to some desired output using flowcharts and cause-and-effect diagrams.

2. Choose promising targets for process improvement, combining the process description with data collection and tools such as Pareto charts.

3. Demonstrate understanding of statistical control, common causes, and special causes by applying these ideas to specific processes.

4. Choose rational subgroups for control charting based on an understanding of the process.

B. VARIABLE CONTROL CHARTS

1. Make an $\bar{x}$ chart using given values of the process μ and σ for monitoring a process that has been in control.

2. Construct $\bar{x}$ and R charts or $\bar{x}$ and s charts or an I chart using estimated values of the process μ and σ.

3. (Optional) Know the distinction between control and capability and apply this distinction in discussing specific processes.

4. (Optional) Know how to estimate the percent of process outcomes meeting specifications.

5. (Optional) Calculate the capability indices C_p, C_{pk}, C_{pu}, and C_{pl} from a process in control. Describe the meaning of these indices with sketches of Normal curves.

C. ATTRIBUTE CONTROL CHARTS

1. Decide when a p chart is appropriate. Make a p chart based on past data.

2. Decide when a c chart is appropriate. Make a c chart based on past data.

CHAPTER 12 Review Exercises

12.52 Enlighten management. A manager who knows no statistics asks you, "What does it mean to say that a process is in control? Is being in control a guarantee that the quality of the product is good?" Answer these questions in plain language that the manager can understand.

12.53 Pareto charts. You manage the customer service operation for a maker of electronic equipment sold to business customers. Traditionally, the most common complaint is that equipment does not operate properly when installed, but attention to manufacturing and installation quality will reduce these complaints. You hire an outside firm to conduct a sample survey of your customers. Here are the percent of customers with each of several kinds of complaints:

Category	Percent
Accuracy of invoices	25
Clarity of operating manual	8
Complete invoice	24
Complete shipment	16
Correct equipment shipped	15
Ease of obtaining invoice adjustments/credits	33
Equipment operates when installed	6
Meeting promised delivery date	11
Sales rep returns calls	4
Technical competence of sales rep	12

(a) Why do the percents not add to 100%?
(b) Make a Pareto chart. What area would you choose as a target for improvement?

12.54 Purchased material. At the present time, about 5 lots out of every 1000 lots of material arriving at a plant site from outside vendors are rejected because they are incorrect. The plant receives about 300 lots per week. As part of an effort to reduce errors in the system of placing and filling orders, you will monitor the proportion of rejected lots each week. What type of control chart will you use? What are the initial center line and control limits?

You have just installed a new system that uses an interferometer to measure the thickness of polystyrene film. To control the thickness, you plan to measure 3 film specimens every 10 minutes and keep $\bar{x}$ and s charts. To establish control you measure 22 samples of 3 films each at 10-minute intervals. Table 12.11 gives $\bar{x}$ and s for these samples. The units are millimeters $\times 10^{-4}$. Exercises 12.55 to 12.57 are based on this process improvement setting.

12.55 s chart. Calculate control limits for s, make an s chart, and comment on control of short-term process variation.

12.56 $\bar{x}$ chart. Interviews with the operators reveal that in Samples 1 and 10 mistakes in operating the interferometer resulted in one high-outlier thickness reading that was clearly incorrect.

TABLE 12.11 $\bar{x}$ and s for samples of film thickness

Sample	$\bar{x}$	s	Sample	$\bar{x}$	s
1	848	20.1	12	823	12.6
2	832	1.1	13	835	4.4
3	826	11.0	14	843	3.6
4	833	7.5	15	841	5.9
5	837	12.5	16	840	3.6
6	834	1.8	17	833	4.9
7	834	1.3	18	840	8.0
8	838	7.4	19	826	6.1
9	835	2.1	20	839	10.2
10	852	18.9	21	836	14.8
11	836	3.8	22	829	6.7

Recalculate $\bar{s}$ after removing Samples 1 and 10. Recalculate UCL for the s chart and add the new UCL to your s chart from the previous exercise. Control for the remaining samples is excellent. Now find the appropriate center line and control limits for an $\bar{x}$ chart, make the $\bar{x}$ chart, and comment on control. 🎞 **FILM**

12.57 Categorizing the output. Previously, control of the process was based on categorizing the thickness of each film inspected as satisfactory or not. Steady improvement in process quality has occurred, so that just 15 of the last 5000 films inspected were unsatisfactory.
(a) What type of control chart discussed in this chapter might be considered for this setting, and what would be the control limits for a sample of 100 films?
(b) Explain why the chart in part (a) would have limited practical value at current quality levels.

12.58 Hospital losses revisited. Refer to Exercise 12.14 (page 654), in which you were asked to construct $\bar{x}$ and s charts for the hospital losses data shown in Table 12.4. 🏥 **HOSPITALLOSS**
(a) Make an R chart and comment on control of the process variation.
(b) Using the range estimate, make an $\bar{x}$ chart and comment on control of process level.

12.59 Bone density revisited. Refer to Exercise 12.26 (page 666), in which you were asked to construct $\bar{x}$ and R charts for the calibration data from a Lunar bone densitometer shown in Table 12.6. 🦴 **BONE**
(a) Make an s chart and comment on control of the process variation.
(b) Based on the standard deviations, make an $\bar{x}$ chart and comment on control of process level.

12.60 Even more signals. There are other out-of-control signals that are sometimes used with $\bar{x}$ charts. One is "15 points in a row within the 1σ level." That is, 15 consecutive points fall between $\mu - \sigma/\sqrt{n}$ and $\mu + \sigma/\sqrt{n}$. This signal suggests either that the value of σ used for the chart is too large or that careless measurement is producing results that are suspiciously close to the target. Find the probability that the next 15 points will give this signal when the process remains in control with the given μ and σ.

12.61 It's all in the wrist. Consider the saga of a professional basketball player plagued with poor free-throw shooting performance. Below are the number of free throws he made out of 50 attempts on 20 consecutive practice days: ⬤ **FREETHROW**

$$
\begin{array}{cccccccccc}
25 & 27 & 31 & 28 & 22 & 21 & 27 & 20 & 25 & 27 \\
23 & 22 & 29 & 34 & 30 & 27 & 26 & 25 & 28 & 25
\end{array}
$$

(a) Construct a p chart for the data. Does the process appear to be in control?

(b) Recognizing that the player needed insight into his free-throw shooting problems, the coach hired an outside consultant to work with the player. The consultant noticed a subtle flaw in the player's technique. Namely, the player was bending back his wrist only 85 degrees when, ideally, the wrist needs to be bent back 90 degrees for proper flick motion. Part of the problem was due to the player's stiff wrist. Over the course of the next week or so, the player was given techniques to loosen his wrist. After implementing a modification to wrist movement, he got the following results on 10 new samples (again out of 50 attempts):

$$
\begin{array}{cccccccccc}
34 & 38 & 35 & 43 & 31 & 35 & 32 & 36 & 28 & 39
\end{array}
$$

Plot the new sample proportions along with the control limits determined in part (a). What are your conclusions? What should be the values of the control limits for future samples?

12.62 Monitoring rare events. In certain SPC applications, we are concerned with monitoring the occurrence of events that can occur at any point within a continuous interval of time, such as the number of computer operator errors per day or plant injuries per month. However, for highly capable processes, the occurrence of events is rare. As a result, the data will plot as many strings of zeros with an occasional nonzero observation. Under such circumstances, a control chart will be fairly useless. In light of this issue, SPC practitioners monitor the time between successive events; for example, the time between accidental contaminated-needle sticks in a health care setting. For this exercise, consider data on the time between fatal commercial airline accidents in the United States between January 1990 and January 2009.[7] ⬤ **AIRFATALITIES**

(a) Construct an individuals chart for the time-between-fatalities data. If the lower control limit computes to a negative number, set it to 0 since negative data values are not possible. Report the lower and upper control limits. Identify any observations flagged as unusual.

(b) Time-between-events data tend to be non-Normal and most often are positively skewed. Construct a histogram for the fatalities data. Is that the case for these data?

(c) For time-between-events data, transforming the data by raising them to the 0.2777 power ($y_i = x_i^{0.2777}$) often Normalizes the data. Apply this transformation to the time-between-fatalities data and construct a histogram for the transformed data. Is this histogram consistent with the Normal distribution?

(d) Construct an individuals (I) chart for the transformed data. If the lower control limit computes to a negative number, set it to 0. Report the lower and upper control limits. Identify the points that are flagged as unusual. With what national tragedy are these observations associated? Are all the observations flagged in the transformed data the same as those flagged in part (a)?

(e) Remove the unusual observations found in part (d), and reestimate and report the control limits. What impressions do you have about the time-between-fatalities process when plotted with the revised limits? Is there evidence of improvement or worsening of the process over the almost 20-year time span?

12.63 Monitoring budgets. Control charts are used for a wide variety of applications in business. In the accounting area, control charts can be used to monitor budget variances. A budget variance is the difference between planned spending and actual spending for a given time period. Often budget variances are measured in percents. For improved budget planning, it is important to identify unusual variances on both the low and high sides. The data file for this exercise includes variance percents for 40 consecutive weeks for a manufacturing work center. ⬤ **BUDGET**

(a) Construct an I chart for the variance percents. Report the lower and upper control limits. Identify any observations that are outside the control limits.

(b) Apply the runs rule based on nine consecutive observations being on one side of the center line. Is there an out-of-control signal based on this rule? If so, what are the associated observations?

(c) Remove all observations associated with out-of-control signals found in parts (a) and (b). Reestimate the control limits and apply them to the remaining observations. Are there any more out-of-control signals? If so, identify them and remove them and reestimate limits. Continue this process until no out-of-control signals are present. Report the final control limits to be used for future monitoring.

12.64 Is it really Poisson? Certain manufacturing environments, such as semiconductor manufacturing and biotechnology, require a low level of environmental pollutants (for example, dust, airborne microbes, and aerosol particles). For such industries, manufacturing occurs in ultraclean environments known as *cleanrooms*. There are federal and international classifications of cleanrooms that specify the maximum number of pollutants of a particular size allowed per volume of air. Consider a manufacturer of integrated circuits. One cubic meter of air is sampled at constant intervals of time, and the number of pollutants of size 0.3 microns or larger is recorded. Here are the count data for 25 consecutive samples: ⬤ **CLEANROOM**

$$
\begin{array}{cccccccccccccccccccccccccc}
7 & 3 & 13 & 1 & 17 & 3 & 6 & 9 & 12 & 5 & 5 & 0 & 6 & 2 & 9 & 1 & 12 & 2 & 3 & 3 & 7 & 5 & 0 & 3 & 13
\end{array}
$$

(a) Construct a c chart for the data. Does the process appear to be in control?

(b) Remove any out-of-control signals found in part (a), and reestimate the c chart limits. Does the process now appear to be in control?

(c) Remove all observations associated with out-of-control signals found in parts (a) and (b). Reestimate the control limits and apply them to the remaining observations. Are there any more out-of-control signals? If so, identify them and remove them and reestimate limits. Continue this process until no out-of-control signals are present. Report the final control limits.

(d) A quality control manager took a look at the data and was suspicious of the numerous rounds of data point removal. Even the final control limits were bothersome to the manager because the variation within the limits seemed too large. The manager made the following statement, "I am not so sure the c chart is applicable here. I have a hunch that the process is not influenced by only Poisson-variation. I suggest we look at the estimated mean and variance of the data values." Calculate the sample variance s^2 of the original 25 values and compare this variance estimate with the mean estimate. Explain how such a comparison can suggest the possibility that a Poisson distribution may not fully describe the underlying process.

CHAPTER 12 Appendix

Using Minitab for Statistics for Quality

Automated procedures for constructing control charts of any sort or calculating process capability indices are not available in standard Excel, but a number of standard control charts are available in the WHFStat Add-In for Excel. Therefore, we focus in this Appendix on using Minitab.

$\bar{x}$ Chart

> Stat ➤ Control Charts ➤ Variables Charts
>
>   for Subgroups ➤ Xbar

The data can be placed either in one column or in separate columns. For example, if the data from Table 12.2 (page 646) were placed all in one column, they would be inputted from left to right, top to bottom; that is, we would input in a column the values 39, 33, 65, 50, 41, 46, and so on. Alternatively, the data values can be placed in five separate columns in the same manner as shown in Table 12.2. Depending on how you have the data in the worksheet, choose the appropriate option from the pull-down menu found at the top of the box. If you selected the option of all the data in one column, then click the data column into the box found below. You will then need to enter the sample size in the **Subgroups sizes** box. If you chose the option of data in more than one column, click all the data columns into the box found below. You will notice that the **Subgroups sizes** box will disappear.

If you know the mean parameter (μ) and the process standard deviation (σ), you should click the **Xbar Options** button and then select the **Parameters** tab. You can then enter the values of μ and σ in the appropriate boxes. If these boxes are left blank, Minitab will estimate the parameters based on the data. In default mode, Minitab estimates the standard deviation using a "pooled estimate"

much like the pooled estimate used with the two-sample t procedure (Section 7.2). If you want to have the limits of the means chart based on the range (R) statistic, then select the **Rbar** option. While you have the **Xbar Options** pop-up box open, you can select the **Tests** tab. Under this tab you can select other out-of-control tests such as nine consecutive points all above or all below the center line. To have Minitab use the standard deviation estimate discussed in this chapter, click the **Xbar Options** button and then the **Estimate** tab. Now select the **Sbar** option. Make sure to leave the checkmark in the **Using unbiasing constant** checkbox. If you want to set aside certain samples because they are found to be associated with special causes, click the **Data Options** button and then choose the option to indicate either which samples to include or which samples to exclude. Subsequently, indicate the rows corresponding to the option chosen (that is, inclusion and exclusion). Once all pop-up option boxes are closed, click **OK** to produce the $\bar{x}$ chart.

R Chart

> Stat ➤ Control Charts ➤ Variables Charts
>
>   for Subgroups ➤ R

Refer to the data entry discussion for the $\bar{x}$ chart, as it applies similarly to the R chart. If you know the process standard deviation (σ), you should click the **R Options** button and then select the **Parameters** tab and enter the value of σ in the **Standard deviation** box. If you don't provide a value for σ, then Minitab will use a pooled estimate based on the sample standard deviations. The use of an estimate based on the sample standard deviations for the construction of R chart limits is a bit unconventional. Typically, when dealing with ranges, the process standard deviation is estimated using the ranges themselves. To base the estimate of the process standard deviation on the ranges, click the **R Options** button and then the **Estimate** tab and finally select the **Rbar** option. While you have the **R Options** pop-up box open, you can select the **Tests** tab. Under this tab you can select other out-of-control tests. As with other control chart routines, you can click the **Data Options** button to set aside specified samples. Once all pop-up option boxes are closed, click **OK** to produce the *R* chart.

s Chart

Stat ➤ Control Charts ➤ Variables Charts

for Subgroups ➤ S

Refer to the data entry discussion for the $\bar{x}$ chart, as it applies similarly to the *s* chart. If you know the process standard deviation (σ), you should click the **S Options** button and then select the **Parameters** tab and enter the value of σ in the **Standard deviation** box. If you want Minitab to estimate the standard deviation and use the estimate discussed in this chapter, click the **S Options** button and then the **Estimate** tab. Now select the **Sbar** option. Make sure to leave the checkmark in the **Using unbiasing constant** checkbox. While you have the **S Options** pop-up box open, you can select the **Tests** tab. Under this tab you can select other out-of-control tests. As with other control chart routines, you can click the **Data Options** button to set aside specified samples. Once all pop-up option boxes are closed, click **OK** to produce the *s* chart.

I Chart

Stat ➤ Control Charts ➤ Variables Charts for

Individuals ➤ Individuals

If you know the process standard deviation (σ), you should click the **I Options** button and then select the **Parameters** tab and enter the value of σ in the **Standard deviation** box. To estimate σ, Minitab uses estimates based on computing the differences between successive observations. If you want Minitab to use the estimate (s/c_4) discussed in this chapter, then you must "trick" Minitab. Calculate s/c_4 separately and then click the **I Options** button and then the **Parameters** tab and enter the value of s/c_4 in the **Standard deviation** box. While you have the **I Options** pop-up box open, you can select the **Tests** tab. Under this tab you can select other out-of-control tests. As with other control chart routines, you can click the **Data Options** button to set aside specified samples. Once all pop-up option boxes are closed, click **OK** to produce the *I* chart.

Process Capability Indices

Stat ➤ Quality Tools ➤ Capability

Analysis ➤ Normal

Data can be placed either in one column or in separate columns. If the data are in one column, select the **Single column** option and then enter the value of "1" into the **Subgroups sizes** box. If you have control data in separate columns (as in Table 12.2, page 646), select the **Subgroups across rows of** option and then click the data columns into the box found below. Now enter the lower and upper specifications in the **Lower spec** and **Upper spec** boxes. Do not place checkmarks in the **Boundary** checkboxes. If you know the mean parameter (μ) and the process standard deviation (σ), enter their values in the **Historical mean** and **Historical standard deviation** boxes. In our discussions about the cautions about capability indices, we noted that it is preferable to use the standard deviation *s* of all measurements as opposed to control-chart-based estimates. In default mode, Minitab reports two sets of capability indices: based on *s* and based on a control chart. To report only the capability indices computed based on *s*, click the **Options** button and remove the checkmark in the **Within subgroup analysis** checkbox. (*Note*: Minitab uses *s* as opposed to the estimate s/c_4 that we used in the chapter. Since capability studies are typically based on large samples, the differences between the two approaches are negligible.) Click **OK** to close the **Options** pop-up box, and click **OK** to produce a histogram of the data with superimposed specification limits along with the capability indices. You will notice that Minitab labels the capability indices P_p and P_{pk} as opposed to C_p and C_{pk}. Minitab uses the labels C_p and C_{pk} to represent the indices that use the control-chart-based estimate of σ.

p Chart

Stat ➤ Control Charts ➤ Attributes Charts ➤ P

This routine works on the basis that there is a worksheet column having the "success" values. For example, for the data of Table 12.8 (page 672), the column would have the values 129, 121, and so on. Click the data column into the **Variables** box. If the sample size (*n*) is the same for all samples, enter its value in the **Subgroups sizes** box. If the sample sizes are different, place the sample sizes for each sample in a worksheet column and click this column into the **Subgroups sizes** box. If you want to specify the value of *p*, click the **P Chart Options** button, select the **Parameters** tab, and then enter the value of p in the **Proportion** box. If you don't specify p, Minitab will estimate the proportion in the manner discussed in Section 12.3. While you have the **P Chart Options** pop-up box open, you can select the **Tests** tab. Under this tab you can select other out-of-control tests. As with other control chart routines, you can click the **Data Options** button to set aside specified samples. Once all pop-up option boxes are closed, click **OK** to produce the p chart.

c Chart

Stat ➤ Control Charts ➤ Attributes Charts ➤ C

Click the data column of counts into the **Variables** box. If you want to specify the value of the Poisson mean μ, click the **C Chart Options** button, select the **Parameters** tab, and then enter the value of μ in the **Mean** box. If you don't specify μ, Minitab will estimate the mean in the manner discussed in Section 12.3. While you have the **C Chart Options** pop-up box open, you can select the **Tests** tab. Under this tab you can select other out-of-control tests. As with other control chart routines, you can click the **Data Options** button to set aside specified samples. Once all pop-up option boxes are closed, click **OK** to produce the c chart.

FIGURE 13.3 Time series plot of daily price changes of Walmart stock, for Example 13.1.

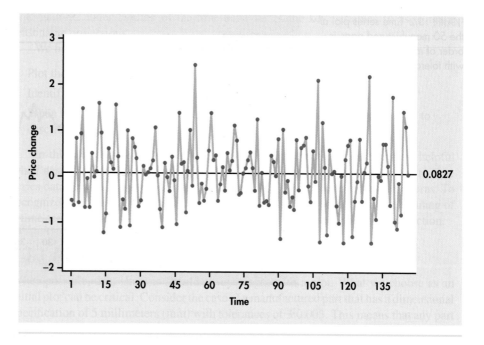

A time series characterized by a constant mean level, no systematic pattern of observations, and a constant level variation define a **random process.** A random process can be represented by the following equation:

random process

$$y_t = \mu + \epsilon_t$$

As with regression, the deviations ϵ_t represent "noise" that prevents us from observing exactly the underlying model, which in this case is simply the constant μ. These deviations are independent, with mean 0 and standard deviation σ. When the deviations are well described by the Normal distribution, we can more specifically refer to the process as a **Normal random process.**

Normal random process

As we have seen with Example 13.1, random processes do occur in practice. It is often the case, however, that time series data exhibit patterns, or systematic departures from random behavior. Here are some overall patterns commonly found in application.

> **Trend, Seasonal, and Autoregressive Variation**
>
> A **trend** in a time series is a persistent, long-term rise or fall.
>
> A pattern in a time series that repeats itself at known regular intervals of time is called **seasonal variation** or **seasonality.**
>
> A time series that shows a persisting tendency for successive observations to be correlated exhibits a form of **autoregressive behavior.**

These patterns can occur in isolation or in combination. Let us now add one more tool in the data analysis toolkit for assessing whether a time series exhibits randomness or not.

The runs test

One simple numerical check for randomness of a time series is a runs test. As a first step, the runs test classifies each observation as being above (+) or below (−) some

tab, and then enter the value of p in the **Proportion** box. If you don't specify p, Minitab will estimate the proportion in the manner discussed in Section 12.3. While you have the **P Chart Options** pop-up box open, you can select the **Tests** tab. Under this tab you can select other out-of-control tests. As with other control chart routines, you can click the **Data Options** button to set aside specified samples. Once all pop-up option boxes are closed, click **OK** to produce the p chart.

c Chart

Stat ➤ Control Charts ➤ Attributes Charts ➤ C

Click the data column of counts into the **Variables** box. If you want to specify the value of the Poisson mean μ, click the **C Chart Options** button, select the **Parameters** tab, and then enter the value of μ in the **Mean** box. If you don't specify μ, Minitab will estimate the mean in the manner discussed in Section 12.3. While you have the **C Chart Options** pop-up box open, you can select the **Tests** tab. Under this tab you can select other out-of-control tests. As with other control chart routines, you can click the **Data Options** button to set aside specified samples. Once all pop-up option boxes are closed, click **OK** to produce the c chart.

EMILY LAI/ALAMY

Time Series Forecasting

Investors around the world check the ups and downs of stock prices. Do these changes reveal patterns or are they the result of a random process? Examples 13.1 and 13.2 examine this question.

Introduction

Many business decisions depend on data tracked over time. Quarterly sales figures, annual health benefits costs, weekly production, monthly product demand, daily stock prices, and changes in market share are all examples of *time series* data.

> **Time Series**
>
> Measurements of a variable taken at regular intervals over time form a **time series.**

We've already encountered time series data in earlier chapters. For example, in Chapter 1, Figure 1.12 (page 20) displays a plot of interest rates for T-bills at regular intervals of four weeks. In Chapter 10, Exercise 10.5 (page 534) considers a small time series of research and development spending, and in Chapter 11, annual corn and soybean yields are analyzed in Exercises 11.104 through 11.110 (page 628). Indeed, all the statistical process control applications of Chapter 12 are based on time series data where the focus is to detect change in an otherwise stable process.

In the modern business environment, large and small companies alike depend on forecasts of numerous time series variables to guide their business decisions and plans. In the short term, forecasting is typically used to predict demand for products or services. Demand forecasting helps in operational decisions such as establishing daily or weekly production levels or staffing. For the longer term, businesses use forecasting to make investment decisions such as determining capacity or deciding where to locate facilities. It is not uncommon for companies to provide forecasts of key quantities in their annual reports; for example, a 2009 annual report may contain forecasts for how the company will perform in 2010. Whether the forecasts are short or long term, the first step is to gain an understanding of the time behavior of the key variables so as to make reasonable projections.

We handle time series data with the same approach used in earlier chapters:

- Plot the data, then add numerical summaries.
- Identify overall patterns and deviations from those patterns.
- When the overall pattern is quite regular, use a compact mathematical model to describe it.

In this chapter, we will focus on the plots and calculations that are most helpful when describing time series data. We will learn to identify patterns common to time series data as well as the models that are commonly used to describe those patterns. To recognize when patterns exist in time series data, it useful to understand the meaning of a time series that lacks pattern. We will explore such a time series in the next section.

13.1 Assessing Time Series Behavior

When possible, data analysis should always begin with a plot. What we choose as an initial plot can be critical. Consider the case of a manufactured part that has a dimensional specification of 5 millimeters (mm) with tolerances of ± 0.003. This means that any part that is less than 4.997 mm or greater than 5.003 mm is considered defective. Suppose 50 parts were sampled from the manufacturing process. It is tempting to start the data analysis with a histogram such as Figure 13.1. The tolerance limits have been added to the plot. The histogram shows a fairly symmetric distribution centered on the specification of 5 mm. Relative to the tolerance limits, the histogram suggests a highly capable process with little chance of producing a defective item. But before handing out a quality award to manufacturing, let us consider a *time series plot*.

> **Time Series Plot**
>
> A **time series plot** of a variable plots each observation against the time at which it was measured. Time is marked on the horizontal scale of the plot, and the variable you are studying is marked on the vertical scale. Connecting sequential data points by lines helps emphasize any change over time.

FIGURE 13.1 Histogram of a sample of the dimensions (millimeters) of 50 manufactured parts, with the tolerance limits indicated.

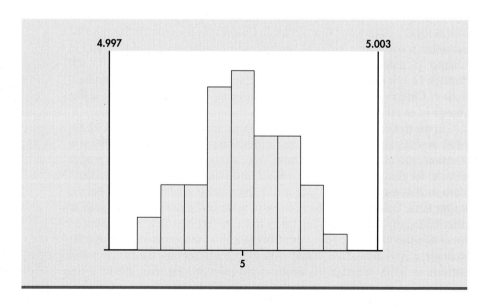

FIGURE 13.2 Time series plot of the 50 manufactured parts in order of manufacture along with tolerance limits.

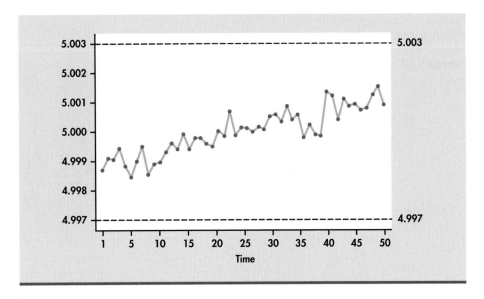

We have previously seen an example of a time series plot in Figure 1.12 (page 20). There we saw that T-bill interest rates show a series in which successive observations are close together in value, and as a result the series exhibits a "meandering" or "snakelike" type of movement. Figure 13.2 is a time series plot of the 50 manufactured parts plotted in order of manufacture. This graph gives a dramatically different conclusion about the process. The process is not stable between the target tolerances but rather exhibits an upward trend. If the trend were to continue, we would expect many defective items to be produced. The histogram in this case is of no predictive value and is misleading. The moral of the story is clear. The first step in the analysis of time series data should always be the construction of a time series plot.

Random processes

EXAMPLE 13.1 Stock Price Changes

WALMART

Whether it is by Blackberry, PC, or cell phone, investors around the world are habitually checking the ups and downs of individual stock prices or of indices (like the Dow Jones or S&P 500). Even the most casual investor knows that stock prices constantly change over time. If the changes are large enough, investors can make a fortune or suffer a great loss.

Figure 13.3 plots the daily stock price changes of Walmart Stores, Inc. from the beginning of January 2008 through the end of July 2008.[1] To add perspective, a horizontal line at the average has been superimposed on the plot. We are left with some impressions:

- Although the price changes exhibit clear variation around the average, they are not trending away from the average over time.

- Sometimes when consecutive observations bounce from one side of the average to the other side, while at other times they do not. Overall, there is no persistent pattern in the sequence of consecutive observations. The fact that an observation is above or below the average provides no insight as to whether future points will fall above or below the average.

- The variation of the observations around the sample average appears about the same throughout the series.

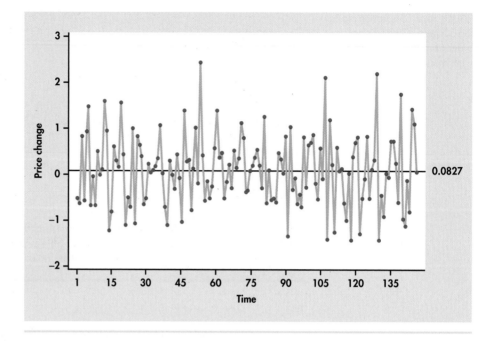

random process

A time series characterized by a constant mean level, no systematic pattern of observations, and a constant level variation define a **random process.** A random process can be represented by the following equation:

$$y_t = \mu + \epsilon_t$$

As with regression, the deviations ϵ_t represent "noise" that prevents us from observing exactly the underlying model, which in this case is simply the constant μ. These deviations are independent, with mean 0 and standard deviation σ. When the deviations are well described by the Normal distribution, we can more specifically refer to the *Normal random process* process as a **Normal random process.**

As we have seen with Example 13.1, random processes do occur in practice. It is often the case, however, that time series data exhibit patterns, or systematic departures from random behavior. Here are some overall patterns commonly found in application.

> **Trend, Seasonal, and Autoregressive Variation**
>
> A **trend** in a time series is a persistent, long-term rise or fall.
>
> A pattern in a time series that repeats itself at known regular intervals of time is called **seasonal variation** or **seasonality.**
>
> A time series that shows a persisting tendency for successive observations to be correlated exhibits a form of **autoregressive behavior.**

These patterns can occur in isolation or in combination. Let us now add one more tool in the data analysis toolkit for assessing whether a time series exhibits randomness or not.

The runs test

One simple numerical check for randomness of a time series is a runs test. As a first step, the runs test classifies each observation as being above (+) or below (−) some reference value such as the sample mean. Based on this classification, a run is a string of consecutive pluses or minuses. To illustrate, suppose that we observe the following sequence of 10 observations:

$$5 \quad 6 \quad 7 \quad 8 \quad 13 \quad 14 \quad 15 \quad 16 \quad 17 \quad 19$$

What do you notice about the sequence? It is distinctly nonrandom because the observations increase in value. Assigning $+$ and $-$ symbols relative to the sample mean of 12, we have the following:

$$- \quad - \quad - \quad - \quad + \quad + \quad + \quad + \quad + \quad +$$

In this sequence, there are two runs: $----$ and $++++++$. The trending values result in runs of longer length, which, in turn, leaves us with a very small number of runs. Imagine now a very different nonrandom sequence of 10 observations. Suppose that the consecutive observations oscillate from one side of the mean to the other side. In such a case, the runs will be of length one (either $+$ or $-$) and we would observe 10 runs. These extreme examples bring out the essence of the runs test. Namely, if we observe too few runs or too many runs, then we should suspect that the process is not random. In the hypothesis-testing framework, we are considering the following competing hypotheses:

H_0: Observations arise from a random process.

H_a: Observations arise from a nonrandom process.

By the phrase "nonrandom process" we are not suggesting that the process is not subject to random variation. The term "nonrandom process" means that a process is not subject to "pure" random variation. Time series with patterns such as trends and/or seasonality are examples of nonrandom processes. We utilize the following facts to use the runs count to test of the hypothesis of a random process.

Runs Test for Randomness

For a sequence of n observations, let n_A be the number of observations above the mean and let n_B be the number of observations below or equal to the mean. If the underlying process generating the observations is random, then the number of runs statistic R has mean

$$\mu_R = \frac{2n_A n_B + n}{n}$$

and standard deviation

$$\sigma_R = \sqrt{\frac{2n_A n_B (2n_A n_B - n)}{n^2(n-1)}}$$

The **runs test** rejects the hypothesis of a random process when the observed number of runs R is far from its mean. For sequences of at least 10 observations, the runs statistic is well approximated by the Normal distribution.

EXAMPLE 13.2   Runs Test and Stock Price Changes

In Example 13.1, we observed $n = 146$ price changes. To make the necessary counts for the runs test, it is convenient to subtract the sample mean from each of the observations. The focus can then be simply on whether the resulting observations are positive or negative. Figure 13.4 shows the sequence of pluses and minuses for the price change series with the first 5 runs identified. Going through the whole sequence, we find 73 runs, and we also find $n_A = 73$ observations above the

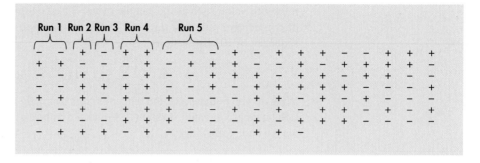

sample mean of 0.0827 and $n_B = 73$ observations below or equal to the sample mean. Given these counts, the number of runs has mean

$$\mu_R = \frac{2n_A n_B + n}{n}$$

$$= \frac{2(73)(73) + 146}{146} = 74$$

and standard deviation

$$\sigma_R = \sqrt{\frac{2n_A n_B (2n_A n_B - n)}{n^2(n-1)}}$$

$$= \sqrt{\frac{2(73)(73)[2(73)(73) - 146]}{146^2(145)}} = \sqrt{36.248} = 6.021$$

The observed number of runs of 73 deviates by only 1 run from the expected number (mean) of runs, which is 74. This tells us that there is no strong evidence against the hypothesis that a random process underlies the changes in stock prices. The results can be summarized with a P-value. The test statistic is given by

$$z = \frac{R - \mu_R}{\sigma_R} = \frac{73 - 74}{6.021} = -0.1661$$

Since evidence against randomness is associated with either too many or too few runs, we have a two-sided test with a P-value of

$$P = P(Z \le -0.1661) + P(Z \ge 0.1661) = 2P(Z \ge 0.1661) = 0.868$$

The P-value indicates that there is no compelling evidence to reject the hypothesis of a random process.

Figure 13.5 shows Minitab output for the runs test. Refer to Example 13.2 above to find all the values counted or computed there in the software output. The runs test serves as a convenient numerical check for the question of randomness. However, as we have learned throughout this book, no single numerical summary will always reveal the whole story. First and foremost, you will want to look carefully at the time series plot for any evidence against randomness. Given the nature of a random process, an intuitive forecast for future values would simply be the sample mean ($\overline{y} = 0.0827$). In terms of the time series plot of Figure 13.3, we would extend the center line into the future as our best guess for future observations.

Forecast

A prediction of a future value of a time series is called a **forecast**.

FIGURE 13.5 Minitab runs test output for the Walmart price change series.

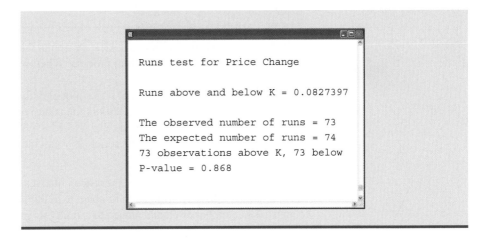

FIGURE 13.6 Histogram of Walmart price changes.

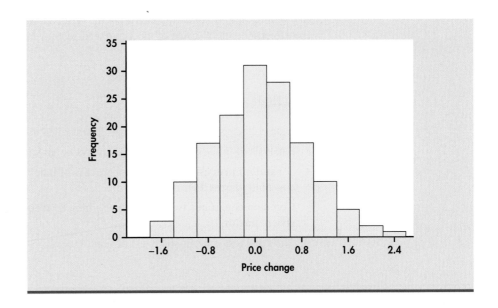

We have found that the price change series is consistent with a random process. What we now can investigate is the distribution of the price changes. Figure 13.6 provides a histogram of the data. We see that the data are compatible with the Normal distribution.[2] This brings up a point that should be emphasized. Randomness and distribution are *distinct* concepts. A process can be random and Normal but it can also be random and not Normal. It is a common mistake to believe that randomness implies Normality.

Our price change data have a standard deviation $s = 0.780$. Assuming Normality for the price changes, we can use the 68–95–99.7 rule to provide an approximate 95% interval for likely future outcomes of the price change series:

$$\bar{y} \pm 2s = 0.0827 \pm 2(0.780) = (-1.4773, 1.6427)$$

prediction interval   As discussed in Chapters 10 and 11, an interval for the prediction of individual observations is known as a **prediction interval.**

APPLY YOUR KNOWLEDGE

 WALMART

13.1 Walmart prices. In Example 13.1, we found the changes in prices of Walmart stock to be consistent with a random process. What can we say about the time series of the actual prices?

(a) Using software, construct a time series plot of daily prices. Is the series consistent with a random process? If not, explain the nature of the inconsistency.

(b) The sample mean of the prices is $\bar{y} = \$53.531$. Without the aid of software, count the number of runs in the series. Is your count consistent with your conclusion in part (a)?

13.2 Spam process. Most email servers keep Inboxes clean by automatically moving incoming mail that is determined to be spam to a "Junk" folder. Below are the weekly counts of the number of emails moved to a Junk folder for 10 consecutive weeks:[3]

| 733 | 905 | 724 | 811 | 760 | 856 | 844 | 681 | 803 | 781 |

(a) Without use of software, determine the observed and expected number of runs.

(b) Determine the P-value of the hypothesis test for randomness.

SECTION 13.1 Summary

- Data collected at regular time intervals are called a **time series.**

- Analysis of time series data should always begin with a **time series plot.**

- A **random process** is a patternless process of independent observations, which vary constantly about their mean over time.

- Common departures from randomness include **trend, seasonality,** and **autoregressive behavior.**

- The **runs test** is a statistical test of randomness.

SECTION 13.1 Exercises

For Exercises 13.1 and 13.2, see page 692.

13.3 Trading volumes. In Chapter 6, Figure 6.2 (page 335) shows the daily trading volumes of AT&T stock from January 2 to May 12, 2009. In Example 6.2 (page 333), it is stated that runs analysis would show that observing a sequencing of observations as patterned as or more patterned than that shown in Figure 6.1 would occur only 2.1% of the time if, in fact, trading volumes are truly random over time. Confirm this statement. TRADINGVOLUMES.

13.4 Yearly changes in personal consumption of services. The U.S. Department of Commerce tracks various measures of personal consumption expenditures by product type. Consider data on the yearly percent changes in personal consumption expenditures on services (for example, transportation, health care, food, and recreation) for the years 1999 to 2008:

Year:	1989	1990	1991	1992	1993	1994	1995
Percent change:	3.0	3.0	1.5	3.6	3.2	3.0	2.5

Year:	1996	1997	1998	1999	2000	2001	2002
Percent change:	2.9	3.1	4.4	4.1	5.0	2.5	1.9

Year:	2003	2004	2005	2006	2007	2008
Percent change:	1.9	2.9	3.0	2.7	2.4	0.7

(a) Determine the values of n_A and n_B and the number of runs around the sample mean.

(b) Assuming the underlying process is random, what is the expected number of runs?

(c) Determine the P-value of the hypothesis test of randomness.

13.5 Runs test output. Below is the runs test output for a time series labeled X:

```
Runs test for X

Runs above and below K = 510.667

The observed number of runs = 46
The expected number of runs = 39.8974
41 observations above K, 37 below
P-value = ???
```

(a) How many observations are in the data series?
(b) Determine the missing *P*-value.

13.6 Randomness versus distribution. In this section, we defined a Normal random process as being a process that generates independent observations that are well described by the Normal distribution. Consider daily data on the average waiting time (minutes) for patients at a health clinic over the course of 50 consecutive workdays. 🕐 CLINIC
(a) Use statistical software to make a time series plot of these data. From your visual inspection of the plot, what do you conclude about the behavior of the process?
(b) Conduct the runs test on the time series. Is there evidence at the 5% level that the process is not random?
(c) Use statistical software to create a histogram and a Normal quantile plot of these data. What do these plots suggest?
(d) Based on what you learned from parts (a), (b), and (c), summarize the overall nature of the clinic waiting-time process.

13.2 Modeling Trend and Seasonality Using Regression

A time series plot along with the runs test provides us with basic tools to assess whether a time series is random or not. With a random time series, our "modeling" of the series trivializes to using a single numerical summary such as the sample mean. However, if we find evidence that the time series is not random, then the challenge is to model the systematic patterns. By capturing the essential features of the time series in a statistical model, we position ourselves for improved forecasting. In this section, we consider two common systematic patterns found in practice: (1) trend and (2) seasonality. We will learn how the regression methods covered in Chapters 10 and 11 can be used to model these time series effects.

Identifying trends

As we saw in Figure 13.2 (page 687), a trend is a steady movement of the time series in a particular direction. Trends are frequently encountered and are of practical importance. A trend might reflect growth in company sales. Continual process improvement efforts might cause a process, such as defect levels, to steadily decline.

EXAMPLE 13.3 Monthly Cable Sales

WIRE

Consider the monthly sales data for festoon cable manufactured by a global distributor of electric wire and cable from January 2005 to December 2007. Festoon cable is a flat cable used in overhead material-handling equipment such as cranes, hoists, and overhead automation systems. The sales data are specifically the total number of linear feet (in thousands of feet) sold in a given month. In the time series plot of Figure 13.7, the upward trend is clearly visible.

   If we ignore the line connections made between successive observations, the plot of Figure 13.7 can be viewed as a scatterplot of the response variable (sales) against the explanatory variable, which is simply the indexed numbers 1, 2, ..., 36. With this perspective we can use the ideas of simple linear regression (Chapter 10) and software to estimate the upward trend. From the Excel regression output on the next page, we estimate the linear trend to be

$$\widehat{SALES}_t = 147.186 + 4.533t$$

where t is a time index of the number of months elapsed since the first month of the time series; that is, $t = 1$ corresponds to January 2005, $t = 2$ corresponds to February 2005, etc. Figure 13.8 is the time series plot of Figure 13.7 with the estimated trend line superimposed.

FIGURE 13.7 Time series plot of monthly sales of festoon cable and excel output, for Example 13.3.

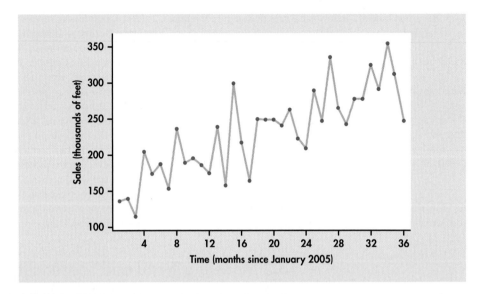

	A	**B**	**C**	**D**	**E**	**F**	**G**
1	SUMMARY OUTPUT						
2							
3	*Regression Statistics*						
4	Multiple R	0.807113898					
5	R Square	0.651432845					
6	Adjusted R Square	0.64118087					
7	Standard Error	35.44415614					
8	Observations	36					
9							
10	ANOVA						
11		*df*	*SS*	*MS*	*F*	*Significance F*	
12	Regression	1	79827.29046	79827.29046	63.5421794	2.74599E-09	
13	Residual	34	42713.79895	1256.288204			
14	Total	35	122541.0894				
15							
16		*Coefficients*	*Standard Error*	*t Stat*	*P-value*		
17	Intercept	147.1864857	12.06523531	12.19922214	5.69731E-14		
18	t	4.532942214	0.56865535	7.971334857	2.74599E-09		

The equation of the line provides a mathematical model for the observed trend in sales. The estimated slope of 4.533 indicates that festoon cable sales increased an average of 4.533 thousand feet per month from January 2005 to December 2007.

In Chapters 10 and 11, we discussed how a plot of the residuals against the explanatory variables can be used to check whether certain conditions of the regression

FIGURE 13.8 Trend line (black)
fitted to monthly sales of
festoon cable, for
Example 13.3.

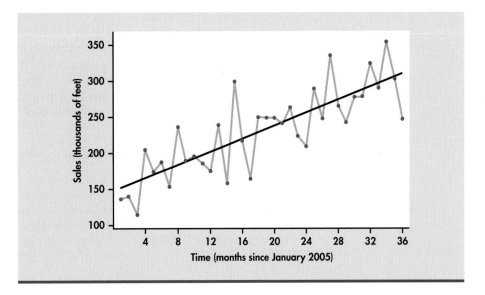

FIGURE 13.8 Trend line (black)
fitted to monthly sales of
festoon cable, for
Example 13.3.

model are being met. When using regression for time series data, we need to recognize that the residuals are themselves time ordered. If we are successful in modeling the systematic component(s) of a time series, then the residuals will behave as a random process. Figure 13.9 shows a time series plot of the residuals along with the runs test

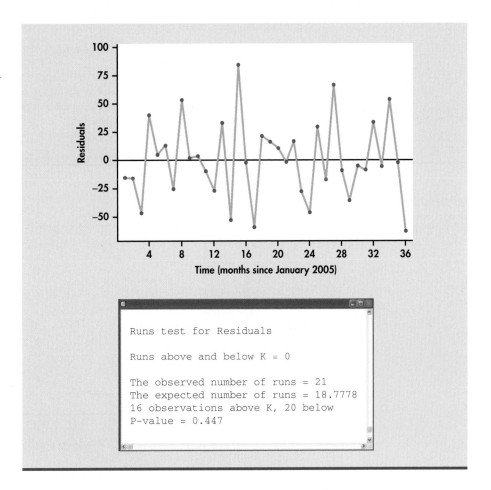

FIGURE 13.9 Randomness
checks of the residuals for the
linear trend fit for festoon
cable sales, for Example 13.3.

for the residuals. Both of these checks indicate that the residuals are consistent with randomness, indicating that the linear trend is a good fit for the time series.

In Chapter 11, we used regression techniques to fit a variety of models to data. For example, we used a polynomial model to predict the prices of houses in Example 11.19 (page 609). Trends in time series data may also be best described by a curved model like a polynomial. The techniques of Chapter 11 help us fit such models to time series data. Example 13.4 illustrates the fitting of a curved model to a nonlinear trend.

EXAMPLE 13.4   Chinese Car Ownership

CHINACARS

China's rapid economic growth can be measured on numerous dimensions. Consider Figure 13.10, which shows the time series plot of the number of passenger cars owned (in tens of thousands) in China from 1990 to 2006.[4] What we see is an increasing rate of growth with time. However, the curved relationship cannot be accommodated by a straight-line model as in the previous example. One possible approach to fitting curved trends is to introduce the square of the time index to the model:

$$\widehat{CARS}_t = b_0 + b_1 t + b_2 t^2$$

quadratic trend model

Such a fitted model is known as a **quadratic trend model.** Figure 13.11 shows the best-fitting quadratic trend model. Even though the quadratic model does attempt to capture the curvature, it is clear that the fits are systematically off and the model is not able to keep pace with the explosive growth.

exponential trend model

An alternative model that is often well suited to rapidly growing (or decaying) data series is an **exponential trend model.** Figure 13.12 shows the China car ownership data with an exponential trend superimposed. Statistical software estimates the exponential trend to be

$$\widehat{CARS}_t = 20.0788 e^{0.266969t}$$

The mathematical constant e was introduced in Chapter 5 on page 320. In comparison to the quadratic trend model, the exponential trend model fits the data series remarkably well.

We can use our trend model to forecast the number of passenger cars owned in a future year. Year 2007 corresponds to $t = 18$, and we get

$$\widehat{CARS}_{18} = 20.0788 e^{0.266969 \times 18} = 2453.0969$$

Since the data are in tens of thousands, the forecasted number is 24,530,969 cars.

FIGURE 13.10 Time series plot of car ownership in China, for Example 13.4.

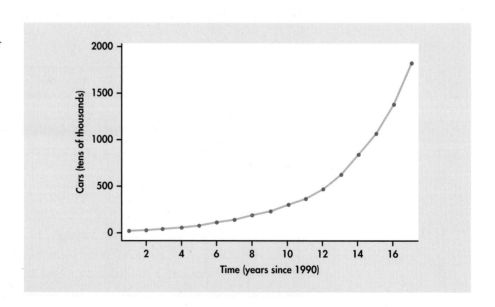

FIGURE 13.11 Quadratic trend (black) fitted to the China car ownership series, for Example 13.4.

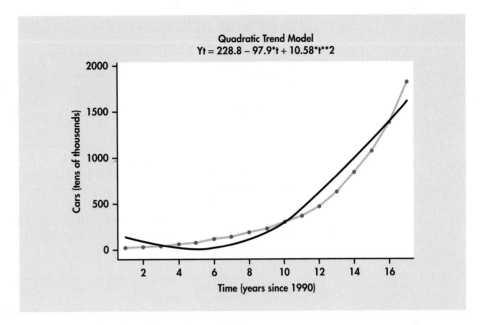

FIGURE 13.12 Exponential trend (black) fitted to the China car ownership series, for Example 13.4.

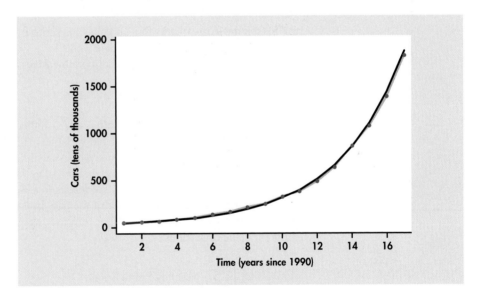

APPLY YOUR KNOWLEDGE

13.7 Monthly cable sales. Refer to the computer output for the linear trend fit of Example 13.3.
(a) Test the null hypothesis that the regression coefficient for the trend term is zero. Give the test statistic, P-value, and your conclusion.
(b) Provide forecasts for festoon cable sales for January 2008 and February 2008.

13.8 Number of portable Macs shipped. Table 13.1 displays the time series of the number of portable Macintosh computers shipped in each of eight consecutive fiscal quarters of fiscal years 2007 and 2008.[5] We usually want a time series longer than just eight time periods, but this short series will give you a chance to do some calculations

TABLE 13.1   Macintosh portable computers shipped per fiscal quarter (thousands of units)

Fiscal year-quarter	Units shipped	Fiscal year-quarter	Units shipped
2007-1st	960	2008-1st	1342
2007-2nd	891	2008-2nd	1433
2007-3rd	1130	2008-3rd	1553
2007-4th	1347	2008-4th	1675

by hand. Hand calculations take the mystery out of what computers do so quickly and efficiently for us.

(a) Make a time series plot of these data.

(b) Using the following summary information, calculate the least-squares regression line for predicting the number of portable Macs shipped (in thousands of units). The variable t simply takes on the values 1, 2, 3, ..., 8 in time order.

Variable	Mean	Std. dev.	Correlation
t	4.5	2.44949	0.969
Portable Macs shipped	1292.5	275.574	

(c) Sketch the least-squares line on your time series plot from part (a). Does the linear model appear to fit these data well?

(d) Interpret the slope in the context of portable Macs shipped.

Seasonal patterns

Variables of economic interest are often tied to other events that repeat with regular frequency over time. Agriculture-related variables will vary with the growing and harvesting seasons. Sales data may be linked to events like regular changes in the weather, the start of the school year, and the celebration of certain holidays. As a result, we find a repeating pattern in the data series that relates to a particular "season," such as month of the year, day of the week, or hour of the day. In the applications to follow, we will see that to improve the accuracy of our forecasts we will need to account for seasonal variation in the time series.

EXAMPLE 13.5   Monthly Retail Sales

USRETAIL

The Census Bureau tracks retail sales using the Monthly Retail Trade Survey.[6] Retail establishments are categorized and tracked by their North American Industry Classification System (NAICS) code. NAICS code 452 corresponds to general merchandise stores (like Walmart). Comparing sales data for a specific retail company with national sales data lets us see how the company is performing compared with its industry as a whole.

Figure 13.13 plots monthly retail sales of general merchandise stores beginning in January 2000 and ending in August 2008 (104 months). The plot reveals interesting characteristics of retail sales for these stores:

• Since January 2000, overall sales have gradually increased. This gradual increase is reflected in the superimposed trend line.

• A distinct pattern repeats itself every 12 months: January sales are lower than during the rest of the year; sales then increase but level off until October, when sales dramatically increase to a peak in December.

FIGURE 13.13 Trend line (black) fitted to U.S. retail sales of general merchandise stores, for Example 13.5.

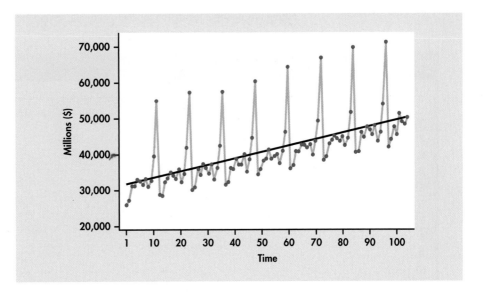

The trend line fitted to the data in Figure 13.13 ignores the seasonal variation in the retail sales time series. Because of this, using a trend model to forecast sales for, say, December 2008 will likely result in a gross underestimate since the line underestimates sales for all the Decembers in the data set. We will need to take the month-to-month pattern into account if we wish to accurately forecast sales in a specific month.

Using indicator variables

Indicator variables were introduced in Chapter 11 (page 611). We can use indicator variables to add the seasonal pattern in a time series to a trend model. Let's look at the details for the monthly retail sales data of Example 13.5.

EXAMPLE 13.6 Monthly Retail Sales

The seasonal pattern in the sales data seems to repeat every 12 months, so we begin by creating $12 - 1 = 11$ indicator variables.

$$Jan = \begin{cases} 1 \text{ if the month is January} \\ 0 \text{ otherwise} \end{cases}$$

$$Feb = \begin{cases} 1 \text{ if the month is February} \\ 0 \text{ otherwise} \end{cases}$$

$$\vdots$$

$$Nov = \begin{cases} 1 \text{ if the month is November} \\ 0 \text{ otherwise} \end{cases}$$

December data are indicated when all 11 indicator variables are 0.

We can extend the trend model with these indicator variables. The new model will capture trend along with the seasonal pattern in the time series.

$$SALES_t = \beta_0 + \beta_1 t + \beta_2 Jan + \cdots + \beta_{12} Nov + \epsilon_t$$

Fitting this multiple regression model to our data, we get the following output:

	A	B	C	D	E	F	G
1	SUMMARY OUTPUT						
2							
3	*Regression Statistics*						
4	Multiple R	0.996931774					
5	R Square	0.993872962					
6	Adjusted R Square	0.993065001					
7	Standard Error	730.4551083					
8	Observations	104					
9							
10	ANOVA						
11		*df*	*SS*	*MS*	*F*	*Significance F*	
12	Regression	12	7876054778	656337898.2	1230.100006	4.52323E-95	
13	Residual	91	48554384.54	533564.6653			
14	Total	103	7924609163				
15							
16		*Coefficients*	*Standard Error*	*t Stat*	*P-value*	*Lower 95%*	*Upper 95%*
17	Intercept	53472.42708	288.7377338	185.1937617	4.4855E-119	52898.88506	54045.96911
18	t	174.6032022	2.391248891	73.0175779	1.42851E-82	169.8532801	179.3531242
19	Jan	-27811.20621	355.1387021	-78.31082911	2.70326E-85	-28516.6456	-27105.76682
20	Feb	-27081.80941	355.0662404	-76.27255517	2.8753E-84	-27787.10487	-26376.51396
21	Mar	-23247.52373	355.0098711	-65.48416148	2.39177E-78	-23952.70721	-22542.34024
22	Apr	-23768.90471	354.9696019	-66.96039487	3.27644E-79	-24474.0082	-23063.80121
23	May	-21335.17458	354.9454381	-60.10832169	4.87875E-75	-22040.23007	-20630.11908
24	June	-22603.55556	354.9373832	-63.68322028	2.86785E-77	-23308.59505	-21898.51606
25	July	-23482.04765	354.9454381	-66.15678109	9.61772E-79	-24187.10314	-22776.99215
26	Aug	-21753.87307	354.9696019	-61.28376333	8.7333E-76	-22458.97656	-21048.76958
27	Sept	-25412.19039	365.2980002	-69.56564334	1.08521E-80	-26137.80998	-24686.57081
28	Oct	-22812.9186	365.2588652	-62.45685121	1.61823E-76	-23538.46044	-22087.37675
29	Nov	-16409.6468	365.2353822	-44.92896252	6.54274E-64	-17135.142	-15684.15159

Figure 13.14 displays the time series plot of sales observations with the trend-and-season model superimposed. You can see the dramatic improvement over the trend-only model by comparing Figure 13.13 with Figure 13.14. The improved model fit is reflected in the R^2-values for the two models: $R^2 = 0.994$ for the trend-and-season model and $R^2 = 0.389$ for the trend-only model.

The regression fit of Example 13.6 implies that the average amount of increase or decrease in sales for a given month around the trend line is the *same* from year to year. As we will see as we explore Case 13.1, not all time series follow such seasonal behavior.

FIGURE 13.14 Trend-and-season model (black) fitted to U.S. retail sales of general merchandise stores, for Example 13.6.

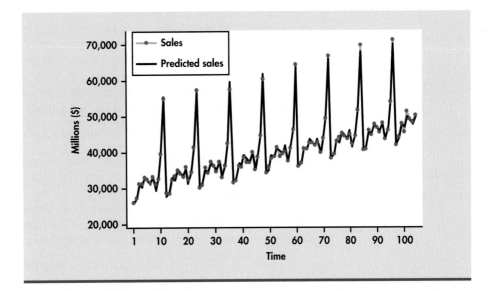

FIGURE 13.15 Time series plot of Amazon quarterly sales from the first quarter of 2000 to the third quarter of 2008.

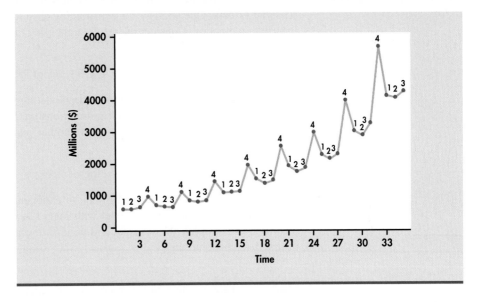

AMAZON

CASE 13.1

Amazon Sales Once just an online bookseller, Amazon continually expands its selection of consumer goods for the online shopper. Figure 13.15 displays a time series plot of Amazon's quarterly sales (in millions of dollars), beginning with the first quarter of 2000 and ending with the third quarter of 2008.[7]

The plot shows that sales are growing at an increasing rate with time. As with the data series of Example 13.4, this suggests the need for a nonlinear trend model. Along with the trend movement, the sales series also shows strong seasonality, with the fourth quarter being the strongest quarter.* Focus on the fourth-quarter spikes. Notice that the amount of increase in sales in the fourth quarter increases each year. One explanation for

---

*The strong fourth quarter seasonality here is clearly related to the holiday shopping season from Thanksgiving to Christmas. However, be aware that companies can follow different fiscal periods. For example, the first quarter of Apple, Inc. spans the holiday shopping season.

this phenomenon is that the seasonal variation is proportional to the general level of sales. Consider, for example, that fourth-quarter sales are on average 40% greater than third-quarter sales. Even though the fourth-quarter percent increase remains fairly constant, the amount of increase in dollars will be greater and greater as the sales series grows.

When we use indicator variables to incorporate seasonality into a trend model, we view the model as a trend component *plus* a seasonal component:

$$y = \text{TREND} + \text{SEASON}$$

additive seasonal effects

You can see this in Example 13.6, where we *added* the indicator variables to the trend model. A time series well represented by such a model is said to have **additive seasonal effects.**

The time series of Case 13.1 suggests that we need to consider the situation where each particular season is some proportion of the trend. Such a perspective views the model for the time series as a trend component *times* a seasonal component:

$$y = \text{TREND} \times \text{SEASON}$$

multiplicative seasonal effects

A time series well represented by such a model is said to have **multiplicative seasonal effects.** One strategy for constructing a multiplicative model is to utilize the logarithmic function. Consider the result of applying the logarithm to the product of the trend and seasonal components:

$$\log(y) = \log(\text{TREND} \times \text{SEASON}) = \log(\text{TREND}) + \log(\text{SEASON})$$

We can see that the logarithm changes the modeling of trend and seasonality from a multiplicative relationship to an additive relationship. When the trend component is an exponential trend $y = ae^{bt}$, we can gain another insight from the application of the natural logarithm:

$$\log(y) = \log(ae^{bt}) = \log(a) + \log(e^{bt}) = \log(a) + bt$$

The above breakdown implies that fitting an exponential model can be accomplished by fitting a simple linear trend model with $\log(y)$ as the response variable. Since the logarithm can simplify the seasonal and trend fitting process, it is commonly used for time series model building.

EXAMPLE 13.7 Fitting and Forecasting Amazon Sales

CASE 13.1

In Case 13.1, we observed that the time series was growing at an increasing rate, which leads us to consider an exponential trend model. Additionally, we saw that the seasonal variation increases with the growth of sales. This indicates that an additive seasonal model is probably not the best choice.

It seems that this data series is a good candidate for logarithmic transformation. Figure 13.16 displays the sales series in logged units. Compare this time series plot with Figure 13.15. In logged units, the trend is now nearly perfectly linear. Furthermore, the seasonal variation from the trend is much more constant. These facts together imply that we can model logged sales as a linear trend with additive seasonal indicator variables:

$$\log(\text{SALES}_t) = \beta_0 + \beta_1 t + \beta_2 Q1 + \beta_3 Q2 + \beta_4 Q3 + \epsilon_t$$

where Q1, Q2, and Q3 are indicator variables for quarters 1, 2, and 3, respectively. Statistical software calculates the fitted model to be

$$\log(\widehat{\text{SALES}_t}) = 6.57356 + 0.061937t - 0.33892Q1 - 0.44316Q2 - 0.44968Q3$$

FIGURE 13.16 Time series plot of Amazon quarterly sales in logged units, for Example 13.7.

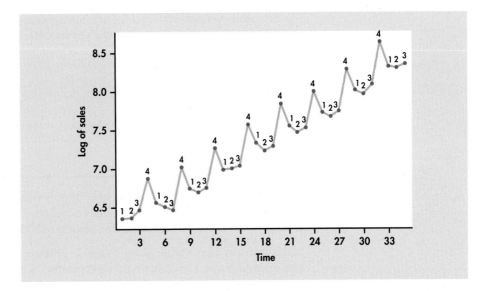

FIGURE 13.17 Trend-and-season model (black) fitted to Amazon sales, for Example 13.7.

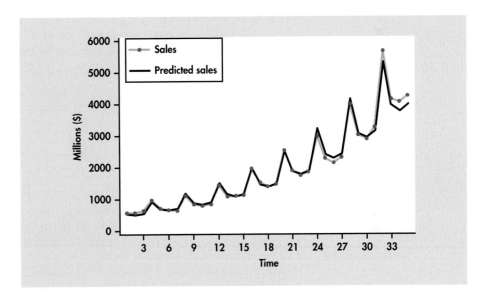

To make predictions of sales in the original units, millions of dollars, we can first predict sales in logged units using the above fitted equation and then untransform the log values by applying the exponential function, which on your calculator is the e^x key. Alternatively, we can apply the exponential function to the log-based regression equation:

$$e^{\log(\widehat{SALES_t})} = e^{6.57356 + 0.061937t - 0.33892Q1 - 0.44316Q2 - 0.44968Q3}$$

$$\widehat{SALES_t} = e^{6.57356} e^{0.061937t} \left(e^{-0.33892Q1}\right)\left(e^{-0.44316Q2}\right)\left(e^{-0.44968Q3}\right)$$

$$= 715.91397 e^{0.061937t} \times \left(e^{-0.33892Q1}\right)\left(e^{-0.44316Q2}\right)\left(e^{-0.44968Q3}\right)$$

Notice that the prediction equation for sales in the original units is indeed of multiplicative form (TREND × SEASON). Figure 13.17 displays the original sales series with the trend-and-season model superimposed. The fitted model not only follows the trend nicely but also does a good job of adapting to the increasing seasonal variation.

We can use our trend-and-season model to forecast sales in a future quarter. The fourth quarter of 2008 corresponds to $t = 36$, which means a forecast of

$$\widehat{SALES}_{36} = 715.91397e^{0.061937 \times 36}(e^{-0.33892 \times 0})(e^{-0.44316 \times 0})(e^{-0.44968 \times 0})$$

$$= 715.91397e^{2.229732}(1)(1)(1)$$

$$= 6656.11997$$

The forecast is in millions of dollars. So, fourth-quarter sales are forecasted to exceed $6.6 billion.

APPLY YOUR KNOWLEDGE

13.9 Monthly retail sales. Consider the monthly retail sales series discussed in Examples 13.5 and 13.6. Using the trend-and-season fitted model shown in Example 13.6, provide forecasts for the four remaining months of 2008.

13.10 Number of portable Macs shipped. In Exercise 13.8, you made a time plot of the data in Table 13.1. With only eight quarters, a strong quarterly pattern is hard to detect. Add indicator variables for the first, second, and third quarters to the linear trend model fitted in Exercise 13.8. Call these indicator variables Q1, Q2, and Q3. Use statistical software to fit this multiple regression model.
(a) Write down the estimated trend-and-season model.
(b) Explain why no indicator variable is needed for the fourth quarters.

13.11 Chinese car ownership. In Example 13.4, we found that an exponential trend model was a good fit to the time series of the number of passenger cars owned in China. Based on the fitted model provided in the example, what is the fitted model for the number of passenger cars owned in logged units? This problem should be done without the use of computer software. Only a calculator is needed.

Using seasonality factors

We have seen how the logarithm can be used to model multiplicative seasonal effects. Another approach does not require the explicit use of the logarithm to account for seasonal variation. This approach calls for ignoring the seasonality at first and just estimating the trend. The trend is then adjusted at each particular season by *multiplying* it by the appropriate **seasonality factor.** One seasonality factor is calculated for each season observed in the data. The idea behind seasonal factors can be seen by rearranging the trend-times-seasonal model:

seasonality factor

$$\frac{y}{\widehat{TREND}} = SEASON$$

Let's see how this works using the monthly retail sales data of Example 13.5.

EXAMPLE 13.8   Monthly Retail Sales

USRETAIL

We have treated each month as a season for the monthly retail sales data, so we will need to calculate 12 seasonality factors. If a linear trend model alone were fitted to the retail sales series, we would find it to be

$$\widehat{SALES}_t = 31{,}616 + 181.24t$$

For each of the 104 months in our time series, we calculate the ratio of actual sales divided by predicted sales, $\widehat{SALES}$. We then average these ratios by month. That is, we compute the average for all the January ratios, then the average for all the February ratios, and so on. These averages become our 12 seasonality factors. The arithmetic needed is straightforward but tedious given that we have 104 months of data. The table below displays the seasonality factors that result.

Month	Seasonality factor	Month	Seasonality factor
January	0.843	July	0.952
February	0.861	August	0.995
March	0.957	September	0.905
April	0.946	October	0.969
May	1.004	November	1.126
June	0.973	December	1.525

Our trend-and-season model is then

$$\widehat{\text{SALES}}_t = (31{,}616 + 181.24t) \times \text{SF}$$

where SF is the seasonality factor for the appropriate month corresponding to the value of t.

The seasonality factors are a snapshot of the typical ups and downs over the course of a year. The factors are used to adjust the trend model to account for the differences from season to season (month to month in this example). The factors themselves have an interpretation if their average is 1—our factors have an average of 1.005, which is very close to 1.* If you think of the factors as percents, then the factors show how each month's sales compare to average monthly sales for all 12 months. For example, December's seasonality factor is 1.525, or 152.5%, indicating that December sales are typically 52.5% above the average for all twelve months. January's seasonality factor is 0.843, or 84.3%, indicating that January sales are typically 15.7% (100 − 84.3) below the annual average. Figure 13.18 plots the seasonality factors in order from January to December. The reference line marked at 1.0 (100%) is a visual aid for interpreting the factors compared to the overall average monthly sales. Notice how the seasonality factors mimic the pattern that is repeated every 12 months in Figure 13.13—dividing the sales data by the estimated trend has isolated the seasonal pattern in the time series.

FIGURE 13.18 Seasonality factors for U.S. retail sales of general merchandise stores plotted against month.

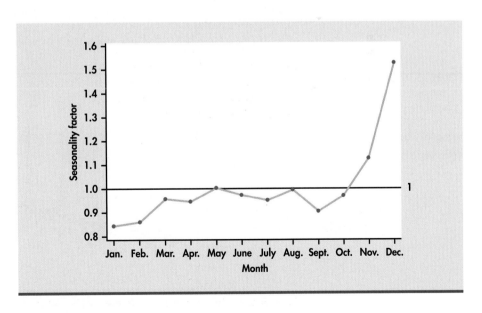

---

*If the factors did not have an average close to 1, we could adjust them so that they did average to 1 by dividing each factor by the average of the 12 factors.

13.12 Number of portable Macs shipped. In Exercise 13.8, you fitted a linear trend-only model to the time series of Macs shipped. Starting with this trend model, incorporate seasonality factors for each quarter. With only eight quarters of data, these calculations can be done by hand (or computer).

(a) Calculate the seasonality factor for each quarter.
(b) Average the four seasonality factors. Is this average close to 1? If so, interpret the seasonality factor for first quarters.
(c) Make a scatterplot of seasonality factor versus quarter with the seasonality factors on the vertical axis and the quarters on the horizontal axis. Connect the points to see the general pattern of seasonal variation. Also, draw a horizontal line at the average of the four seasonality factors.

13.13 Monthly retail sales. Example 13.6 gives an estimated trend-and-season model based on additive seasonality, while Example 13.8 gives an estimated trend-and-season model based on multiplicative seasonality. Given the pattern of seasonal variation in the series as seen in Figure 13.13, explain which model is more likely to be a better fit.

Seasonally adjusted data

Many economic time series are seasonally adjusted to make the overall trend in the numbers more apparent. Government agencies will often release both versions of a time series, so be careful to notice whether you are analyzing seasonally adjusted data or not when using government sources.[8] A seasonally adjusted time series has had each value divided by the seasonality factor corresponding to the appropriate month.

EXAMPLE 13.9   Seasonally Adjusted Monthly Retail Sales

To calculate seasonally adjusted monthly retail sales, we simply divide each actual sales value by the appropriate seasonality factor calculated in Example 13.8. Figure 13.19 displays the original time series and the seasonally adjusted time series. The seasonally adjusted time series, SALES (SA),

FIGURE 13.19 SALES (NSA) is the original U.S. retail sales of general merchandise stores (green). SALES (SA) is the seasonally adjusted data (black).

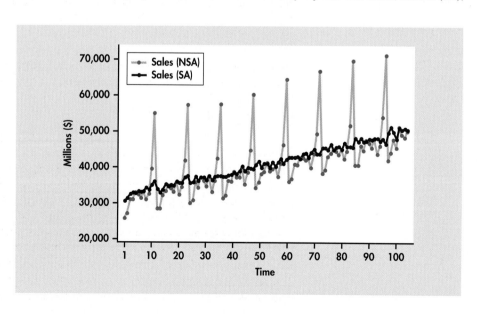

does not show the regular ups and downs of the data that are seen in the original series, SALES (NSA). The seasonally adjusted time series shows the overall trend and some random variation. Seasonally adjusted data are easier to interpret because the technique flattens the regular peaks and valleys that we expect in our data and makes it easier to identify unusual peaks and/or valleys in the data.

There are many different approaches to seasonal adjustment. The indicator variable model of Example 13.6 can be used, although we won't go through the details. Specialized time series models can also be used to seasonally adjust a time series. Different approaches will lead to different estimates and forecasts; however, for many time series the differences will not be great.

APPLY YOUR KNOWLEDGE

AMAZON

CASE 13.1 **13.14 Seasonality factors for Amazon sales data.** Ignoring the seasonal variation, the estimated exponential trend model for the Amazon sales series is

$$\widehat{SALES}_t = 524.1825e^{0.061658t}$$

(a) Calculate the seasonality factor for each quarter. Adjust the factors so that they average out to 1. Excel (or statistical software) can be used to aid the computational effort.
(b) Using software, calculate the seasonally adjusted time series of the Amazon sales data. Make a time series plot of the original Amazon time series with the seasonally adjusted time series superimposed (see Figure 13.19 for an example).
(c) Did seasonally adjusting the Amazon sales data smooth the time series to the degree that seasonally adjusting the sales data in Figure 13.19 did? What does this imply about the strength of the seasonal pattern in these two time series?

CASE 13.1 **13.15 Seasonality factors for Amazon sales data.** Example 13.7 gives the estimated trend-and-seasonal model for the Amazon sales series. Exercise 13.14 gives the estimated trend-only model for the same series. Looking at the exponents of the trend component, you will see that the coefficients for the time variables are quite close in value (0.061937 and 0.061658). However, the constants in front of the two models differ greatly (715.91397 and 524.1825).

(a) What is the multiplicative factor that needs to be multiplied by 524.1825 to obtain 715.91397?
(b) Interpret the multiplicative factor computed in part (a) in the context of seasonality factors. What season does this computed factor relate to?

Looking for autocorrelation

Earlier we said that if the residuals from a regression applied to time series data are consistent with a random process, then the estimated model has captured well the systematic components of the series. In contrast, fitting a trend-only model to a time series influenced by both trend and seasonality will result in residuals that show seasonal variation. Similarly, fitting a seasonal-only model to a time series influenced by both trend and seasonality will result in residuals that show trend behavior.

Modeling the trend and seasonal components correctly does not, however, guarantee that the residuals will be random. What trend and seasonal models do not capture is the influence of past values of time series observations on current values of the same series. A simple form of this dependence on past values is known as *autocorrelation* ("self"-correlation). In Chapter 2, we defined correlation as a measure of the strength of the linear relationship between two variables x and y (page 93). Similarly, autocorrelation measures the strength of the linear relationship between successive values of the same variable.

> **Autocorrelation**
>
> The correlation between successive values y_{t-1} and y_t of a time series is called **first-order autocorrelation** or, simply, **autocorrelation.**

The residuals from a regression model applied to a time series should be examined for signs of autocorrelation. If successive residuals tend to be both negative or both positive, then the residuals have positive autocorrelation. If successive residuals tend to be of opposite sign, then the residuals have negative autocorrelation. In either case, we have evidence that our regression assumption of independent deviations ϵ_t is inappropriate.

Let's examine the residuals that resulted from fitting an exponential trend to the Amazon sales data from Case 13.1.

EXAMPLE 13.10 Amazon Sales

CASE 13.1

positive autocorrelation

Figure 13.20 plots the residuals from the trend-and-season regression model for logged Amazon sales* given in Example 13.7. Notice the pattern of long runs of positive residuals and long runs of negative residuals. The residual series appears to "snake" or "meander" around the mean line. This appearance is due to the fact that successive observations tend to be close to each other. A precise technical term for this behavior is **positive autocorrelation.**

Our visual impression of nonrandomness is confirmed by a runs test:

```
Runs test for Residuals

Runs above and below K = 0

The observed number of runs = 9
The expected number of runs = 18.3714
16 observations above K, 19 below
P-value = 0.001
```

There are only 9 runs, which is much less than the expectation of about 18. The P-value of 0.001 shows that the evidence is strong against the hypothesis of randomness.

An alternative plot for detecting autocorrelation is given in Figure 13.21. Because autocorrelation measures the strength of the linear relationship between successive values in a series, plotting pairs of successive values will show a strong linear pattern when significant autocorrelation exists. Figure 13.21 plots points of the form (e_{t-1}, e_t) where e_t is the residual for time period t. A scatterplot of the residuals e_t against the "lagged" residuals e_{t-1} will show a strong positive association for a series that has positive autocorrelation and a strong negative association for a series that has negative autocorrelation. If the points in a *lagged residual plot* are equally dispersed among the four quadrants of the scatterplot, then we do not have visual evidence of autocorrelation in our residuals.

---

*Since the regression fitting was done on the logged data, the corresponding residuals will have mean 0. This may not be the case for the residuals calculated from the original data relative to the exponential trend-and-season model.

FIGURE 13.20 Time series plot of residuals from the trend-and-season model for logged Amazon sales, for Example 13.10.

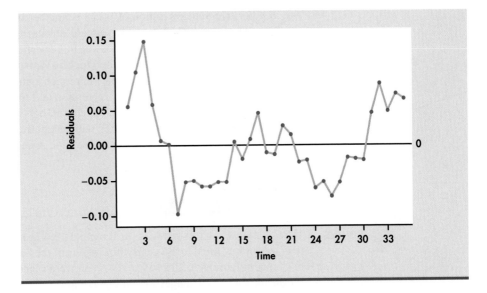

FIGURE 13.21 Lagged residual plot of the residuals from the trend-and-season model for logged Amazon sales.

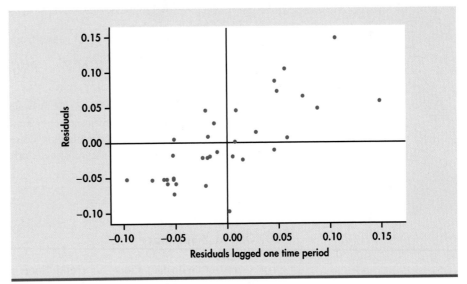

Lagged Residual Plot

A **lagged residual plot** is a scatterplot of residuals e_t against the same residuals lagged one time period; that is, we plot the points $(e_1, e_2), (e_2, e_3), \ldots, (e_{n-1}, e_n)$. The plot is used to detect autocorrelation in residuals from a time series. A positive association in the points (e_{t-1}, e_t) indicates positive autocorrelation, while a negative association indicates negative autocorrelation.

EXAMPLE 13.11 Amazon Sales

CASE 13.1

The linear pattern with positive slope in Figure 13.21 indicates that our residuals may have positive autocorrelation. The correlation for Figure 13.21 is 0.754. This is approximately the autocorrelation for the residuals lagged one time period. Statistical software calculates the autocorrelation to be 0.728 using a modified version of our correlation formula from Chapter 2.

Durbin-Watson test

The positive autocorrelation of 0.728 is another bit of evidence that our residuals are dependent, but is 0.728 a large enough autocorrelation to give us confidence in our conclusion of dependence? To answer this question conclusively, we would use a statistical test for autocorrelation—the **Durbin-Watson test.** The Durbin-Watson test statistic is provided with the multiple regression output of many statistical software packages. You can consult a multiple regression text for the details of this test.[9]

Autocorrelation in the residuals indicates an opportunity to improve the model fit. There are several approaches for dealing with autocorrelation in a time series. In the next section, we explore using past values of the time series as explanatory variables added to the model.

 CHINACARS

APPLY YOUR KNOWLEDGE

13.16 Chinese car ownership. In Example 13.4 (page 696), an exponential trend-only model was fitted to the yearly time series of the number of passenger cars owned in China. In Exercise 13.11 (page 704), you were asked to determine the fitted model for the number of passenger cars in logged units.

(a) Calculate the predicted number of passenger cars in logged units for each year in the time series.

(b) Calculate the residuals using the fitted values from part (a). That is, calculate

$$e_t = \log(CARS_t) - \log(\widehat{CARS_t})$$

for each year in the time series.

(c) Make a time series plot of the residuals. Is there visual evidence of autocorrelation in this plot? If so, describe the evidence.

(d) Make a lagged residual plot and calculate the correlation between successive residuals e_{t-1} and e_t. Is there evidence of autocorrelation? If so, describe the evidence.

SECTION 13.2 Summary

- Time series often display a long-run **trend.** Some time series also display a strong, repeating **seasonal pattern.**

- Regression methods can be used to model the trend and seasonal variation in a time series. **Indicator variables** can be used to model the seasons in a time series. Transformations, such as the logarithm, can simplify the regression-modeling process for trend and seasonal fitting. **Seasonality factors** can be used to adjust trend-only models for seasonal effects.

- **Seasonally adjusted** data have been divided by seasonality factors to remove the effect of seasonal variation on the time series. Government agencies typically release seasonally adjusted data for economic time series.

- Successive residuals may be linearly related. A **lagged residual plot** is a scatterplot of residuals plotted against the residuals lagged by one time period. If residuals exhibit **autocorrelation,** then our usual regression assumptions are not appropriate; in particular, the deviations cannot be assumed to be independent. In this case, a model based on past values of the time series should be considered.

SECTION 13.2 Exercises

For Exercises 13.7 and 13.8, see pages 697–698; for 13.9 to 13.11, see page 704; for 13.12 and 13.13, see page 706; for 13.14 and 13.15, see page 707; and for 13.16, see page 710.

13.17 Consumer loans at commercial banks. The Federal Reserve tracks assets and liabilities of commercial banks in the United States.[10] Consider monthly data on the amount of money (in billions of dollars) held in consumer loans at all commercial banks in the United States, beginning in January 2000 and ending in November 2008. 🔵 **LOANS**

(a) Use statistical software to make a time series plot of these data.

(b) Describe the overall trend present in these data.

(c) Do these data exhibit a regular, repeating pattern (seasonal variation)? If so, describe the repeating pattern.

(d) Some time series exhibit *cycles.* Cycles are clear but irregular up-and-down movements that are not part of the overall trend or seasonal variation present in the time series. Does this time series contain any cycles? If so, describe the cycles by noting when each began and ended and if the cycle was a temporary peak or trough in the series.

13.18 Consumer loans at commercial banks. Use statistical software to fit a linear trend model to the time series described in the previous exercise. Use 1, 2, 3, . . . as the values for the explanatory variable with $t = 1$ corresponding to January 2000, etc.

(a) Record the equation of the fitted linear trend model.

(b) Interpret the slope in the context of the amount of money held in consumer loans.

(c) Indicate stretches of time when the linear trend model consistently underestimates the amount of money in consumer loans. Do the same for stretches of time when the linear trend model consistently overestimates the amount of money in consumer loans.

(d) Forecast the amount of money in consumer loans for December 2008 using the linear trend model. Do you expect this forecast to be less than or more than the actual amount of consumer loans for this month? Explain your reasoning.

13.19 Passenger car production in the United Kingdom. Motor vehicle production in the United Kingdom is tracked monthly. Consider 67 months of production data for the number of passenger cars produced. The data begin with January 2002 and end with July 2007.[11] 🔵 **UKCARS**

(a) Use statistical software to make a time series plot of these data.

(b) Does this time series exhibit a clear trend? If so, describe the trend.

(c) This time series does have clear lows—months when passenger car production was distinctly less than in other months in the series. Identify the months that are clearly lower than the rest of the time series.

13.20 Passenger car production in the United Kingdom. Continue the previous exercise.

(a) Fit a linear trend-only model to the passenger car production time series. Based on this fit, forecast the passenger car production for August 2007. In light of the seasonalities you identified in part (c) of Exercise 13.19, how do you expect the forecast to perform?

(b) Make indicator variables for the months of the year, and fit a linear trend along with 11 indicators to the data series. Based on this fit, make a forecast for August 2007.

(c) Make a time series plot of the residuals. Do they appear consistent with a random process? Which particular month stands out as unusual but did not appear unusual in the original time series plot of the data?

13.21 Gold reserves. Central banks around the world hold gold reserves as financial capital. Consider the total annual world reserves (in metric tons) held by governments around the world from 1981 to 2007.[12] 🔵 **GOLD**

(a) Make a time series plot of these data. Describe the overall movement the data series.

(b) Find the observed and expected number of runs. Determine the P-value for testing the hypothesis of randomness.

(c) Take a closer look at successive observations on the time series plot of part (a). What is happening to the differences (changes) between successive observations over time?

13.22 Gold reserves. Continue the previous exercise.

(a) Compute the changes from year to year in world holdings of gold reserves. Make a time series plot of the change data. Describe the time series behavior seen in the plot.

(b) Fit a linear trend model to the change data. Report your estimated model.

(c) Make a time series plot of the residuals from the linear trend fit and perform the runs test on the residuals. What is your conclusion in terms of how well the model fits the data series?

(d) How do you interpret the linear trend model of changes as it relates to the original data series on world holdings of gold reserves?

13.23 Visitors to Canada. Given the economic implications of tourism on regions, governments and many businesses are keenly interested in tourism at local and national levels. To promote and monitor tourism to Canada, the Canadian national government established the Canadian Travel Commission (CTC).[13] One of the key indicators monitored by the CTC is the number of international visitors to Canada. In this exercise you will explore a data series on the number of monthly visitors to Canada from the United States and other countries. 🔵 **VISITCANADA**

(a) Make two time series plots, one for the number of visitors from the United States and one for the number of visitors from other countries.

(b) In both cases, for what months is visitation highest? During the off-season period, what particular month shows a bit of a surge?

(c) Do there appear to be trends in the two series? If so, are they in the same direction?

13.24 Visitors to Canada. Continue the previous exercise.

(a) Make indicator variables for the months of the year, and fit a linear trend along with 11 indicators to each of the two data series. Report your estimated models.

(b) Based on these fits, make forecasts for the number of visitors from the United States and the number of visitors from countries other than the United States for October 2008.

(c) What is the estimated coefficient for the time trend variable for each model? Interpret these coefficients in the context of the application.

(d) In light of the trend coefficient for the model for number of visitors from the United States, what would you recommend to the CTC and the Canadian government?

13.25 AT&T wireline business. With the continuing growth of the wireless phone market, it is interesting to study the impact on the wireline (landline) phone market. Consider a time series of the quarterly number of AT&T customers (in thousands) who have wireline voice connections.[14] The series begins with the fourth quarter of 2006 and ends with the third quarter of 2008.
ATT

(a) Make a time series plot of these data. Describe the overall movement of the data series.

(b) Does there appear to be seasonality?

(c) Fit the the time series with a linear trend model. Report the estimated model.

(d) Superimpose the fitted linear trend model on the data series. Does the fit seem adequate? If not, explain why not.

13.26 AT&T wireline business. Continue the previous exercise.

(a) Based on the estimated linear trend model from part (c) of Exercise 13.25, compute the residuals and plot them as a time series. Describe the pattern in the residuals.

(b) Now consider a quadratic trend model. Fit a regression model based on two explanatory variables: t and t^2 where $t = 1, \ldots, 8$. Report the estimated model.

(c) Based on the estimated quadratic trend model, compute the residuals and plot them as a time series. Is there a pattern in the residuals?

(d) Superimpose the fitted quadratic trend model on the data series. Explain in what way the quadratic fit is an improvement over the linear trend fit of Exercise 13.25.

13.27 Prediction interval for stock price. In Examples 13.1 and 13.2 (pages 687–690), we found that the daily price changes of Walmart stock were consistent with a random process. We further found that the prices changes were approximately Normal with an estimated mean and standard deviation of 0.0827 and 0.780, respectively.

(a) What is a 90% prediction interval for a future price change?

(b) On July 31, 2008, the closing price for Walmart stock was

$58.38. Given your the prediction interval from part (a), what is a 90% prediction interval for the closing price for the next trading day?

13.28 Seasonally adjusted consumer loans. The consumer loans data used in Exercises 13.17 and 13.18 are not seasonally adjusted (NSA). If we assume that the Federal Reserve used seasonality factors to seasonally adjust the time series, then we can recover approximate seasonality factors. The adjusted series is provided for you in a data file. ADJUSTEDLOANS

(a) Using software, create a new variable called "Approximate SF" by dividing the unadjusted data values by the adjusted data values for each of the 107 months in the time series.

(b) Make a time plot of the Approximate SF series. Do you observe a regular, repeating pattern? If you observe a seasonal pattern, clearly describe the pattern.

13.29 Monthly world oil supply changes. Given the importance of oil supply to the economy and national security, the U.S. government monitors oil production by country and by various aggregates such as the Organization of Petroleum Exporting Countries (OPEC).[15] The data for this exercise are the monthly changes in world oil production (in units of thousands of barrels per day on average) from February 2001 to September 2008.
OIL

(a) Make a time series plot of the monthly changes. Is there any evidence of trend over time?

(b) Superimpose a line at the sample mean of the data. What months are consistently on one side of the sample mean?

(c) Fit a seasonal-only model to the monthly change series by regressing changes on 11 monthly indicator variables. Report your estimated model.

(d) Based on your model from part (c), provide a forecast for the change in oil production from September 2008 to October 2008.

13.30 Autocorrelation in consumer loans? In Exercise 13.18, you fitted a linear trend-only model to the consumer loans data. Using the residuals from this model, look for evidence of autocorrelation.

(a) Make a time series plot of the residuals. Describe any pattern you see in this plot.

(b) Does the time series plot of residuals suggest positive or negative autocorrelation? Explain your response.

(c) Make a lagged residual plot and calculate the correlation between successive residuals e_{t-1} and e_t. Is there evidence of autocorrelation?

13.31 Autocorrelation in monthly world oil supply changes? In Exercise 13.29, you fitted a season-only model to the monthly changes in world oil production.

(a) Obtain the residuals from the estimated seasonal model and plot them as a time series. What conclusions do you draw from visual inspection of the plot?

(b) Perform the runs test on the residuals and report a P-value for the test. What do you conclude?

(c) Make a lagged residual plot and calculate the correlation between successive residuals e_{t-1} and e_t. Is there evidence of autocorrelation?

13.32 Predicting gold reserves. In Exercise 13.22, you fitted a linear trend model to the year-to-year changes in world reserves of gold held by governments.
(a) Let y_t be the amount of world reserves for period t. Reexpress the trend model for changes to give a predictive equation for y_t.
(b) Use the predictive equation from part (a) to get the fitted values for the world reserves series for periods 2, 3, ..., 27.
(c) Superimpose the fitted values from part (b) on the annual world reserve data series plotted in part (a) of Exercise 13.21. (*Note:* When plotting the fitted values, remember that they start with period 2, while the original series starts with period 1.)
(d) Make a time series plot of the residuals from the linear trend fit and perform the runs test on the residuals. What is your conclusion in terms of how well the model fits the data series?
(e) How do you interpret the linear trend model of changes as it relates to the original data series on world holdings of gold reserves?

13.33 Runs test and autocorrelation. Suppose you have three different residual time series S1, S2, and S3. The observed number of runs for series S1 is significantly less than the expected number of runs. The observed number of runs for series S2 is significantly greater than the expected number of runs. The observed number of runs for series S3 is not significantly different from the expected number of runs. If a lagged residual plot was made for each of the residual series, explain what you would likely see.

13.34 Seasonality factors for visitors to Canada. In Exercise 13.24, you fit trend-and-season models to the two series on the number of visitors to Canada. In this exercise, focus on the number of visitors from countries other than the United States.
(a) Fit a linear trend-only model to the series. Report the estimated model.
(b) Calculate the seasonality factors for each month. Adjust the factors so that they average out to 1.
(c) Make a forecast for the number of visitors from countries other than the United States for October 2008.

13.35 A more compact model. Suppose you fit a trend-and-season model to a time series of quarterly sales and you find the following:

$$\widehat{y}_t = 200 + 10t + 30Q1 + 30Q2 + 30Q3$$

Reexpress this fitted model in a more compact form using only one indicator variable.

13.36 Seasonal adjustment for Canadian visitors. In Exercise 13.34, you calculated seasonality factors for the monthly number of visitors to Canada from countries other than the United States. Using these factors, do the following.
(a) Using statistical software, calculate the seasonally adjusted visitor time series. Make a time series plot of the original visitor data with the seasonally adjusted data superimposed (see Figure 13.19, on page 706, for an example).
(b) Did seasonally adjusting the visitor data smooth the time series to the degree that seasonally adjusting the sales data in Figure 13.19 did? What does this imply about the strength of the seasonal pattern in these two time series?
(c) What do you observe in the smoothed series? Is it trending? If so, describe the trend in terms of direction and strength.

13.3 Lag Regression Models

The previous section applied regression methods from Chapters 10 and 11 to time series data. A time period variable and/or seasonal indicator variables were used as the explanatory variables, and the time series was the response variable. If such explanatory variables are sufficient in modeling the patterns in the time series, then the residuals will be independent and consistent with a random process. As we saw for the Amazon sales series in Examples 13.10 and 13.11, residuals can still exhibit nonrandomness even after trend and seasonal components are incorporated in the model. The analysis of the Amazon series suggested that we might consider *past* values of the time series as an explanatory variable.

> **Lag Variable**
> A **lag variable** is a variable based on the past values of the time series.

Notationally, if y_t represents the time series in question, then lag variables are given by $y_{t-1}, y_{t-2}, \ldots$ where y_{t-1} is called a **lag one variable**, y_{t-2} is called a **lag two variable**, and so on. Lag variables can be added to the regression model as explanatory variables. The regression coefficients for the lag variables represent multiples of past values for the prediction of future values.

Autoregressive-based models

Can yesterday's stock price help predict today's stock price? Can last quarter's sales be used to predict this quarter's sales? Sometimes the best explanatory variables are simply past values of the response variable. *Autoregressive time series models* take advantage of the linear relationship between successive values of a time series to predict future values of the series.

> ### First-Order Autoregressive Model
>
> A **first-order autoregressive model** specifies a linear relationship between successive values of the time series. The shorthand for this model is AR(1), and the equation is
>
> $$y_t = \beta_0 + \beta_1 y_{t-1} + \epsilon_t$$

The above model can be expanded to include more lag terms. For example, a time series that is dependent on lag one and lag two variables is referred to as an AR(2) model. Autoregressive models are part of a special class of time-series models known as autoregressive integrated moving-average (ARIMA) models.[16] ARIMA model building strives to find the most compact model for the data. In situations, for instance, where the modeling of a time series requires the use of many lag variables, the ARIMA model-building strategy would attempt to find an alternative model based on fewer explanatory variables. The approach is beyond the scope of this textbook. Without denying the usefulness of the ARIMA approach, it is reassuring that regression-based modeling is an effective approach for modeling most time series in practice.

EXAMPLE 13.12 Corporate Philanthropy

Charitable organizations rely on many sources for contributions, including living individuals, bequests, foundations, and for-profit corporations. Corporate philanthropy is encouraged because it enhances corporate reputation and brand image along with many other business benefits. Indeed, corporate philanthropy is often "advertised" on company Web sites, for example:

> At Bank of America, we are committed to creating meaningful change in the communities we serve through our philanthropic efforts, associate volunteerism, community development activities and investing, support of arts and culture programming and environmental initiatives.

> Our [Sony USA] philanthropic efforts reflect the diverse interests of our key businesses and focus on several distinct areas: arts education; arts and culture; health and human services; civic and community outreach; education; the environment; and volunteerism.

> AT&T is committed to advancing education, strengthening communities and improving lives. Through its philanthropic initiatives and partnerships, AT&T and the AT&T Foundation support projects that create opportunities, make connections and address community needs where we—and our customers—live and work.

PHILANTHROPY

The amount of charitable contributions from all possible sources has grown to a total of more than $300 billion in 2007. The interesting question, however, is whether for-profit corporations have taken on a bigger piece of the pie over time. Figure 13.22 shows the time series of the yearly percent of total U.S. charitable contributions from for-profit corporations from 1967 to 2007.[17] Superimposed on the plot is the 41-year average of 0.05, or 5%. We see that the percents meander about the mean level much like the residual series in Figure 13.20 does. Each time the series drifts

FIGURE 13.22 Time series plot of percent of total charitable contributions by for-profit corporations, for Example 13.12.

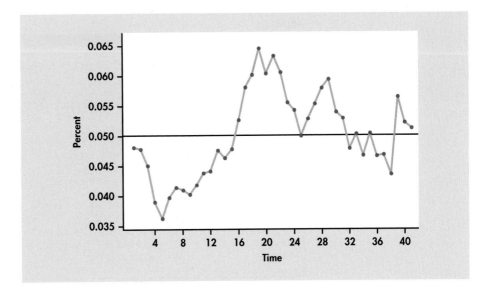

away from the overall mean level (above or below), it reverts back toward the mean. There does not seem to be any strong evidence for a long-term trend.

FIGURE 13.23 Lagged plot of percent of total charitable contributions by for-profit corporations.

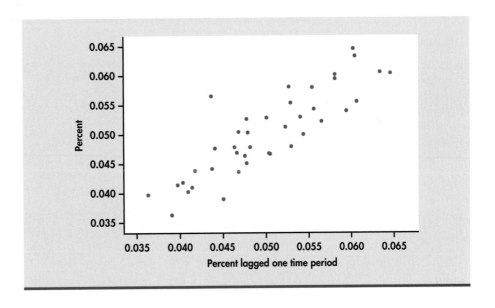

Figure 13.23 plots corporate contribution percents against corporate contribution percents lagged one period. The plot shows a strong positive linear relationship ($r = 0.862$), suggesting that an AR(1) model might be a good fit for the data series.

EXAMPLE 13.13 Fitting Corporate Philanthropy Series

Here is regression output for the estimated AR(1) fitted model:

Minitab

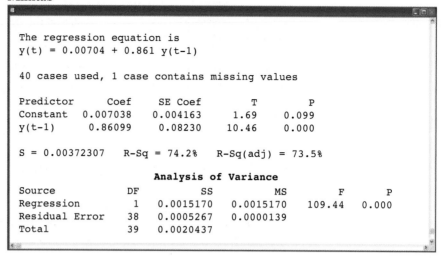

```
The regression equation is
y(t) = 0.00704 + 0.861 y(t-1)

40 cases used, 1 case contains missing values

Predictor        Coef     SE Coef          T          P
Constant      0.007038    0.004163       1.69      0.099
y(t-1)         0.86099    0.08230       10.46      0.000

S = 0.00372307    R-Sq = 74.2%    R-Sq(adj) = 73.5%

                     Analysis of Variance
Source            DF          SS          MS          F          P
Regression         1    0.0015170   0.0015170     109.44      0.000
Residual Error    38    0.0005267   0.0000139
Total             39    0.0020437
```

In summary, the fitted model is given by

$$\hat{y}_t = 0.007038 + 0.86099 y_{t-1}$$

To check the adequacy of the AR(1) fitted model, we turn our attention to the time series plot of residuals in Figure 13.24. The residuals show a mix of short runs and some oscillation with no strong tendency toward one or the other behavior. A runs test reveals 19 observed runs versus the expected number of 20.8 runs, with an associated P-value of 0.560. This means that the runs test provides no evidence against randomness.

FIGURE 13.24 Time series plot of residuals from the AR(1) model for charitable contributions by for-profit corporations, for Example 13.13.

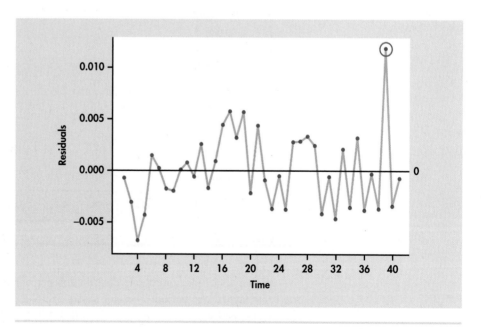

The interesting feature of the residual plot of Figure 13.24 is the unusually large positive residual (see highlighted point), which is associated with the year 2005. The large residual indicates that corporate philanthropic activity increased much greater than predicted. In the time series plot of the original series (Figure 13.22), the unusual movement of year 2005 was buried in the tightly bound snakelike movement of the series. By plotting the residuals, we are able to "tease" out unusual behavior within the overall pattern of the time series.

What explains the unusual involvement in 2005? Looking back, we realize that corporate contributions rose due to the unusual number of natural disasters, including Hurricanes Katrina, Rita, and Wilma and the Asian tsunami at the end of 2004. Since outliers can affect the regression estimates, we might want to consider reestimating the AR(1) after we exclude 2005. We leave this as an exercise for you to try.

EXAMPLE 13.14   Forecasting Corporate Philanthropy for 2008

The corporate philanthropy series ends with the 2007 percent of 5.12092%. Year 2007 is the 41st year of the time series, so our notation is $y_{41} = 0.0512092$. The AR(1) model of Example 13.13 relates the 2008 percent contribution to the 2007 percent contribution, so we can forecast 2008's percent based on the observed 2007 percent. The forecast is

$$\widehat{y_{42}} = 0.007038 + 0.86099 y_{41}$$
$$= 0.007038 + 0.86099 \times 0.0512092$$
$$= 0.0511286$$

Notice that the forecast for 2008 (5.11%) is slightly less than the observed value for 2007 (5.12%). The forecast reflects the tendency of the series to revert back to the overall mean.

Our forecast of 2008's percent was based on an *observed* data value—2007's percent is a known value in our time series. If we wish to forecast 2009's percent, we will have to base our forecast on an *estimated* value because 2008's percent is not a known value in our time series.

EXAMPLE 13.15   Forecasting Corporate Philanthropy for 2009

Because our time series ends with y_{41}, the value of y_{42} is not known. In its place, we will use the value of $\hat{y}_{42}$ calculated in Example 13.14:

$$\widehat{y_{43}} = 0.007038 + 0.86099 \hat{y}_{42}$$
$$= 0.007038 + 0.86099 \times 0.0511286$$
$$= 0.0510592$$

The forecast two periods into the future shows further reversion back to the overall mean.

one-step ahead forecast
two-step ahead forecast

Example 13.14 illustrated a **one-step ahead forecast,** while Example 13.15 illustrated a **two-step ahead forecast.** The process shown in Example 13.15 can be repeated to produce forecasts even further into the future.

In Chapters 10 and 11, we calculated prediction intervals for the response in our regression model. The regression procedures of these chapters can be used to construct prediction intervals for one-step ahead forecasts based on an AR(1) model fit. A difficulty arises when seeking two-step ahead or more prediction intervals. The one-step ahead forecast for the 2008 corporate contribution percent $\hat{y}_{42}$ depends on our estimated values of β_0 and β_1 and the *known* value of y_{41}. However, our forecast of y_{43} depends on estimated values of β_0, β_1, *and* y_{42}. The additional uncertainty involved in estimating y_{42} will make our prediction interval for y_{43} wider than our prediction interval for y_{42}.

The Amazon series ends with third-quarter 2008 sales of $4264 (in millions). The ending quarter is the 35th period in the series, so our notation is $y_{35} = 4264$.

First, use the model to forecast the logarithm of fourth-quarter 2008 sales:

$$\widehat{\log(y_{36})} = 1.90299 + 0.01485(36) - 0.69903(0) - 0.53480(0)$$
$$- 0.46167(0) + 0.76978 \log(y_{35})$$
$$= 2.43759 + 0.76978 \log(y_{35})$$
$$= 2.43759 + 0.76978 \log(4264)$$
$$= 2.43759 + 0.76978(8.35796)$$
$$= 8.87138$$

Second, use your calculator's e^x button (or use software) to calculate the forecast fourth-quarter 2008 sales in the original units:

$$\widehat{\log(y_{36})} = 8.87138$$
$$e^{\widehat{\log(y_{36})}} = e^{8.87138}$$
$$\widehat{y}_{36} = 7125.10684$$

The trend-and-season model with lag one term predicts sales of over $7.1 billion for the fourth quarter of 2008.

In Example 13.7 (page 702), the exponential trend-and-season model predicted sales of $6.6 billion. Our inclusion of the lag one term in the model increased the forecast by half a billion dollars to $7.1 billion.

APPLY YOUR KNOWLEDGE

UNEMPLOYMENT

13.37 Unemployment rate. The Bureau of Labor Statistics tracks national unemployment rates on a monthly and annual basis. Use statistical software to analyze annual unemployment rates from 1948 to 2007.[18]
(a) Make a time series plot of the unemployment rate time series.
(b) Describe any important features of the time series. Is there evidence of a trend in the series over the long run? How would you best describe the time behavior of the series?
(c) Make a lagged time series plot of the unemployment rate time series. Does this plot suggest that an AR(1) model is appropriate for this time series? Why or why not?

13.38 Unemployment rate. Let y_t denote the unemployment rate for time period t.
(a) Use software to fit a simple linear regression model using y_t as the response variable and y_{t-1} as the explanatory variable. Record the estimated regression equation.
(b) Make a lagged residual plot. Does the plot demonstrate that the lag one autocorrelation has been accounted for? Explain.
(c) Use the fitted AR(1) model from part (a) to obtain forecasts for the unemployment rate in 2008, 2009, and 2010. What do you notice about the forecast values? In what way are they similar to the forecast values shown in Figure 13.25?

SECTION 13.3 Summary

- Regression models relating a time series to time and seasonal indicator variables are not always sufficient. This is reflected when the residuals from a trend and/or seasonal model exhibit **autocorrelation.**

- The **first-order autoregressive model AR(1)** is appropriate when successive values of a time series are linearly related. An AR(1) model can be estimated by regressing the time series on a **lag one variable.** In some cases, trend and seasonal variables are needed in addition to the lag variable to fully capture the pattern in the data series.

SECTION 13.3 Exercises

For Exercises 13.37 and 13.38, see page 720.

CASE 13.1 **13.39 Amazon sales.** In Example 13.16 (page 718), the one-step ahead forecast was calculated. Using the model of Example 13.16, determine the two-step ahead forecast in original dollar units.

13.40 Trading volume of shares. Practitioners and financial researchers have found evidence that information on trading volume (number of shares traded in a given period) can be useful for investment decisions. In this exercise, you will explore whether the trading volume of a stock has any predictable pattern. Specifically, consider the daily trading volume for shares of Home Depot, Inc. from January 1 to December 22, 2008.[19] **HOMEDEPOT**
(a) Make a time series plot of the trading volume time series. Describe any important features of the time series.
(b) Make a lagged time series plot of the trading volume time series. Does this plot suggest that an AR(1) model is appropriate for this time series? Why or why not?
(c) Use software to fit a simple linear regression model using y_t as the response variable and y_{t-1} as the explanatory variable. Record the estimated regression equation.
(d) Use the fitted AR(1) model from part (c) to obtain a forecast for trading volume on December 23, 2008.

13.41 Checking model fit for trading volume of shares. In Exercise 13.40, you were asked to fit an AR(1) model to the trading volume time series. Because it is a regression fit, it is important to check the adequacy of the fit.
(a) Obtain the residuals from the AR(1) fit and make a time series plot of the residuals. Do the residuals appear random? Compare this time series plot with the time series plot of the original observations made in part (a) of Exercise 13.40.
(b) Make a lagged residual plot. Compare this plot with the lagged plot made in part (b) of Exercise 13.40. Does it appear that the autocorrelation has been accounted for?
(c) To check the Normality assumption of the regression deviations ϵ_t, make a histogram and Normal quantile plot of the residuals. Are the residuals compatible with the Normal distribution? If not, describe the nature of the distribution.

13.42 Alternative fit for trading volume of shares. In Exercise 13.41, the regression assumption of Normality was checked. When the residuals exhibit non-Normality, a transformation of the response variable can sometimes rectify the issue.
(a) Take the logarithm of the trading volume data. Use software to fit a simple linear regression model using $\log(y_t)$ as the response variable and $\log(y_{t-1})$ as the explanatory variable. Record the estimated regression equation.
(b) Obtain the residuals from the AR(1) fit on the logged data. Make a histogram and Normal quantile plot of these residuals. In comparison to the Normality checks of part (c) of Exercise 13.41, how well do these residuals match up with the Normal distribution?
(c) Use the fitted AR(1) model from part (a) to obtain a forecast for trading volume in logged units on December 23, 2008. Untransform the logged forecast to provide a forecast in the original units.

13.43 Monthly retail sales. Consider the monthly retailed sales series discussed in Examples 13.5 and 13.6 (pages 698–699).
(a) Make a lagged time series plot of the retail sales data using y_t on the vertical axis and y_{t-12} on the horizontal axis (a lag of 12 periods).
(b) What is the correlation between y_t and y_{t-12}?
(c) How well do sales 12 months ago appear to predict this month's sales? Explain your response.

13.44 SA consumer loans. Consider the seasonally adjusted (SA) consumer loan time series of Exercise 13.28 (page 712).
(a) Fit an AR(1) model using y_t as the response variable and y_{t-1} as the explanatory variable. Record the estimated regression equation. Use this model to forecast the seasonally adjusted amount of consumer loans held by commercial banks for December 2008.
(b) Make a time series plot of the residuals. Identify the periods (month and year) that the residuals indicate unusual changes (up or down).

13.4 Moving-Average and Smoothing Models

In the previous sections, we learned that regression is a powerful tool for capturing a variety of systematic effects in time series data. In practice, when we need forecasts that are as accurate as possible, regression and other sophisticated time series methods are commonly employed. However, it is often not practical or necessary to pursue detailed modeling for each and every one of the numerous time series encountered in business. One popular alternative approach used by business practitioners is based on the strategy of "smoothing" out the random variation inherent to all time series. By doing so, one gains a general feel for the longer-term movements in a time series.

Moving-average models

Perhaps the most common method used in practice to smooth out short-term fluctuations is the *moving-average* model. A moving average can be thought of as a rolling average in that the average of the last several values of the time series is used to forecast the next value.

> **Moving-Average Forecast Model**
>
> The **moving-average forecast model** uses the average of the last k values of the time series as the forecast for time period t. The equation is
>
> $$\widehat{y}_t = \frac{y_{t-1} + y_{t-2} + \cdots + y_{t-k}}{k}$$
>
> The number of preceding values included in the moving average is called the **span** of the moving average.

Some care should be taken in choosing the span k for a moving-average forecast model. As a general rule, larger spans smooth the time series more than smaller spans by averaging many ups and downs in each calculation. Smaller spans tend to follow the ups and downs of the time series. With seasonal data, the length of the season is often used for the value of k.

EXAMPLE 13.17 *New York Times* Advertising Revenue

 NYTIMESAD

Advertising revenue is the "lifeblood" of the newspaper industry. Figure 13.27 displays the quarterly advertising revenue (in thousands of dollars) for the *New York Times* beginning with the first quarter of 2005 and ending with the third quarter of 2008.[20] Superimposed on the series are the

FIGURE 13.27 Moving averages based on a span of $k = 4$ (black) superimposed on the *New York Times* quarterly revenue data (green).

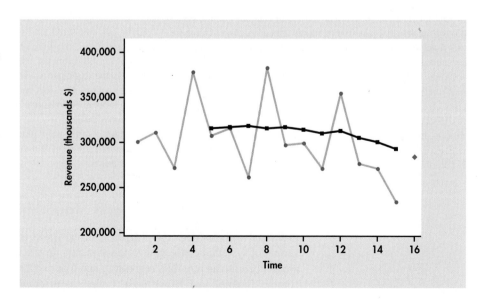

moving-average forecasts based on a span of $k = 4$. Notice that the seasonal pattern in the time series is not present in the moving averages. The moving averages are a smoothed version of the original time series. The moving averages follow the general movements of the time series but not every up and down. From management's perspective, the smoothed forecast line is probably the most important focus point. The line indicates that there is a downward trend in advertising revenues. Even more troubling, the downtrend seems to be nonlinear, which suggests an acceleration in the loss of advertising revenue.

Given that the revenue time series is 15 quarters long, period 16 (fourth quarter of 2008) represents one period into the future. Figure 13.27 includes the forecasted value for the fourth quarter of 2008 (blue point). The forecast is calculated as

$$\widehat{y}_{16} = \frac{y_{15} + y_{14} + y_{13} + y_{12}}{4} = \frac{1,136,644}{4} = 284,161$$

When a strong seasonal pattern is present, moving-average models with the span set equal to the length of the season are similar to the trend-only models presented in Section 13.2. The moving averages will follow the long-run trend of the time series, but the ups and downs of the seasons are ignored. We would not expect the moving-average model of Example 13.17 to forecast fourth-quarter *New York Times* advertising revenue well.

EXAMPLE 13.18   Chicago Cubs Attendance per Game

Major League Baseball's Chicago Cubs have been playing their home games at Wrigley Field ("The Friendly Confines") since 1916. Figure 13.28 displays the average "attendance per home game" time series beginning in 1916 and ending in 2007.[21]

Figure 13.28 also displays moving averages based on a span of 3 years and moving averages based on a span of 15 years. The moving averages based on a span of 15 years highlight the general upward trend in attendance per home game at Wrigley Field without following the ups and downs in the data. The 3-year moving averages are not as smooth as the 15-year moving averages. The 3-year moving averages follow the larger ups and downs while smoothing the smaller changes in the time series.

CUBS

FIGURE 13.28 Time series plot of Chicago Cubs annual attendance per game (blue) with 3-year (green) and 15-year (red) moving averages superimposed.

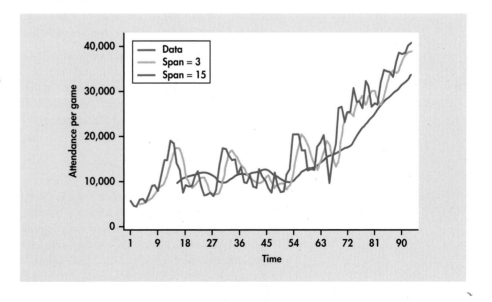

The attendance time series has 92 observations ending with attendance per home game in 2007. Our interest is in forecasting attendance per home game for 2008. We will do this using both the 3-year moving averages and the 15-year moving averages.

$$\widehat{y}_{93} = \frac{y_{92} + y_{91} + y_{90}}{3} = \frac{116,984}{3} = 38,995$$

$$\widehat{y}_{93} = \frac{y_{92} + y_{91} + \cdots + y_{78}}{15} = \frac{505,667}{15} = 33,711$$

The forecast based on the 15-year moving average is less than the forecast based on the 3-year moving average because the 15-year average includes the lower attendance figures of the 1980s. We might guess from looking at Figure 13.28 that the 3-year moving-average forecast will be more accurate than the 15-year moving-average forecast. The actual 2008 observation was 40,743. So, for this series, the 3-year moving-average forecast is more accurate. However, the 3-year moving average could not predict that 2008 would be the most successful regular season for the Cubs since 1945.

Once a new value becomes available (like 2008 average attendance per home game in Example 13.18), we update our time series to include the new observation. We can then calculate a forecast for the next time period (2009 average attendance per home game in Example 13.18).

APPLY YOUR KNOWLEDGE

 BUTTER

13.45 Butter sales. The U.S. Department of Agriculture tracks prices and sales of numerous food commodities. Consider the weekly sales volume (thousands of pounds) of butter sold in the United States starting with the first week of January 2007 and ending with the second week of December 2008.[22]

(a) Use software to make a time series plot of the butter sales time series. Describe the basic features of the time series. Be sure to comment on whether a trend or significant shifts are present or not in the series.

(b) Calculate 1-week, 2-week, 3-week, 4-week, 5-week, and 6-week moving averages for the butter sales time series.

(c) Compute the residuals for the 1-week, 2-week, 3-week, 4-week, 5-week, and 6-week moving-average models fitted in part (b). (*Note:* Given the different spans, you will find a different number of residuals for each of the models. In particular, you will find 101 residuals for the 1-week moving-average model, 100 residuals for the 2-week moving-average model, and so on.)

13.46 Comparing models for butter sales with MAD. When comparing competing forecast methods, a primary concern is the relative accuracy of the methods. Ultimately, how well a forecasting method does is reflected in the residuals ("prediction errors").

mean absolute deviation

(a) One measure of forecasting accuracy is **mean absolute deviation** (MAD). As the name suggests, it is a measure of the average size of the prediction errors. For a given set of residuals (e_t),

$$\text{MAD} = \frac{\sum |e_t|}{n}$$

where e_t is the residual for period t and n is the number of available residuals. Compute MAD for the residuals from the 1-week, 2-week, 3-week, 4-week, 5-week, and 6-week moving-average models determined in Exercise 13.45, part (c).

(b) Plot the MAD values against the moving-average span values of 1, 2, 3, 4, 5, and 6. Describe the behavior of the MAD values versus the span values. For the butter series, at what moving-average span value is MAD smallest?

mean squared error

13.47 Comparing models for butter sales with MSE. In Exercise 13.46, the mean absolute deviation was introduced as a measure of forecasting accuracy. Another measure is **mean squared error*** (MSE), which is a measure of the average size of the prediction errors in squared units:

$$\text{MSE} = \frac{\sum e_t^2}{n}$$

where e_t is the residual for period t and n is the number of available residuals.

(a) Compute MSE for the residuals from the 1-week, 2-week, 3-week, 4-week, 5-week, and 6-week moving-average models determined in Exercise 13.45, part (c).

(b) Plot the MSE values against the moving-average span values of 1, 2, 3, 4, 5, and 6. Describe the behavior of the MSE values versus the span values. For the butter series, at what moving-average span value is MSE smallest?

13.48 Comparing models for butter sales with MAPE. MAD (Exercise 13.46) measures the average magnitude of the prediction errors, and MSE (Exercise 13.47) measures the average squared magnitude of the prediction errors. To put the prediction errors in perspective, it can be useful to measure the errors in terms of percents. One such measure is **mean absolute percentage error** (MAPE):

mean absolute percentage error

$$\text{MAPE} = \frac{\sum \left(\frac{|e_t|}{y_t} \right)}{n} \times 100\%$$

where e_t is the residual for period t, y_t is the actual observation for period t, and n is the number of available residuals.

(a) Compute MAPE to the hundredth place for residuals from the 1-week, 2-week, 3-week, 4-week, 5-week, and 6-week moving-average models determined in Exercise 13.45, part (c).

(b) Plot the MAPE values against the moving-average span values of 1, 2, 3, 4, 5, and 6. Describe the behavior of the MAPE values versus the span values. For the butter series, at what moving-average span value is MAPE smallest?

Exponential smoothing models

Moving-average forecast models appeal to our intuition. Using the average of several of the most recent data values to forecast the next value of the time series is easy to understand conceptually. However, two criticisms can be made against moving-average models. First, our forecast for the next time period ignores all but the last k observations in our data set. If you have 100 observations and use a span of $k = 5$, your forecast will not use 95% of your data! Second, the data values used in our forecast are all weighted equally. In many settings, the current value of a time series depends more on the most recent value and less on past values. We may improve our forecasts if we give the most recent values greater "weight" in our forecast calculation. *Exponential smoothing* models address both of these criticisms.

Several variations exist on the basic exponential smoothing model. We will look at the details of the *simple exponential smoothing model,* which we will refer to as, simply, the *exponential smoothing model*. More complex variations exist to handle time series with specific features, but the details of these models are beyond the scope of this

---

*Minitab refers to this measure as mean squared deviation, or MSD.

chapter. We will only mention the scenarios for which these more complex models are appropriate.

Exponential Smoothing Model

The **exponential smoothing model** uses a weighted average of the observed value y_{t-1} and the forecasted value $\widehat{y}_{t-1}$ as the forecast for time period t. The forecasting equation is

$$\widehat{y}_t = wy_{t-1} + (1-w)\widehat{y}_{t-1}$$

The weight w is called the **smoothing constant** for the exponential smoothing model. The smoothing constant is a value between 0 and 1. Choosing w close to 1 puts more weight on the most recent value.

Choosing the smoothing constant w in the exponential smoothing model is similar to choosing the span k in the moving-average model—both relate directly to the smoothness of the model. Smaller values of w correspond to greater smoothing of the ups and downs in the time series. Larger values of w put most of the weight on the most recent observed value, so the forecasts tend to follow the ups and downs of the series more closely.

EXAMPLE 13.19 Great Lakes Water Level

LAKES

The water levels of the Great Lakes have received much attention in the media. As the world's largest single source of freshwater, more than 40 million people in the United States and Canada depend on the lakes for drinking water. Economies that greatly depend on the lakes include agriculture, shipping, hydroelectric power, fishing, recreation, and water-intensive industries such as steel making and paper and pulp production.

In the first decade of the twenty-first century the water levels of Lakes Superior, Michigan, and Huron were lower than historical averages. The lower levels have caused wetlands along the shore to dry up, which has had serious impacts on the reproductive cycle of numerous fish species and ultimately on the fishing industry. The shipping and steel industries are also feeling the effects in that freighters have to carry lighter loads to avoid running aground in shallower harbors. The lower levels have resulted in "alarming" reports in the media.[23] Contrast this with the higher than normal water levels of the early 1990s, which damaged shoreline properties. The fear that the lake would engulf shoreline properties was so high that cities like Chicago considered building protection systems that would have cost billions of dollars.

Figure 13.29 displays the average annual water levels (in meters) of Lakes Michigan and Huron from 1918 to 2007.[24] Also displayed are the forecasts from the exponential smoothing with $w = 0.1$ and $w = 0.7$.

Using a smoothing constant of 0.7 puts more weight on the most recent values of the time series than using a smoothing constant of 0.1 does. As a result, that curve tends to follow the ups and downs of the time series more closely than the smoother curve for $w = 0.1$ does. With a smoothing constant close to 0, the model will follow only the major changes in the time series. In all the series shown in Figure 13.29, we are left with the impression that water levels meander around a historical average with no obvious evidence of long-term trends (up or down) away from the average.

A little algebra is needed to see that exponential smoothing models address the criticisms against moving-average models. We will start with the forecasting equation for the exponential smoothing model and imagine forecasting the value of the time series for the time period $n + 1$ where n is the number of observed values in the time series. A specific example would be forecasting the 2008 water level of Lakes Michigan and

FIGURE 13.29 Average annual water levels for Lakes Michigan and Huron (blue), with two exponential smoothing models: with $w = 0.1$ (red) and with $w = 0.7$ (green).

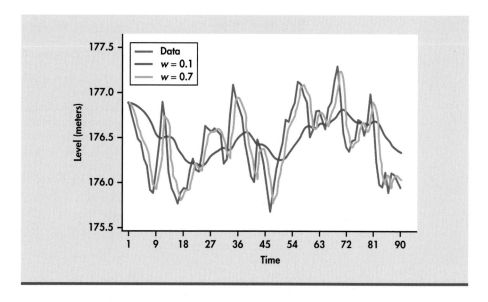

Huron y_{91} using the observed levels for the preceding $n = 90$ months:

$$
\begin{aligned}
\widehat{y}_{n+1} &= wy_n + (1-w)\widehat{y}_n \\
&= wy_n + (1-w)[wy_{n-1} + (1-w)\widehat{y}_{n-1}] \\
&= wy_n + (1-w)wy_{n-1} + (1-w)^2\widehat{y}_{n-1} \\
&= wy_n + (1-w)wy_{n-1} + (1-w)^2[wy_{n-2} + (1-w)\widehat{y}_{n-2}] \\
&= wy_n + (1-w)wy_{n-1} + (1-w)^2wy_{n-2} + (1-w)^3\widehat{y}_{n-2} \\
&\;\;\vdots \\
&= wy_n + (1-w)wy_{n-1} + (1-w)^2wy_{n-2} + \cdots + (1-w)^{n-2}wy_2 \\
&\quad + (1-w)^{n-1}\widehat{y}_1
\end{aligned}
$$

This alternative version of the forecast equation shows exactly how our forecast depends on the values of the time series. Notice that the calculation of $\widehat{y}_{n+1}$ can be tracked all the way back to the initial forecast $\widehat{y}_1$. It is common practice to initiate the exponential smoothing model by using the actual value of the first time period y_1 as the forecast for the first time period $\widehat{y}_1$.* We also learn from the backtracking that the forecast for y_{n+1} uses *all* available values of the time series $y_1, y_2, \ldots, y_n$, not just the most recent k-values like a moving-average model would. The values are not equally weighted. Using a value of w close to 1 puts greater weight on the most recent observation. In this version of the forecast equation, we also see how the model gets its name. The coefficients in the forecast model decrease exponentially in value as you read the equation from left to right (with the exception of the last coefficient, $(1-w)^{n-1}$). Exercise 13.51 (page 730) has you explore what happens to the coefficients as you change the value of the smoothing constant w.

While the second version of our forecasting equation reveals some important properties of the exponential smoothing model, it is easier to use the first version of the equation for calculating forecasts.

---

*Statistical software offers alternative options for initiating the exponential smoothing model. For example, Minitab's default is to use the average of the first six observations as the initial forecast.

EXAMPLE 13.20 Forecasting Great Lakes Water Level

Consider forecasting the 2008 water level using an exponential smoothing model with $w = 0.7$:

$$\widehat{y}_{91} = 0.7y_{90} + (1 - 0.7)\widehat{y}_{90}$$

We need the forecasted value $\widehat{y}_{90}$ to finish our calculation. However, to calculate $\widehat{y}_{90}$ we will need the forecasted value of $\widehat{y}_{89}$! In fact, this pattern continues, and we need to calculate all past forecasts before we can calculate $\widehat{y}_{91}$. We will calculate the first few forecasts here and leave the remaining calculations for software. Taking $\widehat{y}_1$ to be y_1, the calculations begin as follows:

$$\begin{aligned}
\widehat{y}_2 &= 0.7 \times y_1 + (1 - 0.7) \times \widehat{y}_1 \\
&= (0.7)(176.887) + (0.3)(176.887) \\
&= 176.887 \\
\widehat{y}_3 &= 0.7 \times y_2 + (1 - 0.7) \times \widehat{y}_2 \\
&= (0.7)(176.745) + (0.3)(176.887) \\
&= 176.788 \\
\widehat{y}_4 &= 0.7 \times y_3 + (1 - 0.7) \times \widehat{y}_3 \\
&= (0.7)(176.625) + (0.3)(176.788) \\
&= 176.674
\end{aligned}$$

Software continues our calculations to arrive at a forecast for y_{90} of 176.036. We use this value to complete our forecast calculation for y_{91}:

$$\begin{aligned}
\widehat{y}_{91} &= 0.7 \times y_{90} + (1 - 0.7) \times \widehat{y}_{90} \\
&= (0.7)(175.943) + (0.3)(176.036) \\
&= 175.971
\end{aligned}$$

Our model forecasts the average water level to be 175.971 in 2008. Comparing the actual levels to the forecasts, the exponential smoothing model is predicting changes of about 0.1 meters, which translates to about a 4-inch change in water level. This may not seem like much until you realize that a change of 4 inches amounts to more than a three-trillion-gallon change in water volume. The economic impact is many millions of dollars.

With the forecasted value for 2008 from Example 13.20, the forecast for 2009 requires only one calculation:

$$\widehat{y}_{92} = (0.7)(y_{91}) + (0.3)(175.971)$$

Once we observe the actual water level for 2008 (y_{91}), we can enter that value into the forecast equation above. Updating forecasts from exponential smoothing models only requires that we keep track of last period's forecast and last period's observed value. In contrast, moving-average models require that we keep track of the last k observed values of the time series.

The simple exponential smoothing model is best suited for forecasting time series with no strong trend or seasonal variation. Variations on the simple exponential smoothing model have been developed to handle time series with a trend (double exponential smoothing and Holt's exponential smoothing), with seasonality (seasonal exponential smoothing), and with both trend and seasonality (Winters' exponential smoothing). Your software may offer one or more of these smoothing models.

APPLY YOUR KNOWLEDGE

13.49 Domestic average airfare. The Bureau of Transportation Statistics conducts a quarterly survey to monitor domestic and international airfares.[25] Here are the average airfares for U.S. domestic flights for 1995 through 2007:

Year	Airfare	Year	Airfare	Year	Airfare
1995	$292.31	2000	$339.10	2005	$307.21
1996	$276.94	2001	$319.84	2006	$328.33
1997	$287.20	2002	$312.56	2007	$325.54
1998	$309.34	2003	$315.50		
1999	$324.13	2004	$305.91		

(a) Hand calculate forecasts for the time series using an exponential smoothing model with $w = 0.2$. Provide a forecast for the average U.S. domestic airfare for 2008.

(b) Hand calculate forecasts for the time series using an exponential smoothing model with $w = 0.8$. Provide a forecast for the average U.S. domestic airfare for 2008.

(c) Write down the forecast equation for the 2009 average U.S. domestic airfare based on the exponential smoothing model with $w = 0.2$.

13.50 Domestic average airfare. Refer to Exercises 13.46, 13.47, and 13.48 for explanation of the forecast accuracy measures MAD, MSE, and MAPE.

(a) Based on the forecasts calculated in part (a) of Exercise 13.49, calculate MAD, MSE, and MAPE.

(b) Based on the forecasts calculated in part (b) of Exercise 13.49, calculate MAD, MSE, and MAPE.

BEYOND THE BASICS: Spline Fits

smoothing spline

Modern computing capabilities have made possible many statistical tools that would otherwise be impossible or impractical. One such tool is the general-purpose **smoothing spline.** A spline curve can be fitted to any (x, y) data set. For a time series, we take x to be the time variable and y to be the time series values. A spline curve is a single curve consisting of a large number of polynomial curves pieced together in an optimal manner. The more polynomials used, the more flexible and less smooth the spline curve is.

The smoothness of a spline fit is determined by the choice of a positive-valued smoothing constant, similar to the choice of w in the exponential smoothing model. The spline smoothing constant is often denoted by the lowercase Greek letter lambda λ. Choosing a value of λ close to zero results in a very flexible and not very smooth spline curve. As larger values of λ are used, the spline curve becomes less flexible and smoother.

Figure 13.30 displays three spline curves. The red spline curve offers the least amount of smoothing for the data ($\lambda = 0.000001$). With λ nearly zero, the spline curve passes through every point. The green spline curve is based on $\lambda = 10,000$. This spline curve follows the ups and downs of the time series without trying to pass through each point. The black spline curve offers the highest degree of smoothing, with $\lambda = 1,000,000$. The black spline curve reveals the overall pattern in the time series without following the smaller ups and downs.

Spline fits can help you quickly identify overall trends and seasonal variation, as well as less regular patterns, in your time series data. They are a powerful, modern exploratory data analysis tool.

FIGURE 13.30 Spline curve models for three values of the smoothing constant λ: $\lambda = 0.000001$ (red), $\lambda = 10,000$ (green), and $\lambda = 1,000,000$ (black).

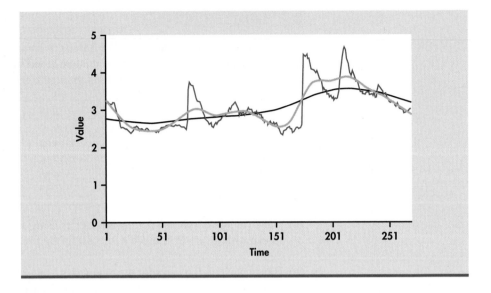

SECTION 13.3 Summary

- **Moving-average forecast models** use the average of the last k observed values to forecast next period's value. The number k is called the **span** of the moving average. Larger values of k result in a smoother model.

- The forecast equation for the **simple exponential smoothing model** is a weighted average of last period's observed value and last period's forecasted value. The degree of smoothing is determined by the choice of a smoothing constant w between 0 and 1. Values of w close to 0 result in a smoother model.

SECTION 13.4 Exercises

For Exercises 13.45 to 13.48, see pages 724–725; and for 13.49 and 13.50, see page 729.

13.51 It's exponential. Exponential smoothing models are so named because the coefficients

$$w, (1 - w)w, (1 - w)^2 w, \ldots, (1 - w)^{n-2} w$$

decrease in value exponentially. For this exercise, take $n = 11$. Use software to do the calculations.
(a) Calculate the coefficients for a smoothing constant of $w = 0.1$.
(b) Calculate the coefficients for a smoothing constant of $w = 0.5$.
(c) Calculate the coefficients for a smoothing constant of $w = 0.9$.
(d) Plot each set of coefficients from parts (a), (b), and (c). The coefficient values should be measured on the vertical axis while the horizontal axis can simply be numbered $1, 2, \ldots, 9, 10$ for the 10 coefficients from each part. Be sure to use a different plotting symbol and/or color to distinguish the three sets of coefficients and connect the points for each set. Also, label the plot so that it is clear which curve corresponds to each value of w used.

(e) Describe each curve in part (d). Which curve puts more weight on the most recent value of the time series when calculating a forecast?
(f) The coefficient of y_1 in the exponential smoothing model is $(1 - w)^{n-1}$. Calculate the coefficient of y_1 for each of the values of w in parts (a), (b), and (c). How do these values compare to the first 10 coefficients you calculated for each value of w? Which value of w puts the greatest weight on y_1 when calculating a forecast?

13.52 *New York Times* monthly revenue. In Example 13.17 (page 722), quarterly advertising revenues of the *New York Times* were studied in conjunction with the moving-average model. For this exercise, consider a time series of monthly total revenues (advertising, circulation, and other revenue in thousands of dollars) starting in January 2005 and ending in October 2008.
🌐 NYTIMESTOTAL
(a) Calculate and plot (on a single time series plot) moving averages for a span of $k = 12$.

(b) What do you observe? Are the moving averages trending? If so, describe the nature of the trend. Is there evidence of a shift? If so, over what period did the moving averages shift?

(c) Calculate a forecast for total monthly revenue for November 2008. Explain why the forecast for November 2008 will most likely be close to the actual observation. For what months will the moving-average model be substantially off in its forecasts?

13.53 Moving averages and linear trend. The moving-average model provides reasonable predictions only under certain scenarios. Use the seasonally adjusted (SA) consumer loan time series of Exercise 13.28 (page 712) to investigate the performance of the moving-average model in the presence of a strong trend.

(a) Calculate moving averages for spans of $k = 20$ and 50. Superimpose the moving averages (on a single time series plot) on a plot of the original time series.

(b) As the span increases, what do you observe about the plot of the moving averages?

(c) At the stock market analysis Web site stockcharts.com, it is stated that moving averages "are best suited for trend identification and trend following purposes, not for prediction." Explain whether or not your results from part (a) are consistent with this claim.

13.54 Exponential smoothing for butter sales. Consider the weekly sales volume of butter (Exercise 13.45, page 724).

(a) Calculate and plot (on a single time series plot) exponential smoothing models using smoothing constants of $w = 0.1, 0.5$, and 0.9.

(b) Comment on the smoothness of each exponential smoothing model in part (a).

(c) The series ended with weekly sales for 12/7/2008 to 12/13/2008. For each model in part (a), calculate forecasts for next week's sales (12/14/2008 to 12/20/2008).

13.55 Exponential smoothing forecast equation. We have learned that the exponential smoothing forecast equation is written as

$$\widehat{y}_t = w y_{t-1} + (1 - w)\widehat{y}_{t-1}$$

(a) Show that the equation can be written as

$$\widehat{y}_t = \widehat{y}_{t-1} + w e_{t-1}$$

where e_{t-1} is the residual, or prediction error, for period $t - 1$.

(b) Explain in words how the reexpressed equation can be interpreted.

STATISTICS IN SUMMARY

Standard regression methods like those described in Chapters 10 and 11 can be used with time series data with "time" as the explanatory variable and the time series as the response variable. These methods can help model both trend and seasonality in a time series, although the models will generally not satisfy the regression assumptions needed for inference. Special time series models relate the current period's value to past values of the time series. These models are often referred to as "smoothing models." We rely on statistical software for fitting time series models and for forecasting future values using these models. Here is a review list of the most important skills you should have acquired from your study of this chapter.

A. Assessing Randomness

1. Recognize whether a process exhibits random-only behavior or not by visual observation of a time series plot.

2. Carry out the runs test to test for randomness.

B. Trends

1. Identify a long-run trend in a time series plot.

2. Use software to fit a regression model to the long-run trend.

3. Use a regression model to forecast future values of a time series based on a trend-only model.

4. Recognize the role of the logarithmic transformation for fitting exponential trend models.

C. Seasons

1. Identify seasonal variation in a time series plot.

2. Model seasonal variation using indicator variables in a regression model.

3. Recognize the role of the logarithmic transformation for fitting multiplicative seasonal effects.

4. Model seasonal variation using seasonality factors.

5. Forecast future values of a time series while taking into account trend and seasonality.

6. Recognize seasonally adjusted data in a time series plot.

D. Lag Regression Models

1. Identify autocorrelation in a lagged residual plot or a lagged time series plot.

2. Use software to calculate the autocorrelation of a time series.

3. Use software to fit a first-order autoregressive model to a time series.

4. Forecast one or more time periods ahead from an autoregressive model.

5. Use software to build a time series model using some combination of lag, trend, and seasonal variables.

E. Moving-Average and Smoothing Models

1. Use software to calculate the moving averages for a time series.

2. Understand how the choice of the span k of a moving-average forecast model relates to the smoothness of the model.

3. Forecast next period's value using a moving-average forecast model.

4. Use software to fit an exponential smoothing model to a time series.

5. Understand how the choice of the smoothing constant value w in an exponential smoothing model relates to the smoothness of the model.

6. Forecast next period's value using an exponential smoothing model.

CHAPTER 13 Review Exercises

13.56 Just use last month's figures! Working with the financial analysts at your company, you discover that, when it comes to forecasting various time series, they often just use last period's value as the forecast for the current period. This strikes you as a very naive approach to forecasting.
(a) If you could pick the estimates of β_0 and β_1 in the AR(1) model, could you pick values such that the AR(1) forecast equation would provide the same forecasts your company's analysts use? If so, specify the values that accomplish this.
(b) What span k in a moving-average forecast model would provide the same forecasts your company's analysts use?
(c) What smoothing constant w in a simple exponential smoothing model would provide the same forecasts your company's analysts use?
(d) Given your responses to parts (a), (b), and (c), are your company's analysts doing the best they can do for forecasting? Explain your response.

13.57 Chicago Cubs attendance trend. Consider the average attendance per Cubs home game beginning in 1916 and ending

in 2007 (92 years). Use statistical software to fit trend models to this time series. 🌀 CUBS
(a) Fit a line to the data. Be sure to use attendance per home game as the response variable and t as the explanatory variable. Report the equation of the least-squares line, the R^2-value, and the regression standard error s.
(b) Now, fit a second-degree polynomial (t and t^2 are explanatory variables) to the data. Report the equation of the quadratic model, the R^2-value, and the regression standard error s.
(c) Finally, fit a third-degree polynomial (t, t^2, and t^3 are explanatory variables) to the data. Report the equation of the cubic model, the R^2-value, and the regression standard error s.
(d) Based on the R^2-values and the regression standard errors, which trend model from parts (a), (b), and (c) would you use for the trend equation for the average attendance per home game time series?

13.58 Chicago Cubs attendance. Continue the analysis of the Chicago Cubs attendance time series.

(a) Obtain the residuals from the third-degree polynomial trend model in Exercise 13.57, part (c). Make a time series plot of the residuals and a lagged residual plot. Is there evidence of autocorrelation?

(b) Fit the attendance time series with two explanatory variables: lag one variable and t^3. Report the fitted equation, the R^2-value, and the regression standard error s.

(c) Obtain the residuals from part (b). Make a time series plot of the residuals and a lagged residual plot. Do the plots suggest that the lag-and-trend model is a good fit for the data series? Explain.

(d) Using the models from parts (a), (b), and (c) of the previous exercise, forecast average attendance per home game for the year 2008. Also, forecast year 2008 average attendance based on the fitted model from part (b) of this exercise.

(e) The actual average attendance per home game in 2008 was 40,743. Which of the four models provided the most accurate forecast? Is this the same model you would have chosen based on R^2-values and the regression standard errors?

13.59 Turkey death rate. More than 270 million turkeys were raised in 2007 for consumption in the United States. To respond to consumer demand, the turkey industry is continually finding ways to produce turkeys at that gain weight faster and have wider breasts (yielding more white meat). The unnaturally rapid weight gains and unnatural anatomy cause a variety of disorders leading to premature death. Given the economic losses associated with premature death, the U.S. Department of Agriculture tracks the rate of death loss of young turkeys. Consider a time series of the annual average rate of death loss from 1976 to 2007.[26] 🦃 **TURKEY**
(a) Make a time series plot of the rate of death data. Describe the time series behavior seen in the plot.
(b) Fit a linear trend model to the data series. Report your estimated model.
(c) On average, by how much is the rate of death loss increasing from year to year?
(d) Estimate the rate of death loss for 2008.

13.60 Annual precipitation. Global temperatures are increasing. Great Lakes water levels meander up and down (see Figure 13.29, page 727). Do all environmental processes exhibit time series patterns? Consider a time series of the annual precipitation (inches) in New Jersey from 1895 to 2007.[27] 📊 **PRECIPITATION**
(a) Make a time series plot of the precipitation series.
(b) Perform the runs test on the series and report a P-value for the test.
(c) Make a lagged time series plot and calculate the correlation between y_t and y_{t-1}.
(d) Based on parts (a), (b), and (c), what do you conclude about the precipitation process?

13.61 Annual precipitation. Continue the analysis of the annual precipitation time series.
(a) Make a histogram and Normal quantile plot of the precipitation data. What do you conclude from these plots?
(b) What is a 90% prediction interval for the annual precipitation for 2008?

13.62 NFL rushing. In the National Football League, the philosophy for winning (rushing, passing, defense) seems to go through cycles. Consider a time series of the average number of rushing yards in the NFL per regular season from 1980 to 2008.[28] 🏈 **NFL**
(a) Make a time series plot. Is there evidence that the average rushing yards is trending in one direction? Describe the general movement of the series.
(b) Make a lagged time series plot and calculate the correlation between y_t and y_{t-1}.
(c) Fit an AR(1) model using y_t as the response variable and y_{t-1} as the explanatory variable. Record the estimated regression equation.
(d) Based on the AR(1) model, forecast the average number of rushing yards in the NFL for the 2009 regular season.

13.63 NFL rushing. Refer to the previous exercise.
(a) Calculate and plot (on a single time series plot) exponential smoothing models using smoothing constants of $w = 0.2, 0.5$, and 0.8.
(b) For each of the exponential smoothing models of part (a), forecast the average number of rushing yards in the NFL for the 2009 regular season.

13.64 NFL rushing. Refer to the previous exercise. For each of the exponential smoothing models in part (a), calculate MAD, MSE, and MAPE (see Exercises 13.46, 13.47, and 13.48, pages 724–725). Which value of w provided the best fit?

13.65 Chicago Cubs attendance per game. The 1993 average attendance per game (y_{78}) was 32,363, and the 2005 average attendance per game (y_{90}) was 38,272. Use *only* these facts and any information found in Example 13.18 (page 711) to calculate 2009 forecasts based on the 3-year moving-average model and the 15-year moving-average model.

13.66 Chinese car ownership. In Example 13.4 (page 696), an exponential trend model was fitted to the yearly time series of the number of passenger cars owned in China. In Exercise 13.16 (page 710), you were asked to investigate whether autocorrelation was present in the residuals associated with the trend fit. 🚗 **CHINACARS**
(a) Fit an AR(1) model using y_t as the response variable and y_{t-1} as the explanatory variable. Record the estimated regression equation.
(b) Based on the AR(1) model, forecast the number of passenger cars owned for 2008. Compare this forecast with the exponential trend forecast given in Example 13.4.

13.67 Chinese car ownership. Continue analyzing the Chinese car ownership data from the previous exercise.
(a) Calculate the residuals for the exponential trend model fit of Example 13.4 (page 696), and calculate MAD, MSE, and MAPE (see Exercises 13.46, 13.47, and 13.48, pages 724–725).
(b) Calculate the residuals for the AR(1) fit of the previous exercise, and calculate MAD, MSE, and MAPE.
(c) Based on the results of parts (a) and (b), which model seems to provide a better fit?

13.68 A special AR(1) model. *Random walk* models for various financial time series are often mentioned in the business literature. A simple random walk model specifies that one-period *differences* in the time series can be modeled as a constant term plus a random deviation term. In other words, period-to-period differences (or changes) will behave as a random process. The equation for this random walk model is

$$y_t - y_{t-1} = \beta_0 + \epsilon_t$$

If we rewrite this equation solving for y_t, we get

$$y_t = \beta_0 + y_{t-1} + \epsilon_t$$

which is our AR(1) model with $\beta_1 = 1$. If we fit an AR(1) model and find that the β_1 estimate is close to 1, then a simple random walk model for the time series is another modeling option.

Consider the monthly closing prices of the Dow Jones Industrial Index beginning with January 2000 and ending with November 2008.[29] **DOWJONES**

(a) Fit an AR(1) model to the Dow Jones time series. Construct a 95% confidence interval for β_1. Does the confidence interval suggest the possibility of a simple random walk model underlying the time series? Explain your answer.

(b) Obtain the one-period differences and plot them as a time series. Perform the runs test on the differences. Based on the plot of the differences and the runs test, can you justify the simple random walk model for the Dow Jones series? Provide appropriate justification.

(c) For a simple random walk model, the estimate of β_0 is simply the average of all one-period differences $y_t - y_{t-1}$ (call this average $\overline{y}_{\text{diff}}$), and the forecast equation is $\widehat{y}_t = y_{t-1} + \overline{y}_{\text{diff}}$. Calculate the one-period differences and their average, and use this to provide a "random walk forecast" for the monthly closing price of December 2008.

13.69 *New York Times* **monthly revenue.** Refer to the time series of the monthly total revenue for the *New York Times* in Exercise 13.52 (page 730).

(a) Fit a regression model based on a linear trend term and monthly seasonal indicator variables.

(b) Make a time series plot of the residuals for the model in part (a). Is there evidence of nonlinearity in the residual plot? If so, describe the pattern.

(c) Fit a regression model based on a linear trend term, a quadratic trend term, and monthly seasonal indicator variables. What is the estimated coefficient for the month of January indicator variable? What is the P-value for the variable?

13.70 **Outlier effect and** *New York Times* **monthly revenue.** Continue the analysis of the *New York Times* revenue time series.

(a) Make a time series plot of the residuals for the model in part (c) of Exercise 13.69. Identify the largest outlier in the residual series. In which month and year did the outlier occur?

(b) When dealing with data sets that are not time series, tossing out an unusual observation from the data set poses no problems. However, with time series applications, deleting an observation can be problematic. In particular, observations that are normally two periods apart around the outlier become adjacent when the intermediate observation is deleted. Especially if lagged variables

are used, this can be a problem. As an alternative, we can create an indicator variable that has the value of 1 for the outlier time period and is 0 otherwise. Using such a variable for the identified outlier in the *New York Times* series, fit a regression model based on a linear trend term, a quadratic trend term, monthly seasonal indicator variables, and the special indicator variable. Report the estimated regression model.

(c) For the model in part (b), what is the estimated coefficient for the month of January indicator variable? What is the P-value for the variable? Compare these results with part (c) of Exercise 13.69. Do you notice any other differences between the two fitted models? What does the comparison tell you about the effect of the outlier on the regression fit?

(d) Make a time series plot of the residuals for the model in part (b). What can you conclude about the fitted model?

(e) Based on the model in part (b), forecast monthly revenue for November 2008.

(f) In Exercise 13.52 (page 730), you were asked to compute the 12-month moving averages for the monthly revenue series. You were also asked if there were any shifts in the moving-average values. Explain any shifts you observed in light of the present exercise.

13.71 **Outlier effect and corporate philanthropy.** In Example 13.13 (page 715) an AR(1) model was fitted to the corporate philanthropy series. Figure 13.24 (page 716) showed that the observation associated with the year 2005 was an outlier due to the unusual events of that period.

(a) Refer to the discussion of Exercise 13.70 and create a special indicator variable for the year 2005. Fit the time series using a lag one variable and the special indicator variable. Report the estimated model.

(b) What is the effect of the outlier on the estimated lag one coefficient?

(c) Using the model fitted in part (a), forecast 2008's corporate philanthropy percent. Compare this forecast with the forecast provided in Example 13.14 (page 717).

13.72 **Best Buy quarterly sales.** Consider a time series of quarterly sales (in millions of dollars) for Best Buy, Inc. starting with the first quarter of fiscal year 2000 and ending with the second quarter of fiscal year 2009.[30] **BESTBUY**

(a) Make a time series plot of quarterly sales. Describe the nature of the series in terms of trend and seasonality. Explain why an additive seasonal model is probably not applicable.

(b) Make a time series plot of sales in logged units. Describe how this plot differs from the plot in part (a).

(c) Fit the logged sales series with a linear trend term and quarterly indicator variables. Report the estimated model and make a forecast for third-quarter sales of fiscal year 2009 in the original units.

13.73 **Best Buy quarterly sales.** Continue the analysis of Best Buy sales in the previous exercise.

(a) Obtain the residuals from the fit on logged sales from Exercise 13.72. Make a time series plot of the residuals. Do the residuals show any pattern? If so, describe the pattern.

(b) Make a lagged residual plot and calculate the correlation between e_t and e_{t-1}. Is there evidence of autocorrelation in the residuals?

(c) Fit the logged sales series with a linear trend term, quarterly indicator variables, and a lag one variable of logged sales.

(d) Obtain the residuals from the fitted model of part (c). Make a time series plot and a lagged residual plot. What do these plots indicate about the adequacy of the fitted model?

(e) Based on the fitted model of part (c), make a forecast for third-quarter sales of fiscal year 2009 in original units. How did the forecast change relative to the forecast made in Exercise 13.72, part (c)?

13.74 Seasonality factors for Best Buy quarterly sales. In the previous exercises, seasonality was directly modeled along with trend in a regression equation. Since the seasonality does not appear to be additive in nature, the trend-and-season modeling was done on sales in logged units. Here you will be asked to separate the estimation of trend and seasonality.

(a) Fit a linear trend model to the sales series in original units. Report the estimated model. Why does a linear trend model seem to be a more reasonable choice for the series than an exponential trend model?

(b) Using the fitted trend model of part (a), calculate the seasonality factors for each quarter. Adjust the factors so that they average out to 1.

(c) Make a forecast for third-quarter sales of fiscal year 2009.

13.75 Walgreens quarterly sales. Walgreens is the nation's largest drugstore chain, with approximately 7000 drugstores as of 2008. Consider a time series of its quarterly sales (in millions of dollars) starting with the first quarter of fiscal year 2000 and ending with the first quarter of fiscal year 2009.[31] 🍃 **WALGREENS**

(a) Make a time series plot of quarterly sales. Describe the nature of the series in terms of trend and seasonality. Explain why an additive seasonal model is probably not applicable.

(b) Fit a linear trend model to the sales series. Report the estimated model.

(c) Fit a quadratic trend model (using t and t^2) to the sales series. Report the estimated model.

(d) Fit an exponential trend model to the sales series. Report the estimated model.

(e) Calculate MAD, MSE, and MAPE (see Exercises 13.46, 13.47, and 13.48, pages 724–725) for each of the three trend fits. Which trend model provides the best fit?

13.76 Seasonality factors for Walgreens quarterly sales. Continue the analysis of Walgreen's sales from the previous exercise.

(a) Using the fitted quadratic trend model from part (c) of Exercise 13.75, calculate the seasonality factors for each quarter. Adjust the factors so that they average out to 1.

(b) Make a forecast for second-quarter sales of fiscal year 2009.

13.77 U.S. air carrier traffic. How much more or less are Americans taking to the air? Consider a time series of monthly total number of passenger miles (in thousands) on U.S. domestic flights starting with January 2003 and ending with September 2008.[32] 🍃 **AIRTRAFFIC**

(a) Make a time series plot of monthly miles. Does the trend appear linear or curved? Describe the trend in the series over times.

(b) Relative to the general trend movement, does the seasonal variation appear to be additive or multiplicative in nature? Justify your answer.

(c) Fit the time series to a quadratic trend (using t and t^2) and monthly indicator variables. Report the estimated model and forecast the total number of passenger miles to be flown in October 2008.

13.78 U.S. air carrier traffic. Continue the analysis of monthly total number of passenger miles on U.S. domestic flights from the previous exercise.

(a) Obtain the residuals from the trend-and-season model of Exercise 13.77, part (c). Make a time series plot of the residuals. Do the residuals show any pattern? If so, describe the pattern.

(b) Make a lagged residual plot and calculate the correlation between e_t and e_{t-1}. Is there evidence of autocorrelation in the residuals?

(c) Fit the passenger miles series to a quadratic trend, monthly indicator variables, and a lag one variable of miles flown. Report the estimated model and forecast the total number of passenger miles to be flown in October 2008.

13.79 U.S. air carrier traffic. Continue the analysis of monthly total number of passenger miles on U.S. domestic flights from the previous exercise.

(a) Obtain the residuals from the fitted model of Exercise 13.78, part (c). Make a time series plot of the residuals. To aid your visual inspection, superimpose a horizontal line at 0 (mean of residuals). Do the residuals show any pattern? If so, describe the pattern.

(b) Make a lagged residual plot of e_t versus e_{t-1}. Also, make a lagged residual plot of e_t versus e_{t-2}. Calculate the correlation between e_t and e_{t-1} and the correlation between e_t and e_{t-2}. Is there evidence of autocorrelation in the residuals that is not accounted for by the lag one variable in the fitted model?

(c) Fit the passenger miles series to a quadratic trend, monthly indicator variables, a lag one variable of miles flown, and a lag two variable of miles flown. Do the P-values for each of the lag variables indicate that the variables should be included in the model?

(d) Forecast the total number of passenger miles to be flown in October 2008. How does this forecast differ from the forecasts from Exercises 13.77 and 13.78?

13.80 Great Lakes water level. Examples 13.19 (page 726) and 13.20 (page 728) explored the annual average level of Lakes Michigan and Huron with the exponential smoothing model. The meandering in the time series (see Figure 13.29, page 727) suggests that an autoregressive model—like an AR(1) model—might fit the data series well. 🍃 **LAKES**

(a) Fit an AR(1) model using y_t as the response variable and y_{t-1} as the explanatory variable. Record the estimated forecast equation, the R^2-value, and the model standard error estimate.

(b) Obtain the residuals from the AR(1) fit and perform a runs test. What do you conclude about the appropriateness of the AR(1) model fit?

(c) Make a lagged residual plot. Does it appear that the autocorrelation has been accounted for by the AR(1) model fit?

13.81 AR(2) and Great Lakes water level. Continue analyzing the annual average water level of Lakes Michigan and Huron.

(a) Introduce y_{t-2} as an additional explanatory variable. This lag variable is a lag two variable representing values of the time series two periods back. Fit an AR(2) model using y_t as the response variable and both y_{t-1} and y_{t-2} as the explanatory variables. Record the estimated forecast equation, the R^2-value, and the model standard error estimate. What improvements in the fit do you notice compared with the AR(1) model of Exercise 13.80?

(b) Obtain the residuals from the AR(2) fit and perform a runs test. What do you conclude about the appropriateness of the AR(2) model fit?

(c) Based on the AR(2) fitted model, forecast the 2008 average water level of Lakes Michigan and Huron.

CHAPTER 13   Appendix

Using Minitab and Excel for Time Series Forecasting

Trend Models

Minitab:

Stat ➤ Time Series ➤ Trend Analysis

With this routine, Minitab can fit different types of trend lines. Click-in the column for the time series variable in the **Series** box. If you wish to fit a linear trend model as done in Example 13.3 (page 693), then choose the **Linear** option. If you wish to fit an exponential trend model as done in Example 13.4 (page 696), then choose the **Exponential Growth** option. The exponential model fitted by this procedure is identical to what has been discussed in the chapter; however, the model is reported in a slightly different form. In particular, if we were to fit the Chinese car ownership data series, we would find that Minitab would report an estimated model of

$$\widehat{CARS}_t = 20.0788(1.30600^t)$$

This is actually the same prediction equation as given in Example 13.4 since $e^{0.266969} = 1.30600$, which implies

$$\widehat{CARS}_t = 20.0788(1.30600^t) = 20.0788(e^{0.266969})^t$$
$$= 20.0788e^{0.266969t}$$

There are several options for whatever type of trend you wish to fit. For example, if you wish to change the time scale (for example, to months or years), then click the **Time** button. By clicking the **Graphs** button, you can produce a variety of residual plots. If you wish to obtain the actual residuals or predictions ("Fits"), then click the **Storage** button. Once you have completed your option selections, click **OK** and the estimated trend model will be outputted in the **Session** window. Minitab will also produce a time series plot of the original data with the trend model superimposed.

An alternative to using Minitab's trend analysis routine for fitting linear and exponential trends is to create a time index variable ($t = 1$, $t = 2$, and so on) and then implement the regression routine in Minitab. To create a time index variable, you can either type its values into the worksheet or perform the following pull-down sequence:

Calc ➤ Make Patterned Data ➤ Simple Set of Numbers

Type the column name in the **Store pattern data in** box to indicate where you want to place the values of the time index variable. Type the value of "1" in the **From first value** box and type in the number of total periods in the **To last value** box. Click **OK**. If you wish to take a logarithm (base e) of the data series so that you can use least-squares regression to estimate an exponential trend model as explained on page 702, then perform the following pull-down sequence:

Calc ➤ Calculator

Type the column name in the **Store result in variable** box to indicate where you want to place the logged values. Now scroll down the function list and select "Natural log (log base e)." Once done, you will find in the **Expression** box the expression "LN(number)." Replace "number" with the column name of the data series you wish to have logged. Click **OK**.

Excel:

Automated trend fitting (linear or exponential) is available in neither standard Excel nor the WHFStat Add-In for Excel. You can, however, create a time index variable ($t = 1$, $t = 2$, and so on) and then implement the regression routine in Excel. These values can be manually typed into a column of Excel. Alternatively, you can set the first cell to the value of "1" and then define the cell below as being equal to the previous cell plus 1 and then copy the formula down the column to a desired ending value. For exponential trend model fitting, you need to take the logarithm (base e) of the data series (see discussion on page 702). To do so, first click on an empty cell in the same row as the first value of the original data series. Now click the **Formulas** tab and then click the **Math & Trig** option found in the **Function Library** group. Scroll down the function list and select the "LN" function. You will then see a **Function Arguments** pop-up box. Click-in the cell location of the first value of the original data series in the **Number** box and then click **OK**. Now copy the formula in the first cell down the column to the row of the last value of the original data series. You can now proceed to the regression routine.

Season Models

Minitab:

Stat ➤ Time Series ➤ Decomposition

With this routine, you can ask Minitab to compute the multiplicative seasonality factors with or without a trend component. Click-in the column for the time series variable in the **Variable** box. In the **Seasonal length** box, type in the number of seasonal periods (for example, type in "12" for monthly data). Select the **Multiplicative** option to obtain seasonality factors that are used as multiplying factors. If you also want to fit a trend as was done in Example 13.8 (page 704), then select the **Trend plus seasonal** option. It should be noted that Minitab first estimates the seasonality factors and then estimates the trend line after the data have been "deseasonalized." For Example 13.8, we estimated the trend line first and then estimated the seasonality factors. The two different approaches result in slightly different estimates, but overall the differences in predictions are inconsequential for practical purposes. If you wish to obtain the actual residuals or predictions ("Fits"), then click the **Storage** button. Click **OK** when finished making selections, and then the estimated seasonal model will be outputted in the **Session** window. Minitab will also produce a number of graphs, including a time series plot of the original data with the seasonal model superimposed.

In our discussions of seasonal models, we illustrated (Example 13.6, page 699) how indicator variables can be added to the prediction model to estimate seasonal variation. The creation of these indicator variables can be done manually by typing 0s and 1s in the worksheet. In the case of Example 13.6, this would mean typing $11 \times 104 = 1144$ values in the worksheet. There is a convenient alternative for creating indicator variables. To begin, we create a coded variable in the worksheet called "Month" that takes the numbers 1, 2, 3, ..., 12 for the twelve months of the year. These numbers are placed in the "Month" column so that we know in what month any value of the data series has occurred. To create this coded variable, perform the following pull-down sequence:

Calc ➤ Make Patterned Data ➤ Arbitrary Set
of Numbers

Type the column name in the **Store patterned data in** box to indicate where you want to place the coded variable. In the **Arbitrary set of numbers** box, if the time series starts in January, type in the values "1," "2," "3," "4," "5," "6," "7," "8," "9," "10," "11," and "12." If the time series starts in another month, cycle through the same values but with a different starting point: for example, for a time series starting in August, type in "8," "9," "10," "11," "12," "1," "2," "3," "4," "5," "6," and "7." Now enter the number of times the sequence of coded values needs to be repeated to cover the entire data series in the **Number of times to list the sequence** box. In the case of Example 13.6, we have 104 data values (which is more than 8 years but less than 9 years), so we would enter "9," which would result in 108 coded values. Click **OK**. If there are any excess values in the coded column, simply go to the worksheet and delete them. You can now proceed to making indicator variables by doing the following pull-down sequence:

Calc ➤ Make Patterned Data ➤ Arbitrary Set
of Numbers

In the **Indicator variable for** box, click-in the column for the coded variable you just created. Minitab will list the distinct values of the coded variable, and it will give column names for the indicator variables. You can either accept the generic names or you can type in your own names: for example, "Jan," "Feb," "Mar," and so on. Click **OK** and then Minitab will create 12 indicator variables in the worksheet. At which point, as explained in Example 13.6, you would only use at most 11 of these indicator variables in the regression analysis.

Excel:

Automated seasonal model fitting is available in neither standard Excel nor the WHFStat Add-In for Excel. Other than typing in all the 0s and 1s, there are a few shortcuts to making indicator variables. For monthly data, you could type in a cell the value of "1" and then "0" into each of the 11 cells beneath it. Thereafter, you can repetitively copy and paste this stack of 12 values into other columns to create all the necessary indicator variables. Another approach is to create a column with 12 distinct coded values so that we know on what month any value of the data series has occurred. Then you can use the "IF" function (found within **Logical** functions) to place 0 or 1 values in other columns depending on the value of the code in a given row (for more information on the "IF" function, click on the **Help on this function** link). Once your indicator variables are created, you can proceed to use them with Excel's regression routine.

Autocorrelation

Minitab:

$$Stat \blacktriangleright Time\ Series \blacktriangleright Autocorrelation$$

Based on a slightly modified formula for computing correlations, the autocorrelation routine in Minitab can be used to report the autocorrelations for a time series variable at various lags. After doing the above pull-down sequence, click-in the column for the time series variable in the **Series** box. You can allow Minitab to determine how many lags it will report, or you can choose the number of lags by inputting a value in the **Number of lags** box. Click **OK**. You will find that Minitab will produce a bar graph of the autocorrelations, and it will also report the autocorrelation values in the **Session** window, along with other test statistics. An alternative for finding autocorrelations is to create a lag variable in the worksheet and then use the standard correlation formula given in Chapter 2. To create a lag variable, perform the following pull-down sequence:

$$Stat \blacktriangleright Time\ Series \blacktriangleright Lag$$

Click-in the column for the time series variable that you wish to lag in the **Series** box. Now type the column name (other than the column of the original variable) in the **Store lags in** box to indicate where you want Minitab to place the lagged values. Finally, indicate how far back you want to lag by typing an integer greater than 0 in the **Lag** box (default lag is "1"). Click **OK**. At this stage, you can use Minitab to compute the correlation between the variable

and its lag using the standard correlation formula (refer to Appendix of Chapter 2).

Excel:

Excel does not have a routine to compute autocorrelations. However, you can create the lagged variable and then use the standard correlation formula. To create the lagged variable, simply highlight the cell range of the column containing the sequence of time series observations and then paste the values into shifted-down cells of an adjacent column. For example, to create a variable lagged one time period, paste the values into cells shifted down by one row. You can now use Excel to compute the correlation between the variable and its lag using the standard correlation formula (refer to Appendix of Chapter 2).

AR(1) Model Estimation

Minitab:

$$Stat \blacktriangleright Time\ Series \blacktriangleright ARIMA$$

We focused on using ordinary regression procedures for fitting autoregressive models like the AR(1) model. This pull-down sequence leads you to a routine for estimating the model coefficients of an AR(1) or a more complicated time series model (known generally as an ARIMA model) using a more sophisticated estimation procedure than used for ordinary regression. Click-in the column for the time series variable in the **Series** box. In the **Autoregressive** box, which is cross-referenced by the **Nonseasonal** label, enter the value of "1." Click **OK**. The AR(1) estimated model will then be outputted in the **Session** window.

Excel:

ARIMA model estimation is available in neither standard Excel nor the WHFStat Add-In for Excel. However, regression procedures can be used in Excel to fit autoregressive models as done in the chapter.

Moving Average Models

Minitab:

$$Stat \blacktriangleright Time\ Series \blacktriangleright Moving\ Average$$

Click-in the column for the time series variable in the **Variable** box. Type the span of the moving average in the **MA length** box. If you wish to change the time scale (for example, to months or years), click the **Time** button. By clicking the **Graphs** button, you can produce a variety of residual plots. If you wish to obtain the actual residuals or predictions ("Fits"), click the **Storage** button. You also have the option of generating forecasts into the future. To do so, place a checkmark next to the **Generate forecasts** option. Then type the number of periods into the future you want to forecast into the **Number of forecasts** box, and finally type the period of origin from which forecasts are made in the **Starting from origin** box. (*Note:* This is typically taken to be the last period of the data series.) Once you have completed your option selections, click **OK** and then Minitab will produce a time series plot of the original data with the moving-average forecasts overlaid. Additionally, if you requested forecasts, Minitab will project out forecasts on the plot along with 95% prediction limits.

Excel:

Select "Moving Average" in the **Data Analysis** menu box and click **OK**. Enter the cell range of the time series data into the **Input Range** box and type the span of the moving average in the **Interval** box. If you want to plot the moving-average forecasts, place a checkmark next to the **Chart Output** option. Click **OK** to have the moving averages outputted to the spreadsheet. If you requested a plot of the moving averages, you will find it to be a bar graph, which you can change to a time series plot by right-clicking on the graph and selecting the **Change Chart Type** option.

Exponential Smoothing Models

Minitab:

Stat ➤ Time Series ➤ Single Exp Smoothing

Click-in the column for the time series variable in the **Variable** box. To use your own smoothing constant (w), choose the **Use** option and type the smoothing constant in the adjacent box. Alternatively, you can let Minitab determine the best choice for the smoothing constant by choosing the **Optimal ARIMA** option. It is important to note that Minitab by default uses the average of the first six observations as the forecast for the first time period. If you wish to use the actual value of the first time period as the forecast for the first time period, as in Example 13.20 (page 728), click the **Options** button and then type "1" in the **K** = box. Minitab allows you to change the "K"-value only if you are using your own smoothing constant. If you wish to change the time scale (for example, to months or years), click the **Time** button. By clicking the **Graphs** button, you can produce a variety of residual plots. If you wish to obtain the actual residuals or predictions ("Fits"), click the **Storage** button. You also have the option of generating forecasts into the future. To do so, place a checkmark next to the **Generate forecasts** option. Then type the number of periods into the future you want to forecast into the **Number of forecasts** box, and finally type the period of origin from which forecasts are made in the **Starting from origin** box. (*Note:* This is typically taken to be the last period of the data series.) Once you have completed your option selections, click **OK** and then Minitab will produce a time series plot of the original data with the exponential smoothing forecasts overlaid, and it will report the smoothing constant used (either your choice or what Minitab has selected). Additionally, if you requested forecasts, Minitab will project out forecasts on the plot along with 95% prediction limits.

Excel:

Select "Exponential Smoothing" in the **Data Analysis** menu box and click **OK**. Enter the cell range of the time series data into the **Input Range** box. In the **Damping factor** box, you should type in 1 minus the smoothing constant, that is, $1 - w$. If you want to plot the exponential smoothing forecasts, place a checkmark next to the **Chart Output** option. Click **OK** to have the exponential smoothing outputted to the spreadsheet. If you requested a plot of the exponential smoothing forecasts, you will find it to be a bar graph, which you can change to a time series plot by right-clicking on the graph and selecting the **Change Chart Type** option.

CHAPTER 14

One-Way Analysis of Variance

Do unskilled workers, skilled workers, and their supervisors have different views on job safety? Example 14.3 begins an in-depth examination of this issue.

Introduction

Many of the most effective statistical studies are comparative, whether we compare the salaries of women and men in a sample of company employees or customer reaction to two advertisements in a completely randomized design. We display comparisons with back-to-back stemplots, side-by-side boxplots, or histograms, and we measure them with five-number summaries or with means and standard deviations.

When only two groups are compared, Chapter 7 provides the tools we need to answer the question "Is the difference between groups statistically significant?" Two-sample t procedures compare the means of two Normal populations, and we saw that these procedures, unlike comparisons of spread, are sufficiently robust to be widely useful. In this chapter, we will compare any number of means by techniques that generalize the two-sample t and share its robustness and usefulness.

14.1 One-Way Analysis of Variance

Which of four advertising offers mailed to sample households produces the highest sales in dollars? Which of 10 brands of automobile tires wears longest? How long do cancer patients live under each of three therapies for their cancer? In each of these settings we wish to compare several treatments. In each case the data are subject to sampling variability. For example, we would not expect the same sales data if we mailed the advertising offers to different sets of households. We therefore pose the question for inference in terms of the *mean* response. We compare, for example, the mean tread lifetimes of different brands of tires. In Chapter 7 we met procedures for comparing the means of two populations. We are now ready to extend those methods to problems involving more than two populations. The statistical methodology for comparing several means is called **analysis of variance,** or simply **ANOVA.** In the sections that follow, we will examine the basic ideas and assumptions that are needed for ANOVA. Although the details differ, many of the concepts are similar to those discussed in the two-sample case.

ANOVA

We will consider two variations of the ANOVA idea. In **one-way ANOVA** we classify the populations of interest according to a single categorical explanatory variable that we call a **factor.** For example, to compare the tread lifetimes of 10 specific brands of tires we use one-way ANOVA with brand as our factor. This chapter presents the details for one-way ANOVA.

one-way ANOVA

factor

In many other comparison studies, there is more than one way to classify the populations. For the advertising study, the mail-order firm might also consider mailing the offers using two different envelope styles. Will each offer draw more sales on the average when sent in an attention-grabbing envelope? Analyzing the effect of advertising offer and envelope style together requires two-way ANOVA. This technique will be discussed in Chapter 15. While adding even more factors necessitates higher-way ANOVA techniques, most of the new ideas in ANOVA with more than two factors already appear in two-way ANOVA.

The ANOVA setting

Do two population means differ? If we have random samples from the two populations, we compute a two-sample t statistic and its P-value to assess the statistical significance of the difference in the sample means. We compare several means in much the same way. We draw a simple random sample (SRS) from each population and use the data to test the null hypothesis that the population means are all equal. Consider the following two examples.

EXAMPLE 14.1 Comparing Magazine Layouts

A magazine publisher wants to compare three different layouts for a magazine that will be offered for sale at supermarket checkout lines. She is interested in whether there is a layout that better catches shoppers' attention and results in more sales. To investigate, she randomly assigns each of 60 stores to one of the three layouts and records the number of magazines that are sold in a one-week period.

EXAMPLE 14.2 Who Shops Where?

How do five bookstores in the same city differ in the demographics of their customers? Are certain bookstores more popular among teenagers? A market researcher asks 50 customers of each store to respond to a questionnaire. One variable of interest is the customer's age.

These two examples are similar in that

- There is a single quantitative response variable measured on many units; the units are the stores in the first example and customers in the second.

- The goal is to compare several populations: stores displaying three magazine layouts in the first example and customers of the five bookstores in the second.

experiment

observational study

There is, however, an important difference. Example 14.1 describes an **experiment** in which stores are randomly assigned to layouts. Example 14.2 is an **observational study** in which customers are selected during a particular time period and not all agree to provide data. We will treat our samples of customers as random samples even though this is only approximately true.

In both examples, we will use ANOVA to compare the mean responses. The response variable is magazines sold in the first example and age of the customer in the second. The same ANOVA methods apply to data from random samples and to data from randomized experiments. Do keep the data production method in mind when interpreting the results, however. A strong case for causation is best made by a randomized experiment.

Comparing means

The question we ask in ANOVA is "Do all groups have the same population means?" We will often use the term *groups* for the populations to be compared in a one-way ANOVA. To answer this question we compare sample means. Figure 14.1 displays the sample means for Example 14.1. Layout 2 has the highest average sales. But is the observed difference among the sample means just the result of chance variation? We do not expect sample means to be equal even if the population means are all identical.

The purpose of ANOVA is to assess whether the observed differences among sample means are *statistically significant*. In other words, could a variation this large be plausibly due to chance, or is it good evidence for a difference among the population means? This question can't be answered from the sample means alone. Because the standard deviation of a sample mean $\bar{x}$ is the population standard deviation σ divided by $\sqrt{n}$, the answer depends upon both the variation within the groups of observations and the sizes of the samples.

Side-by-side boxplots help us see the within-group variation. Compare Figures 14.2(a) and 14.2(b). The sample medians are the same in both figures, but the variation within the groups in Figure 14.2(a) is much larger. Even the boxplots omit essential information, however. To assess the observed differences, we must also know how large the samples are. Nonetheless, boxplots are a good preliminary display of ANOVA data. While ANOVA compares means and boxplots display medians, we expect the data to be approximately Normal and will consider a transformation if they are not. For distributions that are nearly symmetric, these two measures of center will be close together.

FIGURE 14.1 Mean sales of magazines for three different magazine layouts.

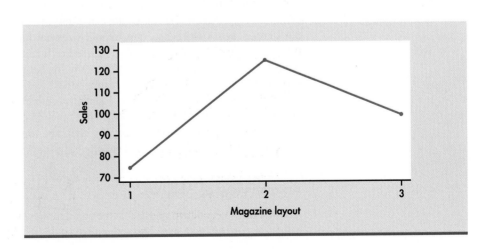

FIGURE 14.2 (a) Side-by-side boxplots for three groups with large within-group variation. The differences among centers may be just chance variation. (b) Side-by-side boxplots for three groups with the same centers as in Figure 14.2(a) but with small within-group variation. The differences among centers are more likely to be significant.

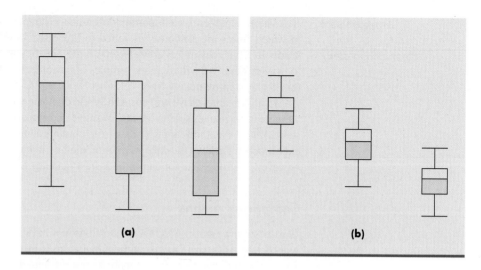

(a)   (b)

The two-sample *t* statistic

Two-sample *t* statistics compare the means of two populations. If the two populations are assumed to have equal but unknown standard deviations and the sample sizes are both equal to *n*, the *t* statistic is (page 430)

$$t = \frac{\overline{x} - \overline{y}}{s_p \sqrt{\dfrac{1}{n} + \dfrac{1}{n}}} = \frac{\sqrt{\dfrac{n}{2}}(\overline{x} - \overline{y})}{s_p}$$

The square of this *t* statistic is

$$t^2 = \frac{\dfrac{n}{2}(\overline{x} - \overline{y})^2}{s_p^2}$$

If we use ANOVA to compare two populations, the ANOVA *F* statistic is exactly equal to this t^2. We can therefore learn something about how ANOVA works by looking carefully at the statistic in this form.

between-group variation

The numerator in the t^2 statistic measures the variation **between** the groups in terms of the difference between their sample means $\overline{x}$ and $\overline{y}$. It includes a factor for the common sample size *n*. The numerator can be large because of a large difference between the sample means or because the sample sizes are large. The denominator measures

within-group variation

the variation **within** groups by s_p^2, the pooled estimator of the common variance. If the within-group variation is small, the same variation between the groups produces a larger statistic and thus a more significant result.

Although the general form of the *F* statistic is more complicated, the idea is the same. To assess whether several populations all have the same mean, we compare the variation *among* the means of several groups with the variation *within* groups. Because we are comparing variation, the method is called *analysis of variance*.

An overview of ANOVA

ANOVA tests the null hypothesis that the population means are *all equal*. The alternative is that they are not all equal. This alternative could be true because all of the means are different or simply because one of them differs from the rest. This is a more complex situation than comparing just two populations. If we reject the null hypothesis, we need to perform some further analysis to draw conclusions about which population means differ from which others.

The computations needed for ANOVA are more lengthy than those for the t test. For this reason we generally use software to perform the calculations. Automating the calculations frees us from the burden of arithmetic and allows us to concentrate on interpretation. The following example illustrates the practical use of ANOVA in analyzing data. Later we will explore the technical details.

EXAMPLE 14.3   Workplace safety

In a study of workplace safety, workers were asked to rate various elements of safety, and a composite score called the Safety Climate Index (SCI) was calculated.[1] The index is the sum of the responses to 10 different questions about safety. The response for each of these questions is an integer ranging from 0 to 10, so the SCI has values from 0 to 100. The workers were classified according to their job category as unskilled, skilled, and supervisor. Here is a summary of the data:

Job category	n	$\overline{x}$	s 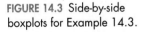
Unskilled workers	448	70.42	18.27
Skilled workers	91	71.21	18.83
Supervisors	51	80.51	14.58

Figure 14.3 gives side-by-side boxplots of the SCI data. Compare Figure 14.3 with the plot of the mean scores in Figure 14.4. The means appear to be different, but there is a large amount of overlap in the three distributions. When we perform an ANOVA on these data, we ask a question about the group means. The null hypothesis is that the population mean SCI scores for the three groups are equal, and the alternative is that they are not all equal. The report on the study states that the ANOVA F statistic is 7.137 with $P < 0.001$. There is very strong evidence that the three groups of workers do not all have the same mean SCI score.

FIGURE 14.3 Side-by-side boxplots for Example 14.3.

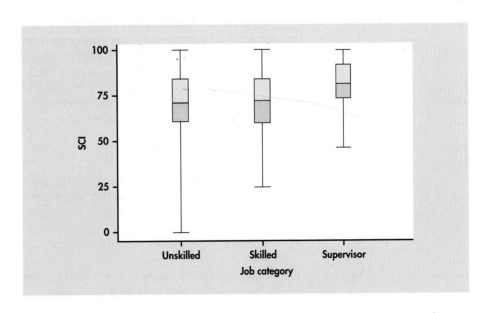

FIGURE 14.4 SCI means for Example 14.3.

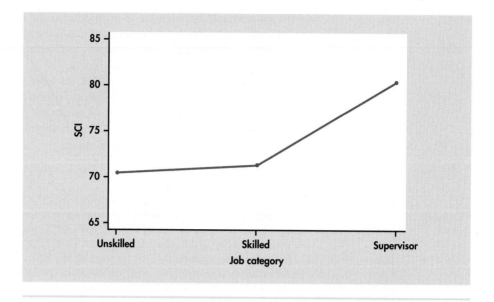

Although we rejected the null hypothesis, we cannot conclude that all three population means are different, only that they are not all the same. Inspection of Figures 14.3 and 14.4 suggests that the means for the skilled workers and the unskilled workers may be the same while the mean for the supervisors is higher. We need to do some additional analysis to compare the three means.

Experts on safety in workplaces have suggested that supervisors face a very different safety environment than other types of workers. Therefore, a reasonable question to ask is whether or not the mean of the supervisors is different from the mean of the others. When there are particular versions of the alternative hypothesis that are of interest before looking at the data, we use **contrasts** to examine them. If we have no specific relations among the means in mind before looking at the data, we instead use a **multiple-comparisons** procedure to determine which pairs of population means differ significantly. Section 14.2 explores both contrasts and multiple comparisons in detail.

contrasts

multiple comparisons

Formulating a clear definition of the populations being compared with ANOVA can be difficult, as in our example. Often some expert judgment is required, and different consumers of the results may have differing opinions. The workers in this study all worked in the same industry in a particular region. They certainly do represent some larger population of similar workers. We are more confident in generalizing our conclusions to similar populations when the results are clearly significant than when the level of significance just barely passes the standard of $P = 0.05$.

APPLY YOUR KNOWLEDGE

14.1 What's wrong? For each of the following, explain what is wrong and why.
(a) ANOVA tests the null hypothesis that the sample means are all equal.
(b) Within-group variation is the variation in the data due to the differences in the sample means.
(c) You use one-way ANOVA when the response variable has only two possible values.
(d) A multiple-comparisons procedure is used to compare a relation among means that was specified prior to looking at the data.

14.2 What's wrong? For each of the following, explain what is wrong and why.

(a) In rejecting the null hypothesis, one can conclude that all the means are different from one another.

(b) A one-way ANOVA can be used only when there are fewer than four means to be compared.

(c) A two-way ANOVA is used when comparing two populations.

The ANOVA model

When analyzing data, we think in terms of an overall pattern and deviations from it. In shorthand form,

$$\text{DATA} = \text{FIT} + \text{RESIDUAL}$$

In the regression model of Chapter 10, the FIT was the population regression line, and the RESIDUAL represented the deviations of the data from this line. We now apply this framework to describe the statistical models used in ANOVA. These models provide a convenient way to summarize the conditions that are the foundation for our analysis. They also give us the necessary notation to describe the calculations needed.

First, recall the statistical model for a random sample of observations from a single Normal population with mean μ and standard deviation σ. If the observations are

$$x_1, x_2, \ldots, x_n$$

we can describe this model by saying that the x_j are an SRS from the $N(\mu, \sigma)$ distribution. Another way to describe the same model is to think of the x's varying about their population mean. To do this, write each observation x_j as

$$x_j = \mu + \epsilon_j$$

The ϵ_j are then an SRS from the $N(0, \sigma)$ distribution. Because μ is unknown, the ϵ's cannot actually be observed. This form more closely corresponds to our

$$\text{DATA} = \text{FIT} + \text{RESIDUAL}$$

way of thinking. The FIT part of the model is represented by μ. It is the systematic part of the model, like the line in a regression. The RESIDUAL part is represented by ϵ_j. It represents the deviations of the data from the fit and is due to random, or chance, variation.

There are two unknown parameters in this statistical model: μ and σ. We estimate μ by $\overline{x}$, the sample mean, and σ by s, the sample standard deviation. The differences $e_j = x_j - \overline{x}$ are the sample residuals and correspond to the ϵ_j in this statistical model.

The model for one-way ANOVA is very similar. We take random samples from each of I different populations. The sample size is n_i for the ith population. Let x_{ij} represent the jth observation from the ith population. The I population means are the FIT part of the model and are represented by μ_i. The random variation, or RESIDUAL, part of the model is represented by the deviations ϵ_{ij} of the observations from the means.

The One-Way ANOVA Model

The data for one-way ANOVA are SRSs from each of I populations. The sample from the ith population has n_i observations, $x_{i1}, x_{i2}, \ldots, x_{in_i}$. The **one-way ANOVA model** is

$$x_{ij} = \mu_i + \epsilon_{ij}$$

for $i = 1, \ldots, I$ and $j = 1, \ldots, n_i$. The ϵ_{ij} are assumed to be from an $N(0, \sigma)$ distribution. The **parameters of the model** are the I population means $\mu_1, \mu_2, \ldots, \mu_I$ and the common standard deviation σ.

FIGURE 14.5 Model for one-way ANOVA with three groups. The three populations have Normal distributions with the same standard deviation.

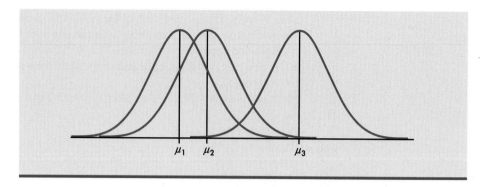

Note that the sample sizes n_i may differ, but the standard deviation σ is assumed to be the same in all the populations. Figure 14.5 pictures this model for $I = 3$. The three population means μ_i are different, but the shapes of the three Normal distributions are the same, reflecting the condition that all three populations have the same standard deviation.

EXAMPLE 14.4   ANOVA Model for the SCI Study

In our worker safety example there are three groups of workers that we want to compare, so $I = 3$. The population means μ_1, μ_2, and μ_3 are the mean SCI values for unskilled workers, for skilled workers, and for supervisors, respectively. The sample sizes n_i are 448, 91, and 51.

Suppose the first unskilled worker sampled is Eve Brogden. The observation x_{11} is the SCI score for Eve. The data for the other unskilled workers are denoted by x_{12}, x_{13}, and so on. Similarly, the data for the other two groups have a first subscript indicating the group and a second subscript indicating the worker in that group.

According to our model, Eve's SCI score is $x_{11} = \mu_1 + \epsilon_{11}$, where μ_1 is the average for *all* unskilled workers and ϵ_{11} is the chance variation due to Eve's opinions. The ANOVA model assumes that the ϵ_{ij} are independent and Normally distributed with mean 0 and standard deviation σ. We have clear evidence that the data are not Normal in our example. The values are numbers ranging from 0 to 100, and we saw some skewness for all three groups in Figure 14.3. However, because our inference is based on the sample means, which will be approximately Normal, we are not overly concerned about this violation of our assumptions.

It is common to use numerical subscripts to distinguish the different means, and some software requires that levels of factors in ANOVA be specified as numerical values. An alternative is to use subscripts that suggest the actual groups. In our example, we could replace μ_1, μ_2, and μ_3 by μ_{UN}, μ_{SK}, and μ_{SU}.

APPLY YOUR KNOWLEDGE

14.3 Magazine layouts. Example 14.1 (page 742) describes a study designed to compare sales based on different magazine layouts. Write out the ANOVA model for this study. Be sure to give specific values for I and the n_i. List all the parameters of the model.

14.4 Ages of customers at different stores. In Example 14.2 (page 742) the ages of customers at different bookstores are compared. Write out the ANOVA model for this study. Be sure to give specific values for I and the n_i. List all the parameters of the model.

Estimates of population parameters

The unknown parameters in the statistical model for ANOVA are the I population means μ_i and the common population standard deviation σ. To estimate μ_i we use the sample mean for the ith group:

$$\bar{x}_i = \frac{1}{n_i} \sum_{j=1}^{n_i} x_{ij}$$

residuals

The **residuals** $e_{ij} = x_{ij} - \bar{x}_i$ reflect the variation about the sample means that we see in the data.

The ANOVA model states that the population standard deviations are all equal. The ANOVA test of equality of means requires this condition. If we have unequal standard deviations, we generally try to transform the data so that they are approximately equal. We might, for example, work with $\sqrt{x_{ij}}$ or $\log x_{ij}$. Fortunately, we can often find a transformation that *both* makes the group standard deviations more nearly equal and also makes the distributions of observations in each group more nearly Normal. If the standard deviations are markedly different and cannot be made similar by a transformation, inference requires different methods.

Unfortunately, formal tests for the equality of standard deviations in several groups share the lack of robustness against non-Normality that we noted in Chapter 7 (page 439) for the case of two groups. Because ANOVA procedures are not extremely sensitive to unequal standard deviations, we do *not* recommend a formal test of equality of standard deviations as a preliminary to the ANOVA. Instead, we will use the following rule of thumb.

> **Rule for Examining Standard Deviations in ANOVA**
>
> If the largest sample standard deviation is less than twice the smallest sample standard deviation, we can use methods based on the condition that the population standard deviations are equal and our results will still be approximately correct.[2]

When we assume that the population standard deviations are equal, each sample standard deviation is an estimate of σ. To combine these into a single estimate, we use a generalization of the pooling method introduced in Chapter 7 (page 429).

> **Pooled Estimator of σ**
>
> Suppose we have sample variances $s_1^2, s_2^2, \ldots, s_I^2$ from I independent SRSs of sizes $n_1, n_2, \ldots, n_I$ from populations with common variance σ^2. The **pooled sample variance**
>
> $$s_p^2 = \frac{(n_1 - 1)s_1^2 + (n_2 - 1)s_2^2 + \cdots + (n_I - 1)s_I^2}{(n_1 - 1) + (n_2 - 1) + \cdots + (n_I - 1)}$$
>
> is an unbiased estimator of σ^2. The **pooled standard error**
>
> $$s_p = \sqrt{s_p^2}$$
>
> is the estimate of σ.

Pooling gives more weight to groups with larger sample sizes. If the sample sizes are equal, s_p^2 is just the average of the I sample variances. Note that s_p is *not* the average of the I sample standard deviations.

EXAMPLE 14.5   Population Estimates for Worker Safety Study

In the worker safety study there are $I = 3$ groups and the sample sizes are $n_1 = 448$, $n_2 = 91$, and $n_3 = 51$. The sample standard deviations are $s_1 = 18.27$, $s_2 = 18.83$, and $s_3 = 14.58$.

Because the largest standard deviation (18.83) is less than twice the smallest ($2 \times 14.58 = 29.16$), our rule indicates that we can use the assumption of equal population standard deviations.

The pooled variance estimate is

$$s_p^2 = \frac{(n_1 - 1)s_1^2 + (n_2 - 1)s_2^2 + (n_3 - 1)s_3^2}{(n_1 - 1) + (n_2 - 1) + (n_3 - 1)}$$

$$= \frac{(447)(18.27)^2 + (90)(18.83)^2 + (50)(14.58)^2}{447 + 90 + 50}$$

$$= \frac{191,745}{587} = 326.7$$

The pooled standard deviation is

$$s_p = \sqrt{326.7} = 18.07$$

This is our estimate of the common standard deviation σ of the SCI scores in the three populations of workers.

APPLY YOUR KNOWLEDGE

14.5 Magazine layouts. Example 14.1 (page 742) describes a study designed to compare sales based on different magazine layouts, and in Exercise 14.3 (page 748) you described the ANOVA model for this study. The three layouts are designated 1, 2, and 3. The following table summarizes the sales data.

Layout	Mean	Standard deviation	Sample size
1	75	85	20
2	125	130	20
3	100	115	20

(a) Is it reasonable to pool the standard deviations for these data? Give a reason for your answer.

(b) For each parameter in your model from Exercise 14.3, give the estimate.

14.6 Ages of customers at different stores. In Example 14.2 (page 742) the ages of customers at different bookstores are compared, and you described the ANOVA model for this study in Exercise 14.4 (page 748). Here is a summary of the ages of the customers:

Store	Mean	Standard deviation	Sample size
A	38	8	50
B	44	10	50
C	23	7	50
D	25	9	50
E	40	13	50

(a) Is it reasonable to pool the standard deviations for these data? Give a reason for your answer.

(b) For each parameter in your model from Exercise 14.4, give the estimate.

Testing hypotheses in one-way ANOVA

Comparison of several means is accomplished by using an F statistic to compare the variation among groups with the variation within groups. We now show how the F statistic expresses this comparison. Calculations are organized in an **ANOVA table,** which contains numerical measures of the variation among groups and within groups.

First we must specify our hypotheses for one-way ANOVA. As usual, I represents the number of populations to be compared.

ANOVA table

> **Hypotheses for One-Way ANOVA**
>
> The **null and alternative hypotheses** for one-way ANOVA are
>
> $$H_0: \mu_1 = \mu_2 = \ldots = \mu_I$$
> $$H_a: \text{not all of the } \mu_i \text{ are equal}$$

The following discussion illustrates how to do a one-way ANOVA for the setting of Case 14.1. The calculations are generally performed using statistical software on a computer, so we focus on interpretation of the output.

CASE 14.1

Do Eyes Affect Ad Response? Research from a variety of fields has found significant effects of eye gaze and eye color on emotions and perceptions such as arousal, attractiveness, and honesty. These findings suggest that a model's eyes may play a role in a viewer's response to an ad. In a recent study, students in marketing and management classes of a southern, predominately Hispanic, university were each presented one of four portfolios.[3] Each portfolio contained a target ad for a fictional product, Sparkle Toothpaste. Students were asked to view the ad and then respond to questions concerning their attitudes and emotions about the ad and product. All questions were from advertising-effects questionnaires previously used in the literature. Each response was on a seven-point scale. Although the researchers investigated nine attitudes and emotions, we will focus on the viewer's "attitudes toward the brand." This response was obtained by averaging 10 survey questions.

The target ads were created using two digital photographs of a model. In one picture the model is looking directly at the camera so the eyes can be seen. This picture was used in three target ads. The only difference was the model's eyes, which were made to be either brown, blue, or green. In the second picture, the model is in virtually the same pose but looking downward so the eyes are not visible. A total of 222 surveys were used for analysis. The table below summarizes the responses for the four portfolios.

Group	n	Mean	Std. dev.
Blue	67	3.19	1.75
Brown	37	3.72	1.73
Down	41	3.11	1.53
Green	77	3.86	1.67

DATA FILE EYES

We want to test the null hypothesis that the four groups represent four populations that all have the same mean score. We should always start an ANOVA with a careful examination of the data using graphical and numerical summaries. Side-by-side boxplots are shown in Figure 14.6. Comparing the first and third quartiles to the median, we can see that the data in each group appear relatively symmetric. The tails, however, suggest some skewness due to the limited range of the response. Groups with scores closer to 1 have much shorter lower tails. With respect to the spread, the ranges and interquartile ranges of the groups are similar.

FIGURE 14.6 Side-by-side boxplots for Case 14.1.

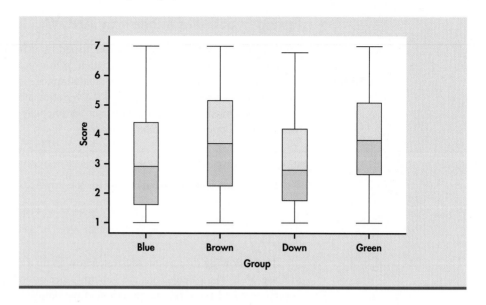

APPLY YOUR KNOWLEDGE

CASE 14.1 **14.7 Histograms.** Display the data of Case 14.1 for the four portfolio groups using four histograms. Are any systematic differences among the groups apparent? Do any groups show strong skewness or clear outliers? Do the data appear reasonably Normal?

EXAMPLE 14.6 Verifying the Conditions for ANOVA

CASE 14.1 If ANOVA is to be trusted, three conditions must hold.

SRSs. Can we regard the four samples as SRSs from four populations? An ideal study would start with an SRS from the population of interest and then randomly assign each participant to a portfolio. This usually isn't practical. The researchers randomly assigned portfolios to students taking marketing and management courses from one university. Can we act as if these students were randomly chosen from the university? People may disagree on the answer.

Normality. Are the attitude scores Normally distributed in each group? Figure 14.7 displays Normal quantile plots for the four groups. The data do look reasonably Normal. There are a limited number of possible values for the score (the average of 10 seven-point questions) so there is some curvature in the plots near 1 and 7. Given the sample sizes, however, this should not cause difficulties in using ANOVA.

Common standard deviation. Because the largest standard deviation (1.75) is less than twice the smallest ($2 \times 1.53 = 3.06$), our rule of thumb tells us that we need not be concerned about violating the condition that the three populations have the same standard deviation.

Because the data look reasonably Normal and meet the condition of equal standard deviations, we proceed with the analysis of variance.

EXAMPLE 14.7 Are the Differences Significant?

CASE 14.1 The ANOVA results produced by Minitab are shown in Figure 14.8. The pooled standard deviation s_p is reported as 1.677. The calculated value of the F statistic appears under the heading F, and its P-value is under the heading P. The value of F is 2.89, with a P-value of 0.036. That is, an F of 2.89 or larger would occur about 3.6% of the time by chance when the population means

FIGURE 14.7 Normal quantile plots of the scores of attitudes toward brand for the four portfolio groups in Case 14.1.

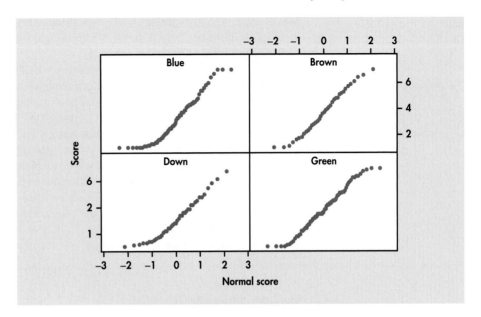

FIGURE 14.8 Minitab analysis of variance output for the brand attitude scores for Case 14.4.

```
One-way ANOVA: Score versus Trt

Source    DF        SS     MS      F        P
Group      3     24.42   8.14   2.89   0.036
Error    218    613.14   2.81
Total    221    637.56

S = 1.677     R-Sq = 3.83%     R-Sq(adj) = 2.51%
```

are equal. We have evidence at the 5% significance level to reject the null hypothesis that the four populations have equal means. There is evidence that the four groups of students do not all have the same mean attitude score.

Examples 14.6 and 14.7 illustrate the basics of ANOVA: state the hypotheses, verify that the conditions for ANOVA are met, look at the F statistic and its P-value, and state a conclusion.

APPLY YOUR KNOWLEDGE

EYES

CASE 14.1 **14.8 An alternative Normality check.** Figure 14.7 displays separate Normal quantile plots for the four groups. An alternative procedure is to make one Normal quantile plot using the *residuals* $e_{ij} = x_{ij} - \bar{x}_i$ for all four groups together. Make this plot and summarize what it shows.

The ANOVA table

Software ANOVA output contains more than simply the test statistic. The additional information shows, among other things, where the test statistic comes from.

ANOVA table

The information in an analysis of variance is organized in an **ANOVA table.** In the software output in Figure 14.8, the columns of this table are labeled Source, DF, SS,

MS, F, and P. The rows are labeled Group, Error, and Total. These are the three sources of variation in the one-way ANOVA. ("Group" was the name used in entering the data to distinguish the four treatment groups.)

variation among groups

The Group row in the table corresponds to the FIT term in our DATA = FIT + RESIDUAL way of thinking. It gives information related to the variation **among** group means. The ANOVA model allows the groups to have different means. The Error row in the table corresponds to the RESIDUAL term in DATA = FIT + RESIDUAL. It gives

variation within groups

information related to the variation **within** groups. The term "error" is most appropriate for experiments in the physical sciences, where the observations within a group differ because of measurement error. In business and the biological and social sciences, on the other hand, the within-group variation is often due to the fact that not all firms or plants or people are the same. This sort of variation is not due to errors and is better described as "residual." Finally, the Total row in the table corresponds to the DATA term in our DATA = FIT + RESIDUAL framework.

For analysis of variance, the idea that

$$DATA = FIT + RESIDUAL$$

translates into an actual equation:

$$total\ variation = variation\ among\ groups + variation\ within\ groups$$

sums of squares

The ANOVA idea is to break the total variation in the responses into two parts: the variation due to differences among the group means and that due to differences within groups. Variation is expressed by **sums of squares.** We use SSG, SSE, and SST for the sums of squares for groups, error, and total. Each sum of squares is the sum of the squares of a set of deviations that expresses a source of variation. SST is the sum of squares of $x_{ij} - \overline{x}$, which measure variation of the responses around their overall mean. Variation of the group means around the overall mean $\overline{x}_i - \overline{x}$ is measured by SSG. Finally, SSE is the sum of squares of the deviations $x_{ij} - \overline{x}_i$ of each observation from its group mean. It is always true that SST = SSG + SSE. This is the algebraic version of the ANOVA idea: total variation is the sum of among-group variation and within-group variation.

EXAMPLE 14.8   Sums of Squares for the Three Sources of Variation

CASE 14.1

The SS column in Figure 14.8 gives the values for the three sums of squares. They are SSG = 24.42, SSE = 613.14, and SST = 637.56. In this example it appears that most of the variation is coming from Error, that is, from within groups.

degrees of freedom

Associated with each sum of squares is a quantity called the **degrees of freedom.** Because SST measures the variation of all N observations around the overall mean, its degrees of freedom are DFT $= N - 1$, the degrees of freedom for the sample variance of the N responses. Similarly, because SSG measures the variation of the I sample means around the overall mean, its degrees of freedom are DFG $= I - 1$. Finally, SSE is the sum of squares of the deviations $x_{ij} - \overline{x}_i$. Here we have N observations being compared with I sample means and DFE $= N - I$.

EXAMPLE 14.9   Degrees of Freedom for the Three Sources

CASE 14.1

The DF column in Figure 14.8 gives the values for the three degrees of freedom. These values are DFT $= 221$, DFG $= 3$, and DFE $= 218$.

mean square

For each source of variation, the **mean square** is the sum of squares divided by the degrees of freedom. Generally, the ANOVA table includes mean squares only for the first two sources of variation.

EXAMPLE 14.10 Mean Squares for the Three Sources

CASE 14.1

The MS column in Figure 14.8 gives values for two of the mean squares. These values are $MSG = 8.14$ and $MSE = 2.81$.

The mean square corresponding to the total source is the sample variance that we would calculate assuming that we have one sample from a single population—that is, assuming that the means of the four groups are the same.

Sums of Squares, Degrees of Freedom, and Mean Squares

Sums of squares represent variation present in the data. They are calculated by summing squared deviations. In one-way ANOVA there are three **sources of variation:** groups, error, and total. The sums of squares are related by the formula

$$SST = SSG + SSE$$

Thus, the total variation is composed of two parts, one due to groups and one due to "error" (variation within groups).

Degrees of freedom are related to the deviations that are used in the sums of squares. The degrees of freedom are related in the same way as the sums of squares:

$$DFT = DFG + DFE$$

To calculate each **mean square,** divide the corresponding sum of squares by its degrees of freedom.

APPLY YOUR KNOWLEDGE

CASE 14.1 *All these exercises use the output in Figure 14.8 (page 753) or the data for Case 14.1.*

14.9 Verify that the SS add. Verify that the sums of squares add, $SST = SSG + SSE$.

14.10 Check the DF. What are I and N for these data? Verify that the DF entries in the output are $DFG = I - 1$, $DFE = N - I$, and $DFT = N - 1$.

14.11 Do the DF add? Verify that degrees of freedom add in the same way that the sums of squares add. That is, $DFT = DFG + DFE$.

14.12 Check the MS. Verify that each mean square in the output is the corresponding sum of squares divided by its degrees of freedom.

14.13 Total mean square. The output does not give the total mean square $MST = SST/DFT$. Calculate this quantity. Then find the mean and variance of all 222 observations and verify that MST is the variance of all the responses.

The F test

The ANOVA table also reports the pooled standard error s_p. It is always true that

$$s_p^2 = MSE = \frac{SSE}{DFE}$$

This fact reinforces the idea that MSE measures variation within groups. From the output in Figure 14.8, we can calculate the variance $s_p^2 = (1.677)^2 = 2.81$. This is the same as MSE up to roundoff error.

The F statistic is the final entry in the ANOVA table. If H_0 is true, there are no differences among the group means. Then MSG will reflect only chance variation and we expect MSG to be about the same as MSE. The F statistic simply compares these two mean squares, $F = $ MSG/MSE. This statistic is near 1 if H_0 is true and tends to be larger if H_a is true. In our example, MSG $= 8.14$ and MSE $= 2.81$, so the ANOVA F statistic is

$$F = \frac{\text{MSG}}{\text{MSE}} = \frac{8.14}{2.81} = 2.89$$

When H_0 is true, the F statistic has an F distribution that depends upon two numbers: the *degrees of freedom for the numerator* and the *degrees of freedom for the denominator*. These degrees of freedom are those associated with the mean squares in the numerator and denominator of the F statistic. For one-way ANOVA, the degrees of freedom for the numerator are DFG $= I - 1$ and the degrees of freedom for the denominator are DFE $= N - I$. We use the notation $F(I - 1, N - I)$ for this distribution.

The ANOVA F Test

To test the null hypothesis in a one-way ANOVA, calculate the **F statistic**

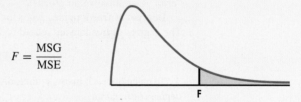

$$F = \frac{\text{MSG}}{\text{MSE}}$$

When H_0 is true, the F statistic has the $F(I - 1, N - I)$ distribution. When H_a is true, the F statistic tends to be large. We reject H_0 in favor of H_a if the F statistic is sufficiently large.

The **P-value** of the F test is the probability that a random variable having the $F(I - 1, N - I)$ distribution is greater than or equal to the calculated value of the F statistic.

Tables of F critical values are available for use when software does not give the P-value. Table E in the back of the book contains the F critical values for probabilities $p = 0.100, 0.050, 0.025, 0.010,$ and 0.001. For one-way ANOVA we use critical values from the table corresponding to $I - 1$ degrees of freedom in the numerator and $N - I$ degrees of freedom in the denominator. We have already seen several examples where the F statistic and its P-value were used to choose between H_0 and H_a.

EXAMPLE 14.11   The Mean Safety Scores Are Not Identical

p	Critical value
0.100	2.33
0.050	3.04
0.025	3.76
0.010	4.71
0.001	7.15

In the study of workplace safety in Example 14.3, $F = 7.137$. There are three populations, so the degrees of freedom in the numerator are DFG $= I - 1 = 2$. The degrees of freedom in the denominator are DFE $= N - I = 590 - 3 = 587$. In Table E first find the column corresponding to 2 degrees of freedom in the numerator. For the degrees of freedom in the denominator, there are entries for 200 and 1000. These entries are very close. To be conservative we use critical values corresponding to 200 degrees of freedom in the denominator since these are slightly larger. Because 7.137 falls between 4.71 and 7.15, we reject H_0 and conclude that the differences in means are statistically significant, with $0.001 < P < 0.01$.

EXAMPLE 14.12 Mean Attitude Scores Differ Significantly

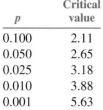

CASE 14.1

In the study of eye color and gaze on attitude towards brand, $F = 2.89$. There are four populations, so the degrees of freedom in the numerator are $DFG = I - 1 = 3$. The degrees of freedom in the denominator are $DFE = N - I = 222 - 4 = 218$. In Table E first find the column corresponding to 3 degrees of freedom in the numerator. For the degrees of freedom in the denominator, there are entries for 200 and 1000. These entries are very close. To be conservative we use critical values corresponding to 200 degrees of freedom in the denominator since these are slightly larger. Because 2.89 is between 2.65 and 3.18, we reject H_0 and conclude that the differences in means are statistically significant, with $0.025 < P < 0.05$.

p	Critical value
0.100	2.11
0.050	2.65
0.025	3.18
0.010	3.88
0.001	5.63

Remember that this F test is always one-sided because any differences among the group means tend to make F large. The ANOVA F test shares the robustness of the two-sample t test. It is relatively insensitive to moderate non-Normality and unequal variances, especially when the sample sizes are similar.

We recommend the *One-Way ANOVA* applet available on the Web site www.whfreeman.com/psbe as an excellent way to see how the value of the F statistic and its P-value depend on both the variability of the data within the groups and the differences among the group means. Exercises 14.16 and 14.17 make use of this applet.

APPLY YOUR KNOWLEDGE

CASE 14.1 **14.14 Pool the variances.** Find the variances for the four samples by squaring the standard deviations given in the table in Case 14.1 (page 751). Then pool the variances using the formula given on page 749. Verify that this calculation gives the MSE entry in Figure 14.8.

14.15 Use Table E. An ANOVA is run to compare three groups. There are 8 subjects in each group.
(a) Give the degrees of freedom for the ANOVA F statistic.
(b) How large would this statistic need to be to have a P-value less than 0.05?
(c) Suppose that we are still interested in comparing the three groups, but we obtain data on 21 subjects per group. How large would the F statistic need to be to have a P-value less than 0.05?
(d) Explain why the answer to part (c) is smaller than what you found for part (b).

14.16 The effect of within-group variation. Go to the *One-Way ANOVA* applet. In the applet display, the black dots are the mean responses in three treatment groups. Move these up and down until you get a configuration with P-value about 0.01. Note the value of the F statistic. Now increase the variation within the groups without changing their means by dragging the mark on the pooled standard error scale to the right. Describe what happens to the F statistic and the P-value. Explain why this happens.

14.17 The effect of among-group variation. Go to the *One-Way ANOVA* applet. Set the pooled standard error near the middle of its scale and drag the black dots so that the three group means are approximately equal. Note the value of the F statistic and its P-value. Now increase the variation among the group means: drag the mean of the second group up and the mean of the third group down. Describe the effect on the F statistic and its P-value. Explain why they change in this way.

Using software

The following display shows the general form of a one-way ANOVA table with the F statistic. The formulas in the sum of squares column can be used for calculations in small problems. There are other formulas that are more efficient for hand or calculator use, but ANOVA calculations are usually done by computer software.

Source	Degrees of freedom	Sum of squares	Mean square	F
Groups	$I - 1$	$\sum_{\text{groups}} n_i(\bar{x}_i - \bar{x})^2$	SSG/DFG	MSG/MSE
Error	$N - I$	$\sum_{\text{groups}} (n_i - 1)s_i^2$	SSE/DFE	
Total	$N - 1$	$\sum_{\text{obs}} (x_{ij} - \bar{x})^2$	SST/DFT	

You should now be able to extract all the ANOVA information we have discussed from output from almost any statistical software. Here is a final example on which to practice this skill.

EXAMPLE 14.13 Advertising and the Quality of a Product

ADQUALITY

Research suggests that customers think that a product is of high quality if it is heavily advertised. An experiment designed to explore this idea collected quality ratings (on a 1 to 7 scale) of a new line of take-home refrigerated entrées based on reading a magazine ad. Three groups were compared. The first group's ad included information that would undermine (U) the expected positive association between quality and advertising; the second group's ad contained information that would affirm (A) the association; and the third group was a control (C).[4] The data are given in Table 14.1. Outputs from SAS, Excel, and Minitab appear in Figure 14.9.

TABLE 14.1 Quality ratings in three groups

Group	Quality ratings
Undermine ($n = 55$)	6 5 5 5 4 5 4 6 5 5 5 5 3 3 5 4 5 5 5 4 5 4 4 5 4 4 4 5 5 5 4 5 5 5 4 4 5 5 4 5 5 4 4 5 4 3 4 5 5 5 3 4 4 4 4
Affirm ($n = 36$)	4 6 4 6 5 5 5 6 4 5 5 5 4 6 6 5 5 7 4 6 6 4 5 4 5 5 6 4 5 5 4 6 4 6 5 5
Control ($n = 36$)	5 4 5 6 5 7 5 6 7 5 7 5 4 5 4 4 6 6 5 6 5 5 4 5 5 6 6 6 5 6 6 7 6 6 5 5

FIGURE 14.9 SAS, Excel, and Minitab outputs for the advertising study of Example 14.13.

SAS

```
General Linear Models Procedure

Dependent Variable: QUALITY
                              Sum of        Mean
Source              DF       Squares       Square    F Value   Pr > F
Model                2     18.828255     9.414127     15.28    0.0001
Error              124     76.384343     0.616003
Corrected Total    126     95.212598

              R-Square          C.V.     Root MSE        QUALITY Mean
              0.197750      15.94832       0.7849              4.9213

Level of           -----------QUALITY-----------
        GROUP        N              Mean                      SD

        affirm      36         5.05555556                0.82615960
        control     36         5.41666667                0.87423436
        under       55         4.50909091                0.69048365
```

Excel

	A	B	C	D	E	F	G
1	Anova: Single Factor						
2	SUMMARY						
3							
4	*Groups*	*Count*	*Sum*	*Average*	*Variance*		
5	affirm	36	182	5.055555556	0.682539683		
6	control	36	195	5.416666667	0.764285714		
7	undermine	55	248	4.509090909	0.476767677		
8							
9	ANOVA						
10							
11	*Source of Variation*	*SS*	*df*	*MS*	*F*	*P-value*	*F crit*
12	Between Groups	18.82825499	2	9.414127495	15.28260579	1.16739E-06	3.069286447
13	Within Groups	76.38434343	124	0.61600277			
14	Total	95.21259843	126				

Minitab

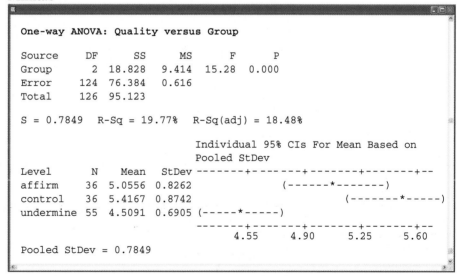

```
One-way ANOVA: Quality versus Group

Source     DF      SS     MS      F      P
Group       2   18.828  9.414  15.28  0.000
Error     124   76.384  0.616
Total     126   95.123

S = 0.7849   R-Sq = 19.77%   R-Sq(adj) = 18.48%

                            Individual 95% CIs For Mean Based on
                            Pooled StDev
Level        N    Mean   StDev -------+-------+-------+-------+--
affirm      36  5.0556  0.8262              (------*-------)
control     36  5.4167  0.8742                       (------*-----)
undermine   55  4.5091  0.6905 (-----*-----)
                                   -------+-------+-------+-------+--
                                    4.55    4.90    5.25    5.60

Pooled StDev = 0.7849
```

FIGURE 14.9 (*Continued*)

14.18 Compare software. Refer to the output in Figure 14.9. Different names are given to the sources of variation in the ANOVA tables.

(a) What are the names given to the source we call Groups?

(b) What are the names given to the source we call Error?

(c) What are the reported P-values for the ANOVA F test?

14.19 Compare software. The pooled standard error for the data in Table 14.1 is $s_p = 0.7849$. Look at the software output in Figure 14.9.

(a) Explain to someone new to ANOVA why SAS labels this quantity as "Root MSE."

(b) Excel does not report s_p. How can you find its value from an Excel output?

SECTION 14.1 Summary

- **One-way analysis of variance (ANOVA)** is used to compare several population means based on independent SRSs from each population. We assume that the populations are Normal and that, although they may have different means, they have the same standard deviation.

- To do an analysis of variance, first examine the data. Side-by-side boxplots give an overview. Examine Normal quantile plots (either for each group separately or for the residuals) to detect outliers or extreme deviations from Normality. Compute the ratio of the largest to the smallest sample standard deviation. If this ratio is less than 2 and the Normal quantile plots are satisfactory, ANOVA can be performed.

- The **null hypothesis** is that the population means are *all equal*. The **alternative hypothesis** is true if there are *any* differences among the population means.

- ANOVA is based on separating the total variation observed in the data into two parts: variation **among group means** and variation **within groups.** If the variation among groups is large relative to the variation within groups, we have evidence against the null hypothesis.

- An **analysis of variance table** organizes the ANOVA calculations. **Degrees of freedom, sums of squares, and mean squares** appear in the table. The **F statistic** and its **P-value** are used to test the null hypothesis.

14.2 Comparing Group Means

The ANOVA F test gives a general answer to a general question: are the differences among observed group means significant? Unfortunately, a small P-value simply tells us that the group means are not all the same. It does not tell us specifically which means differ from each other. Plotting and inspecting the means give us some indication of where the differences lie, but we would like to supplement inspection with formal inference. This section presents two approaches to the task of comparing group means.

Contrasts

The preferred approach is to pose specific questions regarding comparisons among the means before the data are collected. We can answer specific questions of this kind and attach a level of confidence to the answers we give. We now explore these ideas in the setting of Case 14.2.

EDUPRODUCT

CASE 14.2

Evaluation of a New Educational Product Your company markets educational materials aimed at parents of young children. You are planning a new product that is designed to improve children's reading comprehension. Your product is based on new ideas from educational research, and you would like to claim that children will acquire better reading comprehension skills utilizing these new ideas than with the traditional approach. Your marketing material will include the results of a study conducted to compare two versions of the new approach with the traditional method.[5] The standard method is called Basal, and the two variations of the new method are called DRTA and Strat.

VStock/Alamy

Education researchers randomly divided 66 children into three groups of 22. Each group was taught by one of the three methods. The response variable is a measure of reading comprehension called COMP that was obtained by a test taken after the instruction was completed. Can you claim that the new methods are superior to Basal?

We can compare the new with the standard by posing and answering specific questions about the mean responses. First, here is the basic ANOVA.

EXAMPLE 14.14   Are the Comprehension Scores Different?

CASE 14.2   Figure 14.10 gives the summary statistics for COMP computed by SPSS. This software uses only numeric values for the factor, so we coded the groups as 1 for Basal, 2 for DRTA, and 3 for Strat. Side-by-side boxplots appear in Figure 14.11, and Figure 14.12 plots the group means. The ANOVA results generated by SPSS are given in Figure 14.13, and a Normal quantile plot of the residuals appears in Figure 14.14.

Descriptives

COMP

| | | | Std. Deviation | Std. Error | 95% Confidence Interval for Mean | | | |
	N	Mean			Lower Bound	Upper Bound	Minimum	Maximum
1.00	22	41.05	5.636	1.202	38.55	43.54	32	54
2.00	22	46.73	7.388	1.575	43.45	50.00	30	57
3.00	22	44.27	5.767	1.229	41.72	46.83	33	53
Total	66	44.02	6.644	0.818	42.38	45.65	30	57

FIGURE 14.10 Summary statistics for the comprehension scores in the three groups for the new-product evaluation study of Case 14.2.

FIGURE 14.11 Side-by-side boxplots of the comprehension scores in the new-product evaluation study of Case 14.2.

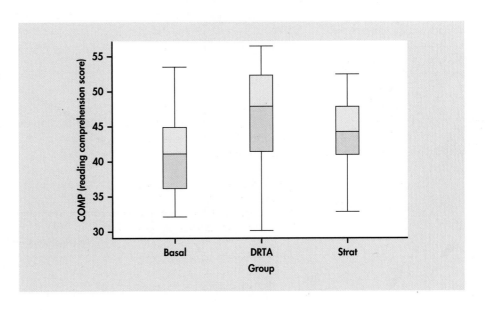

FIGURE 14.12 Comprehension
score group means in the
new-product evaluation study
of Case 14.2.

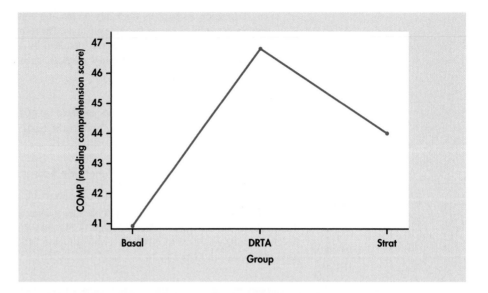

FIGURE 14.13 SPSS analysis of variance output for the comprehension scores in the
new-product evaluation study of Case 14.2.

FIGURE 14.14 Normal quantile
plot of the residuals for the
comprehension scores in the
new-product evaluation study
of Case 14.2.

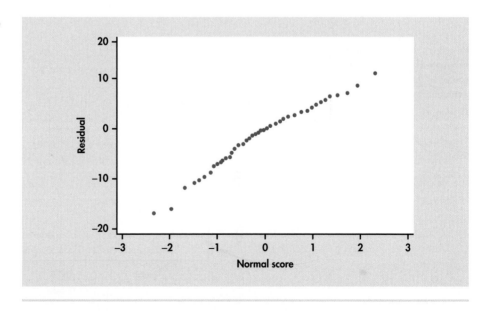

The ANOVA null hypothesis is

$$H_0: \mu_B = \mu_D = \mu_S$$

where the subscripts correspond to the group labels Basal, DRTA, and Strat. Figure 14.13 shows that $F = 4.48$ with degrees of freedom 2 and 63. The P-value is 0.015. We have good evidence against H_0.

What can the researchers conclude from this analysis? The alternative hypothesis is true if $\mu_B \neq \mu_D$ or if $\mu_B \neq \mu_S$ or if $\mu_D \neq \mu_S$ or if any combination of these statements is true. We would like to be more specific.

EXAMPLE 14.15   The Major Question

CASE 14.2   The two new methods are based on the same idea. Are they superior to the standard method? We can formulate this question as the null hypothesis

$$H_{01}: \frac{1}{2}(\mu_D + \mu_S) = \mu_B$$

with the alternative

$$H_{a1}: \frac{1}{2}(\mu_D + \mu_S) > \mu_B$$

The hypothesis H_{01} compares the average of the two innovative methods (DRTA and Strat) with the standard method (Basal). The alternative is one-sided because the researchers are interested in demonstrating that the new methods are better than the old. We use the subscripts 1 and 2 to distinguish two sets of hypotheses that correspond to two specific questions about the means.

EXAMPLE 14.16   A Secondary Question

CASE 14.2   A secondary question involves a comparison of the two new methods. We formulate this as the hypothesis that the methods DRTA and Strat are equally effective,

$$H_{02}: \mu_D = \mu_S$$

versus the alternative

$$H_{a2}: \mu_D \neq \mu_S$$

contrast ψ   Each of H_{01} and H_{02} says that a combination of population means is 0. These combinations of means are called **contrasts.** We use ψ, the Greek letter psi, for contrasts among population means. The two contrasts that arise from our two null hypotheses are

$$\psi_1 = -\mu_B + \frac{1}{2}(\mu_D + \mu_S)$$
$$= (-1)\mu_B + (0.5)\mu_D + (0.5)\mu_S$$

and

$$\psi_2 = \mu_D - \mu_S$$

In each case, the value of the contrast is 0 when H_0 is true. We chose to define the contrasts so that they will be positive when the alternative hypothesis is true. Whenever possible, this is a good idea because it makes some computations easier.

sample contrast   A contrast expresses an effect in the population as a combination of population means. To estimate the contrast, form the corresponding **sample contrast** by using sample means in place of population means. Under the ANOVA assumptions, a sample contrast is a linear combination of independent Normal variables and therefore has a Normal distribution. We can obtain the standard error of a contrast by using the rules for variances given in Section 4.3 (page 253). Inference is based on t statistics. Here are the details.

Contrasts

A **contrast** is a combination of population means of the form

$$\psi = \sum a_i \mu_i$$

where the coefficients a_i have sum 0. The corresponding **sample contrast** is the same combination of sample means,

$$c = \sum a_i \overline{x}_i$$

The **standard error of c** is

$$SE_c = s_p \sqrt{\sum \frac{a_i^2}{n_i}}$$

To test the null hypothesis H_0: $\psi = 0$, use the t **statistic**

$$t = \frac{c}{SE_c}$$

with degrees of freedom DFE that are associated with s_p. The alternative hypothesis can be one-sided or two-sided.

A **level C confidence interval for ψ** is

$$c \pm t^* SE_c$$

where t^* is the value for the t(DFE) density curve with area C between $-t^*$ and t^*.

Because each $\overline{x}_i$ estimates the corresponding μ_i, the addition rule for means (page 248) tells us that the mean μ_c of the sample contrast c is ψ. In other words, c is an unbiased estimator of ψ. Testing the hypothesis that a contrast is 0 assesses the significance of the effect measured by the contrast. It is often more informative to estimate the size of the effect using a confidence interval for the population contrast.

EXAMPLE 14.17 The Coefficients for the Contrasts

CASE 14.2

In our example the coefficients in the contrasts are $a_1 = -1$, $a_2 = 0.5$, $a_3 = 0.5$ for ψ_1 and $a_1 = 0$, $a_2 = 1$, $a_3 = -1$ for ψ_2, where the subscripts 1, 2, and 3 correspond to B, D, and S. In each case the sum of the a_i is 0.

We look at inference for each of these contrasts in turn.

EXAMPLE 14.18 Are the New Methods Better?

CASE 14.2

The sample contrast that estimates ψ_1 is

$$c_1 = -\overline{x}_B + \frac{1}{2}(\overline{x}_D + \overline{x}_S)$$

$$= -41.05 + \frac{1}{2}(46.73 + 44.27) = 4.45$$

with standard error

$$SE_{c_1} = 6.314\sqrt{\frac{(-1)^2}{22} + \frac{(0.5)^2}{22} + \frac{(0.5)^2}{22}}$$
$$= 1.65$$

The t statistic for testing $H_{01}: \psi_1 = 0$ versus $H_{a1}: \psi_1 > 0$ is

$$t = \frac{c_1}{SE_{c_1}}$$
$$= \frac{4.45}{1.65} = 2.70$$

Because s_p has 63 degrees of freedom, software using the $t(63)$ distribution gives the one-sided P-value as 0.0044. If we used Table D, we would conclude that $P < 0.005$. The P-value is small, so there is strong evidence against H_{01}. The researchers have shown that the new methods produce higher mean scores than the old. The size of the improvement can be described by a confidence interval. To find the 95% confidence interval for ψ_1, we combine the estimate with its margin of error:

$$c_1 \pm t^* SE_{c_1} = 4.45 \pm (2.00)(1.65)$$
$$= 4.45 \pm 3.30$$

The interval is $(1.15, 7.75)$. We are 95% confident that the mean improvement obtained by using one of the innovative methods rather than the old method is between 1.15 and 7.75 points.

EXAMPLE 14.19 Comparing the Two New Methods

CASE 14.2 The second sample contrast, which compares the two new methods, is

$$c_2 = 46.73 - 44.27$$
$$= 2.46$$

with standard error

$$SE_{c_2} = 6.314\sqrt{\frac{(1)^2}{22} + \frac{(-1)^2}{22}}$$
$$= 1.90$$

The t statistic for assessing the significance of this contrast is

$$t = \frac{2.46}{1.90} = 1.29$$

The P-value for the two-sided alternative is 0.2020. We conclude that either the two new methods have the same population means or the sample sizes are not sufficiently large to distinguish them. A confidence interval helps clarify this statement. To find the 95% confidence interval for ψ_2, we combine the estimate with its margin of error:

$$c_2 \pm t^* SE_{c_2} = 2.46 \pm (2.00)(1.90)$$
$$= 2.46 \pm 3.80$$

The interval is $(-1.34, 6.26)$. With 95% confidence we state that the difference between the population means for the two new methods is between -1.34 and 6.26.

EXAMPLE 14.20 Using Software

Figure 14.15 displays the SPSS output for the analysis of these contrasts. The column labeled "t" gives the t statistics 2.702 and 1.289 for our two contrasts. The degrees of freedom appear in the column labeled "df" and are 63 for each t. The P-values are given in the column labeled "Sig. (2-tailed)." These are correct for two-sided alternative hypotheses. The values are 0.009 and 0.202. To convert the computer-generated results to apply to our one-sided alternative concerning ψ_1, simply divide the reported P-value by 2 after checking that the value of c is in the direction of H_a (that is, that c is positive).

FIGURE 14.15 SPSS output for contrasts for the comprehension scores in the new-product evaluation study of Case 14.2.

Contrast Coefficients

	Grp1		
Contrast	1.00	2.00	3.00
1	1	−.5	−.5
2	0	−1	1

Contrast		Value of Contrast	Std. Error	t	df	Sig.(2-tailed)
COMP	1	−4.45	1.649	−2.702	63	.009
	2	−2.45	1.904	−1.289	63	.202

Some statistical software packages report the test statistics associated with contrasts as F statistics rather than t statistics. These F statistics are the squares of the t statistics described above. The associated P-values are for the two-sided alternatives.

Questions about population means are expressed as hypotheses about contrasts. A contrast should express a specific question that we have in mind when designing the study. When contrasts are formulated before seeing the data, *inference about contrasts is valid whether or not the ANOVA H_0 of equality of means is rejected.* Because the F test answers a very general question, it is less powerful than tests for contrasts designed to answer specific questions. Specifying the important questions before the analysis is undertaken enables us to use this powerful statistical technique.

APPLY YOUR KNOWLEDGE

14.20 Define a contrast. An ANOVA was run with six groups. Give the coefficients for the contrast that compares the average of the means of the first two groups with the average of the means of the last two groups.

14.21 Find the standard error. Refer to the previous exercise. Suppose that there are 16 observations in each group and that $s_p = 8$. Find the standard error for the contrast.

14.22 Is the contrast significant? Refer to the previous exercise. Suppose that the average of the first two groups minus the average of the last two groups is 7. State an appropriate null hypothesis for this comparison, find the test statistic with its degrees of freedom, and report the result.

14.23 Give the confidence interval. Refer to the previous exercise. Give a 95% confidence interval for the difference between the average of the means of the first two groups and the average of the means of the last two groups.

Multiple comparisons

multiple comparisons

In many studies, specific questions cannot be formulated in advance of the analysis. If H_0 is not rejected, we conclude that the population means are indistinguishable on the basis of the data given. On the other hand, if H_0 is rejected, we would like to know which pairs of means differ. **Multiple-comparisons** methods address this issue. *It is important to keep in mind that multiple-comparisons methods are commonly used only after rejecting the ANOVA H_0.*

Return once more to the reading comprehension study described in Case 14.2. We found in Example 14.14 (page 761) that the means were not all the same ($F = 4.48$, df $= 2$ and 63, $P = 0.015$).

EXAMPLE 14.21 A *t* Statistic to Compare Two Means

CASE 14.2

There are three pairs of population means. We can compare Groups 1 and 2, Groups 1 and 3, and Groups 2 and 3. For each of these pairs, we can write a t statistic for the difference in means. To compare Basal with DRTA (1 with 2), we compute

$$t_{12} = \frac{\overline{x}_1 - \overline{x}_2}{s_p\sqrt{\dfrac{1}{n_1} + \dfrac{1}{n_2}}}$$

$$= \frac{41.05 - 46.73}{6.31\sqrt{\dfrac{1}{22} + \dfrac{1}{22}}} = -2.99$$

The subscripts on t specify which groups are compared.

APPLY YOUR KNOWLEDGE

CASE 14.2 **14.24 Compare Basal with Strat.** Verify that $t_{13} = -1.69$.

CASE 14.2 **14.25 Compare DRTA with Strat.** Verify that $t_{23} = 1.29$. (This is the same t that we used for the contrast $\psi_2 = \mu_2 - \mu_3$ in Example 14.19 (page 765).)

These t statistics are very similar to the pooled two-sample t statistic for comparing two population means, described in Chapter 7 (page 429). The difference is that we now have more than two populations, so each statistic uses the pooled estimator s_p from all groups rather than the pooled estimator from just the two groups being compared. This additional information about the common σ increases the power of the tests. The degrees of freedom for all of these statistics are DFE $= 63$, those associated with s_p.

Because we do not have any specific ordering of the means in mind as an alternative to equality, we must use a two-sided approach to the problem of deciding which pairs of means are significantly different.

> **Multiple Comparisons**
>
> To perform a **multiple-comparisons procedure,** compute t **statistics** for all pairs of means using the formula
>
> $$t_{ij} = \frac{\overline{x}_i - \overline{x}_j}{s_p\sqrt{\dfrac{1}{n_i} + \dfrac{1}{n_j}}}$$

If

$$|t_{ij}| \geq t^{**}$$

we declare that the population means μ_i and μ_j are different. Otherwise, we conclude that the data do not distinguish between them. The value of t^{**} depends upon which multiple-comparisons procedure we choose.

One obvious choice for t^{**} is the upper $\alpha/2$ critical value for the t (DFE) distribution. This choice simply carries out as many separate significance tests of fixed level α as there are pairs of means to be compared. The procedure based on this choice is called the **least-**

LSD method **significant differences method,** or simply LSD. LSD has some undesirable properties, particularly if the number of means being compared is large. Suppose, for example, that there are $I = 20$ groups and we use LSD with $\alpha = 0.05$. There are 190 different pairs of means. If we perform 190 t tests, each with an error rate of 5%, our overall error rate will be unacceptably large. We expect about 5% of the 190 to be significant even if the corresponding population means are the same. Since 5% of 190 is 9.5, we expect 9 or 10 false rejections.

The LSD procedure fixes the probability of a false rejection for each single pair of means being compared. It does not control the overall probability of *some* false rejection among all pairs. Other choices of t^{**} control possible errors in other ways. The choice of t^{**} is therefore a complex problem, and a detailed discussion of it is beyond the scope of this text. Many choices for t^{**} are used in practice. One major statistical package allows selection from a list of over a dozen choices.

Bonferroni method We will discuss only one of these, called the **Bonferroni method.** Use of this procedure with $\alpha = 0.05$, for example, guarantees that the probability of *any* false rejection among all comparisons made is no greater than 0.05. This is much stronger protection than controlling the probability of a false rejection at 0.05 for *each separate* comparison.

EXAMPLE 14.22 Which Means Differ?

CASE 14.2 We apply the Bonferroni multiple-comparisons procedure with $\alpha = 0.05$ to the data from the new-product evaluation study in Example 14.14. The value of t^{**} for this procedure (from software or special tables) is 2.46. The t statistic for comparing Basal with DRTA is $t_{12} = -2.99$. Because $|-2.99|$ is greater than 2.46, the value of t^{**}, we conclude that the DRTA method produces higher reading comprehension scores than Basal.

APPLY YOUR KNOWLEDGE

CASE 14.2 **14.26 Compare Basal with Strat.** The test statistic for comparing Basal with Strat is $t_{13} = -1.69$. For the Bonferroni multiple-comparisons procedure with $\alpha = 0.05$, do you reject the null hypothesis that the population means for these two groups are different?

CASE 14.2 **14.27 Compare DRTA with Strat.** Answer the same question as in the previous exercise for the comparison of DRTA with Strat using the calculated value $t_{23} = 1.29$.

Usually we use software to perform the multiple-comparisons procedure. The formats differ from package to package but they all give the same basic information.

EXAMPLE 14.23   Computer Output for Multiple Comparisons

CASE 14.2

The output from SPSS for Bonferroni comparisons appears in Figure 14.16. The first line of numbers gives the results for comparing Basal with DRTA, Groups 1 and 2. The difference between the means is given as −5.682 with a standard error of 1.904. The P-value for the comparison is given under the heading "Sig." The value is 0.012. Therefore, we could declare the means for Basal and DRTA to be different according to the Bonferroni procedure as long as we are using a value of α that is greater than 0.012. In particular, these groups are significantly different at the *overall* $\alpha = 0.05$ level. The last two entries in the row give the Bonferroni 95% confidence interval. We will discuss this later.

FIGURE 14.16 SPSS Bonferroni multiple-comparisons output for the comprehension scores in the new-product evaluation study of Case 14.2.

Multiple Comparisons

COMP
Bonferroni

(I) Grp1	(J) Grp1	Mean Difference (I-J)	Std. Error	Sig.	95% Confidence Interval Lower Bound	Upper Bound
1.00	2.00	−5.682*	1.904	.012	−10.36	−1.00
	3.00	−3.227	1.904	.285	−7.91	1.46
2.00	1.00	5.682*	1.904	.012	1.00	10.36
	3.00	2.455	1.904	.606	−2.23	7.14
3.00	1.00	3.227	1.904	.285	−1.46	7.91
	2.00	−2.455	1.904	.606	−7.14	2.23

*The mean difference is significant at the 0.05 level.

SPSS does not give the values of the t statistics for multiple comparisons. To compute them, simply divide the difference in the means by the standard error. For comparing Basal with DRTA, we have as before

$$t_{12} = \frac{-5.681}{1.90} = -2.99$$

APPLY YOUR KNOWLEDGE

CASE 14.2 14.28 Compare Basal with Strat. Use the difference in means and the standard error reported in the output of Figure 14.16 to verify that $t_{13} = -1.69$.

CASE 14.2 14.29 Compare DRTA with Strat. Use the difference in means and the standard error reported in the output of Figure 14.16 to verify that $t_{23} = 1.29$.

When there are many groups, the many results of multiple comparisons are difficult to describe. Here is one common format.

EXAMPLE 14.24   Displaying Multiple-Comparisons Results

CASE 14.2

Here is a table of the means and standard deviations for the three treatment groups. To report the results of multiple comparisons, use letters to label the means of pairs of groups that do *not* differ at the overall 0.05 significance level.

Group	Mean	Std. dev.	n
Basal	41.05^A	2.97	22
DRTA	46.73^B	2.65	22
Strat	$44.27^{A,B}$	3.34	22

Label "A" shows that Basal and Strat do not differ. Label "B" shows that DRTA and Strat do not differ. Because Basal and DRTA do not have a common label, they do differ.

The display in Example 14.24 shows that at the overall 0.05 significance level Basal does not differ from Strat and Strat does not differ from DRTA, yet Basal does differ from DRTA. These conclusions appear to be illogical. If μ_1 is the same as μ_3, and μ_3 is the same as μ_2, doesn't it follow that μ_1 is the same as μ_2? Logically, the answer must be "Yes."

This apparent contradiction points out dramatically the nature of the conclusions of tests of significance. A careful statement would say that we found significant evidence that Basal differs from DRTA and failed to find evidence that Basal differs from Strat or that Strat differs from DRTA. *Failing to find strong enough evidence that two means differ doesn't say that they are equal.* It is very unlikely that any two methods of teaching reading comprehension would give *exactly* the same population means, but the data can fail to provide good evidence of a difference. This is particularly true in multiple-comparisons methods such as Bonferroni that use a single α for an entire set of comparisons.

APPLY YOUR KNOWLEDGE

14.30 Which means differ significantly? Here is a table of means for a one-way ANOVA with four groups:

Group	Mean	Std. dev.	n
Group 1	128.4	19.8	30
Group 2	147.8	23.5	30
Group 3	151.3	25.3	30
Group 4	131.5	21.5	30

According to the Bonferroni multiple-comparisons procedure with $\alpha = 0.05$, the means for the following pairs of groups do not differ significantly: 1 and 4, 2 and 3. Mark the means of each pair of groups that do *not* differ significantly with the same letter. Summarize the results.

14.31 The groups can overlap. Refer to the previous exercise. Here is a similar table of means:

Group	Mean	Std. dev.	n
Group 1	128.4	19.8	30
Group 2	140.4	23.5	30
Group 3	151.3	25.3	30
Group 4	131.5	21.5	30

According to the Bonferroni multiple-comparisons procedure with $\alpha = 0.05$, the means for the following pairs of groups do not differ significantly: 1 and 2, 1 and 4, 2 and 3, 2 and 4. Mark the means of each pair of groups that do *not* differ significantly with the same letter. Summarize the results.

Simultaneous confidence intervals

simultaneous confidence intervals

One way to deal with these difficulties of interpretation is to give confidence intervals for the differences. The intervals remind us that the differences are not known exactly. We want to give **simultaneous confidence intervals,** that is, intervals for all the differences among the population means with confidence (say) 95% that *all the intervals at once* cover the true population differences. Again, there are many competing procedures—in this case, many methods of obtaining simultaneous intervals.

Simultaneous Confidence Intervals for Differences between Means

Simultaneous confidence intervals for all differences $\mu_i - \mu_j$ between population means have the form

$$(\overline{x}_i - \overline{x}_j) \pm t^{**} s_p \sqrt{\frac{1}{n_i} + \frac{1}{n_j}}$$

The critical values t^{**} are the same as those used for the multiple comparisons procedure chosen.

The confidence intervals generated by a particular choice of t^{**} are closely related to the multiple comparisons results for that same method. If one of the confidence intervals includes the value 0, then that pair of means will not be declared significantly different, and vice versa.

EXAMPLE 14.25   Software Output for Confidence Intervals

CASE 14.2

For simultaneous 95% Bonferroni confidence intervals, SPSS gives the output in Figure 14.16 for the data in Case 14.2. We are 95% confident that *all three* intervals simultaneously contain the true values of the population mean differences. After rounding the output, the confidence interval for the difference between the mean of the Basal group and the mean of the DRTA group is $(-10.36, -1.00)$. This interval does not include zero, so we conclude that the DRTA method results in higher mean comprehension scores than the Basal method. This is the same conclusion that we obtained from the significance test, but the confidence interval provides us with additional information about the size of the difference.

APPLY YOUR KNOWLEDGE

CASE 14.2 **14.32 Confidence interval for Basal versus Strat.** Refer to the output in Figure 14.16 (page 769). Give the Bonferroni 95% confidence interval for the difference between the mean comprehension score for the Basal method and the mean comprehension score for the Strat method. Be sure to round the numbers from the output in an appropriate way. Does the interval include 0?

CASE 14.2 **14.33 Confidence interval for DRTA versus Strat.** Refer to the previous exercise. Give the interval for comparing DRTA with Strat. Does the interval include 0?

SECTION 14.2 Summary

- The ANOVA F test does not say which of the group means differ. It is therefore usual to add comparisons among the means to basic ANOVA.

- Specific questions formulated before examination of the data can be expressed as **contrasts.** Tests and confidence intervals for contrasts provide answers to these questions.

- If no specific questions are formulated before examination of the data and the null hypothesis of equality of population means is rejected, **multiple-comparisons methods** are used to assess the statistical significance of the differences between pairs of means. These methods are less powerful than contrasts, so use contrasts whenever a study is designed to answer specific questions.

14.3 The Power of the ANOVA Test*

The power of a test is the probability of rejecting H_0 when H_a is in fact true. Power measures how likely a test is to detect a specific alternative. When planning a study in which ANOVA will be used for the analysis, it is important to perform power calculations to check that the sample sizes are adequate to detect differences among means that are judged to be important. Power calculations also help evaluate and interpret the results of studies in which H_0 was not rejected. We sometimes find that the power of the test was so low against reasonable alternatives that there was little chance of obtaining a significant F.

In Chapter 7 (page 444) we found the power for the two-sample t test. One-way ANOVA is a generalization of the two-sample t test, so it is not surprising that the procedure for calculating power is quite similar. Here are the steps that are needed:

1. Specify

   (a) an alternative (H_a) that you consider important; that is, values for the true population means $\mu_1, \mu_2, \ldots, \mu_I$;
   (b) sample sizes $n_1, n_2, \ldots, n_I$; in a preliminary study, these are usually all set equal to a common value n;
   (c) a level of significance α, usually equal to 0.05; and
   (d) a guess at the standard deviation σ.

2. Find the degrees of freedom $DFG = I - 1$ and $DFE = N - I$ and the critical value that will lead to rejection of H_0. This value, which we denote by F^*, is the upper α critical value for the $F(DFG, DFE)$ distribution.

noncentrality parameter   3. Calculate the **noncentrality parameter**[6]

$$\lambda = \frac{\sum n_i (\mu_i - \overline{\mu})^2}{\sigma^2}$$

where $\overline{\mu}$ is a weighted average of the group means,

$$\overline{\mu} = \sum w_i \mu_i$$

---

*This section is optional.

and the weights are proportional to the sample sizes,

$$w_i = \frac{n_i}{\sum n_i} = \frac{n_i}{N}$$

noncentral F distribution

4. Find the power, which is the probability of rejecting H_0 when the alternative hypothesis is true, that is, the probability that the observed F is greater than F^*. Under H_a, the F statistic has a distribution known as the **noncentral F distribution.** This requires special software. SAS, for example, has a function for the noncentral F distribution. Using this function, the power is

$$\text{Power} = 1 - \text{PROBF}(F^*, \text{DFG}, \text{DFE}, \lambda)$$

The noncentrality parameter λ measures how far apart the means μ_i are. If the n_i are all equal to a common value n, $\overline{\mu}$ is the ordinary average of the μ_i and

$$\lambda = \frac{n \sum (\mu_i - \overline{\mu})^2}{\sigma^2}$$

If the means are all equal (the ANOVA H_0), then $\lambda = 0$. Large λ points to an alternative far from H_0, and we expect the ANOVA F test to have high power. Software makes calculation of the power quite easy, but tables and charts are also available.

EXAMPLE 14.26   The Effect of Fewer Subjects

CASE 14.2

The reading comprehension study described in Case 14.2 (page 760) had 22 subjects in each group. Suppose that a similar study has only 10 subjects per group. How likely is this study to detect differences in the mean responses that are similar in size to those observed in the actual study?

Based on the results of the actual study, we will calculate the power for the alternative $\mu_1 = 41$, $\mu_2 = 47$, $\mu_3 = 44$, with $\sigma = 7$. The n_i are equal, so $\overline{\mu}$ is simply the average of the μ_i:

$$\overline{\mu} = \frac{41 + 47 + 44}{3} = 44$$

The noncentrality parameter is therefore

$$\begin{aligned}
\lambda &= \frac{n \sum (\mu_i - \overline{\mu})^2}{\sigma^2} \\
&= \frac{(10)[(41 - 44)^2 + (47 - 44)^2 + (44 - 44)^2]}{49} \\
&= \frac{(10)(18)}{49} = 3.67
\end{aligned}$$

Because there are three groups with 10 observations per group, DFG $= 2$ and DFE $= 27$. The critical value for $\alpha = 0.05$ is $F^* = 3.35$. The power is therefore

$$1 - \text{PROBF}(3.35, 2, 27, 3.67) = 0.3486$$

The chance that we reject the ANOVA H_0 at the 5% significance level is only about 35%.

If the assumed values of the μ_i in this example describe differences among the groups that the experimenter wants to detect, then we would want to use more than 10 subjects per group. Although H_0 is false for these μ_i, the chance of rejecting it at the 5% level is only about 35%. This chance can be increased to acceptable levels by increasing the sample sizes.

EXAMPLE 14.27   Choosing the Sample Size for a Future Study

CASE 14.2

To decide on an appropriate sample size for the experiment described in the previous example, we repeat the power calculation for different values of n, the number of subjects in each group. Here are the results:

n	DFG	DFE	F^*	λ	Power
20	2	57	3.16	7.35	0.65
30	2	87	3.10	11.02	0.84
40	2	117	3.07	14.69	0.93
50	2	147	3.06	18.37	0.97
100	2	297	3.03	36.73	≈ 1

With $n = 40$ the experimenters have a 93% chance of rejecting H_0 with $\alpha = 0.05$ and thereby demonstrating that the groups have different means. In the long run, 93 out of every 100 such experiments would reject H_0 at the $\alpha = 0.05$ level of significance. Using 50 subjects per group increases the chance of finding significance to 97%. With 100 subjects per group, the experimenters are virtually certain to reject H_0. The exact power for $n = 100$ is 0.99989. In most real-life situations the additional cost of increasing the sample size from 50 to 100 subjects per group would not be justified by the relatively small increase in the chance of obtaining statistically significant results.

APPLY YOUR KNOWLEDGE

14.34 Power calculations for planning a study. You are planning a new workplace safety study for a different region of the United States than that studied in Example 14.3. From Example 14.3, we know that the standard deviations for the three groups considered in that study were 18.27, 18.83, and 14.58. In Example 14.5, we found the pooled standard error to be 18.07. Since the power of the F test decreases as the standard deviation increases, use $\sigma = 20$ for the calculations in this exercise. This choice will lead to sample sizes that are perhaps a little larger than we need but will prevent us from choosing sample sizes that are too small to detect the effects of interest. You would like to conclude that the population means are different when $\mu_1 = 70$, $\mu_2 = 75$, and $\mu_3 = 80$.
(a) Pick several values for n (the number of workers that you will select from each group) and calculate the power of the ANOVA F test for each of your choices.
(b) Plot the power versus the sample size. Describe the general shape of the plot.
(c) What choice of n would you choose for your study? Give reasons for your answer.

14.35 Power against a different alternative. Refer to the previous exercise. Repeat all parts for the alternative $\mu_1 = 65$, $\mu_2 = 70$, and $\mu_3 = 80$.

SECTION 14.3 Summary

- The **power** of the F test depends upon the sample sizes, the variation among population means, and the within-group standard deviations. Some software allows easy calculation of power.

STATISTICS IN SUMMARY

Advanced statistical inference often concerns relationships among several parameters. This chapter introduces the ANOVA F test for one such relationship: equality of the means of any number of populations. The alternative to this hypothesis is "many-sided," because it allows any relationship other than "all equal." The ANOVA F test is an overall test that tells us whether the data give good reason to reject the hypothesis that all the population means are equal. Contrasts are used to answer specific questions about group means that can be formulated before the data are collected. When this is not possible, we use a multiple-comparisons procedure to determine which groups differ significantly whenever the overall F test rejects the null hypothesis that all groups have the same mean. You should always accompany the ANOVA by data analysis to see what kind of inequality is present. Plotting the data in all groups side by side is particularly helpful. Here is a review of the most important skills you should have acquired from your study of this chapter.

A. Recognition

1. Recognize when testing the equality of several means is helpful in understanding data.

2. Recognize that the statistical significance of differences among sample means depends on the sizes of the samples and on how much variation there is within the samples.

3. Recognize when you can safely use ANOVA to compare means. Check the data production, the presence of outliers, and the sample standard deviations for the groups you want to compare.

4. Recognize when to use contrasts and multiple comparisons.

5. Recognize when it is appropriate to compute power for a one-way ANOVA.

B. Interpreting ANOVA

1. Explain what null hypothesis F tests in a specific setting.

2. Locate the F statistic and its P-value on the output of a computer ANOVA program.

3. Find the degrees of freedom for the F statistic from the number and sizes of the samples. Use Table E of F critical values to approximate the P-value when software does not give it.

4. If the F test is significant, use graphs and descriptive statistics to see what differences among the means are most important.

5. Formulate contrasts and analyze them using software.

6. Use multiple comparisons to determine pairs of group means that differ significantly.

CHAPTER 14 Review Exercises

For Exercises 14.1 and 14.2, see pages 746–747; for 14.3 and 14.4, see page 748; for 14.5 and 14.6, see page 750; for 14.7, see page 752; for 14.8, see page 753; for 14.9 to 14.13, see page 755; for 14.14 to 14.17, see page 757; for 14.18 and 14.19, see pages 759–760; for 14.20 to 14.23, see page 766; for 14.24 and 14.25, see page 767; for 14.26 and 14.27, see page 768; for 14.28 and 14.29, see page 769; for 14.30 and 14.31, see page 770; for 14.32 and 14.33, see page 771; and for 14.34 and 14.35, see page 774.

14.36 What's wrong? For each of the following, explain what is wrong and why.
(a) The pooled estimate s_p is a parameter of the ANOVA model.
(b) The mean squares in an ANOVA table will add, that is, $MST = MSG + MSE$.
(c) For an ANOVA F test with $P = 0.31$, we conclude that the means are the same.

14.37 Use the F statistic. A study compared 5 groups with 7 observations per group. An F statistic of 2.95 was reported.
(a) Give the degrees of freedom for this statistic and the entries from Table E that correspond to this distribution.
(b) Sketch a picture of this F distribution with the information from the table included.
(c) Based on the table information, how would you report the P-value?
(d) Can you conclude that all pairs of means are different? Explain your answer.

14.38 How large does the F statistic need to be? For each of the following situations, state how large the F statistic needs to be for rejection of the null hypothesis at the 0.05 level.
(a) Compare 5 groups with 3 observations per group.
(b) Compare 5 groups with 6 observations per group.
(c) Compare 5 groups with 10 observations per group.
(d) Summarize what you have learned about F distributions from this exercise.

14.39 Use the F statistic. For each of the following situations, find the F statistic and the degrees of freedom. Then draw a sketch of the distribution under the null hypothesis and shade in the portion corresponding to the P-value. State how you would report the P-value.
(a) Compare 5 groups with 9 observations per group, MSE = 50, and MSG = 77.
(b) Compare 3 groups with 8 observations per group, SSG = 40, and SSE = 140.

14.40 Visualizing the ANOVA model. For each of the following situations, draw a picture of the ANOVA model similar to Figure 14.5 (page 748). Use numerical values for the μ_i. To sketch the Normal curves, you may want to review the 68–95–99.7 rule on page 46.
(a) $\mu_1 = 14$, $\mu_2 = 17$, $\mu_3 = 23$, and $\sigma = 5$.
(b) $\mu_1 = 14$, $\mu_2 = 17$, $\mu_3 = 23$, $\mu_4 = 20$, and $\sigma = 3$.
(c) $\mu_1 = 14$, $\mu_2 = 17$, $\mu_3 = 23$, and $\sigma = 2$.

14.41 The ANOVA framework. For each of the following situations, identify the response variable and the populations to be compared, and give I, the n_i, and N.
(a) Last semester, an alcohol awareness program was conducted for three groups of students at an eastern university. Follow-up questionnaires were sent to the participants two months after each presentation. There were 220 responses from students in an elementary statistics course, 145 from a health and safety course, and 76 from a cooperative housing unit. One of the questions was "Did you discuss the presentation with any of your friends?" The answers were rated on a five-point scale with 1 corresponding to "not at all" and 5 corresponding to "a great deal."

(b) A researcher is interested in students' opinions regarding an additional annual fee to support non-income-producing varsity sports. Students were asked to rate their acceptance of this fee on a five-point scale. She received 94 responses, of which 31 were from students who attend varsity football or basketball games only, 18 were from students who also attend other varsity competitions, and 45 were from students who did not attend any varsity games.
(c) A university sandwich shop wants to compare the effects of providing free food with a sandwich order on sales. The experiment will be conducted from 11:00 A.M. to 2:00 P.M. for the next 20 weekdays. On each day, customers will be offered one of the following: a free drink, free chips, a free cookie, or nothing. Each option will be offered 5 times.

14.42 Describing the ANOVA model. For each of the following situations, identify the response variable and the populations to be compared, and give I, the n_i, and N.
(a) A developer of a virtual-reality (VR) teaching tool for the deaf wants to compare the effectiveness of different navigation methods. A total of 40 children were available for the experiment, of which equal numbers were randomly assigned to use a joystick, wand, dance mat, or gesture-based pinch gloves. The time (in seconds) to complete a designed VR path is recorded for each child.
(b) A waiter designed a study to see the effects of his behaviors on the tip amounts that he received. For some customers, he would tell a joke; for others, he would describe two of the food items as being particularly good that night; and for others he would behave normally. Using a table of random numbers, he assigned equal numbers of his next 30 customers to his different behaviors.
(c) A supermarket wants to compare the effects of providing free samples of cheddar cheese on sales. An experiment will be conducted from 5:00 P.M. to 6:00 P.M. for the next 20 weekdays. On each day, customers will be offered one of the following: a small cube of cheese pierced by a toothpick, a small slice of cheese on a cracker, a cracker with no cheese, or nothing.

14.43 Provide some details. Refer to Exercise 14.41. For each situation, give the following:
(a) Degrees of freedom for the model, for error, and for the total.
(b) Null and alternative hypotheses.
(c) Numerator and denominator degrees of freedom for the F statistic.

14.44 Provide some details. Refer to Exercise 14.42. For each situation, give the following:
(a) Degrees of freedom for the model, for error, and for the total.
(b) Null and alternative hypotheses.
(c) Numerator and denominator degrees of freedom for the F statistic.

14.45 How much can you generalize? Refer to Exercise 14.41. For each situation, discuss the method of obtaining the data and how this would affect the extent to which the results can be generalized.

14.46 How much can you generalize? Refer to Exercise 14.42. For each situation, discuss the method of obtaining the data and how this would affect the extent to which the results can be generalized.

14.47 Pooling variances. An experiment was run to compare four groups. The sample sizes were 40, 32, 360, and 35, and the corresponding estimated standard deviations were 25, 21, 12, and 23.
(a) Is it reasonable to use the assumption of equal standard deviations when we analyze these data? Give a reason for your answer.
(b) Give the values of the variances for the four groups.
(c) Find the pooled variance.
(d) What is the value of the pooled standard deviation?
(e) Explain why your answer in part (c) is much closer to the standard deviation for the third group than to any of the other standard deviations.

14.48 Developing marketing strategies for travel to Hawaii. In 1997 approximately one-third of all tourists to Hawaii were from Japan. Since that time the percent has steadily decreased and is now around 20%.[7] To better understand the reasons for travel to Hawaii, a group of researchers surveyed 315 Japanese tourists who plan to visit Hawaii. The tourists were divide into groups based on their purpose for travel. They were (1) honeymoon, (2) fraternal association, (3) sports, (4) leisure, and (5) business. Their responses to various survey questions were compared across these groups. The responses were on a seven-point scale ranging from 1 (strongly disagree) to 7 (strongly agree). The following table summarizes the mean responses and F test statistics for several questions.

Question	Group 1 $n = 34$	Group 2 $n = 56$	Group 3 $n = 105$	Group 4 $n = 26$	Group 5 $n = 94$	F
I'd like to experience native Hawaiian culture	3.97	4.26	4.25	5.33	4.23	2.46
I'd prefer a group tour to an individual one	3.18	3.38	2.39	2.58	2.98	3.97
I'd like to experience ocean sports	4.71	4.59	4.58	5.33	4.02	2.46
I respect Hawaiian residents' customs	4.88	5.39	5.14	5.83	5.46	1.62

(a) What are the numerator and denominator degrees of freedom for these F tests?
(b) The response variable is not Normally distributed. Explain why this should not cause difficulties in using ANOVA.
(c) Using a significance level of $\alpha = 0.05$ for each question, assess whether there are differences in the group means.
(d) For those questions with a significant F statistic, plot the means and describe their pattern.

14.49 Multiple comparisons. Refer to the previous exercise.
(a) Explain why it is inappropriate to perform a multiple-comparisons analysis for the last question.

(b) For the other questions, use the Bonferroni or another multiple-comparisons procedure to determine which group means differ significantly. The table below gives the MSE for each question.

Question	MSE
I'd like to experience native Hawaiian culture	3.261
I'd prefer a group tour to an individual one	2.841
I'd like to experience ocean sports	4.285
I respect Hawaiian residents' customs	2.905

Summarize your results in a short report.

14.50 Restaurant ambience and consumer behavior. There have been numerous studies investigating the effects of restaurant ambience on consumer behavior. One study investigated the effects of musical genre on consumer spending.[8] At a single high-end restaurant in England over a three-week period, there were a total of 141 participants; 49 of them were subjected to background pop music (for example, Britney Spears, Culture Club, and Ricky Martin) while dining, 44 to background classical music (for example, Vivaldi, Handel, and Strauss), and 48 to no background music. For each participant, the total food bill (in British pounds), adjusted for time spent dining, was recorded. The following table summarizes the means and standard deviations.

Background music	Mean bill	n	s
Pop	21.912	49	2.627
Classical	24.130	44	2.243
None	21.697	48	3.332
Total	22.531	141	2.969

(a) Plot the means versus the type of background music. Does there appear to be a difference in spending?
(b) Is it reasonable to assume that the variances are equal? Explain.
(c) The F statistic is 10.62. Give the degrees of freedom and either an approximate (from a table) or an exact (from software) P-value. What do you conclude?
(d) Refer back to part (a). Without doing any formal analysis, describe the pattern in the means that is likely responsible for your conclusion in part (c).
(e) To what extent do you think the results of this study can be generalized to other settings? Give reasons for your answer.

14.51 Shopping and bargaining in Mexico. Price haggling and other bargaining behaviors among consumers have been observed for a long time. However, research addressing these behaviors, especially in a real-life setting, remains relatively sparse. A group of researchers recently performed a small study to determine whether gender or nationality of the bargainer has an effect in the final price obtained.[9] The study took place in Mexico because of the prevalence of price haggling in informal markets.

Salespersons working at various informal shops were approached by one of three bargainers looking for a specific product. After an initial price was stated by the vendor, bargaining took place. The response was the difference between the initial and the final price of the product. The bargainers were a Spanish-speaking Hispanic male, a Spanish-speaking Hispanic female, and an Anglo non-Spanish-speaking male. The following table summarizes the results.

Bargainer	n	Average reduction
Hispanic male	40	1.055
Hispanic female	40	2.310
Anglo male	40	1.050

(a) To compare the mean reductions in price, what are the degrees of freedom for the ANOVA F statistic?
(b) The reported test statistic is $F = 8.708$. Give an approximate (from a table) or exact (from software) P-value. What do you conclude?
(c) To what extent do you think the results of this study can be generalized? Give reasons for your answer.

14.52 Animals on product labels? Recall Example 7.11 (page 421). This experiment actually involved comparing product preference for a group of consumers that were "primed" and two groups of consumers who served as controls. A bottle of MagicCoat pet shampoo was the product, and participants indicated their attitude toward this product on a seven-point scale (from 1 = dislike very much to 7 = like very much). The bottle of shampoo either had a picture of a collie on the label or just the wording. Also, prior to giving this score, participants were asked to do a word find where four of the words were common across groups (pet, grooming, bottle, label) and four were either related to the image (dog, collie, puppy, woof) or image conflicting (cat, feline, kitten, meow). Here is a summary of the groups:

BRANDPREFERENCE1

Group	Label with dog	Dog "primed"	n
1	Yes	Yes	22
2	Yes	No	20
3	No	Yes	10

(a) Use graphical and numerical methods to describe the data.
(b) Examine the assumptions necessary for ANOVA. Summarize your findings.
(c) Run the ANOVA and report the results.
(d) Use a multiple-comparisons method to compare the three groups. State your conclusions.

14.53 College dining facilities. University and college food service operations have been trying to keep up with the growing expectations of consumers in regard to the overall campus dining experience. Since customer satisfaction has been shown to

be associated with repeat patronage and new customers through word-of-mouth, a public university in the Midwest took a sample of patrons from their eating establishments and asked them about their overall dining satisfaction.[10] The table below summarizes the results for three groups of patrons:

Category	$\bar{x}$	n	s
Student with meal plan	$3.44	489	0.804
Faculty with meal plan	$4.04	69	0.824
Student without meal plan	$3.47	212	0.657

(a) Is it reasonable to use a pooled standard deviation for these data? Why or why not? If yes, compute it.
(b) The ANOVA F statistic was reported as 17.66. Give the degrees of freedom and either an approximate (from a table) or an exact (from software) P-value. Sketch a picture of the F distribution that illustrates the P-value. What do you conclude?
(c) Prior to performing this survey, food service operations thought that satisfaction among faculty would be higher than satisfaction among students. Use the results in the table to test this contrast. Make sure to specify the null and alternative hypotheses, test statistic, and P-value.

14.54 Internet banking. A recent study in Finland looked at consumer perceptions of Internet banking (IB).[11] Data were collected via personal, structured interviews as part of a nationwide consumer study. The sample included 300 active users of IB, aged between 15 and 74. Based on the survey, users were broken down into three groups based on their familiarity with the Internet. For this exercise, we will consider the consumer's perception of status or image in the eyes of other consumers. Standardized scores were used for analysis.

Familiarity	Mean	n
Low	0.21	77
Medium	−0.14	133
High	0.03	90

(a) To compare the mean scores across familiarity levels, what are the degrees of freedom for the ANOVA F statistic?
(b) The MSG = 3.12. If $s_p = 1.05$, what is the F statistic?
(c) Give an approximate (from a table) or exact (from software) P-value. What do you conclude?

14.55 The multiple-play strategy. Multiple play is a bundling strategy in which multiple services are provided over a single network. A common triple-play service these days is Internet, television, and telephone. The market for this service has become a key battleground among telecommunication, cable, and broadband service providers. A recent study compared the pricing (in dollars) among triple-play providers using DSL, cable, or fiber platforms.[12] The following table summarizes the results from 47 providers.

Group	n	$\overline{x}$	s
DSL	19	104.49	26.09
Cable	20	119.98	40.39
Fiber	8	83.87	31.78

(a) Plot the means versus the platform type. Does there appear to be a difference in pricing?

(b) Is it reasonable to assume that the variances are equal? Explain.

(c) The F statistic is 3.39. Give the degrees of freedom and either an approximate (from a table) or an exact (from software) P-value. What do you conclude?

14.56 A contrast. Refer to the previous exercise. Use a contrast to compare the fiber platform with the average of the other two. The hypothesis prior to collecting the data is that the fiber platform price would be smaller. Summarize your conclusion.

14.57 The importance of recreational sports to college satisfaction. The National Intramural-Recreational Sports Association (NIRSA) performed a survey to look at the value of recreational sports on college campuses.[13] One of the questions asked each student to rate the importance of recreational sports to college satisfaction and success. Responses were on a 10-point scale with 1 indicating total lack of importance and 10 indicating very high importance. The following table summarizes the results.

Class	n	Mean score
Freshman	724	7.6
Sophomore	536	7.6
Junior	593	7.5
Senior	437	7.3

(a) To compare the mean scores across classes, what are the degrees of freedom for the ANOVA F statistic?

(b) The MSG = 11.806. If $s_p = 2.16$, what is the F statistic?

(c) Give an approximate (from a table) or exact (from software) P-value. What do you conclude?

14.58 Does sleep deprivation affect your work? Sleep deprivation experienced by physicians during residency training and the possible negative consequences are of concern to many in the health care community. One study of 33 resident anesthesiologists compared their changes from baseline in reaction times on four tasks.[14] Under baseline conditions, the physicians reported getting an average of 7.04 hours of sleep. While on duty, however, the average was 1.66 hours. For each of the tasks the researchers reported a statistically significant increase in the reaction time when the residents were working in a state of sleep deprivation.

(a) If each task is analyzed separately as the researchers did in their report, what is the appropriate statistical method to use? Explain your answer.

(b) Is it appropriate to use a one-way ANOVA with $I = 4$ to analyze these data? Explain why or why not.

14.59 ADA and exams. The Americans with Disabilities Act (ADA) requires that students with learning disabilities (LD) and/or attention deficit disorder (ADD) be given certain accommodations when taking examinations. One study designed to assess the effects of these accommodations examined the relationship between end-of-term grades and the number of accommodations given.[15] The researchers reported the mean grades with sample sizes and standard deviations versus the number of accommodations in a table similar to this:

Accommodations	Mean grade	n	s
0	2.7894	160	0.85035
1	2.8605	38	0.83068
2	2.5757	37	0.82745
3	2.6286	7	1.03072
4	2.4667	3	1.66233
Total	2.7596	245	0.85701

(a) Plot the means versus the number of accommodations. Is there a pattern evident?

(b) A large number of digits are reported for the means and the standard deviations. Do you think that all of these are necessary? Give reasons for your answer and describe how you would report these results.

(c) Should we pool to obtain an estimate of an assumed standard deviation for these data? Explain your answer and give the pooled estimate if your answer is "Yes."

(d) The small numbers of observations with 3 or 4 accommodations lead to estimates that are highly variable in these groups compared with the other groups. Inclusion of groups with relatively few observations in an ANOVA can also lead to low power. We could eliminate these two levels from the analysis or we could combine them with the 37 observations in the group above to form a new group with 2 or more accommodations. Which of these options would you prefer? Give reasons for your answer.

(e) The 245 grades reported in the table were from a sample of 61 students who completed three, four, or five courses during a spring term at one college and who were qualified to receive accommodations. Students in the sample were self-identified, in the sense that they had to request qualification. Even when qualified, some students choose not to request accommodations for some or all of their courses. Based on these facts, would you advise that ANOVA methods be used for these data? Explain your answer. (The authors did not present the results of an ANOVA in their publication.)

(f) To what extent do you think the results of this study can be generalized to other settings? Give reasons for your answer.

(g) Most reasonable approaches to the analysis of these data would conclude that the data fail to provide evidence that the number of accommodations is related to the mean grades. Does this imply that the accommodations are not needed or does it suggest that they are ineffective? Discuss your answer.

14.60 Emotions and cultures of college students. Do people from different cultures experience emotions differently? One study designed to examine this question collected data from 416 college students from five different cultures.[16] The participants were asked to record, on a 1 (never) to 7 (always) scale, how much of the time they typically felt eight specific emotions. These were averaged to produce the global emotion score for each participant. Here is a summary of this measure:

Culture	n	Mean (s)
European American	46	4.39 (1.03)
Asian American	33	4.35 (1.18)
Japanese	91	4.72 (1.13)
Indian	160	4.34 (1.26)
Hispanic American	80	5.04 (1.16)

Note that the convention of giving the standard deviations in parentheses after the means saves a great deal of space in a table such as this.

(a) From the information given, do you think that we need to be concerned that a possible lack of Normality in the data will invalidate the conclusions that we might draw using ANOVA to analyze the data? Give reasons for your answer.

(b) Is it reasonable to use a pooled standard deviation for these data? Why or why not?

(c) The ANOVA F statistic was reported as 5.69. Give the degrees of freedom and either an approximate (from a table) or an exact (from software) P-value. Sketch a picture of the F distribution that illustrates the P-value. What do you conclude?

(d) Without doing any additional formal analysis, describe the pattern in the means that appears to be responsible for your conclusion in part (c). Are there pairs of means that are quite similar?

14.61 A different measure of emotions. Refer to the previous exercise. The experimenters also measured emotions in some different ways. For a period of a week, each participant carried a device that sounded an alarm at random times during a three-hour interval 5 times a day. When the alarm sounded, participants recorded several mood ratings indicating their emotions for the time immediately preceding the alarm. These responses were combined to form two variables: frequency, the number of emotions recorded, expressed as a percent; and intensity, an average of the intensity scores measured on a scale of 0 to 6. At the end of the 1-week experimental period, the subjects were asked to recall the percent of time that they experienced different emotions. This variable was called "recall." Here is a summary of the results:

Culture	n	Frequency mean (s)	Intensity mean (s)	Recall mean (s)
European American	46	82.87 (18.26)	2.79 (0.72)	49.12 (22.33)
Asian American	33	72.68 (25.15)	2.37 (0.60)	39.77 (23.24)
Japanese	91	73.36 (22.78)	2.53 (0.64)	43.98 (22.02)
Indian	160	82.71 (17.97)	2.87 (0.74)	49.86 (21.60)
Hispanic American	80	92.25 (8.85)	3.21 (0.64)	59.99 (24.64)
F statistic		11.89	13.10	7.06

(a) For each response variable state whether or not it is reasonable to use a pooled standard deviation to analyze these data. Give reasons for your answer.

(b) Give the degrees of freedom for the F statistics and find the associated P-values. Summarize what you can conclude from these ANOVA analyses.

(c) Summarize the means, paying particular attention to similarities and differences across cultures and across variables. Include the means from the previous exercise in your summary.

(d) The European American and Asian American subjects were from the University of Illinois, the Japanese subjects were from two universities in Tokyo, the Indian subjects were from eight universities in or near Calcutta, and the Hispanic American subjects were from California State University at Fresno. Participants were paid $25 or an equivalent monetary incentive for the Japanese and Indians. Ads were posted on or near the campuses to recruit volunteers for the study. Discuss how these facts influence your conclusions and the extent to which you would generalize the results.

(e) The percents of female students in the samples were as follows: European American, 83%; Asian American, 67%; Japanese, 63%; Indian, 64%; and Hispanic American, 79%. Use a chi-square test to compare these proportions (see Section 9.2, page 505) and discuss how this information influences your interpretation of the results that you have found in this exercise.

14.62 A natural product can help bones. Kudzu is a plant that was imported to the United States from Japan and now covers over seven million acres in the South. The plant contains chemicals called isoflavones that have been shown to have beneficial effects on bones. One study used three groups of rats to compare a control group with rats that were fed either a low dose or a high dose of isoflavones from kudzu.[17] One of the outcomes examined was the bone mineral density in the femur (in grams per square centimeter). Here are the data: 🌱 **KUDZU**

Treatment	Bone mineral density (g/cm^2)
Control	0.228 0.207 0.234 0.220 0.217 0.228 0.209 0.221 0.204 0.220 0.203 0.219 0.218 0.245 0.210
Low dose	0.211 0.220 0.211 0.233 0.219 0.233 0.226 0.228 0.216 0.225 0.200 0.208 0.198 0.208 0.203
High dose	0.250 0.237 0.217 0.206 0.247 0.228 0.245 0.232 0.267 0.261 0.221 0.219 0.232 0.209 0.255

(a) Use graphical and numerical methods to describe the data.

(b) Examine the assumptions necessary for ANOVA. Summarize your findings.

(c) Run the ANOVA and report the results.

(d) Use a multiple-comparisons method to compare the three groups.

(e) Write a short report explaining the effect of kudzu isoflavones on the femur of the rat.

14.63 Promotions and the expected price of a product. If a supermarket product is frequently offered at a reduced price, do customers expect the price of the product to be lower in the future? This question was examined by researchers in a study conducted on students enrolled in an introductory management course at a large midwestern university. For 10 weeks, 160 subjects read weekly ads for the same product. Students were randomly assigned to read 1, 3, 5, or 7 ads featuring price promotions during the 10-week period. They were then asked to estimate what the product's price would be the following week.[18] Table 14.2 gives the data. 🖱 PRICEPROMO

(a) Make a Normal quantile plot for the data in each of the four treatment groups. Summarize the information in the plots and draw a conclusion regarding the Normality of these data.

(b) Summarize the data with a table containing the sample size, mean, and standard deviation for each group.

(c) Is the assumption of equal standard deviations reasonable here? Explain why or why not.

(d) Carry out a one-way ANOVA. Give the hypotheses, the test statistic with its degrees of freedom, and the P-value. Summarize your conclusion.

14.64 Compare the means. Refer to the previous exercise. Use the Bonferroni or another multiple-comparisons procedure to compare the group means. Summarize the results and support your conclusions with a graph of the means.

14.65 A contrast. Refer to Example 14.3 (page 745). Experts on safety in workplaces suggest that supervisors face a very different safety environment than the other types of workers. It is therefore natural to test the contrast that compares the mean of the supervisors with the average of the two other means. Construct this contrast, perform the significance test, and summarize the results.

14.66 Exercise and healthy bones. Many studies have suggested that there is a link between exercise and healthy bones. Exercise stresses the bones and this causes them to get stronger. One study examined the effect of jumping on the bone density of growing rats.[19] There were three treatments: a control with no jumping, a low-jump condition (the jump height was 30 centimeters), and a high-jump condition (60 centimeters). After eight weeks of 10 jumps per day, five days per week, the bone density of the rats (expressed in milligrams per cubic centimeter) was measured. Here are the data: 🖱 JUMPING

Group	Bone density (mg/cm^3)									
Control	611	621	614	593	593	653	600	554	603	569
Low jump	635	605	638	594	599	632	631	588	607	596
High jump	650	622	626	626	631	622	643	674	643	650

(a) Make a table giving the sample size, mean, and standard deviation for each group of rats. Is it reasonable to pool the variances?

(b) Carry out an analysis of variance. Report the F statistic with its degrees of freedom and P-value. What do you conclude?

14.67 Residuals and multiple comparisons. Refer to the previous exercise.

(a) Examine the residuals. Is the Normality assumption reasonable for these data?

(b) Use the Bonferroni or another multiple-comparisons procedure to determine which pairs of means differ significantly. Summarize your results in a short report. Be sure to include a graph.

14.68 A new material to repair wounds. One way to repair serious wounds is to insert some material as a scaffold for the

---

TABLE 14.2 Price promotion data

Number of promotions	Expected price (dollars)									
1	3.78	3.82	4.18	4.46	4.31	4.56	4.36	4.54	3.89	4.13
	3.97	4.38	3.98	3.91	4.34	4.24	4.22	4.32	3.96	4.73
	3.62	4.27	4.79	4.58	4.46	4.18	4.40	4.36	4.37	4.23
	4.06	3.86	4.26	4.33	4.10	3.94	3.97	4.60	4.50	4.00
3	4.12	3.91	3.96	4.22	3.88	4.14	4.17	4.07	4.16	4.12
	3.84	4.01	4.42	4.01	3.84	3.95	4.26	3.95	4.30	4.33
	4.17	3.97	4.32	3.87	3.91	4.21	3.86	4.14	3.93	4.08
	4.07	4.08	3.95	3.92	4.36	4.05	3.96	4.29	3.60	4.11
5	3.32	3.86	4.15	3.65	3.71	3.78	3.93	3.73	3.71	4.10
	3.69	3.83	3.58	4.08	3.99	3.72	4.41	4.12	3.73	3.56
	3.25	3.76	3.56	3.48	3.47	3.58	3.76	3.57	3.87	3.92
	3.39	3.54	3.86	3.77	4.37	3.77	3.81	3.71	3.58	3.69
7	3.45	3.64	3.37	3.27	3.58	4.01	3.67	3.74	3.50	3.60
	3.97	3.57	3.50	3.81	3.55	3.08	3.78	3.86	3.29	3.77
	3.25	3.07	3.21	3.55	3.23	2.97	3.86	3.14	3.43	3.84
	3.65	3.45	3.73	3.12	3.82	3.70	3.46	3.73	3.79	3.94

body's repair cells to use as a template for new tissue. Scaffolds made from extracellular material (ECM) are particularly promising for this purpose. Because ECMs are biological material, they serve as an effective scaffold and are then absorbed. Unlike biological material that includes cells, ECMs do not trigger tissue rejection reactions in the body. One study compared six types of scaffold material.[20] Three of these were ECMs and the other three were made of inert materials. The subjects were three mice per scaffold type. The response measure was the percent of glucose phosphated isomerase (Gpi) cells in the region of the wound. A large value is good, indicating the presence of many bone marrow cells sent by the body to repair the tissue. Here are the data:

ECM

Material	Gpi (%)		
ECM1	55	70	70
ECM2	60	65	65
ECM3	75	70	75
MAT1	20	25	25
MAT2	5	10	5
MAT3	10	15	10

(a) Make a table giving the sample size, mean, and standard deviation for each of the six types of material. Is it reasonable to pool the variances? Note that the sample sizes are small and the data are rounded.

(b) Run the analysis of variance. Report the F statistic with its degrees of freedom and P-value. What do you conclude?

14.69 Residuals and multiple comparisons. Refer to the previous exercise.

(a) Examine the residuals. Is the Normality assumption reasonable for these data?

(b) Use the Bonferroni or another multiple-comparisons procedure to determine which pairs of means differ significantly. Summarize your results in a short report. Be sure to include a graph.

14.70 A contrast. Refer to the previous two exercises. Use a contrast to compare the three ECM materials with the three other materials. Summarize your conclusions. How do these results compare with those that you obtained from the multiple-comparisons procedure in the previous exercise?

14.71 Exercise and fitness. A study of the effects of exercise on physiological and psychological variables compared four groups of male subjects. The treatment group (T) consisted of 10 participants in an exercise program. A control group (C) of 5 subjects volunteered for the program but were unable to attend for various reasons. Subjects in the other two groups were selected to be similar to those in the first two groups in age and other characteristics. These were 11 joggers (J) and 10 sedentary people (S) who did not regularly exercise.[21] One of the variables measured at the end of the program was a physical fitness score. Here is part of the ANOVA table for these data:

Source	Degrees of freedom	Sum of squares	Mean square	F
Groups	3	104,855.87		
Error	32	70,500.59		
Total				

(a) Fill in the missing entries in the ANOVA table.

(b) State H_0 and H_a for this experiment.

(c) What is the distribution of the F statistic under the assumption that H_0 is true? Using Table E, give an approximate P-value for the ANOVA test. Write a brief conclusion.

(d) What is s_p^2, the estimate of the within-group variance? What is the pooled standard error s_p?

14.72 Exercise and depression. Another variable measured in the experiment described in the previous exercise was a depression score. Higher values of this score indicate more depression. Here is part of the ANOVA table for these data:

Source	Degrees of freedom	Sum of squares	Mean square	F
Groups	3		158.96	
Error	32		62.81	
Total				

(a) Fill in the missing entries in the ANOVA table.

(b) State H_0 and H_a for this experiment.

(c) What is the distribution of the F statistic under the assumption that H_0 is true? Using Table E, give an approximate P-value for the ANOVA test. What do you conclude?

(d) What is s_p^2, the estimate of the within-group variance? What is s_p?

14.73 Weight gain during pregnancy. The weight gain of women during pregnancy has an important effect on the birth weight of their children. If the weight gain is not adequate, the infant is more likely to be small and will tend to be less healthy. In a study conducted in three countries, weight gains (in kilograms) of women during the third trimester of pregnancy were measured.[22] The results are summarized in the following table:

Country	n	$\bar{x}$	s
Egypt	46	3.7	2.5
Kenya	111	3.1	1.8
Mexico	52	2.9	1.8

(a) Find the pooled estimate of the within-country variance s_p^2. What entry in an ANOVA table gives this quantity?

(b) The sum of squares for countries (groups) is 17.22. Use this information and that given above to construct an ANOVA table.

(c) State H_0 and H_a for this study.

(d) What is the distribution of the F statistic under the assumption that H_0 is true? Use Table E to find an approximate P-value for the significance test. Report your conclusion.

14.74 Food intake. In another part of the study described in the previous exercise, measurements of food intake in kilocalories were taken on many individuals several times during the period of a year. From these data, average daily food intake values were computed for each individual. The results for toddlers aged 18 to 30 months are summarized in the following table:

Country	n	$\bar{x}$	s
Egypt	88	1217	327
Kenya	91	844	184
Mexico	54	1119	285

(a) Find the pooled estimate of the within-country variance s_p^2. What entry in an ANOVA table gives this quantity?
(b) The sum of squares for countries (groups) is 6,572,551. Use this information and that given above to construct an ANOVA table.
(c) State H_0 and H_a for this study.
(d) What is the distribution of the F statistic under the assumption that H_0 is true? Use Table E to find an approximate P-value for the significance test. Report your conclusion.

CASE 14.1 14.75 Writing contrasts. Return to the eye study described in Case 14.1 (page 751). Let μ_1, μ_2, μ_3, and μ_4 represent the mean scores for blue, brown, gaze down, and green eyes.
(a) Because a majority of the population is Hispanic (eye color predominantly brown), we want to compare the average score of the brown eyes with the average of the other two eye colors. Write a contrast that expresses this comparison.
(b) Write a contrast to compare the average score when the model is looking at you versus the score when looking down.

14.76 Writing contrasts. You've been asked to help some administrators analyze survey data on textbook expenditures collected at a large public university. Let μ_1, μ_2, μ_3, and μ_4 represent the population mean expenditures on textbooks for the freshmen, sophomores, juniors, and seniors.
(a) Because juniors and seniors take higher-level courses, which might use more expensive textbooks, the administrators want to compare the average of the freshmen and sophomores with the average of the juniors and seniors. Write a contrast that expresses this comparison.
(b) Write a contrast for comparing the freshmen with the sophomores.
(c) Write a contrast for comparing the juniors with the seniors.

CASE 14.1 14.77 Analyzing contrasts. Return to the eyes study in Case 14.1 (page 751). Answer the following questions for the two contrasts that you defined in Exercise 14.75.
(a) For each contrast give H_0 and an appropriate H_a. In choosing the alternatives you should use information given in the description of the problem, but you may not consider any impressions obtained by inspection of the sample means.

(b) Find the values of the corresponding sample contrasts c_1 and c_2.
(c) Using the value $s_p = 2.81$, calculate the standard errors s_{c_1} and s_{c_2} for the sample contrasts.
(d) Give the test statistics and approximate P-values for the two significance tests. What do you conclude?
(e) Compute 95% confidence intervals for the two contrasts.

14.78 High school grades and changing majors. In the computer science department of a large university, many students change their major after the first year. A detailed study of the students enrolled as first-year computer science majors in one year was undertaken to help understand this phenomenon.[23] Students were classified on the basis of their major at the beginning of their second year, and several variables measured at the time of their entrance to the university were obtained. The following table presents data on the high school mathematics grades of the students. The grades have been coded so that $10 = A$, $9 = A-$, and so on.

Second-year major	n	$\bar{x}$	s
Computer science	90	8.77	1.41
Engineering and other sciences	28	8.75	1.46
Other	106	7.83	1.74

Because the first two groups (computer science, engineering and other sciences) are majoring in areas that require mathematics skills, one would want to compare these two groups to see if they are different. In addition, since the third group involves majors requiring less mathematics, it is natural to compare the average of the first two groups with the third. The thought here is that scores will be lower in the third group.
(a) For each of these contrasts state H_0 and an appropriate H_a. In choosing the alternatives you should use the information given in the description of the problem and not consider any impressions obtained by inspection of the sample means.
(b) Find the values of the corresponding sample contrasts c_1 and c_2.
(c) Using the value $s_p = 1.581$, calculate the standard errors s_{c_1} and s_{c_2} for the sample contrasts.
(d) Give the test statistics and approximate P-values for the two significance tests. What do you conclude?
(e) Compute 95% confidence intervals for the two contrasts.

14.79 Contrasts for exercise and fitness. In the exercise program study described in Exercise 14.71, the summary statistics for physical fitness scores are as follows:

Group	n	$\bar{x}$	s
Treatment (T)	10	291.91	38.17
Control (C)	5	308.97	32.07
Joggers (J)	11	366.87	41.19
Sedentary (S)	10	226.07	63.53

The researchers wanted to address the following questions for the physical fitness scores. In these questions "better" means a higher fitness score. (1) Is T better than C? (2) Is T better than the average of C and S? (3) Is J better than the average of the other three groups?

(a) For each of the three questions, define an appropriate contrast. Translate the questions into null and alternative hypotheses about these contrasts.

(b) Test your hypotheses and give approximate P-values. Summarize your conclusions. Do you think that the use of contrasts in this way gives an adequate summary of the results?

(c) You found that the groups differ significantly in the physical fitness scores. Does this study allow conclusions about causation—for example, that a sedentary lifestyle causes people to be less physically fit? Explain your answer.

14.80 Contrasts for exercise and depression. Exercise 14.72 gives the ANOVA table for depression scores from the exercise program study described in Exercise 14.71 (page 782). Here are the summary statistics for the depression scores:

Group	n	$\bar{x}$	s
Treatment (T)	10	51.90	6.42
Control (C)	5	57.40	10.46
Joggers (J)	11	49.73	6.27
Sedentary (S)	10	58.20	9.49

In planning the experiment, the researchers wanted to address the following questions for the depression scores. In these questions "better" means a lower depression score. (1) Is T better than C? (2) Is T better than the average of C and S? (3) Is J better than the average of the other three groups?

(a) For each of the three questions, define an appropriate contrast. Translate the questions into null and alternative hypotheses about these contrasts.

(b) Test your hypotheses and give approximate P-values. Summarize your conclusions. Do you think that the use of contrasts in this way gives an adequate summary of the results?

(c) You found that the groups differ significantly in the depression scores. Does this study allow conclusions about causation—for example, that a sedentary lifestyle causes people to be more depressed? Explain your answer.

14.81 Multiple comparisons for exercise and fitness. Refer to the physical fitness scores for the four groups in the exercise program study discussed in Exercises 14.71 and 14.79. Computer software gives the critical value for the Bonferroni multiple-comparisons procedure with $\alpha = 0.05$ as $t^{**} = 2.81$. Use this procedure to compare the mean fitness scores for the four groups. Summarize your conclusions.

14.82 Multiple comparisons for exercise and depression. Refer to the depression scores for the four groups in the exercise program study discussed in Exercises 14.72 and 14.80. Computer software gives the critical value for the Bonferroni multiple-comparisons procedure with $\alpha = 0.05$ as $t^{**} = 2.81$. Use this

procedure to compare the mean depression scores for the four groups. Summarize your conclusions.

14.83 Power for weight gain during pregnancy (optional). You are planning a study of the weight gains of women during the third trimester of pregnancy similar to that described in Exercise 14.73. The standard deviations given in that exercise range from 1.8 to 2.5. To perform power calculations, assume that the standard deviation is $\sigma = 2.4$. You have three groups, each with n subjects, and you would like to reject the ANOVA H_0 when the alternative $\mu_1 = 2.6$, $\mu_2 = 3.0$, and $\mu_3 = 3.4$ is true. Use software to make a table of powers against this alternative (similar to the table in Example 14.27, page 774) for the following numbers of women in each group: $n = 50, 100, 150, 175,$ and 200. What sample size would you choose for your study?

14.84 More power (optional). Repeat the previous exercise for the alternative $\mu_1 = 2.7$, $\mu_2 = 3.1$, and $\mu_3 = 3.5$. Why are the results the same?

14.85 Planning another emotions study (optional). Scores on an emotional scale were compared for five different cultures in Exercise 14.60 (page 779). Suppose that you are planning a new study using the same outcome variable. Your study will use European American, Asian American, and Hispanic American students from a large university.

(a) Explain how you would select the students to participate in your study.

(b) Use the data from Exercise 14.60 to perform power calculations to determine sample sizes for your study.

(c) Write a report that could be understood by someone with limited background in statistics and that describes your proposed study and why you think it is likely that you will obtain interesting results.

14.86 Planning another restaurant ambience study (optional). Exercise 14.50 (page 777) gave data for a study that examined the effect of background music on total food spending at a high-end restaurant. You are planning a similar study but intend to look at total food spending at a more casual restaurant. Use the results of the study described in Exercise 14.50 to plan your study.

14.87 Transform amounts to percents. Refer to Exercise 14.68 (page 781), where we compared six types of scaffold material to repair wounds. The data are given as percents ranging from 5% to 75%. ECM

(a) Convert these percents into their decimal form by dividing by 100. Calculate the transformed means, standard deviations, and standard errors and summarize them with the sample sizes in a table.

(b) Explain how you could have calculated the table entries directly from the table you gave in part (a) of Exercise 14.68.

(c) Analyze the percents using analysis of variance. Compare the test statistic, degrees of freedom, P-value, and conclusion you obtain here with the corresponding values that you found in Exercise 14.68.

14.88 Changing units and ANOVA. Refer to the previous exercise. Suggest a general conclusion about what happens to the F test statistic, degrees of freedom, P-value, and conclusion when you perform analysis of variance on data after changing the units of measurement. (A change of units is a linear transformation that multiplies each observation by the same positive constant and perhaps adds another constant.)

14.89 The effect of an outlier. Refer to the ECM experiment described in Exercise 14.68 (page 781). Suppose that when entering the data into the computer, you accidentally entered the first observation as 5% rather than 55%. 🔵 ECM
(a) Run the ANOVA with the incorrect observation. Summarize the results.
(b) Compare this run with the results obtained with the correct data set. What does this illustrate about the effect of outliers in an ANOVA?
(c) Compute a table of means and standard deviations for each of the six treatments using the incorrect data. How does this table help you to detect the incorrect observation?

14.90 What colors attract beetles? The presence of harmful insects in farm fields is detected by erecting boards covered with a sticky material and then examining the insects trapped on the board. To investigate which colors are most attractive to cereal leaf beetles, researchers placed six boards of each of four colors in a field of oats in July.[24] The table below gives data on the number of cereal leaf beetles trapped. 🔵 LEAFBEETLES

Color	Insects trapped					
Lemon yellow	45	59	48	46	38	47
White	21	12	14	17	13	17
Green	37	32	15	25	39	41
Blue	16	11	20	21	14	7

The square root transformation is often used for variables that are counts, such as the number of insects trapped in this exercise. In many cases data transformed in this way will conform more closely to the assumptions of Normality and equal standard deviations.
(a) Using the count data, make a table of means and standard deviations for the four colors, and plot the means.
(b) Make a similar table of the data after taking square roots. Are the standard deviations of the transformed data more nearly equal than those of the original data?
(c) Carry out the ANOVA for the raw and transformed data, making sure to check the model assumptions. Compare your conclusions.

14.91 Regression or ANOVA? Refer to the price promotion study that we examined in Exercise 14.63 (page 780). The explanatory variable in this study is the number of price promotions in a 10-week period, with possible values of 1, 3, 5, and 7. ANOVA treats the explanatory variable as categorical—it just labels the groups to be compared. In this study the explanatory variable is in fact quantitative, so we could use simple linear regression rather than one-way ANOVA if there is a linear pattern. 🔵 PRICEPROMO
(a) Make a scatterplot of the responses against the explanatory variable. Is the pattern roughly linear?
(b) In ANOVA, the F test null hypothesis states that groups have no effect on the mean response. What test in regression tests the null hypothesis that the explanatory variable has no linear relationship with the response?
(c) Carry out the regression. Compare your results with those from the ANOVA in Exercise 14.63. Are there any reasons—perhaps from part (a)—to prefer one or the other analysis?

CHAPTER 14 Case Study Exercises

CASE STUDY EXERCISE 1: Compare three products for treating dandruff. Analysis of variance methods are often used to compare different products. In medical settings, studies to assess the effectiveness of one or more treatments for a particular medical condition are often called *clinical trials*. This Case Study is based on W. L. Billhimer et al., "Results of a clinical trial comparing 1% pyrithione zinc and 2% ketoconazole shampoos," *Cosmetic Dermatology,* 9 (1996), pp. 34–39. The study reported in this paper is a clinical trial that compared three treatments for dandruff and a placebo. The products were 1% pyrithione zinc shampoo (PyrI), the same shampoo but with instructions to shampoo two times (PyrII), 2% ketoconazole shampoo (Keto), and a placebo shampoo (Placebo). After six weeks of treatment, eight sections of the scalp were examined and given a score that measured the amount of scalp flaking on a 0 to 10 scale. The response variable was the sum of these eight scores.

An analysis of the baseline flaking measure indicated that randomization of patients to treatments was successful in that

no differences were found among the groups. At baseline there were 112 subjects in each of the three treatment groups and 28 subjects in the Placebo group. During the clinical trial, 3 dropped out from the PyrII group and 6 from the Keto group. No patients dropped out of the other two groups. Summary statistics given in the paper were used to generate random data that give the same conclusions. Here are the data for the first five subjects: 🔵 DANDRUFF

Obs.	Treatment	Flaking
001	PyrI	17
002	PyrI	16
003	PyrI	18
004	PyrI	17
005	PyrI	18

Analyze the data using the methods you learned in this chapter. Be sure to include numerical and graphical summaries, a detailed explanation of the statistical model used, and a summary of all results. Include an analysis of the residuals to examine the assumptions of your model. Include an analysis of the following contrasts: (1) Placebo versus the average of the three treatments; (2) Keto versus the average of the two Pyr treatments; and (3) PyrI versus PyrII.

CASE STUDY EXERCISE 2: Compare three educational products. Jim Baumann and Leah Jones of the Purdue University School of Education conducted a study to compare three methods of teaching reading comprehension. The 66 students who participated in the study were randomly assigned to the methods (22 to each). The standard practice of comparing new methods with a traditional one was used in this study. The traditional method is called Basal and the two innovative methods are called DRTA and Strat.

In the data set the variable Subject is used to identify the individual students. The values are 1 to 66. The method of instruction is indicated by the variable Group, with values B, D, and S, corresponding to Basal, DRTA, and Strat. Two pretests and three posttests were given to all students. These are the variables

Pre1, Pre2, Post1, Post2, and Post3. Data for the first five subjects are given below. READING

Subject	Group	Pre1	Pre2	Post1	Post2	Post3
01	B	4	3	5	4	41
02	B	6	5	9	5	41
03	B	9	4	5	3	43
04	B	12	6	8	5	46
05	B	16	5	10	9	46

Case 14.2 (page 760) compares posttest reading comprehension scores for the three different products. In the actual study there were two pretest measures and three posttest measures. The posttest variable analyzed in Case 14.2 is Post3. Analyze the pretest variables and the other two posttest variables. Use the examples and exercises in the text to guide your analyses. Summarize the results for all five variables in a report. In particular, do the pretests show that the three groups of subjects were not statistically different at the start of the study? Do the posttests point to one or more of the teaching methods as more effective than the others?

CHAPTER 14 Appendix

Using Minitab and Excel for One-Way Analysis of Variance

One-Way ANOVA Analysis
Minitab:

Stat ➤ ANOVA ➤ One-Way

With this pull-down sequence, you should have two worksheet columns. One column should contain all the response data from all the groups. The other column should contain distinct labels (numerical or text) for each group that indicate for a given row which group the corresponding data observation comes from. Given this setup, click the data column into the **Response** box, and click the column of labels into the **Factor** box. Residuals and fitted values from the ANOVA analysis can be stored in the worksheet by selecting the **Store residuals** and **Store fits** options. To make multiple comparisons, click the **Comparisons** button. By means of computing simultaneous confidence intervals, Minitab offers four different methods (which it calls Tukey's, Fisher's, Dunnett's, and Hsu's) for making multiple comparisons. Fisher's method is the least-

significant differences (LSD) method noted in this chapter. Minitab, however, does not offer the Bonferroni method also discussed in this chapter. For making simultaneous confidence intervals for all the differences among the population means, Tukey's method is typically recommended as the best choice among the methods offered by Minitab. You will notice that Tukey's method is the default selection. In the adjacent box, notice that the overall α ("family error rate") is set to 5%. If you wish to change this overall rate, input the rate in percentage units (for example, input "10" for overall $\alpha = 0.10$). Click **OK** to close the pop-up box and then click **OK** to find one-way ANOVA output in the Session window.

Minitab offers another means to obtain one-way ANOVA results using the following pull-down sequence:

Stat ➤ ANOVA ➤ One-Way (Unstacked)

This routine is used if the data for each group are stored in separate columns, as opposed to the data all being in one column. Click all the data columns into the **Responses (in separate columns)** box. All other aspects of this routine are similar to the just-described one-way ANOVA routine.

Excel:

Data from a given group can be either grouped in a column (that is, observations inputted in one column going down the rows) or grouped in a row (that is, observations inputted in one row going across columns). In either case, the data from one group to the next group must be adjacent (that is, in adjacent columns or adjacent rows). To perform one-way ANOVA, select "Anova: Single Factor" in the **Data Analysis** menu box and click **OK.** Enter the cell range of the data into the **Input Range** box. Click **OK.** Multiple-comparisons methods are not available in standard Excel but they are available in the WHFStat Add-In for Excel.

Power (ANOVA Test)

Minitab:

Stat ➤ Power and Sample Size ➤ One-Way ANOVA

Once you have invoked this routine, you can specify values for any two of the following three dialog boxes: **Sample sizes, Values of the maximum difference between means,** or **Power values.** Minitab, in turn, will compute the item not specified. For example, if we wish to compute the power for the ANOVA test, then we would input the group sample sizes and the maximum difference between the smallest and largest means found in the particular alternative of interest. In the case of Example 14.26

(page 773), we only need to input the value "10" in the **Sample sizes** box (since sample sizes are equal) and input the value "6" ($= 47 - 41$) in the **Values of the maximum difference between means** box. This routine can also be used to determine what minimal sample size is required to have a desired power in relation to a particular alternative. In addition to providing values for two of the three dialog boxes, you must input the number of groups in the **Number of levels** box and the estimate for population standard deviation (σ) in the **Standard deviation** box. If you wish to change the value of α from the default value of 0.05, then click the **Options** button and input your desired significance level in the **Significance level** box. Click **OK** to close the pop-up box and then click **OK** to find results reported in the Session window. Additionally, Minitab will produce a "Power Curve," which is simply a graph of the different power values against a range of maximum difference values.

Excel:

Automated power calculations for ANOVA are not available in standard Excel or the WHFStat Add-In for Excel. Furthermore, since the noncentral F distribution is not available in Excel, computing the power by that means is not an option.

NOTES AND DATA SOURCES

CHAPTER 1

1. From the Fatal Accident Reporting System Web site; see `www-fars.nhtsa.dot.gov`.

2. Reported in the May 8, 2008 edition of *GPS Magazine*; see `www.gpsmagazine.com/print/000443.php`.

3. From an American Automobile Association report reported at CNNMoney.com on March 5, 2008; see `money.cnn.com/2008/03/05/news/economy/AAA_study`.

4. From *Traffic Safety Facts 2005*, National Highway Safety Administration, National Center for Statistics and Analysis; see `www.nhtsa.dot.gov`.

5. Pareto charts are named for the Italian economist Vilfredo Pareto (1848–1923). Pareto was one of the first to analyze economic problems with mathematical tools. The Pareto Principle (sometimes called the 80/20 rule) takes various forms, such as "80% of the work is done by 20% of the people." A Pareto chart is a graphical version of the principle—the chart identifies the few important categories (the 20%) that account for most of the responses (the 80%). Of course, in any given setting, the actual percents will vary.

6. From the 2006 Canadian Census; see `www12.statcan.ca/english/census`.

7. See Note 2.

8. Federal Reserve Bank of St. Louis; see `research.stlouisfed.org/fred2/series/WTB6MS`.

9. Our eyes do respond to area but not quite linearly. It appears that we perceive the ratio of two bars to be about the 0.7 power of the ratio of their actual areas. See W. S. Cleveland, *The Elements of Graphing Data*, Wadsworth, 1985, pp. 278–284.

10. Haipeng Shen, "Nonparametric regression for problems involving lognormal distributions," PhD thesis, University of Pennsylvania, 2003. Thanks to Haipeng Shen and Larry Brown for sharing the data.

11. See note 8.

12. From the Color Assignment Web site of Joe Hallock; see `www.joehallock.com/edu/COM498/index.html`.

13. From `news.cnet.com/apple/?tag=rb_content;overviewHead`.

14. From "The Apple iPhone: Successes and challenges for the mobile industry," March 31, 2008, Rubicon Win Markets; see `rubiconconsulting.com`.

15. U.S. Environmental Protection Agency, *Municipal Solid Waste in the United States: 2007 Facts and Figures*, document EPA530-R-08-010, November 2008.

16. September 2008 data from `marketshare.hitslink.com`.

17. See Note 16.

18. See `www.allfacebook.com/2008/08/million-dollar-facebook-application`.

19. See `www.insidefacebook.com`.

20. See Note 19.

21. From the Bureau of Labor Statistics Web site; see `www.bls.gov`.

22. From the 2007 Canadian Labor Force Survey; see `www40.statcan.ca`.

23. Color popularity for 2007 from the Dupont Automotive Color report; see `onlinepressroom.net/DuPont/MultimediaGallery`.

24. Annual reports for Procter and Gamble are available from `www.pg.com/investors/annualreports.shtml`.

25. From the 2006 American Community Survey; see `www.census.gov/acs/www/`.

26. From the 2006 Canadian census; see `www12.statcan.gc.ca/census-recensement/2006/dp-pd/tbt/Index-eng.cfm`. The files at this site provided counts, by province or territory, of the population over 65 and the total population. The percents over 65 were calculated from these quantities.

27. J. Marcus Jobe and Hutch Jobe, "A statistical approach for additional infill development," *Energy Exploration and Exploitation,* 18 (2000), pp. 89–103.

28. See `absc.usgs.gov/research/Fisheries/global_climate_on_salmon.htm`.

29. National Oceanic and Atmospheric Administration; see `www.noaa.gov`.

30. Data from the World Bank, Quick Query option from Key Development Data & Statistics page, http://www.worldbank.org/, Quick Query option on the Key Development Data & Statistics page, 2008.

31. See `www.fueleconomy.gov/`.

32. See Note 23.

33. The Best Countries for Business, forbes.com, June 26, 2008, Jack Gage, from `forbes.com/lists/2008/6/biz_bizcountries08_Best-Countries-for-Business_Rank.html`.

34. From the Interbrand Web site; see `interbrand.com/best_global_brands.aspx?langid=1000`.

35. From `beer100.com/beercalories.htm`, on June 30, 2008.

36. See `www.fueleconomy.gov/`.

37. Information about the Indiana Statewide Testing for Educational Progress program can be found at `www.doe.state.in.us/istep/`.

38. The idea that all distributions are normal in the middle is attributed to Charlie Winsor. See J. W. Tukey. A survey of sampling from contaminated distributions, in I. Olkin, S. G. Ghurye, W. Hoeffding, W. G. Madow, and H. B. Mann (eds.), *Contributions to Probability and Statistics, Essays in Honor of Harold Hotelling,* Stanford University Press, 1960, pp. 448–485.

39. See `www.stubhub.com`.

40. From Matthias R. Mehl et al., "Are women really more talkative than men?" *Science,* 317:5834 (2007), pp. 82.

41. Data provided by the city of Ames, Iowa; see `city.ames.ia.us/assessorweb/assessorpages/CommercialSales.htm`.

42. A subset of data posted at Damodaran Online; see `pages.stern.nyu.edu/~adamodar/`.

43. From `the-numbers.com`.

44. Data tracked by The Conference Board `conference-board.org`.

45. Based on data summaries in G. L. Cromwell et al., "A comparison of the nutritive value of *opaque-2*, *floury-2* and normal corn for the chick," *Poultry Science*, 57 (1968), pp. 840–847.

46. See Note 34.

47. See Note 27.

48. See Note 23.

49. See `cdc.gov/brfss/`. The data set BRFSS described in the data appendix contains several variables from this source.

50. See B. D. McCullough and B. Wilson, "On the accuracy of statistical procedures in Microsoft Excel 2000 and Excel XP," *Computational Statistics and Data Analysis,* 40, No. 4 (2002), pp. 713–721. Also, B. D. McCullough and D. A. Heiser, "On the accuracy of statistical procedures in Microsoft Excel 2007," *Computational Statistics and Data Analysis,* 52, (2008), pp. 4570–4578. Many other articles can be found on the topic of Excel's poor performance in statistical calculations.

CHAPTER 2

1. See "Happy spamiversary! Spam reaches 30," in *NewScientistTech*, April 25, 2008; see `www.NewScientist.com`.

2. Thanks to Doug Crabill, Manager of Computer Systems for the Purdue University Department of Statistics, for providing this background information about spam botnets.

3. See `symantec.com/security_response/writeup.jsp?docid=2007-062007-0946-99`.

4. OECD StatExtracts, Organization for Economic Co-operation Development, downloaded June 29, 2008, from `stats.oecd.org/wbos`.

5. See `forbes.com/lists/2008/6/biz_bizcountries08_Best-Countries-for-Business_Rank.html`.

6. A sophisticated treatment of improvements and additions to scatterplots is W. S. Cleveland and R. McGill, "The many faces of a scatterplot," *Journal of the American Statistical Association*, 79 (1984), pp. 807–822.

7. From the Current Population Survey Web site; see `census.gov/cps`.

8. See `beer100.com/beercalories.htm`.

9. See `www12.statcan.ca/english/census06/analysis/agesex/ProvTerr1.cfm`.

10. See `www.worldbank.org`.

11. A careful study of this phenomenon is W. S. Cleveland, P. Diaconis, and R. McGill, "Variables on scatterplots look more highly correlated when the scales are increased," *Science*, 216 (1982), pp. 1138–1141.

12. Data from a plot in Timothy G. O'Brien and Margaret F. Kinnaird, "Caffeine and conservation," *Science,* 300 (2003), p. 587.

13. From the performance data for the fund presented at the Vanguard Group Web site; see personal.vanguard.com. The EAFE returns differ from those given. For reasons unknown to us, we often find minor variations in EAFE returns reported by different sources. We have elected to use Vanguard's reported values in this exercise.

14. From *The Financial Development Report 2009,* World Economic Forum, 2009; available from www.weforum.org.

15. From a presentation by Charles Knauf, Monroe County (New York) Environmental Health Laboratory.

16. Frank J. Anscombe, "Graphs in statistical analysis," *The American Statistician*, 27 (1973), pp. 17–21.

17. From the Web site oasisnyc.net.

18. See target.com/site/en/corporate.

19. See Gary Taubes, "Magnetic field–cancer link—will it rest in peace?" *Science*, 277 (1997), p. 29.

20. C. M. Ryan, C. A. Northrup-Clewes, B. Knox, and D. I. Thurnham, "The effect of in-store music on consumer choice of wine," *Proceedings of the Nutrition Society,* 57 (1998), p. 1069A.

21. *Education Indicators: An International Perspective, Institute of Education Studies,* National Center for Education Statistics; see nces.ed.gov/surveys/ international/IntlIndicators.

22. For an overview of remote deposit capture, see remotedepositcapture.com/overview/rdc.overview.aspx.

23. From the "Community Bank Competitiveness Survey," 2008, *ABA Banking Journal*. The survey is available at www.nxtbook.com/nxtbooks/sb/ababj-compsurv08/ index.php.

24. The counts reported in Exercise 2.115 were calculated using counts of the numbers of banks in the different regions and the percents given in the ABA report cited in the previous note.

25. From M-Y Chen, et al., "Adequate sleep among adolescents is positively associated with health status and health-related behaviors," *BMC Public Health,* 6:59 (2006); available from biomedicalcentral.com/1471-2458/6/59.

26. See the U.S. Bureau of Census Web site at www.census.gov/ for theses and similar data.

27. Based on *The Ethics of American Youth - 2008,* available from the Josephson Institute at charactercounts.org/programs/reportcard/.

28. From the 2008 edition of the Purdue University Data Digest; see www.purdue.edu/ datadigest.

29. From the *2009 Statistical Abstract of the United States,* available at www.census. gov/compendia/statab/cats/population.html.

30. See Note 4.

31. Information about this procedure was provided by Samuel Flnigan of *U.S. News & World Report.* See `www.usnews.com/usnews/rankguide/rghome.htm` for a description of the variables used to construct the ranks and for the most recent ranks.

32. See `cdc.gov/brfss/`. The data set BRFSS described in the data appendix contains several variables from this source.

33. Oskar Kindvall, "Habitat heterogeneity and survival in a bush cricket metapopulation," *Ecology,* 77 (1996), pp. 207–214.

34. P. Velleman, *ActivStats 2.0*, Addison-Wesley Interactive, 1997.

35. Based on data provided by Professor Michael Hunt and graduate student James Bateman of the Purdue University Department of Forestry and Natural Resources.

36. Reported in the *New York Times*, July 20, 1989, from an article appearing that day in the *New England Journal of Medicine.*

37. Condensed from D. R. Appleton, J. M. French, and M. P. J. Vanderpump, "Ignoring a covariate: An example of Simpson's paradox," *The American Statistician*, 50 (1996), pp. 340–341.

38. Lien-Ti Bei, "Consumers' purchase behavior toward recycled products: An acquisition-transaction utility theory perspective," MS thesis, Purdue University, 1993.

39. From the International Coffee Organization Web site, ico.org

CHAPTER 3

1. See the NORC Web pages at `norc.uchicago.edu`.

2. See `cdc.gov/mmwr/preview/mmwrhtml/mm5839a3.htm`.

3. See Jeffrey G. Johnson et al., "Television viewing and aggressive behavior during adolescence and adulthood," *Science,* 295 (2002), pp. 2468–2471. The authors use statistical adjustments to control for the effects of a number of lurking variables. The association between TV viewing and aggression remains significant.

4. See, for example, `www.absolutebestcare.com/abs-franchise.html`.

5. National Institute of Child Health and Human Development, Study of Early Child Care. The article appears in the July 2003 issue of *Child Development.*

6. The quotation is from the summary on the NICHD Web site; see `nichd.nih.gov`.

7. M. K. Campbell et al., "A tailored multimedia nutrition education pilot program for low-income women receiving food assistance," *Health Education Research,* 14 (1999), pp. 257–267.

8. See, for example, `todayfoundation.org/pdfs/Today_Foundation_White_Paper_2009.pdf`.

9. Based on a study conducted by Tammy Younts directed by Professor Deb Bennett of the Purdue University Department of Educational Studies. For more information about Reading Recovery, see `readingrecovery.org/`.

10. Based on a study conducted by Rajendra Chaini under the direction of Professor Bill Hoover of the Purdue University Department of Forestry and Natural Resources.

11. From the Hot Ringtones list at `www.billboard.com/` on January 5, 2009.

12. From the Hot 100 list at `www.billboard.com/` on January 5, 2009.

13. From the online version of the Bureau of Labor Statistics *Handbook of Methods,* at `bls.gov`. The details of the design are more complicated than the text describes.

14. The nonresponse rate for the CPS can be found at the Bureau of Labor Statistics Web site; see, for example, `www.bls.gov/osmr/pdf/st010080.pdf`. The GSS reports its response rate on its Web site; see `www.norc.org/projects/gensoc.asp`.

15. The Pew Research Center for People and the Press designs careful surveys and is an execllent source of information about nonresponse. See also Special Issue: Non-Response Bias in Household Surveys, *Public Opinion Quarterly,* 70:5 (2006)

16. Robert F. Belli et al., "Reducing vote overreporting in surveys: Social desirability, memory failure, and source monitoring," *Public Opinion Quarterly,* 63 (1999), pp. 90–108.

17. Sex: Tom W. Smith, "The *JAMA* controversy and the meaning of sex," *Public Opinion Quarterly,* 63 (1999), pp. 385–400. Welfare: from a *New York Times*/CBS News poll reported in the *New York Times,* July 5, 1992. Scotland: "All set for independence?" *Economist,* September 12, 1998. Many other examples appear in T. W. Smith, "That which we call welfare by any other name would smell sweeter," *Public Opinion Quarterly,* 51 (1987), pp. 75–83.

18. Giuliana Coccia, "An overview of non-response in Italian telephone surveys," *Proceedings of the 99th Session of the International Statistical Institute, 1993*, Book 3, pp. 271–272.

19. This example is very loosely based on C. J. Schwarz, R. E. Bailey, J. R. Irvine, and F. C. Dalziel, "Estimating salmon spawning escapement using capture-recapture methods," *Canadian Journal of Fisheries and Aquatic Sciences*, 50 (1993), pp. 1181–1197.

20. From `www.gallup.com/home.aspx`.

21. From *CIS Boletin 9, Spaniards' Economic Awareness,* found online at `www.cis.es`.

22. For a full description of the STAR program and its follow-up studies, go to `www.heros-inc.org/star.htm`.

23. Simplified from Arno J. Rethans, John L. Swasy, and Lawrence J. Marks, "Effects of television commercial repetition, receiver knowledge, and commercial length: A test of the two-factor model," *Journal of Marketing Research,* 23 (February 1986), pp. 50–61.

24. Based on an experiment performed by Jake Gandolph under the direction of Professor Lisa Mauer in the Purdue University Department of Food Science.

25. Based on an experiment performed by Evan Whalen under the direction of Professor Patrick Connolly in the Purdue University Department of Computer Graphics Technology.

26. L. L. Miao, "Gastric freezing: An example of the evaluation of medical therapy by randomized clinical trials," in J. P. Bunker, B. A. Barnes, and F. Mosteller (eds.), *Costs, Risks, and Benefits of Surgery,* Oxford University Press, 1977, pp. 198–211.

27. Simplified from David L. Strayer, Frank A. Drews, and William A. Johnston, "Cell phone–induced failures of visual attention during simulated driving," *Journal of Experimental Psychology: Applied,* 9 (2003), pp. 23–32.

28. Based on a study conducted by Brent Ladd, a Water Quality Specialist with the Purdue University Department of Agricultural and Biological Engineering.

29. Based on a study conducted by Sandra Simonis under the direction of Professor Jon Harbor from the Purdue University Earth and Atmospheric Sciences Department.

30. John H. Kagel, Raymond C. Battalio, and C. G. Miles, "Marijuana and work performance: Results from an experiment," *Journal of Human Resources,* 15 (1980), pp. 373–395. A general discussion of failures of blinding is Dean Ferguson et al., "Turning a blind eye: The success of blinding reported in a random sample of randomised, placebo controlled trials," *British Medical Journal,* 328 (2004), p. 432.

31. Simplified from Sanjay K. Dhar, Claudia González-Vallejo, and Dilip Soman, "Modeling the effects of advertised price claims: Tensile versus precise pricing," *Marketing Science,* 18 (1999), pp. 154–177.

32. Joel Brockner et al., "Layoffs, equity theory, and work performance: Further evidence of the impact of survivor guilt," *Academy of Management Journal,* 29 (1986), pp. 373–384.

33. From www.gallup.com/home.aspx on January 5, 2009.

34. From *Drawing the Line: Sexual Harassment on Campus,* a report from the American Association of University Women (AAUW) Educational Foundation published in 2006; see aauw.org.

35. Heather Tait, *Aboriginal Peoples Survey, 2006: Inuit Health and Social Conditions,* (2008) Social and Aboriginal Statistics Division, Statistics Canada. Available from statcan.gc.ca.

36. John C. Bailar III, "The real threats to the integrity of science," *The Chronicle of Higher Education,* (April 21, 1995), pp. B1–B2.

37. The details are available on the Web site of the Office for Human Research Protections of the Department of Health and Human Services; see www.hhs.gov/ohrp.

38. The difficulties of interpreting guidelines for informed consent and for the work of institutional review boards in medical research are a main theme of Beverly Woodward, "Challenges to human subject protections in U.S. medical research," *Journal of the American Medical Association,* 282 (1999), pp. 1947–1952. The references in this paper point to other discussions. Updated regulations and guidelines appear on the OHRP Web site (see Note 2).

39. Quotation from the *Report of the Tuskegee Syphilis Study Legacy Committee,* May 20, 1996. A detailed history is James H. Jones, *Bad Blood: The Tuskegee Syphilis Experiment,* Free Press, 1993.

40. Dr. Hennekens's words are from an interview in the Annenberg/Corporation for Public Broadcasting video series *Against All Odds: Inside Statistics.*

41. See ftc.gov/opa/2009/04/kellogg.shtm.

42. See findarticles.com/p/articles/mi_m0CYD/is_8_40/ai_n13675065/.

43. R. D. Middlemist, E. S. Knowles, and C. F. Matter, "Personal space invasions in the lavatory: Suggestive evidence for arousal," *Journal of Personality and Social Psychology,* 33 (1976), pp. 541–546.

44. For a review of domestic-violence experiments, see C. D. Maxwell et al., *The Effects of Arrest on Intimate Partner Violence: New Evidence from the Spouse Assault Replication Program,* U.S. Department of Justice, NCH188199, 2001. Available online at `ojp.usdoj.gov/nij/pubs-sum/188199.htm`.

45. The report was issued in February 2009 and is available from `ftc.gov/os/2009/02/P085400behavadreport.pdf`.

46. See Note 21.

47. Javier Gimeno et al., "Survival of the fittest? Entrepreneurial human capital and the persistence of under-performing firms," *Administrative Science Quarterly*, 42:4 (1997).

48. Reported in Dupone News, December 10, 2008. See the Dupont Automotive Color Popularity Report at `onlinepressroom.net/DuPont/MultimediaGallery/`.

CHAPTER 4

1. U.S. Census Bureau, Current Population Survey, 2008 Annual Social and Economic Supplement. Available online at `www.census.gov`.

2. Persi Diaconis et al., "Dynamical bias in the coin toss," *SIAM Review*, 49 (2007), pp. 211–235.

3. The Gallup Organization, *Majority of Americans Approve of Labor Unions*, September 1, 2008; see `www.gallup.com/Home.aspx`.

4. You can find a mathematical explanation of Benford's law in Ted Hill, "The first-digit phenomenon," *American Scientist*, 86 (1996), pp. 358–363, and Ted Hill, "The difficulty of faking data," *Chance*, 12:3 (1999), pp. 27–31. Applications in fraud detection are discussed in the second paper by Hill and in Mark A. Nigrini, "I've got your number," *Journal of Accountancy*, May 1999, available online at `www.journalofaccountancy.com/Issues/1999/May/nigrini.htm`.

5. The full 2008 Canadian Medical Association report, *8th Annual National Report Card On Health Care*, can be found at `www.cma.ca`.

6. U.S. Department of Energy, *Annual Energy Review 2007*; see `www.eia.doe.gov`.

7. The Gallup Organization, *U.S. Workers Job Satisfaction Is Relatively High*, August 21, 2008; see `www.gallup.com/Home.aspx`.

8. Reuters Second Life News Center, *Europe takes lead in Second Life Users*, February 9, 2007; see `secondlife.reuters.com`.

9. From the Web site of Statistics Canada; see `www.statcan.gc.ca`.

10. Canadian transportation statistics from Statistics Canada at `www.statcan.gc.ca`. U.S. transportation statistics from U.S. Bureau of Transportation Statistics at `www.bts.gov`.

11. Internet usuage statistics can be found at `www.internetworldstats.com`.

12. From the Dupont Automotive 2006 Color Popularity Survey; see `www2.dupont.com/Automotive/en_US/knowledge_center`.

13. We use $\bar{x}$ both for the random variable, which takes different values in repeated sampling, and for the numerical value of the random variable in a particular sample. Similarly, s stands both for a random variable and for a specific value. This notation is mathematically imprecise but statistically convenient.

14. In this chapter, we consider only the case in which X takes a finite number of possible values. The same ideas, implemented with more advanced mathematics, apply to random variables with an infinite but still countable collection of values.

15. The mean of a continuous random variable X with density function $f(x)$ can be found by integration:

$$\mu_X = \int x f(x) dx$$

This integral is a kind of weighted average, analogous to the discrete-case mean

$$\mu_X = \sum x P(X = x)$$

The variance of a continuous random variable X is the average squared deviation of the values of X from their mean, found by the integral

$$\sigma_X^2 = \int (x - \mu)^2 f(x) dx$$

16. Returns data are available from several sources, including `finance.yahoo.com`.

17. A detailed discussion of risk pooling a supply chain can be found in David Simchi-Levi, Philip Kaminsky, and Edith Simchi-Levi, *Designing & Managing the Supply Chain*, McGraw-Hill, 2007.

18. From the Census Bureau's 2005 American Housing Survey.

19. Pfeiffer Consulting, *Mac Pro 2008 Benchmark Report*; see `www.pfeifferreport.com`.

20. See A. Tversky and D. Kahneman, "Belief in the law of small numbers," *Psychological Bulletin*, 76 (1971), pp. 105–110, and other writings of these authors for a full account of our misperception of randomness.

21. Probabilities involving runs can be quite difficult to compute. That the probability of a run of three or more heads in 10 independent tosses of a fair coin is $(1/2) + (1/128) = 0.508$ can be found by clever counting, as can the other results given in the text. A general treatment using advanced methods appears in Section XIII.7 of William Feller, *An Introduction to Probability Theory and Its Applications*, Vol. 1, 3rd ed., Wiley, 1968.

22. R. Vallone and A. Tversky, "The hot hand in basketball: on the misperception of random sequences," *Cognitive Psychology*, 17 (1985), pp. 295–314. A later series of articles that debate the independence question is A. Tversky and T. Gilovich, "The cold facts about the 'hot hand' in basketball," *Chance*, 2:1 (1989), pp. 16–21; P. D. Larkey, R. A. Smith, and J. B. Kadane, "It's OK to believe in the 'hot hand,'" *Chance*, 2:4 (1989), pp. 22–30; and A. Tversky and T. Gilovich, "The 'hot hand': statistical reality or cognitive illusion?" *Chance*, 2:4 (1989), pp. 31–34.

23. Strictly speaking, the recipe $\sigma/\sqrt{n}$ for the standard deviation of $\bar{x}$ assumes that we draw an SRS of size n from an *infinite* population. If the population has finite size N,

the standard deviation in the recipe is multiplied by $\sqrt{1 - (n-1)/(N-1)}$. This "finite population correction" approaches 1 as N increases. When the population is at least 10 times as large as the sample, the correction factor is between about 0.95 and 1. It is reasonable to use the simpler form $\sigma/\sqrt{n}$ in these settings.

24. From the Web site of U.S. Bureau of Transportation Statistics; see `www.bts.gov`.

25. Compas Inc., *Immense Public Frustration with Politicians over the Global Warming and Climate Change Debate*, October 1, 2008; see `www.compas.ca`.

26. Rapleaf Inc., *Study of Social Network Users vs. Age*, June 18, 2008; see `business .rapleaf.com/company_press.html`.

CHAPTER 5

1. Results from 2006 survey by Society for Human Resource Management; see `www.shrm.org`.

2. Joel Dresang, "1 in 6 executives lie about education credentials in resume, Brookfield firm says," *Journal Sentinel*, July 29, 2008.

3. J. Gollehon, *Pay the Line!* Putnam, 1988.

4. Simon Chadick (Director of the Centre for the International Business of Sport, Coventry University), "Bats and balls the key to economic bounce," from the Reuters global sports blog (`www.dailysportthought.blogspot.com/`), July 8, 2009.

5. K. Badenhausen, M. K. Ozanian, and C. Settimi, "The Business of Football," from the SportsMoney section of Forbes Magazine online site (`www.forbes.com/ business/sportsmoney`), September 10, 2008.

6. Estimated probabilities from National Collegiate Athletic Association (NCAA); see `www.ncaa.org`.

7. Projections by the National Center for Education Statistics; see `nces.ed.gov`.

8. W. D. Witnauer, R. G. Rogers, and J. M. Saint Onge, "Major League Baseball career length in the 20th century," *Population Research and Policy Review*, 26 (2007), pp. 371–386.

9. See Note 8.

10. J. J. Koehler and C. A. Conley, "The hot hand myth in professional basketball," *Journal of Sport and Exercise Psychology*, 25 (2003), pp. 253–259.

11. The Harris Poll, *Americans Satisfied with the Lives They Lead*, August 14, 2007; see `www.harrisinteractive.com`.

12. John Schwartz, "Leisure pursuits of today's young man; forsaking TV for online games and wanton Web sites," *The New York Times*, March 29, 2004.

13. "Ichiro breaks Sisler's all-time hits record," *Associated Press*, October 2, 2004.

14. Office of Technology Assessment, *Scientific Validity of Polygraph Testing: A Research Review and Evaluation*, Government Printing Office, 1983.

15. City of Chicago press release, "Mayor Daley Announces Major Upgrade To Chicago's 911 System," found at `www.cityofchicago.org`, February 19, 2009.

16. Survey results extracted based on 2008 Community College Survey of Student Engagement; see `www.ccsse.org`.

17. From the Web site of Statistics Canada, `www.statcan.gc.ca`.

18. Probabilities are from trials with 2897 people known to be free of HIV antibodies and 673 people known to be infected, reported in J. Richard George, "Alternative specimen sources: Methods for confirming positives," 1998 Conference on the Laboratory Science of HIV, found online at the Centers for Disease Control and Prevention, `www.cdc.gov`.

19. Barbara De Lollis and Barbara Hansen, "Airlines give fliers fewer chances to do the bump," from the Money section of USA Today online site (`www.usatoday.com/money`), December 19, 2009.

20. A collection of documents related to the FDIV bug can be found at `www.mathworks.com/company/pentium/index.shtml`.

CHAPTER 6

1. The 2008 Employer Health Benefits Survey final report can be found at the Kaiser Family Foundation Web site at `ehbs.kff.org`. Data used in the example was provided by Ben Finder of the Kaiser Family Foundation.

2. GAO reports can be found the U.S. Government Accountability Office Web site; see `www.gao.gov`.

3. A standard reference is Bradley Efron and Robert J. Tibshirani, *An Introduction to the Bootstrap,* Chapman Hall, 1993. A less technical overview is in Bradley Efron and Robert J. Tibshirani, "Statistical data analysis in the computer age," *Science,* 253 (1991), pp. 390–395.

4. Seung-Ok Kim, "Burials, pigs, and political prestige in neolithic China," *Current Anthropology,* 35 (1994), pp. 119–141.

5. From L. Pàstor et al., "Entrepreneurial learning, the IPO decision, and the post-IPO drop in firm profitability," *Review of Financial Studies,* 22 (2009), pp. 3005–3046.

6. Based on information reported in "The Graduate Student Credit Card Study," Nellie Mae Corporation (2007); see `www.nelliemae.com`.

7. From Stewart et al., "Exploring the handshake in employment interviews," *Journal of Applied Psychology,* 93 (2008), pp. 1139–1146.

8. From a study by M. R. Schlatter et al., Division of Financial Aid Purdue University.

9. Data provided by Diana Schellenberg, Purdue University School of Health Sciences.

10. Based on D. L. Shankland et al., "The effect of 5-thio-D-glucose on insect development and its absorption by insects," *Journal of Insect Physiology,* 14 (1968), pp. 63–72.

11. From William T. Robinson, "Sources of market pioneer advantages: The case of industrial goods industries," *Journal of Marketing Research,* 25 (1988), pp. 87–94.

12. Warren E. Leary, "Cell phones: Questions but no answers," *New York Times,* October 26, 1999.

13. Robert J. Schiller, "The volatility of stock market prices," *Science,* 235 (1987), pp. 33–36.

14. Sabrina Oesterle et al., "Volunteerism during the transition to adulthood: A life course perspective," *Social Forces,* 83 (2004), pp. 1123–1149.

15. Based on A. M. Garcia et al., "Why do workers behave unsafely at work? Determinants of safe work practices in industrial workers," *Occupational and Environmental Medicine,* 61 (2004), pp. 239–246.

16. Nicholas Stein, "Son of a chicken," *Fortune,* May 13, 2002.

17. Francisco Pèrez-Gonzàlez, "Inherited control and firm performance," *American Economic Review,* 96 (2006), pp. 1559–1588.

18. Data provided by Mugdha Gore and Joseph Thomas, Purdue University School of Pharmacy.

19. Barbara A. Almanza, Richard Ghiselli, and William Jaffe, "Foodservice design and aging baby boomers: Importance and perception of physical amenities in restaurants," *Foodservice Research International,* 12 (2000), pp. 25–40.

CHAPTER 7

1. Information from *Trends in Telephone Service - August 2008.* The report can be downloaded from the Wireline Competition Bureau Statistical Reports Web site; see www.fcc.gov/wcb/iatd/trends.html.

2. From C. Don Wiggins, "The legal perils of 'underdiversification' — a case study," *Personal Financial Planning,* 1:6 (1999), pp. 16–18.

3. Data provided by Joseph A. Wipf, Department of Foreign Languages and Literatures, Purdue University.

4. These recommendations are based on extensive computer work. See, for example, Harry O. Posten, "The robustness of the one-sample t-test over the Pearson system," *Journal of Statistical Computation and Simulation,* 9 (1979), pp. 133–149, and E. S. Pearson and N. W. Please, "Relation between the shape of population distribution and the robustness of four simple test statistics," *Biometrika,* 62 (1975), pp. 223–241.

5. C. Lecavalier. 2006. National and Regional Trends in Business Bankruptcies, 1980 to 2005. Catalogue no. 11-624-MIE. Ottawa: Statistics Canada.

6. 2007 Consumer Expenditures tables were released on November 25, 2008 and are available at www.bls.gov/cex/#tables.

7. Information regarding this consignment sale can be found at www.rerunsrfun.net/Index.aspx.

8. Based on Dan Dauwalter's Master's thesis in the Department of Forestry and Natural Resources at Purdue University. More information is available in Daniel C. Dauwalter et al., "An index of biotic integrity for fish assemblages in Ozark Highland streams of Arkansas," *Southeastern Naturalist,* 2 (2003), pp. 447–468. These data were provided by Emmanuel Frimpong.

9. Based on 2009 information from the USDA Feed Grains Database; see www.ers.usda.gov.

10. Based on results from the February 2004 release of the 2000 National Home and Hospice Care Survey (NHHCS); see www.cdc.gov/nchs/data/nhhcsd/curhomecare00.pdf.

11. Data provided by Timothy Sturm.

12. Data provided by Joseph A. Wipf, Department of Foreign Languages and Literatures, Purdue University.

13. Detailed information about the conservative t procedures can be found in Paul Leaverton and John J. Birch, "Small sample power curves for the two sample location problem," *Technometrics*, 11 (1969), pp. 299–307; in Henry Scheffé, "Practical solutions of the Behrens-Fisher problem," *Journal of the American Statistical Association*, 65 (1970), pp. 1501–1508; and in D. J. Best and J. C. W. Rayner, "Welch's approximate solution for the Behrens-Fisher problem," *Technometrics*, 29 (1987), pp. 205–210.

14. A. A. Labroo et al., "Of frog wines and frowning watches: Semantic priming, perceptual fluency, and brand evaluation," *Journal of Consumer Research,* 34 (2008), pp. 819–831.

15. Extensive simulation studies are reported in Harry O. Posten, "The robustness of the two-sample t test over the Pearson system," *Journal of Statistical Computation and Simulation*, 6 (1978), pp. 295–311; Harry O. Posten, H. Yeh, and D. B. Owen, "Robustness of the two-sample t-test under violations of the homogeneity assumption," *Communications in Statistics*, 11 (1982), pp. 109–126; and Harry O. Posten, "Robustness of the two-sample t-test under violations of the homogeneity assumption, part II," *Journal of Statistical Computation and Simulation*, 8 (1992), pp. 2169–2184.

16. Based on information made available Feb 13, 2009 titled "Wheat Data: Yearbook Tables: Wheat: Average price received by farmers, United States," available at http://www.ers.usda.gov/data/wheat/YBtable18.asp.

17. Based on C. Papoulias and P. Theodossiou, "Analysis and modeling of recent business failures in Greece," *Managerial and Decision Economics*, 13 (1992), pp. 163–169.

18. C. E. Cryfer et al., "Misery is not miserly: Sad and self-focused individuals spend more," *Psychological Science,* 19 (2008), pp. 525–530.

19. Rodney G. Bowden et al., "Lipid levels in a cohort of sedentary university students," *The Internet Journal of Cardiovascular Research,* 2 (2005).

20. The 2008 study can be found at http://www.qsrmagazine.com/reports/drive-thru_time_study/2008/chart-service-time.phtml.

21. M. K. Campbell et al., "A tailored multimedia nutrition education pilot program for low-income women receiving food assistance," *Health Education Research,* 14 (1999), pp. 257–267.

22. B. Bakke et al., "Cumulative exposure to dust and gases as determinants of lung function decline in tunnel construction workers," *Occupational Environmental Medicine,* 61 (2004), pp. 262–269.

23. Samara Joy Nielsen and Barry M. Popkin, "Patterns and trends in food portion sizes, 1977–1998," *Journal of the American Medical Association,* 289 (2003), pp. 450–453.

24. Gordana Mrdjenovic and David A. Levitsky, "Nutritional and energetic consequences of sweetened drink consumption in 6- to 13-year old children," *Journal of Pediatrics,* 142 (2003), pp. 604–610.

25. Data provided by Helen Park; see Helen Park, et al., "Fortifying bread with each of three antioxidants," *Cereal Chemistry*, 74 (1997), pp. 202–206.

26. Based on A. H. Ismail and R. J. Young, "The effect of chronic exercise on the personality of middle-aged men," *Journal of Human Ergology*, 2 (1973), pp. 47–57.

27. Based on a study conducted by Marvin Schlatter, Division of Financial Aid, Purdue University.

28. See the paper of Pearson and Please cited in Note 4 for one example of simulations that demonstrate the lack of robustness of the F test for comparing standard deviations.

29. The problem of comparing spreads is difficult even with advanced methods. Common distribution-free procedures do not offer a satisfactory alternative to the F test because they are sensitive to unequal shapes when comparing two distributions. A good introduction to the available methods is W. J. Conover, M. E. Johnson, and M. M. Johnson, "A comparative study of tests for homogeneity of variances, with applications to outer continental shelf bidding data," *Technometrics*, 23 (1981), pp. 351–361. Modern resampling procedures often work well. See Dennis D. Boos and Colin Brownie, "Bootstrap methods for testing homogeneity of variances," *Technometrics*, 31 (1989), pp. 69–82.

30. G. E. P. Box, "Non-Normality and tests on variances," *Biometrika*, 40 (1953), pp. 318–335. The quote appears on page 333.

31. This city's restaurant inspection data can be found at www.jsonline.com/watchdog/dataondemand/.

32. B. Wansink et al., "Fine as North Dakota wine: Sensory expectations and the intake of companion foods," *Physiology & Behavior,* 90 (2007), pp. 712–716.

33. P. Glick et al., "Evaluations of sexy women in low- and high-status jobs," *Psychology of Women Quarterly*, 29 (2005), pp. 389–395.

34. W. G. Kim and H-B. Kim, "Measuring customer-based restaurant brand equity," *Cornell Hotel and Restaurant Administration Quarterly*, 45 (2004), pp. 115–131.

35. Michael W. Peugh, "Field Investigation of ventilation and air quality in duck and turkey slaughter plants," M.S. thesis, Purdue University, 1996.

36. Ajay Ghei, "An empirical analysis of psychological androgeny in the personality profile of the successful hotel manager," M.S. thesis, Purdue University, 1992.

37. Data from the "wine" database in the archive of machine learning databases at the University of California, Irvine; see `ftp.ics.uci.edu/pub/machine-learning-databases`.

38. See Note 35.

39. Yvan R. Germain, "The dyeing of ramie with fiber reactive dyes using the cold pad-batch method," M.S. thesis, Purdue University, 1988.

40. Refer to Note 39.

41. Data from Wayne Nelson, *Applied Life Data Analysis*, Wiley, 1982, p. 471.

42. G. E. Smith et al., "A cognitive training program based on principles of brain plasticity: Results from the Improvement in Memory with Plasticity-based Adaptive

Cognitive Training (IMPACT) study," *Journal of the American Geriatrics Society,* epub (2009), pp. 1–10.

43. This exercise is based on events that are real. The data and details have been altered to protect the privacy of the individuals involved.

44. Based on G. Salvendy, "Selection of industrial operators: The one-hole test," *International Journal of Production Research*, 13 (1973), pp. 303–321.

45. From J. W. Marr and J. A. Heady, "Within- and between-person variation in dietary surveys: Number of days needed to classify individuals," *Human Nutrition: Applied Nutrition*, 40A (1986), pp. 347–364.

46. Data taken from the Web site `www.assist2sell.com`.

CHAPTER 8

1. From the PriceWaterhouseCooper Web site; see `www.pwc.com/`.

2. From a PEW Internet Project Memo Amanda Lenhart et al. dated December 7, 2008; see `www.pewinternet.org/pdfs/PIP_Adult_gaming_memo.pdf`.

3. From the "Community Bank Competitiveness Survey," 2008, *ABA Banking Journal*; see `www.nxtbook.com/nxtbooks/sb/ababj-compsurv08/index.php`.

4. See A. Agresti and B. A. Coull, "Approximate is better than 'exact' for interval estimation of binomial proportions," *The American Statistician*, 52 (1998), pp. 119–126. A detailed theoretical study is Lawrence D. Brown, Tony Cai, and Anirban Das-Gupta, "Confidence intervals for a binomial proportion and asymptotic expansions," *Annals of Statistics*, 30 (2002), pp. 160–201.

5. Adapted from a survey directed by Professor Joseph N. Uhl of the Department of Agricultural Economics, Purdue University. The survey was sponsored by the Indiana Christmas Tree Growers Association.

6. Heather Tait, *Aboriginal Peoples Survey, 2006: Inuit Health and Social Conditions,* (2008) Social and Aboriginal Statistics Division, Statistics Canada. Available from `statcan.gc.ca/pub`.

7. From a story posted on December 20, 2008 at `stuff.co.nz`. A question based on this survey was used in Michael Feldman's "Whad' Ya Know Quiz" in December 2008; see `notmuch.com`.

8. See `news.teamxbox.com/xbox/18254`.

9. Based on information in "NCAA 2003 national study of collegiate sports wagering and associated health risks," which can be found at the NCAA Web site; see `www.ncaa.org`.

10. From the "National Survey of Student Engagement, The College Student Report 2009," available online at `nsse.iub.edu/index.cfm`.

11. James Otto et al., "Training rats to search and alert on contraband odors," *Applied Animal Behaviour Science,* 77 (2002), pp. 271–232.

12. Information about the survey can be found online at `soc.duke.edu/natcong`.

13. This survey and others that study issues related to college students can be found at `www.nelliemae.com`.

14. From Heeseung Roh Ryu, Roseann M. Lyle, and George P. McCabe, "Factors associated with weight concerns and unhealthy eating patterns among young Korean females," *Eating Disorders*, 11 (2003), pp. 129–141.

15. Data from Roland J. Thorpe Jr. and based on his analysis of "Health ABC," a 10-year longitudinal study of older adults supported by the Laboratory of Epidemiology, Demography, and Biometry of the National Institute on Aging. Additional analyses are given in his PhD dissertation, "Relationship between pet ownership, physical activity, and human health in an elderly population," Purdue University, 2004.

16. Marsha A. Dickson, "Utility of no sweat labels for apparel customers: Profiling label users and predicting their purchases," *The Journal of Consumer Affairs*, 35 (2001), pp. 96–119.

17. From ad JupiterResearch press release dated February 12, 2007; see jupiterresearch.com.

18. Catalina L. Toma et al., "Separating fact from fiction: An examination of deceptive self presentation in online dating profiles," *Personality and Social Psychology Bulletin,* 34:8 (2008), pp. 1023–1036.

19. See Alan Agresti and Brian Caffo, "Simple and effective confidence intervals for proportions and differences of proportions result from adding two successes and two failures," *The American Statistician*, 45 (2000), pp. 280–288. The Wilson interval is a bit conservative (true coverage probability is higher than the confidence level) when p_1 and p_2 are equal and close to 0 or 1, but the traditional interval is much less accurate and has the fatal flaw that the true coverage probability is *less* than the confidence level.

20. Edoardo Botteri et al., "Smoking and colorectal cancer," *Journal of the American Medical Association,* 300:23 (2008), pp. 2765–2778.

21. This information was posted on the Podcast Alley Web site on January 13, 2009; see www.podcastalley.com.

22. From a PEW Internet Project Data Memo by Mary Madden, dated August 2008; see pewinternet.org.

23. From a PEW Internet Project Data Memo by Amanda Lenhart et al., dated December 2008; see pewinternet.org.

24. See S. W. Lagakos, B. J. Wessen, and M. Zelen, "An analysis of contaminated well water and health effects in Woburn, Massachusetts," *Journal of the American Statistical Association*, 81 (1986), pp. 583–596, and the following discussion. This case is the basis for the movie *A Civil Action*.

25. From Monica Macaulay and Colleen Brice, "Don't touch my projectile: Gender bias and stereotyping in syntactic examples," *Language*, 73 (1997), pp. 798–825. The first part of the title is a direct quote from one of the texts.

26. See Note 13.

27. From a PEW Internet report "Teens, video games, and civics," by Amanda Lehnart et al., September 16, 2008; see pewinternet.org.

28. From the Entertainment Software Association Web site; see theesa.com/facts/.

29. Data from the Neilsen Web site; see nielsenmobile.com/documents/WirelessSubstitution.pdf.

30. From M. L. Burr et al., "Effects on respiratory health of a reduction in air pollution from vehicle exhaust emissions," *Occupational and Environmental Medicine,* 61 (2004), pp. 212–218.

31. Lee Rainie et al., "The state of music downloading and file-sharing online," PEW Internet Project and Comscore Media Metrix Data Memo, 2004; see `pewinternet.org/`.

32. From Rick B. van Baaren, "The parrot effect: How to increase tip size," *Cornell Hotel and Restaurant Administration Quarterly*, 46 (2005), pp. 79–84.

33. From Dennis N. Bristow and Richard J. Sebastian, "Holy cow! Wait 'till next year! A closer look at the brand loyalty of Chicago Cubs baseball fans," *Journal of Consumer Marketing*, 18 (2001), pp. 256–275.

34. From Charles T. Clotfelter et al., "State lotteries at the turn of the century: Report to the National Gambling Impact Study Commission," (1999).

35. Based on Robert T. Driescher, "A quality swing with Ping," *Quality Progress*, (August 2001), pp. 37–41.

36. Some details are given in D. H. Kaye and M. Aickin (eds.), *Statistical Methods in Discrimination Litigation*, Marcel Dekker, 1986.

CHAPTER 9

1. From P. Azoulay and S. Shane, "Entrepreneurs, contracts, and the failure of young firms," *Management Science*, 47 (2001), pp. 337–358.

2. When the expected cell counts are small, it is best to use a test based on the exact distribution rather than the chi-square approximation, particularly for 2×2 tables. Many statistical software systems offer an "exact" test as well as the chi-square test for 2×2 tables.

3. The full report of the study appeared in George H. Beaton et al., "Effectiveness of vitamin A supplementation in the control of young child morbidity and mortality in developing countries," United Nations ACC/SCN State-of-the-Art Series, Nutrition Policy Discussion Paper no. 13, 1993.

4. C. M. Ryan, C. A. Northrup-Clewes, B. Knox, and D. I. Thurnham, "The effect of in-store music on consumer choice of wine," *Proceedings of the Nutrition Society*, 57 (1998), p. 1069A.

5. An alternative formula that can be used for hand or calculator computations is:

$$X^2 = \sum \frac{(\text{observed})^2}{\text{expected}} - n$$

6. See, for example, Alan Agresti, *Categorical Data Analysis*, 2nd edition, Wiley, 2002.

7. For an overview of remote deposit capture, see `remotedepositcapture.com/overview/rdc.overview.aspx`.

8. From the "Community Bank Competitiveness Survey," 2008, *ABA Banking Journal*; see `www.nxtbook.com/nxtbooks/sb/ababj-compsurv08/index.php`.

9. The marginal percent yes response in this table does not agree with the corresponding percent from the table in the previous exercise. The counts reported in this exercise

were calculated using counts of the numbers of banks in the different regions and the percents given in the ABA report (see Note 8). The percents match the figures given in the 2008 report.

10. From M-Y Chen, et al., "Adequate sleep among adolescents is positively associated with health status and health-related behaviors," *BMC Public Health,* 6:59 (2006); see biomedicalcentral.com/1471-2458/6/59.

11. Based on *The Ethics of American Youth - 2008,* available from the Josephson Institute at charactercounts.org/programs/reportcard/.

12. See pewinternet.org/About-us.aspx.

13. Data are from the report Home Broadband Adoption 2008 which was prepared by the Pew Internet and American Life Project; see pewinternet.org/Reports/2008/Home-Broadband-2008.aspx.

14. See Note 13.

15. See www.nhcaa.org.

16. These data are a composite based on several actual audits of this type.

17. See, for example, R. Benford amd J. Gess-Newsome, "Factors affecting student adademic succes in gateway courses at Northern Arizona University," 2006; see eric.ed.gov.

18. Data provided by Professor Marcy Towns of the Purdue University Department of Chemistry.

19. See Note 17.

20. From the Survey of Canadian Career College Students Phase II: In-School Student Survey (2008); see hrsdc.gc.ca/eng/publications_resources.

21. Tom Reichert, "The prevalence of sexual imagery in ads targeted to young adults," *Journal of Consumer Affairs,* 37 (2003), pp. 403–412.

22. See George R. Milne, "How well do consumers protect themselves from identity theft?" *Journal of Consumer Affairs,* 37 (2003), pp. 388–402.

23. Based on information in "NCAA 2003 national study of collegiate sports wagering and associated health risks," which can be found at the NCAA Web site; see www.ncaa.org.

24. Christopher Somers et al., "Air pollution induces heritable DNA mutations," *Proceedings of the National Academy of Sciences,* 99 (2002), pp. 15,904–15,907.

25. Ethan J. Temeles and W. John Kress, "Adaption in a plant-hummingbird association," *Science,* 300 (2003), pp. 630–633.

26. Dellavalle et al., "Going, going, gone: Lost Internet references," *Science,* 302 (2003), pp. 787–788.

27. From the National Survey of Student Engagement, 2008 Results; see nsse.iub.edu.

28. From Robert J. M. Dawson, "The 'unusual episode' data revisited," *Journal of Statistics Education*, 3:3 (1995). Electronic journal available at the American Statistical Association Web site; see www.amstat.org.

CHAPTER 10

1. In practice, x may also be a random quantity. Inferences can then be interpreted as *conditional* on a given value of x. If the error in measuring x is large, more advanced inference methods are needed.

2. J. C. Chen and C. Silverthorne, "The impact of locus of control on job stress, job performance and job satifaction in Taiwan," *Leadership and Organizational Development Journal*, 29 (2008), pp. 572–582.

3. National Science Foundation, Division of Science Resources Statistics. 2008. *Academic Research and Development Expenditures: Fiscal Year 2007.* Detailed Statistical Tables NSF 09-303. Arlington, VA; see `www.nsf.gov/statistics/nsf09303/`.

4. As the text notes, the residuals are not independent observations. They also have somewhat different standard deviations. For practical purposes of examining a regression model, we can nonetheless interpret the normal quantile plot as if the residuals were data from a single distribution.

5. Inflation is measured by the December-to-December change in the Consumer Price Index. These data were found at `www.bls.gov/cpi/`. Interest rates for the 6-month secondary market treasury bill were obtained at `www.federalreserve.gov/releases/h15/data.htm`

6. See the essay "Regression toward the mean" in Stephen M. Stigler, *Statistics on the Table*, Harvard University Press, 1999. The quotation from Milton Friedman appears in this essay.

7. In fact, the Excel regression output does not report the sign of the correlation r. The scatterplot in Figure 10.3 shows that r is positive. To get the correlation with the correct sign in Excel, you must use the "Correlation" function.

8. Selling price and assessment value available at `http://php.jconline.com/propertysales/propertysales.php`.

9. Tuition rates for 2000 from the "2000-2001 Tuition and Required Fees Report, University of Missouri. Tuition rates for 2008 are available at `http://colleges.collegetoolkit.com/college/main.aspx`.

10. M. Mondello and J. Maxcy "The impact of salary dispersion and performance bonuses in NFL organizations" *Management Decision,* 47 (2009), pp. 110–123.

11. These data were collected from `www.cbssports.com/nfl/playerrankings/regularseason/` and `http://content.usatoday.com/sports/football/nfl/salaries/`.

12. Based on Dan Dauwakter's master's thesis in the Department of Forestry and Natural Resources at Purdue University. More information is available in Daniel C. Dauwalter et al., "An index of biotic integrity for fish assemblages in Ozark Highland streams of Arkansas," *Southeastern Naturalist,* 2 (2003), pp. 447–468. These data were provided by Emmanuel Frimpong.

13. M. Kuo et al., "The marketing of alcohol to college students: The role of low prices and special promotions," *American Journal of Preventive Medicine*, 25:3 (2003), pp. 204–211.

14. Data were provided by the Ames City Assessor, Ames, Iowa.

15. Net cash flow data from Sean Collins, *Mutual Fund Assets and Flows in 2000*, Investment Company Institute, 2001 and "Trends in Mutual Fund Investing"; see `www.ici.org/stats/mf/arctrends/index.html`. The raw data were converted to real dollars using annual average values of the CPI.

16. These are part of the data from the EESEE story "Blood Alcohol Content," found on the text Web site.

17. Data for the first four years from Manuel Castells, *The Rise of the Network Society*, 2nd ed., Blackwell, 2000, p. 41.

18. Based on summaries in Charles Fombrun and Mark Shanley, "What's in a name? Reputation building and corporate strategy," *Academy of Management Journal*, 33 (1990), pp. 233–258.

19. Data available at `www.ncdc.noaa.gov/oa/climate/sd`.

20. M. J. Braid, K. F. Forbes, and D. W. Moran, "Pharmaceutical 'charge compression' under the Medicare Outpatient Prospective Payment System," *Journal of Health Care Finance,* 30:3 (2004), pp. 21–33.

21. S. E. Kimes and S. K. A. Robson, "The impact of restuarant table characteristics on meal duration and spending," *Cornell Hotel and Restuarant Administration Quarterly,* 45:4 (2004), pp. 333–346.

22. W. G. Kim and H-B. Kim, "Measuring customer-based restaurant brand equity," *Cornell Hotel and Restuarant Administration Quarterly,* 45:2 (2004), pp. 115–131.

23. S. Groschl, "Persons with disabilities, A source of nontraditional labor for Canada's hotel industry," *Cornell Hotel and Restuarant Administration Quarterly,* 46:2 (2005), pp. 258–274.

24. From the Supermarket Facts Web page of the Food Marketing Institute; see www.fmi.org.

25. Summarized annual information of each corporation is available on the Web pages of the Indiana Department of Education; see www.doe.in.gov.

26. Table of values available at `www.ers.usda.gov/Data/AgProductivity/`.

CHAPTER 11

1. Based on the 2007 Space Management Model for Purdue University implemented by Keith Murray, Director of Space Management and Academic Scheduling.

2. The Dow Jones Utility Average is a stock index that follows the performance of 15 prominent utility companies. It was created in 1929 when all utility stocks were removed from the Dow Jones Industrial Average.

3. Annual sales and profits, and assets as of Dec 31, 2008 were obtained from the Markets section at Forbes.com; see `www.forbes.com`.

4. U.S. Federal Deposit Insurance Corp., *Statistics on Banking*, issued annually. Information for 2008 can be found online at `www2.fdic.gov/SDI/SOB/`.

5. N. O. Rule and N. Ambady, "The face of success: Inferences from chief executive officers' appearance predicts company profits" *Psychological Science*, 19 (2008), pp. 109–111.

6. R. M. Smith and P. A. Schumacher, "Predicting success for actuarial students in undergraduate mathematics courses," *College Student Journal*, 39:1 (2005), pp. 165–177.

7. Available at *www.consumerreports.org*. Latest summary posted January 2008.

8. From Alan Levinsohn, "Online brokerage, the new core account?" *ABA Banking Journal*, (September 1999), pp. 34–42.

9. Data provided by the owners of Duck Worth Wearing, Ames, Iowa.

10. From a table entitled "Largest Indianapolis-Area Architectural Firms," *Indianapolis Business Journal,* December 16, 2003.

11. The data were graciously provided by Bruce Nash, founder of `www.the-numbers.com` and Nash Information Services.

12. KISS refers to the empirical principle "Keep it simple, stupid." In regression, this refers to keeping the models simple and avoiding unnecessary complexity.

13. From Kiyoshi Takahashi, "Effects of wage and promotion incentives on the motivation levels of Japanese employees," *Career Development International*, 11 (2006), pp. 193–203.

14. From Stephanie Aaronson and Julie Coronado, "Are firms or workers behind the shift away from DB pension plans?" Federal Reserve Board, (2005); see `www.federalreserve.gov/pubs/feds`.

15. From Michael E. Staten et al., "Information costs and the organization of credit markets: A theory of indirect lending," *Economic Inquiry*, 28 (1990), pp. 508–529.

16. The summary information taken from "FINAL REPORT : Canada Small Business Financing Program (CSBFP) Awareness and Satisfaction Study," prepared for Industry Canada by Phoenix Strategic Perspectives Inc., July 2007.

17. From Susan Stites-Doe and James J. Cordeiro, "An empirical assessment of the determinants of bank branch manager compensation," *The Journal of Applied Business Research*, 15 (1999), pp. 55–66.

18. The data were collected from the Web site `www.realtor.com` on October 8, 2001.

19. Tom Reichert, "The prevalence of sexual imagery in ads targeted to young adults," *Journal of Consumer Affairs,* 37 (2003), pp. 403–412.

20. For more information on logistic regression, see Chapter 17.

21. Bill Merrilees and Tino Fenech, "From catalog to Web: B2B multi-channel marketing strategy," *Industrial Marketing Management,* 36 (2007), pp. 44–49.

22. Based on M. U. Kalwani and C. K. Yim, "Consumer price and promotion expectations: An experimental study," *Journal of Marketing Research,* 29 (1992), pp. 90–100.

23. Based on Dan Dauwalter's master's thesis in the Department of Forestry and Natural Resources at Purdue University. More information is available in Daniel C. Dauwalter et al., "An index of biotic integrity for fish assemblages in Ozark Highland streams of Arkansas," *Southeastern Naturalist,* 2 (2003), pp. 447–468. These data were provided by Emmanuel Frimpong.

24. R. East et al., "Measuring the impact of positive and negative word of mouth on brand purchase probability," *International Journal of Research in Marketing,* 25 (2008), pp. 215–224.

25. A description of this case, as well as other examples of the use of statistics in legal settings, is given in Michael O. Finkelstein, *Quantitative Methods in Law*, Free Press, 1978.

26. This data set was provided by Joanne Lasrado, Purdue University, Department of Foods and Nutrition. More information regarding PCBs can be found at www.epa.gov/waterscience/fishstudy.

27. Yield data can be obtained at www.nass.usda.gov/Data_and_Statistics/Quick_Stats.

28. Darlene Gordon of the Purdue University School of Education provided the data for this case study exercise.

29. See G. P. McCabe, "Regression analysis in discrimination cases," in D. H. Kaye and M. Aickin (eds.), *Statistical Methods in Discrimination Litigation,* Marcel Dekker, 1986, for an explanation of how multiple regression is used in this setting.

CHAPTER 12

1. As of 2009, the American Society for Quality (ASQ) has honored twenty-three individuals by conferring on them the status of honorary member. A detailed summary of the background and contributions of the individuals noted here along with other pioneers can be found on the ASQ Web site; see www.asq.org/about-asq/who-we-are/honorary-members.html.

2. The cause-and-effect diagram was prepared by S. K. Bhat of the General Motors Technical Center as part of a course assignment at Purdue University.

3. Simulated data based on information appearing in Arvind Salvekar, "Application of six sigma to DRG 209," found at the Smarter Solutions Web site; see www.smartersolutions.com.

4. The exact formula for c_4 is given by

$$c_4 = \sqrt{\frac{2}{n-1}} \frac{(\frac{k}{2} - 1)!}{(\frac{k-1}{2} - 1)!}$$

where k is the number of observations. If the argument of the factorial is a noninteger it is computed as follows:

$$\left(\frac{k}{2}\right)! = \left(\frac{k}{2}\right)! \left(\frac{k}{2} - 1\right)! \left(\frac{k}{2} - 2\right)! \cdots \left(\frac{1}{2}\right) \sqrt{\pi}$$

5. Data provided by Linda McCabe, Purdue University.

6. Data provided by Colleen O'Brien, Team Leader Quality Resource and Privacy and Safety Officer, Bellin Health.

7. Data on aviation accidents can be found at the Federal Aviation Administration (FAA) statistics page; see www.faa.gov/data_statistics/accident_incident.

CHAPTER 13

1. Stock prices (including Wal-Mart) can be found at `finance.yahoo.com`.

2. Many studies have shown that stock price changes (and returns) sometimes follow a symmetric distribution with tails a bit thicker than the normal distribution. For our discussions, the Normal distribution is sufficient.

3. These data come from a weekly count of mails in spam folder of Layth Alwan's (co-author) University of Wisconsin-Milwaukee e-mail account during the Fall of 2008.

4. Data from the National Bureau of Statistics of China Web site; see `www.stats.gov.cn/english/`.

5. The number of Macintoshes shipped per fiscal quarter can be found in Apple's quarterly earnings press release. Follow the "Earnings Releases" link at `www.apple.com/investor/` to find the documents.

6. The home page for the Monthly Retail Trade Survey is found at `www.census.gov/mrts/www/mrtshist.html`.

7. Amazon quarterly net sales data were extracted from quarterly reports found by following the link "Investor Relations" at `www.amazon.com`.

8. There is an informative "Frequently Asked Questions" section about seasonal adjusting at the following U.S. government site: `www.census.gov/const/www/faq2.html`.

9. For example, details of the Durbin-Watson test for autocorrelation can be found in M. H. Kutner, C. J. Nachtschiem, J. Neter, and W. Li, *Applied Linear Statistical Models*, 5th ed., McGraw-Hill, 2005.

10. The consumer loans data is maintained at the Federal Reserve Web site. The Web address for monthly consumer loans data is `www.federalreserve.gov/releases/h8/data.htm`. Values in this series are sometimes revised after their initial posting to the site.

11. The motor vehicle production data are provided at the National Statistics Web site for the UK at `www.statistics.gov.uk`. The passenger car production time series is coded JCYM.

12. A variety of historical data on gold can be found at the World Gold Council Web site; see `www.gold.org`. Follow the "Research and Statistics" link.

13. The home page (in English) for the Canadian Tourism Commission is `www.corporate.canada.travel/en/ca/index.html`. Follow the "Research and Statistics" link.

14. For data on the number of ATT wireline customers, follow the link "About Us" at `www.att.com`, and then follow the link "Investor Relations" to find the documents under "Financials."

15. Oil supply and a variety of energy-related statistics are found at `www.eia.doe.gov`.

16. The authoritative reference book on ARIMA modeling is G. E. P. Box, G. M. Jenkins, and G. C. Reinsel, *Time Series Analysis: Forecasting and Control*, 4th ed., Wiley, 2008.

17. Data kindly provided by The Giving USA Foundation; see `www.givingusa.org`. The foundation publishes an annual report *Giving USA* which provides a comprehensive annual study of philanthropy in the United States.

18. The unemployment data can be found at Bureau of Labor Statistics database site; see `www.bls.gov/data/`.

19. Revenue data for the *NY Times* can be found at `www.nytco.com/investors/financials/revenue.html`.

20. Attendance data for the Cubs and other teams can be found by following the link "Teams" at `www.baseball-reference.com`.

21. Butter sales along with a wide variety of other agricultural statistics can be found at `www.nass.usda.gov/QuickStats`.

22. Two recent articles about the water level of the Great Lakes include "Water Levels in 3 Great Lakes Dip Far Below Normal," *NY Times*, August 14, 2007 and "Great Lakes' Lower Water Levels Propel a Cascade of Hardships," *Washington Post*, January 27, 2008.

23. Data on the Great Lakes can be found at a Web site of U.S. Army Corps of Engineering; see `www.lre.usace.army.mil/greatlakes`. Follow the link "Hydraulics and Hydrology" to find a link to historical data.

24. Transportation statistics can be found under "Data and Statistics" at `www.bts.gov`.

25. See Note 1.

26. See Note 1.

27. See Note 21.

28. Data found at the Office of the New Jersey State Climatologist Web site; see `climate.rutgers.edu/stateclim`.

29. Data found at `www.pro-football-reference.com`.

30. Quarterly data for Best Buy Inc. are found in its annual reports. Follow the link "For Our Investors" at `www.bestbuy.com`, and then follow the link "Annual Reports."

31. Quarterly data for Walgreens are found in its annual reports at `investor.walgreens.com/annual.cfm`. However, Walgreens posts data only going back 5 years. To obtain data for years prior, go to the U.S. government's SEC Filings Web site; see `www.sec.gov/edgar.shtml`.

32. See Note 24.

CHAPTER 14

1. Based on A. M. Garcia et al., "Why do workers behave unsafely at work? Determinants of safe work practices in industrial workers," *Occupational and Environmental Medicine,* 61 (2004), pp. 239–246.

2. This rule is intended to provide a general guideline for deciding when serious errors may result by applying ANOVA procedures. When the sample sizes in each group are very small, the sample variances will tend to vary much more than when the sample sizes are large. In this case, the rule may be a little too conservative. For unequal sample sizes, particular difficulties can arise when a relatively

small sample size is associated with a population having a relatively large standard deviation. Careful judgment is needed in all cases. By considering P-values rather than fixed level α testing, judgments in ambiguous cases can more easily be made; for example, if the P-value is very small, say 0.001, then it is probably safe to reject H_0 even if there is a fair amount of variation in the sample standard deviations.

3. Penny M. Simpson et al., "The eyes have it, or do they? The effects of model eye color and eye gaze on consumer ad response," *The Journal of Applied Business and Economics,* 8 (2008), pp. 60–71.

4. This example is based on Amna Kirmani and Peter Wright, "Money talks: Perceived advertising expense and expected product quality," *Journal of Consumer Research*, 16 (1989), pp. 344–353.

5. Case 14.2 is based on data from a study conducted by Jim Baumann and Leah Jones of the Purdue University School of Education.

6. Several different definitions for the noncentrality parameter of the noncentral F distribution are in use. When $I = 2$, the λ defined here is equal to the square of the noncentrality parameter δ that we used for the two-sample t test in Chapter 7. Many authors prefer $\phi = \sqrt{\lambda/I}$. We have chosen to use λ because it is the form needed for the SAS function PROBF.

7. Samuel S. Kim and Jerome Agrusa, "Segmenting Japanese tourists to Hawaii according to tour purpose," *Journal of Travel and Tourism Marketing,* 24 (2008), pp. 63–80.

8. Adrian C. North et al., "The effect of musical style on restaurant consumers' spending," *Environment and Behavior,* 35 (2003), pp. 712–718.

9. Jesus Tanguma et al., "Shopping and bargaining in Mexico: The role of women," *The Journal of Applied Business and Economics,* 9 (2009), pp. 34–40.

10. Woo Gon Kim et al., "Influence of institutional DINESERV on customer satisfaction, return intention, and word-of-mouth," *International Journal of Hospitality Management,* 28 (2009), pp. 10–17.

11. Katariina Mäenpää et al., "Consumer perceptions of Internet banking in Finland: The moderating role of familiarity," *Journal of Retailing and Consumer Services,* 15 (2008), pp. 266–276.

12. Sangwon Lee and Seonmi Lee, "Multiple play strategy in global telecommunication markets: An empirical analysis," *International Journal of Mobile Marketing,* 3 (2008), pp. 44–53.

13. "Value of recreational sports on college campuses," National Intramural-Recreational Sports Association (2002).

14. P. Bartel et al., "Attention and working memory in resident anaesthetists after night duty: Group and individual effects," *Occupational and Environmental Medicine,* 61 (2004), pp. 167–170.

15. Based on Jack K. Trammel, "The impact of academic accommodations on final grades in a post-secondary setting," *Journal of College Reading and Learning,* 34 (2003), pp. 76–90.

16. Christie N. Scollon et al., "Emotions across cultures and methods," *Journal of Cross-cultural Psychology*, 35 (2004), pp. 304–326.

17. The experiment was performed in Connie Weaver's lab in the Purdue University Department of Foods and Nutrition. The data were provided by Berdine Martin and Yong Jiang.

18. Based on M. U. Kalwani and C. K. Yim, "Consumer price and promotion expectations: An experimental study," *Journal of Marketing Research*, 29 (1992), pp. 90–100.

19. Data provided by Jo Welch of the Purdue University Department of Foods and Nutrition.

20. See S. Badylak et al., "Marrow-derived cells populate scaffolds composed of xenogeneic extracellular matrix," *Experimental Hematology*, 29 (2001), pp. 1310–1318.

21. Data provided by Dennis Lobstein, from his Ph.D. dissertation, "A multivariate study of exercise training effects on beta-endorphin and emotionality in psychologically normal, medically healthy men," Purdue University, 1983.

22. These data were taken from Collaborative Research Support Program in Food Intake and Human Function, *Management Entity Final Report*, University of California, Berkeley, 1988.

23. Results of the study are reported in P. F. Campbell and G. P. McCabe, "Predicting the success of freshmen in a computer science major," *Communications of the ACM*, 27 (1984), pp. 1108–1113.

24. Modified from M. C. Wilson and R. E. Shade, "Relative attractiveness of various luminescent colors to the cereal leaf beetle and the meadow spittlebug," *Journal of Economic Entomology*, 60 (1967), pp. 578–580.

TABLES

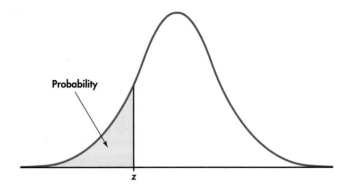

Table entry for z is the area under the standard normal curve to the left of z.

TABLE A Standard normal probabilities

z	.00	.01	.02	.03	.04	.05	.06	.07	.08	.09
−3.4	.0003	.0003	.0003	.0003	.0003	.0003	.0003	.0003	.0003	.0002
−3.3	.0005	.0005	.0005	.0004	.0004	.0004	.0004	.0004	.0004	.0003
−3.2	.0007	.0007	.0006	.0006	.0006	.0006	.0006	.0005	.0005	.0005
−3.1	.0010	.0009	.0009	.0009	.0008	.0008	.0008	.0008	.0007	.0007
−3.0	.0013	.0013	.0013	.0012	.0012	.0011	.0011	.0011	.0010	.0010
−2.9	.0019	.0018	.0018	.0017	.0016	.0016	.0015	.0015	.0014	.0014
−2.8	.0026	.0025	.0024	.0023	.0023	.0022	.0021	.0021	.0020	.0019
−2.7	.0035	.0034	.0033	.0032	.0031	.0030	.0029	.0028	.0027	.0026
−2.6	.0047	.0045	.0044	.0043	.0041	.0040	.0039	.0038	.0037	.0036
−2.5	.0062	.0060	.0059	.0057	.0055	.0054	.0052	.0051	.0049	.0048
−2.4	.0082	.0080	.0078	.0075	.0073	.0071	.0069	.0068	.0066	.0064
−2.3	.0107	.0104	.0102	.0099	.0096	.0094	.0091	.0089	.0087	.0084
−2.2	.0139	.0136	.0132	.0129	.0125	.0122	.0119	.0116	.0113	.0110
−2.1	.0179	.0174	.0170	.0166	.0162	.0158	.0154	.0150	.0146	.0143
−2.0	.0228	.0222	.0217	.0212	.0207	.0202	.0197	.0192	.0188	.0183
−1.9	.0287	.0281	.0274	.0268	.0262	.0256	.0250	.0244	.0239	.0233
−1.8	.0359	.0351	.0344	.0336	.0329	.0322	.0314	.0307	.0301	.0294
−1.7	.0446	.0436	.0427	.0418	.0409	.0401	.0392	.0384	.0375	.0367
−1.6	.0548	.0537	.0526	.0516	.0505	.0495	.0485	.0475	.0465	.0455
−1.5	.0668	.0655	.0643	.0630	.0618	.0606	.0594	.0582	.0571	.0559
−1.4	.0808	.0793	.0778	.0764	.0749	.0735	.0721	.0708	.0694	.0681
−1.3	.0968	.0951	.0934	.0918	.0901	.0885	.0869	.0853	.0838	.0823
−1.2	.1151	.1131	.1112	.1093	.1075	.1056	.1038	.1020	.1003	.0985
−1.1	.1357	.1335	.1314	.1292	.1271	.1251	.1230	.1210	.1190	.1170
−1.0	.1587	.1562	.1539	.1515	.1492	.1469	.1446	.1423	.1401	.1379
−0.9	.1841	.1814	.1788	.1762	.1736	.1711	.1685	.1660	.1635	.1611
−0.8	.2119	.2090	.2061	.2033	.2005	.1977	.1949	.1922	.1894	.1867
−0.7	.2420	.2389	.2358	.2327	.2296	.2266	.2236	.2206	.2177	.2148
−0.6	.2743	.2709	.2676	.2643	.2611	.2578	.2546	.2514	.2483	.2451
−0.5	.3085	.3050	.3015	.2981	.2946	.2912	.2877	.2843	.2810	.2776
−0.4	.3446	.3409	.3372	.3336	.3300	.3264	.3228	.3192	.3156	.3121
−0.3	.3821	.3783	.3745	.3707	.3669	.3632	.3594	.3557	.3520	.3483
−0.2	.4207	.4168	.4129	.4090	.4052	.4013	.3974	.3936	.3897	.3859
−0.1	.4602	.4562	.4522	.4483	.4443	.4404	.4364	.4325	.4286	.4247
−0.0	.5000	.4960	.4920	.4880	.4840	.4801	.4761	.4721	.4681	.4641

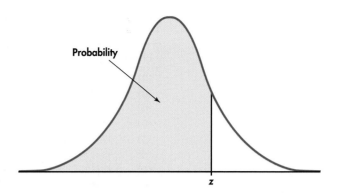

Probability

Table entry for z is
the area under the
standard normal curve
to the left of z.

TABLE A Standard normal probabilities (*continued*)

z	.00	.01	.02	.03	.04	.05	.06	.07	.08	.09
0.0	.5000	.5040	.5080	.5120	.5160	.5199	.5239	.5279	.5319	.5359
0.1	.5398	.5438	.5478	.5517	.5557	.5596	.5636	.5675	.5714	.5753
0.2	.5793	.5832	.5871	.5910	.5948	.5987	.6026	.6064	.6103	.6141
0.3	.6179	.6217	.6255	.6293	.6331	.6368	.6406	.6443	.6480	.6517
0.4	.6554	.6591	.6628	.6664	.6700	.6736	.6772	.6808	.6844	.6879
0.5	.6915	.6950	.6985	.7019	.7054	.7088	.7123	.7157	.7190	.7224
0.6	.7257	.7291	.7324	.7357	.7389	.7422	.7454	.7486	.7517	.7549
0.7	.7580	.7611	.7642	.7673	.7704	.7734	.7764	.7794	.7823	.7852
0.8	.7881	.7910	.7939	.7967	.7995	.8023	.8051	.8078	.8106	.8133
0.9	.8159	.8186	.8212	.8238	.8264	.8289	.8315	.8340	.8365	.8389
1.0	.8413	.8438	.8461	.8485	.8508	.8531	.8554	.8577	.8599	.8621
1.1	.8643	.8665	.8686	.8708	.8729	.8749	.8770	.8790	.8810	.8830
1.2	.8849	.8869	.8888	.8907	.8925	.8944	.8962	.8980	.8997	.9015
1.3	.9032	.9049	.9066	.9082	.9099	.9115	.9131	.9147	.9162	.9177
1.4	.9192	.9207	.9222	.9236	.9251	.9265	.9279	.9292	.9306	.9319
1.5	.9332	.9345	.9357	.9370	.9382	.9394	.9406	.9418	.9429	.9441
1.6	.9452	.9463	.9474	.9484	.9495	.9505	.9515	.9525	.9535	.9545
1.7	.9554	.9564	.9573	.9582	.9591	.9599	.9608	.9616	.9625	.9633
1.8	.9641	.9649	.9656	.9664	.9671	.9678	.9686	.9693	.9699	.9706
1.9	.9713	.9719	.9726	.9732	.9738	.9744	.9750	.9756	.9761	.9767
2.0	.9772	.9778	.9783	.9788	.9793	.9798	.9803	.9808	.9812	.9817
2.1	.9821	.9826	.9830	.9834	.9838	.9842	.9846	.9850	.9854	.9857
2.2	.9861	.9864	.9868	.9871	.9875	.9878	.9881	.9884	.9887	.9890
2.3	.9893	.9896	.9898	.9901	.9904	.9906	.9909	.9911	.9913	.9916
2.4	.9918	.9920	.9922	.9925	.9927	.9929	.9931	.9932	.9934	.9936
2.5	.9938	.9940	.9941	.9943	.9945	.9946	.9948	.9949	.9951	.9952
2.6	.9953	.9955	.9956	.9957	.9959	.9960	.9961	.9962	.9963	.9964
2.7	.9965	.9966	.9967	.9968	.9969	.9970	.9971	.9972	.9973	.9974
2.8	.9974	.9975	.9976	.9977	.9977	.9978	.9979	.9979	.9980	.9981
2.9	.9981	.9982	.9982	.9983	.9984	.9984	.9985	.9985	.9986	.9986
3.0	.9987	.9987	.9987	.9988	.9988	.9989	.9989	.9989	.9990	.9990
3.1	.9990	.9991	.9991	.9991	.9992	.9992	.9992	.9992	.9993	.9993
3.2	.9993	.9993	.9994	.9994	.9994	.9994	.9994	.9995	.9995	.9995
3.3	.9995	.9995	.9995	.9996	.9996	.9996	.9996	.9996	.9996	.9997
3.4	.9997	.9997	.9997	.9997	.9997	.9997	.9997	.9997	.9997	.9998

TABLE B Random digits

Line							
101	19223	95034	05756	28713	96409	12531	42544	82853
102	73676	47150	99400	01927	27754	42648	82425	36290
103	45467	71709	77558	00095	32863	29485	82226	90056
104	52711	38889	93074	60227	40011	85848	48767	52573
105	95592	94007	69971	91481	60779	53791	17297	59335
106	68417	35013	15529	72765	85089	57067	50211	47487
107	82739	57890	20807	47511	81676	55300	94383	14893
108	60940	72024	17868	24943	61790	90656	87964	18883
109	36009	19365	15412	39638	85453	46816	83485	41979
110	38448	48789	18338	24697	39364	42006	76688	08708
111	81486	69487	60513	09297	00412	71238	27649	39950
112	59636	88804	04634	71197	19352	73089	84898	45785
113	62568	70206	40325	03699	71080	22553	11486	11776
114	45149	32992	75730	66280	03819	56202	02938	70915
115	61041	77684	94322	24709	73698	14526	31893	32592
116	14459	26056	31424	80371	65103	62253	50490	61181
117	38167	98532	62183	70632	23417	26185	41448	75532
118	73190	32533	04470	29669	84407	90785	65956	86382
119	95857	07118	87664	92099	58806	66979	98624	84826
120	35476	55972	39421	65850	04266	35435	43742	11937
121	71487	09984	29077	14863	61683	47052	62224	51025
122	13873	81598	95052	90908	73592	75186	87136	95761
123	54580	81507	27102	56027	55892	33063	41842	81868
124	71035	09001	43367	49497	72719	96758	27611	91596
125	96746	12149	37823	71868	18442	35119	62103	39244
126	96927	19931	36089	74192	77567	88741	48409	41903
127	43909	99477	25330	64359	40085	16925	85117	36071
128	15689	14227	06565	14374	13352	49367	81982	87209
129	36759	58984	68288	22913	18638	54303	00795	08727
130	69051	64817	87174	09517	84534	06489	87201	97245
131	05007	16632	81194	14873	04197	85576	45195	96565
132	68732	55259	84292	08796	43165	93739	31685	97150
133	45740	41807	65561	33302	07051	93623	18132	09547
134	27816	78416	18329	21337	35213	37741	04312	68508
135	66925	55658	39100	78458	11206	19876	87151	31260
136	08421	44753	77377	28744	75592	08563	79140	92454
137	53645	66812	61421	47836	12609	15373	98481	14592
138	66831	68908	40772	21558	47781	33586	79177	06928
139	55588	99404	70708	41098	43563	56934	48394	51719
140	12975	13258	13048	45144	72321	81940	00360	02428
141	96767	35964	23822	96012	94591	65194	50842	53372
142	72829	50232	97892	63408	77919	44575	24870	04178
143	88565	42628	17797	49376	61762	16953	88604	12724
144	62964	88145	83083	69453	46109	59505	69680	00900
145	19687	12633	57857	95806	09931	02150	43163	58636
146	37609	59057	66967	83401	60705	02384	90597	93600
147	54973	86278	88737	74351	47500	84552	19909	67181
148	00694	05977	19664	65441	20903	62371	22725	53340
149	71546	05233	53946	68743	72460	27601	45403	88692
150	07511	88915	41267	16853	84569	79367	32337	03316

TABLE B Random digits *(continued)*

Line							
151	03802	29341	29264	80198	12371	13121	54969	43912
152	77320	35030	77519	41109	98296	18984	60869	12349
153	07886	56866	39648	69290	03600	05376	58958	22720
154	87065	74133	21117	70595	22791	67306	28420	52067
155	42090	09628	54035	93879	98441	04606	27381	82637
156	55494	67690	88131	81800	11188	28552	25752	21953
157	16698	30406	96587	65985	07165	50148	16201	86792
158	16297	07626	68683	45335	34377	72941	41764	77038
159	22897	17467	17638	70043	36243	13008	83993	22869
160	98163	45944	34210	64158	76971	27689	82926	75957
161	43400	25831	06283	22138	16043	15706	73345	26238
162	97341	46254	88153	62336	21112	35574	99271	45297
163	64578	67197	28310	90341	37531	63890	52630	76315
164	11022	79124	49525	63078	17229	32165	01343	21394
165	81232	43939	23840	05995	84589	06788	76358	26622
166	36843	84798	51167	44728	20554	55538	27647	32708
167	84329	80081	69516	78934	14293	92478	16479	26974
168	27788	85789	41592	74472	96773	27090	24954	41474
169	99224	00850	43737	75202	44753	63236	14260	73686
170	38075	73239	52555	46342	13365	02182	30443	53229
171	87368	49451	55771	48343	51236	18522	73670	23212
172	40512	00681	44282	47178	08139	78693	34715	75606
173	81636	57578	54286	27216	58758	80358	84115	84568
174	26411	94292	06340	97762	37033	85968	94165	46514
175	80011	09937	57195	33906	94831	10056	42211	65491
176	92813	87503	63494	71379	76550	45984	05481	50830
177	70348	72871	63419	57363	29685	43090	18763	31714
178	24005	52114	26224	39078	80798	15220	43186	00976
179	85063	55810	10470	08029	30025	29734	61181	72090
180	11532	73186	92541	06915	72954	10167	12142	26492
181	59618	03914	05208	84088	20426	39004	84582	87317
182	92965	50837	39921	84661	82514	81899	24565	60874
183	85116	27684	14597	85747	01596	25889	41998	15635
184	15106	10411	90221	49377	44369	28185	80959	76355
185	03638	31589	07871	25792	85823	55400	56026	12193
186	97971	48932	45792	63993	95635	28753	46069	84635
187	49345	18305	76213	82390	77412	97401	50650	71755
188	87370	88099	89695	87633	76987	85503	26257	51736
189	88296	95670	74932	65317	93848	43988	47597	83044
190	79485	92200	99401	54473	34336	82786	05457	60343
191	40830	24979	23333	37619	56227	95941	59494	86539
192	32006	76302	81221	00693	95197	75044	46596	11628
193	37569	85187	44692	50706	53161	69027	88389	60313
194	56680	79003	23361	67094	15019	63261	24543	52884
195	05172	08100	22316	54495	60005	29532	18433	18057
196	74782	27005	03894	98038	20627	40307	47317	92759
197	85288	93264	61409	03404	09649	55937	60843	66167
198	68309	12060	14762	58002	03716	81968	57934	32624
199	26461	88346	52430	60906	74216	96263	69296	90107
200	42672	67680	42376	95023	82744	03971	96560	55148

TABLE C Binomial probabilities

Entry is $P(X = k) = \binom{n}{k} p^k (1 - p)^{n-k}$

						p				
n	k	.01	.02	.03	.04	.05	.06	.07	.08	.09
2	0	.9801	.9604	.9409	.9216	.9025	.8836	.8649	.8464	.8281
	1	.0198	.0392	.0582	.0768	.0950	.1128	.1302	.1472	.1638
	2	.0001	.0004	.0009	.0016	.0025	.0036	.0049	.0064	.0081
3	0	.9703	.9412	.9127	.8847	.8574	.8306	.8044	.7787	.7536
	1	.0294	.0576	.0847	.1106	.1354	.1590	.1816	.2031	.2236
	2	.0003	.0012	.0026	.0046	.0071	.0102	.0137	.0177	.0221
	3				.0001	.0001	.0002	.0003	.0005	.0007
4	0	.9606	.9224	.8853	.8493	.8145	.7807	.7481	.7164	.6857
	1	.0388	.0753	.1095	.1416	.1715	.1993	.2252	.2492	.2713
	2	.0006	.0023	.0051	.0088	.0135	.0191	.0254	.0325	.0402
	3			.0001	.0002	.0005	.0008	.0013	.0019	.0027
	4									.0001
5	0	.9510	.9039	.8587	.8154	.7738	.7339	.6957	.6591	.6240
	1	.0480	.0922	.1328	.1699	.2036	.2342	.2618	.2866	.3086
	2	.0010	.0038	.0082	.0142	.0214	.0299	.0394	.0498	.0610
	3		.0001	.0003	.0006	.0011	.0019	.0030	.0043	.0060
	4						.0001	.0001	.0002	.0003
	5									
6	0	.9415	.8858	.8330	.7828	.7351	.6899	.6470	.6064	.5679
	1	.0571	.1085	.1546	.1957	.2321	.2642	.2922	.3164	.3370
	2	.0014	.0055	.0120	.0204	.0305	.0422	.0550	.0688	.0833
	3		.0002	.0005	.0011	.0021	.0036	.0055	.0080	.0110
	4					.0001	.0002	.0003	.0005	.0008
	5									
	6									
7	0	.9321	.8681	.8080	.7514	.6983	.6485	.6017	.5578	.5168
	1	.0659	.1240	.1749	.2192	.2573	.2897	.3170	.3396	.3578
	2	.0020	.0076	.0162	.0274	.0406	.0555	.0716	.0886	.1061
	3		.0003	.0008	.0019	.0036	.0059	.0090	.0128	.0175
	4				.0001	.0002	.0004	.0007	.0011	.0017
	5								.0001	.0001
	6									
	7									
8	0	.9227	.8508	.7837	.7214	.6634	.6096	.5596	.5132	.4703
	1	.0746	.1389	.1939	.2405	.2793	.3113	.3370	.3570	.3721
	2	.0026	.0099	.0210	.0351	.0515	.0695	.0888	.1087	.1288
	3	.0001	.0004	.0013	.0029	.0054	.0089	.0134	.0189	.0255
	4			.0001	.0002	.0004	.0007	.0013	.0021	.0031
	5							.0001	.0001	.0002
	6									
	7									
	8									

TABLE C Binomial probabilities *(continued)*

Entry is $P(X = k) = \binom{n}{k} p^k (1 - p)^{n-k}$

n	k	.10	.15	.20	.25	.30	.35	.40	.45	.50
2	0	.8100	.7225	.6400	.5625	.4900	.4225	.3600	.3025	.2500
	1	.1800	.2550	.3200	.3750	.4200	.4550	.4800	.4950	.5000
	2	.0100	.0225	.0400	.0625	.0900	.1225	.1600	.2025	.2500
3	0	.7290	.6141	.5120	.4219	.3430	.2746	.2160	.1664	.1250
	1	.2430	.3251	.3840	.4219	.4410	.4436	.4320	.4084	.3750
	2	.0270	.0574	.0960	.1406	.1890	.2389	.2880	.3341	.3750
	3	.0010	.0034	.0080	.0156	.0270	.0429	.0640	.0911	.1250
4	0	.6561	.5220	.4096	.3164	.2401	.1785	.1296	.0915	.0625
	1	.2916	.3685	.4096	.4219	.4116	.3845	.3456	.2995	.2500
	2	.0486	.0975	.1536	.2109	.2646	.3105	.3456	.3675	.3750
	3	.0036	.0115	.0256	.0469	.0756	.1115	.1536	.2005	.2500
	4	.0001	.0005	.0016	.0039	.0081	.0150	.0256	.0410	.0625
5	0	.5905	.4437	.3277	.2373	.1681	.1160	.0778	.0503	.0313
	1	.3280	.3915	.4096	.3955	.3602	.3124	.2592	.2059	.1563
	2	.0729	.1382	.2048	.2637	.3087	.3364	.3456	.3369	.3125
	3	.0081	.0244	.0512	.0879	.1323	.1811	.2304	.2757	.3125
	4	.0004	.0022	.0064	.0146	.0284	.0488	.0768	.1128	.1562
	5		.0001	.0003	.0010	.0024	.0053	.0102	.0185	.0312
6	0	.5314	.3771	.2621	.1780	.1176	.0754	.0467	.0277	.0156
	1	.3543	.3993	.3932	.3560	.3025	.2437	.1866	.1359	.0938
	2	.0984	.1762	.2458	.2966	.3241	.3280	.3110	.2780	.2344
	3	.0146	.0415	.0819	.1318	.1852	.2355	.2765	.3032	.3125
	4	.0012	.0055	.0154	.0330	.0595	.0951	.1382	.1861	.2344
	5	.0001	.0004	.0015	.0044	.0102	.0205	.0369	.0609	.0937
	6			.0001	.0002	.0007	.0018	.0041	.0083	.0156
7	0	.4783	.3206	.2097	.1335	.0824	.0490	.0280	.0152	.0078
	1	.3720	.3960	.3670	.3115	.2471	.1848	.1306	.0872	.0547
	2	.1240	.2097	.2753	.3115	.3177	.2985	.2613	.2140	.1641
	3	.0230	.0617	.1147	.1730	.2269	.2679	.2903	.2918	.2734
	4	.0026	.0109	.0287	.0577	.0972	.1442	.1935	.2388	.2734
	5	.0002	.0012	.0043	.0115	.0250	.0466	.0774	.1172	.1641
	6		.0001	.0004	.0013	.0036	.0084	.0172	.0320	.0547
	7				.0001	.0002	.0006	.0016	.0037	.0078
8	0	.4305	.2725	.1678	.1001	.0576	.0319	.0168	.0084	.0039
	1	.3826	.3847	.3355	.2670	.1977	.1373	.0896	.0548	.0313
	2	.1488	.2376	.2936	.3115	.2965	.2587	.2090	.1569	.1094
	3	.0331	.0839	.1468	.2076	.2541	.2786	.2787	.2568	.2188
	4	.0046	.0185	.0459	.0865	.1361	.1875	.2322	.2627	.2734
	5	.0004	.0026	.0092	.0231	.0467	.0808	.1239	.1719	.2188
	6		.0002	.0011	.0038	.0100	.0217	.0413	.0703	.1094
	7			.0001	.0004	.0012	.0033	.0079	.0164	.0312
	8					.0001	.0002	.0007	.0017	.0039

TABLE C Binomial probabilities *(continued)*

Entry is $P(X = k) = \binom{n}{k} p^k (1 - p)^{n-k}$

						p				
n	k	.01	.02	.03	.04	.05	.06	.07	.08	.09
9	0	.9135	.8337	.7602	.6925	.6302	.5730	.5204	.4722	.4279
	1	.0830	.1531	.2116	.2597	.2985	.3292	.3525	.3695	.3809
	2	.0034	.0125	.0262	.0433	.0629	.0840	.1061	.1285	.1507
	3	.0001	.0006	.0019	.0042	.0077	.0125	.0186	.0261	.0348
	4			.0001	.0003	.0006	.0012	.0021	.0034	.0052
	5						.0001	.0002	.0003	.0005
	6									
	7									
	8									
	9									
10	0	.9044	.8171	.7374	.6648	.5987	.5386	.4840	.4344	.3894
	1	.0914	.1667	.2281	.2770	.3151	.3438	.3643	.3777	.3851
	2	.0042	.0153	.0317	.0519	.0746	.0988	.1234	.1478	.1714
	3	.0001	.0008	.0026	.0058	.0105	.0168	.0248	.0343	.0452
	4			.0001	.0004	.0010	.0019	.0033	.0052	.0078
	5					.0001	.0001	.0003	.0005	.0009
	6									.0001
	7									
	8									
	9									
	10									
12	0	.8864	.7847	.6938	.6127	.5404	.4759	.4186	.3677	.3225
	1	.1074	.1922	.2575	.3064	.3413	.3645	.3781	.3837	.3827
	2	.0060	.0216	.0438	.0702	.0988	.1280	.1565	.1835	.2082
	3	.0002	.0015	.0045	.0098	.0173	.0272	.0393	.0532	.0686
	4		.0001	.0003	.0009	.0021	.0039	.0067	.0104	.0153
	5				.0001	.0002	.0004	.0008	.0014	.0024
	6							.0001	.0001	.0003
	7									
	8									
	9									
	10									
	11									
	12									
15	0	.8601	.7386	.6333	.5421	.4633	.3953	.3367	.2863	.2430
	1	.1303	.2261	.2938	.3388	.3658	.3785	.3801	.3734	.3605
	2	.0092	.0323	.0636	.0988	.1348	.1691	.2003	.2273	.2496
	3	.0004	.0029	.0085	.0178	.0307	.0468	.0653	.0857	.1070
	4		.0002	.0008	.0022	.0049	.0090	.0148	.0223	.0317
	5			.0001	.0002	.0006	.0013	.0024	.0043	.0069
	6						.0001	.0003	.0006	.0011
	7								.0001	.0001
	8									
	9									
	10									
	11									
	12									
	13									
	14									
	15									

TABLE C Binomial probabilities *(continued)*

						p				
n	k	.10	.15	.20	.25	.30	.35	.40	.45	.50
9	0	.3874	.2316	.1342	.0751	.0404	.0207	.0101	.0046	.0020
	1	.3874	.3679	.3020	.2253	.1556	.1004	.0605	.0339	.0176
	2	.1722	.2597	.3020	.3003	.2668	.2162	.1612	.1110	.0703
	3	.0446	.1069	.1762	.2336	.2668	.2716	.2508	.2119	.1641
	4	.0074	.0283	.0661	.1168	.1715	.2194	.2508	.2600	.2461
	5	.0008	.0050	.0165	.0389	.0735	.1181	.1672	.2128	.2461
	6	.0001	.0006	.0028	.0087	.0210	.0424	.0743	.1160	.1641
	7			.0003	.0012	.0039	.0098	.0212	.0407	.0703
	8				.0001	.0004	.0013	.0035	.0083	.0176
	9						.0001	.0003	.0008	.0020
10	0	.3487	.1969	.1074	.0563	.0282	.0135	.0060	.0025	.0010
	1	.3874	.3474	.2684	.1877	.1211	.0725	.0403	.0207	.0098
	2	.1937	.2759	.3020	.2816	.2335	.1757	.1209	.0763	.0439
	3	.0574	.1298	.2013	.2503	.2668	.2522	.2150	.1665	.1172
	4	.0112	.0401	.0881	.1460	.2001	.2377	.2508	.2384	.2051
	5	.0015	.0085	.0264	.0584	.1029	.1536	.2007	.2340	.2461
	6	.0001	.0012	.0055	.0162	.0368	.0689	.1115	.1596	.2051
	7		.0001	.0008	.0031	.0090	.0212	.0425	.0746	.1172
	8			.0001	.0004	.0014	.0043	.0106	.0229	.0439
	9					.0001	.0005	.0016	.0042	.0098
	10							.0001	.0003	.0010
12	0	.2824	.1422	.0687	.0317	.0138	.0057	.0022	.0008	.0002
	1	.3766	.3012	.2062	.1267	.0712	.0368	.0174	.0075	.0029
	2	.2301	.2924	.2835	.2323	.1678	.1088	.0639	.0339	.0161
	3	.0852	.1720	.2362	.2581	.2397	.1954	.1419	.0923	.0537
	4	.0213	.0683	.1329	.1936	.2311	.2367	.2128	.1700	.1208
	5	.0038	.0193	.0532	.1032	.1585	.2039	.2270	.2225	.1934
	6	.0005	.0040	.0155	.0401	.0792	.1281	.1766	.2124	.2256
	7		.0006	.0033	.0115	.0291	.0591	.1009	.1489	.1934
	8		.0001	.0005	.0024	.0078	.0199	.0420	.0762	.1208
	9			.0001	.0004	.0015	.0048	.0125	.0277	.0537
	10					.0002	.0008	.0025	.0068	.0161
	11						.0001	.0003	.0010	.0029
	12								.0001	.0002
15	0	.2059	.0874	.0352	.0134	.0047	.0016	.0005	.0001	.0000
	1	.3432	.2312	.1319	.0668	.0305	.0126	.0047	.0016	.0005
	2	.2669	.2856	.2309	.1559	.0916	.0476	.0219	.0090	.0032
	3	.1285	.2184	.2501	.2252	.1700	.1110	.0634	.0318	.0139
	4	.0428	.1156	.1876	.2252	.2186	.1792	.1268	.0780	.0417
	5	.0105	.0449	.1032	.1651	.2061	.2123	.1859	.1404	.0916
	6	.0019	.0132	.0430	.0917	.1472	.1906	.2066	.1914	.1527
	7	.0003	.0030	.0138	.0393	.0811	.1319	.1771	.2013	.1964
	8		.0005	.0035	.0131	.0348	.0710	.1181	.1647	.1964
	9		.0001	.0007	.0034	.0116	.0298	.0612	.1048	.1527
	10			.0001	.0007	.0030	.0096	.0245	.0515	.0916
	11				.0001	.0006	.0024	.0074	.0191	.0417
	12					.0001	.0004	.0016	.0052	.0139
	13						.0001	.0003	.0010	.0032
	14								.0001	.0005
	15									

TABLE C Binomial probabilities *(continued)*

						p				
n	k	.01	.02	.03	.04	.05	.06	.07	.08	.09
20	0	.8179	.6676	.5438	.4420	.3585	.2901	.2342	.1887	.1516
	1	.1652	.2725	.3364	.3683	.3774	.3703	.3526	.3282	.3000
	2	.0159	.0528	.0988	.1458	.1887	.2246	.2521	.2711	.2818
	3	.0010	.0065	.0183	.0364	.0596	.0860	.1139	.1414	.1672
	4		.0006	.0024	.0065	.0133	.0233	.0364	.0523	.0703
	5			.0002	.0009	.0022	.0048	.0088	.0145	.0222
	6				.0001	.0003	.0008	.0017	.0032	.0055
	7						.0001	.0002	.0005	.0011
	8								.0001	.0002
	9									
	10									
	11									
	12									
	13									
	14									
	15									
	16									
	17									
	18									
	19									
	20									

						p				
n	k	.10	.15	.20	.25	.30	.35	.40	.45	.50
20	0	.1216	.0388	.0115	.0032	.0008	.0002	.0000	.0000	.0000
	1	.2702	.1368	.0576	.0211	.0068	.0020	.0005	.0001	.0000
	2	.2852	.2293	.1369	.0669	.0278	.0100	.0031	.0008	.0002
	3	.1901	.2428	.2054	.1339	.0716	.0323	.0123	.0040	.0011
	4	.0898	.1821	.2182	.1897	.1304	.0738	.0350	.0139	.0046
	5	.0319	.1028	.1746	.2023	.1789	.1272	.0746	.0365	.0148
	6	.0089	.0454	.1091	.1686	.1916	.1712	.1244	.0746	.0370
	7	.0020	.0160	.0545	.1124	.1643	.1844	.1659	.1221	.0739
	8	.0004	.0046	.0222	.0609	.1144	.1614	.1797	.1623	.1201
	9	.0001	.0011	.0074	.0271	.0654	.1158	.1597	.1771	.1602
	10		.0002	.0020	.0099	.0308	.0686	.1171	.1593	.1762
	11			.0005	.0030	.0120	.0336	.0710	.1185	.1602
	12			.0001	.0008	.0039	.0136	.0355	.0727	.1201
	13				.0002	.0010	.0045	.0146	.0366	.0739
	14					.0002	.0012	.0049	.0150	.0370
	15						.0003	.0013	.0049	.0148
	16							.0003	.0013	.0046
	17								.0002	.0011
	18									.0002
	19									
	20									

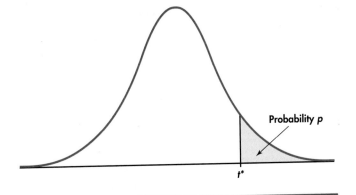

Table entry for p and C is the critical value t^* with probability p lying to its right and probability C lying between $-t^*$ and t^*.

Probability p

t^*

TABLE D t distribution critical values

df	.25	.20	.15	.10	.05	.025	.02	.01	.005	.0025	.001	.0005
						Upper tail probability p						
1	1.000	1.376	1.963	3.078	6.314	12.71	15.89	31.82	63.66	127.3	318.3	636.6
2	0.816	1.061	1.386	1.886	2.920	4.303	4.849	6.965	9.925	14.09	22.33	31.60
3	0.765	0.978	1.250	1.638	2.353	3.182	3.482	4.541	5.841	7.453	10.21	12.92
4	0.741	0.941	1.190	1.533	2.132	2.776	2.999	3.747	4.604	5.598	7.173	8.610
5	0.727	0.920	1.156	1.476	2.015	2.571	2.757	3.365	4.032	4.773	5.893	6.869
6	0.718	0.906	1.134	1.440	1.943	2.447	2.612	3.143	3.707	4.317	5.208	5.959
7	0.711	0.896	1.119	1.415	1.895	2.365	2.517	2.998	3.499	4.029	4.785	5.408
8	0.706	0.889	1.108	1.397	1.860	2.306	2.449	2.896	3.355	3.833	4.501	5.041
9	0.703	0.883	1.100	1.383	1.833	2.262	2.398	2.821	3.250	3.690	4.297	4.781
10	0.700	0.879	1.093	1.372	1.812	2.228	2.359	2.764	3.169	3.581	4.144	4.587
11	0.697	0.876	1.088	1.363	1.796	2.201	2.328	2.718	3.106	3.497	4.025	4.437
12	0.695	0.873	1.083	1.356	1.782	2.179	2.303	2.681	3.055	3.428	3.930	4.318
13	0.694	0.870	1.079	1.350	1.771	2.160	2.282	2.650	3.012	3.372	3.852	4.221
14	0.692	0.868	1.076	1.345	1.761	2.145	2.264	2.624	2.977	3.326	3.787	4.140
15	0.691	0.866	1.074	1.341	1.753	2.131	2.249	2.602	2.947	3.286	3.733	4.073
16	0.690	0.865	1.071	1.337	1.746	2.120	2.235	2.583	2.921	3.252	3.686	4.015
17	0.689	0.863	1.069	1.333	1.740	2.110	2.224	2.567	2.898	3.222	3.646	3.965
18	0.688	0.862	1.067	1.330	1.734	2.101	2.214	2.552	2.878	3.197	3.611	3.922
19	0.688	0.861	1.066	1.328	1.729	2.093	2.205	2.539	2.861	3.174	3.579	3.883
20	0.687	0.860	1.064	1.325	1.725	2.086	2.197	2.528	2.845	3.153	3.552	3.850
21	0.686	0.859	1.063	1.323	1.721	2.080	2.189	2.518	2.831	3.135	3.527	3.819
22	0.686	0.858	1.061	1.321	1.717	2.074	2.183	2.508	2.819	3.119	3.505	3.792
23	0.685	0.858	1.060	1.319	1.714	2.069	2.177	2.500	2.807	3.104	3.485	3.768
24	0.685	0.857	1.059	1.318	1.711	2.064	2.172	2.492	2.797	3.091	3.467	3.745
25	0.684	0.856	1.058	1.316	1.708	2.060	2.167	2.485	2.787	3.078	3.450	3.725
26	0.684	0.856	1.058	1.315	1.706	2.056	2.162	2.479	2.779	3.067	3.435	3.707
27	0.684	0.855	1.057	1.314	1.703	2.052	2.158	2.473	2.771	3.057	3.421	3.690
28	0.683	0.855	1.056	1.313	1.701	2.048	2.154	2.467	2.763	3.047	3.408	3.674
29	0.683	0.854	1.055	1.311	1.699	2.045	2.150	2.462	2.756	3.038	3.396	3.659
30	0.683	0.854	1.055	1.310	1.697	2.042	2.147	2.457	2.750	3.030	3.385	3.646
40	0.681	0.851	1.050	1.303	1.684	2.021	2.123	2.423	2.704	2.971	3.307	3.551
50	0.679	0.849	1.047	1.299	1.676	2.009	2.109	2.403	2.678	2.937	3.261	3.496
60	0.679	0.848	1.045	1.296	1.671	2.000	2.099	2.390	2.660	2.915	3.232	3.460
80	0.678	0.846	1.043	1.292	1.664	1.990	2.088	2.374	2.639	2.887	3.195	3.416
100	0.677	0.845	1.042	1.290	1.660	1.984	2.081	2.364	2.626	2.871	3.174	3.390
1000	0.675	0.842	1.037	1.282	1.646	1.962	2.056	2.330	2.581	2.813	3.098	3.300
z^*	0.674	0.841	1.036	1.282	1.645	1.960	2.054	2.326	2.576	2.807	3.091	3.291
	50%	60%	70%	80%	90%	95%	96%	98%	99%	99.5%	99.8%	99.9%
						Confidence level C						

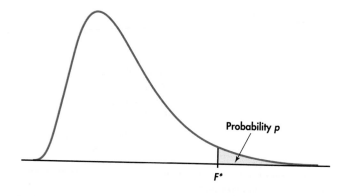

Table entry for *p* is the
critical value *F** with
probability *p* lying to
its right.

Probability *p*

F*

TABLE E *F* critical values

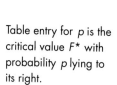

	p	\multicolumn{9}{c}{Degrees of freedom in the numerator}
		1	2	3	4	5	6	7	8	9
1	.100	39.86	49.50	53.59	55.83	57.24	58.20	58.91	59.44	59.86
	.050	161.45	199.50	215.71	224.58	230.16	233.99	236.77	238.88	240.54
	.025	647.79	799.50	864.16	899.58	921.85	937.11	948.22	956.66	963.28
	.010	4052.2	4999.5	5403.4	5624.6	5763.6	5859.0	5928.4	5981.1	6022.5
	.001	405284	500000	540379	562500	576405	585937	592873	598144	602284
2	.100	8.53	9.00	9.16	9.24	9.29	9.33	9.35	9.37	9.38
	.050	18.51	19.00	19.16	19.25	19.30	19.33	19.35	19.37	19.38
	.025	38.51	39.00	39.17	39.25	39.30	39.33	39.36	39.37	39.39
	.010	98.50	99.00	99.17	99.25	99.30	99.33	99.36	99.37	99.39
	.001	998.50	999.00	999.17	999.25	999.30	999.33	999.36	999.37	999.39
3	.100	5.54	5.46	5.39	5.34	5.31	5.28	5.27	5.25	5.24
	.050	10.13	9.55	9.28	9.12	9.01	8.94	8.89	8.85	8.81
	.025	17.44	16.04	15.44	15.10	14.88	14.73	14.62	14.54	14.47
	.010	34.12	30.82	29.46	28.71	28.24	27.91	27.67	27.49	27.35
	.001	167.03	148.50	141.11	137.10	134.58	132.85	131.58	130.62	129.86
4	.100	4.54	4.32	4.19	4.11	4.05	4.01	3.98	3.95	3.94
	.050	7.71	6.94	6.59	6.39	6.26	6.16	6.09	6.04	6.00
	.025	12.22	10.65	9.98	9.60	9.36	9.20	9.07	8.98	8.90
	.010	21.20	18.00	16.69	15.98	15.52	15.21	14.98	14.80	14.66
	.001	74.14	61.25	56.18	53.44	51.71	50.53	49.66	49.00	48.47
5	.100	4.06	3.78	3.62	3.52	3.45	3.40	3.37	3.34	3.32
	.050	6.61	5.79	5.41	5.19	5.05	4.95	4.88	4.82	4.77
	.025	10.01	8.43	7.76	7.39	7.15	6.98	6.85	6.76	6.68
	.010	16.26	13.27	12.06	11.39	10.97	10.67	10.46	10.29	10.16
	.001	47.18	37.12	33.20	31.09	29.75	28.83	28.16	27.65	27.24
6	.100	3.78	3.46	3.29	3.18	3.11	3.05	3.01	2.98	2.96
	.050	5.99	5.14	4.76	4.53	4.39	4.28	4.21	4.15	4.10
	.025	8.81	7.26	6.60	6.23	5.99	5.82	5.70	5.60	5.52
	.010	13.75	10.92	9.78	9.15	8.75	8.47	8.26	8.10	7.98
	.001	35.51	27.00	23.70	21.92	20.80	20.03	19.46	19.03	18.69
7	.100	3.59	3.26	3.07	2.96	2.88	2.83	2.78	2.75	2.72
	.050	5.59	4.74	4.35	4.12	3.97	3.87	3.79	3.73	3.68
	.025	8.07	6.54	5.89	5.52	5.29	5.12	4.99	4.90	4.82
	.010	12.25	9.55	8.45	7.85	7.46	7.19	6.99	6.84	6.72
	.001	29.25	21.69	18.77	17.20	16.21	15.52	15.02	14.63	14.33

Degrees of freedom in the denominator

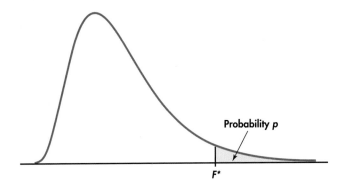

Table entry for p is the critical value F^* with probability p lying to its right.

Probability p

F^*

TABLE E F critical values (continued)

	Degrees of freedom in the numerator										
	10	12	15	20	25	30	40	50	60	120	1000
	60.19	60.71	61.22	61.74	62.05	62.26	62.53	62.69	62.79	63.06	63.30
	241.88	243.91	245.95	248.01	249.26	250.10	251.14	251.77	252.20	253.25	254.19
	968.63	976.71	984.87	993.10	998.08	1001.4	1005.6	1008.1	1009.8	1014.0	1017.7
	6055.8	6106.3	6157.3	6208.7	6239.8	6260.6	6286.8	6302.5	6313.0	6339.4	6362.7
	605621	610668	615764	620908	624017	626099	628712	630285	631337	633972	636301
	9.39	9.41	9.42	9.44	9.45	9.46	9.47	9.47	9.47	9.48	9.49
	19.40	19.41	19.43	19.45	19.46	19.46	19.47	19.48	19.48	19.49	19.49
	39.40	39.41	39.43	39.45	39.46	39.46	39.47	39.48	39.48	39.49	39.50
	99.40	99.42	99.43	99.45	99.46	99.47	99.47	99.48	99.48	99.49	99.50
	999.40	999.42	999.43	999.45	999.46	999.47	999.47	999.48	999.48	999.49	999.50
	5.23	5.22	5.20	5.18	5.17	5.17	5.16	5.15	5.15	5.14	5.13
	8.79	8.74	8.70	8.66	8.63	8.62	8.59	8.58	8.57	8.55	8.53
	14.42	14.34	14.25	14.17	14.12	14.08	14.04	14.01	13.99	13.95	13.91
	27.23	27.05	26.87	26.69	26.58	26.50	26.41	26.35	26.32	26.22	26.14
	129.25	128.32	127.37	126.42	125.84	125.45	124.96	124.66	124.47	123.97	123.53
	3.92	3.90	3.87	3.84	3.83	3.82	3.80	3.80	3.79	3.78	3.76
	5.96	5.91	5.86	5.80	5.77	5.75	5.72	5.70	5.69	5.66	5.63
	8.84	8.75	8.66	8.56	8.50	8.46	8.41	8.38	8.36	8.31	8.26
	14.55	14.37	14.20	14.02	13.91	13.84	13.75	13.69	13.65	13.56	13.47
	48.05	47.41	46.76	46.10	45.70	45.43	45.09	44.88	44.75	44.40	44.09
	3.30	3.27	3.24	3.21	3.19	3.17	3.16	3.15	3.14	3.12	3.11
	4.74	4.68	4.62	4.56	4.52	4.50	4.46	4.44	4.43	4.40	4.37
	6.62	6.52	6.43	6.33	6.27	6.23	6.18	6.14	6.12	6.07	6.02
	10.05	9.89	9.72	9.55	9.45	9.38	9.29	9.24	9.20	9.11	9.03
	26.92	26.42	25.91	25.39	25.08	24.87	24.60	24.44	24.33	24.06	23.82
	2.94	2.90	2.87	2.84	2.81	2.80	2.78	2.77	2.76	2.74	2.72
	4.06	4.00	3.94	3.87	3.83	3.81	3.77	3.75	3.74	3.70	3.67
	5.46	5.37	5.27	5.17	5.11	5.07	5.01	4.98	4.96	4.90	4.86
	7.87	7.72	7.56	7.40	7.30	7.23	7.14	7.09	7.06	6.97	6.89
	18.41	17.99	17.56	17.12	16.85	16.67	16.44	16.31	16.21	15.98	15.77
	2.70	2.67	2.63	2.59	2.57	2.56	2.54	2.52	2.51	2.49	2.47
	3.64	3.57	3.51	3.44	3.40	3.38	3.34	3.32	3.30	3.27	3.23
	4.76	4.67	4.57	4.47	4.40	4.36	4.31	4.28	4.25	4.20	4.15
	6.62	6.47	6.31	6.16	6.06	5.99	5.91	5.86	5.82	5.74	5.66
	14.08	13.71	13.32	12.93	12.69	12.53	12.33	12.20	12.12	11.91	11.72

TABLE E *F* critical values *(continued)*

| | *p* | \multicolumn{9}{c}{Degrees of freedom in the numerator} |
		1	2	3	4	5	6	7	8	9
8	.100	3.46	3.11	2.92	2.81	2.73	2.67	2.62	2.59	2.56
	.050	5.32	4.46	4.07	3.84	3.69	3.58	3.50	3.44	3.39
	.025	7.57	6.06	5.42	5.05	4.82	4.65	4.53	4.43	4.36
	.010	11.26	8.65	7.59	7.01	6.63	6.37	6.18	6.03	5.91
	.001	25.41	18.49	15.83	14.39	13.48	12.86	12.40	12.05	11.77
9	.100	3.36	3.01	2.81	2.69	2.61	2.55	2.51	2.47	2.44
	.050	5.12	4.26	3.86	3.63	3.48	3.37	3.29	3.23	3.18
	.025	7.21	5.71	5.08	4.72	4.48	4.32	4.20	4.10	4.03
	.010	10.56	8.02	6.99	6.42	6.06	5.80	5.61	5.47	5.35
	.001	22.86	16.39	13.90	12.56	11.71	11.13	10.70	10.37	10.11
10	.100	3.29	2.92	2.73	2.61	2.52	2.46	2.41	2.38	2.35
	.050	4.96	4.10	3.71	3.48	3.33	3.22	3.14	3.07	3.02
	.025	6.94	5.46	4.83	4.47	4.24	4.07	3.95	3.85	3.78
	.010	10.04	7.56	6.55	5.99	5.64	5.39	5.20	5.06	4.94
	.001	21.04	14.91	12.55	11.28	10.48	9.93	9.52	9.20	8.96
11	.100	3.23	2.86	2.66	2.54	2.45	2.39	2.34	2.30	2.27
	.050	4.84	3.98	3.59	3.36	3.20	3.09	3.01	2.95	2.90
	.025	6.72	5.26	4.63	4.28	4.04	3.88	3.76	3.66	3.59
	.010	9.65	7.21	6.22	5.67	5.32	5.07	4.89	4.74	4.63
	.001	19.69	13.81	11.56	10.35	9.58	9.05	8.66	8.35	8.12
12	.100	3.18	2.81	2.61	2.48	2.39	2.33	2.28	2.24	2.21
	.050	4.75	3.89	3.49	3.26	3.11	3.00	2.91	2.85	2.80
	.025	6.55	5.10	4.47	4.12	3.89	3.73	3.61	3.51	3.44
	.010	9.33	6.93	5.95	5.41	5.06	4.82	4.64	4.50	4.39
	.001	18.64	12.97	10.80	9.63	8.89	8.38	8.00	7.71	7.48
13	.100	3.14	2.76	2.56	2.43	2.35	2.28	2.23	2.20	2.16
	.050	4.67	3.81	3.41	3.18	3.03	2.92	2.83	2.77	2.71
	.025	6.41	4.97	4.35	4.00	3.77	3.60	3.48	3.39	3.31
	.010	9.07	6.70	5.74	5.21	4.86	4.62	4.44	4.30	4.19
	.001	17.82	12.31	10.21	9.07	8.35	7.86	7.49	7.21	6.98
14	.100	3.10	2.73	2.52	2.39	2.31	2.24	2.19	2.15	2.12
	.050	4.60	3.74	3.34	3.11	2.96	2.85	2.76	2.70	2.65
	.025	6.30	4.86	4.24	3.89	3.66	3.50	3.38	3.29	3.21
	.010	8.86	6.51	5.56	5.04	4.69	4.46	4.28	4.14	4.03
	.001	17.14	11.78	9.73	8.62	7.92	7.44	7.08	6.80	6.58
15	.100	3.07	2.70	2.49	2.36	2.27	2.21	2.16	2.12	2.09
	.050	4.54	3.68	3.29	3.06	2.90	2.79	2.71	2.64	2.59
	.025	6.20	4.77	4.15	3.80	3.58	3.41	3.29	3.20	3.12
	.010	8.68	6.36	5.42	4.89	4.56	4.32	4.14	4.00	3.89
	.001	16.59	11.34	9.34	8.25	7.57	7.09	6.74	6.47	6.26
16	.100	3.05	2.67	2.46	2.33	2.24	2.18	2.13	2.09	2.06
	.050	4.49	3.63	3.24	3.01	2.85	2.74	2.66	2.59	2.54
	.025	6.12	4.69	4.08	3.73	3.50	3.34	3.22	3.12	3.05
	.010	8.53	6.23	5.29	4.77	4.44	4.20	4.03	3.89	3.78
	.001	16.12	10.97	9.01	7.94	7.27	6.80	6.46	6.19	5.98
17	.100	3.03	2.64	2.44	2.31	2.22	2.15	2.10	2.06	2.03
	.050	4.45	3.59	3.20	2.96	2.81	2.70	2.61	2.55	2.49
	.025	6.04	4.62	4.01	3.66	3.44	3.28	3.16	3.06	2.98
	.010	8.40	6.11	5.19	4.67	4.34	4.10	3.93	3.79	3.68
	.001	15.72	10.66	8.73	7.68	7.02	6.56	6.22	5.96	5.75

Degrees of freedom in the denominator

TABLE E *F critical values (continued)*

	Degrees of freedom in the numerator									
10	12	15	20	25	30	40	50	60	120	1000
2.54	2.50	2.46	2.42	2.40	2.38	2.36	2.35	2.34	2.32	2.30
3.35	3.28	3.22	3.15	3.11	3.08	3.04	3.02	3.01	2.97	2.93
4.30	4.20	4.10	4.00	3.94	3.89	3.84	3.81	3.78	3.73	3.68
5.81	5.67	5.52	5.36	5.26	5.20	5.12	5.07	5.03	4.95	4.87
11.54	11.19	10.84	10.48	10.26	10.11	9.92	9.80	9.73	9.53	9.36
2.42	2.38	2.34	2.30	2.27	2.25	2.23	2.22	2.21	2.18	2.16
3.14	3.07	3.01	2.94	2.89	2.86	2.83	2.80	2.79	2.75	2.71
3.96	3.87	3.77	3.67	3.60	3.56	3.51	3.47	3.45	3.39	3.34
5.26	5.11	4.96	4.81	4.71	4.65	4.57	4.52	4.48	4.40	4.32
9.89	9.57	9.24	8.90	8.69	8.55	8.37	8.26	8.19	8.00	7.84
2.32	2.28	2.24	2.20	2.17	2.16	2.13	2.12	2.11	2.08	2.06
2.98	2.91	2.85	2.77	2.73	2.70	2.66	2.64	2.62	2.58	2.54
3.72	3.62	3.52	3.42	3.35	3.31	3.26	3.22	3.20	3.14	3.09
4.85	4.71	4.56	4.41	4.31	4.25	4.17	4.12	4.08	4.00	3.92
8.75	8.45	8.13	7.80	7.60	7.47	7.30	7.19	7.12	6.94	6.78
2.25	2.21	2.17	2.12	2.10	2.08	2.05	2.04	2.03	2.00	1.98
2.85	2.79	2.72	2.65	2.60	2.57	2.53	2.51	2.49	2.45	2.41
3.53	3.43	3.33	3.23	3.16	3.12	3.06	3.03	3.00	2.94	2.89
4.54	4.40	4.25	4.10	4.01	3.94	3.86	3.81	3.78	3.69	3.61
7.92	7.63	7.32	7.01	6.81	6.68	6.52	6.42	6.35	6.18	6.02
2.19	2.15	2.10	2.06	2.03	2.01	1.99	1.97	1.96	1.93	1.91
2.75	2.69	2.62	2.54	2.50	2.47	2.43	2.40	2.38	2.34	2.30
3.37	3.28	3.18	3.07	3.01	2.96	2.91	2.87	2.85	2.79	2.73
4.30	4.16	4.01	3.86	3.76	3.70	3.62	3.57	3.54	3.45	3.37
7.29	7.00	6.71	6.40	6.22	6.09	5.93	5.83	5.76	5.59	5.44
2.14	2.10	2.05	2.01	1.98	1.96	1.93	1.92	1.90	1.88	1.85
2.67	2.60	2.53	2.46	2.41	2.38	2.34	2.31	2.30	2.25	2.21
3.25	3.15	3.05	2.95	2.88	2.84	2.78	2.74	2.72	2.66	2.60
4.10	3.96	3.82	3.66	3.57	3.51	3.43	3.38	3.34	3.25	3.18
6.80	6.52	6.23	5.93	5.75	5.63	5.47	5.37	5.30	5.14	4.99
2.10	2.05	2.01	1.96	1.93	1.91	1.89	1.87	1.86	1.83	1.80
2.60	2.53	2.46	2.39	2.34	2.31	2.27	2.24	2.22	2.18	2.14
3.15	3.05	2.95	2.84	2.78	2.73	2.67	2.64	2.61	2.55	2.50
3.94	3.80	3.66	3.51	3.41	3.35	3.27	3.22	3.18	3.09	3.02
6.40	6.13	5.85	5.56	5.38	5.25	5.10	5.00	4.94	4.77	4.62
2.06	2.02	1.97	1.92	1.89	1.87	1.85	1.83	1.82	1.79	1.76
2.54	2.48	2.40	2.33	2.28	2.25	2.20	2.18	2.16	2.11	2.07
3.06	2.96	2.86	2.76	2.69	2.64	2.59	2.55	2.52	2.46	2.40
3.80	3.67	3.52	3.37	3.28	3.21	3.13	3.08	3.05	2.96	2.88
6.08	5.81	5.54	5.25	5.07	4.95	4.80	4.70	4.64	4.47	4.33
2.03	1.99	1.94	1.89	1.86	1.84	1.81	1.79	1.78	1.75	1.72
2.49	2.42	2.35	2.28	2.23	2.19	2.15	2.12	2.11	2.06	2.02
2.99	2.89	2.79	2.68	2.61	2.57	2.51	2.47	2.45	2.38	2.32
3.69	3.55	3.41	3.26	3.16	3.10	3.02	2.97	2.93	2.84	2.76
5.81	5.55	5.27	4.99	4.82	4.70	4.54	4.45	4.39	4.23	4.08
2.00	1.96	1.91	1.86	1.83	1.81	1.78	1.76	1.75	1.72	1.69
2.45	2.38	2.31	2.23	2.18	2.15	2.10	2.08	2.06	2.01	1.97
2.92	2.82	2.72	2.62	2.55	2.50	2.44	2.41	2.38	2.32	2.26
3.59	3.46	3.31	3.16	3.07	3.00	2.92	2.87	2.83	2.75	2.66
5.58	5.32	5.05	4.78	4.60	4.48	4.33	4.24	4.18	4.02	3.87

TABLE E F critical values (continued)

				Degrees of freedom in the numerator					
	p	1	2	3	4	5	6	7	8	9
18	.100	3.01	2.62	2.42	2.29	2.20	2.13	2.08	2.04	2.00
	.050	4.41	3.55	3.16	2.93	2.77	2.66	2.58	2.51	2.46
	.025	5.98	4.56	3.95	3.61	3.38	3.22	3.10	3.01	2.93
	.010	8.29	6.01	5.09	4.58	4.25	4.01	3.84	3.71	3.60
	.001	15.38	10.39	8.49	7.46	6.81	6.35	6.02	5.76	5.56
19	.100	2.99	2.61	2.40	2.27	2.18	2.11	2.06	2.02	1.98
	.050	4.38	3.52	3.13	2.90	2.74	2.63	2.54	2.48	2.42
	.025	5.92	4.51	3.90	3.56	3.33	3.17	3.05	2.96	2.88
	.010	8.18	5.93	5.01	4.50	4.17	3.94	3.77	3.63	3.52
	.001	15.08	10.16	8.28	7.27	6.62	6.18	5.85	5.59	5.39
20	.100	2.97	2.59	2.38	2.25	2.16	2.09	2.04	2.00	1.96
	.050	4.35	3.49	3.10	2.87	2.71	2.60	2.51	2.45	2.39
	.025	5.87	4.46	3.86	3.51	3.29	3.13	3.01	2.91	2.84
	.010	8.10	5.85	4.94	4.43	4.10	3.87	3.70	3.56	3.46
	.001	14.82	9.95	8.10	7.10	6.46	6.02	5.69	5.44	5.24
21	.100	2.96	2.57	2.36	2.23	2.14	2.08	2.02	1.98	1.95
	.050	4.32	3.47	3.07	2.84	2.68	2.57	2.49	2.42	2.37
	.025	5.83	4.42	3.82	3.48	3.25	3.09	2.97	2.87	2.80
	.010	8.02	5.78	4.87	4.37	4.04	3.81	3.64	3.51	3.40
	.001	14.59	9.77	7.94	6.95	6.32	5.88	5.56	5.31	5.11
22	.100	2.95	2.56	2.35	2.22	2.13	2.06	2.01	1.97	1.93
	.050	4.30	3.44	3.05	2.82	2.66	2.55	2.46	2.40	2.34
	.025	5.79	4.38	3.78	3.44	3.22	3.05	2.93	2.84	2.76
	.010	7.95	5.72	4.82	4.31	3.99	3.76	3.59	3.45	3.35
	.001	14.38	9.61	7.80	6.81	6.19	5.76	5.44	5.19	4.99
23	.100	2.94	2.55	2.34	2.21	2.11	2.05	1.99	1.95	1.92
	.050	4.28	3.42	3.03	2.80	2.64	2.53	2.44	2.37	2.32
	.025	5.75	4.35	3.75	3.41	3.18	3.02	2.90	2.81	2.73
	.010	7.88	5.66	4.76	4.26	3.94	3.71	3.54	3.41	3.30
	.001	14.20	9.47	7.67	6.70	6.08	5.65	5.33	5.09	4.89
24	.100	2.93	2.54	2.33	2.19	2.10	2.04	1.98	1.94	1.91
	.050	4.26	3.40	3.01	2.78	2.62	2.51	2.42	2.36	2.30
	.025	5.72	4.32	3.72	3.38	3.15	2.99	2.87	2.78	2.70
	.010	7.82	5.61	4.72	4.22	3.90	3.67	3.50	3.36	3.26
	.001	14.03	9.34	7.55	6.59	5.98	5.55	5.23	4.99	4.80
25	.100	2.92	2.53	2.32	2.18	2.09	2.02	1.97	1.93	1.89
	.050	4.24	3.39	2.99	2.76	2.60	2.49	2.40	2.34	2.28
	.025	5.69	4.29	3.69	3.35	3.13	2.97	2.85	2.75	2.68
	.010	7.77	5.57	4.68	4.18	3.85	3.63	3.46	3.32	3.22
	.001	13.88	9.22	7.45	6.49	5.89	5.46	5.15	4.91	4.71
26	.100	2.91	2.52	2.31	2.17	2.08	2.01	1.96	1.92	1.88
	.050	4.23	3.37	2.98	2.74	2.59	2.47	2.39	2.32	2.27
	.025	5.66	4.27	3.67	3.33	3.10	2.94	2.82	2.73	2.65
	.010	7.72	5.53	4.64	4.14	3.82	3.59	3.42	3.29	3.18
	.001	13.74	9.12	7.36	6.41	5.80	5.38	5.07	4.83	4.64
27	.100	2.90	2.51	2.30	2.17	2.07	2.00	1.95	1.91	1.87
	.050	4.21	3.35	2.96	2.73	2.57	2.46	2.37	2.31	2.25
	.025	5.63	4.24	3.65	3.31	3.08	2.92	2.80	2.71	2.63
	.010	7.68	5.49	4.60	4.11	3.78	3.56	3.39	3.26	3.15
	.001	13.61	9.02	7.27	6.33	5.73	5.31	5.00	4.76	4.57

Degrees of freedom in the denominator

TABLE E *F* critical values *(continued)*

| Degrees of freedom in the numerator | | | | | | | | | | |
10	12	15	20	25	30	40	50	60	120	1000
1.98	1.93	1.89	1.84	1.80	1.78	1.75	1.74	1.72	1.69	1.66
2.41	2.34	2.27	2.19	2.14	2.11	2.06	2.04	2.02	1.97	1.92
2.87	2.77	2.67	2.56	2.49	2.44	2.38	2.35	2.32	2.26	2.20
3.51	3.37	3.23	3.08	2.98	2.92	2.84	2.78	2.75	2.66	2.58
5.39	5.13	4.87	4.59	4.42	4.30	4.15	4.06	4.00	3.84	3.69
1.96	1.91	1.86	1.81	1.78	1.76	1.73	1.71	1.70	1.67	1.64
2.38	2.31	2.23	2.16	2.11	2.07	2.03	2.00	1.98	1.93	1.88
2.82	2.72	2.62	2.51	2.44	2.39	2.33	2.30	2.27	2.20	2.14
3.43	3.30	3.15	3.00	2.91	2.84	2.76	2.71	2.67	2.58	2.50
5.22	4.97	4.70	4.43	4.26	4.14	3.99	3.90	3.84	3.68	3.53
1.94	1.89	1.84	1.79	1.76	1.74	1.71	1.69	1.68	1.64	1.61
2.35	2.28	2.20	2.12	2.07	2.04	1.99	1.97	1.95	1.90	1.85
2.77	2.68	2.57	2.46	2.40	2.35	2.29	2.25	2.22	2.16	2.09
3.37	3.23	3.09	2.94	2.84	2.78	2.69	2.64	2.61	2.52	2.43
5.08	4.82	4.56	4.29	4.12	4.00	3.86	3.77	3.70	3.54	3.40
1.92	1.87	1.83	1.78	1.74	1.72	1.69	1.67	1.66	1.62	1.59
2.32	2.25	2.18	2.10	2.05	2.01	1.96	1.94	1.92	1.87	1.82
2.73	2.64	2.53	2.42	2.36	2.31	2.25	2.21	2.18	2.11	2.05
3.31	3.17	3.03	2.88	2.79	2.72	2.64	2.58	2.55	2.46	2.37
4.95	4.70	4.44	4.17	4.00	3.88	3.74	3.64	3.58	3.42	3.28
1.90	1.86	1.81	1.76	1.73	1.70	1.67	1.65	1.64	1.60	1.57
2.30	2.23	2.15	2.07	2.02	1.98	1.94	1.91	1.89	1.84	1.79
2.70	2.60	2.50	2.39	2.32	2.27	2.21	2.17	2.14	2.08	2.01
3.26	3.12	2.98	2.83	2.73	2.67	2.58	2.53	2.50	2.40	2.32
4.83	4.58	4.33	4.06	3.89	3.78	3.63	3.54	3.48	3.32	3.17
1.89	1.84	1.80	1.74	1.71	1.69	1.66	1.64	1.62	1.59	1.55
2.27	2.20	2.13	2.05	2.00	1.96	1.91	1.88	1.86	1.81	1.76
2.67	2.57	2.47	2.36	2.29	2.24	2.18	2.14	2.11	2.04	1.98
3.21	3.07	2.93	2.78	2.69	2.62	2.54	2.48	2.45	2.35	2.27
4.73	4.48	4.23	3.96	3.79	3.68	3.53	3.44	3.38	3.22	3.08
1.88	1.83	1.78	1.73	1.70	1.67	1.64	1.62	1.61	1.57	1.54
2.25	2.18	2.11	2.03	1.97	1.94	1.89	1.86	1.84	1.79	1.74
2.64	2.54	2.44	2.33	2.26	2.21	2.15	2.11	2.08	2.01	1.94
3.17	3.03	2.89	2.74	2.64	2.58	2.49	2.44	2.40	2.31	2.22
4.64	4.39	4.14	3.87	3.71	3.59	3.45	3.36	3.29	3.14	2.99
1.87	1.82	1.77	1.72	1.68	1.66	1.63	1.61	1.59	1.56	1.52
2.24	2.16	2.09	2.01	1.96	1.92	1.87	1.84	1.82	1.77	1.72
2.61	2.51	2.41	2.30	2.23	2.18	2.12	2.08	2.05	1.98	1.91
3.13	2.99	2.85	2.70	2.60	2.54	2.45	2.40	2.36	2.27	2.18
4.56	4.31	4.06	3.79	3.63	3.52	3.37	3.28	3.22	3.06	2.91
1.86	1.81	1.76	1.71	1.67	1.65	1.61	1.59	1.58	1.54	1.51
2.22	2.15	2.07	1.99	1.94	1.90	1.85	1.82	1.80	1.75	1.70
2.59	2.49	2.39	2.28	2.21	2.16	2.09	2.05	2.03	1.95	1.89
3.09	2.96	2.81	2.66	2.57	2.50	2.42	2.36	2.33	2.23	2.14
4.48	4.24	3.99	3.72	3.56	3.44	3.30	3.21	3.15	2.99	2.84
1.85	1.80	1.75	1.70	1.66	1.64	1.60	1.58	1.57	1.53	1.50
2.20	2.13	2.06	1.97	1.92	1.88	1.84	1.81	1.79	1.73	1.68
2.57	2.47	2.36	2.25	2.18	2.13	2.07	2.03	2.00	1.93	1.86
3.06	2.93	2.78	2.63	2.54	2.47	2.38	2.33	2.29	2.20	2.11
4.41	4.17	3.92	3.66	3.49	3.38	3.23	3.14	3.08	2.92	2.78

TABLE E *F* critical values *(continued)*

			Degrees of freedom in the numerator								
		p	1	2	3	4	5	6	7	8	9
Degrees of freedom in the denominator	28	.100	2.89	2.50	2.29	2.16	2.06	2.00	1.94	1.90	1.87
		.050	4.20	3.34	2.95	2.71	2.56	2.45	2.36	2.29	2.24
		.025	5.61	4.22	3.63	3.29	3.06	2.90	2.78	2.69	2.61
		.010	7.64	5.45	4.57	4.07	3.75	3.53	3.36	3.23	3.12
		.001	13.50	8.93	7.19	6.25	5.66	5.24	4.93	4.69	4.50
	29	.100	2.89	2.50	2.28	2.15	2.06	1.99	1.93	1.89	1.86
		.050	4.18	3.33	2.93	2.70	2.55	2.43	2.35	2.28	2.22
		.025	5.59	4.20	3.61	3.27	3.04	2.88	2.76	2.67	2.59
		.010	7.60	5.42	4.54	4.04	3.73	3.50	3.33	3.20	3.09
		.001	13.39	8.85	7.12	6.19	5.59	5.18	4.87	4.64	4.45
	30	.100	2.88	2.49	2.28	2.14	2.05	1.98	1.93	1.88	1.85
		.050	4.17	3.32	2.92	2.69	2.53	2.42	2.33	2.27	2.21
		.025	5.57	4.18	3.59	3.25	3.03	2.87	2.75	2.65	2.57
		.010	7.56	5.39	4.51	4.02	3.70	3.47	3.30	3.17	3.07
		.001	13.29	8.77	7.05	6.12	5.53	5.12	4.82	4.58	4.39
	40	.100	2.84	2.44	2.23	2.09	2.00	1.93	1.87	1.83	1.79
		.050	4.08	3.23	2.84	2.61	2.45	2.34	2.25	2.18	2.12
		.025	5.42	4.05	3.46	3.13	2.90	2.74	2.62	2.53	2.45
		.010	7.31	5.18	4.31	3.83	3.51	3.29	3.12	2.99	2.89
		.001	12.61	8.25	6.59	5.70	5.13	4.73	4.44	4.21	4.02
	50	.100	2.81	2.41	2.20	2.06	1.97	1.90	1.84	1.80	1.76
		.050	4.03	3.18	2.79	2.56	2.40	2.29	2.20	2.13	2.07
		.025	5.34	3.97	3.39	3.05	2.83	2.67	2.55	2.46	2.38
		.010	7.17	5.06	4.20	3.72	3.41	3.19	3.02	2.89	2.78
		.001	12.22	7.96	6.34	5.46	4.90	4.51	4.22	4.00	3.82
	60	.100	2.79	2.39	2.18	2.04	1.95	1.87	1.82	1.77	1.74
		.050	4.00	3.15	2.76	2.53	2.37	2.25	2.17	2.10	2.04
		.025	5.29	3.93	3.34	3.01	2.79	2.63	2.51	2.41	2.33
		.010	7.08	4.98	4.13	3.65	3.34	3.12	2.95	2.82	2.72
		.001	11.97	7.77	6.17	5.31	4.76	4.37	4.09	3.86	3.69
	100	.100	2.76	2.36	2.14	2.00	1.91	1.83	1.78	1.73	1.69
		.050	3.94	3.09	2.70	2.46	2.31	2.19	2.10	2.03	1.97
		.025	5.18	3.83	3.25	2.92	2.70	2.54	2.42	2.32	2.24
		.010	6.90	4.82	3.98	3.51	3.21	2.99	2.82	2.69	2.59
		.001	11.50	7.41	5.86	5.02	4.48	4.11	3.83	3.61	3.44
	200	.100	2.73	2.33	2.11	1.97	1.88	1.80	1.75	1.70	1.66
		.050	3.89	3.04	2.65	2.42	2.26	2.14	2.06	1.98	1.93
		.025	5.10	3.76	3.18	2.85	2.63	2.47	2.35	2.26	2.18
		.010	6.76	4.71	3.88	3.41	3.11	2.89	2.73	2.60	2.50
		.001	11.15	7.15	5.63	4.81	4.29	3.92	3.65	3.43	3.26
	1000	.100	2.71	2.31	2.09	1.95	1.85	1.78	1.72	1.68	1.64
		.050	3.85	3.00	2.61	2.38	2.22	2.11	2.02	1.95	1.89
		.025	5.04	3.70	3.13	2.80	2.58	2.42	2.30	2.20	2.13
		.010	6.66	4.63	3.80	3.34	3.04	2.82	2.66	2.53	2.43
		.001	10.89	6.96	5.46	4.65	4.14	3.78	3.51	3.30	3.13

TABLE E F critical values (continued)

				Degrees of freedom in the numerator						
10	12	15	20	25	30	40	50	60	120	1000
1.84	1.79	1.74	1.69	1.65	1.63	1.59	1.57	1.56	1.52	1.48
2.19	2.12	2.04	1.96	1.91	1.87	1.82	1.79	1.77	1.71	1.66
2.55	2.45	2.34	2.23	2.16	2.11	2.05	2.01	1.98	1.91	1.84
3.03	2.90	2.75	2.60	2.51	2.44	2.35	2.30	2.26	2.17	2.08
4.35	4.11	3.86	3.60	3.43	3.32	3.18	3.09	3.02	2.86	2.72
1.83	1.78	1.73	1.68	1.64	1.62	1.58	1.56	1.55	1.51	1.47
2.18	2.10	2.03	1.94	1.89	1.85	1.81	1.77	1.75	1.70	1.65
2.53	2.43	2.32	2.21	2.14	2.09	2.03	1.99	1.96	1.89	1.82
3.00	2.87	2.73	2.57	2.48	2.41	2.33	2.27	2.23	2.14	2.05
4.29	4.05	3.80	3.54	3.38	3.27	3.12	3.03	2.97	2.81	2.66
1.82	1.77	1.72	1.67	1.63	1.61	1.57	1.55	1.54	1.50	1.46
2.16	2.09	2.01	1.93	1.88	1.84	1.79	1.76	1.74	1.68	1.63
2.51	2.41	2.31	2.20	2.12	2.07	2.01	1.97	1.94	1.87	1.80
2.98	2.84	2.70	2.55	2.45	2.39	2.30	2.25	2.21	2.11	2.02
4.24	4.00	3.75	3.49	3.33	3.22	3.07	2.98	2.92	2.76	2.61
1.76	1.71	1.66	1.61	1.57	1.54	1.51	1.48	1.47	1.42	1.38
2.08	2.00	1.92	1.84	1.78	1.74	1.69	1.66	1.64	1.58	1.52
2.39	2.29	2.18	2.07	1.99	1.94	1.88	1.83	1.80	1.72	1.65
2.80	2.66	2.52	2.37	2.27	2.20	2.11	2.06	2.02	1.92	1.82
3.87	3.64	3.40	3.14	2.98	2.87	2.73	2.64	2.57	2.41	2.25
1.73	1.68	1.63	1.57	1.53	1.50	1.46	1.44	1.42	1.38	1.33
2.03	1.95	1.87	1.78	1.73	1.69	1.63	1.60	1.58	1.51	1.45
2.32	2.22	2.11	1.99	1.92	1.87	1.80	1.75	1.72	1.64	1.56
2.70	2.56	2.42	2.27	2.17	2.10	2.01	1.95	1.91	1.80	1.70
3.67	3.44	3.20	2.95	2.79	2.68	2.53	2.44	2.38	2.21	2.05
1.71	1.66	1.60	1.54	1.50	1.48	1.44	1.41	1.40	1.35	1.30
1.99	1.92	1.84	1.75	1.69	1.65	1.59	1.56	1.53	1.47	1.40
2.27	2.17	2.06	1.94	1.87	1.82	1.74	1.70	1.67	1.58	1.49
2.63	2.50	2.35	2.20	2.10	2.03	1.94	1.88	1.84	1.73	1.62
3.54	3.32	3.08	2.83	2.67	2.55	2.41	2.32	2.25	2.08	1.92
1.66	1.61	1.56	1.49	1.45	1.42	1.38	1.35	1.34	1.28	1.22
1.93	1.85	1.77	1.68	1.62	1.57	1.52	1.48	1.45	1.38	1.30
2.18	2.08	1.97	1.85	1.77	1.71	1.64	1.59	1.56	1.46	1.36
2.50	2.37	2.22	2.07	1.97	1.89	1.80	1.74	1.69	1.57	1.45
3.30	3.07	2.84	2.59	2.43	2.32	2.17	2.08	2.01	1.83	1.64
1.63	1.58	1.52	1.46	1.41	1.38	1.34	1.31	1.29	1.23	1.16
1.88	1.80	1.72	1.62	1.56	1.52	1.46	1.41	1.39	1.30	1.21
2.11	2.01	1.90	1.78	1.70	1.64	1.56	1.51	1.47	1.37	1.25
2.41	2.27	2.13	1.97	1.87	1.79	1.69	1.63	1.58	1.45	1.30
3.12	2.90	2.67	2.42	2.26	2.15	2.00	1.90	1.83	1.64	1.43
1.61	1.55	1.49	1.43	1.38	1.35	1.30	1.27	1.25	1.18	1.08
1.84	1.76	1.68	1.58	1.52	1.47	1.41	1.36	1.33	1.24	1.11
2.06	1.96	1.85	1.72	1.64	1.58	1.50	1.45	1.41	1.29	1.13
2.34	2.20	2.06	1.90	1.79	1.72	1.61	1.54	1.50	1.35	1.16
2.99	2.77	2.54	2.30	2.14	2.02	1.87	1.77	1.69	1.49	1.22

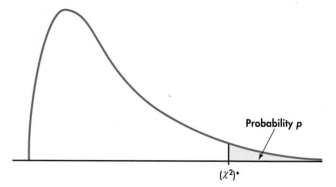

Table entry for *p* is the critical value $(\chi^2)^*$ with probability *p* lying to its right.

Probability *p*

$(\chi^2)^*$

TABLE F χ^2 distribution critical values

df	Tail probability *p*											
	.25	.20	.15	.10	.05	.025	.02	.01	.005	.0025	.001	.0005
1	1.32	1.64	2.07	2.71	3.84	5.02	5.41	6.63	7.88	9.14	10.83	12.12
2	2.77	3.22	3.79	4.61	5.99	7.38	7.82	9.21	10.60	11.98	13.82	15.20
3	4.11	4.64	5.32	6.25	7.81	9.35	9.84	11.34	12.84	14.32	16.27	17.73
4	5.39	5.99	6.74	7.78	9.49	11.14	11.67	13.28	14.86	16.42	18.47	20.00
5	6.63	7.29	8.12	9.24	11.07	12.83	13.39	15.09	16.75	18.39	20.51	22.11
6	7.84	8.56	9.45	10.64	12.59	14.45	15.03	16.81	18.55	20.25	22.46	24.10
7	9.04	9.80	10.75	12.02	14.07	16.01	16.62	18.48	20.28	22.04	24.32	26.02
8	10.22	11.03	12.03	13.36	15.51	17.53	18.17	20.09	21.95	23.77	26.12	27.87
9	11.39	12.24	13.29	14.68	16.92	19.02	19.68	21.67	23.59	25.46	27.88	29.67
10	12.55	13.44	14.53	15.99	18.31	20.48	21.16	23.21	25.19	27.11	29.59	31.42
11	13.70	14.63	15.77	17.28	19.68	21.92	22.62	24.72	26.76	28.73	31.26	33.14
12	14.85	15.81	16.99	18.55	21.03	23.34	24.05	26.22	28.30	30.32	32.91	34.82
13	15.98	16.98	18.20	19.81	22.36	24.74	25.47	27.69	29.82	31.88	34.53	36.48
14	17.12	18.15	19.41	21.06	23.68	26.12	26.87	29.14	31.32	33.43	36.12	38.11
15	18.25	19.31	20.60	22.31	25.00	27.49	28.26	30.58	32.80	34.95	37.70	39.72
16	19.37	20.47	21.79	23.54	26.30	28.85	29.63	32.00	34.27	36.46	39.25	41.31
17	20.49	21.61	22.98	24.77	27.59	30.19	31.00	33.41	35.72	37.95	40.79	42.88
18	21.60	22.76	24.16	25.99	28.87	31.53	32.35	34.81	37.16	39.42	42.31	44.43
19	22.72	23.90	25.33	27.20	30.14	32.85	33.69	36.19	38.58	40.88	43.82	45.97
20	23.83	25.04	26.50	28.41	31.41	34.17	35.02	37.57	40.00	42.34	45.31	47.50
21	24.93	26.17	27.66	29.62	32.67	35.48	36.34	38.93	41.40	43.78	46.80	49.01
22	26.04	27.30	28.82	30.81	33.92	36.78	37.66	40.29	42.80	45.20	48.27	50.51
23	27.14	28.43	29.98	32.01	35.17	38.08	38.97	41.64	44.18	46.62	49.73	52.00
24	28.24	29.55	31.13	33.20	36.42	39.36	40.27	42.98	45.56	48.03	51.18	53.48
25	29.34	30.68	32.28	34.38	37.65	40.65	41.57	44.31	46.93	49.44	52.62	54.95
26	30.43	31.79	33.43	35.56	38.89	41.92	42.86	45.64	48.29	50.83	54.05	56.41
27	31.53	32.91	34.57	36.74	40.11	43.19	44.14	46.96	49.64	52.22	55.48	57.86
28	32.62	34.03	35.71	37.92	41.34	44.46	45.42	48.28	50.99	53.59	56.89	59.30
29	33.71	35.14	36.85	39.09	42.56	45.72	46.69	49.59	52.34	54.97	58.30	60.73
30	34.80	36.25	37.99	40.26	43.77	46.98	47.96	50.89	53.67	56.33	59.70	62.16
40	45.62	47.27	49.24	51.81	55.76	59.34	60.44	63.69	66.77	69.70	73.40	76.09
50	56.33	58.16	60.35	63.17	67.50	71.42	72.61	76.15	79.49	82.66	86.66	89.56
60	66.98	68.97	71.34	74.40	79.08	83.30	84.58	88.38	91.95	95.34	99.61	102.7
80	88.13	90.41	93.11	96.58	101.9	106.6	108.1	112.3	116.3	120.1	124.8	128.3
100	109.1	111.7	114.7	118.5	124.3	129.6	131.1	135.8	140.2	144.3	149.4	153.2

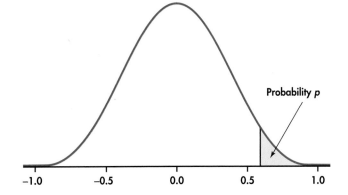

Table entry for *p* is
the value of *r* with
probability *p* lying
above it.

TABLE G Critical values of the correlation *r*

n	Tail probability *p*									
	.20	.10	.05	.025	.02	.01	.005	.0025	.001	.0005
3	0.8090	0.9511	0.9877	0.9969	0.9980	0.9995	0.9999	1.0000	1.0000	1.0000
4	0.6000	0.8000	0.9000	0.9500	0.9600	0.9800	0.9900	0.9950	0.9980	0.9990
5	0.4919	0.6870	0.8054	0.8783	0.8953	0.9343	0.9587	0.9740	0.9859	0.9911
6	0.4257	0.6084	0.7293	0.8114	0.8319	0.8822	0.9172	0.9417	0.9633	0.9741
7	0.3803	0.5509	0.6694	0.7545	0.7766	0.8329	0.8745	0.9056	0.9350	0.9509
8	0.3468	0.5067	0.6215	0.7067	0.7295	0.7887	0.8343	0.8697	0.9049	0.9249
9	0.3208	0.4716	0.5822	0.6664	0.6892	0.7498	0.7977	0.8359	0.8751	0.8983
10	0.2998	0.4428	0.5494	0.6319	0.6546	0.7155	0.7646	0.8046	0.8467	0.8721
11	0.2825	0.4187	0.5214	0.6021	0.6244	0.6851	0.7348	0.7759	0.8199	0.8470
12	0.2678	0.3981	0.4973	0.5760	0.5980	0.6581	0.7079	0.7496	0.7950	0.8233
13	0.2552	0.3802	0.4762	0.5529	0.5745	0.6339	0.6835	0.7255	0.7717	0.8010
14	0.2443	0.3646	0.4575	0.5324	0.5536	0.6120	0.6614	0.7034	0.7501	0.7800
15	0.2346	0.3507	0.4409	0.5140	0.5347	0.5923	0.6411	0.6831	0.7301	0.7604
16	0.2260	0.3383	0.4259	0.4973	0.5177	0.5742	0.6226	0.6643	0.7114	0.7419
17	0.2183	0.3271	0.4124	0.4821	0.5021	0.5577	0.6055	0.6470	0.6940	0.7247
18	0.2113	0.3170	0.4000	0.4683	0.4878	0.5425	0.5897	0.6308	0.6777	0.7084
19	0.2049	0.3077	0.3887	0.4555	0.4747	0.5285	0.5751	0.6158	0.6624	0.6932
20	0.1991	0.2992	0.3783	0.4438	0.4626	0.5155	0.5614	0.6018	0.6481	0.6788
21	0.1938	0.2914	0.3687	0.4329	0.4513	0.5034	0.5487	0.5886	0.6346	0.6652
22	0.1888	0.2841	0.3598	0.4227	0.4409	0.4921	0.5368	0.5763	0.6219	0.6524
23	0.1843	0.2774	0.3515	0.4132	0.4311	0.4815	0.5256	0.5647	0.6099	0.6402
24	0.1800	0.2711	0.3438	0.4044	0.4219	0.4716	0.5151	0.5537	0.5986	0.6287
25	0.1760	0.2653	0.3365	0.3961	0.4133	0.4622	0.5052	0.5434	0.5879	0.6178
26	0.1723	0.2598	0.3297	0.3882	0.4052	0.4534	0.4958	0.5336	0.5776	0.6074
27	0.1688	0.2546	0.3233	0.3809	0.3976	0.4451	0.4869	0.5243	0.5679	0.5974
28	0.1655	0.2497	0.3172	0.3739	0.3904	0.4372	0.4785	0.5154	0.5587	0.5880
29	0.1624	0.2451	0.3115	0.3673	0.3835	0.4297	0.4705	0.5070	0.5499	0.5790
30	0.1594	0.2407	0.3061	0.3610	0.3770	0.4226	0.4629	0.4990	0.5415	0.5703
40	0.1368	0.2070	0.2638	0.3120	0.3261	0.3665	0.4026	0.4353	0.4741	0.5007
50	0.1217	0.1843	0.2353	0.2787	0.2915	0.3281	0.3610	0.3909	0.4267	0.4514
60	0.1106	0.1678	0.2144	0.2542	0.2659	0.2997	0.3301	0.3578	0.3912	0.4143
80	0.0954	0.1448	0.1852	0.2199	0.2301	0.2597	0.2864	0.3109	0.3405	0.3611
100	0.0851	0.1292	0.1654	0.1966	0.2058	0.2324	0.2565	0.2786	0.3054	0.3242
1000	0.0266	0.0406	0.0520	0.0620	0.0650	0.0736	0.0814	0.0887	0.0976	0.1039

SOLUTIONS TO ODD-NUMBERED EXERCISES

CHAPTER 1

1.1 Answers will vary.

1.3 Exam1 = 71; Exam2 = 80; Final = 79.

1.5 Cases are available apartments. Monthly rent, number of bedrooms, and distance to campus are quantitative. Free cable and pets allowed are categorical.

1.7 **(a)** is a bar graph. **(b)** Garmin has a higher percentage of the market share in the United States. A company's market share would be higher in the area where it is located.

1.9 All < signs should be replaced with ≤ signs, all ≤ signs should be replaced with < signs, and an interval should be added for 50 < score ≤ 60.

1.11 Stemplots will vary but should show data that is skewed to the right.

1.13 **(a)** The cases represent the employees in the company. **(b)** Employee identification number, last name, first name, and middle initial are labels. Department and education are categorical. The remaining variables are quantitative. **(c)** The spreadsheet columns should represent the variables and labels from part (b); the spreadsheet rows should represent employees in the company.

1.15 Quantitative variables are (a), (d), and (e) as these take on numerical values. Categorical variables are (b), (c), and (f) as these place the students in one of a number of categories.

1.17 Answers will vary. Examples of quantitative variables might include "average number of hours studied per week" or "number of classes requiring more than one hour of studying per week." Examples of categorical variables might include "do you study on weekends? (yes/no)" or "do you watch television while studying? (yes/no)," among others.

1.19 Bar graph will show the tallest bar for orange, followed by brown. Purple, yellow, and gray make up the next lowest group of bars at approximately the same level, followed by green and white, and then red. Black and blue will show no bars as these were not selected by anyone.

1.21 Bar graph will show the tallest bars for "other phone" and "Motorola Razr." Windows Mobile and Blackberry have the next highest bars, followed by Nothing, Palm, Sidekick, and Symbian.

1.23 The bar graph shows that Google – Global has by far the largest market share. In distant second place is Yahoo Global, followed by MSN Global, AOL Global, Microsoft Live, Ask Global, and finally Other.

1.25 The bar graph shows the highest bars for the United Kingdom followed by Canàda. Quite a bit below these are the bars for Turkey, Australia, Columbia, Chile, ánd France. The remaining countries show bars half the size of France's or smaller with the shortest bars for Israel and Italy.

1.27 **(a)** shows a histogram that is skewed to the right with a range from 3.3% to 8.9%. There is a peak around the upper 4% range. **(b)** shows the stemplot of the same data with a peak in the 4% to 5% range. **(c)** The histogram breaks up the intervals into different intervals than used in the stemplot, allowing you to see a finer breakdown of the data for this problem. Answers may vary for preference but should reflect an understanding of the features of both plots.

1.29   **(a)** and **(b)** are bar graphs. **(c)** One way to combine the graphs is to use the "clustered bar graph" option on SPSS.

1.31   **(a)** Alaska has 7.0% and Florida has 17.0% older residents. **(b)** The distribution is unimodal and symmetric with a peak around 13%. Without Alaska and Florida, the range is 8.8% to 15.5%.

1.33   **(a)** The stemplot or histogram is unimodal and skewed left. **(b)** A histogram might combine the intervals, especially for the higher percentage values, causing a loss in the information regarding the shape of the graph.

1.35   The graph shows a unimodal, skewed right distribution with a mean of 48.28 and a range from 2.0 to 204.9 barrels. There are 4 high outliers at 118.2, 156.5, 196.0, and 204.9 barrels.

1.37   **(a)** Many more readers who completed the survey owned Brand A than Brand B. **(b)** It would be better to consider the proportion of readers owning each brand who required service calls. In this case 22% of the Brand A owners required a service call while 40% of the Brand B owners required a service call.

1.39   When you change the scales, some extreme changes on one scale will be barely noticeable on the other. The addition of white space in the graph also changes visual impressions.

1.41   The mean service time is 196.575 seconds.

1.43   The median is 40 days. Without the outlier, the median is 36.5 days. The median moves to the next largest value with the addition of Suriname.

1.45   The median exam score is 82.5.

1.47   The boxplot shows a distribution that is slightly skewed to the right. The distance from the minimum value to Q_1 is slightly smaller than the distance from Q_3 to the maximum. The distance from Q1 to the median is slightly larger than the distance from Q_3 to the maximum. This puts the median just slightly below the midpoint of the range from the minimum to the maximum value. The boxplot does not include as much detail about the features of the data as the stemplot does.

1.49   The standard deviation is 18.6 with Suriname excluded and 132.6 with Suriname included.

1.51   **(a)** Mean = 196.575; standard deviation = 342.022. **(b)** Minimum = 1, Q_1 = 54.5, median = 103.5, Q_3 = 200, maximum = 2631. **(c)** The five number summary is better because the distribution is skewed with outliers.

1.53   **(a)** Median = 5.5%; Q_1 = 4.0%, and Q_3 = 6.9%. **(c)** Excluding the outliers, the median = 5.5%, Q_1 = 4.0%, and Q_3 = 6.5%.

1.55   **(a)** The graphical summaries will show a unimodal distribution that is skewed to the right with many outliers. **(b)** Low outliers are the United States, Spain, United Kingdom, Italy, Australia, Greece, Turkey, France, Romania, South Africa, India, Portugal, and Poland. High outliers are Taiwan, Malaysia, Canada, Sweden,

Algeria, United Arab Emirates, Singapore, Kuwait, Norway, Netherlands, Switzerland, Russia, Saudi Arabia, Germany, Japan, and China. **(c)** Without the outliers, the interior of the data set can be seen much more clearly in a histogram of the data. Again, the graph shows a skewed-right distribution. **(d)** The distribution is skewed to the right with 29 outliers. The median is $0.5 billion with a range of −$747.1 billion to $363.3 billion. Removing the outliers slightly shifts the median to −$0.6 billion.

1.57   **(a)** Mean = 6.93%; standard deviation = 3.18%. **(b)** 3.5%, 4.2%, 6.8%, 8.0%, 13.6%. **(c)** is a boxplot. **(d)** The distribution is skewed to the right with no outliers. The data would be best described by the five number summary because the distribution is skewed.

1.59   **(a)** Mean = 48.25; standard deviation = 40.24. **(b)** Five number summary is 2.0, 21.5, 37.8, 60.1, 204.9. **(c)** is a boxplot. **(d)** The distribution is skewed to the right with outliers. The five number summary would be the best numerical summary of the data.

1.61   **(a)** is a histogram that is strongly skewed to the right. **(b)** Mean = $12,143.95; standard deviation = $12,421.48. Five number summary is $3338, $4589, $7558.50, $13416, $66667. The five number summary gives the best numerical summary of the data. **(c)** The distribution is right-skewed with 9 high outliers. Other information from numerical and graphical summaries above should also be included in the paragraph.

1.63   **(a)** Median is 4.70% with or without the outlier. Mean = 4.76% with the outlier included and 4.81% with the outlier excluded. **(b)** Standard deviation is 0.752; without the outlier, it decreases to 0.586. Q_1 = 4.3; 4.35 when the outlier is excluded. Q_3 = 5.0 with or without the outlier. **(c)** Removing the outlier affects the mean and standard deviation but does not change the median or the third quartile and only slightly affects the third quartile.

1.65   Answers will vary.

1.67   **(a)** The five number summary is 7.0%, 12.15%, 13.10%, 13.55%, 17.0%. **(b)** Any low outliers would have to be less than 10.05%, and any high outliers would have to be greater than 15.65%. Therefore, Alaska and Florida are outliers, but so are Utah with 8.8%, Georgia with 9.9%, and Texas with 10.0% older residents.

1.69   You will learn about the effects of outliers on the mean and median by interactively using the applet.

1.71   Both sets of data have a mean of 7.5 and a standard deviation of 2.03. Data A has a distribution that is left-skewed while Data B has a distribution that is fairly symmetric except for one high outlier. Thus, we see two distributions with quite different shapes but with the same mean and standard deviation.

1.73   **(a)** Mean = $79,375. All except the owner have salaries less than the mean. Median = $30,000. **(b)** Mean = $91,875; median does not change.

1.75   **(a)** 10, 10, 10, 10. **(b)** 10, 10, 20, 20. **(c)** There are several answers for (a), but (b) is unique.

1.77   Mean = $12,143.95; trimmed mean = $10,262.12.

1.79   You can use the Uniform distribution for an example of a symmetric distribution.

1.81   (a) Mean is C; median is B. (b) Mean is A; median is A. (c) Mean is A; median is B.

1.83   (a) 0.025. (b) 64 inches to 74 inches. (c) 16%.

1.85   (419, 725).

1.87   The proportion who scored less than 600 is 0.7088. The proportion who scored greater than or equal to 600 is 0.2912. A sketch with the relationship between three Normal curves should be shown.

1.89   A score of 656 is needed.

1.91   (a) The plot shows a curve that is concave up instead of one that follows a straight line. (b) Figure 1.28 has a curve with an upper portion that follows a straight line. Figure 1.29 shows a curve with an upper portion that is concave down. When the upper portion of the curve does not follow a straight line, the curve does not have a Normal distribution. For right-skewed data, the Normal quantile plot will show the highest points below the diagonal. For left skewed data, the Normal quantile plot will show the highest points above the diagonal. For symmetric data, the points will follow the diagonal fairly closely, even at the ends.

1.93   (a) and (b) show sketches of two overlapping Normal curves, the curve showing the distribution with a mean of 10 is to the left of the curve showing the distribution with the mean of 20. (c) Changing the mean causes the curve to shift along the x-axis.

1.95   Sketches will vary.

1.97   (a) The 68% interval is (5232, 23362), the 95% interval is (-3833, 32427), and the 99.7% interval is (-12898, 41492). (b) The negative values in the intervals would not make sense in terms of words spoken. (c) The 68% interval is (5004, 23116), the 95% interval is (-4052, 32172), and the 99.7% interval is (-13108, 41228). These intervals also contained substantially negative numbers of words. (d) The numbers in these intervals were close for men compared to women. The fact that the women's intervals were slightly higher may not be enough to show a substantial difference between the two groups.

1.99   (a) Standardized score are -0.2, -1.6, 2.2, 0.5, 0.3, 2.8, -0.6, -1.5, 1.0, 0.0. (b) The cut-off is the standardized score of 1.04. (c) Students with scores of 92 and 98 earned As.

1.101   (a) shows histograms and Normal quantile plots for both sales price and building area. (b) The graphs are not Normally distributed. The histograms are not symmetric, and the Normal quantile plots show that neither distribution follows a straight line. (c) The outlier has a selling price of $7,900,000 and a building area of 114,412 sq. ft.

1.103   (a) Mean = $65.30; standard deviation = $16.29. (b) $x \pm s$ give the interval (49.01, 81.59). $x \pm 2s$ give the interval (32.72, 97.88), and $x \pm 3s$ give the interval (16.43, 114.17). (c) Tables may vary. (d) Normality is indicated.

1.105   (a) Mean = 579,338; standard deviation = 1,496,814. (b) (-3911104, 5069779). (c) There are no negative values in the data, even though this interval indicates that a portion should spread into the negative numbers.

1.107   (a) is a histogram that is strongly right skewed. (b) is a sketch of a Normal quantile plot, showing that the upper tail of the distribution definitely does not follow the pattern of a Normal distribution. **(c)** should compare the sketch to the computer generated plot.

1.109   Answers may vary, but the standard deviations should be close to 0.2 for the taller curve and 0.5 for the shorter curve.

1.111   **(a)** 0.0107. **(b)** 0.9893. **(c)** 0.0446. **(d)** 0.9447.

1.113   **(a)** Subtracting the area to the right of 90 from the area to the right of 105 gives an area of 0.5328. **(b)** 85% of the data values will be between 85.6 and 114.4.

1.115   **(a)** The first and third quartiles for the standard Normal distribution are approximately ±0.67. **(b)** The first quartile is about 255 days, and the third quartile is about 277 days.

1.117   The histograms and Normal quantile plots will differ for each student.

1.119   These pairs go together: (1) and (c), (2) and (b), (3) and (d), (4) and (a).

1.121   **(a)** Iowa. **(b)** The small slices would be indistinguishable from each other as they are too small compared to the size of the largest slice. **(c)** is a pie chart.

1.123   **(a)** The five number summary is −$1.5, −$4.80, −$6, −$10, −$78. **(b)** 75.73%. **(c)** is a boxplot. **(d)** The boxplot shows that the distribution is strongly skewed to the left.

1.125   **(a)** 0.6316. **(b)** 0.15% **(c)** 0%. 0.15% of 22 companies is 0.033 cases, which rounds to 0.

1.127   Gender and automobile preference are categorical. Age and household income are quantitative.

1.129   Hatchbacks have a right-skewed distribution with a five number summary of 16, 20, 21, 5, 25, and 30. Sedans have a left-skewed distribution with a five number summary of 13, 16, 17, 17, 19; two high outliers at 19 mpg; and two low outliers at 13 mpg. The range for hatchbacks is larger than for sedans as shown in a side-by-side boxplot. The hatchbacks have much greater miles per gallon than the sedans, both on average and for the entire range of values.

1.131   There are four outliers at 118.2, 156.5, 196, and 204.9.

1.133   Reports will vary based on the year of data selected. Information should be included on the mean, standard deviation, and five number summaries of both the "business starts" and "business failures." The shape of the distribution from either a histogram or a stemplot should be discussed as well as the presence of any outliers. A comparison of the two graphs and/or numerical summaries between these two variables should also be made.

1.135   Reports will vary based on data selected but should include graphical summaries and numerical summaries.

CHAPTER 2

2.1   **(a)** Cases: individual employees. **(b)** Labels: employee identifier. **(c)** Variables may vary. Quantitative variable examples are hours of sleep, units produced, and so on. Categorical variables include type of job. **(d)** If the study uses sleep to predict employee effectiveness, then there is an explanatory/response relationship.

2.3　　**(a)** is a spreadsheet. **(b)** 10. **(c)** and **(d)** Label: botnet. Variables: number of bots measured in thousands (quantitative) and number of spams in billions (quantitative).

2.5　　**(a)** and **(b)** show a scatterplot like Figure 2.1 with the point at an x-value of 185 labeled as Bobax.

2.7　　The debts for 2007 are higher than those for 2006 by an average of approximately 19 million dollars.

2.9　　The top third of companies have high log GDP per capita. The middle third generally fall in the mid-range of GDP per capita. The bottom third generally have the lowest log GDP per capita.

2.11　Sketches will vary.

2.13　**(a)** A bar graph of the data would be appropriate. The highest bars are for "25 to 34 years" and for "45 to 64 years." The "65 years and older" group has the smallest number. The remaining three categories are similar at a height of about $3/4$ the height of the tallest two. **(b)** Percents from the youngest age group to the oldest are 18.4%, 17.7%, 22.8%, 17.1%, 22.8%, and 1.2%. **(c)** The highest bar for the percent of uninsured is for the "18 to 24 years" group. This is followed in decreasing order by "25 to 34 years," "35 to 44 years," 35 to 44 years," "45 to 64 years," "Under 18 years," and finally, with a very short bar, "65 years and older." **(d)** The graph in part (b) shows the relationship between the number uninsured and the size of the population for each group. **(e)** People 65 years and older have a small number as well as a small portion that are uninsured. Those under 18 years and those between 45 and 64 years have the next smallest percent uninsured, but their counts are high. This is due to the large size of these two populations. The 18 to 24 age group has the largest percent uninsured, but the number of uninsured in this group is not high compared to the other groups. This is due to the small number of people in this group.

2.15　When looking at the number who are uninsured compared to the total population, we can draw conclusions regarding the likelihood that an unemployed person would come from a particular age group. We can also discuss at what ages the greatest number or percent of the uninsured are located. When we compare the number of uninsured in each age group with the total size of that age group, we see how deep the uninsured problem is within each group. We can draw conclusions about how likely it is for a person to be uninsured within an age group.

2.17　**(a)** A strong linear relationship does not tell us that the numbers remained approximately the same; however, the fact that these points lined up on an approximate 45-degree angle line would tell us that the numbers are very similar between 2006 and 2007. **(b)** There was an approximate 3% decrease in the number of uninsured from 2006 to 2007. This change was not the same across all age groups. The "65 years and older" group saw a 27% increase. However, the number of the increase or decrease from each group was small relative to the total number of people studied, so even a 27% increase does not show up as a huge change when looking at a comparison of the numbers of uninsured. When we graph the percent changes by age group, however, we do see large differences between the age groups.

2.19　**(a)** and **(b)** The scatterplot shows a moderately strong, positive linear relationship between percent alcohol and number of calories.

2.21　**(a)** The scatterplot shows that the three territories have much lower percents of people over age 65 and greater percents of people under age 15. **(b)** The provinces still appear to have a negative relationship but not as strong of a relationship as previously suggested when these data were not separated into territories vs. provinces.

2.23   There still appears to be a positive, moderate relationship between Internet users and life expectancy. While the relationship is not precisely linear, it is much closer to a linear relationship than what is seen when all 181 countries are included. In Europe, there are very few countries with close to 0 Internet users per 100 people while there were quite few in the 181 countries scatterplot. The life expectancy is higher overall for the European countries as well.

2.25 $r = .884$.

2.27   The correlation between 2006 debt and 2007 debt is near 1. Because low values of debt occur together for both years, and high values from both years occur together, the correlation will be positive. Because the points on the scatterplot closely follow the model as shown by the line through the scatterplot, the correlation will be strong, and, therefore, close to one.

2.29   (a) The scatterplot shows points increasing from $x = 10$ to $x = 40$ and decreasing from $x = 40$ to $x = 60$. (b) There is a very strong relationship between y and x; however, it is definitely not linear throughout the relationship. There is a positive linear relationship from $x = 10$ through $x = 40$, and there is a negative linear relationship from $x = 40$ through $x = 60$. (c) $r = 0.162$. (d) Correlation only makes sense when looking at data that has a single linear relationship throughout the data set.

2.31   (a) Correlation should not be used to describe the strength of a relationship that is not linear. (b) -0.479. (c) Because this relationship is linear above a log unemployment rate of 1.5, the correlation would be a good measure of strength for this range of data.

2.33   (a) A student who scores well on the second exam would demonstrate an understanding of statistical concepts and an ability to think analytically. This type of student would also probably score well on the final exam (and vice versa). (b) The scatterplot shows a weak to moderate positive relationship between the variables. (c) $r = 0.519$. (d) Answers may vary. Some possibilities may include additional factors, such as prior knowledge, that might also affect exam 1. By the second exam, students have also figured out the level of studying and effort that they intend to apply to a particular course, and this would likely flow through to the final exam.

2.35   The correlation is 0.287.

2.37   The correlation is -0.839. This is not a good measure of the strength of the relationship because the relationship between the variables is not linear.

2.39   (a) See 2.21a. (b) The correlation for the territories is -0.967. (c) The correlation for the provinces is -0.331. (d) The negative relationship that is seen in the data is very strong when only the territories are considered but rather weak when only the provinces are considered.

2.41   (a) See the scatterplot in Figure 2.14. (b) $r = 0.686$. (c) The correlations are almost identical between the two problems, but the scatterplots are quite different. The scatterplot for Exercise 2.40 does not show a linear relationship, so the correlation is not a good measure of strength for this data. In this exercise, the data do appear to have closer to a linear relationship, so the correlation would be a better measure of strength here.

2.43 Correlation is not affected by changes in units of measurement, so there would be no difference in the correlation values.

2.45 Applet.

2.47 The scatterplot shows an exact linear relationship between the data points. The correlation is 1.00. When the points of the scatterplot line up exactly on a line (in this case, the line $y = x + 2$, the correlation will be precisely 1.00.

2.49 (a) The relationship between x and y is moderate, positive, and linear. (b) $r = 0.523$.

2.51 (a) The outlier is at the far right of the data set. The x-value makes this point an outlier, even though the y-value is not different from that seen in the rest of the data. (b) The correlation here is $r = 0.072$, a much weaker value than was seen in the original data set. (c) Outliers that extend the relationship already seen in the data (outliers in both the x and y directions) can make the relationship between the x and y variables appear stronger. Outliers in only one direction, either x or y, can make the relationship appear weaker than it otherwise would.

2.53 In a perfectly correlated relationship, the points in the scatterplot of the relationship between two variables would line up exactly on a line. In this case, if Fund A is plotted on the x-axis, and Fund B is plotted on the y-axis, the points would lie exactly on a line with a slope of $1/2$.

2.55 The estimated net assets per capita is $200. This gives a prediction error of $80.

2.57 (a) The slope of the regression line is 20. (b) The intercept of the regression line is 10. (c) When x is 10, the predicted value of y is 210. When x is 20, the predicted values is 410. When x is 30, the predicted value of y is 610. (d) The plot shows a line increasing from a point on the y-axis at 10 to a y value of 1010 when x is 50.

2.59 (a) The percent is 35.5% because $r^2 = (0.596)^2 = 0.355$. (b) $\hat{y} = 6.08\% + 1.707x$. (c) $\hat{y} = 9.08\%$. The least-squares regression line passes through the point $(\overline{x}, \overline{y})$. Thus, we would predict $\hat{y} = \overline{y} = 9.07\%$ when $x = \overline{x} = 1.75\%$.

2.61 (a) The predicted value for Bobax is $\hat{y} = -3.4192 + .17048(185) = 28.1196$. (b) The residual is $y - \hat{y} = 9 - 28.1196 = -19.1196$. (c) Bobax has the greater deviation. Srizbi's observed y value is 9.5 units away from the regression line while Bobax's observed y value is 19.1196 units away.

2.63 (a) shows a labeled scatterplot. (b) Points can be found by matching the residual plot x-values to the data set x-values.

2.65 $y = 5.00 + 1.10x$.

2.67 (a) The number y in inventory after x weeks must be $y = 96 - 6x$. (b) The graph shows a negatively sloping line. (c) No. After 16 weeks, the equation predicts that there will be $y = -4$.

2.69 $\hat{y} = 3.379 + 1.615x$, where x is percent alcohol and $\hat{y}$ is the predicted carbohydrates.

2.71 (a) The scatterplot shows points that closely follow a positive linear model. (b) $\hat{y} = 318.594 + 0.935x$. (c) The residual is -399.319. (d) The residual plot shows a steadily decreasing pattern from 1998 to 2002, then sharp increases through 2006 and a sharp decrease again in 2007. This indicates that the fund returns may have a somewhat cyclical pattern compared to their predicted return over time. (e) 56.0% of the variation in EAFE is explained by year.

2.73 **(a)** The new point is a high outlier in both the x and y directions. The slope of the regression line is steeper but still positive. **(b)** $\hat{y} = 1.470 + 1.443x$. **(c)** The residual plot also indicates that the new point is an outlier. **(d)** 71.1%. **(e)** Adding this point to the data causes the relationship between the data point to appear stronger. It affects the regression model and decreases the percent of variation that is explained by something other than the regression model.

2.75 **(a)** $r = 0.999993$, so the calibration does not need to be done again. **(b)** $\hat{y} = 1.6571+0.113301x$. For $x = 500$, $\hat{y} = 58.3$. We would expect the predicted absorbance to be very accurate based on the plot and the correlation.

2.77 Experiment with the applet on the Web and comment.

2.79 $r^2 = 0.64$. Smaller raises tended to be given to those who missed more days, and so the variables are negatively associated. Thus, $r = -0.8$.

2.81 **(a)** The plot shows a negatively sloping regression line that does not fit the data very well. **(b)** It is not appropriate to use the least-squares regression line here as the relationship was not linear. **(c)** There is a large grouping of points below the residual line at the far left of the plot, a more spread out grouping above the residual line at the far left, and a group of points trailing off to the right. The group of points going toward the right shows us that the points in the scatterplot do not decrease as fast as the linear model would suggest. The points to the left of the plot show a wide range of residuals around a small range of unemployment values, indicating that the regression model is not fitting this data very well.

2.83 **(a)** This plot shows a decreasing linear relationship between the log unemployment and the log GDP per capita. There appears to be a moderately strong relationship between these two variables. **(b)** Because the relationship between the log unemployment and log GDP per capita is linear, it is appropriate to use the least-squares regression line to describe this relationship. **(c)** The residual plot shows a little greater concentration toward the lower log unemployment rates but otherwise a good scattering of points around this residual line. As there is no pattern to these residuals, it appears that the assumption that the relationship here was linear is justified.

2.85

City	Residual
Los Angeles	5994.83
Washington, D.C.	2763.74
Minneapolis	2107.59
Philadelphia	169.41
Oakland	27.91
Boston	20.96
San Francisco	−75.78
Baltimore	−131.55
New York	−283.04
Long Beach	−1181.71
Miami	−2129.21
Chicago	−7283.28

2.87   **(b)** It would not be reasonable to fit a straight line to the data because the relationship is not linear. **(c)** $\hat{y} = 8.739 - 0.00042x$. **(d)** 8.7%. The relationship here is not linear, so the least-squares regression line is not an appropriate model for this data.

2.89   The sketch should look similar to Figure 2.26 with the cluster closest to the x-axis representing academia, and the cluster closest to the y-axis representing businesses.

2.91   No. Correlation does not imply causation. The most seriously ill patients may be sent to larger hospitals rather than smaller hospitals, and this could account for the observed correlation.

2.93   **(a)** If all residuals are positive, this just means the points in the scatterplot are all above the regression line. It does not mean the slope of the line is positive. A negatively sloped line can be drawn with points above this line to give a counterexample. **(b)** The direction of the relationship does not tell anything about causation. If a positive change in the explanatory variable causes the response variable to decrease, this would give us a negative relationship. **(c)** A lurking variable may help explain changes in the response variable, but it is not the response variable itself.

2.95   Another factor that might explain this relationship is the wealth of the countries. Wealthier, developed countries would have longer life expectancies as well as more Internet usage.

2.97   Answers will vary but may include age, years of education, years of experience in job, and location.

2.99   Artificial sweeteners tend to be used by people trying to watch their calories. People who are overweight would be more likely to diet, therefore using more artificial sweeteners.

2.101 Intelligence or family background may be lurking variables. Families with a tradition of more education and high-paying jobs may encourage children to follow a similar career path.

2.103 The marginal distribution shows 34.6% of the bottles of wine were bought when Italian music was playing, 30.9% when French music was playing, and 34.6% when no music was playing. The bar graph of this data shows that slightly fewer bottles of wine were purchased when French music was playing than when Italian or no music was playing, which resulted in an equal number of bottles.

2.105 **(a)** Totals for Field of Study are 1697 for "social sciences, business, law"; 911 for "science, mathematics, engineering"; 801 for "arts and humanities"; 319 for "education"; and 857 for "other" fields. Totals for Country are 176 for Canada, 672 for France, 218 for Germany, 321 for Italy, 654 for Japan, 475 for United Kingdom, and 2069 for United States **(b)** Following is the marginal distribution by country. A bar graph would be appropriate for displaying this data.

Canada	France	Germany	Italy	Japan	U.K.	U.S.
3.8%	14.7%	4.8%	7.0%	14.3%	10.4%	45.1%

(c) Following is the marginal distribution by field of study. A bar graph would be appropriate for displaying this data.

Social sciences, business, law	37.0%
Science, mathematics, engineering	19.9%
Arts and humanities	17.5%
Education	7.0%
Other	18.7%

2.107 (a)

	Italian music
French wine	30
Italian wine	19
Other wine	35
Total	84

(b)

Wine type	French	Italian	Other
Percent when Italian music is playing	35.7%	22.6%	41.7%

(c) A bar graph shows "Other" has the highest percent, followed by "French" and then "Italian." (d) Italian wine was sold more often when Italian music was playing.

2.109 (a) See the conditional distributions for each country in the following chart.

Field of study	Canada	France	Germany	Italy	Japan	U.K.	U.S.
Social sciences, business, law	36.4%	22.8%	30.3%	38.9%	39.6%	32.0%	42.4%
Science, mathematics, engineering	19.9%	16.5%	30.3%	24.9%	20.8%	26.9%	17.2%
Arts and humanities	15.3%	11.0%	15.1%	13.1%	18.8%	22.1%	19.2%
Education	11.4%	6.7%	8.3%	5.0%	6.0%	2.9%	8.1%
Other	17.0%	43.0%	16.1%	18.1%	14.8%	16.0%	13.1%

(b) A combined bar graph would be appropriate for displaying this data. (c) The bar graph shows that most countries have a greater percent of their students in social sciences, business and law; however, the most popular field of study in France is "other." In most countries, science, math, and engineering is the second most common field of study; however, in the United States, Arts and Humanities is the second most common choice. Education has the smallest percent for all of the countries.

2.111 (a) Both of these approaches allow us to study different things (see part c). There are some items that stand out in both cases, like the large number of students in France in "other" areas of study. **(b)** When looking at the conditional distribution by field of study, we are primarily seeing the difference in the sizes of college populations between the countries. If we are interested in studying the popularity of the different fields of study, the conditional distributions of field of study given the different countries would be more instructive. **(c)** The first of these approaches allows us to see which fields are most/least popular across the different countries while the second approach allows us to see where the greatest number of students are within each field.

2.113 **(a)** 3.8%, 4%, Hospital A. **(b)** 1.0%, 1.3%, Hospital A. **(c)** Hospital B only loses 2% of patients compared with 3% at Hospital A. **(d)** Those in "poor" condition make up a greater percentage of Hospital A's surgeries, and these have a lower survival rate than those in "good" condition.

2.115 On average, 37% of banks offer RDC. However, the region with the largest number of banks, the Midwest with 31%, has the smallest percent (25.8%) of banks that offer RDC. In the Northeast, Southeast, and West regions, over 40% of banks offer this service. The Southeast leads the group with 48.3% offering RDC.

2.117 **(a)** Of those who get adequate sleep, 56.77% are high exercisers and 43.23% are low exercisers. **(b)** Of those who do not get adequate sleep, 37.95% are high exercisers and 62.05% are low exercisers. **(c)** Boys and girls in the 13- to 18-year-old range are more likely to get high amounts of exercise when they get adequate amounts of sleep while those who do not get adequate amounts of sleep are much less likely to be high exercisers.

2.119 **(a)**

	Full time	Part time
15–19	89.7%	10.3%
20–24	81.8%	18.2%
25–34	50.1%	49.9%
35 and over	27.1%	72.9%

(b) Use a bar graph that shows percents for each group in different colors. **(c)** Students who are younger that 24 are much more likely to be full-time students than part-time students. Students age 35 and up are much more likely to be part-time students. Students in the 25- to 35-year-old group are equally as likely to be full or part-time students.

2.121 **(a)**

	Male	Female	Total
Agree	11724	14169	25893
Disagree	1163	746	1909
Total	12887	14915	27802

(b) 95% of females and 91% of males agreed that honesty and trust were essential. These percents were very similar. An overall 93% of students agreed with this statement. **(c)** When asked whether students agreed or disagreed about whether trust and honesty were essential in the workplace, most (93%) of them agreed that they were. A slightly smaller proportion of males agreed with this statement (91%) than females (95%).

2.123 **(a)**

	Hired	Not hired
Applicants < 40	6.82%	93.18%
Applicants ≥ 40	0.61%	99.39%

(b) A graph shows results similar to those in (a). **(c)** Only a small percent of all applicants are hired, but the percent (6.44%) of applicants who are less than 40 who are hired is more than 10 times the percent (0.61%) of applicants who are 40 or older who are hired. **(d)** Lurking variables that might be involved are past employment history (why are older applicants without a job and looking for work?) or health.

2.125 **(a)** Never married: 25,262; married: 65,128; widowed: 11,208; divorced: 13,208. **(b)** The marginal distribution for marital status is Never married: 22.0%; married: 56.7%; widowed: 9.8%; divorced: 11.5%.

Following is the joint distribution.

	Never married	Married	Widowed	Divorced
18 to 24	9.9%	2.1%	0.0%	0.2%
25 to 39	7.3%	16.8%	0.1%	2.2%
40 to 64	4.2%	29.8%	2.0%	7.4%
$\geq$ 65	0.7%	8.0%	7.6%	1.7%

Each row of the following table shows the conditional distribution for a given age.

	Never married	Married	Widowed	Divorced
18 to 24	81.3%	17.3%	0.1%	1.3%
25 to 39	27.5%	63.7%	0.6%	8.3%
40 to 64	9.6%	68.6%	4.7%	17.1%
$\geq$ 65	3.7%	44.5%	42.2%	9.6%

Each column of the following table shows the conditional distribution for a given marital status.

	Never married	Married	Widowed	Divorced
18 to 24	45.0%	3.7%	0.2%	1.4%
25 to 39	33.0%	29.7%	1.5%	19.0%
40 to 64	19.0%	52.6%	20.8%	64.6%
$\geq$ 65	3.0%	14.1%	77.5%	15.0%

(c) More young women are unmarried, more women above age 65 are widowed, and most women in the 25 to 64 year age range are either married or divorced. Divorce is highest among the 40 to 64 age range. The largest group of women fall in the 40 to 64 and married group.

2.127 **(a)** Never married: 30,856; married: 64,590; widowed: 2693; divorced: 9610. **(b)** The marginal distribution for marital status is Never married: 28.6%; married: 59.9%; widowed: 2.5%; divorced: 8.9%. Never married: 25,262; married: 62,128; widowed: 11,208; divorced: 13,208.

Following is the joint distribution.

	Never married	Married	Widowed	Divorced
18 to 24	12.0%	1.2%	0.0%	0.1%
25 to 39	10.7%	15.8%	0.1%	1.7%
40 to 64	5.4%	32.1%	0.6%	6.1%
≥ 65	0.6%	10.8%	1.9%	1.1%

Each row of the following table shows the conditional distribution for a given age.

	Never married	Married	Widowed	Divorced
18 to 24	89.9%	9.3%	0.1%	0.7%
25 to 39	37.8%	56.0%	0.2%	6.0%
40 to 64	12.2%	72.8%	1.3%	13.8%
≥ 65	4.0%	75.4%	13.1%	7.5%

Each column of the following table shows the conditional distribution for a given marital status.

	Never married	Married	Widowed	Divorced
18 to 24	42.0%	2.1%	0.4%	1.0%
25 to 39	37.3%	26.4%	2.0%	18.9%
40 to 64	18.7%	53.5%	22.7%	68.0%
≥ 65	2.0%	18.0%	74.9%	12.1%

(c) More young men are unmarried, more men above age 65 are widowed, and most men in the 25 to 64 year age range are either married or divorced. Divorce is highest among the 40 to 64 age range. The largest group of men fall in the 40 to 64 and married group.

2.129 Answers will vary, but the relationships described should hold. Try answers where the percent of smokers who are overweight is less than the percent of nonsmokers who are overweight.

2.131 (a) There is a moderately weak positive linear relationship between the two variables with no outliers or influential observations present. (b) $\hat{y} = 109.82 + 0.126x$. (c) 129.98. (d) Sales for the Netherlands is 107, so the residual is -22.98. (e) 10.3%.

2.133 (a) There is a weak positive linear relationship between the two variables with no outliers or influential observations present. (b) $\hat{y} = 112.257 + 0.115x$. (c) 126.632. (d) Sales for Finland are 136, so the residual is 9.368. (e) 2.3%.

2.135 (a) The residual plot shows a curve that has a pattern other than a linear form. The salaries are increasing more slowly in the beginning years and faster in later years, as indicated by the decreasing and then increasing pattern to the residuals. (b) The residual plot makes the pattern in the data more obvious than it was in the scatterplot.

2.137 **(a)** \$139,579. **(b)** \$158,856. **(c)** The prediction from the log salary model will be better because the relationship here is actually linear while the salary model is not. **(d)** For 0 to 20 years, both models will give fairly similar results as the points are very close to the linear model in both cases. However, the original salary model does not take into account that the changes in salary are increasing in the later years. The log salary model does account for this and would be the better prediction model. **(e)** Extrapolation can lead to inaccurate predictions, and when the model does not precisely show the pattern of the data, there can be even greater discrepancies between the prediction and the actual value. In this case, if we do not use the log salary model, the prediction will not take into account the increasing salaries in the later years, and, as a result, the prediction will not be accurate.

2.139 **(a)** $\hat{y} = 5403 + 0.982x$. **(b)** The residual shows a random scattering of points around the 0 residual line. There do not appear to be any outliers or patterns to the residual plot.

2.141 The residual will tell whether the school's graduation rate is better or worse than what is predicted. So a school with a positive residual would have a higher graduation rate than expected and vice versa.

2.143 **(a)** The scatterplot shows a moderately strong, positive, linear relationship between the proportion of adults who eat at least five servings of fruits and vegetables each day and the proportion of adults who have completed college. **(b)** The line does give an overall picture of the trend of the data. The points do not follow the line exactly, but it does show the general form of the data. **(c)** Answers will vary based on states chosen. **(d)** There is no reason to believe that a causation relationship exists between these two variables. There are other variables that could help explain why this relationship exists such as financial well-being and education of parents.

2.145 Answers will vary depending on teams chosen.

2.147 Based on this formula, there is a negative relationship between population variation and diversity. So, assuming the relationship is linear, the greater the diversity, the less variation there would be in population size. The correlation would be $r = -0.583$, so the strength is only moderately strong.

2.149 **(a)** The amps of the saw can be used to predict the weight of the saw, so amps are the explanatory variable. **(b)** shows a scatterplot with a weak positive linear relationship between the variables. **(c)** $\hat{y} = 5.8 + 0.4x$, $s = 0.782$, $r^2 = 45.7\%$. **(d)** When amps increase by 1, weight increases by 0.4 pounds on average. **(e)** 2.5 amps. **(f)** The residual plot shows a general upward trend.

2.151 Suppose the return on Fund A is always twice that of Fund B. Then if Fund B increases by 10%, Fund B increases by 20%, and if Fund B increases by 20%, Fund A increases by 40%. This is a case where there is a perfect straight-line relation between Funds A and B.

2.153 **(a)** and **(b)** The overall pattern shows strength decreasing as length increases until length is 9, then the pattern is relatively flat with strength decreasing only very slightly for lengths greater than 9. **(c)** The least-squares line is Strength $= 488.38 - 20.75 \times$ Length. A straight line does not adequately describe these data because it fails to capture the "bend" in the pattern at Length = 9. **(d)** The equation for the lengths of 5 to 9 inches is $283.1 - 3.4 \times$ Length. The equation for the lengths of 9 to 14 inches is $667.5 - 46.9 \times$ Length. The two lines describe the data much better than a single line. It would be interesting to know how the strength of the wood product compares to solid wood of various lengths. For what lengths is it stronger, and are these common lengths that are used in building?

2.155 **(a)** 76% of smokers and 69% of nonsmokers were alive after 20 years.

	Smoker	Not
Dead	139	230
Alive	443	502

(b) For 18–44 year olds, 93% of smokers and 96% of nonsmokers were still alive. For 46- to 64-year-olds, 68% of smokers and 74% of nonsmokers were still alive. For the 65+ age group, 14% of smokers and 15% of nonsmokers were still alive. **(c)** The percentage of smokers in the three age groups were 46%, 55%, and 20%.

CHAPTER 3

3.1 This is anecdotal evidence and cannot be generalized to the population.

3.3 This is anecdotal evidence. Ashley's preference may be due to being a runner and would not necessarily generalize to all young people.

3.5 Answers will vary.

3.7 This is an experiment as the injections were deliberately imposed on the adults. The explanatory variable is whether the adult received the vaccine. Response variables include the presence of the various adverse affects and whether there was a presence of the antibodies in the adult.

3.9 This is an observational study. The researcher is merely asking a question, not imposing a treatment. The explanatory variable is the consumer's gender, and the response variable is which health plan the consumer prefers.

3.11 Many variables (lurking variables) could have changed over the five years to increase the unemployment rate, such as a recession or factories leaving the area. In addition, those who were trained well may be able to complete their work efficiently enough so that not as many workers are needed.

3.13 To give a complete answer, we would need to know whether all forest owners from the population were sent a survey or whether the 772 forest owners represented a sample from the overall population. The population would include all forest owners from this region. If the survey was mailed to only some of the forest owners, then the sample includes the 772 forest owners who were asked to complete the survey. The response rate is 45.1%.

3.15 **(a)** All adult U.S. residents. **(b)** All U.S. households. **(c)** All voltage regulators from the supplier.

3.17 Answers will vary.

3.19 The numbers 29, 07, 22, and 10 would be selected for junior associates; numbers 05 and 09 would be selected for senior associates.

3.21   (a) Any households with no phones or those with phones that are not listed would not be included (examples may vary). (b) Unlisted numbers would be included.

3.23   (a) To make this an experiment, randomly select a number of poker players to play a game of poker with the friend. Record the results. Biases that could have occurred in the original scenario include the friend only selecting to play games with very inexperienced players and/or misrepresenting the actual results. (b) To make this an experiment, randomly assign students to seats in the classroom from the first day of class. The unseen bias here is that individuals who are self-motivated or who care about the class and their grades might be more likely to choose particular seats in the classroom, usually those closer to the front of the room.

3.25   (a) Passages in text should be randomly selected. The current method does not account for differences in readability throughout the text. (b) Students should be randomly selected from a list of all students. Early morning students may not be representative of all students. (c) Subjects should be randomly selected from a list of all subjects. Not all subjects have a potential opportunity to be selected.

3.27   (a) All U.S. adults aged 18; 1001. (b) Advantages and disadvantages depend on the use for the information obtained. For popularity with the members of this population, the overall group gives a better measure. For quality of work, the use of those who knew the person would be better.

3.29   Using two-digit numbers from 01 to 33 and numbering the complexes in order alphabetically, the randomly selected sample using line 137 of Table B is 12 – Country View, 14 – Crestview, 11 – Country Square, 16 – Fairington, 08 – Burberry.

3.31   Applet. Answers will vary.

3.33   Various labels are possible. We labeled tracts numbered in the 1000s from 01 to 06, in the 2000s from 07 to 18, and in the 3000s from 19 to 44. The labels and blocks selected were 04 (block 1003), 17 (block 2010), 32 (block 3013), 22 (block 3003), and 09 (block 2002).

3.35   Answers will vary depending on the line number used. Within each group, number the blocks; use 1 through 6 for Group 1, 01 through 12 for Group 2, and 01 through 26 for Group 3. Select a line on Table B. Starting at the beginning of the line, select the first number on the first line between 1 and 6 for Group 1. Then select the next two nonrepeating two-digit numbers between 01 and 12. Then select the next three nonrepeating two-digit numbers between 01 and 26. The corresponding blocks will be selected.

3.37   The first number selected in the range of 1 to 25 defines a group of 4 numbers: x, $x+25$, $x+50$, and $x+75$. There are 25 groups, each of which has a 1 in 25 chance of being selected. Therefore, each number has a 1 in 25 chance of being selected. However, any group that is not defined in this way has a 0 chance of being selected.

3.39   One method for assigning numbers would be to give each parcel in Climax 1 a number from 01 to 36. Then give each parcel in Climax 2 a number between 01 and 72. Each in Climax 3 would be given a number from 01 to 31. Then the Secondary parcels would be numbered from 01 to 42. Find the random numbers from Climax 1, then starting immediately after the last number used for this selection, begin selecting the parcels for Climax 2, and so forth. By this method, the following parcels would be selected: for Climax 1 – 05, 16, 17, and 20; for Climax 2 – 19, 72, 45, 05, 71, 66, and 32; for Climax 3 – 19, 04, and 25; and for Secondary – 29, 20, 16, and 37. Other methods would yield other solutions.

3.41   (a) The question is slanted because it gives the impression that a relationship exists between brain cancer and cell phones. (b) This question encourages the responder to agree with the statement and gives reasons why agreement is the correct response. (c) This question uses words that sound negative and may not be fully understood by the average responder to encourage a desired response.

3.43   This sample is biased because the teacher is only sampling from those families who have children of school age. Young families with just one baby or preschooler at home would be left out of the study.

3.45   The wording of questions has the most important influence on the answers to a survey. Leading questions can introduce strong bias, and both of these questions lead the respondent to answer "Yes," which results in contradictory responses.

3.47   This study compares two treatments: instruction in drawing using paper-based instruction and instruction in drawing using Web-based instruction. Because these treatments are imposed on the individuals and they are not allowed to choose their own method of instruction, this is an experiment. Two drawing tests are taken, so the response variable would be the improvement in the drawing skills from one test to the other. The factor being studied is the type of drawing instruction, which has two levels: paper-based and Web-based. The experimental units are the students in the course. Because the students are interested in computer applications and graphics, judged by their enrollment in this class, the results of this experiment might not generalize to other populations where this interest and/or experience is not present.

3.49   Because the students chose for themselves which group to be in, there would be some bias. Students who are interested in trying new things and those who are good with computer applications would likely make up a good portion of those selecting to use the new software. Those who are less confident with computer applications would be more likely to stick with the previous version of the software. These characteristics may also have an effect on the final exam scores of these groups.

3.51   There is only one factor with two different levels: paper-based and Web-based. Lay these out in a diagram like that in Figure 3.5 with one column and two rows.

3.53   To assign the 20 pairs to the four treatments, give each of the pairs a two-digit number from 01 to 20. The first 5 selected are assigned to Group 1, the next 5 to Group 2, and the next 5 to Group 3. The remaining 5 are assigned to Group 4. Group 1 consists of pairs numbered 16, 04, 19, 07, and 10; Group 2 contains pairs 13, 15, 05, 09, and 08; Group 3 includes pairs 18, 03, 01, 06, and 11. The remaining 5 pairs are assigned to Group 4. Use a diagram like those in Figures 3.4 and 3.5 to describe the design.

3.55   If charts or indicators are introduced in the second year, and the electric consumption in the first year is compared with the second year, you won't know if the observed differences are due to the introduction of the chart or indicator or due to lurking variables.

3.57 A statistically significant result means that it is unlikely that the salary differences we are observing are just due to chance.

3.59 If the experimenter thought that meditation would help reduce anxiety, this could have been communicated to the subjects through tone of voice, body language, and so on. Even if the experimenter did not try to influence the subjects, he or she could have done so without knowing it. In a blind experiment, the experimenter would not have had access to the subjects before the treatment.

3.61 For a completely randomized design, students are randomly assigned to one of two groups. One of the groups will have access to charts, one will not. The amount of money that the groups make on average will be compared. For a matched pairs design, each student will participate in two sessions of currency trading. Half of the students will be randomly assigned to do this with charts in the first session and without charts in the second session. The other half will trade without charts first and with charts second. Each students' results will be compared between the two sessions.

3.63 **(a)** There is no random assignment here. To randomize this, each subject should be assigned a number from 01 through 50, and a random number table or software should be used to select the random groups. **(b)** All eight tosses could come up heads, or all eight could come up tails. This may not result in two groups of four. Assign each subject a number and randomly assign them to groups based on these numbers. **(c)** This is assuming that all rats are essentially identical. This may not be true in the different batches. If there were any differences in the way the rats were treated or fed, or if there were an illness in one of the groups, and so on, it would make that group sufficiently different from the other groups and lead to different results. Each batch should be randomly assigned with five rats in each of the treatment groups.

3.65 Answers will vary. Some variables might include rank within company (manager, employee, etc.), gender, previous team building experiences, and years working at the company.

3.67 Answers will vary. Design should include using a control group of trained professionals and comparing the results with a volunteer group.

3.69 Applet.

3.71 **(a)** This is a completely randomized design with two treatment groups. **(b)** Answers will vary using software, but using line 131 on Table B, Group 1 will contain: Dubois (05), Travers (19), Chen (04), Ullman (20), Quinones (16), Thompson (18), Fluharty (07), Lucero (13), Afifi (02), and Gerson (08). Group 2 will then contain the remaining subjects: Abate, Brown, Engel, Gutierrez, Hwang, Iselin, Kaplan, McNiell, Morse, and Rosen.

3.73 **(a)** Randomly assign your subjects to either Group 1 or Group 2. Each group will taste and rate both the regular and the light mocha drink. However, Group 1 will drink them in the regular/light order, and Group 2 will drink them in the light/regular order. For each group, the taste ratings of the regular and light drinks will be compared, and then the results of the two groups will be compared to see if the order of tasting made a difference to the ratings. To properly blind the subjects, both mocha drinks should be in identical opaque (we are only measuring taste, not appearance) cups with no labels on them. **(b)** Using line 141 on Table B, the regular/light group will use subjects with labels: 12, 16, 02, 08, 17, 10, 05, 09, 19, 06. The light/regular group will use the remaining 10 subjects.

3.75 This is a comparative experiment with two treatments: steady price and price cuts. The explanatory variable is the price history that is viewed. The response variable is the price the student would expect to pay for the detergent.

3.77 (a) The subjects are the 210 children. (b) The factor is the set of choices that are presented to each subject. The levels correspond to the three sets of choices. The response variable is whether they chose a milk drink or a fruit drink. (c) Use a diagram like those in Figures 3.4 and 3.5 to describe the design. (d) The first 5 children to be assigned to receive Set 1 have labels 119, 033, 199, 192, and 148.

3.79 (a) Use a diagram like those in Figures 3.4 and 3.5 to describe the design. (b) As a practical issue, it may take a long time to carry out the study. Some people might object because some participants will be required to pay for part of their health care, while others will not. People who have to pay a greater portion of their costs may choose not to get needed medical care, which could result in ethical issues.

3.81 (a) Use a diagram like those in Figures 3.4 and 3.5 to describe the design. (b) Each subject will do the task twice, once under each temperature condition in a random order. The difference in the number of correct insertions at the two temperatures is the observed response.

3.83 (a) Diagram shows random assignment into three groups, one with each treatment as described. (b) Using line 123, subjects 102, 063, 035 and 090 would be the first four selected.

3.85 This section talks about using a statistic to estimate a parameter. You would not use information from a voluntary response sample to predict population percentages because these samples usually have quite a bit of bias associated with them and would therefore not give accurate results for the population.

3.87 The center of the sampling distribution would be the same with either sample size; however, the spread of the sampling distribution would be smaller with the sample of size 400 than the sample with size 200.

3.89 The firm wants to decrease the variability of the sample. Selecting a larger sample will accomplish this as well as make the *margin of error* smaller.

3.91 (a) The population is all students in U.S. four-year colleges. The sample is the 17,096 students selected for the survey. (b) The population consists of all restaurant workers while the sample contains the 100 restaurant workers in the study. (c) The population here is the 584 longleaf pines in the forest tract, and the sample is the 40 trees that were measured.

3.93 (a) Variability is based on the sample size, not the population size. It is a measure of how different one sample of 2000 people in one state would be from another sample of 2000 people in that same state. Variability will not change from state to state based on different population sizes. (b) The variability will change from state to state as the number of people sampled will vary from state to state and as variability is based on the sample size.

3.95 Answers will vary. (a) One attempt of this problem resulted in an average $\hat{p}$ value of 0.585 with a standard deviation of 0.06747 and a range from 0.38 to 0.72. (b) One set of generated data resulted in an average $\hat{p}$ value of 0.294 with a standard deviation of 0.05857 and values ranging from 0.12 to 0.42. The average $\hat{p}$ value is lower when the population parameter p is lower and vice versa.

The spread of the data was also a little lower in this case when we lowered the p value. **(c)** One result gave an average $\hat{p}$ value of 0.595 with a standard deviation of 0.03295 and a range from 0.52 to 0.66. Increasing the sample size made only a slight change to the mean (due to a different set of generated values) but reduced the spread of the data quite a bit.

3.97 Applet.

3.99 **(a)** High variability, high bias. **(b)** Low variability, low bias. **(c)** High variability, low bias. **(d)** Low variability, high bias.

3.101 **(a)** Answers will vary. **(b)** The center of the distribution should be close to 0.50.

3.103 Applet.

3.105 **(a)** This one has "minimal risk." **(b)** The risk here is probably minimal as drawing blood is a part of routine physical exams, but this would be fall in a gray area under this vague definition. **(c)** This is not "minimal risk."

3.107 Answers will vary.

3.109 Answers will vary.

3.111 **(a)** Clinical trials should benefit the subjects, not just the future of science. Benefits and risks to the subjects should be balanced. **(b)** The board should also follow up on each experiment at least once a year. **(c)** Other considerations are important as well, such as personal rights of the individual, whether or not they give their informed consent, and so on.

3.113 Answers will vary.

3.115 Answers will vary.

3.117 This is not anonymous because it takes place in the person's home. However, it is confidential if the name/address is then separated from the response before the results are publicized.

3.119 **(a)** The pollsters must tell the potential respondents what type of questions will be asked and how long it will take to complete the survey. **(b)** This is required so the respondents can make sure the polling group is legitimate or so the respondent can issue a complaint if necessary. **(c)** Revealing the sponsor would bias the poll as people with strong party affiliation would look for the "correct" answer to support their party. The sponsor should be announced so that readers know whether to question the motivation of the poll, the wording of the question, or the legitimacy of the polling group.

3.121 Psychology 001 uses dependent subjects, which does not seem ethical. The other two courses have acceptable alternatives that make the use of the students more ethical; however, if students feel their grade may be somehow tied to participation, there may be ethical issues here as well.

3.123 Answers will vary.

3.125 Answers will vary.

3.127 Information here may vary. Some ideas to include are that experiments include treatments applied to the individuals or units while surveys are based on observations of the way things happen in a natural setting. Experiments have an advantage in that they can be used to explore the possibility of a causation relationship by holding all variables fixed except the variable believed to cause the response. However, experiments often do not show situations in a real-life setting. Observational studies show things in a more realistic light but have a disadvantage that lurking variables cannot be controlled as they are in experiments.

3.129 Summaries should mention the confidentiality assurances that the NORC gives as well as the pledge not to use the data for any other than its intended purpose.

3.131 This is an experiment because the students are put into treatment groups. The treatments are (1) view a steady price history and (2) review a variable price history with temporary price cuts. The explanatory variable is the price history viewed; the response variable is the price the student would expect to pay for the detergent.

3.133 Answers will vary.

3.135 (a) This is an experiment because students are put into an environment to observe these ads and respond instead of responding naturally in a normal environment. (b) The explanatory variable is whether coupon offerings are seen or not. The response variable is the price the student would expect to pay.

3.137 (a) This is a completely randomized design with two treatment groups (20 men in each group). One group will receive a calcium supplement, and one will receive a placebo. At the end of the experiment, a blood pressure measurement will be taken and the blood pressures compared between the two groups. (b) Using Table B at line 131, the group receiving the calcium supplement will consist of Chen (05), Rodriguez (32), Hruska (19), Bikalis (04), Liang (25), O'Brian (29), Imrani (20), Guillen (16), Tullock (37), Willis (39), Plochman (31), Howard (18), Cranston (07), Fratianna (13), Rosen (33), Asihiro (02), Townsend (36), Krushchev (23), Marsden (27), and Tompkins (35).

3.139 (a) Without a control group, there is no reason to believe that the difference seen is due to the variable being studied. Other lurking variables might also be responsible for the change. For example, changes in job conditions, time away from work to meditate, and the act of doing something for yourself may also play a roll in the levels of job satisfaction. (b) The experimenter's knowledge may make him or her expect to see results from the meditation, and therefore they may see results that aren't really present. (c) In a proper experiment, a control group would be set up. This group could also have consideration from the company in terms of their health, such as a time set aside for daily exercise or personal time. A test or survey should be standardized to determine job satisfactions, and this should be given both before and after the experiment. The researcher giving the survey should not know which subjects were in which groups.

3.141 This is an experiment because there is a treatment imposed on the subjects. They are to taste the two types of muffins.

3.143 (a) An experiment to test the effectiveness of daytime running lights would compare the accident rate for cars with and cars without this feature. Because there are some areas where daytime running lights are more prevalent than others, multiple locations might be studied. (b) When new features are added to vehicles, they are very noticeable at first. However, after time they are not noticed as much. Therefore, even if the results show a decreased accident rate when daytime running lights are used, this result might not continue after people become used to this feature on the vehicles.

3.145 Label the students from 0001 to 3478. The first six students selected using line 125 of Table B would be 1868, 1844, 2351, 1962, 1033, and 3136.

3.147 **(a)** This study has a population that includes all students registered for classes at your college. Because part-time students would also purchase these clothes, they should also be included. **(b)** Different variables may be used to create a stratified sample. These might include gender, bachelors vs. masters or doctoral students, students involved in school activities or fraternities vs. those not involved, and/or full-time vs. part-time students. Any of these variables might show groups that are more or less likely to purchase college brand apparel that might influence their attitudes toward the labor practices. A student's major might also be an indicator about whether he or she would be more or less sympathetic regarding the labor practices in factories. **(c)** Answers will vary. Issues may include nonresponse rate, difficulty in contacting students, and so on.

3.149 The average return on the first three Mondays is "not significantly different from zero" means that the return on these Mondays is close enough to zero, and any difference from zero is just as likely to be due to random chance. These weeks are "significantly higher" than the last two weeks means that the last two weeks means that there is a decrease in the returns of these last two weeks that is too large to occur just by chance.

3.151 An experiment here would have two factors: time of day and ZIP code. The originating post office and the receiving address should be the same for all letters mailed. A selection of times of day should be made. For each time of day, a selected number of letters should be mailed, half with and half without ZIP codes. Variability may result due to the day of the week the letter is mailed, so this process may be repeated for each day of the week.

3.153 Answers will vary.

3.155 Simulations will give different results. For $p = 0.1$, the average should be close to 0.1, and the standard deviation should be close to 0.0387. For $p = 0.3$, the average should be close to 0.3, and the standard deviation should be close to 0.0592. For $p = 0.5$, the average should be close to 0.5, and the standard deviation should be close to 0.0645.

3.157 Subjects would more likely admit that they had used illegal drugs in a CASI. People are less likely to admit to something that might make them feel embarrassed or under disapproval, so they would be more likely to lie in a CAPI situation.

CHAPTER 4

4.1 We spun a nickel 50 times and got 22 heads. We estimate the probability of heads to be 0.44.

4.3 Answers will vary.

4.5 **(a)** The first 200 digits contain 21 0s. The proportion of 0s is 0.105. **(b)** Answers will vary. The proportion should be close to 0.1.

4.7 We tossed a thumbtack 100 times and it landed with the point up 39 times. The approximate probability of landing point up is .39.

4.9 **(a)** 0. **(b)** 1. **(c)** 0.01. **(d)** 0.6.

4.11 **(a)** The number of "buys" (1s) in our simulation was 50. The percent of simulated customers who bought a new computer is 50%. **(b)** The longest run of "buys" (1s) was 4. The longest run of "not buys" was 5.

4.13 **(a)** In our 100 draws, the number in which at least 14 people approved of labor unions was 16. The approximate probability is 0.16. **(b)** The shape of our histogram was roughly symmetric and bell-shaped, the center appeared to be about 59%, and the values ranged from 30% to 85%. **(c)** The shape of our histogram was roughly symmetric and bell-shaped, the center was about 59%, and the values ranged from 52% to 65%. **(d)** Both distributions are roughly symmetric with similar centers. The histogram in (b) is more spread out.

4.15 **(a)** Answers will vary but should be close to 6%. **(b)** This simulation represents 100 randomly selected items that have been purchased. A success is an item that is returned. Over the long run, there is a 0.06 probability that a purchased item will be returned.

4.17 **(a)** $S = \{$any number (including fractional values) between 0 and 24 hours$\}$. **(b)** $S = \{$any integer value between 0 and 11,000$\}$. **(c)** $S = \{0, 1, 2, 3, 4, 5, 6, 7, 8, 9, 10, 11, 12\}$. **(d)** One possibility is $S = \{$any number 0 or larger$\}$. **(e)** $S = \{$any whole number 0 or larger$\}$.

4.19 **(a)** P(death was either agriculture-related or manufacturing related) $= 0.177$. **(b)** P(death was related to some other occupation) $= 0.823$.

4.21 Model 1: Not legitimate. The probabilities do not sum to 1. Model 2: Legitimate. Model 3: Not legitimate. The probabilities do not sum to 1. Model 4: Not legitimate. Some of the probabilities are greater than 1.

4.23 **(a)** P(completely satisfied) $= 0.48$ because probabilities sum to 1. **(b)** P(dissatisfied) $= 0.09$.

4.25 **(a)** The area of a triangle is $(1/2) \times$ base $\times$ height $= (1/2) \times 1 \times 2 = 1$. **(b)** The probability that T is less than 1 is 0.5. **(c)** The probability that T is less than 0.5 is 0.125.

4.27 **(a)** $P(Y > 515) = 0.50$. **(b)** $P(Y > 631) = 0.16$.

4.29 **(a)** For "Small Business," the probability would be 0.09. For "Big Business," the probability would be 0.37. **(b)** Subtracting the answers in part (a) from 1.00, the probability of some confidence in small businesses is 0.91. The probability for big businesses is 0.63.

4.31 **(a)** All probabilities are between 0 and 1, and all probabilities sum to 1.00. **(b)** P(somewhat successful or very successful) $= 0.41 + 0.48 = 0.89$.

4.33 **(a)** The probability that a Canadian employee who works out of the home uses an automobile is 0.8. The same probability for a U.S. employee is 0.902. **(b)** The probability that a Canadian employee uses "sustainable" transportation is 0.187. The probability that a U.S. employee does the same is 0.082.

4.35 **(a)** $P(Y > 1) = 0.69$. **(b)** 0.28. **(c)** 0.65.

4.37 **(a)** P(not farmland) $= 0.08$. **(b)** P(either farmland or forest) $= 0.93$. **(c)** P(something other than farmland or forest) $= 0.07$.

4.39 **(a)** P(red) $= 0.12$. **(b)** P(M&M is green, yellow, or orange) $= 0.5$.

4.41   **(a)** Legitimate. **(b)** Not legitimate. The sum of the probabilities is greater than 1. **(c)** Not legitimate. The probabilities of the outcomes do not sum to 1.

4.43   **(a)** $P(A) = 0.29. P(B) = 0.18.$ **(b)** "A does not occur" means the farm is 50 acres or more. $P(A$ does not occur$) = 0.71.$ **(c)** "A or B" means the farm is less than 50 acres or 500 or more acres. $P(A$ or $B) = 0.47.$

4.45   **(a)** The eight arrangements of preferences are NNN, NNO, NON, ONN, NOO, ONO, OON, OOO. Each must have probability 0.125. **(b)** $P(X = 2) = 0.375.$ **(c)**

X	0	1	2	3
$P(X)$	0.125	0.375	0.375	0.125

4.47   **(a)** All the probabilities given are between 0 and 1 and they sum to 1. **(b)** $P(X \geq 5) = 0.11.$ **(c)** $P(X > 5) = 0.04.$ **(d)** $P(2 < X \leq 4) = 0.32.$ **(e)** $P(X = 1) = 0.75.$ **(f)** $P($a randomly chosen household contains more than two persons$) = P(X > 2) = 0.43.$

4.49   **(a)** $P(X > 550) = 0.3821.$ **(b)** $P(600 < X < 700) = 0.1768.$

4.51   **(a)** Continuous. All times greater than 0 are possible without any separation between values. **(b)** Discrete. It is a count that can take only the values 0, 1, 2, 3, or 4. **(c)** Continuous. Any number 0 or larger is possible without any separation between values. **(d)** Discrete. Household size can take only the values 1, 2, 3, 4, 5, 6, or 7.

4.53   **(a)** All the probabilities given are between 0 and 1, and they sum to 1. **(b)** $\{X \geq 1\}$ means the household owns at least 1 car. $P(X \geq 1) = 0.91.$ **(c)** Households that have more cars than the garage can hold have 3 or more cars. 20% of households have more cars than the garage can hold.

4.55   **(a)** The probability that a tire lasts more than 50,000 miles is $P(X > 50,000) = 0.5.$ **(b)** $P(X > 60,000) = 0.0344.$ **(c)** The normal distribution is continuous, so $P(X = 60,000) = 0$ and $P(X \geq 60,000) = P(X > 60,000) = 0.0344.$

4.57 $\mu = 169.5.$ If we record the size of the hard drive chosen by many, many customers in the 60-day period and compute the average of these sizes, the average will be close to 169.5. Knowing μ is not very helpful, as it is not one of the possible choices and does not indicate which choice is most popular.

4.59   **(a)** $\mu_X = 1.9.\ \mu_Y = 3.05.$ **(b)** The mean cost for the product development process is \$15,200. The mean cost for the design of the manufacturing process is \$91,500. **(c)** The mean completion time is 4.95 weeks. The mean cost is \$106,700.

4.61 $\mu_Y = 445, \sigma_Y^2 = 19,225.00,$ and $\sigma_Y = 138.65.$

4.63   **(a)** The variance for the number of weeks to complete the development process is 0.49. The standard deviation is 0.7. **(b)** The variance for the number of weeks to complete the design of the manufacturing process is 1.048. The standard deviation is 1.023.

4.65 $\mu_{X-Y} = 100.\ \sigma_{X-Y}^2 = 10,000.\ \sigma_{X-Y} = 100.$ If the correlation between two variables is positive, then large values of one tend to be associated with large values of the other, resulting in a relatively small difference. Also, small values of one tend to be associated with small values of the other, again resulting in a relatively small difference. This suggests that when two variables are positively associated, they vary together, and the difference tends to stay relatively small and varies little.

4.67　(a) 720.

(b)

No. of transactions	0	1	2	3	4	5
% of clients	0.08	0.17	0.25	0.19	0.18	0.13

(c) Yes, if for 5 transactions the 0.125 is rounded to 0.13, then the probabilities add up to 1. If the 0.125 is rounded down to 0.12, the probabilities add up to 0.99.

4.69　(a) 0.0010. (b) 0.1131. (c) 394.24 bets per second.

4.71　(a) −1280.

(b)

X	218,720	−1280
probability	0.003	0.997

4.73　(a) Two important differences between the histograms are (1) the renter-occupied rooms has a distribution that is skewed to the right with a center that is to the left of the center of the distribution of the number of rooms of owner-occupied units, and (2) the spread of the distribution of the number of rooms of owner-occupied units is slightly larger than the spread of the distribution of the number of rooms of renter-occupied units. (b) $\mu_{owned} = 6.389$. $\mu_{rented} = 4.386$. The mean number of rooms for owner-occupied units is larger than the mean number of rooms for renter-occupied units. This reflects the fact that the center of the distribution of the number of rooms of owner-occupied units is larger than the center of the distribution of the number of rooms of renter-occupied units.

4.75　The histogram for renter-occupied units is more peaked and less spread out than the histogram for owner-occupied units. $\sigma^2_{owned} = 2.8398$, $\sigma^2_{rented} = 2.0232$, $\sigma_{owned} = 1.6852$, $\sigma_{rented} = 1.4224$.

4.77　Answers will vary.

4.79　There are 1000 three-digit numbers (000 to 999). If you pick a number with three different digits, your probability of winning is 6/1000, and your probability of not winning is 994/1000. Your expected payoff is $0.49998.

4.81　(a) We would expect X and Y to be independent because they correspond to events that are widely separated in time. (b) Experience suggests that the amount of rainfall on one day is not closely related to the amount of rainfall on the next. We might expect X and Y to be independent or, because they are not widely separated in time, perhaps slightly dependent. (c) Orlando and Disney World are close to each other. Rainfall usually covers more than just a very small geographic area. We would not expect X and Y to be independent.

4.83　(a) If X is the time to bring the part from the bin to its position on the automobile chassis, and Y is the time required to attach the part to the chassis, then the total time for the entire operation is $X + Y$. $\mu_{X+Y} = 31$ seconds. (b) It will not affect the mean. (c) The answer in (a) will remain the same in both cases.

4.85　$\sigma_{X+Y} = 4.47$. If X and Y have correlation 0.3, $\sigma_{X+Y} = 4.98$. If X and Y are positively correlated, then large values of X and Y tend to occur together, resulting in a very large value of $X + Y$. Likewise, small values of X and Y tend to occur together, resulting in a small value of $X + Y$. Thus, $X + Y$ exhibits larger variation when X and Y are positively correlated than if they are not.

4.87 $\mu_X = (\mu + \sigma)(0.5) + (\mu - \sigma)(0.5) = \mu$. $\sigma_X^2 = (\mu + \sigma - \mu_X)^2(0.5) + (\mu - \sigma - \mu_X)^2(0.5) = \sigma^2$. Thus, $\sigma_X = \sigma$.

4.89 **(a)** $\sigma_{X+Y}^2 = 7{,}819{,}225$ and $\sigma_{X+Y} = 2796.29$. **(b)** $\sigma_Z^2 = \sigma_{2000X+3500Y}^2 = 3.14355 \times 10^{13}$. Thus, $\sigma_Z = 5{,}606{,}739$.

4.91 **(a)** The two students are selected at random, and we expect their scores to be unrelated or independent. **(b)** $\mu_{\text{female}-\text{male}} = 15$. $\sigma_{\text{female}-\text{male}}^2 = 2009$ and $\sigma_{\text{female}-\text{male}}^2 = 44.82$. **(c)** We cannot find the probability that the woman-chosen scores are higher than the man-chosen scores because we do not know the probability distribution for the scores of women or men.

4.93 **(a)** The mean of the difference between Y and X is 0.87. The standard deviation is 11.77. **(b)** The probability that the difference between Y and X will be positive, and Vegas will win the bet, is 0.5279.

4.95 $\sigma_{0.8W+0.2Y}^2 = 0.001122$. Thus, $\sigma_{0.8W+0.2Y} = 0.0335$. This is smaller than the result in Exercise 4.94 because we no longer include the positive term $2\rho\sigma_{0.8W}\sigma_{0.2Y}$. The mean return remains the same.

4.97 If $\rho_{XY} = 1$, $\sigma_{X+Y}^2 = \sigma_X^2 + \sigma_Y^2 + 2\rho_{XY}\sigma_X\sigma_Y = \sigma_X^2 + \sigma_Y^2 + 2\sigma_X\sigma_Y = (\sigma_X + \sigma_Y)^2$. So $\sigma_{X+Y} = \sigma_X + \sigma_Y$.

4.99 **(a)** $\mu_X = 550$. $\sigma_X = 5.7$. **(b)** $\mu_{X-550} = 0$. $\sigma_{X-550}^2 = 32.5$ and $\sigma_{X-550} = 5.7$. **(c)** $\mu_Y = \mu_{(9X/5)+32} = 1022$. $\sigma_Y^2 = \sigma_{(9X/5)+32}^2 = 105.3$ and $\sigma_Y = 10.26$.

4.101 These are statistics. Pfeiffer undoubtedly tested only a sample of all the models produced by Apple, and these means are computed from these samples.

4.103 The law of large numbers says that in the long run, the average payout to Joe will be $0.60. However, Joe pays $1.00 to play each time, so in the long run his average winnings are −$0.40. Thus, in the long run, if Joe keeps track of his net winnings and computes the average per bet, he will find that he loses an average of $0.40 per bet.

4.105 That is not right. The law of large numbers tells us that the long-run average will be close to 34%. Six of seven at-bats is hardly the "long run." Furthermore, the law of large numbers says nothing about the next event. It only tells us what will happen if we keep track of the long-run average.

4.107 **(a)** Using statistical software, we compute the mean of the 10 sizes to be $\mu = 69.4$. **(b)** Using line 120, our SRS is companies 3, 5, 4, 7. $\bar{x} = 67.25$. **(c)** The center of our histogram appears to be at about 69, which is close to the value of $\mu = 69.4$ computed in part (a).

4.109 **(a)** To say that $\bar{x}$ is an unbiased estimator of μ means that $\bar{x}$ will neither systematically overestimate or underestimate μ in repeated use and that if we take many, many samples, calculate $\bar{x}$ for each, and compute the average of these $\bar{x}$-values, this average would be close to μ. **(b)** If we draw a large sample from a population, compute the value of some statistic (such as $\bar{x}$), repeat this many times, and keep track of our results, these results will vary less from sample to sample than the results we would obtain if our samples were small.

4.111 The sampling distribution of $\bar{x}$ should be approximately $N\,(1.6,\ 0.085)$. $P(\bar{x} > 2) = P(Z > 4.76)$, which is approximately 0.

4.113 500,000,000 is a parameter, and 5.6 is a statistic.

4.115 19 is a parameter. 14 is a statistic.

4.117 **(a)** The mean $\bar{x}$ of $n = 3$ weighings will follow an $N\,(123, 0.0462)$ distribution. **(b)** $P(\overline{X} \ge 124) = P(Z \ge 21.65)$, which is approximately 0.

4.119 $P(\overline{X} > 210.53) = 0.0052$.

4.121 0.0125.

4.123 The range 0.11 to 0.19 will contain approximately 95% of the many $\bar{x}$'s.

4.125 **(a)** $P(\bar{x} > 400) = P(Z > 10.94)$, which is approximately 0. **(b)** We would need to know the distribution of weekly postal expenses to compute the probability that postage for a particular week will exceed \$400. In part (a), we applied the central limit theorem, which does not require that we know the distribution of weekly expenses.

4.127 **(a)** $E(X) = -620$ and $\sigma_X = 12{,}031.808$. **(b)** $E(Y) = 219{,}340$ and $\sigma_Y = 12{,}031.808$. **(c)** The standard deviation is the same for both.

4.129 **(a)** 0.0866. **(b)** 9. **(c)** There is always sampling variability, but this variability is reduced when an average is used instead of an individual measurement. The larger the sample size, the smaller the variability.

4.131 **(a)** 0.3544. **(b)** 0.8558. **(c)** 0.9990. **(d)** The probabilities using the central limit theorem are more accurate as the sample size increases. The 150-bag probability calculation is probably fairly accurate, but the 3-bag probability calculation is probably not.

4.133 P(Chavez promoted) $= 0.55$.

4.135 **(a)** All probabilities must add to 1.00, so the probability that a person does not believe global warming is taking place would be 0.01. **(b)** No, because there is a probability of 0.02 that the person will have no opinion and 0.08 that the person will think that it is debatable whether global warming is taking place or not.

4.137 **(a)** All the probabilities listed are between 0 and 1, and they sum to 1. **(b)** P(worker is female) $= 0.43$. **(c)** P(not in occupation F) $= 0.96$. **(d)** P(occupation D or E) $= 0.28$. **(e)** P(not in occupation D or E) $= 0.72$.

4.139 The probability distribution is:

Y	1	2	3	4	5	6	7	8	9	10	11	12
$P(Y)$	1/12	1/12	1/12	1/12	1/12	1/12	1/12	1/12	1/12	1/12	1/12	1/12

4.141 The weight of a carton varies according to an $N(780, 17.32)$ distribution. Letting Y denote the weight of a carton, $P(750 < Y < 825) = P(-1.73 < Z < 2.60) = 0.9535$.

4.143 **(a)** All the probabilities listed are between 0 and 1, and they sum to 1. **(b)** $\mu = 2.45$.

4.145 $\rho = 0$.

4.147 $P(\overline{X} > \text{mean monthly return}) = 0.90$ results in a mean monthly return of -0.0411 for Dell, -0.0383 for Apple, and -0.0306 for the portfolio. 90% of the time the mean monthly return would be expected to be greater than these values.

4.149 If you sold only 10 policies, you would probably lose a lot of money. Even though the average loss per person is only $400, that average comes from a distribution that has some losses equal to $0 and a very few losses that are greater, perhaps $200,000 or more. One loss would be much more than the profit from 9 policies. If, instead, you sold thousands of policies, the gain from the many, many policies that paid out $0 would be much more than the few losses incurred.

4.151 $P(\text{age at death} \geq 26) = 0.99058. \mu_X = 303.35.$

4.153 $\sigma_X^2 = 94,236,826.6. \sigma_X = 9707.57.$

4.155 (a) $S = \{0, 1, 2, 3, \ldots, 100\}$ is one possibility (but assuming a person could be employed for 100 years is extreme). (b) Because we are allowing only a finite number of possible values, X is discrete. (c) We included 101 possible values (0 to 100).

4.157 (a) $S = \{$ all numbers between 0 and 35 ml with no gaps$\}$. (b) S is a continuous sample space. All values between 0 and 35 ml are possible with no gaps. (c) We included an infinite number of possible values.

CHAPTER 5

5.1 (a) The probability that the rank of the second student falls into the five categories is unaffected by the rank of the first student selected. (b) 0.1681. (c) 0.0041.

5.3 (a) P(all are truthful) $= (0.841)^5 = 0.421.$ (b) P(at least one has a misrepresentation) $= 1 - $ P(all are truthful) $= 0.579$

5.5 0.66761.

5.7 (a) 15% drink only cola. (b) 20% drink none of these beverages.

5.9 (a) $W = \{0, 1, 2, 3\}.$

W	Arrangements	Part (b) Probability of each arrangement	Part (c) Probability for W
$W = 0$	DDD	$(0.73)^3 = 0.389$	0.389
$W = 1$	DDF	$(0.27)(0.73)^2 = 0.144$	0.432
	DFD	$(0.27)(0.73)^2 = 0.144$	
	FDD	$(0.27)(0.73)^2 = 0.144$	
$W = 2$	DFF	$(0.27)^2(0.73) = 0.0532$	0.160
	FDF	$(0.27)^2(0.73) = 0.0532$	
	FFD	$(0.27)^2(0.73) = 0.0532$	
$W = 3$	FFF	$(0.27)^3 = 0.0197$	0.197

5.11   **(a)** $(0.09)^6 = 0.0000005$. **(b)** $(0.91)^6 = 0.568$. **(c)** 0.337.

5.13   0.09608.

5.15   **(a)** 0.2746. **(b)** The probability of the price being down in any given year is $1 - 0.65 = 0.35$. Because the years are independent, the probability of the price being down in the third year is 0.35. **(c)** 0.5450.

5.17   0.8.

5.19   Yes. A and B are independent if $P(A \text{ and } B) = P(A)P(B)$ and $0.3 = (0.6)(0.5)$.

5.21   This is not correct. If $P(A \text{ or } B) = P(A) + P(B)$, then this means $P(A \text{ and } B) = 0$. If A and B were independent, then $P(A \text{ and } B) = P(A)P(B)$. But this cannot be equal unless one of $P(A)$ and $P(B)$ is zero. Because neither are zero, A and B must not be independent.

5.23   **(a)** The probability of an 11 is 2/36. The probability of three 11s in three independent throws is 0.000171. **(b)** The writer's first statement is correct. The odds against throwing three straight 11s, however, are $\dfrac{1-p}{p} = \dfrac{1-(2/36)^3}{(2/36)^3} = 5831$ to 1. When computing the odds for the three tosses, the writer multiplied the odds, which is not the correct way to compute the odds for the three throws.

5.25   0.1472.

5.27   **(a)** 0.6946. **(b)** 0.7627. **(c)** 0.2921. **(d)** 0.0158. **(e)** 0.1884.

5.29   **(a)** 0.5833. **(b)** 0.3621. **(c)** If the events A and B were independent, then $P(B|A) = P(B)$, which is not the case.

5.31   Using Bayes's rule, the probability is 0.7660.

5.33   33.70%.

5.35   0.8989. Given that the customer defaults on the loan, there is an 89.89% chance that the customer overdraws the account. This percent changes to 81.63% when there is a 25% chance that the customer will overdraw his account.

5.37   **(a)** 0.25. **(b)** 0.3333.

5.39   0.0043.

5.41 $P(Y < 1/2 \text{ and } Y > X) = 1/8$. Drawing a diagram should help.

5.43   Unemployment rates are 0.0714 for "Did not finish High School," 0.0436 for "High School," 0.0355 for "Some College," and 0.0202 for "Bachelor's or higher." The unemployment rate decreases for higher levels of education. If the results were independent, then the probability of a person being unemployed would not change given a particular level of education; $P(B \mid A) = P(B)$.

5.45   **(a)** 0.20. **(b)** 0.62. These answers are simpler to see if you first draw a tree diagram.

5.47   Using Bayes's rule, the probability is 0.323.

5.49   **(a)** Using Bayes's rule, the probability is 0.064. **(b)** Expect 94 not to have defaulted. **(c)** With the credit manager's policy, the vast majority of those whose future credit is denied would not have defaulted.

5.51   Yes. A and B are independent if $P(A \text{ and } B) = P(A)P(B)$ and $0.3 = (0.6)(0.5)$.

5.53   Using Bayes's rule, the probability is 0.1930.

5.55   Because the number of trials is not fixed, X does not have a binomial distribution.

5.57   (a) X can take any of the values 0, 1, 2, 3, 4, or 5. (b) $P(X = 0) = 0.4207$, $P(X = 1) = 0.3977$, $P(X = 2) = 0.1504$, $P(X = 3) = 0.0284$, $P(X = 4) = 0.0027$, and $P(X = 5) = 0.0001$.

5.59 $P(X = 11) = 0.0074$.

5.61 $\mu = 8$, and $\sigma = \sigma = \sqrt{np(1-p)} = \sqrt{20(0.4)(0.6)} = 2.191$.

5.63   (a) $\mu = 16$. (b) $\sigma = \sigma = \sqrt{np(1-p)} = \sqrt{20(0.8)(0.2)} = 1.789$. (c) When $p = 0.9$, $\sigma = 1.342$; and when $p = 0.99$, $\sigma = 0.455$. As the value of p gets closer to 1, there is less variability in the values of X.

5.65   (a) Using the Normal approximation, $P(X \geq 100) = 0.0019$. This probability is extremely small and suggests that p may be greater than 0.4. (b) $P(X \geq 10)$ was evaluated for a sample of size 20 and found to be 0.2447. The proportion in the sample will be closer to 40% for larger sample sizes, so the chance of the proportion in the sample being as large as 50% decreases.

5.67   (a) $\mu_x = np = 6$, $\mu_{\hat{p}} = p = 0.5$. (b) $\mu_x = 60$ if $n = 120$, and $\mu_x = 600$ if $n = 1200$. $\mu_{\hat{p}}$ stays the same regardless of sample size.

5.69   (a) X is binomial with $n = 1555$ and $p = 0.20$. (b) Using the Normal approximation to the binomial for proportions, $\mu_{\hat{p}} = 0.20$, $\sigma_{\hat{p}} = 0.0101$, and the $P(\hat{p} \leq 0.193) = 0.2451$. Using software for the binomial distribution, $P(X \leq 300) = 0.254$.

5.71   (a) The 20 machinists selected are not an SRS from a large population of machinists and could have different success probabilities. (b) We know that the count of successes in an SRS from a large population containing a proportion p of successes is well approximated by the binomial distribution. This description fits this setting.

5.73   (a) $n = 10$, and $p = 0.25$. (b) $P(X = 2) = 0.2816$. (c) $P(X \leq 2) = P(X = 0) + P(X = 1) + P(X = 2) = 0.5256$. (d) $\mu = 2.5$, and $\sigma = \sqrt{np(1-p)} = \sqrt{10(0.25)(0.75)} = 1.37$.

5.75   (a) $n = 5$, and $p = 0.65$. (b) The possible values of X are 0, 1, 2, 3, 4, 5. (c) $P(X = 0) = 0.0053$, $P(X = 1) = 0.0488$, $P(X = 2) = 0.1811$, $P(X = 3) = 0.3364$, $P(X = 4) = 0.3124$, and $P(X = 5) = 0.1160$. (d) $\mu = 3.25$, and $\sigma = \sqrt{np(1-p)} = \sqrt{5(0.65)(0.35)} = 1.067$. The value of 3.25 should be included in the histogram from part (c) and should be a good indication of the center of the distribution.

5.77   (a) Using the Normal approximation, $P(X \leq 70) = 0.1251$. (b) $P(X \leq 175)$ corresponds to Jodi scoring 70% or lower. Using the Normal approximation, $P(X \leq 175) = 0.0344$.

5.79   (a) The binomial distribution with $n = 150$, and $p = 0.5$ is reasonable for the count of successes in an SRS of size n from a large population, as in this case. (b) We expect 75 businesses to respond. (c) Using the Normal approximation, $P(X \leq 70) = 0.2061$. (d) n must be increased to 200.

5.81   (a) $\mu = 180$, and $\sigma = \sqrt{np(1-p)} = \sqrt{1500(0.12)(0.88)} = 12.586$. (b) Using the Normal approximation, $P(X \leq 170) = 0.2148$.

5.83 $P(X \geq 261.5) = 0.0069$. This value is closer to the exact probability than the original approximation.

5.85 (a) Poisson distribution with a mean of $12 \times 7 = 84$ accidents. (b) $P(X \le 66) = 0.0248$.

5.87 (a) The employees are independent and each equally likely to be hospitalized. (b) 0.0620. (c) 0.938. (d) 0.256.

5.89 (a) 0.0821. (b) 0.2424.

5.91 (a) Poisson distribution with a mean of 48.7. $P(X \ge 50) = 1 - P(X \le 49) = 0.4450$. (b) For a 15-minute period, $\sigma = \sqrt{\mu} = \sqrt{48.7} = 6.98$. For a 30-minute period, $\sigma = \sqrt{\mu} = \sqrt{97.4} = 9.87$. (c) Poisson with mean $2 \times 48.7 = 97.4$. $P(X \ge 100) = 1 - P(X \le 99) = 1 - 0.5905 = 0.4095$.

5.93 (a) $\sigma = \sqrt{\mu} = \sqrt{15} = 3.873$. (b) $P(X \le 10) = 0.1185$. (c) $P(X > 30) = 1 - P(X \le 30) = 1 - 0.9998 = 0.0002$.

5.95 (a) $P(X \ge 5) = 1 - P(X \le 4) = 1 - 0.0018 = 0.9982$. (b) Poisson with mean $1/2 \times 14 = 7$. $P(X \ge 5) = 1 - P(X \le 4) = 1 - 0.1730 = 0.8270$. (c) Poisson with mean $1/4 \times 14 = 3.5$. $P(X \ge 5) = 1 - P(X \le 4) = 1 - 0.7254 = 0.2746$.

5.97 (a) $\sigma = \sqrt{\mu} = \sqrt{2.3} = 1.52$. (b) $P(X > 5) = 1 - P(X \le 5) = 1 - 0.9700 = 0.0300$. (c) $k = 3$.

5.99 (14632.58, 15367.42).

5.101 (a) 0.1416. (b) 0.1029. (c) The people are independent from each other, and each person has a 70% chance of being male.

5.103 0.75.

5.105 0.84.

5.107 (a) 0.5649. (b) 0.2377. (c) 0.0483. (d) 0.6677.

5.109 (a) Drawing the tree diagram will be helpful in solving the remaining parts. (b) 0.01592. (c) 0.627.

5.111 (a) $(0.99) \times (0.01)$. (b) $(0.99)^2 (0.01)$. (c) $P(\text{first defective is } k\text{th item}) = (0.99)^{k-1}(0.01)$.

5.113 (a) 0.799. (b) 0.921.

5.115 (a) 0.151. (b) 0.0465.

5.117 Using Bayes's rule, the probability is 0.0404.

5.119 (a) $\mu = 3.75$. (b) $P(X \ge 10) = 1 - P(X \le 9) = 1 - 0.9992 = 0.0008$. (c) Using the Normal approximation, $P(X \ge 275) = 0.0336$.

5.121 (a) $\mu = 1250$. (b) Using the Normal approximation, $P(X \ge 1245) = 0.5596$.

5.123 (a) The probability of a success p should be the same for each observation. To ensure that this is true, it is important to take the observations under similar conditions, such as the same location and time of day. (b) The probability of the driver being male for observations made outside a church on Sunday morning may differ from the probability for observations made on campus after a dance. (c) $P(X \le 8) = 0.4557$. (d) Using software, $P(X \le 80) = 0.1065$.

5.125 The Poisson distribution requires that the probability of a success occurring in any 10-minute period will be the same for all 10-minute periods. Because that is not true here, the Poisson distribution does not apply.

5.127 0.0334.

5.129 0.1805.

CHAPTER 6

6.1   The standard deviation for x is $22.

6.3 $2 \times$ (standard deviation for x) = $44.

6.5   ($394.70, $415.34).

6.7 $n = 1244.68$. Use $n = 1245$.

6.9   **(a)** The standard deviation should be divided by the square root of the sample size. **(b)** The confidence interval gives a range for the population mean, not the sample mean. **(c)** The confidence interval does not estimate where a portion of the population falls, only where the population mean falls. **(d)** The central limit theorem says that the distribution of the sample mean will be approximately normal when the sample size is large enough. It does not indicate anything about the distribution of the population.

6.11

Sample size	10	20	40	100
Margin of error	3.10	2.19	1.55	0.98

As sample size increases, the width of the confidence interval (or the size of the margin of error) decreases.

6.13   The students in your major will have a smaller standard deviation because many of them will be taking the same classes that require the same textbooks. The smaller standard deviation leads to a smaller margin of error.

6.15   **(a)** In general, a 100% confidence interval for the mean would extend from negative infinity to positive infinity. In other words, it would contain every possible value, however unlikely, for the population mean. **(b)** The 0% confidence interval would just be the sample mean. There is a 0% chance, or immeasurably small chance, that the population mean and the sample mean would be exactly equal to each other in a continuous distribution.

6.17   The given interval can be used to find the sample mean. The z^* value can then be exchanged for the desired z^* value, resulting in the interval (327.32, 359.01).

6.19   The high nonresponse rate could result in a biased result if the customers who did not respond would have given different answers from those who did respond. A small margin of error around an incorrect value will not accurately reflect the desired population value.

6.21 9.29 ± 0.92.

6.23   **(a)** 115 ± 13.25 minutes, or (101.75 minutes, 128.25 minutes). **(b)** No. The confidence coefficient of 95% is the probability that the method we used will give an interval containing the correct value of the population mean study time. It does not tell us about individual study times.

6.25   **(a)** The mean weight of the runners in kilograms is $\overline{x} = 61.863$. The mean weight of runners in pounds is 136.099. **(b)** The standard deviation of the mean weight in kilograms is 0.9186. The standard deviation of the mean weight in pounds is 2.0208. **(c)** A 95% confidence interval for the mean weight, in kilograms, of the population is (60.062, 63.664). In pounds, we get (132.137, 140.061).

6.27 11.8 ± 0.77 years or (11.03 years, 12.57 years).

6.29   The sample size should be 75 shoppers.

6.31   **(a)** No. We are 95% confident that the true population percent falls in this interval. **(b)** This particular interval was produced by a method that will give an interval that contains the true percent of the population who like their job 95% of the time. When we apply the method once, we do not know if our interval correctly includes the population percent or not. Because the method yields a correct result 95% of the time, we say we are 95% confident that this is one of the correct intervals. **(c)** 1.531. **(d)** No, the margin of error only covers random variation.

6.33   Answers will vary. Some possibilities include blue collar vs. white collar jobs, and management vs. entry- and mid-level employees.

6.35   **(a)** 95% confidence means that this particular interval was produced by a method that will give an interval that contains the true percent of the population who will vote for Ringel 95% of the time. When we apply the method once, we do not know if our interval correctly includes the true population percent or not. Thus, 95% refers to the method, not to any particular interval produced by the method. **(b)** The margin of error is 3%, so the 95% confidence interval is 52% $\pm$ 3% = (49%, 55%). The interval includes 50%, so we cannot be "confident" that the true percent is 50% or even slightly less than 50%. Hence, the election is too close to call from the results of the poll.

6.37   The results are not trustworthy. The formula for a confidence interval is relevant if our sample is an SRS (or can plausibly be considered an SRS). Phone-in polls are not SRSs.

6.39 $H_0: \mu = 0$, $H_a: \mu < 0$.

6.41   **(a)** $z = -1.58$. **(b)** P-value $= 0.1142$. This would not be considered strong evidence that Cleveland differs from the national average.

6.43   If the homebuilders have no idea whether Cleveland residents spend more or less than the national average, then they are not sure whether μ is larger or smaller than 31%. The appropriate hypotheses are $H_0: \mu = 31\%$, $H_a: \mu \neq 31\%$.

6.45 z-values that are significant at the $\alpha = 0.005$ level are $z > 2.807$ and $z < -2.807$.

6.47   The significance level corresponding to $z^* = 2$ is 0.0456, and the significance level corresponding to $z^* = 3$ is 0.0026.

6.49   **(a)** P-value $= 0.0287$. **(b)** P-value $= 0.9713$. **(c)** P-value $= 0.0574$.

6.51   **(a)** With a P-value of 0.07, we would not reject $H_0: \mu = 15$, and thus 15 would not fall outside the 95% confidence interval. **(b)** With a P-value of 0.07, we would reject $H_0: \mu = 15$, and thus 15 would fall outside the 90% confidence interval.

6.53   **(a)** Yes. $0.049 < 0.05$. **(b)** No. $0.049 > 0.01$. **(c)** Reject the null hypothesis whenever the P-value is $\leq \alpha$.

6.55   **(a)** The null hypothesis should be $\mu = 0$. The alternative hypothesis could be $\mu > 0$. **(b)** The standard deviation of the sample mean should be $18/\sqrt{30}$. **(c)** The hypothesis test uses the population parameter μ, not the sample statistic x.

6.57   If a significance level of 0.05 is used, do not reject the null hypothesis. There is not enough evidence to say that students in the West region have a significantly different amount of credit card debt than those in the South region.

6.59 (a) $H_0: \mu = 31$, $H_a: \mu > 31$. (b) $H_0: \mu = 4$, $H_a: \mu = 4$. (c) $H_0: \mu = 1400$, $H_a: \mu < 1400$.

6.61 (a) $H_0: p_M = p_F$, $H_a: p_M > p_F$, where the parameters of interest are the percent of males, p_M, and the percent of females, p_F, in the population who name economics as their favorite subject. (b) $H_0: \mu_A = \mu_B$, $H_a: \mu_A > \mu_B$, where μ_A is the mean score on the test of basketball skills for the population of all sixth-grade students if all were treated as those in group A, and μ_B is the mean score on the test of basketball skills for the population of all sixth-grade students if all were treated as those in group B. (c) $H_0: \rho = 0$, $H_a: \rho > 0$, where the parameter of interest is the correlation ρ between income and the percent of disposable income that is saved by employed young adults.

6.63 When comparing the percent of customers who planned to purchase a new vehicle of the same make, the percents were close for customers who never called or who had a positive resolution to their complaints. Any difference between these two percents was small enough that it could be just due to the selection of people who were chosen. However, when comparing either of these groups to the group of people who complained and had a negative resolution to their problem, the percent that would purchase a new vehicle was so much smaller than the other two groups that it would only happen in the most rare circumstance. With a P-value lower than 0.001, this indicates that we can be extremely confident that there is a significantly lower percent of those with a negative resolution to their complaint who plan to purchase the same make again.

6.65 (a) Answers may vary, but one possibility is a two-sample comparison of means test with $H_0: \mu_A = \mu_B$, $H_a: \mu_A \neq \mu_B$ where group A consists of students who exercise regularly and group B consists of students who do not exercise regularly. (b) No, the P-value is large. There is not enough evidence to say that exercise significantly affects how students perform on their final exam in statistics. (c) It would be good to know how the sample was selected and if this was an observational study or an experiment.

6.67 $z = 3.56$. The P-value is $P(Z \geq 3.56) < 0.002$. We would conclude that there is strong evidence that these 5 sonnets come from a population with a mean number of new words that is larger than 6.9, and thus we have evidence that the new sonnets are not by our poet.

6.69 The P-value is 0.0164. This is reasonably strong evidence against the null hypothesis that the population mean corn yield is 135. Our results are based on the mean of a sample of 50 observations, and such a mean may vary approximately according to a Normal distribution (by the central limit theorem) even if the population is not Normal, so our conclusions are probably still valid.

6.71 (a) The P-value $= 0.2040$, and the result is not significant at the 5% level. (b) The P-value is 0.2040, and the result is not significant at the 1% level.

6.73 The approximate P-value is <0.001.

6.75 (a) $z > 1.645$. (b) $z > 1.96$ or $z < -1.96$. (c) In part (a), we reject H_0 only for large values of z because such values are evidence against H_0 in favor of $H_a: \mu > 0$. In part (b), we reject H_0 in favor of $H_a: \mu \neq 0$ if either z is too large or z is too small because both extremes are evidence against H_0 in favor of $H_a: \mu \neq 0$.

6.77   **(a)** (99.041, 109.225). **(b)** The hypotheses are $H_0: \mu = 105$, $H_a: \mu \neq 105$. The 95% confidence interval in part (a) contains 105, so we would not reject H_0 at the 5% level.

6.79   **(a)** $H_0: \mu = 7$, $H_a: \mu \neq 7$. The 95% confidence interval does not contain 7, so we would reject H_0 at the 5% level. **(b)** 5 is in the 95% confidence interval, so we would not reject the hypothesis that $\mu = 5$ at the 5% level.

6.81   **(a)** $z = 1.64$. We reject H_0 at the 5% level if $z > 1.645$. The result is not significant at the 5% level. **(b)** $z = 1.65$. The result is significant at the 5% level.

6.83   Any convenience sample (phone-in or write-in polls, surveys of acquaintances only) will produce data for which statistical inference is not appropriate. Poorly designed experiments also provide examples of data for which statistical inference is not valid.

6.85   Statistical significance and practical significance are not necessarily the same thing. With a significance level this high, we would be willing to reject the null hypothesis even when no practical effect exists. A P-value of 0.5 corresponds to a Z value of 0, which means that our sample statistic would be 0 standard deviations away from the null hypothesis mean, but we would still reject the null hypothesis.

6.87   Answers will vary.

6.89   **(a)** No. **(b)** Yes. **(c)** No.

6.91   **(a)** P-value = 0.3821. **(b)** P-value = 0.1711. **(c)** P-value = 0.0013.

6.93   The conclusion is not justified. Statistical inference is not valid for badly designed surveys or experiments. This is a call-in poll, and such polls are not random samples.

6.95   Using the Bonferroni procedure for $k = 6$ tests with $\alpha = 0.05$, we should require a P-value of $\alpha/k = 0.05/6 = 0.0083$ (or less) for statistical significance for each test. Of the six P-values given, only two, 0.008 and 0.001, are below $0.05/6 = 0.0083$.

6.97   **(a)** X has a binomial distribution with $n = 77$ and $p = 0.05$. **(b)** The probability that 2 or more are significant is 0.90266.

6.99   As sample size increases, power increases.

6.101 Because 80 is farther away from 50 than 70 is, the power will be higher than 0.5.

6.103 **(a)** A sample size of 35 is required and will give a power of 0.905. **(b)** The power curve only shows positive differences because the alternative hypothesis was one-sided. The power when the sample size is 25 is 0.80. **(c)** Alternative mean values above about 1.5% give an almost certain detection of the alternative.

6.105 To decrease the probability of a Type II error, either (1) increase the significance level, (2) increase the sample size, (3) decrease σ, or look at an alternative that is farther away from μ_0.

6.107 **(a)** $0.25n$. **(b)** With a smaller sample size, there will be an increased variability in the sampling distribution. This could lead to, among other things, a greater chance of not detecting a true alternative hypothesis or confidence intervals that are too large to provide useful information.

6.109   (a) The hypotheses are H_0: patient does not need to see a doctor and H_a: patient does need to see a doctor. The program can make two types of error: (1) The patient is told to see a doctor when, in fact, the patient does not need to see a doctor. This is a "false-positive." (2) The patient is diagnosed as not needing to see a doctor when, in fact, the patient does need to see one. This is a "false-negative." (b) The error probability one chooses to control usually depends on which error is considered more serious. In most cases, a false-negative is considered more serious. If you have an illness, and it is not detected, the consequences can be serious.

6.111

Age	18	19	20	21	22	23	24	25	26
Months employed, $\bar{x}$	2.9	4.2	5.0	5.3	6.4	7.4	8.5	8.9	9.3
95% CI, $\bar{x} \pm 0.32$	(2.58, 3.22)	(3.88, 4.52)	(4.68, 5.32)	(4.98, 5.62)	(6.08, 6.72)	(7.08, 7.72)	(8.18, 8.82)	(8.58, 9.22)	(8.98, 9.62)

There is a strong, positive, linear relationship between age and months employed. All the widths of all the confidence intervals are exactly the same.

6.113   Answers will vary.

6.115 $(-4.774, -3.806)$.

6.117   (a) The stemplot shows that the data are roughly symmetric. (b) (26.06 μg/l, 34.74 μg/l). (c) Let μ denote the mean DMS odor threshold among all beginning oenology students. We test the hypotheses $H_0: \mu = 25$, $H_a: \mu > 25$. $z = 2.44$. The P-value is 0.0073. This is strong evidence that the mean odor threshold for beginning oenology students is higher than the published threshold of 25 μg/l.

6.119 $963.43 \pm 4.41.

6.121   (a) The authors probably want to draw conclusions about the population of all adult Americans. The population to which their conclusions most clearly apply is all people listed in the Indianapolis telephone directory.
(b)

Store type	95% confidence interval
Food stores	18.67 ± 3.45
Mass merchandisers	32.38 ± 4.61
Pharmacies	48.60 ± 4.92

(c) None of these intervals overlap, which suggests that the observed differences are likely to be real.

6.123   (a) Increasing the size n of a sample will decrease the width of a level C confidence interval. (b) Increasing the size n of a sample will decrease the P-value. (c) Increasing the sample size n will increase the power.

6.125   No. The null hypothesis is either true or false. Statistical significance at the 0.05 level means that if the null hypothesis is true, the probability that we will obtain data that lead us to incorrectly reject H_0 is 0.05.

6.127   (a) Assume that in the population of all mothers with young children, those who would choose to attend the training program and those who would not choose to attend actually remain on welfare at the same rate. The probability is less than 0.01 that we would observe a difference as or more extreme than that actually observed. **(b)** 95% confidence means that the method used to construct the interval $21\% \pm 4\%$ will produce an interval that contains the true difference 95% of the time. Because the method is reliable 95% of the time, we say that we are 95% confident that this particular interval is accurate. **(c)** The study is not good evidence that requiring job training of all welfare mothers will greatly reduce the percent who remain on welfare for several years. Mothers chose to participate in the training program. They were not assigned using randomization. Thus, the effect of the program is confounded with the reasons why some women chose to participate and some didn't.

6.129   Only in 5 cases did we reject H_0. Thus, the proportion of times was 0.05. This is consistent with the meaning of a 0.05 significance level. In 100 trials where H_0 was true, we would expect the proportion of times we reject to be about 0.05.

6.131   **(a)** Simulations will vary. **(b)** The central limit theorem states that the sample mean will have an approximately normal distribution, regardless of the distribution it was sampled from, as long as the sample size is large enough. **(c)** 29.1842. **(d)** The calculations agree with our simulation result up to roundoff error. **(e)** 13 of the 25, or 52%, of our simulations contained $\mu = 1000$. If we repeated the simulations, we would not expect to get exactly the same number of intervals to contain $\mu = 1000$. This is because each simulation is random, so results will vary from one simulation to the next. The probability that any given simulation contains $\mu = 1000$ is 0.50, and so in a very large number of simulations, we would expect about 50% to contain $\mu = 1000$.

CHAPTER 7

7.1   **(a)** SE = 19.5. **(b)** There are 15 degrees of freedom.

7.3   The 95% confidence interval for the mean monthly rent is ($466.45, $549.55).

7.5   Hypotheses are $H_0: \mu = 550$ and $H_a: \mu > 550$. $t = -2.15$ with 15 df, and using the t-table, we get P-value between $= 0.95$ and 0.975. We conclude that there is not much evidence that the mean rent of all advertised apartments exceeds $550.

7.7   **(a)** Two-sided, because we are interested in whether the average sales are different from last month (no direction of the difference is specified). Hypotheses are $H_0: \mu = 0\%$ and $H_a: \mu \neq 0\%$. **(b)** $t = -1.687$ with 39 df, and $0.10 < P$-value < 0.20. There is not enough evidence that the average sales have changed. **(c)** With a mean of -3.2% and a standard deviation of 12%, it is reasonable to conclude that some stores will have positive changes, and some will have negative changes, so a 0% change is within the range of reasonable values.

7.9   The P-value from SPSS is 0.024.

7.11   The confidence interval for Drink B – Drink A is $(-7.7440, 5.7440)$.

7.13   The sample size is quite large here, much bigger than $n = 40$, so the t procedures would be fine for this problem.

7.15   The power of the t test against the alternative $\mu = 2.4$ is 0.9678.

7.17   **(a)** $t^* = 2.179$ using degrees of freedom $n - 1 = 12$. **(b)** $t^* = 2.060$ using degrees of freedom $n - 1 = 25$. **(c)** $t^* = 2.787$ using degrees of freedom

$n - 1 = 25$. **(d)** As the sample size increases, the t^* decreases for the same confidence level. As the confidence level increases, t^* increases for the same sample size.

7.19   **(a)** Degrees of freedom $= 17$. **(b)** 1.740 and 2.110. **(c)** 0.05 and 0.025. **(d)** $0.025 < P$-value < 0.05. **(e)** Significant at 5%, not significant at 1%. **(f)** Excel gives 0.042421.

7.21   **(a)** Degrees of freedom $= 149$, but use df $= 100$ on Table D to be conservative. **(b)** P-value < 0.0005. **(c)** Excel gives 9.4145×10^{-5} using 149 degrees of freedom.

7.23   (25.57, 32.03).

7.25   (6.50, 11.46).

7.27   **(a)** and **(b)** The histogram shows that the distribution is somewhat skewed to the right. The boxplot confirms this. The normal probability plot indicates that the distribution is not exactly normal. There is a pattern to the dots that shows a curve rather than a straight line. There are no outliers to the data. **(c)** Because the sample size of 81 is greater than 40, even though there is some skewness to the data and the distribution is not Normal, the t procedures would still be appropriate based on the robustness guidelines.

7.29   **(a)** H_0: $\mu = 1550$, and H_a: $\mu > 1550$. The test statistic is $t = 1.800$, and the P-value is 0.038. At a 5% significance level, reject the null hypothesis. There is evidence that the average number of seeds in a one-pound scoop is greater than 1550. **(b)** H_0: $\mu = 1560$, and H_a: $\mu > 1560$. The test statistic is $t = 1.218$, and the P-value is 0.1135. At a 5% significance level, do not reject the null hypothesis. There is not enough evidence that the average number of seeds in a one-pound scoop is greater than 1560. **(c)** The 90% confidence interval says that, using a procedure that will capture the true population mean 90% of the time, we expect the population mean to be above 1552.34. So a test value below this number should only happen in 5% of cases if the null hypothesis is true. Thus, using a test value of 1550 results in a rejection of the null hypothesis but a test value of 1560 does not.

7.31   **(a)** The histogram shows a unimodal distribution that is skewed to the left. The boxplot shows the presence of eight outliers. The Normal probability plot shows that the distribution is not Normal. The t-distribution is not robust against the presence of outliers; however, with such a large sample size, the t-procedure will be quite accurate. **(b)** The mean is 35,288.25 with a standard deviation of 10362.0092. The standard error is 1092.2517, and the margin of error is 1817.507. **(c)** The 90% confidence interval is (33,470.74, 37,105.76). **(d)** H_0: $\mu = 33000$, and H_a: $\mu > 33000$. The test statistic is $t = 2.095$, and the P-value is 0.0195. At a 5% significance level, reject the null hypothesis. There is evidence that the average pounds of product treated in one hour is greater than 33000.

7.33   **(a)** 6 streams are classified very poor or poor out of 49 total, so the sample proportion is 0.122. **(b)** No, sample proportions use z^*, not t^*; Sample proportions are for categorical data; sample means are for quantitative data. We have converted quantitative data into categorical data by sorting the streams into very poor, poor, and other.

7.35   **(a)** The stem-and-leaf plot shows a distribution that is skewed to the right. It also has more observations in the tails of the distribution than would be seen in a

Normal distribution. **(b)** The mean is 31.134, the standard deviation is 21.24836, and the standard error is 3.35966. **(c)** (27.3384, 40.9296).

7.37   ($3.465, $3.815).

7.39   (3.69, 3.81).

7.41   **(a)** 17.14% of the bags are below the advertised target. **(b)** The histogram shows a unimodal distribution with outliers on both sides of the distribution. The boxplot confirms that there are a number of outliers. The normal probability plot shows that the distribution is not Normal as there are more values in the tails of the distributions than expected. **(c)** If we test H_0: $\mu_{\text{diff}} = 500$, H_a: $\mu_{\text{diff}} > 500$, we will get a test statistic of $t = 0.372$ resulting in a P-value of 0.3555. This does not give us enough evidence that the average difference is above 500. But, we just want to make sure that it is at least 500. So we test H_0: $\mu_{\text{diff}} = 500$, H_a: $\mu_{\text{diff}} < 500$. The test statistic is $t = 0.372$, and the P-value $= 0.6445$. There is not enough evidence that average difference between the number of seeds in the bag and the target number of seeds is less than 500. This is enough to substantiate the claim based on the criteria given. **(d)** Excluding a low outlier would cause the test statistic to increase and the area below the test statistic to increase, so the P-value would increase. **(e)** The new P-value $= 0.8695$, which does increase as predicted.

7.43   The 90% confidence interval for the mean time advantage is $(-21.17, -5.47)$. The ratio of mean time for right-hand threads as a percent of mean time for left-hand threads is $\bar{x}_R / \bar{x}_L = 88.7\%$, so those using the right-hand threads complete the task in about 90% of the time it takes those using the left-hand threads.

7.45   **(a)** H_0: $\mu_{\text{diff}} = 0$, and H_a: $\mu_{\text{diff}} \neq 0$. **(b)** $t = 4.358$. P-value $= 0$. These results show that there is a difference between the driver's calculations and the car's computer estimates of miles per gallon.

7.47   Taking the differences (Variety A – Variety B) to determine if there is evidence that Variety A has the higher yield corresponds to the hypotheses H_0: $\mu = 0$ and H_a: $\mu > 0$ for the mean of the differences. $t = 1.414$ with 9 df, and $0.05 < P$-value < 0.10. Do not reject the null hypothesis. There is not enough evidence to support that Variety A tomatoes have a higher mean yield than Variety B tomatoes.

7.49   **(a)** Single sample. **(b)** The best answer is two independent samples. Even though we are looking at changes of opinion over time from the same basic group, in a SRS, the same group would not necessarily be selected each year. However, if the exact same sample was used both years with a stable population of customers, matched pairs would also be an acceptable answer.

7.51   **(a)** $t^* = 2.403$. **(b)** $x \geq 287.89$. **(c)** The power is $P(Z \geq -0.1011) = .5398$. A difference of $300 will produce significance at the 1% level in 53.98% of all possible samples. A larger sample size would be required to be quite certain of detecting this increase.

7.53   **(a)** H_0: population median $= 0$, and H_a: population median > 0, or the population of differences (time to complete task with left-hand thread) $-$ (time to complete task with right-hand thread). If p is the probability of completing the task faster with the right-hand thread, the hypotheses would be H_0: $p = \frac{1}{2}$ and H_a: $p > \frac{1}{2}$. **(b)** The number of pairs with a positive difference in our data is 19, and $n = 24$

because the zero difference is dropped. The P-value is $P(X \geq 19) = 0.0021$ using the Normal approximation or 0.0033 using the binomial distribution.

7.55 $H_0\colon p = 0.5$, $H_a\colon p > 0.5$. $P(X \geq 8) = 0.0547$. This is not very likely, but we would not reject the null hypothesis at the 5% level. The results for Exercise 7.44 produced a stronger result in this case.

7.57   (a) Use a two-sided significance test because you just want to know if there is evidence of a difference in the two designs. (b) 44 df, but use 40 on Table D to be conservative. (c) To reject the null hypothesis, the t-statistic must be outside the range from -2.021 to 2.021.

7.59   This is a matched pairs experiment with the pairs being the two measurements on each day of the week.

7.61   Randomization makes the two groups similar except for the treatment and is the best way to ensure that no bias is introduced.

7.63   (a) Excel, Minitab, SAS, and SPSS all report both means. JMP just reports the difference between the two means. (b) Excel reports the two variances; SPSS, Minitab, and SAS report both the standard deviation and the standard error of the mean for each group. JMP reports just the Standard Error of the Difference. (c) SPSS and SAS provide the t with Satterthwaite degrees of freedom as well as the pooled t, and Excel, Minitab, and JMP provide the t with Satterthwaite degrees of freedom. JMP shows the equivalent upper-tail t value only. All report P-values to various accuracies. (d) With the exception of Excel, all report the confidence interval for the mean difference. (e) Excel has the least information, while SAS seems to provide the most information because it includes more information about the two groups individually.

7.65   The pooled t test statistic is 5.678 with 133 degrees of freedom. This results in a P-value less than 0.001. The same conclusion holds as for Example 7.13; however, with larger degrees of freedom, this result will be even stronger.

7.67   First write pooled variance as the average of the individual variances, and then use this expression in the pooled t to see that it gives the same result for equal sample sizes.

7.69   (a) Reject the null hypothesis because 0 is not inside the confidence interval. $H_0\colon \mu_A = \mu_B$ can also be written as $H_0\colon \mu_A - \mu_B = 0$. (b) As sample size increases, the margin of error decreases.

7.71   95% CI $= (5.35, 14.65)$ using 40 degrees of freedom as the conservative estimate from the t-table. A 99% CI will be wider than a 95% CI because the t^* value increases as the confidence level increases.

7.73   (a) The t procedures are quite robust, especially when the sample sizes are large. The sample sizes in this problem are large enough to make these procedures appropriate. (b) $H_0\colon \mu_F = \mu_M$, and $H_a\colon \mu_F \neq \mu_M$. (c) $t = 0.2759$, df $= 36$, P-value > 0.50. There is not enough evidence of a difference between the total cholesterol levels of males and females. (d) $(-12.0957, 15.8757)$. This interval shows that the females could have total cholesterol levels lower or higher than the males which agrees with the conclusion from the hypothesis test. (e) If the students had modified their behavior based on taking the health class, such as eating healthier or exercising regularly, then the results might not generalize to the rest of the population. However, because we know these students were sedentary, they were likely not adjusting their activities based on the health class information, and the information would not be affected by knowing that these students were taking a health class.

7.75 **(a)** The histograms and boxplots indicate that Plant 1748 has a skewed distribution with gaps in the data. Plant 1746 is more symmetric according to the boxplot but has its lowest count in the interval from 1600 to 1650, right in the center of the distribution. Neither plant has a seed count with a Normal distribution. **(b)** The sample sizes are large enough to use the t procedures despite the nonnormality. **(c)** $(-412.845, -82.315)$. The observed seed count for Plant 1746 is lower on average than for Plant 1748. **(d)** $H_0: \mu_{1746} - \mu_{1748} = 0$, $H_a: \mu_{1746} - \mu_{1748} \neq 0$, $t = -4.172$, P-value ≈ 0. **(e)** There is evidence that there is a difference in the average seed counts for these two plants.

7.77 **(a)** Females gave a mean rating of 4.079 with a standard deviation of 0.986. Males gave an average rating of 3.833 with a standard deviation of 1.068. **(b)** With sample sizes this large (689 all together), the t procedures are appropriate. **(c)** $H_0: \mu_F = \mu_M$, and $H_a: \mu_F \neq \mu_M$. $t = 2.898$. SPSS give a P-value of 0.004. There is evidence that there is a difference in how satisfied males and females were with the drive-thru window speaker quality. **(d)** $(0.079, 0.414)$ using SPSS. **(e)** The estimated average difference in ratings is between 0.079 units and 0.414 units. A difference of 0.25 units is in the middle of this range. There is evidence that there is a difference in the male and female ratings, but there is not evidence that this difference is over 0.25 units.

7.79 **(a)** The data are probably not exactly normally distributed because there are only five discrete answer choices. **(b)** Yes, the large sample sizes would compensate for uneven distributions. The two-sample comparison of means t test is fairly robust. **(c)** $H_0: \mu_I = \mu_C$, and $H_a: \mu_I > \mu_C$. (A "$\neq$" would also be appropriate in the alternative hypothesis, but be sure to double the P-value when checking your answer with this one.) **(d)** $t = 3.57$, P-value < 0.0005. Reject the null hypothesis. There is strong evidence the average self-efficacy score for the intervention group is significantly higher than the average self-efficacy score for the control group. **(e)** $(0.191, 0.669)$.

7.81 **(b)** $(4.374, 5.426)$. **(c)** $H_0: \mu_{D/B} = \mu_C$, and $H_a: \mu_{D/B} > \mu_C$. $t = 18.49$, P-value is close to 0. Reject the null hypothesis. There is very strong evidence the average exposure to respirable dust is significantly higher for the drill and blast workers than it is for the concrete workers. **(d)** The sample sizes are so large that a little skewness will not affect the results of the two-sample comparison of means test.

7.83 **(a)** No. If we use the 68–95–99.7% rule, then 68% of the younger kids would consume between -2.5 and 18.9 oz. of sweetened drinks every day. The same problem with negative consumption for the older kids (starting with 95%) as well. **(b)** $H_0: \mu_{\text{older}} = \mu_{\text{younger}}$, and $H_a: \mu_{\text{older}} \neq \mu_{\text{younger}}$. $t = 1.44$, $0.2 < P$-value < 0.3. Do not reject the null hypothesis. There is not enough evidence to say that there is a significant difference between the average consumption of sugary drinks between the older and younger groups of children. **(c)** $(-5.85, 18.45)$. **(d)** The sample sizes are very uneven and fairly small. The sample data is not normally distributed either. The t procedures are not particularly appropriate here. **(e)** How were these children selected to participate? Where they chosen because they consume such large quantities of sweetened drinks?

7.85 **(a)** The hypotheses are $H_0: \mu_0 = \mu_3$ and $H_a: \mu_0 > \mu_3$, where 0 corresponds to immediately after baking and 3 corresponds to three days after baking. $t = 22.16$, df $= 1.47$, and P-value $= 0.014$, which indicates there is strong evidence that the vitamin C content has decreased after three days. **(b)** The 90% confidence interval is (19.26, 34.57).

7.87 **(a)** The hypotheses are H_0: $\mu_0 = \mu_3$ and H_a: $\mu_0 > \mu_3$, where 0 corresponds to immediately after baking and 3 corresponds to three days after baking. $t = -0.32$, and P-value > 0.5, which indicates no evidence of a loss of vitamin E. **(b)** The 90% confidence interval is $(-11.29, 10.2)$.

7.89 **(a)** The hypotheses are H_0: $\mu_{\text{Low}} = \mu_{\text{High}}$ and H_a: $\mu_{\text{Low}} \neq \mu_{\text{High}}$. $t = -8.23$, df $= 21.77$, and the P-value is approximately 0. You reject at both the 1% and 5% levels of significance. **(b)** The individuals in the study are not random samples from low-fitness and high-fitness groups of middle-aged men, so the possibility of bias is definitely present. **(c)** These are observational data.

7.91 **(a)** Using 50 df to be conservative and using Table D, the 95% confidence interval is $(-0.9, 6.9)$. **(b)** The negative values correspond to a decrease in sales. If the true difference in means was -1, which is included in the confidence interval, then the percent sales would have decreased.

7.93 **(a)** The boxplots and histograms show no outliers for either distribution; however, there is some skewness in the scores for both the men and the women. Both distributions are somewhat skewed to the right. Because the skewness is not strong, the t procedures are appropriate with the combined sample size of 38. **(b)** H_0: $\mu_{\text{women}} = \mu_{\text{men}}$, H_a: $\mu_{\text{women}} > \mu_{\text{men}}$, $t = 2.223$, P-value $= 0.0165$. This gives fairly strong evidence that the mean SSHA score is lower for men than for women at this college. **(c)** (4.907, 35.981).

7.95 The approximate degrees of freedom are 39.535.

7.97 **(a)** $t^* = 2.776$. **(b)** $t^* = 2.306$. **(c)** The increase in degrees of freedom means that the value of the t statistic needs to be less extreme for the pooled t statistic to find a statistically significant difference.

7.99 **(a)** The upper 5% critical value is $F^* = 2.59$. **(b)** Significant at the 10% level but not at the 5% level.

7.101 For differences smaller than 1.15, you would expect the power to decrease. The noncentrality parameter would decrease, $t^* - \delta$ would increase, and $P(Z > t^* - \delta)$ would decrease.

7.103 **(a)** $F = 3.59$. **(b)** 3.77. **(c)** Do not reject the null hypothesis. There is no evidence of a difference in the two population standard deviations.

7.105 **(a)** The value in the table is 647.79. **(b)** $F = 3.96$, so there is little evidence of a difference in variances between the two groups.

7.107 **(a)** The hypotheses are H_0: $\sigma_M = \sigma_W$ and H_a: $\sigma_M > \sigma_W$. **(b)** $F = 1.74$. **(c)** Using an $F(19, 17)$ distribution, P-value > 0.10, so there is little evidence that the males have larger variability in their scores.

7.109 The power is 0.966003 for $n = 25$, 0.986689 for $n = 30$, 0.994990 for $n = 35$, and 0.998175 for $n = 40$. The power increases as the number of scoops increases.

7.111 **(a)** Using SAS, POWER $= 1 - \text{PROBT}(t^*, \text{DF}, \delta) = 0.3390$. **(b)** Using SAS, POWER $= 1 - \text{PROBT}(t^*, \text{DF}, \delta) = 0.7765$, which is greater than the power found in (a).

7.113 (a) The stemplot using split stems shows the data are clearly skewed to the right and have several high outliers. (b) The 95% confidence interval is (20.00, 27.12), which agrees quite well with the bootstrap intervals. The lesson is that for large sample sizes, the t procedures are very robust.

7.115 (a) The P-value for the one-sided alternative hypothesis would be 0.04, so you would reject the null hypothesis. (b) The P-value for the one-sided hypothesis would be 0.96, so you would not reject the null hypothesis.

7.117 The amount of wine a person drinks and the amount of food they eat are related to whether they are eating alone or in a group. This should be taken into account when performing this study.

7.119

Perceived quality	$\overline{x}_H - \overline{x}_L$	$\sqrt{\frac{s_H^2}{170} + \frac{s_L^2}{224}}$	t	P-value	Conclusion
Food served in promised time	0.45	0.139	3.24	Between 0.001 and 0.002	Reject H_0
Quickly corrects mistakes	0.16	0.125	1.28	Between 0.2 and 0.3	Do not reject H_0
Well-dressed staff	0.39	0.136	2.87	≈ 0.005	Reject H_0
Attractive menu	0.21	0.143	1.47	Between 0.1 and 0.2	Do not reject H_0
Serving accurately	0.37	0.123	3.01	Between 0.002 and 0.005	Reject H_0
Well-trained personnel	0.06	0.125	0.48	>0.5	Do not reject H_0
Clean dining area	0.08	0.127	0.630	>0.5	Do not reject H_0
Employees adjust to needs	0.14	0.123	1.14	>Between 0.2 and 0.3	Do not reject H_0
Employees know menu	0.29	0.119	2.44	Between 0.01 and 0.02	Reject H_0
Convenient hours	0.24	0.129	1.86	Between 0.05 and 0.1	Do not reject H_0

7.121 (a) Study was done in South Korea; results may not apply to other countries. Only selected QSRs were studied; results may not apply to other QSRs. Response rate is low (394/950); we would trust the results more if the rate were higher. The fact that no differences were found when the demographics of this study were compared with the demographics of similar studies suggests that we do not have a serious problem with bias based on these characteristics. (b) Answers will vary.

7.123 $H_0: \mu = 4.88$, and $H_a: \mu > 4.88$. $t = 21.98$; P-value is close to 0 so we can reject the null hypothesis. There is strong evidence that hotel managers have a significantly higher average masculinity score than the general male population.

7.125 (a) No outliers, slightly skewed right but fairly symmetric. (b) (12.9998, 13.3077).

7.127 $\overline{x}_C = 48.9513$, $s_C = 2.1537$, $\overline{x}_R = 41.6488$, $s_R = 0.39219$. Side-by-side box-plot shows cotton much higher than ramie. $H_0: \mu_C = \mu_R$, and $H_a: \mu_C > \mu_R$. $t = 46.162$; P-value is very close to 0 so reject the null hypothesis. There is strong evidence that cotton has a significantly higher mean lightness than ramie.

7.129 (1.0765, 7.6035).

7.131 5.8836 ± 0.1434.

7.133 The 95% confidence interval on percentage of lower priced products at the alternate supplier is (64.55, 92.09). This suggests that more than half of the products at the alternate supplier are priced lower than the original supplier.

7.135 (a) This study used a matched pairs design; therefore, they used a single sample t test. (b) The average weight loss in this program was significantly different from zero, and we can conclude that the program was effective. **(c)** The *P*-value is approximately equal to zero.

7.137   **(a)** $(0.207, 0.573)$. **(b)** $(-0.312, 0.012)$. **(c)** Based on a process that gives accurate results 90% of the time, the average daily alcohol consumption of London double-decker bus conductors is between 0.207 and 0.573 grams. Using a process that would result in capturing the true population difference in alcohol consumption between drivers and conductors 80% of the time, on average conductors consume between 0.312 grams more of alcohol and 0.012 grams less of alcohol daily than conductors.

7.139   As the degrees of freedom increase for small values of *n*, the value of t^* rapidly approaches the z^* value. As sample sizes get larger, the t^* values are close to $z^* = 1.96$ but never become greater than 1.96.

CHAPTER 8

8.1   **(a)** The sample size is 780. **(b)** $X = 283$. This count represents those who expect to acquire another bank within five years. **(c)** $\hat{p} = 0.3724$.

8.3   **(a)** The standard error of $\hat{p}$ is 0.0175. **(b)** The confidence interval is 0.3724 ± 0.0344. **(c)** $(33.80\%, 40.68\%)$.

8.5   The plus four confidence interval is the same as that in Example 8.3: $(0.50, 0.54)$.

8.7   Answers may vary. Smaller sample sizes will make a bigger difference when the sample proportion is close to zero.

8.9   The confidence interval is $(44.10\%, 85.90\%)$. When looking at the population of people using sun protection, we are 95% confident that the proportion who get better protection from your product is between 44.10% and 85.90%. Since this interval includes 50% as well as values above and below 50%, we can not conclude that your product provides protection to a substantially greater percentage of the population than your competitor's.

8.11   When the sample size is increased to 40, the new test statistic is $z = 1.90$. The *P*-value is 0.0574. In a 5% significance test, we still do not have enough evidence to reject the null hypothesis. The data are compatible with the hypothesis that there is no difference between the proportion who get more protection with your product and the proportion who get more protection with your competitor's product.

8.13   Assuming every choice is equally likely, we would use a p^* of 0.4. 93 people would need to be sampled to get a margin of error of 0.1.

8.15   **(a)** Hypothesis tests give information about population parameters, in this case *p*, not sample parameters like $\hat{p}$. **(b)** The test statistic for one-sample proportion hypothesis tests is based on the Normal distribution, not the *t*-distribution. **(c)** For a 95% confidence interval, you would need to use a z^* value of 1.96 multiplied by the standard error to get the margin of error.

8.17   **(a)** $\mu_{\hat{p}} = 0.4$, $\sigma_{\hat{p}} = 0.0693$. **(b)** The normal distribution will be centered at a *p* value of 0.4. **(c)** Using the 68-95-99.7% rule, the middle 95% of the distribution will be within two standard deviations of the mean, or between $\mu_{\hat{p}} \pm 0.1386$.

8.19   **(a)** $\hat{p} = 0.161$; confidence interval $= (0.1508, 0.1712)$. **(b)** $\hat{p} = 16.1\%$; confidence interval $= (15.08\%, 17.12\%)$.

8.21 **(a)** $(0.6412, 0.6988)$. **(b)** $(0.6534, 0.6866)$. **(c)** The larger the group is that does not currently play a musical instrument, the smaller the margin of error will be. In this case, the intervals are rather similar, even with the difference in the margins of error.

8.23 **(a)** $(0.4007, 0.4335)$. **(b)** The proportion of success and the sample sizes were not the same for these two problems, and the margin of error depends on these numbers.

8.25 **(a)** 67179 students plan to study abroad. **(b)** $(41.68\%, 42.32\%)$, which translates to an interval of $(66670, 67687)$ students.

8.27 **(a)** $(0.300, 0.354)$. **(b)** $(0.301, 0.355)$. The methods have the same margin of error $= 0.027$, but the plus four method shifts the interval slightly higher. **(c)** Nonresponse rate is only 3.64%, which is small, so we can trust these results. **(d)** Yes, the person delivering the sermon might think the sermon is shorter than it actually is, and the congregation might think the sermon is longer than it actually is.

8.29 $(0.7380, 0.7823)$.

8.31 The 95% confidence interval would be narrower than the 99% interval. The interval is $(41.76\%, 42.24\%)$, which is between 66792 and 67565 students.

8.33 $(0.635, 0.745)$.

8.35 $(0.218, 0.251)$.

8.37 $(0.766, 0.887)$.

8.39 $(0.642, 0.768)$.

8.41 H_0: $p = 0.36$, H_a: $p \neq 0.36$. $Z = 0.9317$. P-value $= 0.3524$. There is no evidence to believe the sample does not represent the population with respect to rural versus urban residence.

8.43 A sample of at least 764 is needed.

8.45 **(a)** The test statistic is $z = 1.34$. The P-value $= 0.1802$, and because this is larger than 0.05, we would not reject the null hypothesis that the probability that Kerrich's coin comes up heads is 0.5. **(b)** $(0.4969, 0.5165)$.

8.47 $n = 288.12$. Round up to get $n = 289$.

8.49 We round up to get $n = 201$. The margin of error for the 90% confidence interval is 0.0691.

8.51

$\hat{p}$:	0.1	0.2	0.3	0.4	0.5	0.6	0.7	0.8	0.9
m:	0.048	0.064	0.073	0.078	0.080	0.078	0.073	0.064	0.048

8.53 The mean is -0.1, and the standard deviation is 0.1339.

8.55 **(a)** For $\hat{p}_1$: mean $\mu_{\hat{p}_1} = p_1$, standard deviation $\sigma_{\hat{p}_1} = \sqrt{\frac{p_1(1-p_1)}{n}}$. For $\hat{p}_2$: mean $\mu_{\hat{p}_2} = p_2$, standard deviation $\sigma_{\hat{p}_2} = \sqrt{\frac{p_2(1-p_2)}{n}}$. **(b)** $\mu_D = \mu_{p_1-p_2} = \mu_{p_1} - \mu_{p_2} = p_1 - p_2$. **(c)** $\sigma_D^2 = \sigma_{\hat{p}_1-\hat{p}_2}^2 = \sigma_{\hat{p}_1}^2 + \sigma_{\hat{p}_2}^2 = \frac{p_1(1-p_1)}{n} + \frac{p_2(1-p_2)}{n}$.

8.57 **(a)** $\hat{p}_m = 0.600$; $\hat{p}_w = 0.575$. **(b)** $D = 0.025$. **(c)** $SE_D = 0.1100$. **(d)** $(-0.191, 0.241)$.

8.59   Plus four interval: $(-0.125, 0.267)$, z interval: $(-0.112, 0.272)$.

8.61   **(a)** H_0: $p_m = p_w$, H_a: $p_m \neq p_w$. This tests whether there is a difference between the proportion of women and the proportion of men who lied. Because the hypotheses should be stated before the data is gathered, we would only use a one-sided alternative hypothesis if we wanted to test whether one group had a greater proportion than the other group. **(b)** $z = 1.12$ (or -1.12 depending on the order of subtraction). P-value $= 0.2628$. **(c)** With a P-value this large, we would not reject the null hypothesis. There is not significant evidence that there is a difference in the population proportion of men who lied about their height and women who lied about their height.

8.63   **(a)**

Population	Population proportion	Sample size	Count of successes	Sample proportion
Feb – April 2006	p_1	2822	198	0.0702
May 2008	p_2	1553	295	0.1900

(b) $D = 0.1198$. **(c)** The number of successes and failures in each sample are more than 10, so it is appropriate to use the large sample confidence interval is appropriate. **(d)** $(0.0981, 0.1415)$. **(e)** $D = 11.98\%$; confidence interval: $(9.81\%, 14.15\%)$. **(f)** The difference in the dates could definitely make a difference if more people tend to download podcasts at different times of the year. This could happen in May if people download more podcasts when school is not in session.

8.65   **(a)** There will be variability in these numbers depending on the sample selected, so we cannot conclude that this ratio holds true for the populations. We can only conclude that there will be a difference between the proportions for these two time periods. **(b)** A one-sided hypothesis test would show that the 2008 time period has a significantly higher proportion of downloads than the 2006 period. However, we would need to assume that the time periods selected are representative of the entire years to apply our conclusion to the entire years.

8.67 H_0: $p_{adult} = p_{teen}$, H_a: $p_{adult} \neq p_{teen}$. $z = 17.88$. P-value $= 0$. There is evidence of a significant difference in the proportion of teens and the proportion of adults who played games on game consoles.

8.69 H_0: $p_{adult} = p_{teen}$, H_a: $p_{adult} \neq p_{teen}$. $z = 1.60$. P-value $= 0.1096$. There is no evidence of a difference in the proportion of teens and the proportion of adults who played games on computers.

8.71   **(a)** $\mu_{\hat{p}_1 - \hat{p}_2} = -0.1$, and $\sigma_{\hat{p}_1 - \hat{p}_2} = 0.0947$. **(b)** This should be a sketch of the Normal curve with center at -0.1. **(c)** $(-0.289, 0.089)$.

8.73   **(a)** The proportion of women who worked last summer has an approximately Normal distribution with a mean of 0.83 and a standard deviation of 0.0188. The equivalent distribution for men is Normal with a mean of 0.85 and a standard deviation of 0.0179. **(b)** The distribution of the difference is Normal with a mean of 0.02 and a standard deviation of 0.0259.

8.75   **(a)** H_0: $p_1 = p_2$, H_a: $p_1 \neq p_2$. **(b)** $z = 1.23$, P-value $= 0.2186$. This is not strong evidence that there is a difference in preference for natural trees versus artificial trees between urban and rural households. **(c)** $(-0.0209, 0.1389)$.

8.77 **(a)** For juvenile males, the sample proportion is 0.394. For juvenile females, the sample proportion is 0.800. **(b)** The 95% confidence interval of (0.275, 0.537) indicates that textbooks more often refer to females in the juvenile sense than they do males. Because the confidence interval does not contain 0, there is a significant difference in the population proportion of juvenile references that are male versus female.

8.79 **(a)** The test statistic is $z = 0.73$. This results in a P-value of 0.4654. There is not a statistically significant difference between the proportion of men who are employed and the proportion of women who are employed. **(b)** When the sample sizes are smaller, we are less likely to observe a difference in the population proportions.

8.81 (0.0747, 0.1453).

8.83 **(a)** The 95% confidence interval is (0.0713, 0.1487). For the hypotheses H_0: $p_{2000} = p_{2004}$, H_a: $p_{2000} \neq p_{2004}$, the test statistic is $z = 5.48$, which results in a P-value of approximately 0. Reject the null hypothesis. There is significant evidence that the population proportion of student with four or more credit cards in 2004 is different from the proportion in 2000. **(b)** The 95% confidence interval is (0.0772, 0.1428). For the hypotheses H_0: $p_{2000} = p_{2004}$, H_a: $p_{2000} \neq p_{2004}$, the test statistic is $z = 6.59$, which results in a P-value of approximately 0. Reject the null hypothesis. There is significant evidence that the population proportion of student with four or more credit cards in 2004 is different from the proportion in 2000. **(c)** Changing the sample size does change the confidence interval the test statistic and the P-value of the hypothesis test. In this case, the results of the hypothesis test were the same, although there was even stronger evidence against the null hypothesis with the larger sample size.

8.85 When the sample size increases, the center of the distribution remains at $\hat{p} = 0.6951$, but the margin of error decreases. When the sample size decreases, the margin of error increases.

8.87 **(a)** The chance of more than one missing the proportion would be equal to $1 - P$ (all intervals capture the true proportion). If the events "the interval misses the proportion" are independent, then the chance of more than one missing the proportion would be equal to $1 - 0.95^6 = 0.2649$. This is a much greater chance than 5%. **(b)** $z = 2.64$.
(c)

Genre	Racing	Puzzle	Sports	Action	Adventure	Rhythm
Confidence Interval	(0.705, 0.775)	(0.684, 0.756)	(0.643, 0.717)	(0.633, 0.707)	(0.622, 0.698)	(0.571, 0.649)

8.89 **(a)** 0.164. 2460 households. **(b)** (0.158, 0.17). **(c)** $16.4\% \pm 0.6\%$. **(d)** $D = 0.122$. **(e)** 0.0067.

8.91 H_0: $p_1 = p_2$, H_a: $p_1 \neq p_2$, $z = 20.18$, P-value is approximately 0, so reject the null hypothesis. There is evidence of a significant difference between the proportion of male athletes who admit to cheating and the proportion of female athletes who admit to cheating. The 95% confidence interval (male − female) is (0.1963, 0.2377). Someone who gambles will be less likely to respond to the survey. Do you think men or women are more likely to report that they do not gamble when, in fact, they do gamble?

8.93   **(a)** and **(b)**

Category	$\hat{p}$ (in %)	n	m (in %)
Download less	38	247	3.09
Peer-to-peer	33.3	247	3.00
E-mail and IM	24	247	2.72
Web sites	20	247	2.55
iTunes	17	247	2.39
Overall use of new services	7	1371	0.69
Overall use paid services	3	1371	0.46

(c) Argument for (A): Readers should understand that the population percent is not guaranteed to be at the sample percent; there is variability involved in taking a sample. Argument for (B): Listing each individual margin of error does seem excessive, so you could summarize by saying that the margin of error was no greater than 3.09% for each of these questions. You could also separate out the last two questions by saying their margin of error was less than 1%.

8.95   **(a)** $\hat{p}_{repeat} = 0.783$, $\hat{p}_{norepeat} = 0.517$, 95% CI (repeat − no repeat) is (0.102, 0.430). **(b)** H_0: $p_1 = p_2$, H_a: $p_1 \neq p_2$, $Z = 3.05$, P-value = 0.0022, so reject the null hypothesis. There is strong evidence of a significant difference in the proportion of tips received between servers who repeat the customer's order and those who do not repeat the order. **(c)** Cultural differences, personalities of the servers, and gender differences could all play a role in interpreting these results. Did one server only do the repeating while the other server did no repeating, or did they switch off? **(d)** Answers will vary.

8.97   Let p_1 be the proportion of all die-hard fans who attend a Cubs game at least once a month and p_2 the proportion of less loyal fans who attend a Cubs game at least once a month. We test H_0: $p_1 = p_2$, H_a: $p_1 > p_2$. $z = 9.04$, P-value = approximately 0. This is strong evidence that the proportion of die-hard Cubs fans who attend a game at least once a month is larger than the proportion of less loyal Cubs fans who attend a game at least once a month. A 95% confidence interval for the difference in the two proportions is (0.3755, 0.5645), so the magnitude of the difference is large.

8.99   The 95% confidence interval is (0.5524, 0.6736). This does not include 0.485, the proportion of U.S. adults who are male, so we would conclude that the proportion of heavy lottery players who are male is different from the proportion of U.S. adults who are male.

8.101   Let p be the proportion of products that will fail to conform to specifications in the modified process. We test the hypotheses H_0: $p = 0.11$, H_a: $p < 0.11$. $z = -3.16$, P-value = 0.0008. This is strong evidence that the proportion of nonconforming items is less than 0.11, the former value. This conclusion is justified provided the trial run can be considered a random sample of all (future) runs from the modified process and that items produced by the process are independently conforming or nonconforming.

8.103

n	10	25	50	100	150	200	400	500
m	0.438	0.277	0.196	0.139	0.113	0.098	0.069	0.062

As sample size increases, the margin of error decreases.

8.105   Starting with a sample size of 25 for the first sample, it is not possible to guarantee a margin of error of 0.15 or less. $m = 1.960 \sqrt{\frac{(0.5)(0.5)}{25} + \frac{(0.5)(0.5)}{n_2}}$. This leads to a negative value for n_2, which is not feasible.

8.107   **(a)** $p_0 = 0.791$. **(b)** $\hat{p} = 0.390$, $z = -29.09$, P-value $=$ approximately 0. We would conclude that there is strong evidence that the probability that a randomly selected juror is Mexican American is less than the proportion of eligible voters who are Mexican American. **(c)** $z = -29.20$, P-value $=$ approximately 0. The z statistic and P-value are almost the same as in part (b).

CHAPTER 9

9.1   **(a)** Gender would be the explanatory variable, and lying about height (yes/no) would be the response variable. Gender could logically help explain whether a person would lie about their height while the reverse would not make sense. **(b)** r and c would both be 2 because there are two options for each variable. **(c)** 4. **(d)**

	Men	Women	Total
Lied	22	17	39
Did not lie	18	23	41
Total	40	40	80

9.3   87.8% of successful firms and 72.3% of unsuccessful firms offer exclusive territories.

9.5   142 firms have exclusive territories. 27.6% of all firms are unsuccessful. If there is no association between these two variables, 27.6% of exclusive-territory clause companies and 27.6% of no-exclusive-territory clause companies would be unsuccessful.

9.7   df $= 15$.

9.9   **(a)** Minitab calculates $\chi^2 = 10.949$. $z^2 = (3.31)^2 = 10.95$. **(b)** $(3.291)^2 = 10.83$. **(c)** No relation says that whether or not a person is a label user has no relationship to gender, which is the same as H_0: $p_1 = p_2$.

9.11   **(a)** See Exercise 2.115. **(b)** H_0: There is no association between region and whether or not a community bank offers RDC; H_a: There is an association between region and whether or not a community bank offers RDC. $\chi^2 = 24.059$, df $= 5$. **(c)** P-value $= 0$. **(d)** There is substantial evidence that there is an association between region and whether or not a community bank offers RDC. On average, 37% of banks have RDC. The Northeast, Southeast, and West have a greater percentage of banks with RDC than the other regions.

9.13   **(a)**

	Male	Female	Total
Lied	3228	10295	13523
Didn't Lie	9659	4620	14279
Total	12887	14915	27802

(b) Only 25% of males lied at least once to a teacher. 69% of females lied at least once to a teacher. **(c)** Approximately half of all students have lied at least once to a teacher. However, a much higher percentage of females have lied at least once to a teacher than males. More males have not lied to their teachers. **(d)** H_0: There is no association between gender and whether or not a student lied to a teacher; H_a: There is an association between gender and whether or not a student lied to a teacher. $\chi^2 = 5351.94$, df $= 1$, P-value $= 0$. There is substantial evidence that there is an association between gender and whether or not a student lied to a teacher. A student is more likely to have lied to a teacher at least once if the student was a female.

9.15   **(a)** through **(d)** See Exercise 2.123. **(e)** H_0: There is no association between age and whether an applicant was hired or not; H_a: There is an association between age and whether an applicant was hired or not. $\chi^2 = 9.106$, df $= 1$, P-value $= 0.003$. There is evidence that there is an association between age and whether or not an applicant was hired.

9.17   **(a)**

	Admit	Deny
Male	490	210
Female	280	220

(b) 70% of male applicants are admitted, and 56% of female applicants are admitted. **(c)** For the business school: Male = 80%, Female = 90%. For the law school: Male = 10%, Female = 33%. **(d)** Without considering the variable of the professional school to which the applicant has applied, the percentages do not reflect a clear picture. More men apply to Wabash's professional schools overall; therefore, based solely on applications to professional school, more men are being admitted than women. After we separate the data by type of school, we see that Wabash is actually accepting a higher percentage of their women applicants. **(e)** The χ^2 value for this test is 24.863 with a P-value of 0. There is strong evidence that there is an association between gender and whether or not a student is admitted to Wabash Tech. **(f)** In testing the business school data, $\chi^2 = 10.390$, and the P-value is 0.001. We can conclude that there is an association between gender and whether or not the applicant is admitted for the business school. In testing the law school data, $\chi^2 = 20.481$, and the P-value is 0. We can conclude that there is even stronger evidence of an association between gender and whether or not the applicant is admitted for the law school.

9.19   **(a)** When $a = 0$, V will have observations of zero 50% of the time when $U = 0$. When $U = 1$, V will also have observations of zero 50% of the time. The null hypothesis would say there is no association between U and V, and when a is equal to zero, we can see that the distribution of V will be the same regardless of the value of U.
(b)

a	0	5	10	15	20	25
% zeros for $V \mid U = 1$	50%	52.6%	55.6%	58.8%	62.5%	66.7%
χ^2		0.000	0.501	2.020	4.604	8.333	13.333
P-value		1.000	0.479	0.155	0.032	0.004	0.000

(c) As the percent of zeros for V increases, the Chi-squared value increases at a greater rate the farther a gets away from zero. The P-value decreases as the percent of zeros for V increases. **(d)** When the percentages within the group change, the P-value reacts very quickly to the change. The chi-squared value reacts a little less quickly. The greater the change in the percentages in the table, the greater the change in the chi-squared statistic. The P-value also continues to decrease, but not as quickly as with the initial change.

9.21 **(a)**

	Apr 2001	Apr 2004	Mar 2007	Apr 2008
Broadband	113	540	1080	1237
No broadband	2137	1710	1170	1013

(b) H_0: There is no association between the date of the survey and whether or not a home has broadband; H_a: There is an association between the date of the survey and whether or not a household accesses the Internet using broadband. $\chi^2 = 1599.503$, df = 3, P-value = 0. There is substantial evidence that there is an association between the date of the survey and whether or not a household had broadband. **(c)** The test statistic in this case is 21.930 with one degree of freedom. The P-value is still 0, so we would still reject the null hypothesis, but the result is not nearly as strong as that seen when all four dates were considered.

9.23 Answers will vary depending on changes made in selected scenarios.

9.25 **(a)**

	Claim allowed	Claim not allowed	Total claims
Small	51	6	57
Medium	12	5	17
Large	4	1	5
Total	67	12	79

(b) 10.53% of small claims, 29.41% of medium claims, and 20% of large claims were not allowed. **(c)** H_0: There is no association between the size of the claim and whether or not the claim was allowed; H_a: There is an association between the size of the claim and whether or not the claim was allowed. **(d)** $\chi^2 = 3.721$, df = 2, P-value = 0.156. There is not enough evidence to indicate that an association exists between the size of the claim and whether or not it was allowed.

9.27 In testing whether there is an association between the changes in the gateway course and the DFW rate, a chi-square test was performed which resulted in a test statistic of $\chi^2 = 1932.003$ with two degrees of freedom. The P-value for this test was 0, indicating that the changes did have an impact on the DFW rate.

9.29 **(a)**

	Entered after HS	Entered later
Trades	320	622
Design	274	310
Health	2034	3051
Media/IT	976	2172
Service	486	864
Other	1173	1082

(b) Using a chi-square test of association with five degrees of freedom to determine whether there is an association between time of entry and field of study with five degrees of freedom, the test statistic is $\chi^2 = 276.110$, and the P-value is 0. **(c)** There is evidence that there is an association between the field of study and the time of entry of Canadian students enrolled in private career colleges. A bar graph shows that those entering the Media/IT field were the most likely to enter later whereas those entering Other fields and Design were the groups most likely to enter right after leaving high school.

9.31 **(a)**

	Family funded	Not family funded
Trades	188	754
Design	222	377
Health	1361	3873
Media/IT	518	2720
Service	248	1130
Other	943	1357

(b) H_0: There is no association between field of study and whether or not the student received funding from family; H_a: There is an association between field of study and whether or not the student received funding from family. $\chi^2 = 544.787$, df $= 5$, P-value $= 0$. There is evidence that there is an association between the field of study and whether or not the student received funding from parents, spouse, or family. **(c)** A bar graph shows that those entering the Media/IT, Service, or Trades fields were the least likely to used funds from family while those entering Other fields and Design were the groups most likely to use family funding.

9.33 Answers will vary, but one example would be:

	X	Y	Z	Totals
A	10	10	10	30
B	10	10	10	30
C	10	10	10	30
Totals	30	30	30	90

9.35 The percent of Department A's classes that are small is $32/52 = 61.54\%$. The percent of Department B's classes that are small is $42/106 = 39.62\%$. The percent of Department A's classes that are for third, and fourth-year students is $40/52 = 76.92\%$. The percent of Department B's classes that are for third- and fourth-year students is $36/106 = 33.96\%$. Department A teaches a much larger percentage of small classes, but Department A teaches more than twice the percentage of third- and fourth-year students as Department B.

9.37 $\chi^2 = 2.591$, P-value $= 0.107$ from SPSS. Do not reject the null hypothesis. There is not enough evidence to say that there is an association between model dress and magazine readership age group. The marginal percentages for model dress were 73.6% of the ads had models dressed not sexually, and 26.4% had models dressed sexually. The marginal percentages for age group were 33.3% in the mature adult category and 66.7% in the young adult category.

9.39 **(a)** $\chi^2 = 76.675$, P-value is close to 0 from SPSS. Reject the null hypothesis. There is very strong evidence that there is an association between collegiate sports division and report of cheating. **(b)** Answers will vary, but here is one possibility: If we change all the sample sizes to 1000 but keep the percentages the same, $\chi^2 = 15.713$, and the *P*-value remains very close to 0. If we change all the "yes" answer counts to 100 but keep the percentages the same, then $\chi^2 = 7.749$, and the *P*-value increases to 0.021. **(c)** The people most likely not to respond are those who gamble, so the results may be biased toward "no" answers. **(d)** If one member of a team is cheating, it may be more likely that others are cheating too. The teammates also may have had a discussion about how to fill out the form.

9.41 $\chi^2 = 12, P\text{-value} = 0.001$ from SPSS. Reject the null hypothesis. There is strong evidence to say that there is an association between gender and visits to the *H. bihai* flowers. The 0 cell count does not invalidate the significance test. It is the expected counts that need to be 5 or greater for a 2×2 table for the chi-square test to be appropriate.

		Visits H. bihai	
Gender	Yes	No	Total
Female	20	29	49
Male	0	21	21
Total	20	50	70

9.43   **(a)**

	Responded yes	Responded no
First-year students	79317	105140
Seniors	97429	97429

(b) H_0: There is no association between grade level of students and whether or not they think statistical data needs to be explained; H_a: There is an association between grade level of students and whether or not they think statistical data needs to be explained. **(c)** $\chi^2 = 1865.752, P\text{-value} = 0$. There is substantial evidence that there is an association between the class of a student and whether or not they think that an explanation is required for statistical or numerical information. **(d)** The data does show that seniors are more likely to believe that an explanation is required for statistical or numerical data. However, instructors might wish for a greater change in the percent who believe this than what is seen between these two groups of students.

9.45 $\chi^2 = 50.81, df = 2, P\text{-value} = 0.000$. The older employees (over 40) are almost twice as likely to fall into the lowest performance category but are only one-third as likely to fall into the highest category.

9.47   Exercise 9.16 used model 1, a comparison of several populations based on separate samples. The remaining problems used model 2, data from a single sample.

9.49   **(a)**

	NoRDC	YesRDC
Asset size under 100	257.5	114.5
Asset size 101 to 200	132.2	58.8
Asset size 201 or More	136.3	60.7

(b) $\chi^2 = 96.305$. **(c)** $df = 2$. **(d)** $P\text{-value} < 0.0005$.

CHAPTER 10

10.1   Predicted stress level is 2.3752. The residual is 0.3048.

10.3   **(a)** $\beta_0 = 0.3$. When the U.S. market is flat, the average overseas return will be 0.3. **(b)** $\beta_1 = 0.85$. For each 1% increase in return on common stocks in the United States, the overseas return will increase 0.85% on average. **(c)** Mean Overseas Return $= 0.3 + 0.85 \times$ U.S. Return $+ \varepsilon$. ε allows the overseas returns to vary for a fixed value of U.S. returns.

10.5   **(a)** Yes, there is a strong, positive, linear relationship between year and spending. **(b)** $\hat{y} = -4566.24 + 2.3x$.
(c)

2003	2004	2005	2006	2007
−0.56	0.34	0.54	0.14	−0.46

(d) $SPENDING = \beta_0 + \beta_1 \cdot YEAR + \varepsilon$, with estimates $\beta_1 = 2.3$, $\beta_0 = -4566.242.3$, $\varepsilon = 0.563$. **(e)** For 2001, the predicted spending is $36.06. The residual is $3.26. This is an extrapolation. The trend might not have been the same prior to 2003.

10.7 $b_1 = 0.6269$. $SE_{b1} = 0.0889$. $H_0: \beta_1 = 0$, $H_a: \beta_1 > 0$. $t = 7.052$ with 49 degrees of freedom. P-value < 0.0005.

10.9   The mutual funds that had the highest returns last year did so well partly because they were a good investment and partly because of good luck (chance).

10.11 **(a)** $r = 0.7096$. $t = 7.05$ with 49 degrees of freedom. P-value < 0.0005, and there is strong evidence that the population correlation $\rho > 0$. **(b)** The output gives $t = 7.0501$.

10.13 **(a)** 22 out of 30 have an assessed value less than the selling price. As of Spring 2010, this would not be true. Home prices have dropped dramatically, causing selling prices to fall below assessed values much more frequently. **(b)** There is a strong, positive linear relationship between the assessed value and selling price of real estate. **(c)** $\hat{y} = 21.499 + .947x$. **(d)** There don't appear to be any major problems with the residual plot. There is one point that has an assessed value that is quite a bit bigger than the others; however, it is not large enough to be an outlier. **(e)** A Normal probability plot of the residuals shows no major deviations from normality. **(f)** The assumptions of linear regression are met. The residuals appear to be normally distributed, and there is a linear relationship between the response variable and the explanatory variable.

10.15 **(a)** The scatterplot shows a strong positive linear relationship between the tuitions for these two years. **(b)** $\hat{y} = 1132.75 + 1.692x$. **(c)** There is nothing unusual that stands out about the residual plot. The dots appear to have random scatter in the residual plot. **(d)** The Normal probability plot does not show Normality. The points in the plot show a stair-step type of pattern where the observed probability has occasional larger increases than would be expected in the Normal distribution, followed by a few smaller increases, and then a larger increase again. **(e)** H_0: $\beta_1 = 0$, H_a: $\beta_1 \neq 0$. **(f)** $t = 10.552$. P-value $= 0$. There is evidence that there is a linear relationship between the 2000 and 2008 tuitions.

10.17 (a) Both "percentage" and "rating" are strongly skewed to the right. Rating has a median of 6.31 and a mean of 7.759. The range is from 0 to 27.88. Percentage has a median of 1.4286 and a mean of 14.2205. The range is from 0 to 85.015. (b) The values of y for any fixed x need to have a Normal distribution, but the individual variables do not. (c) The scatterplot does not support a linear relationship. (d) $\hat{y} = 6.247 + 0.106x$. (e) The Normal probability plot has a curved shape and does not support normality of the residuals.

10.19 (a) Area: $x = 28.29$, $s = 17.714$, min $= 2$, max $= 70$, fairly symmetric and normally distributed except for two high outliers. IBI: $x = 65.94$, $s = 18.280$, min $= 29$, max $= 9$, skewed left but no outliers. (b) Scatterplot looks weak, positive, and linear. It's very hard to tell if there are any outliers or unusual patterns because the relationship is so weak. (c) IBI $= \beta_0 + \beta_1 \cdot$ AREA $+ \varepsilon$. (d) $H_0: \beta_1 = 0$, $H_a: \beta_1 = 0$. (e) $\hat{y} = 52.923 + 0.460x$. From SPSS, the hypothesis test in part (d) has $t = 3.415$ and a P-value of 0.001, so we can reject the null hypothesis. There is strong evidence that Area and IBI have a linear relationship. $R^2 = 19.9\%$, and the regression standard error is $s = 16.535$. (f) The residual plot shows that the residuals get slightly less spread out as Area increases, but overall it doesn't look too bad. (g) Yes, the normal probability plot looks good. (h) The assumptions are reasonable—we are assuming a simple random sample from the population, there is a fairly linear (although weak) relationship between IBI and Area, there is approximately the same spread above and below the regression line on the scatterplot that is fairly uniform for all Area values on the plot, and the Normal probability and residual plots look good.

10.21 Area is the better explanatory variable for regression with IBI. Forest has a much weaker relationship with IBI than Area does.

10.23 $H_0: \rho = 0$, $H_a: \rho \neq 0$, P-value $= 0.061$, so do not reject the null hypothesis. There is not enough evidence to say that the correlation is significantly different from 0 if a significance level of 0.05 is used. This agrees with what we saw in the test using the slope in the 10.20. The correlation is a numerical description of the linear relationship between Area and Forest, which is very weak.

10.25 (a) Each of the vertical stacks corresponds to an integer value of age. (b) Older men might earn more than younger men because of seniority or experience. Younger men might earn more than older men because they are better trained for certain types of high-paying jobs (technology) or have more education. The correlation is positive, so there is a positive association between age and income in the sample. $r^2 = 0.03525$, so that age explains only about 3% of the variation in income. The relationship is weak. (c) Income $= 24{,}874.3745 + 892.1135 \times$ Age. The slope tells us that for each additional year in age, the predicted income increases by \$892.1135.

10.27 (a) We see the skewness in the plot by looking at the vertical stacks of points. There are many points in the bottom (lower income) portion of each stack. At the upper (higher income) portion of each vertical stack, the points are more dispersed, with only a very few at the highest incomes. (b) Regression inference is robust against a moderate lack of Normality for larger sample sizes.

10.29 (a) The intercept tells us what the T-bill percent will be when inflation is at 0%. No one will purchase T-bills if they do not offer a rate greater than 0. (b) The estimate of the intercept is 2.9645; the standard error is 0.4452. (c) Because the

95% confidence interval only contains positive values, we have evidence that $\beta_0 > 0$. **(d)** (2.06, 3.86).

10.31 The correlation tells us that the binge drinking rate decreases as the average price for a bottle of beer near campus increases. Because the correlation is closer to zero than to one, we can also tell that this relationship is rather weak, meaning that the relationship is not perfect and has quite a bit of variation from a linear model through the data while still having a generally negative trend.

10.33 **(a)** There is a moderate, positive, linear relationship between size and price. $r^2 = 0.431$, indicating that square footage is helpful for predicting selling price. **(b)** *Selling price* $= 21.398 + 0.077 \times$ *Square footage*. *P*-value < 0.0001. There is a statistically significant straight-line relationship between selling price and square footage.

10.35 H_0: $\rho = 0$, H_a: $\rho > 0$, $r = 0.656$, $t = 4.601$ with 48 degrees of freedom, *P*-value < 0.0001. We conclude that there is strong evidence that $\rho > 0$, and, therefore, larger houses have higher prices.

10.37 The r^2 is 38.0%, $t = 4.068$ with a very small *P*-value, and the equation of the line is $\hat{y} = 40.647 + 0.058x$. The conclusions are the same as they were before, so this outlier is not influential.

10.39 **(a)** Growth is much faster than linear. **(b)** The plot looks very linear. **(c)** (0.432, 0.461).

10.41 **(a)** (2.751, 3.037). **(b)** (2.774, 3.014).

10.43 **(a)** (87.87%, 108.60%). **(b)** (8543, 13354). **(c)** (13085, 18628). **(d)** The prediction interval is wider for Moneypit U because we have less information at this extreme of a value for the 2000 tuition.

10.45 **(a)** (69.34, 86.21). **(b)** (43.46, 112.09). **(c)** A 95% confidence interval for mean response is the interval for the average IBI for all areas of 54 km². A 95% prediction interval for a future response estimates a single IBI for a 54 km² area. **(d)** This would depend on the type of terrain and the location. Mountain regions, latitude, and proximity to industrial areas might make a difference.

10.47 The two confidence intervals do not overlap. The confidence intervals, however, are intended to give a location for the mean, not for an individual value. The prediction intervals do overlap, and a future measurement from a similar area would likely result in an IBI in the overlapping range of these two intervals.

10.49 **(a)** $\hat{y} = 24{,}874.3745 + 892.1135 \times$ Age $= 51{,}637.78$. **(b)** Under the column labeled 95.0% CI, we see that the desired interval is (49,780, 53,496). **(c)** Under the column labeled 95.0% PI, we see that the desired interval is $(-41{,}735, 145{,}010)$. It is too wide to be very useful.

10.51 (58,915, 62,203).

10.53 **(a)** (44,446, 55,314). **(b)** This interval is wider because the regression output uses all 5712 men in the data set, while the one-sample *t* procedure uses only the 195 men of age 30.

10.55 A 90% prediction interval for the BAC of someone who drinks five beers is (0.03996, 0.11428). This interval includes values larger than 0.08, so Steve can't be confident that he won't be arrested if he drives and is stopped.

10.57 $SE_{b1} = 0.0889$.

10.59 The hypotheses are $H_0: \beta_1 = 0$, $H_a: \beta_1 \neq 0$. $F = 49.705$, and $t = 7.0501$. $\sqrt{F} = t$. P-value $= 5.55 \times 10^{-9}$.

10.61 **(a)** $r^2 = \dfrac{\text{Regression SS}}{\text{Total SS}} = 0.5036$. **(b)** $s = \sqrt{\text{Residual MS}} = 1.8761$.

10.63 $r^2 = 0.533$; $s = 14.899$.

10.65 **(a)** $H_0: \beta_1 = 0$, $H_a: \beta_1 \neq 0$, $t = 6.041$, and P-value $= 0.0001$. There is strong evidence that the slope of the population regression line of profitability on reputation is nonzero. **(b)** 19.36% of the variation in profitability among these companies is explained by regression on reputation. **(c)** Part (a) tells us that the test of whether the slope of the regression line of profitability on reputation is 0 is statistically significant, but the value of r^2 in part (b) tells us that only 19.36% of the variation in profitability among these companies is explained by regression on reputation.

10.67 $(0.111822, 0.140586)$.

10.69 The F statistic (36.492) is the square of the t statistic (6.041) for the slope. P-value for F statistic $= 0.0001 = P$-value for t statistic for the slope.

10.71 $r^2 = 0.1936$, which equals the value of the Sum of Squares for the Model divided by the Sum of Squares for the Error.

10.73 **(a)** The slope describes the change in y for a unit change in x. **(b)** The population regression line is $\mu_y = \beta_0 + \beta_1 x$. **(c)** The confidence interval for a mean response is based on the value of x.

10.75 **(a)** There does appear to be a positive linear relationship between the driver's and computer's calculations of mpg based on a scatterplot of these two variables. **(b)** $\hat{y} = 11.812 + 0.775x$. **(c)** The regression model would be similar to $\hat{y} = x$ if the calculations were the same on average, so in this case they appear not to be the same.

10.77 Answers will vary.

10.79 **(a)** For more expensive items, the pharmacy would be charging less of a markup. **(b)** $\hat{y} = 2.885 - 0.295x$ where $\hat{y}$ is the predicted (log) markup, and x is the (log) cost. **(c)** df $= 137$, but using the t table we would use df $= 100$ to be conservative. For testing $H_0: \beta_1 = 0$, $H_a: \beta_1 < 0$, the P-value is less than 0.005, so we can reject the null hypothesis. There is strong evidence that (log) markup and (log) cost have a negative linear relationship (evidence that charge compression is taking place).

10.81 **(a)** Some questions to consider: Is this a national chain or an independent restaurant? What kind of food is prepared? Do the restaurants have similar staffing and experience? **(b)** Answers will vary.

10.83 **(a)** The relationship looks positive, linear, and moderately strong except for potential outliers for the two largest hotels (1388 and 1590 rooms each). **(b)** The moderately strong linear relationship is good, but the outliers are not. **(c)** $\hat{y} = 101.981 + 0.514x$, where $\hat{y}$ is the predicted number of employees, and x is the number of rooms. **(d)** $H_0: \beta_1 = 0$, $H_a: \beta_1 \neq 0$ with $t = 4.287$ and P-value from SPSS of 0.001, so reject the null hypothesis. There is strong evidence that there is a significant linear relationship between the number of rooms and the number of employees working at hotels in Toronto. **(e)** $(0.253, 0.775)$.

10.85   **(a)** Hotel 1 (1388 rooms) and Hotel 11 (1590 rooms) are the two outliers. **(b)** The R^2 drops from 60.5% in the original model to just 26.3% in the model without the outliers. The s drops from 188.774 in the original model to 129.151 in the model without the outliers. The new equation of the line is $\hat{y} = 72.526 + 0.589x$. The new t test statistic is 1.891, and the P-value is now 0.088, so we cannot reject the null hypothesis. There is not enough evidence to say that the slope is significantly different from 0. The 95% confidence interval for the slope is $(-0.105, 1.284)$, which contains 0. The 95% confidence interval for the slope from the original model did not contain 0.

10.87   The scatterplot shows a strong, linear, and positive relationship between the number of students and the total yearly expenditures. The least squares regression line is $\hat{y} = 0.526 + 0.530x$, where $\hat{y}$ is the predicted total yearly expenditure, and x is the number of students. R^2 is very good at 98.4%, and the regression standard error is $s = 1.70$. A two-sided test of the slope gives a $t = 41.623$ and a P-value close to 0. These results sound very promising; however, the normal probability plot does not look good, and the residual plot has a distinct funnel shape. The assumptions for regression may not be appropriate here.

10.89 $t = 2.16$, and $0.02 < P$-value < 0.04.

CHAPTER 11

11.1   PredSpace $= 210 + 160(45.25) + 160(15.50) + 150(41.52) + 65(411.88) + 120(25.24) = 45{,}959 \text{ ft}^2$.

11.3   **(a)** The response variable is bank assets. **(b)** The explanatory variables are the number of banks and deposits. **(c)** $p = 2$. **(d)** $n = 52$.

11.5   Multiple regression requires that the residuals have a Normal distribution, but the variables themselves do not need to be normally distributed.

11.7   The relationship between Assets and Number is weakly positive with a correlation of 0.018. The outliers of North Carolina and Ohio stand out for Assets. Nevada also shows higher assets than most other banks. Texas and Illinois show a greater number of banks than the other states. Assets and Deposits have a strong, positive linear relationship with a correlation of 0.998. Again, North Carolina and Ohio show much higher values for both Assets and Deposits than the other locations. Number and Deposits are weakly linear with a correlation of 0.049 with the same four states distinguished again.

11.9 $\widehat{\text{Assets}} = 8.090 - 0.108 \times \text{Number} + 1.566 \times \text{Deposits}$.

11.11   The residual plots highlight the outliers again. For the remaining points, the plots show a scattering of points that is close to the zero line with a few points straying away from the line. The residual plot against Number shows a denser scattering below the zero line than above. The histogram of the residuals shows that the residuals are denser toward the center of the distribution than would be expected of a Normal distribution. The Normal probability plot echoes this by showing a curved pattern to the residuals.

11.13   The residual plots show points that are more spread out along the zero residual line, not clustered toward the lower x values as was seen in the original data. We can still see the presence of the outliers, but the residuals are more randomly scattered. The Normal probability plot and the histogram of the residuals still show that the residuals are not normally distributed, but they are much closer to Normal than the original set of residuals was.

11.15   For Excel, the unrounded values are $s = 0.354000516$ and $s^2 = 0.12531637$. The name given to s in the output is Standard Error. For Minitab, the unrounded values are $s = 0.354001$ and $s^2 = 0.1253$. The name given to s in the output is S. For SPSS, the unrounded values are $s = 0.354001$ and $s^2 = 0.125$. The name given to s in the output is Std. Error of the Estimate. For SAS, the unrounded values are $s = 0.35400$ and $s^2 = 0.12532$. The name given to s in the output is Root MSE.

11.17   (a) Overall math GPA. (b) 106. (c) 4. (d) SAT math score, SAT verbal score, class rank, Bryant College's mathematics placement score.

11.19   (a)

	Mean	Median	Std. deviation
Price	116.58	120	47.142
Weight	12.32	12	1.945
Amps	13.711	15	1.7104
Depth	2.41137	2.44	0.066938
Speed	4.05	4	1.026
Power	4.26	4	0.806
Ease	3.63	4	0.597
Construction	4.26	5	0.872

(b) and (c) Most of the distributions are skewed to the left with the exception of Weight. Weight also has three observations that are right on the border of being considered outliers. Also of note is that Amps has over half of its values equal to its maximum value. Otherwise, nothing stands out as unusual in terms of completing the multiple regression.

11.21   (a) $\widehat{\text{Price}} = -467.715 + 11.567 \times \text{Weight} + 4.767 \times \text{Amps} + 82.935 \times \text{Depth} - 13.275 \times \text{Speed} - 1.186 \times \text{Power} + 18.751 \times \text{Ease} + 39.233 \times \text{Construction}$. (b) The regression standard error is 26.642. (c) The Normal probability plot shows that the points do not line up precisely on the diagonal line, but there is nothing that stands out as a major problem, so the model could be used with caution.

11.23   (a)

Variable	Mean	St. dev.	Median	Min.	Max.	Q_1	Q_3
Share	8.96	7.74	8.85	1.30	27.50	2.80	11.60
Accounts	794	886	509	125	2500	134	909
Assets	48.9	76.2	15.35	1.3	219.0	5.9	38.8

(b) Use split stems for stemplot. (c) All three distributions appear to be skewed to the right, with high outliers.

11.25   (a) Share $= 5.16 - 0.00031$ Accounts $+ 0.0828$ Assets. (b) $s = 5.488$.

11.27   All summaries are smaller except minimums. Largest effect is on the mean and standard deviation.

Variable	Mean	St. dev.	Median	Min.	Max.	Q_1	Q_3
Share	6.60	4.63	6.00	1.30	12.90	2.50	10.80
Accounts	392	293	316.5	125	909	132	602.5
Assets	13.76	12.27	9.00	1.30	38.80	5.70	20.30

11.29 Share $= 1.85 + 0.00663$ Accounts $+ 0.157$ Assets, and $s = 3.501$.

11.31 **(a)** Stemplots show that all four distributions are right-skewed and that gross sales, cash items, and credit items have high outliers.

Variable	Mean	Median	St. dev.
Gross	320.3	263.3	180.1
Cash	20.52	19.00	11.80
Credit	20.04	15.00	14.07
Check	7.68	5.00	7.98

(b) There is a strong linear trend between gross sales and both cash and credit card items. The association is less strong with check items, but this is due to an outlier. **(c)** Gross Total Sales $= 0.3412 + 7.1103 \times$ Cash Items $+ 6.987 \times$ Check Items $+ 4.4579 \times$ Credit Card Items. **(d)** Plot the residuals versus the three explanatory variables, and give a Normal quantile plot for the residuals. There are no obvious problems in the plots. **(e)** This model is not intended to create exact values for every potential combination of values of the explanatory variables. To use this model to predict the Gross Total Sales when all the explanatory variables are zero would be an extrapolation and, therefore, not an accurate use of the model.

11.32 Opening weekend revenue is significant in predicting U.S. revenue when using a model including Budget, number of Theaters and Opening weekend revenue.

Package	Coefficient	Standard error	t-statistic	df	P-value
Excel	3.2417285	0.296467651	10.93451	39	5.42E-13
Minitab	3.2417	0.2965	10.93	39	0.000
SPSS	3.242	0.296	10.935	39	0.000

11.35 $\widehat{\text{USRevenue}} = 50.071 - 0.499 \times \text{Budget} + 3.230 \times \text{Opening}$. There is very little change between this model and the previous one. There was a slight change in the coefficient of Opening and a slight change in the significance of Budget. The fact that removing Theaters from the model caused little change indicates that it was not significant to the model when these two other variables were included.

11.37 **(a)** (76.62, 253.51). **(b)** (79.78, 251.34). **(c)** When theaters is included in the model, the prediction interval is wider than when only opening-weekend revenue and budget are included.

11.39

Variables	Model	R^2
Budget, Opening	$\widehat{Revenue} = 50.071 - 0.499$ Budget $+ 3.230$ Opening	0.872
Budget, Theaters	$\widehat{Revenue} = -360.505 + 0.433$ Budget $+ 0.134$ Theaters	0.446
Opening, Theaters	$\widehat{Revenue} = 21.602 - 0.003$ Theaters $+ 2.932$ Opening	0.850
Budget	$\widehat{Revenue} = 18.792 + 1.136$ Budget	0.169
Opening	$\widehat{Revenue} = 12.268 + 2.901$ Opening	0.850
Theaters	$\widehat{Revenue} = -368.389 + 0.149$ Theaters	0.426

The R^2 value is high only when opening-weekend revenue is contained in the model. This indicates that opening-weekend revenue is very important in predicting total revenue. Including Budget as well results in a model with a slightly better R^2 value, indicating that Budget does help explain some of the variation in the model, but Budget by itself is a poor predictor of overall revenue.

11.41 $R_1^2 = 0.872$, and $R_2^2 = 0.446$. $F = 119.8125$ with 1 and 36 degrees of freedom. P-value ≈ 0. Opening-weekend revenue contributes significantly to explaining total revenue, even when Budget and number of Theaters are included in the model. F-statistic $= t^2$ from Example 11.13.

11.43 Because b_1 and SE_{b_1} do not change in these four cases, the test statistic stays the same at $t = 1.986$. The P-value will change due to different degrees of freedom; however, using t-tables, our estimate is the same. The P-value in each case is between 0.05 and 0.10.

11.45 **(a)** The squared multiple correlation, R^2, gives the proportion of the variation in the response variable that is explained by the explanatory variables. **(b)** The null hypothesis should be H_0: $\beta_2 = 0$. **(c)** A small P-value for the ANOVA F test only implies that at least one explanatory variable is statistically different from zero.

11.47 **(a)** DFR $= p = 3$, so MSR $= 16/3 = 5.3333$. DFE $= n - p - 1 = 23$, so MSE $= 120/23 = 5.2174$. $F = $ MSR/MSE $= 1.022$. **(b)** $R^2 = $ SSR/SST $= 16/136 = 0.1176$.

11.49 **(a)** *Finding Nemo* obtained \$109.49 million more than predicted, and *Lord of the Rings: The Return of the King* made \$139.50 more than predicted. **(b)**

	Coefficient	Std Error	t statistic	P-value
Budget	−0.311	0.139	−2.230	0.032
Opening	2.896	0.208	13.942	0.000
Theaters	0.005	0.013	0.346	0.731

(c) Compared to Example 11.13, Theaters now has a positive impact instead of a negative impact, although it is still not a significant variable. Opening and Budget are still significant, but their coefficients have decreased in size. **(d)** The residuals do not appear to be precisely Normal. The positive residuals seem to follow the normal distribution fairly well, but the negative residuals do not. In particular, there are more movies with residuals between -25 and -12.5 than would be expected in a Normal distribution.

11.51 **(a)** $t = -8.917$, df $= 40 - 30 - 1 = 9$, P-value < 0.001. There is strong evidence that the coefficient of P5 is not 0. **(b)** $(-1.487, -0.885)$. Because we rejected the null hypothesis in part (a), we did not expect the confidence interval to include 0.

11.53 The difference in signs makes sense because employees generally sign up for either the DC or the DB plan. The total number of dual-earner couples available to sign up for a plan probably stays about the same from year to year; however, as more dual-earner couples sign up for DB plans, fewer would be signing up for DC plans.

11.55 **(a)** The hypotheses about the jth explanatory variable are H_0: $\beta_j = 0$ and H_a: $\beta_j \neq 0$. The degrees of freedom for the t statistics are 2215. At the 5% level, values that are less than -1.96 or greater than 1.96 will lead to rejection of the null hypothesis. **(b)** The significant explanatory variables are loan size, length of loan, percent down payment, cosigner, unsecured loan, total income, bad credit report, young borrower, own home, and years at current address. If a t is not significant, conclude that the explanatory variable is not useful in prediction when all other variables are available to use for prediction. **(c)** The interest rate is lower for larger loans, lower for longer length loans, lower for a higher percent down payment, lower when there is a cosigner, higher for an unsecured loan, lower for those with higher total income, higher when there is a bad credit report, higher when there is a young borrower, lower when the borrower owns a home, and lower when the number of years at current address is higher.

11.57   **(a)** The hypotheses about the jth explanatory variable are H_0: $\beta_j = 0$ and H_a: $\beta_j \neq 0$. The degrees of freedom for the t statistics are 5650. At the 5% level, values that are less than -1.96 or greater than 1.96 will lead to rejection of the null hypothesis. **(b)** The statistically significant explanatory variables are loan size, length of loan, percent down payment, and unsecured loan. **(c)** The interest rate is lower for larger loans, lower for longer length loans, lower for a higher percent down payment, and higher for an unsecured loan.

11.59   **(a)** F has 8 degrees of freedom in the numerator and 494 in the denominator. **(b)** 68%. **(c)** $F = 20.58$. P-value < 0.001. There is evidence that at least one of the other variables is helpful in predicting satisfaction when quality of service received and the possibility of negotiating the terms of financing are included in the model.

11.61   From the stemplot, six of the excluded homes are the six most expensive homes. Three of these are clearly outliers (the three most expensive) for this location.

11.63   1000 sq. ft. homes, predicted price = $79,621.62. For 1500 sq. ft. homes, predicted price = $96,783.43.

11.65   For 1000 sq. ft. homes, predicted price = $78,253.47. For 1500 sq. ft. homes, predicted price = $97,041.71. Overall, the two sets of predictions are fairly similar.

11.67   The pooled t test statistic is $t = 2.875$ with df $= 35$ and $0.005 < P$-value < 0.01. These results agree (up to roundoff) with the results for the coefficient of Bed3 in Example 11.21.

11.69   The trend is increasing and roughly linear as one goes from zero to two garages and then levels off.

11.71   With an extra half bath, predicted price = $99,719. Without an extra half bath, predicted price = $84,585. The difference in price is $15,134.

11.73   Because we have no data on smaller homes with an extra half bath, our regression equation is probably not trustworthy.

11.75   **(a)** The relationship between μ_y and x is gently curved and increasing. **(b)** The relationship between μ_y and x is more strongly curved and increasing. **(c)** The relationship between μ_y and x is curved and decreasing. **(d)** The relationship between μ_y and x is less strongly curved and decreasing.

11.77   For part **(a)**, the difference in means (Group B $-$ Group A) $= 4 =$ the coefficient of x. This can be verified directly for the other parts and shown to be true in general.

11.79   For part **(a)**, the difference in slopes (coefficient x_2 for Group B $-$ coefficient x_2 for Group A) $= 12 - 6 = 6 =$ the coefficient of $x_1 x_2$. The difference in the intercepts (Group B $-$ Group A) $= 100 - 70 = 30 =$ coefficient of x_1. This can be verified directly for the other parts and shown to be true in general.

11.81   **(a)** Assets $= 7.61 - 0.0046$ Account $+ 0.000034$ Account^2. **(b)** 0.000034 ± 0.000021. **(c)** $t = 3.76$, df $= 7$, and $P = 0.007$. The quadratic term is useful for predicting assets in a model that already contains the linear term. **(d)** The variables Account and the square of Account are highly correlated.

11.83   **(a)** $F(2,37) = 16.223$ with a P-value of 0, so at least one of the variables is significant in the model that includes both Theaters and the square of Theaters. However, the test statistics for the individual variables ($t = -1.226$ with P-value of 0.228 for Theaters and $t = 1.690$ with P-value of 0.099 for Theaters^2) are not significant at the 5% level, indicating that the combination is significant but not the individuals. This result looks strange and is due to the dependence of the variables. **(b)** $R^2 = 0.467$ for the model with Theaters and Theaters^2. $R^2 = 0.426$ for the model with only Theaters. $F = 2.846$ with a P-value is 0.100, indicating that Theaters does not contribute significantly to explaining U.S. box office revenue when Theaters^2 is already in the model. **(c)** These are approximately equal.

11.85

Variables	Constant	Opening	Budget	Theaters	Sequel	R^2	s
Opening, Budget, Theaters, Sequel	23.951	2.856 (*)	−0.317 (*)	0.004	5.207	0.930	28.673
Opening, Budget, Theaters	22.050	2.896 (*)	−0.311 (*)	0.005	0	0.930	28.299
Opening, Budget, Sequel	36.691	2.896 (*)	−0.317 (*)	0	5.523	0.964	28.291
Opening, Theaters, Sequel	0.811	2.681 (*)	0	0.004	0.758	0.920	30.298
Opening, Budget	35.641	2.942 (*)	−0.310 (*)	0	0	0.930	27.941
Opening, Theaters	0.600	2.688 (*)	0	0.004	0	0.920	29.863
Opening, Sequel	12.289	2.717 (*)	0	0	1.051	0.919	29.894
Opening	12.190	2.727 (*)	0	0	0	0.919	29.478

COEFFICIENTS

The preferred model is still the model with both Opening and Budget included but neither Theaters nor Sequel included.

11.87   **(a)** $\mu_y = \beta_0 + \beta_1 x_{LOC} + \beta_2 x_{age} + \beta_3 x_{gender}$. **(b)** β_0, β_1, β_2, β_3, and σ are the parameters. **(c)** The estimate for β_0 is 3.284. The estimate for β_1, is 0.028. The estimate for β_2 is -0.023. The estimate for β_3 is -0.150. The estimate for σ is 0.414. **(d)** $F(3, 96) = 11.008$ with a P-value of 0.000. There is evidence that at least one variable is significant for predicting Stress when LOC, age, and gender are included in the model. **(e)** $R^2 = 0.256$. **(f)** LOC has a test statistic of 2.271 with df $= 96$, and the P-value $= 0.025$. Age has a test statistic of -4.348 with df $= 96$, and the P-value $= 0.000$. Gender has a test statistic of -1.480 with df $= 96$, and the P-value $= 0.142$. For the model including all three variables, LOC and Age are both significant for predicting Stress while Gender is not significant.

11.89   **(a)** The constant in the single-variable model is smaller (2.2555 as compared to 3.284), and the coefficient of LOC is larger in the single-variable model (0.0399 as compared to 0.028). The estimate of the standard deviation is also larger for the model with only LOC (0.4513 as compared to 0.414). **(b)** Predicted stress level is 2.598 for the full model and 2.655 for the single-variable model. **(c)** A one-unit change in LOC changes Stress by 0.028 in the full model and 0.0399 in the single-variable model. **(d)** R^2 for the full model is 0.256. Because this is quite a bit higher than the R^2 in the single-variable model, we would conclude that age and gender are helpful predictors of stress when LOC is already in the model.

11.91 **(a)** Price vs. Promotions: negative, moderate, linear. Price vs. Discount: negative, weak, fairly linear. **(b)** The following table shows the mean and standard deviation for expected price for each combination of promotion and discount.

Promotions	Discount			
	10	20	30	40
1	4.920, 0.1520	4.689, 0.2331	4.225, 0.3856	4.423, 0.1848
3	4.756, 0.2429	4.524, 0.2707	4.097, 0.2346	4.284, 0.2040
5	4.393, 0.2685	4.251, 0.2648	3.89, 0.1629	4.058, 0.1760
7	4.269, 0.2699	4.094, 0.2407	3.76, 0.2618	3.780, 0.2144

(b) and **(c)** At every promotion level, the 10% discount yields the highest expected price, then 20%, then 40%, and then 30%. For every discount level, the 1 promotion yields the highest expected price, then 3, then 5, and then 7.

11.93 In the previous exercise, the residual plot for promotions looks okay but the residual plot for discount seems to have a curved shape. Therefore, try the quadratic term for discount and the interaction of discount and promotion. If the quadratic term for discount is used, the R^2 increases to 61.1%, the s decreases to 0.2511, the F test statistic is 81.809, and the P-value is still very close to 0. All coefficients are significantly different from 0. If the quadratic term is removed and the interaction of discount and promotion is used instead, the coefficient for the interaction term is not significant, and R^2 and s are very close to what they were in the original model. The best model is the one with the quadratic term for discount: $\hat{y} = 5.540 - 0.102$ Promotion $- 0.060$ Discount $+ 0.001$ Discount^2.

11.95 **(a)**

	Area	Forest	IBI
Mean	28.29	39.39	65.94
St. dev.	17.714	32.204	18.280
Description of stemplot	Right skewed, 2 high outliers	Right skewed, no outliers	Left skewed, no outliers

(b) All relationships are linear and moderately weak. Area and Forest have a negative relationship, but the others are positive. Data point #40 looks like a potential outlier on the Area/Forest scatterplot but not anyplace else. **(c)** $y_i = \beta_0 + \beta_{Area}x_{Areai} + \beta_{Forest}x_{Foresti} + \varepsilon_i$. **(d)** H_0: $\beta_{Area} = \beta_{Forest} = 0$, H_a: The coefficients are not both 0. **(e)** $R^2 = 35.7\%$, $s = 14.972$, $F = 12.776$, P-value $= 0.000$. All coefficients are significant. $\hat{y} = 40.629 + 0.569$ Area $+ 0.234$ Forest. **(f)** Residual plots look good—random and no outliers. **(g)** A histogram of the residuals looks slightly skewed left, but the Normal probability plot looks good. **(h)** Yes. We had general linear relationships between the variables, and the residuals look good; however, only 35.7% of the variation in IBI is explained by Area and Forest.

11.97 **(a)** 20%. **(b)** Room for change (PPP) with a test statistic of -16.82, strength of expression of WOM with a test statistic of -3.385, and amount of WOM given with a test statistic of -3.636 are statistically significant. **(c)** The more strongly the consumer states their negative opinion about a certain brand, the more negative the impact will be. **(d)** When the NWOM is regarding the main brand, the impact will be more negative than when the NWOM is about another brand.

11.99 **(a)** Based on these results, neither variable is significant in predicting y when both variables are included in the model. **(b)** When only $x1$ is included in the model, the P-value of this coefficient is 0.032, indicating that it is significant in predicting y. $X1$ and y are positively correlated with $R = 0.393$. When only $x2$ is included, the P-value of this coefficient is 0.025, indicating that it is significant in predicting y. $X2$ is also positively correlated with y with $R = 0.409$. **(c)** Multiple regression results are not sufficient for describing relationships between variables. The relationships between the individual variables should also be considered.

11.101 **(a)**

	Mean	St. dev.	Description of distribution
PCB	68.4674	59.3906	Skewed right, 5 high outliers
PCB52	0.9580	1.5983	Skewed right, 6 high outliers
PCB118	3.2563	3.0191	Skewed right, 5 high outliers
PCB138	6.8268	5.8627	Skewed right, 5 high outliers
PCB180	4.1584	4.9864	Skewed right, 7 high outliers

(b) All the variables are positively correlated with each other. All the explanatory variables are significantly correlated with PCB, although PCB52 is the most weakly correlated with PCB and with the other explanatory variables. Only the correlation between PCB52 and PCB180 is not significant (P-value $= 0.478$). **(c)** $\hat{y} = 0.937 + 11.873 \cdot PCB52 + 3.761 \cdot PCB118 + 3.884 \cdot PCB138 + 4.182 \cdot PCB180$. $R^2 = 98.9\%$, $s = 6.3821$, $F = 1456.178$, P-value is very close to 0. All the coefficients are significant. **(d)** There does appear to be a slight pattern to the residuals. The histogram confirms that the residual values are more concentrated around the zero line than would be seen if these were normally distributed. **(e)** #50 and #65 are the two potential outliers. #50 (residual $= -22.0864$) is the overestimate. **(f)** $\hat{y} = 1.628 + 14.442 PCB52 + 2.600 PCB118 + 4.054 PCB138 + 4.109 PCB180$. All coefficients are significant again. $R^2 = 99.4\%$, $s = 4.555$, $F = 2628.685$, P-value $= 0$. The residuals look fairly normally distributed, and the residual plots look random. There are two new potential outliers though at #44 and #58.

11.103 **(a)** Results will vary with software. **(b)** Most software should ignore these data points, but ignoring data is not a good idea. **(c)** The following Table uses Base 10 logs.

Log of variable	Mean	St. dev.	Description of stemplot
*PCB*138	0.7009	0.3494	Fairly symmetric, 1 low outlier
*PCB*153	0.7397	0.3914	Fairly symmetric, no outliers
*PCB*180	0.4235	0.4028	Fairly symmetric, 1 high outlier
*PCB*28	−0.5793	0.4918	Fairly symmetric, 1 low and 1 high outlier
*PCB*52	−0.3354	0.5167	Fairly symmetric, 1 low and 2 high outliers
*PCB*126	2.1044	0.3325	Right skewed, 16 high outliers
*PCB*118	0.3717	0.3592	Fairly symmetric, 1 low and 1 high outlier
PCB	1.701	0.3483	Fairly symmetric, no outliers
TEQ	0.3495	0.2591	Right skewed, no outliers

11.105 **(a)** Corn yield $= -41.817 + 4.614 \times$ Soybean yield, $R^2 = 0.9109$, $t = 22.151$, P-value $= 0$. **(b)** The residuals appear fairly Normal. **(c)** There is an increasing and then decreasing effect when looking at the residuals plotted against Year. It makes sense to include Year in the regression model.

11.107 **(a)** Corn yield $= -1508.127 + 0.767 \times$ Year $-0.014 \times$ Year$^2 + 2.954 \times$ Soybean Yield. **(b)** H_0: All coefficients are equal to zero. H_a: At least one coefficient is not equal to zero. $F = 277.563$, df $= 3$ and 47, P-value $= 0$. This indicates that the model provides significant prediction of corn yield. **(c)** $R^2 = 0.947$ compared to 0.939 from the previous model. **(d)** In the order they appear in the model, the t statistics for the coefficients are $t = 4.078$ with a P-value $= 0$, $t = -2.501$ with a P-value $= 0.016$, and $t = 6.436$ with a P-value $= 0$. **(e)** The residuals still show somewhat of a pattern when compared to Year. The residuals do not appear completely random when compared to Year^2 either.

11.109 For the linear model, predicted yield for 2008 was 151.37 with a 95% prediction interval of (131.45, 171.28). For the quadratic model, predicted yield for 2008 was 150.66 with a 95% prediction interval of (129.54, 171.79). The linear model gave a closer prediction to actual.

11.111 The scatterplots do not show strong relationships. The plot of SATM vs. GPA shows two possible outliers on the left side. The plot of SATV vs. GPA does not show any obvious outliers.

11.113 The residuals do not show any obvious patterns or problems.

11.115 **(a)** GPA $= 0.666 + 0.193$ HSM $+ 0.00061$ SATM. **(b)** H_0: All coefficients are equal to 0. H_a: At least one coefficient is not equal to zero. This means that we assume the model is not a significant predictor of GPA and let the data provide evidence that the model is a significant predictor. $F = 26.63$ and P-value $= 0$. This model provides significant prediction of GPA. **(c)** HSM: (0.1295, 0.2565), SATM: (−0.00059, 0.001815). Yes, the interval describing SATM coefficient does contain zero. **(d)** HSM: $t = 5.99$, P-value $= 0$. SATM: $t = 0.999$, P-value $= 0.319$. Based on the large P-value associated with the SATM coefficient, one should conclude that SATM does not provide significant prediction of GPA. **(e)** $s = 0.703$. **(f)** $R^2 = 0.1942$.

11.117 Looking at the sample of males only shows the same results when compared to the sample of all students. $R^2 = 0.184$ (compared to 0.20), and while HSM is a significant predictor, HSS and HSE are not.

11.119 **(a)** GPA = $0.582 + 0.155$ HSM $+ 0.050$ HSS $+ 0.044$ HSE $+ 0.067$ Gender $+ 0.05$ GHSM $- 0.05$ GHSS $- 0.012$ GHSE. The t statistics for each coefficient have P-values greater than 0.10 for all except the explanatory variable HSM. **(b)** Verify. **(c)** Verify. **(d)** $F = 0.143$, P-value $= 0.966$. This indicates there is no reason to include gender and the interactions.

CHAPTER 12

12.1 Include items like getting out of the car, checking baggage outdoors or indoors if bringing baggage to check, checking in, receiving boarding pass, determining departure gate, going through security checkpoints, and purchasing souvenirs or snacks.

12.3 Include items like your knowledge of and experience with the process, the time to find the picture you want to upload, the speed of your computer, the quality of Internet connection, and the source of the upload (camera, hard drive, etc.).

12.5 Common cause variation might include a longer line at the security checkpoint or at the ticket counter or crowds in the airport corridor. Special cause variation might include forgetting something in the car and having to go back for it, an electrical malfunction on internal airport shuttle on which you are riding, a power outage at the airport, or an increased security alert requiring all passengers' bags to be searched prior to boarding.

12.7 5:45 A.M., alarm rings $\rightarrow$ 5:45 A.M., get out of bed $\rightarrow$ 5:46 A.M., start coffeemaker $\rightarrow$ 5:47 A.M., shower, shave, and dress $\rightarrow$ 6:15 A.M., breakfast $\rightarrow$ 6:30 A.M., brush teeth $\rightarrow$ 6:35 A.M., leave home $\rightarrow$ 6:50 A.M., park at school $\rightarrow$ 6:55 A.M., arrive at office.

12.9 My main reasons for late arrivals at work are "don't get out of bed when the alarm rings" (responsible for about 40% of late arrivals), "too long eating breakfast" (responsible for about 25% of late arrivals), "too long showering, shaving, and dressing" (responsible for about 20% of late arrivals), and "slow traffic on the way to work" (responsible for about 15% of late arrivals).

12.11 When a point falls above the upper limit of the R chart, this means that the particular subgroup in question has greater variability than is generally seen in the other subgroups. A point above the upper control limit of the $\bar{x}$ chart indicates that the process level for the group is above the expected range, and the process itself is out of control.

12.13 Using the data with the two subgroups removed as stated in the text results in an $\bar{R}$ of 0.00899 and an $\bar{\bar{x}}$ of 2.60755. For the R chart, $D_4 = 2.282$, $D_3 = 0.000$, the UCL $= 2.282 \times 0.00899 = 0.02052$, the LCL $= 0 \times 0.00899 = 0$, and the CL $= \bar{R} = 0.00899$. For the $\bar{x}$ chart, $A_2 = 0.729$, the UCL $= 2.60755 + 0.729 \times 0.00899 = 2.6141$, the CL $= \bar{\bar{x}} = 2.60755$, and the LCL $= 2.60755 - 0.729 \times 0.00899 = 2.60100$.

12.15 For the s chart, $B_4 = 2.089$, $B_3 = 0.000$, the UCL $= 2.089 \times 9.2015 = 19.2219$, the LCL $= 0 \times 9.2015 = 0$, and the CL $= \bar{s} = 9.2015$. For the $\bar{x}$ chart, $A_3 = 1.427$, the UCL $= 40.62 + 1.427 \times 9.2015 = 53.7505$, the CL $= \bar{\bar{x}} = 40.62$, and the LCL $= 40.62 - 1.427 \times 9.2015 = 27.4895$.

12.17 Answers will vary. For me, some examples might be dollars spent on weekly grocery shopping trips or number of minutes it takes to drive/walk/bike to work.

12.19 LSL and USL define the acceptable set of values for *individual* measurements on the output of a process. These limits represent the *desired performance* of the process. LCL and UCL for x are the lower and upper control limits for the *means of samples* of several individual measurements on the output of a process. They represent the *actual performance* of the process when it is in control.

12.21 $1 - P(2.592 \leq \text{diameter} \leq 2.632) = 0.000086$. This is equivalent to 86 defective parts per million parts.

12.23 We want to compute the probability that $\bar{x}$ will be beyond the control limits; that is, $\bar{x}$ will be either larger than UCL = 713 or smaller than LCL = 687. The desired probability is 0.1591.

12.25 For the R chart, LCL = 0, CL = 0.372, and UCL = 0.7864. For the $\bar{x}$ chart, LCL = 10.3486, CL = 10.5632, and UCL = 10.7778.

12.27 When selecting a group of nine consecutive observations, 1.953 times out of 1000, all nine of the observations would be on the same side of the center line. This is computed using the probability of 0.5 that an individual observation would fall on a particular side of the center line.

12.29 Both the sample ranges and the sample means stay within the three-sigma limits; however, the last nine observations have sample means below the CL, indicating that the process is out of control based on the nine-in-a-row rule. This affects subgroups 27 through 35.

12.31 For the s chart, the LCL = 2.1318, the CL = 7.50638, and the UCL = 12.8809. For the $\bar{x}$ chart, LCL = 23.3646, CL = 30.6833, and UCL = 38.0020.

12.33 Answers will vary.

12.35 **(a)** 17.78% of clip openings will meet specifications. **(b)** $\hat{C}_{pk} = 0.06$.

12.37 Answers will vary.

12.39 $C_p = (\text{USL} - \text{LSL})/6\sigma$, so if $C_p \geq 2$, we must have USL $-$ LSL $\geq 12\sigma$. If the process is properly centered, then its mean μ will be halfway between USL and LSL, and hence USL and LSL will be at least 6σ from μ. Thus, this is called six-sigma quality.

12.41 **(a)** The total number of opportunities for unpaid invoices = 128,750, $\bar{p} = 0.0334$. **(b)** UCL = 0.0435, CL = 0.0334, LCL = 0.0233.

12.43 The center line is given by $\bar{c} = 6.304$. The UCL = 13.837, and the LCL = $-1.228 \rightarrow 0$. There are no points above the UCL or below the UCL after the original out-of-control observation has been removed.

12.45 UCL = 0.01307, CL = 0.00599, LCL = 0.

12.47 **(a)** $\bar{p} = 0.356$, $\bar{n} = 921.8$. **(b)** For the p chart, UCL = 0.4033, CL = 0.356, LCL = 0.3087. The process (proportion of students per month with three or more unexcused absences) appears to be in control. That is, there are no months with unusually high or low proportions of absences, and no obvious trends. **(c)** For October, UCL = 0.4027 and LCL = 0.3093. For June, UCL = 0.4043 and LCL = 0.3077. These exact limits do not affect our conclusions in this case.

12.49 UCL = 26.619, and LCL = 3.381.

12.51 With attribute control charts, if an out-of-control point is on the above the upper control limit, an investigation should be launched with the goal of removing and protecting against any special causes of poor quality. However, if an out-of-control point is below a lower control limit, an investigation should also be conducted to promote any special causes of high quality.

12.53 **(a)** The percents add up to larger than 100% because customers can have more than one complaint. **(b)** The category with the largest number of complaints is the ease of obtaining invoice adjustments/credits, so we might target this area for improvement.

12.55 $CL = 7.65$, $UCL = 19.642$, $LCL = 0$. The first sample is out of control, perhaps reflecting an initial lack of skill or initial problems in using the new system. After the first sample, the process is in control with respect to short-term variation, although sample 10 is very close to the upper control limit.

12.57 **(a)** We would use a p chart. The control limits would be $UCL = 0.019$, $CL = 0.003$, and $LCL = 0$. **(b)** If the proportion of unsatisfactory films is 0.003, then in a sample of 100 films, the expected number of unsatisfactory films is 0.3. Thus, most of our samples will have no unsatisfactory films, and plotting the sample values (most of which will be 0) will not be very informative.

12.59 **(a)** Using a B_3 of 0 and a B_4 of 3.267, the UCL is 0.00916, and the LCL is 0. No standard deviation is outside the range of the LCL and UCL. **(b)** The limits for the $\bar{x}$ chart are $UCL = 1.269$ and $LCL = 1.254$. The process is in control.

12.61 **(a)** With $\bar{p} = 0.522$, $UCL = 0.734$, and $LCL = 0.310$, the process does appear to be in control. **(b)** The new sample proportions are not in control according to the original control limits from part (a). Attempts 2, 4, and 10 are above the UCL. With new $\bar{p} = 0.702$, the new limits for future samples should be $UCL = 0.896$, and $LCL = 0.508$.

12.63 **(a)** Using an approximation for c_4 of 0.99363, an $\bar{x}$ of 0.75625, and an s of 3.8876, the $UCL = 12.494$ and the $LCL = -10.9813$. There is one observation below the LCL: variance $= -11.98$. **(b)** There is an out of control signal based on the nine-in-a-row rule. These observations are 4.28, 1.63, 1.43, 2.92, 4.56, 3.04, 2.25, 2.42, and 3.86, which all fall above the mean of 0.75625. **(c)** The sample mean is now 0.528 with standard deviation $s = 3.5865$. c_4 is approximated as 0.99145. The new UCL is 11.380 and the new LCL is -10.324. The largest variance still remaining in the data set is 7.69, and the smallest is -6.01. There are no further occurrences of nine points in a row above or below the sample mean, so these are the final control limits.

CHAPTER 13

13.1 **(a)** There is an upward trend to the actual prices, indicating that this time series is not a random process. **(b)** Two runs are present—the first with prices below the mean and the second with prices above the mean. This is consistent with the upward trend noted.

13.3 $R = 35$, $n_A = 42$, $n_B = 48$, $\mu_R = 45.80$, $\sigma_R = 4.70$. $Z = -2.30$ giving a P-value of 0.0214.

13.5 **(a)** $R = 41 + 37 = 78$. **(b)** 0.1630.

13.7 **(a)** Based on the output, $t = 7.97$. P-value $= 2.745E-09$. There is evidence that the regression coefficient for the trend line is not zero. **(b)** Predicted sales are

314.91 thousand feet for January 2008 and 319.44 thousand feet for February 2008.

13.9   Forcasted sales are $46,393.57 million for September 2008, $49,167.45 million for October 2008, $55,745.32 million for November 2008 and $72,329.57 million for December 2008.

13.11 Taking the log of both sides of the equation in Example 13.4 results in this equation in logged units: $\log(\widehat{Cars_t}) = 3.00 + 0.266969t$.

13.13 Additive seasonality would be more appropriate for this data. The amount of increase or decrease appears to be relatively constant from year to year for a given month.

13.15 **(a)** 1.3658. **(b)** When Q1, Q2, and Q3 are set to zero, we are left with the constant 715.91397 times the trend component which is the prediction equation for the fourth quarter. Since the trend components for the trend-only model and the trend-and-seasonal model are nearly equal, the factor of 1.3658 implies fourth quarter's sales are 1.3658 times the general trend of the series.

13.17 **(a)** and **(b)** There is an upward trend to these data that appears to be somewhat curved rather than linear. **(c)** While there is not a strong, obvious seasonal pattern as seen in the previous examples, there does appear to be some seasonality. Notice particularly that January of each year is at a peak, followed by a decrease of the next couple of months, then a fluctuating increase up to the next January. **(d)** There are three groups of points seen in this data. The first is through approximately January 2004. The second is from January 2004 to January 2007, and the third begins at January 2007. In the first of these three groups, the level of consumer loans is increasing gently. Then there is a sizeable increase at the end of 2003. At this time, we see a higher level of loans with more fluctuation in the loan amounts. Finally, starting at the beginning of 2007, we can see much steeper increases occurring throughout the two years shown in the data.

13.19 **(a)** and **(b)** The time series shows a slightly decreasing process with a number of clearly low points compared to the rest of the series. **(c)** The clear lows occur in August and December of each year.

13.21 **(a)** There is a downward trend to this series. From 1981 to 1991, the trend is slightly downward, almost flat. From 1991 to 1996, the trend decreases more sharply. From 1996 to 2007, there is a very sharp decreasing trend with the exception of 1998 to 1999 where the reserves are almost level. **(b)** The expected number of runs is 14.04, but the observed number of runs is only 2. The P-value for testing randomness is approximately 0, indicating that this process is not random. **(c)** The distances between the observations show an increasing pattern over time.

13.23 **(a)** Both time series show three large peaks during the summer months and much lower values during the winter months. **(b)** July is the highest point for the non-U.S. visitors. July and August are both high for U.S. visitors. December shows a small peak in the middle of the off-peak season. January has the smallest number of visitors every year for U.S. visitors, and November and January are both low for non-U.S. visitors. **(c)** Visually, the U.S. visitor series seems to be trending down slightly while the non-U.S. visitor series is trending up.

13.25 **(a)** The data has a negative trend with the number of AT&T wireline customer decreasing more rapidly toward the end of the time series. **(b)** Seasonality is not apparent in this series. **(c)** $\hat{y}_t = 33683 - 604.643t$. **(d)** The model does not provide a good fit for the actual data seen in the time plot and will overestimate the number of customers in any future period because it does not decrease as rapidly as the actual data does.

13.27 **(a)** $(-1.20, 1.37)$. **(b)** $(57.18, 59.75)$.

13.29 **(a)** There is no evidence of a trend in this series. **(b)** There is no completely consistent pattern of one month appearing above or below the line every time, but there are months that tend to fall on one side or the other of the sample mean: June is below 7 out of 8 times, July is above 7 out of 8 times, January is below 5 out of 7 times, December is below 5 out of 7 times, and October is above 5 out of 7 times. **(c)** $\hat{y}_t = -188.6 + 493.9Feb + 230.7Mar - 36.3Apr + 329.5May + 72.8Jun + 975.1Jul - 73.2Aug + 129.5Sep + 798.9Oct + 293.9Nov - 115.3Dec$. **(d)** $\hat{y}_{93} = -59.1$, $\hat{y}_{94} = 610.3$.

13.31 **(a)** The residuals exhibit a random pattern with some short runs. **(b)** The observed number of runs is 45, and the expected number of runs is 46.9130. The P-value of the runs test is 0.678. There is not enough evidence against randomness to say that the residuals are nonrandom. **(c)** The correlation is approximately zero, and the lagged residual plot shows random scatter in all four quadrants, so there is no evidence of autocorrelation.

13.33 For S1, when the observed number of runs is significantly less than expected, the series will have positive autocorrelation. A lagged residual plot will show points with a strong positive association. For S2, when the observed number of runs is significantly greater than expected, the series will have negative autocorrelation. A lagged residual plot will show points with a strong negative association. For S3, when the series does not have a significantly different number of runs than expected, the lagged residual plot will show points equally dispersed among the four quadrants of the scatterplot.

13.35 Because the coefficients are the same for the first three quarters, we can just write a model based on whether we are looking at the fourth quarter or not. $\hat{y}_t = 230 + 10t - 30Q_4$.

13.37 **(a)** and **(b)** The time series plot follows a meandering path with a slight upward trend. **(c)** An AR(1) model is probably appropriate because the lagged time series plot shows a positive linear relationship between Rate and Rate lagged one period.

13.39 $\widehat{\log y_{37}} = 1.90299 + 0.01485(37) - 0.69903(1) + 0.76978\log(8.87138) = 8.58242$. $\hat{y}_{37} = 5337.01$.

13.41 **(a)** The residual series exhibits less meandering behavior than the original series. **(b)** There is a relatively random scattering of points in the lagged residual plot with a slightly denser grouping in the third quadrant and a looser grouping in the first quadrant, but no strong indication that autocorrelation has not been accounted for. **(c)** The residuals have a right-skewed distribution instead of a Normal distribution as seen through the probability plot and the histogram.

13.43 **(a)** The lagged time series plot shows a very strong, positive linear relationship between the lagged variable and the original data. **(b)** The correlation is 0.995. **(c)** Based on the relationship between these variables, using a 12-month lagged variable as a predictor of retail sales seems reasonable.

13.45 **(a)** There is no significant trend seen in this model. **(b)** The final two moving averages are 2834.18 and 2835.91 for one-week moving average, 2583.04 and 2835.04 for the two-week moving average, 2732.13 and 2667.33 for the three-week moving average, 2887.93 and 2758.07 for the four-week moving average,

2966.92 and 2877.52 for the five-week moving average, and 3036.53 and 2945.09 for the six-week moving average. Following are the moving average residuals for one point. The remaining residuals should be calculated equivalently.

	1-wk	2-wk	3-wk	4-wk	5-wk	6-wk
Residual for 12/13/08	1.73	252.87	103.78	−52.02	−131.02	−200.62

13.47 **(a)** The MSE are 1929978 for a span of 1, 1465070 for a span of 2, 1319551 for a span of 3, 1316952 for a span of 4, 1292744 for a span of 5, and 1342976 for a span of 6. **(b)** The MSE decreases sharply at first and then gradually to a low value at a span of 5. At a span of 6, the MSE increases.

13.49 **(a)**

year	1995	1996	1997	1998	1999	2000	2001
$\hat{y}_{year}$	292.31	292.31	289.24	288.83	292.93	299.17	307.16

year	2002	2003	2004	2005	2006	2007
$\hat{y}_{year}$	309.69	310.27	311.31	310.23	309.63	313.37

The forecast for 2008 is 315.80.

(b)

year	1995	1996	1997	1998	1999	2000	2001
$\hat{y}_{year}$	292.31	292.31	280.01	285.76	304.62	320.23	335.33

year	2002	2003	2004	2005	2006	2007
$\hat{y}_{year}$	322.94	314.64	315.33	307.79	307.33	324.13

The forecast for 2008 is 325.26.
(c) $\hat{y}_{2009} = 0.2(y_{2008}) + (1 - 0.2)(\hat{y}_{2008}) = .2(y_{2008}) + .8(315.80)$.

13.51 **(a)** When $w = 0.1$, the coefficients are 0.1, 0.09, 0.081, 0.0729, 0.06561, 0.059049, 0.0531441, 0.0478297, 0.0430467, and 0.0387420. **(b)** When $w = 0.5$, the coefficients are 0.5, 0.25, 0.125, 0.0625, 0.03125, 0.015625, 0.0078125, 0.0039062, 0.0019531, and 0.0009766. **(c)** When $w = 0.9$, the coefficients are 0.9, 0.09, 0.009, 0.0009, 0.00009, 0.000009, 0.0000009, 0.0000001, 0.0000000, and 0.0000000. **(d)**, **(e)** The curves for $w = 0.9$ and $w = 0.5$ show a more rapid exponential decrease than the curve for $w = 0.1$, which decreases more slowly. The curve for $w = 0.9$ puts more weight on the most recent value of the time series in the computation of a forecast. **(f)** The coefficients of y_1 are 0.348678, 0.000977, and 0.0000000 for the values $w = 0.1$, $w = 0.5$, and $w = 0.9$, respectively. The value of $w = 0.1$ puts the greatest weight on y_1 in the calculation of a forecast. Note that as indicated in the text, the values of the coefficients in the forecast model decrease exponentially in value with the exception of this last coefficient.

13.53 **(a)** All three series are increasing. The graph shows the actual time series above the moving average with a span of 20, which is in turn above the moving average of span 50. **(b)** The larger the span, the farther the moving average is below the actual values of the series. This is due to the fact that the series is increasing, and longer spans will include more of the lower values of the series. **(c)** The quote is confirmed by this exercise which shows the increasing trend clearly but does not provide information of use in predicting future values. Any future values would be substantially underestimated by these models.

13.55 **(a)** $\hat{y}_t = wy_{t-1} + (1-w)\hat{y}_{t-1} = wy_{t-1} + \hat{y}_{t-1} - w\hat{y}_{t-1} = \hat{y}_{t-1} + w(y_{t-1} - \hat{y}_{t-1}) = \hat{y}_{t-1} + we_{t-1}$. **(b)** The forecasted value is equal to the predicted value for the previous observation plus a portion of how far this predicted value was from being accurate.

13.57 **(a)** The equation of the least-squares line is $\hat{y}_t = 2964 + 298.26t$ with an R^2 of 0.685 and an s of 5429.71. **(b)** The regression equation is $\hat{y}_t = 11438 - 242.65t + 5.8163t^2$ with an R^2 of 0.832 and an s of 3987.92. **(c)** The least-squares regression line is given by $\hat{y}_t = 6775 + 343.2t - 9.847t^2 + 0.11228t^3$ with an R^2 of 0.862 and an s of 3637.94. **(d)** The third model is the best choice as it has the largest R^2 and the smallest s. Notice that all three explanatory variables, t, t^2, and t^3, are significant in this model as well.

13.59 **(a)** The time series plot shows a definite increasing trend. **(b)** $\hat{y}_t = 8.064 + 0.113t$. **(c)** There is an increase in the death rate of 0.113 per year on average. **(d)** $\hat{y}_{32} = 11.68$.

13.61 **(a)** The histogram and the probability plot both show data that has an approximately Normal distribution. **(b)** Using a mean of 45.015 and a standard deviation of 6.063, the 90% prediction interval is (35.04, 54.99).

13.63 **(a)** The prediction models are smoother as w gets smaller. For $w = 0.8$, the prediction model follows the data closely. The model with $w = 0.2$ is much smoother and gives the general trend of the data but does not follow the actual data points closely enough to give a good prediction of future values. **(b)** For 2009, the predictions are 113.73 when $w = 0.2$, 133.36 when $w = 0.5$, and 112.02 when $w = 0.8$.

13.65 The 3-year moving average is $\hat{y}_{94} = \dfrac{y_{93} + y_{92} + y_{91}}{3}$. Example 13.18 tells us that $y_{92} + y_{91} + y_{90} = 116,984$. Since we know that $y_{90} = 38,272$ and $y_{93} = 40,743$, $\hat{y}_{94} = \dfrac{40743 + 116984 - 38272}{3} = 39,818$. The 15-year moving average is likewise $\hat{y}_{94} = \dfrac{40743 + 505667 - 32363}{15} = 34,270$.

13.67 **(a)** For the exponential trend model, MAD = 16.919, MSE = 550.54, and MAPE = 5.773%. **(b)** For the AR(1) model, MAD = 12.661, MSE = 272.26, and MAPE = 4.589%. **(c)** Because we prefer lower values for MAD, MSE, and MAPE, the AR(1) model would provide the best fit.

13.69 **(a)** $\hat{y}_t = 298728 - 511.8t - 6616\,Jan - 41143\,Feb - 41041\,Mar + 17280\,Apr - 42361\,May - 46355\,Jun - 28947\,Jul - 70970\,Aug - 21806\,Sep + 36377\,Oct - 5378\,Nov$. **(b)** The time series plot shows that the residuals have an increasing trend followed by a decreasing trend instead of random scatter, indicating that there is evidence of nonlinearity. **(c)** $\hat{y}_t = 281883 + 1224.4\,t - 36.941\,t^2 - 2774\,Jan - 37597\,Feb - 37716\,Mar + 20457\,Apr - 39258\,May - 43252\,Jun - 25771\,Jul - 67646\,Aug - 18260\,Sep + 40219\,Oct - 5378\,Nov$. The indicator variable for January has an estimated coefficient of -2774 with a P-value of 0.721.

13.71 **(a)** $\hat{y}_t = 0.00476 + 0.90026y_{t-1} + 0.012436\,(\mathrm{Ind}_{2005})$. **(b)** The coefficient of the lag one variable has increased from the previous value of 0.86099. **(c)** $\hat{y}_t = 0.05086$. This estimate is a little lower than the one in Example 13.14, which was 0.0511286.

13.73 **(a)** The time series of the residuals shows an increasing trend in the residual through the fourth quarter of 2002, followed by a decreasing trend. It does not appear to show just random variations, so there is an indication that we haven't found the best model yet. **(b)** We see evidence of autocorrelation in the lagged residual plot because there is a positive relationship with a correlation of 0.766 between the residuals and the lag one residuals. **(c)** $\widehat{\mathrm{Log}(y_t)} = 2.4843 + 0.008629t - 0.69311\mathrm{Q}1 - 0.29161\mathrm{Q}2 - 0.25568\mathrm{Q}3 + 0.7338\,\mathrm{Log}(y_{t-1})$.
(d) With the addition of the lagged sales variable, we now have residuals with a random time series and no pattern to the lagged residual plot. This model now looks much better. **(e)** The forecast for third quarter 2009 is 11036. This is much closer to what we would expect to see for this point than the previous forecast of 11545.

13.75 **(a)** The time series has an increasing trend with obvious seasonality. The second quarter has a sizable jump each year from the first quarter. Sales then decrease in the third quarter and again in the fourth quarter, rising slightly in the first quarter. **(b)** $\hat{y}_t = 4158 + 296t$. **(c)** $\hat{y}_t = 4706.0 + 211.64t + 2.2174t^2$. **(d)** $\hat{y}_t = 5071.88e^{0.031605t}$.
(e)

	MAD	MSE	MAPE
Linear model	362	203414	4.014%
Quadratic model	318	152408	3.315%
Exponential trend model	373	233990	3.826%

All three measures are lowest for the quadratic model, so this one provides the best fit.

13.77 **(a)** There is a curved trend to the series. It increases and then appears to level off. At the very end of the series, there seems to be a slight decrease in the trend.
(b) Because the increases and decreases do not appear to change for particular seasons throughout the series, an additive model would be appropriate. **(c)** $\hat{y}_t = 35400927 + 364870t - 2671.2t^2 - 3706652\,Jan - 5175343\,Feb + 3752082\,Mar + 1168197\,Apr + 2214362\,May + 5242785\,Jun + 7849675\,Jul + 6118787\,Aug - 4219674\,Sep + 139004\,Oct - 1816936\,Nov.\ \hat{y}_{82} = 47499467$.

13.79 **(a)** The time series has an oscillating pattern, bouncing from a positive residual to a negative residual and back again. The number of runs here is greater than what we would expect to see. **(b)** The lagged residual plot shows a negative relationship between the residual and the lag one residual. The correlation between e_t and e_{t-1} is -0.24. **(c)** $\hat{y}_t = 10146180 + 120623t - 1049.1t^2 - 3558944\,Jan - 4783694\,Feb + 6455376\,Mar + 1860862\,Apr - 174651\,May + 3656518\,Jun + 4916885\,Jul + 1111092\,Aug - 9836686\,Sep - 1908367\,Oct - 675931\,Nov + 0.2959y_{t-1} + 0.4307y_{t-2}$. The P-value for y_{t-1} is 0.012, and the P-value for y_{t-2} is 0.000, indicating that both of these should be included in the final model.
(d) $\hat{y}_{82} = 45844197$, which is smaller than either of the other predictions.

13.81 **(a)** $\hat{y}_t = 39.216 + 1.21151y_{t-1} - 0.4338y_{t-2}.\ R^2 = 0.763$ and $s = 0.1853$. This model has a higher R^2 value and a lower standard error estimate. Both lag variables are significant with P-values of approximately zero. **(b)** The observed number of runs is 42, and the expected number of runs is 45. The P-value of the runs test is 0.52. There is no evidence against randomness. **(c)** $\hat{y}_{91} = 176.017$.

CHAPTER 14

14.1 **(a)** "Sample" should be "population." **(b)** "Within-groups" should say "Between-groups." **(c)** "One-Way ANOVA" should say "a one-sample proportion analysis." **(d)** "Multiple comparisons" should read "Contrasts."

14.3 $I = 3$, $N = 60$, $n_1 = 20$, $n_2 = 20$, and $n_3 = 20$. The ANOVA model is $x_{ij} = \mu_i + \varepsilon_{ij}$. The ε_{ij} are assumed to be from an $N(0, \sigma)$ distribution where σ is the common standard deviation. The parameters of the model are the I population means μ_1, μ_2, and μ_3, and the common standard deviation σ.

14.5 **(a)** The largest standard deviation is 130, and the smallest is 85. The ratio of these is $130/85 = 1.53$, which is less than 2, so it is reasonable to pool the standard deviations for these data. **(b)** $I = 3$, $\bar{x}_1 = 75$, $\bar{x}_2 = 125$, $\bar{x}_3 = 100$. We estimate σ by $s_p = 111.58$.

14.7 None of the groups have Normal distributions. The blue and brown groups are skewed to the right with a peak toward the left of the distribution. The green and brown groups have a somewhat uniform distribution on the left side of the graph, a peak toward the center, and decreasing frequencies to the right.

14.9 $SSG + SSE = 24.42 + 613.14 = 637.56 = SSG$.

14.11 $DFG + DFE = 3 + 218 = 221 = DFT$.

14.13 $MST = SST / DFT = 637.56 / 221 = 2.885$. The variance of all observations is also 2.885 with a mean of 3.497.

14.15 **(a)** The ANOVA F statistic has 2 numerator degrees of freedom and 21 denominator degrees of freedom. **(b)** The F statistic would need to be larger than 3.47 to have a P-value less than 0.05. **(c)** With 2 degrees of freedom in the numerator and 60 in the denominator, the F statistic would need to be larger than 3.15. **(d)** As the degrees of freedom in the denominator increases, F gets smaller, making it less likely that we will reject the null hypothesis with a smaller sample size.

14.17 Try the applet. You should find that the F statistic increases, and the P-value decreases, if the pooled standard error is kept fixed and the variation among the group means increases.

14.19 **(a)** It is always true that $s_p^2 = MSE$. Hence, $s_p = \sqrt{MSE}$. Because of this, SAS calls s_p "Root MSE." **(b)** $s_p = \sqrt{MSE}$. In Excel, MSE is found in the ANOVA table in the row labeled "Within Groups" and under the column labeled "MS." Take the square root of this entry (0.616003 in this case) to get s_p.

14.21 The standard error for the contrast $SE_c = 2$.

14.23 The 95% confidence interval is $7 \pm 3.98 = (3.02, 10.98)$.

14.25 $t_{23} = \dfrac{46.7273 - 44.2727}{6.31\sqrt{\frac{1}{22} + \frac{1}{22}}} = 1.29$.

14.27 The value of t^{**} for the Bonferroni procedure when $\alpha = 0.05$ is $t^{**} = 2.46$ (see Example 14.22). $t_{23} = 1.29$. Because $|1.29| < t^{**}$, using the Bonferroni procedure, we would not reject the null hypothesis that the population means for groups 2 and 3 are different.

14.29 From the output in Figure 14.16, the difference between x_2 and x_3 is 2.455, and the standard error for this difference is 1.904. Thus, $t_{23} = 1.2894$.

14.31 Groups 1, 2, and 4 should be marked "A," and groups 2 and 3 should be marked "B." There is not a significant difference between groups 1, 2, and 4. Group 3 does not differ significantly from group 2, but it does differ significantly from groups 1 and 4.

14.33 The Bonferroni 95% confidence interval is $(-2.23, 7.14)$. This interval includes 0.

14.35 **(a)**

Sample size	10	20	30	40	50	100
Power	0.275	0.532	0.726	0.851	0.923	0.998

(b) As the sample size increases, the power increases. **(c)** Of the sample sizes selected, 50 would be the best because the power is relatively high; the sample size would have to increase greatly to make a substantial increase in the power.

14.37 **(a)** $F(4, 30)$.

p	0.100	0.050	0.025	0.010	0.001
F	2.14	2.69	3.25	4.02	6.12

(b) and **(c)** The P-value would be between 0.025 and 0.05. **(d)** You can never conclude from an F test that all the means are different. The alternative hypothesis is that "not all the means are the same." In this case, we would reject the null hypothesis that all the means are equal.

14.39 **(a)** $F(4, 40) = 1.54$, P-value $= 0.2091$ (from software). **(b)** $F(2, 21) = 3$, P-value $= 0.0714$ (from software).

14.41 **(a)** $I = 3$, $n_1 = 220$, $n_2 = 145$, $n_3 = 76$, $N = 441$, response variable $= 1$ to 5 rating of "Did you discuss the presentation with any of your friends?" **(b)** $I = |, 3$, $n_1 = 31$, $n_2 = 18$, $n_3 = 45$, $N = 94$, response = acceptance rating from 1 to 5. **(c)** $I = 4$, $n_i = 5$, $N = 20$, response = sales.

14.43 **(a)** $H_0: \mu_1 = \mu_2 = \mu_3$, and H_a: Not all the means are the same. DFG $= 2$, DFE $= 438$, DFT $= 440$, $F(2, 438)$. **(b)** $H_0: \mu_1 = \mu_2 = \mu_3$, and H_a: Not all the means are the same. DFG $= 2$, DFE $= 91$, DFT $= 93$, $F(2, 91)$. **(c)** $H_0 : \mu_1 = \mu_2 = \mu_3 = \mu_4$, and H_a: Not all the means are the same. DFG $= 3$, DFE $= 16$, DFT $= 19$, $F(3, 16)$.

14.45 Answers will vary.

14.47 **(a)** 2×12 is not greater than 25, so it is not reasonable to assume that the standard deviations are equal. **(b)** $s_1^2 = 625$, $s_2^2 = 441$, $s_3^2 = 144$, $s_2^2 = 529$. **(c)** $s_p^2 = 232.67$. **(d)** $s_p = 15.25$. **(e)** The third group had the largest sample size, so it has the heaviest weight in the pooled standard deviation.

14.49 **(a)** For the last question, we would not reject the null hypothesis that the means are equal because an F of 1.62 results in a P-value of 0.169, so no test would be needed to find which mean is significantly different. **(b)** Excel gives a t^{**} value of 2.25. For the question "I'd like to experience native Hawaiian culture," the test statistics are $t_{12} = -0.74$, $t_{13} = -0.79$, $t_{14} = -2.89$, $t_{15} = -0.72$, $t_{23} = 0.03$, $t_{24} = -2.50$, $t_{25} = 0.10$, $t_{34} = -2.73$, $t_{35} = 0.08$, and $t_{45} = 2.75$. For this question, there is evidence that group 4 differs significantly from all of the other groups, but no other groups have significant differences from each other. For the question "I'd prefer a group tour to an individual one," the test statistics are $t_{12} = -0.55$, $t_{13} = 2.38$, $t_{14} = 1.37$, $t_{15} = 0.59$, $t_{23} = 3.55$, $t_{24} = 2.00$, $t_{25} = 1.41$, $t_{34} = -0.51$, $t_{35} = -2.47$, and $t_{45} = -1.07$. For this question, there is evidence that group 3 differs significantly from groups 1, 2, and 5, but no other groups have significant differences from each other. For the question "I'd like to experience ocean sports," the test statistics are $t_{12} = 0.27$, $t_{13} = 0.32$, $t_{14} = -1.15$, $t_{15} = 1.67$, $t_{23} = 0.03$, $t_{24} = -1.51$, $t_{25} = 1.63$, $t_{34} = -1.65$, $t_{35} = 1.91$, and $t_{45} = 2.86$. For this question, groups 4 and 5 are the only pair that have a significant difference between their means.

14.51 **(a)** $F(2, 117)$. **(b)** P-value $= 0.0003$. There is evidence that at least one of the groups has a different population average price reduction than the others. **(c)** It is not clear whether the shoppers were asked to bargain or whether bargaining behavior was observed. If shoppers were asked to bargain, then we may be measuring their tenacity without the risk of losing the object they were bargaining for. More information is needed to determine whether the results would generalize to other populations.

14.53 **(a)** It is reasonable to use a pooled standard deviation because twice the smallest standard deviation is larger than the largest s. $s_p = 0.768$. **(b)** $F(2, 767)$. The P-value is less than 0.001. There is significant evidence that at least one of the groups rated the dining satisfaction differently on average. **(c)** Numbering the groups in the order presented in the table, $\psi = \frac{1}{2}(\mu_1 + \mu_3) - \mu_2$. $H_0: \frac{1}{2}(\mu_1 + \mu_3) = \mu_2$, $H_a: \frac{1}{2}(\mu_1 + \mu_3) < \mu_2$. The test statistic is $t = -5.98$ with 767 degrees of freedom. The P-value is less than 0.0001. There is evidence that the faculty satisfaction rating is higher on average.

14.55 **(a)** The means do appear to be quite different, but a means plot alone cannot determine whether the pricing difference is significant. **(b)** $2(26.09) > 40.39$, so it is reasonable to assume the population standard deviations are equal. **(c)** With 2 and 44 degrees of freedom, the P-value of the F statistic is 0.0427.

14.57 **(a)** 3 and 2286. **(b)** MSE $= 4.6656$, so $F = 2.5304$. **(c)** P-value $= 0.056$. There is not a significant difference between the mean scores for the different classes at the 5% level.

14.59 **(a)** In general, as the number of accommodations increased, the mean grade decreased, but it is not a steady decline. **(b)** So many decimal places are not necessary. No additional information is gained by having so many. Using two decimal places would be fine. **(c)** The biggest s is not exactly less than two times the smallest s (1.66233 vs. 2×0.82745), but it is close. Pooling should not be used (s_p would be 0.86). **(d)** It is never a good idea to eliminate data points without a good reason. Because the mean grade is affected similarly for two, three, and four accommodations, it is not unreasonable to group these data together. **(e)** These data do not represent 245 independent observations. Some students were measured multiple times. **(f)** Answers will vary. University policies, teacher policies, and student backgrounds might not be the same from school to school. **(g)** There is no control group, so it is impossible to comment on the effectiveness of the accommodations.

14.61 (a) Checking whether the largest $s < 2 \times$ smallest s for each group indicates that it is appropriate to pool the standard deviations for intensity and recall but not frequency. (b) $F(4, 405)$. The P-value for all three one-way ANOVA tests is < 0.001, so for frequency, intensity, and recall, there is strong evidence that not all the means are the same.

P	0.100	0.050	0.025	0.010	0.001
F	1.95	2.42	2.85	3.41	4.81

(c) Hispanic Americans were highest in each of the three groups. For Emotion score, the European, Asian, and Indian scores were low but close together. For Frequency and Intensity and Recall, Asian and Japanese were both low, and European and Indian are close together in the middle. (d) Answers will vary. People near a university might not be representative of all citizens of those countries. (e) The chi-square test has a test statistic of 11.353 and a P-value of 0.023, so at the 5% significance level, there is enough evidence to say that there is a relationship between gender and culture. In other words, there is evidence that the proportion of men and women is not the same for all of the cultural groups. This may affect how broadly the results can be applied.

14.63 (a) There no serious departures from Normality. The assumption that the data are approximately Normal is not unreasonable.
(b)

Number of promotions	Average	Standard deviation	Sample size
1	4.56	0.36	40
3	4.42	0.34	40
5	4.15	0.29	40
7	3.98	0.32	40

(c) The ratio of the largest to the smallest is $0.36/0.29 = 1.24$. This is less than 2, so it is not unreasonable to assume that the population standard deviations are equal. (d) $H_0: \mu_1 = \mu_2 = \mu_3 = \mu_4$; H_a: not all of the μ_i are equal. Following is the ANOVA table.

Source	DF	Sum of squares	Mean square	F	P-value
Groups	3	8.361	2.787	25.655	≤ 0.001
Error	156	16.946	0.109		
Total	159	25.307			

We would conclude that there is strong evidence that the population mean expected prices associated with the different numbers of promotions are not all equal.

14.65 The contrast is $\psi = \mu_{\text{supervisor}} - \frac{1}{2}(\mu_{\text{skilled}} + \mu_{\text{unskilled}})$. The sample contrast is $c = 9.695$ with a standard error of 2.736. The test statistic for $H_0: \psi = 0$ vs. $H_a: \psi \neq 0$ is $t = 3.543$. The P-value for this test is less than 0.001. There is evidence that the average safety index is significantly different for the average of skilled and unskilled workers than for supervisors. If a one-sided test had been performed, the P-value would have been less than 0.0005, and we would conclude that supervisors have a significantly higher SCI than the average of the workers.

14.67 **(a)** Plots show no serious departures from Normality, so the Normality assumption is reasonable.

(b)

	Bonferroni Tests		
Group $i - j$	Difference	Std. error	P-value
$2 - 1$	11.4000	9.653	0.574592
$3 - 1$	37.6000	9.653	0.001750
$3 - 2$	26.2000	9.653	0.033909

At the $\alpha = 0.05$ level, we see that Group 3 (the high-jump group) differs from the other two. The other two groups (the control group and the low-jump group) are not significantly different. It appears that the mean density after eight weeks is different (higher) for the high-jump group than for the other two.

14.69 **(a)** Plots show no serious departures from Normality (but note the granularity of the Normal probability plot, which shows that values are rounded to the nearest 5%), so the Normality assumption is not unreasonable.

(b)

	Bonferroni		
Tests Group $i - j$	Difference	Std. error	P-value
ECM2 − ECM1	−1.66667	3.600	1.00000
ECM3 − ECM1	8.33333	3.600	0.450687
ECM3 − ECM2	10.0000	3.600	0.223577
MAT1 − ECM1	−41.6667	3.600	0.000001
MAT1 − ECM2	−40.0000	3.600	0.000002
MAT1 − ECM3	−50.0000	3.600	0.000000
MAT2 − ECM1	−58.3333	3.600	0.000000
MAT2 − ECM2	−56.6667	3.600	0.000000
MAT2 − ECM3	−66.6667	3.600	0.000000
MAT2 − MAT1	−16.6667	3.600	0.008680
MAT3 − ECM1	−53.3333	3.600	0.000000
MAT3 − ECM2	−51.6667	3.600	0.000000
MAT3 − ECM3	−61.6667	3.600	0.000000
MAT3 − MAT1	−11.6667	3.600	0.101119
MAT3 − MAT2	5.00000	3.600	0.957728

At the $\alpha = 0.05$ level, we see that none of the ECMs differ from each other, that all the ECMs differ from all the other types of materials (the MATs), and that MAT1 and MAT2 differ from each other. The most striking differences are those between the ECMs and the other materials.

14.71 **(a)**

Source	Degrees of freedom	Sum of squares	Mean square	F
Groups	3	104,855.87	34,951.96	15.86
Error	32	70,500.59	2,203.14	
Total	35	175,356.46	5,010.18	

(b) H_0: $\mu_1 = \mu_2 = \mu_3 = \mu_4$, H_a: not all of the μ_i are equal. **(c)** The F statistic has the $F(3, 32)$ distribution. The P-value is smaller than 0.001. **(d)** $s_p^2 = 2{,}203.14$, $s_p = 46.94$.

14.73 **(a)** $s_p^2 = 3.90$. This quantity corresponds to MSE in the ANOVA table.

(b)

Source	Degrees of freedom	Sum of squares	Mean square	F
Groups	2	17.22	8.61	2.21
Error	206	803.4	3.90	
Total	208	175,356.46	5010.18	

(c) H_0: $\mu_1 = \mu_2 = \mu_3$; H_a: not all of the μ_i are equal. **(d)** The F statistic has the $F(2, 206)$ distribution. The P-value is greater than 0.100. We conclude that these data do not provide evidence that the mean weight gains of pregnant women in these three countries differ.

14.75 **(a)** $\psi_1 = \mu_{\text{brown}} - \frac{1}{2} (\mu_{\text{blue}} - \mu_{\text{green}})$. **(b)** $\psi_2 = \mu_{\text{down}} - \frac{1}{3} (\mu_{\text{blue}} - \mu_{\text{green}} + \mu_{\text{brown}})$.

14.77 **(a)** For contrast ψ_1, it would be reasonable to test the hypotheses H_0: $\psi_1 = 0$, H_a: $\psi_1 > 0$. For the contrast ψ_2, it would be reasonable to test the hypotheses H_0: $\psi_2 = 0$, H_a: $\psi_2 \neq 0$. **(b)** $c_1 = 0.195$. $c_2 = -0.48$. **(c)** $SE_{c1} = 0.310$, $SE_{c2} = 0.293$. **(d)** The test statistic for H_0: $\psi_1 = 0$, H_a: $\psi_1 > 0$, is $t = 0.629$. P-value > 0.25. We conclude that there is not enough evidence that the responses were greater for the models with brown eyes than the average scores for those with blue or green eyes. For H_0: $\psi_2 = 0$, H_a: $\psi_2 \neq 0$, the test statistic is $t = -1.636$. $0.05 < P$-value < 0.10. We conclude that there is not strong evidence that the average response for models with their eyes looking downward is different from the mean of the responses for those whose eyes were visible. **(e)** Using 100 df to be more conservative, a 95% confidence interval for ψ_1 is $(-0.42, 0.81)$, and a 95% confidence interval for ψ_2 is $(-1.062, 0.102)$.

14.79 **(a)** Question 1. Contrast: $\psi_1 = \mu_1 - \mu_2$. Hypotheses: H_0: $\psi_1 = 0$, H_a: $\psi_1 > 0$. Question 2. Contrast: $\psi_2 = \mu_1 - (0.5)\mu_2 - (0.5)\mu_4$. Hypotheses: H_0: $\psi_2 = 0$, H_a: $\psi_2 > 0$. Question 3. Contrast: $\psi_3 = \mu_3 - (1/3)\mu_1 - (1/3)\mu_2 - (1/3)\mu_4$. Hypotheses: H_0: $\psi_3 = 0$, H_a: $\psi_3 > 0$. **(b)** Question 1: P-value > 0.25. We do not have evidence that T is better than C. Question 2: $0.10 < P$-value < 0.15. There is at best weak evidence that T is better than the average of C and S. Question 3: P-value < 0.005. There is strong evidence that J is better than the average of the other three groups. **(c)** This is an observational study. Males were not assigned at random to treatments. Thus, although the researchers tried to match those in the groups with respect to age and other characteristics, there are reasons why people choose to jog or choose to be sedentary that may affect other aspects of their health. It is always risky to draw conclusions of causality from a single (small) observational study, no matter how well designed it is in other respects.

14.81 For each pair of means, we get the following: Group 1 (T) vs. Group 2 (C): $t_{12} = -0.66$. $|-.66|$ is not larger than $t^{**} = 2.81$, so we do not have strong evidence that T and C differ. Group 1 (T) vs. Group 3 (J): $t_{13} = -3.65$. $|-3.65|$ is larger than $t^{**} = 2.81$, so we have strong evidence that T and J differ. Group 1 (T) vs. Group 4 (S): $t_{14} = 3.14$. $|3.14|$ is larger than $t^{**} = 2.81$, so we have strong evidence that T and S differ. Group 2 (C) vs. Group 3 (J): $t_{23} = -2.29$. $|-2.29|$ is not larger than $t^{**} = 2.81$, so we do not have strong evidence that C and J differ. Group 2 (C) vs. Group 4 (S): $t_{24} = 3.22$. $|3.22|$ is larger than $t^{**} = 2.81$, so we have strong evidence that C and S differ. Group 3 (J) vs. Group 4 (S): $t_{34} = 6.86$. $|6.86|$ is larger than $t^{**} = 2.81$, so we have strong evidence that J and S differ.

14.83

n	DFG	DFE	F^*	λ	Power
50	2	147	3.0576	2.78	0.2950
100	2	297	3.0261	5.56	0.5453
150	2	447	3.0158	8.34	0.7336
175	2	522	3.0130	9.73	0.8017
200	2	597	3.0108	11.12	0.8548

A sample size of 175 gives reasonable power. The gain in power by using 200 women per group may not be worthwhile unless it is easy to get women for the study. If it is difficult or expensive to include more women in the study, one might consider a sample size of 150 per group.

14.85 **(a)** Answers will vary. **(b)** A sample of 75 gives a power of 0.836. A sample size of 100 results in a power of 0.935. A sample size of 200 gives a power of 0.999. A good sample size for this study would be 100. **(c)** Answers will vary.

14.87 **(a)**

Level	Sample size	Mean	Standard dev.
ECM1	3	0.6500	0.0866
ECM2	3	0.6333	0.0289
ECM3	3	0.7333	0.0289
MAT1	3	0.2333	0.0289
MAT2	3	0.6667	0.0289
MAT3	3	0.1167	0.0289

(b) We get the same result by dividing the means and standard deviations by 100 as well. **(c)** The results are identical to those in Exercise 14.68 where $F(5, 12) = 137.94$ with a P-value of 0.

14.89 **(a)** With the incorrect observation, the test statistic changes from 137.94 to 9.65, and the P-value changes from 0 to 0.001. **(b)** Outliers can have a substantial effect on the ANOVA test results. **(c)** As the following table shows, the standard deviation is quite different for the group with the incorrect observation. This helps us identify the problem.

Level	Sample size	Mean	Standard dev.
ECM1	3	48.33	37.53
ECM2	3	63.33	2.89
ECM3	3	73.33	2.89
MAT1	3	23.33	2.89
MAT2	3	66.67	2.89
MAT3	3	11.67	2.89

14.91   **(a)** The pattern in the scatterplot is a roughly linear, decreasing trend. **(b)** The test in regression that tests the null hypothesis that the explanatory variable has no linear relationship with the response variable is the t test of whether or not the slope is 0. **(c)** Using software we obtain the following:

Variable	Coefficient	Std. error of coeff.	t ratio	P-value
Constant	4.36453	0.0401	109	≤ 0.0001
Number of promotions	−0.116475	0.0087	−13.3	≤ 0.0001

The t statistic for testing whether the slope is 0 is −133, and we see that the P-value is ≤ 0.0001. Thus, there is strong evidence that the slope is not 0. The ANOVA in Exercise 14.63 showed that there is strong evidence that the mean expected price is different for the different numbers of promotions. This is consistent with our regression results here because if the slope is different from 0, the mean expected price is changing as the number of promotions changes. In this example, the regression is more informative. It not only tells us that the means differ but also gives us information about how they differ.

INDEX

Table entry for p and C is the critical value t^* with probability p lying to its right and probability C lying between $-t^*$ and t^*.

Probability p

t^*

TABLE D t distribution critical values

df	.25	.20	.15	.10	.05	.025	.02	.01	.005	.0025	.001	.0005
						Upper tail probability p						
1	1.000	1.376	1.963	3.078	6.314	12.71	15.89	31.82	63.66	127.3	318.3	636.6
2	0.816	1.061	1.386	1.886	2.920	4.303	4.849	6.965	9.925	14.09	22.33	31.60
3	0.765	0.978	1.250	1.638	2.353	3.182	3.482	4.541	5.841	7.453	10.21	12.92
4	0.741	0.941	1.190	1.533	2.132	2.776	2.999	3.747	4.604	5.598	7.173	8.610
5	0.727	0.920	1.156	1.476	2.015	2.571	2.757	3.365	4.032	4.773	5.893	6.869
6	0.718	0.906	1.134	1.440	1.943	2.447	2.612	3.143	3.707	4.317	5.208	5.959
7	0.711	0.896	1.119	1.415	1.895	2.365	2.517	2.998	3.499	4.029	4.785	5.408
8	0.706	0.889	1.108	1.397	1.860	2.306	2.449	2.896	3.355	3.833	4.501	5.041
9	0.703	0.883	1.100	1.383	1.833	2.262	2.398	2.821	3.250	3.690	4.297	4.781
10	0.700	0.879	1.093	1.372	1.812	2.228	2.359	2.764	3.169	3.581	4.144	4.587
11	0.697	0.876	1.088	1.363	1.796	2.201	2.328	2.718	3.106	3.497	4.025	4.437
12	0.695	0.873	1.083	1.356	1.782	2.179	2.303	2.681	3.055	3.428	3.930	4.318
13	0.694	0.870	1.079	1.350	1.771	2.160	2.282	2.650	3.012	3.372	3.852	4.221
14	0.692	0.868	1.076	1.345	1.761	2.145	2.264	2.624	2.977	3.326	3.787	4.140
15	0.691	0.866	1.074	1.341	1.753	2.131	2.249	2.602	2.947	3.286	3.733	4.073
16	0.690	0.865	1.071	1.337	1.746	2.120	2.235	2.583	2.921	3.252	3.686	4.015
17	0.689	0.863	1.069	1.333	1.740	2.110	2.224	2.567	2.898	3.222	3.646	3.965
18	0.688	0.862	1.067	1.330	1.734	2.101	2.214	2.552	2.878	3.197	3.611	3.922
19	0.688	0.861	1.066	1.328	1.729	2.093	2.205	2.539	2.861	3.174	3.579	3.883
20	0.687	0.860	1.064	1.325	1.725	2.086	2.197	2.528	2.845	3.153	3.552	3.850
21	0.686	0.859	1.063	1.323	1.721	2.080	2.189	2.518	2.831	3.135	3.527	3.819
22	0.686	0.858	1.061	1.321	1.717	2.074	2.183	2.508	2.819	3.119	3.505	3.792
23	0.685	0.858	1.060	1.319	1.714	2.069	2.177	2.500	2.807	3.104	3.485	3.768
24	0.685	0.857	1.059	1.318	1.711	2.064	2.172	2.492	2.797	3.091	3.467	3.745
25	0.684	0.856	1.058	1.316	1.708	2.060	2.167	2.485	2.787	3.078	3.450	3.725
26	0.684	0.856	1.058	1.315	1.706	2.056	2.162	2.479	2.779	3.067	3.435	3.707
27	0.684	0.855	1.057	1.314	1.703	2.052	2.158	2.473	2.771	3.057	3.421	3.690
28	0.683	0.855	1.056	1.313	1.701	2.048	2.154	2.467	2.763	3.047	3.408	3.674
29	0.683	0.854	1.055	1.311	1.699	2.045	2.150	2.462	2.756	3.038	3.396	3.659
30	0.683	0.854	1.055	1.310	1.697	2.042	2.147	2.457	2.750	3.030	3.385	3.646
40	0.681	0.851	1.050	1.303	1.684	2.021	2.123	2.423	2.704	2.971	3.307	3.551
50	0.679	0.849	1.047	1.299	1.676	2.009	2.109	2.403	2.678	2.937	3.261	3.496
60	0.679	0.848	1.045	1.296	1.671	2.000	2.099	2.390	2.660	2.915	3.232	3.460
80	0.678	0.846	1.043	1.292	1.664	1.990	2.088	2.374	2.639	2.887	3.195	3.416
100	0.677	0.845	1.042	1.290	1.660	1.984	2.081	2.364	2.626	2.871	3.174	3.390
1000	0.675	0.842	1.037	1.282	1.646	1.962	2.056	2.330	2.581	2.813	3.098	3.300
z^*	0.674	0.841	1.036	1.282	1.645	1.960	2.054	2.326	2.576	2.807	3.091	3.291
	50%	60%	70%	80%	90%	95%	96%	98%	99%	99.5%	99.8%	99.9%
						Confidence level C						